Geological Society of America
Centennial Articles

Reprinted from Volume 100
Geological Society of America Bulletin

Edited by

R. D. Hatcher, Jr.
Department of Geological Sciences
University of Tennessee
Knoxville, Tennessee 37996-1410

and

William A. Thomas
Department of Geology
University of Alabama
Tuscaloosa, Alabama 35487

Special Paper 253

A Reprint Volume

© 1990 The Geological Society of America, Inc.
All rights reserved.

All materials subject to this copyright and included
in this volume may be photocopied for the noncommercial
purpose of scientific or educational advancement.

Copyright is not claimed on any material prepared
by government employees within the scope of their
employment.

Published by The Geological Society of America, Inc.
3300 Penrose Place, P.O. Box 9140, Boulder, Colorado 80301

GSA Books Science Editor Richard A. Hoppin

Printed in U.S.A.

Library of Congress Cataloging-in-Publication Data
Centennial articles : reprinted from volume 100, Geological Society of
  America bulletin / Geological Society of America ; edited by
  R.D. Hatcher, Jr., and W.A. Thomas.
    p.   cm. — (Special paper ; 253)
  Includes bibliographical references.
  ISBN 0-8137-2253-5
  1. Geology. 2. Geology—United States. I. Hatcher, Robert. D.,
1940- . II. Thomas, William A., 1946- . III. Geological
Society of America. IV. Geological Society of America bulletin.
V. Series: Special papers (Geological Society of America) ; 253.
QE35.C46 1990
550—dc20   90-34478
                                               CIP
                                               Rev.

Contents

Foreword v

Landscape analysis and the search for geomorphic unity 1
    Dale F. Ritter

The Geological Society of America Bulletin and the development of quantitative geomorphology 13
    Marie Morisawa

Geological fluvial geomorphology 21
    Victor R. Baker

Dynamic changes and processes in the Mississippi River delta 33
    James M. Coleman

Forty years of sequence stratigraphy 51
    L. L. Sloss

Tectonics of sedimentary basins 57
    Raymond V. Ingersoll

Evolution of thought on passive continental margins from the origin of geosynclinal theory (~1860) to the present 73
    Gerard C. Bond and Michelle A. Kominz

Alpine serpentinites, ultramafic magmas, and ocean-basin evolution: The ideas of H. H. Hess 99
    E. M. Moores and F. J. Vine

Impact of earthquake seismology on the geological community since the Benioff zone 107
    Keiiti Aki

Plate tectonics and island arcs 113
    Warren B. Hamilton

Paleozoic paleogeography of North America, Gondwana, and intervening displaced terranes: Comparisons of paleomagnetism with paleoclimatology and biogeographical patterns 139
    Rob Van der Voo

Paleoceanography: A review for the GSA Centennial 153
    William W. Hay

The mechanical paradox of large overthrusts 177
    Raymond A. Price

Material balance in Alpine orogeny 189
    Hans Laubscher

Strike-slip faults 205
    Arthur G. Sylvester

Basin and Range extensional tectonics at the latitude of Las Vegas, Nevada 243
Brian Wernicke, Gary J. Axen, and J. Kent Snow

Seismic imaging of extended crust with emphasis on the western United States 263
Jill McCarthy and George A. Thompson

Significance of past and recent heat-flow and radioactivity studies in the Southern Rocky Mountains region 277
Edward R. Decker, Henry P. Heasler, Kenneth L. Buelow, Keith H. Baker, and James S. Hallin

Progress in understanding jointing over the past century 313
David D. Pollard and Atilla Aydin

Low-temperature deformation mechanisms and their interpretation 337
Richard H. Groshong, Jr.

The formation of continental crust: Part 1. A review of some principles; Part 2. An application to the Proterozoic evolution of southern North America 369
M. E. Bickford

The origin of granite: The role and source of water in the evolution of granitic magmas 387
James A. Whitney

Crystal capture, sorting, and retention in convecting magma 399
Bruce D. Marsh

The development of gravity and magnetic studies, emphasizing articles published in the Geological Society of America Bulletin 417
G. R. Keller

Three decades of geochronologic studies in the New England Appalachians 427
Robert E. Zartman

100 years of economic geology and GSA 441
J. M. Guilbert

*The **Bulletin of the Geological Society of America** and Charles Doolittle Walcott* 449
Ellis L. Yochelson

Paleontology and The Geological Society of America: The first 100 years 459
J. Thomas Dutro, Jr.

Foreword

This volume brings together the Centennial Articles that were published in the twelve monthly issues of Volume 100 of the *Geological Society of America Bulletin*. All of these articles were invited, and the topic of each was specified in the invitation. The series of articles was intended to span the entire field of geological science, in the tradition of GSA publications. We believe that this is an outstanding set of papers that deserves publication as a group. Many have already proved useful as readings in graduate courses and seminars, one of the intended uses of the articles when they were invited. Several of the authors have told us of unusually numerous requests for reprints.

The conceptual framework of the Centennial Articles for the *Bulletin* was designed by an advisory committee composed of Arden Albee, Arthur L. Bloom, Jack Oliver, A. R. Palmer, and the two *Bulletin* editors. Planning the Centennial Articles as a special component of the celebration of the GSA Centennial, the committee recommended that some emphasis should be placed on the scientific achievements of GSA publications during the past century. It was also decided that the articles should provide comprehensive reviews, and might also include appropriate new data and ideas.

Authors were provided with information about the guiding philosophy of the series of articles, and were encouraged to develop their topics as they wished. Some articles have a greater emphasis on review and history, whereas others are primarily research articles that provide new data and ideas. Most authors chose the middle ground, and their articles involve both review and the introduction of new data and ideas. We believe that many of the Centennial Articles will become landmark papers, and we hope that this reprint volume will provide convenient access for many readers.

Robert D. Hatcher, Jr.
William A. Thomas

CENTENNIAL ARTICLE

Landscape analysis and the search for geomorphic unity

DALE F. RITTER *Department of Geology, Southern Illinois University, Carbondale, Illinois 62901*

ABSTRACT

Geomorphology has been an integral part of geological science since the inception of the Geological Society of America, even though different investigative goals have existed continuously in the discipline. The dichotomy of purpose began with the vastly different perception of landscape analysis embraced by William Morris Davis and G. K. Gilbert. Historical and physical geomorphic studies have always been conducted simultaneously, but one of these research activities has dominated the field at any given time. During the first half of this century, most geomorphic work was devoted to the interpretation of long-term, evolutionary history of regional landscapes. In the past four decades, however, research has concentrated on the study of geomorphic processes. This revised direction placed greater emphasis on the physical component of geomorphology and provided a more complete understanding of the time factor in landscape analysis. Historical geomorphology is now less concerned with theories of cyclic-time landscape development and more involved in determining the time and sequence of shorter episodes of geomorphic disequilibrium caused by tectonism and/or climate change.

The tenor of future research in surficial geology will require input from both physical and historical geomorphology, and therefore a greater unity of purpose among geomorphologists can be expected. Geomorphologists will have the opportunity to provide geologists and engineers with useful data about major scientific and technical problems related to plate-tectonic theory, interpretation of the stratigraphic record, and prediction of landform stability for environmental planning.

INTRODUCTION

The accumulation of knowledge in any science inevitably results in the fragmentation of that science into distinct subparts. The Geological Society of America, for example, presently recognizes ten divisions which derive identity because their members share mutual scientific interests. Underlying this commonality are intellectual threads which, woven together, provide the philosophic basis for each subdiscipline and manifest the scientific goals that make a particular field unique. It is also true, however, that recognition of common goals does not insure agreement about how to achieve those goals or what investigative thrust should be first among all others. Nonetheless, in most disciplines, the intellectual bonds are stronger than differences which arise concerning goals or methodology.

The Geological Society of America is the only organization that recognizes geomorphology as an integral part of geology and correctly joins it with Quaternary Geology as a major Society division. Geomorphology is the scientific study of landforms, and because landforms constitute the building blocks of regional landscapes, the science has historically been associated with the development of landscapes. The question, of course, is how does one analyze a landscape? William Morris Davis provided the answer by stressing that landforms (and therefore landscapes) are a function of the combined effect of structure, process, and time, where structure refers to all aspects of geology (lithology, stratigraphy, tectonic characteristics) and time refers to a relative stage of development rather than absolute years. Thus, in simple terms, the form or forms of a landscape reflect some unique accommodation between a particular geologic framework being acted on by both endogenic and exogenic processes over an indistinct period of time. There is nothing inherently wrong with Davis' generalized statement; it is, in fact, true and certainly eloquent in its simplicity. It does, however, give geomorphologists the option to emphasize one variable to the exclusion of the others, thereby leading to the problem of diverse approach and purpose. Such difficulties are especially apparent when either time or process becomes the primary target of study because each has a different goal and each requires a different scientific approach. Therefore, the very nature of landscapes requires geomorphology to assume the dual nature of being both a historical and physical science.

When geomorphology is adopted as a historical science, landscapes (and their component landforms) are used to discern the sequence of geological events that are reflected and preserved in their present condition. In this approach, erosional and depositional landforms are used as evidence for tectonic or climatic events, and the time and sequence of those events are the research goals. When geomorphology is considered to be a physical science, research analysis revolves around process and is totally different. This approach attempts to explain *why* rather than *when* landforms develop. Process geomorphologists observe physical phenomena over short time spans, with the ultimate goal being prediction of what will occur if factors controlling process (climate, tectonics, human activity) are altered. Thus, the study of process often concentrates on the mechanics of agents producing erosion or deposition rather than the landforms which they create.

The dichotomy of approach in geomorphology is not new. It developed in North America at the same time that seminal observations were leading to the creation of geomorphology as a distinct science. Such multiplicity of purpose is not necessarily bad, especially when it leads to cutting-edge discoveries. It can be argued, however, that diversity of goal and approach, and

Geological Society of America, v. 100, p. 160–171, 3 figs., 3 tables, February 1988.

the vagueness of paradigms, have given geomorphology the undeserved reputation of having less scientific rigor than other segments of geoscience. Geomorphologists know that this is not true, but many in other branches of geology hold lingering doubts and are confused as to what it is that geomorphologists actually do.

This paper is an attempt to trace how the primary emphasis in geomorphic thinking has changed since the inception of the Geological Society of America and, wherever possible, stress how the Society has functioned as a vehicle for that change. It has not been written for geomorphologists; they know what they do and why they do it. Instead, the paper is directed toward other geoscientists who may perpetuate misconceptions concerning geomorphology because they cling to ideas which lend themselves to easy pedagogy but are possibly wrong and certainly do not reflect geomorphic science as it is today or what it will be like in the future.

APPROACHES OF LANDSCAPE ANALYSIS

The Geomorphology of G. K. Gilbert

One of the most honored American geologists, G. K. Gilbert (1843–1918), is generally recognized as our first truly great geomorphologist. During his illustrious career, Gilbert provided us with the scientific approach that has become the basis for much of process-oriented geomorphology being employed today. Gilbert was an amalgamation of geologist, physicist, and engineer, who viewed all scientific problems as a balance between force and resistance; the force being applied by some geological agent and resistance stemming from the geological framework. His treatment of landforms was no different, and his study of the Henry Mountains of Utah (Gilbert, 1877) stands as a classic example of how he examined landscapes. His goal was not to place landforms in a developmental sequence but to understand how processes functioned to create the features spread before him. Gilbert visualized geomorphology as a physical science and accepted the principles of thermodynamics as his basic paradigm (Baker and Pyne, 1978). For example, Gilbert's landscapes are analogous to chemical systems in equilibrium. When equilibrium is upset by a change in thermodynamic variables, the systemic response will be to establish a new but different equilibrium condition adjusted to the changed values of the controlling variables. Similarly, in his approach, rivers in equilibrium must adjust to changes in load or discharge to re-establish equilibrium with new values of those independent variables. An increase in load or particle size would require greater erosional and transportational energy, which Gilbert believed would be provided by a change in declivity (channel slope). This, of course, was the first expression of grade in geomorphic analysis and represents the quintessence of Gilbert's equilibrium approach in all aspects of his varied geological work.

Gilbert's preoccupation with equilibrium also filtered into his thinking about the relationship between process and landform; in fact, he postulated physical "laws" to explain topography. His laws of structure (geology), divides, and declivities stemmed from his belief that the interaction of each determined the landscape character. Terms such as "dynamic adjustment" and "balanced condition" permeate his writing and illustrate his perception that landforms reflect some unique balance between process and geology. For example, he stated (Gilbert, 1877, p. 115–116):

> Erosion is most rapid where the resistance is least; and hence as the soft rocks are worn away the hard are left prominent. The differentiation continues until an equilibrium is reached through the law of declivities. When the ratio of erosive action as dependent on declivities becomes equal to the ratio of resistances as dependent on rock character, there is equality of action.

In simple terms, this statement suggests that landforms and landscapes reflect a continuing equilibrium as long as erosive action or resistance does not change. When those variables are changed, the landscape character will be brought into a new equilibrium configuration by a change in the slope.

Gilbert's perception of time as a factor in geomorphology followed logically from the scientific approach he espoused. He resisted any concept that required continuous unidirectional change (Pyne, 1980; Baker and Pyne, 1978), and therefore time was simply a framework within which events occurred that upset the condition of equilibrium. In that sense, Gilbert's time was not an independent variable of landscape form, nor could time ever obliterate the influence of geology at any scale. For example, he stated (Gilbert, 1877, p. 115):

> It is evident that if steep slopes are worn more rapidly than gentle, the tendency is to abolish all differences of slope and produce uniformity. The law of uniform slope thus opposes diversity of topography, and if not complemented by other laws, would reduce all drainage basins to plains. But in reality it is never free to work out its full results; for it demands a uniformity of conditions which nowhere exists. Only a water sheet of uniform depth, flowing over a surface of homogeneous material would suffice; and every inequality of water depth or of rock texture produces a corresponding inequality of slope and diversity of form.

Gilbert's work on Lake Bonneville (Gilbert, 1890) demonstrated that he considered geomorphic history as nothing more than episodes of equilibrium being periodically disrupted by geologic events (Baker and Pyne, 1978). Even though he suggests that one goal of the Bonneville study was local Pleistocene history, his main emphasis was placed on shoreline processes. After he understood what features resulted from those processes, the landforms (for example, bars, strandlines) were used to reconstruct the positions of various lake levels. Thus, his "geological history" relied totally on the study of physical processes and certainly did not consist of the multidisciplinary approaches used by modern Quaternary geologists. Furthermore, his precise leveling of shoreline features led him to important insights about crustal rebound associated with the removal of stress as Lake Bonneville disappeared. Gilbert, therefore, considered landform history primarily because it preserved the record of both endogenic and exogenic processes. He had little interest in determing when, or in what sequence, events occurred, and thus his work often strayed from our usual understanding of historical geology.

In 1905, Gilbert was given the task of assessing the effects of increased sediment loads caused by hydraulic mining in the Sierra Nevada and determining whether such mining could be reinitiated without irreversible damage to the environments in downstream areas. The results were published in two studies which once again demonstrated his focus on process. In contrast to the Henry Mountains and Lake Bonneville reports, however, the study of transportation by running water (Gilbert, 1914) and his analysis of effects of hydraulic mining on river systems (Gilbert, 1917) also revealed Gilbert as a pragmatist. His approach was not a total concentration of effort in the Sierra foothills. Instead, Gilbert chose to examine the mechanics of stream transportation in flumes constructed on the Berkeley campus specifically to provide insight to the Sierra problem. Thus, he intuitively understood the difference between the process analysis that he provided for stream behavior in the Henry Mountains and the detailed, quantitative understanding of fluvial action required to solve the question of hydraulic mining. Actually, his flume studies were a frustration to him because they did not result in clear-cut laws that could be expressed in precise mathematical terms. Instead, he was the first geologist to realize that relationships between variables involved in the phenomenon of sediment transportation were so complex that they defied deterministic solution.

A discussion of Gilbert's complete analysis of the hydraulic mining problem is beyond the scope of this discussion. Suffice it to say that it stands as a primer for all environmental geologists and engineers. It clearly contains all of the basic principles of modern process geomorphology, including systems analysis, thresholds, complex response, and process links. It therefore stands as a testament to the premise that physical aspects of environmental geology are to a large extent applied process geomorphology.

The Davisian Model

Gilbert was highly honored by both national and international scientific societies; in fact, he alone served twice as president of the Geological Society of America. It seems remarkable then that Gilbert's solid scientific base for physical geomorphology was somehow pushed aside by a historical approach which proved to be a more fashionable mode of landscape analysis. The fledgling Geological Society of America had its beginnings at about the same time that William Morris Davis (1850–1934) was formulating his grand scheme of landscape evolution.

Davis rose quickly from a rather insecure position as an instructor of physical geography at Harvard University to become a world-renowned physiographer. It is ironic that Davis had no formal training in physical geography, but, in fact, earned degrees in geology and mining engineering, and his early publications were devoted more to meteorology than either geology or geography (Chorley and others, 1973). It is fair to say that the vehicle of his ascent to world prominence was one great idea called the "geographical cycle" or the cycle of erosion. This theory of landscape evolution was introduced in his early papers (Davis, 1889, 1890, 1899a) and was woven into his writings in one form or another throughout his life.

As most geologists know, Davis' ideal cycle of erosion begins with rapid uplift of a landmass and is followed by a long period of tectonic quiescence during which the landscape progresses through a sequence of stages which he called "youth, maturity, and old age." As a region evolved from one stage into another, its characteristic landforms also changed. When nearly all of the original mass had been removed by erosion, what remained had the form of a featureless plain, called a peneplain. The key point here is that the properties of landforms were considered to be diagnostic of stage and, therefore, relative time. Thus, Davis thought of landforms in the same sense as strata; that is, both are tools in the analysis of geologic history.

A region in a youthful stage is characterized by a small number of poorly integrated streams which are actively downcutting in rather narrow valleys, and the topography is dominated by high, broad divides. As the master stream nears its base level and drainage lines increase into well-defined networks, the region enters the mature stage. Main streams begin to erode laterally, and flood plains develop in their lower reaches. The increased numbers of rivers tend to dissect the region so that divides become narrow and most of the area is occupied by valley-side slopes. Rivers become graded, and the entire system is coordinated to erode and transport debris efficiently from the area. Gradually the valleys widen by lateral erosion, slope angles decrease, and relief is lowered until the region evolves into old age. In that stage, the original landmass has been virtually consumed. A few very large rivers separated by low, broad divides flow sluggishly across the region. By applying this conceptual scheme, any landscape could be described by its physiographic characteristics and, more important, placed in a relative time framework on the basis of its developmental stage.

The concept of the cycle of erosion is a masterpiece of synthesis. As Davis readily admitted, the ideas needed to formulate a cyclic thesis were already available to him in the writings of Powell, Gilbert, Dutton, and T. C. Chamberlain. Davis viewed his own contribution as that of integrating diffuse landforms into a model for the evolution of regional landscapes. He stated in a 1922 paper published in the *Geological Society of America Bulletin* (Davis, 1922, p. 594–595):

The essence and object of the scheme of the cycle does not lie in its terminology, but in its capacity to set forth the reasonableness of land forms and to replace the arbitrary, empirical methods of description formerly in universal use, by a rational, explanatory method in accord with the evolutionary philosophy of the modern era. All the older descriptions of land forms treated each form by and for itself. The idea that certain groups of forms may be arranged in a genetic sequence based upon structure, process and stage, and the further idea that the different form-elements of a given structural mass are at each stage of its physiographic evolution systematically related to one another, were not then recognized.

The intellectual climate at the end of the 19th century was perfect for a concept like the cycle of erosion. The idea of biological evolution was permeating all disciplines, and "time" became almost synonymous with continuous and irreversible change. Davis's model was in vogue and, as seen above, he made the analogy with organic evolution in his presentations of the cycle.

Although North American geomorphologists generally accepted the cyclic model, the concept was criticized from its very inception. In fact, it soon became apparent that the ideal model envisioned by Davis was extremely difficult to prove and often led to applications which were based on rather tenuous assumptions. Nevertheless, the cyclic concept spurred a wave of studies using an approach known as denudation chronology (Beckinsale and Chorley, 1968): the practice of reconstructing the evolution of landforms. Denudation chronology flourished between 1900 and 1939, and most geomorphologists in this era were dedicated to determining how many cycles could be documented in any area and when they existed.

In denudation chronology, the present landscape is explained by deducing a sequence of evolutionary changes in topography until the landscape achieves its present form (Chorley and others, 1973); that is, the past becomes the key to the present. The basic assumption in this approach is that the configuration of the topography is known at the start of any cycle; this is a spurious assumption at best, because erosion in one cycle would remove parts of the landscape developed in any previous cycle. In fact, the completion of any cycle would remove all vestiges of prior topography, and thus gaps in the evolutionary sequence were inevitable. This left room for creative thinking rather than direct observation, and geological events, including peneplanation, were imagined as occurring within the intervals of observational gaps in order to explain the denudational history (see Johnson, 1931).

One of the more appealing aspects of Davisian geomorphology to geologists was the concept of the peneplain, even though Davis readily admitted that we have no modern peneplains to study (Davis, 1899b, p. 232–233);

While it may be true that there are today no extensive peneplains still standing close to the sea level with respect to which they were denuded, the examples given in this and in the preceding section seem to me to prove that the Earth contains many approximations to the peneplain condition, inasmuch as it preserves some excellent fossil peneplains; and that the stratigraphic as well as the physiographic method of investigation yields abundant and accordant evidence of their occurrence.

Davis never defined a peneplain in succinct terms but rather chose to hint at its meaning in his early papers treating the cycle of erosion. What he gave us were statements such as "erosional plain near base level," or "a broad surface of very gentle undulation," or "a nearly featureless plain." Because of this loose definition, every geologist was free to establish his own meaning for the term. Unfortunately, many failed to recognize that all erosion surfaces are not peneplains even though Davis clearly intended that certain properties must exist before an erosion surface can be called a peneplain.

These properties are as follows: (1) the feature represents the ultimate stage of a Davisian cycle and must be formed as a result of one totally complete cycle, (2) it is subcontinental in size, (3) it is a totally erosional landform developed primarily by fluvial action with ancillary weathering and mass wasting, (4) it truncates all rock types of differing resistance, (5) the surface is cut near base level, (6) the surface has extremely low relief, and (7) erosion forming the peneplain is entirely subaerial.

In 1922, Davis made it clear that the base level he was talking about was sea level and suggested that a peneplain surface may stand well above sea level in the interior parts of the landscape. This was presumably included to strengthen the case that the high plains area of Montana (where he claims to have first developed the idea) was a peneplain graded to sea level.

It is important to note at this juncture that Davis' ideal model represents a different type of historical geomorphology because his stages of development are not Gilbert-type punctuated events of disequilibrium but passive transitions with time as the landscape progresses from youth to peneplain without interruption by uplift, base-level lowering, or significant climate change. This ideal case represents a truly closed system where mass and energy provided during the uplift that initiated the cycle were gradually and irreversibly lost as the cycle ran its course. In the end, little or no mass and energy remained as the system approached total entropy.

Actually, even denudation chronologists knew that factors controlling landforms could rarely, if ever, remain constant long enough for any cycle to be totally complete. In fact, all geologists recognized that certain landforms were out of phase with the present cycle; that is, landscapes are palimpsets of forms presumed to be remnants of different incomplete cycles. The evidence cited for incomplete cycles was erosion surfaces of limited areal extent, commonly referred to as "partial peneplains." The preservation of such features suggested that cycles were not free to go to completion, but were rejuvenated (made young) by uplift or eustatic changes of base level. Thus, any landscape could have youthful characteristics superimposed on areas containing mature or old-age characteristics. The result was interpretive confusion because denudation chronologists were trying to fit Gilbert's disequilibrium events into a model that was not suited to accommodate such events, and therefore much of Davis' professional life was devoted to explaining exceptions to or interruptions of the ideal case. In fact, in defense of the concept he gave us the following remarkable statement about a model that served as the cornerstone of geomorphic thought for nearly a half century (Davis, 1905, p. 152):

> the scheme of the cycle is not meant to include any actual examples at all, because it is by intention a scheme of the imagination and not a matter for observation.

My purpose for this extensive discussion of Davisian geomorphology is simply an attempt to convince the geological community that the cyclic concept may not be valid and that peneplains are indeed a rare commodity if, in fact, they exist or ever existed. Davis, perhaps unintentionally, gave us stringent guidelines as to what constitutes a true peneplain. There is no question that widespread erosion surfaces exist, but identifying these as peneplains or partial peneplains immediately endorses the cyclic concept when, in fact, such features may not meet the requisites or may not have evolved through the distinct stages envisioned by Davis (Hack, 1960; Flint, 1963; Flemal, 1971). Every aspect of the cyclic theory has been challenged (see Flemal, 1971 for review), and almost every major proposed peneplain has been explained in a different way (Denny, 1956; Hack, 1960; Steven, 1968). Thus, it has been clear to geomorphologists for decades that the cyclic concept is fraught with difficulties and is impossible to prove.

Other Models of Landscape Development

In retrospect, the opposition to Davis' theory of landscape evolution and its eventual demise created a large philosophical vacuum in geomorphology with no comprehensive model that might serve as a fundamental guideline for geomorphic research. As a result, other theories of landscape development were suggested. In general, the most prominent of these accepted Davis' basic idea that landform development was cyclic and controlled by structure, process, and time. They deviated from Davis, however, on whether slope declivity changed with time in the manner suggested by the geographical cycle, and on the importance of endogenic processes. Walther Penck (1924, 1953) stressed how different rates of crustal movement relative to erosion rates would produce characteristic slope segments and forms. Regional doming, for example, would theoretically begin with accelerating uplift rates that resulted in steepened slopes and a general convexity of slope forms (Chorley and others, 1984). With time, however, uplift rates would decline and rapid parallel retreat of the upper-slope segments would destroy the initial convexity. Thus, a cycle of landform development was directly related to the change in uplift rates through time. The Penck model received little support in the United States, mainly because it is based on speculative tectonics, and because Penck's premature death in 1923 left his classic work (Penck, 1924) incomplete and very difficult to understand. In fact, a correct English translation did not appear until 1953.

L. C. King (1953, 1962) also supported the concept of evolutionary development of landscapes; however, he rejected Davis' argument that slopes progressively decline as a landscape proceeds through its various evolutionary stages, and he believed that the "normal" climate was semiarid. King viewed landscapes as resulting from parallel slope retreat associated with ever-expanding pediments. His pediplanation cycle therefore begins with rapid diastrophism which is followed by a long period of tectonic quiescence. During tectonic inactivity, the landscape changes through stages to an end product called a "pediplain." The basic model is very similar to Davis' cycle. It differs in that whereas Davis' landscape wears down, King's landscape wears back by headward scarp erosion. His cycle, therefore, involves progressive integration of pediments into an expansive surface (Higgins, 1976) that is covered by debris and that has a subdued configuration, broken only by residual knobs of uneroded bedrock.

The one theory of landscape development that attained considerable interest was given to us by Hack (1960, 1976). Hack suggested that in the Appalachian region of the eastern United States the present topography manifests an equilibrium condition between structure and process. More important, he proposed the hypothesis that the evolutionary scheme in Davis' cycle was not necessary to explain the landscape. Instead, it is equally likely that the components of the topography were rapidly adjusted to one another and thereafter maintained in a condition of dynamic equilibrium. After this equilibrium was established, and assuming that erosional energy remained constant, all elements of the landscape downwasted at the same rate. This preserved the topographic configuration through time.

In Hack's dynamic equilibrium model, (1) relief and form are explained in spatial rather than temporal terms, (2) landscapes become essentially time independent, and (3) the interaction between process and structure (lithology) dominates the system. Hack envisioned the "maturely dissected peneplain" of Davis and denudation chronologists as representing the equilibrium form in the central Appalachians rather than simply being a stage along the way in the progression toward Davisian old age. This model is a significant diversion from those of Davis, Penck, and King because it has no cyclic implication nor does it assume any conditions of continuous or changing uplift. In fact, Hack suggests the possibility that the present landscape may have been formed or inherited through one continuous period of dying orogeny or isostatic adjustment after the formation of the Appalachian Mountains. This interpretation, of course, removes the need for multicyclic geomorphic history in the Appalachian region, and with it all the peneplains and partial peneplains.

Hack believed that his dynamic-equilibrium model returned us to Gilbert's perception of landscape analysis. He is certainly partly correct because Gilbert viewed landscapes as being neatly adjusted between rocks and process. This is especially evident in Gilbert's Henry Mountains work where he viewed topographic form as being in a "dynamic adjustment." The missing constituent, and the difference between Hack and Gilbert, however, is that Gilbert never thought of landscape development over the time spans being considered by Hack, Davis, or any other progenitor of general theories of landscape development. In fact, Gilbert had very little interest in long-term landscape development, an aspect of his thinking that greatly puzzled Davis. Instead, as stated above, Gilbert was more concerned about how landforms changed in response to punctuated episodes of disequilibrium, and his geomorphic history was a linkage of these spasms rather than the long-term evolution or dynamic equilibrium of the landscape. He never seemed to consider what the landscape would be like given infinite amounts of time for erosion to work out its ultimate end result.

THE TIME FRAMEWORK

In light of the above, it appears that one major difference between Gilbert and any general theory of landscape development is that each model considers landforms over a time scale that is significantly longer than that being used in the approach of Gilbert and that adopted by most present geomorphologists. The implication here is that geomorphic analyses must be conditioned by the amount of time involved. Schumm and Lichty (1965) first suggested that time intervals used in geomorphic studies can be divided into three categories called cyclic, graded, and steady. Cyclic time involves intervals of time that are truly geologic in magnitude (Fig. 1). During cyclic time spans, we can expect exponential change in landscape components such as stream gradients, valley-floor altitudes, and perhaps slope angles. Thus, this temporal scale is applied best to analyses of regional landscape denudation, and it represents the time involved in any general theory of landscape development whether it is proposed as being evolutionary or as downwasting in dynamic equilibrium. The importance here is that cyclic time can be subdivided into the shorter graded and steady time periods which are not suited for theories of long-term landscape development but are appropriate for different types of geomorphic investigations, including Gilbert's episodes of disequilibrium.

During graded time, the character of landscape components will attain some average condition. Within this time interval, values of

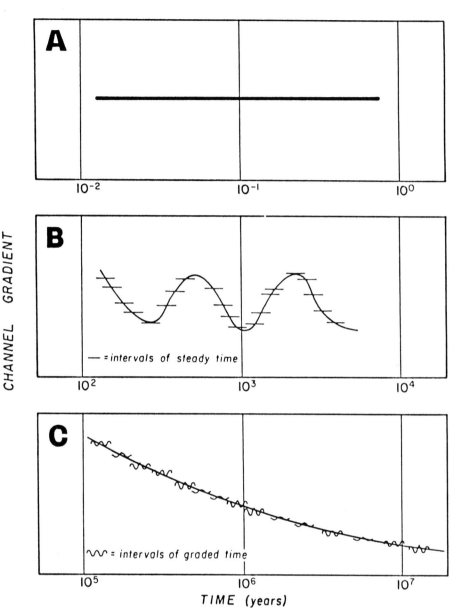

Figure 1. Diagram representing the time framework used in geomorphic studies. (A) Steady time. No change in the geomorphic variable (channel gradient) over short time interval. (B) Graded time. Channel gradient varies around an average value. Time interval is longer. (C) Cyclic time. Time interval is very long. Channel gradient experiences a continuous and irreversible decline, even though many episodes of graded time fluctuations occur within the cyclic time interval (after Ritter, 1986).

any parameter will fluctuate above or below the mean (Fig. 1). Thus, climate change or diastrophism will temporarily displace landform parameters away from their average condition, but these events will initiate responses within the landscape system that tend to offset the effect generated by the change. The landscape components, therefore, will simply vary around the average condition which is considered to represent a form of equilibrium. A prime example might be the concept of a graded river as expressed by Mackin (1948) and refined by Knox (1976).

Steady time is so short (days to months) that no change is detectable in the landscape variable during this interval (Fig. 1). Studies such as the hydraulic geometry of rivers (Leopold and Maddock, 1953) are reflecting events occurring during steady time, but they are not applicable for analyses over longer time intervals. Thus, the static condition demonstrated by hydraulic geometry cannot be used as evidence that a river is in equilibrium over a graded or cyclic time interval.

The above discussion is meant to emphasize that the concept of equilibrium is clearly dependent on the time factor. For example, dynamic equilibrium as expressed by Hack suggests that landscapes simply downwaste with time while maintaining an unchanging configuration. The major assumption, of course, is that factors controlling landforms (process and structure) remain constant in rate and character. In reality (and certainly recognized by Hack), external factors that affect process (climate, tectonics) are subject to repeated alterations. Landforms, therefore, assuredly change with time, and the true meaning of equilibrium depends on the time interval over which the balanced condition is being considered. Over cyclic time, for example, landscapes might continuously lose mass and energy as they presumably approach, but perhaps never attain, some equilibrium condition. During this march to total entropy, however, numerous graded-time alternations of equilibrium and disequilibrium may occur which mask the type of landform development occurring over cyclic time. It is even possible, as suggested by Schumm (1976), that landscape components such as stream gradients and valley floor altitudes do not progressively decrease during cyclic time. Instead, functions inherent in fluvial systems prevent their reduction with time, and the geomorphically normal denudation history is composed of brief episodes of instability and incision separated by long periods of stability (dynamic metastable equilibrium; Schumm, 1977). Notwithstanding these arguments, intuition may still suggest that given enough time most (but not necessarily all) landscapes will progress to the lowest possible relief, and some regions could even attain an eroded surface similar in form to a Davisian peneplain, although it may be impossible to totally remove interfluve hills (Horton, 1945). Landscapes may attain this ultimate form in different ways, but in a closed system seeking complete entropy over vast time spans any region might reach a greatly subdued topographic condition.

It is incorrect, however, to suggest that all landscapes will be reduced to low relief given enough time. Considerable evidence exists to suggest that large portions of the Earth's surface are very old and have changed little since middle Tertiary or even before (King, 1950; Twidale, 1976, 1985; Twidale and others, 1976; Ollier, 1979; Sevon, 1985; Bloom, 1985). Some of these paleosurfaces have been exhumed during the Cenozoic. Some of great antiquity (see Twidale, 1976), however, have apparently survived long exposures to weathering and erosion and, although slightly modified, have retained their original configuration throughout cyclic-time spans.

Twidale (1976) suggests that the survival of paleosurfaces, essentially intact with respect to relief and position of ridges and valleys, is probably a function of several factors that create extremely low rates of denudation. First, spatial variations of diastrophism may place paleosurfaces in the cores of continents or mountain ranges where the land surface is less vulnerable to erosion. Second, river erosion within any large catchment may be localized to the extent that upland areas or ridges are significantly less affected by erosion than are the valley floors. Third, where slope retreat is the dominant process of landscape development, upland paleosurfaces bounded by abrupt scarps become protected. This occurs because the rate of scarp retreat is slowed when coarse debris derived from the bluffs collects in gullies crossing the slopes. Because gullying is primarily responsible for regrading slopes and continued scarp retreat, the coarse debris effectively protects channels from further gully erosion. Therefore, the system becomes self regulating, especially where coarse debris survives for extended periods under arid-climate weathering. Thus, in contrast to closed systems seeking total entropy, some parts of the Earth's surface have been essentially unaltered during cyclic time, even though the systems are open and commonly subjected to repeated uplift. This happens because unique geologic controls isolate and protect these areas from rapid denudation.

Accepting the probability that some relatively high geomorphic surfaces are very old, the single remaining difficulty with models of long-term landscape development is demonstrating how landforms change during shorter intervals of disequilibrium set within the cyclic-time span. Topography as we see it may be dominated by adjustments to the graded-time episodes of disequilibrium envisioned by Gilbert. Indeed, these perturbations may only be a "blip" in terms of cyclic geomorphology. We must entertain the possibility that geomorphic history is nothing more than a continuum of "blips," however, and that they occur with enough frequency to rule out prolonged and progressive evolution of a landscape. In light of this, the suggestion can be made that no model of long-term landscape development can ever be totally satisfactory to all geomorphologists because evidence needed to prove or disprove such models is lost when landscapes adjust during or following periods of disequilibrium. The search for evidence of cyclic-time landscape development, whether in equilibrium or in stages of evolution, may therefore be an exercise in futility. No one can correctly piece together the effects of graded-time events into a model of landscape development considered over cyclic time scales. It may be that the imprint of tectonism and climatic change creates such diversity in the landscape that placing all of these elements into one comprehensive theory of landscape development is extremely difficult, if not impossible (Higgins, 1976).

THE EMPHASIS IN MODERN GEOMORPHOLOGY

It now seems certain that the philosophical vacuum created by the disenchantment with Davisian geomorphology was not replaced by another general theory of landscape development but instead was filled by a dramatic shift of emphasis in geomorphic analyses. This change in direction represents the real return to Gilbert, because geomorphologists turned away from discussions concerning the viability of various landscape-development models and espoused Gilbert's preoccupation with process and how processes and landforms adjust to bursts of disequilibrium within graded-time intervals. The philosophical shift did not occur because the Davisian or any other model was wrong. Instead, the approach couched in the models was simply not germane to the concepts and questions associated with the new wave of geomorphology (Baker and Pyne, 1978; Sevon and others, 1983).

The above discussion is not meant to imply that studies of geological processes were totally abandoned during the Davisian era; on the contrary, numerous process studies were conducted in the first half of this century. Examples can be cited for the analysis of river mechanics and channel form (Rubey, 1938; Hjulstrom, 1939; Friedkin, 1945), wind processes (Bagnold, 1941), glacial mechanics (Perutz, 1950), slope processes (Horton, 1945; Terzaghi, 1950), coastal and beach processes (Keulegan and Krumbein, 1949; Bascom, 1951), and periglacial mechanics (Taber, 1930; Grawe, 1946). Compared to the sheer volume of work concerning cyclic geomorphology, however, the research dedicated to process analysis was, at best, minor and at worst, a geological disgrace. In fact, much of the process-oriented research was left by default to civil engineers, and very little was conducted by geologists. What geomorphologists lacked during the period between the waning influence of Davis and the rise of process geomorphology was a clear reason to change emphasis and a philosophical framework or paradigm that demonstrated the common bond between the variety of process studies being conducted.

The philosophical framework of process geomorphology was not established precipitously; rather, the basic premises involved in this type of geomorphology were gradually put in place over a period of several decades. One reason for the shift to process geomorphology may have been that many geologists serving in the armed forces during World War II were placed in a milieu of pragmatic science (Sevon and others, 1983; Tinkler, 1985). Questions being asked about the Earth's surface during the war effort were more concerned with how processes worked rather than when landforms developed. Geomorphologists conducting this type of research continued along those lines after the war. This thrust, coupled with the frustration that research based on landscape-development models resulted in interpretations that were always open to debate (Bishop, 1980; Tinkler, 1985) and had little utility in geoscience, may have provided the impetus for the change in geomorphic emphasis.

Studies documenting the establishment of process geomorphology as a distinct subdiscipline are too numerous to mention in this short discussion. Table 1 is presented as my own interpretation of some critical steps taken in developing the framework of process-oriented geomorphology. It is not intended to be all inclusive, but to show, through selected examples, how the conceptual basis of process geomorphology evolved during a three-decade period. For the early stages of this transition, two papers (Horton, 1945; Strahler, 1952) are noteworthy in this review because they were published in the *Geological Society of America Bulletin.*

In 1945, R. E. Horton used hydrophysical principles to examine how infiltration, runoff, and slope-surface characteristics were combined to initiate stream channels and ultimately lead to an orderly drainage network that was amenable to statistical analysis. Therefore, this paper appears to be the initial attempt at quantitative geomorphology. Although his analysis of slope hydrology was later shown to be inappropriate for humid climate situations, Horton's work probably represents the first clear attempt after Gilbert to treat regional phenomena as part of physical geomorphology. In fact, Horton suggested that landscapes cannot proceed to a Davisian peneplain. Instead, hydrophysical laws require a definite endpoint to the development of streams and valleys that demands the survival of interfluve hills or divides in a regional landscape.

A. N. Strahler (1952) appealed to geologists to adopt a dynamic basis for geomorphology. In Strahler's dynamic model, processes are analagous to stresses acting on different geological materials to produce characteristic varieties of strain or failure. The failures manifest responses to stress and ultimately express the mechanics of weathering, erosion, transportation, and deposition as accommodations between stress and strain. Furthermore, he suggested that such a dynamic approach requires that processes be analyzed in terms of open systems which tend to achieve a steady-state condition and may be self-regulating. These systems and process mechanics should be expressable in mathematical models derived by both rational and empirical analyses of observed data. Strahler's work clearly demonstrates the dramatic shift in geomorphology from a historical to a physical science. It represents the first real attempt to establish a philosophical framework for process studies, and therefore his paper can reasonably be cited as the beginning of process geomorphology as a distinct subdiscipline.

The analysis of processes has now matured to the point where we are again asking the critical questions about how processes determine the specific character of landforms. In fact, many geomorphologists are looking at the relationship between landforms and processes on larger scales than the micro-spatial studies that typified our early attempts to understand process. For example, the realization that some landforms reflect an accommodation between process and structural control (and also tectonic history) is the basis of a relatively new subdiscipline, called "tectonic geomorphology," that is already entrenched in modern geomorphic practice and stands on the brink of enormous geologic utility (Bull and McFadden, 1977; Keller and others, 1982; Colman and Watson, 1983; Bull, 1984). Carried to an even greater scale, the expanding comprehension of process geomorphology has helped to provide a firm scientific basis for planetary geology. Much of the geological interpretation of extraterrestrial bodies relies heavily on analyses of surficial landscapes. It follows, therefore, that reasonable interpretations demand a detailed understanding of the genetic relationship between process and form.

Space technology has also worked in the opposite direction by giving us a perception of the process/form relationships on Earth that may not be easily derived by normal land-based investigations. The ongoing NASA program devoted to analysis of Mars has generated an enormous volume of research that revealed insights about a variety of Earth processes and landforms, including wind action, catastrophic flooding, thermokarst, sapping, mass movements, drainage and valley evolution, and volcanic landscapes (for reviews see Sharp, 1980; Baker, 1981, 1982). The multitude of papers resulting from this program established a solid basis for comparative process analyses. In addi-

TABLE 1. SELECTED PAPERS THAT DEMONSTRATE THE DEVELOPMENT OF A FRAMEWORK FOR PROCESS-ORIENTED FLUVIAL GEOMORPHOLOGY

Author, year	Significance
Horton, 1945	Use of hydrophysical laws to explain origin and development of river networks on a regional basis. Statistical analysis of physical parameters.
Strahler, 1952	Suggestion that geomorphic processes can be treated as stress-strain phenomena operating in open systems. First attempt to provide a physical paradigm for process geomorphology.
Leopold and Maddock, 1953	Analysis of hydraulic geometry and the demonstration of equilibrium between fluvial parameters.
Chorley, 1962	Argues the importance of the open-system approach in geomorphology because it emphasizes tendency toward adjustment between process and form. Stresses utility of open-system analysis for nonprogressive landscape changes with time.
Schumm and Lichty, 1965	Introduces a time framework for geomorphic analyses and stresses the importance of time in the study of fluvial behavior.
Schumm, 1969	Attempt to demonstrate how rivers might respond to changes in discharge and load.
Schumm, 1973	Introduction of the threshold concept and complex response in fluvial systems.

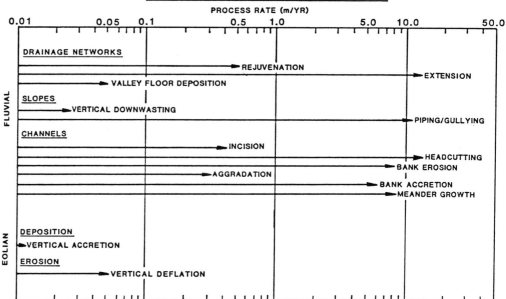

Figure 2. Magnitudes of geomorphic processes which pose a hazard to the long-term stability of uranium tailings disposal sites in northwestern New Mexico. Processes having the highest rates pose the greatest threat (from Wells and Gardner, 1985).

tion, Landsat photography and other spaceborne techniques allowed us to directly examine the Earth's surface and make cogent interpretations about processes and landforms on a regional basis.

Geomorphology today, therefore, is attempting to analyze processes over a wide spectrum of temporal and spatial scales and to utilize a variety of investigative techniques. It is incorrect, however, to think that the shift toward process geomorphology means that historical geomorphology has been abandoned in the discipline as it now stands. Nothing could be farther from the truth. Landforms do reflect the effects of diastrophism and climate and, more important, episodic bursts of tectonic activity and climate change. In fact, much of the topography as we see it today may be dominated by adjustments to Quaternary climate change and/or diastrophism. These events represent important ingredients of geological history and, as such, deservedly receive the attention of geomorphologists. In conjunction with soils studies, palynology, and dating techniques, landform characteristics are still being used to interpret Pleistocene and Holocene history, in addition to increasing our understanding of how process mechanics change in response to tectonism or climatic alteration. It seems clear, however, that the primary goal of historical geomorphology has changed from one of theoretical landscape evolution over cyclic time to one that is similar to Gilbert's analyses of punctuated events of disequilibrium.

A LOOK IN THE CRYSTAL BALL

Assuming that we understand what geomorphology is today and how it became that way, it is fair to ask where the discipline will be going in the future? Each geomorphologist will view that journey in a different way. In fact, no individual possesses a blueprint for this field, nor can anyone begin to espouse all of the divergent opinions that exist concerning future developments in geomorphology. The following, therefore, merely represents my own, admittedly biased, perception of what lies ahead.

Tinkler (1985) presented a convincing argument that the rise of computer technology has brought us to the brink of a new wave of theorizing about landscapes. The rationale, of course, is that the computer can assume all possible environmental controls for landscape systems and reveal any discontinuities in landscape development that exist as a function of those controls. An array of simulation models can therefore be produced that might be closer to the truth about landform development than those derived from short-term data or those conceived in the minds of geomorphologists who preceded sophisticated electronics. I agree with Tinkler that such computer-generated models are inevitable. The tools needed to create the models are available, and it is unrealistic to expect that they will not be used in such an endeavor. It is important, however, that geomorphologists use computers to develop insights about significant problems and concerns (for example see Craig, in press). Computer-generated theories about cyclic-time landscape evolution are still theories about landscape evolution, and emphasis on that goal will probably leave us as scientifically sterile as we were in the Davisian era. The next generation of geomorphologists thus will probably utilize the power of computer technology to address different types of landscape problems and to conduct different types of geomorphic analyses.

If geomorphology is to remain as a viable part of geoscience, it is mandatory that future research in this discipline provide geologists and geophysicists with useful information. That aim can be accomplished in several ways. First, geologists should become more cognizant that the traits of many large surficial systems are direct reflections of plate tectonics; that is, the geomorphic framework and process mechanics are conditioned by plate-tectonic style. Each particular type of plate margin is likely to engender a unique set of surface processes and distinct landforms and geomorphic history (for example, uplift and erosion rates, terrace sequences, regional erosion surfaces, raised and tilted shoreline features). Geomorphic analyses, combined with geophysical and geological data, should therefore be useful in identifying plate-boundary conditions and how the mechanics operating at those boundaries have functioned through time. Research linking geomorphology and plate tectonics has already begun (Judson, 1975; Ollier, 1979; Gilpin and others, 1981; Strecker and others, 1984; Tosdal and others, 1984; Taylor

TABLE 2. EXPECTED CHANGES IN RUNOFF AND SEDIMENT YIELD PRODUCED BY CLIMATE CHANGE, TECTONISM, OR HUMAN ACTIVITY

Climate change							
Variables	a / b	Arid to semiarid / Semiarid to arid	Semiarid to subhumid / Subhumid to semiarid	Subhumid to humid / Humid to subhumid	Nonglacial to glacial	Nonglacial to periglacial	Uniform to seasonal
Runoff	a	+	+	+	+	+	−
	b	−	−	−			
Sediment yield	a	+	−	−	+	+	+
	b	−	+	+			

Tectonics							
	Upstream		On Site		Downstream	
Variables	Up	Down	Up	Down	Up	Down
Runoff	0	0	0	0	0	0
Sediment yield	+	−	+	+	0	0

Land use and vegetation change									
				Dam construction		Gravel mining		Channelization	
Variables	Fire	Increased agriculture	Timbering	Upstream	Downstream	Upstream	Downstream	Upstream	Downstream
Runoff	+	+	+	−	0	0	0	+	0
Sediment yield	+	+	+	−	0	− or +	0	+	0

Note: Symbols (+, −, 0) indicate increase, decrease, or no change, respectively. Data adapted from Schumm and others, 1982.

and others, 1985), but much more can be expected in the future.

Second, geomorphologists should stress the fact that details contained in the stratigraphic record reflect changes of sediment type produced in source areas or, in the case of gaps in the terrestrial record, erosion spurred by base-level decline or diastrophism (Kraus and Middleton, 1987). In either case, it is important to note that evidence of erosional or depositional events preserved in the stratigraphic record is filtered through a geomorphic screen, regardless of the primary climatic or tectonic cause.

Although geologists have always utilized tectonism or climate change to explain changes in

TABLE 3. GEOMORPHIC HAZARDS AND THEIR PROCESS RATES WHICH MAY IMPACT THE STABILITY OF URANIUM TAILINGS DISPOSAL SITES IN NORTHWESTERN NEW MEXICO

Geomorphic hazard	Material property	Rates of geomorphic processes		
		Historic (<200 yr)	Holocene (>200, <10,000 yr)	Pleistocene (>10,000 yr)
Fluvial Activity				
A. Drainage networks				
1. Erosion				
a. rejuvenation	bedrock	..	0.3–0.5 cm/yr	0.01–0.04 cm/yr
b. extension	alluvium	up to 50.0 cm/yr	up to 0.2 cm/yr	..
2. Deposition (valley floor)	alluvium	0.5 cm/yr	0.25–0.30 cm/yr	..
3. Pattern change		Development of drainage networks in valley floors	Increased drainage density	Reduction to watershed size
B. Slopes				
1. Erosion				
a. Vertical downwasting	sandstone/sand alluvium	0.3–2.5 cm/yr	..	..
	mudtone/shale	0.1–0.5 cm/yr	..	..
	bedrock	..	0.3–0.5 cm/yr	..
b. Gullying/piping (lateral extension)	alluvium	up to 10.5 m/yr	..	..
C. Channels				
1. Erosion				
a. Incision	alluvium	6.5–50.0 cm/yr	0.25–0.45 cm/yr	..
b. Headcutting	alluvium	1.0–12.5 m/yr	..	..
c. Erosion	alluvium	1.5–8.0 m/yr	..	..
2. Deposition				
a. Channel floor	..	4.0–30.0 cm/yr	0.025–0.400 cm/yr	..
b. Bank accretion	..	1.5–6.0 m/yr	..	..
3. Pattern Change				
a. Meander growth	alluvium	2.0–8.0 m/yr	..	..
b. Cutoff activity	alluvium	3 per 50 yr	..	..
4. Metamorphosis				
a. Complete pattern change		braided to meandering	..	..
b. Channel width-depth ratio		decrease	decrease	..

Note: adapted from Wells and Gardner, 1985.

sedimentary sequences or to reconstruct paleogeography, too many nagging questions remain about how the geomorphic screens function in response to climate change or tectonism for us to assume that these interpretive leaps are easy or correct. The truth is that sudden changes in a sedimentary sequence cannot be confidently attributed to a specific geological cause until we know more about the intermediate phase of geomorphic response. In fact, abrupt change in the depositional sequence is reflecting thresholds, complex response and episodic behavior within geomorphic systems, and we are only beginning to understand how those phenomena are reflected in a sedimentary sequence (Schumm, 1981). Geomorphologists must therefore provide the information needed to bridge the interpretive gap between what remains in the stratigraphic record and what geological events occurred to produce that record.

Notwithstanding the essential nature of the above, there exist numerous other research avenues that geomorphology might follow. Therefore, rather than attempting to consider each possibility, a brief discussion of one example may epitomize the kinds of activities looming in the future. This potential direction was chosen because it is both theoretical and applied in scope; furthermore, work already accomplished along this line indicates that it contains both a physical and historical component in the research involved. Therefore, it is possible that the geomorphic unity we have been seeking will eventually be found in this or similar types of efforts.

Landform Stability, Thresholds, and Event Effectiveness

Geologists and engineers are increasingly aware that identification of geomorphic hazards over various temporal and spatial scales is an important consideration in environmental planning. Geomorphic hazards can be defined as any change which disrupts landform stability in a manner that produces an adverse effect on living things (Chorley and others, 1984). Because stable landforms are those that are unthreatened by geomorphic hazards (Scheidegger, 1975), identification of hazards is directly linked to prediction of landform stability.

Hazard identification is desirable in a variety of societal concerns such as (1) slope and river channel stability related to highway and bridge construction, (2) drainage network development associated with mined lands reclamation, and (3) storage of hazardous wastes. Each analysis is unique because the potential hazards differ according to the area being considered (regional versus local scale) and the time over which land-

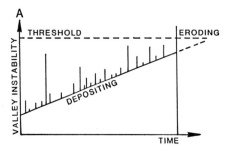
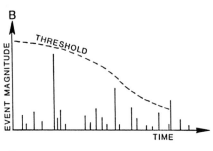
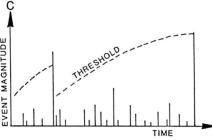

Figure 3. Different perceptions of geomorphic thresholds. (A) Constant threshold values with time (after Schumm, 1973); (B) decreasing threshold values; (C) increasing threshold values. (From Beven, 1981.)

form stability must be maintained. Nonetheless, the predictive methodology is similar in most cases because identifying hazards requires direct application of knowledge concerning geomorphic processes, especially how those processes might change in rate or character when they are affected by tectonism, climate change, or human activity. In fluvial systems, for example, changes in those controlling factors result in altered runoff or sediment yield which are the precursors of threshold-related phases of disequilibrium that require a response in processes or landforms (Table 2). Identifying a hazard, therefore, involves a systematic appraisal of geomorphic processes and their rates operating within the area of concern. Hydrologic regime (flood frequency and magnitude) is usually an important component of the analysis. Data are commonly tabulated to provide the basis for prediction of landform stability (Table 3) and the most likely type of hazard that exists over the time span involved (Fig. 2).

Long-term landform stability (>10,000 yr) is a critical factor in decisions involving safe storage of high-level nuclear wastes and mine tailings. In fact, guidelines for those decisions have already been developed (Schumm and others, 1982; Wells and Gardner, 1985), and continuing interest in the topic is suggested by a workshop at the 1987 Geological Society of America national meeting (Site Characterization for High-Level Nuclear Waste Disposal; R. G. Craig, organizer). It is important to recognize, however, that prediction of stability over millennia cannot be totally based on measurements of modern process rates. Instead, these analyses require a detailed understanding of geomorphic history in the area being considered for a potential storage site. In fact, dating the sequence of landform development during the Holocene and Pleistocene provides the link needed to estimate long-term process rates and the rigor of systemic responses to tectonism and/or climate change.

Identification of hazards is clearly based on the tacit assumption that we know the conditions at which a significant change in process or landform will occur; that is, geomorphic thresholds (Schumm, 1973; Coates and Vitek, 1980) are predictable for all systems and under all temporal and spatial scales. This assumption is simply not valid at present, and therefore a major research effort is needed before we can realize the enormous potential resting in the threshold concept.

Thresholds exist in all geomorphic systems, but we are still uncertain as to which of the many threshold crossings will produce responses that signal landform instability. For example, we might ask whether an extreme or catastrophic precipitation and runoff event will cause a lasting effect on fluvial processes or landforms? Such events normally produce scouring of channels and erosion of banks or valley-side slopes, but we cannot be certain if those threshold crossings portend the onset of a graded-time episode of degradation or landform destruction. Our ability to predict suffers because we do not understand the effectiveness of geomorphic events in terms of the time needed for channels or slopes to recover their original character after a disruptive event (recovery time of Wolman and Gerson, 1978). In some cases, recovery time may be short enough to erase the effect of the threshold crossing before another event of similar magnitude can occur. The system, therefore, will experience no long-term change, as it will revert to its original condition. Longer recovery times may result in the opposite effect. Unfortunately, different components of a drainage basin system are not on the same threshold clock, and therefore the analysis of effectiveness is more complex than it seems. For example, Anderson and Calver (1977) found that channel deepening produced in the Cannon Hill valley of southwest England would probably survive longer than the recurrence interval of the flood that produced the entrenchment; however, channel-side erosion scars that formed in the same event were healed quickly. Clearly, different geomorphic components were disrupted by the same event, but responses to the threshold crossings involved varied with regard to their ephemeral or lasting effect on the landscape.

Effectiveness analyses are further complicated because statistical recurrence intervals of events may be less significant than the actual time interval between events of the same magnitude. In fact, studies are now showing that successive events of similar magnitude may produce different geomorphic responses because the systems had not totally recovered from the first event and were therefore in a different environmental condition when the second event occurred (Newson, 1980; Kochel and others, 1987). The complexities involved prompted Beven (1981) to suggest that geomorphic effectiveness and its control on landform stability is partly dependent on the order in which various events occur and the interarrival time between events of the same magnitude. Furthermore, Beven raised the possibility that threshold limits themselves may not be constant but are subject to change with time (Fig. 3).

Superimposed on the temporal problems surrounding thresholds and effectiveness is the growing awareness that responses in surficial systems may be strongly controlled by spatial characteristics existing within the geomorphic setting. In a recent Presidential address presented at a Geological Society of America Annual Meeting, M. G. Wolman revitalized Gilbert's concept that structure (in the broad sense) is pervasive at all geomorphic scales. In addition, he stressed that structurally controlled units of the Earth's surface have preferred rather than random dimensions and that the interaction between structural boundaries and external processes define these landform scales. A collage of different-sized landforms therefore probably exists within any regional landscape, and these landforms can be analyzed according to their scalar properties.

Spatial analysis lends itself to space technology because large segments of the Earth's surface can be viewed as single structural entities (Hayden, 1985). It also presents a unique opportunity to couple mega-geomorphic characteristics with various plate-tectonic settings which, as suggested above, is a probable topic for future geomorphic research. The influence of structural fabric on thresholds, effectiveness of events, and process mechanics has yet to be determined. Clearly, these relationships must be detailed in future studies before we can hope to confidently predict geomorphic hazards and landform stability.

SUMMARY AND CONCLUSIONS

In this paper, I have attempted to show that multiplicity of approach and purpose is not new in American geomorphology. It stems from the very different perceptions of geomorphology espoused by Gilbert and Davis in the latter part of the 19th century. Both historical and physical geomorphic studies have always been conducted simultaneously, but one approach or the other has dominated geomorphology at different times. The most common historical approach has been to interpret landscapes and their component landforms as evidence for the sequence of geological evolution within large areas. Landscape development models, created as a basis for such historical interpretations, are very difficult to prove or disprove because they encompass cyclic-time intervals. During those intervals, shorter term episodes of tectonism and climate change upset the balance in surficial systems and make it difficult, if not impossible, to know how any landscape developed over the long cyclic period. For this and other reasons, a dramatic shift toward physical geomorphology has occurred in the past three or four decades. It is now clear, therefore, that arguments concerning models of landscape evolution no longer represent the essence of modern geomorphology, nor will they do so in the coming years.

Future developments in geomorphology will probably necessitate a greater unity of purpose than ever before because the questions and problems facing the discipline will require both a physical and historical understanding of geomorphic systems. In addition, it is no longer acceptable for geomorphologists and other geologists to isolate their efforts from one another. Many scientific investigations can benefit from an interdisciplinary approach that fosters an exchange of data and ideas, and the Geological Society of America will continue to be an important vehicle in facilitating those collaborative efforts.

ACKNOWLEDGMENTS

Discussions with R. C. Kochel and J. Miller and constructive reviews by J. Costa, A. Bloom, J. Utgaard, J. Miller, S. Miller, and R. C. Kochel were instrumental in providing focus to the manuscript. Their help is greatly appreciated, and any value found in the paper is largely due to their efforts. I also thank M. G. Wolman and R. G. Craig for sharing thoughts on their unpublished work. Interpretations and all shortcomings found in the paper are my own.

REFERENCES CITED

Anderson, M. G., and Calver, A., 1977, On the persistence of landscape failures formed by a large flood: Institute of British Geographers Transactions, new ser., v. 2, p. 243–254.

Bagnold, R. A., 1941, Physics of blown sand and desert dunes: London, Methuen, 265 p.

Baker, V. R., 1981, The geomorphology of Mars: Progress in Physical Geography, v. 5, p. 453–513.

——— 1982, The channels of Mars: Austin, Texas, University of Texas Press, 198 p.

Baker, V. R., and Pyne, S., 1978, G. K. Gilbert and modern geomorphology: American Journal of Science, v. 278, p. 97–123.

Bascom, W., 1951, The relationship between sand size and beachface slope: American Geophysical Union Transactions, v. 32, p. 866–874.

Beckinsale, R. P., and Chorley, R. J., 1968, History of geomorphology, in Fairbridge, R., ed., Encyclopedia of geomorphology: New York, Reinhold Corporation, p. 410–416.

Beven, K., 1981, The effect of ordering on the geomorphic effectiveness of hydrologic events: International Association of Hydrological Sciences, Publication no. 132, p. 510–526.

Bishop, P., 1980, Popper's Principle of Falsifiability and the irrefutability of the Davisian Cycle: Professional Geographer, v. 32, p. 310–315.

Bloom, A. L., 1985, Andean examples of mega-geomorphic themes, in Hayden, R., ed., Global mega-geomorphology: Washington, D.C., NASA Conference Publication 2312, p. 44–45.

Bull, W. B., 1984, Tectonic geomorphology: Journal of Geological Education, v. 32, p. 310–324.

Bull, W. B., and McFadden, L. D., 1977, Tectonic geomorphology north and south of the Garlock fault, California, in Doehring, D., ed., Geomorphology in arid regions: Binghamton, New York, State University of New York Publications in Geomorphology, p. 115–138.

Chorley, R. J., 1962, Geomorphology and general systems theory: U.S. Geological Survey Professional Paper 500-B, 10 p.

Chorley, R. J., Beckinsale, R. P., and Dunn, A. J., 1973, The history of the study of landforms, Volume 2, The life and work of William Morris Davis: London, Methuen and Co.

Chorley, R. J., Schumm, S. A., and Sugden, D. E., 1984, Geomorphology: London, Methuen, 605 p.

Coates, D. R., and Vitek, J. D., eds., 1980, Thresholds in geomorphology: London, Allen and Unwin, 498 p.

Colman, S., and Watson, K., 1983, Ages estimated from a diffusion model for scarp degradation: Science, v. 221, p. 263–265.

Craig, R. G., in press, Portraits of a Missoula flood: Two dimensional analysis, in Mayer, L., and Nash, D., eds., Catastrophic flooding: London, Allen and Unwin.

Davis, W. M., 1889, The rivers and valleys of Pennsylvania: National Geographic Magazine, v. 1, p. 183–253.

——— 1890, The rivers of northern New Jersey with notes on the classification of rivers in general: National Geographic Magazine, v. 2, p. 81–110.

——— 1899a, The geographical cycle: Geographical Journal, v. 14, p. 481–504.

——— 1899b, The peneplain: American Geologist, v. 23, p. 207–239.

——— 1905, Complications of the geographical cycle: Report of the 8th Geographical Congress, 1904, p. 150–163.

——— 1922, Peneplains and the geographical cycle: Geological Society of America Bulletin, v. 23, p. 587–598.

Denny, C. S., 1956, Surficial geology and geomorphology of Potter County, Pennsylvania: U.S. Geological Survey Professional Paper 288.

Flemal, R. C., 1971, The attack on the Davisian system of geomorphology: A synopsis: Journal of Geological Education, v. 19, p. 3–13.

Flint, R. F., 1963, Altitude, lithology and the Fall Zone Peneplain in Connecticut: Journal of Geology, v. 71, p. 683–697.

Friedkin, J. F., 1945, A laboratory study of the meandering of alluvial rivers: Vicksburg, Mississippi, U.S. Waterways Experimental Station, 40 p.

Gilbert, G. K., 1877, Geology of the Henry Mountains (Utah): U.S. Geographical and Geological Survey of the Rocky Mountain region, 160 p.

——— 1890, Lake Bonneville: U.S. Geological Survey Monograph 1, 438 p.

—— 1914, The transportation of debris by running water: U.S. Geological Survey Professional Paper 86, 263 p.
—— 1917, Hydraulic-mining debris in the Sierra Nevada: U.S. Geological Survey Professional Paper 105, 154 p.
Gilpin, L., Isacks, B., Bloom, A., Jouannic, C., Mallet, P., and Taylor, F., 1981, Geomorphic effects of subduction of D'Entrecasteaux fracture zone under Santa Island, Vanuatu (New Hebrides): Geological Society of America Abstracts with Programs, v. 13, no. 7, p. 459.
Grawe, R., 1936, Ice as an agent in rock weathering: A discussion: Journal of Geology, v. 44, p. 173–182.
Hack, J. T., 1960, Interpretation of erosional topography in humid temperate regions: American Journal of Science, v. 258-A, p. 80–97.
—— 1976, Dynamic equilibrium and landscape evolution, in Melhorn, W., and Flemal, R., eds., Theories of landform development: Binghamton, New York, State University of New York Publications in Geomorphology, p. 87–102.
Hayden, R. S., ed., 1985, Global mega-geomorphology: Washington, D.C., NASA Conference Publication 2312, 126 p.
Higgins, C. G., 1976, Theories of landscape development: A perspective, in Melhorn, W., and Flemal, R., eds., Theories of landform development: Binghamton, New York State University of New York Publications in Geomorphology, p. 1–28.
Hjulstrom, F., 1939, Transportation of detritus by moving water, in Trask, P., ed., Recent marine sediments: A symposium: Tulsa, Oklahoma, American Association of Petroleum Geologists.
Horton, R. E., 1945, Erosional development of streams and their drainage basins: Hydrophysical approach to quantitative morphology: Geological Society of America Bulletin, v. 56, p. 275–370.
Johnson, D. W., 1931, Stream sculpture on the Atlantic slope: New York, Columbia University Press, 142 p.
Judson, S., 1976, Evolution of Appalachian topography, in Melhorn, W., and Flemal, R., eds., Theories of landform development: Binghamton, New York, State University of New York Publications in Geomorphology, p. 29–44.
Keller, E. A., Bonkowski, M. S., Korsch, R. J., and Shlemon, R. J., 1982, Tectonic geomorphology of the San Andreas fault zone in the southern Indio Hills, Coachella Valley, California: Geological Society of America Bulletin, v. 93, p. 46–56.
Keulegan, G. H., and Krumbein, W. C., 1949, Stable configuration of bottom slope in a shallow sea and its bearing on geological processes: American Geophysical Union Transactions, v. 30, p. 855–861.
King, L. C., 1950, The study of the world's plainlands: Geological Society of London Journal, v. 106, p. 101–131.
—— 1953, Canons of landscape evolution: Geological Society of America Bulletin, v. 64, p. 721–752.
—— 1962, The morphology of the Earth: Edinburgh, Oliver and Boyd, 699 p.
Knox, J. C., 1976, Concept of the graded stream, in Melhorn, W., and Flemal, R., eds., Theories of landform development: Binghamton, New York, State University of New York Publications in Geomorphology, p. 169–198.

Kochel, R. C., Ritter, D. F., and Miller, J., 1987, Role of tree dams in the construction of pseudo-terraces and variable geomorphic response to floods in Little River valley, Virginia: Geology, v. 15, p. 718–721.
Kraus, M. J., and Middleton, L. T., 1987, Dissected paleotopography and base-level changes in a Triassic fluvial sequence: Geology, v. 15, p. 18–23.
Leopold, L. B., and Maddock, T., 1953, The hydraulic geometry of stream channels and some physiographic implications: U.S. Geological Survey Professional Paper 252, 57 p.
Mackin, J. H., 1948, Concept of the graded river: Geological Society of America Bulletin, v. 59, p. 463–511.
Newson, M., 1980, The geomorphological effectiveness of floods. A contribution stimulated by two recent events in mid-Wales: Earth Surface Processes, v. 5, p. 1–16.
Penck, W., 1924, Die morphologische Analyse: Geographische Abhandlungen: 2 Reihe, Heft 2: Stuttgart, Germany, 283 p.
—— 1953, Morphological analysis of landforms: Translated and edited by H. Czech and K. Boswell: London, Macmillan, 429 p.
Perutz, M. F., 1950, Direct measurement of the velocity distribution in a vertical profile through a glacier: Journal of Glaciology, v. 1, p. 382–383.
Pyne, S. J., 1980, Grove Karl Gilbert: A great engine of research: Austin, Texas, University of Texas Press, 306 p.
Ritter, D. F., 1986, Process geomorphology (2nd edition): Dubuque, Iowa, Wm. C. Brown, 579 p.
Rubey, W. W., 1938, The force required to move particles on a stream bed: U.S. Geological Survey Professional Paper 189-E, p. 121–141.
Scheidegger, A. E., 1975, Physical aspects of natural catastrophes: Amsterdam, The Netherlands, Elsevier.
Schumm, S. A., 1969, River metamorphosis: American Society of Civil Engineers, Journal of Hydraulics Division, HY 1, p. 255–273.
—— 1973, Geomorphic thresholds and complex response of drainage systems, in Morisawa, M., ed., Fluvial geomorphology: Binghamton, New York, State University of New York Publications in Geomorphology, p. 299–310.
—— 1976, Episodic erosion: A modification of the geomorphic cycle, in Melhorn, W., and Flemal, R., eds., Theories of landform development: Binghamton, New York, State University of New York Publications in Geomorphology, p. 69–85.
—— 1981, Evolution and response of the fluvial system, sedimentologic implications, in Ethridge, F. G., and Flores, R. M., eds., Recent and ancient nonmarine depositional environments: Models for exploration: Tulsa, Oklahoma, Society of Economic Paleontologists and Mineralogists Special Publication 31, p. 19–29.
Schumm, S. A., and Lichty, R. W., 1965, Time, space, and causality in geomorphology: American Journal of Science, v. 263, p. 110–119.
Schumm, S. A., Costa, J. E., Toy, T., Knox, J., Warner, R., and Scott, J., 1982, Geomorphic assessment of uranium mill tailings disposal sites: Uranium Mill Tailings Management, Fort Collins, Colorado, Colorado State University, OECD Nuclear Energy Agency, and U.S. Department of Energy, p. 69–87.

Sevon, W. D., 1985, Pennsylvania's polygenetic landscape: Harrisburg, Pennsylvania, Harrisburg Area Geological Society, 4th annual field trip guidebook, p. 1–35.
Sevon, W. D., Potter, N., Jr., and Crowl, G. H., 1983, Appalachian peneplains: An historical review: Earth Science History, v. 2, p. 156–164.
Sharp, R. P., 1980, Geomorphological processes on terrestrial planetary surfaces: Annual Reviews of Earth and Planetary Science, v. 8, p. 231–261.
Steven, T. A., 1968, Critical review of the San Juan peneplain, southwestern Colorado: U.S. Geological Survey Professional Paper 594-I, 19 p.
Strahler, A. N., 1952, Dynamic basis of geomorphology: Geological Society of America Bulletin, v. 63, p. 923–938.
Strecker, M., Bloom, A., and Allmendinger, R., 1984, Neotectonic activity in northwest Argentina: Santa Maria valley: Geological Society of America Abstracts with Programs, v. 15, p. 700.
Taber, S., 1930, Mechanics of frost heaving: Journal of Geology, v. 38, p. 303–317.
Taylor, F. W., Jouannic, C., and Bloom, A. L., 1985, Quaternary uplift of the Torres Islands, northern New Hebrides frontal arc: Comparison with Santo and Malekula islands, central New Hebrides: Journal of Geology, v. 93, p. 419–438.
Terzaghi, K., 1950, Mechanism of landslides, in Paige, S., ed., Application of geology to engineering practice: Geological Society of America, Berkey Volume, p. 83–123.
Tinkler, K. J., 1985, A short history of geomorphology: Totowa, New Jersey, Barnes and Noble, 317 p.
Tosdal, T. M., Clark, A. H., and Farrar, E., 1984, Cenozoic polyphase landscape and tectonic evolution of the Cordillera Occidental, southernmost Peru: Geological Society of America Bulletin, v. 95, p. 1318–1332.
Twidale, C. R., 1976, On the survival of paleoforms: American Journal of Science, v. 276, p. 77–95.
—— 1985, Ancient landscapes—Their nature and significance for the question of inheritance, in Hayden, R., ed., Global mega-geomorphology: Washington, D.C., NASA Conference Publication 2312, p. 29–40.
Twidale, C. R., Bourne, J. A., and Smith, D. M., 1976, Age and origin of paleosurfaces on Eyre Peninsula and in the southern Gawler Ranges, south Australia: Zeitschrift für geomorphologie, v. 20, p. 28–55.
Wells, S. G., and Gardner, T. W., 1985, Geomorphic criteria for selecting stable uranium tailings disposal sites in New Mexico: Santa Fe, New Mexico, New Mexico Energy Research and Development Institute, Technical Report 2-69-1112, v. 1, 353 p.
Wolman, M. G., and Gerson, R., 1978, Relative scales of time and effectiveness in watershed geomorphology: Earth Surface Processes, v. 3, p. 189–208.

MANUSCRIPT RECEIVED BY THE SOCIETY JULY 5, 1987
REVISED MANUSCRIPT RECEIVED SEPTEMBER 9, 1987
MANUSCRIPT ACCEPTED SEPTEMBER 21, 1987

CENTENNIAL ARTICLE

The *Geological Society of America Bulletin* and the development of quantitative geomorphology

MARIE MORISAWA *Department of Geological Sciences and Environmental Studies, State University of New York at Binghamton, Binghamton, New York 13901*

ABSTRACT

The influence of the *Geological Society of America Bulletin* on the development of quantitative geomorphology is demonstrated by three papers: Horton (1945), Mackin (1948), and Strahler (1952a), all published in the *Bulletin*. Horton (1945) advocated a move from qualitative to quantitative approaches in the study of landforms. He proposed quantitative tools and techniques by which we could analyze the landscapes around us. His paper not only stimulated statistical and mathematical approaches to geomorphic research and description but raised fundamental questions such as the randomness or determinism of surface processes. Mackin's (1948) paper was influential in the development of quantitative geomorphology because it prompted the work of Leopold and many others on the hydraulic geometry of river channels. In that paper, he also proposed several concepts fundamental to the dynamic geomorphology system of Strahler. Strahler's (1952a) paper set forth some guiding principles underlying quantitative methods of investigating landforms. He advocated the use of statistical techniques in landscape analyses of all geomorphic processes. In that paper, Strahler also integrated Mackin's (1948) concept of system and that of a dynamic equilibrium into a new "dynamic" geomorphology. These three papers were fundamental in the origin and growth of quantitative, process-oriented geomorphology.

Future trends lie in the fields of environmental geomorphology, global and planetary studies, and morpho-tectonics with accent on applied problems and cooperation with other disciplines.

INTRODUCTION

Understanding the landscape involves a description of the landform, determining its relationship to the underlying rock and structure, and an analysis of the process or processes which created, developed, and are still changing the surface features. These are the concerns of the geomorphologist.

Significant contributions to the basic tenets and terms of geomorphology were made by American geologists exploring the western United States during the latter part of the nineteenth century. Viewing the magnificent scenery laid bare before them, Newberry, Dutton, Gilbert, and Powell gave graphic descriptions of the landforms they saw and the processes which they thought created them, processes which they observed as still in operation. All of these men contributed to the basic concepts and terminology which are familiar to geomorphologists today, emphasizing the relations between landscape, structure, and process. In particular, Gilbert showed great insight into the dynamic processes involved in shaping the Earth's surface features in terms of slope forms and river action. He emphasized the balance he saw in the erosive, transportive, and depositional processes and the dynamic adjustment made to that equilibrium as changes took place in the environment.

The concepts and approaches manifest in the work of these geologists, however, were overshadowed by the writings of William Morris Davis. Davis took the ideas of these men and others and integrated them into the grand theme of the "erosion cycle." In using the cyclic concept to explain the development of landforms, Davis lost sight of the dynamism of the system and ignored the importance of process (mechanics), adopting instead the historical approach, tracing the sequence of events that formed the surface features.

Although Davis advocated the trilogy of structure, process, and stage in landform development, he suppressed the importance of the first two to the dominance of the last. Davis used the word "process" to mean the type of geomorphic agent at work (that is, running water, ice, wind, and so on) rather than the present-day usage of process in process-form or process-geomorphology to mean mechanics, or the way in which a given geomorphic agent works. His primary approach to geomorphology thus emphasized qualitative description and denudation chronology—stage, rather than process and structure.

We must admit the importance of Davis in the development and growth of geomorphic thought.

he synthesized the scattered elements of geomorphology and made it into a coherent subject for the first time . . . His aim was . . . to enable the trained reader to understand the descriptions of the trained observer of landforms, as both would speak the same language and have the same concepts. To this extent he succeeded magnificently. His work has proved to be a simple and effective vehicle for teaching and it has had a profound effect on the development of geomorphology in the English-speaking world, firing the imaginations of generations of students and accounting in no small measure for the popularity of the subject as a field of study (Hart, 1986, p. 15).

Overlapping in time the work of Davis, but essentially disregarded by the mainstream of geomorphic studies at the time, there appeared papers by Gilbert (1914), Rubey (1938), Hjulstrom (1935), and Bagnold (1941)

TABLE 1. QUANTITATIVE MEASURES AND CONCEPTS, HORTON, 1945

Stream order	o (later changed to u)
Drainage density	$D_d = \frac{\Sigma L}{A}$
Stream frequency	$F_s = N/A$
Overland flow	
length	$l_o = \frac{1}{2D_d}\sqrt{1-(S_c/S_g)^2}$
wave trains	
flow equation	$q_s = k_s \delta^{5/3}$
belt of no erosion	$x_c = \frac{65}{q_s n}\left(\frac{R_i}{f(s)}\right)^{5/3}$
slope function	$f(s) = \frac{\sin\alpha}{\tan^{0.3}\alpha}$
erosion force	$F_1 = \frac{w_1}{12}\left(\frac{q_s n x}{1020}\right)^{3/5} \cdot f(s)$
erosion rate	$e_r = k_e\, w_1 \frac{\delta_x}{12}\sin\alpha$
erosion/slope/length	$E = B l_o (l_o^{3/5} - x_c^{3/5})$
erosion/rainfall intensity	
Laws of drainage composition	
Playfair's law quantified	$D_d = \frac{l_1 r_b^{s-1}}{A} \cdot \frac{\rho^3 - 1}{\rho - 1}$
Infiltration theory	
Rill development	
Tributary entrance angles	$\cos z_c = \frac{\tan S_c}{\tan S_g}$

where ΣL is total length of streams, A is basin area, N is number of streams, S_c is main channel slope, S_g is ground surface slope of tributary valley, q_s is surface runoff intensity, δ is depth of overland sheet flow at foot of a slope, n is Manning roughness factor, R_i is initial surface resistance to sheet erosion, α is slope angle, w_1 is weight of runoff, x is distance from the watershed crest, ρ is r_f/r_b and k_s, k_e, and B are empirically derived constants.

which laid the groundwork for the later development of quantitative geomorphology. It was not until after World War II that a significant change in the approach of geomorphologists to the study of landscapes took place. The new approach was not only quantitative in collection and treatment of data but also included the application of physical, chemical, and engineering principles to the analysis of landforms (Morisawa, 1985).

This paper is concerned with the influence of the *Geological Society of America Bulletin* on the development of quantitative geomorphology. I will examine this theme from the effects of three papers: Horton (1945), Mackin (1948), and Strahler (1952a). The discussion refers primarily to the quantification of fluvial geomorphology, drainage-basin analysis, and rill formation. It has been argued that drainage basins represent the fundamental geomorphic unit and that rivers are the primary molders of the landscape (Leopold and others, 1964; Chorley, 1969; Selby, 1985). Moreover, both the Horton and Strahler papers have ramifications beyond fluvial processes to the entire range of geomorphic analysis and thought.

HORTON'S 1945 PAPER

The most influential paper in the quantification of geomorphology has been the paper by R. E. Horton, published in the *Bulletin* in 1945 (Leopold and others, 1964; Salisbury, 1971; Morisawa, 1985).

Modern quantitative fluvial geomorphology . . . begins . . . with R. E. Horton's 1945 paper (Kennedy, 1978, p. 219).

It is difficult to say when quantification began because many early studies used numbers in one way or another. To many geomorphologists, however, the classic work of Horton (1945) marks the beginning of the new era (Hart, 1986, p. 113).

In that paper, Horton explained that physiography (a term that included studies of vegetation and surface hydroclimatology as well as landforms) was qualitative because of the lack of adequate tools for measuring and operating.

In spite of the general renaissance of science in the present century, physiography . . . still remains largely qualitative. Stream basins . . . described as "youthful", "mature", "old" . . . without specific information as to how, how much or why. This is probably the result largely of lack of tools with which to work . . . One purpose of this paper is to describe two sets of tools which permit an attack on the problems of the development of landforms . . . along quantitative lines (Horton, 1945, p. 281).

His approach was hydrophysical, using overland runoff and infiltration to attack the problems of fluvial erosion and slope formation. To analyze drainage, he offered the following quantitative tools of measurement and description: stream order (u), drainage density (D_d), stream frequency (F), and other parameters leading to his "laws of drainage composition" which he stated was a quantification of Playfair's law of accordant junctions (Table 1). He deliberately set out to change the basis for the analysis of landforms:

Horton's goal was to replace the qualitative description of drainage basins (e.g., youthful, mature, well-drained) with quantitative ones (Smart, 1978, p. 130).

The operating tool with a hydrophysical basis for explaining drainage development and slope erosion, as well as overland flow, was set forth as the infiltration theory of runoff. Equations for overland flow, erosion force on a slope, erosion rate, and the relation of amount of erosion to slope length and rainfall intensity were formulated. All explanations were in terms of the hydraulics of flow—not a general practice at that time in geomorphic literature. From overland flow, Horton moved into the development of rills and stream channels, explaining the orderly development of drainage networks using hydrophysical (hydraulic) principles. Finally, he quantified the angle of junction of tributaries and the topography of basin surface and interfluves.

If we accept the previous statements that Horton's 1945 paper was the start of the "quantitative revolution," then the *Geological Society of America Bulletin* was, indeed, influential in the reformation of geomorphology, both in concept and methodology. To substantiate this, let us examine some effects of the paper.

Horton's stated goal was to change the basis for geomorphic analysis from a qualitative to a quantitative one, and he supplied the tools to do so. The first tool was the technique of ordering stream reaches.

Quantitative analysis of drainage network composition began with the well known paper by Horton (1945) and practically all of the early work in this field was based on the Horton or Strahler (1952) ordering procedures (Smart, 1978, p. 129).

As indicated in the above quote, Strahler was the first to modify the Horton ordering procedure. Later a number of papers were published criticizing the Horton-Strahler ordering system (Bowden and Wallis, 1964; Shreve, 1966; Smart, 1978; Werrity, 1972, for example). Other terminologies and ways of classifying stream links were proposed: link magnitude by Shreve (1966), ambilateral classes by Smart (1969), and cis and trans links by James and Krumbein (1969). Still other classifications were introduced by Mock (1971) and Flint (1980). More topologic parameters were supplied by Smart in a series of papers (1969, 1970, 1972, and 1978). The concepts of topologic distance (Jarvis, 1972), path length (Werner and Smart, 1973), network diameter (Jarvis, 1976), and network volume (Gregory, 1977) were proposed for topologic analysis. Horton's ordering tool thus spawned numerous substitutes and stimulated topologic analysis of network drainage patterns.

Using order as a basis, various stream and watershed characteristics (Table 1) were measured in a "Horton analysis" that gave a quantitative description of the composition of the drainage network (Table 2). This method was widely accepted and used both in geomorphic and hydrologic studies (Schumm, 1956a and 1956b, Schumm and others, 1987; Melton, 1958a, 1958b; Brush, 1961; Morisawa, 1959, 1962; Eyles, 1968).

Shreve (1966, 1967) and Smart (1967, 1969) disputed the validity of Horton's "laws" of drainage composition (Table 2). Smart (1968) developed a random link length model and derived an expression for the stream length of a given order which was more accurate than Horton's law of stream numbers. The question also arose: are drainage networks randomly generated or are their growths orderly (determined)? Shreve (1966, 1967) stated that network topology was random in the absence of geologic constraints and developed topologically random network models. Shreve's work was supported by Smart (1969) as well as by random walk models (Leopold and Langbein, 1962). Abrahams (1984) pointed out that the apparent randomness in network topology was a result of the variations of a large number of factors controlling the net. A probabilistic approach to geomorphology was proposed by Scheidegger and Langbein, 1966. Using central place theory, Woldenberg (1966, 1969) upheld Horton's deterministic viewpoint. Werrity (1972) criticized Horton's ordering scheme as insensitive to structure and lithology. He also stated that Woldenberg could not relate central place theory and drainage nets because the two were not dimensionally isomorphic and that using convergent means was not justified. Research by James and Krumbein (1969), Flint (1980), and Abrahams (1975, 1977) all indicate inherent space filling or other controls on network growth, in support of Horton's premise. As stated by Schumm (1977, p. 63),

the addition of a tributary to a growing drainage network occurs where there is sufficient runoff to permit erosion of the material underlying the drainage basin . . . the pattern develops in relation to hydrologic, geomorphic, and erodibility characteristics of the basin . . . Such a relation is evidence that the drainage network develops . . . in a deterministic manner.

The usefulness of a Horton analysis in relating quantitative basin parameters to hydrology was demonstrated by Brush (1961), Morisawa (1962), Carlston (1963), Rodda (1967), Stall and Fok (1967), Chorley and Morgan (1962), and Kirkby (1976).

Horton's paper also prompted studies of network topology using angles of junction and stream orientation. According to Howard (1971, p. 863),

Horton was apparently the first to offer a quantitative model and rational explanation for angles of junction.

Both Lubowe (1964) and Howard (1971) examined the validity of Horton's equation

$$\cos\theta j = \tan S_m / \tan S_t$$

where θj is angle of junction and S_m and S_t are gradients of the main stream and tributary, respectively. Junction angles were studied in flume experiments by Mosley (1976) to determine the effects of entrance angle, discharges, widths, and sediment loads on scour and fill in the channel below the confluence. Quantitative investigations of orientation of stream flow have been made by Judson and Andrews (1955), Morisawa (1963), Jarvis (1976) and Cox and Harrison (1979).

Horton's concept of overland flow and the development of rills and channels has not generated the heated controversies of his drainage laws. It is generally agreed that Hortonian overland flow occurs in semi-arid and arid climates; another type, called "saturation overland flow," takes place under humid conditions. Subsequent studies, motivated by Horton's paper, have substantiated and extended his observations of hydraulics of overland flow (Smith and Wischmeier, 1962; Emmett, 1970; and Pearce, 1973).

Systematic changes in developing rill and gully systems formed by overland flow have been studied in the field by a number of investigators,

TABLE 2. COMPONENTS OF A HORTON ANALYSIS

Component	Mathematical equation	Source
Number of streams of order u	$N_u = R_b^{s-u}$	Horton, 1945
Number of streams	$N = \dfrac{R_b^s - 1}{R_b - 1}$	Horton, 1945
Average length of all streams	$L_a = L_1 R_L^{u-1}$	Horton, 1945
Total length of streams of order u	$L_u = L_1 R_b^{s-u} R_L^{u-1}$	Horton, 1945
Average basin area	$A_u = A_1 R_a^{u-1}$	Horton, 1945, Schumm, 1956
Average gradient	$S_u = S_1 R_g^{s-u}$	Horton, 1945, Morisawa, 1962
Basin relief	$H_u = H_1 R_R^{u-1}$	Horton, 1945, Morisawa, 1962

where symbols are as follows:
- u   a given order
- S   the highest order
- N   number of streams, and so on
- N_u number of streams of order, u
- R_b bifurcation ratio (N_u/N_{uti})
- L_a average length of all streams
- L_1 average length of first-order streams
- L_u total length of streams of order, u
- R_L stream length ratio
- A_u average area of a basin of order, u
- A_1 average area of a first-order basin
- R_a basin area ratio
- S_u average gradient of a stream of order, u
- S_1 average gradient of a first-order stream
- R_g stream gradient ratio
- H_u average relief of a basin of order, u
- H_1 average relief of first-order basin
- R_R relief ratio

including Schumm (1956b), Morisawa (1964), and Wells and others (1985), and in the laboratory by Schumm and others (1987). Both Ruhe (1952) and Hack (1965) have used Hortonian tools and techniques to examine chronologic development of drainage systems.

It was some time before Horton's ideas penetrated to the geomorphic community. Some thirty years after his paper was published in the *Bulletin*, his concepts and methodology have permeated the literature. They have been tested and expanded upon; some have been rejected, but his imprint is deeply inscribed in geomorphic thought.

Since the first (Horton) systematic treatment of landform geometry . . . geomorphology has experienced a revolution in both aim and technique . . . changed scope from denudation chronology to processes (Eyles, 1968, p. 702).

MACKIN'S 1948 PAPER

Perhaps it may seem strange to some that I have included the paper on the graded river by J. H. Mackin published in the *Geological Society of America Bulletin* (1948) as a significant publication in the development of quantitative geomorphology, especially in view of his criticism of the quantitative approach (Mackin, 1963). His 1948 discussion of the concept of the graded river, however, laid the theoretical foundation needed for the development of quantitative approaches and spurred the quantification of stream-channel morphology, an aspect disregarded by Horton.

A highly simplified and yet . . . fruitful approach to quantitative fluvial morphology is that suggested by Robert E. Horton (1945) . . . it barely touches upon the morphology of the individual stream channel (Nemenyi, 1952).

Although the term and basic concept of grade was elucidated by Gilbert (1877) in his discussion of the relation of amount of load carried by stream erosion and deposition, it was Davis who defined the term:

TABLE 3. HYDRAULIC GEOMETRY OF RIVER CHANNELS

$$w = aQ^b$$
$$D = cQ^f$$
$$V = kQ^m$$
$$S = tQ^z$$
$$n = rQ^y$$
$$G_s = pQ^j$$

where w, channel width
D, mean channel depth
V, mean velocity
S, water surface slope
n, Manning roughness
G_s, suspended-load transport rate
Q, discharge

The original excess of ability over work will thus . . . be corrected and when an equality of these two quantities is brought about, the river is graded (1899, *in* Davis, 1909, p. 258).

'grade', meaning balance, always implies an equality of two quantities . . . grade is a condition of an essential balance between corrosion and deposition (1902, *in* Davis, 1909, p. 390).

Davis equated the condition of grade to the mature or old river stage in the sequential development of landscapes. He related the balance of load and ability to do work to the river gradient or slope of equilibrium. The development of the concept of grade has been thoroughly discussed elsewhere (see especially Dury, 1966, and Knox, 1975). There were dissenters to the concept of grade. Kesseli (1941) believed that Davis' condition of balance was impossible. Dury (1966) stated

The uncertainty here . . . (Leopold and Wolman, 1956) throws more doubt than ever on efforts to equate transporting power with total load . . . the simple connections required by the concept of grade do not obtain in nature (p. 229).

Although he discussed grade in qualitative terms, Mackin based much of his analysis on papers by Gilbert (1914) and Rubey (1933, 1938, 1952) which were quantitative. Moreover, he prophesied that (p. 465)

future advances in knowledge of stream processes will certainly be based increasingly on quantitative measurement and mathematical analysis.

and advocated the use of engineering principles in geomorphic studies.

In the 1948 paper, Mackin used concepts which later became fundamental in quantitative analyses: the concept of a fluvial *system* (p. 476).

It is useful and necessary to consider a graded stream as a system in equilibrium.

and the adaptation of Le Chatelier's principle of compensation for a change in chemical equilibria. Mackin's fluvial system in equilibrium was dynamic and self-regulating, adjusting its morphology to changing conditions. It was what we now call a "process-response system."

Mackin defined a graded river as

A graded stream is one in which, over a period of years, slope is delicately adjusted to provide, with available discharge and with prevailing channel characteristics, just the velocity required for the transportation of the load supplied from the drainage basin. The graded stream is a system in equilibrium; its diagnostic characteristic is that any change in any of the controlling factors will cause a displacement of the equilibrium in a direction that will tend to absorb the effect of the change (p. 471).

Although in this definition he accented adjustment of the river gradient (slope), Mackin later (p. 484) amended the concept:

Both slope *and* channel characteristics vary from segment to segment, and any change in external controls usually results in changes in both of these variables.

A question arises, then, as to whether the foregoing definition . . . should not be revised to read, 'in which *slope and channel characteristics* are delicately adjusted, etc.'

He emphasized that river morphology is adjusted to enable the stream to transport debris supplied to it. He qualitatively analyzed the effects of increased discharge on width, depth, velocity, and slope. He discussed the effect on velocity as channel roughness, width, depth, wetted perimeter, slope, channel pattern, and frictional resistance changed. Above all, he insisted that discharge and load were the independent variables upon which all other stream-channel characteristics were dependent. These concepts became a second major focus in the quantification of fluvial geomorphology.

As Horton quantitatively expressed Playfair's law of accordant junctions, so Leopold and Maddock (1953) and Leopold and Miller (1956) quantitatively expressed Mackin's definition of a graded stream with their concept of hydraulic geometry. The hydraulic geometry equations represented the empirical morphology of a river in equilibrium: they showed, mathematically, the relationship of width, depth, and velocity to discharge (Table 3). The number of hydraulic geometry characteristics have been increased to include the relation of slope and roughness to discharge, as well as sediment load to discharge (Table 3).

Mackin's paper thus prepared the way for studies of hydraulic geometry of stream channels which have flooded the geomorphic literature since Leopold and Maddock's seminal publication. A bibliography of hydraulic geometry studies compiled by Williams in 1979 (handout at the 10th Binghamton Geomorphology Symposium) consisted of 86 entries. Such investigations still continue to be a popular way to examine channel morphology, and they have greatly increased our understanding of the adjustment that a river in equilibrium makes as discharge and load change. Although most approaches have been empirical (for example, Knighton, 1974; Richards, 1973; Andrews, 1979; Osterkamp and Hedman, 1982), the theoretical aspects have also been explored (Smith, 1974; Williams, 1978; Hey, 1978; Osterkamp and others, 1983). The hydraulic geometry of alluvial channels has been said to be indeterminate by Kennedy and Brooks (1963), Maddock (1970), and Williams (1978) but determinate by Hey (1978). Control of hydraulic geometry of specific fluvial systems has been related to sediment size, peak discharge, and bankfull discharge (Brush, 1961; Carlston, 1965; Leopold and Maddock, 1953; Leopold and Miller, 1956; Leopold and Wolman, 1957; Andrews, 1979). Changes in hydraulic geometry in response to human alterations of the watershed or channel have been documented by Leopold (1973), Morisawa and LaFlure (1979), Petts (1979), Richards (1979), and Williams and Wolman (1984). Later studies have shown that hydraulic geometry relationships are not necessarily constant (straight-line relationships) (Knighton, 1974). Bennett (1987) attributed such nonlinear relationships to form drag caused by changing flow resistance.

Mackin also discussed the problem of time scales: shifts in the equilibrium which occurred over the short term and those taking place over long (geologic) time. These concepts have since been elucidated by Schumm and Lichty (1965), who suggested three time spans: long-term, cyclic time; intermediate length of time as graded time; and short-term, steady-state time.

These time concepts have been related to river equilibrium and stream-channel gradient (Chorley and Kennedy, 1971; Chorley and others, 1984; Schumm and Lichty, 1965). The idea of time is so basic and important in geomorphic studies that recently three books have been devoted to the subject: Thornes and Brunsden (1977), Cullingford and others (1980), and Thorn (1982).

Finally, as pertinent to this discussion, Mackin suggested that the principle of least work was effective in the mutual adjustment of morphologic factors to changing load and discharge (p. 485). This concept was applied in a statistical approach to explain adjustments in the hydraulic geometry by Leopold and Langbein (1962) and Leopold and others (1964). It has also been applied to meandering (Langbein and Leopold, 1966).

Mackin's paper, therefore, although qualitative, gave rise to a number of ideas used later in quantitative analyses and led directly to the quantification of the characteristics of graded stream-channel morphology.

STRAHLER'S 1952 PAPER

It was several years before Horton's paper showed an effect in the geomorphic literature. The groundwork was laid, however; the paper was read and heeded. In 1950, A. N. Strahler and one of his students (Smith, 1950) began what was to be a significant series of papers in geomorphology, using a quantitative approach. In these papers, published in various geological journals, Strahler and his students at Columbia University did two things to advance the cause of quantitative geomorphology. First, they used Horton's measuring tools and techniques to analyze landforms. In so doing, they corroborated and expanded on his methods and concepts. Secondly, they applied statistical techniques in their studies.

Of particular importance in this discussion is the paper by Strahler, published in the *Geological Society of America Bulletin* in 1952a. Here Strahler set forth "the guiding principles underlying the quantitative investigation of erosional landforms" (Strahler, 1952a, p. 923). Moreover, he stated (p. 923):

The aim of this paper is to outline a system of geomorphology grounded in basic principles of mechanics and fluid dynamics.

He insisted that geomorphology must have a basis in soil mechanics and fluid dynamics if we are to fully comprehend the behavior of streams, slope movements, glaciers, wind, and waves. He thus also set the stage for the movement toward process geomorphology.

Moreover, he pointedly drew a distinction between "timeless" (dynamic) and "timebound" (historical) viewpoints. The timeless, dynamic approach was later applied by Hack (1960), who completely rejected the Davisian stages and proposed in their place a dynamic equilibrium concept wherein landforms become independent of time.

Strahler also adapted and expanded the ideas of system and equilibrium from Mackin's (1948) paper discussed above. Whereas Mackin used these terms to describe streams, Strahler (p. 935) proposed treating all geomorphic processes and forms as systems, both open and closed.

Many of the geomorphic processes operate in clearly defined systems . . . A drainage system . . . within the geographical confines of a watershed represents such a dynamic system. A cross-sectional belt of unit width across a shore line or sand dune, or down a given slope . . . would constitute another . . . type of dynamic system.

Such systems, as material and energy are imported and exported, achieve a time-independent steady state which is self-regulatory. He clearly envisioned the concept of "feedback" as the system adjusted to a change with an opposite reaction, that is, a response to offset the effects of the initial change. The system concept in geomorphology was thoroughly explored by Chorley (1962) and Chorley and Kennedy (1971).

Finally, in that paper, Strahler strongly advocated the use of statistical (empirical) methods of analysis and mathematical (rational) modeling in geomorphic research.

. . . the pioneer in the application of statistical techniques was Strahler (Hart, 1986).

He suggested that morphometric variables could be objectively related to a force by empirical equations (regressions) to determine a cause/effect association. An empirical relation, when enough data are gathered, may result in a generalized mathematical model.

In closing, Strahler pointed out a program for the future development of geomorphology on a dynamic, quantitative basis: (1) a study of geomorphic processes and landforms as responses of different types of materials to shear stresses, (2) quantification of landforms and causative factors, (3) formulation of statistical regression equations, (4) treating geomorphic processes and forms as open systems in a steady state, and (5) deduction of quantitative laws and mathematical models. Now, 25 years later, all elements of his program have been implemented by geomorphologists.

The initial key to carrying out this program was, in great part, the work of Strahler's students at Columbia University who applied statistical methods in their analyses. They used tests of significance, frequency histograms, and other simple statistical techniques, as well as regression equations. By the middle 1960s, more complicated and advanced statistics, such as factor analysis, cluster analysis, principal components analysis, and discriminant analysis came into use in geomorphic studies. Mathematical modeling, used as a predictive tool by Morisawa (1962), has become commonplace. For example, mathematical models are applied in describing slope development (Ahnert, 1973; Kirkby, 1971, 1985; Scheidegger, 1961; Young, 1972). Although the analysis of scarp retreat by Wallace (1978) was descriptive, more recent literature has quantified changes in scarp morphology using the diffusion equation (Nash, 1980; Colman and Watson, 1983) adapted from geochemistry. Computer modeling has been used in coastal studies to examine shoreline shape (Price and others, 1973), delta growth (Komar, 1973), headland erosion (Komar, 1985), and spiral spit bars (LeBlond, 1979; Yasso, 1964), to mention a few. Equilibrium beach planforms have been quantitatively described by Hoyle and King (1958), Dicken (1961), and McLean (1967). Monitoring systems have greatly increased quantitative studies of coastal forms and processes (Morisawa, 1985).

In glacial studies, quantification took place early in glaciology as research on the mechanism of glacial flow turned to application of concepts from physics and material science to ice (Nye, 1951, 1952). Morphometric studies of glacial landforms include those by Chorley (1959) and Reed and others (1962) on drumlin shapes and by Svensson (1959) on cross-valley profiles of glacial troughs. Boulton (1974) developed a mathematical theory of erosion relating process and form.

The influence of Strahler's paper is shown in a recent geomorphology textbook where Selby (1985, p. 11) stated that

Since about 1950 detailed process studies . . . are more firmly based in the physical sciences and especially upon detailed measurements . . . secondly, analysis is more rigorous and makes full use of statistical techniques.

RECENT AND FUTURE TRENDS

In analyzing the contents of these three papers and their effects, we see that the *Bulletin* of the Geological Society of America has been significant in the development of modern geomorphic thought and methodology as the change from a descriptive, qualitative science to a quantitative, process-oriented one took place. The new era began with Horton's 1945 paper, which prompted topological studies of networks and network growth. Recent trends have been to relate network topology to structure (Abrahams and Flint, 1983). Although studies such as those of Schumm (1956b), Schumm and others (1987), Morisawa (1964), and Wells and others (1985) furthered our understanding of the development of networks, much more work needs to be done in this area. Perhaps an awareness of the processes taking place will arise out of current research on piping, zero-order basins, mechanics of soil erosion, and drainage evolution in tectonically active areas (Hansen, 1986; Oberlander, 1985).

Statistical and mathematical techniques were championed and a dynamic, process approach to geomorphology was advocated by Strahler

(1952a). From a beginning of the application of simple statistical techniques to the analysis of morphology, geomorphologists have moved to sophisticated mathematical and computer modeling. This methodology has often, but not always, helped in understanding the mechanics of geomorphic processes. We need to advance from descriptive morphometry to explanation of form to process. Moreover, to accomplish this, geomorphologists must have backgrounds in other fields such as physics, biology, chemistry, mathematics, soil mechanics, engineering hydraulics and hydrology, and computers, or they must associate with those who do. This is becoming more and more the case.

Mackin's (1948) definition and discussion of the graded river led to the hydraulic geometry of Leopold and his co-workers and to concepts expanded by Strahler and integrated into his "dynamic" geomorphology. The trend after the initial papers on hydraulic geometry was to verify the concepts proposed by analyzing data from other rivers. Knighton (1974) found that the family of curves for different flow frequencies was not a system of parallel sloping lines. Control of hydraulic geometry exponents by channel shape (Richards, 1973), resistance of bed and banks (Heede, 1972; Morisawa, 1972), discharge (Dury, 1974; Selby, 1974; Wolman and Miller, 1960), and vegetation (Heede, 1976) were investigated. Many questions remain unanswered as to the influence of bed and bank materials on channel morphometry, the geometry of bedrock channels and gravel-bed rivers, and the response of channel geometry to catastrophic flooding and changing hydraulics. Incidentally, there is still debate about the definition of channel-forming discharge, which must be settled.

As a result, there has been a fundamental change in the study of landscapes. Morphological and process aspects of landforms are now routinely analyzed in quantitative terms by statistical techniques and mathematical modeling. More precise descriptions of landforms have resulted. From simple beginnings to more sophisticated methods of treatment, quantitative geomorphology has grown and has effected a change in geomorphic thinking. Morphometric variables are viewed as parts of geomorphic systems in dynamic equilibrium (Chorley and Kennedy, 1971). Landscapes are no longer timebound (Hack, 1960). Geomorphologists have borrowed and adapted principles from physics, chemistry, and biology, as well as from soil mechanics, hydrology, and hydraulics to the study of surface features and processes (Costa and Baker, 1981). Emphasis is on process (mechanics) rather than stage (Ritter, 1978). Quantification also allows prediction of some geomorphic events and other practical applications.

In addition to future trends already posed, my crystal ball indicates other booming areas for geomorphologists. We are already well into studies of environmental geomorphology (Costa and Baker, 1981, for example). In particular, geologic hazards are often a result of geomorphic processes such as mass movements, subsidence, flooding, erosion, and weathering, or they can be related to landforms (fault scarps, terraces, and so on). Secondly, geomorphologists are beginning to realize the relations of landforms to tectonics in the developing field of morphotectonics (Bull and Knuepfer, 1987; King, 1972; Mayer, 1985; Morisawa and Hack, 1985; Ouchi, 1985).

Thirdly, since early in the space program, geomorphologists have been involved in the study of lunar and planetary geomorphology (Mason and Nordmeyer, 1969; Schumm, 1970; Eppler and others, 1983; Whitney, 1979, for example). This will continue to be a fertile field of investigation. And lastly, geomorphologists must integrate their studies into current research on the global level. Surely, nothing is more important than the surface of the Earth on which we live.

ACKNOWLEDGMENT

I gratefully acknowledge a review of this paper by Donald Coates and helpful comments by Will Graf.

REFERENCES CITED

Abrahams, A. D., 1975, Initial bifurcation process in natural channel networks: Geology, v. 3, p. 307–308.
——1977, The factor of relief in the evolution of channel networks in mature drainage basins: American Journal of Science, v. 277, p. 626–646.
——1984, Channel networks: A geomorphological perspective: Water Resources Research, v. 20, p. 161–188.
Abrahams, A. D., and Flint, J., 1983, Geological controls on the topological properties of some trellis channel networks: Geological Society of America Bulletin, v. 94, p. 80–91.
Ahnert, F., 1973, COSLOPE 2-A comprehensive model program for simulating slope profile development: Geocommunication Programs 8, p. 99–119.
Andrews, E. D., 1979, Hydraulic adjustment of the East Fork River, Wyoming, to the supply of sediment, *in* Rhoades, D., and Williams, G. P., eds., Adjustments of the fluvial system: London, England, Allen and Unwin, p. 69–74.
Bagnold, R. A., 1941, The physics of blown sand and desert dunes: London, England, Methuen & Co., 264 p.
Bennett, S. J., 1987, Temporal variations in channel morphology and hydrology of the eastern Susquehanna River in N.Y. state [M.A. thesis]: Binghamton, New York, State University of New York, 221 p.
Boulton, G. S., 1974, Processes and patterns of glacial erosion, *in* Coates, D. R., ed., Glacial geomorphology: State University of New York at Binghamton, Publications in Geomorphology, p. 41–87.
Bowden, K. L., and Wallis, J. R., 1964, Effect of stream-ordering technique on Horton's law of drainage composition: Geological Society of America Bulletin, v. 75, p. 767–774.
Broscoe, A. J., 1959, Quantitative analysis of longitudinal stream profiles of small watersheds: Office of Naval Research, Technical Report No. 18, Contract N6271-30, New York, Columbia University, 73 p.
Brush, L. M., 1961, Drainage basins, channels and flow characteristics of selected streams in central Pennsylvania: U.S. Geological Survey Professional Paper 282 F, 180 p.
Bull, W. L., and Knuepfer, P.L.K., 1987, Adjustments by the Charwell River, New Zealand, to uplift and climatic changes: Geomorphology, v. 1, p. 15–32.
Carlston, C. W., 1963, Drainage density and stream flow: U.S. Geological Survey Professional Paper 282 F, 180 p.
——1965, The relation of free meander geometry to stream discharge and its geomorphic implications: American Journal of Science, v. 263, p. 864–885.
Chorley, R. J., 1957, Illustrating the laws of morphometry: Geological Magazine, v. 94, p. 140–150.
——1957, Climate and morphometry: Journal of Geology, v. 65, p. 628–638.
——1958, Group operator variance in morphometric work with maps: American Journal of Science, v. 256, p. 208–218.
——1959, The shape of drumlins: Journal of Glaciology, v. 3, p. 339–344.
——1962, Geomorphology and general systems theory: U.S. Geological Survey Professional Paper 500-B, 10 p.
——1969, The drainage basin as a fundamental geomorphic unit, *in* Chorley, R. J., ed., Water, earth and man: London, England, Methuen.
Chorley, R. J., and Kennedy, B. A., 1971, Physical geography, a systems approach: London, England, Prentice-Hall, 370 p.
Chorley, R. J., and Morgan, U. A., 1962, Comparison of morphometric features, Unaka Mts., Tennessee and North Carolina and Dartmoor, England: Geological Society of America Bulletin, v. 73, p. 17–34.
Chorley, R. J., Schumm, S. A., and Sugden, D. E., 1984, Geomorphology: London, England, Methuen & Co., Ltd., 589 p.
Coates, D. R., 1958, Quantitative geomorphology of small drainage basins of southern Indiana: Office of Naval Research, Technical Report No. 10, Contract NR271-30. New York, Columbia University, 67 p.
Colman, S. M., and Watson, K., 1983, Ages estimated from a diffusion-equation model for scarp degradation: Science, v. 221, p. 263–265.
Costa, J. E., and Baker, V. R., 1981, Surficial geology: Building with the earth: New York, John Wiley & Sons, 498 p.
Cox, J. C., and Harrison, S. S., 1979, Fracture-trace influenced stream orientation in glacial drift, northwestern Pennsylvania: Canadian Journal of Earth Sciences, v. 16, p. 1511–1514.
Cullingford, R. A., Davidson, D. A., and Lewin, J., 1980, Time-scales in geomorphology: Chichester, England, John Wiley and Sons, 360 p.
Davis, W. M., 1899, The geographical cycle: Geographical Journal, v. 14, p. 481–504.
——1902, Base level, grade and peneplain: Journal of Geology, v. 10, p. 77–111.
——1909, Geographical essays: New York, Ginn, 777 p.
Dicken, S. S., 1961, Some recent physical changes of the Oregon coast: Office of Naval Research, Final Technical Report, Contract 2771(04), Project NR 388-062, Eugene, University of Oregon, 151 p.
Dury, G. H., 1966, The concept of grade, *in* Dury, G. H., ed., Essays in geomorphology: New York, Elsevier, p. 211–234.
——1974, Magnitude-frequency analysis and channel morphometry, *in* Morisawa, M., ed., Fluvial geomorphology: London, England, Allen & Unwin, p. 9–21.
Emmett, W. W., 1970, Hydraulics of overland flow on hillslopes: U.S. Geological Survey Professional Paper No. 662A, 68 p.
Eppler, D. T., Ehrlich, R., Nummedal, D., and Schultz, P. H., 1983, Sources of shape variation in lunar impact craters: Fourier shape analysis: Geological Society of America Bulletin, v. 94, p. 274–291.
Eyles, R. J., 1968, Stream net ratios in West Malaysia: Geological Society of America Bulletin, v. 79, p. 701–712.
Flint, J. J., 1980, Tributary arrangements in fluvial systems: American Journal of Science, v. 280, p. 26–45.
Gilbert, G. K., 1877, Report on the geology of the Henry Mts.: Washington, D.C., U.S. Geological Survey, 170 p.
——1914, The transportation of debris by running water: U.S. Geological Survey Professional Paper 86, 259 p.
Gregory, K. G., 1977, Stream network volume: An index of channel morphometry: Geological Society of America Bulletin, v. 88, p. 1075–1080.
Hack, J. T., 1960, Interpretation of erosional topography in humid temperate regions: American Journal of Science, Bradley Volume, v. 258-A, p. 80–97.
——1965, Post-glacial drainage evolution and stream geometry in the Ontonagan area, Michigan: U.S. Geological Survey Professional Paper 524 C, 40 p.
——1973, Stream-profile analysis and stream-gradient index: U.S. Geological Survey Journal of Research, v. 1, p. 421–429.
Hansen, W. R., 1986, Neogene tectonics and geomorphology of the Eastern Uinta Mountains in Utah, Colorado, and Wyoming: U.S. Geological Survey Professional Paper 1356, 78 p.
Hart, M. G., 1986, Geomorphology pure and applied: London, England, Allen and Unwin, 228 p.
Heede, B. H., 1972, Influences of a forest on the hydraulic geometry of two mountain streams: Water Resources Bulletin, v. 8 (3), p. 523–530.
Hey, R. D., 1978, Determinate hydraulic geometry of river channels: American Society of Civil Engineers, Journal of the Hydraulics Division, v. 104 (H46), p. 869–885.
Hjulstrom, F., 1935, The morphological activity of rivers as illustrated by River Fyris: University of Uppsala Geological Institute Bulletin No. 25, p. 221–527.
Horton, R. E., 1945, Erosional development of streams and their drainage basins: Hydrophysical approach to quantitative morphology: Geological Society of America Bulletin, v. 56, p. 275–370.
Howard, A. D., 1971, Optimal angles of stream junction: Geometric, stability to capture, and minimum power criteria: Water Resources Research, v. 7, p. 863–873.
Hoyle, J. W., and King, G. T., 1958, Origin and stability of beaches, *in* Coastal Engineering Conference, 6th, Palm Beach, Florida, Dec. 1957, Proceedings, p. 281–301.
James, W. R., and Krumbein, W. C., 1969, Frequency distribution of stream link lengths: Journal of Geology, v. 77, p. 544–565.
Jarvis, R. S., 1972, New measure of the topologic structure of dendritic drainage networks: Water Resources Research, v. 8, p. 1265–1271.
——1976, Link length organization and network scale dependencies in the network diameter model: Water Resources Research, v. 12, p. 1215–1225.
Judson, S., and Andrews, G. W., 1955, Pattern and form of some valleys in the driftless area, Wisconsin: Journal of Geology, v. 63, p. 328–336.
Kennedy, B. A., 1978, After Horton: Earth Surface Processes, v. 3, p. 328–336.
Kennedy, John F., and Brooks, Norman H., 1963, Laboratory study of an alluvial stream at constant discharge, *in* Federal Inter-Agency Sedimentation Conference, Jan. 28–Feb. 1, 1963, p. 320–330.
Kesseli, J. E., 1941, The concept of the graded river: Journal of Geology, v. 49, p. 561–588.
King, Lewis, 1972, Relation of plate tectonics to geomorphic evolution of Canadian Atlantic provinces: Geological Society of America Bulletin, v. 83, p. 3083–3090.
Kirkby, M. J., 1971, Hillslope process-response models based on the continuity equation, *in* Institute of British Geographers, Special Publication 3, p. 15–30.

——— 1976, Tests of the random network model and its application to basin hydrology: Earth Surface Processes, v. 1, p. 197–213.
——— 1985, A model for the evolution of regolith-mantled slopes, *in* Woldenberg, M. J., ed., Models in geomorphology: London, England, Allen and Unwin, p. 213–237.
Knighton, A. D., 1974, Variation in width-discharge relations and some implications for hydraulic geometry: Geological Society of America Bulletin, v. 85, p. 1069–1076.
Knox, J. C., 1975, Concept of the graded stream, *in* Melhorn, W. and Flemal, R., eds., Theories of landform development: State University of New York at Binghamton, Publications in Geomorphology, p. 169–198.
Komar, P. D., 1973, Computer models of delta growth due to sediment input from rivers and longshore transport: Geological Society of America Bulletin, v. 84, p. 2217–2226.
——— 1985, Computer models of shoreline configuration: Headland erosion and the graded beach revisited, *in* Woldenberg, M. J., ed., Models in geomorphology: London, England, Allen and Unwin, p. 155–170.
Langbein, W. B., and Leopold, L. B., 1966, River meanders—Theory of minimum variance: U.S. Geological Survey Professional Paper 422H, 15 p.
LeBlond, P. H., 1979, An explanation of the logarithmic spiral plan shape of headland-bay beaches: Journal of Sedimentary Petrology, v. 49, p. 1093–1100.
Leopold, L. B., 1973, River channel change with time: Geological Society of America Bulletin, v. 84, p. 1845–1860.
Leopold, L. B., and Langbein, W. B., 1962, Concept of entropy in landscape evolution: U.S. Geological Survey Professional Paper 500-A, 20 p.
Leopold, L. B., and Maddock, T., 1953, The hydraulic geometry of stream channels and some physiographic implications: U.S. Geological Survey Professional Paper 252, 57 p.
Leopold, L. B., and Miller, J. P., 1956, Ephemeral streams-hydraulic factors and their relation to the drainage net: U.S. Geological Survey Professional Paper 282-A, 36 p.
Leopold, L. B., and Wolman, M. G., 1957, River channel patterns: Braided, meandering, straight: U.S. Geological Survey Professional Paper 282-B, p. 39–85.
Leopold, L. B., Wolman, M. G., and Miller, J. P., 1964, Fluvial processes in geomorphology: San Francisco, Freeman and Co., 522 p.
Lubowe, J. K., 1964, Stream junction angles in the dendritic drainage pattern: American Journal of Science, v. 262, p. 325–339.
Mackin, J. J., 1948, Concept of the graded river: Geological Society of America Bulletin, v. 59, p. 463–512.
——— 1963, Rational and empirical methods of investigation in geology, *in* Albritton, C., ed., The fabric of geology: Stanford, California, Freeman and Co., p. 135–163.
Maddock, T., 1970, Indeterminate hydraulics of alluvial channels: American Society of Civil Engineers, Journal of Hydraulics Division, v. 96, p. 2309–2323.
Mason, C. C., and Nordmeyer, E. F., 1969, An empirically derived erosion law and its application to lunar module landing: Geological Society of America Bulletin, v. 80, p. 1783–1788.
Maxwell, J. C., 1960, Quantitative geomorphology of the San Dimas Experimental Forest, Calif.: Office of Naval Research Technical Report 19, Contract NR271-30, New YOrk, Columbia University, 95 p.
Mayer, L., 1985, Tectonic geomorphology of the Basin and Range–Colorado Plateau boundary in Arizona, *in* Morisawa, M., and Hack, J. T., eds., Tectonic geomorphology: London, England, Allen and Unwin, p. 235–259.
McLean, R., 1967, Plan shape and orientation of beaches along the east coast, South Island, New Zealand: New Zealand Geographer, v. 23, p. 16–22.
Melton, M. A., 1957, An analysis of the relation among elements of climate, surface properties, and geomorphology: Office of Naval Research Technical Report No. 11, contract N60NR271-30, New York, Columbia University, 102 p.
——— 1958a, List of sample parameters of quantitative properties of landforms: Their use in determining the size of geomorphic experiments: Office of Naval Research Technical Report No. 16, Contract N60NR271-30, New York, Columbia University, 17 p.
——— 1958b, Correlation structure of morphometric properties of drainage systems and their controlling agents: Journal of Geology, v. 66, p. 442–460.
——— 1959, A derivation of Strahler's channel ordering system: Journal of Geology, v. 67, p. 345–356.
——— 1960, Intravalley variation in slope angles related to micro-climate and erosional environment: Geological Society of America Bulletin, v. 71, p. 133–144.
Miller, V. C., 1953, A quantitative geomorphic study of drainage basin characteristics in the Clinch Mt. area, Virginia and Tennessee: Office of Naval Research Technical Report No. 3, Contract NR271-30, New York, Columbia University, 30 p.
Mock, S. J., 1971, A classification of channel links in stream networks: Water Resources Research, v. 7, p. 1558–1566.
Morisawa, M. E., 1957, A classification of channel links in stream lengths from topographic maps: American Geophysical Union Transactions, v. 38, p. 86–88.
——— 1958, Measurement of drainage basin outline form: Journal of Geology, v. 66, p. 587–591.
——— 1959, Relation of morphometric properties to runoff in Little Mill Creek, Ohio, drainage basin: Office of Naval Research Technical Report 17, Contract NR271-30, New York, Columbia University, 10 p.
——— 1962, Quantitative geomorphology of some watersheds in the Appalachian Plateau: Geological Society of America Bulletin, v. 73, p. 1025–1046.
——— 1963, Distribution of stream-flow direction in drainage patterns: Journal of Geology, v. 71, p. 528–529.
——— 1964, Development of drainage systems on an upraised lake floor: American Journal of Science, v. 262, p. 3340–3354.
——— 1972, Hydrogeology of the Green and Wind Rivers, Wyoming [abs.]: Geological Society of America Abstracts with Programs, v. 4, p. 396–397.
——— 1985, Development of quantitative geomorphology: Geological Society of America Centennial Special Volume 1, p. 79–107.
Morisawa, M., and Hack, J. T., 1985, Tectonic geomorphology: London, England, Allen and Unwin, 390 p.
Morisawa, M., and LaFlure, E., 1979, Hydraulic geometry, stream equilibrium and urbanization, *in* Rhodes, D., and Williams, G. P., eds., Adjustments of the fluvial system: London, England, Allen and Unwin, p. 333–350.
Mosley, M. P., 1976, An experimental study of channel confluences: Journal of Geology, v. 84, p. 535–562.
Nash, D. B., 1980, Morphologic dating of degraded normal fault scarps: Journal of Geology, v. 88, p. 353–360.
Nemenyi, P. F., 1952, Annotated and illustrated bibliographic material on the morphology of rivers: Geological Society of America Bulletin, v. 63, p. 595–644.
Nye, J. F., 1951, The flow of glaciers and ice-sheets as a problem in plasticity: Royal Society of London Proceedings, ser. A, v. 207, p. 554–572.
——— 1952, The mechanics of glacier flow: Journal of Glaciology, v. 2, p. 82–93.
Oberlander, T. M., 1985, Origin of drainage transverse to structures in orogens, *in* Morisawa, M., and Hack, J. T., eds., Tectonic geomorphology: London, England, Allen and Unwin, p. 155–181.
Osterkamp, W. R., and Hedman, E. R., 1982, Perennial streamflow characteristics related to channel geometry and sediment in Missouri River basin: U.S. Geological Survey Professional Paper 1242, 37 p.
Osterkamp, W. R., Lane, L. J., and Foster, G. R., 1983, An analytical treatment of channel-morphology relations: U.S. Geological Survey Professional Paper 1288, 21 p.
Ouchi, S., 1985, Response of alluvial rivers to slow active tectonic movement: Geological Society of America Bulletin, v. 96, p. 504–515.
Pearce, A. J., 1973, Mass and energy flux in physical denudation; defoliated areas, Sudbury, Ontario: McGill University, Department of Geological Sciences, Technical Report No. 75-1, 235 p.
Petts, G. E., 1979, Complex response of river channel morphology subsequent to reservoir construction: Progress in Physical Geography, v. 3, p. 329–362.
Playfair, J., 1802, Illustrations of the Huttonian Theory of the Earth: Edinburgh, Scotland, William Creech, 528 p.
Price, W. A., Tomlinson, K. W., and Willis, D. H., 1973, Predicting the changes in the plan shape of beaches, *in* 16th Conference on Coastal Engineering, Proceedings, p. 1321–1329.
Reed, B., Calvin, C. J., Jr., and Miller, J. P., 1962, Some aspects of drumlin geometry: American Journal of Science, v. 260, p. 200–210.
Richards, K. S., 1973, Hydraulic geometry and channel roughness—a non-linear system: American Journal of Science, v. 273, p. 877–896.
——— 1979, Channel adjustment to sediment pollution by the china-clay industry in Cornwall, England, *in* Rhodes, D., and Williams, G. P., eds., Adjustments of the fluvial system: London, England, Allen and Unwin, p. 309–332.
Ritter, D. F., 1978, Process geomorphology: Dubuque, Iowa, W. C. Brown Publishers, 579 p.
Rodda, J. C., 1967, The significance of characteristics of basin rainfall and morphometry in a study of floods in the United Kingdom: International Association of Scientific Hydrology Publication 85, p. 834–845.
Rubey, W. W., 1933, Settling velocities of gravel, sand and silt: American Journal of Science, v. 225, p. 325–338.
——— 1938, The force required to move particles on a stream bed: U.S. Geological Survey Professional Paper 189-E, p. 121–141.
——— 1952, Geology and mineral resources of the Hardin and Brussels quadrangles (Illinois): U.S. Geological Survey Professional Paper 218, 179 p.
Ruhe, R. V., 1952, Topographic discontinuities of the Des Moines lobe: American Journal of Science, v. 250, p. 46–56.
Salisbury, N., 1971, Threads of inquiry in quantitative geomorphology, *in* Morisawa, M., ed., Quantitative geomorphology: Some aspects and applications: State University of New York at Binghamton, Publications in Geomorphology, p. 9–40.
Scheidegger, A. E., 1961, Mathematical models of slope development: Geological Society of America Bulletin, v. 72, p. 37–50.
——— 1967, On the topology of river nets: Water Resources Research, v. 3, p. 103–106.
——— 1968, Horton's law of stream numbers: Water Resources Research, v. 4, p. 655–658.
Scheidegger, A. E., and Langbein, W. B., 1966, Probability concepts in geomorphology: U.S. Geological Survey Professional Paper No. 500-C, 14 p.
Schumm, S. A., 1956a, The role of creep and rainwash on the retreat of badland slopes: American Journal of Science, v. 254, p. 693–706.
——— 1956b, Evolution of drainage systems and slopes in badlands at Perth Amboy, New Jersey: Geological Society of America Bulletin, v. 67, p. 597–646.
——— 1962, Erosion of miniature pediments in Badlands National Monument, South Dakota: Geological Society of America Bulletin, v. 73, p. 719–724.
——— 1970, Experimental studies on the formation of lunar surface features of fluidization: Geological Society of America Bulletin, v. 81, p. 2539–2552.
——— 1977, The fluvial system: New York, John Wiley and Sons, 338 p.
Schumm, S. A., and Lichty, R. W., 1965, Time, space and causality in geomorphology: American Journal of Science, v. 263, p. 110–119.
Schumm, S. A., Mosley, M. P., and Weaver, W. E., 1987, Experimental fluvial geomorphology: New York, John Wiley and Sons, 413 p.
Selby, M. J., 1974, Dominant geomorphic events in landform evolution: International Association of Engineering Geology Bulletin, v. 9, p. 85–89.
——— 1985, Earth's changing surface: Oxford, England, Clarendon Press, 607 p.
Shreve, R. L., 1966, Statistical law of stream numbers: Journal of Geology, v. 74, p. 17–37.
——— 1967, Infinite topologically random channel networks: Journal of Geology, v. 75, p. 179–186.
Smart, J. S., 1967, A comment on Horton's law of stream numbers: Water Resources Research, v. 3, p. 773–776.
——— 1968, Statistical properties of stream length: Water Resources Research, v. 4, p. 1001–1014.
——— 1969, Topological properties of channel networks: Geological Society of America Bulletin, v. 80, p. 1757–1773.
——— 1970, Use of topologic information in processing data for channel networks: Water Resources Research, v. 6, p. 932–936.
——— 1972, Channel networks: Advances in Hydroscience, v. 8, p. 305–346.
——— 1978, The analysis of drainage network composition: Earth Surface Processes, v. 3, p. 129–170.
Smith, D. D., and Wischmeier, W. H., 1962, Rainfall erosion: Advances in Agronomy, v. 14, p. 109–148.
Smith, K. G., 1950, Standards for grading texture of erosional topography: American Journal of Science, v. 248, p. 655–668.
——— 1958, Erosional processes and landforms in Badlands National Monument, South Dakota: Geological Society of America Bulletin, v. 69, p. 975–1008.
Smith, T. R., 1974, A derivation of the hydraulic geometry of steady state channels from conservation principles and sediment transport laws: Journal of Geology, v. 82, p. 98–104.
Stall, J. B., and Fok, Y. S., 1967, Discharge as related to stream system morphology: International Association of Scientific Hydrology Publication 75, p. 224–235.
Strahler, A. N., 1950, Equilibrium theory of erosional slopes approached by frequency distribution analysis: American Journal of Science, v. 248, p. 673–696 and 800–814.
——— 1952a, Dynamic basis of geomorphology: Geological Society of America Bulletin, v. 63, p. 923–938.
——— 1952b, Hypsometric (area-altitude) analysis of erosional topography: Geological Society of America Bulletin, v. 63, p. 1117–1142.
——— 1954, Statistical analysis in geomorphic research: Journal of Geology, v. 62, p. 1–25.
——— 1956, Quantitative slope analysis: Geological Society of America Bulletin, v. 67, p. 571–596.
——— 1957, Quantitative analysis of watershed geomorphology: American Geophysics Union Transactions, v. 38, p. 913–920.
——— 1958, Dimensional analysis applied to fluvially eroded landforms: Geological Society of America Bulletin, v. 69, p. 279–300.
——— 1964, Quantitative geomorphology of drainage basins and channel networks, *in* Ven te Chow, ed., Handbook of applied hydrology: New York, McGraw Hill, p. 4–39.
Strahler, A. N., and Koons, D., 1960, Objective and quantitative field methods of terrain analysis: Office of Naval Research, Final Report, Contract 266-50, New York, Columbia University, 51 p.
Svensson, H., 1959, Is the cross-section of a glacial valley a parabola?: Journal of Glaciology, v. 3, p. 362–363.
Thorn, C. E., ed., 1982, Space and time in geomorphology: London, England, Allen and Unwin, 379 p.
Thornes, J. B., and Brunsden, D., 1977, Geomorphology and time: London, England, Methuen, 208 p.
Wallace, R. E., 1978, Geometry and rates of change of fault-generated range fronts, north-central Nevada: U.S. Geological Survey Journal of Research, v. 6, no. 5, p. 637–650.
Wells, S. G., Dohrenwend, J. C., McFadden, L. D., Turrin, B. D., and Mahrer, K. D., 1985, Late Cenozoic landscape evolution on lava surfaces of the Cima volcanic field, Mojave Desert, California: Geological Society of America Bulletin, v. 96, p. 1518–1529.
Werner, C., and Smart, J. S., 1973, Some new methods of topologic classification of channel networks: Geological Analysis, v. 5, p. 271–295.
Werrity, A., 1972, The topology of stream networks, *in* Chorley, R. J., ed., Spatial analyses in geomorphology: New York, Harper and Row, p. 167–196.
Whitney, M. L., 1979, Aerodynamic and vorticity erosion of Mars: The formation of channels: Geological Society of America Bulletin, v. 90, p. 1111–1127.
Williams, G. P., 1978, Hydraulic geometry of river cross sections—Theory of minimum variance: U.S. Geological Survey Professional Paper 1029, 47 p.
Williams, G. P., and Wolman, M. G., 1984, Downstream effects of dams on alluvial rivers: U.S. Geological Survey Professional Paper 1286, 83 p.
Woldenberg, M. J., 1966, Horton's laws justified in terms of allometric growth and steady state in open systems: Geological Society of America Bulletin, v. 77, p. 431–434.
——— 1969, Spatial order in fluvial systems: Horton's laws derived from mixed hexagonal hierarchies of drainage basin areas: Geological Society of America Bulletin, v. 80, p. 97–112.
Wolman, M. G., and Miller, J., 1960, Magnitude and frequency of forces in geomorphic processes: Journal of Geology, v. 68, p. 54–74.
Yasso, W. E., 1964, Geometry and development of spit-bar shorelines at Horseshoe Cove, Sandy Hook, New Jersey: Office of Naval Research Technical Report No. 5, Contract 266-08, New York, Columbia University, 104 p.
Young, A., 1972, Slopes: Edinburgh, Scotland, Oliver and Boyd, 288 p.

Manuscript Received by the Society July 20, 1987
Revised Manuscript Received January 25, 1988
Manuscript Accepted January 27, 1988

Printed in U.S.A.

Geological fluvial geomorphology

CENTENNIAL ARTICLE

VICTOR R. BAKER *Department of Geosciences, University of Arizona, Tucson, Arizona 85721*

ABSTRACT

The history of American fluvial geomorphology over the past century is viewed as one of conflict and crises. From 1888 to 1938, a controversy arose between (1) a rational approach to understanding landscape genesis and history, with its roots in geology, and (2) a spatial-analytical approach to landscape classification and description, with its roots in geography. By the 1960s, geomorphology, led by fluvial studies, had changed its emphasis from historical studies to process studies, and the geology/geography dispute became irrelevant. Since the 1960s, a new conflict has arisen between (1) problem-oriented studies of landform genesis and (2) method-oriented studies. The latter emphasize useful predictions and a methodology that generates respect from other scientific and engineering disciplines. In extreme cases, approach 2 may bypass the understanding of phenomena in order to generate useful predictions of systems assumed to embody the behavior of those phenomena. In order to achieve its goal of intellectually satisfying understanding of phenomena, approach 1 may require the stimulus of the occasional outrageous hypothesis, thereby posing a seeming anathema to an existing scientific program. The identification and explanation of anomalies is critical to approach 1. Because of the inherent conflict in these approaches to fluvial geomorphology, there is a need to balance opposing tendencies.

INTRODUCTION

Geomorphology is one science divided by arbitrary classifications that contribute nothing to the understanding of landscapes but much to the impediment of such understanding. Fluvial geomorphologists are divided into geographers and geologists by arbitrary academic convention, into process or historical geomorphologists by their emphasis on time scales of study, and into rational or empirical geomorphologists by their scientific methodologies. Members of each group have their own sets of problems and may disdain the approaches and/or problems of others. Moreover, there are fundamental dichotomies in the basic assumptions made by individual geomorphological researchers as they pursue various studies. Views such as gradualism in process operation, randomness in nature, and dynamic equilibrium in landscape development may be held in personal preference to contrary assumptions.

A century ago, geomorphology was in its golden age. It was a science filled with wonder and excitement. The source of scientific stimulus was not the establishment of new methodologies for landscape study. Rather, it was the discovery of new phenomena in landscapes that seemed as bizarre and alien to the residents of humid-temperate academia as the surfaces of other planets seem to most of today's terrestrial geomorphologists. For American geological geomorphologists, the most important scientific trinity was not structure, process, and stage, emerging from the heuristic synthesis of a Harvard scholar. Rather the critical trinity was Gilbert, Powell, and Dutton, who were stimulated by studies of the arid western United States.

The American character of geological fluvial geomorphology derives directly from the scientific example of Grove Karl Gilbert and the organization of the United States Geological Survey by John Wesley Powell. Considerable history has been written on this era of fluvial geomorphology (Chorley and Beckinsale, 1980; Pyne, 1980; Tinkler, 1985), and so, rather than again describing its inspiration, let suffice Dutton's (1885) acknowledgment of his two colleagues. "If I were to attempt payment, I would be bankrupt" (p. 198).

Probably the only great revolutionary change that ever occurred in geomorphology was the one that Hubbert (1967) recognized for all geology and which he termed "Huttonian-Lyellian-Darwinian." The Darwinian influence was paramount by the early 1900s, when Russell (1904) described geomorphology as a science "vivified by evolution." As noted by Judson (1958), the scientific program established by William Morris Davis merely followed in this trend. The Davisian emphasis on time-directed evolution of landscapes overshadowed the emphasis that Gilbert had placed on timeless, equilibrium processes of landform genesis (Chorley and Beckinsale, 1980; Ritter, 1988).

There are parallels between the ways scientists organize their own programs of study and the ways that historians may organize the analyses of such programs through time. Thus, one obvious approach to the history of American geological fluvial geomorphology over the past century might be an evolutionary one, perhaps punctuated with the occasional major change in scientific program. Thus, Morisawa's (1985) history recognizes a progressive increase in the use of quantitative approaches to problems, punctuated by Horton's (1945) seminal paper on drainage-basin analysis. Whether or not quantification constituted a true scientific revolution for geomorphology, it certainly engendered a new program of effort based on quantitative measurements of processes, statistical treatment of data, and predictive models with practical applications. As pointed out by Morisawa (1985), much of the methodology for the new program was borrowed directly from engineering disciplines, such as hydraulics and hydrology, as well as from basic physics, chemistry, and mathematics. Moreover, the kinds of geomorphological problems studied under this new program tended to be those that were most easily analyzed by the new methodology.

Perhaps even more substantive than mere quantification has been the move toward a dynamical basis of geomorphology, inspired by Strahler (1952). It would be heartening to record the progressive march of geomorphology toward understanding modern surficial Earth processes in a physical-quantitative fashion. As a geological science, however, geomorphology must strive not only to understand modern Earth processes, but to place this understanding within the broader context of Earth history. As stated by Bryan (1950): "the essence of geomorphology is the discrimination of the ancient from the modern." Thus, another important theme is that of tension between historical and dynamical emphases in geomorphological research (Ritter, 1988).

There are many sources of conflict in geomorphological research directions (Table 1).

TABLE 1. SOME CONFLICTING RESEARCH DIRECTIONS OR PROGRAMS IN FLUVIAL GEOMORPHOLOGY

Geography	Geology
Process	Historical
Empirical	Rational
Applied	Basic
Experimental	Field
Quantitative	Qualitative

Much has been written on these approaches to research by their adherents. Another area of conflict is the fundamental assumptions that underlie individual geomorphological investigations (Table 2). Whether termed "fundamental concepts" (Thornbury, 1969), "philosophical assumptions" (Twidale, 1977), "paradigms" (Ollier, 1981), or "basic postulates" (Pitty, 1982), these ideas are the conventional wisdom for the science. Whereas the analysis of research directions must emphasize the backgrounds and methodologies of groups, the analysis of fundamental assumptions must emphasize the basic philosophies of individuals.

The view that science advances by the interaction of incompatible alternative hypotheses is discussed at length in the book *Against Method* (Feyeraband, 1975). In the view of Feyerabend (1975), it is the proliferation of mutually inconsistent theories that is most healthy for scientific advancement. In this spirit, therefore, my review will emphasize the role of problems and questions. I will argue that several fundamental crises have developed during the past century of fluvial geomorphological research. Moreover, I believe that the issues in these crises transcend the simple classifications of Tables 1 and 2.

From 1888 to 1938, a controversy evolved over the purpose for studies of terrestrial landforms. A rational approach to the genetic understanding of landscapes developed from roots in geology, but found its most eloquent expression in the writings of a geographer, William Morris Davis. More basic to its roots in geography was an empirical approach to landforms as spatial elements of the Earth's surface, where objective measurement was more important than subjective interpretations of genesis. This conflict disappeared after World War II, when both

TABLE 2. SOME CONFLICTING FUNDAMENTAL ASSUMPTIONS IN FLUVIAL GEOMORPHOLOGICAL RESEARCH

Randomness or indeterminacy	Causality
Gradualism	Catastrophism
Disorder	Order
Complexity	Simplicity
Dominance of present-day processes	Importance of ancient conditions for relict landscapes
Youthfulness of topography	Antiquity of relict landscapes
Morphoclimatic zonation of landscapes	Azonal landforms controlled by structure

geological and geographical geomorphologists turned their attention to process studies.

As stated by Thornes and Brunsden (1977): "The current paradigm is one in which process studies prevail effected principally and increasingly by mathematical and stochastic models." Indeed, fluvial geomorphology has been so infused by new research methods that those methods are determining the choice of appropriate geomorphological research problems. This has resulted in the paradox described by Church (1980): "Contemporary process studies are of little worth in evaluating landscape evolution." The process studies most amenable to the new method-dominated approach to geomorphology may be largely irrelevant to the most important scientific questions in problem-dominated genetic geomorphology. The issues in this new crisis for geomorphology are further complicated by attempts by geomorphologists to make their science more relevant in technical applications and more respectable in comparison to sister basic sciences. The resulting choices of research direction (Table 1) and fundamental assumptions (Table 2) by the new method-dominated geomorphologists reflect a new crisis of purpose that will be described in this review.

GEOMORPHOGRAPHY OR GEOMORPHOGENY?

In reviewing the status of American geomorphology to a Polish audience, John P. Miller (1959) wrote: "One curious aspect of geomorphology in North America is its alliance with geology rather than geography, as is the case in other parts of the world." Miller's Ph.D. advisor had been Kirk Bryan, who became Professor of Physiography at Harvard in 1943. Bryan preferred the term "geomorphology" to describe his scientific activity. This caused much dismay to Bryan's predecessor at Harvard, William Morris Davis, who reportedly told him (Bryan, 1941, p. 6): "Bryan, I am afraid that you will always teach more Geomorphogeny than Geomorphography." In its use of terminology to stand for complex ideas, this quote is typical of Davis, but it also conveys the extreme irony of his argument. Davis was not in the evolving mainstream of geography in his definition of "geomorphography" as the explanatory description of the land. Explanation, in terms of genesis, was the purpose of "geomorphogeny," as originally defined by Lawson (1894). Davis believed that a genetic description of landforms was more understandable and heuristically preferable to an empirical one. This view made Davis a cherished member of the geological community, who regarded his geographical assertions as mere idiosyncrasy (Bryan, 1941). Davis pleased his geological colleagues with statements such as that at his Presidential Address to the Geological Society of America (Davis, 1912, p. 121): "all geography belongs under geology, since geography is neither more nor less than the geology of today."

Such statements did not reflect the mainstream of thinking in geography, where one could argue that empirical landscape description was preferable to hypothesizing genetic explanations. Questions of landform genesis can be considered irrelevant when concern is with the human use of landscapes. Davis' explanatory description was subjective and qualitative. By concentrating on objective, quantitative measures of slopes, spatial relationships, and so on, geomorphography could become more appropriate as a branch of geography. In being what Cotton (1956) terms a "utilitarian art," geomorphography would divorce itself from rational argument over landform genesis and concentrate on an empirical approach that would be useful for social purposes. It was this evolving view of geomorphography that caused Douglas Johnson (1929) to conclude: "Geomorphology itself has suffered, and will continue to suffer, from attempts to include it in the geographic realm. In the history of its development, in its methods, and in its affiliations it is a part of geology."

Fortunately for terminology explosion, the term "geomorphology" entered America from Europe (Cotton, 1956), displacing both "geomorphogeny" and "physiography" about the time of the 50th anniversary of the Geological Society of America (Bryan, 1941). Fortunately for geographer-geomorphologists, the empirical view of geomorphography did not prevail into the modern era. By the 1960s, geomorphology seemed to have finally accepted what Dury (1972) termed "its somewhat ambivalent geographic-geologic location as a research field." In part this was achieved by the shift in emphasis from the historical-based explanatory description of Davis to modern process studies (Chorley, 1962). A new geography arose as both human and physical geographers adapted Baconian scientific methods to their disciplines (Johnston, 1983). New tools such as statistical treatment of data, spatial analysis, and systems theory were found to be useful in both physical and human geography.

The American concerns over geomorphogeny and geomorphography were best expressed by successive Harvard geomorphologists: William Morris Davis, Kirk Bryan, and John P. Miller. These giants of our science inspired successive generations of geomorphologists, but ultimately they seem to have inspired neither their geographical nor their geological colleagues. Geomorphology was abolished during the 1960s as an academic enterprise at Harvard, and that example of arbitrary definition has been followed by other universities that purport to cover the subject matter of the earth sciences in their curricula. Such decisions, in effect, constitute value judgments on geomorphology and/or geomorphologists. Because these value judgments have been made and continue to be made by others, there is a need for geomorphologists themselves to assess the basic values of their discipline.

RATIONAL OR EMPIRICAL GEOMORPHOLOGY

The *Fiftieth Anniversary Volume* of the Geological Society of America contained a single address exploring the issue of geomorphogeny versus geomorphography (Bryan, 1941). The 75th anniversary volume, entitled *The Fabric of Geology*, contained two papers, both relying on fluvial geomorphological examples, which explored a new dichotomy in the field. The papers, one by J. Hoover Mackin (1963), and the other by Luna B. Leopold and Walter B. Langbein (1963), are generally assumed to represent a debate over methodology in fluvial process studies (Thornes, 1979). I shall argue that these papers presaged the predominant conflict for the most recent generation of fluvial geomorphologists.

Mackin's (1963) paper was derived from an oral address given at a banquet of the Branner Club during the meeting of the Cordilleran Section of the Geological Society of America in Los Angeles, April 17, 1962. In an unpublished review of this after-dinner talk, Mackin included several examples of "shotgun empiricism" that did not survive editing into the 1963 paper. He specifically analyzed a geomorphological study that related channel slope to bed-material size, width-depth ratios, and drainage areas. Scatter diagrams of these parameters illustrated interesting trends. Analyzed without concern for individual data points, however, these trends also concealed an important distinction described by Mackin as follows:

in any graded segment of a river, the channel is shaped by the river itself in alluvial materials; the outstanding characteristic of this self-modeled channel is that the slope is adjusted—it is precisely that slope which provides the transporting power required to move the load with the available discharge, and with the prevailing channel characteristics, whatever these may be. This equilibrium, or steady-state condition, is timeless; the graded condition is maintained without change as long as controlling conditions remain the same. . . . The slope of a rapid or waterfall, on the other hand, is a function of (a) *rock erodibility*, which depends on resistance of the rock to abrasion, spacing and orientation of joints, and so on; (b) *discharge*—how big the river is; and (c) *time*—how long the river has been operating at that place. . . . I suggest that while there is much profit in further investigation of the slope of a graded river segment, and some, perhaps, in investigation of the slope or height of waterfalls, there cannot possibly be any profit in analyzing a slope that averages these two unlike types of slope. Indeed, the averaging of these basically different types of slopes is certain to defeat any attempt to understand either of them.

Thus, Mackin clearly distinguished two aspects of slope in relation to the operation of rivers: (1) a *timeless* aspect in which equilibrium adjustments prevailed and (2) a *timebound* aspect in which time, the history of the system, was of paramount importance. The relationship of time to fluvial analysis was subsequently stated more explicitly by Schumm and Lichty (1965) and elaborated upon *in extenso* by Thornes and Brunsden (1977).

Mackin's concern, however, was not merely with timeless versus timebound geomorphology. Rather, he made it clear that the appropriate goal was the *understanding of fluvial systems*. His worry was that an emphasis on method, specifically the empirical studies that appeared in the 1950s, would overshadow this goal. He wrote (Mackin, 1963, p. 148–149):

If this empirical approach—this blind probing—were the only way of quantifying geology, we would have to be content with it. But it is not; the quantitative approach is associated with the empirical approach, but it is not wedded to it. If you will list mentally the best papers in your own field, you will discover that most of them are quantitative and rational. In the study of rivers I think of Gilbert's field and laboratory studies of Sierra Nevada mining debris (1914, 1917), and Rubey's analysis of the force required to move particles on a stream bed (1938). These geologists, and many others that come to mind, have (or had) the happy faculty of dealing with numbers without being carried away by them—of quantifying without, in the same measure, taking leave of their senses.

Mackin emphasized the use of rational scientific thinking because that thinking sometimes became clouded in the use of certain new investigative tools that streamlined analysis. Despite his misgivings, however, Mackin clearly respected the new methodologies, as long as they had what he considered to be a geological purpose. He concluded the unpublished version of his talk as follows:

I am not sure that I have made these points clear, and I think that I will be damned by some in the groups on both sides. . . . But if this talk starts some discussion—I mean, of course, rational discussion—it will have served its purpose. There are many more things to be said for my position, and there are surely other points of view that are just as valid or perhaps more valid than mine. The things that I *am* sure of are that geology is a science, with different sorts of problems and methods, but not in any sense inferior to, or less mature than, any other science; and that anyone who hires out as a geologist, whether in practice, or in research, or in teaching, and then thinks like a physicist or a chemist or a statistician or an engineer, is not living up to his contract. A verse keeps running through my head:

I wonder what the vinters buy,
one half so precious as the stuff they sell.

In the Leopold and Langbein (1963) paper, progress in science is viewed as a transition from qualitative description to quantitative prediction:

A continued interest in classification, during the first third of the present century, took the form of assigning names to features of the landscape. Streams were designated as subsequent, superimposed, etc., and each such designation carried with it appropriate inference about both operative processes and historical sequence. Little attention was paid to the study of process, which, looking back at the record, now appears to have led to a neglect of field studies as the foundation of geomorphic science. As a result, the subject became one of decreasing interest to other workers in geology. An important aspect of this growing disinterest was that geomorphology, as practiced, seemed to lose its inherent usefulness.

In science usefulness is measured in part by ability to forecast, i.e., to predict relations postulated by reasoning about associations and subsequently subject to verification by experiment or field study. With this in mind, it is apparent that preoccupation with description could lead to decreasing usefulness because classification and description are usually insufficient bases for extrapolation and thus for prediction.

Note that the emphasis is on usefulness. Leopold and Langbein (1963) describe a revitalization of geomorphology through its concentration on process studies, increased use of quantitative data, and mathematical expression. In addition, given this new emphasis on prediction, Leopold and Lanbein (1963) also foresaw the emergence of stochastic principles as a dominating theme in fluvial geomorphology. Somewhat ironically, however, at the end of their paper, Leopold and Langbein (1963, p. 192), like Mackin (1963), concluded that problems in fluvial geomorphology are more important than methods:

The measure of a research man is the kind of question he poses. So, also, the vitality of a branch of science is a reflection of the magnitude or importance of the questions on which its students are applying their effort. Geomorphology is an example of a field of inquiry rejuvenated not so much by new methods as by recognition of the great and interesting questions that confront the geologist.

Because both of the *Fabric of Geology* papers cited above ultimately espoused geomorphogeny, both were seriously attempting to resolve a major conflict that was emerging in their science. Other fluvial investigators, however, have exacerbated the issues in this conflict beyond the simple divisions of rational and empirical, qualitative and quantitative, or pure and applied.

GEOMORPHOGENY OR GEOMORPHOTECHNICS?

To streamline the discussion of the modern crises in fluvial geomorphology, I will resurrect Lawson's (1894) old term "geomorphogeny." This will be defined as the *study of the origin, development, and changes in the landscapes and landforms of Earth and the Earth-like planets*. It will be contrasted with a hypothetical subdiscipline, "geomorphotechnics," defined as *the use of scientific and engineering methodology to acquire, interpret, and apply knowledge of the Earth's landscape and the processes operating upon it*. The emphasis of geomorphotechnics is on methodology and usefulness of results in the conduct of geomorphological research. The emphasis of geomorphologeny is on identifying critical questions that lead to scientific understanding. It is my belief that the recent growth of the geomorphotechnical approach to landform study has produced an imbalance of emphases in modern fluvial geomorphological research. The arbitrary device of defining subdisciplines is intended to facilitate discussion of this imbalance, not to introduce new terminology.

In their extreme forms, geomorphogeny and geomorphotechnics have numerous distinctive elements (Table 3). Their most divisive aspect, however, is in the views with which extremists in each camp hold the extremists of the opposite camp. Extremist geomorphogenists are viewed as prone to qualitative generalization, deductive reasoning, and emphasis on the "big picture" rather than the detailed evidence (Fig. 1). Extremist geomorphotechnicians, on the other hand, are viewed either as so immersed in predictive mathematical modeling that nature is ignored in order to simulate, or so tied to detailed measurement of processes that their inductive methodology ignores the relevance of their measurements to the genesis of landforms and landscapes (Fig. 2).

The infusion of mathematical rigor into geomorphotechnics may be viewed as both a blessing and a curse. Similar trends are apparent in modern hydrology, where, as noted by Klemeš (1987) "paradoxically, it has become almost axiomatic that to be a good hydrologist means to learn how to be a mediocre mathematician or statistician." By concentrating on mathematical modeling, facilitated by the "computer revolution," it has become easier to bypass understanding to achieve elegant quantitative predictions. Klemeš (1986) warns that an overemphasis on such models has the danger of transforming hydrology from a science into a type of dilettantism.

TABLE 3. CONTRAST BETWEEN EXTREME SCIENTIFIC APPROACHES IN FLUVIAL GEOMORPHOTECHNICS AND GEOMORPHOGENY

	Geomorphotechnics	Geomorphogeny
Emphasis	Use of scientific and engineering methodology (method dominated)	Study of landscape origins, development, and changes (problem dominated)
Theme	Utility of answers	Quality of questions
Goal	Prediction	Understanding
Measure of success	Objective tests of predictive capability	Intellectual satisfaction (multiple working hypotheses)
Quantification	Essential in all cases	Desirable where appropriate to problem
Optimum scales of study (temporal and spatial)	Small	Large
Process studies	Measure modern processes	Reconstruct ancient processes
Causality	Indeterminate or irrelevant	Deterministic by assumption
Examples of appropriate problems for study	Rill development on a hillslope Sediment transport in alluvial channels	Origin of pediments Origin of the Channeled Scabland Origin of channels and valleys on Mars
Landmark papers	Horton, 1945 (a hydrologist introducing a quantitative, predictive theory to the study of common fluvial landscapes)	Bretz and others, 1956 (geologists using the method of multiple working hypotheses and the principle of uniformity to understand an anomalous fluvial landscape)

Sparks (1971) observed that classifications are merely arbitrary constructions, designed to facilitate the discussion of diverse phenomena at the risk of some distortion of the truth. This undoubtedly applies to my arbitrary division of modern fluvial geomorphologists. Nevertheless, the distinction will serve to illustrate some interesting issues of conflict. Figure 3 makes some of the relevant points using the systems-terminology so in vogue among geomorphotechnicians. Because engineering prediction must be achieved, associations and analogies are extrapolated from observation, often with the aid of mathematical "short cuts." Science, however,

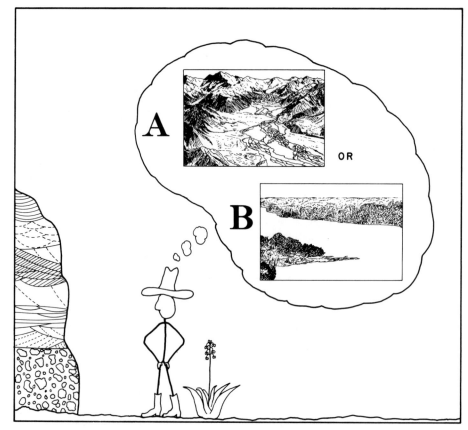

Figure 1. **Uncomplimentary view of a geomorphogenist studying an outcrop of fluvial sediments. Alternative genetic hypotheses for the sediments are envisioned simplistically as a mountain environment of braided streams (A) or as a lowland environment of a meandering river (B).**

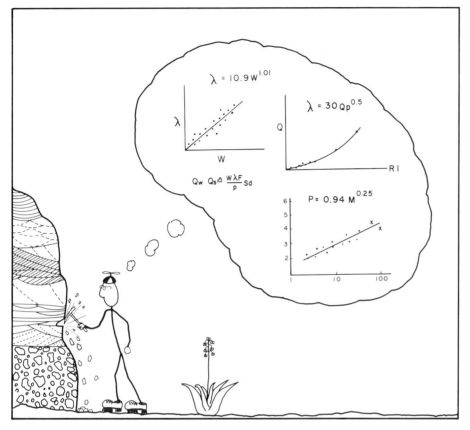

Figure 2. Uncomplimentary view of a geomorphotechnician studying same sediments as in Figure 1. Quantitative-engineering methodology results in many useful predictive relationships, but their relevance to this field problem is unclear.

views were those of Johnson (1932), who hypothesized lateral planation by streams; Lawson (1915) and Rich (1935), who hypothesized the parallel retreat of slopes; and Bryan (1922), Gilluly (1937), and Sharp (1940), who hypothesized complex combinations of planation, weathering, and wash in pedimentation processes.

In his last paper, published posthumously, Mackin (1970) reflected on the conflict that arose over the geomorphology of pediments. He noted the existence of two end members in an isomorphous series: corrasion (lateral planation) pediments and weathering-washing (sheet wash) pediments. In reflecting on the controversy over pediment genesis, Mackin (1970, p. 85) stated:

much of the controversy in the United States regarding pediments has arisen from the fact that people working in different places have seen different things. If a man is familiar with pediments of only one origin, and thinks that all pediments are the same in origin, he is likely to disagree with those who have worked with pediments elsewhere. Many of our problems are of this nature; the disagreements between different schools of thought in the United States, over the years, would have been greatly reduced had the men exchanged visits in the field.

When new field studies turn up phenomena inconsistent with prevailing views, one has anomalies. These introduce conflict in the existing scientific program. In the above case, it has become obvious that the same geomorphic form (the pediment) can be arrived at by multiple genetic paths. Moreover, many pediments may be relict forms, on which the presently active processes are irrelevant for their genesis. Here then is a geomorphogenetic problem that is particularly unsuited to geomorphotechnics. Indeed, Schumm (1985) lists the phenomenon of different processes or causes yielding similar effects, which he termed "convergence" or "equifinality," as one of seven reasons for geologic uncertainty for extrapolation in geomorphology. He views extrapolation as based on the use of analogy and the reliance on the concept of geological uniformity.

Extrapolation can be abused in either geomorphogeny or geomorphotechnics. King (1953) generalized concepts of parallel slope retreat into a model of regional pedimentation producing globally correlated planation surfaces. Because the model was inconsistent with abundant evidence that planation surfaces formed by other mechanisms, the persistence of its author in holding to it can only be considered an extravagance. Similarly, a computer simulation model, despite its mathematical elegance, is no more than an intellectual game if it fails to be consistent with detailed evidence in the field.

A good geomorphotechnical extrapolation must have the respect of those who will use it.

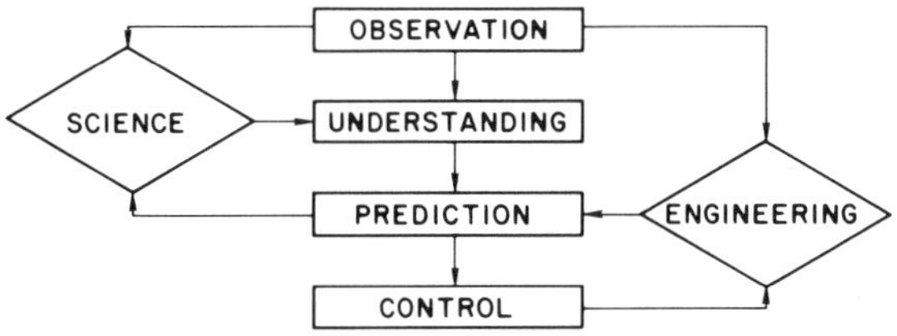

involves feedback loops so that temporary understandings (hypotheses) can be tested by their implications (predictions). Prediction for both science and engineering is a tool, not a goal. In science it serves to test hypotheses in order to generate a more satisfactory understanding. This is an endless process with a utopian goal (truth). In engineering, prediction serves a utilitarian purpose. It must be achieved quickly in order to control critical systems important to humankind.

A well-known hydrologist recently told me, "one does not do science unless all of the relevant equations can be written down and programmed into a computer." Such an attitude might serve well in geomorphotechnical studies, but it would not be much use in relation to a problem identified by Gilbert (1877). Surrounding the diorite porphyry intrusions of the Henry Mountains of Utah, there are slopes cut across relatively weak sedimentary rocks. These are veneered with cap rocks of gravel resting on surfaces that truncate the rock structure. The gravel caps are fluvial in origin but only several meters thick. Gilbert termed these mountain-bounding surfaces "slopes of planation." McGee (1897) subsequently provided the inevitable descriptive term for such landforms, "pediments," which are common geomorphological features in arid and semiarid regions (Fig. 4).

This review cannot do full justice to the prolonged debate over the origin of pediments and related concerns over the role of desert stream processes. At one extreme, Keyes (1912) argued that pediments were produced by wind erosion. This extravagant hypothesis was simply inconsistent with the field evidence. More rational

Figure 3. Systems diagram illustrating some differences between scientific and engineering methodologies (after Dooge, 1986).

Figure 4. Contrasting pediments in arid environments. (A) Dissected pediment surfaces of probable lateral planation origin south of the James Ranges in the Amadeus Basin in central Australia. The lateritic weathering profiles developed on the pediment gravel indicate that these surfaces probably formed during a Tertiary period of tropical weathering long before dissection during the modern arid climatic regime. (B) Exhumed rock pediment surface (right center) developed on granite west of the Catalina Mountains of south-central Arizona. Late Tertiary fan gravel at left probably correlates to the former mantle that once overlaid this ancient landform. (Photograph by Peter Kresan.)

This means that phenomena must be predicted according to certain standards and that the reasoning employed in the analysis must have general acceptance in the scientific community. Of itself this is merely a path to applying existing knowledge, not a means of discovery that will lead to new understanding. If the goal is discovery, then it is necessary to focus on anomalies, the failures of extrapolations, thereby bringing into question the prevailing understanding of the phenomenon under investigation. Because such questions are so vital to science, researchers may even employ in their analysis reasoning or methods that are not respectable to many contemporaries.

OUTRAGEOUS OR EXTRAVAGANT HYPOTHESES

In describing what he perceived to be a lack of scientific excitement at the meetings of the Geological Society of America, William Morris Davis (1926) wrote: "We shall be indeed fortunate if geology is so marvelously enlarged in the next thirty years as physics has been in the last thirty. But to make such progress violence must be done to our accepted principles." In this view, it is not respectable new methodologies that will advance geomorphology, but rather advancement will come from outrageous hypotheses that eventually are proven correct.

There is no more sacred principle in geomorphology than that of gradualism. Its antithesis, that of cataclysms, epitomizes disruptive activity in both science and its history. Cataclysmic geomorphological processes are events of unusual suddenness and magnitude that generate exceptional change in the landscapes of a planetary surface. The study of such processes follows in a long tradition of somewhat disreputable scientific activity. This lack of respect from other scientists derives from centuries of mistaken views as to what constituted proper scientific pursuits. Through the 17th and 18th centuries, it was common scientific practice to try to reconcile the surface features of the Earth with cataclysmic events, such as the Noachian flood (Davies, 1969). By the 19th century, the efforts of James Hutton, John Playfair, and Charles Lyell had replaced the biblical-catastrophist view of Earth history with a concept of gradualism. The new dogma held that fluvial landform development, like science itself, proceeded slowly and with order. Catastrophist views came into disrepute.

A great misconception among many earth scientists is that the above transition had something to do with uniformitarianism. As reviewed by Albritton (1967), Gould (1965), Hubbert

(1967), and Shea (1982), uniformitarianism in modern earth science holds merely that among competing hypotheses, the simple hypothesis often tends to prevail. This principle, also known as "Occam's razor," leads to hypotheses such as the following: the basic laws of nature remain invariant with time (or at least over the time period of interest). Uniformitarianism in its 20th century form has absolutely nothing to do with whether or not a process is catastrophic. Rather, it has everything to do with whether a hypothesized cataclysm obeys the laws of physics and is consistent with the field evidence.

One of the greatest of geological controversies arose in the 1920s and 1930s because of the absolutely erroneous belief that hypotheses involving cataclysmic origins of features could be rejected merely because catastrophic processes were inconsistent with uniformitarianism. J Harlen Bretz, in a series of a dozen papers, documented the cataclysmic flood origin of the Channeled Scabland in the finest tradition of uniformitarianism. In their righteous defense of an anachronistic, Victorian concept of scientific dignity, it was Bretz's critics who were the nonuniformitarians.

Despite this legacy, there remain vestiges of concern that the immense power and energy of extraordinary floods are something to be downplayed. How are such processes to be reconciled with the orderly, slow progression of landscape change? In science such paradoxes are known as anomalies, and it is in the study of such anomalies that major advances in knowledge can occur (Kuhn, 1962).

If the "Great Scablands Debate," as it was dubbed by Gould (1978), did not concern uniformity, why was it so significant for geological fluvial geomorphology? Bretz (1928, p. 701) provided the answer at the conclusion of his detailed descriptive paper on scabland bars:

Ideas without precedent are generally looked on with disfavor and men are shocked if their conceptions of an orderly world are challenged. A hypothesis earnestly defended begets emotional reaction which may cloud the protagonist's view, but if such hypotheses outrage prevailing modes of thought the view of antagonists may also become fogged.

On the other hand, geology is plagued with extravagant ideas which spring from faulty observation and misinterpretation. They are worse than "outrageous hypotheses," for they lead nowhere. The writer's Spokane Flood hypothesis may belong to the latter class, but it cannot be placed there unless errors of observation and direct inference are demonstrated. The writer insists that until then it should not be judged by the principles applicable to valley formation, for the scabland phenomena are the product of river channel mechanics. If this is in error, inherent disharmonies should establish the fact, and without adequate ac-

Figure 5. Giant current ripples in the Channeled Scabland of eastern Washington. The ripples are composed of gravel and boulders. Their size proved to be incontrovertible evidence consistent only with the cataclysmic flood hypothesis for the origin of the Channeled Scabland (Bretz and others, 1956). (A) Gravel bar adjacent to modern Snake River immediately downstream of junction with Palouse River (top center). Spacing of giant current ripples averages about 60 m. Note railway bridge for scale. (B) Giant current ripples on divide between Crab Creek (top left) and Canniwai Creek, Washington. Ripples with a spacing of about 60 m developed beneath flood water that innundated both valleys and spilled over divides between them.

quaintance with the region, this is the logical field for critics.

When Bretz and others (1956) documented the cataclysmic flood origin of the Channeled Scabland, they used meticulous field evidence to destroy the hypotheses of those who had disputed Bretz's outrageous hypothesis (Fig. 5). In essence the concept of the outrageous hypothesis espoused by Davis (1926) was made respectable. The methods employed by Bretz in his scabland studies were those dictated by what was important to achieve fundamental geomorphological understanding. Moreover, those methods had to be adapted from their original intended uses and modified to be geomorphological tools. This point was made most clearly by J. Hoover Mackin, who was trained both as an engineer and geologist. Mackin is quoted by Bretz and others (1956, p. 960) as stating, "to understand the scablands one must throw away textbook treatments of river work."

A new debate, in many ways comparable to that over the origin of the Channeled Scabland, has recently arisen with the discovery of channels and valleys on Mars (Fig. 6). Despite the current consensus about an aqueous origin for Martian channels (Mars Channel Working Group, 1983), it is fascinating that in the scientific literature nearly every conceivable fluid has been invoked to explain the Martian channels. The list includes low-viscosity turbulent-flow lava, wind, glacial ice, liquefaction of crustal materials, debris flow, and water. Some fluids were proposed without reasonable analogs to their geomorphic effects (for example, liquid alkanes and liquid CO_2). Some models achieved theoretical elegance (for example, the eolian hypothesis of Cutts and Blasius, 1981) but failed in their consistency with the available evidence, argued, not in the field, but from imagery of the planetary surface.

The rapid pace of hypothesis formulation to explain Martian channels illustrates the fine line that exists between the outrageous hypothesis that goes beyond the bounds of existing theory to explain startling new facts, and the extravagant hypothesis that ignores important facts merely to present speculation as a startling new theory. The former can serve as a stimulating source of scientific advancement, as in the famous Spokane Flood Debate over the origin of the Channeled Scabland. The latter, however, tends to suppress facts that are inconsistent with the favored hypothesis.

Clearly one may need to throw out the textbooks to advance scientific understanding, but one cannot throw out the necessity of testing hypotheses against the field data. To establish a hypothesis, the investigator cannot simply test the analogies (or models) that relate to that hypothesis; rather, the investigator must test the rival hypotheses. If the hypotheses are too closely identified with the hypothesizer, then this may be viewed as a regrettable, even an offensive, way to proceed. Hypotheses (models) that survive tests while their rivals are disproven are never fully established. They merely begin to seem more probable than other explanations. At some point, a model may begin to abstract from raw data the facts that its inventor perceives to be fundamental and controlling, placing these in relation to each other in ways that were not understood before, and thereby generating predictions of surprising new facts (Judson, 1980). At this point, the model has the qualities of a theory. True theories bind diverse consequences together in such an elegant manner that they compel belief by the scientific community. Geomorphology has little in the way of true scientific theories as defined in this manner. Its recent activity has largely consisted of the development of models (hypotheses) that move from one level of analogic reasoning to another.

Note that there are elements of the above process, such as surprise, elegance, and belief, that do not program well into a computer. If geomorphology is to develop scientific theories, it must cultivate these elements. It must continually question its basic tenets, even causality itself.

RANDOMNESS OR CAUSALITY?

The complexity of operation of fluvial systems has for centuries been a source of frustration for engineers and one of fascination for scientists. Albert Einstein, who published on river meandering, is said to have been impressed with the difficulties in explaining fluvial phenomena. One manifestation of this complexity has been termed "complex response" (Schumm, 1977). Complex response appears in river terraces (Womack and Schumm, 1977), which otherwise might be presumed to have certain genetic significance.

The concept of complexity is also central to arguments about the statistical (random) or deterministic (causal) nature of fluvial phenomena. For example, Leopold and Langbein (1963) argue for a basic indeterminacy of fluvial phenomena:

Where a large number of interacting factors are involved in a large number of individual cases or examples, the possibilities of combination are so great that physical laws governing forces and motions are not sufficient to determine the outcome of these interactions in an individual case. The physical laws may be completely fulfilled by a variety of combinations of the interrelated factors. The remaining statements are stochastic in nature rather than physical. These stochastic statements differ from deterministic physical laws in that the former carry with them the idea of an irreducible uncertainty. As more is known about the processes operating and as more is learned about the factors involved, the range of uncertainty will decrease, but it never will be entirely removed.

In essence, the orderliness and causality of nature are matters of faith. This is no better illustrated than in the classic exchange between Albert Einstein and Niels Bohr. Paraphrasing its recounting by Bohr (1949), Einstein's question was, "Do you really believe that God plays dice?" Bohr's reply was, in effect, "It is presumptuous of us to say what God does." Heisenberg's famous "uncertainty principle" was applied to the indeterminancy of specifying the position and momentum of electrons (Heisenberg, 1958). Bohr extended that principle to a philosophical one: "There is no such thing as an electron with definite position and momentum." To this, Einstein's rational mind rebelled. For him the uncertainty principle was not an inherent feature of reality, but rather it was merely a shortcoming of the current theory for reality.

The Einstein-Bohr debate was never resolved. Indeed it may be the ultimate dilemma of all science. The need to predict certain phenomena often requires assumptions of randomness to predict at least probability distributions of large populations of those phenomena. This expedient, however, has also been termed "the philosophy of scientific desperation." Sometimes it is the individual phenomena that require scrutiny.

Many scientists might rebel at this view that causality and determinism are matters of faith. Adherents of Bridgman's (1936, 1959) logical positivism would hold that a proposition has scientific meaning only if it can be tested for validity by accepted scientific principles. Smart (1979), for example, concluded that claims of either inherent macroscopic randomness or determinism for fluvial processes are both propositions without scientific content because neither can be adequately tested.

A pragmatic view of scientific understanding, one that is testable, is that it is achieved when the properties of the system under study can be predicted to a satisfactory level of accuracy. There is no question that empirical geomorphological studies, such as those of drainage networks or river-channel hydraulic geometry, are made immensely more useful when fitted to predictive models. Indeed, Shreve (1979) argued that such models constitute an appropriate goal for modern fluvial geomorphology. Models can be based on randomness, as is the probabilistic-topologic approach to drainage-basin geomorphology (Shreve, 1975), or they can be deterministic, as are diffusion-equation-based models of hillslope evolution (Kirby, 1971). The quality of such studies is that they provide predictions of phenomena that are useful in many applications.

Success at prediction is certainly a goal that is

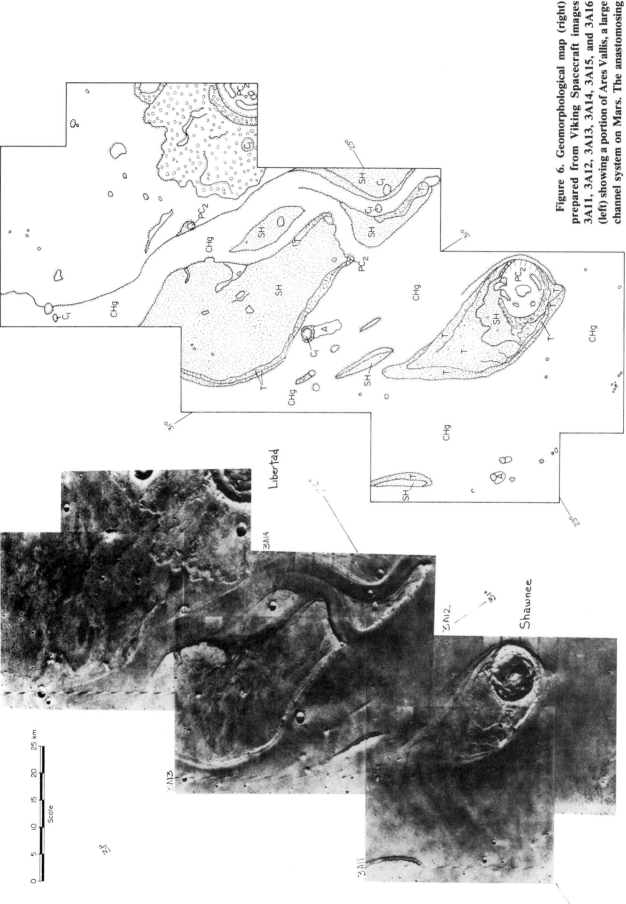

Figure 6. Geomorphological map (right) prepared from Viking Spacecraft images 3A11, 3A12, 3A13, 3A14, 3A15, and 3A16 (left) showing a portion of Ares Vallis, a large channel system on Mars. The anastomosing pattern of channelized zones (CHg) indicates flow to the north (from lower right to upper left). The large crater at the bottom center (named "Shawnee") acted as an obstacle to large-scale fluid flows, resulting in preservation of the terraced upland (T) on its downstream margin. Streamlined hills (SH) are uplands that were modified by erosion in the large-scale fluid flows. Scarps bounding these uplands are up to a few hundred meters high. Relatively fresh craters are designated C_1, and craters modified by fluid flows are designated PC_2. Wind streaks (A) show local effects of winds blowing from north (top) to south (bottom).

considered paramount to many clients of scientific knowledge. Fluvial geomorphology will exhibit soundness and prestige through its quanitative predictions. Although we scientists must concern ourselves with this public perception of scientific success, we must not let the scientific image conflict with the scientific process.

Kuhn (1962, 1978) argued that laboratory measurements in science generally refine principles that are presupposed by the prevailing consensus of "normal science," that is, the paradigm. In geology and hydrology, the "laboratory" is often the Earth itself, but the principle remains that the quantitative regularity sought is conditioned by the regularity that is expected. Stated another way, Kuhn's argument holds that prediction with "loaded dice" is far more prevalent in science than frustrating attempts to fathom the unknown. Thus, one can demonstrate that the predictive capabilities of many models arise because the tests of those models are severely constrained.

A story about Alexander the Great may illustrate the choice that must be made. King Gordius of Phrygia supposedly tied an immensely intricate knot and stated that the knot could be undone only by the future ruler of Asia. Faced with the knot's complexity, Alexander summarily cut it with his sword.

There are also shortcuts in science. Some especially powerful methodologies can achieve useful results, but in doing so, they may bypass the fundamental understanding of the natural world. Faced with the complexity of the geological record, many geologists have tried compromise on this issue through a rational method of working hypotheses (Gilbert, 1886; Chamberlin, 1897). Gilbert (1886) even tried to avoid the pitfalls of causality by conceiving of "antecedent and consequent" relations constituting a "plexus" that pervades nature. Hypotheses are used to penetrate this plexus, but it is not their predictive ability that the scientist must seek. Rather it is the points at which models (hypotheses) fail to predict accurately. Science will progress only by concern with the failings of its own accomplishments. Fluvial geomorphology is fated to this same dilemma.

CONCLUSIONS

Niels Bohr believed that there were two kinds of scientific truth: trivialities, where opposites are obviously absurd, and profound truths, recognized by the fact that the opposite is also a profound truth. There is much good to be said of both geomorphogeny and geomorphotechnics. I have chosen a theme of conflict to provoke thought rather than offense. In the first half of this century, geomorphology split into one camp that held explanatory genetic description to be all important and another camp that argued for empirical description and objective analysis that was geographically useful. The geography/geology dichotomy of the past has now been replaced by another dichotomy, more subtle yet more divisive. One activity seeks to satisfy the rational mind by searching for ultimate origins or causes, while the other uses objective criteria, including tests and predictability to wed technological advances into a geomorphology that is both useful and respectable. These are both worthy goals, but the advantages and limitations of each must be kept in mind by their practitioners.

Consistent with the view of fluvial geomorphology argued in this paper, one might define a turning point of study, one based on a change of view related to geomorphological problems rather than methods. This change need not have affected the majority of geomorphologists, but it must carry the essential element of anomaly that creates conflict and drives a science toward developing its own methodologies to achieve understanding. For this reason, I do not view the critical turning point in the history of fluvial geomorphological thought to be defined by the paper of a hydrologist (Horton, 1945). Horton (1945) introduced very significant quantitative approaches to principles of stream junctions, slopes, and divides that were already appreciated, at least in part, by earlier masters of metaphorical description such as William Morris Davis. Nor would I pick papers by geographers, introducing the rigor of spatial statistical treatment of the extant landscape. Nor do I propose that the important contributions of regime theory or sediment-transport hydraulics, the two conflicting approaches in river engineering (Leliavsky, 1955), are the critical inputs. Instead, I propose that geological fluvial geomorphology has been and will continue to be most enriched by the demonstration by a geologist that his outrageous hypothesis for the cataclysmic flood origin of a landscape was indeed the appropriate route to understanding. Published after his last season of field work in the Channeled Scabland, J Harlen Bretz's detailed refutation of all criticisms of his cataclysmic flood hypothesis (Bretz and others, 1956) is meticulous on detail but profound in its lasting lesson of how to pursue fluvial studies as a geologist.

The reverence with which geological fluvial geomorphologists hold the work of G. K. Gilbert might be considered ironic in view of the conflicts described in this paper. Gilbert's focus on equilibrium emphasized the timeless over the timebound aspects of geomorphology. He used physical principles in analyzing geomorphological processes. I think that what most impresses geologists, however, is that Gilbert carried the principle of balance into his scientific philosophy. His purpose was always clearly geomorphogenetic, and he chose the geomorphotechnical approach that was most appropriate to the task. It is true that he worked in a simpler time, before the advent of computers, geostatistics, and simulation modeling. Nevertheless, these developments merely make the need today even more acute for the type of philosophical balance exemplified by Gilbert's scientific work.

In a very perceptive review on how much hydraulic engineering has contributed to the understanding of rivers, J. F. Kennedy (1983) found that the history of the subject illustrated the operation of a nonphysical conservation law. The law states that the anguish of the river researcher over the status of his subject is balanced only by his joy that so much remains to be elucidated and reliably formulated.

ACKNOWLEDGMENTS

I thank William L. Graf, Jim E. O'Connor, Stanley A. Schumm, and C. R. Twidale for review comments and discussions. This paper was completed while the author was Visiting Research Fellow, Department of Geology and Geophysics, The University of Adelaide, South Australia.

REFERENCES CITED

Albritton, C. C., 1967, Uniformity, the ambiguous principle, *in* Albritton, C. C., ed., Uniformity and simplicity: Geological Society of America Special Paper 89, p. 1–2.
Bohr, N., 1949, Discussion with Einstein on epistemological problems in atomic physics, *in* Schilpp, P. A., ed., Albert Einstein: Philosopher-scientist: London, England, Cambridge University Press, p. 199–241.
Bretz, J H., 1928, Bars of the Channeled Scabland: Geological Society of America Bulletin, v. 39, p. 643–702.
Bretz, J H., Smith, H.T.U., and Neff, G. E., 1956, Channeled Scabland of Washington: New data and interpretations: Geological Society of America Bulletin, v. 67, p. 957–1049.
Bridgman, P. W., 1936, The nature of physical theory: Princeton, New Jersey, Princeton University Press.
—— 1959, The way things are: Cambridge, Massachusetts, Harvard University Press.
Bryan, Kirk, 1922, Erosion and sedimentation in the Papago Country, Arizona: U.S. Geological Survey Bulletin 730, p. 19–90.
—— 1941, Physiography, *in* Geology, 1888–1938, Fiftieth Anniversary Volume: New York, Geological Society of America, p. 1–15.
—— 1950, The place of geomorphology in the geographic sciences: Annals of the Association of American Geographers, v. 40, p. 196–208.
Chamberlin, T. C., 1897, The method of multiple working hypotheses: Journal of Geology, v. 5, p. 837–848.
Chorley, R. J., 1962, Geomorphology and general systems theory: U.S. Geological Survey Professional paper 500-B, 10 p.
Chorley, R. J., and Beckinsale, R. P., 1980, G. K. Gilbert's geomorphology, *in* Yochelson, E. L., ed., The scientific ideas of G. K. Gilbert: Geological Society of America Special Paper 183, p. 129–142.

Church, M., 1980, Records of recent geomorphological events, *in* Cullingford, R. A., Davidson, D. A., and Lewin, J., eds., Timescales in geomorphology: New York, John Wiley, p. 13–29.

Cotton, C. A., 1956, Geomorphology, geomorphography, geomorphogeny and geography: New Zealand Geographer, v. 12, p. 89–90.

Cutts, J. A., and Blasius, K. R., 1981, Origin of Martian outflow channels: The eolian hypothesis: Journal of Geophysical Research, v. 86, p. 5075–5102.

Davies, G. L., 1969, The Earth in decay: London, England, MacMillan.

Davis, W. M., 1912, Relation of geography to geology: Geological Society of America Bulletin, v. 23, p. 93–124.

—— 1926, The value of outrageous geological hypotheses: Science, v. 63, p. 463–468.

Dooge, J.C.I., 1986, Looking for hydrologic laws: Water Resources Research, v. 22, p. 46S–58S.

Dury, G. H., 1972, Some current trends in geomorphology: Earth-Science Reviews, v. 8, p. 45–72.

Dutton, C. E., 1885, Mount Taylor and the Zuni Plateau: U.S. Geological Survey Annual Report, no. 6, p. 113–198.

Fenneman, N. M., 1939, The rise of physiography: Geological Society of America Bulletin, v. 50, p. 349–360.

Feyeraband, P., 1975, Against method: London, England, NLB, 339 p.

Gilbert, G. K., 1877, Geology of the Henry Mountains: Washington, D.C., U.S. Geological and Geographical Survey, 160 p.

—— 1886, The inculcation of scientific method by example: American Journal of Science (3rd ser.), v. 31, p. 284–299.

—— 1914, The transportation of debris by running water: U.S. Geological Survey Professional Paper 86, 263 p.

—— 1917, Hydraulic-mining debris in the Sierra Nevada: U.S. Geological Survey Professional Paper 105, 154 p.

Gilluly, James, 1937, Physiography of the Ajo region, Arizona: Geological Society of America Bulletin, v. 48, p. 323–348.

Gould, S. J., 1965, Is uniformatarianism necessary?: American Journal of Science, v. 263, p. 223–228.

—— 1978, The great scablands debate: Natural History, v. 87, no. 7, p. 12–18.

Heisenberg, W., 1958, Physics and philosophy: New York, Harper, 206 p.

Horton, R. E., 1945, Erosional development of streams and their drainage basins: Hydrophysical approach to quantitative morphology: Geological Society of America Bulletin, v. 56, p. 275–370.

Hubbert, M. K., 1967, Critique of the principle of uniformity, *in* Albritton, C. C., ed., Uniformity and simplicity: Geological Society of America Special paper 89, p. 3–33.

Johnson, D. W., 1929, The geographic prospect: Association of American Geographers Annals, v. 19, p. 168–231.

—— 1932, Rock fans of arid regions: American Journal of Science, 5th ser., v. 23, p. 389–416.

Johnston, R. J., 1983, Geography and geographers: Anglo-American human geography since 1945: London, England, Edward Arnold, 264 p.

Judson, H. F., 1980, The search for solutions: New York, Holt, Rinehart and Winston, 212 p.

Judson, Shelton, 1958, Geomorphology and geology: New York Academy of Sciences Transactions, ser. II, v. 20, p. 305–315.

Kennedy, J. F., 1983, Reflections on rivers, research, and Rouse: American Society of Civil Engineers, Journal of Hydrological Engineering, v. 109, p. 1258–1260.

Keyes, C. R., 1912, Deflative systems of the geographical cycle in an arid climate: Geological Society of America Bulletin, v. 23, p. 537–562.

King, L. C., 1953, Canons of landscape evolution: Geological Society of America Bulletin, v. 64, p. 721–752.

Kirby, M. J., 1971, Hillslope process-response models based on the continuity equation, *in* Brunsden, D., ed., Slopes: Form and process: Institute of British Geographers Special Publication 3, p. 15–30.

Klemeš, V., 1986, Dilettantism in hydrology: Transition or destiny?: Water Resources Research, v. 22, p. 177S–188S.

—— 1987, Hydrological and engineering relevance of flood frequency analysis, *in* Singh, V. P., ed., Hydrologic frequency modeling: Boston, Massachusetts, Dordrecht, p. 1–18.

Kuhn, T. S., 1962, The structure of scientific revolutions: Chicago, Illinois, University of Chicago Press, 172 p.

—— 1978, The essential tension: Chicago, Illinois, University of Chicago Press, 366 p.

Lawson, A. C., 1894, Geomorphogeny of the coast of northern California: University of California Publications, Bulletin of the Department of Geology, v. 1, p. 241–272.

—— 1915, The epigene profiles of the desert: University of California Publications, Bulletin of the Department of Geology, v. 9, p. 23–48.

Leliavsky, Serge, 1955, An introduction to fluvial hydraulics: London, England, Constable and Company, 257 p.

Leopold, L. B., and Langbein, W. B., 1963, Association and indeterminancy in geomorphology, *in* Albritton, C. C., ed., The fabric of geology: Reading, Massachusetts, Addison-Wesley, p. 184–192.

Leopold, L. B., Wolman, M. G., and Miller, J. P., 1964, Fluvial processes in geomorphology: San Francisco, California, Freeman, 522 p.

Mackin, J. H., 1963, Rational and empirical methods of investigation in geology, *in* Albritton, C. C., ed., The fabric of geology: Reading, Massachusetts, Addison-Wesley, p. 135–163.

—— 1970, Origin of pediments in the western United States, *in* Problems of relief planation: Studies in Hungarian geography, Volume 8: Budapest, Hungary, Akademia Kiado, p. 85–105.

Mars Channel Working Group, 1983, Channels and valleys on Mars: Geological Society of America Bulletin, v. 94, p. 1035–1054.

McGee, W. J., 1897, Sheetflood erosion: Geological Society of America Bulletin, v. 8, p. 87–112.

Miller, J. P., 1959, Geomorphology in North America: Przeglad Geograficzny (Polish Geographical Review), v. 31, p. 567–587.

Morisawa, M., 1985, Development of quantitative geomorphology, *in* Drake, E. T., and Jordan, W. M., eds., Geologists and ideas: A history of North American geology: Geological Society of America, The Geology of North America, Centennial Special Volume 1, p. 79–107.

Ollier, C. D., 1981, Tectonics and landforms: London, England, Longman, 324 p.

Pitty, A. F., 1982, The nature of geomorphology: London, England, Methuen, 161 p.

Pyne, S. J., 1980, Grove Karl Gilbert: A great engine of research: Austin, Texas, University of Texas Press, 306 p.

Rich, J. L., 1935, Origin and evolution of rock fans and pediments: Geological Society of America Bulletin, v. 46, p. 999–1024.

Ritter, D. F., 1988, Landscape analysis and the search for geomorphic unity: Geological Society of America Bulletin, v. 100, p. 160–171.

Rubey, W. W., 1938, The force required to move particles on a stream bed: U.S. Geological Survey Professional Paper 189-E, p. 120–140.

Russell, I. C., 1904, Physiographic problems of today: Journal of Geology, v. 12, p. 524–550.

Schumm, S. A., 1977, The fluvial system: New York, Wiley, 338 p.

—— 1985, Explanation and extrapolation in geomorphology: Seven reasons for geologic uncertainty: Japanese Geomorphological Union, Transactions, v. 6, p. 1–18.

Schumm, S. A., and Lichty, R. W., 1965, Time, space, and causality in geomorphology: American Journal of Science, v. 263, p. 110–119.

Sharp, R. P., 1940, Geomorphology of the Rubey–East Humbolt Range, Nevada: Geological Society of America Bulletin, v. 51, p. 337–372.

Shea, J. H., 1982, Twelve fallacies of uniformitarianism: Geology, v. 10, p. 455–460.

Shreve, R. L., 1975, The probabilistic-topologic approach to drainage-basin geomorphology: Geology, v. 3, p. 527–529.

—— 1979, Models for prediction in fluvial geomorphology: Mathematical Geology, v. 11, p. 165–174.

Smart, J. S., 1979, Determinism and randomness in fluvial geomorphology: EOS (American Geophysical Union Transactions), v. 60, p. 651–655.

Sparks, B. W., 1971, Rocks and relief: London, England, Longman, 404 p.

Strahler, A. N., 1952, Dynamic basis of geomorphology: Geological Society of America Bulletin, v. 63, p. 923–938.

Thornbury, W. D., 1969, Principles of geomorphology: New York, Wiley, 594 p.

Thornes, J. B., 1979, Fluvial processes, *in* Embleton, C., and Thornes, J., eds., Process in geomorphology: New York, Wiley, p. 213–271.

Thornes, J. B., and Brunsden, Denys, 1977, Geomorphology and time: London, England, Methuen, 208 p.

Tinkler, K. J., 1985, A short history of geomorphology: London, England, Croom Helm, 317 p.

Twidale, C. R., 1977, Fragile foundations: Some methodological problems in geomorphological research: Revue de Geomorphologie Dynamique, v. 26, p. 81–95.

Womack, W. R., and Schumm, S. A., 1977, Terraces of Douglas Creek, northwestern Colorado: An example of episodic erosion: Geology, v. 5, p. 72–76.

Manuscript Received by the Society December 17, 1987
Revised Manuscript Received April 18, 1988
Manuscript Accepted April 19, 1988

Dynamic changes and processes in the Mississippi River delta

CENTENNIAL ARTICLE

JAMES M. COLEMAN *Coastal Studies Institute, School of Geoscience, Louisiana State University, Baton Rouge, Louisiana 70803*

ABSTRACT

Research in the modern delta of the Mississippi River has revealed short-term changes and processes that are of significant magnitude. Deltaic lobes, each lobe covering an area of 30,000 sq km and having an average thickness of 35 km, switch sites of deposition on an average of every 1,500 yr. Through short periods of geologic time, this process results in a relatively thick accumulation of stacked deltaic cycles covering extremely large areas. Within a single delta lobe, and operating on an even higher frequency, are bay fills and overbank splays. Bay fills, having areas of 250 sq km and thickness of 15 m, require only 150 yr to accumulate. Four major events have taken place in the modern Balize delta since 1838. Overbank splays are much smaller, covering areas of less than 2 sq km and having thicknesses of 3 m, but are associated with high floods on the river. At the river mouth, continued progradation of the distributary channel can form distributary mouth sand bodies that have dimensions of 17 km long, 8 km wide, and a thickness of 80 m in a period of only 200 yr. Differential sedimentary loading at the river mouth results in formation of diapirs that display vertical movements in excess of 100 m in a period of 20 yr. On the subaqueous delta platform, sediment instabilities operate nearly continuously, mass-moving large quantities of shallow-water deposits to deeper-water environments via arcuate rotational slides and mudflow gullies and depositional lobes. All of these changes and processes operate at differing spatial and temporal scales, but all result in deposition of large volumes of sediment over extremely short periods of time.

INTRODUCTION

Deltas are extremely dynamic environments, a fact that was recognized early in the geologic literature. Lyell (1847) and Riddell (1846) commented on the extremely rapid changes in the sites of sedimentation at the mouths of the Mississippi River, but they had little information on the styles of sedimentation or on the reasons responsible for rapid changes. Credner (1878), in a study of numerous large world deltas, alluded to rapid changes in sites of deposition through time, but he had little knowledge of sedimentation patterns. Prestwich (1885) speculated on the age and dynamics of several deltas, including the Nile delta, "it would have required, at the present rate of deposit, 13,500 years to accumulate the 39 feet of Nile sediment which underlies the statue of Rameses." He recognized that deltas were subject to dynamic sedimentation processes: rapid progradation and infilling of the depositional basins; and deposition of extensive, thick volumes of sediment over relatively short periods of geologic time. Johnson (1891), in an article in the *Geological Society of America Bulletin*, described the Nita Crevasse along the Mississippi River, which broke off the main river channel and deposited a relatively large volume of sediment in the adjacent interdistributary basin within only a few years. Utilizing historic maps, he showed that such "crevassing" was a common process of the Mississippi River. Barrell (1917), speculating on depositional rates throughout geologic time, indicated that present depositional rates (especially in river deltas) probably prevailed in the past. These and other geologists who speculated on the dynamic nature of deltas had few or no methods with which to quantify the nature of the changes or to evaluate the sedimentary framework of the modern delta deposits and their rate of accumulation.

The true nature of the dynamic tendency of deltas to change through short periods of time was not fully realized until the pioneering research on the Mississippi River delta by Russell, Fisk, and other contemporaries in the 1930s, '40s, and '50s (Trowbridge, 1930, 1954; Russell, 1936, 1940, 1958; Russell and Russell, 1939; Fisk, 1944, 1947, 1952, 1955, 1956, 1961; Fisk and others, 1954; Fisk and McFarlan, 1955; Bates, 1953; Shepard, 1955, 1956; Scruton, 1960; Kolb and Van Lopik, 1966; Welder, 1959). These classical studies formed the foundation for subsequent delta research, not only on the Mississippi River delta but also on a wealth of other modern deltas world-wide. In addition, they provided the basis for application of observations on modern sedimentation to the study of ancient delta sedimentary sequences, both on outcrops and in the subsurface. Since these pioneering works, other researchers have added details concerning the sedimentary patterns in the subaerial delta and, with the development of marine remote sensing techniques (high-resolution seismic and side-scan sonar), have made significant observations on the subaqueous delta platform of the Mississippi River delta. This paper will attempt to summarize some of the more dynamic aspects of the sedimentation processes that have been documented on the Mississippi River delta since these pioneering studies.

MISSISSIPPI RIVER DELTA

The Mississippi River system, the largest in North America, drains an area of 3,344,560 sq km. This river system has been active since at least Late Jurassic times and has significantly influenced depositional patterns in the northern Gulf of Mexico. The modern river has an average water discharge of 15,360 cu m/sec, and average maximum and minimum discharges are 57,900 and 2,830 cu m/sec, respectively. Annual sediment discharge is estimated at 6.21×10^{11} kg; the bedload consists of 90% fine sand, and the suspended load is characterized by 65% clay and 35% silt and very fine sand. Thus the Mississippi River carries a substantial sediment load annually, and a high percentage consists of fine-grained clays and silts.

The drainage basin and alluvial valley (Fig. 1) cover a large area of the North American continent. A wide variety of igneous, metamorphic, and sedimentary rocks are present in the basin, and the sediment load is composed of a highly varied mineralogical suite. The limits of the drainage basin have existed since at least Cretaceous times, consisting of the Rockies

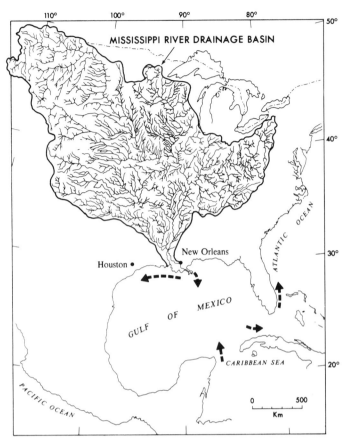

Figure 1. Drainage system of the Mississippi River.

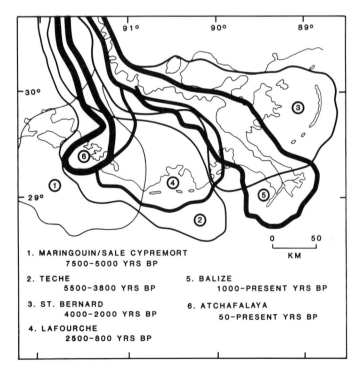

1. MARINGOUIN/SALE CYPREMORT
   7500-5000 YRS BP
2. TECHE
   5500-3800 YRS BP
3. ST. BERNARD
   4000-2000 YRS BP
4. LAFOURCHE
   2500-800 YRS BP
5. BALIZE
   1000-PRESENT YRS BP
6. ATCHAFALAYA
   50-PRESENT YRS BP

Figure 2. Shifting sites of deltaic sedimentation during the past 7,500 yr. Note the newly emerging Atchafalaya delta lobe (no. 6), which began forming only 50 yr ago.

on the west, the Appalachians on the east, and the Precambrian Shield on the north. The modern delta area forms the coastal wetlands of Louisiana and covers some 28,568 sq km. Through time, both the alluvial valley and the delta system have shifted sites of deposition. The bulk of the sediments comprising the Gulf Coast geosyncline have been derived in part from ancestral Mississippi River systems.

The Gulf of Mexico is the receiving basin for this large fluvial sediment load. The basin is a semi-enclosed body of water characterized by relatively low-energy marine conditions. Tidal range averages 0.43 m; littoral currents are minimal and generally set from east to west on the shallow shelf. Wave energy is extremely low, averaging 0.034×10^7 ergs/sec/m of coast. Compared to many other modern-day major rivers, the Mississippi delta displays extremely low wave-energy levels. Regional subsidence in the basin is high, and areas of localized subsidence, caused primarily by loading and compaction, can be extreme. The delivery of high sediment load to the Gulf basin has resulted in building a thick sequence of alluvial, deltaic, shelf, slope, and basinal deposits, which, since Cretaceous time, has prograded the coastal-plain shoreline seaward.

SWITCHING SITES OF DELTA DEPOSITION

Since at least Cretaceous times, the drainage basin of the Mississippi River system has been delivering sediments to the Gulf of Mexico basin. Throughout this long period of geologic time, the sites of maximum deposition or depocenters have shifted within the Gulf coastal plain, resulting in continuous outbuilding or progradation of the shoreline. During the Quaternary, the river was significantly influenced by changing sea levels associated with the advance and retreat of continental glaciers. During the last low sea-level stand (some 18,000 yr ago), the Mississippi River entrenched its valley, numerous channels scoured across the continental shelf, and deltas were depositing sediments near the shelf edge. This resulted in rapid seaward building of the shelf edge. As sea level began to rise, the site of maximum sedimentation shifted to the alluvial valley, and few deltas can be found on the shelf from the period 18,000 to 9,000 yr B.P. By 9,000 yr B.P., a significant part of the alluvial valley had been filled, and the river commenced to construct its modern delta plain.

In these more recent times, the shifting deltas of the Mississippi River have constructed a delta plain that has a total area of 28,568 sq km, of which 23,900 sq km is subaerial in nature. This type of delta switching, combined with a continuously subsiding basin, results in vertically stacked cyclic sequences that operate at extremely high frequencies. Frazier (1967), in a classical paper, illustrated the complexity of these overlapping delta lobes in the Holocene deltaic plain of the Mississippi River. The various sites of the major delta lobes during the past 7,000 yr are illustrated in Figure 2. One of the earlier deltas shown, the Maringouin/Sale Cypremort, prograded and constructed a delta lobe along the western flank of the present Mississippi River delta plain during the period ~7,500 to 5,000 yr ago. This delta lobe spread far out onto the present shelf, and thicknesses are on the order of 10 to 20 m. After ~2,500 yr of active progradation, the bifurcating and shifting channel pattern became extremely complex; the river channel abandoned this site of deposition, and a new delta lobe, the Teche delta lobe, began to be constructed along the central Louisiana coast. A similar sequence of events ensued, and within a period of 1,700 yr, this delta site was abandoned, and a new delta lobe (St. Bernard) began a period of active build-out. This process of switching sites of delta lobes continued, each delta lobe requiring a period of 1,200 to 1,700 yr to construct a single lobe.

The modern delta (Balize or Birdfoot) began its progradation ~800 to 1,000 yr ago; its rate of progradation has diminished with time, and the river is presently seeking a new site of deposition. Within the past 100 yr, a new distributary, the Atchafalaya, has begun to take more of the river's flow. Without the interference of man, this channel would have captured

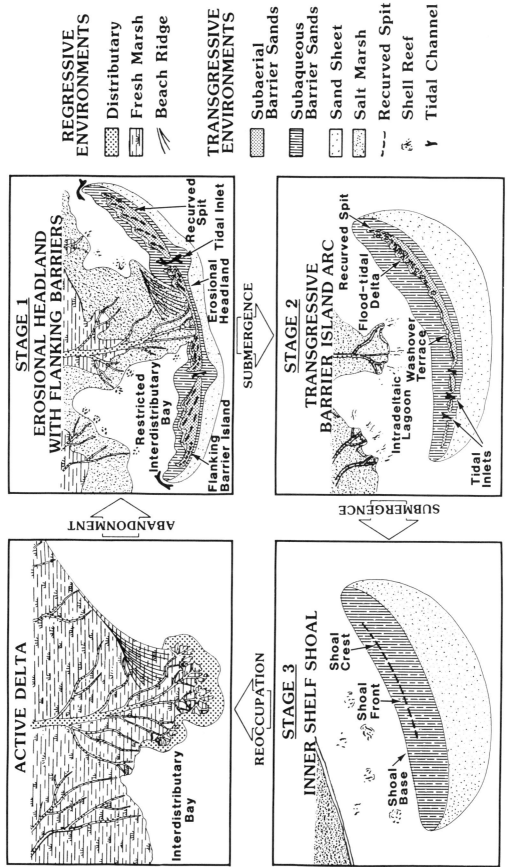

Figure 3. Stages in a deltaic cycle. After Penland and Boyd, 1981.

the flow, and the modern Balize delta would have begun a period of deterioration. Even with man's control, the new Atchafalaya River delta (Fig. 2) has begun a period of active delta construction within Atchafalaya Bay (Roberts and others, 1980a). This orderly repetition of depositional events and shifting sites of sedimentation results in the formation of overlapping and laterally displaced deltaic sequences extending ~400 km along the Louisiana coast and nearly 200 km in a dip direction, with sediment thickness ranging from only a few tens of meters to a maximum of 200 m. The spatial and temporal magnitude of these overlapping delta lobes would make it very difficult to decipher similar sequences in ancient sedimentary rocks; it would be particularly difficult to utilize fossil remains or even radiometric techniques to resolve the time intervals involved in such rapidly developing sequences.

Although Russell (1936) recognized that each delta lobe underwent a series of events from its original phase of progradation to its eventual transgression by marine waters, it was Scruton (1960) who formulated this process into a well-defined concept. The phase of build-out or progradation was termed the "constructional phase," and the period of abandonment was called the "destructional phase." Scruton described this process verbally and used deltas of various ages and states of decay to illustrate this cycle. Coleman and Gagliano (1964), using a model developed from the modern Mississippi delta, set this delta switching process into a sedimentological framework. Penland and Boyd (1981) illustrated this cyclic nature graphically, as shown in Figure 3.

During the period of active progradation, distributaries rapidly prograde seaward, and overbanking during floods fills the innumerable shallow interdistributary bays (active delta, Fig. 3). Thus sediments accumulate at a faster rate than subsidence or rising sea level can bury them. When a new site of deposition is formed, the distributary channels of the old delta lobe, no longer receiving river sediment, cease to prograde; wave processes commence to rework the distributary mouth bar sands, and initial transgressive processes commence. When annual flooding does not occur, sediments are not deposited overbank, and the subaerial delta, deprived of sediment, begins to be inundated with marine waters. This sequence is shown in stage 1 in Figure 3. The process of deterioration continues, and soon the reworked sands are separated from the mainland (old delta), and the barrier island develops more fully by alongshore spreading of sediments. Transgression continues, and a thin sand sheet is left seaward of the landward-migrating barrier. This sequence is shown in stage 2 in Figure 3. As greater amounts of sand and reworked lagoon deposits accumulate, the low wave-energy conditions cannot maintain the previous rate of sediment reworking; landward shoreline migration decreases, and subsidence begins to become dominant. The barrier island becomes a submerged shoal on which normal marine processes spread sediments laterally. These shoals are extremely rich in carbonate shell debris (stage 3, Fig. 3). The deterioration process, to this point, generally requires 2,000 to 4,000 yr to complete. Eventually another delta lobe begins its progradation over the deteriorated delta lobe.

Thus each major delta lobe or cycle is composed of a detrital lens (regressive or progradational phase) that is bounded on all sides by essentially nondetrital sediments indigenous to the basin of deposition (transgressive or abandonment phase, Coleman and Gagliano, 1964). The regressive phase is characterized by high percentages of relatively coarse clastics, abrupt facies changes, and rapid accumulation and burial rates. The bounding sediments (or transgressive phase) tend to be rich in organic constituents and chemical precipitants, to have a slower accumulation rate, and to be tabular or blanket accumulations with considerable lateral continuity. In the Mississippi River delta, each delta cycle covers an area of ~30,000 sq km and has an average thickness of 35 m. Within even short periods of geologic time, these types of stacked cyclic deltaic sequences can attain thicknesses of several thousands of meters and cover areas in excess of 150,000 sq km. In the Pleistocene depocenter of the Gulf Coast, similar deltaic sequences exceed 3,600 m in thickness and were deposited in a time framework of less than 2.5 m.y. The cyclic concept provides a framework for organizing the complex environmental relationships and facies distributions resulting from delta building and abandonment.

OVERBANK SPLAYING AND BAY-FILLING PROCESSES

Large interdistributary bays exist between the major distributaries of the Mississippi River. These areas normally receive only fine-grained suspended sediment load during high floods or through small breaks in the

Figure 4. Photograph showing an overbank splay in the lower Mississippi River delta. The original break (A) in the levee occurred in 1975.

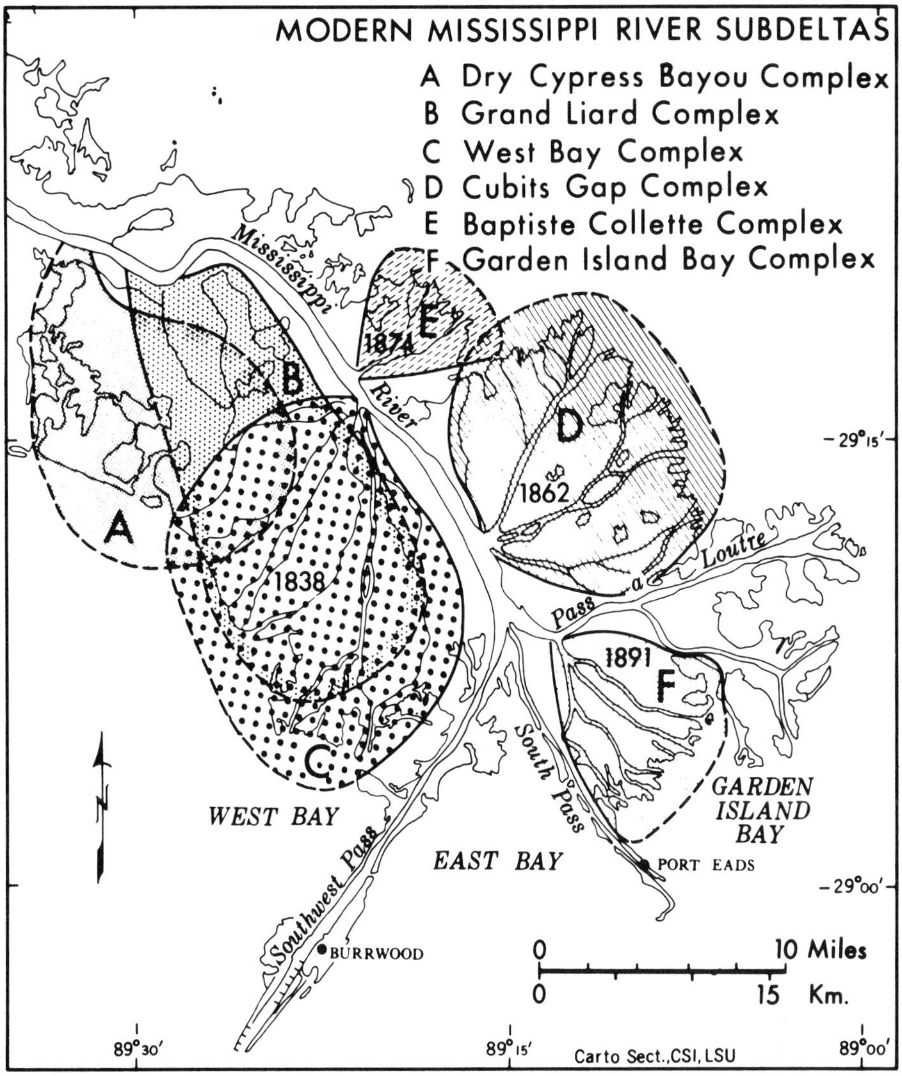

Figure 5. Bay fills of the modern Balize delta, Mississippi River. Filling episodes A and B occurred prior to historic maps in the delta. Filling episodes C–F occurred during the period 1838 to present, and the dates indicated are the period of the break. After Coleman and Gagliano, 1964.

MODERN MISSISSIPPI RIVER SUBDELTAS
A Dry Cypress Bayou Complex
B Grand Liard Complex
C West Bay Complex
D Cubits Gap Complex
E Baptiste Collette Complex
F Garden Island Bay Complex

channel banks which form overbank splays (Fig. 4). The Nita crevasse, described by Johnson (1891), was one of the major crevasse splays that occurred along the banks of the Mississippi River. This process of overbank deposition forms the major subaerial land (marshes) found in the lower Mississippi River delta. River banks in the lower delta rarely exceed 0.5 m in elevation, and major floods on the river have stage heights that exceed the elevation of the channel banks. Overbanking does not occur each year, but it is normally associated with high flood events. Small channels, especially in low areas along the channel margin, will scour the channel bank and provide a point source for suspended sediment to move from the main channels into the adjacent interdistributary bays. Such morphologic features are called "overbank splays" or "crevasse splays" (Fig. 4). The splay shown in Figure 4 scoured the distributary channel bank in 1975, and within a period of 10 yr, had nearly filled an interdistributary bay 4 km long, 3 km wide, and 3 m deep. These splays are rarely active for more than 10 to 15 yr and, during this period of time, will deposit a sedimentary sequence that is generally less than 3 m thick and that covers an area of 12 to 15 sq km. Thus in a period of only a few years, a depositional unit ~3 m thick and covering a considerable area can form by this overbanking process. Through longer periods of time, stacked overbank sediments can attain a thickness of several tens of meters and cover considerable areas. This relatively episodic continuous process of overtopping the channel banks results in deposition of sediment along the margin of the distributary channel and leads to slightly higher elevations of land bordering the channels. The coalescing of numerous overbank splays through time results in the formation of what is morphologically referred to as "natural levees"; this is especially true in the lower delta plain.

Crevassing of small splays off distributaries is a relatively high frequency process that is associated with large floods on the river. In areas where relatively large interdistributary bays exist between the major distributaries, a channel of a crevasse splay can scour deeply enough to allow a major depositional episode to take place. In these instances, overbank deposition will fill relatively large bays that have areal extents of 300 to 400 sq km and depths of 10 to 15 m; this process is referred to as "bay filling." Figure 5 illustrates the historic sequence of bay fills that have formed within the modern delta during the past few hundred years (Coleman and Gagliano, 1964). Of the six bay fills, four have been dated historically, and much of their development can be traced on historic maps. Note that the bay fills are quite large features, some 10 to 15 km wide and 20 to 25 km long; their thicknesses vary as a function of the depth of the interdistributary bay but may range from 5 to 20 m. The vertical sequence forms a depositional cycle which requires 100 to 150 yr to complete.

Cubits Gap is one of the major bay fills that has occurred in historic times, and the history of its development can be easily traced and illustrated with historic maps. Figure 6, which illustrates only a few of the numerous historic maps that are available, shows the sequential development of the Cubits Gap bay fill during the period 1838–1971. The 1838 map was surveyed prior to the break and shows a narrow natural levee bordering the main channel of the Mississippi River adjacent to Bay Rondo. In 1862, along a low section of the levee, an oyster fisherman named Cubits excavated a small ditch across the levee. The high flood of 1862 enlarged this small channel, and by 1868 the channel width had increased to 740 m, and the depth was on the order of several meters. Bay Rondo, the adjacent interdistributary bay, had maximum water depths on the order of 30 m. The remaining maps in Figure 6 show the development of this bay fill for the next 100 yr. By 1884, some 22 yr after the initial break, a complicated bifurcated channel system had formed; note that subaqueous deposition had resulted in shallowing of the original water depths in the bay. By 1905, a well-developed channel pattern had been established, and progradation had been rapid during the past 21 yr. In 1922, progradation was continuing on some of the distributaries, but already small bays were opening up in the areas between the main channels. By 1946, many of the bays had enlarged significantly, and only a few of the major distributaries continued to prograde seaward. By 1971, compaction had caused substantial submergence, and only a few of the channels were still building seaward. Today, in the late 1980s, about 75% of the original land area has been lost to subsidence and compaction.

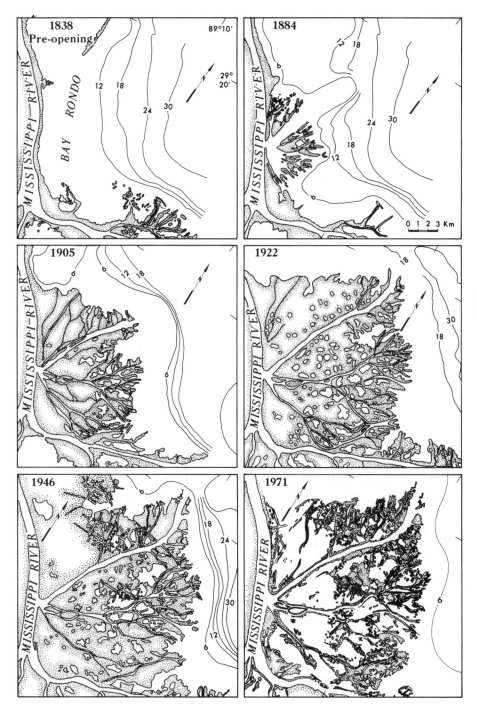

Figure 6. Historic maps illustrating the development of Cubits Gap bay fill during the period 1838 to 1971.

The pattern of land gain (progradation) and land loss (deterioration) for the four major historic bay fills is graphically illustrated in Figure 7. The x-axis represents the number of years after the first appearance of subaerial land within the bay fill, and the y-axis is percent growth. Note that it takes ~70 yr to attain full development of the progradational or build-out phase, whereas the deterioration stage is much more variable, requiring from 30 to 100 yr to complete. This variation in accumulation rates within a single sedimentary bay fill is quite common and shows the error in computing average accumulation rates.

Thus, each bay fill forms initially as a break in the major distributary channel during flood stage, gradually increases in flow through successive floods, reaches a peak of discharge and deposition, wanes, and becomes inactive. As a result of continued subsidence, the bay-fill deposits are inundated by marine waters, reverting to a bay environment, thus completing a relatively high frequency sedimentary cycle. The mass of sediment resulting from this cyclic process varies in thickness from 5 to 30 m and requires 100 to 150 yr. The bay-fill sedimentary cycle is similar to the main delta lobes in many respects but is of less magnitude and requires considerably less time to complete (Coleman and Gagliano, 1964).

Figure 8 is a block diagram illustrating the sedimentary fill of West Bay and a portion of the sedimentary fill of an older bay-fill cycle. This diagram is based on more than 200 cored borings and 300 drill holes penetrating at least one depositional cycle and often two complete bay-fill cycles (Coleman, 1976). Note the extreme complexity of the lithofacies (generally well correlated with particular depositional environments). The most widespread units within the sedimentary sequence consist of shallow-marine bay clays (ST & CL, interdistributary bay) and marsh peats (peat, marsh). These units cap and separate the detrital or progradational units. They form the bounding deposits that encase the regressive wedge of the active bay fill. At the point of the original channel break, the sands of the detrital lens are extremely thick and generally tend to erode down into earlier depositional cycles. In the central and more distal parts of the bay fill, the cycle begins with a marine clay (or marsh peat) which represents the original surface of the bay bottom or the old marsh surface that has subsided below sea level. This unit grades upward into the prodelta clays and delta-front deposits, which consist of silts and clays. This, in turn, is overlain by the coarsest deposits of the regressive sequence, the distributary mouth bars (SD, Fig. 8). Finally, the capping deposits of the marsh cover the regressive unit, forming one complete cycle of deposition. Figure 9 summarizes the major characteristics of the sedimentary sequence associated with the bay fills. The vertical sequence displays a coarsening-upward trend, commencing with highly bioturbated (and often highly calcareous) clays and grading upward into well-laminated clays and silts of the initial subaqueous fill. Above this, alternating sands, silts, and clays form the delta-front, each sand layer becoming progressively thicker and more prominent stratigraphically upward. The major coarse detritus unit, the distributary mouth bar, consists primarily of sand with thin stringers of silt and a few clay laminations. Often, the top of the sand is highly bioturbated and rich in iron compounds. Immediately above this coarse detritus unit, there is a sequence of silts and clays (representing localized overbanking from the distributary channels within the bay fill), and the unit is finally capped by high organic clays alternating with peat stringers.

Through longer periods of time, bay fills stack one upon another, often forming a rather thick accumulation of cyclic sequences. A schematic diagram of this process is illustrated in Figure 10, a reproduction of the original diagram by Coleman and Gagliano (1964), who originally recognized the nature of these high-frequency cycles. In this instance, three cycles of deposition are illustrated, and the variations in the lithofacies at any one vertical profile can be seen. This cyclic model has been particularly significant in the use of the Mississippi River delta as a model for ancient analogues.

RIVER-MOUTH PROGRADATION

The river mouth is the point at which the seaward-flowing water leaves the confines of the channel banks and spreads and mixes with ambient waters of the receiving basin. It is the dynamic dissemination point for sediments that contribute to continued delta progradation and is responsible for forming one of the major sand bodies associated with deltaic sequences, the distributary mouth bar. The Mississippi River debouches into the Gulf of Mexico, a basin containing nearly open marine salinities. The sediment-laden, fresh-water river effluent has a density that approximates 1.000 g/cu cm, whereas the saline Gulf of Mexico waters have a density of about 1.028 g/cu cm (Wright and Coleman, 1974). The river effluent is therefore lighter than the ambient basin water, and the buoyancy of the lighter river water becomes of paramount importance in

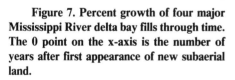

Figure 7. Percent growth of four major Mississippi River delta bay fills through time. The 0 point on the x-axis is the number of years after first appearance of new subaerial land.

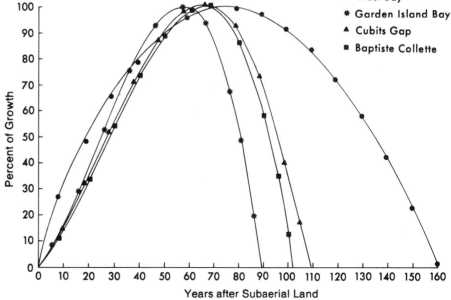

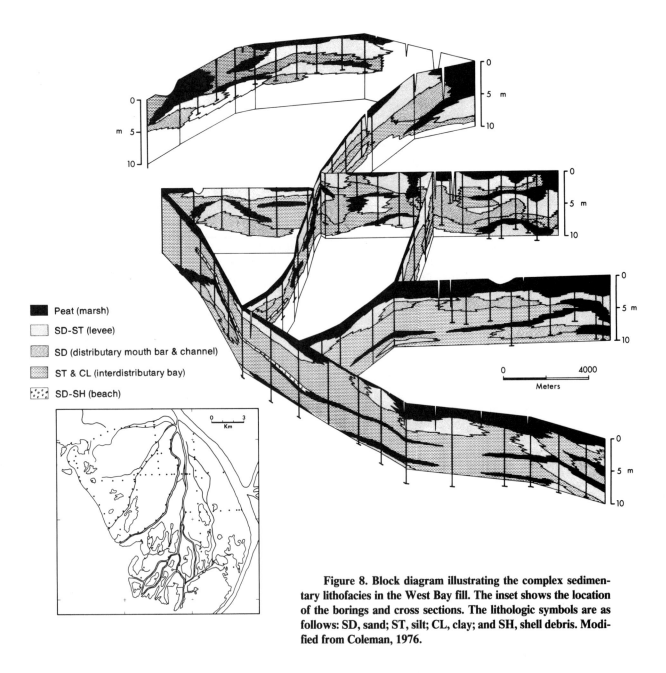

Figure 8. Block diagram illustrating the complex sedimentary lithofacies in the West Bay fill. The inset shows the location of the borings and cross sections. The lithologic symbols are as follows: SD, sand; ST, silt; CL, clay; and SH, shell debris. Modified from Coleman, 1976.

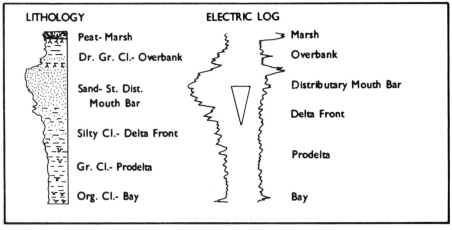

Figure 9. Vertical sequence of a bay fill deposit. Symbols are as follows: Dr., dark; Gr., gray; St., silt; Cl., clay; Org., organic. From Coleman, 1976.

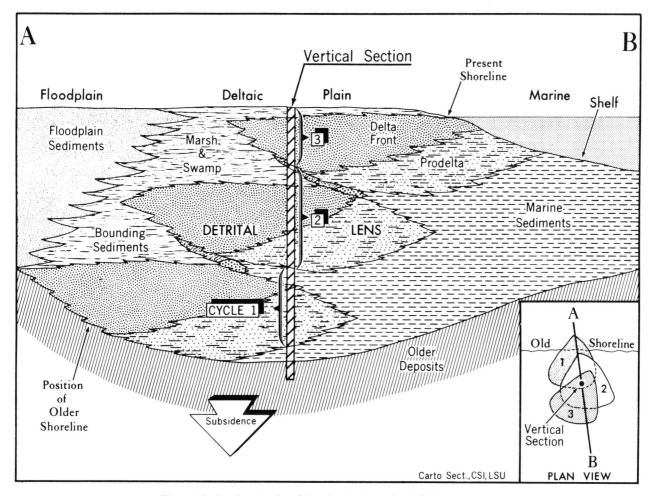

Figure 10. Overlapping bay-fill cycles in a hypothetical delta complex.

controlling the outflow pattern. Low-density, turbid, fresh water flows via the distributary over denser saline Gulf waters. Although the fresh-water effluent carries a heavy sediment load, the concentration (generally less than 2,000 ppm) is not great enough to influence or increase the density of the effluent over that of the saline basin waters. This lighter water initially floats on and over the denser saline waters; gravity, however, results in spreading of the plume. As the effluent plume spreads, velocity is reduced, and thus a hydraulic sorting process begins whereby the coarser sediments are deposited near the river mouth and the finer-grained sediments are deposited progressively farther offshore. The sands being deposited nearer the river mouth accumulate rather rapidly, but the finer-grained sediments, being carried in an ever-expanding effluent plume, become more widespread, and depositional rates are significantly lower. This concept was originally defined by Scruton (1960) and is illustrated diagrammatically in a modified version of the original diagram in Figure 11. More detailed descriptions of the processes active at a river mouth can be found in Coleman (1976), Wright and Coleman (1971), Borichansky and Mikhailov (1966), and Mikhailov (1966).

Thus the hydraulic sorting of sediments at a river mouth results in deposition of a sedimentary sequence that can be sedimentologically divided into three environments: the distributary mouth bar, the delta-front, and the prodelta (Fig. 11). The distributary mouth bar consists primarily of the coarsest detrital sediments being transported by the river and is concentrated nearest the active river mouth. The delta-front sediments are composed of alternating beds of well-sorted sand alternating with silt and clay, whereas the prodelta sediments consist of fine-grained clays alternating with thin silt laminations.

As long as the distributary channel is actively receiving sediment, the river mouth and these sedimentary units prograde seaward through time.

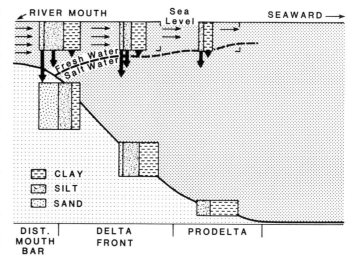

Figure 11. Sediment dispersal from an effluent river mouth plume. Modified from Scruton, 1960.

The progradation results in building the classical vertical sequence that is illustrated in virtually every geology textbook. This is a classic example illustrating Walther's Law of lateral and vertical facies relationships. The river mouths of the Mississippi River prograde seaward at differing rates, depending on the volume of sediment carried by each distributary and the slope of the adjacent continental shelf. The higher the sediment yield and the lower the offshore slope, the faster the rate of progradation. Figure 12, modified and updated from Gould (1970), shows the progradation of Southwest Pass, Mississippi River delta, as determined from historic maps (1764 to 1979) and from borings acquired along the channel. In a period of only 215 yr, the river mouth (10 m contour) prograded a distance of 17 km. The distributary mouth bar at Southwest Pass varies in width but is ~8 km wide; in thickness, it ranges from 30 to 90 m. With continued subsidence, this deposition has a relatively high probability of being preserved in the geologic record. In a period slightly greater than 200 yr, a sand body some 8–10 km wide, 17 km long, and in excess of 80 m thick has formed. The dynamics of deltaic deposition in a large river system is illustrated by this example. Similar progradation rates are observed at the mouths of the other major distributary channels of the active Mississippi River.

DELTA-FRONT SEDIMENT INSTABILITY

The early classical research on the Mississippi River delta conducted by Russell, Fisk, and colleagues was concentrated primarily on the subaerial part of the delta plain. Techniques such as high-resolution geophysics and side-scan sonar were not available for extensive investigations on the submerged platform fronting the active delta. As a result, the continental shelf and upper continental slope fronting the delta were thought to be an area of "quiet suspension sedimentation" settling from the effluent plume. This concept had been repeatedly published in numerous scientific journals and was the sedimentological model most often cited. Shepard (1955), in a classical paper in the *Geological Society of America Bulletin*, reported on a series of radial gullies that creased the delta-front of the active river delta. His observations were made from newly constructed bathymetric maps of the subaqueous segment of the delta. He continued his research on these "delta-front gullies" and concluded that they represented massive subaqueous landslides. He faced a problem, however, in explaining how such massive slumping of sediments could occur on such low-angle slopes as exist in the delta-front. Offshore slopes in this region of the delta rarely exceed 0.5°, and most of the slopes are less than 0.2°.

By the early 1970s, the offshore petroleum industry had begun to drill and construct bottom-fixed platforms in water depths of 100 to 150 m. Following a major hurricane (Camille) in 1969, several platforms in water depths of 110 m were severely damaged or lost. Post-hurricane analysis indicated that the damage was not the result of hurricane-driven wind, wave, or current forces, but of bottom sediment instability (Bea, 1971). Massive slides on a slope of less than 0.3° had been responsible for the platform damage. Interestingly, the type of failure that caused this structural damage had been discussed in the classical paper by Shepard (1955) some 15 yr earlier. Unfortunately, little attention was given to this paper when it was published; it was thought that massive sediment failure just could not be present on such low-angle slopes. It required some 20 yr and the aid of marine remote-sensing tools (side-scan sonar and high-resolution seismic), *in situ* testing instruments, and soil-mechanic models to prove that Shepard had been correct in his original assessment of the cause of major changes he had observed in the offshore Mississippi River delta.

Recent detailed marine geological investigations on the subaqueous parts of the continental shelves and upper continental slopes seaward of many river deltas experiencing high depositional rates and large quantities of fine-grained sediment have revealed that contemporary recurrent subaqueous gravity-induced mass movements are common phenomena (Fisk and McClelland, 1959; Coleman and others, 1974; Coleman and Garrison, 1977; Prior and Coleman, 1978a, 1978b; Coleman and Prior, 1980; Roberts and others, 1976, 1980b). These subaqueous sediment instabilities are an integral component of the normal deltaic process and offshore marine-sediment transport. The sediment instabilities generally display the following characteristics: (a) instability occurs on very low slope angles (generally less than 2°), and (b) large quantities of sediment are transported from shallow waters offshore to deeper marine waters along well-defined transport paths and in a variety of modes. The major factors that influence the stability of bottom sediments are as follows.

1. Rapid sedimentation at the river mouths results in widespread sedimentary loading of the upper delta-front slopes.

2. Coarse-grained sands and silts, which comprise the distributary mouth bars, differentially load the underlying prodelta clays (Morgan, 1961).

3. Fine-grained delta deposits, because of their rapid deposition, are generally underconsolidated, with large excess pore-fluid pressures causing low sediment strengths.

4. Rapid biochemical degradation of organic material leads to the formation of large volumes of methane gas, which in the bubble phase contributes to the generation of excess pore pressures within the sediments (Whelan and others, 1975).

5. The offshore region experiences winter storms and hurricanes, which cause cyclic loading on the sea floor; this loading imparts downslope stresses and contributes to pore-water pressure generation.

The main types of subaqueous slope instabilities that can be recognized in water depths of 5 to 200 m are illustrated schematically in Figure 13. This figure illustrates the instabilities distributed around a single distributary and part of an interdistributary bay, but similar features can be found at each of the main distributaries within the Mississippi River delta (Coleman and others, 1980). In the immediate vicinity of the river mouths, differential weighting caused by dense distributary mouth bar sands causes vertical diapiric uplift and the formation of clay diapirs (mud lumps). In the adjacent shallow-water bays, a variety of small-scale features such as

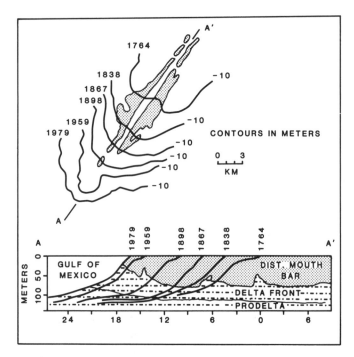

Figure 12. Progradation of Southwest Pass, Mississippi River, during the period 1764–1979. Updated and modified from Gould, 1970.

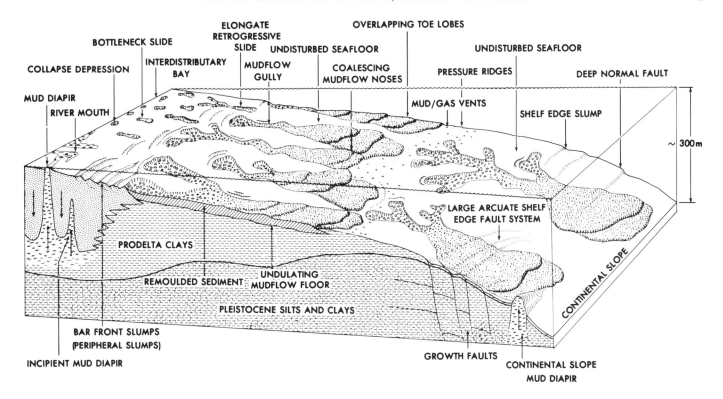

Figure 13. Schematic diagram illustrating the types of subaqueous sediment instabilities in the Mississippi River delta offshore. From Coleman and Prior, 1980.

collapse depressions and bottleneck slides occur (Coleman and Prior, 1980; Prior and Coleman, 1978a). Seaward of the river mouth and on the peripheral edge of the distributary mouth bar, a wide variety of arcuate peripheral slides are found. Immediately seaward of the bar, commencing in water depths of 10 m or less, there are a large number of elongate retrogressive slides (mudflow gullies and depositional lobes). Near the shelf edge, extremely large sediment instabilities, such as contemporaneous faults, submarine canyons, and shelf-edge slumps, are common (Coleman and others, 1983).

Differential Weighting and Mud-Lump Formation

Rapid deposition of localized dense distributary mouth-bar sands over less dense, plastic prodelta and marine clays results in a major type of sediment instability. The loading stress results in diapiric intrusion of the weak delta-front and prodelta sediments from beneath and into the overlying sand body, the distributary mouth bar. The major mappable feature that results from this differential weighting is thin spines of diapirically intruded mud which form islands or "mud lumps" (Fig. 14) at the mouths of the major distributaries (Morgan, 1961; Morgan and others, 1968). During the period 1876–1973, some 105 individual diapirs were mapped at the mouth of South Pass in an area covering only 13 sq km. These diapiric spines form rapidly, some rising 3–5 m over a single flood cycle. During much of the year (during low-river stage), diapirs do not display movement activity; with the initiation of flood stage and deposition of large quantities of sediment at the river mouth, vertical movement commences over the short flood period. In 1961 a hole was drilled on one of the mud diapirs at the mouth of South Pass. Three strings of drilling casing were inserted to a depth of 314 m and anchored in late Pleistocene sediments. By 1985, some 145 m of the casing had been extruded through

Figure 14. Photograph of a "mud lump" or diapiric spine formed at the mouth of South Pass, Mississippi River.

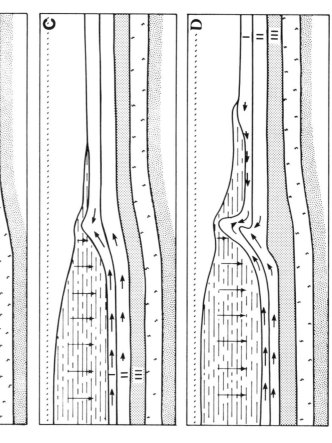

Figure 16. Schematic diagram illustrating the development of river-mouth "mud lumps" or diapirs. Modified from Morgan and others, 1968.

Figure 15. Photograph of mud lump at South Pass, Mississippi River, showing the extruded drilling casing.

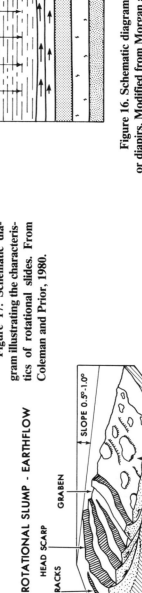

Figure 17. Schematic diagram illustrating the characteristics of rotational slides. From Coleman and Prior, 1980.

vertical movement of the diapiric spine in the 24 yr since its emplacement. Figure 15 is a photograph of the drilling casing attitude after only a few years of vertical movement.

Figure 16 diagrammatically illustrates the process of diapiric intrusion associated with differential weighting of the distributary mouth bar on the underlying prodelta clays. Stage A depicts the initial loading and compaction of underlying prodelta and shelf deposits by the progradation of massive sands of the river mouth. During Stage B, continued accumulation of delta-front and distributary mouth-bar sediment accelerates lateral flowage of the underlying clays. The river mouth, however, is not static, and the river mouth continues to prograde seaward. This isolates an area that begins to form the locus for the diapiric spine; this is shown in Stage C. In Stage D, continued loading results in vertical movement of the intruded mass, and movement will occur as long as the loading process continues or until most of the underlying clay has been involved in the vertical movement. With time, the diapir breaks the sea surface and forms an island or "mud lump." In this type of mass wasting, large volumes of older and deeper water prodelta clays are intruded into younger delta-front deposits. Wave reworking of the intruded marine clays results in incorporating deep-water microfauna into the shallow-water sands. In many cases, therefore, the shallow-water distributary mouth-bar sands take on a deep-water marine faunal characteristic because of this process.

Rotational Slumps

Downslope movement of large sediment masses often begins high on the upper delta-front slope near the distributary mouth bar (Coleman and others, 1974). Bottom slopes range from 0.2° to 1.0°, but in many instances major scarps associated with rotational slides, such as schematically illustrated in Figure 17, scar this rather smooth slope. The shear planes display distinctive curved or curvilinear plan view and are concave upward in the subsurface. Often they grade into bedding planes with depth. Scarp heights vary from 3 to 7 m, and the shear planes cut to depths of 30 to 35 m. The surface of the slump block normally has extensive hummocky and irregular bottom topography and displaced clasts of sediment.

A high-resolution seismic line run across one of these rotational slumps is shown in Figure 18. The two main shear planes (A), which display surface relief, mark the zone of weakness along which the rotational movement is taking place. As the sediment is removed, the upslope area becomes unstable and a new shear plane (B) begins to form. Its shear plane will be somewhat shallower than the previous ones. This process will continue until a stable slope is once again established. This process of slumping moves large quantities of highly fractured and partially remolded sediment downslope. Most of this sediment has been deposited in shallow-water delta-front and distributary mouth-bar environments and is being mass-moved into deeper waters at the shelf edge or onto the upper continental slope.

Mudflow Gullies

Extending radially seaward from each distributary, there are major elongate systems of gullies that have been referred to as "delta-front gullies." These features were first described from hydrographic maps by Shepard (1955), who indicated that he thought that they represented massive slumps on the bar front. These gullies are well displayed on the bathymetric map shown in Figure 19, modified from the original presented by Shepard (1955). Recent research using high-resolution seismic techniques and drill-hole information has indicated that these "gullies" represent a type of mass movement that is generally referred to as "retrogressive slides" (Prior and Coleman, 1978a, 1978b). The major characteristics of an elongate retrogressive slide (also commonly known as mudflow gullies) are schematically illustrated in Figure 20. The gullies emerge from within an extremely disturbed area of slump topography high on the delta-front. Each gully has a clearly recognizable area of rotational instability or shear slumps at its upslope margin; each also has a long, often sinuous, narrow chute that links a depressed hummocky source area on the upslope end to composite overlapping depositional lobes. As shown in the diagram, the instability is bounded by a bowl-shaped depression that serves as the source area for the blocky debris. Multiple head scarps and crown cracks are present, indicating upslope retrogression. Within the source area, hummocky, irregular, distinctive blocks of various sizes and arrangements can be discerned. Downslope is an essentially elongate narrow chute bounded by sharp linear escarpments and numerous sidewall rotational slumps. The gully floors are 3 to 20 m below the adjacent intact bottom. These gullies extend downslope at approximately right angles to the depth contours for distances of 7 to 10 km. Highly sinuous patterns, alternating narrow constrictions, and wide bulbous sections exist along their length. At the downslope end of the gully systems are found massive, often overlapping depositional lobes composed of debris discharged from the gully system.

The side-scan sonar image (Fig. 21) illustrates two major gully systems (A) that commence high on the delta-front. This image is composed of adjacent side-scan sonar lines that have been merged to form a mosaic 1.2 km by 1.8 km. Off one gully, a major zone of upslope retrogressive failure has taken place (B) and is feeding debris to one of the major gullies. This area of retrogression has advanced nearly to the edge of the adjacent gully. The narrow gullies (C) are relatively deeply incised, and evidence of

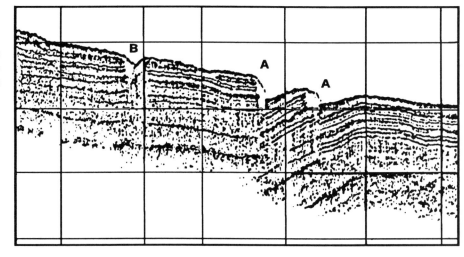

Figure 18. High-resolution seismic line run across several rotational slides. The navigation fixes are 152 m apart, and timing lines are 10 m apart. A. Rotational slides. B. Newly formed rotational slide.

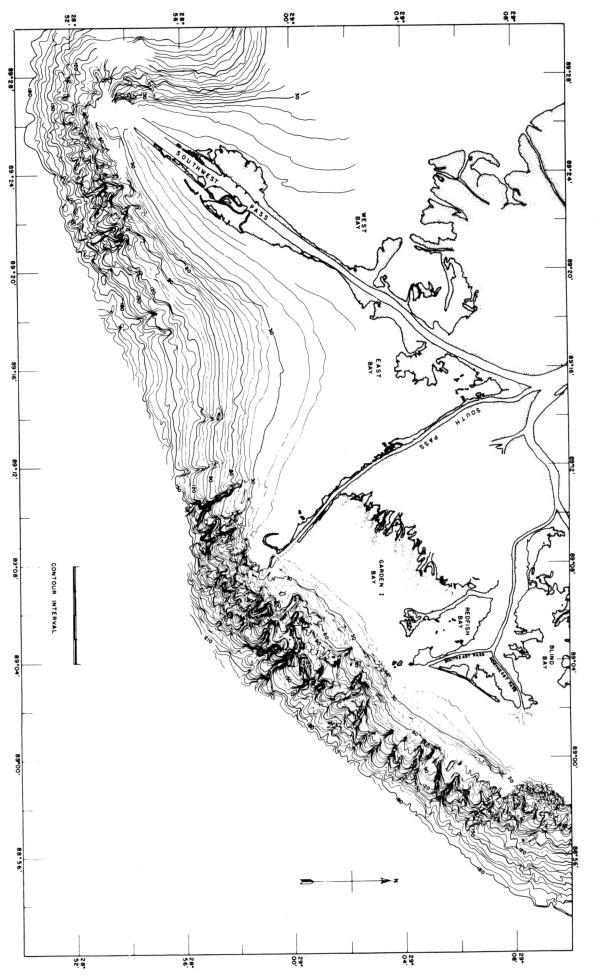

19. Submarine topography of the delta-front platform, Mississippi River delta. Modified from Shepard, 1955.

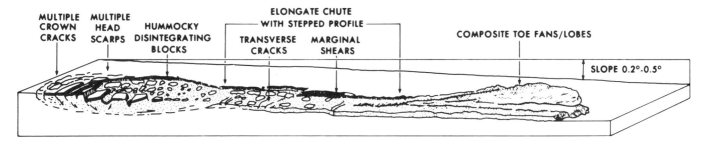

Figure 20. Schematic diagram illustrating the major characteristics of an elongate slide or "mudflow gully." From Coleman and Prior, 1980.

sidewall instability is indicated by small slumps along the gully margins (D) and by the alternations of bulbous and restricted regions. The widths of the individual gullies range from 18–150 m at the narrow points to 370–550 m at the widest. The floors of the gullies are characterized by large erratic blocks of different sizes (E). The sidewalls of the gullies are often subject to major instability; this process is responsible for the widening of the gully in a seaward direction.

At the seaward or downslope ends of the mudflow gullies, there are extensive areas of irregular bottom topography that are composed of discharged, blocky, disturbed debris. Figure 22 is a side-scan sonar image illustrating some of the characteristics of this discharged debris; the mosaic covers an area 1.5 by 2.2 km. Three overlapping lobes (A, B, C) have been discharged from a single upslope gully system. Large erratic blocks (D) are often incorporated in the depositional lobes and may be up to 1 km in width. The discharged debris often plows into the underlying undisturbed marine sediments, and a series of pressure ridges (E) form in front of the

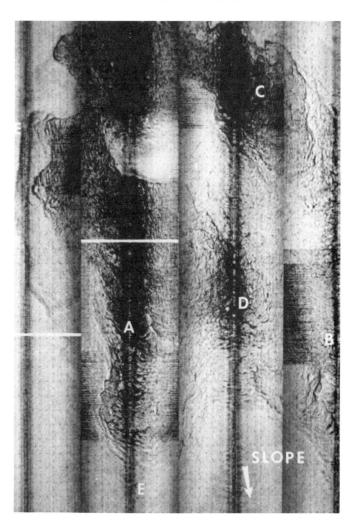

Figure 21. Side-scan sonar mosaic illustrating elongate slides or "mudflow gullies."

Figure 22. Side-scan sonar mosaic illustrating the depositional lobes of an elongate slide. A–D show individual depositional lobes; E illustrates pressure ridges formed in front of prograding depositional mudflow lobes.

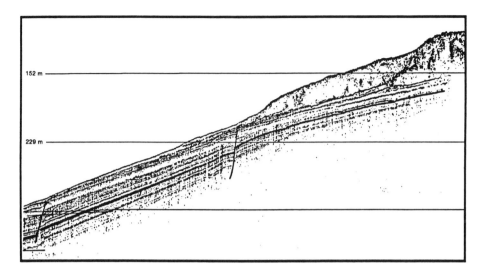

Figure 23. High-resolution seismic profile (dip section) run across the seaward edge of a depositional lobe.

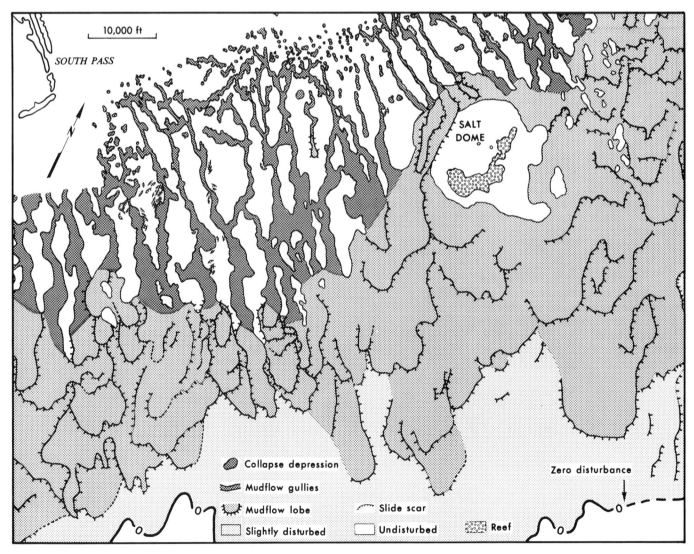

Figure 24. Map illustrating the distribution of elongate slides on the delta-front of the Mississippi River delta.

advancing lobe. Figure 23 is a high-resolution seismic profile run due south or in a dip direction off South Pass, Mississippi River. The large sediment mass (upper chaotic seismic sequence lacking coherent reflectors) is a recent mudflow depositional lobe that is ~60 m thick and has moved downslope from near the mouth of South Pass to water depths in excess of 150 m near the continental-shelf edge. Emplacement of this mudflow lobe has taken place since 1947 (the seismic line was run in 1978); thus in a period of less than 31 yr, a shallow-water deposit ~60 m thick has been deposited in water depths of 150 m.

Surveying of most of the delta by side-scan sonar and high-resolution geophysical techniques has been completed, and virtually the entire delta-front is scarred by this type of mass-movement process (Coleman and others, 1980). Figure 24 illustrates the distribution of retrogressive slides and arcuate slides that scar the delta-front area seaward of Southwest Pass. From the various studies conducted, it is estimated that as much as 40% of the sediment that is annually deposited on the shallow part of the delta-front platform is transported downslope by mass-movement processes. This is indeed a dynamic short-term process that is responsible for transporting large volumes of shallow-water sediment into deeper-water environments, another example of the extremely dynamic nature of delta environments. Early researchers had no way of knowing the magnitude and dynamic nature of the subaqueous part of the delta; yet it is just as dynamic as the subaerial delta.

CONCLUSIONS

Deltas are indeed dynamic environments, and research in the Mississippi River reveals the magnitude and rapidity of these changes. Delta lobes switch alongshore on a frequency on the order of every 1,500 yr; each delta lobe covers an area of nearly 30,000 sq km and has an average thickness of 35 m. Through even short periods of geologic time, on the order of 8,000 yr, these complex interfingering delta lobes cover an area in excess of 80,000 sq km and accumulate to a total thickness in excess of 200 m. Bay fills operate on an even higher frequency, filling an interdistributary bay with dimensions of 250 sq km and depths of 10 to 20 m within a period of 150 yr. Progradation of the river-mouth distributaries develops distributary mouth-bar sand bodies that have dimensions 17 km long, 8 km wide, and 80 m thick within a period of 200 yr.

The subaqueous delta-front platform, perceived by early researchers in the delta to be an area of "quiet suspension sedimentation," has been shown to be an area characterized by extremely dynamic mass-movement processes. Differential weighting of prodelta clays by denser distributary mouth-bar sands results in formation of diapirs or mud lumps at the mouths of the distributaries. These diapirs display vertical movements in excess of 100 m over a period of 20 yr. Subaqueous sediment instabilities, such as arcuate rotational slides and elongate slides (mudflow gullies), displace large volumes of sediment from shallow-water, delta-front environments to deeper-water environments at the shelf edge. It is estimated that ~40% of the annual sediment load of the river is displaced downslope annually on the delta-front platform of the modern Mississippi River delta. These processes lead to rapid accumulation of sediments in the deeper-water areas seaward of the river mouths.

All of the processes described operate over short periods of geologic time, ranging from a few years to a few thousand years, but result in deposition of significant volumes of sediment over extensive areas. Continued accumulation rates of these magnitudes develop a sequence of thick cyclic deposits within a basin-filling episode and result in complex vertical and lateral facies changes. Knowledge of the rapidity of changes and processes in modern delta settings will aid in the interpretation and understanding of similar ancient sedimentary sequences.

REFERENCES CITED

Barrell, J., 1917, Rhythms and measurement of geologic time: Geological Society of America Bulletin, v. 28, p. 745–904.
Bates, C. C., 1953, Rational theory of delta formation: American Association of Petroleum Geologists Bulletin, v. 37, p. 2119–2162.
Bea, R. G., 1971, How sea floor slides affect offshore structures: Oil and Gas Journal, v. 69, no. 48, p. 88–92.
Borichansky, L. S., and Mikhailov, V. N., 1966, Interaction of river and sea water in the absence of tides, in Problems of the humid tropic zone deltas and their implications: UNESCO, Dacca Symposium Proceedings, p. 175–180.
Coleman, J. M., 1976, Deltas: Processes of deposition and models for exploration: Champaign, Illinois, Continuing Education Publishing Company; (now available from Burgess Publishing Company, 7108 Ohms Lane, Minneapolis, Minnesota 55435), 102 p.
Coleman, J. M., and Gagliano, S. M., 1964, Cyclic sedimentation in the Mississippi River deltaic plain: Gulf Coast Association of Geological Societies Transactions, v. 14, p. 67–80.
Coleman, J. M., and Garrison, L. E., 1977, Geological aspects of marine slope instability, northwestern Gulf of Mexico: Marine Geotechnology, v. 2, p. 9–44.
Coleman, J. M., and Prior, D. B., 1980, Deltaic sand bodies: Continuing Education Course Note Series no. 15: Tulsa, Oklahoma, American Association of Petroleum Geologists, 171 p.
Coleman, J. M., Suhayda, J. N., Whelan, T., III, and Wright, L. D., 1974, Mass movements of Mississippi River delta sediments: Gulf Coast Association of Geological Societies Transactions, v. 24, p. 49–68.
Coleman, J. M., Prior, D. B., and Garrison, L. E., 1980, Subaqueous sediment instabilities in the offshore Mississippi River delta: New Orleans, Louisiana, Bureau of Land Management Open-File Report 80-01.
Coleman, J. M., Prior, D. B., and Lindsay, J. F., 1983, Deltaic influences on shelf edge instability processes, in Stanley, D. J., and Moore, G. T., eds., The shelf break; Critical interface on continental margins: Society of Economic Paleontologists and Mineralogists Special Publication 33, p. 121–137.
Credner, G. R., 1878, Die deltas, ihre morphologie, geographische, verbreitung, und entstehungs bedingungen: Petermans Geographische Mittheilungen (Erganzungsland), v. 12, p. 1–74.
Fisk, H. N., 1944, Geological investigation of the alluvial valley of the lower Mississippi River: Vicksburg, Mississippi, U.S. Army Corps of Engineers, Mississippi River Commission.
—— 1947, Fine-grained alluvial deposits and their effects on Mississippi River activity: Vicksburg, Mississippi, U.S. Waterways Experiment Station, Volumes 1 and 2.
—— 1952, Geological investigation of the Atchafalaya basin and the problem of Mississippi River diversion: Vicksburg, Mississippi, Mississippi River Commission, 145 p.
—— 1955, Sand facies of Recent Mississippi Delta deposits: 4th World Petroleum Congress, Proceedings, Section 1, p. 377–398.
—— 1956, Nearshore sediments of the continental shelf off Louisiana: 8th Texas Conference on Soil Mechanics and Foundation Engineering, p. 1–23.
—— 1961, Bar-finger sands of the Mississippi delta, in Peterson, J. A., and Osmond, J. C., eds., Geometry of sandstone bodies: Tulsa, Oklahoma, American Association of Petroleum Geologists, p. 29–52.
Fisk, H. N., and McClelland, B., 1959, Geology of continental shelf and its influence on offshore foundation design: Geological Society of America Bulletin, v. 70, p. 1369–1394.
Fisk, H. N., and McFarlan, E., Jr., 1955, Late Quaternary deltaic deposits of the Mississippi River—Local sedimentation and basin tectonics, in Poldervaart, A., ed., Crust of the earth, a symposium: Geological Society of America Special Paper 62, p. 279–302.
Fisk, H. N., McFarlan, E., Jr., Kolb, C. R., and Wilbert, L. J., Jr., 1954, Sedimentary framework of the modern Mississippi Delta: Journal of Sedimentary Petrology, v. 24, p. 79–99.
Frazier, D. E., 1967, Recent deltaic deposits of the Mississippi River, their development and chronology: Gulf Coast Association of Geological Societies Transactions, v. 17, p. 287–315.
Gould, H. R., 1970, The Mississippi delta complex, in Morgan, J. P., ed., Deltaic sedimentation: Modern and ancient: Society of Economic Paleontologists and Mineralogists Special Publication 15, p. 3–30.
Johnson, L. C., 1891, The Nita crevasse: Geological Society of America Bulletin, v. 2, p. 20–25.
Kolb, C. R., and Van Lopik, J. R., 1966, Depositional environments of the Mississippi River deltaic plain, southeastern Louisiana, in Shirley, M. L., and Ragsdale, J. A., eds., Deltas: Houston, Texas, Geological Society, p. 17–62.
Lyell, C., 1847, On the Mississippi delta: American Journal of Science, 2nd ser., v. 3, p. 118–119.
Mikhailov, V. N., 1966, Hydrology and formation of river mouth bars, in Scientific problems of the humid tropic zone deltas and their implications: UNESCO, Dacca Symposium Proceedings, p. 59–64.
Morgan, J. P., 1961, Mudlumps at the mouths of the Mississippi River, in Genesis and paleontology of the Mississippi River mudlumps: Louisiana Department of Conservation Geological Bulletin 35, 116 p.
Morgan, J. P., Coleman, J. M., and Gagliano, S. M., 1968, Mudlumps: diapiric structures in Mississippi delta sediments, in Diapirism and diapirs: American Association Petroleum Geologists Memoir No. 8, p. 145–161.
Penland, Shea, and Boyd, R., 1981, Shoreline changes on the Louisiana barrier coast: IEEE Oceans, v. 81, p. 209–219.
Prestwich, Joseph, 1885, Geology: Chemical, physical, and stratigraphical: Vol. 1: Oxford, Clarendon Press, 477 p.
Prior, D. B., and Coleman, J. M., 1978a, Disintegrating retrogressive landslides on very-low-angle subaqueous slopes, Mississippi Delta: Marine Geotechnology, v. 3, p. 37–60.
—— 1978b, Submarine landslides on the Mississippi River delta-front slope: Geoscience and Man: School of Geoscience, Louisiana State University, Baton Rouge, v. 19, p. 41–53.
Riddell, J. L., 1846, Deposits of the Mississippi and changes in its mouth: De Bow's Review, v. 2, p. 433–448.
Roberts, H. H., Cratsley, D., and Whelan, T., III, 1976, Stability of Mississippi delta sediments as evaluated by analysis of structural features in sediment borings: 8th Offshore Technology Conference, Houston, Texas, p. 9–28.
Roberts, H. H., Adams, R. D., and Cunningham, R. W., 1980a, Evolution of sand-dominant subaerial phase: Atchafalaya Delta, Louisiana: American Association of Petroleum Geologists Bulletin, v. 64, p. 264–279.
Roberts, H. H., Suhayda, J. N., and Coleman, J. M., 1980b, Sediment deformation and transportation on low-angle slopes: Mississippi River delta, in Coates, D. R., and Vitek, J. D., eds., Thresholds in geomorphology: London, George Allen and Unwin, p. 131–167.
Russell, R. J., 1936, Physiography of the lower Mississippi River delta, in Reports on the geology of Plaquemines and St. Bernard Parishes: Louisiana Department of Conservation Geological Bulletin 8, p. 3–193.
—— 1940, Quaternary history of Louisiana: Geological Society of America Bulletin, v. 51, p. 1199–1234.
—— 1958, Geological geomorphology: Geological Society of America Bulletin, v. 69, p. 1–22.
Russell, R. J., and Russell, R. D., 1939, Mississippi River delta sedimentation, in Recent marine sediments: American Association of Petroleum Geologists, p. 153–177.
Scruton, P. C., 1960, Delta building and the deltaic sequence, in Shepard, F. P., and others, eds., Recent sediments, northwest Gulf of Mexico: Tulsa, Oklahoma, American Association of Petroleum Geologists, p. 82–102.
Shepard, F. P., 1955, Delta-front valleys bordering the Mississippi distributaries: Geological Society of America Bulletin, v. 66, p. 1489–1498.
—— 1956, Marginal sediments of the Mississippi delta: American Association of Petroleum Geologists Bulletin, v. 40, p. 2537–2623.
Trowbridge, A. C., 1930, Building of Mississippi delta: American Association of Petroleum Geologists Bulletin, v. 14, p. 867–901.
—— 1954, Mississippi River and Gulf Coast terraces and sediments as related to Pleistocene history—A problem: Geological Society of America Bulletin, v. 65, p. 793–813.
Welder, F. A., 1959, Processes of deltaic sedimentation in the Lower Mississippi River: Baton Rouge, Louisiana, Louisiana State University, Coastal Studies Institute Technical Report 12, 90 p.
Whelan, Thomas, III, Coleman, J. M., Suhayda, J. N., and Garrison, L. E., 1975, The geochemistry of Recent Mississippi River delta sediments: Gas concentration and sediment stability: 7th Offshore Technology Conference, Houston, Texas, p. 71–84.
Wright, L. D., and Coleman, J. M., 1971, Effluent expansion and interfacial mixing in the presence of a salt wedge, Mississippi River delta: Journal of Geophysical Research, v. 76, p. 8649–8661.
—— 1974, Mississippi River mouth processes: Effluent dynamics and morphologic development: Journal of Geology, v. 82, p. 751–778.

Manuscript Received by the Society January 19, 1988
Revised Manuscript Received March 11, 1988
Manuscript Accepted March 14, 1988

Forty years of sequence stratigraphy

CENTENNIAL ARTICLE

L. L. SLOSS *Department of Geological Sciences, Northwestern University, Evanston, Illinois 60208*

ABSTRACT

The principles and practice of sequence stratigraphy are of ancient heritage. In North America, the separation of the Carboniferous into two systems and the identification of Ozarkian and Comanchean and other indigenous chronostratigraphic entities were efforts toward making the segmentation of the geologic column more representative of stratigraphic observations, particularly on cratons and their margins. Growing awareness of the influence of tectonics on sedimentation, as well as recognition that unconformity-bounded sedimentary packages identified in Montana were craton-wide, led to this writer's 1948 GSA paper which formally applied Native American tribe names to *sequences* as lithostratigraphic units.

Acceptance of the sequence philosophy beyond the Northwestern campus lacked early manifestations of fervor and enthusiasm, although isolated pockets of true believers were known to exist. Except for papers by me, my students, and the late, lamented H. E. Wheeler, sequences did not appear in the public prints for more than a decade—and then through misappropriation by the U.S. Geological Survey. The list of Indian-tribe sequences was emended and completed at another GSA presentation in 1959.

Meanwhile the sequence concept was alive and well in a research facility of a company later to be known as Exxon. Here, Peter Vail and a cohort of preconditioned colleagues seized upon the stratigraphic imagery made available by multichannel, digitally recorded, and computer-massaged reflection seismography to establish the discipline of *seismic stratigraphy*. The "Vail curve," recording relative change of coastal onlap, defines successive "sequences" which are the third-order bottom rungs of an elaborate hierarchy that is topped by "megacycle sets," themselves subdivisions of the ancestral sequences.

The taxonomy of sequence stratigraphy is not at issue (although it has been complicated by the sanctification of "synthem" by the International Subcommission on Stratigraphic Classification). Interpretation is important, however; the Exxon people, their alumni, and adherents see sea-level change as the "be-all" and "end-all" of coastal onlap and its erosional and depositional concomitants. No one will question the influence of sea-level change on many of the observations that make sequence stratigraphy viable, but there are distinctions among and within major-scale unconformity-bounded stratal packages that can be explained only by tectonic change. Whether tectonic influence can be extended to Exxonian sequences remains to be demonstrated. Meanwhile, it is worth considering the proposition that sea-level change is a second- or third-order response to some more significant global phenomenon.

INTRODUCTION

The philosophy guiding sequence stratigraphy is much older than the title of this paper might suggest. In the late decades of the 18th century, geologists were striving toward a stratigraphic taxonomy within which their observations could find organization and structure. Some of the early schemes of classification were largely descriptive and relatively free of the taint of genetic implication (for example, "Primitive," "Secondary," and so on), whereas others, such as "Flötzgebirge" and "Aufgeschwemmte," applied by the Wernerians, were explicitly genetic artifacts of the Plutonist School. By the middle of the 19th century, the gross elements of geochronology and chronostratigraphy, the periods and corresponding systems (Cambrian, Cretaceous, and so on) were widely recognized and accepted in a form that has required only modest change over the ensuing years. Most of the classical systems established in Great Britain and western Europe were defined at base or top or both by major unconformities (such as Hutton's post-Caledonide "Great Unconformity") and thus are "unconformity-bounded units" in the modern sense. Several system boundaries were placed at positions of marked paleontologic change and were assumed to represent times of catastrophic extinction and replacement without physical evidences of erosion or depositional hiatus. Parenthetically, it is interesting to note that early students of the Paris Basin seized upon what would now be called "seismic-stratigraphy criteria" (truncation, onlap, facies progradation, and so on) to establish the boundaries of Tertiary stages.

In any case, transplanting classical chronostratigraphic units to the New World, whether defined by unconformities and other physical changes or by paleontologic changes, was not a simple or wholly satisfying operation. The Permian System, established on grounds that mixed geopolitics with stratigraphy, was not a popular import, but North Americans were early converts to the acceptance of a biostratigraphically defined Ordovician erected to separate Cambrian and Silurian. The Carboniferous System, which makes uncomfortable bedfellows of the "Mountain limestone" and the Coal Measures in Britain, became the Mississippian and Pennsylvanian Systems on this continent, the common boundary placed at the position of a pervasive cratonic unconformity. Meanwhile, other revisionist attempts were launched in efforts to bring North American chronologic and chronostratigraphic units into greater conformity with observations. A prime example is E. O. Ulrich's excision of Canadian (Early

Ordovician) time and strata from the Ordovician—a precursive recognition of the separation of the Sauk and Tippecanoe sequences. Similarly, the isolation of Lower Cretaceous strata as the "Comanchean System" identified the stratal discontinuity now employed to separate Zuni II and III (more or less the division point of Upper Zuni A-1 and A-2 of Haq and others, 1987).

As the 20th century advanced, stratigraphers were made increasingly aware of the necessity of distinguishing between what are now termed "lithostratigraphic" and "chronostratigraphic" units. The practice of demanding a stratal discontinuity at each major chronostratigraphic (system or series) boundary—a practice that led to the erection of a confusing multiplicity of continent- or craton-limited system/period and series/epoch boundaries—slowly but perceptibly declined in community regard. At a lesser scale, the study and naming of cyclothems (for example, Wanless and Weller, 1932) anticipated the recognition of "genetic units" and other assemblages of strata identified in terms of the sedimentary processes represented. In the same decades, the three-dimensional view of stratified rocks provided by subsurface exploration and the practical requirements of subsurface stratigraphic nomenclature in the service of industry and government produced an environment within which nonclassical approaches were fostered and developed.

FORMALIZATION OF THE SEQUENCE CONCEPT

The Age of Stratigraphic Enlightenment dawned in near synchronism with this investigator's emplacement among the abundant outcrops of western and central Montana. Here, the stratigraphic succession was arrayed in discrete packages defined at base and top by unconformities that evinced no hard-and-fast relationship to the established hierarchical system. Further, each major package appeared to have been deposited within a tectonic framework that commonly featured previously unrepresented positive and negative elements along with some, but not all, of such elements active during earlier (and later) depositional episodes. For example, the Belt Supergroup, which forms the earliest discernible stratigraphic package in the Montana area, rests on an unconformity cut on basement crystallines and projects eastward into the craton along an aulacogen-like trough defined along its southern margin by syndepositional faulting. The overlying package, Middle Cambrian to Early Ordovician in age, rests on a second major unconformity representing a lacuna of more than 700 m.y. Erosion during this time span removed all vestiges of an intervening Windermere succession, and post-Windermere (latest Proterozoic and Early Cambrian) nondeposition prevailed until Middle Cambrian time, when sedimentary accumulation began anew under conditions of extreme tectonic stability that continued until the Early Ordovician. In regions northeast of central Montana (Williston Basin and the Manitoba lakes), southeast and south (Black Hills and the uplifts along the Montana-Wyoming border), and west into Idaho, the Cambrian–Lower Ordovician package is overlain disconformably by upper Middle and Upper Ordovician strata that widely overstep the truncated edges of underlying units. Over the greater part of western and central Montana, extending north into southern Alberta, the Cambrian–Lower Ordovician package is succeeded by Upper Devonian beds deposited on an unconformity that is commonly obscure in individual outcrops but along which all Middle and Upper Ordovician and Silurian strata have been stripped. Clearly, a new and different tectonic geography emerged before the erosion surface subsided below depositional base level.

It is not necessary to complete here a review of the sedimentary and tectonic evolution of Montana; rather, it is hoped that two points have been made: (1) *the stratigraphy of a cratonic region is divisible into rational and useful packages by reference to major regional unconformities,* and (2) *the history of such a region is divisible into tectonically characterized segments representing the times of accumulation of the succession of stratigraphic packages.*

Transfer of the seat of my activities to the cratonic heartland brought me to intensive contact with E. C. Dapples and W. C. Krumbein. Among our major concerns was the development of a practical and readily transferrable technology for facies mapping, an effort that yielded some success and which attracted numbers of impressionable graduate students. Preparation of a useful facies map is dependent on the identification and correlation of a significant stratigraphic unit; the need to establish the correlation of significant units at regional and interregional scales led, in turn, to the recognition of *operational* units, that is, stratigraphic entities on which operations such as facies mapping could be performed. Such operational units included bodies of strata defined by lithologic or geophysical-log markers, by key beds, and by unconformities. The availability of willing student labor made it possible to establish, correlate, and map stratigraphic units, including unconformity-bounded units, over wide areas of the craton.

Early belief in the craton-wide universality of the major stratigraphic packages recognized in the Montana area was greatly supported and strengthened by the interregional correlations demanded by facies mapping. When Chester Longwell organized a symposium on facies for the 1948 GSA meeting, therefore, a paper (Sloss and others, 1949) on our facies-mapping technique was presented, incorporating four major Paleozoic stratigraphic packages now identified as the Sauk, Tippecanoe, Kaskaskia, and Absaroka *sequences* (using Native American names to emphasize the North American derivation of the units). This, the first explicit reference to the sequence concept burst upon an unready stratigraphic community with all of the impact of a failed soap bubble.

In a closely following paper (Sloss, 1950), the four named sequences, identified as unconformity-bounded packages representing "depositional cycles," were applied in a regional study. The effort to gain acceptance for the concept was renewed in a textbook (Krumbein and Sloss, 1951) which termed a sequence to be "the rock record of a major tectonic cycle" and which stressed the defining character of sequence-bounding, craton-wide unconformities. Nevertheless, except for H. E. Wheeler and his associates, certain oil-industry research groups infected by our students, and certain authors of historical geology texts (notably from McGill and Wisconsin), the concept and its applications were seldom aired publicly beyond the Northwestern campus. In fairness, it should be said that sequence stratigraphy did indeed make early entry into the thinking and instruction at some enlightened universities and to the practices of certain advanced state surveys.

Meanwhile, back on the Northwestern campus, P. R. Vail was struggling to reconcile his observations on the Upper Mississippian of the Cumberland Plateau (his dissertation topic) with the common boundary of the Kaskaskia and Absaroka sequences, as this was first defined. In the Montana area and in the Illinois Basin, the ubiquitous Lower and middle Mississippian carbonate blankets are succeeded by younger Mississippian detritals (Kibbey, Aux Vases, and so on), commonly involving an intervening unconformity. These evidences of hiatus and change in petrology were interpreted as a fundamental shift in the tectonic framework of the craton worthy of designation as a sequence boundary, ignoring other Mississippian discontinuities and other incursions of Shield-derived clastics (for example, Borden, Bedford). Vail's studies made clear what should have been apparent all along: the cyclical detrital/carbonate couplets of the Illinois Basin Chester beds pass to the east to be replaced without significant interruption by a largely carbonate succession. Vail (1959) further demonstrated, employing an ancestral "Vail curve," that the major marine regression and the logical geochronologic position of the close of Kaskaskia sedimentation occurred very late in Mississippian time.

As noted, the term "sequence" (as applied to a package of strata) as well as the concept represented were not commonly adopted by stratigraphers and tectonists in the decade of the '50s. Some kind of change was heralded in a paper by Silberling and Roberts (1962); here, avoiding the cluttering of the page with attributions to previous usage, sequences are defined as "lithologically and *geographically* [emphasis added] discrete units of major rank . . . that are set apart from underlying or overlying sequences by unconformities." Sequences constrained by narrow geographic limitations have not found wide application.

More-or-less concurrently with the above-noted foray into sequence stratigraphy, the formalization of sequences denoted by Native American names was finalized in a paper before the Pittsburgh meeting of the Society in 1959; publication, unaccountably delayed (Sloss, 1963), recorded the aforementioned revision of the position of the pre-Absaroka lacuna and completed the roster by the addition of the Zuni and Tejas sequences. It is this paper that most workers tend to quote as *the* earliest exposition of the modern-era sequence concept.

The 1963 paper insists that "stratigraphic sequences are rock-stratigraphic [= lithostratigraphic] units of higher rank than group, megagroup, or supergroup." Chang (1975) has recommended that unconformity-bounded stratal units be separated from the lithostratigraphic hierarchy and identified as *synthems*, a position now sanctified by the International Subcommission on Stratigraphic Classification (Salvador, chairman, 1987). It is too early to tell whether the stratigraphic community will accept the ruling with a significant level of unanimity; the North American Commission on Stratigraphic Nomenclature, for example, has approached the question with (pointed?) restraint.

Regardless of the eventual choice between *sequence* and *synthem*, the wisdom of erecting a new category of stratigraphic classification to accommodate unconformity-bounded units, thereby excluding these from the company of lithostratigraphic assemblages of strata, is by no means agreed upon by all interested workers. It is claimed that because unconformity-bounded units have chronostratigraphic applications, such units represent something unique in the hierarchy of stratigraphic taxonomy. This position would seem to deny or ignore the chronostratigraphic significance of stratal units defined by time-parallel surfaces such as those identified by bentonites and other lithologic evidences of short-term events. The chairman of the International Subcommission and I have had previous debate on these nonvital matters (A. Salvador, 1987, personal commun.) without reaching agreement.

ENTER SEISMIC STRATIGRAPHY

While these momentous events were transpiring, the same Peter Vail noted above had moved on to the research department of The Carter Oil Company in Tulsa. A major function of the laboratory was the integration of stratigraphy as observed in outcrops, cuttings and cores, and electric or gamma/neutron logs with geophysical observations displayed as reflection-seismograph profiles. Early on (P. Dickey and R. Sarmiento, 1988, personal commun.), the relatively poor seismic records of the time could be interpreted to clarify truncation and onlap relationships at unconformities—as in the Tertiary of the Maracaibo fields of Venezuela, for example. As digital recording and computer processing of reflection-seismic data advanced, profiles were produced that increasingly came to resemble stratigraphic cross sections. With the addition of interval-velocity information, time sections were translatable to depth sections that aided in bridging the intellectual gap between stratigraphy and geophysics. Continuing evolution in the technologies of acquisition, processing, and display of seismic data was accompanied by evolving sophistication of stratigraphic analysis of seismic profiles.

Vail and colleagues (Vail and Wilbur, 1966) were particularly impressed by evidences of onlap over unconformity surfaces revealed by the progressive landward termination of reflectors, especially (but not exclusively) as these terminations were made apparent in dip sections across passive margins. Sea-level change was identified as the most logical choice among potential driving mechanisms in the development of certain unconformities that appeared to have world-wide distribution at the same chronostratigraphic positions.

By the mid-'60s, the exploration and production research efforts of the "Jersey companies" (notably Carter and Humble) were consolidated at the Esso (soon Exxon) lab in Houston. The transported Tulsa group survived the inevitable distress accompanying a shotgun wedding and, buoyed by success in application of their principles to the emerging complexities of the North Sea Basin, pushed ahead with the elaboration and codification of the doctrinal precepts of seismic stratigraphy. The most visible product in the public realm is Memoir 26 of the American Association of Petroleum Geologists: *Seismic stratigraphy: Applications to hydrocarbon exploration* (Payton, editor, 1977). Whereas my earlier attempts to introduce new approaches to the organization and analysis of stratigraphic successions had found little acceptance beyond former students and close acquaintances, seismic stratigraphy became an instantly and widely popular topic among subequal sets as an object of either praise or abuse. A recent paper by Cross and Lessenger (1988) presents an excellent review of seismic stratigraphy after a decade in the public domain, giving freedom here to explore the influence of seismic stratigraphy on the "new stratigraphy" and on the resulting revisions of concept required.

SEISMIC STRATIGRAPHY AND THE SEQUENCE CONCEPT

Much of the stratigraphic framework within which Exxonian seismic stratigraphy operates derives directly from the sequence concept and the recognition of unconformity-bounded stratal units; indeed, Memoir 26 defines a *depositional sequence* as a "stratigraphic unit . . . bounded at its top and base by unconformities *or their correlative conformities*" [emphasis added] (Mitchum and others, 1977). Here, the addition of the phrase italicized above does no particular violence to the original sequence definition; rather, it is the natural product of the continuum of observations made possible by seismic exploration. Note, however, that *sequence* in the context of Memoir 26 and subsequent writings from the Exxon group refers to stratal packages at a much smaller scale than the original sequences (the latter were referred to as *supersequences* in the terminology of Memoir 26).

Where a profile is constructed across a shelf, slope, rise, and beyond (or from the margin to the interior of a sedimentary basin), the geometric discontinuities (terminations of reflectors) that identify unconformities pass seaward (or basinward) into seemingly conformable successions. Some such appearances of conformity are artifacts created by the low resolving power of reflections in the normally employed frequency range; that is, the spacing of potential reflectors (bedding planes or other sonic-impedance discontinuities) is too small to be "seen" by the seismic detectors. In other commonly encountered situations, however, it is rational to assume, in the absence of biostratigraphic data to the contrary, that uninterrupted sedimentation is represented and that the position of the sequence-bounding unconformity is reasonably approximated by the equivalent observable reflector, especially in "deeper water" pelagic or turbidite successions. Such a "conformity" represents the lithostratigraphic and chronostratigraphic position of the surface of zero lacuna, that is, the position where both the erosional vacuity and/or nondepositional hiatus (in the sense of Wheeler, 1958) present at an unconformity are replaced basinward by uninterrupted deposition.

The earliest published "Vail curve" for the totality of Phanerozoic time (Vail and others, 1977) shows "relative changes of sea level" which describe "second-order cycles (global supercycles)." The supercycles are the chronostratigraphic equivalents of the supersequences of Mitchum and others (1977) and fit with reasonable comfort into the sequence chronology of Sloss (1963)—serious exception being taken only to the boundary between supercycles O–S (= Tippecanoe) and D–M (= Kaskaskia); this choice by Vail and colleagues is clearly in error. Expanded versions of the "relative sea level" curves for Mesozoic and Cenozoic time appear in the same chapter of Memoir 26 (Vail and others, 1977). Publication of the curves aroused great interest and a fair share of controversy, some justifiable and some not. A common complaint concerned the lack of supporting data and, indeed, a large measure of faith was required of the reader. Complaining parties failed to recognize that the 1977 curves represent two decades of analysis of data, much of it proprietary, including thousands of kilometers of seismic lines, hundreds of subsurface records, and untold man-hours of biostratigraphic work. Further, curves showing a higher level of detail in the Cretaceous were not released for publication, and this omission added to the malaise of an ungrateful segment of the public. Many of the deficiencies of the 1977 product have been corrected in subsequent publications; progress continues with activity now spread over a broad spectrum of industrial, academic, and governmental agencies.

A more meaningful criticism of the work of Vail and his colleagues is based on their dependence on sea-level change as the driving mechanism controlling the freeboard of land masses and the distribution of sedimentary environments, their petrologies, and syndepositional structures. In the original publications (Vail and others, 1977), the excursions and retreats of the curves defining "second- and third-order cycles" were described as illustrating *relative change of sea level* and were interpreted to represent "global cycles of sea-level change." In a decade of continuing publication, there has been a relaxation of insistence that the landward overstep of shallow-marine and littoral deposits across a surface of erosion or nondeposition be ascribed directly and quantitatively to sea-level change; instead the rubric *relative change of coastal onlap* is employed.

The most recent available publication by the Exxon group (Haq and others, 1987) adds a wealth of post-Paleozoic biostratigraphic, magnetostratigraphic, and chronostratigraphic detail integrated with an updated "relative change of coastal onlap" curve (the now-familiar sawtooth curve) derived from observations at identified outcrop sites as well as from subsurface and seismic record-section data. The paper should go a long way toward stilling complaints about documentation and the reliance on proprietary information. At the same time, however, a continuing degree of obsession with sea level as a primary control is revealed by appended long-term and short-term "eustatic curves" expressed in meters above current mean sea level; the short-term curve represents a smoothing of the onlap curve such that implied sea-level highstands are shown approximately midway in the time span of an individual onlap event, and progressive onlap is indicated to continue to the close of the onlap cycle.

Postponing further discussion of the relationship between depositional cycles and sea level, it is appropriate to examine other examples of the continuing evolution of principle and practice as put forth by the founding fathers of seismic stratigraphy in the decade between Memoir 26 (Payton, editor, 1977) and the paper in *Science* (Haq and others, 1987) now before us. The Native-American sequences of 1949 and 1963, termed "supersequences" in Memoir 26 (Mitchum and others, 1977), are represented by major subdivisions, *megacycles* (for example, Upper Zuni), retaining the original tribal names. Megacycle boundaries are placed at the positions of prominent recessions of the coastal-onlap (sawtooth) curve and, of the two boundaries present in the stratigraphic and chronologic span considered, one (the Absaroka-Zuni boundary) is entirely satisfactory (meaning it agrees with the time-honored placement at the close of Early Jurassic time, separating Toarcian and older strata from Aalenian and younger Mesozoic beds).

The close of the Upper Zuni megacycle is placed near the end of the Cretaceous (late in the Maastrichtian) some 8 m.y. earlier and, commonly, tens to hundreds of meters below the mid-Paleocene position that right-thinking stratigraphers would prefer. The pick by Haq and others (1987) coincides with a modest recession of the onlap curve interpreted as a sharply defined lowstand of sea level, ignoring more impressive events marked by the onlap curve at the close of Danian and early in Thanetian time. It is not made clear whether the end of Exxon's Zuni megacycle reflects a gesture of conformity with the Mesozoic-Cenozoic boundary and the K-T extinction event or an urge to cultivate the perception of regular periodicity in the spacing of high-amplitude sea-level events. In any case, reassignment of the Zuni-Tejas boundary performs no useful service for workers concerned with a rational segmentation of the stratigraphic record, whether the segments be termed sequences, megacycles, or synthems.

The history of stratigraphic classification and nomenclature suggests that it is perilous to apply an ordinal system (for example, Primary, Secondary, Tertiary, Quaternary) to successive units; inevitably, advancing knowledge demands further subdivisions and these are identified by lower ordinal levels of letters, numbers, subscripts, and so on. Nevertheless, in recognition of the complexity of the Indian-tribe sequences as originally defined, I have been impelled (most recently in Sloss, 1988) to recognize *subsequences* labeled by Roman numerals (for example, Zuni III). The hierarchy proposed by Haq and others (1987) goes much farther, however, dividing megacycles (for example, Upper Zuni) among *supercycle sets* (for example, Upper Zuni A or UZA, the lowest supercycle set in the Upper Zuni megacycle) which are made up of *supercycles* (for example, UZA-3, the third supercycle above the base of the supercycle set) these in turn are composed of third-order cycles called *sequences* identified by a number separated by a decimal point following the number of the supercycle concerned (for example, UZA-3.5 to indicate the fifth sequence in the third supercycle of the Upper Zuni A supercycle set. It does not stretch the imagination to picture the numeric and alphabetic bouillabaisse that will appear with the addition of new elements to the system. In fact, packages of strata at the scale of a few meters and bearing genetic implications are already in use (for example, the "PACs" of Goodwin and Anderson, 1985) and, as was inevitable, local and regional variants of the Exxonian coastal-onlap curve are appearing. It will be difficult to accommodate these within the proposed system.

DOES SEQUENCE STRATIGRAPHY REFLECT MORE THAN SEA-LEVEL CHANGE?

Stratigraphers and tectonists will derive little benefit from prolonged debate on the language and semantic structure employed in sequence stratigraphy, and it is not profitable to expend significant effort in defending this or that lithostratigraphic or chronostratigraphic choice as a critical sequence-stratigraphic boundary. Rather, the community of concerned scholars would be more usefully engaged seeking a response to the question, "What does it all mean?" As was noted above, the founders of seismic stratigraphy and its applications to sequence stratigraphy are possessed by a belief in global-eustatic sea-level control of coastal onlap and its sequelae as observed in unconformities and documented by the distribution of sedimentary facies. By extension, in the absence of a more acceptable mechanism for high-frequency eustatic fluctuations, the founders are forced to an unspoken reliance on glacio-eustatics even for times apparently lacking continents with appropriate latitudinal qualifications and/or evidences of inequitable climatic regimes.

These questions have been voiced repeatedly over the past decade, and it requires no breakthrough to state that "[the Vail/Haq curves] should not be used as templates for . . . correlation" (Miall, 1986). Still, in spite of all the nay-sayers and the clear case to be made for influence by local and regional uplift and subsidence, by shifts in sediment-distribution routes and volumes, and so on, the fact remains that the published "global" onlap curves present a remarkably faithful framework for a first cut at correlation and a baseline from which to study local or regional departures from the "template," at least in regions free of the effects of syndepositional active-margin tectonism. Further, it remains evident that, although there are important exceptions, many of the observations upon which the Exxon people (and Exxon alumni) base their coastal onlap charts do indeed reflect sea-level changes. What mechanism, then *is* in control if it is not change in the volume of liquid-phase water available to increase or decrease the freeboard of continents? Anything that would alter the volume capacity of the world ocean would be effective even if continents maintained a constant elevation with respect to Earth's center. Oceanic volcanism and related lithospheric inflation at divergent margins and at intraplate sites has been a popular resort in this regard; a volcanic edifice of significant magnitude can be quickly erected, but the response time of cooling and subsidence is an order of magnitude too slow to account for short-period sea-level cycles.

It is useful to recall that one of the primary factors promoting the initial development of sequence stratigraphy was the need to recognize secular change in the tectonic framework governing the distribution, character, and thickness of sedimentary accumulation. The increasing availability of lithostratigraphic syntheses covering parts of several cratons in the '60s and '70s made it clear that the Indian-tribe sedimentary-tectonic sequences of the '40s and '50s were not limited to the North American craton (for example, Sloss, 1987). This realization led to re-examination of cratonic behavior in the light of global-tectonic theory and to the suggestion that tectonic episodes of near pancratonic effectiveness are responsible for the Indian-tribe sequences and their equivalents beyond North America (Sloss and Speed, 1974). It was further postulated that the timing and mode of cratonic uplift, subsidence, and deformation were related to the accumulation and discharge of thermally generated products trapped under cratons; these thoughts receive a degree of support from accumulating evidence of secular change in radial heat flow.

It would be an exaggeration to state that there is widespread belief in a global-tectonic basis for the observations on which sequence stratigraphy is founded (see, for example, the demurrer entered by Miall, 1986). Undeniably, episodes of continental glaciation and deglaciation affect global sea levels and may be represented by some of the later Cenozoic third-order sequences of Haq and others (1987). Other interregional and intercratonic examples of coastal onlap may well prove to be the results of geoidal perturbations and other unidentified manifestations. Nevertheless, I persist in the claim that cratons, their margins, and their interior basins "do not just lie there" passively waiting to be encroached upon by rising sea levels or laid bare to erosion as sea levels fall. Students and practitioners of sequence stratigraphy are, for better or worse, recorders and interpreters of tectonic evolution.

REFERENCES CITED

Chang, K. H., 1975, Unconformity-bounded stratigraphic units: Geological Society of America Bulletin, v. 86, p. 1544–1552.
Cross, T. A., and Lessenger, M. A., 1988, Seismic stratigraphy, *in* Wetherell, G. W., Albee, A. A., and Stehli, F. G., eds., Annual Review of Earth and Planetary Sciences, v. 16, p. 319–354.
Goodwin, P. W., and Anderson, E. J., 1985, Punctuated aggradational cycles: A general hypothesis of episodic stratigraphic accumulation: Journal of Geology, v. 93, p. 515–533.
Haq, B. U., Hardenbol, J., and Vail, P. R., 1987, Chronology of fluctuating sea levels since the Triassic (250 m.y. to present): Science, v. 235, p. 1156–1167.
Krumbein, W. C., and Sloss, L. L., 1951, Stratigraphy and sedimentation (1st edition): San Francisco, W. H. Freeman and Co., 497 p.
Miall, A. D., 1986, Eustatic sea level changes interpreted from seismic stratigraphy: A critique of the methodology with particular reference to the North Sea Jurassic record: American Association of Petroleum Geologists Bulletin, v. 70, p. 131–137.
Mitchum, R. M., Jr., Vail, P. R., and Thompson, S., III, 1977, The depositional sequence as a basic unit for stratigraphic analysis, *in* Payton, C. E., ed., Seismic stratigraphy—Applications to hydrocarbon exploration: American Association of Petroleum Geologists Memoir 26, p. 53–62.
Payton, C. E., ed., 1987, Seismic stratigraphy—Applications to hydrocarbon exploration: American Association of Petroleum Geologists Memoir 26, VII + 516 p.
Salvador, A., chairman, International Subcommission on Stratigraphic Classification, 1987, Unconformity-bounded stratigraphic units: Geological Society of America Bulletin, v. 98, p. 232–237.
Silberling, N. J., and Roberts, R. J., 1962, Pre-Tertiary stratigraphy and structure of northwestern Nevada: Geological Society of America Special Paper 72, 58 p.
Sloss, L. L., 1950, Paleozoic stratigraphy in the Montana area: American Association of Petroleum Geologists Bulletin, v. 34, p. 423–451.
—— 1963, Sequences in the cratonic interior of North America: Geological Society of America Bulletin, v. 74, p. 93–114.
—— 1987, The Williston Basin in the family of cratonic basins: Rocky Mountain Association of Geologists, 1987 Symposium, p. 1–8.
—— 1988, Tectonic evolution of the craton in Phanerozoic time, *in* Sloss, L. L., ed., Sedimentary cover of the craton: U.S.: Geological Society of America, Geology of North America, v. D-2, p. 25–52.
Sloss, L. L., and Speed, R. C., 1974, Relationship of cratonic and continental-margin tectonic episodes, *in* Dickinson, W. R., ed., Tectonics and sedimentation: Society of Economic Paleontologists and Mineralogists Special Publication 22, p. 38–55.
Sloss, L. L., Krumbein, W. C., and Dapples, E. C., 1949, Integrated facies analysis, *in* Longwell, C. R., chairman, Sedimentary facies in geologic history: Geological Society of America Memoir 39, p. 91–124.
Vail, P. R., 1959, Stratigraphy and lithofacies of Upper Mississippian rocks in the Cumberland Plateau [Ph.D. dissert.]: Evanston, Illinois, Northwestern University.
Vail, P. R., and Wilbur, R. O., 1966, Onlap, key to worldwide unconformities and depositional cycles [abs.]: American Association of Petroleum Geologists Bulletin, v. 50, p. 638–639.
Vail, P. R., Mitchum, R. M., Jr., and Thompson, S., III, 1987, Relative changes of sea level from coastal onlap, and Global cycles of relative changes of sea level, *in* Payton, C. E., ed., Seismic stratigraphy—Applications to hydrocarbon exploration: American Association of Petroleum Geologists Memoir 26, p. 63–98.
Wanless, H. R., and Weller, J. M., 1932, Correlation and extent of Pennsylvanian cyclothems: Geological Society of America Bulletin, v. 43, p. 1003–1016.
Wheeler, H. E., 1958, Time stratigraphy: American Association of Petroleum Geologists Bulletin, v. 42, p. 1047–1063.

MANUSCRIPT RECEIVED BY THE SOCIETY JUNE 28, 1988
REVISED MANUSCRIPT RECEIVED JULY 20, 1988
MANUSCRIPT ACCEPTED JULY 20, 1988

CENTENNIAL ARTICLE

Tectonics of sedimentary basins

RAYMOND V. INGERSOLL *Department of Earth and Space Sciences, University of California, Los Angeles, California 90024-1567*

ABSTRACT

Simultaneous breakthroughs in our understanding of plate-tectonic processes, depositional systems, subsidence mechanisms, chronostratigraphy, and basin-exploration methods have resulted in rapidly improving actualistic models for sedimentary basins. Basin analysis has become a true science with the development of quantitatively testable models based on modern basins of known plate-tectonic setting. Major subdivisions of basin settings include divergent, convergent, transform, and hybrid; 23 basin categories occur within these settings. Basins are classified according to primary plate-tectonic controls on basin evolution: (1) type of substratum, (2) proximity to plate boundary, and (3) type of nearest plate boundary(s). Sedimentary basins subside primarily owing to (1) attenuation of crust as a result of stretching and erosion, (2) contraction of lithosphere during cooling, and (3) depression of lithosphere by sedimentary and tectonic loads. The first two processes dominate in most divergent settings, whereas the third process dominates in most convergent settings. Intraplate, transform, and hybrid settings experience complex combinations of processes. Several basin types have low preservation potential, as predicted by their susceptibilities to erosion and uplift during orogeny and as confirmed by their scarcity in the very ancient record.

Key references concerning actualistic plate-tectonic models for each type of basin form the basis for reviewing the present state of the science. The key references come from many sources, with diverse authorship, including several publications of the Geological Society of America. The further development and refinement of actualistic basin models will lead to improved testable paleotectonic reconstructions.

INTRODUCTION

The 1970s and 1980s have been a dynamic era in the science of basin analysis. Several breakthroughs in our understanding of modern and ancient processes, at large and small scales, have resulted in major advances in several disciplines central to basin analysis. Best known to geoscientists, of course, is the "revolution in the earth sciences" brought about by the widespread acceptance of the paradigm of plate tectonics (for example, Cox and Hart, 1986). Equally important to basin analysts has been the revolution in our understanding of modern depositional systems and the consequent major advances in sophistication of actualistic depositional models (for example, Davis, 1983; Walker, 1984; Reading, 1986). Actualistic petrologic models relating sediment composition, especially sand and sandstone, to plate-tectonic settings have been developed (for example, Dickinson and Suczek, 1979). Exploration techniques, especially seismic stratigraphy and detailed mapping of ocean floors, have provided new avenues for investigation of both ancient and modern basins. In addition, refinement in analytical methods, including chronostratigraphy, subsidence analysis, microanalytical investigation of thermal history, paleomagnetism, and paleoclimatology, to name just a few, has revolutionized basin analysis, as summarized by Miall (1984) in a landmark textbook. Miall pointed out that a "new stratigraphy" is upon us. The past 20 years have been the beginning of the "golden age of the new stratigraphy" owing to the concurrent revolutions affecting the earth sciences. In fact, stratigraphy has become a true science only with the development of testable actualistic models based on a combination of theory, observation, and experiment. These actualistic models are the key to basin analysis from the micro- to the megascale. Dott (1978) reviewed the development of pre–plate-tectonic (primarily nonactualistic) models for megascale basins (geosynclines).

The emphasis of this article is on actualistic plate-tectonic models for sedimentary basins. Models dealing solely with plate-tectonic processes, without application to sedimentary basins, have been excluded. Also excluded are models dealing with details of depositional systems, basin architecture, petrology, or other aspects of the basins themselves. Several excellent case studies of specific basins are not mentioned, not because they are deficient, but because they do not develop actualistic plate-tectonic models for a class of basins. To paraphrase Walker (1984, p. 6), a basin model should act as (1) a norm, for purposes of comparison; (2) a framework and guide for future observations; (3) a predictor in other situations; and (4) an integrated basis for interpretation of the class of basin it represents.

Selection of key papers for each type of basin ranged from obvious and easy to arbitrary and difficult; selection was limited to the English literature. Each paper discussed herein captures the essence of the important controls on basin formation and evolution for that specific plate-tectonic setting. Some of these papers are reviews that summarize years of both the authors' and others' work. Other papers are breakthrough insights concerning the development of some basin types. Each selected paper is not necessarily the first or the final statement on a particular type of basin; rather,

Figure 1. Sketch map showing sites of key sediment accumulations (basins) in relation to plate boundaries, continental margins, and associated sources of detrital sediment. Unpatterned areas represent oceanic crust. Continental margins and tectonic features are indicated by solid lines; other basin margins are indicated by dashed lines. Stipples indicate areas of sediment accumulation; "smoking" triangles represent magmatic arc. Solid barbs represent subduction zones; open barbs represent foreland fold-thrust belts. An intracratonic basin is shown between "drainage divide" (peripheral bulge) and "rifted-margin sediment prism" (continental rise and terrace). Modified from Dickinson (1980).

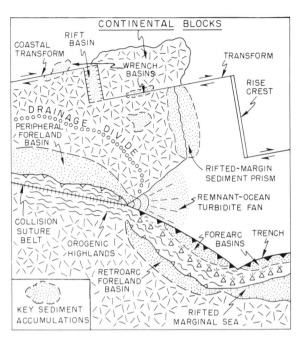

each paper represents (in my opinion) the best over-all statement and/or model at present. "Truth emerges more readily from error than from confusion" [Francis Bacon (Kuhn, 1970, p. 18)].

As is appropriate for a field as diverse as basin analysis, the authors of the papers range from sedimentologists to geophysicists. One of the most exciting aspects of this field of research is the integration of diverse disciplines.

One of the purposes of this Geological Society of America Centennial article is to highlight contributions of publications of the Society. The key papers selected come from diverse sources. As discussed below, however, publications of the Society have had a large impact on the science.

OVERVIEW AND BASIN CLASSIFICATION

Dickinson (1974, 1976) has provided the most comprehensive actualistic classification of basin types related to plate-tectonic processes. This classification is the basis for the organization of the remainder of this paper. Dickinson (1974) is a more formal publication, whereas Dickinson (1976) provides more detail. The true-scale diagrams of orogenic basins are an important addition to the latter publication.

To paraphrase Dickinson (1974, 1976), plate tectonics emphasizes horizontal movements of the lithosphere, which induce vertical movements due to changes in crustal thickness, thermal character, and isostatic adjustment. These vertical movements cause the formation of sedimentary basins, uplift of sediment source areas, and reorganization of dispersal paths. Primary controls on basin evolution (and the basis for classification) are (1) type of substratum, (2) proximity to plate boundary, and (3) type of nearest plate boundary(s). Types of substratum include continental crust, oceanic crust, transitional crust, and anomalous crust. Primary types of plate boundaries are divergent, convergent, and transform; hybrid boundaries also occur. Throughout this discussion, the distinction between continental margins and plate boundaries is important; they may or may not correspond. [For this reason, usage such as "collision of plates" should be avoided. Plates "converge"; only nonsubductable (buoyant) crustal components can "collide."] "The evolution of a sedimentary basin thus can be viewed as the result of a succession of discrete plate-tectonic settings and plate interactions whose effects blend into a continuum of development" (Dickinson, 1974, p. 1).

Subsidence of the crust to form sedimentary basins is induced by the following processes (Dickinson, 1974, 1976): (1) attenuation of crust due to stretching and erosion; (2) contraction of lithosphere during cooling; and (3) depression of both crust and lithosphere by sedimentary or tectonic loads, which are isostatically compensated either locally or regionally.

The latter type of regional compensation of loads results in lithospheric flexure that can result in both subsidence and uplift far from applied loads. The first two processes dominate along most divergent plate boundaries, whereas the third process dominates along most convergent boundaries. Intraplate (including rifted continental margins), transform, and hybrid settings experience complex combinations of processes. Ancillary effects such as paleolatitude, paleogeography, and eustatic changes may provide modifying influences.

Tectonic controls on basin formation are diverse; therefore, the basin analyst must assimilate an enormous literature. Gross subdivision of basins into compressional and extensional types, or active versus passive margins, is even less useful than geosynclinal classification, which if nothing else, recognized the diversity and complexity of basin types (for example, Kay, 1951).

The first step in identifying essential components controlling basin development commonly is the construction of accurate maps and cross sections (preferably at true scale) of modern plate-tectonic systems. Figure 1 illustrates a courageous attempt to summarize almost every basin type in one diagram. Each of these basin types is discussed in terms of a key reference in the following sections.

Table 1 summarizes the types of basins discussed below. Dickinson (1974, 1976) discussed the over-all controls on basin development of most of these types. The key references discussed below provide either more complete reviews of controlling processes, further work detailing these processes, or more refined and quantitative models. Some of these basin models have been brought to high levels of quantitative sophistication, whereas others remain general and qualitative. Significant additional work is necessary before quantitatively testable models are available for all basin types.

Bally and Snelson (1980), Miall (1984), and Klein (1987), among others, provide useful reviews of additional aspects of basin classification, kinematics, and dynamics.

DIVERGENT SETTINGS

Sequential Rift Development and Continental Separation

Kinsman (1975) reviewed the sequential evolution of terrestrial rift valleys into juvenile ocean basins and ultimately, into mature rifted continental margins. Figure 2 illustrates the post-rift stages, using somewhat different nomenclature from that of Kinsman (1975). Kinsman's model is based on isostatic adjustment to lithospheric temperature and density distributions, and subsidence is analogous to the thermal subsidence of sea floor formed at spreading centers, with additional subsidence due to sedimentary loading. Early domal uplifts related to mantle plumes are postulated to precede crustal rifting, without consideration of the possibility (in fact, probability) that domal uplifts are responses to lithospheric extension rather than its cause (Sengor and Burke, 1978; Morgan and Baker, 1983). Kinsman also assumed "normal" continental crust prior to rifting, which is at odds with theoretical studies showing that thickened and heated crust is weaker and, thus, is more likely to be the locus of extension when tensile stresses are applied (Kusznir and Park, 1987; Lynch and Morgan, 1987). In any case, elevated rift shoulders (arch rims of Veevers, 1981) form on either side of the terrestrial rift and persist through the proto-oceanic phase (also, see Hellinger and Sclater, 1983) (Fig. 2A). Supracrustal thinning (erosion) occurs while these rims are elevated, resulting in subsidence below sea level during subsequent cooling as the continental edges move away from the spreading center. Kinsman suggested that the width of attenuated continental crust seaward of the rim is relatively narrow (60–80 km), although other workers have suggested that wider attenuated margins are common (for example, Cochran, 1983a; Lister and others, 1986). Rifting and continental separation commonly occur along alternating divergent and transform margins linking hot spots (Burke and Dewey, 1973), resulting in interplume, transcurrent, and plume margins, respectively. Kinsman's (1975) model emphasized simple isostatic compensation at a depth of 100 km, whereas regional lithospheric flexure (for example, Walcott, 1972) is a major control on subsidence, especially in terms of the seaward tilting of the shallow-marine (shelf) prism (for example, Pitman, 1978). Nonetheless, Kinsman's model successfully predicts the maximum stratigraphic thickness found in continental embankments (16–18 km) based on the loading capacity of oceanic lithosphere.

Terrestrial Rift Valleys

Surprisingly little work has been published on the tectonic development of terrestrial rift valleys in terms of basin models. Many volumes have been published on geophysics, geochemistry, volcanism, geomorphology, and other aspects of modern rifts, but few attempts have been made to synthesize the temporal evolution of rifts in terms of general models for basin development. Rifts form in many plate-tectonic settings: (1) craton interiors unrelated to orogeny (East Africa), (2) intracontinental zones related to continental collision (Rhine graben and Baikal rift), (3) transtensional rifts along transform faults (Dead Sea and Salton Sea), and (4) settings related to complex plate interactions at subduction zones and transform margins (Rio Grande rift). Orogenic activity immediately following rifting may deform rifts and convert them into complex basins. If no orogenic activity occurs following a finite amount of rifting, the rifts may "fail," and intracratonic basins may develop subsequently. In contrast, if rifting is "successful," then continental separation occurs and the rift basins are pulled apart to form new ocean basins. Thus, ancient rift basins commonly are buried under younger sediments, deposited either in intracratonic basins (overly-

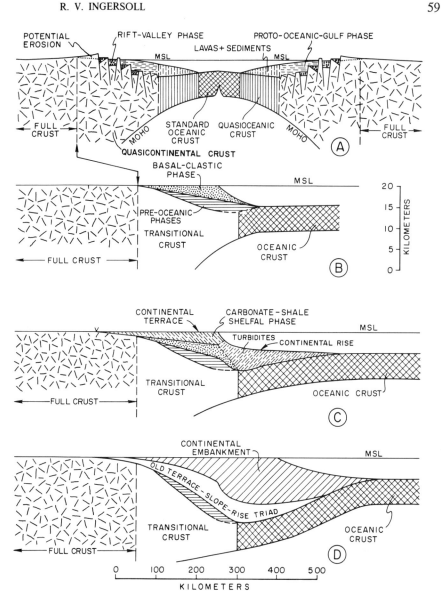

Figure 2. Schematic diagrams (vertical exaggeration 10×) to illustrate general evolution of rifted-margin prism along rifted continental margin: (A) proto-oceanic stage showing rift-valley depositional phases on top of attenuated continental (quasi-continental) crust, adjacent to thickened basaltic (quasi-oceanic) crust; (B) end of proto-oceanic stage when thermal subsidence is nearing completion; quasi-continental and quasi-oceanic crust are combined into transitional crust, underlying the subsiding continental margin; (C) continental terrace-slope-rise configuration during open-ocean stage, during which sediment loading is predominant subsidence mechanism; (D) continental-embankment stage, reached only where sediment delivery is voluminous enough to cause progradation of shoreline over oceanic crust (in areas of major deltas, usually at open ends of failed rifts). Modified from Dickinson (1976).

TABLE 1. BASIN CLASSIFICATION

Divergent settings
  Terrestrial rift valleys: rifts within continental crust, commonly associated with bimodal volcanism
  Proto-oceanic rift troughs: incipient oceanic basins floored by new oceanic crust and flanked by young rifted continental margins
  Continental rises and terraces: mature rifted continental margins in intraplate settings at continental-oceanic interfaces
  Continental embankments: progradational sediment piles constructed off edges of rifted continental margins
  Failed rifts and aulacogens: inactive terrestrial rift valleys, which may be reactivated during convergent tectonics and become aulacogens at high angles to orogenic belts
  Intracratonic basins: broad cratonic basins floored by failed rifts in axial zones
  Oceanic basins: basins floored by oceanic crust formed at divergent plate boundaries unrelated to arc-trench systems
  Oceanic islands, aseismic ridges, and plateaus: sedimentary aprons and platforms formed in intraoceanic settings other than magmatic arcs
Convergent settings
  Trenches: deep troughs formed by subduction of oceanic lithosphere
  Trench-slope basins: local structural depressions developed on subduction complexes
  Forearc basins: basins developed between subduction complexes and magmatic arcs
  Intra-arc basins: local basins within magmatic arcs
  Interarc and backarc basins: oceanic basins between and behind intraoceanic magmatic arcs, and continental basins behind continental-margin magmatic arcs without foreland fold-thrust belts
  Retroarc foreland basins: foreland basins on continental sides of continental-margin arc-trench systems
  Remnant ocean basins: shrinking ocean basins caught between colliding continental margins and/or arc-trench systems, and ultimately subducted or deformed within suture belts
  Peripheral foreland basins: foreland basins above rifted continental margins that have been pulled into subduction zones during crustal collisions
  Piggyback basins: basins formed and carried atop moving thrust sheets
  Foreland intermontane basins: basins formed among basement-cored uplifts in foreland settings
Transform settings
  Transtensional basins: basins formed by extension along strike-slip fault systems
  Transpressional basins: basins formed by compression along strike-slip fault systems
  Transrotational basins: basins formed by rotation of crustal blocks about vertical axes within strike-slip fault systems
Hybrid settings
  Intracontinental wrench basins: diverse basins formed within and on continent crust due to distant collisional processes
  Successor basins: basins formed in intermontane settings following cessation of local orogenic activity

Note: table modified after Dickinson (1974, 1976).

ing failed rifts) or along rifted continental margins (overlying successful rifts and adjacent failed rifts).

Leeder and Gawthorpe (1987) have provided the latest synthesis of sedimentary models for extensional tilt-block/half-graben basins. They outlined the primary control that faulting exerts on dispersal paths, depositional systems, and basin architecture and presented four qualitative models relating surface morphology, fault activity, and facies. In all of the models, coarse-grained steep fans of small volume are derived from footwall uplands, whereas finer-grained broad cones are derived from hanging-wall dip slopes, except in the case of marine inundation. Lacustrine, fluvial, or deep-marine sediments may accumulate in areas immediately adjacent to fault scarps, depending upon over-all paleogeography. Tilt-block/half-graben basins are common in the ancient record, but it is not clear how many of them represent primarily high-angle faulting, listric faulting, or low-angle detachment faulting, as envisioned by Wernicke (1985) and Kusznir and others (1987). Development of refined depositional models based on studies of young extensional basins is needed to provide constraints for interpreting fault geometry and tectonics in ancient settings (for example, Frostick and Reid, 1987).

Proto-oceanic Rift Troughs

The Red Sea is the type proto-oceanic rift trough. Caution is needed, however, in constructing a general model for such settings because the Red Sea (including the Gulf of Aden) is the only modern proto-ocean on Earth. The Gulf of California is primarily a transtensional feature, although it shares many characteristics with the Red Sea. It is not surprising, therefore, that a summary of the history of the Red Sea is the basis for a proto-oceanic model.

Cochran (1983a) presented convincing geophysical evidence for active sea-floor spreading in the southern Red Sea, whereas north of 25°N, the center of the trough appears to be attenuated continental crust. Plate reconstructions dictate that significant divergence has occurred between Africa and Arabia since the end of the Oligocene. This necessitates the presence of quasicontinental and quasioceanic crust under most of the Red Sea (for example, Fig. 2A). Cochran (1983a) argued that the northern and southern Red Sea represent the earlier and later stages, respectively, of the transition from terrestrial rifting to sea-floor spreading. "An initial period of diffuse extension by rotational faulting and dike injection over an area perhaps 100 km wide is followed by concentration of extension at a single axis and the initiation of sea-floor spreading" (Cochran, 1983a, p. 41).

An extended phase of rifting and diffuse extension must be accounted for in thermal models for post-rifting subsidence (Cochran, 1983b). Specifically, horizontal heat flow causes additional cooling within the rift and uplift of the rift shoulders. Subsequent thermal subsidence of the continental margins is less than would be expected following "instantaneous" subsidence (for example, McKenzie, 1978).

Continental Rises and Terraces

Subsidence mechanisms at rifted continental margins include (1) thinning of crust due to stretching and erosion during doming; (2) thermal subsidence following rifting; (3) loading by sediment, resulting in both local isostatic and regional lithospheric flexure; and (4) lower-crustal and sub-crustal flow and densification following rifting.

Pitman (1978) elegantly demonstrated how transgressive/regressive sequences are formed on shelves along rifted continental margins. He showed that subsidence rates at shelf edges are normally greater than rates at which eustatic sea level changes. Shelves can be modeled as platforms rotating about a landward hinge line. Thus, shorelines seek locations that reflect a balance among sea-level change, subsidence rate, and sedimentation rate. The net result is that transgressive/regressive sequences reflect changes in rates of sea-level change, rather than changes in sea level. All modern rifted continental margins are younger than Paleozoic (post-breakup of Pangea), and most seismic "sea-level curves" (for example, Vail and others, 1977) are constructed primarily from Cretaceous and Cenozoic sequences along Atlantic margins. Sea level has been generally falling since its high point in the Cretaceous, owing to the combination of increase of the average age of oceanic crust (for example, Heller and Angevine, 1985) and the initiation of continental glaciation during the middle Cenozoic. The record of Cenozoic transgressive/regressive sequences along Atlantic margins is consistent with continually falling sea level, but at varying rates. Diverse tectonic processes provide mechanisms for second-order changes in sea level (for example, Cloetingh and others, 1985; Karner, 1986).

Pitman's model shows how the seaward-thickening wedge of the continental terrace can form (for example, Fig. 2C) in response to the combined effects of tilting about a landward hingeline, changes in sea level, and changes in sedimentation rates. The shoreline would be at the shelf edge only during times of unusually rapid sea-level fall (for example, glaciation or sudden flooding of a formerly isolated dry ocean basin as in the case of, for example, the Messinian Mediterranean) or in areas of very high sedimentation (see below).

Continental Embankments

Burke (1972) summarized depositional processes leading to the development of the Niger Delta (the surficial expression of a continental embankment), as schematically represented in Figure 2D. Prevailing southwest winds impinge symmetrically on the Niger Delta and neighboring coasts, thus setting up converging longshore drifts, which meet at the delta corners. Submarine canyons at these corners (and at the center of the delta during low sea level) feed voluminous sediment to submarine fans at the delta foot. The net result is a five-layer structure as the continental margin progrades seaward over oceanic crust and older pelagic sediments. From bottom to top, these layers are (1) deep-sea-fan sands; (2) transitional sand and mud; (3) slope mud, with abundant diapirs and slumps; (4) transitional shoreline and shelf sand and mud; (5) continental (primarily fluvial) sand. Continental embankments are gravitationally unstable owing to rapid burial of low-density, water-saturated sediments, and the common occurrence of deeply buried evaporite deposits inherited from the proto-oceanic stage. Therefore, salt diapirs, mud diapirs, and growth faulting are common.

Continental embankments such as the Niger Delta and the Mississippi Delta form in areas of unusually rapid sedimentation, primarily at the mouths of failed rifts (Burke and Dewey, 1973). Drainage from large cratons commonly is directed toward the mouths of failed rifts and

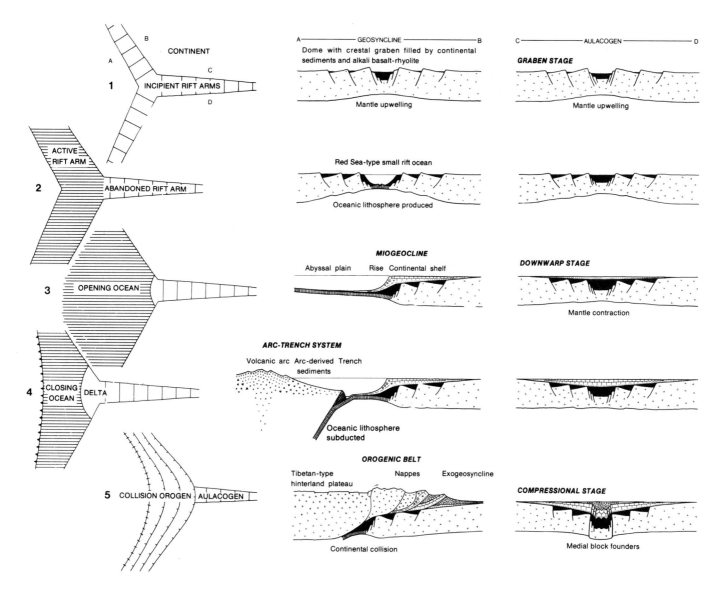

Figure 3. Model for evolution of an aulacogen and related orogenic belt. (1) A three-armed radial rift system accompanied by crustal doming; strike-slip component (transform fault) may be important along any of the rift arms. (2) Two rift arms spread to produce narrow ocean similar to Red Sea; rifting of third arm insufficient for continental separation. (3) Spreading of two active rift arms produces large ocean basin. Downwarping and terrace-rise sedimentation occur along new aseismic continental margins. Third arm fails and remains as transverse trough located at re-entrant on continental margin. The failed rift arm evolves from incipient rift to broad downwarp (intracratonic basin). (4) Ocean is closed by subduction along trench, producing adjacent magmatic arc. History of ocean closure may take many paths, only one of which is shown here. (5) Closing of ocean ultimately results in continental collision and development of collision orogen. Abandoned rift arm is preserved as an aulacogen located where orogen makes re-entrant into its foreland. Aulacogen is further loaded with peripheral-foreland sediments derived from advancing orogen, and its medial block founders, resulting in final stage of compressional deformation and faulting. Reproduced by permission of Society of Economic Paleontologists and Mineralogists (from Hoffman and others, 1974).

away from adjacent "normal" rifted continental margins. Similar massive outbuilding of continental margins due to deltaic progradation occurs in remnant ocean basins (for example, Bengal and Indus fans; see below); however, these latter deposits form in tectonically active settings related to continental suturing. Therefore, provenance, dispersal orientations, sediment types, related petrotectonic assemblages, and final deformational features contrast with continental embankments formed at failed rifts, even though gross structural cross sections of these progradational continental margins are similar. Kinsman (1975) predicted that the maximum possible stratal thickness of such margins is 16–18 km.

Failed Rifts and Aulacogens

During continental rifting, three rifts commonly form at approximately 120°, probably because this is a least-work configuration (Burke and Dewey, 1973). Regardless of whether initiating processes are "active" or "passive" (Sengor and Burke, 1978; Morgan and Baker, 1983), in the majority of cases, two rift arms proceed through the stages of continental separation (outlined above), whereas one rift arm tends to fail (Fig. 3). Hoffman and others (1974) discussed the resulting sedimentary accumulations, with emphasis on a Proterozoic example. They outlined five stages in the development of the Athapuscow aulacogen, which with slight modification, provide a model applicable to most aulacogens (linear sedimentary troughs at high angles to orogens): (1) rift stage, (2) transitional stage, (3) downwarping stage, (4) reactivation stage, (5) post-orogenic stage.

Although the actualistic model developed by Hoffman and others (1974) may be applied to several failed rifts and aulacogens, Hoffman (1987) has recently questioned whether it is the best model for the Athapuscow "aulacogen." Also, Thomas (1983, 1985) has discussed the possibility that the southern Oklahoma "aulacogen" originated as a transform boundary rather than as a failed rift. Neither of these reinterpretations discredits Hoffman and others' (1974) original model; rather, both examples illustrate the difficulty in applying any model to the complexity of the real world. Progress results from the application and testing of models; the real world is always more complex.

Successful rifts evolve into shelf-slope-rise margins, with continental embankments at reentrants (for example, Niger Delta). Failed rifts grade from embankments at their mouths to terrestrial rifts within cratons. As lithospheric extension ceases (for any reason), the rift areas cool and flexurally subside to form intracratonic basins, especially where three failed rifts meet. Upon activation or collision of the continental margin, the rifted-margin sedimentary prisms are intensely deformed, especially at continental promontories (Dewey and Burke, 1974; Graham and others, 1975). As orogeny proceeds, failed rifts become aulacogens, which may experience compressional, extensional, or translational deformation.

Impactogens (Sengor and others, 1978) resemble aulacogens, but without the pre-orogenic stages (1–3) (see below).

Intracratonic Basins

With increased exploration of deeper levels of well-known intracratonic basins (for example, the Michigan basin), it has become clear that most intracratonic basins have formed above ancient failed rifts (see Klein and Hsui, 1987).

Derito and others (1983) developed a lithospheric-flexure model with a nonlinear Maxwell viscoelastic rheology that helps explain how intracratonic basins can have long histories of synchronous subsidence over broad areas of continents. They pointed out that predicted subsidence due to stretching and cooling following cessation of rifting (for example, McKenzie, 1978; Sclater and Christie, 1980) is insufficient to explain the magnitude or timing of subsidence for most intracratonic basins. Many intracratonic basins (for example, Michigan, Illinois, and Williston basins) have experienced renewed subsidence during times of orogeny in adjacent orogenic belts. These periods of reactivation may be due to the reduction of effective viscosity of the lithosphere resulting from applied stress (due to adjacent orogeny) (Derito and others, 1983). Due to flexural rigidity of the lithosphere,

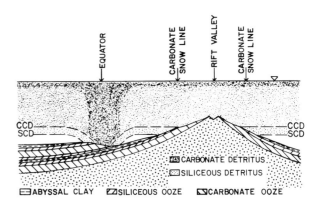

Figure 4. Model of axially accreting oceanic sedimentation (extreme vertical exaggeration, especially for stratal thicknesses). Deposition of carbonate ooze dominates on oceanic crust (solid-triangle pattern) shallower than the CCD (carbonate compensation depth); siliceous ooze accumulates between the CCD and SCD (silica compensation depth), and only abyssal clay accumulates below the SCD. Both the CCD and the SCD are depressed near equator and in other areas of high biologic productivity (zones of upwelling). Predictable stratigraphic sequences result as oceanic crust formed at spreading center (rift valley) cools and subsides as it moves away from spreading center (see text). After Heezen and others (1973).

dense loads emplaced in the crust during rifting (for example, basaltic dikes) remain isostatically uncompensated for geologically long periods of time. Any stress applied to the lithosphere results in a geologically instantaneous relaxation and increased subsidence. Repeated changes of stress due to orogenic activity decrease the effective viscosity beneath the basin, and the basin subsides more rapidly than during nonorogenic periods. Thus, the nearby orogenic activity allows the basin (underlain by a failed rift) to approach isostatic compensation more rapidly than it would without orogenic activity. Cumulative subsidence over hundreds of millions of years approaches that predicted by simple thermal models (for example, McKenzie, 1978). The model of Derito and others (1983) presents an elegant explanation for continent-wide synchronism of depositional sequences due to orogenic activity along nearby continental margins.

Quinlan (1987) has provided a thorough review of mechanisms and models for intracratonic basins.

Oceanic Basins

Heezen and others (1973) presented a kinematic model to explain the distribution of Cretaceous and Cenozoic sediment in the modern Pacific Ocean. The model is based on the systematic increase in depth of oceanic crust with age, combined with the dependence of sediment type on depth of water (Fig. 4). Biogenic oozes are volumetrically predominant away from continental margins, with generally more rapid production of calcareous pelagic detritus than siliceous pelagic detritus. Virtually all of this biogenic material is produced within the photic zone (approximately the upper 150 m of water),

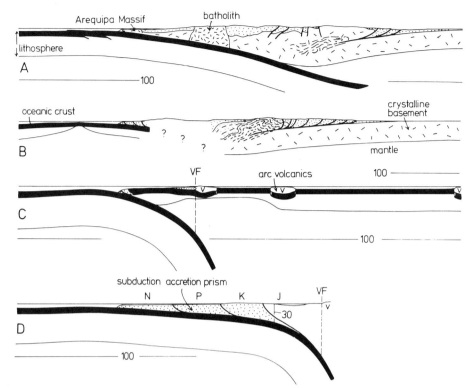

Figure 5. True-scale sections across Pacific plate margins (scale in kilometers). (A) Central Peru. (B) Western Canada. (C) Marianas. (D) Alaska (age of subduction-accretion prism: J = Jurassic, K = Cretaceous, P = Paleogene, N = Neogene). See text for discussion. Modified from Dewey (1980).

and upon the death of the secreting organisms, these tests settle slowly through the water column, dissolving as they fall. The depth below which no carbonate remains is the carbonate compensation depth (CCD). This depth is determined by the balance of biologic productivity in the photic zone and of the rates at which tests fall and dissolve in the water column (Berger and Winterer, 1974). Changes in any of these rates can change the CCD, which presently averages –3,700 m in the open ocean (Heezen and others, 1973) and –5,000 m near the equator (Berger, 1973). The CCD is depressed near the equator (and in other areas of upwelling) owing to higher biologic productivity. The silica compensation depth (SCD) is less well defined than is the CCD, but it is universally deeper than the CCD owing to the slower dissolution of silica tests. Below the SCD, only nonbiogenic clay accumulates (Fig. 4).

Oceanic crust formed at the East Pacific Rise south of the equator has the potential to subside below both the CCD and the SCD before crossing the equator; given the right depth (age) of crust and depth of equatorial CCD and SCD, an equatorial sequence of siliceous and calcareous oozes also may be deposited. Ancient oceanic crust with this history should, therefore, be overlain by the following vertical lithologic sequence: carbonate ooze, siliceous ooze, abyssal clay, siliceous ooze, carbonate ooze, siliceous ooze, abyssal clay. Heezen and others (1973) developed this model to illustrate possible combinations of ridge location and equatorial-transit history. The result is a predictive model that is in excellent agreement with the Cretaceous and Cenozoic stratigraphy of the Pacific plate. They discussed the formation of time-transgressive lithofacies (see the interesting discussion of oceanic stratigraphic principles by Cook, 1975) including pyroclastic material derived from western Pacific magmatic arcs and turbidite fans derived from the North American margin. Winterer (1973) has provided an additional example of paleotectonic reconstructions using Pacific plate stratigraphy.

Heezen and others' (1973) kinematic model is a potentially powerful tool for interpreting depositional histories of ophiolite sequences preserved in subduction complexes and suture zones. Caution is advised, however, because most ophiolites probably formed in ancient marginal basins rather than open-ocean settings, in which case, models for interarc and backarc basins are more useful for reconstructing paleotectonic settings. In addition, biologic controls on carbonate and siliceous sedimentation have changed markedly owing to evolutionary processes (Berger and Winterer, 1974); therefore, pre-Cenozoic oceanic models cannot be truly actualistic.

Oceanic Islands, Aseismic Ridges, and Plateaus

The model discussed above for the systematic subsidence of oceanic lithosphere as it travels away from divergent boundaries has important implications for intraoceanic sedimentation in general. All islands, aseismic ridges, and plateaus constructed of unusually thick basaltic material (for example, Hawaiian-Emperor chain) experience subsidence equal to the subsidence of oceanic lithosphere on which they ride. During growth due to active basaltic volcanism, only minor fringing sediments are likely to be deposited.

Clague (1981) divided the post-volcanic history of seamounts into three stages: subaerial, shallow water, and deep water or bathyal. Following cessation of volcanism, seamounts are likely to accumulate pelagic and shallow-marine carbonate sediments, with rate and type of sedimentation controlled largely by latitude (Clague, 1981). If carbonate sedimentation is equal to or greater than thermal subsidence, then fringing reefs and atolls will form atop the submerged islands and plateaus. Some of the depositional and migratory behavior of ancient islands, ridges, and plateaus accreted to continental margins at subduction zones may be reconstructed based on this model.

CONVERGENT SETTINGS

Arc-Trench Systems

Dewey (1980) classified arc-trench systems as extensional, neutral, or compressional, analogous to Dickinson and Seely's (1979) migratory-detached, noncontracted-stationary, and contracted categories (Fig. 5). Extensional arcs are intraoceanic owing to the formation of oceanic crust within and behind magmatic arcs as a result of trench rollback being faster than trenchward migration of the overriding plate. The western Pacific intraoceanic arcs are typical modern examples, with steep Benioff-Wadati zones dipping westward and subduction of old oceanic lithosphere (for example, Molnar and Atwater, 1978). Compressional arcs occur where a continental margin advances trenchward faster than trench rollback. The Andes typify these arcs, with shallow Benioff-Wadati zones dipping eastward and subduction of young oceanic lithosphere. Neutral arcs result where trench rollback is approximately equal to the trenchward advance of the overriding plate. The Aleutian and Indonesian arcs typify these arcs, with intermediate-dip Benioff-Wadati zones dipping northward and subduction of intermediate-age oceanic lithosphere.

Whether facing direction, age of subducted lithosphere, or a linked combination of both processes is the controlling factor in arc behavior is yet to be resolved (Dickinson, 1978). Dewey's (1980) kinematic model for arc behavior is useful, however, whatever the dynamic controls. Second-order effects include the probability that extensional arcs experience primarily basaltic magmatism and have low relief, thin sediments, and deep trenches. In contrast, compressional arcs experience silicic magmatism and have high relief, abundant sediments, and shallow trenches. Most arc-trench systems have intermediate characteristics, commonly including transform motion along arc trends; the superposition of several complex convergent zones results in great complexity (for example, Hamilton, 1979).

Further development of predictive models for the evolution of continental margins during the opening and closing of ocean basins (the Wilson cycle; see below) will come primarily from ancient orogenic belts, owing to the limited variety of young continental collisions. Dewey's (1980) kinematic model based on actualistic examples of magmatic arcs is a step in the direction of constraining the possibilities.

Trenches

Thornburg and Kulm (1987) have provided one of the most detailed studies of sedimentation in a modern trench. The Chile Trench is especially interesting because it provides markedly different climatic and sedimentologic conditions along the studied length from 18° to 45° S latitude. Subducted crust is younger to the south, but subduction is orthogonal and rapid everywhere. Thus, plate-tectonic processes are relatively constant along the trench, and sedimentary processes can be isolated as causes and effects.

Karig and Sharman (1975) and Schweller and Kulm (1978), among others, discussed the dynamic nature of sedimentation and accretion at trenches. The sediment wedge of a trench is in dynamic equilibrium when subduction rate and angle, sediment thickness on oceanic plate, rate of sedimentation, and distribution of sediment within the trench are constant. Schweller and Kulm (1978) presented an empirical model relating convergence rate and sediment supply to types of trench deposits. Thornburg and Kulm (1987) provided documentation for the dynamic interaction of longitudinally transported material (trench wedge with axial channel) and transversely fed material (trench fan). With increasing transverse supply of sediment to the trench, the axial channel of the trench wedge is forced seaward, and the trench wedge widens. Contrasts in dynamic trench-fill geometries help determine not only trench bathymetry and depositional systems but also accretionary geometry (Fig. 6). This dynamic model may be useful for reconstruction of sedimentary and tectonic processes in trenches, as expressed in ancient subduction complexes.

Trench-Slope Basins

Moore and Karig (1976) developed a model for sedimentation in small ponded basins along inner trench walls. Deformation within and on the subduction complex results in irregular bathymetry, commonly including ridges subparallel to the trench. Turbidites are ponded behind these ridges, and trench-slope basins form. Average width, sediment thickness, and age of basins increase up slope due to the progressive

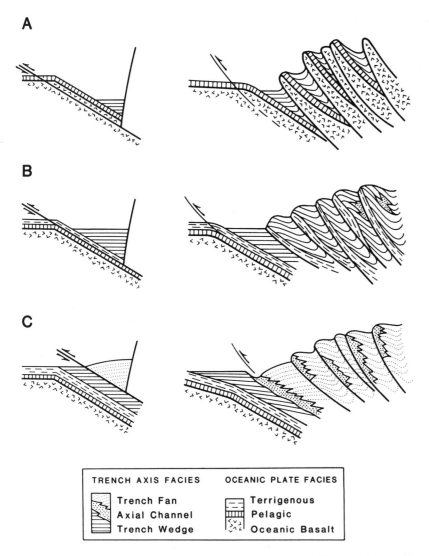

Figure 6. Accretion and preservation of trench deposits; stratigraphic control on position of décollement zone (thrust symbol) within the trench basin (Chile Trench). The lithofacies succession (left) determines the stratigraphic assemblage that is preferentially accreted to the adjacent continental slope (right). Sediment supply varies from meager (A) to abundant (C) along a margin, resulting in subduction complexes that include slivers of oceanic crust (ophiolite) with thin stratal packages (A), mixtures of trench and oceanic facies (B), or predominantly trench facies (C). Sediments that are not accreted at the trench presumably are underplated and/or melted at greater depths. From Thornburg and Kulm (1987).

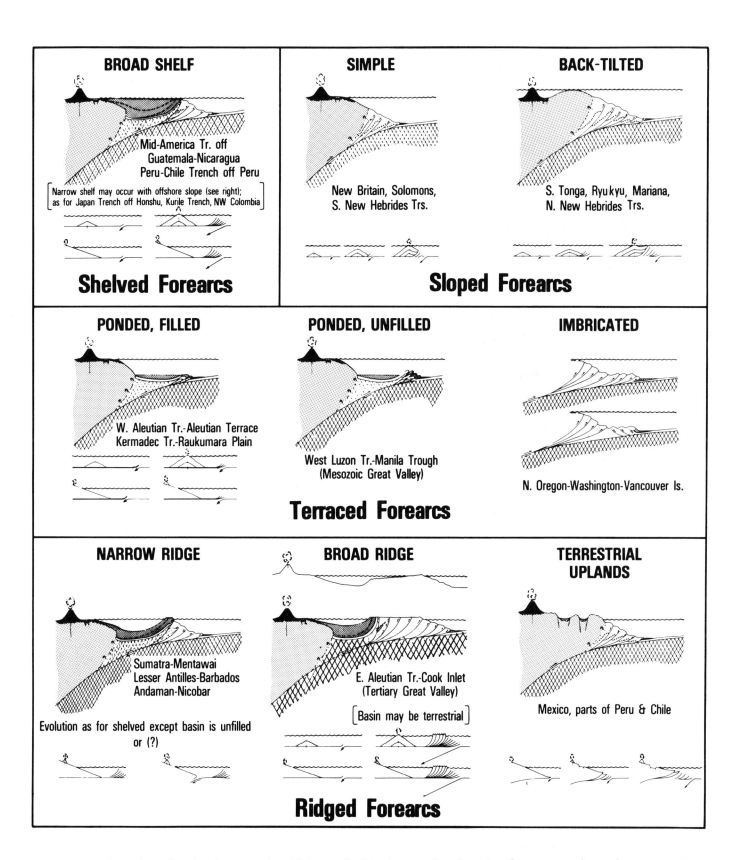

Figure 7. Configurations of modern forearc regions. Volumes of subduction complexes in modern forearcs are unknown in most areas (note question marks on diagrams), as are subsurface extent and nature of forearc igneous and metamorphic rocks. Where volumes of subduction complexes are small, arc massifs may be nearer trench than shown. Modified from Dickinson and Seely (1979).

uplift of deformed material and the widening of fault spacing during dewatering and deformation of offscraped sediment. In ancient subduction complexes, trench-slope basins are filled with relatively undeformed, locally derived turbidites surrounded by highly deformed accreted material of variable origin. Contacts between trench-slope basins and accreted material are both depositional and tectonic.

Moore and Karig's (1976) model was developed for Nias Island near Sumatra, an area of rapid accretion of thick sediments. Their model is less useful for sediment-starved forearc areas. Nonetheless, their general principles governing the development of sedimentary basins on the lower trench slope are fundamental to reconstructing ancient subduction settings.

Forearc Basins

Dickinson and Seely (1979) provided a classification of arc-trench systems, similar to Dewey's (1980), and outlined plate-tectonic controls governing subduction initiation and forearc development. Figure 7 illustrates the variability of forearc types. Factors controlling forearc geometry include (1) initial setting, (2) sediment thickness on subducting plate, (3) rate of sediment supply to trench, (4) rate of sediment supply to forearc area, (5) rate and orientation of subduction, and (6) time since initiation of subduction. Arc-trench gaps tend to widen through time (Dickinson, 1973) owing to prograde accretion at trenches and retrograde migration of magmatic arcs following subduction initiation. Prograde accretion is especially rapid where thick sequences of sediment are accreted. The net result of widening of the arc-trench gap is the general tendency for forearc basins to enlarge through time (for example, Great Valley forearc basin; Ingersoll, 1979, 1982).

Forearc basins include the following types (Dickinson and Seely, 1979): (1) intramassif (one type of intra-arc), (2) accretionary (trench-slope), (3) residual (lying on oceanic or transitional crust trapped behind the trench when subduction began), (4) constructed (lying across the boundary of arc massif and subduction complex), and (5) composite (combination of above settings). Residual and constructed basins tend to evolve into composite basins; commonly, this evolutionary trend is accompanied by filling and shallowing of the forearc basins.

Intra-arc Basins

Intra-arc basins are dominated by volcanic flows and volcaniclastic deposits, owing to their intimate association with active volcanoes. Depositional environments are varied but are determined, in large part, by whether the arc-trench system is extensional (oceanic), neutral (shelfal), or compressional (mountainous) (Dickinson, 1974, 1976; Dewey, 1980). Composition of magma, and therefore, style of eruption and volcanism, is determined, in large part, by type of underlying crust, oceanic (basaltic), transitional (andesitic), or continental (rhyodacitic) (Hamilton, 1979). Basins form primarily owing to tectonomagmatic collapse within eruptive centers, especially within silicic systems. Thus, "tectonic" processes are dominated by local extension above rising plutons, with little regard to regional stresses. Intra-arc basins may evolve into interarc basins during backarc spreading in extensional systems (see below), they may collapse as calderas and be engulfed in plutons in neutral systems, or they may be uplifted and destroyed in compressional systems. Thus, intra-arc basins have low preservation potential, and ancient examples are scarce.

Few studies of modern intra-arc basins have been published. This dearth of studies is due largely to the lack of integration of volcanology, sedimentology, and basin analysis. Publication of two recent textbooks (Fisher and Schmincke, 1984; Cas and Wright, 1987) suggests that this neglect is ending.

Busby-Spera (1984a, 1984b) provided a model to explain marine intra-arc sedimentation along the early Mesozoic continental margin of California. This basin model is not rigorously actualistic owing to the lack of modern studies. The sedimentologic and volcanologic aspects of the study, however, are actualistic. There is a risk in developing "actualistic" basin models based on ancient examples, especially in structurally complex, metamorphosed terranes such as Sierra Nevada "roof pendants" (for example, Christensen, 1963). Nonetheless, Busby-Spera (1984a, 1984b) outlined broad characteristics of intra-arc basins formed within neutral arc-trench systems and developed a model with predictive value in other studies.

Busby-Spera (1984a, 1984b) has demonstrated that complex facies changes are to be expected in the dynamic situation typified by silicic-andesitic volcanism near sea level along an open-ocean coast. Voluminous eruptions of rhyolite ash-flow tuff, construction of small andesite stratocones, and growth of submarine fans and debris aprons occur during times of volcanic activity; quiescence results in deposition of fine-grained epiclastic sediment and ash and/or progradational shelf sequences. All of the variability of modern shelves, shorelines, and nearby basins occurs within intra-arc basins near sea level. High sedimentation rates, complex facies interfingering, and complex interaction of magmatic, tectonic, eustatic, and sedimentary processes are expected. Development of more predictive actualistic models awaits study of modern systems.

Interarc and Backarc Basins

Carey and Sigurdsson (1984) have proposed a model for volcaniclastic sedimentation behind intraoceanic magmatic arcs that experience backarc spreading. Their model is a refinement of the model of Karig and Moore (1975), which in turn, is based on the tectonic model of Karig (1971). This model is applicable to intraoceanic arcs (for example, Marianas and Lesser Antilles), where trench rollback exceeds trenchward motion of the overriding plate (see Dewey, 1980); the magmatic arc is split to form a backarc basin, bounded by the still-active part of the magmatic arc and an inactive remnant arc (hence, this type of backarc basin also is called an "interarc basin"). The model needs modification in order to be applicable to backarc basins formed by the rifting of a continental-margin arc (for example, Japan Sea, where a rifted continental margin replaces the oldest remnant arc) or to backarc basins formed by the trapping of old oceanic crust (for example, Bering Sea, where no backarc spreading has occurred). Royden and others (1982) discussed an example of failed backarc spreading closely associated with continental collision. Also, Klein (1985) discussed modifications in sedimentation patterns in intraoceanic settings, resulting from climatic and oceanographic effects.

In Carey and Sigurdsson's (1984) model, the relative importance of sediment sources are (1) magmatic arc (primarily volcaniclastites, resulting from subaerial and subaqueous eruptions, and erosion of arc complex), (2) backarc spreading center (hyaloclastites and hydrothermally derived sediments), and (3) remnant arc (minor subaqueous gravity-flow deposits). Deposition of magmatic-arc–derived volcaniclastites results in the development of a sedimentary apron on the back side of the arc, with pronounced asymmetry of basin fill. Backarc basins experience four stages of evolution: (1) early rifting, with rapid influx of volcaniclastites; (2) basin widening, with active volcanism and spreading, and development of an asymmetric apron; (3) basin maturity, with waning volcaniclastic input and increased pelagic and hemipelagic deposition; and (4) basin inactivity, with cessation of spreading, and continued pelagic and hemipelagic deposition. A new cycle initiates with splitting of the arc and formation of a new backarc (interarc) basin.

Retroarc Foreland Basins

Compressional arc-trench systems commonly develop foreland basins behind the arcs owing to partial subduction of continental crust beneath the arc orogens (Figs. 5A and 5B). "Foreland basin" is a pre–plate-tectonic term used to describe a basin between an orogenic belt and a craton (Allen and others, 1986). Dickinson (1974) proposed that the term "retroarc" be used to describe foreland basins formed behind compressional arcs, in contrast to peripheral foreland basins formed during continental collisions (see below). Thus, although "backarc" and "retroarc" literally are synonymous, the former is used for extensional and neutral arcs, whereas the latter is used for compressional arcs.

Jordan (1981) presented an analysis of the asymmetric Cretaceous foreland basin associated with the Idaho-Wyoming thrust belt. She used a two-dimensional elastic model to show how thrust loading and sedimentary loading resulted in broad flexure of the lithosphere. The location of maximum flexure migrated eastward as thrusting migrated eastward. The area of subsidence was broadened owing to the erosional and depositional redistribution of part of the thrust load. Comparison of modeled basin and basement geometries with isopach maps provides tests of possible values of flexural rigidity of the lithosphere. The modern sub-Andean thrust belt and foreland basin have similar topography to that proposed for the Cretaceous of the Idaho-Wyoming system. Topography is controlled by thrust-fault geometry and isostatic subsidence.

The model presented by Jordan (1981) is broadly applicable to other retroarc foreland basins. It demonstrates how tectonic activity in the foreland fold-thrust belt is the primary cause of subsidence in associated foreland basins. Sedimentary redistribution, autocyclic sedimentary processes, and eustatic sea-level changes are important modifying factors in terms of regressive-transgressive sequences, but compressional tectonics behind the arc-trench system is the driving force. The Cretaceous seaway of North America was largely the result of this compressional tectonic activity (combined with high eustatic sea level). Details concerning timing of thrusting and initial sedimentary response to thrusting within the Idaho-Wyoming thrust belt are debated (for example, Heller and others, 1986), but the essential role of compressional tectonics in creating retroarc foreland basins is clear.

Remnant Ocean Basins and Suture Belts

Intense deformation occurs in suture belts during the attempted subduction of nonsubductable, buoyant continental or magmatic-arc crust. Suture belts can involve rifted continental margins and continental-margin magmatic arcs (terminal closing of an ocean basin) or various combinations of arcs and continental margins. Figure 8 illustrates stages in the development of suture belts, either in time or in space. Colliding continents tend to be irregular, and great variability of timing, structural deformation, sediment dispersal patterns, and preservability occurs along strike (Dewey and Burke, 1974).

Graham and others (1975) used the Cenozoic development of the Himalayan-Bengal system as

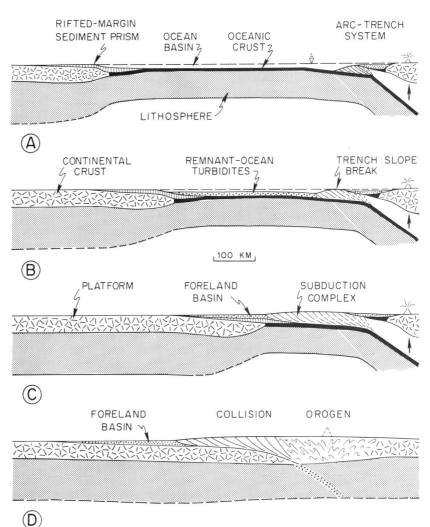

Figure 8. Idealized true-scale diagrams showing inferred evolution (A to D) of sedimentary basins associated with crustal collision to form cryptic intercontinental suture belt within collision orogen. Diagrams represent a sequence of events in time at one place along a developing collision orogen or coeval events at different places along a suture belt marked by diachronous closure. Hence, erosion in one segment (D), where suture has formed, provides sediment which is dispersed longitudinally past a migrating transition point (B to C), to feed subsea fans in a remnant ocean basin (B) along tectonic strike. Also, see Figure 9. Reproduced by permission of American Association of Petroleum Geologists (from Dickinson, 1976).

an analogue for the late Paleozoic development of the Appalachian-Ouachita system and proposed a general model for sediment dispersal related to sequentially suturing orogenic belts (Fig. 9). "Most sediment shed from orogenic highlands formed by continental collisions pours longitudinally through deltaic complexes into remnant ocean basins as turbidites that are subsequently deformed and incorporated into the orogenic belts as collision sutures lengthen" (Graham and others, 1975, p. 273). This model provides a general explanation for many syn-orogenic flysch and molasse deposits associated with suture belts (Fig. 8), although many units called "flysch" and "molasse" have different tectonic settings. For this reason, use of these geosynclinal terms is not recommended outside of their type areas in the Alps.

Peripheral Foreland Basins

As continental collision occurs between a rifted continental margin and the subduction zone of an arc-trench system, a tectonic load is placed on the rifted margin, first below sea level and later subaerially (Fig. 8). A peripheral foreland basin forms as the elastic lithosphere flexes under the encroaching dynamic load. The discrimination of ancient retroarc and peripheral forelands is difficult, but it may be possible based on the following characteristics: (1) polarity of magmatic arc, (2) presence of oceanic subduction complex associated with earliest phases of peripheral foreland, (3) asymmetry of suture belt (closer to peripheral foreland), (4) protracted development of retroarc (long-term arc evolution) versus discrete development of peripheral foreland (terminal ocean closure without precursor), and (5) possible volcaniclastic input to retroarc throughout history versus minimal volcaniclastic input to peripheral foreland.

Stockmal and others (1986) have provided a dynamic two-dimensional model for the development of peripheral foreland basins, following finite times of rifting. They modified the model of Speed and Sleep (1982) and demonstrated the effects of rifted-margin age and topography on lithospheric flexure and basin development. The primary effect of age shows up as a higher flexural forebulge and thicker trench fill during the earlier stages for an old (120 m.y.) margin. Subsequent development is relatively insensitive to margin age. Foreland-basin subsidence is strongly sensitive to overthrust load, with depths possibly exceeding 10 km. Crustal thickness may reach 70 km during the compressional phase (for example, Himalayas). Tens of kilometers of uplift and erosion, of both the allochthon and the foreland basin, are predicted during and after deformation. Most of this eroded material is deposited elsewhere owing to uplift within the foreland; longitudinal transport into remnant ocean basins results (Graham and others, 1975). Thick overthrusts with low topographic expression are to be expected where broad, attenuated rifted continental margins have been pulled into subduction zones.

Piggyback Basins

Ori and Friend (1984) defined "piggyback basins" as basins that have formed and been

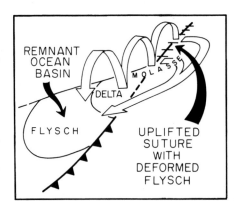

Figure 9. Conceptual diagram to illustrate progressive incorporation of synorogenic flysch within an orogenic suture belt by sequential closure of remnant ocean basin. Figure 8B illustrates the "flysch phase," Figure 8C illustrates the "delta phase" (transition point), and Figure 8D illustrates the "molasse phase." From Graham and others (1975).

filled while being carried on top of moving thrust sheets. They discussed examples from the Apennine and Pyrenean fold-thrust systems. Piggyback basins are dynamic settings for sediment accumulation; most sediment, if not all, is derived from associated fold-thrust belts. The fold-thrust belts can be in peripheral, retroarc, or transpressional settings. Piggyback basins share characteristics with foreland basins and trench-slope basins. They have low preservation potential owing to their formation on top of growing thrust belts; therefore, they are generally found only in young orogenic systems (for example, Burbank and Tahirkheli, 1985).

Foreland Intermontane Basins

Low-angle subduction beneath compressional arc-trench systems may result in basement-involved deformation within retroarc foreland basins (Dickinson and Snyder, 1978). The Rocky Mountain region of the western United States is the best-known example of this style of deformation, although similar modern provinces have been documented in the Andean foreland (Jordan and others, 1983). Overthrusting and wrench deformation, similar to processes related to intracontinental wrench basins, are likely.

Chapin and Cather (1981) discussed controls on Eocene sedimentation and basin formation of the Colorado Plateau and Rocky Mountain area. Types of associated uplifts include (1) Cordilleran thrust-belt uplifts, (2) basement-cored Rocky Mountain uplifts, and (3) monoclinal uplifts of the Colorado Plateau. Resulting basins can be classified into three types: (1) Green River type (large, equidimensional to elliptical, bounded on three or more sides by uplifts, and commonly containing lakes), (2) Denver type (elongate, open, asymmetric synclinal downwarps with uplift on one side), and (3) Echo Park type (narrow, highly elongate, fault-bounded, with through drainage, and strike-slip origin). Composite uplifts and basins also formed. Green River–type basins have quasi-concentric facies zonation, in contrast to wedge-shaped, unidirectional facies distribution in Denver-type basins. Echo Park–type basins have complex facies, with common sedimentary breccias and sheetwash deposits, associated with active faulting, erosion, and stream diversion typical of strike-slip basins. Chapin and Cather (1981) also discussed geomorphic and climatic effects on basin evolution and used the occurrence and timing of different basin types to constrain interpretations of the Laramide orogeny. They proposed a two-stage model for basin formation, which can be related to changes in North American plate interactions both in the Atlantic and the Pacific oceanic basins. Studies in similar modern settings (for example, Andean foothills; Jordan and Allmendinger, 1986) should improve these models.

TRANSFORM SETTINGS

Strike-Slip Systems

The complexity and variety of sedimentary basins associated with strike-slip faults are almost as great as for all other types of basins. Transform faults in oceanic lithosphere generally behave according to the plate-tectonic model, whereas strike-slip faults in continental lithosphere are extremely complex and difficult to fit into a model involving rigid plates. Simple mechanical models based on homogeneous media have little application to the heterogeneous media of continental crust.

The Reading cycle (for example, Reading, 1980) predicts that any strike-slip fault within continental crust is likely to experience alternating periods of extension and compression as slip directions adjust along major crustal faults. Thus, opening and closing of basins along strike-slip faults (Reading cycle) is analogous, at smaller scale, to the opening and closing of ocean basins (Wilson cycle).

Basins related to strike-slip faults can be classified into end-member types, although most basins are hybrids. Transtensional (pull-apart) basins form near releasing bends, and transpressional basins form at constraining bends (Crowell, 1974b). Basins associated with crustal rotations about vertical axes within the rotating

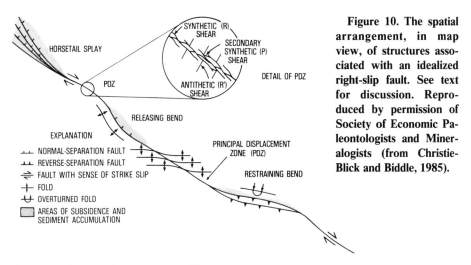

Figure 10. The spatial arrangement, in map view, of structures associated with an idealized right-slip fault. See text for discussion. Reproduced by permission of Society of Economic Paleontologists and Mineralogists (from Christie-Blick and Biddle, 1985).

blocks (herein termed "transrotational") may experience any combination of extension, compression, and strike slip.

Christie-Blick and Biddle (1985) have provided the most comprehensive summary of the structural and stratigraphic development of strike-slip basins, based, in large part, on the pioneering work of Crowell (1974a, 1974b). They illustrated the structural complexity likely along strike-slip faults (Fig. 10) and the implications for associated basins. Primary controls on structural patterns are (1) degree of convergence and divergence of adjacent blocks, (2) magnitude of displacement, (3) material properties of deformed rocks, and (4) pre-existing structures (Christie-Blick and Biddle, 1985, p. 1). Subsidence in sedimentary basins results from crustal attenuation, thermal subsidence during and following extension, flexural loading due to compression, and sedimentary loading. Thermal subsidence is less important than in elongate orthogonal rifts due to lateral heat conduction in narrow pull-apart basins. Distinctive aspects of sedimentary basins associated with strike-slip faults include (Christie-Blick and Biddle, 1985, p. 1) (1) mismatches across basin margins; (2) longitudinal and lateral basin asymmetry; (3) episodic rapid subsidence; (4) abrupt lateral facies changes and local unconformities; and (5) marked contrasts in stratigraphy, facies geometry, and unconformities among different basins in the same region.

Transtensional Basins

Pull-apart basins form at left-stepping sinistral fault junctures and at right-stepping dextral fault junctures. Mann and others (1983) proposed a model for such basins based on a comparative study of well-studied pull-apart basins at various stages of development. Pull-apart basins evolve through the following stages: (1) nucleation of extensional faulting at releasing bends of master faults; (2) formation of spindle-shaped basins defined and commonly bisected by oblique-slip faults; (3) further extension, producing "lazy-S" or "lazy-Z" basins; (4) development into rhombochasms, commonly with two or more subcircular deeps; and (5) continued extension, resulting in the formation of oceanic crust at short spreading centers offset by long transforms. Basaltic volcanism and intrusion may become important in the transition from stages 3 through 5 (for example, Crowell, 1974b). Most pull-apart basins have low length-to-width ratios, owing to their short histories in changing strike-slip regimes (Mann and others, 1983).

Transpressional Basins

Transpressional basins include two types: (1) severely deformed and overthrust margins along sharp restraining bends that result in flexural subsidence due to tectonic load (for example, south side of San Gabriel Mountains, southern California) and (2) fault-wedge basins at gentle restraining bends that result in uplift of one or two margins and downdropping of a basin as one block moves past the restraining bend (for example, Ridge basin, southern California) (Crowell, 1974b). A basin model for type 1 would involve flexural loading similar to that of the foreland models discussed above, although at smaller scale.

Ridge basin is one of the most elegantly exposed and carefully studied transpressional basins in the world, as summarized by Crowell and Link (1982). Crowell (1982) presented a dynamic model for the evolution of Ridge basin (12–5 Ma), a narrow crustal sliver caught between the San Gabriel fault to the southwest and a set of northwest-trending faults that became active sequentially in a northeast direction (Fig. 11). Ridge basin became inactive when motion was transferred completely to the modern San Andreas fault. As a result of movement on the San Gabriel fault, the southwest side of the basin was uplifted and the Violin Breccia was deposited along the basin margin. The depressed floor of the basin moved past this uplifted margin, as

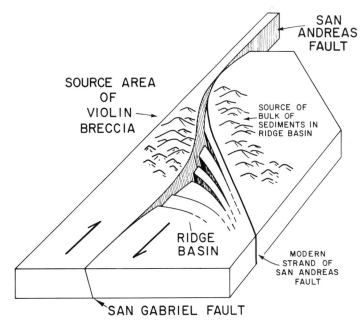

Figure 11. Block diagram (looking toward the northwest) illustrating origin of Ridge basin at a sigmoidal bend in the San Andreas fault. See text for discussion. Reproduced by permission of the Pacific Section, Society of Economic Paleontologists and Mineralogists (from Crowell, 1982).

TABLE 2. SUMMARY OF KEY REFERENCES AND PUBLISHERS

Topic	Key reference	Publisher
Overview and basin classification	Dickinson (1974, 1976)	SEPM and AAPG
Sequential rift development and continental separation	Kinsman (1975)	Princeton U. Press
Terrestrial rift valleys	Leeder and Gawthorpe (1987)	Geol. Soc. London
Proto-oceanic rift troughs	Cochran (1983a)	AAPG
Continental rises and terraces	Pitman (1978)	GSA
Continental embankments	Burke (1972)	AAPG
Failed rifts and aulacogens	Hoffman and others (1974)	SEPM
Intracratonic basins	Derito and others (1983)	Tectonophysics
Oceanic basins	Heezen and others (1973)	Nature
Oceanic islands, aseismic ridges, and plateaus	Clague (1981)	SEPM
Arc-trench systems	Dewey (1980)	Geol. Assoc. Canada
Trenches	Thornburg and Kulm (1987)	GSA
Trench-slope basins	Moore and Karig (1976)	GSA
Forearc basins	Dickinson and Seely (1979)	AAPG
Intra-arc basins	Busby-Spera (1984a, 1984b)	AGU and Pac. Sec. SEPM
Interarc and backarc basins	Carey and Sigurdsson (1984)	Geol. Soc. London
Retroarc foreland basins	Jordan (1981)	AAPG
Remnant ocean basins and suture belts	Graham and others (1975)	GSA
Peripheral foreland basins	Stockmal and others (1986)	AAPG
Piggyback basins	Ori and Friend (1984)	GSA
Foreland intermontane basins	Chapin and Cather (1981)	Arizona Geol. Soc.
Strike-slip systems	Christie-Blick and Biddle (1985)	SEPM
Transtensional basins	Mann and others (1983)	Journal of Geology
Transpressional basins	Crowell (1982)	Pac. Sec. SEPM
Transrotational basins	Luyendyk and Hornafius (1987)	Prentice-Hall
Intracontinental wrench basins	Sengor and others (1978)	Am. Jour. of Science
Successor basins	Eisbacher (1974)	SEPM

it received abundant sediment from the northeast. Previous depocenters moved southeastward past the restraining bend, after receiving sediment in conveyor-belt fashion, with uplift and tilting following deposition. The result is a stratigraphic thickness of more than 11 km in outcrop, although vertical thickness of the basin fill is approximately one-third of this. Many extraordinarily thick coarse clastic units in ancient, narrow fault-bounded basins likely were deposited in similar settings.

Transrotational Basins

Paleomagnetic data from southern California have documented extensive clockwise rotation of several crustal blocks, beginning in the Miocene and continuing today (Luyendyk and others, 1980; Hornafius and others, 1986). Luyendyk and Hornafius (1987) further developed their geometric model in order to make testable predictions concerning amount and direction of slip on faults bounding rotated and nonrotated blocks and concerning areas of gaps (basins) and overlap (overthrusts) among blocks. Within this setting, all of the complexity of transtensional and transpressional basins is likely. The unique aspect of Luyendyk and Hornafius' (1987) geometric model is that it successfully predicts the positions, shapes, areas, and ages of most southern California Neogene basins. Similar geometric models may be possible along other complex transform boundaries involving significant crustal rotations.

HYBRID SETTINGS

Intracontinental Wrench Basins

The collision of continents of varying shapes and sizes can lead to bewildering complexity in ancient orogenic belts (for example, Dewey and Burke, 1974; Graham and others, 1975; Molnar and Tapponnier, 1975; Sengor, 1976; Tapponnier and others, 1982). As Tapponnier and others (1982) demonstrated through the use of plasticine models, the collision of India and Asia has resulted in major intracontinental strike-slip faults, with associated transtensional, transpressional, and transrotational basins, including the formation of impactogens.

Sengor and others (1978) developed criteria for distinguishing failed rifts formed during the opening of oceans that are later closed (aulacogens) from intracontinental rifts formed owing to crustal collisions (impactogens). Both types of rift valleys trend at high angles to orogenic belts; however, aulacogens have a rifting history coincident with formation of a new ocean basin prior to collision, whereas impactogens have no pre-collisional history. A third category of rifts at high angles to orogenic belts is random rifts unrelated to either ocean formation or collision orogenesis. All of these rift basins are deformed during suturing in associated orogens. Tests for distinguishing them must come from the stratigraphic record because temporal correlation of initial rifting (or lack thereof) is the primary test for the geodynamic origin for these ancient rifts located at high angles to orogenic belts. One suggestive type of evidence for discrimination is that aulacogens tend to form at re-entrants along rifted continental margins (Dewey and Burke, 1974), whereas impactogens are more likely to form opposite coastal promontories, where deformation of colliding continents is more intense (Sengor, 1976). This criterion must be applied cautiously, however, due to the difficulty of definitively reconstructing pre-collision geometry (for example, Thomas, 1983, 1985).

Successor Basins

The original definition of successor basins (King, 1966) as "deeply subsiding troughs with limited volcanism associated with rather narrow uplifts, and overlying deformed and intruded eugeosynclines" (Kay, 1951, p. 107; Eisbacher, 1974) needs modification; "deeply subsiding" and "eugeosynclines" should be replaced by "intermontane" and "terranes," respectively. Within the context of plate tectonics, successor basins form primarily in intermontane settings on top of inactive fold-thrust belts, suture belts, transform belts, and noncratonal failed rifts. The presence of successor basins indicates the end of orogenic activity, and so their ages constrain interpretations of timing of suturing and deformation. Thus, they have special significance in "terrane analysis"; they represent overlap assemblages which provide minimum ages for terrane accretion (Howell and others, 1985).

Little work has been published on actualistic models for such basins; Eisbacher (1974) summarized models based on work in the Canadian Cordillera. This dearth of work may reflect the diversity of successor basins and their tectonic settings. In a sense, all basins are successor basins because they form following some orogenic event represented in the basement of the basin. In fact, one of Kay's (1951) examples of epieugeosynclines (successor basins) is the post-Nevadan basin of central California, which is now interpreted as a forearc basin, overprinted in the Tertiary by transform tectonics (Ingersoll, 1982; Ingersoll and Schweickert, 1986; Graham, 1987). Modern use of the term "successor basin" should be restricted to post-orogenic basins that do not fall into any other plate-tectonic framework during their development. Additional work on modern successor basins is needed before an actualistic model is available.

DISCUSSION

This review discusses actualistic models for basins in all plate-tectonic settings. It should be obvious that some of the basin types are common and volumetrically important, whereas others are rare and volumetrically minor. An important, but seldom discussed, factor in basin analysis and paleotectonic reconstruction is the preservability of tectonostratigraphic assemblages. Several basin types are rarely preserved in the very ancient record (for example, intra-arc, trench-slope, and successor basins); their absence in ancient orogenic belts is predicted by their locations and susceptibilities to erosion and deformation. Thus, their absence is not a valid test of plate-tectonic models. Veizer and Jansen (1979, 1985) have provided an empirical method of determining the "half life" of tectonostratigraphic elements. They estimated the half lives of active-margin basins at 30 m.y., oceanic sediments at 40 m.y., oceanic crust at 55 m.y., passive margins at 80 m.y., immature orogenic belts at 100 m.y., and mature orogenic belts and platforms at 380 m.y. The application of Veizer and Jansen's type of analysis to all of the basin types discussed herein will provide additional quantitative constraints on paleotectonic reconstructions.

Table 2 lists the key references and where they were published. Not surprisingly, more than half of these references were published by three societies with strong emphases on basin analysis and tectonics: American Association of Petroleum Geologists, Society of Economic Paleontologists and Mineralogists, and Geological Society of America. Table 2, however, also demonstrates the diversity of information sources and the interdisciplinary nature of the field; staying current in the field of sedimentary tectonics is a challenge.

It is hoped that this review will focus attention on important references, as well as suggest where work is needed most. Certain tectonic settings (for example, continental rises, continental terraces, and foreland basins) have received much attention in the last decade, so that actualistic models are becoming quantitatively sophisticated. In contrast, other settings (for example, successor basins, intra-arc basins, and continental embankments) have received relatively little attention. As all of these models improve, it is likely that more integration of successive orogenic processes into a continuum of models will provide strong predictive capabilities for paleotectonic reconstructions.

ACKNOWLEDGMENTS

I thank W. R. Dickinson and S. A. Graham for years of rewarding interaction, and R. D. Hatcher and W. A. Thomas for inviting me to write this. I thank the following for critically reviewing the manuscript: W. Cavazza, D. S. Diamond, W. R. Dickinson, P. L. Heller, T. E. Jordan, L. D. Kulm, A. Linn, K. M. Marsaglia, and L. J. Suttner. The following provided copies of original figures: K. T. Biddle, W. R. Dickinson, P. F. Hoffman, L. D. Kulm, and D. R. Seely. The following organizations provided permission to reproduce original illustrations: American Association of Petroleum Geologists, Geological Association of Canada, and Society of Economic Paleontologists and Mineralogists. Also, I thank students at the University of New Mexico and at the University of California, Los Angeles (UCLA), for helping to keep me honest, especially participants in Earth and Space Sciences 244 (Tectonics of Sedimentary Basins) at UCLA.

REFERENCES CITED

Allen, P. A., Homewood, P., and Williams, G. D., 1986, Foreland basins: An introduction: International Association of Sedimentologists Special Publication 8, p. 3–12.
Bally, A. W., and Snelson, S., 1980, Realms of subsidence: Canadian Society of Petroleum Geologists Memoir 6, p. 9–75.
Berger, W. H., 1973, Cenozoic sedimentation in the eastern tropical Pacific: Geological Society of America Bulletin, v. 84, p. 1941–1954.
Berger, W. H., and Winterer, E. L., 1974, Plate stratigraphy and the fluctuating carbonate line: International Association of Sedimentologists Special Publication 1, p. 11–48.
Burbank, D. W., and Tahirkheli, R.A.K., 1985, The magnetostratigraphy, fission-track dating, and stratigraphic evolution of the Peshawar intermontane basin, northern Pakistan: Geological Society of America Bulletin, v. 96, p. 539–552.
Burke, K., 1972, Longshore drift, submarine canyons, and submarine fans in development of Niger delta: American Association of Petroleum Geologists Bulletin, v. 56, p. 1975–1983.
Burke, K., and Dewey, J. F., 1973, Plume-generated triple junctions: Key indicators in applying plate tectonics to old rocks: Journal of Geology, v. 81, p. 406–433.
Busby-Spera, C. J., 1984a, The lower Mesozoic continental margin and marine intra-arc sedimentation at Mineral King, California, in Crouch, J. K., and Bachman, S. B., eds., Tectonics and sedimentation along the California margin: Pacific Section, Society of Economic Paleontologists and Mineralogists Publication 38, p. 135–155.
——— 1984b, Large-volume rhyolite ash flow eruptions and submarine caldera collapse in the lower Mesozoic Sierra Nevada, California: Journal of Geophysical Research, v. 89, p. 8417–8427.
Carey, S., and Sigurdsson, H., 1984, A model of volcanogenic sedimentation in marginal basins: Geological Society of London Special Publication 16, p. 37–58.
Cas, R.A.F., and Wright, J. V., 1987, Volcanic successions modern and ancient: Boston, Massachusetts, Allen and Unwin, 528 p.
Chapin, C. E., and Cather, S. M., 1981, Eocene tectonics and sedimentation in the Colorado Plateau–Rocky Mountain area: Arizona Geological Society Digest, v. 14, p. 173–198.
Christensen, M. N., 1963, Structure of metamorphic rocks at Mineral King, California: University of California Publications in Geological Sciences, v. 42, p. 159–198.
Christie-Blick, N., and Biddle, K. T., 1985, Deformation and basin formation along strike-slip faults: Society of Economic Paleontologists and Mineralogists Special Publication 37, p. 1–34.
Clague, D. A., 1981, Linear island and seamount chains, aseismic ridges and intraplate volcanism: Results from DSDP: Society of Economic Paleontologists and Mineralogists Special Publication 32, p. 7–22.
Cloetingh, S., McQueen, H., and Lambeck, K., 1985, On a tectonic mechanism for regional sealevel variations: Earth and Planetary Science Letters, v. 75, p. 157–166.
Cochran, J. R., 1983a, A model for development of Red Sea: American Association of Petroleum Geologists Bulletin, v. 67, p. 41–69.
——— 1983b, Effects of finite rifting times on the development of sedimentary basins: Earth and Planetary Science Letters, v. 66, p. 289–302.
Cook, H. E., 1975, North American stratigraphic principles as applied to deep-sea sediments: American Association of Petroleum Geologists Bulletin, v. 59, p. 817–837.
Cox, A., and Hart, R. B., 1986, Plate tectonics: How it works: Palo Alto, California, Blackwell Scientific Publications, 392 p.
Crowell, J. C., 1974a, Sedimentation along the San Andreas fault: Society of Economic Paleontologists and Mineralogists Special Publication 19, p. 292–303.
——— 1974b, Origin of late Cenozoic basins in California: Society of Economic Paleontologists and Mineralogists Special Publication 22, p. 190–204.
——— 1982, The tectonics of Ridge basin, southern California, in Crowell, J. C., and Link, M. H., eds., Geologic history of Ridge basin southern California: Pacific Section, Society of Economic Paleontologists and Mineralogists Publication 22, p. 25–41.
Crowell, J. C., and Link, M. H., eds., 1982, Geologic history of Ridge basin southern California: Pacific Section, Society of Economic Paleontologists and Mineralogists Publication 22, 304 p.
Davis, R. A., Jr., 1983, Depositional systems: A genetic approach to sedimentary geology: Englewood Cliffs, New Jersey, Prentice-Hall, 669 p.
Derito, R. F., Cozzarelli, F. A., and Hodge, D. S., 1983, Mechanism of subsidence of ancient cratonic rift basins: Tectonophysics, v. 94, p. 141–168.
Dewey, J. F., 1980, Episodicity, sequence, and style at convergent plate boundaries: Geological Association of Canada Special Paper 20, p. 553–573.
Dewey, J. F., and Burke, K., 1974, Hot spots and continental break-up: Implications for collisional orogeny: Geology, v. 2, p. 57–60.
Dickinson, W. R., 1973, Widths of modern arc-trench gaps proportional to past duration of igneous activity in associated magmatic arcs: Journal of Geophysical Research, v. 78, p. 3376–3389.
——— 1974, Plate tectonics and sedimentation: Society of Economic Paleontologists and Mineralogists Special Publication 22, p. 1–27.
——— 1976, Plate tectonic evolution of sedimentary basins: American Association of Petroleum Geologists Continuing Education Course Notes Series 1, 62 p.
——— 1978, Plate tectonic evolution of north Pacific rim: Journal of the Physics of the Earth, v. 26, Supplement, p. S1–S19.
——— 1980, Plate tectonics and key petrologic associations: Geological Association of Canada Special Paper 20, p. 341–360.
Dickinson, W. R., and Seely, D. R., 1979, Structure and stratigraphy of forearc regions: American Association of Petroleum Geologists Bulletin, v. 63, p. 2–31.
Dickinson, W. R., and Snyder, W. S., 1978, Plate tectonics of the Laramide orogeny: Geological Society of America Memoir 151, p. 355–366.
Dickinson, W. R., and Suczek, C. A., 1979, Plate tectonics and sandstone compositions: American Association of Petroleum Geologists Bulletin, v. 63, p. 2164–2182.
Dott, R. H., Jr., 1978, Tectonics and sedimentation a century later: Earth-Science Reviews, v. 14, p. 1–34.
Eisbacher, G. H., 1974, Evolution of successor basins in the Canadian Cordillera: Society of Economic Paleontologists and Mineralogists Special Publication 19, p. 274–291.
Fisher, R. V., and Schmincke, H.-U., 1984, Pyroclastic rocks: New York, Springer-Verlag, 472 p.
Frostick, L. E., and Reid, I., 1987, Tectonic control of desert sediments in rift basins ancient and modern: Geological Society of London Special Publication 35, p. 53–68.
Graham, S. A., 1987, Tectonic controls on petroleum occurrence in central California, in Ingersoll, R. V., and Ernst, W. G., eds., Cenozoic basin development of coastal California (Rubey Volume 6): Englewood Cliffs, New Jersey, Prentice-Hall, p. 47–63.
Graham, S. A., Dickinson, W. R., and Ingersoll, R. V., 1975, Himalayan-Bengal model for flysch dispersal in Appalachian-Ouachita system: Geological Society of America Bulletin, v. 86, p. 273–286.
Hamilton, W., 1979, Tectonics of the Indonesian region: U.S. Geological Survey Professional Paper 1078, 345 p.
Heezen, B. C., and 11 coauthors, 1973, Diachronous deposits, a kinematic interpretation of the post-Jurassic sedimentary sequence on the Pacific plate: Nature, v. 241, p. 25–32.
Heller, P. L., and Angevine, C. L., 1985, Sea-level cycles during the growth of Atlantic-type oceans: Earth and Planetary Science Letters, v. 75, p. 417–426.
Heller, P. L., and 7 coauthors, 1986, Time of initial thrusting in the Sevier orogenic belt, Idaho-Wyoming and Utah: Geology, v. 14, p. 388–391.
Hellinger, S. J., and Sclater, J. G., 1983, Some comments on two-layer extensional models for the evolution of sedimentary basins: Journal of Geophysical Research, v. 88, p. 8251–8269.
Hoffman, P. F., 1987, Continental transform tectonics: Great Slave Lake shear zone (ca. 1.9 Ga), northwest Canada: Geology, v. 15, p. 785–788.
Hoffman, P., Dewey, J. F., and Burke, K., 1974, Aulacogens and their genetic relation to geosynclines, with a Proterozoic example from Great Slave Lake, Canada: Society of Economic Paleontologists and Mineralogists Special Publication 19, p. 38–55.
Hornafius, J. S., Luyendyk, B. P., Terres, R. R., and Kamerling, M. J., 1986, Timing and extent of Neogene tectonic rotation in the western Transverse Ranges, California: Geological Society of America Bulletin, v. 97, p. 1476–1487.
Howell, D. G., Jones, D. L., and Schermer, E. R., 1985, Tectonostratigraphic terranes of the circum-Pacific region, in Howell, D. G., ed., Tectonostratigraphic terranes of the circum-Pacific region: Circum-Pacific Council for Energy and Mineral Resources Earth Sciences Series 1, p. 3–30.
Ingersoll, R. V., 1979, Evolution of the Late Cretaceous forearc basin, northern and central California: Geological Society of America Bulletin, v. 90, Part I, p. 813–826.
——— 1982, Initiation and evolution of the Great Valley forearc basin of northern and central California, U.S.A.: Geological Society of London Special Publication 10, p. 459–467.
Ingersoll, R. V., and Schweickert, R. A., 1986, A plate-tectonic model for Late Jurassic ophiolite genesis, Nevadan orogeny and forearc initiation, northern California: Tectonics, v. 5, p. 901–912.
Jordan, T. E., 1981, Thrust loads and foreland basin evolution, Cretaceous, western United States: American Association of Petroleum Geologists Bulletin, v. 65, p. 2506–2520.

Jordan, T. E., and Allmendinger, R. W., 1986, The Sierras Pampeanas of Argentina: A modern analogue of Rocky Mountain foreland deformation: American Journal of Science, v. 286, p. 737-764.

Jordan, T. E., Isacks, B. L., Allmendinger, R. W., Brewer, J. A., Ramos, V. A., and Ando, C. J., 1983, Andean tectonics related to geometry of subducted Nazca plate: Geological Society of America Bulletin, v. 94, p. 341-361.

Karig, D. E., 1971, Origin and development of marginal basins in the western Pacific: Journal of Geophysical Research, v. 76, p. 2542-2561.

Karig, D. E., and Moore, G. F., 1975, Tectonically controlled sedimentation in marginal basins: Earth and Planetary Science Letters, v. 26, p. 233-238.

Karig, D. E., and Sharman, G. F., III, 1975, Subduction and accretion in trenches: Geological Society of America Bulletin, v. 86, p. 377-389.

Karner, G. D., 1986, Effects of lithospheric in-plane stress on sedimentary basin stratigraphy: Tectonics, v. 5, p. 573-588.

Kay, M., 1951, North American geosynclines: Geological Society of America Memoir 48, 143 p.

King, P. B., 1966, The North American Cordillera: Canadian Institute of Mining and Metallurgy Special Volume 8, p. 1-25.

Kinsman, D.J.J., 1975, Rift valley basins and sedimentary history of trailing continental margins, in Fischer, A. G., and Judson, S., eds., Petroleum and global tectonics: Princeton, New Jersey, Princeton University Press, p. 83-126.

Klein, G. deV., 1985, The control of depositional depth, tectonic uplift, and volcanism on sedimentation processes in the back-arc basins of the western Pacific Ocean: Journal of Geology, v. 93, p. 1-25.

—— 1987, Current aspects of basin analysis: Sedimentary Geology, v. 50, p. 95-118.

Klein, G. deV., and Hsui, A. T., 1987, Origin of cratonic basins: Geology, v. 15, p. 1094-1098.

Kuhn, T. S., 1970, The structure of scientific revolutions (2nd edition): Chicago, Illinois, University of Chicago Press, 210 p.

Kusznir, N. J., and Park, R. G., 1987, The extensional strength of the continental lithosphere: Its dependence on geothermal gradient, and crustal composition and thickness: Geological Society of London Special Publication 28, p. 35-52.

Kusznir, N. J., Karner, G. D., and Egan, S., 1987, Geometric, thermal and isostatic consequences of detachments in continental lithosphere extension and basin formation: Canadian Society of Petroleum Geologists Memoir 12, p. 185-203.

Leeder, M. R., and Gawthorpe, R. L., 1987, Sedimentary models for extensional tilt-block/half-graben basins: Geological Society of London Special Publication 28, p. 139-152.

Lister, G. S., Etheridge, M. A., and Symonds, P. A., 1986, Detachment faulting and the evolution of passive continental margins: Geology, v. 14, p. 246-250.

Luyendyk, B. P., and Hornafius, J. S., 1987, Neogene crustal rotations, fault slip, and basin development in southern California, in Ingersoll, R. V., and Ernst, W. G., eds., Cenozoic basin development of coastal California (Rubey Volume 6): Englewood Cliffs, New Jersey, Prentice-Hall, p. 259-283.

Luyendyk, B. P., Kamerling, M. J., and Terres, R., 1980, Geometric model for Neogene rotations in southern California: Geological Society of America Bulletin, v. 91, Part I, p. 211-217.

Lynch, H. D., and Morgan, P., 1987, The tensile strength of the lithosphere and the localization of extension: Geological Society of London Special Publication 28, p. 53-65.

Mann, P., Hempton, M. R., Bradley, D. C., and Burke, K., 1983, Development of pull-apart basins: Journal of Geology, v. 91, p. 529-554.

McKenzie, D., 1978, Some remarks on the development of sedimentary basins: Earth and Planetary Science Letters, v. 40, p. 25-32.

Miall, A. D., 1984, Principles of sedimentary basin analysis: New York, Springer-Verlag, 490 p.

Molnar, P., and Atwater, T., 1978, Interarc spreading and Cordilleran tectonics as alternates related to age of subducted oceanic lithosphere: Earth and Planetary Science Letters, v. 41, p. 330-340.

Molnar, P., and Tapponnier, P., 1975, Cenozoic tectonics of Asia: Effects of a continental collision: Science, v. 189, p. 419-426.

Moore, G. F., and Karig, D. E., 1976, Development of sedimentary basins on the lower trench slope: Geology, v. 4, p. 693-697.

Morgan, P., and Baker, B. H., 1983, Introduction—Processes of continental rifting: Tectonophysics, v. 94, p. 1-10.

Ori, G. G., and Friend, P. F., 1984, Sedimentary basins formed and carried piggyback on active thrust sheets: Geology, v. 12, p. 475-478.

Pitman, W. C., III, 1978, Relationship between eustacy and stratigraphic sequences of passive margins: Geological Society of America Bulletin, v. 89, p. 1389-1403.

Quinlan, G., 1987, Models of subsidence mechanisms in intracratonic basins, and their applicability to North American examples: Canadian Society of Petroleum Geologists Memoir 12, p. 463-481.

Reading, H. G., 1980, Characteristics and recognition of strike-slip fault systems: International Association of Sedimentologists Special Publication 4, p. 7-26.

Reading, H. G., ed., 1986, Sedimentary environments and facies (2nd edition): New York, Elsevier, 615 p.

Royden, L. H., Horvath, F., and Burchfiel, B. C., 1982, Transform faulting, extension, and subduction in the Carpathian Pannonian region: Geological Society of America Bulletin, v. 93, p. 717-725.

Schweller, W. J., and Kulm, L. D., 1978, Depositional patterns and channelized sedimentation in active eastern Pacific trenches, in Stanley, D. J., and Kelling, G., eds., Sedimentation in submarine canyons, fans, and trenches: Stroudsburg, Pennsylvania, Dowden, Hutchinson & Ross, p. 311-324.

Sclater, J. G., and Christie, P.A.F., 1980, Continental stretching: An explanation of the post-mid-Cretaceous subsidence of the central North Sea basin: Journal of Geophysical Research, v. 85, p. 3711-3739.

Sengor, A.M.C., 1976, Collision of irregular continental margins: Implications for foreland deformation of Alpine-type orogens: Geology, v. 4, p. 779-782.

Sengor, A.M.C., and Burke, K., 1978, Relative timing of rifting and volcanism on Earth and its tectonic implications: Geophysical Research Letters, v. 5, p. 419-421.

Sengor, A.M.C., Burke, K., and Dewey, J. F., 1978, Rifts at high angles to orogenic belts: Tests for their origin and the upper Rhine Graben as an example: American Journal of Science, v. 278, p. 24-40.

Speed, R. C., and Sleep, N. H., 1982, Antler orogeny and foreland basin: A model: Geological Society of America Bulletin, v. 93, p. 815-828.

Stockmal, G. S., Beaumont, C., and Boutilier, R., 1986, Geodynamic models of convergent margin tectonics: Transition from rifted margin to overthrust belt and consequences for foreland-basin development: American Association of Petroleum Geologists Bulletin, v. 70, p. 181-190.

Tapponnier, P., Peltzer, G., LeDain, A. Y., Armijo, R., and Cobbold, P., 1982, Propagating extrusion tectonics in Asia: New insights from simple experiments with plasticine: Geology, v. 10, p. 611-616.

Thomas, W. A., 1983, Continental margins, orogenic belts, and intracratonic structures: Geology, v. 11, p. 270-272.

—— 1985, The Appalachian-Ouachita connection: Paleozoic orogenic belt at the southern margin of North America: Annual Review of Earth and Planetary Sciences, v. 13, p. 175-199.

Thornburg, T. M., and Kulm, L. D., 1987, Sedimentation in the Chile Trench: Depositional morphologies, lithofacies, and stratigraphy: Geological Society of America Bulletin, v. 98, p. 33-52.

Vail, P. R., and others, 1977, Seismic stratigraphy and global changes of sea level: American Association of Petroleum Geologists Memoir 26, p. 49-212.

Veevers, J. J., 1981, Morphotectonics of rifted continental margins in embryo (East Africa), youth (Africa-Arabia), and maturity (Australia): Journal of Geology, v. 89, p. 57-82.

Veizer, J., and Jansen, S. L., 1979, Basement and sedimentary recycling and continental evolution: Journal of Geology, v. 87, p. 341-370.

—— 1985, Basement and sedimentary recycling—2: Time dimension to global tectonics: Journal of Geology, v. 93, p. 625-643.

Walcott, R. E., 1972, Gravity, flexure, and the growth of sedimentary basins at a continental edge: Geological Society of America Bulletin, v. 83, p. 1845-1848.

Walker, R. G., ed., 1984, Facies models (2nd edition): Geoscience Canada Reprint Series 1, 317 p.

Wernicke, B., 1985, Uniform-sense normal simple shear of the continental lithosphere: Canadian Journal of Earth Sciences, v. 22, p. 108-125.

Winterer, E. L., 1973, Sedimentary facies and plate tectonics of equatorial Pacific: American Association of Petroleum Geologists Bulletin, v. 57, p. 265-282.

MANUSCRIPT RECEIVED BY THE SOCIETY JANUARY 20, 1988
REVISED MANUSCRIPT RECEIVED JUNE 8, 1988
MANUSCRIPT ACCEPTED JUNE 9, 1988

Evolution of thought on passive continental margins from the origin of geosynclinal theory (~1860) to the present

CENTENNIAL ARTICLE

GERARD C. BOND
MICHELLE A. KOMINZ } *Lamont-Doherty Geological Observatory of Columbia University, Palisades, New York 10964*

ABSTRACT

Most of the current views on the evolution of passive margins have roots in ideas that were developed before 1930 in the context of continental drift and geosynclinal theory. These ideas include the concept of an Atlantic type of margin formed by rifting and continental drift; the presence of a thick sedimentary deposit beneath the continental shelves; and subsidence in response to such mechanisms as crustal thinning, igneous underplating, thermal contraction, flexure, and sediment loading. As large amounts of new surface and subsurface data were acquired from modern passive margins after World War II, owing to significant advances in technology for geological and geophysical exploration of the ocean basins, these early ideas were strengthened and modified. With the development of plate-tectonic theory, the origin of passive margins as rifted trailing edges of continents became widely accepted, and significant changes in thinking involved the role of passive margins and their implications for large horizontal displacements in the evolution of geosynclines. Within the past decade, a large number of geophysical models have been developed for passive margins that focus once again on the problem of the mechanisms of vertical movements of the Earth's crust. At the same time, new developments in the acquisition and processing of data from ocean basins, especially deep-reflection data, have resulted in major new concepts about the deep structure of passive margins, including recognition of the importance of underplating and plutonic activity in the thinned rifted crust and the unexpected degree of faulting and formation of horizontal reflectors in the lower continental crust and subcrustal lithosphere.

INTRODUCTION

The purpose of this paper is to give an overview of the evolution of thought on passive or Atlantic-type continental margins, emphasizing important contributions from Geological Society of America publications where appropriate. The ideas about passive margins are traced from early concepts that were formulated in the context of geosynclines and continental drift through more than 100 years of progressively more sophisticated understanding of the structure and evolution of these major features of the Earth's surface (for example, Fig. 1). This long span of time is divided into four periods, in each of which the development of ideas and acquisition of new data have a broad common theme. The first period extends from the origin of geosynclinal theory to World War II, when most of the contributions to thinking about passive margins came, largely indirectly, from studies within the continents. The second period extends from World War II to the widespread acceptance of plate-tectonic theory, when geological and geophysical exploration of ocean basins was undertaken for the first time on a large scale, leading to many new data that bore directly on the nature and evolution of modern passive margins. The third corresponds to the short period of time during which plate-tectonic theory was developed, confirming the role of passive margins in continental drift and significantly revising concepts of passive margins in the evolution of geosynclines. During the last period, extending to the present, rapid technological advances in data gathering and processing have led to major advances in knowledge about passive margins, especially their deeper structure, and have provided a basis for a remarkably large number of diverse and sophisticated computer models of passive-margin evolution. Many of the important contributions to stratigraphy and sedimentology that have been made in the evolution of ideas about passive margins are beyond the scope of this paper, and the emphasis is on the significance of passive margins as fundamental, lithospheric-scale structures in the context of extension, vertical movements, and modification of continental crust.

ORIGINS OF CURRENT CONCEPTS OF PASSIVE MARGINS: EARLY VIEWS FROM A PREDOMINANTLY CONTINENTAL PERSPECTIVE (~1857–1944)

Much of the current thinking about passive margins can be traced back several decades to concepts that were formulated in early studies of the continents and that evolved within the framework of controversy over continental drift and the origin of geosynclines. The following is a brief summary of the origin of these early ideas and the context in which they are presently applied to passive margins. With a few exceptions, no attempt is made to compare these early views with the many other contemporaneous ideas that have been largely abandoned.

Identification of an Atlantic (Passive) Type of Continental Margin

In his synthesis of global geology, first published in 1885, Suess recognized two fundamentally different types of continental margins, based on subaerial coastal features. One of these was an "Atlantic coast," characterized by structural evidence of subsidence, minor volcanism and seismicity, and a discordance between the outline of the adjacent ocean and the outer sides of the folded mountain ranges along the continental edge.

> With the exception of the cordillera of the Antilles and the mountain fragment of Gibraltar, which form respectively the boundary of the two mediterranean seas, the outer side of a folded range nowhere determines the outline of the Atlantic Ocean.... The inner sides of folded ranges, jagged rias coasts which indicate the subsidence of mountain chains, fractured margins of horsts, and fractured table-land form the diversified boundary of the Atlantic Ocean. (Suess, 1906, v. 2, p. 203)

Geological Society of America Bulletin, v. 100, p. 1909–1933, 14 figs., December 1988.

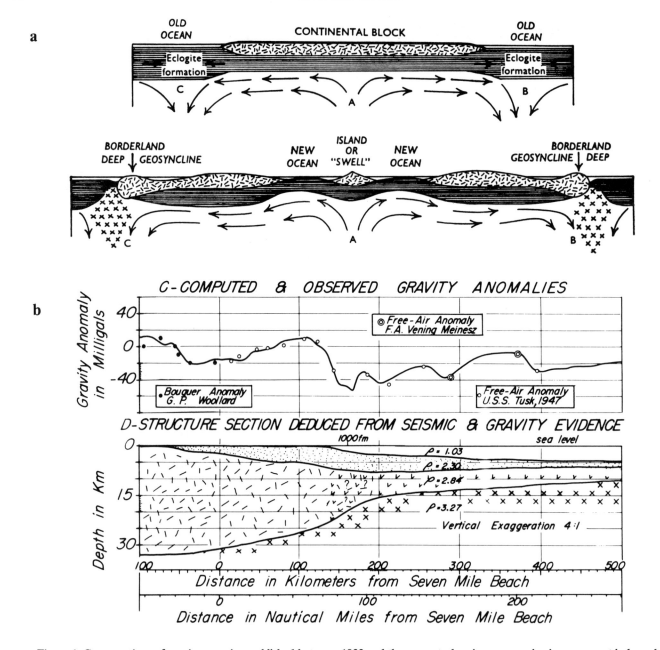

Figure 1. Cross sections of passive margins, published between 1933 and the present, showing progressive improvement in knowledge of passive-margin structure and stratigraphy.

(a) An early concept of convection-driven continental drift from Holmes (1930), showing crustal thinning beneath passive margins and a fundamental compositional difference between continents and ocean basins. The cross section was largely conceptual and derived from Holmes' concept of convection in the mantle and his belief in continental drift.

(b) A cross section of the modern passive margin off the Virginia coast from Worzel (1965), based on one of the earliest refraction experiments in a passive margin (Ewing and others, 1939) and gravity measurements acquired at sea. The refraction and gravity data do not resolve faults at depth but suggest the presence of thinned crust and a thick sedimentary wedge in the margin. The refraction velocity data clearly support a fundamental difference between continental and oceanic crust.

Suess regarded most of the Atlantic, Indian, and Australian margins as belonging to the Atlantic category. The concept of an Atlantic-type coast, therefore, included most of the continental margins that were viewed as "passive" in the plate-tectonic sense nearly 80 years later. (The term "passive margin" was coined by Michell and Reading in 1969.)

Suess' definition of Atlantic-type coasts was later incorporated in the theory of continental drift by Wegener (1929), and the discordance between the trends of the coasts and the folds of the coastal mountain ranges was regarded as particularly strong support for that concept. DuToit (1937) published one of the first maps showing the global distribution of Atlantic-type coasts (his continental fault-line coasts), which included more than half of the margins now classified as passive (compare Figs. 2a and 2b).

Early Ideas on the Cross-Sectional Structure of Passive Margins

In the late 1800s and early 1900s, the continental shelves were considered by some to be essentially erosional features, underlain by

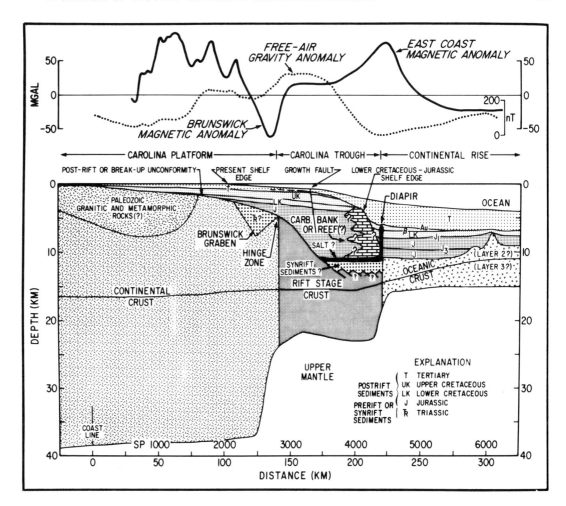

Figure 1. (c) A cross section from Hutchinson and others (1982) of the modern passive margin along U.S. Geological Survey line 32 off the coast of North Carolina about 350 km south of the cross section in b (above). The greater detail of structure and stratigraphy compared with those of b (above) reflects the improved quality of seismic reflection data and increase in deep drilling data in the late 1970s.

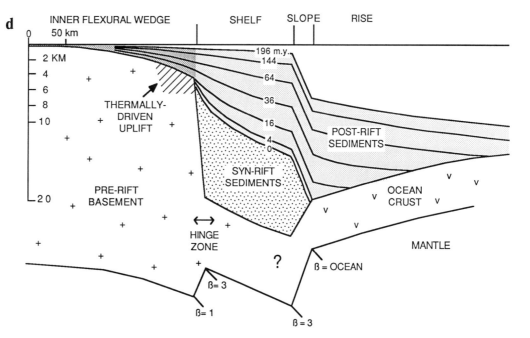

(d) An example from Steckler (1981) of the many sophisticated two-dimensional thermo-mechanical models of passive margins that were developed in the early 1980s. β refers to the amount of stretching of the crust; the hinge zone marks the boundary between stretched and unstretched crust, the area of thermally driven uplift results from the lateral flow of heat during rifting and cooling of the margin, and the inner flexural wedge is a region that subsides as a result of flexural bending of the unrifted basement as the margin cools, becomes more rigid, and fills with sediment.

continental crust, and by others to be the surfaces of a thick sedimentary deposit. The prevailing view as argued in the *Bulletin of the Geological Society of America* by Barrell in 1912, following earlier discussions by Chamberlin (1898) and Gulliver (1899), favored the depositional model (Fig. 3a). Direct evidence supporting the depositional model had been reported by Upham (1894) and Dahl (1925), who identified Tertiary and Cretaceous fossils in rocks brought to the surface in fishermen's nets. Later, Stetson (1936) published in the *Bulletin* the results of a systematic program of offshore dredging, which indicated that Upper Cretaceous rocks occur in submarine canyons in Georges Bank, implying that older rocks must underlie the shelves and crop out in the canyons.

The first compelling support for the depositional origin came from

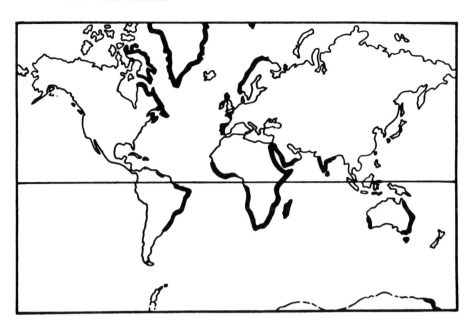

Figure 2. Comparison of an early map (a) of the world's Atlantic-type coastlines from DuToit (1937), based on mainly coastal morphology and discordance between coastline and onshore structures, with the present distribution of passive margins (b) as determined from various subsurface data (Bally, 1979).

the earliest refraction measurements in continental margins. These refraction experiments were supported in part by the Geological Society of America. The results of these first measurements were described in the *Bulletin* by Ewing and others (1937, 1939). Similar measurements were made a short time later off the coast of Britain by Bullard and Gaskill (1941). The results clearly showed that the continental crust dipped seaward beneath the shelves and was buried by a thick layer of sedimentary rock (for example, Fig. 4a).

Inferences regarding the nature of the deeper structure of the margins were necessarily indirect and mostly linked to the controversies over continental drift and the Pratt and Airy-Heiskanen isostatic models for the Earth. Daly (1926), Wegener (1929), and Holmes (1930), among others, thought that both isostatic models were partly correct and that the deep structure of the margins reflected the transition between two fundamentally different types of crust, sial and sima (for example, Fig. 1a), terms originally coined by Suess. Wegener concluded that ocean basins and continents are fundamentally different because the two occupy fundamentally different levels on the Earth's surface and because the gravity anomalies across continental margins reported by Helmert (1909) appear to contain evidence of a strong edge effect. The prevailing view, however, held even by DuToit (1937), was that only parts of the Pacific Ocean and north polar basins were underlain by sima and that the other ocean basins bounded by most of the Atlantic-type margins were floored by thinned sial.

The fault origin of Atlantic-type margins advocated by Suess (1906), Wegener (1929), and DuToit (1937) also was disputed. Umbgrove (1947) opposed the fault interpretation on the basis of the absence of subsurface faults in the early refraction measurements of Ewing and others (1937, 1939) and Bullard and Gaskill (1941). Although the limitations of using refraction data to identify subsurface faulting were known, Umbgrove adhered to the marginal-flexure concept developed by Bourcart (1938), who believed that subsidence of the margin below sea level was caused by warping of the continental edge. Even as late as the mid-1960s, many cross sections of passive margins indicated only a gradual thinning of the crust beneath the sediments, which was modeled to fit the available gravity data (for example, Fig. 1b; Vening-Meinesz, 1941).

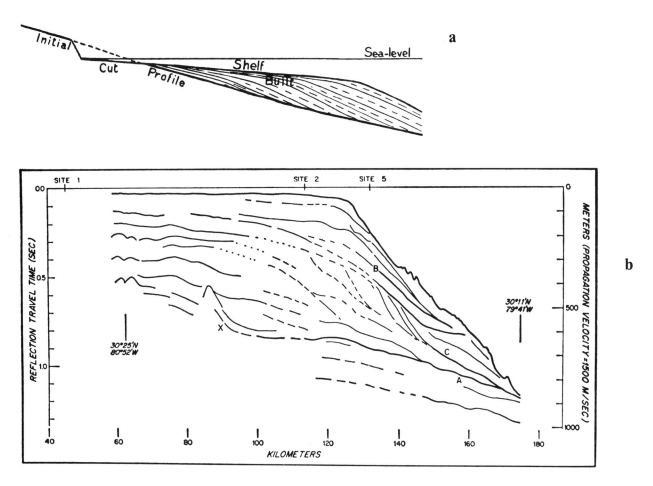

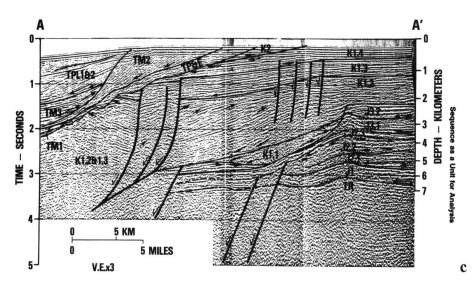

Figure 3. Examples of increasing understanding of the internal structure of passive margins, beginning with (a) hypothetical concept based on analogy with terrestrial deltas (Cotton, 1918) that was common in the early 1900s, (b) evidence from an early seismic reflection profile across the Blake Plateau off the southeast coast of the United States (JOIDES, 1965), supporting the over-all progradational nature of some passive margins as predicted by the delta analogue, and finally (c) the complex internal stratigraphy and structure in the passive margin offshore of western Africa, revealed by seismic stratigraphic methods developed in the late 1970s (Vail and others, 1977). Arrows indicate sequence boundaries that are presumed to have chronological significance as defined by Vail and others (1977).

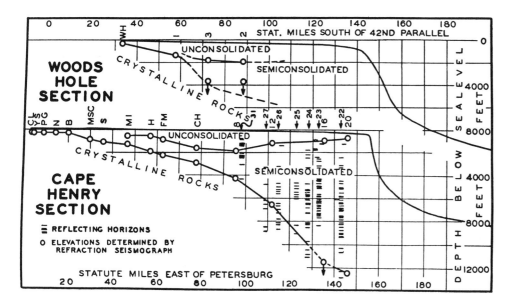

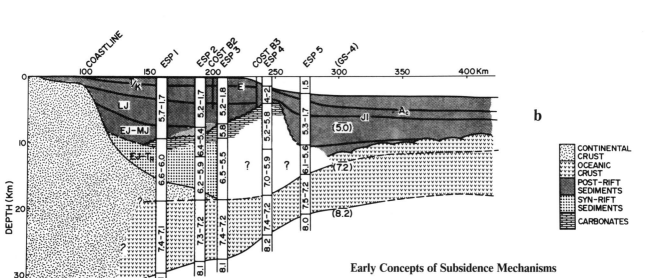

Figure 4. Comparison of results of the earliest refraction experiments conducted in a passive margin (a) from Ewing and others (1939) with results of one of the recent deep-reflection experiments in a passive margin (LASE Study Group, 1986). Sections in a are from off the coast of Virginia (below) and off the coast of Massachusetts (above), and the section in b is from off the coast of New Jersey. ESP in b refers to locations of expanding spread profiles with velocities in kilometers per second indicated, and the acronym COST gives locations of deep commercial wells.

It is now clear that sedimentary deposits of passive margins consist of a continentward tapering wedge reaching thicknesses of as much as 12 km beneath the outer shelf and rise (Fig. 1c). According to one estimate, more than half of the sedimentary sequences younger than 200 m.y. were deposited in passive margins since the breakup of Pangea (Bally, 1979). The continental crust beneath the sedimentary wedge is generally regarded as thinned, modified by igneous material, and complexly faulted (Fig. 1c), although direct evidence has been difficult to obtain from the deeply buried basement in most margins.

Early Concepts of Subsidence Mechanisms

The fundamental question of how the Earth's crust subsides to accommodate thick successions of shallow-marine deposits, such as occur in passive margins, had been clearly defined in the late 1800s during the debate over the origin of geosynclines and mountain belts, especially between James Hall and J. D. Dana. Hall (1859) had shown that the thickness of lower to middle Paleozoic strata in the Appalachian Mountains of New York (~4.9 km) was about six times greater than that of equivalent strata far to the west in the upper Mississippi Valley.

> We have evidence of this subsidence in the great amount of material accumulated; for we cannot suppose that the sea has been originally as deep as the thickness of these accumulations. On the contrary, the evidences from ripplemarks, marine plants, and other conditions, prove that the sea in which these deposits have been successively made was at all times shallow, or of moderate depth. The accumulation, therefore, could only have been made by a gradual or periodical subsidence of the ocean bed. (Hall, 1859, p. 69–70)

About 100 years later, after the predominantly shallow-marine origin of the thick sedimentary fill in the passive margin of the eastern United States was thoroughly documented by dredging and drilling (Heezen and Sheridan, 1966), the same concept was rediscovered and became one of the important criteria incorporated in geological and geophysical analyses of the evolution of the margin (for example, Sheridan, 1976; Rona, 1974; Watts and Ryan, 1976).

After Hall (1859) presented the evidence for crustal subsidence in geosynclines, a number of hypotheses, including one of his own, were developed to explain how such large magnitudes of subsidence occurred. By the early 1900s, the roots of most of the modern thinking about mechanisms of subsidence had appeared in print. Few of these concepts pertained directly to passive margins, however, and it was not uncommon for each subsidence mechanism to be regarded as the principal control of subsidence.

Sediment Loading. In his publication documenting subsidence in the New York Appalachians, Hall (1859) attempted to account for the process, together with the linear shape of the subsiding or "synclinal" trough, solely as a result of the weight of sediment accumulating beneath a persistent marine current (Hall, 1859, p. 73). Although it can be demonstrated from isostasy (defined later by Dutton in 1871) that this process alone cannot produce the geosynclinal thicknesses of shallow-marine strata such as in New York, Hall's ideas were among the earliest suggestions of a relation between the strength of the crust and the importance of sediment loading to the subsidence of sedimentary basins. It is now thought that the weight of accumulating sediment amplifies subsidence caused by other mechanisms by as much as 300% (for example, Sleep, 1971; Watts, 1981).

Wegener (1912, p. 188) recognized that sediment loading alone is not sufficient to produce the geosynclinal thicknesses of shallow-marine deposits such as described by Hall (1859) because the sediments are less dense than basement rock and a sedimentary basin will gradually tend to fill. Bowie (1927) appears to be one of the first to have described this relationship quantitatively and in terms of different initial water depths in the basin.

But the density of sediments in place is not over 2.4 or 2.5 until after they have been consolidated. The density of the material at the bottom of the crust must be 3.0 or close to that.... With this large difference in density a layer of sediments 1000 feet in thickness would require a layer of matter at the bottom of the crust of from 800 to 900 feet in thickness to move away to restore equilibrium.... It would be impossible to have many thousands of feet of sediment deposited over an area, all in shoal water, when the only force causing the downward movement is the weight of the sediment. (Bowie, 1927, p. 260)

Similar discussions quantifying the relation between sediment loading and crustal subsidence in geosynclines were common in the early 1900s, and the need for a tectonic mechanism of subsidence to account for deposition of thick shallow-water sediments was clearly recognized.

Lateral and Vertical Thermal Contraction. J. D. Dana, who originated the term "geosyncline" and applied it to Hall's synclinal trough in the Appalachians (Dana, 1873), was critical of the mechanism of subsidence proposed by Hall (1859). He did not recognize the isostatic problem posed by Hall's subsidence mechanism and argued instead that subsidence could not proceed with such a "fine reaction" to incremental deposition of sediment as required by the sediment-loading model. He believed that geosynclinal subsidence (and mountain building) was caused by buckling of the crust under lateral pressure arising from cooling and thermal contraction of the Earth, especially at the boundary between continents and ocean basins (Dana, 1873). Crustal subsidence thus was viewed as the result of strong horizontal compressive stresses induced by cooling of the Earth.

In a remarkably short and strongly worded paragraph, Dana dismissed outright the possibility that geosynclinal subsidence could be controlled by cooling and thermal contraction of the continental crust *beneath* the thick sedimentary deposits, apparently because he believed that cooling of the continents was essentially completed early in the history of the Earth.

Another cause of local subsidence is local cooling beneath, accompanying the increasing accumulation of sediments. But this idea is too obviously absurd to require remark. (Dana, 1873, p. 427)

Dana's opinion contrasts markedly with the current view that after rifting ends, subsidence in passive margins is controlled mainly by cooling and thermal contraction of the subjacent heated lithosphere. It is thought that as much as 6 km of sediments has been deposited during the cooling phase in some modern and ancient passive margins (for example, Sleep, 1971; Watts, 1981; Bond and Kominz, 1984).

One of the earliest suggestions that subsidence could be caused by an isostatic response to thermal contraction of the crust beneath the strata appears to have been made by W. Bowie.

There is another cause of the sinking independent of the weight of the sediment and that is the contraction of the crustal material beneath the sedimentary strata. That contraction seems to result from processes set up by the change in temperature of the crustal matter as a result of the upward movement of that matter when the surface above it was undergoing erosion. (Bowie, 1927, p. 260)

This concept was first published in 1921 (Bowie, 1921) and later appeared in a series of papers on isostasy published by the *Bulletin* (Bowie, 1922). Bowie did not believe that the Earth's crust had sufficient strength to transmit horizontal stresses over the distances required to cause uplift of mountains, and his explanation of the mechanism of subsidence was part of an effort to explain the Earth's major changes in relief in terms of purely vertical motions. A highly modified form of this concept is part of recent two-phase models for passive margins in which part of the crustal thinning is caused by erosion during uplift accompanying rifting and is followed by subsidence after rifting as the thermal anomaly decays (for example, Beaumont and others, 1982). The uplift is assumed to be caused by an elevation of the temperature of the lithosphere by a shallowing of the hotter asthenosphere, however, and most of the thermal contraction results from decay of that thermal anomaly.

A strikingly modern view of the relation between temperature and subsidence in ocean basins was held by Wegener (1929, p. 208). He suggested, as a plausible (but not favored) hypothesis, that the difference in depth of ocean basins could reflect an isostatic response to a difference in age and, consequently, in temperature, reasoning that the shallowest portions of the ocean floors would be the youngest and least dense.

Flexure. Dutton (1889) recognized that although the Earth's crust is weak enough to respond to the emplacement or removal of loads, the crust also has sufficient strength to partly support loads. The magnitude of that strength, as given by estimates of the difference between regional and local (Airy) isostatic compensation, was highly debated, however (for example, Bowie, 1927; Barrell, 1914). Vening-Meinesz (1931) incorporated the concept of regional isostatic compensation into a model in which crustal loads were partly supported by bending of the adjacent rigid crust, giving rise to deep depressions adjacent to the mountain chains (Fig. 5a). This concept was incorporated into a mechanism of subsidence of Atlantic-type continental margins by Gunn (1944) in an attempt to quantitatively explain the variation between the theoretical and observed gravity anomalies at these margins. Gunn concluded that the gravity data supported the concept of a lithosphere of finite strength that had deformed by flexural bending in response to the growth of sediment load, producing a deep sediment-filled trough at the edge of the continent. Like Umbgrove (1947),

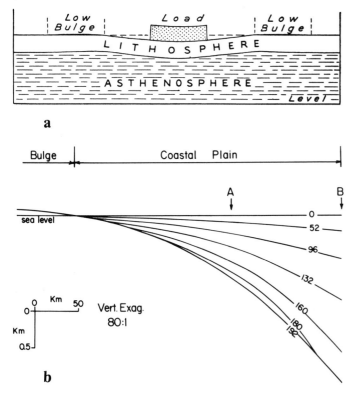

Figure 5. (a) Figure from Daly (1940), showing flexural response of the lithosphere to loads, based on concepts of regional compensation developed by Vening-Meinesz (1931).

(b) A recent application of the concept of regional compensation for lithospheric loading to a passive margin (from Steckler and Watts, 1982). The coastal plain corresponds to the inner flexural wedge in Figure 1d where the rigid, unrifted edge of the craton bends in response to the loads forming in the passive margin.

Gunn did not believe that extensional faulting was the main cause of subsidence of Atlantic-type margins. Gunn's work, however, pioneered the modern quantitative studies of flexure in passive margins. These studies have shown that flexure occurs predominantly in the post-rift or cooling phase of subsidence and, therefore, is compatible with extensional faulting during rifting of the margins (Watts and Ryan, 1976; Beaumont and others, 1982). In addition, flexure is thought to cause progressive bending and subsidence of the craton edge landward of the hinge zone to account, in large part, for deposition of coastal-plain strata along continental margins (Figs. 1d and 5b; Watts, 1981).

Crustal Thinning by Extension. Between the latter part of the 19th century and the early part of the 20th century, various attempts were made to explain the rift structures and subsidence such as observed along Atlantic-type coastlines and in rift valleys. These efforts contained some of the earliest inferences that passive margins formed as a result of regional extension and that subsidence was an isostatic response to extensional thinning of the crust.

Suess (1909) was one of the first to emphasize the importance of regional extension in the formation of graben structures in continental rifts. He viewed this process as the result of stresses caused by contraction of the Earth and concluded that grabens formed by collapse of crustal blocks between steeply dipping normal faults. Bucher (1924, 1933) was a strong advocate of horizontal extension as a mechanism for the origin and subsidence in rifts and geosynclines. He compared the process with the extension and ductile thinning of metals under tension and illustrated the process with figures from contemporaneous physics textbooks.

Wegener (1929, p. 181–205) argued that horizontal crustal extension accompanied rifting and continental breakup, producing the structures and subsidence of Atlantic-type coasts and eventually forming the ocean basins. As indicated in a simple sketch from his book (Fig. 6a; Wegener, 1929, p. 185), he clearly viewed the process as involving substantial amounts of crustal thinning. He further suggested that in the earlier stages of rifting of continental blocks,

a gaping fissure would form only in the uppermost, more brittle layers, while the lower, plastic layers would stretch. (Wegener, 1929, p. 191)

Wegener's view of the rifting process thus also anticipated the vertical change from brittle to ductile behavior that has been incorporated in many of the more recent extension models.

Increase in Crustal Density by Basalt Injection and Phase Changes. Mechanisms of subsidence in passive margins involving perturbations in the density of the lower crust and upper mantle, which have been invoked by some to minimize or eliminate the importance of crustal extension, also have roots in ideas that were developed well before the 1940s. LeConte (1872) proposed that the subsidence of geosynclines was caused by a progressive increase in density of the sediments with burial, first by compaction (lithification) and by heating and metamorphism at greater depths. A somewhat similar concept based on deep-seated thermal metamorphism (producing a change from greenschist to amphibolite grade in rocks of the lower crust) has been incorporated in one of the recent models for subsidence during heating and rifting in modern passive margins (for example, Falvey, 1974). DuToit (1937) incorporated ideas of Holmes (1926) into a concept of a world-wide "paramorphic zone" within the crust in which subsidence due to sediment loading causes a basalt-to-eclogite phase transformation. He thought that the accompanying increase in density would provide a mechanism for maintaining the subsidence. Basalt-eclogite phase transformations have been considered recently in models of the subsidence of passive margins (for example, Neugebauer and Spohn, 1982; Artemjev and Artyushkov, 1971), but are controversial because of the widely held view that the Moho is not a phase transition.

Barrell (1927) was one of the first to suggest that intrusion of basic magma into the crust would cause subsidence, based on the concave-upward shape of many basic intrusive bodies. Bucher (1933) apparently was the first to apply this concept on a much larger scale to the "tensile phase" of crustal deformation, during which a new mobile belt comes into existence. He believed that although subsidence of deep furrows (areas of thick sediments in geosynclines) occurred in part as a result of regional extension and thinning of the crust, most of the subsidence occurred in response to injection of large amounts of basic material into the extending crust. Interpretation of recent deep-reflection profiling suggests that basaltic injection into the lower crust or basaltic underplating has occurred, apparently on a large scale, beneath some modern passive margins (for example, Mutter and others, 1982; LASE Study Group, 1986; Keen and de Voogd, 1988).

Crustal Thinning by Gravitationally Induced Body Forces. The idea that rock could deform and flow under gravitational stress was developed in the late 1800s and early 1900s by Reyer (1888), who used the concept as an alternative to the contraction theory of folding and orogeny. The concept appears to have been first applied to the continental margins by Chamberlin (1913, p. 578), who recognized that crustal blocks bounded by free inclined surfaces, such as along continental margins, cannot be in isostatic equilibrium. He argued that the continental margins must creep, in a manner analogous to movement of a glacier, outward over the oceans. The process would cause crustal thinning (possibly associated

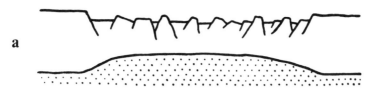

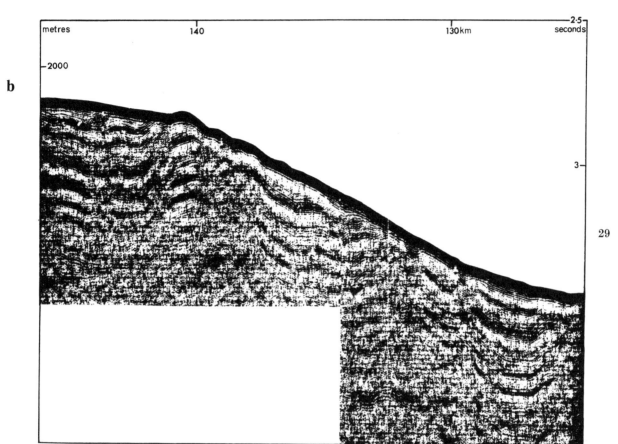

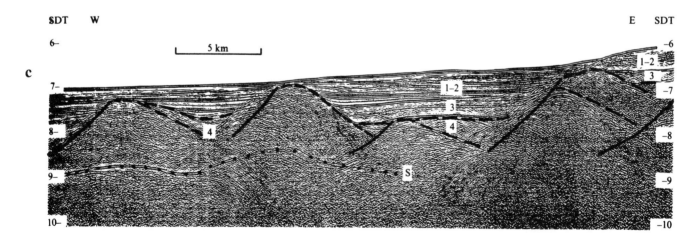

Figure 6. Examples of the development of evidence supporting rifting and crustal thinning in passive margins.
(a) Conceptual view of extension in a rift belt from Wegener (1929), indicating his view of the relation of tilted fault blocks and crustal thinning to regional extension. He also thought that the lower crust beneath the fault blocks would extend by ductile deformation.
(b) One of the earliest seismic reflection profiles showing evidence of tilted strata and faulting in a modern passive margin. The profile is from the eastern flank of Galicia Bank off western Spain (Stride and others, 1969).
(c) Recent seismic profile from immediately west of Galicia Bank (de Charpal and others, 1978), not far from the line in b (above), acquired with substantially improved equipment and processing, showing tilted basement blocks above normal faults and possible Moho (dotted lines), a result supporting the relation between normal faulting and crustal thinning pictured by Wegener in a (above).

with normal faulting) and subsidence, which would create space for sedimentation and growth of the shelf. This concept in nearly the same form was applied quantitatively to modern passive margins by Bott and Dean (1972), although they emphasized that creep must be concentrated in the lower crust and normal faults will occur in the upper crust. A more recent view has tended to minimize the importance of this mechanism, and it is regarded as mostly a secondary process responsible for the formation of local troughs or basins within passive margins (Bott, 1981). The process may also be partly responsible for growth faulting in the thick sedimentary sections on passive margins.

Red Sea and East African Rift Systems as Classic Analogues of Initiation of Passive Margins

The Red Sea and the East African rift systems have become two of the most frequently cited examples of the early stages of rifting and initial drift in passive margins, appearing in many textbooks as general illustrations of the stages of plate divergence leading to oceanic lithosphere. The analogy perhaps can be traced to Suess, who concluded that the Red Sea was a continuation and advanced stage of rifting of the East African rift belt (Suess, 1904, p. 374–376). In a discussion closer to the context of passive margins, Wegener (1929, p. 191) described the evolution of Atlantic-type coasts in terms of the East African rifts. DuToit (1937, p. 258–259) compared the geomorphology of Atlantic or fault-line coasts with that of the East African rift valleys and inferred a similar, extensional origin. The most direct analogies between rifts and passive margins, stage by stage, were not drawn until much later, however, by Carey (1958) and Heezen (1959).

The extensional origin of the African rifts and other rift belts of the world is a fundamental concept in the analogue with passive margins, but it received strong opposition in the first half of the 19th century. One of the most influential geologists of the early 1900s, Bailey Willis, favored a compressional origin of rifts. Willis believed that rifts were bounded by steeply dipping reverse faults that had been obscured by slumping of the rift valley walls (Willis, 1936). Bullard (1936) interpreted the strong negative Bouguer anomalies over the rift valley as evidence supporting, in part, the compressional model. The controversy between these two schools of thought persisted into the 1960s, even as new paleomagnetic data were reviving and strengthening the continental-drift theory. The extensional origin was finally confirmed after computed gravity values were found to be smaller beyond the edges of the rift valleys than was predicted by the compressional or reverse-fault model (Girdler, 1964) and after earthquake mechanism studies indicated that the present-day motion along the faults was a combination of dip slip and strike slip (Sykes, 1968; Fairhead and Girdler, 1971).

GEOLOGICAL AND GEOPHYSICAL EXPLORATION OF THE OCEANS PRIOR TO DEVELOPMENT OF PLATE-TECTONIC THEORY (~1944 TO 1970)

Between the end of World War II and the development of plate-tectonic theory in the early 1970s, the principal advances in studies of passive margins resulted from steady improvements in marine geophysical technology coupled with commercial and scientific deep-ocean drilling. The new data acquired from the ocean basins and their margins not only further emphasized existing questions about passive margins, particularly with respect to their internal structure and subsidence mechanisms, but also raised new questions about the nature of the transition from continental to oceanic crust. The major deep-sea drilling program, Joint Oceanographic Institutions for Deep Earth Sampling (JOIDES), sponsored by the National Science Foundation, was initiated in 1964 in order to extend the knowledge that had been obtained from dredging and piston coring. This program initially focussed mainly on the deep-ocean basins, and it produced critical information on only certain passive margins. JOIDES drilling on passive margins did not contribute greatly until 1975–1983. The most important conceptual advances in passive-margin studies were an outgrowth of the growing support for sea-floor spreading, leading to a renewed emphasis on the significance of rift belts and rifted continental margins in the theory of continental drift.

New Evidence on the Shallow to Deep Structure of Passive Margins

Seismic Refraction Measurements and Airborne Magnetic Surveys. The offshore refraction experiments begun by Ewing and his co-workers were resumed after World War II (with continued financial support from the Geological Society of America), and similar experiments were begun by other groups in ocean basins and continental margins (for example, Hersey and others, 1959). By the mid-1950s, the offshore refraction measurements, combined with Rayleigh and Love wave dispersion data, had documented beyond doubt the fundamental differences in thickness and composition between continental and oceanic crust (for example, Fig. 1b). A series of papers summarizing the early arguments and inferences about the nature of the continent-ocean transition was published in 1955 in the Geological Society of America Special Paper 62 (A. Poldervaart, ed.).

By the end of the 1960s, a significant amount of offshore subsurface structural and stratigraphic detail had been added to the fairly limited data base for passive margins that existed prior to World War II. Papers given at the third Symposium of Continental Margins and Island Arcs in Zurich (Keen and Loncarevic, 1968) and in Volume 4, Parts 1 and 2, of *The sea* (Maxwell, 1970) give a good summary of the status of knowledge about passive margins just before the development of plate-tectonic theory. Refraction data clearly indicated that the crystalline basement in the margins was buried by several kilometers of sedimentary rock (for example, Drake and others, 1959; Worzel, 1965; McConnell and others, 1966; Fenwick and others, 1968). In addition, refraction measurements revealed the presence of a deep crustal layer with an anomalous velocity of 7.2–7.7 km/s beneath many margins, including passive margins, and orogenic belts. Drake and Nafe (1968) and Drake and Kosminskaya (1969) suggested that this anomalous crust was derived from the mantle and possibly of a transient nature related to changes in elevation associated with tectonic activity. Another important result of the refraction measurements was the discovery of significant relief on the crystalline basement beneath the margins (Fig. 7).

The most notable presumed basement structure was a distinct outer high or ridge underlying the shelf break and part of the slope (Drake and others, 1959). A basement contour map, of the depth to the >5.6 km/s velocities, constructed by Drake and others (1968) showed that this outer high separated two thick sedimentary basins, one beneath the shelf and the other beneath the rise. The complexity of the basement structure was further suggested by magnetic anomaly data that were compiled for the first time on regional scales from airborne surveys over the passive margins of eastern North America, South Africa, arctic North America, and South America (summarized in Drake and others, 1963; Taylor and others, 1967; Fenwick and others, 1968). These magnetic anomaly maps indicated the presence of a nearly continuous magnetic anomaly, averaging about 600 gammas over many passive margins, close to the continental slope and flanked on each side by smaller, discontinuous anomalies. Along the eastern North American margin, this anomaly is known as the "East Coast magnetic anomaly" or "ECMA" (Fig. 1c).

Various explanations were proposed to account for the magnetic data and structures revealed by the refraction measurements. The basement

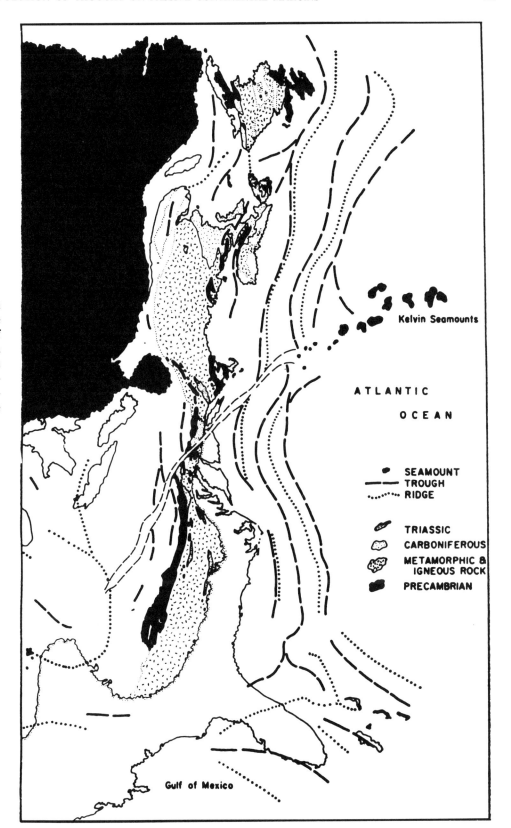

Figure 7. Summary figure from Drake and others (1968), showing the first evidence on a regional scale of complex basement topography in a passive margin, interpreted as trough and ridges at the time and now regarded as largely a result of extensional faulting enhanced by subsidence and sediment loading.

structures seen in the United States–Canadian Atlantic margin and subsequently in other margins were generally referred to as troughs and arches resulting from regional warping (for example, Drake and others, 1959). The refraction data and magnetic anomaly patterns did not provide unambiguous evidence for a fault origin of the basement relief, and although some basement faults were proposed, they were speculative and controversial (Drake and Woodward, 1963; Taylor and others, 1967; Sheridan and others, 1969). The ECMA was regarded by Taylor and others (1967) as evidence of a felsic intrusion paralleling the edge of the pre-Paleozoic landmass, based on model calculations. Drake and others (1959) suggested

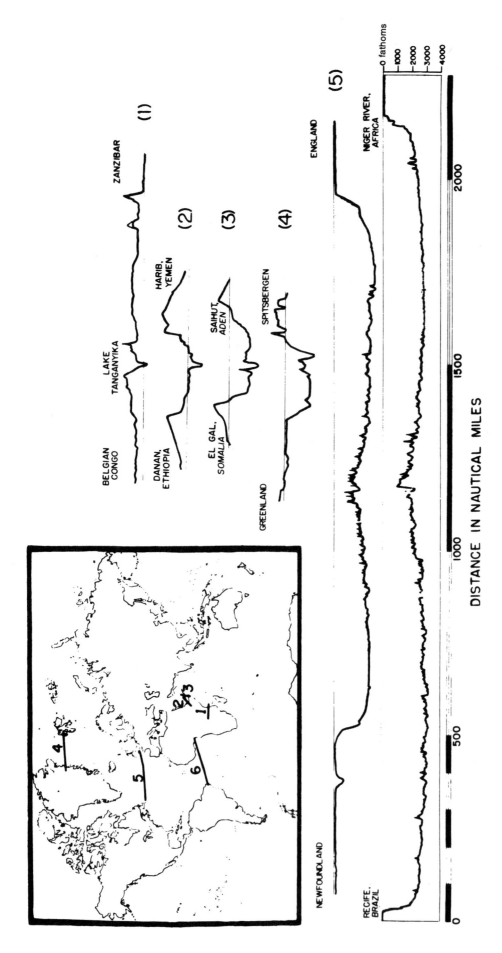

Figure 8. Early ideas about the evolution of passive margins, using the East African rift system and Red Sea as analogues. (a) Figure from Heezen (1959), comparing the topographic profiles across rifts in East Africa and elsewhere with the results of PDR profiling of the Atlantic Ocean floor. The presence of a ridge culminating in an axial trough was considered especially compelling evidence in support of the common origin of the structures in the ocean and in East Africa.

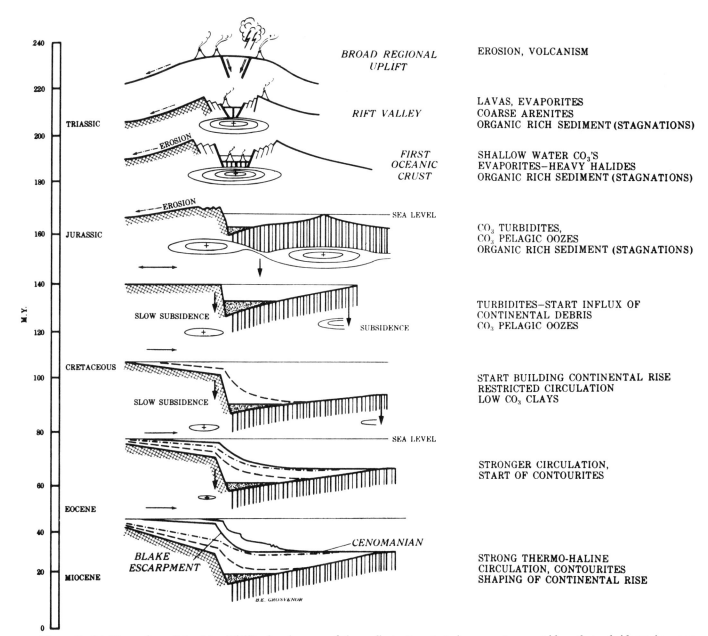

Figure 8. (b) Figure from Schneider (1969), showing one of the earliest attempts to incorporate recent ideas that subsidence in ocean basins and passive margins is controlled by thermal contraction resulting from decay of thermal anomalies (pictured as concentric ellipses) formed during rifting.

that it is related to the continuous basement ridge identified in the refraction measurements, and Keen (1969) considered the anomaly to be an edge effect arising from juxtaposition of different crustal types at the continent-ocean transition. Modern seismic reflection data strongly suggest that the outer basement high is actually the edge of a Mesozoic carbonate bank (Grow and others, 1988).

Continuous Seismic Profiling. Beginning in the late 1950s, deep-penetration echo sounders (Beckmann and others, 1959), later improved and called "seismic profilers" (Ewing and Tirey, 1961), were employed to probe the subsurface in ocean basins and continental margins. The first published seismic profile (as opposed to deep echo soundings) near a passive margin was from the Gulf of Mexico (Ewing and Tirey, 1961). A few years later, seismic-profiling experiments demonstrated the presence of faults and rotated blocks in the deep subsurface of a few passive margins, providing the first direct evidence in support of the rift origin of Atlantic-type margins. Among the earliest of these profiles were a series of lines off western Europe between the Faeroe Isles and Lisbon (Stride and others, 1969). Here, the records indicated a widespread occurrence of disrupted and tilted strata bounded by faults (Fig. 6b), in many places lying unconformably beneath flat-lying, presumably post-rift strata. The evidence for basement faulting could be observed here because in contrast to other margins such as off the United States and Canada, the margin is sediment starved and lacks the thick sedimentary cover that severely attenuates seismic signals. This area, particularly in the Bay of Biscay and off the coast of Iberia, has been re-examined a number of times since the 1970s with improved methods of seismic data acquisition and processing and

JOIDES drilling. It has become a classic example of rotated fault blocks and thinned continental crust in the rift phase of a passive margin (for example, Fig. 6c).

The early seismic profiling experiments also provided much new information that was calibrated with existing commercial well data and limited JOIDES drilling data to reconstruct and interpret the internal stratigraphy of passive margins. It soon became evident that although the delta models proposed much earlier had some applicability, the origin of strata in passive margins was much more complex. One of the earliest examples, published in 1965, was from the Blake Plateau (Fig. 3b), where the results were regarded as evidence of progradation of the shelf through the Tertiary and erosional thinning of supra-Paleocene strata by the Gulf Stream (JOIDES, 1965; Drake and others, 1968). By 1970, single-channel seismic profiling combined with drill data had shown that in some passive margins, strata grew seaward by a delta-like progradation; in others, the strata were terminated seaward by erosion and faulting, and in some, the strata appeared to terminate against a reef that had grown upward beneath the shelf edge (Hedberg, 1970; Emery and others, 1970). Early in the next decade (1975–1985), the technology of multichannel seismic profiling used in petroleum exploration was applied routinely by academic institutions to the studies of passive margins.

Passive Margins in the Context of Continental Drift and Sea-Floor Spreading

After publication of the classic paper by Hess in 1962 by the Geological Society of America, new ideas on the tectonic implications of passive margins appeared in the context of sea-floor spreading theory as it gained increasing acceptance in the light of ocean-floor magnetics, paleomagnetic data from continents, and further development of convection-current models that provided a plausible mechanism. Carey (1958), arguing for continental drift on an expanding Earth, and later Heezen (1959) developed the widely cited evolutionary model for passive margins based on stages of rifting observed in East Africa and the Red Sea (Fig. 8a).

In the late 1960s, a series of papers appeared relating new data on heat flow, age, and bathymetry of ocean floors to the cooling of lithosphere moving away from a spreading-ridge axis (Langseth and others, 1966; McKenzie, 1967; Vogt and Ostenso, 1967; McKenzie and Sclater, 1969; Sleep, 1969). These papers marked the first appearance of quantitative thermal contraction models for the subsidence of ocean floor that were to have a major influence on concepts of subsidence in passive margins in the next two decades. Vogt and Ostenso (1967) appear to have been the first to suggest that modern passive margins should have a subsidence history comparable to that of ocean floors and were also the first to compare the subsidence history of commercial wells in passive margins with the early thermal contraction models for ocean floor. In an obscure journal, Schneider (1969) incorporated the ideas of Vogt and Ostenso in a modification of Heezen's and Carey's sequential stages in formation of a passive margin, that although qualitative, has much in common with current thermo-mechanical models for the evolution of passive margins (Fig. 8b).

Passive Margins and Geosynclines

A major contribution to the continually evolving ideas about the geosyncline and what it represents resulted from the merging of two unrelated lines of research underway between about 1940 and 1960. One of these was the program of refraction studies off the east coast of the United States and Canada, and the other appeared in a series of papers on geosynclines by Marshall Kay, culminating in a classic publication by the Geological Society of America in 1951 (Kay, 1951). In the late 1940s, Kay developed an elaborate classification of geosynclines in an attempt to reconcile the remarkably diverse and conflicting views on their origin that had appeared since the publications of Hall and Dana. Building on ideas that originated with Stille (1941), Kay termed the most common geosyncline an "orthogeosyncline" (true geosyncline on a continental edge, as opposed to an intracratonic geosyncline or basin) and proposed that it consisted of two fundamental divisions, a "miogeosyncline" and "eugeosyncline." He defined the eugeosyncline as a surface that subsided deeply in a belt of voluminous active volcanism and the miogeosyncline as a belt of less subsidence and much less active volcanism. Kay regarded the miogeosynclinal stratigraphy as similar to that of the adjacent craton, but thicker. His most cited example is the "restored" Appalachian miogeosyncline and eugeosyncline of Cambrian and Ordovician age between New York and Maine (Fig. 9b).

In the same landmark paper containing the results of seismic refraction measurements from the western Atlantic margin, Drake and his co-workers (Drake and others, 1959) emphasized what they regarded as a striking similarity between the results of the refraction experiments and the miogeosynclinal-eugeosynclinal cross section from Kay (1951; compare Figs. 9a and 9b). They suggested that a geosyncline was, in part at least, the deformed remnants of a passive-margin shelf, slope, and rise. The enigmatic outer basement high in the refraction data was considered analogous to Kay's Vermontia, as inferred subaerial "tectonic land" or geanticline that shed coarse detritus eastward into part of the miogeosyncline.

A few years later, following the lead of Drake and his colleagues, Dietz (1963) used the geosynclinal cross section of Kay and his own interpretation of the refraction data from passive margins to support what he termed an "actualistic concept" of geosynclines (compare Figs. 9b and 9c). Dietz thought that most of the sediment delivered to the passive margins accumulated at the base of the slope, producing a thick continental-rise prism which he regarded as the modern analogue of the eugeosyncline. Following the earlier ideas of Gunn (1944), Dietz suggested that the weight of sediment in the rise was sufficient to cause an isostatic depression of the shelf and create the space for thinner, shallow-water shelf strata, which he considered analogous to the miogeosyncline. He argued that the eugeosyncline was subsequently decoupled from the miogeosyncline and thrust beneath it by sea-floor spreading, giving rise to compressional orogenesis and the mountain-building phase of the geosynclinal cycle. The models of Dietz and Drake and others, although subsequently shown to be largely incorrect, represent the earliest efforts to recast the geosynclinal concept in the context of new subsurface data from modern continental margins.

Later, other comparisons between miogeosynclinal stratigraphy and modern passive margins were published. Rodgers (1968) compared the major and abrupt eastward transition from shallow-marine carbonate-bank facies to deeper-water pelitic rocks in Cambrian to Lower Ordovician strata of the Appalachian belt with the abrupt seaward edge of the present Bahama Banks. He concluded that this facies change marked the eastern edge of the North American continent at that time. A similar concept was applied much later by Churkin (1974) to explain the carbonate–to–graptolitic-shale transition in lower Paleozoic miogeoclinal strata of the North American Cordillera, although he viewed the transition as occurring within a backarc basin.

One of the most stimulating and prophetic papers published in this period was by J. T. Wilson in 1966. He suggested that the confusing distribution of two early Paleozoic faunal realms in the Appalachian-Caledonide orogen on both sides of the North Atlantic could be explained simply by the opening and subsequent closing of a proto-Atlantic Ocean. His interpretation implied that the lower Paleozoic strata containing the two faunal realms were remnants of the margins of the ancient ocean, a concept that as Wilson noted, was anticipated much earlier by Grabau (1936). In this paper, Wilson clearly and succinctly laid much of the groundwork for the major revision of the geosynclinal concept and the role of ancient passive margins in geosynclinal evolution that was to occur in the next decade.

It is interesting to note that at the end of this period, just before the widespread application of plate-tectonic theory to geosynclines, there was a general skepticism about all geosynclinal models (Schwab, 1982), perhaps best reflected in a significant contribution by one of Marshall Kay's students, Robert Dott, Jr. (1964). Dott emphasized the great diversity of geosynclines and the difficulty of explaining all in terms of any single model. In particular, he pointed out important differences between the Atlantic passive margin and conventional orthogeosynclines as defined by Stille and Kay and questioned the validity of the actualistic geosynclinal models of Drake and Dietz.

PLATE TECTONICS (~1970–1975)

The general theory of plate tectonics was developed rapidly in the early 1970s as new, compelling evidence was acquired through an intensive program of geophysical and geological exploration of the world's ocean basins. The history of this revolution and its major impact on tectonics and geosynclinal theory are well known and can be found in several excellent publications (for example, Schwab, 1982; Dennis, 1982; Shea, 1985; Hsü, 1973; Dott, 1974, 1978, 1979).

Although a remarkably large amount of new data was gathered from the deep ocean floor in order to test the sea-floor spreading concept, passive margins were generally of secondary interest, and the principal advances in thinking related to these structures were inferences drawn from the developing plate-tectonic theory. By the mid-1970s, there was general acceptance of the view, proposed many years earlier by Wegener and DuToit, that modern passive margins were the deeply buried edges of continental fragments that had rifted and drifted apart.

Passive Margins and Plate-Tectonic Models of Geosynclines

One of the most significant implications of plate-tectonic theory for passive margins concerned their role in continental orogenesis and geosynclinal evolution. In a series of papers on plate tectonics and geosynclines published in the short span of time between 1970 and 1972 (for example, Mitchell and Reading, 1969; Dewey, 1969; Dewey and Bird, 1970; Dewey and Horsfield, 1970; Dickinson, 1971), the basic concept of the geosyncline and its role in orogenesis was completely recast in the framework of actualistic models derived from modern plate tectonics. Two of the earliest of these papers were published in the *Bulletin* by Bird and Dewey (1970), who applied plate-tectonic models to the Appalachian geosyncline near the cross section of Kay (compare Figs. 9b and 10), and by Coney (1970), who questioned the fundamental premises of the geosynclinal theory. In this new view of geosynclines, ancient passive margins that had been incorporated into mountain belts were thought to occur only within the miogeosynclines as defined by Stille/Kay. Kay's eugeosynclines were interpreted as complex structures produced by plate convergence that postdated and terminated the passive-margin phase. The early actualistic models of Drake and others (1959) and Dietz (1963) that envisioned the miogeosyncline and eugeosyncline as paired and essentially contemporaneous elements of a passive margin thus were superseded by a much more mobile concept in which the passive margin played a less dramatic but important role.

This significantly revised view of geosynclines resulted in a remarkably intensive effort to compare the stratigraphy, structure, and petrology of orogenic belts with the various plate-tectonic elements that were emerging from studies of modern ocean basins and continents. Most of his effort centered on dramatic features such as island arcs, obducted oceanic crust, and sutures between continents and other continents or island arcs. Passive margins received attention mainly as criteria for recognizing and dating periods of plate divergence and, through their destruction, as sensitive recorders of the initiation of plate convergence and an active continental margin. Passive margins thus were regarded primarily as indicators of the timing and nature of horizontal crustal displacements. In contrast to the emphasis in previous decades, relatively little attention was given to the mechanisms responsible for the large magnitudes of vertical displacements required by the immense thicknesses of strata in passive margins.

The earliest attempts to apply plate-tectonic models to geosynclines were in the Appalachian and Cordilleran orogenic belts of North America and the Caledonian belt of Britain. Dewey (1969), Bird and Dewey (1970), and Williams and Stevens (1974) interpreted the upper Proterozoic to Lower Ordovician strata in the Appalachian miogeosyncline of Kay and in correlative miogeosynclinal strata in Britain as an ancient passive margin formed during opening of a proto-Atlantic Ocean (Fig. 10), following and amplifying the original concept of Wilson (1966). Later, Stewart (1972), in an article in the *Bulletin*, along with Burchfiel and Davis (1972), Stewart and Poole (1974), and Gabrielse (1972), interpreted the extensive miogeosynclinal wedge of Late Proterozoic to Devonian age in the Cordillera of western North America as evidence of a passive margin that as in the Appalachian belt, was initiated by continental rifting in the Late Proterozoic. A short time later, Stewart (1976) suggested in an article published in *Geology* that North America is rimmed by broadly correlative passive-margin deposits ranging from Late Proterozoic to early Paleozoic in age, and he compared the continent and its relation to plate boundaries with that of Africa after breakup of Gondwana. He further emphasized evidence that appeared to document widespread rifting and initiation of the passive margins in Late Proterozoic time and concluded that North America must have been an interior piece of a much larger continent that rifted apart in Late Proterozoic time. A significant contribution in this paper, which was anticipated in some respects by Valentine and Moores (1972), was the use of the distribution and age of ancient passive-margin deposits to document the presence and subsequent breakup history of supercontinents for the vast majority of geologic time in which direct evidence from sea-floor spreading histories is not preserved.

Development of Actualistic Models for Recognizing Ancient Passive Margins

As Schwab (1982) has emphasized, the early papers that attempted to reinterpret ancient geosynclines in terms of plate tectonics met with mixed success, underscoring the need to develop better criteria for identifying plate-tectonic mechanisms. These criteria were developed largely from the ongoing studies of modern plate-tectonic regimes within the continents and the ocean basins. By the mid-1970s, largely descriptive stratigraphic, sedimentologic, and petrologic analogues of ancient passive margins had been developed and were being widely applied, along with analogues of other types of plate-tectonic regimes, to studies of orogenic belts. Many of the best early summaries of the sedimentologic and stratigraphic indicators of ancient passive margins were published in 1974 in Special Publications 19 and 22 of the Society of Economic Paleontologists and Mineralogists. At about the same time, Pearce and Cann (1973) laid the groundwork for using major- and trace-element criteria as magmatic indicators of plate settings.

The identification of an ancient passive margin by analogy with these modern examples was (and still is) far from straightforward. It was commonly assumed that all syn-rift materials underlying mature modern passive margins, and, by analogy, underlying their ancient equivalents, should be broadly similar to the volcanic-rich, predominantly arkosic examples from East Africa and the Triassic-Jurassic syn-rift strata of eastern North America. There is very little direct evidence supporting this assumption as a general rule because of the difficulty of thoroughly sampling the deeply buried rift sequences in modern passive margins. In fact, presumed syn-rift strata in some classic examples of ancient passive margins, such as the early Paleozoic miogeosynclinal wedge of the North American Cordillera, lack large amounts of immature sediments in many places and contain limited amounts of volcanic material (Stewart, 1972; Gabrielse, 1972; Bond and others, 1985). Moreover, the large amounts of uplift in the East African analogues did not occur during formation of some modern margins, such as in the Bay of Biscay and Galicia (Montadert and others, 1977). Although wedge-shaped deposits of shallow-marine to paralic strata that compose the shelf portion of a typical passive margin are a fundamental element of passive-margin analogues, they occur in other settings as well. For example, wedge-shaped deposits tapering toward the craton and composed mostly of shallow-marine carbonate rocks are present beneath the modern Persian Gulf. These deposits occur in a foreland basin associated with the modern convergent boundary between the Asian and African plates. An especially significant problem is whether an ancient passive margin prograded into a true Atlantic type of ocean or into a backarc basin. This question has been raised in reconstructions of the early Paleozoic continental margins in the Appalachian orogen by Hatcher in an article published by the *Bulletin* (Hatcher, 1972) and in the Cordilleran orogen by Monger and Price (1979). So far, there are no reliable criteria from the passive margins themselves to resolve the issue. These issues, along with similar difficulties in developing unambiguous analogues for other plate-tectonic elements lead to lively disagreements over the appropriate plate-tectonic models for evolution of specific orogenic belts.

The Ocean-Floor Cooling Curve and a Thermal Contraction Model for Subsidence of Passive Margins

While the intensive efforts to reinterpret geosynclines in terms of plate-tectonic models were underway, studies of the relation between heat flow, age, and bathymetry in ocean basins (for example, Sclater and Francheteau, 1970; Sclater and Detrick, 1973, in the *Bulletin*; Davis and Lister, 1974) were producing results that would later have a major impact on concepts of subsidence in passive margins. Two models of the subsidence of ocean floor moving away from ridge crests had been proposed, a simple cooling half-space model in which subsidence decays with respect to $t^{1/2}$ (t = time) and a cooling-plate model in which the lithospheric thickness is finite and the subsidence decays exponentially to a constant value. Both models seemed to approximately fit the available age-depth data for ocean floor. In 1977, Parsons and Sclater compiled new data on subsidence of ocean floor out to 160 m.y. and found that subsidence of young ocean floor decayed as a function of $t^{1/2}$ as predicted by the half-space model and that subsidence of older ocean floor decayed exponentially as predicted by the plate model. They inverted these empirical data and concluded that a cooling half-space model adequately accounts for the first 60 to 80 m.y. of ocean-floor subsidence and that a plate model with a constant thermal-equilibrium thickness of 125 km, maintained by a constant influx of heat to its base, accounts for subsidence of ocean floor older than ~20 m.y. By the late 1970s, there was general agreement that subsidence of ocean floor was controlled by cooling and thermal contraction of hot lithosphere moving away from ridge crests, but the validity of the plate model for older ocean floor was questioned by some (for example, Heestand and Crough, 1981).

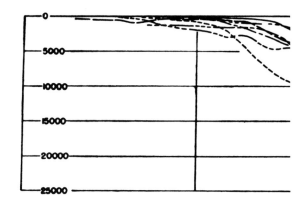

NEW YORK TO MAINE
MIOGEOSYNCLINAL AND
EUGEOSYNCLINAL BELTS
CAMBRIAN THROUGH MIDDLE ORDOVICIAN
WEST   130 MIL. YRS.

Hedreocraton   *Adirondack Line*   *Champlain Belt*

ORDOVICIAN
T—TRENTONIAN
-BOLARIAN
Ch—CHAZYAN
C—CANADIAN
CAMBRIAN
S—ST. CROIXAN
A—ALBERTAN
W—WAUCOBAN

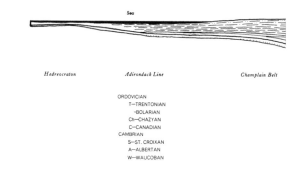

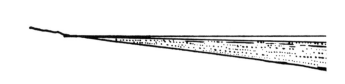

Figure 9. (c) Hypothetical cross section of a passive margin from Dietz (1963). Both Dietz (1963) and Drake and others (1959) attempted to interpret the entire geosyncline in terms of modern passive margins, assuming that the "tectonic land" separating the mio- and eugeosynclines corresponded to the outer basement highs of modern passive margins, which appear as the prominent basement ridges between the shelf and slope in a (above).

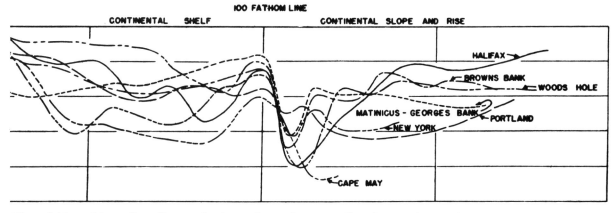

Figure 9. Two of the earliest efforts to develop analogues for geosynclines from modern passive margins.
(a) Cross sections from off eastern North America, showing depth to basement across the modern passive margin (from Drake and others, 1959).

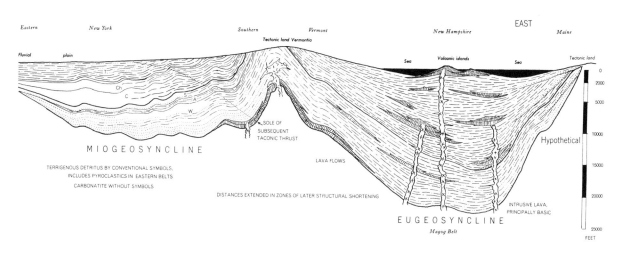

Figure 9. (b) "Restored" cross section of the Appalachian orthogeosyncline from New York to Maine, showing the location of the miogeosyncline and eugeosyncline from Kay (1951).

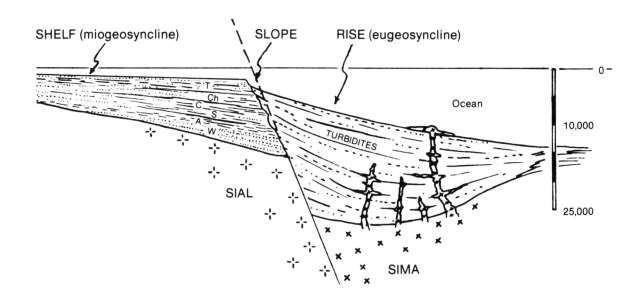

Figure 10. One of the earliest views of the role of passive margins in the evolution of geosynclines in terms of plate-tectonic theory (from Bird and Dewey, 1970).
(a) The passive-margin phase of the geosyncline in Cambrian and Early Ordovician time.
(b) and (c) Progressive destruction of the passive margin by plate convergence and development of a cross section comparable to that of Kay in Figure 9b.

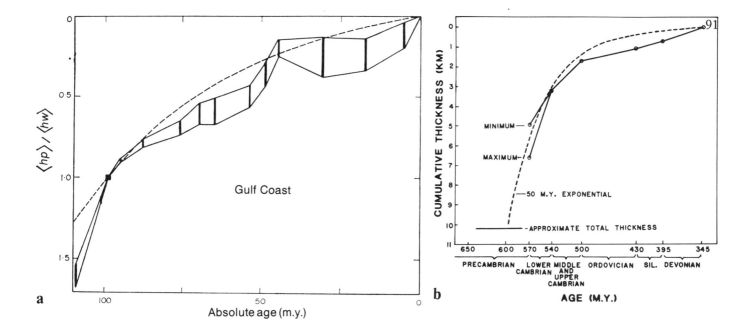

Figure 11. Comparisons of subsidence history in modern and ancient passive margins with the exponential subsidence curve that is predicted by cooling-plate models for ocean floor and the post-rift subsidence of passive margins.

(a) One of the first curves produced by a "backstripping" procedure. From Sleep (1971), for a number of wells on the Gulf Coast of the United States. The dashed curve is a 50-m.y. exponential curve fitted to the average subsidence in the wells, corrected for the component of subsidence caused by the loading of the sediments. The depth is normalized to the base of the Woodbine Formation. The solid vertical bars are the 90% confidence limits on the parameter. Sleep (1971) interpreted the deviations from the exponential as due to eustatic sea-level changes.

(b) The first effort to compare the subsidence of an ancient passive margin with an exponential curve, from Stewart and Suczek (1977). The curve is for upper Proterozoic and lower Paleozoic strata in the White-Inyo Mountains of southeastern California. Although Stewart and Suczek did not correct for sediment loading and compaction, the subsidence appeared to fit that of modern passive margins remarkably well.

(c) Example of subsidence curve from Bond and Kominz (1984) for Cambrian and Ordovician strata in the ancient passive margin in the southern Canadian Rocky

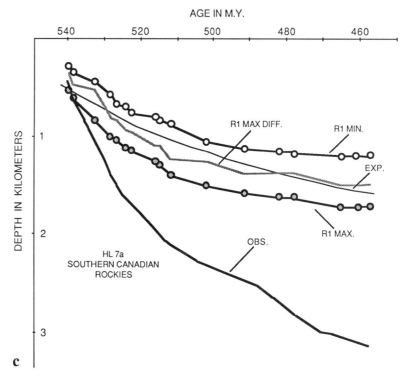

Mountains. OBS refers to the measured cumulative thickness curve. R1 is the reduced cumulative curve after correcting for sediment loading and compaction. R1 minimum and R1 maximum are limits for assumed extremes in the corrections for compaction. R1 maximum difference is the maximum difference between the two extremes. EXP is a best-fit exponential curve with a decay constant of 62.8 m.y. to the R1 maximum difference curve.

Curves such as those in b and c for lower Paleozoic strata in a number of localities around the world are taken as evidence that thermal contraction was a principal mechanism of subsidence in the passive-margin phases of geosynclinal evolution and that the early Paleozoic passive margins were initiated by breakup of a supercontinent near the end of the Proterozoic (Bond and others, 1984).

In a pioneering paper that has not received the recognition it deserves, Sleep (1971) quantitatively applied an early form of the ocean-floor thermal contraction model to passive margins. Following ideas originally developed by Vogt and Ostenso (1967), Sleep suggested that as continents split apart and the rifted edges move away from the ridge crests, the subsidence of heated and rifted lithosphere has the same age-depth curve as that of ocean floor. In his 1971 paper, Sleep developed this concept analytically for Atlantic-type continental margins, assuming that ocean floor subsides exponentially with a decay constant of 50 m.y., a value Sleep derived from ocean-floor age-depth data in an earlier paper (Sleep, 1969). He showed that subsidence curves from wells in the Atlantic and Gulf Coast margins fit the exponential form reasonably well after correcting for sediment loading, and even suggested that deviations from these curves were caused by eustatic sea-level changes during late Mesozoic and Cenozoic time (Fig. 11a). Although he was primarily concerned with the post-rift subsidence, he clearly recognized that if the pre-rift continental surface is elevated by an increase in temperature of the continental lithosphere, that surface will subside to its original elevation as the lithosphere cools. He concluded that in order to account for deposition of the thick shallow-marine sediments in mature passive margins, the crust must have been thinned by some means during rifting, and he considered erosion, extension, or a combination of both as viable mechanisms. In addition to developing one of the earliest analytical models for thermal contraction and subsidence in passive margins, Sleep laid the groundwork for the "backstripping" techniques of subsidence analysis that have been widely applied to modern and ancient passive margins and to many other types of basins as well. The thermal contraction model was not immediately accepted, partly because Sleep (1971) had overemphasized the role of erosion in crustal thinning. In addition, Drake and others (1968) considered that a proto-Atlantic Ocean had existed, so that no breakup was necessary, and Emery and others (1970) considered the breakup age of the Atlantic margin to be Permian and, therefore, too old for thermal contraction to account for the late Mesozoic and Cenozoic subsidence.

MODELS AND MORE NEW DATA (~1978–PRESENT)

Development of Geophysical Models for Passive Margins— A Return to the Problem of Vertical Motions

Between the late 1970s and early 1980s, a series of papers appeared that focussed attention once again on the mechanisms of vertical motions in passive margins. In contrast to the largely geologic approach in earlier

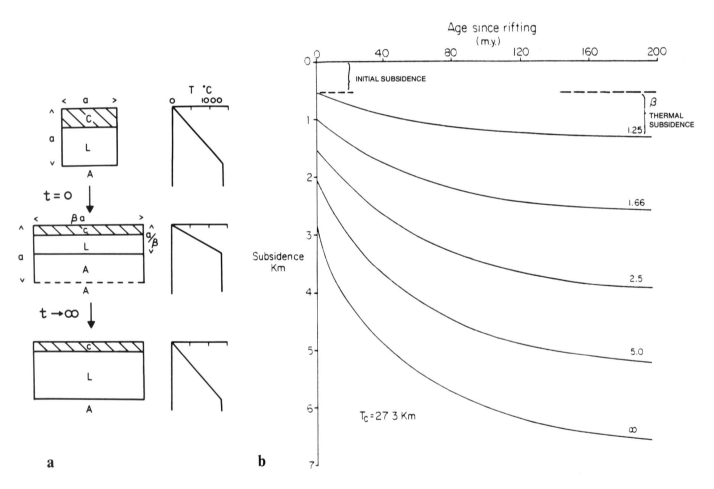

Figure 12. (a) The stages in the uniform and instantaneous stretching model as shown by McKenzie (1978). C is crust, L is lithosphere, and A is asthenosphere. The lengths a are unit dimensions. β is the amount of stretching. The crust and lithosphere are stretched, causing compression of the isotherms and passive heating of the lithosphere as shown in the cartoons on the right.

(b) Subsidence curves calculated from the stretching model (from Watts, 1981). Initial subsidence is the subsidence that occurs during rifting. T_c is the initial thickness of the crust.

studies, the problem was attacked in these papers with geophysical models of the processes assumed to control the rifting and post-rift phases of passive margins. One of the most influential of these models was published by Dan McKenzie in 1978. McKenzie (1978), following Sleep (1971) and Kinsman (1975), was concerned with the problem of how the crust is thinned during rifting. He rejected erosional thinning and phase changes as mechanisms and suggested instead that the problem of space for sediments and the problem of heating can be avoided by assuming that the entire lithosphere is stretched. The subsidence after rifting was constrained to have the same form as the average age-depth curve of ocean floor described earlier by Parsons and Sclater (1977). This became the highly popular instantaneous uniform-stretching model for passive margins (Fig. 12). It is interesting to recall that the idea of subsidence as a simple isostatic response to crustal thinning by extension was advocated much earlier by Wegener (1929) and Bucher (1933).

Shortly before the stretching model of McKenzie was published, the method for analyzing subsidence in wells originally described by Sleep (1971) was refined with the objective of further testing the mechanisms of subsidence in passive margins. Watts and Ryan (1976) informally termed the procedure "backstripping." They concluded from the forms of the subsidence curves that the post-rift subsidence was exponential as predicted by the thermal contraction models. A significant contribution was recognition that the form of subsidence corrected for sediment loading through time differed significantly depending on whether an Airy or regional compensation model was assumed for sediment and water loads. They suggested that an Airy model is a good approximation of the crustal response early in the subsidence when the lithosphere is hot and weak and that a flexural model is the best approximation of the later history when the margin has cooled and stiffened. This paper contains the earliest concept of a model combining time-dependent flexure and thermal contraction for subsidence of passive margins.

Later, Steckler and Watts (1978) modified the analytical method of subsidence to include corrections for sediment compaction and applied the procedure to the COST B-2 well off the coast of New York. They concluded that the total subsidence in the well was best explained by a simple thermal model of cooling of an initially thinned lithosphere and crust that was subsequently loaded by sediments. Watts and Steckler (1979) further developed and substantiated these ideas and also recognized evidence of Mesozoic and Cenozoic sea-level changes in the curves, as had been suggested earlier by Sleep (1971). By the late 1970s, there was widespread acceptance of a two-stage model for the tectonic or driving component of subsidence that consisted of crustal thinning during rifting and thermal contraction after rifting. There was much debate, however, about the precise mechanisms and magnitudes of crustal thinning.

Within a remarkably short period of time after publication of the stretching model by McKenzie in 1978, an overwhelming number of different geophysical models for the initiation and subsidence of passive margins appeared, many of them incorporating variations on the lithospheric stretching model and constrained by new data acquired from modern passive margins. These models include such diverse mechanisms as dike injection and underplating accompanying uniform stretching (as envisioned much earlier by Bucher); stretching by different amounts above and below an arbitrary reference level (non-uniform or depth-dependent extension), with or without dike injection and underplating; stretching over a finite length of time; and stretching incorporating both brittle and ductile processes (as suggested by Wegener). Good summaries of these models can be found in Watts (1981), Scrutton (1982), and Beaumont and others (1982, 1987).

In the early 1980s, the growing evidence that much of the Tertiary to Recent extension in the Great Basin of the United States occurred along low-angle normal or detachment faults (Wernicke, 1985) led to the development of detachment or simple-shear models for passive margins as alternatives to the pure-shear stretching models and their variations. In some detachment models, extension occurs by simple shear along a gently dipping detachment surface that is assumed to extend through the entire lithosphere (Wernicke, 1985), whereas in others, extension is assumed to occur by a combination of simple shear along detachments in the upper crust and pure shear in the lower crust and subcrustal lithosphere, following ideas published in an article in *Geology* by Lister and others (1986).

With the rapid development of high-speed and relatively inexpensive computers, most of the passive-margin models have been incorporated in elaborate two-dimensional and three-dimensional computer simulations that include the effects of lateral heat flow, time-dependent flexure, different assumptions about the rheology of the lithosphere, and different mechanisms of extension. These computer models have reproduced many of the first-order structural features of passive margins with a reasonable degree of success and have been used to predict heat flow, thermal maturation history of the sediments, and the development of sedimentary facies within the margin (for example, Beaumont and others, 1982; Steckler and Watts, 1982; Watts and Thorne, 1984). Their main shortcomings are in accounting for the increasing evidence of the variability and complexity of the rifting processes that can be observed directly in the upper crust and the limited direct evidence for the mechanisms of extension in the deep crust and subcrustal lithosphere. If these difficulties can be overcome, modeling offers the substantial advantage of quantitatively tracking the thermo-mechanical processes by which continental crust is modified and replaced by oceanic crust step by step through sophisticated computer simulations.

Application of Geophysical Models to Ancient Passive Margins— Refining the Passive-Margin Analogue

The successful identification of the thermal component of subsidence in modern passive margins using simple one-dimensional procedures encouraged attempts to apply similar methods to exposures of what were presumed to be ancient analogues of passive margins in the miogeosynclinal portions of orogenic belts. The first of these efforts was made by Stewart and Suczek in 1977. Following the ideas of Sleep (1971), they compared an exponential curve (50-m.y. decay constant) with a cumulative subsidence curve for lower Paleozoic strata in the White-Inyo Mountains of the western United States. Although they did not correct for sediment loading and compaction, the subsidence of the lower Paleozoic strata fit the exponential curve reasonably well (Fig. 11b). Moreover, they noted that the exponential curve implied ages of 600 to 650 m.y. for initiation of sedimentation in the margin. They suggested that the ages of between 800 and 900 m.y. previously assumed for rifting might reflect early rifting phases predating the main rifting event that initiated the early Paleozoic passive margin. This paper marked the first effort to use the thermal form of post-rift subsidence to estimate the ages of breakup in an ancient passive margin.

Subsequently, the analytical methods for analyzing subsidence were refined to include corrections for sediment loading and lithification of the strata (Kominz and Bond, 1982) and applied to several exposed sections of the lower Paleozoic passive-margin strata in the Great Basin of the United States and in the fold-and-thrust belt of the southern Canadian Rockies. The results of these analyses were published in two outlets of the Geological Society of America, by Armin and Mayer (1983) in *Geology* and by Bond and Kominz (1984) in the *Bulletin*, and elsewhere (Bond and Kominz, 1982; Bond and others, 1983). The results confirmed the exponential form of the post-rift subsidence along more than 2,000 km of the early Paleozoic passive margin (for example, Fig. 11c). Bond and others (1983) and Bond and Kominz (1984) further showed that the breakup age could be inferred from the slopes of the post-rift subsidence curves (an approximation of the rate of cooling) without projecting the best-fit exponential curves into older, poorly dated strata or making assumptions about the specific rifting process. The breakup ages for strata from the Canadian Rockies to the southern Great Basin were found to lie between 600 and 555 m.y., confirming the results of Stewart and Suczek (1977) and leading to new efforts to identify the complex rifting events that initiated this ancient margin. The subsidence methods were later applied to lower Paleozoic passive-margin strata around the world (Bond and others, 1984). The results suggested that a large number of early Paleozoic passive margins were initiated between 625 and 555 Ma, further documenting the existence of a controversial late Proterozoic supercontinent and refining its breakup history.

Apart from constraining rifting and breakup histories in orogenic belts, the subsidence analyses have provided a quantitative basis for identifying parts of the thick stratigraphic successions in geosynclines that correspond to plate divergence and growth of a passive margin. For example, in an article published recently in the *Bulletin*, Marshak (1986) analyzed a stratigraphic section not far from the location of Kay's cross section, shown in Figure 9, and found that the passive-margin phase is characterized by a smoothly decaying, concave-upward curve, whereas the destruction of the passive margin and subsequent tectonic events tend to be marked by abrupt deflections and convex-upward curves (Fig. 13).

A Quantum Jump in Data Acquisition

At the same time that geophysical models for passive margins were being developed, the quality of subsurface data acquired from passive margins substantially improved as a result of major technological developments, including large energy sources for multichannel profiling, digital streamers with as many as 240 channels, and new experimental designs such as ESP (expanding spread profiles), for refining the velocities at great depths, and multiship wide-aperture CDP (common depth point) profiling, for resolving structures at great depths. One of the most exciting results

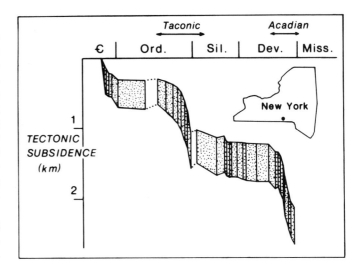

Figure 13. Subsidence curve from Marshak (1986), corrected for sediment loading and compaction from the inner part of the miogeosyncline near the cross section of Kay shown in Figure 9. The Cambrian to Early Ordovician portion of the curve corresponds to the cooling and subsidence of the ancient passive margin, and the rapid accelerations in the Late Ordovician and Middle Devonian correspond to subsequent phases of convergence during the Taconic and Acadian orogenies.

of the new seismic studies has been the identification of deep interfaces beneath passive margins in the lower continental crust, at the Moho, and in the mantle. For example, in the Grand Banks region off eastern Canada (Keen and de Voogd, 1988) and around the Exmouth Plateau (J. C. Mutter, 1988, personal commun.), upper-crustal extensional faults appear to be truncated along mid-crustal shear zones or detachment faults. These data are providing support for detachment mechanisms of extension in passive margins. Deep faults, possibly related to Caledonide orogenesis,

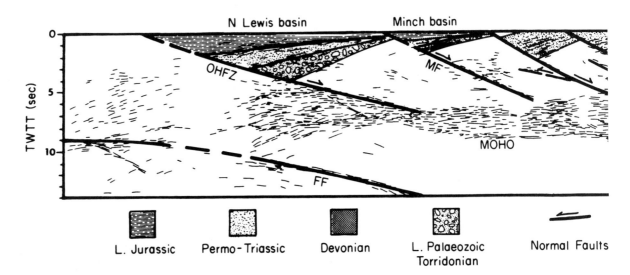

Figure 14. A deep-reflection profile acquired by the BIRPS group from north of Scotland (from Cheadle and others, 1987). FF, OHFZ, and MF identify the Flannan fault, the Outer Hebrides fault, and the Minch fault, respectively. The profile clearly reveals reflection in the lower crust and in the subcrustal lithosphere. Many of the faults, such as the OHFZ, have been mapped as Caledonian thrust faults at the surface and are recognized in the subsurface as reactivated extensional faults above which lie rift basins with tilted sediments, such as in the N. Lewis basin. The vertical to horizontal scale is 1:1 for 5 km/s velocities.

have been observed cutting through the entire continental crust and extending well into the mantle off the British Isles (McGeary and Warner, 1985; Smythe and others, 1982). Some of these are inferred to have been reactivated during post-Paleozoic extension (Cheadle and others, 1987). Reactivation of Paleozoic thrust faults as Mesozoic extensional faults also has been inferred for parts of the passive margin off the east coast of the United States (Hutchinson and others, 1985, in *Geology*; Ratcliffe and others, 1986, in *Geology*). It is becoming apparent from these studies that the lower crust typically is strongly reflective and that deformation of the subcrustal lithosphere is much more extensive than had been thought previously (Fig. 14).

In addition, deep-reflection profiling across the continent-ocean boundary, especially in the Grand Banks area of Canada off the east coast of the United States and in the Greenland-Norwegian conjugate margins, is revealing evidence that significant amounts of mafic igneous material have intruded or underplated the rifted and thinned lower continental crust adjacent to the boundary. In the conjugate margin of northern East Greenland-Norway and southern Greenland-Rockall Plateau, as much as 20 km of mafic igneous material occurs at the ocean-continent boundary, leading to the suggestion that in passive margins with large amounts of volcanic material, a convective partial-melting process may have accompanied the onset of drift (Mutter and others, 1988). Mutter and others (1982) have suggested that the outer highs of some passive margins may be thick accumulations of igneous material at the continent-ocean boundary, such as those seen in the North Atlantic conjugate margins. The thick volcanic complexes were drilled in various JOIDES legs in the North Atlantic (Roberts and others, 1984; Eldholm and others, 1986).

In an article published in *Geology*, the Large Aperture Seismic Experiment (LASE) group identified a deep crustal layer with a nearly uniform p-wave velocity of 7.2 km/s and about 6 to 9 km thick lying beneath nearly all of the coast off New Jersey, about 250 km north of the first refraction lines shot across a passive margin (LASE Study Group, 1986; Fig. 4b). This is presumably part of the same crustal layer with anomalous velocity of 7.2–7.7 km/s that had been recognized two decades earlier by Drake and his co-workers (Drake and Nafe, 1968) beneath the continental rise off eastern North America. A significant interpretation of the LASE experiment is that this layer is continuous from the shelf to the deep ocean basin, implying a continuous progression between plutonic emplacement into thinned lower continental crust and plutonic construction of the lower oceanic crust. The data further underscore the possibility of underplating and injection of mafic rock into thinned continental crust during rifting. Deeper commercial and JOIDES drilling on passive margins has contributed significantly to these seismic advances. Drilling has aided identification of carbonate-bank complexes along their outer edges (Mattick and Libby-French, 1988), penetrated some of the volcanic complexes at the ocean-continent transition (Roberts and others, 1984), and sampled the early-rift–stage sediments and volcanic rocks in a sediment-starved margin (de Graciansky and others, 1985).

The development of high-resolution multichannel seismic techniques has also led to major advances in understanding the stratigraphic framework in continental margins, leading to what many regard as a revolution in stratigraphy. This has been accomplished using a new methodology, termed "seismic stratigraphy," that was developed mainly by geologists from the Exxon Production Research Company (Vail and others, 1977). The underlying principle of this methodology is that primary seismic reflectors in sedimentary strata are caused by physical contrasts across bedding and unconformities. Therefore, rocks above continuous reflectors are assumed to be everywhere younger than those below the reflectors. The resulting seismic section is interpreted to be a record of chronostratigraphic depositional and structural patterns rather than a record of the time-transgressive lithostratigraphy. The development of the seismic stratigraphic method has led to a new set of principles of stratigraphic analysis and correlation, based on predominantly physical criteria and an elaborate terminology that has been applied mainly to seismic sections, many of which come from the thick strata that can be observed with reflection profiling in passive margins (for example, Fig. 3c). One of the most intriguing and controversial results of seismic stratigraphy is a global sea-level curve that is thought to contain evidence of multiple orders of sea-level change ranging from more than 100 m.y. to 1 to 2 m.y. (Vail and others, 1977; Haq and others, 1987).

SUMMARY AND CONCLUSIONS

Much of the modern thinking about passive margins has origins in ideas about rifts, geosynclines, and continental drift that were formulated decades ago, mostly from studies within the continents. These ideas were steadily strengthened and modified as progressively larger and better data sets were gathered from modern ocean basins after World War II, and most have become widely accepted only relatively recently. This is perhaps best illustrated by the reluctance to accept the idea of passive margins as products of continental drift, which had been proposed in the early 1900s, until compelling evidence in support of sea-floor spreading appeared in the mid-1970s. A similar history can be traced for other ideas such as extensional faulting, crustal thinning, magmatic underplating, and thermal contraction and flexure, most of which originated before the mid-1930s and were gradually accepted much later as the quality and quantity of refraction and especially deep-reflection data from ocean basins steadily increased.

A number of current lines of research seem to hold the promise of major advances and even substantial revisions in concepts of passive margins in the future. Studies of the remarkable deep interfaces and detachments in the lower crust and upper mantle are suggesting that pre-existing thrust faults are reactivated as major extensional faults during rifting and continental breakup. These ideas may lead to explanations of why the breakup of continents typically (but not always) occurs near or on the site

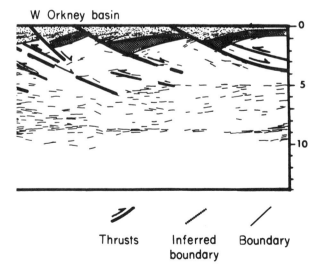

Figure 14. (*Continued*).

of a pre-existing compressional orogen. The growing evidence for large volumes of mafic plutonic rocks beneath passive margins is raising the possibility that magmatic underplating may play a major role in modifying the composition of continental crust during extension (for example, Dewey, 1986). Deeper-penetrating vertical seismic reflection profiling across modern passive margins is needed to document the magmatic underplating and detachment hypotheses.

There is also growing interest in understanding how a passive margin is converted to a convergent margin, and ideas concerning this issue incorporate the steadily increasing observational data on deep crustal and subcrustal structures in both extensional and compressional regimes (Dewey, 1982; Cloetingh and others, 1982). Recently, sophisticated computer simulations, building on the existing models of both passive and convergent margins, have been developed to quantitatively address questions of how these processes operate on a lithospheric scale (for example, Stockmal and others, 1986). Finally, as the structure, stratigraphy, and subsidence history of modern passive margins are becoming better understood, there is increasing interest in refining criteria for identifying their ancient analogues. In particular, efforts are underway to search the Proterozoic record for ancient passive margins in order to compare their evolution with that of modern margins and to better constrain the existence and breakup history of supercontinents over the past 2 to 2.5 b.y. of Earth history (Hoffman, 1980, and Grotzinger, 1986, published in the *Bulletin*).

ACKNOWLEDGMENTS

We thank John Mutter and John Diebold for helpful review and criticisms of the manuscript. Preparation of this paper was supported in part by National Scientific Foundation Grants EAR 85-18644 and EAR 84-17439 to G. Bond.

REFERENCES CITED

Armin, R. A., and Mayer, L., 1983, Subsidence analysis of the Cordilleran miogeocline: Implications for timing of Late Proterozoic rifting and amount of extension: Geology, v. 11, p. 702–705.

Artemjev, M. E., and Artyushkov, E. V., 1971, Structure and isostasy of the Biakal Rift and the mechanism of rifting: Journal of Geophysical Research, v. 76, p. 1197–1211.

Bally, A. W., chairman, 1979, Continental margins, geological and geophysical research needs and problems: Washington, D.C., National Academy of Sciences, National Research Council Report, 302 p.

Barrell, J., 1912, Criteria for the recognition of ancient delta deposits: Geological Society of America Bulletin, v. 23, p. 377–446.

——— 1914, The strength of the earth's crust—Part II. Regional distribution of isostatic compensation: Journal of Geology, v. 22, p. 145–165.

——— 1927, On continental fragmentation and the geologic bearing of the moon's surficial features: American Journal of Science, 5th series, v. 13, p. 283–314.

Beaumont, C., Keen, C. E., and Boutillier, R., 1982, On the evolution of rifted continental margins: Comparison of models and observations for the Nova Scotian margin: Royal Astronomical Society Geophysical Journal, v. 70, p. 667–715.

Beckmann, W. C., Roberts, A. C., and Luskin, B., 1959, Sub-bottom depth recorder: Geophysics, v. 24, p. 749–760.

Bird, J. M., and Dewey, J. F., 1970, Lithosphere plate–continental margin tectonics and the evolution of the Appalachian orogen: Geological Society of America Bulletin, v. 81, p. 1031–1059.

Bond, G. C., and Kominz, M. A., 1982, Evidence of thermal subsidence and sea level change for the lower Paleozoic stratigraphic succession in the southern Canadian Rocky Mountains: Geological Society of America Abstracts with Programs, v. 14, p. 447.

——— 1984, Construction of tectonic subsidence curves for the early Paleozoic miogeocline, southern Canadian Rocky Mountains: Implications for subsidence mechanisms, age of breakup, and crustal thinning: Geological Society of America Bulletin, v. 95, p. 155–173.

Bond, G. C., Kominz, M. A., and Devlin, W. J., 1983, Thermal subsidence and eustasy in the lower Paleozoic miogeocline of western North America: Nature, v. 316, p. 742–745.

Bond, G. C., Nickeson, P. A., and Kominz, M. A., 1984, Breakup of a supercontinent between 625 Ma and 555 Ma: New evidence and implications for continental histories: Earth and Planetary Science Letters, v. 70, p. 325–345.

Bond, G. C., Christie-Blick, N., Kominz, M. A. and Devlin, W. J., 1985, An early Cambrian rift to post-rift transition in the Cordillera of western North America: Nature, v. 316, p. 742–745.

Bott, M.H.P., 1981, Stress based tectonic mechanisms at passive continental margins, *in* Scrutton, R. A., ed., Dynamics of passive margins: American Geophysical Union Geodynamics Series, v. 6, p. 147–153.

Bott, M.H.P., and Dean, D. S., 1972, Stress systems at young continental margins: Nature, v. 235, p. 23–25.

Bourcart, J., 1938, La Marge continentale: Bull. Soc. Geol. France, v. 8.

Bowie, W., 1921, The relation of isostasy to uplift and subsidence: American Journal of Science, v. 2.

——— 1922, Theory of isostasy—A geological problem: Geological Society of America Bulletin, v. 33, p. 273–286.

——— 1927, Isostasy: New York, E. P. Dutton and Company, 275 p.

Bucher, W. H., 1924, The pattern of the earth's mobile belts: Journal of Geology, v. 32, p. 265–290.

——— 1933, Deformation of the earth's crust: New York, Hafner, 518 p.

Bullard, E. C., 1936, Gravity measurements in East Africa: Royal Society Philosophical Transactions, v. 235 A, p. 445–531.

Bullard, E. C., and Gaskill, T. F., 1941, Submarine seismic investigations, *in* Royal Society Proceedings, Series A, v. 177, p. 476–499.

Burchfiel, B. C., and Davis, G. A., 1972, Structural framework and evolution of the southern part of the Cordilleran orogen, western United States: American Journal of Science, v. 272, p. 97–118.

Carey, S. W., 1958, A tectonic approach to continental drift, *in* Carey, S. W., convener, Continental drift—A symposium: Hobart, Tasmania, University of Tasmania Geology Department, p. 177–355.

Chamberlin, T. C., 1898, The ulterior basis of time divisions and the classification of geologic history: Journal of Geology, v. 6, p. 449–462.

——— 1913, Diastrophism and the formative process: Part III: Journal of Geology, v. 21, p. 577–587.

Cheadle, M. J., McGeary, S., Warner, M. R., and Matthews, D. H., 1987, Extensional structures on the western UK continental shelf: A review of evidence from deep seismic profiling, *in* Coward, M. P., Dewey, J. F., and Hancock, P. L., eds., Continental extensional tectonics: Geological Society of London Special Publication 28, p. 445–465.

Churkin, M., Jr., 1974, Paleozoic marginal ocean basin–volcanic arc systems in the Cordilleran foldbelt, *in* Dott, R. H., Jr., and Shaver, R. H., Modern and ancient geosynclinal sedimentation: Society of Economic Paleontologists and Mineralogists Special Publication 19, p. 174–192.

Cloetingh, S.A.P.L., Wortel, M.J.R., and Vlaar, N. J., 1982, Evolution of passive continental margins and initiation of subduction zones: Nature, v. 297, p. 139–142.

Coney, P. J., 1970, The geotectonic cycle and the new global tectonics: Geological Society of America Bulletin, v. 81, p. 739–747.

Cotton, C. A., 1918, Conditions of deposition on the continental shelf and slope: Journal of Geology, v. 24, p. 135–160.

Dahl, W. H., 1925, Tertiary fossils dredged off the northeastern coast of North America: American Journal of Science, ser. 5, v. 10, p. 215.

Daly, R. A., 1926, Our mobile earth: New York, Scribner's, 342 p.

——— 1940, Strength and structure of the earth: New York, Hafner Publishing Company, 434 p.

Dana, J. D., 1873, On some results of the earth's contraction from cooling, including a discussion of the origin of mountains and the nature of the earth's interior: American Journal of Science, ser. 3, v. 5, p. 423–443; v. 6, p. 6–14, 104–115, 161–171.

Davis, E. E., and Lister, C.R.B., 1974, Fundamentals of ridge crest topography: Earth and Planetary Science Letters, v. 21, p. 405–413.

de Charpal, O., Guennoc, P., Montadert, L., and Roberts, D. G., 1978, Rifting, crustal attenuation and subsidence in the Bay of Biscay: Nature, v. 275, p. 706–711.

de Graciansky, P. C., Poag, W. C., and OPD Leg 81 scientists, 1985, The Goban Spur transect: Geologic evolution of a sediment-starved passive continental margin: Geological Society of America Bulletin, v. 96, p. 58–76.

Dennis, J. G., 1982, Orogeny (Benchmark Papers in Geology): Stroudsburg, Pennsylvania, Hutchinson Ross Publishing Company, 411 p.

Dewey, J. F., 1969, Evolution of the Appalachian/Caledonian orogen: Nature, v. 222, p. 124–129.

——— 1982, Plate tectonics and the evolution of the British Isles: Geological Society of London Journal, v. 139, p. 371–412.

——— 1986, Diversity in the lower continental crust, *in* Dawson, J. B., and others, eds., The nature of the lower continental crust: Geological Society Special Publication 24 (Blackwell Scientific Publications, Oxford, p. 71–78.

Dewey, J. F., and Bird, J. M., 1970, Plate tectonics and geosynclines: Tectonophysics, v. 10, p. 625–638.

Dewey, J. F., and Horsfield, B., 1970, Plate tectonics, orogeny and continental growth: Nature, v. 225, p. 521–525.

Dickinson, W. R., 1971, Plate tectonic models of geosynclines: Earth and Planetary Science Letters, v. 10, p. 165–174.

Dietz, R. S., 1963, Collapsing continental rises: An actualistic concept of geosynclines and mountain building: Journal of Geology, v. 71, p. 314–333.

Dott, R. H., Jr., 1964, Mobile belts, sedimentation and orogenesis: New York Academy of Sciences Transactions, v. 27, p. 135–143.

——— 1974, The geosynclinal concept, *in* Dott, R. H., Jr., and Shaver, R. H., Modern and ancient geosynclinal sedimentation: Society of Economic Paleontologists and Mineralogists Special Publication 19, p. 1–13.

——— 1978, Tectonics and sedimentation a century later: Earth Science Reviews, v. 14, p. 1–34.

——— 1979, The geosyncline—A first major geosynclinal concept made in America (Two Hundred Years of Geology in America): Hanover, New Hampshire, University Press of New England, p. 239–264.

Drake, C. L., and Kosminskaya, I. P., 1969, The transition from continental to oceanic crust: Tectonophysics, v. 7, p. 363–384.

Drake, C. L., and Nafe, J. E., 1968, The transition from ocean to continent from seismic refraction data, *in* Knopoff, L., and others, eds., The crust and upper mantle of the Pacific area: Geophysical Monograph 12, p. 174–186.

Drake, C. L., and Woodward, H. P., 1963, Appalachian curvature, wrench faulting and offshore structures: New York Academy of Science Transactions, v. 26, p. 48–63.

Drake, C. L., Ewing, M., and Sutton, G. H., 1959, Continental margins and geosynclines: The east coast of North America north of Cape Hatteras: Physics and Chemistry of the Earth, v. 3, p. 110–198.

Drake, C. L., Heirtzler, J., and Hirshman, J., 1963, Magnetic anomalies off eastern North America: Journal of Geophysical Research, v. 68, p. 5259–5275.

Drake, C. L., Ewing, J. I., and Stockard, H., 1968, The continental margin of the eastern United States: Canadian Journal of Earth Sciences, v. 5, p. 993–1010.

DuToit, A. L., 1937, Our wandering continents: Westport, Connecticut, Greenwood Press, 366 p.

Dutton, C. E., 1889, On some of the greater problems of physical geology: Washington Philosophical Society Bulletin, ser. B, v. 11, p. 51–64.

Eldholm, O., and ODP Leg 104 scientists, 1986, Formation of the Norwegian Sea: Nature, v. 319, p. 360–361.

Emery, K. O., Uchupi, E., Phillips, J. D., Bowin, C. O., Bunce, E. T., and Knott, S. T., 1970, Continental rise of eastern North America: American Association of Petroleum Geologists Bulletin, v. 54, p. 44–108.

Ewing, M., and Tirey, G. B., 1961, Seismic profiler: Journal of Geophysical Research, v. 66, p. 2917–2927.

Ewing, M., Crary, A. P., and Rutherford, H. M., 1937, Geophysical investigations in the emerged and submerged Atlantic coastal plain, Part 1. Methods and results: Geological Society of America Bulletin, v. 48, p. 753–802.

Ewing, M., Woollard, G. P., and Vine, A. C., 1939, Geophysical investigations in the emerged and submerged Atlantic coastal plain, Part III, Barnegat Bay, N.J., section: Geological Society of America Bulletin, v. 50, p. 257–296.

Fairhead, J. D., and Girdler, R. W., 1971, The seismicity of Africa: Royal Astronomical Society Geophysical Journal, v. 24, p. 271–301.

Falvey, D. A., 1974, The development of continental margins in plate tectonic theory: Australian Petroleum Exploration Association Journal, v. 14, p. 95–106.

Fenwick, D.K.B., Keen, M. J., Keen, C., and Lambert, A., 1968, Geophysical studies of the continental margin northeast of Newfoundland: Canadian Journal of Earth Sciences, v. 5, p. 483–500.

Gabrielse, H., 1972, Younger Precambrian of the Canadian Cordillera: American Journal of Science, v. 272, p. 521–536.

Girdler, R. W., 1964, Geophysical studies of rift valleys: Physics and Chemistry of the Earth, v. 5, p. 121–156.

Grabau, A. W., 1936, Paleozoic formations in the light of the pulsation theory: National University of Peking University Press, 680 p.

Grotzinger, J. P., 1986, Cyclicity and paleoenvironmental dynamics, Rocknest Formation, Wopmay orogen, northwest Canada: Geological Society of America Bulletin, v. 97, p. 1208–1231.

Grow, J. A., Klitgord, K. D., and Schlee, J. S., 1988, Structure and evolution of the Baltimore Canyon Trough, *in* Sheridan, R. E., and Grow, J. A., eds., The Atlantic continental margin: U.S.: Boulder, Colorado, Geological Society of America, The Geology of North America, v. 1-2, p. 269–290.

Gulliver, F. P., 1899, Shoreline topography: American Academy of Arts and Sciences Proceedings, v. 34, 176 p.

Gunn, R., 1944, A quantitative study of the lithosphere and gravity anomalies along the Atlantic Coast: Franklin Institute Journal, v. 237, p. 139–154.

Hall, J., 1859, Description and figures of the organic remains of the Lower Helderberg Group and the Oriskany Sandstone, *in* Natural history of New York (Volume 3): New York, van Benthusen, 532 p.

Haq, B. U., Hardenbol, J., and Vail, P. R., 1987, Chronology of fluctuating sea levels since the Triassic: Science, v. 235, p. 1156–1167.

Hatcher, R. D., 1972, Developmental model for the southern Appalachians: Geological Society of America Bulletin, v. 83, p. 2735–2760.
Hedberg, H. D., 1970, Continental margins from viewpoint of the petroleum geologist: American Association of Petroleum Geologists Bulletin, v. 54, p. 3–43.
Heestand, R. L., and Crough, S. T., 1981, The effect of hot spots on the oceanic age-depth relation: Journal of Geophysical Research, v. 86, p. 6107–6114.
Heezen, B., 1959, Paleomagnetism, continental displacements and the origin of submarine topography, *in* Sears, M., ed., Preprints, 1st International Oceanographic Congress: Washington, D.C., American Association for the Advancement of Science, p. 26–28.
Heezen, B. C., and Sheridan, R. E., 1966, Lower Cretaceous rocks (Neocomian-Albian) dredged from Blake Escarpment: Science, v. 154, p. 1644–1647.
Helmert, F. R., 1909, Die Tiefe Der Ausgleichflache Bei Des Prattschen Hypothese Fur Das Gleicagewicht Der Erdkruste Und Der Verlauf Der Schwerestorung vom Innern Der Kontinente Und Ozeane Nach Den Kusten: Sitzber. Deut. Kgl. Preusz. Akad. Wiss., v. 18, p. 1192–1198.
Hersey, J. B., Bunce, E. T., Wyrick, R. F., and Dietz, F. T., 1959, Geophysical investigations of the continental margin between Cape Henry, Virginia and Jacksonville, Florida: Geological Society of America Bulletin, v. 70, p. 437–465.
Hess, H. H., 1962, History of ocean basins, *in* Petrologic studies: A volume to honor A. F. Buddington: New York, Geological Society of America, p. 599–620.
Hoffman, P. F., 1980, Wopmay orogen: A Wilson cycle of Early Proterozoic age in the northwest of the Canadian Shield, *in* Strangeway, D. W., ed., The continental crust and its mineral deposits: Geological Association of Canada Special Paper 20, p. 523–549.
Holmes, A., 1926, Contributions to the theory of magmatic cycles: Geological Magazine, v. 306–329.
——— 1930, Radioactivity and earth movements: Geological Society of Glasgow Transactions (1928–1929), v. 18, p. 559–606.
Hsü, K. J., 1973, The odyssey of geosyncline: Evolving concepts in sedimentology: Baltimore, Maryland, Johns Hopkins University Press, p. 66–92.
Hutchinson, D. R., Grow, J. A., Klitgord, K. D., and Swift, B. A., 1982, Deep structure and evolution of the Carolina Trough, *in* Watkins, J. S., and Drake, C. L., eds., Studies in continental margin geology: American Association of Petroleum Geologists Memoir 34, p. 129–152.
Hutchinson, D. R., Klitgord, K. D., and Detrick, R. S., 1985, Block Island fault: A Paleozoic crustal boundary on the Long Island platform: Geology, v. 13, p. 875–879.
JOIDES, 1965, Ocean drilling on the continental margin: Science, v. 150, p. 709–716.
Kay, M., 1951, North American geosynclines: Geological Society of America Memoir 48, 143 p.
Keen, C. E., and de Voogd, B., 1988, The continent-ocean boundary at the rifted margin off eastern Canada: New results from deep seismic reflection studies: Tectonics, v. 7, p. 107–124.
Keen, M. J., 1969, Possible edge effect to explain magnetic anomalies off the eastern seaboard of the United States: Nature, v. 222, p. 72–74.
Keen, M. J., and Loncarevic, B. D., eds., 1968, Third Symposium on Continental Margins and Island Arcs: Canadian Journal of Earth Sciences, v. 5, p. 963–1125.
Kinsman, D.J.J., 1975, Rift valley basin and sedimentary history of trailing continental margins, *in* Fischer, A.G., and Judson, S., eds., Petroleum and global tectonics: Princeton, New Jersey, Princeton University Press, p. 83–126.
Kominz, M. A., and Bond, G. C., 1982, Tectonic subsidence calculated from lithified basin strata: Geological Society of America Abstracts with Programs, v. 14, p. 534.
Langseth, M., LePichon, X., and Ewing, M., 1966, Crustal structure of the mid-ocean ridges, 5, Heat flow through the Atlantic Ocean floor and convection currents: Journal of Geophysical Research, v. 71, p. 5321–5355.
LASE Study Group, 1986, Deep structure of the US East Coast passive margin from large aperture seismic experiments (LASE): Marine and Petroleum Geology, v. 3, p. 234–242.
LeConte, J., 1872, On the formation of the features of the earth's surface: American Journal of Science, ser. 3, v. 4, p. 345–355.
Lister, G. S., Etheridge, M. A., and Symonds, P. A., 1986, Detachment faulting and the evolution of passive continental margins: Geology, v. 14, p. 246–250.
Marshak, S., 1986, Structure and tectonics of the Hudson Valley fold-thrust belt, eastern New York State: Geological Society of America Bulletin, v. 97, p. 354–368.
Mattick, R. E., and Libby-French, J., 1988, Petroleum geology of the United States Atlantic continental margin, *in* Sheridan, R. E., and Grow, J. A., eds., The Atlantic continental margin: U.S.: Boulder, Colorado, Geological Society of America, The Geology of North America, v. 1–2, p. 445–462.
Maxwell, A. R., ed., 1970, New concepts of sea floor evolution, *in* The sea (Volume 4, Parts 1 and 2): New York, Wiley-Interscience, pt. 1, 791 p.; pt. 2, 664 p.
McConnell, R. K., Jr., Gupta, R. N., and Wilson, J. T., 1966, Compilation of deep crustal seismic profiles: Reviews of Geophysics, v. 4, p. 41–100.
McGeary, S., and Warner, M. R., 1985, Seismic profiling of the continental lithosphere: Nature, v. 317, p. 795–797.
McKenzie, D. P., 1967, Some remarks on heat flow and gravity anomalies: Journal of Geophysical Research, v. 72, p. 6261–6271.
——— 1978, Some remarks on the development of sedimentary basins: Earth and Planetary Science Letters, v. 40, p. 25–32.
McKenzie, D. P., and Sclater, J. G., 1969, Heat flow in the eastern Pacific and sea floor spreading: Bulletin Volcanologique, v. 33-1, p. 101–118.
Mitchell, A. H., and Reading, H. G., 1969, Continental margins, geosynclines and ocean floor spreading: Journal of Geology, v. 77, p. 629–646.
Monger, J.W.H., and Price, R. A., 1979, Geodynamic evolution of the Canadian Cordillera—Progress and problems: Canadian Journal of Earth Sciences, v. 16, p. 770–791.
Montadert, L., Roberts, D. G., Auffret, G., Bock, W., DuPeuble, P. A., Hailwood, E. A., Harrison, W., Kagami, H., Lumsden, D. N., Muller, C., Schnitker, D., Thompson, R. W., Thompson, T. L., and Timofeev, P. P., 1977, Rifting and subsidence on passive continental margins in the North East Atlantic: Nature, v. 268, p. 305–309.
Mutter, J. C., Talwani, M., and Stoffa, P. L., 1982, Origin of seaward dipping reflectors in oceanic crust off the Norwegian margins by "subaerial sea-floor spreading": Geology, v. 10, p. 353–357.
Mutter, J. C., Buck, W. R., and Zehnder, C. M., 1988, Convective partial melting. 1. A model for the formation of thick basaltic sequences during the initiation of spreading: Journal of Geophysical Research, v. 93, p. 1031–1048.
Neugebauer, H. J., and Spohn, T., 1982, Metastable phase transitions and progressive decline of gravitational energy: Aspects of Atlantic type margin dynamics, *in* Scrutton, R. A., ed., Dynamics of passive margins: American Geophysical Union Geodynamics Series, v. 6, p. 166–183.
Parsons, B., and Sclater, J. G., 1977, An analysis of the variation of ocean floor bathymetry and heat flow with age: Journal of Geophysical Research, v. 32, p. 803–827.
Pearce, J. A., and Cann, J. R., 1973, Tectonic setting of basic volcanic rocks determined using trace element analyses: Earth and Planetary Science Letters, v. 19, p. 290–300.
Ratcliffe, N. M., Burton, W. C., Burton, W. C., D'Angelo, R. M., and Costain, J. K., 1986, Low-angle extensional faulting, reactivated mylonites, and seismic reflection geometry of the Newark basin margin in eastern Pennsylvania: Geology, v. 14, p. 766–770.
Reyer, E., 1888, Theoretische Geologie: Stuttgart, Germany.
Roberts, D. G., Morton, A. C., and Backman, J., 1984, Late Paleocene–early Eocene volcanic events in the northern North Atlantic Ocean, *in* Roberts, D. G., and Schnitker, D., eds., Initial reports of the Deep Sea Drilling Project: Washington, D.C., U.S. Government Printing Office, v. 81, p. 913–923.
Rodgers, J., 1968, The eastern edge of the North American continent during the Cambrian and Early Ordovician, *in* Zen, E-An, and others, eds., Studies of Appalachian geology, northern and maritime: New York, Interscience Publishers, p. 141–151.
Rona, P. A., 1974, Subsidence of Atlantic continental margins: Tectonophysics, v. 22, p. 283–299.
Schneider, E. D., 1969, The deep sea—A habitat for petroleum: Undersea Technology, v. 10, p. 32–57.
Schwab, F. L., 1982, Geosyncline—Concept and place within plate tectonics (Benchmark Papers in Geology 64: Stroudsburg, Pennsylvania, Hutchinson Ross Publishing Company, 411 p.
Sclater, J. G., and Detrick, R., 1973, Elevation of mid-ocean ridges and the basement age of JOIDES deep sea drilling sites: Geological Society of America Bulletin, v. 84, p. 1547–1554.
Sclater, J. G., and Francheteau, J., 1970, The implications of terrestrial heat flow observations on current tectonic and geochemical models of the crust and upper mantle of the earth: Royal Astronomical Society Geophysical Journal, v. 20, p. 509–542.
Scrutton, R. A., 1982, ed., Dynamics of passive margins: American Geophysical Union Geodynamics Series, v. 6, 200 p.
Shea, J. H., 1985, Plate tectonics (Benchmark Papers in Geology 89): Stroudsburg, Pennsylvania, Hutchinson Ross Publishing Company, 411 p.
Sheridan, R. E., 1976, Sedimentary basins of the Atlantic margin of North America: Tectonophysics, v. 36, p. 113–132.
Sheridan, R. E., Houtz, R. E., Drake, C. L., and Ewing, M., 1969, Structure of continental margin off Sierra Leone, West Africa: Journal of Geophysical Research, v. 74, p. 2512–2530.
Sleep, N. H., 1969, Sensitivity of heat flow and gravity to the mechanism of sea floor spreading: Journal of Geophysical Research, v. 72, p. 542–549.
——— 1971, Thermal effects of the formation of Atlantic continental margins by continental breakup: Royal Astronomical Society Geophysical Journal, v. 24, p. 325–350.
Smythe, D. K., Dobinson, A., McQuillin, R., Brewer, J. A., Matthews, D. H., Blundell, D. J., and Kelk, B., 1982, Deep structure of the Scottish Caledonides revealed by the MOIST reflection profile: Nature, v. 299, p. 338–340.
Steckler, M. S., 1981, The thermal and mechanical evolution of Atlantic-type continental margins [Ph.D. thesis]: Palisades, New York, Columbia University, 261 p.
Steckler, M. S., and Watts, A. B., 1978, Subsidence of the Atlantic-type continental margin off New York: Earth and Planetary Science Letters, v. 41, p. 1–13.
——— 1982, Subsidence history and tectonic evolution of Atlantic-type continental margins, *in* Scrutton, R. A., ed., Dynamics of passive margins: American Geophysical Union Geodynamics Series, v. 6, p. 184–196.
Stetson, H. C., 1936, Geology and paleontology of Georges Bank canyons: Geological Society of America Bulletin, v. 47, p. 339–366.
Stewart, J. H., 1972, Initial deposits in the Cordilleran geosyncline: Evidence of a late Precambrian (<850 m.y.) continental separation: Geological Society of America Bulletin, v. 83, p. 1345–1360.
——— 1976, Late Precambrian evolution of North America: Plate tectonics implication: Geology, v. 4, p. 11–15.
Stewart, J. H., and Poole, F. G., 1974, Lower Paleozoic and uppermost Precambrian Cordilleran miogeocline, Great Basin, western United States, *in* Dickinson, W. R., ed., Tectonics and sedimentation: Society of Economic Paleontologists and Mineralogists Special Publication 22, p. 28–58.
Stewart, J. H., and Suczek, C. A., 1977, Cambrian and latest Precambrian paleogeography and tectonics in the western United States, *in* Stewart, J. H., Stevens, C. H., and Fritche, A. E., eds., Paleozoic paleogeography of the western United States: Society of Economic Paleontologists and Mineralogists, Pacific section, Pacific Coast Paleogeography Symposium, 1st, p. 1–17.
Stille, H., 1941, Einführung in den Bau Amerikas: Berlin, Germany, Borntraeger, 717 p.
Stockmal, G. S., Beaumont, C., and Boutilier, R., 1986, Geodynamic models of convergent margin tectonics: Transition from rifted margin to overthrust belt and consequences for foreland basin development: American Association of Petroleum Geologists Bulletin, v. 70, p. 181–190.
Stride, A. H., Curray, J. R., Moore, D. G., and Belderson, R. H., 1969, Marine geology of the Atlantic continental margin of Europe: Royal Society Philosophical Transactions, v. 264, p. 31–75.
Suess, E., 1904, The face of the earth (Volume 1): Oxford, United Kingdom, 604 p. (English translation; first edition in German in 1885).
——— 1906, The face of the earth (Volume 2): Oxford, United Kingdom, 556 p. (English translation).
——— 1909, Das Antlitz der Erde (Volume 3): Wien, Austria, Tempsky, 789 p.
Sykes, L. R., 1968, Seismological evidence for transform faults, sea floor spreading and continental drift, *in* Phinney, R. A., ed., The history of the Earth's crust: Princeton, New Jersey, Princeton University Press, p. 120–150.
Taylor, P. T., Zietz, I., and Dennis, L. S., 1968, Geologic implications of aeromagnetic data for the eastern continental margin of the United States: Geophysics, v. 33, p. 755–780.
Umbgrove, J.H.F., 1947, The pulse of the earth: The Hague, the Netherlands, Martinus Nijhoff, 358 p.
Upham, W., 1894, The fishing banks between Cape Cod and Newfoundland: American Journal of Science, 3rd series, v. 47, p. 123–129.
Vail, P. R., Mitchum, R. M., Jr., and Thompson, S., III, 1977, Seismic stratigraphy and global changes of sea level, Part 4: Global cycles of relative changes of sea level, *in* Payton, C. E., ed., Seismic stratigraphy—Applications to hydrocarbon exploration: American Association of Petroleum Geologists Memoir 26, p. 83–97.
Valentine, J. W., and Moores, E. M., 1972, Global tectonics and the fossil record: Journal of Geology, v. 80, p. 167–184.
Vening-Meinesz, F. A., 1931, Une nouvelle methode pour la reduction isostatique regionale de l'intensite de la pesanteur: Bulletin Geod., no. 29.
——— 1941, Gravity over the continental edges: Koninkl. Ned. Akad. Wetenschap. Proc., v. 44.
Vogt, P., and Ostenso, N., 1967, Steady state crustal spreading: Nature, v. 215, p. 810–817.
Watts, A. B., 1981, The U.S. Atlantic continental margin: Subsidence history, crustal structure and thermal evolution, *in* Geology of passive continental margins: History, structure and sedimentologic record: American Association of Petroleum Geologists Education Course Note Series, no. 19, chapt. 2, 75 p.
Watts, A. B., and Ryan, W.B.F., 1976, Flexure of the lithosphere and continental margin basins: Tectonophysics, v. 36, p. 25–44.
Watts, A. B., and Steckler, M. S., 1979, Subsidence and eustasy at the continental margin of eastern North America, *in* Maurice Ewing Symposium Series 3: Washington, D.C., American Geophysical Union, p. 218–234.
Watts, A. B., and Thorne, J., 1984, Tectonics, global changes in sea level and their relationship to stratigraphical sequences at the US Atlantic continental margin: Marine and Petroleum Geology, v. 1, p. 319–339.
Wegener, A., 1912, The origin of continents: Petermanns Geog. Mitteilungen, v. 58, p. 185–309.
——— 1966, The origin of continents and oceans: New York, Dover, 246 p. (translated from the fourth revised German edition).
Wernicke, B., 1985, Uniform-sense normal simple shear of the continental lithosphere: Canadian Journal of Earth Sciences, v. 22, p. 108–125.
Williams, H., and Stevens, R. K., 1974, The ancient continental margin of eastern North America, *in* Burk, C. A., and Drake, C. L., eds., The geology of continental margins: New York, Springer-Verlag, p. 781–796.
Willis, B., 1936, East African plateaus and rift valleys: Carnegie Institute of Washington Publication 470, 358 p.
Wilson, J. T., 1966, Did the Atlantic close and then reopen?: Nature, v. 211, p. 676–681.
Worzel, J. L., 1965, Pendulum gravity measurements at sea: 1936–1959: New York, John Wiley & Sons, 422 p.

MANUSCRIPT RECEIVED BY THE SOCIETY APRIL 28, 1988
REVISED MANUSCRIPT RECEIVED AUGUST 9, 1988
MANUSCRIPT ACCEPTED AUGUST 10, 1988

Printed in U.S.A.

Alpine serpentinites, ultramafic magmas, and ocean-basin evolution: The ideas of H. H. Hess

CENTENNIAL ARTICLE

E. M. MOORES *Department of Geology, University of California, Davis, California 95616*
F. J. VINE *School of Environmental Sciences, University of East Anglia, Norwich, NR4 7TJ, United Kingdom*

ABSTRACT

Hess' ideas related to the tectonic significance of ultramafic rocks, ultramafic magmas, evolution of oceanic crust, and H_2O outgassing, all emanating from his abiding interest in serpentinite. Hess recognized the prime importance of ultramafic rocks in understanding the tectonic development of collisional (Alpine-type) mountain belts. They are not magmatic intrusions, as he proposed, but tectonic remnants of fossil plate boundaries. Hess was the formulator of the concept of sea-floor spreading. His ideas on the nature of oceanic crust changed through the years, but he repeatedly argued that oceanic crust was partly or predominantly serpentinite. Despite the lack of subsequent popularity of this view, he was at least partly right. Serpentinite is a part of oceanic crust, notably in tectonically thinned regions along rifted ridges, near transform faults, and along obliquely rifted continental margins, such as the Mesozoic western Alpine continental margin.

INTRODUCTION

It is fitting that one of the centennial articles be devoted to the work and ideas of H. H. Hess, one of the giants of twentieth-century geology. Throughout his research career, which began in the early 1930s, Hess was continually at the center of some of the most exciting developments in the earth sciences. The variety of fields his career spanned is indicated by the subdivisions listed in the Geological Society of America (GSA) Memoir 132, the Hess Memorial Volume: geotectonics, ultrabasic rocks, Caribbean geology, petrology, mineralogy, deformation of materials, environmental geology, and space science. It was said by his contemporaries that he was one of the people, if not the last person, to gain mastery of these diverse fields. Among other achievements, he was a member of the National Academy of Sciences, Chairman of its Space Science Board, and President of GSA (1963). He was awarded the Feltrinelli prize by the Accademia Nazionale dei Lincei in 1966, the GSA Penrose medal in 1966, and, posthumously, the National Atmospheric and Space Administration Distinguished Public Service award.

Hess is perhaps best known in the geologic community for two principal contributions: his many publications on mafic and ultramafic rocks and minerals and his seminal paper on the evolution of ocean basins (Hess, 1962) that precipitated the plate-tectonic revolution. Both contributions emphasize his keen interest throughout his career in ultramafic rocks, their tectonic environment, and oceanic tectonics.

After receiving his bachelor's degree from Yale, Hess spent 2 years as a geologist working for a mining company in Northern Rhodesia (now Zambia), conducting traverses spaced ¼ mile apart. He stated that "at 17 miles a day, I developed leg muscles, a philosophical attitude toward life, and a profound respect for field work" (Hess, 1968, p. 85). He once related to one of us (EMM) that after a year had gone by without result, he and his partner requested that the spacing be increased to 1 mile. When their request was refused, Hess quit and returned to academia.

His having worked in economic geology, it may have seemed natural to do a thesis in that field. At any rate, Hess began to work under the economic geologist at Princeton, Edward Sampson, on a soapstone deposit near Schuyler, Virginia. Why he did so can only be conjectured, but rumor has it that the steps in the geology building at Princeton, Guyot Hall, which are made of soapstone, actually come from the same quarry. His work included the hydrothermal alteration not only of the Schuyler ultramafic rocks but also of numerous other localities in the Appalachians. In the course of this work, Hess noted that ultramafic rocks in the Appalachians appeared to be symmetrically disposed about the center of the orogenic belt along its entire length.

Also during his graduate years, with the encouragement of Professor Richard Field, Hess began to work with F. A. Vening Meinesz on gravity measurements in a submarine in the Caribbean region, thereby beginning his long interest in and involvement with marine geology and geophysics. In order to be able to give orders to members of the crew, Hess found it expedient to obtain a reserve commission in the U.S. Navy. Hess served in the Navy in World War II, and he maintained his Navy connections after the war, spending 2 weeks each summer on active duty in Washington, D.C., as a reserve officer, eventually rising to the rank of Rear Admiral.

From the first, Hess was one to "think big" (a characteristic that he inherited, perhaps, as did Tuzo Wilson and Maurice Ewing, from Richard Field). The earliest manifestation of this was his article at the International Geological Congress in Moscow (Hess, 1939) in which he attempted to combine the results of Vening Meinesz's marine geological and geophysical work, his own observations on ultramafic rocks in the Appalachians, and the distribution of ultramafic rocks around the world (Benson, 1926) into a synthesis. Hess argued that orogenic belts were formed by downbuckles or "tectogenes" caused by convection currents in the Earth's mantle. Peridotites or serpentinites were "intruded in the first great deformation of the belt when the downbuckle is formed, and they do not accompany later deformations."

Hess' ideas at the time received less attention than they merited, partly because World War II intervened and also because he was a young scientist trying to synthesize. He once observed to EMM that he had learned a hard if valuable lesson—that a young scientist must first make a reputation in an established field if he/she wants to synthesize credibly in a controversial field.

Geological Society of America Bulletin, v. 100, p. 1205–1212, 6 figs., August 1988.

Hess retreated from tectonics and concentrated on a study of the Stillwater complex, eventually published as a GSA memoir (Hess, 1960). In the course of this work, Hess published a number of pioneering studies applying first optical techniques and later X-ray diffraction to the problem of compositional change of pyroxenes during fractional crystallization of mafic and ultramafic magma.

When the United States entered World War II, Hess at first was assigned to the antisubmarine warfare office, where he had spectacular success in predicting the activity of U-boats (Buddington, 1972) (Hess once expressed the opinion that geologists made better intelligence analysts than did other physical scientists because they were accustomed to making decisions on faulty and incomplete information).

During the latter part of World War II, Hess was Executive Officer and Captain of the attack transport *U.S.S. Cape Johnston* in the Pacific. He kept his depth sounder going essentially at all times, even under battle conditions. He may have occasionally sailed in regions where he otherwise would not have done in order to explore interesting topographic features (James, 1973). (A member of Hess' crew, the paleobiologist James W. Valentine, once recounted that Hess was a popular Captain, but the crew was amused by his eagerness to get off the ship and run around banging on rocks at any island they put into.)

The result of the thousands of miles of new ship tracks of the *Cape Johnston*, the *U.S.S. Massachusetts*, and many others was a vastly improved understanding of the topography of the western Pacific basin.

The most famous of Hess' contributions to result from the war was his paper on flat-topped seamounts, or guyots, in the Pacific basin (Hess, 1946). His original idea that they were Precambrian was soon disproven, but the problem of their erosion and subsidence below wave base occupied his thoughts for years and figured in many of his subsequent papers.

In addition to his guyot work, Hess collected the new soundings and published them as two landmark articles, one in the GSA *Bulletin* (Hess, 1948; Hess and Maxwell, 1953). The 1948 GSA paper presented a new bathymetric chart of a vast area, roughly 10 million km², between Japan on the north, the Philippines on the west, the Mariana trench on the east, and Indonesia on the south, as well as a brief discussion of its principal tectonic features. Hess compiled evidence of earthquakes along "45 degree shear zones" beneath the island arcs of the region (see Fig. 1). He thus postdated Wadati (1940) but slightly predated Benioff (1949) in his recognition of the importance of these seismic zones.

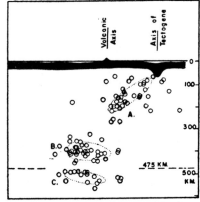

A. Composite profile Guam to Honshu

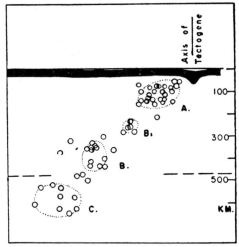

B. Composite profile Kurile Islands

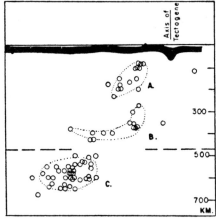

C. Composite profile Tonga-Kermadoc Islands

FIGURE 6.—*Composite profiles showing relation of deep-focus earthquakes to the axis of the trench or tectogene in each of three island arc groups*

Figure 1. Hess' (1948) figure and caption showing his compilation of the earthquakes beneath (A) the Marianas, (B) the Kuriles, and (C) the Tonga-Kermadec Islands.

Hess' major *GSA* contributions, however, are embodied in two papers (Hess, 1955, 1962). The first article, "Serpentines, orogeny and epeirogeny," updated and restated his synthesis of 1939. The second article, "History of ocean basins," laid out the idea of sea-floor spreading. The principal subjects treated in these articles lie in three categories: (1) the tectonic significance of ultramafic rocks and the origin of Alpine-style mountain belts, (2) the existence of ultramafic

Figure 2. Hess' (1955) figure and caption showing serpentine belts of North America.

magmas, and (3) the evolution of oceanic crust and of H_2O outgassing from the Earth's mantle with time. Through these diverse subjects runs the common theme of serpentinization and its possible role in tectonics, uplift, and oceanic crust.

TECTONIC SIGNIFICANCE OF ULTRAMAFIC ROCKS AND ORIGIN OF ALPINE MOUNTAIN SYSTEMS

Hess updated in 1955 his observations about serpentine belts in mountain systems (Fig. 2). He also argued that the ophiolite association, as defined by Steinmann (1905, 1927), of serpentine, pillow lava, and chert was a misleading grouping. He maintained that the most important of these three rocks was serpentinite, "found in every Alpine mountain system." He reiterated his 1939 argument that they are intruded into continental rocks in the initial stages of orogeny. By "Alpine mountain system," Hess meant orogenic belts such as the Alps or Appalachians with outer overthrust belts of sedimentary rocks and an inner "core" of crystalline nappes, high-grade metamorphic rocks, and/or intrusive rocks of granitic and ultramafic composition.

Hess also argued, both in 1939 and 1955, that island arcs represented early stages in the development of Alpine-style mountain systems. This idea gave rise to an ambitious years-long research project, the Caribbean research project, that ultimately involved some 35 doctoral theses, many of which were published in two complete issues of the GSA *Bulletin* (January 1953 and March 1960) and a GSA memoir (Memoir 98). During this work, the simple model of a gradation between island arcs and continental orogenic belts gradually fell by the wayside under the sheer weight of a myriad of local stratigraphic, structural, and petrologic details. Hess persevered and learned many new facts and skills (even becoming conversant with foram biostratigraphy).

Hess also suggested that thick sedimentary sequences or "geosynclines" were less important for understanding the evolution of mountain systems than were the serpentine belts. In addition, he argued that the time of emplacement of serpentines was more important in timing the development of an orogenic belt than was the age of sediment deformation. With regard to the Appalachians, he maintained that the Taconic emplacement of ultramafic rocks (serpentines), as indicated in Figure 2, was a more profound event than the late Paleozoic "Alleghanyian."

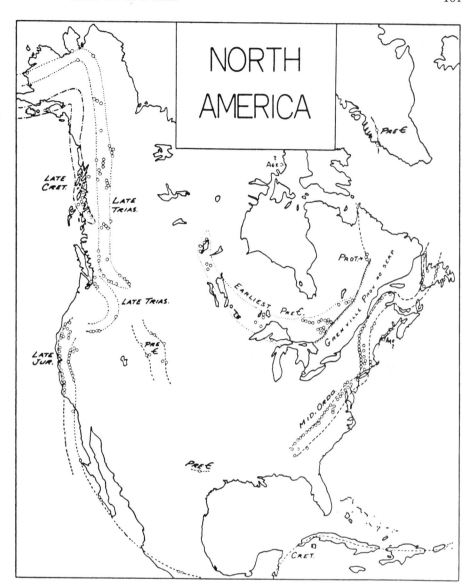

FIGURE 1.—*Serpentine belts of North America*

The Taconic was when the tectonic action began that produced the mountain system. He observed (1955, p. 395), "We do not give a man's age in years since he graduated from high school or since he was married, but from his birth. Similarly mountain systems should be dated from birth, not from fortuitous events which may occur as much as 50 million years or more later. Thus dating mountain systems by means of radioactive minerals in granites and pegmatites has a serious disadvantage in that these may be intruded in relation to recurring episodes of compression for a very long period after the birth of the system. There may be granitic igneous rocks following only the first deformation, or there may be several such episodes over a long time, as in the Appalachians."

Hess knew that he was onto something important in the emplacement of ultramafic rocks in orogenic belts, and with the benefit of hindsight, we can say that he was right. We now recognize generally that Alpine-type mountain belts represent collisional belts where two former continents have converged and collided. The ultramafic rocks predominantly represent remnants of oceanic lithosphere involved in this collision (for example, Moores, 1970, 1982; Dewey and Bird, 1970; Dewey, 1976; Gealey, 1980; Church and Stevens, 1971; Malpas and Stevens, 1977).

The emplacement of mantle rocks is predominantly by thrusting of oceanic crust and mantle (ophiolite) over continental margins. This process most commonly seems to be by collision of a continental margin with a subduction zone, that is, by aborted subduction (for example, Moores and others, 1984; Moores, 1970, 1982; Casey and Dewey, 1984). Subsequent disruption and solid emplacement of these rocks can give rise to the isolated serpentinite bodies that concerned Hess.

Hess' observations that these rocks are "intruded in the initial stages of a mountain belt" can easily be understood in plate-tectonic terms. The oldest rocks in a collisional mountain belt are those deposited on or off the edge of a continental margin or in an ocean basin prior to its closure. These deposits will generally remain flat lying until deformed in a subduction zone, particularly by one that collides with the continental margin. The first deformation of continental-margin rocks thus coincides with ophiolite emplacement.

Hess dropped his opposition to the ophiolite suite (Hess, 1965) as a meaningful entity after seeing field work by his associates in Italy and Greece (Maxwell and Azzaroli, 1962; Moores, 1969; see Fig. 3). He maintained that the oceanic crust was serpentinite and overlying basalt, as implied by Steinmann's trinity. He was skeptical about the wisdom of our proposal in 1966 to go to Cyprus to study the Troodos massif as a slice of oceanic crust and mantle formed by spreading because it contained a thick gabbro section that Hess thought should not be in the oceanic crust. Part of this skepticism was based on his observations in the Appalachians, where few complete ophiolite sequences are preserved (Malpas and Stevens, 1977), and on Steinmann's bias toward the western Mediterranean, where most ophiolites contain little or no mafic section and probably are derived from narrow oceanic rifts near transform faults (for example, Lemoine and others, 1987).

It might be said that the importance of ophiolite belts and their emplacement is still not adequately appreciated. These rocks reflect the emplacement of oceanic crust and mantle rocks along thrust faults rooted in the mantle. These mantle thrusts are more profound than the more famous examples of thrusts involving only crustal rocks, but they are less amenable to analysis than are the more familiar structures. Mantle thrust complexes cannot be balanced; mantle thrust displacements are indeterminate but very large. They represent the traces of former oceans and of fossil plate boundaries; fabrics along their basal contacts may reflect fossil plate motions.

Hess was right in a sense about the idea that island arcs represent an early stage in the development of an Alpine mountain belt. If the latter represents the convergence and collision of two formerly separate continents, they must bulldoze before them all the island arcs and other anomalously thick oceanic regions (for example, Nur and Ben-Avraham, 1982). These separate anomalous regions will collide first with one continental margin or each other, then finally with the second continent, in the process forming the composite accretionary terranes now so popular

Figure 3. Photograph of Harry H. Hess standing on the mafic-ultramafic contact, Vourinos complex, northern Greece, August 1964. Photograph by E. M. Moores.

in all orogenic analyses (Williams and Hatcher, 1983; Coney and others, 1980). Hess' idea on island arcs was essentially stabilist, that is, he predicted that an island arc would develop *in situ* into an Alpine-style mountain belt. One needs to add the mobility that the consequences of sea-floor spreading gave him in order to get it right.

ULTRAMAFIC MAGMAS

Hess entered the debate on the existence of ultramafic magmas with an article published in the *American Journal of Science* (1938). As evidence, he cited the presence of large "batholith-sized" bodies of ultramafic rocks in such places as Cuba, Newfoundland, and California. The current reigning model was origin of peridotite by fractional crystallization of basaltic magma as argued by Bowen (1927) and Bowen and Schairer (1935). Hess noted that in the places in question, the large amounts of basaltic material required by the Bowen fractional crystallization process were lacking. He suggested that the lack of high-temperature effects could have resulted from depression of the high melting temperature of olivine determined by Bowen and Schairer by the presence of water (that is, a serpentine magma). Hess doggedly held onto his ideas for nearly 20 years, despite the experimental results of Bowen and Tuttle (1949), who found no indication of a magma of serpentine composition.

Instead, their results indicated that serpentine dissociated into olivine plus water at approximately 500 °C. In 1955, Hess drew on the possibility that the isolated SiO_4^{-4} tetrahedral structure of olivine could make it possible for it to flow at lower stresses than other minerals, and that serpentine might flow at still lower stresses. Hess rejected, however, such solid flow as a general solution to the problem of the occurrence of ultramafic rocks because of the abundance of serpentines "along the margins of Alpine systems where the deformation has not been intense. They form extensive concordant sills in low dipping sedimentary rocks—shales, graywackes, sandstones and phyllites" (p. 401–402). Hess did not know how to resolve the problem and observed, "some vital piece of evidence is still missing."

After publication of his 1955 paper, events began to erode Hess' confidence in his peridotite magma hypothesis. De Roever (1957) published an article suggesting that peridotites were fault slices of the Earth's mantle. Bailey and McCallien (1953) revived the mélange concept of Greenly (1919), and work in the Caribbean by Hess' students indicated that many serpentinites were exotic blocks rather than intrusions. In addition, work on several peridotites indicated intrusion as hot solid masses at considerable depths (Green, 1964), and work in southwest Puerto Rico suggested that the peridotite there was the basement upon which the island arc developed. Thus confronted, Hess "reluctantly" withdrew his hypothesis in his introduction to the Caribbean memoir (Hess, 1966).

After this reversal, Hess was understandably skeptical about the first reports of the discovery of komatiites in the Barberton Mountain Land, South Africa (Viljoen and Viljoen, 1970). He was stunned when he saw them for himself on a field trip just before his death in the summer of 1969 and wrote to one of us (FJV) that it was the greatest field trip of his life.

In his early writings on the subject, Hess made the same mistake that so many had made about ultramafic rocks, beginning with Brongniart (1827), in assuming that they were igneous because they were coarse grained. It was not until the careful experimental and petrographic work on olivine and peridotite (for example, MacKenzie, 1972) that the true nature of the large "batholith-sized" masses of ultramafic rock became clear. It now seems clear that they are exposures of the Earth's mantle that preserve textural and mineralogic evidence of high-temperature flow and crystallization in the Earth's mantle, mostly at spreading centers. Practically all peridotites thus exposed are probably remnants of the mantle. For a number of years in the 1960s and 1970s, a number of workers proposed the existence of "high-temperature Alpine-type ultramafic intrusions" with thermal aureoles, thought to represent hot intrusions into the crust (see Moores and MacGregor, 1972, for summary). Most of these occurrences are now recognized as allochthonous slabs of suboceanic or subcontinental mantle. The metamorphic rocks associated with the margins of these bodies are either associated with normal processes at a spreading center or are thin garnet amphibolite slices along the base of the complexes that result from fracture zone metamorphism or hot slab emplacement (for example, Malpas, 1979; Parrot and Whitechurch, 1978).

Rocks formed by crystallization of true ultramafic magmas, komatiites, are widespread in Archean terranes (for example, Arndt and Nisbet, 1982; Nisbet and others, 1987), and there is at least one Phanerozoic example (Echeverria, 1980). These rocks are present as flows with quench-textured olivines, as dikes and sills, perhaps as sheeted dikes (M. de Wit, 1986, oral commun.), and they may be the parent magmas of some Archean ultramafic-mafic plutonic complexes. These magmas differ in composition from the average mantle tectonite, particularly in having less MgO, more CaO, and different trace-element compositions. Their formation may require melting of as much as 60% of the mantle.

More problematic is the question of whether primary ultramafic lavas exist in the modern world. The most abundant magma of the modern world—oceanic tholeiite—probably is not primary (for example, Engel and others, 1965) as it apparently represents a low-density derivative magma (for example, Stolper and Walker, 1980; O'Hara, 1968) derived by a two-or-more-stage differentiation of a more ultramafic magma (for example, Duncan and Green, 1980). Many ophiolite complexes contain dunites in their ultramafic rocks. These dunites may be refractory residua, but some seem to be of igneous cumulate origin (for example, Harkins and others, 1980), suggesting fractionation and olivine settling during ascent of ultramafic magma to produce mafic magma at the surface (Elthon, 1979). These data thus suggest that modern primary magmas may well be ultramafic at depth, but they fractionate to mafic compositions during ascent. Many of the intimate "intrusive" field relations used by Hess to suggest the existence of ultramafic magma apparently formed by other processes. In particular, J. P. Lockwood (1972), one of Hess' Caribbean students, summarized the evidence for sedimentary deposition of serpentinite layers, previously thought to be igneous, and for solid protrusion or diapiric emplacement of serpentinite. In addition, serpentinite-matrix mélanges are known from many orogenic regions and may result from shearing of oceanic lithosphere along transform or consuming plate margins.

NATURE OF THE OCEANIC CRUST

In his early writings, particularly in his 1939 paper, Hess assumed, as did others, that the crust of the oceans differed little in composition from that of the continents. His 1939 tectogene diagram showed downbuckling of a 25-km-thick granitic layer above a 35-km-thick basaltic layer. The early seismic refraction work at sea (for example, Raitt, 1956; Ewing and Ewing, 1959) changed the picture drastically by indicating that the oceanic crust was uniformly thin and of a velocity compatible with that of basalt. In 1955, thus, Hess portrayed the oceanic crust as a 4-km-thick layer of basalt over the mantle. A thin layer of sediment (0.7 km thick) overlay the basalt. Hess still implicitly considered ocean basins as permanent features.

The development of paleomagnetism and of evidence for "polar wandering" in the late 1950s clearly was troubling to Hess. In his lectures in late 1959, however, he remained publicly skeptical about the validity of the results. He particularly emphasized the possibilities of self-induced reversals and anisotropy (for example, Balsey and Buddington, 1960) to give rise to the apparent polar wander paths.

In his 1962 paper, however, he wrote, "paleomagnetic data . . . strongly suggest that the continents have moved by large amounts in geologically recent times. One may quibble over the details, but the general picture on paleomagnetism is sufficiently compelling that it is much more reasonable to accept it than to disregard it" (p. 608). This statement must have been the result of quite a mental wrench, but it enabled Hess to go on to propound his sea-floor spreading hypothesis (see Fig. 4). He suggested the existence of "a dynamic situation in the present Earth whereby the continents move to positions dictated by a fairly regular system of convection cells in the mantle. . . . The mid-ocean ridges could represent the traces of the rising limbs of convection cells, while the circum-Pacific belt of deformation and vulcanism represents descending limbs. The mid-Atlantic Ridge is median because the continental areas on each side of it have moved away from it at the same rate—1 cm/year. This is not exactly the same as continental drift. The continents do not plow through oceanic crust impelled by unknown forces; rather they ride passively on mantle material as it comes to the surface at the crest of the ridge and then moves laterally away from it. On this

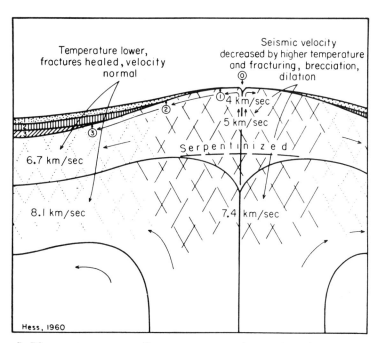

Figure 7. Diagram to represent (1) apparent progressive overlap of ocean sediments on a mid-ocean ridge which would actually be the effect of the mantle moving laterally away from ridge crest, and (2) the postulated fracturing where convective flow changes direction from vertical to horizontal. Fracturing and higher temperature could account for the lower seismic velocities on ridge crests, and cooling and healing of the fractures with time, the return to normal velocities on the flanks.

Figure 4. Hess' (1962) figure and caption showing his concept of sea-floor spreading, as of 1960. Note the progressive onlap of sediments, a fact subsequently confirmed by the Deep Sea Drilling Project. In addition, the diagram shows his concept of a crust of serpentinized peridotite.

basis the crest of the ridge should have only recent sediments on it, and recent and Tertiary sediments on its flanks; the whole Atlantic Ocean and possibly all of the oceans should have little sediment older than Mesozoic" (p. 608–609). Hess argued that the postulated convection systems involved the whole mantle with an "approximate diameter of 3000 to 6000 km in cross-section (the other horizontal dimension might be 10,000–20,000 km, giving them a banana-like shape)" (1962, p. 608). He subsequently (1965) proposed a "tennis-ball" pattern of convection but postulated a convective system limited to the uppermost 750 km, suggesting that below this depth, radiative transfer of heat was adequate to dissipate any thermally induced stresses.

Hess' schematic cross section of a mid-oceanic ridge (Fig. 4) shows his idea of the onlap of progressively younger sediments toward the crest of the ridge, a relation subsequently confirmed in all the oceans. It also shows his concept of the formation of oceanic layer 3 by serpentinization of peridotite. His oceanic crust idea was based upon Shand's (1949) report of serpentinized peridotite on the mid-Atlantic ridge, his reasoning about the depth to the 500 °C isotherm beneath ridge crests, and the effect of serpentinization on seismic velocities in peridotite.

Hess' idea of oceanic crust as serpentinized peridotite has never gained much acceptance. Rapid accumulation of seismic data and dredge hauls from the oceanic crust indicated a three- or four-layer crust (Sutton and others, 1970) and abundant amounts of basalt and gabbro (Engel and Engel, 1970; Fox and Stroup, 1981). In addition, by 1970–1971, the ophiolite analogue for oceanic crust had gained acceptance. As mentioned above, although Hess adopted the ophiolite analogue for oceanic crust in 1965, he never considered mafic plutonic rocks important.

The correlation of ophiolite pseudostratigraphy with oceanic crustal layers has been repeated many times. Most workers have adopted an interpretation whereby extrusive and/or zeolite-facies rocks are equivalent to seismic layers 2A and 2B; greenschist-facies rocks, predominantly dikes, to layer 2C; and mafic plutonic rocks and/or amphibolite-facies rocks, layer 3 (Christensen and Salisbury, 1975; Vine and Moores, 1972; Coleman, 1977; Gass and Smewing, 1975).

Two problems have arisen with the ophiolite analogy with oceanic crust. (1) The composition of extruded and inferred magma in many ophiolites is unlike those in modern oceans and is more reminiscent of modern island arcs (for example, Miyashiro, 1973, 1975; Robinson and others, 1983; Schmincke and others, 1983; Rautenschlein, 1988); and (2) the thickness of the magmatic section of most ophiolites is considerably less than that of oceanic crust (Coleman, 1971; Moores, 1982; Vine and Moores, 1972).

The magma geochemistry of most ophiolites has caused a number of workers to argue that they represent "supra-subduction zone" oceanic crust and mantle (for example, Pearce, 1975) or that they originated in an environment such as that currently found in the Andaman Sea (Moores and others, 1984) or in the Woodlark Basin (Scott and others, 1988; Dilek and Moores, 1988). The presence of dike complexes and the thin mafic section in many ophiolite complexes argues against any simple origin in an island-arc complex, however.

The presence of a too-thin mafic section in most ophiolite complexes may result from the fact that many ophiolites are derived from lithosphere adjacent to and including transform faults. In a major review paper, Fox and Gallo (1984) documented the nature of slow-, intermediate-, and fast-spreading transform faults and the variation of crustal structure near transform fault zones. The thickness of oceanic crust apparently thins markedly near transform fault zones, and in these regions, a continuous mafic section may be lacking. In addition, as discussed below, the crust in these regions may be partly serpentinite.

In formulating the concept of sea-floor spreading, Hess invoked whole-mantle convection, equating the ridge crests with upwelling

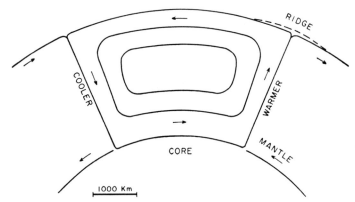

Figure 8. Possible geometry of a mantle convection cell

Figure 5. Hess' (1962) figure and caption showing his concept then of a convection cell involving the whole mantle.

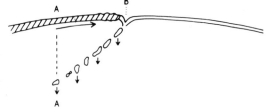

Figure 121. Continent overrides trench forcing it to the right. Settling, heavier, cooler masses move vertically downwards. Trench was at A when heavier mass beneath it was at the surface. Island arcs commonly override their trenches in a similar manner.

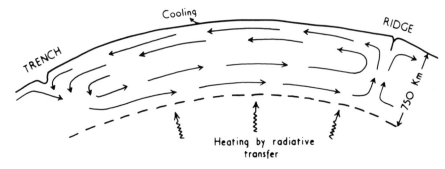

Figure 122. Postulated convecting system in the upper mantle.

Figure 6. Two diagrams with captions from Hess' (1965) paper, illustrating his revised concept of a convection cell limited to the upper 750 km of the Earth, with upwelling under the ridges and heavier, cooler material moving downward beneath a continental or island-arc margin.

and the trenches with downwelling (see Fig. 5). Modern ideas on the nature of mantle convection are in a state of flux. A seismic discontinuity at 650 km is well established. The current debate centers on whether convection is whole mantle, as argued by Hess in 1962; limited to the upper 650–750 km, as Hess proposed in 1965 (see Fig. 6); or 2 stage (for example, Alvarez, 1982; Ringwood and Irifune, 1988). Recent results from seismic tomography suggest that high-velocity slabs penetrate to depths of at least 2,000 km, suggesting whole-mantle convection. On the other hand, isotopic relations of magmas and possible mantle sources suggest chemical zonation of the mantle with resultant two-stage convection (Ringwood and Irifune, 1988). Recently, Ringwood and Irifune (1988) have proposed a compromise model whereby young lithosphere is trapped in a gravitationally stable layer at 600–700 km depth, whereas old lithospheric plates may descend to the core-mantle boundary. In addition, an increasing number of workers argue for coupling of core and upper-mantle processes (for example, Courtillot and Besse, 1987). It seems clear, however, that there is no simple correlation between ridges and upwelling convective limbs or subduction zones and downwelling limbs, as Hess assumed (McKenzie, 1983).

Hess' model for oceanic crust formed largely of partially serpentinized peridotite included addition of juvenile water from the mantle. Hess thought that this meant progressive increase in the volume of ocean water, but that the constant continental elevation would be maintained by the erosive action of water. He thus argued that as ocean water grew in volume, continents grew in thickness. He regarded the 5-km-thick oceanic crust as "axiomatic," implying an invariant oceanic geotherm throughout geologic time.

The problem of continent-ocean elevation difference, water volume, and average sea level is still unresolved. Hess' proposal for constant addition of juvenile water has never been accepted, however, and it is not part of any current consensus.

SERPENTINIZATION

Throughout his research career, Hess clearly was fascinated by the possible role of serpentinization in explaining certain tectonic processes, as well as the properties of oceanic crust, not least its remarkably constant thickness. He was probably the first to realize the dramatic effects of progressive serpentinization on the physical properties of pre-existing peridotite. The most obvious of these was the change in density and volume that could potentially be a mechanism for epeirogenic uplift. The effect of serpentinization extends beyond the effects on density and elastic properties to all other physical properties, for example, magnetic and electric properties, which are also possibly of great relevance to the interpretation of geophysical data.

The basic constituents of serpentine—water and peridotite—are very abundant at and near the Earth's surface, although admittedly, the largest reservoirs are typically separated by the Earth's crust. It thus is perhaps surprising that serpentinization and serpentinites are rarely invoked today to explain tectonic processes or geophysical results. Hess obviously had less conceptual difficulty in bringing the constituents together near the Earth's surface (at temperatures below 500 °C) in that he believed in steady and continuing outgassing of large volumes of water from the mantle throughout geologic time. Isotopic studies have shown, however, that the water of serpentinization is derived from the surface, that is, is of meteoric or sea-water origin (for example, Spooner and others, 1977).

Ten years ago, the lower crust in both continental and oceanic areas was thought to be dry and to have essentially zero porosity and permeability. More recently, however, this assumption has been questioned. Variable seismic velocities, including, in places, a low-velocity layer in the lowermost oceanic crust, coupled with a possible slight increase in the thickness of oceanic crust with time, led Lewis (1983) to postulate varying degrees of serpentinization in a basal crustal layer. Several authors (for example, Lee and others, 1983; Gough, 1987) have invoked the presence of small amounts of free water, at least in the uppermost part of the lower continental crust, to explain an inferred highly electrically conducting layer at these depths. Serpentinite itself, however, has rarely been invoked to explain regions of high electrical conductivity in continental areas except in the case of presumed suture zones (Camfield and Gough, 1977).

Potentially, serpentinized peridotite in the crust could explain regions or layers of anomalously low seismic velocity, high electrical conductivity, and high intensities of magnetization. Moreover, the temperature-dependent phase change water + peridotite = serpentine might explain the "mobile" Moho of some authors in both continental and oceanic areas (Barton, 1986; Matthews and Smith, 1987; Lewis, 1983).

In his last writing, Hess (in Vine and Hess, 1970) listed a total of 40 dredge sites where serpentinites were recovered in the oceans. In addition, serpentinite was cored by Deep Sea Drilling Project leg 37. Serpentinite is in fact commonly recovered from rifted ridges, particularly near fracture zones, probably as a result of tectonic thinning of the oceanic crust because of starved magma supply in these settings. Also, Francis (1981) has pointed out that the presence of earthquakes at depths of 8 km below the sea floor along active slow-spreading ridges indicates that ridge-parallel faults penetrate into the mantle, where they would be likely loci of serpentinization. Locally, serpentinite is thus a major constituent of oceanic crust (Karson, 1988). Similarly, tectonic thinning is a major characteristic of many rifted continental margins (for example, Lemoine and others, 1987; Lister and others, 1986), with the potential, in the more extreme cases, for partial melting and hydration of the upwelling mantle. Within the framework of plate tectonics, such settings are ultimately involved in continental collision and suturing, and they might well supply many of the Alpine-type peridotites that concerned Hess.

Modern ideas on tectonic thinning in oceanic areas thus are very pertinent to the formation of serpentinite in specific tectonic settings and begin to provide a more detailed understanding of Hess' prescient observations on the occurrence of serpentinite in the oceanic crust and in mountain belts.

EPILOGUE

We have touched on the outcome of Hess' tendency to "think big" in geology. He extended this tendency to planetary science and to science funding as well, despite the predilection of his geological peers to regard such indulgences as being rather less than respectable.

In planetary science, Hess was probably the first person to suggest spreading on another planet. He wrote in October 1968, "Mars, judging from the few Mariner photos, resembles the Moon and has not so far shown the curvo-linear features that characterize spreading on Earth. Mercury (does) not look unlike Mars. . . . Radar observations (of Venus) indicate areas of very rough topography but much less differential relief than Mars. The high surface temperature . . . —ca 700 °K—favors a mantle capable of convection. The lower differential relief suggests isostatic adjustment, erosion by the atmosphere, or both, and the rough areas perhaps indicate mountain building that is analogous to the Earth's. Venus, therefore, seems to be the only candidate in the solar system in which another example of spreading might be found" (Vine and Hess, 1970, p. 616). Thus, two decades ago, Hess anticipated recent suggestions for divergent plate activity on Venus (Head and Crumpler, 1987).

Hess was also intimately linked with project Mohole, a proposal first made in 1957 to drill a hole through the oceanic crust into the mantle. Partly conceived as a geologic response to the Soviet Sputnik challenge, this project was an ambitious, even outlandish, proposal to answer a question of true multidisciplinary interest—the composition of the Earth mantle—by obtaining a direct sample. Support for the project was by no means unanimous among geologists, and after 8 years of preliminary drilling, professional infighting, and partisan politics, the project was terminated in 1966. It did serve as the forerunner, however, of the enormously successful Deep Sea Drilling Project.

Harry Hess was a gentle person of dry wit, who maintained throughout both a sense of humility and of enthusiasm. For example, in his Penrose medal response, Hess noted, "As a geologist who has often guessed wrong, I deeply appreciate the generosity of the Society in balancing my errors against deductions of mine not yet proven incorrect. I am pleased to come out with a positive balance" (1968, p. 85).

Every student and colleague has a catalogue of Hess stories, the total of which could fill a large volume. He also had a way with children. He was above all a lifelong inspiration to those fortunate enough to become acquainted with him, not only for his science, but for his qualities as a human being.

REFERENCES CITED

Alvarez, W. S., 1982, Geological evidence for the geographical pattern of mantle return flow and the driving mechanism of plate tectonics: Journal of Geophysical Research, v. 87, p. 6697–6710.

Arndt, N. T., and Nisbet, E. G., eds., 1983, Komatiites: London, United Kingdom, Allen and Unwin, 526 p.

Bailey, E. B., and McCallien, 1953, Serpentine lavas, the Ankara melange, and the Anatolian thrust: Royal Society of Edinburgh, Transactions, v. 62, Part II, p. 403–442.

Balsey, J. R., and Buddington, A. F., 1960, Magnetic susceptibility, anisotropy and fabric of some Adirondack granites and orthogneisses: American Journal of Science, v. 258A, 6 p.

Barton, P., 1986, Deep reflections on the Moho: Nature, v. 323, p. 392–393.

Benioff, H., 1949, Seismic evidence for the fault origin of oceanic deeps: Geological Society of America Bulletin, v. 60, p. 1837–1856.

Benson, W. N., 1926, The tectonic conditions accompanying the intrusion of basic and ultrabasic igneous rocks: National Academy of Sciences Memoir 1, p. 1–90.

Bowen, N. L., 1927, The origin of ultrabasic and related rocks: American Journal of Science, v. 14, p. 89–108.

Bowen, N. L., and Schairer, J. F., 1935, The system MgO-FeO-SiO_2: American Journal of Science, v. 29, p. 151–217.

Bowen, N. L., and Tuttle, O. F., 1949, The system MgO-SiO_2-H_2O: Geological Society of America Bulletin, v. 60, p. 939–960.

Brongniart, A., 1827, Classification et Characteres Mineralogiques des Roches Homogenes et Heterogenes: Paris, France, F. G. Levrault.

Buddington, A. F., 1972, Memorial to Harry Hammond Hess, 1906–1969: Geological Society of America Memorials for 1969–1972, p. 18–26.

Camfield, P. A., and Gough, D. I., 1977, A possible Proterozoic plate boundary in North America: Canadian Journal of Earth Sciences, v. 14, p. 1229–1238.

Casey, J. F., and Dewey, J. F., 1984, Initiation of subduction zones along transform and accreting plate boundaries, triple-junction evolution, and forearc spreading centers—Implications for ophiolitic geology and obduction, *in* Gass, I. G., Lippard, S. J., and Shelton, A. W., eds., Ophiolites and oceanic lithosphere: Geological Society of America Special Publication 13, p. 269–290.

Christensen, N. I., and Salisbury, M. H., 1975, Structure and constitution of the lower oceanic crust: Reviews of Geophysics and Space Physics, v. 13, p. 57–86.

Church, W. R., and Stevens, R. K., 1971, Early Paleozoic ophiolite complexes of the Newfoundland Appalachians as mantle-oceanic crustal sequences: Journal of Geophysical Research, v. 76, p. 1460–1466.

Coleman, R. G., 1971, Plate tectonic emplacement of upper mantle peridotites along continental edges: Journal of Geophysical Research, v. 76, p. 1212–1222.

——— 1977, Ophiolites: New York, Springer-Verlag, 229 p.

Coney, P., Jones, D. L., and Monger, J.W.H., 1980, Cordilleran suspect terranes: Nature, v. 288, p. 329–333.

Courtillot, V., and Besse, J., 1987, Magnetic field reversals, polar wander, and core-mantle coupling: Science, v. 237, p. 1140–1147.

De Roever, W. P., 1957, Sind die alpinotypen Peridotitmassen vielleicht tektonisch Verfrachtete Bruchstücke der Peridotitschale?: Geologische Rundschau, v. 46, p. 137–146.

Dewey, J. F., 1976, Ophiolite obduction: Tectonophysics, v. 31, p. 93–120.

Dewey, J. F., and Bird, J., 1970, Origin and emplacement of the ophiolite suite: Appalachian ophiolites in Newfoundland: Journal of Geophysical Research, v. 76, p. 3179–3206.

Dilek, Y., and Moores, E. M., 1988, Regional setting of the Troodos ophiolite, Cyprus: Troodos 1987 Conference, Proceedings (in press).

Duncan, R., and Green, D. H., 1980, Role of multistage melting in the formation of oceanic crust: Geology, v. 8, p. 22–26.

Echevarria, L. M., 1980, Tertiary komatiites of Gorgona Island, Colombia: Carnegie Institute of Washington Yearbook 79, p. 340–344.

Elthon, D., 1979, High magnesia liquids as the parental magma for ocean floor basalts: Nature, v. 278, p. 514–518.

Engel, A.E.J., and Engel, C. G., 1970, Mafic and ultramafic rocks, *in* Maxwell, A. E., ed., The sea, Volume 4, Part 1, New concepts of sea floor evolution: New York, Wiley, p. 465–520.

Engel, A.E.J., Engel, C. G., and Havens, R. G., 1965, Chemical characteristics of oceanic basalts and the upper mantle: Geological Society of America Bulletin, v. 76, p. 719–734.

Ewing, J., and Ewing, M., 1959, Seismic refraction profiles in the Atlantic Ocean basins, in the Mediterranean Sea, on the Mid-Atlantic Ridge and in the Norwegian Sea: Geological Society of America Bulletin, v. 70, p. 291–318.

Fox, P. J., and Gallo, D. G., 1984, A tectonic model for ridge-transform plate boundaries: Implications for the structure of oceanic lithosphere: Tectonophysics, v. 104, p. 205–242.

Fox, P. J., and Stroup, J., 1981, The plutonic foundation of the oceanic lithosphere, *in* Emiliani, C., ed., The sea, Volume 7: New York, J. Wiley p. 119–218.

Francis, T.J.G., 1981, Serpentinization faults and their role in the tectonics of slow-spreading ridges: Journal of Geophysical Research, v. 86, Part B12, p. 11616–11622.

Gass, I. G., and Smewing, J. D., 1975, Intrusion, extrusion and metamorphism at constructive margins: Evidence from the Troodos massif, Cyprus: Nature, v. 242, p. 26–29.

Gealey, W., 1980, Ophiolite obduction mechanisms, *in* Geological Survey Department, Cyprus, International Ophiolite Symposium, Proceedings, p. 228–243.

Gough, D. I., 1987, Seismic reflectors, conductivity, water and stress in the continental crust: Nature, v. 323, p. 143–144.

Green, D. H., 1964, The petrogenesis of the high-temperature peridotite intrusion in the Lizard area, Cornwall: Journal of Petrology, v. 5, p. 134–188.

Greenly, E., 1919, The geology of Anglesey, Great Britain: Geological Survey Memoir, 980 p.

Harkins, M. E., Green, H. W., and Moores, E. M., 1980, Multiple intrusive events documented from the Vourinos complex, northern Greece: American Journal of Science, v. 280A, p. 284–295.

Head, James W., III, and Crumpler, L. S., 1987, Evidence for divergent plate-boundary characteristics and crustal spreading on Venus: Science, v. 238, p. 1380–1385.

Hess, H. H., 1938, A primary peridotite magma: American Journal of Science, v. 35, p. 321–344.

——— 1939, Island arcs, gravity anomalies, and serpentine intrusions: International Geological Congress, Moscow, 1937, Report 17, v. 2, p. 263–283.

——— 1946, Drowned ancient islands of the Pacific basin: American Journal of Science, v. 244, p. 772–791.

——— 1948, Major structural features of the western north Pacific, an interpretation of H.O. 5485, bathymetric chart, Korea to New Guinea: Geological Society of America Bulletin, v. 59, p. 417–446.

——— 1955, Serpentines, orogeny and epeirogeny: Geological Society of America Special Paper 62, p. 391–408.

——— 1960, Stillwater Igneous Complex, Montana, a quantitative mineralogical study: Geological Society of America Memoir 80, 230 p.

——— 1962, History of ocean basins: Geological Society of America Buddington Volume, p. 599–620.

——— 1965, Mid-oceanic ridges and tectonics of the sea floor: Colston Research Society, Symposium, 17th, Butterworths Scientific Publication, Colston Papers, Proceedings, p. 317–333.

——— 1966, Caribbean research project, 1965 and bathymetric chart: Geological Society of America Memoir 98, p. 1–10.

——— 1968, Response, in Presentation of the 1966 Penrose Medal to Harry Hammond Hess: Geological Society of America Proceedings Volume for 1966, p. 83–86.

Hess, H. H., and Maxwell, J. C., 1953, Major structural features of the southwest Pacific: A preliminary interpretation of H. O. 5484 bathymetric chart, New Guinea to New Zealand: Pacific Science Congress, 7th, New Zealand, Proceedings, v. 27, p. 14–17.

James, H. L., 1973, Harry Hammond Hess: National Academy of Sciences Biographical Memoirs, v. 43, p. 109–128.

Karson, J. A., 1988, Sea floor spreading on the mid-Atlantic ridge: Implication for the structures of ophiolites and oceanic lithosphere produced in slow-spreading environments: Troodos 1987 Conference, Proceedings (in press).

Lee, C. D., Vine, F. J., and Ross, R. G., 1983, Electrical conductivity models for the continental crust based on laboratory measurements on high-grade metamorphic rocks: Geophysical Journal, v. 72, p. 353–371.

Lemoine, M., Tricart, P., and Boillot, G., 1987, Ultramafic and gabbroic ocean floor of the Ligurian Tethys (Alps, Corsica, Apennines): In search of a genetic model: Geology, v. 15, p. 622–625.

Lewis, B.T.R., 1983, The process of formation of oceanic crust: Science, v. 220, p. 151–157.

Lister, G. S., Etheridge, M. A., and Symonds, P. A., 1986, Detachment faulting and evolution of passive continental margins Geology, v. 14, p. 246–250.

Lockwood, J. P., 1972, Possible mechanisms for the emplacement of Alpine-type serpentinite: Geological Society of America Memoir 132, p. 273–288.

MacKenzie, D. B., 1972, Peridotite fabrics and velocity anisotropy in the Earth's mantle: Geological Society of America Memoir 132, p. 593–604.

Malpas, J., 1979, The dynamothermal aureole of the Bay of Islands ophiolite suite: Canadian Journal of Earth Sciences, v. 16, p. 2086–2101.

Malpas, J., and Stevens, R. K., 1977, Origin and emplacement of the ophiolite suite, with examples from western Newfoundland: Geotectonics, v. 11, p. 453–466.

Matthews, D., and Smith, C., 1987, Deep seismic reflection profiling of the continental lithosphere, Preface to special issue: Geophysical Journal, v. 89, p. vii–xiii.

Maxwell, J. C., and Azzaroli, A., 1962, Submarine extrusion of ultramafic magma: Geological Society of America Abstracts with Programs, v. 103A.

McKenzie, D. P., 1983, The Earth's mantle: Scientific American, v. 249, p. 50–63.

Miyashiro, A., 1973, The Troodos complex was probably formed in an island arc: Earth and Planetary Science Letters, v. 18, p. 218–224.

——— 1975, Classification, characterization, and origin of ophiolites: Journal of Geology, v. 83, p. 249–281.

Moores, E. M., 1969, Petrology and structure of the Vourinos ophiolite complex: Geological Society of America Special Paper 118, 69 p.

——— 1970, Ultramafics and orogeny, with models for the United States Cordillera and the Tethys: Nature, v. 228, p. 837–842.

——— 1982, Origin and significance of ophiolites: Reviews of Geophysics and Space Physics, v. 20, p. 735–760.

Moores, E. M., and MacGregor, I. D., 1972, Types of Alpine ultramafic rocks and their implications for fossil plate interactions: Geological Society of America Memoir 132, p. 209–223.

Moores, E. M., Robinson, P. T., Malpas, J., and Xenophontos, C., 1984, Model for the origin of the Troodos massif, Cyprus, and other Middle East ophiolites: Geology, v. 12, p. 500–503.

Nisbet, E. G., Arndt, N. T., Bickle, M. J., Cameron, W. E., Chauvel, C., Cheadle, M., Hegner, E., Kyser, T. R., Martin, A., Renner, R., and Roedder, E., 1987, Uniquely fresh 2.7 Ga komatiites from the Belingwe greenstone belt, Zimbabwe: Geology, v. 15, p. 1147–1150.

Nur, A., and Ben-Avraham, Z., 1982, Oceanic plateaus, the fragmentation of continents, and mountain building: Journal of Geophysical Research, v. 87, p. 3661–3694.

O'Hara, M. J., 1968, Are ocean floor basalts primary magmas?: Nature, v. 220, p. 683–686.

Parrot, J. F., and Whitechurch, H., 1978, Subduction anterieur au charraiges nord-sud de la croute Tethysienne: Review de Geographic Physique et Geologie Dynamique, v. 20, p. 153–170.

Pearce, J. A., 1975, Basalt geochemistry used to investigate past tectonic environments on Cyprus: Tectonophysics, v. 25, p. 41–67.

Raitt, R. W., 1956, Seismic refraction studies of the Pacific ocean basin: Geological Society of America Bulletin, v. 67, p. 1623–1640.

Rautenschlein, M., 1988, Geochemistry and origin of Troodos extrusive sequence, Akaki Canyon, Cyprus: Troodos 1987 Conference, Proceedings (in press).

Ringwood, E., and Irifune, T., 1988, Nature of the 650 km seismic discontinuity: Implication for mantle dynamics and differentiation: Nature, v. 331, p. 131–136.

Robinson, P. T., Nelson, W. G., O'Hearn, T., and Schmincke, H.-U., 1983, Volcanic glass compositions of the Troodos ophiolite, Cyprus: Geology, v. 11, p. 400–404.

Schmincke, H., Rautenschlein, M., Robinson, P. T., and Mehegan, J. M., 1983, Troodos extrusive series of Cyprus: A comparison with oceanic crust: Geology, v. 11, p. 405–409.

Scott, S. D., Chase, R. L., Hannington, M. D., Michael, P. J., McConaky, T. F., and Shea, G. P., 1988, Sulfide deposits, tectonics and petrogenesis of the southern Explorer ridge: Troodos 1987 Conference, Proceedings (in press).

Shand, S. J., 1949, Rocks of the mid-Atlantic ridge: Journal of Geology, v. 57, p. 89–92.

Spooner, E.T.C., Chapman, H. J., and Smewing, J. D., 1977, Strontium isotopic contamination and oxidation during ocean floor hydrothermal metamorphism of the ophiolitic rocks of the Troodos massif, Cyprus: Geochimica et Cosmochimica Acta, v. 41, p. 873–890.

Steinmann, G., 1905, Geologische Beobachtungen in den Alpen, II. Die Schart'sch Überfaltungstheorie und die geologische Bedeutung der Tiefseeabsätze und der ophiolitischen Massengesteine: Berichte Naturwissenschaftische gesellschaft Freiburg, i, v. 16, p. 44–65.

——— 1927, Die ophiolitischen Zonen in der Mediterranean Kettengebirgen: International Geological Congress, 14th, Madrid, Compte Rendu 2, p. 639–667.

Stolper, E., and Walker, E., 1980, Melt density and the average composition of basalt: Contributions to Mineralogy and Petrology, v. 74, p. 7–12.

Sutton, G. H., Maynard, G. L., and Hussong, D. M., 1970, Widespread occurrence of a high-velocity basal crustal layer in the Pacific crust found with repetitive sources and sonobuoys: American Geophysical Union Monograph 14, p. 193–209.

Viljoen, M. J., and Viljoen, R. P., 1970, The geology and geochemistry of the lower ultramafic unit of the Onverwacht group and a proposed new class of igneous rocks: Geological Society of South Africa Special Publication 2, p. 55–85.

Vine, F. J., and Hess, H. H., 1970, Sea floor spreading, *in* Maxwell, A. E., ed., The sea, Volume 4, Part 2: New York, Wiley, p. 587–622.

Vine, F. J., and Moores, E. M., 1972, A model for the gross structure, petrology, and magnetic properties of oceanic crust: Geological Society of America Memoir 132, p. 195–208.

Wadati, K., 1940, Deep focus earthquakes in Japan and its vicinity: Pacific Science Congress, 6th, Proceedings, v. 1, p. 139–148.

Williams, H., and Hatcher, R. D., Jr., 1983, Appalachian suspect terranes: Geological Society of America Memoir 158, p. 33–53.

MANUSCRIPT RECEIVED BY THE SOCIETY FEBRUARY 10, 1988
REVISED MANUSCRIPT RECEIVED MAY 11, 1988
MANUSCRIPT ACCEPTED MAY 16, 1988

Printed in U.S.A.

Impact of earthquake seismology on the geological community since the Benioff zone

CENTENNIAL ARTICLE

KEIITI AKI *Department of Geological Sciences, University of Southern California, Los Angeles, California 90089-0740*

ABSTRACT

For a small segment of a scientific community to make a significant impact on the broader community, the segment needs to gain strength by developing a consensus within itself. Earthquake seismologists reached a consensus about the existence of deep-focus earthquakes about 1930 and accepted the first quantitative measure of earthquake magnitude in the 1940s. Benioff (1949, 1954) was able to present this consensus to the geological community by publishing a coherent picture of deep tectonic processes in the *Bulletin* of the Geological Society of America.

The next major impact from earthquake seismology after the Benioff zone came from the final acceptance of Reid's elastic rebound theory in the early 1960s after two controversies were resolved. The mathematical framework for determining the earthquake fault process from observed seismograms was firmly established and has been extensively used in the past two decades. One outcome was the new measure of the size of an earthquake, seismic moment, which directly links geological observations of faults with the seismic ground motions.

The latest impact which has begun to be felt by the geological community comes from the consensus about "seismic tomography" which represents the commitment of the seismological community to go beyond the classic one-dimensional earth model to search for a three-dimensional image of Earth's interior.

INTRODUCTION

A seismologist looks at the Earth through a seismograph. The view through a seismograph is quite limited, but when it is properly focused on a subject of geological significance, it gives a sharp understanding of the subject.

The approach of seismologists to the understanding of geologic processes may be compared to Galileo's. Unlike Aristotle, who offered explanations for everything in nature, Galileo studied a very limited subject, such as the falling of a body. Galileo's approach was, however, quantitative; he made measurements of natural phenomena and compared them with calculations. For example, he measured the location of a ball rolling down on a slightly tilted table top at equal time intervals (judged by musical beats), and discovered that the displacement is exactly proportional to the square of the elapsed time, and that the acceleration is therefore a constant. This experience must have given him a true taste of attaining knowledge with a perfect understanding. At the same time, he was struck by the utility of mathematics as a tool in the study of physics.

Like Galileo, seismologists make measurements and compare them with calculations based on so-called "models" abstracted from nature. It is relatively easier for seismologists to follow Galileo's approach than for other workers in geological sciences, because of the very limited view that can be seen through seismographs. For the same reason, unfortunately, it does not happen very often that a discovery from seismology makes a deep and broad impact on the thinking of geologists. The discovery of the subduction zone was such a rare case.

In papers published in 1949 and 1954 in the *Bulletin* of the Geological Society of America, Benioff presented a global picture of the broad zone of shearing produced by the underthrusting of ocean under continent using the results of the first truly quantitative study of seismicity of the Earth made by Gutenberg and Richter (1941, 1949, 1954). In these papers, Benioff attempted a quantitative description of a tectonic process based on the observed locations and magnitudes of earthquakes as well as the results from laboratory experiments on rock failure.

In the present paper, we shall first inquire why the Benioff papers made such a far-reaching and profound impact on the geological community. We shall find that the main reason is the strong consensus reached among seismologists about the extraordinary phenomena called "deep-focus earthquakes" as well as the development and broad acceptance of a quantitative measure of earthquakes called "magnitude" by the seismological community. Benioff was able to present a coherent picture of a deep global tectonic process on the basis of the consensus developed in the seismological community.

We shall then describe the impact made by seismologists since the mid-1960s after they finally resolved two major controversies about the fault origin of earthquakes and established a mathematical framework for determining the earthquake fault process from observed seismograms.

We shall finally describe a new major impact which comes from the consensus and commitment of the seismological community to go beyond the classic 1-D earth model to search for 3-D images of Earth's interior.

DISCOVERY OF DEEP-FOCUS EARTHQUAKES

The greatest impact that seismology has ever made to solid earth sciences is probably the discovery of deep-focus earthquakes. One of the important roles that the Benioff papers played was to have presented the global distribution of deep-focus earthquakes as a well-established result to the broader community of earth scientists.

When first reported in the early 1920s, deep-focus earthquakes received strong objections from the geophysical establishment, which was then enjoying the broad confirmation of the concept of isostasy from analyses of gravity surveys. For example, at the second Pan-Pacific Science Congress held in Australia in 1923, William Bowie, then Chief of Division of Geology, U.S. Coast and Geodetic Survey, spoke about the maximum possible focal depth of an earthquake. He concluded that since we have a crust ~100 km in thickness in isostatic equilibrium, it is most probable that the material below the crust is highly plastic to long-continued stress, and therefore, it would not subject to rupture, fracture, or fissuring. If we replace the word "crust" by the term "lithosphere," Bowie's conclusion is still valid today except for the subduction zone.

The main objective of Bowie's paper was to refute Turner's reports about focal depths as deep as 1,000 km. Turner (1922) reported a great range of focal depths of earthquakes from the variability of travel time observed near the antipode of the epicenter. His evidence was, however, not convincing because some of his earthquakes had so-called "high focus"; that is, they were located up in the air.

In order to establish the existence of a deep-focus earthquake, Wadati (1928, 1929, 1931) took full advantage of the improved high-density seismograph network in Japan, which had then already expanded to about 50 stations. He measured the arrival times of P waves at these stations and was able to map the wave front spreading from the epicenter. He found two events which shared the same epicenter. The wave front for one event spread twice as fast as the wave front for the other. The only explanation for this is a difference in focal depth of several hundred kilometres. To convince people who were bothered by large absolute clock errors at that time, he also measured the S-P times and plotted constant time lines on the map. As expected, the deep-focus earthquake showed a very long S-P time at the epicenter. Wadati's work was followed by Stoneley's (1931) paper on the lack of surface waves from deep-focus earthquakes, and by Scrase's (1932) paper on so-called "depth phases" such as pP, which leave the focus upward to the surface and reflect back to follow the first P wave. Subsequently, the existence of deep-focus earthquakes became widely accepted among seismologists.

Discovery of deep-focus earthquakes must have had a great disruptive and disconcerting effect on the harmonious world of isostasy. It took about 40 years to reconstruct a new harmony, namely, plate tectonics, which embraced both isostasy and deep-focus earthquakes.

The distribution of deep-focus earthquakes presented by Benioff (1949, 1954) is based on the earthquake locations determined by Gutenberg and Richter (1949). The typical seismic zone along a plane dipping under the continent from the oceanic trench was first delineated under Japan by Wadati (1935). For this reason, the Benioff zone is sometimes called the Wadati-Benioff zone.

ESTIMATION OF STRAIN RELEASE USING EARTHQUAKE MAGNITUDE

Benioff (1949) acknowledges the generosity of Gutenberg and Richter, who made available to him a list of magnitudes, hypocenters, and origin times of earthquakes from the manuscript of their book on seismicity of the Earth (1954). Magnitude, the first quantitative measure of earthquake size, was introduced by Richter (1935) and extended by Gutenberg and Richter (1942). Benioff used magnitudes to estimate strain release along an earthquake fault on the basis of Reid's (1910) elastic-rebound theory. In estimating strain release, he used Gutenberg-Richter's (1942) relation between magnitude and energy, and assumed that the effective volume of fault rock participating in an earthquake is common to all earthquakes occurring on a given fault. The cumulative strain release plotted as a function of the time of earthquake occurrence was then compared with the strain-time relation for creep in laboratory experiments for identifying the type of creep on the earthquake fault.

This part of Benioff's idea is no longer acceptable to us. The main problem is his assumption of the common effective volume for all earthquakes from a fault zone. His assumption was challenged by Tsuboi (1956), who believed that the effective volume rather than the magnitude of failure strain varied from one earthquake to another. Tsuboi's view has prevailed. A large earthquake is large not because of a large strain release but because of a large effective volume. This result, sometimes referred to

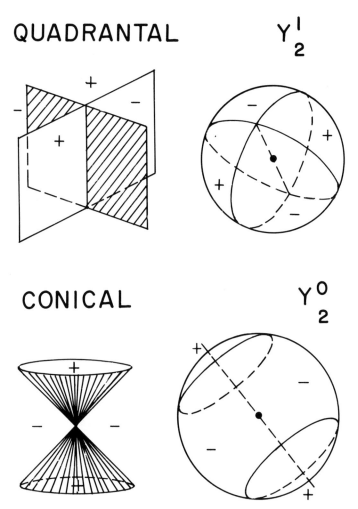

Figure 1. Quadrantal and conical distribution of compressions and dilatations of first P waves. The former can be generated by a source of pressure distributed as the spherical harmonics Y_2^1 and the latter by Y_2^0.

self-similarity, is supported broadly by various observations for a vast range of earthquake sizes (Aki, 1967; Kanamori and Anderson, 1975; Hanks, 1979). Thus Benioff's arguments based on the comparison of the strain-release curve with laboratory results are no longer valid.

FAULT ORIGIN OF EARTHQUAKES AND FAULT-PLANE SOLUTIONS

Geologists are usually skeptical about a quantitative modeling approach because the Earth and geologic processes are infinitely complicated, and many relevant parameters may be missing in a mathematically tractable model. It seems to me that the earthquake may be an exception, judging from the success of the mathematical modeling approach in earthquake seismology. The reason for success may be the relative simplicity of the phenomena. An earthquake is primarily a mechanical process, and the Earth, except in the earthquake source region, behaves as an elastic body during the short time span of the phenomena.

At the time when Benioff's papers (1949, 1954) were written, however, the mathematical modeling of an earthquake as a dynamic fault rupture had not been well established. The elastic-rebound theory of Reid

(1910) was severely objected to by Ishimoto and his colleagues in Japan about 1930 and had yet to survive another controversy called "single-couple or double-couple?" ca. 1960. The first controversy is about the radiation pattern of P-wave first motions, and the second one is about the radiation pattern of S waves.

As the number of seismograph stations increased, it was recognized in the late 1910s that the sense of first motion (compression or dilatation) shows a regular quadrantal pattern as shown in Figure 1. The sense of motion changed across the orthogonal planes, called nodal planes, and the pattern was immediately interpreted by seismologists who believed in Reid's theory as due to slip along a fault oriented along one of the nodal planes.

Then, a controversy began in the late 1920s in Japan, where the world's best local station network was located (for a detailed historical sketch, see Kawasumi, 1937). Ishimoto, who was one of the most influential seismologists at that time, argued that faulting is merely a result of an earthquake, that is, damage to the Earth just like damage to buildings by the shaking of an earthquake. The same argument was revived as late as 1963 by Evison. Ishimoto (1935) proposed that the real origin of an earthquake is a sudden intrusion of magma. A forced intrusion of magma along a narrow channel would generate a radiation pattern of P waves with nodes along the surface of two cones as shown in Figure 1. This pattern was called "conical type" and is rather difficult to distinguish from the quadrantal pattern when the station coverage is incomplete.

At the time of this controversy in Japan, the general solution of the elastic equation of motion in spherical coordinates was already known. Thus, a large number of papers were written about two spherical sources corresponding to the conical and quadrantal radiation patterns, namely, the sources with pressure distribution in spherical harmonics Y_2^0 and Y_2^1 as shown in Figure 1, and theoretical results were compared with observations. The conical source Y_2^0 was later revived by Knopoff and Randall (1970) under the new name of "Compensated Linear-Vector Dipole," and the difficulty of distinguishing between Y_2^0 and Y_2^1 sources still exists even with the modern seismograph network, as witnessed in the recent controversy about the magma intrusion in the Mammoth Lakes area, California. Such cases are, however, exceptions, and we know now that almost all earthquakes, including deep-focus earthquakes, show the quadrantal radiation pattern for P waves.

The fault-plane solutions of deep-focus earthquakes under Japan were studied extensively by Honda and Masatsuka (1952) and were considered by Benioff (1954) to be "in substantial agreement with the assumptions and finding of this investigation." It is, however, now clear that it was not the agreement that was substantial, but the difference. According to Benioff's thrust-fault model, the slip must be parallel to the plane of the Benioff zone. The directions of slip observed by Honda and Masatsuka, however, are ~45° away from the plane of the Benioff zone. The observed direction can be explained if the shear stress vanishes along the plane of the Benioff zone, as expected if the lithospheric slab is subducting in the asthenosphere that cannot sustain shear stress for a long time. Thus, the observed regularity of fault-plane solutions of deep-focus earthquakes was used as evidence supporting plate tectonics by Isacks and others (1968) in their landmark paper titled "Seismology and the new global tectonics."

MATHEMATICAL FRAMEWORK FOR EARTHQUAKE STUDIES

In the early 1960s, seismologists experienced a new understanding of an earthquake when they finally resolved the controversy called "single-couple or double-couple?" mentioned above.

Most seismologists in North America, who believed in Reid's elastic-rebound theory, thought that an earthquake was a slip parallel to the fault plane; therefore, it must be equivalent to a couple with the force in the direction of slip and the arm directed normal to the fault plane. On the other hand, Honda (1962) from Japan showed convincing evidence that although the observed radiation pattern for P waves agrees with the single-couple source, that for S waves does not. The observed pattern of S waves required an additional couple orthogonal to the initial one with the moment of identical magnitude but opposite sign. This model was called a "double couple," and Honda attributed its physical basis to a catastrophic process in which a volume of the Earth originally under shear stress suddenly collapses due to the vanishing of rigidity.

This controversy was resolved by the consistent support of the double-couple source from observations that included improved data on North American earthquakes and by the mathematical proof that the point force system equivalent to a fault slip must be a double couple, and cannot be a single couple. The proof was given for a special case of a homogeneous isotropic body by Maruyama (1963) and for a general case by Burridge and Knopoff (1964).

The consensus finally reached among seismologists about the fault origin of an earthquake might not have impacted the broader geological community as immediately as the discovery of deep-focus earthquakes, but the impact has been equally profound and far reaching. It gave seismologists a mathematical framework which accurately relates the dynamic process at the earthquake source and the structure of the Earth with the observed seismograms. This framework, together with the advent of digital computers and the availability of high-quality data from the Worldwide Standardized Seismograph Station Network (WWSSN) and later networks of digital seismographs, led to two decades of productive work in quantitative seismology.

It is interesting to note that the beginning of plate tectonics came after the above consensus was reached among seismologists. In other words, seismologists were ready to accept earthquakes as a slip along plate boundaries, when Isacks and others (1968) made a comprehensive survey of seismological evidence supporting plate tectonics. Thus, applying the method of determining fault-plane solution developed by Byerly (1938) and Hodgson (1957) to records from the WWSSN, Sykes (1967) was able to turn many seismologists into accepting the plate-tectonic theory by proving Wilson's (1965) idea of transform fault.

The success of plate tectonics in turn definitely supported the fault origin of earthquakes. Slip vectors obtained from fault-plane solutions of numerous earthquakes became one of the basic data set for determining the present-day plate motion by Minster and Jordan (1978). The discovery of the high Q region along the Benioff zone by Oliver and Isacks (1967) and Utsu (1967) was another major contribution to plate tectonics from earthquake seismology.

A fundamental development made in earthquake seismology in the past two decades was the acceptance of seismic moment as the measure of an earthquake source size. The seismic moment offered a direct link between causative faults and the effect of earthquakes generated by them, and its widespread use was possible only after the broad acceptance of the fault origin of earthquakes by the seismological community.

SEISMIC MOMENT

Seismic moment is the moment of component couple of the double-couple point source equivalent to a fault slip as mentioned above. The equivalence is given quantitatively by the relation $M_o = \mu \bar{u} S$, where M_o is the seismic moment, μ is the rigidity of elastic medium containing the fault, S is the area of fault plane, and $\bar{u}$ is the slip averaged over the fault plane. The seismic moment is proportional to the amplitude of long-period waves with wave lengths much longer than the fault length. The seismic source behaves like a point for such long waves. Since the contribution from all elements of the fault would be in phase, their amplitude would be proportional both to the fault area and amount of slip.

The seismic moment was first measured for the Niigata earthquake of 1964 using the records of the WWSSN. The moment was shown to be consistent with the slip and fault area determined from near-field data such as the geodetic and geologic data on deformation, the tsunami source area, and the aftershock area (Aki, 1966). Since then, the seismic moment has served as a direct link between seismological, geological, and geodetical observations. In particular, Kanamori (1971) demonstrated that the seismic moment is a crucial parameter for understanding the behavior of giant earthquakes in subduction zones.

The cumulative seismic moment of earthquakes that occurred in a fault zone was translated to the cumulative slip in the zone by Brune (1968) assuming an appropriate area of the fault, and has replaced Benioff's strain-release curve discussed above. The cumulative seismic-moment curve for a given plate boundary can be directly compared with the slip rate estimated from plate tectonics, and the deficiency of slip at a given instant has been used as a basis for long-term predictions of earthquakes.

The seismic-moment tensor, which includes the orientation of the double couple in addition to the magnitude of moment (or, in general, including possible deviations from a double couple), is now being determined routinely using data from the global digital network (Dziewonski and others, 1983).

In addition to the seismic moment, the accurate mathematical framework relating the earthquake source with observed seismograms enabled seismologists to improve the accuracy of location of earthquakes using wave forms of surface waves and body waves. Samples of some of the results of geological significance, from the mid-oceanic ridge to the Benioff zone, include the following: the extreme shallowness of mid-oceanic ridge, dip-slip earthquakes (Weidner and Aki, 1973), the increase of the maximum focal depth with the age of the oceanic lithosphere (Chen and Molnar, 1983; Wiens and Stein, 1983), the state of stress in the bending part of the subduction plate from precise determinations of focal depth (Forsyth, 1982), and fine details of the modes of deformation within the Benioff zone by identifying the lineations in seismicity pattern with planes of shear failure (Giardini and Woodhouse, 1984), among others.

Through extensive analyses of seismograms using the above mathematical framework, fault-rupture processes from nucleation to stopping have been studied for many earthquakes (Archuleta, 1984). The nucleation occurs either as the failure of a relatively stronger patch (asperity) by stress concentration due to aseismic slip surrounding the patch, or as the growth of a crack originating in a weaker patch. The rupture propagation is fast with velocity usually in the range from 2 to 3 km/sec. The slip velocity (rate of slip on the fault plane) is another stable parameter and is usually about 100 cm/sec. The stopping of rupture occurs either by "running out of gas" (entering the area where there is no stress accumulated), or by encountering barriers, as in the case of the Parkfield earthquake of 1966 (Lindh and Boore, 1981). Sometimes barriers can be left unbroken behind a slip front, as in the case of the Dasht-e-Bayaz earthquake of 1968 (Tchalenko and Berberian, 1975).

Asperities and barriers are important for estimating the rupture length of a potential earthquake fault because they may define the starting and stopping points on the fault and offer a physical basis for the "characteristic earthquakes" proposed by Schwartz and Coppersmith (1984). Asperities and barriers also play important roles in earthquake prediction, because precursory phenomena, such as foreshocks and preslips, are anticipated there.

SEISMIC TOMOGRAPHY: A NEW CONSENSUS

About 50 years after the deep-focus earthquake and 20 years after the seismic moment, the seismological community recently reached another consensus, namely, the value of seismic tomography, which is beginning to make a new major impact on the broader geological community. Although the word "tomography" was borrowed from the medical technology for imaging the internal organs of a human body using X-rays, the method of determining 3-D images of the Earth's interior was independently discovered by seismologists, and the first result on the deep structure of the San Andreas fault was reported at the 1974 Fall meeting of the American Geophysical Union (Aki and others, 1974).

Such a development was very natural for seismologists because they had been well prepared by that time for attacking inverse problems for a large number of model parameters by Backus and Gilbert (1967). A large amount of high-quality data accumulated at newly built seismic arrays such as the Large Aperture Seismic Array in Montana for monitoring underground nuclear explosions and the U.S. Geological Survey Central California array for monitoring regional seismicity also pressured seismologists to seek a new Earth modeling approach beyond the classic laterally homogeneous Earth model.

In fact, the 3-D imaging method first reported at the 1974 Fall AGU meeting was developed at an array center called NORSAR in southern Norway by Aki, Christofferson, and Husebye in the summer of 1974. Their paper, which dealt with the inversion of arrival-time data from distant earthquakes, was published in 1977. The method was extended by Aki and Lee (1976) to the data from local earthquakes occurring within the Earth's volume, of which the seismic image is being sought. This problem presents a unique case never encountered in medical tomography, and various methods have been used to simultaneously determine the 3-D seismic image and the location of earthquake hypocenters (Pavlis and Booker, 1980; Spencer and Gubbins, 1980).

Figure 2 shows an example of a seismic image of the upper mantle under northeastern Honshu obtained by Hasemi and others (1984) using the data from the Tohoku University, Japan, seismic network. The contour drawn in the east-west vertical cross section along 39.8°N shows

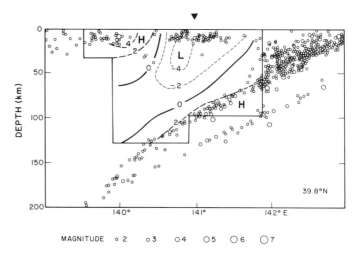

Figure 2. Distribution of P-wave velocity (fractional deviation from a laterally homogeneous initial model in percent) in a vertical cross section along the latitude 39.8°N. A closed triangle on the surface points to the location of volcanic front.

the fractional anomaly of seismic velocity relative to a laterally homogeneous initial model in percent. A high-velocity zone (marked H) coincides with the subducting Pacific plate, and a low-velocity column (marked L) appears beneath the volcanic belt bounded by the volcanic front marked by a solid triangle in the figure.

A similar seismic image of the Benioff zone has been obtained by Hirahara (1981) under Japan, by Michaelson and Weaver (1986) under Washington and Oregon, by Roecker and others (1987) under Taiwan, and by Kuge and Satake (1987) under New Zealand.

Looking at Figure 2, one wonders if the plate-tectonic theory could have been invented solely from the seismic imaging. Figure 2 also shows the remarkable double-seismic zone (Umino and Hasegawa, 1975) which has been attributed to various causes, including unbending of the slab, stresses associated with phase changes, sagging of the slab, and thermoelastic stresses (see, for example, Kawakatsu, 1986).

Numerous applications of the 3-D seismic imaging in local and regional scales have been made to various parts of the Earth in the past decade as summarized by Aki (1982) and Thurber and Aki (1987).

The seismic tomography was extended to the global scale by Dziewonski and others (1977) using the ISC (International Seismological Center) data and has grown into a very exciting area of seismology where the 3-D seismic image of the deep interior of the Earth is related to the pattern of convection current in the mantle (Dziewonski and Anderson, 1984; Tanimoto and Anderson, 1984; Woodhouse and Dziewonski, 1984).

CONCLUSION

The present writer followed the history of Earthquake seismology in the past 80 years to discover a young productive branch of solid earth sciences making occasional but powerful impacts on the broader community. We saw repeated developments of consensus after vigorous controversies, each time contributing to the deeper understanding of structure and processes in the Earth's interior.

As mentioned in the introduction, the view of the Earth through seismograph is limited. The very limitation, however, makes it easier to abstract simple and useful mathematical models from nature. As long as we follow Galileo faithfully in making tedious measurements and comparing them with scrupulous calculations based on models, earthquake seismology will continue to make important contributions to the understanding of structure and processes in the Earth's interior.

ACKNOWLEDGMENT

The present work was supported by the United States Department of Energy under Grant DE-FG03-85ER13336.

REFERENCES CITED

Aki, K., 1966, Generation and propagation of G waves from the Niigata earthquake of June 16, 1964, 2. Estimation of earthquake moment, released energy, and stress-strain drop from G wave spectrum: Tokyo University, Earthquake Research Institute Bulletin, v. 44, p. 23–88.

——— 1967, Scaling law of seismic spectrum: Journal of Geophysical Research, v. 72, p. 1212–1231.

——— 1982, Three-dimensional seismic inhomogeneities in the lithosphere and asthenosphere: Evidence for decoupling in the lithosphere and flow in the asthenosphere: Review of Geophysics and Space Physics, v. 20, p. 161–170.

Aki, K., and Lee, W.H.K., 1976, Determination of three-dimensional velocity anomalies under a seismic array using first P arrival times from local earthquakes, 1. A homogeneous initial model: Journal of Geophysical Research, v. 81, p. 4381–4399.

Aki, K., Christoffersson, A., Husebye, E., and Powell, C., 1974, Three-dimensional seismic-velocity anomalies in the crust and upper-mantle under the U.S.G.S. California Seismic array: EOS (American Geophysical Union Transactions), v. 56, p. 1145.

Aki, K., Christoffersson, A., and Husebye, E., 1977, Determination of the three-dimensional seismic structures of the lithosphere: Journal of Geophysical Research, v. 82, p. 277–296.

Archuleta, R. J., 1984, A faulting model for the 1979 Imperial Valley earthquake: Journal of Geophysical Research, v. 89, p. 4559–4585.

Backus, G., and Gilbert, F., 1967, Numerical applications of a formalism for geophysical inversion problems: Royal Astronomical Society Geophysical Journal, v. 13, p. 247–276.

Benioff, H., 1949, Seismic evidence for the fault origin of oceanic deeps: Geological Society of America Bulletin, v. 60, p. 1837–1856.

——— 1954, Orogenesis and deep crustal structure—Additional evidence from seismology: Geological Society of America Bulletin, v. 65, p. 386–400.

Brune, J. N., 1968, Seismic moment, seismicity, and rate of slip along major fault zones: Journal of Geophysical Research, v. 73, p. 777–784.

Burridge, R., and Knopoff, L., 1964, Body force equivalents for seismic dislocation: Seismological Society of America Bulletin, v. 54, p. 1875–1888.

Byerly, P., 1938, The earthquake of July 6, 1934: Amplitudes and first motion: Seismological Society of America Bulletin, v. 28, p. 1–13.

Dziewonski, A. M., and Anderson, D. L., 1984, Seismic tomography of the earth's interior: American Scientist, v. 721, p. 483–494.

Dziewonski, A. M., Hager, B. H., and O'Connell, R. J., 1977, Large-scale heterogeneities in the lower mantle: Journal of Geophysical Research, v. 82, p. 239–255.

Dziewonski, A. M., Friedman, A., Giardini, D., and Woodhouse, J. H., 1983, Global seismicity of 1982: Centroid-moment tensor solutions for 308 earthquakes: Physics of the Earth and Planetary Interior, v. 33, p. 76–90.

Evison, F. F., 1963, Earthquakes and faults: Seismological Society of America Bulletin, v. 53, p. 873–892.

Forsyth, D. W., 1982, Determination of the focal depths of earthquakes associated with the bending of oceanic plates at trenches: Physics of the Earth and Planetary Interior, v. 28, p. 141–160.

Giardini, D., and Woodhouse, J. H., 1984, Deep seismicity and modes of deformation in Tonga subduction zones: Nature, v. 307, p. 505–509.

Gutenberg, B., and Richter, C. F., 1942, Earthquake magnitude, intensity, energy and acceleration: Seismological Society of America Bulletin, v. 32, p. 163–191.

——— 1954, Seismicity of the Earth and associated phenomena (2nd edition): Princeton, New Jersey, Princeton University Press.

Hanks, T. C., 1979, b values and ω-γ seismic source models: Implications for tectonic stress variations along active crustal fault zones and the estimation of high-frequency strong ground motion: Journal of Geophysical Research, v. 84, p. 2235–2242.

Hasemi, A. H., Ishii, H., and Takagi, A., 1984, Fine structure beneath the Tohoku district, northeastern Japan arc, as derived by the inversion of P-wave arrival times from local earthquakes: Tectonophysics, v. 101, p. 245–265.

Hirahara, K., 1981, Three-dimensional seismic structure beneath southwest Japan: The subducting Philippine Sea plate: Tectonophysics, v. 79, p. 1–44.

Hodgson, J. H., 1957, Nature of faulting in large earthquakes: Seismological Society of America Bulletin, v. 68, p. 611–643.

Honda, H., 1962, Earthquake mechanism and seismic waves: Journal of Physics of the Earth, v. 10, p. 1–97.

Honda, H., and Masatsuka, A., 1952, On the mechanism of the earthquakes and the stresses producing them in Japan and its vicinity: Tohoku University Science Reports, Ser. 5, Geophysics, v. 4, p. 42–60.

Isacks, B., Oliver, J., and Sykes, L. R., 1968, Seismology and the new global tectonics: Journal of Geophysical Research, v. 73, p. 5855–5899.

Ishimoto, M., 1935, Study of earthquakes: Tokyo, Kokin-Shoin, (in Japanese).

Kanamori, H., 1971, Great earthquakes at island arcs and the lithosphere: Tectonophysics, v. 12, p. 187–198.

Kanamori, H., and Anderson, D. L., 1975, Theoretical basis of some empirical relations in seismology: Seismological Society of America Bulletin, v. 65, p. 1073–1096.

Kawasumi, H., 1937, A historical sketch of the development of knowledge concerning the initial motion of an earthquake: Publication Bureau of the Center for International Seismology, Ser. A, Travaux Scientifiques, v. 15, p. 1–76.

Knopoff, L., and Randall, M., 1970, The compensated linear-vector dipole: A possible mechanism for deep earthquakes: Journal of Geophysical Research, v. 75, p. 4957–4963.

Kuge, K., and Satake, K., 1987, Lateral segmentation within the subducting lithosphere: Three-dimensional structure beneath the North Island, New Zealand: Tectonophysics, v. 139, p. 223–237.

Lindh, A. G., and Boore, D. M., 1981, Control of rupture by fault geometry during the 1966 Parkfield earthquake: Seismological Society of America Bulletin, v. 71, p. 95–117.

Maruyama, T., 1963, On the force equivalent of dynamic elastic dislocations with reference to the earthquake mechanism: Tokyo University Earthquake Research Institute Bulletin, v. 41, p. 467–486.

Michaelson, C. A., and Weaver, C. W., 1986, Upper mantle structure from teleseismic P wave arrivals in Washington and northern Oregon: Journal of Geophysical Research, v. 91, p. 2077–2094.

Minster, J. B., and Jordan, T. M., 1978, Present-day plate motions: Journal of Geophysical Research, v. 91, p. 2077–2094.

Oliver, J., and Isacks, B., 1967, Deep earthquake zones, anomalous structures in the upper mantle, and the lithosphere: Journal of Geophysical Research, v. 83, p. 5331–5354.

Pavlis, G. L., and Booker, J. R., 1980, The mixed discrete continuous inverse problem: Application to the simultaneous determination of earthquake hypocenters and velocity structure: Journal of Geophysical Research, v. 85, p. 4801–4810.

Reid, H. F., 1910, The mechanism of the earthquake, in the California earthquake of April 18, 1906: Report of the state Investigation Commission, Volume 2, Carnegie Institution of Washington.

Richter, C. F., 1935, An instrumental earthquake magnitude scale: Seismological Society of America Bulletin, v. 25, p. 1–32.

Roecker, S. W., Yeh, Y. W., and Tsai, Y. B., 1987, Three-dimensional P and S wave velocity structures beneath Taiwan: Deep structure beneath an arc-continent collision: Journal of Geophysical Research, v. 92, p. 10547–10580.

Schwartz, D. P., and Coppersmith, K. J., 1984, Fault behavior and characteristic earthquakes: Examples from the Wasatch and San Andreas fault zones: Journal of Geophysical Research, v. 89, p. 5681–5698.

Scrase, F. J., 1932, The characteristics of a deep focus earthquake; A study of the disturbance of February 20, 1931: Philosophical Transactions, ser. A, v. 231, p. 207–234.

Spencer, C., and Gubbins, D., 1980, Travel time inversion for simultaneous earthquake location and velocity structure determination in laterally varying media: Royal Astronomical Society Geophysical Journal, v. 63, p. 95–116.

Stoneley, R., 1931, On deep-focus earthquakes: Beitrage Geophysics, v. 29, p. 417–435.

Sykes, L. R., 1967, Mechanism of earthquakes and nature of faulting on the mid-oceanic ridges: Journal of Geophysical Research, v. 72, p. 2131.

Tanimoto, T., and Anderson, D. L., 1984, Mapping convection in the mantle: Geophysical Research Letters, v. 11, p. 287–290.

Tchalenko, J. S., and Berberian, M., 1975, Dasht-e-Bayaz fault, Iran: Earthquake and earlier related structure in bedrock: Geological Society of America Bulletin, v. 86, p. 703–709.

Thurber, C. H., and Aki, K., 1987, Three-dimensional seismic imaging: Annual Review of Earth and Planetary Science, v. 15, p. 115–139.

Tsuboi, C., 1956, Earthquake energy, earthquake volume, aftershock area, and strength of the earth's crust: Journal of the Physics of the Earth, v. 4, p. 63–66.

Turner, H. H., 1922, On the arrival of earthquake waves at the antipodes and on the measurement of the focal depth of an earthquake: Royal Astronomical Society Geophysical Supplement, Monthly Notice, v. 1, p. 1–13.

Umino, N., and Hasegawa, A., 1975, On the two-layered structure of deep seismic plane in northeastern Japan arc: Seismological Society of Japan Journal, v. 28, p. 125–139 (in Japanese).

Utsu, T., 1967, Anomalies in seismic wave velocity and attenuation associated with a deep earthquake zone: Hokkaido University, Faculty of Science Journal, Ser. 7, Geophysics, v. 3, p. 1–25.

Wadati, K., 1928, 1929, 1931, On shallow and deep earthquakes: Geophysics Magazine, Tokyo, v. 1, p. 162–202; v. 2, p. 1–36; v. 4, p. 231–285.

——— 1935, On the activity of deep-focus earthquakes in the Japan Islands and neighborhoods: Geophysics Magazine, Tokyo, v. 8, p. 305–325.

Wilson, J. T., 1965, A new class of faults and their bearing on continental drift: Nature, v. 207, p. 343.

Woodhouse, J. H., and Dziewonski, A. M., 1984, Mapping the upper mantle: Three dimensional modeling of earth structure by inversion of seismic wave forms: Journal of Geophysical Research, v. 89, p. 5953–5986.

MANUSCRIPT RECEIVED BY THE SOCIETY OCTOBER 13, 1987
REVISED MANUSCRIPT RECEIVED JANUARY 12, 1988
MANUSCRIPT ACCEPTED JANUARY 13, 1988

Printed in U.S.A.

Plate tectonics and island arcs

CENTENNIAL ARTICLE

WARREN B. HAMILTON   *U.S. Geological Survey, Denver, Colorado 80225*

ABSTRACT

The plate-tectonic concepts that developed rapidly in the late 1960s made possible the understanding of island arcs. Before that time, mobilistic concepts evolved slowly, hindered, particularly in the United States, by an obstructionist geoscience establishment.

The volcanic belts of island arcs form about 100 km above subducting plates. Convergent-plate boundaries evolve complexly with time and, at any one time, vary greatly along their lengths. Seismicity defines positions, but not trajectories, of descending slabs, which sink more steeply than they dip and are overridden by advancing upper plates. Subduction occurs beneath only one side at a time of an internally rigid plate, and the common regime in an overriding plate, behind a surficial accretionary wedge, is extensional, except where a collision is underway. Back-arc-basin lithosphere is built behind, or by, migrating island arcs, which lengthen and increase their curvatures. A collision can involve two active arcs, in which case, intervening lithosphere sinks beneath both of them, or an active margin and a passive one. Either type of collision generally is followed by the breaking through of new subduction, beneath the composite mass of light crust, from a new trench on the outside of the aggregate; conversely, a new subduction system commonly is a by-product of collision. A strip of back-arc-basin crust is in many cases left attached to the aggregate, in front of the new trench, and becomes the basement for a fore-arc basin, the leading edge of which is raised as mélange is stuffed under it.

Sedimentation in trenches is predominantly longitudinal and can be from distant sources. Accretionary wedges are dynamic, being thickened at both toes and bottoms by tectonic accretion and thinned by gravitational forward flow; mélange is largely a product of tectonic imbrication and flowage driven by these conflicting processes, not of submarine sliding. High-pressure metamorphic rocks form beneath overriding plates, not within wedges in front of them.

Arc magmas incorporate much material from the lithosphere through which they rise and vary correspondingly with the evolving composition of that lithosphere. Arc crust is inflated into geanticlines by intrusive rocks and thermal expansion. Submarine island-arc volcanic rocks are widely spilitized, with Na enrichment and Ca depletion, by hydrothermal reaction with sea water. The lower crust of mature island arcs consists of granulite-facies rocks of mafic, intermediate, and felsic-intermediate compositions. The Mohorovičić discontinuity may be primarily a constructional boundary, representing the shallow limit of crystallization of voluminous rocks of ultramafic composition or plagioclase-free mineralogy.

INTRODUCTION

Arc systems develop where oceanic plates sink beneath overriding plates that can be continental, transitional, or oceanic. Many individual arcs are continuous across the diverse crustal types; continental and oceanic arcs belong to a continuum and should be considered together. Arcs are not steady-state tectonic systems but instead evolve and change complexly and rapidly, and different parts of a single, continuous arc can have grossly different histories and characteristics. Arcs commonly are inaugurated by subduction reversals consequent on collisions between other arcs and light crustal masses, and collision histories vary greatly along trend. Oceanic sectors of arcs migrate and lengthen with time, and one sector of a continuous arc can have been inaugurated tens of millions of years later than another sector. Petrologic and crustal features evolve as activity continues in a given sector.

Such characteristics can be illustrated from many modern arc systems. I use the arcs of the Indonesian–southern Philippine–western Melanesian region for my major examples in the following discussion. This is both the region I know best and the region of the greatest modern variety and complexity, and hence of the most informative examples.

In the first section of this report, I review the development of the mobilist concepts that made preliminary comprehension of island arcs possible by 1970. The remainder of the essay is a summary of the characteristics and behavior of island arcs, both as oceanic features and as assemblages accreted to continents.

DEVELOPMENT OF CONCEPTS

Progress that finally permitted our present modest comprehension of island arcs was slow and erratic before the late 1960s and was encumbered by the rejection by most of the geoscience community, particularly in the United States, of large-scale lateral mobilism. My review here of this slow progress emphasizes, where appropriate, publications of the Geological Society of America (GSA) and also emphasizes my own viewpoint and experiences as a pro-drift continental geologist. Menard (1986) presented a superb insider's review of the development of concepts of sea-floor spreading, and then of plate tectonics, from marine-geophysical data in the 1960s. Glen (1982) detailed the development of the paleomagnetic time scale which provided the key component in that evolution.

Mobilists and Stabilists

The first broadly important proposal of a continental-drift theory was Frank Taylor's (1910) paper in the GSA *Bulletin*. Taylor proposed that

---

Geological Society of America Bulletin, v. 100, p. 1503–1527, 9 figs., October 1988.

the Tethyan and circum-Pacific orogenic belts are being crumpled in front of drifting Atlantic and northern continents, which are sliding away from the Mid-Atlantic Ridge and Arctic Ocean. Trenches are depressed by the weight of arcs thrust over them. The 90° inflection in tectonic trends in southern Alaska is an orocline (in Carey's later terminology). Nares Strait is a strike-slip fault, and the geometry of the Canadian Arctic region requires that Greenland, Baffin, the Arctic Islands, and mainland Canada moved as separate plates. Taylor (1860–1938) published many papers on the Pleistocene geology of the Great Lakes region, and analogy with spreading ice sheets provided the impetus for his insightful foray into continental drift.

Occasional major works on the road to plate tectonics appeared outside the United States before 1960. Meteorologist Alfred Wegener (1915, and subsequent revisions) recognized that many paleoclimatic and paleontologic features of the Gondwana continents required prior juxtaposition of those continents, and he deduced that the oceans were underlain by dense material, the continents by light. Emile Argand (1924) saw that orogenic belts within continents were products of continental collisions. He dimly perceived sea-floor spreading and subduction and regarded island arcs as migrating overfolds. Argand recognized that the North Atlantic Ocean had closed during Paleozoic time, producing the Appalachians and Caledonides by collision, then reopened later, and he coined the term "Proto-Atlantic Ocean" for the early ocean. (Wilson, 1966, mistakenly claimed credit for this concept four decades later.) Arthur Holmes (1931 and other papers) added other geologic evidence for drift, which he explained as due to mantle convection currents rising and diverging beneath spreading ocean basins and converging and sinking at migrating trenches—30 years ahead of the next clear statements of this model. A. L. Du Toit (1937) systematized and added great detail to the analysis of geologic ties between the Gondwana continents. S. W. Carey (1958) published a global analysis of mobilistic continental tectonics that has proved correct in many aspects, although he was by then mired in an expanding-Earth concept.

Clegg, Almond, and Stubbs (Clegg and others, 1954) proposed that the magnetization directions they measured in Triassic strata were evidence for post-Triassic rotation and latitudinal shift of Britain. Soon other British (as, Creer and others, 1957; Runcorn, 1959) and other groups were generating continental paleomagnetic data which provided, as they emphasized, strong evidence for drift. Much such evidence had been amassed when Cox and Doell (1960) reviewed global paleomagnetic data for the GSA *Bulletin*; although they then sought stabilist explanations, they were tentatively advocating drift within a few years. A contemporary broad pro-drift synthesis of paleomagnetic data (Deutsch, 1963; written in 1960) was published in a rare symposium, convened by Arthur Munyan, that included pro-drift participants. Although only a few were then writing and lecturing about it, paleomagnetic latitudes were compatible with those deduced from paleoclimatic and paleobiogeographic data, and the complementary data sets required not only continental drift but also the aggregation of continents with orogenic belts between collided lesser continents. (A limited modern review of such relationships was given by Van der Voo, 1988.) Opdyke and Runcorn (1960) argued that late Paleozoic paleowind directions in the western United States fit trade-wind orientations as predicted from paleomagnetic latitudes.

Mobilism, however, came slowly to GSA publications, which before 1969 were filled primarily with papers describing, in stabilist terms, the geology of bits of North America. Gutenberg (1936) proposed that the Atlantic Ocean had opened by gravitational flattening and spreading of the flanking continents, which overrode the Pacific basin; he misinterpreted teleseismic data to indicate the Atlantic Ocean to have a thin continental crust, whereas Wegener had deduced that it does not. Gutenberg (1954) summarized his evidence for a low-velocity asthenosphere, which we now recognize as the zone of decoupling of lithosphere plates from the deeper

mantle. Benioff (1949, 1954) defined the inclined seismic zones that dip arcward from trenches (and that had been recognized earlier by K. Wadati in Japan and by H. H. Turner in South America) and the thrust character of shallow coseismic slip in those zones.

The infrequent broad syntheses of orogenesis in GSA reports mostly were variants of themes of collapsing geosynclines, contraction, thermal uplift and subsidence, and gravitational sliding. The GSA Presidential Addresses on megatectonics by geologist Billings (1960) and geophysicist Birch (1965), both then exceedingly influential, dismissed mobilism; the earlier megatectonic address by petrologist Knopf (1948) ignored it. Gilluly (1949) also neglected drift in his address, but he became a tentative advocate for it in the 1950s and an explicit one in the 1960s.

My own acceptance of a mobilistic view of the Earth had come with my reading in graduate school, about 1949, of Du Toit's (1937) "Our wandering continents." Although the reality of continental drift was demonstrated by evidence such as Du Toit, Holmes, and others had by then assembled, most American geologists and geophysicists dismissed the subject. I began writing and lecturing on pro-drift topics (as, Hamilton, 1963c, 1963d; papers written in 1960 and 1961, respectively) after my 1958 field season in Antarctica, when I realized that the geology predicted by Du Toit's reconstructions was indeed present there. It was generally difficult in those years to get pro-drift materials published outside of the rare mobilistic symposia, whereas it was both easy and commendable to publish anti-drift papers. A 1962 manuscript by L. W. Morley, correctly interpreting the magnetic lineations of oceanic crust as due to sea-floor spreading during alternating periods of normal and reversed geomagnetic fields, was rejected by both *Nature* and the *Journal of Geophysical Research* (*JGR*; Glen, 1982). On the other hand, G.J.F. MacDonald as a young man received widespread acclaim for his repeatedly published calculations (for example, MacDonald, 1964), based on invalid assumptions regarding the Earth's rigidity and heat loss, that continental drift was impossible. (The trend continues in that much published geophysical modeling incorporates bad assumptions, but nowadays the assumptions are mobilistic.) F. G. Stehli published in major journals many anti-drift presentations of his misconceptions of the distribution of Permian fossils indicative of water temperatures (as, Stehli, 1957, 1970, and many between). The *JGR* published an anti-drift rationalization of paleofloras by Axelrod (1963; he has since published important biogeographic evidence for drift). I wrote a detailed pro-drift refutation, but the editor would accept only an undocumented note (Hamilton, 1964). A long review of global late Paleozoic and younger paleontologic, paleoclimatic, and paleomagnetic evidence for continental drift and aggregation that I wrote in the early 1960s for U.S. Geological Survey monographic publication was basically correct in its content but was cycled in series to hostile reviewers who collectively delayed it for 2 years before I gave up on it; only small parts of it were published, within short papers (as, Hamilton, 1964 and 1968, the latter written in 1965).

Very large strike-slip offset on the San Andreas fault was demonstrated in a GSA paper by Hill and Dibblee (1953). My first mobilistic GSA paper (Hamilton, 1961) built on this to link the San Andreas fault to oblique opening of the Gulf of California. (My manuscript had previously been rejected, as foolish speculation, by the *Bulletin of the American Association of Petroleum Geologists*; indeed, my proposed mechanism was foolish.) The distribution of upper Paleozoic tillites in the Gondwana continents had long been recognized by geologists in those continents as powerful evidence for continental drift, although American geologists tended to assume that the deposits at issue were not glacial; Hamilton and Krinsley (1967) reiterated field evidence for glacial origin, added petrographic and electron-micrographic evidence, and argued for drift. J. C. Crowell and associates, beginning with Frakes and Crowell (1967), applied modern sedimentological methods to Gondwana glacial strata and soon thereafter (Frakes and Crowell, 1969, in a GSA paper; papers published elsewhere in 1968) argued that continental drift was required to explain the distribution of the glacial materials.

Spreaders and Subducters

The early evidence for continental drift came from the continents, and it was unclear how the ocean floors fit into the picture. Some advocates of continental drift, from Wegener on, had visualized continental rafts floating across dense oceanic material, whereas others, from Taylor on, had visualized sea-floor spreading. Direct evidence for spreading came with the oceanographic data accumulated during the 1950s (Glen, 1982). Bruce Heezen and Marie Tharp (as, Heezen and others, 1959) presented GSA bathymetric maps of the oceans, from which Heezen argued in other papers for the globe-girdling character of ridges that spread because the Earth is expanding, whereas Maurice Ewing (as, Ewing and others, 1964) for a while argued that the ridges were not spreading. Raff and Mason (1961) presented a map of sea-floor magnetic lineations west of the northwest United States, for which Vacquier, Raff, and Warren (Vacquier and others, 1961) recognized strike-slip, but not spreading, significance.

Holmes (1931) visualized what we would now term sea-floor spreading, subduction, and migrating plate boundaries, all of which were ignored by most of the earth-science community. The suggestion by Griggs (1939) that ocean floors underthrust continents from trenches also met general indifference. U.S. Navy bathymetry of the trenches, island arcs, and marginal basins of the west-central Pacific was presented by Hess (1948), who then regarded trenches as formed by local downbuckling. Dietz (1954) published a Japanese Navy map of about the same region and proposed, correctly, that the Japan and Okhotsk Seas had opened as island arcs and continental fragments migrated away from Asia. Coats (1962) had perhaps the first clear visualization of subduction and postulated that magmatic arcs formed by melting of sedimentary rocks thrust beneath arcs along Benioff seismic zones.

The lasting recognition that sea floor produced by spreading might disappear beneath arcs and continents by subduction probably came first to Hess (1962; Glen, 1982), but Dietz (1961) had it in more sophisticated form. Their initial views were essentially two-dimensional and were in that sense more primitive than those of Holmes, whereas Wilson (1961) saw that spreading ridges must themselves migrate and change shapes and lengths. Wilson was a major contributor to mobilist concepts in the early 1960s, although he had been a militant stabilist in the 1950s.

Plate Tectonicists

The short period from 1963 to 1968 saw the demonstrations from geophysical data that the Earth's lithosphere is fragmented into plates, all moving relative to all others—pulling apart at ridges, sinking beneath one another at trenches, sliding past one another on transform faults. This drama played in journals, especially *JGR*, *Nature*, and *Science*, which previously had been bastions of stabilism. The history of this development has been discussed in detail by Glen (1982), Menard (1986), and others.

Vine and Matthews (1963), unlike Morley, were able to publish their proposal that the magnetic anomalies parallel to oceanic ridges recorded conveyor-belt crystallization during spreading in alternating periods of normal and reversed magnetic polarity. Their suggestion generated mostly a mixture of disinterest and hostile responses for 3 years but then was proved correct by various groups, notably by the well-organized Lamont group (for example, Heirtzler and others, 1968), which demonstrated the magnetic symmetry of ridges and integrated the dating of oceanic crust by deep-sea drilling with an extrapolated geomagnetic time scale. Coode (1965) and Wilson (1965) proposed simultaneously that the fracture zones known to mark offsets of the ridges did not postdate the ridges but rather were "transform faults" (Wilson's term) formed by stepping of spreading between ridge segments that were perpendicular to the spreading direction; Sykes (1967) demonstrated that the slip sense of fracture-zone earthquakes accorded with this concept. Other geophysicists added more confirming data for the evolving mobilistic concepts. So many independent lines of evidence required the same conclusions that the general validity of those conclusions was quickly established.

Euler-plate geometry was used implicitly by Carey (1958), who made reconstructions on a transparent hemisphere moved about a globe and drew reconstructions incorporating spherical geometry. Bullard and others (1965) used a computer to fit Atlantic continents together and specified the Euler pole required. First to formalize the global Euler-plate behavior required by the spherical geometry of spreading ridges and transform faults probably was Morgan (1968). Only a few months behind were McKenzie and Parker (1967; their paper was written after Morgan's), and behind them Le Pichon (1968), who had been publishing stabilist papers several years before. McKenzie and Morgan (1969) analyzed the geometric behavior of evolving triple junctions between plates.

Plate tectonics (at that time, "the new global tectonics") was a demonstrated reality. Most geophysicists who were paying attention were quickly convinced, whereas most geologists lagged behind. (My own conversion, from a previously muddled view that had incorporated little awareness of marine geophysics, came in 1968, a year or two behind the involved geophysicists.) It remained to apply the concepts to global geology.

Continental and Island-Arc Geologists

Geologists at last had a framework within which to place the empirically related features of island arcs and continents. Davis (1969) discussed the Klamath Mountains in terms of Mesozoic subduction imbrication, which he had in part perceived earlier (Davis, 1968). Interpretations of island arcs in plate-tectonic terms first appeared in the GSA *Bulletin* in 1969 (Isacks and others, 1969; Molnar and Sykes, 1969; Rodolfo, 1969), although rearguards Von Huene and Shor (1969) argued that the Aleutian Trench recorded downwarping, not subduction. The same year came my analysis (Hamilton, 1969a) of California as a product of Jurassic tectonic accretion and Cretaceous Andean-style tectonics; this paper was the first clear statement of the assembly of a broad orogenic tract from island arcs and other bits conveyor-belted in from far away. (The manuscript was held up for half a year by U.S. Geological Survey reviewers and a supervisor who considered it too radical for public display, and an abstract based on it was one of the few volunteered papers rejected for presentation at a 1970 regional GSA meeting.) Dickinson (1969, 1970c) and Hamilton (1969a, 1969b) argued that batholiths such as the Sierra Nevada were the roots of continental volcanic arcs and were not products of anatexis in "geosynclines." This expanded on the analysis by Hamilton and Myers (1967) that batholiths in general are overlain by silicic volcanic complexes and underlain by migmatites; our concept was widely rejected at the time. The relationship of belts of high-pressure, low-temperature metamorphic rocks such as blueschists to likely trenches was recognized by Miyashiro (1961), who, like Ernst (1965), invoked depression by "downbuckling," whereas Blake and others (1969) and Coleman (1967) appealed to "tectonic overpressures." Blueschists were put in the context of subduction by Ernst (1970) and Hamilton (1969a). Hsü (1968, and other papers) began to make sense of the Franciscan mélanges, which he at that time attributed to gravity sliding, of coastal California; Hamilton (1969a) put them in an accretionary-wedge context.

In 1969, only a dozen papers on mobilist topics were presented at all seven GSA meetings—and half of those papers took stabilist positions. Late 1969, however, saw the important GSA Penrose Conference on "The meaning of the new global tectonics for magmatism, sedimentation, and metamorphism in orogenic belts," convened by William R. Dickinson at

Asilomar, California (Dickinson, 1970a, 1970b). The 90 attendees included not only most of the few geologists who were already active in the new field but also most of those who would make important plate-tectonic geologic contributions during the 1970s. Dickinson's conference produced an abrupt dissemination of awareness that convergent-plate tectonics controls much of the evolution of continents.

The year 1970 brought a large increase in mobilist publication by the GSA. Bracey and Vogt (1970), Grow and Atwater (1970), and Luyendyk (1970) presented important papers on the tectonics of island arcs. Atwater (1970) put the Cenozoic geology of western North America into the essential framework of evolving triple junctions. Bird and Dewey (1970) explained the Appalachians, and I (Hamilton, 1970) the Uralides, as products of continental collisions as intervening oceans were subducted beneath flanking continents and beneath intervening arcs that were accreted to the continents. Coney (1970) summarized what the synthesizers were learning. There were other plate-tectonic and pro-drift papers—and also rearguard papers arguing that subduction had nothing to do with trenches or continental tectonics. Among particularly important 1970 papers on plate tectonics and continental geology in other journals was that by Dewey and Bird (1970), on broad aspects of orogenic systems, and that by Dickinson (1970c), integrating volcanism, plutonism, and sedimentation into a plate framework.

Papers dealing with plate tectonics, subduction, and island arcs have been numerous in GSA publications since 1970. I mention here a few of the papers from the early 1970s that advanced understanding of continental and arc geology. Island-arc migration and back-arc spreading were documented by Karig (1971, 1972) and Sclater and others (1972). Grow (1973) presented the best geophysical analysis to that time of an accretionary wedge and fore-arc basin, those of the Aleutian Islands. Silver (1971a, 1971b) applied marine geophysics to analysis of the Mendocino triple junction, critical for comprehension of California tectonics. Barbat (1971) and Page (1972) much advanced understanding of the nature of the contact along which Cretaceous California had overridden oceanic materials. Broad syntheses were attempted in plate-tectonic terms by Dewey and others (1973) and Ernst (1973) for the Alpine system, James (1971) for the Andes, Hatcher (1972) for the southern Appalachians, and Malfait and Dinkelman (1972) for the Caribbean region.

Although plate tectonics provided the framework within which the behavior of island arcs could potentially be understood, the requisite geologic data had long been accumulating. Hess (1948) and Dietz (1954), among others, focused attention on the tectonic bathymetry of arc systems. Kay (1951) recognized that island arcs are important components of continental orogenic belts—a major advance, although he explained their presence with stabilist geosynclinal theory. Hess (1955) recognized the continuity of belts of mantle peridotites—mostly accreted ophiolites, in modern parlance—in orogenic belts but also sought an explanation in geosynclinal theory and vertical tectonics. Dietz (1963, 1966) made early attempts to relate continental drift to conveyor-belting atop convecting mantle and continental-margin tectonics to convergence and subduction. I showed (Hamilton, 1963a, 1963b) that the metavolcanic rocks of western Idaho were of oceanic island-arc petrology and had been overthrust from the east by continental-crustal rocks. In Hamilton (1963d; written in 1961), I proposed that the north and south flanks of the Caribbean and Scotia arc systems had been plated out on the sides as the arcs migrated eastward, and in Hamilton (1966), I suggested that western Pacific arcs were migrating eastward faster than were the continents behind them. In my 1966 paper, I argued on petrologic grounds that both oceanic island arcs and the floors of marginal seas were incorporated in continental "eugeosynclines." Krause (1965, 1966) presented moderately mobilistic explanations for Indonesian and Melanesian arcs and marginal seas. Burk (1965) sought vertical-tectonics explanations for the transition he described from offshore Aleutian arc to onshore Alaskan one. Dickinson and Hatherton (1967) and Kuno (1966 and earlier papers) showed how cross-strike variations in island-arc volcanoes correlate with depth to the inclined seismic zone beneath, although not until later (Dickinson, 1969, 1970c; Hatherton and Dickinson, 1969) did they perceive the relationships in terms of subduction.

Vening Meinesz (1954) summarized for GSA his pioneer gravity work with Indonesian arcs. The "isostatic" anomalies he calculated were very strongly negative along the fore-arc ridges, and he proposed that trenches are held down dynamically, far out of gravitational equilibrium, as "tectogenes." His gravity anomalies were calculated with the invalid assumption that all material beneath the sea floor has the same density. In their analogous report on the gravity anomalies of the West Indies, Ewing and Worzel (1954) recognized that thick low-density material, not dynamic imbalance, produced the negative anomalies; they did not attempt an explanation for the trenches in that paper, although in other papers of the period, they argued for extensional origins. We now know that (as anticipated by Ewing and Worzel) the maximum thickness of accretionary wedges lies along the fore-arc ridges and that Vening Meinesz's anomalies were dominated by the thicknesses of those wedges. Free-air anomalies of the ridges are positive and broadly correlative with bathymetry (Watts and others, 1978), and so the accretionary-wedge load is in part supported by beam strength of the subducting plate. Karig and others (1976) evaluated quantitatively the depression of subducting plates by accretionary-wedge loading.

Current Status

Plate tectonics has given us the framework within which to begin to comprehend the geology of continents and island arcs. Relationships between the tectonic and magmatic components of modern convergent-plate systems are so systematic that derived generalizations now allow predictions, the testing of which refines our understanding. The obvious success of the plate-tectonic paradigm has, however, produced a complacency that has cluttered the geological and geophysical literature with invalid convergent-plate models that reflect naive assumptions rather than understanding of actual plate systems. The problems are being passed on to the next generation; I have just examined eight current physical-geology textbooks, all of which incorporate gross misconceptions regarding plate convergence, and most of which are little better regarding plate divergence.

PLATE TECTONICS

Seven very large lithospheric plates, and numerous mid- and small-sized ones (the concept of coherent plates breaks down at the small-scale end), are now all moving relative to all others. All plate boundaries—divergent, convergent, strike-slip, and oblique—are also moving, at widely varying rates, and most boundaries change greatly in length and shape with time. Although plates tend to be internally rigid and to interact mostly at their boundaries, parts of many plates undergo severe internal deformation. Relative velocities between adjacent plates presently range up to about 13 cm/yr.

Mechanism

"Absolute" velocities of present large plates—their relative velocities in an approximate zero-sum frame, with or without qualifications regarding true polar wander (Davis and Solomon, 1985) or rationalizations regarding semifixed hot spots—correlate positively with the lengths of ridges and of trenches along their perimeters and negatively with the proportion of continental lithosphere within them (Carlson, 1981). It appears from the quantitative correlations between these parameters that plates are propelled primarily by gravitational forces and that on average, pull by the descending slab is about 2.5 times as important in moving plates as is slide of plates away from ridges, whereas thick continental lithosphere retards motion by drag (Carlson, 1981). The 80 or 100 km of relief of the base of an oceanic-lithosphere plate, between lithosphere and less-dense asthenosphere, is much more important in producing ridge slide than is the 3 or 4 km of bathymetric relief of the top of the plate. Major plate motions thus apparently are controlled by large lateral variations in lithosphere density and thickness that result primarily from cooling (Carlson, 1981; Hager and O'Connell, 1981), although much of the negative buoyancy, mechanical behavior, and seismicity of subducting slabs is due to density-phase changes (Pennington, 1983; Rubie, 1984). Many complications are discussed by Jarrard (1986). Velocity of lithosphere is in general greater at low latitudes than at high latitudes, and so the Earth's rotation likely is an additional factor in driving forces (Solomon and others, 1975), perhaps by a gyroscopic feedback mechanism. Motions of small plates are primarily by-products of motions of adjacent large plates.

Convection in the upper mantle is largely a complex product, not a major cause, of plate motion (Alvarez, 1982). Spreading ridges form where plates move apart and hot mantle wells into the gap, and ridges migrate and change shape and length at widely varying rates. The return flow that compensates for lithosphere motions probably occurs mostly in the asthenosphere, where it likely is pervasive beneath oceanic plates (Chase, 1979) but may be concentrated in channels beneath thin lithosphere, as in the Scotia and Caribbean gaps, where continents are involved (Alvarez, 1982; as he emphasized, this process was anticipated by Hamilton, 1963d).

Hot spots—sites of long-continuing asthenospheric upwelling beneath moving plates, shown at the surface as migrating zones of volcanism—figure in many analyses and explanations of plate kinematics and commonly are assumed to represent sources of heat fixed in the mantle. An alternative explanation is that hot spots are products of propagating rifts in the lithosphere and hence are primarily responses to cooling at the top rather than to heating at the bottom. Hot-spot volcanism is controlled by upper-lithosphere fractures which can be explained as related to interacting regional-plate and volcano-loading stresses. (For discussions and citations of some of the relevant references, see Clague and Dalrymple, 1987, and ten Brink and Brocher, 1987.) Even the best behaved of the oceanic hot spots are moving relative to one another with velocities of 1–2 cm/yr (Molnar and Stock, 1987), perhaps more. Many linear volcanic chains proposed as hot-spot tracks in fact lack systematic age progressions (as, Turner and Jarrard, 1982), and the best examples display much irregularity. The apparent requirement of the hot-spot concept that lava productivity not be a function of plate velocity is not met (McNutt, 1988).

Heat and Variations with Time

Plate motions are responsible for most of the Earth's heat loss. Of the total heat lost by the Earth, about 60% is lost by magmatism at spreading ridges and by the subsequent cooling of new oceanic lithosphere as it moves away from ridges (Sclater and others, 1981). Because the rate of heat loss by the Earth has probably decreased with time and because ancient crustal nonmagmatic thermal gradients demonstrable by petrologic thermobarometry were little if any steeper than modern ones, it seems likely that plate motions have become slower on average with time. There may have been fluctuations within this progression, representing variations of 10% or 20% in the rates of plate generation and consumption (compare Parsons, 1982), and there certainly have been major unidirectional changes in the petrologic evolution of crust and mantle. Nevertheless, plate tectonics appears to have operated, broadly as it does now, at least during Proterozoic and Phanerozoic time. Archean crust displays the effects of much more voluminous, and in part higher temperature, magmatism than

that of younger time and of expulsion of light, continent-forming elements directly from a little-differentiated mantle, and specific processes that then operated are much debated. Many of us see Archean geology as likely recording the more-rapid motions of more and smaller plates than those of later time.

Subduction

Much published tectonic speculation and geophysical modeling have been built on the false assumption that a subducting plate rolls over a hinge and slides down a slot that is fixed in the mantle, and that overriding plates commonly are shortened compressively across their magmatic arcs and belts of foreland deformation. These assumptions are disproved both by the characteristics of modern convergent-plate systems in which the subducting plate is of normal oceanic lithosphere and the Benioff seismic zone has a moderate to steep inclination, and by analyses of "absolute" plate motions. Hinges commonly retreat—roll back—into incoming oceanic plates as overriding plates advance, even though at least most subducting plates are also advancing in "absolute" motion. Subducting slabs sink more steeply than the inclinations of Benioff seismic zones, which mark positions, not trajectories, of slabs. Perhaps the most obvious evidence for rollback comes from the fact that the Pacific Ocean is becoming smaller with time as flanking continents and marginal-sea plates advance trenchward over ocean-floor plates, but many other types of evidence for the phenomenon have been presented by, among others, Carlson and Melia (1984), Chase (1978), Dewey (1980), Garfunkel and others (1986), Hamilton (1979), Hawkins and others (1984), Kincaid and Olson (1987), Malinverno and Ryan (1986), Molnar and Atwater (1978), and Uyeda and Kanamori (1979). As most of these authors emphasized, the typical regime in an overriding plate above a sinking slab is one of extension, not shortening.

A corollary, often overlooked by geologists producing palinspastic cartoons, is that subduction can occur beneath only one side at a time of an internally rigid plate. Retrograde motion could occur only if a dense slab could push light mantle forward and upward out of its way—an impossibility in a gravity-dominated system.

Exceptions that can be argued to counter these interpretations are of doubtful validity. Plate-motion analyses which deduce retrograde motion for the Mariana arc and trench are no stronger than their poorly constrained estimates of internal motions within eastern Asia and its marginal seas. Subduction now occurs inward beneath both sides of the Caribbean region—Antilles on the east, Central America on the west—but poorly understood plate boundaries intervene. Subduction now occurs at both east and west sides of southern Mindanao, but the trajectories, and in part even boundaries, of the many small plates in that region are so poorly constrained that this cannot yet be evaluated properly.

Arc Migration and Back-Arc Spreading

Karig (as, 1972, 1975) demonstrated that the Mariana island arc has migrated Pacificward as new back-arc–basin oceanic crust formed behind it. Karig and many others since (as, Taylor and Karner, 1983) have found that island arcs generally migrate in such fashion. Some migration is accomplished by the splitting of the magmatic arc and the migration of the forward half away from the rear half, and some, by irregular sea-floor spreading behind the entire arc. The magmatic welt can move forward with the advancing part of the overriding plate, can be abandoned as a remnant arc on the relatively retreating part, or can be split longitudinally between them. Oceanic island arcs do not bound rigid plates of old lithosphere but, instead, mark the fronts of plates of young lithosphere that are widening in the extensional regimes above sinking slabs. Oceanic arcs commonly are not inaugurated by the breaking of subduction through old oceanic crust but, rather, break through near boundaries between thin and thick crust and migrate over the plates of thin crust (Hamilton, 1979; Karig, 1982). In any one system, periods of back-arc spreading may alternate irregularly with periods of magmatism along a volcanic-arc welt (Crawford and others, 1981; some authorities disagree).

An island arc should be viewed as a product of a subducting slab rather than as a fixture of an overriding plate. A belt of arc-magmatic rocks forms above that part of a subducting slab, the top of which is 100 km or so deep—and migrates to track that contour as the slab falls away. Mechanisms of back-arc spreading are still debated, but it appears to me, as to some others (perhaps including Hawkins and others, 1984, and Shervais and Kimbrough, 1985), that although some oceanic back-arc–basin lithosphere forms by regular or irregular spreading behind an arc, much forms instead by the rapid migration of a magmatic arc which plates out a variable-thickness sheet of arc crust rather than forming a full island-arc welt of thick crust.

Arc Festoons

Arcs increase in curvature as they migrate. A migrating arc becomes pinned where it encounters thick crust in the subducting plate that either is nonsubductible or that forms a stiffening girder, and festoons and sharply curving arcs result where migration continues away from such obstructions (McCabe, 1984). Pinning against the Caroline Ridge may explain the Yap-Mariana syntaxis, and pinning against the Emperor Seamount Ridge may explain the Kamchatka-Aleutian one.

Ophiolites

On-land ophiolites are sections of upper oceanic lithosphere that were long assumed to be samples of spreading mid-ocean–ridge materials. Many researchers now believe that instead, most, perhaps all, large sheets of ophiolite incorporated tectonically into the continents are products of arc magmatism, back-arc spreading, or both together (Bloomer and Hawkins, 1983; Coleman, 1984; Hawkins and others, 1984; Pearce and others, 1984; Shervais and Kimbrough, 1985). Much has yet to be learned about the pre-collision evolution of these complexes, but the mechanisms of irregular spreading and fast-migrating arcs appear capable of explaining many relationships.

The Eocene Acoje ophiolite of western Luzon was described, as a "nascent island arc," by Hawkins and Evans (1983). The moderately dipping Acoje section exposes the entire crust, about 9 km thick, and about 10 km of the underlying mantle. All but the top 1 km of the mantle section consists of serpentinized and tectonized residual harzburgite and subordinate dunite and chromite; clinopyroxene-rich pods that increase in abundance downsection crystallized from late melts that may have been either introduced or segregated nearby. The upper 1 km or so of the geophysical mantle, but the basal 1 km of the arc-magmatic section, consists of undeformed cumulates of olivine and clinopyroxene. These are intercalated, over a thickness of several hundred meters, with the basal part of the gabbroic rocks that make up the lower 7 km or so of the overlying crust, which has a total thickness of about 9 km. Most of this gabbroic section consists of layered-cumulate two-pyroxene gabbro. The cumulates grade upward into massive gabbro and norite, about 1 km thick, in the upper part of which are abundant small plutons and dikes of plagiogranites (hornblende tonalite and leucotonalite). The top 1 or 2 km of the crustal section consists of dikes, sills, and pillow flows of basalt compositionally like that of modern primitive island arcs rather than like spreading-ridge lava. The crustal section is much thicker than that formed at mid-ocean ridges. Formation in a steady-state magma chamber nevertheless seems likely, and rapid migration of a belt of arc magmatism in a spreading-marginal-basin setting can be inferred.

Little in the description just given of the Acoje ophiolite, except for irregular variations in thickness, would have to be changed to apply to many sections of ophiolites in continental accreted terranes around the world. The Cretaceous Oman ophiolite of the Arabian Peninsula (Lippard and others, 1986) and the Jurassic Coast Range ophiolite of California (Hopson and others, 1981) are well-studied examples that are dimensionally and petrologically similar to the Acoje complex, although Lippard, Hopson, and their associates favored explanations in terms of spreading-ridge magmatism. Fragments of ophiolites of these types have been dredged from the arcward slopes of trenches—the leading edges of overriding plates—in open-ocean settings in which there is little development of masking accretionary wedges (Bloomer and Hawkins, 1983).

There appear to me to be two major processes of emplacement of ophiolites within orogenic belts, and neither of them represents capture of random bits of spreading-ridge lithosphere. One mode is in the collision of an advancing arc with a continent or other island arc, the thin ophiolitic leading edge of the overriding plate ramping onto thick-crustal parts of subducting plates. Such thrusting is in the sense of subduction. (The hypothetical process of "obduction," whereby a great sheet of oceanic lithosphere is split from a subducting slab and shoved, in the opposite sense of thrusting, atop the thick crust of an overriding island-arc or continental plate, is invoked by many writers—but the process defies mechanical analysis and has yet to be proved to have operated anywhere. Confusion is introduced by those writers who misuse the term "obduction" to imply onramping in the sense of subduction, which is opposite to the original definition of the term.)

The second major process of ophiolite emplacement is a subset of the first and is a common by-product of an arc collision. A new subduction system of opposite dip breaks through behind the collided arc and leaves attached to it a strip of back-arc–basin crust that becomes raised as accretionary-wedge materials are stuffed beneath it. Such an ophiolitic strip may remain at the leading edge of a plate, or may become part of a suture system after other crustal masses collide with it. Examples are noted in subsequent sections.

TECTONICS: ARCS OF INDONESIA AND VICINITY

Introduction

The complex characteristics and histories of island arcs are exemplified by the arcs of Indonesia and surrounding regions. The active tectonism and magmatism there record the complex interactions of the Asian, Pacific, and Indian-Australian lithosphere megaplates and of dozens of lesser plates. In various reports and maps, culminating in a monograph (Hamilton, 1979) and an accompanying tectonic map (also published separately: Hamilton, 1978a), I integrated onshore and offshore geological and geophysical data for Indonesia, southeast Asia, the southern Philippines, western Melanesia, and the adjacent seas into a synthesis of modern plate behavior and of the evolution of plate-tectonic features. My concepts have evolved since I completed the book, but interpretations not otherwise credited herein come primarily from that monograph, which contains the relevant documentation both as voluminous data newly reported there and as synthesis of the findings of others. Some reports published since the completion of the book are cited herein and also in a more detailed update (Hamilton, 1988b); the new data require modification of details of my synthesis but have in general substantiated it. Figures 1 and 2 show locations of features discussed, and Figures 3 and 4 illustrate some of the concepts. The bathymetric map by Mammerickx and others (1976) is more detailed than the map used as the base for my maps. Map summaries of offshore geophysical data, in part incorporating data and interpretations from Hamilton (1974a, 1974b), were presented by Anderson and others (1978, thermal properties), Hayes and Taylor (1978, earthquakes), Hayes and others (1978, crustal structure), Mrozowski and Hayes (1978, sediment isopachs), Watts and others (1978, free-air gravity), and Weissel and Hayes (1978, magnetic anomalies).

The diverse subduction systems of the Indonesian region record the interactions between three megaplates and many smaller plates. Relative to internally stable northwestern Eurasia, the India–Indian Ocean–Australia megaplate is moving approximately northward in this region, whereas the Pacific megaplate is moving west-northwestward. The Asian continental megaplate is fragmented into dozens of internally deformed subplates. Many small oceanic and continental plates intervene between parts of the megaplates, and many of these small plates also have been much deformed internally. Southeast Asia was crowded eastward out of the way of the indenting Indian subcontinent and has since been rotating clockwise over the oceanic Bay of Bengal (Hamilton, 1979; Tapponnier and others, 1986). Convergence between Indian and Asian megaplates is now being taken up primarily by the continuous Burma-Andaman-Sunda-Banda subduction system, whereas that between Pacific and Asian megaplates is taken up on many subduction systems that trend mostly northerly into the Philippines and along boundaries farther east. Complex subduction and strike-slip systems separate the plates in the zone of interaction between Indian and Pacific megaplates, along and north of New Guinea and in the tectonic knot in northeastern Indonesia and surrounding regions.

If present gross plate motions continue for another 50 m.y. or so, the continental scraps, composite island arcs, and accretionary wedges of much of the Indonesian–Philippine–northern Melanesian region likely will be squashed between Australia and Asia. The result will be another broad orogenic terrane akin to those we elsewhere term Tethyan, Hercynian, Caledonian, Pan-African, and so on.

Sunda Subduction System

A great subduction system is continuous from Burma around the Banda Arc. In this section, I discuss the central 3,000-km Sunda sector, along Sumatra, Java, Bali, and Sumbawa, of this plate boundary. This sector consists of concentric, arcuate tectonic features that typify those along other active margins of continents, mature island arcs, and the transitions between them. In the south is the trench, and northward from it rises the surface of the accretionary wedge, at the front of the overriding plate, to a culmination at a fore-arc ridge.[1] Islands stand on the ridge along Sumatra, but the ridge is wholly submarine south of Java, Bali, and Lombok. Between the ridge and the magmatic arc is the submarine fore-arc basin. Along the Sunda sector, Indian Ocean lithosphere is being subducted, at high to moderate convergence angles, beneath an arc system that changes along strike from continental in Sumatra through transitional in Java to oceanic in Bali and Sumbawa. This sector of the subduction system has been active only since middle Tertiary time.

Trench. The Sunda Trench, like trenches that mark traces of subduction systems along continental margins and mature island arcs elsewhere, has inner and outer "walls" that slope only 7° or so. The trench does not mark either an abrupt hinge in the subducting Indian Ocean lithosphere or the contact between lithosphere plates, but rather is the dihedral angle between a surficial accretionary wedge, in front of the overriding plate, and oceanic lithosphere depressed by that wedge. An outer rise on the oceanward side of the trench is an elastic response to that depression. The tectonic hinge, where the subducting plate tips downward into the mantle, lies 100–200 km arcward from the bathymetric trench. Trenches of oceanic island-arc systems commonly are illustrated with reflection profiles with vertical exaggerations of 25× or so, giving the visual impression of very steep slopes, but actual slopes of these also commonly are gentle.

Clastic sedimentation in trenches is primarily in the form of longitudinal turbidites, and the long profiles of trench-floor fills slope gently away from the sources. Sunda Trench sediments as far southeast as Java came largely from the Ganges and Brahmaputra Rivers, 3,000 km away (see also Ingersoll and Suczek, 1979, and Moore and others, 1982); the supply has recently been cut off by collision of the Ninetyeast Ridge with the trench in the Andaman sector. Aleutian Trench turbidites sluice a similar distance from Alaskan rivers. Source terranes thus need bear little similarity to nearby parts of the overriding plate against which trench turbidites are plated in an accretionary wedge. Dickinson (1982) demonstrated this to be true for various fossil accretionary wedges around the Pacific Ocean. Continental detritus sluiced along trenches, or abyssal-fan materials from continents, can be accreted to, and subducted beneath, oceanic sectors of arc systems.

Accretionary Wedge. Sediments and other materials scraped from subducting Indian Ocean lithosphere accumulate, snowplow fashion, in the accretionary wedge at the front of the overriding Sunda plate. The surface of the wedge is furrowed by longitudinal ridges and basins defined by imbricate thrust faults (Karig and others, 1980b). Trench fill can be seen on reflection profiles to be scraped off at the front of a wedge, the shallowest materials being accreted against the toe, the deeper being accreted beneath the wedge farther back. Quaternary coral reefs are raised high above sea level on islands along the fore-arc ridge—the crest of the accretionary wedge—and the rapid uplift presumably is a result of thickening of the wedge by underplating. The base of the Sunda wedge—the top of the subducting plate—dips very gently at least as far arcward as the crest of the fore-arc ridge; the wedge is a thin, dynamic debris pile only 15 km or so thick 75–150 km from the trench. Reflection profiles showing internal structure of this and analogous wedges elsewhere generally display semiconstant imbrication angles, dipping 30° or so arcward, independent of position in the wedge, discordant to the gently dipping décollement atop the subducting plate. (Steeper planes may also be present, for they would not be imaged in reflection profiles.) Surface slopes of most accretionary wedges define broadly similar convex-upward curves regardless of the widths and thicknesses of the wedges and hence presumably are profiles of dynamic equilibrium.

Such features indicate to me that an accretionary wedge is simultaneously thickened, by underplating and by dragging back of its base, and thinned, by flowing forward by gravitational spreading. The result is internal imbrication of the wedge and maintenance of a dynamic profile, akin to that of an ice sheet, as the wedge grows both laterally and vertically. Some other observers see wedges as more static features, broadened by imbrication and offscraping at their toes but relatively stable in their arcward portions.

Reflection profiles across the accretionary wedge indicate that ratios of imbricated, disrupted, and coherently folded materials vary widely but relate systematically to the convergence rates and directions and to the thickness and character of sedimentary sections being accreted (Moore and others, 1980b). No wells have yet been drilled into the Sunda wedge. Drillholes in other similar sediment-rich modern wedges show them to be typified near their toes by broken formations with scaly-clay matrices and by highly variable proportions of broken formations and coherently imbricated strata farther back in the wedges. Island exposures of the top of the Sunda wedge have been studied primarily on Nias (Moore and Karig, 1980; Moore and others, 1980a), where coherent, unmetamorphosed lower Miocene to lower Pliocene strata structurally overlie on the northeast, and elsewhere are imbricated with, polymict mélange composed of undated materials. The mélange is slightly metamorphosed, is extremely

---

[1] I have previously (as, 1979) used the term "outer-arc ridge" for this feature because of the conflict of the more widely used term "fore-arc ridge" with the classical term "foreland." I here yield to the more common usage. Similarly, the "fore-arc basin" of most recent literature and this essay is the "outer-arc basin" of my previous reports. Note that a "foreland basin" is on the same side of an arc as is a "back-arc basin" and on the side opposite to a "fore-arc basin."

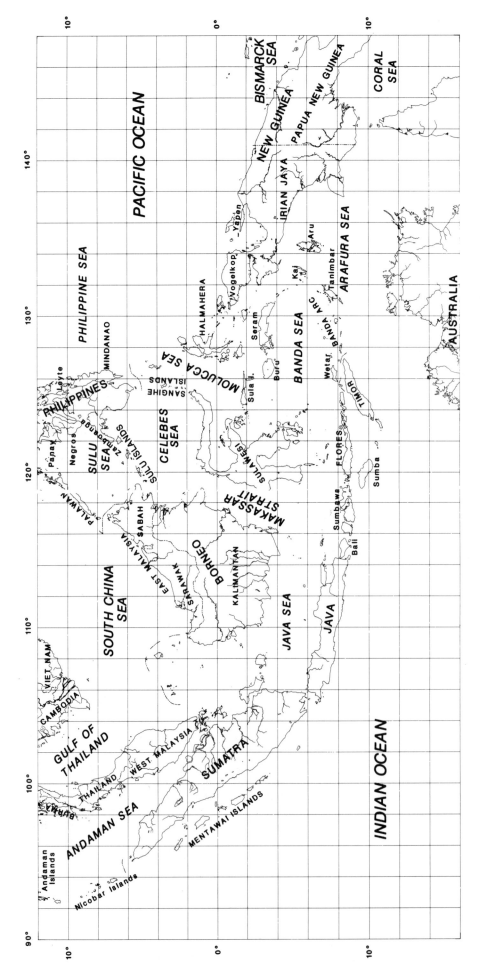

Figure 1. Index map of the Indonesian region. For bathymetry and more detail, see Hamilton (1978a or 1979, Plate 1) or Mammerickx and others (1976).

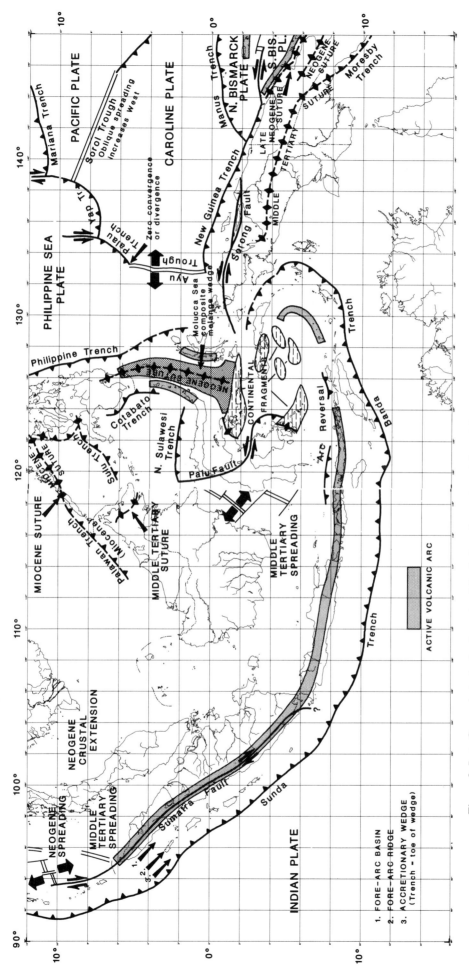

Figure 2. Late Cenozoic tectonic elements of the Indonesian region. Adapted from 1981 edition of Hamilton (1978a) and other sources. Plate boundaries are not complete in the poorly understood region between Sulawesi and New Guinea. The composite mélange wedge of the collided Sangihe and Halmahera arcs is darkly shaded.

sheared and disrupted, and is dominated by terrigenous clastic sediments of deep-water origin but contains abundant fragments of chert and basalt from the upper oceanic crust, and sparse fragments of mafic and ultramafic deeper-seated oceanic rocks. Fossils indicate the coherent strata to have been deposited in water depths that decreased with time. Although no depositional contacts of strata on mélange were found, Moore and his associates inferred that the Neogene strata were deposited atop the wedge and imbricated into it. Alternative possibilities are that at least the older Neogene materials were scraped off the subducting plate and imbricated into the wedge with little internal deformation or that they were deposited on the crystalline oceanic basement of what was then the outer part of the fore-arc basin, that basement having been since removed by tectonic erosion by the subducting plate and itself subducted. In the latter terms, the fore-arc ridge would have migrated arcward with time, not seaward as Moore and associates, and Karig (1982), inferred, and the formation of the undated polymict mélange might largely overlap in time the deposition of the older Neogene strata. Similar inferences seem to me to be plausible for the analogous Cretaceous systems of coastal California, where geometric relationships are much better known but where also interpretations are disputed.

Some investigators (for example, Silver and Reed, 1988) interpret ambiguous reflection profiles to indicate that the overriding-plate backstops for accretionary wedges commonly slope trenchward under the wedges rather than arcward. The apparent lack of on-land exposures of such phenomena is one of many lines of evidence I see against the general validity of such interpretations.

Many students of fossil accretionary wedges assume most broken formation and mélange in them to have formed as thick olistostromes (submarine slumps). Minor slumps may be abundant on the surfaces of wedges, but only one major slump (Moore and others, 1976) has yet been documented on reflection profiles across the floors of the Sunda or other modern trenches, and so I regard sedimentary-mélange interpretations of accretionary wedges as in general unlikely. Any sedimentary mélanges that are formed in trench settings must be imbricated into wedges and tectonized in the process. Broken formation in wedges is produced primarily by shear related to subduction, not by downslope sliding. The ratios within exposed accretionary wedges of polymict mélange, broken formation, and strata coherent at outcrop scale vary greatly, as does the proportion of soft-sediment to brittle deformation. These variables reflect differences in convergence rates, in amount and character of sedimentary strata being added to the wedges, and in positions within wedges. Consolidated sediments and shreds of the subducting lithosphere plate are scraped off against the bottom of the wedge or are carried beneath the overriding plate.

Oceanic lithosphere disappears beneath overriding plates at rates typically near 50 or 100 km/m.y., and most light material on subducting plates is destined for tectonic accretion against overriding plates. Something like 3,000 km of Indian Ocean lithosphere have disappeared beneath Sumatra and Java in the 30 m.y. that the Sunda subduction system has been operating, and so much far-traveled material must be incorporated in the accretionary wedge. Far more subduction is recorded in complexes, as along western North America and many other parts of the world, that incorporate the products of series of such subduction systems.

Arcward slopes of active intraoceanic island arcs, in settings in which little sediment is present on the subducting plate so that little accretionary wedge is present, have been sampled in the Mariana and Tonga Trenches. The Mariana slope consists mostly of igneous rocks of arc origin—calc-alkalic, tholeiitic, and high-magnesium basalts, andesites and dacites, cumulate and massive two-pyroxene gabbros, and serpentinized ultramafic rocks including both cumulate and residual types (Bloomer, 1983;

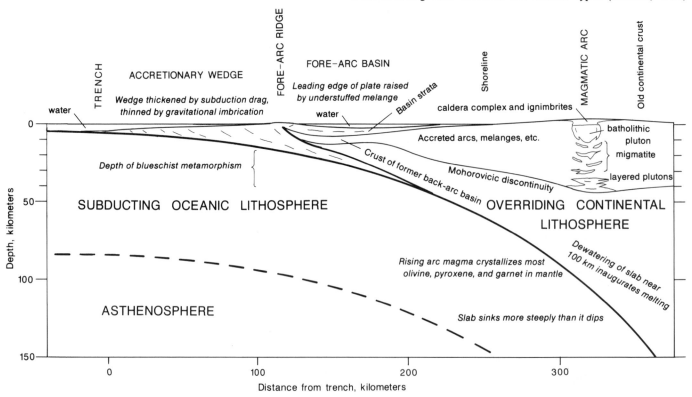

Figure 3. Section across a continental-margin subduction system. The diagram is scaled to modern Sumatra (Indian Ocean to left, southwest; Sumatra on the right, northeast), following constraints of surface dimensions and geology and of seismicity and refraction seismology (mostly from Hamilton, 1979). The dimensions and geology greatly resemble those of middle Cretaceous California also (compare with Hamilton, 1978b, 1988a). The deep erosion of parts of the California analogues exposes variations with depth which are integrated.

Bloomer and Hawkins, 1983; Natland and Tarney, 1981). The Tonga slope yielded what appears to be a nearly complete crustal and upper-mantle section of primitive arc rocks (Bloomer and Fisher, 1987). Tectonic erosion of the bases of the overriding Mariana and Tonga plates is inferred. Analogous complexes would be termed "ophiolitic" if they were encountered in ancient accreted terranes, but they clearly are of arc origins (Bloomer and Fisher, 1987; Bloomer and Hawkins, 1983).

Fore-Arc Basin. Between the fore-arc ridge and the shoreline of the Sunda system is the bathymetric and structural fore-arc basin, which is 150–200 km wide and contains at least 5 km of strata in the Sumatra sector (Beaudry and Moore, 1981; Hamilton, 1979; Karig and others, 1980a). On the arcward side of the basin, undeformed lower Miocene and higher strata lap progressively farther landward onto the basement. On the oceanic side, strata become increasingly deformed toward the fore-arc ridge, and that deformation includes both arcward-directed thrusts and folds and the diapiric rise of shale into folds; basement generally is not defined by reflection data. Depocenters of successively younger stratal packages are displaced arcward. This displacement is better documented by published data for the fore-arc basins of Peru and Chile (Coulbourn and Moberly, 1977) and Luzon (Lewis and Hayes, 1984). I have seen proprietary reflection profiles on which it appears that deep strata now tilted arcward on the oceanic side of the Sunda fore-arc basin were deposited as units prograded trenchward in deep water, before their basement was raised to define the basin. The Aleutian fore-arc basin similarly evolved as its front was raised (Harbert and others, 1986). The basement beneath the outer parts of both the Sumatra (Kieckhefer and others, 1980) and Java (Naomi Benaron, 1982, written commun.) basins has velocities typical of oceanic, not continental, crust, although the thickness of crust of such velocity in the Sumatra sector is much greater than is typical of oceanic lithosphere.

I integrate these features for this and other modern fore-arc basins, the features of the fore-arc ridge noted previously, and characteristics of some ancient analogs to infer that the fill of a fore-arc basin is deposited across the boundary between continental crust and a narrow strip of oceanic upper lithosphere that is attached to the front of the overriding continental plate. The basin is formed primarily by the raising of the thin, oceanic leading edge of the overriding plate as accretionary-wedge mélange and packets of sediments are stuffed under it. Depth of the basin is augmented by elastic downflexure behind that raised leading edge. The fore-arc ridge is the crest of accretionary-wedge debris accumulated in snowplow fashion in front of the overriding leading edge. Debris that overtops the leading edge is imbricated gravitationally arcward over the shallow strata of the basin. As tectonic erosion trims the leading edge of the overriding plate, the fore-arc ridge migrates arcward relative to that plate, and the fore-arc basin is narrowed.

Fore-arc basins of similar characteristics are common along subduction-system margins of continents and mature island arcs. The ridge and basin may be displayed in bathymetry as well as structure (as in the modern Sunda system and in the paleobathymetry and longitudinal deposition of the Lower Cretaceous part of the "Valley Facies" of California) or may appear as a bathymetric shelf underlain by structural ridge and basin (as, modern Chile and southern Alaska and much of the Upper Cretaceous and Paleogene parts of the "Valley Facies"). Basal strata in such fore-arc basins are generally pelagic sediments and abyssal-fan strata that predate the inauguration of the basins and of the subduction systems that bound them.

Exposed basement of the outer parts of fore-arc basins consists of oceanic crust [for example, Cretaceous California (Hamilton, 1978b; Ingersoll and Schweickert, 1986) and middle Tertiary Luzon (Bachman and others, 1983; Karig, 1982)] and is arguably in most cases of marginal-basin origin. A similar origin of the basement of modern basins, including the Sunda system, accords with geophysical data. The leading edge of overriding continental plates may commonly be a strip, 100 km or so wide, of oceanic lithosphere. As noted below, such a strip likely forms behind a migrating oceanic island arc which then collides with a continent or another arc, the strip then being left attached to the enlarged crustal mass when reversal of subduction polarity took place.

Lack of Shortening. Sunda and other fore-arc basin fills and their thin upper-plate–lithosphere basements are not commonly shortened compressively across their width, although they are subjected to tectonic erosion and rumpling at their trenchward sides. Thick and undisturbed basin-filling strata can be seen on reflection profiles across many fore-arc basins in Indonesian and other active subduction systems. This lack of deformation disproves the common assumption (for example, Hutchinson, 1980) that the leading edges of overriding plates are crumpled against subducting plates. Extreme shear imbricates the surficial accretionary wedge pushed in front of an overriding plate, but that plate itself commonly is not shortened. Slight to severe extension, not shortening, occurs across most modern magmatic arcs, perhaps because the steeply sinking subducting slabs displace underlying mantle downward, resulting in extension of the mantle, asthenosphere, and lithosphere above the slabs.

Relation to Arc Reversal. Subduction systems typically are inaugurated by reversal of subduction polarity following a collision between thick crustal masses. Subduction can no longer occur within the newly enlarged crustal mass, and as convergence continues, a new subduction system breaks through on an oceanic side of the enlarged landmass. The break commonly occurs not at the boundary between thick and thin crust, but within oceanic lithosphere 100 km or so oceanward of that boundary, so that a strip of oceanic lithosphere thus becomes the thin leading edge of the newly defined overriding plate. In the case of a reversal following an island-arc collision, this oceanic strip is the youngest part of the back-arc–basin lithosphere formed by the migrating arc, hence is only slightly older than the collision itself. Such an explanation is best documented for the case of latest Jurassic California (Ingersoll and Schweickert, 1986) but accords with data from many other arcs, including Sumatra and Java (Hamilton, 1988b). (Contrary views are expressed by Karig, 1982.)

High-Pressure Metamorphism. The only high-pressure metamorphic rocks yet known within Neogene mélange in the Sunda system are blocks of garnet amphibolite found on Nias by Moore and Karig (1980). (Glaucophane schist is known in Neogene mélange farther east, in the Banda sector.) High-pressure metamorphic rocks, of blueschist and locally eclogite or garnet amphibolite facies, are widely known in pre-Neogene Phanerozoic subduction complexes in the Indonesian region and elsewhere about the world. The petrology of such rocks requires that they have been metamorphosed mostly at depths of 25–45 km at relatively low to moderate temperatures, then returned to shallow depths before equilibration of geothermal gradients to normal values for such depths; this apparently occurs as a return-flow by-product of subduction (for example, Cloos, 1985, and Wang and Shi, 1984). I infer from the geologic relationships of many occurrences around the world and from the geometry of modern wedges that such metamorphic rocks never form within an accretionary wedge between trench, fore-arc ridge, and subducting lithosphere but, rather, that they form only where crustal and supracrustal materials have been subducted beneath the overriding plate.

Sediment on subducting plates can partly bypass the accretionary wedge and ride far beneath the overriding plate. This is shown directly where anticlinal windows in southern California broadly expose metamorphosed oceanic sedimentary and crustal rocks (termed Pelona, Orocopia, and Rand Schists) that were subducted in latest Cretaceous time beneath lower continental crust.

Reasoning by analogy with Mesozoic California and other deeply eroded ancient systems of accretionary wedges and fore-arc basins, I infer that beneath the sub-basin leading edge of the overriding Sunda plate, mélange is now being metamorphosed at blueschist facies, and perhaps eclogite facies, and that beneath this metamélange is metamorphosing crust of the subducting oceanic lithosphere. The thick zone with oceanic crustal velocities beneath the basin fill, as defined by Kieckhefer and others (1980), may represent a sandwich of thick arc-type ophiolitic overriding-plate basement to the basin, metasedimentary rocks beneath that, and crust of the subducting Indian Ocean plate still deeper.

Magmatic Arc. Sunda volcanoes now erupt in a belt about 100 km above the top of the inclined Benioff zone of mantle earthquakes, or about 130 km above the midplane (Hamilton, 1974a, 1978a; Hayes and Taylor, 1978). This magmatic arc changes along strike from continental in Sumatra to transitional in Java to a mature oceanic island arc in Bali, Lombok, and Sumbawa. Sunda-system volcanism did not begin until early Miocene time in Sumatra. Middle Tertiary volcanic rocks are widespread but poorly dated on land, but the inception and subsequent continuity of major silicic magmatism are defined, in sections drilled in the Gulf of Thailand, by voluminous volcanogenic mixed-layer clays in middle lower Miocene and higher shales. Volcanism, dated by the paleontologic age of intercalated strata in drill holes, was active by late Oligocene time in offshore southern Java; whether this magmatism records the Sunda system or an oceanic island arc that collided with the pre–Sunda-system continent is unclear, but the continuity of upper Oligocene and higher strata across central and western Java and the continental shelves to the north and northwest shows that region to have been a coherent part of Southeast Asia by then. The Paleogene of mainland Sumatra, inland from the collided arc noted subsequently, records pre-arc sedimentation across a low and stable landmass from Southeast Asian cratonic sources.

The volcanoes of the magmatic arc rise above a geanticline, within which are most exposures of pre-Miocene rocks of Java and Sumatra. Presumably, this geanticline is a product of magmatic inflation and thermal uplift of pre-existing crust. Continental Sumatra has much the higher geanticline of pre-volcanic rocks, and I infer that there, a crustal column was heated by intrusions to near-magmatic temperatures, with formation of voluminous migmatites, before much magma reached the surface to form volcanoes.

The compositions of the volcanic rocks vary systematically with the character of the crust through which their magmas have been erupted. The crust of Sumatra was continental by late Paleozoic time, when silicic, radiogenic granites were formed, and likely was so during the Precambrian, although no rocks of that age have been identified. The modern magmatic-arc rocks atop this continental crust are mostly intermediate to silicic in composition. They approximate rhyodacite (granodiorite) in bulk composition; there is little basalt. Lake Toba caldera, produced by collapse accompanying voluminous late Pleistocene silicic ignimbritic eruptions, is the largest caldera known anywhere and is about the same size and shape as what is perhaps the largest upper-crustal granitic pluton yet mapped, the Late Cretaceous Mount Whitney pluton of the Sierra Nevada of California. In Java, where the pre-Neogene crust is of near-continental thickness but consists of mélanges and mafic to intermediate magmatic rocks, young volcanic rocks are mafic to intermediate—mostly pyroxene andesite and high-alumina basalt, with subordinate dacite. Similar mafic and intermediate rocks characterize the mature oceanic island arc of Bali and Sumbawa, where exposed rocks are entirely of Neogene age. Farther east, in the Banda Arc sector discussed subsequently, the volcanic arc is younger and consists mostly of more primitive basalts, rocks much less evolved petrologically than those even of the mature oceanic part of the Sunda sector. Comparable transitions, from evolved and silicic magmatic rocks to more primitive and mafic ones, can be seen wherever about the Pacific continuous magmatic arcs cross from continental to oceanic lithosphere.

Indian Ocean lithosphere is being subducted beneath all of the Sunda sector, and presumably, the deep proto-magmas generated by subduction-related processes—melting of mantle consequent on dehydration of subducted hydrous rocks?—are similar olivine-rich basaltic melts along the entire length of the sector. The volcanic rocks which reach the surface have been profoundly modified by reactions in and with the crust through which they have passed. Even the primitive rocks farther east record magmas equilibrated at shallow depths; no deep-mantle magmas reach the surface without great modification.

Volcanoes north of the main Sunda magmatic belt show the marked variations, including increased potassium relative to silicon, that characterize eruptions above deep parts of subducting slabs.

Pre-Neogene Tectonics of Sumatra. The modern Sunda system, involving subduction of Indian Ocean lithosphere beneath Sumatra and Java, was inaugurated only in middle Tertiary time. Much of the older geology records subduction in quite different tectonic systems. Most of Sumatra has been continental at least since late Paleozoic time and belongs to the same system of late Paleozoic and early Mesozoic sutures and magmatic arcs as does the Malay Peninsula. Sumatra may have been rifted from what is now medial New Guinea in Middle Jurassic time, and a rifted-margin stratal wedge can be inferred from meager data to be present in Sumatra. Java, on the other hand, has been constructed entirely by post-Jurassic subduction-related processes of magmatism and tectonic accretion. Many reconnaissance data regarding the pre-Neogene geology of Sumatra have been released since the completion of my 1979 book, as 1:250,000 photogeologic maps constrained by sparse field traverses and brief rock descriptions (as, Bennett and others, 1981; Cameron and others, 1982; Rock and others, 1983). I interpret these works to show that the pre–Late Jurassic–age rocks of the old continental crust are bounded on the southwest by a broad belt of polymict subduction mélange and broken formation of late Mesozoic and(?) Paleogene age. This accretionary-wedge complex includes not only the small areas identified as mélange and serpentinite by these authors but also most of the larger terranes they designated as eastern Woyla Group and as Babahrot and Belok Gadang Formations, the brief descriptions of which indicate the presence of widespread broken formation and polymict mélange. (Rock-unit names applied in these reports have little lithostratigraphic significance.) This broad accretionary-wedge tract lies within the medial part of far northern Sumatra, where its distribution is complicated by the active right-slip Sumatran fault system, but closer to the southwest coast in central Sumatra; southern Sumatra lacks exposures of pre-Neogene rocks in the relevant coastal belt. To the west of the broad belt of probable mélange is a belt of volcanic, volcaniclastic, and sedimentary rocks, of island-arc type and of Late Jurassic and Early Cretaceous age at the few localities where dated paleontologically, which was assigned to the western Woyla Group by the mappers.

I interpret these relationships to indicate that a northward-migrating oceanic island arc collided with the margin of Sumatra, which had been a trailing edge since its mid-Jurassic separation from New Guinea, in Paleogene time. Convergence of Sumatra and Indian Ocean continued, and the subduction system that is now active broke through south of the continent as enlarged by the collision, leaving a narrow strip of marginal-sea lithosphere, which had been formed behind the advancing arc, as the leading edge of the new upper plate. (Bennett and others, 1981, and Rock and others, 1983, recognized the island-arc character of the southwestern rocks but interpreted them in terms quite different from mine.)

Pre-Neogene Tectonics of Java. The modern subduction system in Java was inaugurated no earlier than late Oligocene time. Exposures of pre-Neogene–age rocks are limited to small areas, in central and southwest Java, of polymict mélange of Late Cretaceous and early Paleogene age and of overlying middle or late Eocene through Oligocene quartzose clastic strata and shallow-water carbonates. Much more information has come from the subsurface of northern Java and the Java Sea shelf. Mélange of Cretaceous and early Paleogene age dominates the basement in a broad belt trending northeastward from Java across the shelf to southeast Borneo, where it is widely exposed. This mélange may be paired to wide-

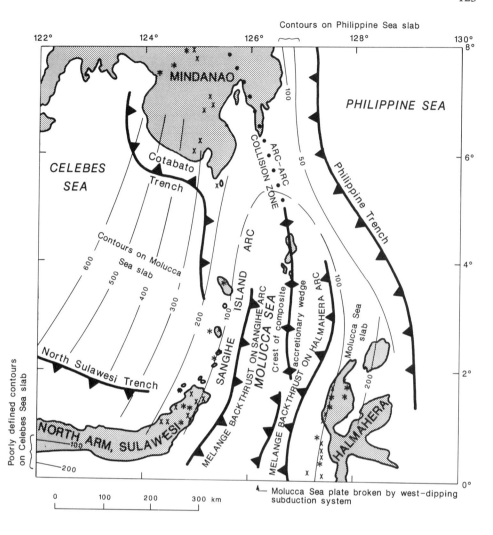

Figure 4. Plate-tectonic features of the Molucca Sea region. A collision between east-facing Sangihe island arc and west-facing Halmahera arc has progressed southward with time as the Molucca Sea plate has sunk (subducted) beneath them. The opposed accretionary wedges merged and backflowed onto the advancing arcs. The young Cotabato Trench has broken through on the west side of the old part of the aggregate; thrust earthquakes, but not a clearly defined Benioff seismic zone, are associated with it. The Philippine Trench, on the east side of the aggregate, has only a shallow west-dipping seismic zone, the relation of which to the east-dipping zone beneath Halmahera is unclear. The marked "collision zone" within Mindanao and the northern Molucca Sea may still be active as a left-slip transpressive plate boundary. Contours show depths to tops of subducting slabs, in kilometers; *, historically active volcano; x, Quaternary volcano. Adapted mostly from data and interpretations of Hamilton (1974b, 1979), but with modifications for data and interpretations of Cardwell and others (1980), McCaffrey (1982), McCaffrey and others (1980), and Moore and Silver (1982).

spread Cretaceous granitic and volcanic rocks to the northwest in Borneo and the Java Sea basement. During late Paleogene time, western and central Java and the Java Sea were tectonically and magmatically dormant and were fused to the subcontinent that included most of Sumatra and all of the Malay Peninsula, and subduction was beneath the opposite side of the small continent; South China Sea lithosphere was then being subducted southward beneath what is now northwest Borneo. If a northward-migrating arc collided with Java in Paleogene time, as might be expected from the interpretation just made for Sumatra, then it now lies offshore in the subsurface, where the upper Oligocene volcanic rocks drilled south of central Java may belong to such an arc. Eastern Java, Bali, Lombok, Sumbawa, and Flores project east of all known pre-Neogene complexes and expose only oceanic Neogene island-arc rocks.

Neogene Deformation. Popular conjecture, residual from geosynclinal theory, assumes great crustal shortening to be a precursor of arc magmatism. Such deformation is not recorded in the Sunda system or other modern magmatic arcs. In Java, middle Tertiary strata are openly folded; deformation decreases in intensity away from magmatic centers, about which structures tend to be concentric (as, Djuri, 1975), and gravitational spreading of magmatic chambers and edifices is likely a major cause of deformation. In Sumatra, middle Tertiary pre-magmatic strata within the modern volcanic belt but distant from local centers are subhorizontal or gently dipping and display normal faulting. Gravitational spreading related to magmatic crustal thickening can be inferred for Sumatra also. Normal faulting, not compressional deformation, is commonly seen in the old parts of mature island arcs. Where collisions of light crustal masses are involved, however, or where convergence is so rapid that an overriding continental plate drags on a gently dipping subducted plate, severe shortening and major crustal thrusting can result.

Banda Subduction System

The Banda Arc continues the great subduction system eastward from the Sunda sector. The character of the system changes greatly along strike. Oceanic lithosphere is being subducted beneath a continental plate in Sumatra, beneath transitional lithosphere in Java, and beneath another oceanic plate in the Bali-Sumbawa-Flores sector; in the Banda Arc, an oceanic arc is now colliding with the continent of Australia and New Guinea. That collision becomes progressively younger in the Neogene eastward, and as the arc complex has accreted to the continent, subduction has reversed beneath the south limb of the arc to become southward beneath it. The south limb of the Banda Arc is wholly of Neogene age, and it becomes progressively younger in age of inception of magmatism along its eastward trend; the arc has lengthened with time. In the eastern Banda Arc, trench, fore-arc ridge and basin, and volcanic arc all trend concentrically around a tight curve. A well-defined Benioff zone of earthquakes dips northward deep into the mantle from the accretionary wedge of the Sunda Arc and the south limb of the Banda Arc. The seismic zone curves in the east, concentric to the bathymetric features, to define a spoon-shaped zone that plunges gently westward but that can be traced unambiguously only to a little north of the geometric axis of the Banda Arc. Various conclusions of my 1979 and earlier reports have been replicated and expanded by others from much new geophysical data, some of which is cited herein.

Trench. Whereas the trench in the Sunda sector overlies oceanic lithosphere, the shallow trench in the Banda sector overlies continental crust around the entire curve of the arc. The continuity around the arc of the distinctive tectonic morphology of trench and accretionary wedge is shown by scores of reflection profiles. The trench marks the gentle dihedral angle between shallow-water strata bowed down from the Australia-Arafura-New Guinea continental shelf on one side and the toe of the accretionary wedge on the other. The accretionary front advances discontinuously as new thrust slices develop within the shelf strata (Karig and others, 1987). Continental crust is demonstrated by refraction data to extend beneath the accretionary wedge at least to the inner edge of the fore-arc ridge (Bowin and others, 1980; Jacobson and others, 1979). McCaffrey and others (1985) inferred that the thin leading edge of the continent has been subducted to a depth of 150 km in the Timor sector and that still deeper subducted oceanic lithosphere is detached and sinking independently.

Fore-Arc Ridge. The top of the accretionary wedge is wholly submarine from Java to Flores, but where it stands upon continental crust, it forms the large, high island of Timor; lower, smaller, and later-starting islands around the tight eastern curve of the Banda Arc; and large, high Seram on the north limb of the system. Continuity of the wedge around the arc as a thick aggregate of low-density material is indicated by its continuous gravity anomaly (Bowin and others, 1980). Descriptions of island geology that postdate Hamilton (1979) add details to my accretionary-wedge descriptions. The wedge consists of polymict mélange and broken formations imbricated, with generally arcward dips, with variably coherent strata that include strata from the continental shelf onto which the wedge has been ramped, strata deposited atop the wedge, abyssal pelagic sediments, and slices and fragments of both ophiolitic and continental crystalline rocks. Fore-arc–basin materials may have been imbricated into the wedge after tectonic removal of their overriding-plate basement. Quaternary reefs have been elevated as high as 1,000 m above sea level as the top of the wedge has been raised both by thickening of the wedge by accretion and imbrication and by ramping farther onto continental crust.

Berry and Grady (1981) described metamorphism of sedimentary rocks that decreases from uppermost amphibolite facies to greenschist facies away from an ophiolite mass at the north edge of central Timor. Potassium-argon ages of hornblende show the metamorphism to be of about late middle Miocene age. I infer from the relationships mapped by Berry that the temperature of metamorphism decreased downward beneath the ophiolite sheet, which I regard as the hot leading edge of the onramping island arc. (Berry and Grady inferred vertical or strike-slip tectonics and suggested no heat source.) The Tethyan region has many analogous ophiolite sheets that were emplaced while hot. Farther west in north-coastal Timor, upper Miocene tholeiitic and calc-alkalic basalt are thrust southward onto the wedge (Abbott and Chamalaun, 1981); again, I infer onramping of the advancing arc.

Fore-Arc Basin. The fore-arc basin is continuous (except at Sumba) around the Banda Arc. Little-deformed basin strata lap onto the fore-arc ridge on the outside of the basin and grade into volcaniclastic aprons of the magmatic arc on the inside. The bathymetric basin deepens symmetrically along both limbs of the Banda Arc toward the axis of its tight horseshoe curve to define the Weber Deep, the depth of which reaches 7.5 km precisely at that axis. I interpret the basin as formed by the elastic deflection of the thin leading part of the overriding lithosphere as its edge has been ramped up by the stuffing beneath it both of accretionary-wedge mélange and of continental crust. This depression is focused at the Weber Deep from three sides.

The basin is markedly narrower along northern and eastern Timor than elsewhere around the south limb and eastern curve of the Banda Arc. I infer tectonic erosion of the leading edge of the overriding plate, and the imbrication into the Timor wedge of what were strata deposited on that leading edge. There is no suggestion on reflection profiles of subduction within this or other sectors of the basin; narrowing (or, in the Weber Deep, deepening) by subduction cannot be proposed.

The concentricity of the Banda Arc deteriorates in the Buru–western Seram sector of the north limb. No fore-arc basin is present inward from that part of the fore-arc ridge. Islands of Pliocene volcanic rocks, which presumably represent the extinct magmatic arc and were erupted through silicic continental rocks (Abbott and Chamalaun, 1981), are separated from the ridge only by narrow straits. Tectonic erosion of the overlying plate may here also be part of the explanation.

The large island of Sumba rises within what is otherwise the fore-arc basin, and its almost undeformed Miocene to Quaternary strata are continuous with those of the basin; the island is a raised part of the basin, domed above a poorly understood complex of pre-Miocene-age crystalline and sedimentary rocks. I suggested (Hamilton, 1979) that the old rocks represent a crustal fragment rifted from the Java shelf, whereas Silver and others (1983c) suggested that they represent one of the crustal fragments in front of Australia, subducted beneath the basin.

Magmatic Arc. Although the magmatic arc is continuous around the south limb and eastern curve of the Banda Arc, its history varies systematically with position. The width and volume of the magmatic edifice decrease eastward along the south limb of the arc and correspond to a decreasing age of inception of magmatism, from early Miocene in the west to Pliocene in the east part of the south limb, and probably Quaternary within the tight eastern arc. (Data postdating my monograph include those of Abbott and Chamalaun, 1981, and Suwarna and others, 1981.) Around the sharp eastern curve, the magmatic arc is represented only by small, active-volcano islands atop a narrow and poorly continuous ridge. The volcanic rocks change correspondingly from andesites and evolved basalts in the older sector to more primitive basalts in the young sector. Volcanic rocks on the short, irregular north limb of the arc are of Pliocene age, but here, tectonic relationships are poorly understood. Volcanoes are currently active all along the south limb of the magmatic arc and around the eastern curve of the arc, except for a length of about 500 km, to the north and northeast of eastern Timor, and along the short north limb in the Buru–western Seram sector, in both of which activity ended in Pliocene time, apparently following the cessation of subduction beneath the Banda Sea as a consequence of the arc-continent collision.

Arc Reversal. Two sectors, each about 500 km long, of the south limb of the Banda Arc are now marked by trenches, the tectonic geometry of which indicates subduction relatively southward, at the north base of the volcanic arc. This polarity is opposite to that of the main Banda system. I (Hamilton, 1979) identified the trenches on reflection profiles and argued for arc reversal following collision of arc with continent. Breen and others (1986), Karig and others (1987), McCaffrey and Nabelek (1984, 1987), Reed and others (1986), and Silver and others (1983c, 1986) further defined the character and extent of the frontal and reversed trenches and the accretionary wedge from reflection profiles, side-scan mapping, seismicity, and other data. The eastern of these new trenches is north of central and eastern Timor and coincides with that part of the volcanic arc in which magmatism ceased in late Pliocene time. The western of the new trenches lies north of Flores, Sumbawa, and Lombok, where magmatism apparently belonging to the north-dipping subduction system is still active but appears to have decreased within late Quaternary time.

Banda Sea. The small but complex Banda Sea, enclosed by the Banda Arc, consists of the oceanic North and South Banda Basins and an intervening group of submarine ridges. These ridges are known from dredging to be fragments of continental crust (Silver and others, 1985).

Minicontinental fragments are exposed on the partly submerged platforms around the northern part of the Banda Sea—Buton in the west, Banggai-Sula in the northwest, and Buru–Ambon–western Seram in north-center (Hamilton, 1979; Pigram and Panggabean, 1983; Silver and others, 1983b; Silver and others, 1985). The age of formation of the oceanic crust of the two major Banda Sea basins is not yet constrained by drilling. I suggested (Hamilton, 1979) that the basins formed behind a migrating Banda Arc and are of Cenozoic age. Bowin and others (1980), Lee and McCabe (1986), Pigram and Panggabean (1983), and Silver and others (1985), by contrast, have regarded both basins as trapped bits of Mesozoic lithosphere. Their interpretation is plausible for parts of the North Banda Basin, which is discontinuously rimmed by fragments of pre-Cenozoic continental crust, although reconnaissance heat-flow measurements from the southern subbasin of the North Banda Basin are so high that Neogene rifting is likely there also (Van Gool and others, 1987). The old-crust interpretation is implausible for the South Banda Basin, for the Banda Arc, which defines the south edge of this basin, has lengthened during late Neogene time.

Interpretation. The age of inception of the Banda magmatic arc becomes progressively younger eastward along the arc, from early Miocene to Pliocene and probably to Quaternary; the arc has lengthened with time. The collision of the arc with the Australia–New Guinea continent also has progressed eastward with time, occurring earlier at Timor than around the axis of the tight curve in the east. Timor has not slid past Australia on strike-slip faults but has remained attached to it since the collision in that sector; Banda Sea lithosphere is beginning to be subducted southward beneath the continent, as enlarged by the accreted arc, at a new trench, even as subduction at the axis of curvature of the arc is relatively westward beneath the Banda Sea. Such relationships, to me, require that the crust of the South Banda Basin has formed by spreading behind a rapidly migrating Banda Arc or has been plated out by the fast-migrating arc itself. The Banda Sea does not represent an internally rigid plate neatly pre-shaped to slide into the Arafura concavity between Australia and New Guinea; rather, the Banda plate expanded as needed to fill a concavity which likely has itself changed shape as Jurassic oceanic crust attached to the continent sank in front of the Banda plate.

This much of the story is analogous to that in my 1979 book, but clearly, I erred there in picturing the entire Banda Arc and Banda Sea as a simple migrating arc paired to an extensional back-arc basin. The north limb of the arc (Seram and Buru), the North Banda Basin, and the submarine ridges require much more complex explanations. All observers agree that the continental fragments must have been torn from New Guinea, but details remain highly ambiguous. A viable solution must incorporate rapid northward motion of New Guinea and westward motion of Pacific plates, and probably southward motion of the Sunda system, and must account for the bewildering array of diversely oriented tectonic elements in and north of the Banda Sea, as well as for migration and lengthening of at least the southern and eastern parts of the Banda Arc.

Caribbean, Scotian, and Carpathian Arcs. Each of these three east-facing, horseshoe-shaped arcs is dimensionally and geometrically so like the Banda Arc and displays so many analogous features that similar origins are likely. Each can in my view (but not in the views of most local experts) be explained in terms primarily of eastward-migrating oceanic arcs. Caribbean and Scotia arcs collided with the Pacific sides of Central and South America and West Antarctica in late Mesozoic time but continued to migrate through the oceanic gaps between those landmasses, beaching arc material against north and south sides progressively eastward with time. The initial frontal collisions were followed by reversals of subduction polarity which inaugurated the Andean systems that have operated subsequently along the continental margins. The Carpathian arc migrated into a continental concavity during Tertiary time and also beached its flanks successively eastward with time; minicontinental fragments trailed behind.

Northern Indonesia and the Southern Philippines

Molucca Sea Collision Zone. A collision between inward-facing island arcs is under way in the Molucca Sea region, where the collision between the east-facing Sangihe island arc and west-facing Halmahera arc has progressed southward with time (Fig. 4). The suture zone is fully closed in the north and is exposed on land in Mindanao. In the central sector, the northern Molucca Sea region, the accretionary wedges of the two arcs have been joined by collision and have been thickened to at least 15 km, and the composite surface raised to near sea level and locally above it, in the medial zone. The overthickened composite wedge has flowed gravitationally across the inward-facing trenches and onto the arcs on both sides, so that surficial thrusting of mélange has the sense opposite to that of subduction. Following the collision, arc magmatism ceased in this central sector, and subduction polarity of the Sangihe arc was reversed. In the southern Molucca Sea region, the two accretionary wedges have met in the center, but subduction and arc magmatism are still active in their pre-collision sense.

This system of colliding arcs is important for comprehension of plate behavior and has been studied extensively, particularly by Eli Silver and his associates, since my work on it. Recent information on this collision system that in general documents the conclusions just summarized was reported by Cardwell and others (1980), Hall (1987), McCaffrey (1982), McCaffrey and others (1980), Moore and Silver (1982), and Silver and others (1983a). Weissel (1980) identified sea-floor–spreading magnetic anomalies of the southwestern Celebes Sea (the marginal basin opened behind the Sangihe arc but now being subducted beneath northern Sulawesi, southwestern Mindanao, and the northern Sangihe arc) as probably Eocene in age, whereas Lee and McCabe (1986) regarded them as of latest Cretaceous age.

The Molucca Sea plate is being subducted simultaneously relatively westward beneath the Sangihe arc and eastward beneath the Halmahera arc. A well-defined Benioff seismic zone dips westward beneath the Sangihe arc to a depth of about 650 km beneath the Celebes Sea, and another zone dips eastward beneath Halmahera to a depth of about 250 km. The active volcanoes of both arcs are concentrated about 100 km above the tops of the respective seismic zones, which merge beneath the Molucca Sea composite mélange wedge. This two-sided subduction cannot be explained in terms of subduction as a process of injection down fixed slots and requires that the subducting Molucca Sea plate fell away on both sides as overriding plates advanced over it.

Relationships around the southern Molucca Sea region are exceedingly complex and still poorly understood. Data and synthesis postdating mine were presented by Silver and others (1983b).

Aggregation of the Southern Philippines. The Philippine Islands are a collage of variably collided, reversed, oroclinally twisted, and magmatically overprinted components of island arcs—magmatic arcs, accretionary wedges, large and small ophiolitic masses, and sedimentary assemblages. The collided Sangihe and Halmahera arcs and intervening wedge come ashore in southern Mindanao, where suturing was completed during middle Tertiary time (Hawkins and others, 1985), and their products are being overprinted by arc magmatism paired to subduction relatively eastward from the Cotabato Trench, which was inaugurated by arc reversal following the collision. The Benioff zone inclined westward from the Philippine Trench extends only to shallow depth and has no obviously associated arc volcanoes south of Leyte (Cardwell and others, 1983); kinematic relationships of this Philippine Trench system to those of the rest of Mindanao are not yet clear. Farther west, the recently inactivated, northwest-facing Sulu island arc comes ashore as the Zamboanga Peninsula of western Mindanao, and the slow-subduction, southwest-facing Negros arc system apparently is closing southward against the Sulu-Zamboanga arc, colliding with its own earlier projection. Still farther west,

Figure 5. Volcanic rocks of the modern volcanic arc of North Island, New Zealand. A. Ruapehu stratovolcano of basaltic andesite, and Lower Tama Lake explosion craters, at nonextending south end of arc, which stands on geanticline. B. Normal faulted and rotated middle Quaternary ignimbrite sheets of silicic biotite rhyodacite, in the rapidly extending northern part of the arc, which is little above sea level. Paeroa Range, 25 km south of Lake Rotorua.

the Palawan island arc, beneath which South China Sea lithosphere was subducted during middle Tertiary time, comes ashore in the west-central Philippines, much complicated by a collision with a North Palawan minicontinent rifted from from China as the South China Sea was opened.

Six different middle and late Cenozoic subduction systems are thus clearly recorded in the southern Philippines aggregate. As arc-type materials are as old as Cretaceous in the southern Philippines, many additional complexities have yet to be understood. Other aspects of this accretionary history have been discussed by Hawkins and others (1985), Karig and others (1986), McCabe and others (1987), and Sarewitz and Karig (1986). It appears that many far-traveled arcs and fragments have here been aggregated entirely in an intra-oceanic setting; it is the ultimate fate of this composite mass to be accreted to a continent.

Collisions and Subduction

Examples such as those noted briefly in this section make it obvious that long-continuing, steady-state subduction systems are atypical; that complex sequences of collision, aggregation, reversal, rifting, and internal deformation are the rule; and that aggregates of collided bits can be assembled far from their final resting places. Histories and kinematics can vary dramatically along strike in continuous complexes. Collisions and reversals progress along strike with time, and strike-slip and oroclinal deformation are common. Collisions do not occur between neatly matched shapes; irregular masses meet, and highly variable deformation occurs before they are fully jostled together.

Large plates commonly continue to converge after a collision, and the result is the inauguration of a new subduction system on an oceanic side of the new aggregate; often, this represents a reversal of polarity of subduction as well as a jump in position. Subduction of oceanic lithosphere beneath a continental plate commonly begins as a consequence of a plate collision. Convergence between megaplates continues, but the light crust on the subducting plate is too low in density to be subducted, and so a new subduction system breaks through oceanward of the continental plate as enlarged by the subduction. Such post-collision reversals are now under way in the Timor and Molucca regions; dozens of others are recorded in circum-Pacific geology. The Solomon-Admiralty arc complex displays two reversals, one of which presently is progressing along strike as the arc slides past a trench-trench-transform triple junction (Hamilton, 1979).

Major plate-convergence complexes record subduction at rates on the order of 10 cm/yr, or 100 km/m.y. Large motions and great complexity are the common case. Subduction systems are as likely to be multiple as single and commonly vary greatly along strike and are linked by diverse boundaries of other types.

ROCKS OF ISLAND ARCS

Volcanic Rocks

Volcanic rocks of oceanic island arcs typically show a progression with time from primitive to evolved compositions. Rocks erupted in young arcs of small crustal volume are predominantly tholeiitic basalts, many of which differ from spreading-ridge basalts primarily in their lower contents of the high-field-strength elements titanium, zirconium, and hafnium. Rocks erupted in mature arcs of large crustal volume are typically calc-alkalic basalt, andesite, and dacite—rocks relatively rich in aluminum and calcium. Two-pyroxene basalt and andesite with plagioclase phenocrysts, with or without olivine, and pyroxene dacite are common types, although many andesites and dacites are hornblendic. The rocks of oceanic arcs mostly are primitive in isotopic compositions, and the significance of moderate departures from primitive isotopes is much debated. Arcs erupted through continental crust, or through thick terrigenous sedimentary rocks proxying for such crust, commonly are much more silicic in bulk composition and more evolved in their isotopes.

Some oceanic arcs contain also early rocks of a richly magnesian series, of which the mafic members lack plagioclase but the most common

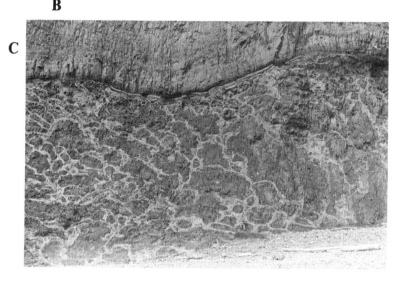

Figure 6. Paleogene submarine island-arc rocks of Unalaska, Aleutian Islands. Photographs by George L. Snyder. A. Dome (lower half of view, right) and bulbous masses (lower half, left) of keratophyre underlie partly fragmented bedded argillite (across center) and an unbroken sill (upper cliffs). Cliff is 350 m high. B. Large pillows of lithoidal latite with thin rims of black-weathering glass. Exposure is 10 m high. C. Large detached pillows in altered argillite, topped by thin, contorted argillite and overlain by an undeformed sill. Exposure is 25 m high.

member is magnesian andesite, boninite (Bloomer and Hawkins, 1987). These rocks are very low in the high–field-strength cations, and as these elements go preferentially from solids into partial melts, the boninitic and arc-tholeiitic magmas likely are derived largely from harzburgitic mantle that has previously undergone partial melting to yield ridge basalt (Bloomer and Hawkins, 1987; see also Fisk, 1986).

Magmatic arcs developed in continental or transitional crust contain more highly evolved volcanic rocks, as noted in the prior discussion of the Sunda system and the subsequent one of New Zealand.

Petrologic modeling, mostly cantilevered from the compositions of the final volcanic rocks, of various combinations of major and minor elements and of isotopes has in recent years yielded an array of hypotheses of origin of the mainline calc-alkalic rocks, involving partial melting of diverse mantle and subducted materials, fractionation at various levels, contamination, and magma mixing. Recent papers include those by Brophy and Marsh (1986), Crawford and others (1987), Davidson (1987), DeLong and others (1985), Gill (1981, 1987), Hawkins and others (1984), Kay and Kay (1985), Myers and Marsh (1987), Nye and Reid (1986), Wheller and others (1987), and White and Dupre (1986). The major point of agreement in these explanations is that the melting is somehow due to subduction.

Much of this mathematic-petrologic modeling incorporates such simplistic notions of one- or two-stage processes that it has little chance of achieving correct solutions. Melts rising in the mantle cannot be stable partial melts in equilibrium with specific wall rocks (O'Hara, 1985). Melts reaching the crust are already highly evolved, and complications such as those pictured by O'Hara and Mathews (1981, p. 237) likely are the rule as the melts evolve further in crustal chambers.

If periodically replenished, periodically tapped, continuously fractionated magma chambers exist, they will evolve products whose phase petrology and trace element chemistry appear (in hitherto conventional interpretations) to require variable degrees of partial melting of inhomogeneous source regions for their petrogenesis. When the additional effects arising from assimilation by the magma of the roof of the chamber and from variation of mineralogy at the solidus of even a chemically homogeneous peridotite mantle are added, the scope for confusion is greatly increased. Moreover, these relationships cannot be inverted in order to deduce uniquely the magma chamber parameters or mantle source compositions from a knowledge of the erupted products.

O'Hara and Mathews were discussing the simplest of systems, that of spreading-ridge basaltic magmatism. Arc-magmatic systems add the great additional complications of progressively changing crust-and-mantle col-

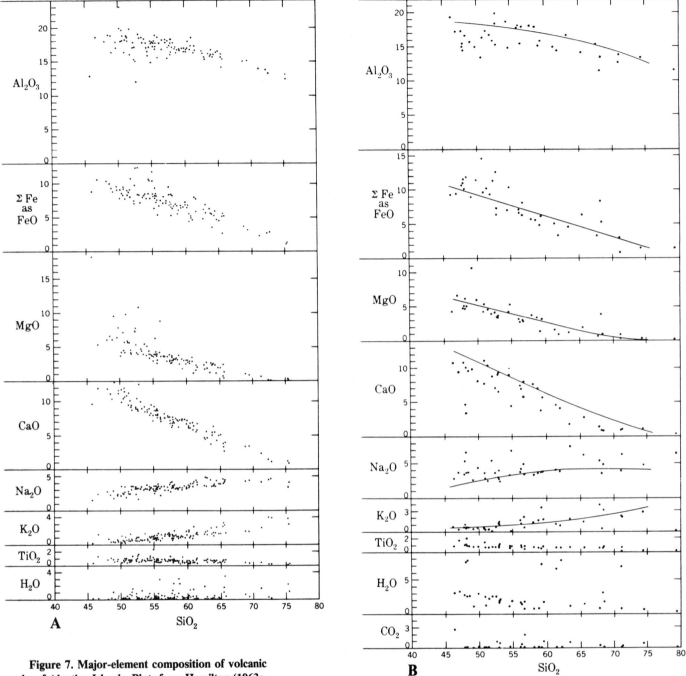

Figure 7. Major-element composition of volcanic rocks of Aleutian Islands. Plots from Hamilton (1963a, Figs. 65–67); data from U.S. Geological Survey reports, as cited by Hamilton. A. Silica-variation diagram of weight-percent analyses of subaerial volcanic rocks. B. Silica-variation diagram of weight-percent analyses of submarine volcanic rocks and contemporaneous intrusive rocks. Lines indicate variation trends in subaerial rocks (from A).

umns through which later magmas rise and evolve, with resultant systematic changes in composition of volcanic rocks along continuous arcs crossing crust of different type.

New Zealand System

A continuous, straight magmatic arc, 3,000 km long, trends north-northeastward from a terminus in southern North Island, New Zealand, across the continental shelf, and along the oceanic Kermadec and Tonga Islands. The arc illustrates not only the contrast between continental- and oceanic-arc magmas but the invalidity of some popular concepts regarding distinctions between arc and rift magmatism.

The modern arc lies above a continuous west-dipping slab of subducting Pacific lithosphere. Velocity of convergence and subduction decreases southward along the system, which gives way within New Zealand to a transpressive plate boundary. The New Zealand part of the arc is devel-

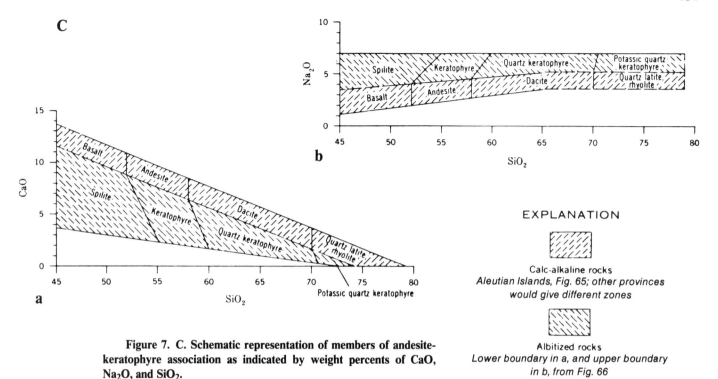

Figure 7. C. Schematic representation of members of andesite-keratophyre association as indicated by weight percents of CaO, Na_2O, and SiO_2.

oped on a crust of continental thickness but formed of Mesozoic accretionary-wedge materials, mostly variably metamorphosed terrigenous clastic sedimentary rocks. Along the oceanic Kermadec-Tonga part of the system, small volcanic islands rise from a submarine ridge. An oceanic back-arc basin has opened behind the migrating oceanic sector. The extensional zone rises southward onto the New Zealand continental shelf, and northern North Island is undergoing rapid, southward-decreasing extension and eastward-migrating, southward-lengthening arc magmatism (Stern, 1985).

Kermadec-Tonga lavas are typical of a mature oceanic arc—basalt, basaltic andesite, andesite, and subordinate dacite, all petrologically primitive and mostly of high-alumina types (Ewart and others, 1977).

Magmatism has only recently reached the south end of the modern arc in southern North Island, where there is little if any extension under way, and where large stratovolcanoes (Fig. 5A) are forming atop a low geanticlinal ridge that presumably is rising because of magmatic heating and inflation. The volcanic rocks are predominantly high-alumina andesites and basaltic andesites that display incorporation of continental materials—from the terrigenous strata of the accretionary wedge—in their minor elements and radiogenic isotopes (Cole, 1979; Ewart and others, 1977).

In north-central North Island, the arc is undergoing rapid extension synchronous with magmatism (Fig. 5B), and the predominant volcanic rocks are ignimbrites and flows of high-silica rhyodacite and quartz latite, with subordinate true rhyolite and much-subordinate basalt and dacite (Cole, 1979; Ewart and others, 1977). The isotopic similarity of the high-silica rocks to the underlying accretionary-wedge sedimentary rocks is indicative of a high degree of partial melting of clastic strata, likely as a result of heating of the deep crust by rising mantle diapirs and relatively primitive arc magmas. Extension is proceeding faster than the addition of magma from the mantle into the crustal column and the region is subsiding, and the northern part of the magmatic belt is below sea level on the subsiding continental shelf. Were this strongly bimodal magmatic assemblage met in an ancient setting, it would be regarded by most petrologists as evidence against a subduction setting—yet, it is in fact forming in a magmatic arc. Both the extending and nonextending parts of the onshore magmatic arc are at about the same height, 100 km, above the top of the subducting slab (Adams and Ware, 1977).

Submarine Volcanic Rocks

Oceanic island arcs are submarine ridges, built by magma, on which stand subaerial volcanoes that comprise a very small proportion of the crustal volume. The submarine rocks are erupted with compositions like those of the subaerial rocks but are variably altered by hydrothermal sea water to quite different compositions. Our petrologic data on arcs come overwhelmingly from the volumetrically minor subaerial rocks. Gill (1981) did not even mention submarine rocks in his generally excellent monograph on andesites. Ancient arcs accreted tectonically to continents are now exposed almost entirely at what were submarine levels, and so it is comparisons with submarine rocks that are relevant for paleotectonic analysis—but most geologists working with ancient accreted arcs mistakenly compare the petrology of ancient submarine rocks with modern subaerial rocks, and many reach invalid paleotectonic conclusions as a result. Submarine rocks commonly are exposed on some islands of mature island arcs, where they are raised, presumably, by magmatic inflation of the ridges. The submarine rocks commonly are altered to brown or green assemblages of fine-grained secondary minerals, and most petrologists and mapping geologists do little with them.

Much of our information on submarine rocks of still-active arcs comes from the reconnaissance study of the Aleutian Islands by U.S. Geological Survey geologists during the field seasons of 1946–1954, as reported particularly by Byers (1959), Drewes and others (1961), Fraser and Snyder (1959), Gates and others (1971), and Snyder and Fraser (1963). Subsequent Aleutian studies include those by Hein and others (1984) and McLean and Hein (1984). Eocene, Oligocene, and Miocene

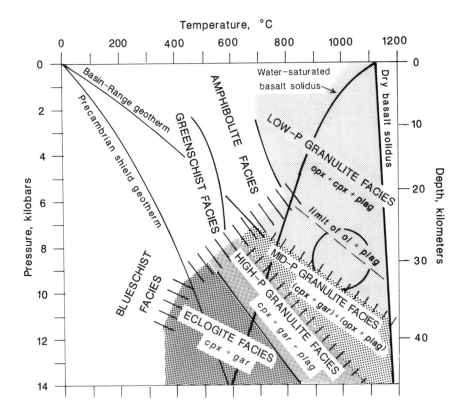

Figure 8. Generalized pressure-temperature diagram of mineral assemblages relevant to the crust of continents and mature island arcs. Boundaries approximate those for mafic and intermediate rocks but vary with bulk composition; coexisting minerals vary in composition across each facies. The boundary between amphibolite and granulite facies at pressures greater than about 5 kb shifts greatly with activity of H_2O, and a garnet amphibolite facies (not shown) of much-varying P/T width commonly intervenes. Abbreviations: cpx, clinopyroxene; gar, garnet; ol, olivine; opx, orthopyroxene; plag, plagioclase. Rocks of similar bulk compositions become progressively more dense going from low-pressure granulite to eclogite as plagioclase reacts with ferromagnesian minerals to produce successively denser phases; plagioclase reacts out successively with olivine, orthopyroxene, and clinopyroxene, although albite is stable in the higher T/P part of the blueschist facies, and sanidine is stable in high-T eclogite. Most igneous rocks in exposures of the upper part of the lower crust crystallized in or near the field indicated by the circle. From Hamilton (1988a, Fig. 2), where references are given.

submarine rocks are superbly exposed in the high seacliffs of the islands (Fig. 6). Near-source complexes are dominated by basaltic to dacitic lavas, pillow lavas, breccias including pillow breccias, and large and small sheets and nodular masses, intruded by gabbroic to granodioritic stocks and small batholiths. More-distal materials include volcaniclastic breccias and wackes, argillites, and cherts.

Severe alteration broadly affected the submarine volcanic rocks. Much of the alteration was diagenetic and much was hydrothermal, driven by heat from nearby plutons and by heat from the cooling flows and small intrusions themselves. Chlorite, epidote, albite, calcite, quartz, zeolites, clays, and oxides are widely developed. In addition to variable hydration, oxidation, and carbonation, much of the volcanic and hypabyssal assemblage underwent variable and often extreme change in bulk composition. The major-element change is defined particularly by enrichment in Na and depletion in Ca. (Systematic minor-element studies have not been made on interior submarine complexes of this or any other still-active arc so far as I am aware.) Contents of Na and Ca in the submarine rocks define a spectrum from amounts like those in the unaltered subaerial rocks of the island chain to extreme enrichment in Na and depletion in Ca (Fig. 7; Hamilton, 1963a). The rocks are compositionally basalt, andesite, and dacite and their Na-enriched, Ca-depleted equivalents, spilite, keratophyre, and quartz keratophyre. (The latter terms are often misapplied to greenschist-facies metavolcanic rocks of little-changed calc-alkalic bulk compositions in which plagioclase has been converted to albite, epidote, and other secondary minerals.) The aberrant, now-sodic rocks crystallized from calc-alkalic magmas, like those of the modern subaerial volcanoes, not from hypothetical sodic or hydrous melts. Relic clinopyroxenes are ordinary augitic varieties, relic high-temperature plagioclase is normal labradorite, and the voluminous albite has low-temperature crystal structure; the aberrant compositions are products of fluid exchanges under conditions comparable to those of low greenschist facies (Byers, 1959; Drewes and others, 1961; Wilcox, 1959). The reacting fluid must have had high activities of Na and CO_2 and a low activity of Ca, and must have affected submarine rocks selectively. Sea water is the obvious candidate, as Wilcox (1959) and others have emphasized. Brine concentrated from sea water produces the required albitization at low pressure and greenschist-facies temperature, provided other reactions release silica to the fluid (Rosenbauer and others, 1988).

Spreading-ridge basalts commonly display variable hydrothermal alteration by sea water, but the changes so far defined in bulk composition are far less severe than in many of the Aleutian rocks (Alt and others, 1986; Thompson, 1983). On the other hand, many on-land ophiolites—products of nascent island arcs?—display variable severe spilitization (Hawkins and Evans, 1983; Hopson and others, 1981; Lippard and others, 1986). Water depth may be a factor; contact between still-hot magmatic rocks and circulating sea water at the water depths, greater than 2.5 km, of spreading ridges may be inadequate to produce the observed arc-rock reactions. A possible explanation for the severe alteration of arc assemblages is that violent hydrothermal systems, involving great concentration of brines by boiling, are set up as submarine arc magmas cool in water shallower than the 2-km critical depth of water or are induced around plutons emplaced in shallow-water settings. The extreme variability of mineral-fluid reactions near the critical point of water may also be important. Further, the abundance of fragmental rocks in arc assemblages makes them highly permeable.

I showed that an ancient island arc now part of western Idaho had

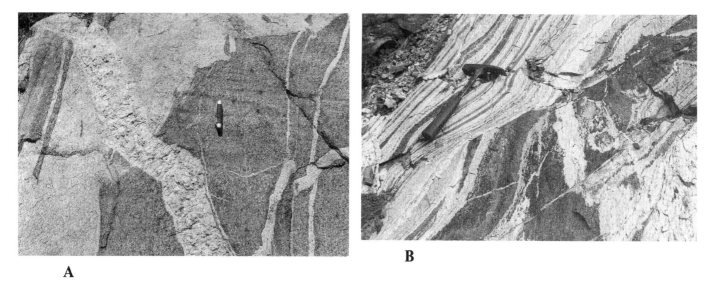

Figure 9. Middle-crust rocks of Mesozoic island arcs accreted tectonically to western North America. A. Dike of sodic pegmatite cutting light trondhjemite which migmatized tonalite gneiss. Sixmile Creek, Riggins quadrangle, west-central Idaho. B. Amphibolite migmatized by trondhjemite, which probably was derived by partial melting consequent on metamorphic dehydration of amphibolite. Near Diablo Lake, North Cascades, northwest Washington.

compositional variations, from calc-alkalic to spilitic-keratophyric, quantitatively like those of the submarine Aleutian rocks (Hamilton, 1963a). Various investigators (as, Roobol and others, 1983) have defined similar spectra in ancient submarine arc assemblages but then have argued that the sodic divergence from the compositions of modern subaerial volcanoes indicates alkaline magmatic associations.

Island-Arc Crust

Oceanic island-arc magmatism builds crust of continent-like thickness in mature arcs. Although this crust can be seen in exposure in various regions, its character has been minimally integrated in most petrologic modeling of arc magmas, which has been based primarily on the composition of subaerial volcanic rocks. The upper-crust substrates of volcanoes are widely exposed within still-active mature oceanic arcs and include abundant intrusive masses—dikes, sills, inflated pods, stocks, small batholiths—of gabbro, diorite, tonalite, and granodiorite and less commonly of more sodic granitic rocks. The plutonic rocks probably are on average more felsic than are the volcanic rocks. References to publications describing Indonesian and Melanesian examples were given by Hamilton (1979). Aleutian examples were described by Byers (1959), Drewes and others (1961), and others.

Rocks formed deeper in the crust of oceanic island arcs are exposed in some arcs accreted tectonically to continents and there deeply eroded. Figure 8 is a compilation of relevant crystallization facies for mafic rocks, applicable both to metamorphic and magmatic rocks. Facies designations herein accord with this figure, and not necessarily with the terminology of authors cited. Exposures of island-arc middle crust, beneath the levels of crosscutting mafic and intermediate plutons, in many cases are dominated by isotopically primitive amphibolitic, tonalitic, and trondhjemitic gneisses, the amphibolites becoming increasingly garnetiferous or pyroxene-bearing with depth; west-central Idaho and the North Cascades of Washington State provide excellent examples (Fig. 9). Trondhjemite (sodic leucotonalite; the term is misused by some geologists to include leucogranodiorite and andesine leucotonalite) may form primarily by partial melting, under lower-crustal conditions, of amphibolite of spilitic composition (Rapp and Watson, 1988).

Layered ultramafic and gabbroic complexes, fractionated both gravitationally and by fluid flow and contorted diapirically, are found in the roots of some arcs (as, Burns, 1985; Himmelberg and others, 1986; Irvine, 1974; Murray, 1972; Snoke and others, 1981). These complexes represent broad ranges of depth of formation, some having crystallized within the stability field of olivine plus plagioclase, others deeper (see Fig. 8). Clinopyroxene is the predominant pyroxene in most of these complexes, although orthopyroxene is abundant in many. Some of these mafic and ultramafic assemblages are associated with more felsic plutons.

An obliquely eroded north-dipping crustal section through a probable island-arc complex, of Cretaceous and early Tertiary age, which ramped southward onto northwest India in Eocene(?) time, has been studied in reconnaissance in Kohistan, northernmost Pakistan (Bard, 1983; Coward and others, 1982; Dietrich and others, 1983; Jan and Howie, 1981; D. E. Karig, 1988, written commun.; Tahirkheli, 1982). The much-deformed crustal section is perhaps 40 km thick, and mantle rocks extend as much as 5 km deeper to a truncation at the structural base of the section atop blueschist and mélange. Mafic and intermediate volcanic and volcaniclastic rocks, with abundant intercalated turbidite (Karig regards this part of the complex as of back-arc–basin origin) and downward-increasing stocks and small batholiths mostly of massive to gneissic diorite and tonalite, form the upper and middle crust, within which contact metamorphism on a regional scale increases downward from low greenschist through lower and upper amphibolite to garnet amphibolite facies. The lower crust consists of mafic granulites and mafic plutonic rocks, within which the grade of metamorphism, which is syn-plutonic with regard to some intrusions and post-plutonic with regard to others, increases downward from low-through middle- to high-pressure granulite facies. The variably metamorphosed plutonic rocks were fractionated from basaltic magmas and include norite, gabbro, and, in thin layers, anorthosite; magmatic olivine and plagioclase crystallized together in the upper part of the lower crust but not in the lower part. Rocks of intermediate composition are more abundant high in the lower crust than low in it. The mantle rocks at the base of the section consist of interlayered and injected residual, cumulate, and magmatic clinopyroxenite, peridotite, dunite, and subordinate olivine-free norite and gabbro, variably deformed and re-equilibrated at high-pressure granulite facies. Bard (1983) regarded the metamorphism as having occurred at much higher pressures than did the magmatism, but the facies relationships permit a contrary inference of isobaric magmatism and metamorphism.

A deep-crustal section of an isotopically primitive oceanic island arc, which probably was both crystallized and metamorphosed mostly within Early Cretaceous time, is exposed in far southwestern New Zealand (Mattinson and others, 1986), and includes Paleozoic components (Gibson and others, 1988). The rocks have been studied in reconnaissance by Blattner (1978), Gibson (1982), Gibson and others (1988), Mattinson and others (1986), Oliver (1980), and Williams and Smith (1983). The following synthesis represents my inference from their petrologic and structural data; they disagree variably with each other and with me. The crustal section was ramped up westward in Neogene time, as part of the transpressive deformation along the Alpine fault, and has been eroded obliquely. Gabbro, diorite, and tonalite dominate the deep, western part of the section, within which lenses of ultramafic rocks increase downward in abundance; leucogabbro, calcic anorthosite, and granodiorite are subordinate. Magmatic crystallization was in the high-pressure part of the low-pressure granulite facies (two pyroxenes; plagioclase stable with orthopyroxene but not with olivine; no garnet). At the deepest structural levels, these rocks were widely retrograded to gneisses in the middle- and high-pressure granulite facies and locally to eclogite facies; somewhat shallower rocks widely preserve igneous fabrics or were retrograded at garnet amphibolite facies. The facies relationships permit the inference that magmatism and retrogression were essentially isobaric and, for the deepest rocks exposed, occurred at a depth of about 35 km. Elsewhere in the complex, olivine and plagioclase crystallized together in mafic plutonic rocks, metavolcanic and calc-silicate gneisses are present, and retrogression occurred at amphibolite and garnet amphibolite facies; I infer isobaric magmatism and retrogression at depths of 20–25 km. Both massive and layered-differentiated plutonic rocks were present at both lower- and mid-crustal levels.

The character of crust in a nascent island arc was noted in the previous section on ophiolites.

Crust and Mantle

The lower crust of the two mature island arcs described above is dominated by mafic rocks in the Kohistan example, but by mafic, intermediate, and felsic-intermediate rocks in the Fiordland one. The high acoustic velocity and density of lower-crust rocks are due primarily to their granulite-facies mineralogy—to the presence of much of what would be plagioclase at lower pressure as pyroxene and garnet of granulite—and not necessarily to gabbroic bulk compositions. Similarly, mantle rocks include high-pressure plagioclase-free rocks as well as ultramafic rocks.

The Mohorovičić discontinuity exposed in the Kohistan section appears to be a gradational boundary within fractionated magmatic rocks, which are predominantly ultramafic beneath and predominantly granulitic and olivine-free noritic and gabbroic above. The discontinuity was constructed by arc magmatism and is not a fossil lithologic boundary. I have argued elsewhere (Hamilton, 1981) that this is the general character of the base of the crust in magmatic arcs—that the Mohorovičić discontinuity of continents and mature island arcs represents primarily the shallow limit of crystallization of voluminous arc-magmatic rocks of ultramafic composition or of plagioclase-free mineralogy. An example of similar relationships across the Mohorovičić discontinuity of a continental magmatic arc is given by the Ivrea zone of the northwest Italian Alps (Rivalenti and others, 1981). Arc magmas that reach the base of the crust have basaltic or even intermediate compositions, yet the protomelts generated deeper in the mantle likely are there in approximate equilibrium with olivine-rich rocks and hence likely are olivine-rich basalts, and so it follows that likely most of the ultramafic component of the primary magmas is crystallized within the mantle. Plagioclase-free rocks, formed at pressures too high for plagioclase to be stable, also are limited to the mantle. The mantle-crust boundary is a self-perpetuating density filter for rising melts, much evolution occurs within the mantle, melts that reach the crust are already highly fractionated, and those that reach the surface are more so. O'Hara (1985), Quick (1981), and Stolper and Walker (1980) made related points.

ACCRETION TO CONTINENTS

Island arcs migrate by back-arc spreading and are conveyor-belted toward subduction zones, so that island arcs sooner or later collide with one another and with continents. All island arcs older than middle Mesozoic have been accreted to continents, as have many much younger arcs. Collisions between continents commonly are preceded by long periods of subduction with complexly changing patterns, and collided arcs commonly are major components of the broad tracts of tectonic flotsam crunched between collided continents. Accreted arcs have now been demonstrated to occur within such tracts of all ages from the Archean onward. Among many examples are those discussed by Burchfiel and Davis (1981), Condie (1986), Dickinson (1981), Hamilton (1978b, 1979), Hanson and Schweickert (1986), Shervais and Kimbrough (1985), Silver and Smith (1983), Stoeser (1986), Sylvester and others (1987), and Windley (1984).

The complex histories of collisions, subduction reversals, rifting, and strike-slip and oroclinal deformation of the modern arc systems discussed earlier in this paper presumably have analogues, however difficult they are to decipher, in ancient accreted-arc terranes. Paleotectonic analysis of island arcs and other subduction-related complexes should, but too often does not, incorporate awareness of the complex variations and behavior of modern arc systems. Departures from actualistic models should record intent, not ignorance. The study of modern arcs, and the testing of predictions implicit in paleotectonic analyses in terms of modern analogues, are urged upon anyone who would interpret ancient arcs.

ACKNOWLEDGMENTS

This review draws on the published work of hundreds of geologists and geophysicists, only a relative few of whom can be cited, and on my discussions over the years with scores more. The manuscript was much improved as a result of helpful criticism by W. R. Dickinson, A. B. Ford, William Glen, J. W. Hawkins, and D. E. Karig.

REFERENCES CITED

Abbott, M. J., and Chamalaun, F. H., 1981, Geochronology of some Banda Arc volcanics: Indonesia Geological Research and Development Centre Special Publication 2, p. 253–268.

Adams, R. D., and Ware, D. E., 1977, Subcrustal earthquakes beneath New Zealand; locations determined with a laterally inhomogeneous velocity model: New Zealand Journal of Geology and Geophysics, v. 20, p. 59–83.

Alt, J. C., Honnorez, J., Laverne, C., and Emmermann, R., 1986, Hydrothermal alteration of a 1 km section through the upper oceanic crust, Deep Sea Drilling Project hole 504B—Mineralogy, chemistry, and evolution of seawater-basalt interactions: Journal of Geophysical Research, v. 91, p. 10309–10335.

Alvarez, W., 1982, Geological evidence for the geographical pattern of mantle return flow and the driving mechanism of plate tectonics: Journal of Geophysical Research, v. 87, p. 6697–6710.

Anderson, R. N., Langseth, M. G., Hayes, D. E., Watanabe, T., and Yasui, M., 1978, A geophysical atlas of the east and southeast Asian seas—Heat flow, thermal conductivity, thermal gradient: Geological Society of America Map and Chart Series MC-25, scale 1:6,442,194.

Argand, E., 1924, La tectonique de l'Asie: Congres Geologique International, Comptes Rendus de la XIIIe Session, en Belgique, 1922, v. 1, p. 171–372.

Atwater, T., 1970, Implications of plate tectonics for the Cenozoic tectonic evolution of western North America: Geological Society of America Bulletin, v. 81, p. 3513–3536.

Axelrod, D. I., 1963, Fossil floras suggest stable, not drifting, continents: Journal of Geophysical Research, v. 68, p. 3257–3263.

Bachman, S. B., Lewis, S. D., and Schweller, W. J., 1983, Evolution of a forearc basin, Luzon Central Valley, Philippines: American Association of Petroleum Geologists Bulletin, v. 67, p. 1143–1162.

Barbat, W. F., 1971, Megatectonics of the Coast Ranges, California: Geological Society of America Bulletin, v. 82, p. 1541–1562.

Bard, J. P., 1983, Metamorphism of an obducted island arc—Example of the Kohistan sequence (Pakistan) in the Himalayan collided range: Earth and Planetary Science Letters, v. 65, p. 133–144.

Beaudry, D., and Moore, G. F., 1981, Seismic-stratigraphic framework of the forearc basin off central Sumatra, Sunda Arc: Earth and Planetary Science Letters, v. 54,, p. 17–28.

Benioff, H., 1949, Seismic evidence for the fault origin of oceanic deeps: Geological Society of America Bulletin, v. 60, p. 1337–1356.

——— 1954, Orogenesis and deep crustal structure—Additional evidence from seismology: Geological Society of America Bulletin, v. 65, p. 385–400.

Bennett, J. D., and 10 others, 1981, Geologic map of the Banda Aceh quadrangle, North Sumatra: Indonesia Geological Research and Development Centre, 19 p. + map, scale 1:250,000.

Berry, R. F., and Grady, A. E., 1981, Deformation and metamorphism of the Aileu Formation, north coast, East Timor and its tectonic significance: Journal of Structural Geology, v. 3, p. 143–167.

Billings, M. P., 1960, Diastrophism and mountain building: Geological Society of America Bulletin, v. 71, p. 363–398.

Birch, F., 1965, Speculations on the Earth's thermal history: Geological Society of America Bulletin, v. 76, p. 133–154.

Bird, J. M., and Dewey, J. F., 1970, Lithosphere plate–continental margin tectonics and the evolution of the Appalachian orogen: Geological Society of America Bulletin, v. 81, p. 1031–1060.

Blake, M. C., Jr., Irwin, W. P., and Coleman, R. G., 1969, Blueschist-facies metamorphism related to regional thrust faulting: Tectonophysics, v. 8, p. 237–246.

Blattner, P., 1978, Geology of the crystalline basement between Milford Sound and the Hollyford Valley, New Zealand: New Zealand Journal of Geology and Geophysics, v. 21, p. 33–47.

Bloomer, S. H., 1983, Distribution and origin of igneous rocks from the landward slopes of the Mariana Trench—Implications for its structure and evolution: Journal of Geophysical Research, v. 88, p. 7411–7428.

Bloomer, S. H., and Fisher, R. L., 1987, Petrology and geochemistry of igneous rocks from the Tonga Trench—A non-accreting plate boundary: Journal of Geology, v. 95, p. 469–495.

Bloomer, S. H., and Hawkins, J. W., 1983, Gabbroic and ultramafic rocks from the Mariana Trench—An island-arc ophiolite: American Geophysical Union Geophysical Monograph 27, p. 294–317.

——— 1987, Petrology and geochemistry of boninite series volcanic rocks from the Mariana trench: Contributions to Mineralogy and Petrology, v. 97, p. 361–377.

Bowin, C., Purdy, G. M., Johnston, C., Shor, G., Lawver, L., Hartono, H.M.S., and Jezek, P., 1980, Arc-continent collision in Banda Sea region: American Association of Petroleum Geologists Bulletin, v. 64, p. 868–915.

Bracey, D. R., and Vogt, P. R., 1970, Plate tectonics in the Hispaniola area: Geological Society of America Bulletin, v. 81, p. 2855–2860.

Breen, N. A., Silver, E. A., and Hussong, D. M., 1986, Structural styles of an accretionary wedge south of the island of Sumba, Indonesia, revealed by SeaMARC II side-scan sonar: Geological Society of America Bulletin, v. 97, p. 1250–1261.

Brophy, J. G., and Marsh, B. D., 1986, On the origin of high-alumina arc basalt and the mechanics of melt extraction: Journal of Petrology, v. 27, p. 763–789.

Bullard, E., Everett, J. E., and Smith, A. G., 1965, The fit of continents around the Atlantic: Royal Society of London Philosophical Transactions, ser. A, v. 258, p. 41–51.

Burchfiel, B. C., and Davis, G. A., 1981, Triassic and Jurassic tectonic evolution of the Klamath Mountains–Sierra Nevada geologic terrane, *in* Ernst, W. G., ed., The geotectonic development of California: Englewood Cliffs, New Jersey, Prentice-Hall, p. 50–70.

Burk, C. A., 1965, Geology of the Alaska Peninsula—Island arc and continental margin: Geological Society of America Memoir 99, 250 p.

Burns, L. E., 1985, The Border Ranges ultramafic and mafic complex, south-central Alaska—Cumulate fractionates of island-arc volcanics: Canadian Journal of Earth Sciences, v. 22, p. 1020–1038.

Byers, F. M., Jr., 1959, Geology of Umnak and Bogoslof Islands, Aleutian Islands, Alaska: U.S. Geological Survey Bulletin 1028, p. 267–369.

Cameron, N. R., and 10 others, 1982, The geology of the Tapaktuan quadrangle, Sumatra: Indonesia Geological Research and Development Centre, 18 p. + map, scale 1:250,000.

Cardwell, R. K., Isacks, B. L., and Karig, D. E., 1980, The spatial distribution of earthquakes, focal mechanism solutions, and subducted lithosphere in the Philippine and northeastern Indonesian islands: American Geophysical Union Geophysical Monograph 23, p. 1–35.

Carey, S. W., 1958, The tectonic approach to continental drift, *in* Carey, S. W., ed., Continental drift—A symposium: Geology Department, University of Tasmania, p. 177–374.

Carlson, R. L., 1981, Boundary forces and plate tectonics: Geophysical Research Letters, v. 8, p. 958–961.

Carlson, R. L., and Melia, P. J., 1984, Subduction hinge migration: Tectonophysics, v. 102, p. 399–411.

Chase, C. G., 1978, Extension behind island arcs and motions relative to hot spots: Journal of Geophysical Research, v. 83, p. 5385–5387.

——— 1979, Asthenospheric counterflow—A kinematic model: Royal Astronomical Society Geophysical Journal, v. 56, p. 1–18.

Clague, D. A., and Dalrymple, G. B., 1987, The Hawaiian-Emperor volcanic chain, Part I, Geologic evolution: U.S. Geological Survey Professional Paper 1350, p. 5–54.

Clegg, J. A., Almond, M., and Stubbs, P.H.S., 1954, The remanent magnetism of some sedimentary rocks in Britain: Philosophical Magazine, ser. 7, v. 45, p. 583–598.

Cloos, M., 1985, Thermal evolution of convergent plate margins—Thermal modeling and reevaluation of isotopic Ar-ages for blueschists in the Franciscan complex of California: Tectonics, v. 4, p. 421–433.

Coats, R. R., 1962, Magma type and crustal structure in the Aleutian arc: American Geophysical Union Monograph 6, p. 92–109.

Cole, J. W., 1979, Structure, petrology, and genesis of Cenozoic volcanism, Taupo Volcanic Zone, New Zealand—A review: New Zealand Journal of Geology and Geophysics, v. 22, p. 631–657.

Coleman, R. G., 1967, Glaucophane schists from California and New Caledonia: Tectonophysics, v. 4, p. 479–498.

——— 1984, The diversity of ophiolites: Geologie en Mijnbouw, v. 63, p. 141–150.

Condie, K. C., 1986, Geochemistry and tectonic setting of Early Proterozoic supracrustal rocks in the southwestern United States: Journal of Geology, v. 94, p. 845–864.

Coney, P. J., 1970, The geotectonic cycle and the new global tectonics: Geological Society of America Bulletin, v. 81, p. 739–748.

Coode, A. M., 1965, A note on oceanic transcurrent faults: Canadian Journal of Earth Sciences, v. 2, p. 400–401.

Coulbourn, W. T., and Moberly, R., 1977, Structural evidence of the evolution of fore-arc basins of South America: Canadian Journal of Earth Sciences, v. 14, p. 102–116.

Coward, M. P., Jan, M. Q., Rex, D., Tarney, J., Thirlwall, M., and Windley, B. F., 1982, Geo-tectonic framework of the Himalaya of N Pakistan: Geological Society of London Journal, v. 139, p. 299–308.

Cox, A., and Doell, R. R., 1960, Review of paleomagnetism: Geological Society of America Bulletin, v. 71, p. 645–768.

Crawford, A. J., Beccaluva, L., and Serri, G., 1981, Tectono-magmatic evolution of the West Philippine–Mariana region and the origin of boninites: Earth and Planetary Science Letters, v. 54, p. 346–356.

Crawford, A. J., Falloon, T. J., and Eggins, S., 1987, The origin of island arc high-alumina basalts: Contributions to Mineralogy and Petrology, v. 97, p. 417–430.

Creer, K. M., Irving, E., and Runcorn, S. K., 1957, Geophysical interpretation of palaeomagnetic directions from Great Britain: Royal Society of London Philosophical Transactions, ser. A, v. 250, p. 144–156.

Davidson, J. P., 1987, Crustal contamination *versus* subduction zone enrichment—Examples from the Lesser Antilles and implications for mantle source compositions of island arc volcanic rocks: Geochimica et Cosmochimica Acta, v. 51, p. 2185–2198.

Davis, D. M., and Solomon, S. C., 1985, True polar wander and plate-driving forces: Journal of Geophysical Research, v. 90, p. 1837–1841.

Davis, G. A., 1968, Westward thrust faulting in the south-central Klamath Mountains, California: Geological Society of America Bulletin, v. 79, p. 911–934.

——— 1969, Tectonic correlations, Klamath Mountains and western Sierra Nevada, California: Geological Society of America Bulletin, v. 80, p. 1095–1108.

DeLong, S. E., Perfit, M. R., McCulloch, M. T., and Ach, J., 1985, Magmatic evolution of Semisopochnoi Island, Alaska—Trace-element and isotopic constraints: Journal of Geology, v. 93, p. 609–618.

Deutsch, E. R., 1963, Polar wandering and continental drift—An evaluation of recent evidence: Society of Economic Paleontologists and Mineralogists Special Publication 10, p. 4–46.

Dewey, J. F., 1980, Episodicity, sequence and style at convergent plate boundaries: Geological Association of Canada Special Paper 20, p. 553–573.

Dewey, J. F., and Bird, J. M., 1970, Mountain belts and the new global tectonics: Journal of Geophysical Research, v. 75, p. 2625–2647.

Dewey, J. F., Pitman, W. C., III, Ryan, W.B.F., and Bonnin, J., 1973, Plate tectonics and the evolution of the Alpine system: Geological Society of America Bulletin, v. 84, p. 3137–3180.

Dickinson, W. R., 1969, Evolution of calc-alkaline rocks in the geosynclinal system of California and Oregon: Oregon Department of Geology and Mineral Industries Bulletin 65, p. 151–156.

——— 1970a, 2d Penrose Conference—The new global tectonics: GeoTimes, v. 15, no. 4, p. 18–22.

——— 1970b, Meetings—Global tectonics: Science, v. 168, p. 1250–1259.

——— 1970c, Relations of andesites, granites, and derivative sandstones to arc-trench tectonics: Reviews of Geophysics and Space Physics, v. 8, p. 813–860.

——— 1981, Plate tectonics and the continental margin of California, *in* Ernst, W. G., ed., The geotectonic development of California: Englewood Cliffs, New Jersey, Prentice-Hall, p. 1–28.

——— 1982, Compositions of sandstones in circum-Pacific subduction complexes and fore-arc basins: American Association of Petroleum Geologists Bulletin, v. 66, p. 121–137.

Dickinson, W. R., and Hatherton, T., 1967, Andesitic volcanism and seismicity around the Pacific: Science, v. 157, p. 801–803.

Dietrich, V. J., Frank, W., and Honegger, K., 1983, A Jurassic-Cretaceous island arc in the Ladakh-Himalayas: Journal of Volcanology and Geothermal Research, v. 18, p. 405–433.

Dietz, R. S., 1954, Marine geology of northwestern Pacific—Description of Japanese bathymetric chart 6901: Geological Society of America Bulletin, v. 65, p. 1199–1224.

——— 1961, Continent and ocean basin evolution by spreading of the sea floor: Nature, v. 190, p. 854–857.

——— 1963, Collapsing continental rises—An actualistic concept of geosynclines and mountain building: Journal of Geology, v. 71, p. 314–333.

——— 1966, Passive continents, spreading sea floors, and collapsing continental rises: American Journal of Science, v. 265, p. 177–193.

Djuri, M., 1975, Geologic map of the Purwokerto and Tegal quadrangles, Java: Geological Survey of Indonesia, scale 1:100,000.

Drewes, H., Fraser, G. D., Snyder, G. L., and Barnett, H. F., Jr., 1961, Geology of Unalaska Island and adjacent insular shelf, Aleutian Islands, Alaska: U.S. Geological Survey Bulletin 1028, p. 583–676.

Du Toit, A. L., 1937, Our wandering continents: Edinburgh, United Kingdom, Oliver and Boyd, 366 p.

Ernst, W. G., 1965, Mineral parageneses in Franciscan metamorphic rocks, Panoche Pass, California: Geological Society of America Bulletin, v. 76, p. 879–914.

——— 1970, Tectonic contact between the Franciscan melange and the Great Valley Sequence—Crustal expression of a late Mesozoic Benioff zone: Journal of Geophysical Research, v. 75, p. 886–901.

——— 1973, Interpretative synthesis of metamorphism in the Alps: Geological Society of America Bulletin, v. 84, p. 2053–2078.

Ewart, A., Brothers, R. N., and Mateen, A., 1977, An outline of the geology and geochemistry, and the possible petrogenetic evolution of the volcanic rocks of the Tonga–Kermadec–New Zealand island arc: Journal of Volcanology and Geothermal Research, v. 2, p. 205–250.

Ewing, M., and Worzel, J. L., 1954, Gravity anomalies and structure of the West Indies, Part I: Geological Society of America Bulletin, v. 65, p. 165–174.

Ewing, M., Ewing, J. I., and Talwani, M., 1964, Sediment distribution in the oceans—The Mid-Atlantic Ridge: Geological Society of America Bulletin, v. 75, p. 17–36.

Fisk, M. B., 1986, Basalt magma interaction with harzburgite and the formation of high-magnesium andesites: Geophysical Research Letters, v. 13, p. 467–470.

Frakes, L. A., and Crowell, J. C., 1967, Facies and paleogeography of late Paleozoic Lafonian diamictite, Falkland Islands: Geological Society of America Bulletin, v. 78, p. 37–58.

——— 1969, Late Paleozoic glaciation, I, South America: Geological Society of America Bulletin, v. 80, p. 1007–1042.

Fraser, G. D., and Snyder, G. L., 1959, Geology of southern Adak Island and Kagalaska Island, Alaska: U.S. Geological Survey Bulletin 1028, p. 371–408.

Garfunkel, Z., Anderson, C. A., and Schubert, G., 1986, Mantle circulation and the lateral migration of subducted slabs: Journal of Geophysical Research, v. 91, p. 7205–7223.

Gates, O., Powers, H. A., and Wilcox, R. E., 1971, Geology of the Near Islands, Alaska: U.S. Geological Survey Bulletin 1028, p. 709–822.

Gibson, G. M., 1982, Stratigraphy and petrography of some metasediments and associated intrusive rocks from central Fiordland, New Zealand: New Zealand Journal of Geology and Geophysics, v. 25, p. 21–43.

Gibson, G. M., McDougall, I., and Ireland, T.R., 1988, Age constraints on metamorphism and the development of a metamorphic core complex in Fiordland, southern New Zealand: Geology, v. 16, p. 405–408.

Gill, J. B., 1981, Orogenic andesites and plate tectonics: Berlin, Springer-Verlag, 390 p.

——— 1987, Early geochemical evolution of an oceanic island arc and backarc—Fiji and the South Fiji Basin: Journal of Geology, v. 95, p. 589–615.

Gilluly, James, 1949, Distribution of mountain building in geologic time: Geological Society of America Bulletin, v. 60, p. 561–590.
Glen, William, 1982, The road to Jaramillo: Stanford, California, Stanford University Press, 459 p.
Griggs, D. T., 1939, A theory of mountain building: American Journal of Science, v. 237, p. 611–650.
Grow, J. A., 1973, Crustal and upper mantle structure of the central Aleutian arc: Geological Society of America Bulletin, v. 84, p. 2169–2192.
Grow, J. A., and Atwater, T., 1970, Mid-Tertiary tectonic transition in the Aleutian arc: Geological Society of America Bulletin, v. 81, p. 3715–3722.
Gutenberg, B., 1936, Structure of the Earth's crust and the spreading of the continents: Geological Society of America Bulletin, v. 47, p. 1587–1610.
———— 1954, Low-velocity layers in the Earth's mantle: Geological Society of America Bulletin, v. 65, p. 337–348.
Hager, B. H., and O'Connell, R. J., 1981, A simple global model of plate dynamics and mantle convection: Journal of Geophysical Research, v. 86, p. 4843–4867.
Hall, Robert, 1987, Plate boundary evolution in the Halmahera region, Indonesia: Tectonophysics, v. 144, p. 337–352.
Hamilton, W. B., 1961, Origin of the Gulf of California: Geological Society of America Bulletin, v. 72, p. 1307–1318.
———— 1963a, Metamorphism in the Riggins region, western Idaho: U.S. Geological Survey Professional Paper 436, 95 p.
———— 1963b, Overlapping of late Mesozoic orogens in western Idaho: Geological Society of America Bulletin, v. 74, p. 779–788.
———— 1963c, Antarctic tectonics and continental drift: Society of Economic Paleontologists and Mineralogists Special Publication 10, p. 74–93.
———— 1963d, Tectonics of Antarctica: American Association of Petroleum Geologists Memoir 2, p. 4–15.
———— 1964, Discussion of paper by D. I. Axelrod, 'Fossil floras suggest stable, not drifting, continents': Journal of Geophysical Research, v. 69, p. 1666–1668.
———— 1966, Origin of the volcanic rocks of eugeosynclines and island arcs: Geological Survey of Canada Paper 66-15, p. 348–356.
———— 1968, Cenozoic climatic change and its cause: American Meteorological Society Meteorological Monographs, v. 8, no. 30, p. 128–133.
———— 1969a, Mesozoic California and the underflow of Pacific mantle: Geological Society of America Bulletin, v. 80, p. 2409–2430.
———— 1969b, The volcanic central Andes—A modern model for the Cretaceous batholiths and tectonics of western North America: Oregon Department of Geology and Mineral Industries Bulletin 65, p. 175–184.
———— 1970, The Uralides and the motion of the Russian and Siberian Platforms: Geological Society of America Bulletin, v. 81, p. 2553–2576.
———— 1974a, Map of sedimentary basins of the Indonesian region: U.S. Geological Survey Miscellaneous Investigations Series Map I-875-B, scale 1:5,000,000.
———— 1974b, Earthquake map of the Indonesian region: U.S. Geological Survey Miscellaneous Investigations Series Map I-875-C, scale 1:5,000,000.
———— 1978a, Tectonic map of the Indonesian region: U.S. Geological Survey Miscellaneous Investigations Series Map I-875-D, scale 1:5,000,000; reprinted with corrections, 1981.
———— 1978b, Mesozoic tectonics of the western United States: Society of Economic Paleontologists and Mineralogists, Pacific Section, Pacific Coast Paleogeography Symposium, 2nd, p. 33–70.
———— 1979, Tectonics of the Indonesian region: U.S. Geological Survey Professional Paper 1078, 345 p.; reprinted with corrections, 1981 and 1985.
———— 1981, Crustal evolution by arc magmatism: Royal Society of London Philosophical Transactions, ser. A, v. 301, p. 279–291.
———— 1988a, Tectonic setting and variations with depth of some Cretaceous and Cenozoic structural and magmatic systems of the western United States, in Ernst, W. G., ed., Metamorphism and crustal evolution of the western United States: Englewood Cliffs, New Jersey, Prentice-Hall, p. 1–40.
———— 1988b, Convergent-plate tectonics viewed from the Indonesian region, in Sengor, A.M.C., ed., Tectonic evolution of the Tethyan domain: Amsterdam, the Netherlands, Reidel.
Hamilton, W. B., and Krinsley, D., 1967, Upper Paleozoic glacial deposits of South Africa and southern Australia: Geological Society of America Bulletin, v. 78, p. 783–800.
Hamilton, W. B., and Myers, W. B., 1967, The nature of batholiths: U.S. Geological Survey Professional Paper 554-C, 29 p.
Hanson, R. E., and Schweickert, R. A., 1986, Stratigraphy of mid-Paleozoic island-arc rocks in part of the northern Sierra Nevada, Sierra and Nevada Counties, California: Geological Society of America Bulletin, v. 97, p. 986–998.
Harbert, W., Scholl, D. W., Vallier, T. L., Stevenson, A. J., and Mann, D. M., 1986, Major evolutionary phases of a forearc basin of the Aleutian terrace—Relation to North Pacific tectonic events and the formation of the Aleutian subduction complex: Geology, v. 14, p. 757–761.
Hatcher, R. D., 1972, Developmental model for the southern Appalachians: Geological Society of America Bulletin, v. 83, p. 2735–2760.
Hatherton, T., and Dickinson, W. R., 1969, The relationship between andesitic volcanism and seismicity in Indonesia, the Lesser Antilles, and other island arcs: Journal of Geophysical Research, v. 74, p. 5301–5310.
Hawkins, J. W., and Evans, C. A., 1983, Geology of the Zambales Range, Luzon, Philippine Islands—Ophiolite derived from an island arc-back arc basin pair: American Geophysical Union Geophysical Monograph 27, p. 95–123.
Hawkins, J. W., Bloomer, S. H., Evans, C. A., and Melchior, J. T., 1984, Evolution of intra-oceanic arc-trench systems: Tectonophysics, v. 102, p. 174–205.
Hawkins, J. W., Moore, G. F., Villamor, R., Evans, C., and Wright, E., 1985, Geology of the composite terranes of east and central Mindanao: Circum-Pacific Council for Energy and Mineral Resources, Earth Science Series, v. 1, p. 437–463.
Hayes, D. E., and Taylor, B., 1978, A geophysical atlas of the east and southeast Asian seas—Tectonics: Geological Society of America Map and Chart Series MC-25, scale 1:6,442,194.
Hayes, D. E., Houtz, R. E., Jarrard, R. D., Mrozowski, C. L., and Watanabe, T., 1978, A geophysical atlas of east and southeast Asian seas—Crustal structure: Geological Society of America Map and Chart Series MC-25, scale 1:6,442,194.
Heezen, B. C., Tharp, M., and Ewing, M., 1959, The floors of the oceans. I. The North Atlantic: Geological Society of America Special Paper 65, 122 p.
Hein, J. R., McLean, H., and Vallier, T., 1984, Reconnaissance geology of southern Atka Island, Aleutian Islands, Alaska: U.S. Geological Survey Bulletin 1609, 19 p.
Heirtzler, J. R., Dickson, G. O., Herron, E. M., Pitman, W. C., III, and Le Pichon, X., 1968, Marine magnetic anomalies, geomagnetic field reversals, and motions of the ocean floor and continents: Journal of Geophysical Research, v. 73, p. 2119–2136.
Hess, H. H., 1948, Major structural features of the western North Pacific, an interpretation of H.O. 5485, bathymetric chart, Korea to New Guinea: Geological Society of America Bulletin, v. 59, p. 417–446.
———— 1955, Serpentines, orogeny, and epeirogeny: Geological Society of America Special Paper 62, p. 391–408.
———— 1962, History of ocean basins, in Engel, A.E.J., James, H. L., and Leonard, B. F., eds., Petrologic studies, A volume in honor of A. F. Buddington: Boulder, Colorado, Geological Society of America, p. 599–620.
Hill, M. L., and Dibblee, T. W., Jr., 1953, San Andreas, Garlock, and Big Pine faults, California: Geological Society of America Bulletin, v. 64, p. 443–458.
Himmelberg, G. R., Loney, R. A., and Craig, J. T., 1986, Petrogenesis of the ultramafic complex at the Blashke Islands, southeastern Alaska: U.S. Geological Survey Bulletin 1662, 14 p.
Holmes, A., 1931, Radioactivity and earth movements: Geological Society of Glasgow Transactions, v. 18, p. 559–606.
Hopson, C. A., Mattinson, J. W., and Pessagno, E. A., Jr., 1981, Coast Range ophiolite, western California, in Ernst, W. G., ed., The geotectonic development of California: Englewood Cliffs, New Jersey, Prentice-Hall, p. 418–510.
Hsü, K. J., 1968, Principles of melanges and their bearing on the Franciscan-Knoxville paradox: Geological Society of America Bulletin, v. 79, p. 1063–1074.
Hutchinson, R. W., 1980, Massive base metal sulphide deposits as guides to tectonic evolution: Geological Association of Canada Special Paper 20, p. 659–694.

Ingersoll, R. V., and Schweickert, R. A., 1986, A plate-tectonic model for Late Jurassic ophiolite genesis, Nevadan orogeny and forearc initiation, northern California: Tectonics, v. 5, p. 901–912.
Ingersoll, R. V., and Suczek, C. A., 1979, Petrology and provenance of Neogene sand from Nicobar and Bengal fans, DSDP sites 211 and 218: Journal of Sedimentary Petrology, v. 49, p. 1217–1228.
Irvine, T. N., 1974, Petrology of the Duke Island ultramafic complex, southeastern Alaska: Geological Society of America Memoir 138, 240 p.
Isacks, B., Sykes, L. B., and Oliver, Jack, 1969, Focal mechanisms of deep and shallow earthquakes in the Tonga-Kermadec region and the tectonics of island arcs: Geological Society of America Bulletin, v. 80, p. 1443–1470.
Jacobson, R. S., Shor, G. G., Jr., Kieckhefer, R. M., and Purdy, G. M., 1979, Seismic refraction and reflection studies in the Timor-Aru trough system and Australian continental shelf: American Association of Petroleum Geologists Memoir 29, p. 209–222.
James, D. E., 1971, Plate tectonic model for the evolution of the central Andes: Geological Society of America Bulletin, v. 82, p. 3325–3346.
Jan, M. Q., and Howie, R. A., 1981, The mineralogy and geochemistry of the metamorphosed basic and ultrabasic rocks of the Jijal complex, Kohistan, NW Pakistan: Journal of Petrology, v. 22, p. 85–126.
Jarrard, R. D., 1986, Relations among subduction parameters: Reviews of Geophysics, v. 24, p. 217–284.
Karig, D. E., 1971, Structural history of the Mariana arc system: Geological Society of America Bulletin, v. 82, p. 323–344.
———— 1972, Remnant arcs: Geological Society of America Bulletin, v. 83, p. 1057–1068.
———— 1975, Basin genesis in the Philippine Sea: Initial reports of the Deep Sea Drilling Project, v. 31, p. 857–879.
———— 1982, Initiation of subduction zones—Implications for arc evolution and ophiolite development: Geological Society of London Special Publication 10, p. 563–576.
Karig, D. E., Caldwell, J. G., and Parmentier, E. M., 1976, Effects of accretion on the geometry of the descending lithosphere: Journal of Geophysical Research, v. 81, p. 6281–6291.
Karig, D. E., Lawrence, M. B., Moore, G. F., and Curray, J. R., 1980a, Structural framework of the fore-arc basin, NW Sumatra: Geological Society of London Journal, v. 137, p. 77–91.
Karig, D. E., Moore, G. F., Curray, J. R., and Lawrence, M. B., 1980b, Morphology and shallow structure of the lower trench slope off Nias Island, Sunda Arc: American Geophysical Union Geophysical Monograph 23, p. 179–208.
Karig, D. E., Sarewitz, D. R., and Haeck, G. D., 1986, Role of strike-slip faulting in the evolution of allochthonous terranes in the Philippines: Geology, v. 14, p. 852–855.
Karig, D. E., Barber, A. J., Charlton, T. R., Klemperer, S., and Hussong, D. M., 1987, Nature and distribution of deformation across the Banda Arc–Australian collision zone in Timor: Geological Society of America Bulletin, v. 98, p. 18–32.
Kay, M., 1951, North American geosynclines: Geological Society of America Memoir 48, 143 p.
Kay, S. M., and Kay, R. W., 1985, Role of crystal cumulates and the oceanic crust in the formation of the Aleutian arc: Geology, v. 13, p. 461–464.
Kieckhefer, R. M., Shor, G. G., Jr., Curray, J. R., Sugiarta, W., and Hehuwat, F., 1980, Seismic refraction studies of the Sunda Trench and forearc basin: Journal of Geophysical Research, v. 85, p. 863–889.
Kincaid, C., and Olson, P., 1987, An experimental study of subduction and slab migration: Journal of Geophysical Research, v. 92, p. 13832–13840.
Knopf, A., 1948, The geosynclinal theory: Geological Society of America Bulletin, v. 59, p. 649–670.
Krause, D. C., 1965, Submarine geology north of New Guinea: Geological Society of America Bulletin, v. 76, p. 27–42.
———— 1966, Tectonics, marine geology, and bathymetry of the Celebes Sea–Sulu Sea region: Geological Society of America Bulletin, v. 77, p. 813–832.
Kuno, H., 1966, Lateral variation of basalt magma across continental margins and island arcs: Geological Survey of Canada Paper 66-15, p. 317–335.
Lee, C.-S., and McCabe, R., 1986, The Banda-Celebes-Sulu basin—A trapped piece of Cretaceous-Eocene oceanic crust?: Nature, v. 322, p. 51–54.
Le Pichon, X., 1968, Sea-floor spreading and continental drift: Journal of Geophysical Research, v. 73, p. 3661–3697.
Lewis, S. D., and Hayes, D. E., 1984, A geophysical study of the Manila Trench, Luzon, Philippines. 2. Fore arc basin structural and stratigraphic evolution: Journal of Geophysical Research, v. 89, p. 9196–9214.
Lippard, S. J., Shelton, A. W., and Gass, I. G., 1986, The ophiolite of northern Oman: Geological Society of London Memoir 11, 178 p.
Luyendyk, B. P., 1970, Dips of downgoing lithospheric plates beneath island arcs: Geological Society of America Bulletin, v. 81, p. 3411–3416.
MacDonald, G.J.F., 1964, The deep structure of continents: Science, v. 143, p. 921–929.
Malfait, B. T., and Dinkelman, M. G., 1972, Circum-Caribbean tectonic and igneous activity and the evolution of the Caribbean plate: Geological Society of America Bulletin, v. 83, p. 251–272.
Malinverno, A., and Ryan, W.B.F., 1986, Extension in the Tyrrhenian Sea and shortening in the Apennines as result of arc migration driven by sinking of the lithosphere: Tectonics, v. 5, p. 227–245.
Mammerickx, J., Fisher, R. L., Emmel, F. J., and Smith, S. M., 1976, Bathymetry of the east and southeast Asian seas: Geological Society of America Map and Chart Series MC-17, scale 1:6,442,194.
Mattinson, J. M., Kimbrough, D. L., and Bradshaw, J. Y., 1986, Western Fiordland orthogneiss—Early Cretaceous arc magmatism and granulite facies metamorphism, New Zealand: Contributions to Mineralogy and Petrology, v. 92, p. 383–392.
McCabe, R., 1984, Implications of paleomagnetic data on the collision related bending of island arcs: Tectonics, v. 3, p. 409–428.
McCabe, R., Kikawa, E., Cole, J. T., Malicse, A. J., Baldauf, P. E., Yumul, J., and Almasco, J., 1987, Paleomagnetic results from Luzon and the central Philippines: Journal of Geophysical Research, v. 92, p. 555–580.
McCaffrey, R., 1982, Lithospheric deformation within the Molucca Sea arc-arc collision—Evidence from shallow and intermediate earthquake activity: Journal of Geophysical Research, v. 87, p. 3663–3678.
McCaffrey, R., and Nabelek, J., 1984, The geometry of back arc thrusting along the eastern Sunda Arc, Indonesia—Constraints from earthquake and gravity data: Journal of Geophysical Research, v. 89, p. 6171–6179.
———— 1987, Earthquakes, gravity, and the origin of the Bali Basin—An example of a nascent continental fold-and-thrust belt: Journal of Geophysical Research, v. 92, p. 441–460.
McCaffrey, R., Silver, E. A., and Raitt, R. W., 1980, Crustal structure of the Molucca Sea collision zone, Indonesia: American Geophysical Union Geophysical Monograph 23, p. 161–178.
McCaffrey, R., Molnar, P., Roecker, S. W., and Joyodiwiryo, Y. S., 1985, Microearthquake seismicity and fault plane solutions related to arc-continent collision in the eastern Sunda arc, Indonesia: Journal of Geophysical Research, v. 90, p. 4511–4528.
McKenzie, D. P., and Morgan, W. J., 1969, Evolution of triple junctions: Nature, v. 224, p. 125–133.
McKenzie, D. P., and Parker, R. L., 1967, The North Pacific—An example of tectonics on a sphere: Nature, v. 216, p. 1276–1280.
McLean, H., and Hein, J. R., 1984, Paleogene geology and chronology of southwestern Umnak Island, Aleutian Islands, Alaska: Canadian Journal of Earth Sciences, v. 21, p. 171–180.
McNutt, M., 1988, Thermal and mechanical properties of the Cape Verde Rise: Journal of Geophysical Research, v. 93, p. 2784–2794.
Menard, H. W., 1986, The ocean of truth—A personal history of global tectonics: Princeton University Press, 353 p.
Miyashiro, A., 1961, Evolution of metamorphic belts: Journal of Petrology, v. 2, p. 277–311.
Molnar, P., and Atwater, T., 1978, Interarc spreading and Cordilleran tectonics as alternates related to the age of subducted oceanic lithosphere: Earth and Planetary Science Letters, v. 41, p. 330–340.
Molnar, P., and Stock, J., 1987, Relative motions of hotspots in the Pacific, Atlantic, and Indian Oceans since late Cretaceous time: Nature, v. 327, p. 587–591.
Molnar, P., and Sykes, L. R., 1969, Tectonics of the Caribbean and Middle America regions from focal mechanisms and seismicity: Geological Society of America Bulletin, v. 80, p. 1639–1684.
Moore, D. G., Curray, J. R., and Emmel, F. J., 1976, Large submarine slide (olistostrome) associated with Sunda Arc subduction zone, northeast Indian Ocean: Marine Geology, v. 21, p. 211–226.
Moore, G. F., and Karig, D. E., 1980, Structural geology of Nias Island, Indonesia—Implications for subduction zone

tectonics: American Journal of Science, v. 280, p. 193–223.
Moore, G. F., and Silver, E. A., 1982, Collision processes in the northern Molucca Sea: American Geophysical Union Geophysical Monograph 27, p. 360–372.
Moore, G. F., Billman, H. G., Hehanussa, P. E., and Karig, D. E., 1980a, Sedimentology and paleobathymetry of Neogene trench-slope deposits, Nias Island, Indonesia: Journal of Geology, v. 88, p. 161–180.
Moore, G. F., Curray, J. R., Moore, D. G., and Karig, D. E., 1980b, Variations in geologic structure along the Sunda fore arc, northeastern Indian Ocean: American Geophysical Union Geophysical Monograph 23, p. 145–160.
Moore, G. F., Curray, J. R., and Emmel, F. J., 1982, Sedimentation in the Sunda Trench and forearc region: Geological Society of London Special Publication 10, p. 245–258.
Morgan, W. J., 1968, Rises, trenches, great faults, and crustal blocks: Journal of Geophysical Research, v. 73, p. 1959–1982.
Mrozowski, C. L., and Hayes, D. L., 1978, A geophysical atlas of east and southeast Asian seas—Sediment isopachs: Geological Society of America Map and Chart Series MC-25, scale 1:6,442,194.
Murray, C. G., 1972, Zoned ultramafic complexes of the Alaskan type—Feeder pipes of andesitic volcanoes: Geological Society of America Memoir 132, p. 313–335.
Myers, J. D., and Marsh, B. D., 1987, Aleutian lead isotopic data—Additional evidence for the evolution of lithospheric plumbing systems: Geochimica et Cosmochimica Acta, v. 51, p. 1833–1842.
Natland, J. H., and Tarney, J., 1981, Petrologic evolution of the Mariana arc and back-arc basin system—A synthesis of drilling results in the Philippine Sea: Initial reports of the Deep Sea Drilling Project, v. 60, p. 877–908.
Nye, C. J., and Reid, M. R., 1986, Geochemistry of primary and least fractionated lavas from Okmok volcano, central Aleutians—Implications for arc magma genesis: Journal of Geophysical Research, v. 91, p. 10271–10287.
O'Hara, M. J., 1985, Importance of the 'shape' of the melting regime during partial melting of the mantle: Nature, v. 314, p. 58–62.
O'Hara, M. J., and Mathews, R. E., 1981, Geochemical evolution in an advancing, periodically replenished, periodically tapped, continuously fractionated magmatic chamber: Geological Society of London Journal, v. 138, p. 237–277.
Oliver, G.J.H., 1980, Geology of the granulite and amphibolite facies gneisses of Doubtful Sound, Fiordland, New Zealand: New Zealand Journal of Geology and Geophysics, v. 23, p. 27–41.
Opdyke, N. D., and Runcorn, S. K., 1960, Wind direction in the western United States in the late Paleozoic: Geological Society of America Bulletin, v. 71, p. 959–972.
Page, B. M., 1972, Oceanic crust and mantle fragment in subduction complex near San Luis Obispo, California: Geological Society of America Bulletin, v. 83, p. 957–972.
Parsons, B., 1982, Causes and consequences of the relation between area and age of the ocean floor: Journal of Geophysical Research, v. 87, p. 289–302.
Pearce, J. A., Lippard, S. J., and Roberts, S., 1984, Characteristics and tectonic significance of supra-subduction zone ophiolites: Geological Society of London Special Publication 16, p. 77–94.
Pennington, W. D., 1983, Role of shallow phase changes in the subduction of oceanic crust: Science, v. 220, p. 1045–1047.
Pigram, C. J., and Panggabean, H., 1983, Age of the Banda Sea, eastern Indonesia: Nature, v. 301, p. 231–234.
Quick, J. E., 1981, The origin and significance of large, tabular dunite bodies in the Trinity peridotite, northern California: Contributions to Mineralogy and Petrology, v. 78, p. 413–422.
Raff, A. D., and Mason, R. G., 1961, Magnetic survey off the west coast of North America, 40° N. latitude to 52° N. latitude: Geological Society of America Bulletin, v. 72, p. 1267–1270.
Rapp, R. P., and Watson, E. B., 1988, Partial melting of amphibolite/eclogite and the origin of tonalitic-trondhjemitic magmas [abs.]: EOS (American Geophysical Union Transactions), v. 69, p. 521.
Reed, D. L., Silver, E. A., Prasetyo, H., and Meyer, A. W., 1986, Deformation and sedimentation along a developing terrane suture—Eastern Sunda forearc, Indonesia: Geology, v. 14, p. 1000–1003.
Rivalenti, G., Garuti, G., Rossi, A., Siena, F., and Sinigoi, S., 1981, Existence of different peridotite types and of a layered igneous complex in the Ivrea zone of the Western Alps: Journal of Petrology, v. 22, p. 127–153.
Rock, N.M.S., and 8 others, 1983, The geology of the Lubuksikaping quadrangle, Sumatra: Indonesia Geological Research and Development Centre, 60 p. + map, scale 1:250,000.
Rodolfo, K. S., 1969, Bathymetry and marine geology of the Andaman Basin, and tectonic implications for southeast Asia: Geological Society of America Bulletin, v. 80, p. 1203–1230.
Roobol, M. J., Jackson, N. J., and Darbyshire, D.F.P., 1983, Late Proterozoic lavas of the central Arabian shield—Evolution of an ancient arc system: Geological Society of London Journal, v. 140, p. 185–202.
Rosenbauer, R. J., Bischoff, J. L., and Zierenberg, R. A., 1988, The laboratory albitization of mid-ocean ridge basalt: Journal of Geology, v. 96, p. 237–244.
Rubie, D. C., 1984, The olivine-spinel transformation and the rheology of subducting lithosphere: Nature, v. 308, p. 505–508.
Runcorn, S. K., 1959, Rock magnetism: Science, v. 129, p. 1002–1012.
Sarewitz, D. R., and Karig, D. E., 1986, Processes of allochthonous terrane evolution, Mindoro Island, Philippines: Tectonics, v. 5, p. 525–552.
Sclater, J. G., Hawkins, J. W., Mammerickx, J., and Chase, C. G., 1972, Crustal extension between the Tonga and Lau Ridges—Petrologic and geophysical evidence: Geological Society of America Bulletin, v. 83, p. 505–518.
Sclater, J. G., Parsons, B., and Jaupart, C., 1981, Oceans and continents—Similarities and differences in the mechanisms of heat loss: Journal of Geophysical Research, v. 86, p. 11535–11552.
Shervais, J. W., and Kimbrough, D. L., 1985, Geochemical evidence for the tectonic setting of the Coast Range ophiolite—A composite island arc–oceanic crust terrane in western California: Geology, v. 13, p. 35–38.
Silver, E. A., 1971a, Transitional tectonics and late Cenozoic structure of the continental margin off northernmost California: Geological Society of America Bulletin, v. 82, p. 1–22.
——— 1971b, Tectonics of the Mendocino triple junction: Geological Society of America Bulletin, v. 82, p. 2965–2978.
Silver, E. A., and Reed, D. L., 1988, Backthrusting in accretionary wedges: Journal of Geophysical Research, v. 93, p. 3116–3126.
Silver, E. A., and Smith, R. B., 1983, Comparison of terrane accretion in modern Southeast Asia and the Mesozoic North American Cordillera: Geology, v. 11, p. 198–202.
Silver, E. A., McCaffrey, R., Joyodiwiryo, Y., and Stevens, S., 1983a, Ophiolite emplacement by collision between the Sula Platform and the Sulawesi Island Arc, Indonesia: Journal of Geophysical Research, v. 88, p. 9419–9435.
Silver, E. A., McCaffrey, R., and Smith, R. B., 1983b, Collision, rotation, and the initiation of subduction in the evolution of Sulawesi, Indonesia: Journal of Geophysical Research, v. 86, p. 11535–11552.
Silver, E. A., Reed, D. L., McCaffrey, R., and Joyodiwiryo, Y., 1983c, Back arc thrusting in the eastern Sunda Arc, Indonesia—A consequence of arc-continent collision: Journal of Geophysical Research, v. 88, p. 7429–7448.
Silver, E. A., Gill, J. B., Schwartz, D., Prasetyo, H., and Duncan, R. A., 1985, Evidence for a submerged and displaced continental borderland, north Banda Sea, Indonesia: Geology, v. 13, p. 687–691.
Silver, E. A., Breen, N. A., Prasetyo, Hardi, and Hussong, D. M., 1986, Multibeam study of the Flores backarc thrust belt, Indonesia: Journal of Geophysical Research, v. 91, p. 3489–3500.
Snoke, A. W., Quick, J. E., and Bowman, H. R., 1981, Bear Mountain igneous complex, Klamath Mountains, California—An ultrabasic to silicic calc-alkaline suite: Journal of Petrology, v. 22, p. 501–552.
Snyder, G. L., and Fraser, G. D., 1963, Pillowed lavas, I—Intrusive layered lava pods and pillowed lavas, Unalaska Island, Alaska: U.S. Geological Survey Professional Paper 454-B, 23 p.
Solomon, S. C., Sleep, N. H., and Richardson, R. M., 1975, On the forces driving plate tectonics—Inferences from absolute plate velocities and intraplate stress: Royal Astronomical Society Geophysical Journal, v. 42, p. 769–801.
Stehli, F. G., 1957, Possible Permian climatic zonation and its implications: American Journal of Science, v. 255, p. 607–718.
——— 1970, A test of the Earth's magnetic field during Permian time: Journal of Geophysical Research, v. 75, p. 3325–3342.
Stern, T. A., 1985, A back-arc basin formed within continental lithosphere—The Central Volcanic Region of New Zealand: Tectonophysics, v. 112, p. 385–409.
Stoeser, D. B., 1986, Distribution and tectonic setting of plutonic rocks of the Arabian Shield: Journal of African Earth Sciences, v. 4, p. 21–46.
Stolper, E., and Walker, D., 1980, Melt density and the average composition of basalt: Contributions to Mineralogy and Petrology, v. 74, p. 7–12.
Suwarna, N., Koesoemadinata, S., and Santosa, S., 1981, Peta geologi lembar ende Nusatenggara Timur: Indonesia Geological Research and Development Centre, 23 p. + map, scale 1:250,000.
Sykes, L. R., 1967, Mechanisms of earthquakes and nature of faulting on the mid-oceanic ridges: Journal of Geophysical Research, v. 72, p. 2131–2153.
Sylvester, P. J., Attoh, K., and Schulz, K. J., 1987, Tectonic setting of late Archean bimodal volcanism in the Michipicoten (Wawa) greenstone belt, Ontario: Canadian Journal of Earth Sciences, v. 24, p. 1120–1134.
Tahirkheli, R.A.K., 1982, Geology of the Himalaya, Karakoram and Hindukush in Pakistan: University of Peshawar Geological Bulletin, v. 15, 51 p.
Tapponnier, P., Peltzer, G., and Armijo, R., 1986, On the mechanics of the collision between India and Asia: Geological Society of London Special Publication 19, p. 115–157.
Taylor, B., and Karner, G. D., 1983, On the evolution of marginal basins: Reviews of Geophysics and Space Physics, v. 21, p. 1727–1741.
Taylor, F. B., 1910, Bearing of the Tertiary mountain belt on the origin of the Earth's plan: Geological Society of America Bulletin, v. 21, p. 179–226.
ten Brink, U. S., and Brocher, T. M., 1987, Multichannel seismic evidence for a subcrustal intrusive complex under Oahu and a model for Hawaiian volcanism: Journal of Geophysical Research, v. 92, p. 13687–13707.
Thompson, G., 1983, Basalt-seawater interaction, *in* Rona, P. A., Bostrom, K., Laubier, L., and Smith, K. L., Jr., eds., Hydrothermal processes at seafloor spreading centers: New York, Plenum Press, p. 225–278.
Turner, D. L., and Jarrard, R. D., 1982, K-Ar dating of the Cook-Austral island chain—A test of the hot-spot hypothesis: Journal of Volcanology and Geothermal Research, v. 12, p. 187–220.
Uyeda, S., and Kanamori, H., 1979, Back-arc opening and the mode of subduction: Journal of Geophysical Research, v. 84, p. 1049–1061.
Vacquier, V., Raff, A. D., and Warren, R. E., 1961, Horizontal displacements in the floor of the northeastern Pacific Ocean: Geological Society of America Bulletin, v. 72, p. 1251–1258.
Van der Voo, R., 1988, Paleozoic paleogeography of North America, Gondwana, and intervening terranes—Comparisons of paleomagnetism with paleoclimatology and biogeographical patterns: Geological Society of America Bulletin, v. 100, p. 311–324.
Van Gool, M., Huson, W. J., Prawirasasra, R., and Owen, T. R., 1987, Heat flow and seismic observations in the northwestern Banda Arc: Journal of Geophysical Research, v. 92, p. 2581–2586.
Vening Meinesz, F. A., 1954, Indonesian Archipelago—A geophysical study: Geological Society of America Bulletin, v. 65, p. 143–164.
Vine, F. J., and Matthews, D. H., 1963, Magnetic anomalies over oceanic ridges: Nature, v. 199, p. 947–949.
Von Huene, R., and Shor, G. G., 1969, The structure and tectonic history of the eastern Aleutian Trench: Geological Society of America Bulletin, v. 80, p. 1889–1902.
Wang, C.-Y., and Shi, Y.-L., 1984, On the thermal structure of subduction complexes—A preliminary study: Journal of Geophysical Research, v. 89, p. 7709–7719.
Watts, A. B., Bodine, J. H., and Bowin, C. O., 1978, A geophysical atlas of the east and southeast Asian seas—Free air gravity field: Geological Society of America Map and Chart Series MC-25, scale 1:6,442,194.
Wegener, A., 1915, Die Enstehung der Kontinente und Ozeane: Braunschweig, Vieweg, 94 p.
Weissel, J. K., 1980, Evidence for Eocene oceanic crust in the Celebes Basin: American Geophysical Union Geophysical Monograph 23, p. 37–48.
Weissel, J. K., and Hayes, D. E., 1978, A geophysical atlas of the east and southeast Asian seas—Magnetic anomalies: Geological Society of America Map and Chart Series MC-25, scale 1:6,442,194.
Wheller, G. E., Varne, R., Foden, J. D., and Abbott, M. J., 1987, Geochemistry of Quaternary volcanism in the Sunda-Banda arc, Indonesia, and three-component genesis of island-arc basaltic magmas: Journal of Volcanology and Geothermal Research, v. 32, p. 137–160.
White, W. M., and Dupre, B., 1986, Sediment subduction and magma genesis in the Lesser Antilles—Isotopic and trace element constraints: Journal of Geophysical Research, v. 91, p. 5927–5941.
Wilcox, R. E., 1959, Igneous rocks of the Near Islands, Aleutian Islands, Alaska: International Geological Congress, 20th, Mexico City, sec. 11-A, p. 365–378.
Williams, J. G., and Smith, I.E.M., 1983, The Hollyford gabbronorite—A calcalkaline cumulate: New Zealand Journal of Geology and Geophysics, v. 26, p. 345–357.
Wilson, J. T., 1961, Untitled discussion: Nature, v. 192, p. 125–128.
——— 1965, A new class of faults and their bearing on continental drift: Nature, v. 207, p. 343–347.
——— 1966, Did the Atlantic close and then reopen?: Nature, v. 211, p. 676–681.
Windley, Brian, 1984, The evolving continents (2nd edition): Chichester, England, John Wiley & Sons, 399 p.

Manuscript Received by the Society February 29, 1988
Revised Manuscript Received May 5, 1988
Manuscript Accepted May 5, 1988

Paleozoic paleogeography of North America, Gondwana, and intervening displaced terranes: Comparisons of paleomagnetism with paleoclimatology and biogeographical patterns

CENTENNIAL ARTICLE

ROB VAN DER VOO *Department of Geological Sciences, University of Michigan, Ann Arbor, Michigan 48109-1063*

ABSTRACT

New paleomagnetic data have become available in the past 5 yr that require modifications in previously published paleogeographic reconstructions for the Silurian and Devonian. In this paper, the new paleopoles are compared to published paleogeographic models based on paleoclimatologic and biogeographic data. The data from the three fields of paleomagnetism, paleoclimatology, and biogeography are generally in excellent agreement, and an internally consistent paleogeographic evolutionary picture of the interactions between North America, Gondwana, and intervening displaced terranes is emerging. During the interval of the Ordovician, Silurian, and Devonian, North America stayed in equatorial paleoposition, while rotating counterclockwise. The northwest African part of Gondwana was in high southerly latitudes during the Late Ordovician and was fringed by peri-Gondwanide terranes, such as southern Europe (Armorica) and Avalonian basement blocks now found in eastern Newfoundland, Nova Scotia, the Boston Basin, the Appalachian Piedmont, and northern Florida. Subsequently, Gondwana and the peri-Gondwanide terranes displayed rapid drift with respect to the pole. This drift translates into the following pattern of movement for northwest Africa. During the latest Ordovician–Early Silurian, this area moved rapidly northward from polar to subtropical latitudes, followed by equally rapid southward motion from subtropical to intermediate (about 50°S) paleolatitudes during the Late Silurian–Middle Devonian. It is likely that significant east-to-west motion accompanied the latter shift in paleolatitudes, with the Caledonian-Acadian orogeny the result of Silurian to Early Devonian convergence and collision between Gondwana and North America. This collision sandwiched several of the intervening displaced terranes between Gondwana and North America. Subsequent to this collision, Gondwana was separated in the Late Devonian by a medium-width ocean from North America and the Avalonian and southern European blocks which were left behind adjacent to North America. This new ocean closed during the Carboniferous, and the resulting convergence and collision were the cause of the Hercynian-Alleghanian orogenic belt. Problems remaining for future research, besides the further gathering of reliable paleopoles, involve the uncertain pre-Devonian position of the southern British Isles in this scenario and the very rapid velocity with respect to the pole that results from the rapid Late Ordovician–Silurian apparent polar wander for Gondwana.

INTRODUCTION

Paleomagnetism, paleoclimatology, and biogeography, as subdisciplines of the geosciences, each contribute significantly to our collective paleogeographic knowledge, because the data from each field show a strong dependence on latitude. Given that Paleozoic oceans have left little evidence other than that found in highly deformed sequences of accreted terranes, the continental record that is contained in the magnetization of rocks, their latitude-dependent lithofacies, and their fauna forms the only data base from which the construction of Paleozoic paleogeographic maps can be attempted. Each of these subdisciplines, however, has its drawbacks, uncertainties, and controversies, and often the ambiguities involve widely divergent paleolatitude options for the same geological period in all of the three fields. An example of such ambiguities can be found in published Devonian reconstructions, where paleomagnetism, paleoclimatology, and biogeography have each been used to indicate paleolatitude patterns different by up to 50°, as will be discussed below.

In recent years, however, some additional paleomagnetic results have become available that seem to narrow the occasional interpretation gaps between the three fields. It is the purpose of this paper to illustrate the convergence of paleogeographic models for Ordovician through Devonian times, where most of the above-mentioned ambiguities could previously be found. When Editors Hatcher and Thomas invited me to write an article for the GSA Centennial *Bulletin*, they asked me to integrate my insights with those gained from other recent Geological Society of America publications, as well as the general literature, and to discuss the effects of these papers on subsequent thought, concepts, and ideas. In looking through the publications by the Geological Society of America of the past five years, I was struck by the fact that although papers discussing Paleozoic paleogeography have been published at regular intervals in each of the three fields of paleomagnetism, paleoclimatology, and biogeography, not much cross-referencing took place. Although this paper, given the length constraints, cannot possibly be a comprehensive review, I hope that the following sections will illustrate how improved paleomagnetic-analysis techniques yield results that are, within the margins of error, in good agreement with our understanding of paleolatitudes derived from paleoclimatology and biogeography. Some previously existing ambiguities appear to be resolved, whereas others persist.

The following sections will discuss, first, the strong and weak points of each of the methodologies, and second, the paleogeographic data base for the major continents North America and Gondwana and the interven-

Geological Society of America Bulletin, v. 100, p. 311–324, 8 figs., 2 tables, March 1988.

ing displaced terranes, followed by a discussion of the tectonic implications. The paleogeographic data base is organized by paleogeographic units, using subsections dealing with the temporal aspects of the Ordovician, Silurian, and Devonian periods. Because I will not discuss data from other continental elements, such as the Baltic Shield–Russian Platform, Siberia, or China, the paleogeographic maps presented in this paper do not include these elements.

PALEOLATITUDE METHODOLOGIES

Paleomagnetism

Granted the assumption of a geocentric, co-axial geomagnetic dipole field (on average) in the geological past, paleomagnetic inclinations with respect to paleohorizontal yield direct information about geographic paleolatitude. The choice between northern and southern hemisphere can usually be well documented if a sufficiently continuous record of paleopole positions exists to work back in time from the present-day pole. For North America and Gondwana, such records, in the form of more-or-less continuous apparent polar wander paths, do indeed exist, whereas for many of the intervening displaced terranes, a choice between hemispheres must be based on minimum motion and coherent plate interactions.

Are these assumptions and choices justified? The question as to whether the geomagnetic field was geocentric has been evaluated for late Mesozoic and Cenozoic time, with the result that only small departures have been documented (Merrill and McElhinny, 1983). These departures, causing pole errors generally less than $8°$, are insufficient to cause us concern for the mid-Paleozoic, given the imprecision of the entire data base. The question as to whether the field was co-axial is, of course, critical to a successful comparison between paleomagnetic and paleoclimatological data, and this paper will confirm previous affirmative answers (for example, Briden, 1968a) by showing that, in general, paleomagnetic poles and polar regions defined by other techniques appear to agree very well. The choice of northern versus southern hemisphere for the displaced terranes seems likewise to cause no disagreement between the various techniques and therefore, does not present a concern.

The determination of paleohorizontal, on the other hand, constitutes one of the fundamental reliability criteria for paleomagnetic results, but is readily made only for stratified rocks. For many of the recent paleopole determinations discussed in the following section, this criterion has been met. Difficult to evaluate for nonpaleomagnetists, but nevertheless critical, is the quality of the paleomagnetic analysis itself, as judged from the thoroughness of the demagnetization techniques applied to isolate the characteristic remanence from unwanted overprints. Furthermore, a sufficient quantity of sites and samples is needed from the rock sections to average out errors and secular variation of the geomagnetic field, and the presence of near-antipodal reversals constitutes another reliability criterion (Van der Voo, 1988). Lastly, most important is the age of the rock, or better yet, the age of the magnetization. Although the danger of remagnetization has long been recognized (Creer, 1968), it has become quite clear in recent years that many apparent tectonic inconsistencies in the available paleomagnetic data are the result of magnetizations that were not recognized as being significantly younger than the rocks themselves. As an example, one can cite results from some restudied Late Devonian and early Carboniferous rocks of North America, which have now been shown to possess late Carboniferous–Early Permian overprints as well as, occasionally, some primary magnetizations (Irving and Strong, 1984; Kent and Opdyke, 1985; Miller and Kent, 1986). Consequently, in the past few years, the inferred paleolatitude positions of North America for the Late Devonian and early Carboniferous have become more southerly by about $15°$ to $20°$.

Tests are available to constrain the age of magnetization, and in recent years, these have taken on ever greater importance. It has, therefore, become imperative to redo several of the earlier studies, but with better field (fold or conglomerate) tests and with better dating techniques to determine the age of the rock or that of its alteration or cooling from elevated metamorphic temperatures. As such, paleomagnetic interpretation has some advantages over the techniques from paleoclimatology and biogeography: not only is a paleomagnetic paleolatitude determination generally more quantitative, but the technique also allows one to test the results again and again by carefully designed experiments that enable us to unambiguously accept or reject the hypothesis that a given magnetization was acquired in a given time interval. It must therefore be emphasized that the paleomagnetic data base is not static, an aspect often underestimated by paleogeographers who express puzzlement or skepticism over the sometimes drastic changes in interpretation from one to the next paper in a series of successive paleomagnetic publications (for example, Boucot and Gray, 1983, p. 574–575; Young, 1986). Several recent paleomagnetic results have resolved previous ambiguities or revealed the secondary nature of magnetizations previously regarded as primary. If remagnetizations are not (yet) recognized as secondary, the paleopoles for a given period often are inconsistent. All else being equal, it would be my preference in such a case to give greater weight to those poles for which the location least resembles poles for any younger period and thus are least likely to be based on remagnetizations (see also Tarling, 1985).

Because of the axially symmetric nature of the dipole field model used to determine paleolatitude, absolute longitude cannot be determined, and relative longitudes can be determined only if accurate apparent polar wander path segments exist for two blocks that presumably drifted together for the duration of the path segments. For pre-Carboniferous comparisons between North America, the other major continents or smaller displaced terranes, the presently available paths are not sufficient to determine relative paleolongitudes.

Paleoclimatology

As noted by Robinson (1973), "Paleoclimatology involves the study of the motion of all three complex systems of hydrosphere, atmosphere, and lithosphere, and hence it is the most difficult of the Earth sciences." Ocean circulation and, hence, climatic zonation on Earth are strongly influenced by the presence of (usually irregular) land barriers, and topographic relief on the continents has pronounced influence on precipitation, as illustrated by Ziegler and co-workers (1979). Hence, assertions about paleoclimatic conditions require a rather complete paleogeographic knowledge about the distribution of land, shallow seas, and oceans. If the Earth's topography (above, as well as below, sea level) were to show axial symmetry, then a more or less zonal (latitudinal) climatological pattern would develop, but obviously the Paleozoic world was far from zonally arranged. Correlations of coal swamps, thick clastic sequences, and glacial tillites with wet belts, however, are usually excellent; conversely, dry climates are indicated by evaporites.

What is needed, therefore, is a model that "translates" dry or wet conditions (combined with warm- or cold-temperature indicators) into latitudinal position. The difficulty in achieving this renders the use of paleoclimatology a very complex subject for making inferences about latitude. Warm-water carbonates (aragonite, high Mg-calcite) may develop up to mid-latitudes ($45°$), but so can continental glaciations during alternating long-term climatic cycles. For relatively stationary continents

TABLE 1. SELECTED PALEOMAGNETIC POLES FOR THE ORDOVICIAN THROUGH DEVONIAN PERIODS FOR LAURENTIA, GONDWANA, AND THE INTERVENING DISPLACED TERRANES

Rock unit, location	Age	Pole position		k	α_{95}	Reliability							Q	Reference
		Lat.	Long.			1	2	3	4	5	6	7		
North American Craton and Cratonic Margins (south poles)														
Catskill red beds, south, PA	Du	26°S,	304°E	16	16	x		x			x	x	4	Miller and Kent (1986)
Catskill red beds, north, PA	Du	33°S,	270°E	165	7	x	x	x	x		x	x	5	Miller and Kent (1986)
Peel Sound Fm., NWT, Canada	Dl	25°S,	279°E	66	9	x	x	x	x	x	x	x	6	Dankers (1982)
Bloomsburg Fm., PA	Su	32°S,	282°E	35	9	x	x	x	x		x	x	6	Roy and others (1967)
Rose Hill Fm., PA-VA	Sm	19°S,	309°E	18	6	x	x	x	x		x	x	6	French and Van der Voo (1979)
Wabash Reef Ls., IN	Sm	17°S,	305°E	74	5	x	x	x	x	x	x	x	7	McCabe and others (1985)
Ringgold Gap sediments, GA	Sl/Ou	28°S,	322°E	62	7	x	x	x			x	x	5	Morrison and Ellwood (1986)
Cordova secondary, Ontario	446	31°S	282°E	18	11	x	x	x		x	x	x	5	Dunlop and Stirling (1985)
Juniata Fm., PA-VA	Ou	32°S,	294°E	53	5	x	x	x	x		x		5	Van der Voo and French (1977)
Steel Mtn. Secondary, W. Newf.	451	23°S,	319°E	22	13		x	x		x		x	4	Murthy and Rao (1976)
Chapman Ridge Fm., TN	Om	27°S,	292°E	38	15	x		x	x				3	Watts and Van der Voo (1979)
Moccasin-Bays Fms., TN	Om	33°S,	327°E	135	6	x	x	x	x		x	x	6	Watts and Van der Voo (1979)
St. George Fm., Newfoundland	Ol	26°S,	306°E	202	7	x	x	x		x			4	Beales and others (1974)
Oneota Dolomite, IA, MN, WI	Ol	10°S,	346°E	18	12	x	x	x		x	x	x	6	Jackson and Van der Voo (1985)
Gondwana Craton (African coordinates; south poles)														
Canning Basin Reef Ls., Austr.	Du	8°N,	23°E	62	8	x	x	x			x	x	6	Hurley and Van der Voo (1987)
Bokkeveld Grp., S. Africa	Dm	10°N,	15°E	33	7					x	x	x	4	Bachtadse and others (1987)
Mereenie Ss., Central Austr.	D?	15°N,	25°E	15	11			x	x		x	x	4	Embleton (1972)
Air Ring Complexes, Niger	435	43°S,	9°E	50	6	x	x	x			x	x	5	Hargraves and others (1987)
Pakhuis/Cedarberg, S. Africa	Sl/Ou	25°N,	343°E	9	18	x		x			x	x	5	Bachtadse and others (1987)
Stairway Ss., Central Austr.	Om	43°N,	31°E	25	9	x		x		x		x	4	Embleton (1972)
Taylor Valley dikes, Antarct.	470	36°N,	15°E	75	11	x		x			x	x	4	Manzoni and Nanni (1977)
Sør Rondane Intr., Antarctica	480	12°N,	14°E		5	x		x			x		3	Zijderveld (1968)
Jinduckin Fm., N. Australia	Ol	31°N,	1°E	10	11	x	x	x		x	x	x	6	Luck (1972)
Graafwater Fm., S. Africa	Ol	28°N,	14°E	25	9	x	x	x		x	x	x	6	Bachtadse and others (1987)
Ntonya Ring Str., Malawi	474	28°N,	345°E	1054	2	x	x	x	x		x		5	Briden (1968b)
Hook Intrusives, Zambia	500	14°N,	336°E	13	36	x			x			x	3	Brock (1967)
Central Mobile Belt of Newfoundland, New Brunswick, Maine (south poles)														
Compton Fm., Quebec	Du	28°S,	257°E	30	7	x	x	x	x		x	x	6	Seguin and others (1982)
Traveler Felsite, ME	Dl	29°S,	262°E	16	11	x	x	x			x	x	5	Spariosu and others (1983)
Eastport Fm., ME	Dl	24°S,	294°E	19	9	x	x	x	x		x		5	Kent and Opdyke (1980)
Hersey Fm., ME	Dl	20°S,	309°E	36	6	x	x	x	x		x		5	Kent and Opdyke (1980)
Botwood Volc., C. Newfoundland	Sm	13°S,	305°E	177	9	x		x		x	x	x	4	Lapointe (1979)
Wigwam red beds, C. Newf.	Sm	25°S,	280°E	9	14	x		x			x	x	4	Lapointe (1979)
Avalon Basement Terranes of Nova Scotia, New Brunswick, Boston Basin (south poles?)														
Metamorphics, Massachusetts	Du	23°S,	306°E	11	10		x	x				x	3	Schutts and others (1976)
Mascarene Fm., C, New Brunsw.	Su	28°S,	265°E	666	4	x	x	x			x	x	5	Roy and Anderson (1981)
Mascarene Fm., A-B, New Brunsw.	Su	5°N,	267°E	33	7	x		x	x		x	x	5	Roy and Anderson (1981)
Dunn Point Fm., Nova Scotia	Sl/Ou	2°N,	316°E	79	4	x	x	x	x	x	x	x	6	Van der Voo and Johnson (1985)
Nahant Gabbro, Massachusetts	O	34°N,	282°E			x	x	x			x	x	5	Weisse and others (1985)
Delaware Piedmont (south poles?)														
Metamorphic rocks, DE	Ou	48°N,	288°E	15	16		x	x				x	4	Brown and Van der Voo (1983)
Arden Pluton, DE	Ou	16°N,	303°E	32	14	x	x	x				x	4	Rao and Van der Voo (1980)
Southern Ireland, Southern England, and Wales (south poles)														
Old Red Sandst., Wales	Dl/Su	3°N,	298°E	4	13	x		x	x		x		5	Chamalaun and Creer (1964)
N. Wales Intrusives	O	68°N,	288°E	17	5		x	x	x				5	Thomas and Briden (1976)
Builth Inlier Intr., Wales	Ou	2°N,	2°E	37	11	x	x	x			x	x	5	Piper and Briden (1973)
Builth Volcanics, Wales	Ol	3°S,	355°E	22	4	x	x	x	x		x	x	6	Briden and Mullan (1984)
Tramore Volcanics, Ireland	Om	11°N,	18°E			x	x	x		x	x	x	6	Deutsch (1984)
Central and Southern Europe (south poles)														
Sediments, Volcanics, Germany	Du	30°S,	9°E			x	x	x	x		x	x	6	Bachtadse and others (1983)
San Pedro Red beds, Spain*	Dl	22°S,	319°E*	30	10	x	x	x	x		x	x	6	Perroud and Bonhommet (1984)
Cabo de Peñas, Spain	Ou	30°N,	330°E	50	6	x	x	x			x		4	Perroud and Bonhommet (1981)
Buçaco, Portugal*	Ou	25°N,	335°E*	83	6	x	x	x			x		4	Perroud and Bonhommet (1981)
Thouars Massif, France	Ou	34°N,	5°E	27	5	x	x	x			x	x	5	Perroud and Van der Voo (1985)
M. de Chateaupanne Fm., France	Ou	34°N,	343°E	65	6	x	x	x		x		x	5	Perroud and others (1986)
Erquy Volcanics, France	479	35°N,	344°E	47	11	x		x				x	3	Duff (1979)
Arenig Sandst., Poland	Ol	22°N,	9°E		10	x		x	x		x		4	Lewandowski (1987)

Explanation: D = Devonian, S = Silurian, O = Ordovician, u = upper, m = middle, l = lower; k and α_{95} are the statistical parameters associated with the means; Q = quality factor (maximum 7), comprised of the following reliability criteria: 1 = well-determined age, 2 = sufficient number of samples (greater than 25) and high enough precision (k greater than 10, α_{95} lower than 16°), 3 = demagnetization published in sufficient detail, 4 = positive field (fold-, conglomerate-, contact-) tests, 5 = tectonic coherence with block or craton and sufficient structural control, 6 = presence of antipodal reversals, 7 = lack of similarity with poles for younger times (see Van der Voo, 1988).
*Poles recalculated after restoration of an 80° counterclockwise rotation during Hercynian deformation of the Ibero-Armorican arc (Perroud and Bonhommet, 1981). Poles have been selected for quality (Q); the poles for the southern British Isles have been taken from Briden and Duff (1981); poles with their quality classification of C or D have not been used.

(for example, North America in the Paleozoic), averaging over longer time spans may yield satisfactory results, but for rapidly drifting continents (for example, Paleozoic Gondwana) conditions may change not only because of short-term (secular) climatic variations, but also because of rapid and systematic, yet imprecisely dated, shifts in latitude.

A further complication arises from the fact that not every diamictite is a glacial deposit and that not all authors agree on the interpretation of some questionable tillites. It should also be kept in mind that climatic features of the Mesozoic and Cenozoic have been shown to depart significantly from latitudinal predictions based on paleomagnetism (Barron, 1984) and that this could play a role by analogy in the Paleozoic. As we will see below, however, there is generally good agreement between paleopoles and glacial deposits, as long as they are abundant in space or long-lasting in time.

Biogeography

Faunal provinces may provide a useful test of predicted paleolatitudes or relative longitudinal proximity. Examples can be found in the Early Devonian faunas, such as those of the Malvinokaffric realm (cold-water) and the Eastern Americas and Old World realms (warm-water), because the organisms appear to have had different tolerances to such variables as temperature or light (Boucot and Gray, 1983). Although different provinces in time periods of noncosmopolitan faunas may still span large geographical distances, their boundaries usually must be interpreted in terms of geographic barriers or other reproductive isolation mechanisms. At the same time, a faunal province found to occur in tectonic elements now far apart implies that reproductive communication must have been possible.

Biogeographic relationships, therefore, provide excellent constraints on relative positioning of individual tectonic elements but are by themselves less well able to position the continents in a latitude framework. The major difficulty in Paleozoic biogeography is that ocean currents, water depth, the positions of islands, and climatic or sea-level fluctuations all are incompletely known, and yet they can profoundly influence the reproductive isolation and communication. Objections to paleogeographic reconstructions may always be overcome by *ad hoc* hypotheses about barriers or seaways.

In a qualitative way, biogeography can indicate changes in relative paleolongitudes, as shown by McKerrow and Cocks (1976) and noted by Boucot and Gray (1983), who wrote that "distance itself may be a reproductively isolating factor for organisms whose propagules have limited viability and low dispersal potential." This important aspect is unique to biogeography, as neither paleomagnetism nor paleoclimatology has any ability to predict longitudinal separation.

With the current interest in Phanerozoic extinction mechanisms, correlations between times of mass extinctions and climatic changes have recently been inferred (Stanley, 1984; Copper, 1986). Stanley argues that tropical faunas become trapped without escape, whereas cold-water biota move equatorward during times of global refrigeration. Although several mass extinctions coincided with recognized intervals of climatic cooling, thus providing the ultimate testing of such paleogeographically controlled extinction models, there is the danger of circular (circum-global?) reasoning without quantitative data on ocean temperatures. Oxygen isotopic signatures of Paleozoic carbonates can possibly provide us with such quantitative data, but at this time, it does not yet appear to be possible to derive paleolatitudinal information from isotopic studies (K. C. Lohmann, 1987, personal commun.). On the other hand, trace elements in carbonates have been inferred (Rao, 1981) to provide paleotemperature information correlative with paleolatitudes.

NORTH AMERICAN PALEOLATITUDES

Ordovician

North America, or more appropriately, Laurentia, includes the craton, Greenland, and Scotland and northern Ireland north of the Caledonian suture (Ziegler, 1984). Some new high-quality paleomagnetic data have become available in recent years for the Ordovician of Laurentia, but these new data do not change appreciably its paleolatitude models. For a recent review of the paleomagnetic data base for North America, the reader is referred to Van der Voo (1988). The available Ordovician paleopoles (Table 1) fall along a track between 10°S, 346°E and about 30°S, 280°E, which indicates equatorial paleolatitudes for the craton throughout the Ordovician. The endpoints of this track are defined by an earliest

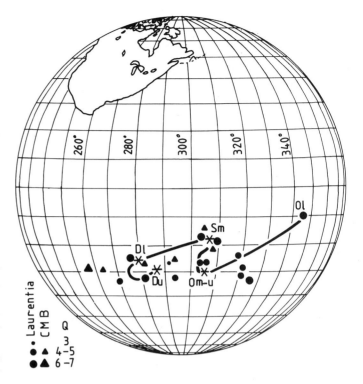

Figure 1. Ordovician through Late Devonian apparent polar wander path for Laurentia (dots) and the Central Mobile Belt (CMB) of the northern Appalachians (triangles). The symbols for individual paleopoles are plotted with different sizes according to the quality factor (Q, as listed in Table 1); the mean paleopoles used to make paleogeographic reconstructions (asterisks) are listed in Table 2. O = Ordovician, S = Silurian, D = Devonian, l = lower, m = middle, and u = upper. Note the agreement between the Laurentian and CMB paleopoles for Middle Silurian through Devonian time; there are no Ordovician paleopoles for the CMB. All available paleopoles are about 90° away from the continent, indicating equatorial paleopositions for North America throughout the Ordovician to Late Devonian interval.

Ordovician paleopole for the Oneota Dolomite (Jackson and Van der Voo, 1985), which falls close to reliable Late Cambrian poles, and a new Late Ordovician pole based on ^{40}Ar/^{39}Ar cooling ages of 446 m.y. in the Grenville province (Dunlop and Stirling, 1985), which confirms a previously determined paleopole for the Juniata Formation. Middle Ordovician poles are available for the Moccasin, Bays, and Chapman Ridge Formations of East Tennessee (Table 1). The Ordovician apparent polar wander path segment (Fig. 1) indicates that Laurentia rotated, while remaining on the equator, in a counterclockwise sense by about 45°. Paleomagnetic data from Scotland (Briden and others, 1984) also show this rotation, but the paleolatitudes are slightly more southerly (by about 15°) than one would expect from the cratonic North American data. Whether this means that conventional fitting of northern Scotland against southeast Greenland is inappropriate for the Ordovician, thereby implying that northern Scotland is a displaced terrane, remains very controversial (Briden and others, 1984).

Paleoclimatological indicators for the Middle Ordovician of North America (Bambach and others, 1980) consist of evaporites symmetrically distributed at about 15° latitude north and south of the equator (for details, see Cook and Bally, 1975), thus confirming the equatorial paleomagnetic

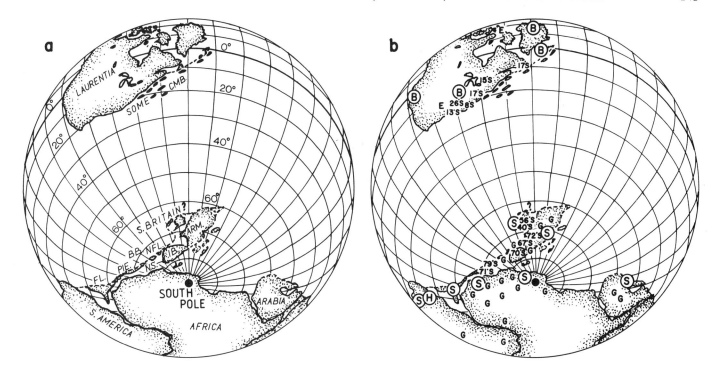

Figure 2. Middle to Late Ordovician paleogeographic reconstruction. a. Map showing the names of the separate tectonic elements discussed in this paper: CMB = Central Mobile Belt of the northern Appalachians, ARM = Central Europe, and IB = Iberian Peninsula, together constituting Armorica; NFL = eastern Newfoundland, NS = Nova Scotia, BB = Boston Basin, PIE = Avalonian part of the Piedmont province (Williams and Hatcher, 1983), and FL = northern Florida. The continental elements are positioned according to the mean paleomagnetic poles of Table 2.

Figure 2b. The same map as in Figure 2a, but with paleomagnetic paleolatitudes (numbers), biogeographical (encircled letters), and paleoclimatological indicators (letters). B = the Bathyurid fauna, S = *Selenopeltis* fauna, and H = Hungaiid-Calymenid trilobite fauna (from Whittington and Hughes, 1972). G = glacial relicts, and E = evaporite occurrences.

latitudes. Early to Late Ordovician phosphates in Nevada have been interpreted to mark the west-facing margin of Laurentia, with paleolatitudes between 50° and the equator (Coles and Snyder, 1985).

Whittington and Hughes (1972), in their study of Ordovician trilobite faunas, described a Bathyurid fauna for North America and Scotland. This fauna is also found in northern Siberia and north China, but not in the Baltic Shield, southern England and Wales, southern Europe, south China and Gondwana, suggesting no connection between Laurentia and Gondwana or these other areas. Middle Ordovician brachiopods define an American realm (Williams, 1973) which coincides with the Bathyurid fauna, although it extends also to south China and Australia. Coral and graptolite associations do not negate the evidence from trilobites and brachiopods (Kaljo and Klaamann, 1973; Boucek, 1972).

In summary, paleoclimatological and biogeographical data appear to confirm paleolatitudinal separation of Laurentia from Gondwana and smaller blocks such as found in England, Wales, and southern Europe. North America was situated on the equator throughout the Ordovician, but rotated counterclockwise. North America's paleomagnetic determinations for Late Ordovician times have not changed significantly during the past decade, and my paleolatitude reconstruction of North America (Fig. 2)

TABLE 2. MEAN PALEOPOLES USED TO POSITION THE MAJOR CONTINENTS AND SOME TERRANES IN FIGURES 2 THROUGH 5

Continental element	Ordovician (Om-u)			Silurian (Sm)			Devonian (Dl)			Devonian (Du)			Comments
	Pole	k	α_{95}	Pole	k	α_{95}	Pole	k	α_{95}	Pole	k	α_{95}	
Laurentia	30°S, 306°E	24	14	18°S, 307°	..	..	25°S, 279°E	..	..	29°S, 287°E	..	..	Om-u mean of 6 poles
Gondwana	34°N, 7°E	..	..	43°S, 9°E	..	..	30°S, 10°E	..	..	8°N, 23°E	..	..	Om-u mean of 2 poles E. Dev. by interpolation
Central Mobile Belt	..			19°S, 293°E	..	..	26°S, 289°E	14	35	28°S, 257°E	..	..	Sm mean of 2, Dl mean of 3
Avalon	34°N, 282°E	..	..	..			25°S, 285°E	..	..	23°S, 306°E	..	..	Dl pole by interpolation
Delaware Piedmont	32°N, 296°E	..	..	..			–			..			Ou mean of 2 poles
S. Europe (Armorica)	31°N, 343°E	35	16	..			22°S, 319°E	..	..	30°S, 9°E	..	..	Ou mean of 4 poles

Explanation: Period abbreviations, k and α_{95} are the same as in Table 1. Means have been calculated from the poles listed in Table 1; if more than one pole is used, then this is noted under Comments; if no poles were available, then sometimes a pole was calculated by interpolation, as noted under Comments.

Poles are given in the coordinates of the individual blocks; in general, the pole longitudes are similar to the longitudes of the present-day block locations, with the primary variation in pole location in the pole latitude. Given this variation primarily in pole latitude, note the agreement for the Ou poles for Gondwana, Armorica (both African/European coordinates) and Avalon, and the Delaware Piedmont (North American coordinates) and the contrast with Laurentia. Note also the agreement for the Du poles of Laurentia, the Central Mobile Belt, Avalon and Armorica, and the contrast with Gondwana.

is similar to previous ones (Bambach and others, 1980; Cocks and Fortey, 1982; Neuman, 1984; Perroud and others, 1984; Scotese, 1984; Scotese and others, 1979; Ziegler and others, 1979).

Silurian

A new paleopole from Indiana (McCabe and others, 1985) has confirmed previously existing results from the Rose Hill Formation in the Appalachians (see Table 1). The corresponding equator runs through north-central Greenland and most of Canada toward Vancouver Island. Results from the Bloomsburg Formation yield a southerly low-latitude position in agreement with the other Silurian studies, but the Bloomsburg declination and, hence, the paleopole may be slightly displaced because of *in situ* rotations in the Pennsylvania salient (Kent, 1986). Paleopoles are shown in Figure 1.

Evaporites and reefs in the Michigan Basin confirm the paleomagnetic results paleoclimatologically, with a paleolatitude of about 25°S (Fig. 3). Silurian brachiopod faunas are mostly cosmopolitan, but with important Gondwana exceptions that will be discussed below. A group of Silurian agnathans (fishes), the Thelodonts, shows a faunal province that is distinct for North America and the Russian Platform, whereas other Thelodont faunas in Siberia, the Canadian Maritime provinces, England and Wales, and the Scottish-Norwegian Caledonides have different characteristics (Turner and Tarling, 1982).

Silurian paleolatitudes determined for Laurentia by paleomagnetic means again have not changed appreciably from previous ones (Bambach and others, 1980; McKerrow and Ziegler, 1972; Scotese and others, 1979,

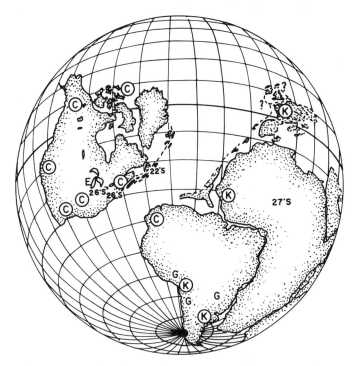

Figure 3. Middle Silurian paleogeographic reconstruction. For an explanation of the use of letters and numbers, see Figure 2. The continental elements are positioned according to the mean paleomagnetic poles of Table 2. No poles are available for Armorica, Avalon, and the Piedmont. C = cosmopolitan and K = *Clarkeia* brachiopod fauna (Cocks, 1972). The position of the southern British Isles is very uncertain.

1985; Ziegler and others, 1977). The Silurian paleogeographic map of Turner and Tarling (1982) places North America too far south and with an orientation that is too "canted."

Devonian

A paleopole for the Early Devonian of the Canadian Arctic (Dankers, 1982) and new paleopoles from the Upper Devonian Catskill red beds (Miller and Kent, 1986) have significantly changed North America's paleolatitude position from those published previously, especially those for the Late Devonian (Bambach and others, 1980; Perroud and others, 1984; Scotese, 1984; Scotese and others, 1979, 1985; Ziegler and others, 1979). The new paleopoles are plotted in Figure 1. Results from the Central Mobile Belt in New England and the Canadian Maritime provinces (Spariosu and Kent, 1983; Seguin and others, 1982) now appear to agree with those for the Laurentian craton, although local rotations about a vertical axis could account for a longitudinal spread in the paleopoles (Table 1; Fig. 1). The hypothesis of a Devonian-Carboniferous "Acadia" displaced terrane (Kent and Opdyke, 1978) is thus no longer tenable with the newer results. It appears that Devonian and Silurian paleolatitudes are not very different from each other, placing the Early or Late Devonian equator through southern Greenland, Hudson Bay, and British Columbia (Fig. 4), or Montana (Fig. 5). The ancient meridians for the Devonian of eastern North America are thus more or less parallel to the present ones, although the possibility of thrust sheet rotations in the Appalachians causes some uncertainty in the paleomagnetic declinations (but not the inclinations) of the Catskill results (Miller and Kent, 1986). Although the Devonian paleolatitudes of Turner and Tarling (1982) are approximately correct, the orientation of North America is again rotated too much (clockwise) with respect to the paleomeridian.

Paleoclimatologically, Lower to Upper Devonian evaporites in the Yukon, Hudson Bay, and Williston Basin (Cook and Bally, 1975) are located between 15°N and 15°S in agreement with the new paleomagnetic results.

Previous Early Devonian reconstructions have been particularly problematic for biogeographers (Boucot and Gray, 1983). Lower Devonian brachiopods reveal an eastern Americas realm which is found in the American mid-continent region, but also in the western United States and, importantly, in northern South America, including the Amazon Basin (Fig. 4). This realm, on the other hand, is not found in Europe, the northeastern Appalachians, or the rest of Gondwana. The endemism of Lower Devonian faunas is usually ascribed to regressive sea levels and barriers caused by Caledonian mountain building. The brachiopod similarities between North America and South America require reproductive communication and some degree of proximity.

Another Lower Devonian brachiopod province, called the "Old World realm," is found in Europe, the Canadian Maritime provinces, northern Africa, and, importantly, in Oklahoma and the western Cordillera of North America (Fig. 4). A third province, the Malvinokaffric realm, persists in the cold-water regions of southern South America, South Africa, and Antarctica (Fig. 4). The Old World realm requires reproductive communication between Europe, Africa, and the Appalachians down to the Gulf area of the southern United States. Conodont and coral distributions (Fahraeus, 1976; Oliver, 1976) generally follow the provinces outlined above and also require a Caledonian barrier land mass.

The Late Devonian was, by contrast to the Early Devonian, characterized by a highly cosmopolitan marine fauna, with no biogeographic realms to be distinguished (Boucot and Gray, 1983). There are therefore few biogeographic constraints on reconstructions for this time.

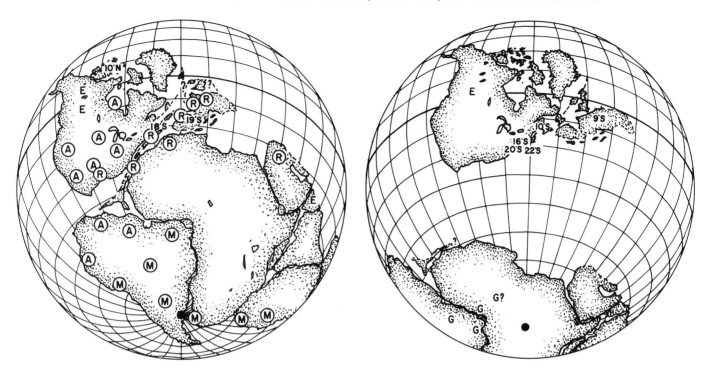

Figure 4. Early Devonian paleogeographic reconstruction. For an explanation of the use of letters and numbers, see Figure 2. A = Eastern Americas realm, R = Rhenish-Bohemian subprovince of Old World realm, and M = Malvinokaffric realm of brachiopod provincialities (from Boucot and Gray, 1983). As in the other figures, the continental elements are positioned according to the mean paleopoles of Table 2; note that the position for Gondwana is based on an interpolated paleopole location between the Middle Silurian and the Late Devonian, but new results (Schmidt and others, 1987) for the Early Devonian Snowy River Volcanics of eastern Australia support this paleopole location.

Figure 5. Late Devonian paleogeographic reconstruction. For an explanation of the use of letters and numbers, see Figure 2. The continental elements are positioned according to the mean paleomagnetic poles of Table 2. The Piedmont has no paleopoles available for this time, and its position is uncertain.

GONDWANA

Ordovician

Ordovician paleopoles have long been available from Africa, Australia, South America, and Antarctica; although they are of variable quality, they cluster well around a mean (in African coordinates) of about 30°N, 5°E (Fig. 6). Recently, a new Early Ordovician pole of good quality has been obtained from the Graafwater Formation in the Table Mountain Group of South Africa, as well as a Late Ordovician–earliest Silurian paleopole of lesser quality from the Pakhuis and Cedarberg Formations in the same Group (Bachtadse and others, 1987). The newer paleopoles (Table 1) also fall in the vicinity of northwestern Africa and support the earlier determinations. In terms of Gondwana paleolatitudes for the Ordovician, therefore, there is little change from previous publications (Bambach and others, 1980; Cocks and Fortey, 1982; Neuman, 1984; Perroud and others, 1984; Scotese and others, 1979; Ziegler and others, 1979).

The presence of Late Ordovician glaciation in northern Africa has been similarly well established (Spjeldnaes, 1961; Beuf and others, 1971; Caputo and Crowell, 1985; Robardet and Doré, 1988; Van Houten, 1985; Van Houten and Hargraves, 1987; Whittington and Hughes, 1972) and agrees very well with the paleomagnetic poles (Fig. 2).

Biogeographically, trilobites reveal two faunas in Gondwana: the *Selenopeltis* and Hungaiid-Calymenid faunas. The latter is also found in south China, whereas the former is apparently found in colder water areas and occurs also in southern Europe, England and Wales, Turkey, and Iran (Whittington and Hughes, 1972). Brachiopods (Williams, 1973) of the Anglo-French-Bohemian fauna occur in northern Africa and southern Europe, thus further linking these two areas during the Ordovician. An anomaly in the brachiopod record occurs in the American Realm, which is found in North America, as well as Australia, which are both located on the Ordovician equator. Whether this implies longitudinal proximity or widespread equatorial distribution is an ambiguous matter that cannot be resolved with paleomagnetic or paleoclimatological techniques.

In summary, Ordovician paleogeographic models for Gondwana remain relatively unchanged in the last decade.

Silurian

Until recently, essentially no paleomagnetic poles have been available for the Silurian of cratonic Gondwana, and most reconstructions used interpolations between Ordovician and Devonian paleopoles. Siluro-Devonian paleopoles from the Tasman mobile belt of Australia have occasionally been used for alternative models of positioning Gondwana (for example, Path Y of Morel and Irving, 1978) but always with the caveat that these results may have come from displaced terranes. These eastern Australian results fall on a loop in the apparent polar wander path between central Africa and the area west of Chile, whereas other reconstructions have been based on an interpolated Silurian pole location in

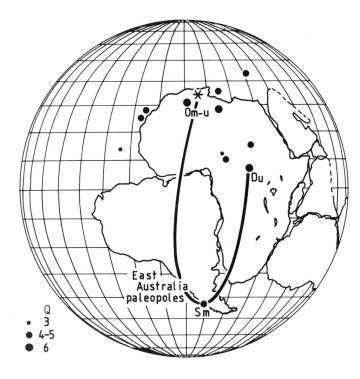

Figure 6. Ordovician through Late Devonian apparent polar-wander path for Gondwana. Symbols and abbreviations are explained in Figure 1. Silurian paleopoles from eastern Australia are located west of Chile, but they are from the mobile (noncratonic) part of Gondwana and have not been used in this paper, other than to indicate support for the Sm paleopole of Hargraves and others (1987).

central Africa. A new Silurian paleopole (Table 1), however, is now available for cratonic Gondwana (Hargraves and others, 1987) which *grosso modo* confirms this loop (Fig. 6) and drastically changes the mid-Paleozoic paleolatitudes for Gondwana. This Silurian result, which is obtained from well-dated ring complexes in Niger, has been thoroughly documented in terms of paleomagnetic demagnetizations and is not subject to uncertainties about tectonic tilting (Hargraves and others, 1987).

Paleoclimatologically, this new Silurian pole agrees with the presence and absence of mid-Paleozoic glacial relicts reported by Caputo and Crowell (1985), who proposed a path of glacial centers moving across Gondwana that agrees well with the paleomagnetically determined loop discussed above. The apex of this loop, located to the west of Chile, occurs in the Silurian-Devonian according to Caputo and Crowell, whereas paleomagnetic results from South Africa (Bachtadse and others, 1987) and the Niger pole bracket the loop to the period between latest Ordovician and Middle Devonian. Evaporites in western Australia (Bambach and others, 1980) are in agreement with any of the paleopole locations, because Australia's location is insensitive to the range of poles discussed that are all about 70 degrees away.

The consequences of this new Silurian pole for the Siluro-Devonian paleogeographic locations of Gondwana and for mid-Paleozoic tectonics in the Atlantic realm are large (Hargraves and others, 1987; Van Houten and Hargraves, 1987). Most of the previously published models for these times must be changed (Bambach and others, 1980; Perroud and others, 1984; Scotese, 1984; Scotese and others, 1979, 1985; Ziegler and others, 1977, 1979; see also Rodgers and Sougy, 1984, and Ziegler, 1984). With the new Silurian paleopole, the north African part of Gondwana is in subtropical latitudes and is much more northerly than in previous models (Fig. 3).

The new paleomagnetic result may also remove several objections to these previous models made on biogeographic grounds, as discussed below in the subsection for the Devonian, and may allow us to return to an early model for the Acadian orogeny (McKerrow and Ziegler, 1972) by implying a Silurian-Devonian collision between Gondwana and North America, as discussed below. Silurian brachiopod faunas are mostly cosmopolitan and warm-water, but with important exceptions for a colder water Malvinokaffric province in South America and a separate province in Siberia. Cocks (1972) recognized a shallow-water *Clarkeia* community within the Malvinokaffric province, which is also found in southern Europe (Fig. 3), thus linking this area to Gondwana through Silurian times.

Devonian

A new paleomagnetic result from the Late Devonian (Frasnian-Famennian) of western Australia yields a pole in central Africa (Hurley and Van der Voo, 1987), thereby requiring that the apparent polar wander path for Gondwana loops from an Ordovician location in northern Africa through the Silurian pole of Hargraves and colleagues (1987) in Chile back up to the equator in Africa (Fig. 6; Table 1). The Australian result is from well-dated limestones in the Canning basin, resting on the Precambrian of the Kimberley block in West Australia, and represents an intracratonic setting. The magnetizations are dual polarity, with the reversals stratigraphically controlled (Hurley and Van der Voo, 1986), and they do not resemble directions for younger periods. New Late Devonian to earliest Carboniferous paleomagnetic results from the Hervey Group in New South Wales, Australia, support the location of the Canning Basin pole (Schmidt, 1987). A previous Late Devonian result from the Msissi Norite in Morocco does not meet modern reliability criteria and must be discarded (Salmon and others, 1986). In addition, a mid-Devonian paleopole of lesser quality has been published for the Bokkeveld Group of South Africa (Bachtadse and others, 1987), which also falls in central Africa.

I should briefly discuss two other recent Devonian poles for Gondwana, which, in my opinion, should not be used. One pole, for the Gneiguira Formation of Mauritania (Kent and others, 1984), resembles younger (late Paleozoic) paleopoles and could well represent a remagnetization. No field test results are available to test this possibility. The other pole is obtained from the late Lower to early Upper Devonian Comerong Volcanics of eastern Australia (Schmidt and others, 1986) and is of excellent quality. The field tests, however, show that rotation of parts of the sampling area has occurred, and although the magnetization is older than this megakinking, the problem is that with the possibility of rotation no reference declination can be established. Thus, in a comparison between the Australian Devonian results, it appears that the Canning Basin pole (Hurley and Van der Voo, 1987) is least likely to be subject to uncertainties due to local rotations and should therefore be preferred for the Late Devonian positioning of cratonic Gondwana.

The path of glacial centers migrating across Gondwana published by Caputo and Crowell (1985) shows the Devonian pole moving through Peru and Brazil into an early Carboniferous location in central Africa. This is in agreement with the paleomagnetic polar wander path (Fig. 6), except that the ages of the paleomagnetic results require that by Late Devonian time the pole is already near the African equatorial location. After a mid-Silurian to mid-Devonian period with a warmer climate in Gondwana, strong evidence for Late Devonian glaciations can be found in northern Brazil (Fig. 5), and weak evidence exists for glacial relicts in west Africa near Accra and Agades, Niger (Caputo and Crowell, 1985). Veevers and Powell (1987) also discussed glacial episodes in Gondwana but limited themselves to Late Devonian through Permian time. Although they used a different paleomagnetic pole based on work by Schmidt and others (1986), their marine and nonmarine glacial relict data lend support to a central African pole location for Late Devonian time.

The Early Devonian endemic faunas of Gondwana include a continuation of the Malvinokaffric realm for the brachiopods (Boucot and Gray, 1983) and corals (Oliver, 1976), which grades into the Eastern Americas Realm of northwestern South America as well as North America, requiring continuity between the two Americas (Fig. 4). In northern Africa, Europe, and Nova Scotia, the Rhenish Bohemian subprovince of the Old World Realm is found.

A model for a low climatic gradient has been proposed by Boucot and Gray (1983) on the basis of warm-water goniatite cephalopods and brachiopods. On the other hand, the Late Devonian glacial relicts and paleopole location in central Africa, and the carbonate deposits in Morocco appear to give a high climatic gradient. This presents difficulties for the interpretation of climatic conditions in Gondwana. Carbonates and patchy reefs appear on the northern fringes of Gondwana in the late Early to Middle Devonian (Scotese and others, 1985; Wendt, 1985), suggesting that the shelf environment here must be warming up just when this area is thought to be moving southward with the new paleomagnetic results. The limited development of shelf carbonates, with only small mud mounts and biostromes, however, suggests a latitude for Morocco (which would be placed at about 50°S with the new paleomagnetic data; see Fig. 5) close to the southern limit of actual shelf carbonate deposition (Wendt, 1985). Moreover, Stanley (1984) and Copper (1986) argue that climatic cooling and/or equatorial diversion of cold waters in the Late Devonian are the prime causes for the Late Devonian extinction. Copper's mechanism of a Gondwana-Laurentia collision, however, seems to be hindered by the wrong timing, as is discussed below. The climatic cooling models contrast with the model of Boucot and Gray (1983). This controversy is not easily resolved with the presently available data, and it presents a challenge for further research.

THE INTERVENING DISPLACED TERRANES

Ordovician

A review of the geology and biogeography of islands in the Ordovician Iapetus Ocean (Neuman, 1984) suggests locations for such smaller terranes as central Newfoundland, central Maine and New Brunswick, southeast Ireland and Anglesey, the Avalonian basement blocks of Nova Scotia and eastern Newfoundland, southern England and Wales, and the Meguma terrane. To this must be added the Cadomian massifs of central and southern Europe (the Armorican and Bohemian Massifs, the Massif Central, the Iberian Meseta, and poorly known blocks that are now buried or reworked in the Alpine orogeny), the Avalonian basement block of the Boston Basin, the central and southern Appalachian exotic terranes of the Piedmont province (Secor and others, 1983), and northern Florida.

Paleomagnetic data for the Ordovician and older rocks of the Avalonian basement areas of Newfoundland, Nova Scotia, and the Boston Basin indicate high paleolatitudes (Fig. 7) that suggest a significant distance from the equatorial North American craton and proximity to the northwest African margin of Gondwana (Johnson and Van der Voo, 1985, 1986; Van der Voo and Johnson, 1985; Weisse and others, 1985; Fang Wu and others, 1986), in agreement with the locations indicated by Neuman (1984) and others before him. Similarly, paleolatitudes for southern Ireland, southern England, Wales, and the blocks of southern Europe are higher than would be expected if they were adjacent to North America or northern Scotland and Ireland (Deutsch, 1984; Briden and Mullan, 1984; Perroud and others, 1984; Perroud and Van der Voo, 1985; Van der Voo and others, 1985a; Lewandowski, 1987). The paleopoles for these blocks are listed in Table 1. Consequently, it seems likely that many, if not all, of these areas were also in the vicinity of northwest Africa. Lastly, high Late Ordovician paleolatitudes are also suggested for the Wilmington area in Delaware in the central Appalachian Piedmont province and for northern Florida (Rao and Van der Voo, 1980; Brown and Van der Voo, 1983; Jones and others, 1983).

The similarity of the Avalonian, Cadomian, and Pan-African basement in these high-latitude terranes and in Africa led Van der Voo (1982) to propose an Armorica plate, which was associated with Gondwana in the early Paleozoic. As is shown below, these areas are thought to be separate from Gondwana by Late Devonian time (but now adjoined to North America and northern Europe), thus arguably outlining separate "plate" status for Armorica at some time in the mid-Paleozoic. The drift history of Armorica was initially based on paleomagnetic data from the Armorican massif only, thus requiring further work to outline the limits of the Armorica plate. Now that Ordovician paleopoles have been determined for southern Poland (Lewandowski, 1987), the Boston Basin (Weisse and others, 1985), and the Iberian Meseta (Perroud and Bonhommet, 1981), the extent of Armorica has become better delineated. The northern margin of Armorica is still ill-defined, however: on the one hand, faunal, geological, and pre-Ordovician paleomagnetic evidence for southern Ireland, southern England and Wales, and the Belgian Ardennes seems to argue for inclusion of these areas in Armorica (Neuman, 1984; André and others, 1986; Perigo and others, 1983; Deutsch, 1984; Piper, 1982), but on the other hand, Ordovician paleolatitudes for southern England and Wales are ambiguous (see Thomas and Briden, 1976; Briden and Mullan, 1984) and not uniformly as high as those from the Armorican Massif (Burrett, 1985; Van der Voo and others, 1985a).

In light of these paleolatitudinal uncertainties for the southern part of the British Isles, aggravated by the possibility of apparent polar wander during the Ordovician, it is probably better to think of Armorica as a loosely comprised mosaic of elements, with Ordovician paleolatitudes ranging between a high mean of 70°S for France and Spain, and a mean possibly as low as about 40°S for the southern British Isles. A further complication results from the subsequent deformation of central and southern Europe during the Hercynian orogeny, which produced significant local rotations of France and Spain with respect to the stable foreland of northern Europe (Bachtadse and Van der Voo, 1986; Edel and others, 1981; Perroud, 1986). Thus, Armorica certainly did not resemble the geographic configuration its elements have today. Matte (1986) has discussed the plate-tectonic evolution of Hercynian Europe and has suggested possible locations for intra-Armorican ocean basins on the basis of ophiolite remains and tectono-metamorphic characteristics.

The other terranes for which paleomagnetic evidence of high latitudes exists (Table 1; Delaware Piedmont and Florida) were also located close to northwestern Africa (Fig. 2). In response to a comment by Palmer and Secor (1983), I proposed to call these (as well as the Armorican) elements "peri-Gondwanide terranes." Not all of these peri-Gondwanide terranes belonged to Armorica, however, because an essential element of the Armorica definition is comprised in their subsequent Paleozoic history: Armorica, by hypothesis, is thought to have separated from Gondwana before the Late Devonian. It is a matter of speculation how and when the Piedmont terranes became adjoined to North America, whereas for Florida it is very likely that it remained part of Gondwana until Pangea was formed in the Carboniferous.

The Ordovician paleoclimatology of Armorican terranes includes strong evidence for glaciomarine conditions (Spjeldnaes, 1961; Fortuin, 1984; Robardet and Doré, 1988). Thus the long-known climatic evidence

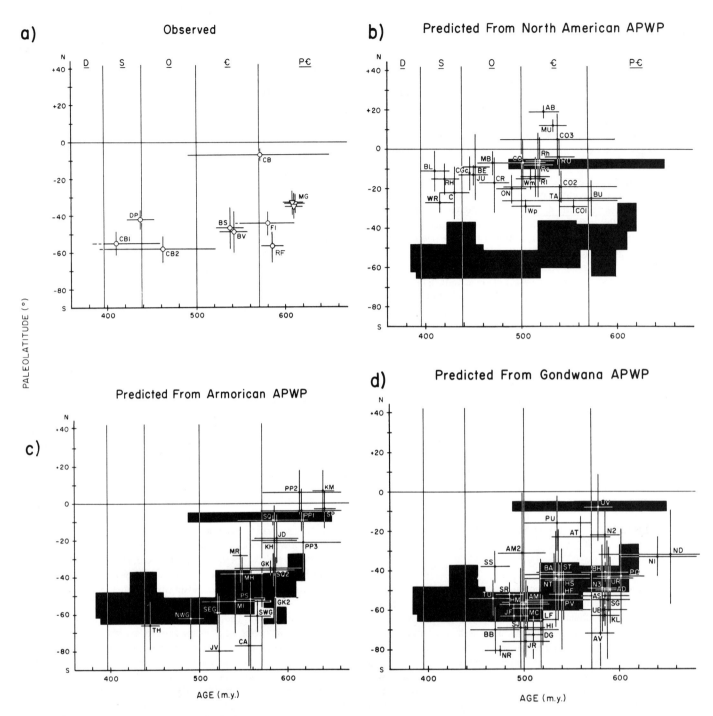

Figure 7. Paleolatitudes observed for the Avalonian basement terranes of the northern Appalachians compared with paleolatitudes for Nova Scotia predicted on the basis of paleomagnetic data from other tectonic elements in a Pangea configuration (from Johnson and Van der Voo, 1986; and Van der Voo, 1988). Error bars in time and space are calculated from the age uncertainties and the paleomagnetic statistics; data and abbreviations are explained in Johnson and Van der Voo's (1986) paper. a. Observed paleolatitudes plotted as a function of early Paleozoic time. b. Paleolatitudes predicted for Nova Scotia from the cratonic paleomagnetic data of North America, on the assumption that Nova Scotia had always been in today's relative position. The envelope of the data with their error bars of Figure 7a is repeated in black for easy comparison. c. Same as Figure 7b, but with predictions based on the paleopoles for the Armorican Massif. d. Same as Figure 7b, but with predictions based on Gondwana paleopoles. The three last frames show that Avalon was separated by an ocean from North America, and that Avalon, Armorica, and Gondwana display the same paleolatitudes during the early Paleozoic, in support of the hypothesis that they drifted together.

from Spain, France, Germany (Thuringia), Nova Scotia, and eastern Newfoundland (Fig. 2), is confirmed by the paleomagnetically determined high paleolatitudes for the Late Ordovician. These areas are likely to have been in proximity to the glaciated areas of northern Africa (Robardet and Doré, 1988).

For the southern British Isles, the occurrence of the Lower Ordovician *Selenopeltis* trilobite fauna (Whittington and Hughes, 1972), which is found also in southern Europe, Florida, and northern Africa, suggests that southern England and Wales must have been in proximity to Gondwana and Armorica. Brachiopod faunas, however, provide only weak support for this colder water faunal link based on trilobites, as a Baltic-Celtic brachiopod province, found in northern Europe and Wales (Williams, 1973), contrasts with an Anglo-French-Bohemian fauna found in Armorica and Morocco. Graptolite province characteristics (Boucek, 1972) do not contradict or amplify the inconclusive evidence obtained from the trilobite and brachiopod provincialities.

In summary, Ordovician paleomagnetic data have now become available for many of the terranes that will later become sandwiched between the major Gondwana and Laurentia continents in a Pangea configuration, and these data indicate high-latitude peri-Gondwanide locations in agreement with paleoclimatological and biogeographical evidence, and with the locations of the map of Neuman (1984, Fig. 1B). No paleomagnetic data are available for the Central Mobile Belt of the northern Appalachians, and so the locations of the latter are relatively unconstrained other than by the biogeographic considerations enumerated by Neuman (1984). It remains unresolved whether the Ordovician location of the southern British Isles was adjacent to Armorica or whether it was in mid-paleolatitudes between Armorica and Laurentia.

Silurian

Few paleomagnetic data exist for the Silurian of the displaced terranes, with the exception of several low-quality data for the southern British Isles (Briden and Duff, 1981). The latter are thought to be rapidly converging with the Laurentian (Scottish) margin during the Silurian, and the paleomagnetically determined paleolatitudes (Briden and others, 1984) and faunal distance arguments (McKerrow and Cocks, 1976) support this. Paleomagnetic results (Table 1) from the Botwood volcanics and red beds of the Central Mobile Belt in Newfoundland (Lapointe, 1979) and from New Brunswick (Roy and Anderson, 1981) also appear to indicate proximity to Laurentia.

Biogeographically, the presence of a *Clarkeia* (brachiopod) community in Armorica as well as in the Malvinokaffric realm of Gondwana argues for a continued association between the two areas (Fig. 3). Most of the Silurian fish are marine, and, even for those genera found in nonmarine beds, it is suspected that they had a marine stage in their ontogeny, thus precluding strong biogeographic conclusions.

The Late Silurian was a time in which marine sedimentation in many areas of the Caledonian-Acadian orogenic belt gave way to continental deposits. Basins in eastern Greenland, Norway, Great Britain, and the northern Appalachians were beginning to receive the Old Red Sandstone deposits in the Late Silurian. These basins show lateral gradation into continued marine conditions in the central and southern Appalachians. The lack of marine deposits in the northern half of this long belt indicates the onset of continent-continent collision.

In summary, it is clear that the Silurian was a time of transition. During the Ordovician Period, Laurentia was clearly separated from Gondwana (combined with Armorica and other paleogeographic fringe areas), whereas in the latest Silurian–Early Devonian, such separation is not generally indicated by the presently available data. Rapid northward drift of Gondwana apparently swept up the intervening displaced terranes, which ended up in proximity to Laurentia.

Devonian

Devonian paleomagnetic data exist for Great Britain, Spain, Germany, and for coastal Maine and the Boston Basin (Briden and others, 1984; Perroud and Bonhommet, 1984; Bachtadse and others, 1983; Kent and Opdyke, 1980; Spariosu and Kent, 1983; Schutts and others, 1976). These data require further corroboration, as not all results are of high quality; the possibility of remagnetization (Roy, 1982) needs to be further investigated. Equatorial to low-southerly paleolatitudes are uniformly indicated, but with one exception for southern Portugal (Perroud and others, 1985). The latter result, from the Upper Devonian Beja Gabbro, indicates a paleolatitude of about 35°S, which is higher than the Late Devonian paleolatitudes predicted for Armorica. Perroud and co-workers (1985) argued that the Beja area was therefore not part of Armorica in the Late Devonian. Armorica appears to have been separated at this time from Gondwana, and it is inferred that the Avalonian basement terranes in the northern Appalachians have similarly become attached to North America and are by now separated from Gondwana (Perroud and others, 1984).

Whereas the Lower Devonian brachiopod realms still show similarities between northern Africa and the southern Gulf States of the United States, Nova Scotia, and Europe (Fig. 4), the Upper Devonian faunas were too cosmopolitan to indicate paleogeographic relationships (Boucot and Gray, 1983). Thus there are no biogeographic constraints to the amount of latitudinal separation between these areas in the Late Devonian.

In summary, the available paleomagnetic results indicate the opening of a new ocean between Gondwana and Armorica during the Devonian (Fig. 5), whereas the previously existing Ordovician and Silurian ocean between Laurentia and Armorica has closed at the latest by mid-Devonian time. Biogeographic and paleoclimatological evidence does not negate such a conclusion.

TECTONIC IMPLICATIONS AND CONCLUSIONS

The two major continents, Gondwana and Laurentia, show largely different drift scenarios during the Ordovician-Devonian interval. Laurentia remained on the equator, rotating counterclockwise during the Ordovician by about 45°, and again by a smaller amount during the Late Silurian–Early Devonian. Gondwana, on the other hand, traveled a large distance, such that the pole was located in northern Africa during the Ordovician, then moved to the west of southern Chile during the Silurian, and finally, moved back again to central Africa by Middle to Late Devonian time (Fig. 6). The corresponding motion of northwestern Africa, with respect to the pole, is first northward from polar to subtropical latitudes during the Early Silurian, then southward from subtropical to the intermediate latitudes of about 50°S of the Middle to Late Devonian (Figs. 2–5). In this drift scenario, possible east-west motions constitute the big unknown; Hargraves and co-workers (1987) have suggested that northern Africa moved westward over thousands of kilometres with respect to Laurentia while not changing much in paleolatitude (Figs. 3–4). Although untestable paleomagnetically, this hypothesis is quite reasonable, because the northern African continental fringe, which is located in very low latitudes during the Silurian (Fig. 3), should not overlap with the more southerly location of southern Laurentia. The east-to-west motion of northern Africa allowed Hargraves and co-workers (1987) to resurrect the model of McKerrow and Ziegler (1972) for the Caledonian-Acadian orogenies in modified form.

McKerrow and Ziegler (1972) proposed that the Acadian orogeny was the result of ocean closure between northwestern South America and the northern Appalachians, with the Avalonian basement terranes constituting a prong (connected to Great Britain and the Baltic Shield) in between. The oceans on either side of this prong were called the "Theic" and "Proto-Atlantic" (= Iapetus).

The high southerly paleolatitudes of Armorica in the early Paleozoic imply a location close to northwest Africa (Fig. 2). No other location with respect to Gondwana (For example, near Colombia or Peru) matches the observed Armorican paleolatitudes (Hartnady, 1985; Van der Voo and others, 1985b). If Armorica (including the Avalonian basement terranes) moved with Gondwana throughout the Ordovician and Silurian, as seems likely (Hargraves and others, 1987), then it cannot have been in the prong location of McKerrow and Ziegler (1972). This aspect requires a modification of their model, in that the Acadian continent-continent collision must have been between Laurentia and northwest Africa with Avalonian terranes and cannot have been between South America and the Appalachian margin, because in the latter model Armorica would not end up adjacent to Laurentia.

Another major aspect of the relative drift history between Gondwana and Laurentia lies in the Devonian retreat of Gondwana away from Laurentia, after the Acadian collision took place (Figs. 4–5). The Devonian paleomagnetic data for Armorica suggest that it stayed behind with the northern continents, and that the new ocean opened up between Gondwana and Armorica. Thus, instead of Copper's (1986) model of a Late Devonian ocean closure, the new data suggest an opening ocean. How this affects climatic models for the Late Devonian merits further investigation.

In more detail, the orogenic phases in the Appalachian-Caledonian mountain belt (Penobscot-Grampian, Taconic, Caledonian, Acadian, and Alleghanian-Hercynian) need to be explained by the overall plate-tectonic movements of the major continents and the displaced terranes (Williams and Hatcher, 1983). Little agreement exists on the synchroneity or the timing of individual phases, however; for this discussion, therefore, I suggest to group them in three major intervals: the Ordovician, the Silurian–Early Devonian, and the Carboniferous–Early Permian. When the Ordovician orogenic phases occur, we also have a time of wide oceanic separation between Laurentia, on the one hand, and Gondwana and Armorica (including Avalon), on the other. Consequently, the Ordovician deformation in North America cannot have been caused by collisions involving Gondwana or the Ordovician peri-Gondwanide terranes. As it is likely that a collision between Laurentia and some continental fragments nevertheless took place, these continental fragments must be sought in the Central Mobile Belt in the northern Appalachians. Paleogeographic data are not available at this time to delineate these fragments, although an island-arc setting has often been inferred (Bird and Dewey, 1970; for recent literature see: Leo, 1985; Neuman, 1984; Ludman, 1986; van der Pluijm, 1987; Williams and Hatcher, 1983). Silurian–Early Devonian deformation has been widespread along the Appalachian mountain chain and is ascribed to a collision between the Appalachian margin of Laurentia and Gondwana's margin in northwest Africa, with the peri-Gondwanide terranes sandwiched in between. A newly opened ocean was forming between the Gondwana craton and Laurentia (with Avalonian/Armorican accreted terranes left behind, adjacent to Laurentia) during the Middle and Late Devonian, and this ocean subsequently closed in the Early to Late Carboniferous to cause the Alleghanian-Hercynian orogeny.

Following Young's (1986) method of using area cladograms to display the timing of collisions between the elements involved, I have constructed a cladogram for the above-mentioned scenario. This cladogram (Fig. 8) does not include the continental block of Baltica, comprised of the Baltic Shield and the Russian Platform, because insufficient paleomagnetic data are available for this block to make meaningful paleogeographic reconstructions. The major new elements in my cladogram are the Acadian collision between Gondwana and Laurentia, and the subsequent retreat of Gondwana.

REMAINING PROBLEMS AND FUTURE WORK

The most surprising aspect of the new paleopoles for Gondwana is the rapid northward motion that this large continent must have undergone during the latest Ordovician and Silurian. Because the Ordovician paleopoles for Gondwana are well established and the new pole from Niger (Hargraves and others, 1987) is well dated and paleomagnetically reliable, there either must have been rapid motion of Gondwana with respect to the pole (apparent polar wandering), or there must have been rapid motion of the dipole axis with respect to the lithosphere (true polar wandering). The North American polar-wander path shows a rotation of the continent during the mid-Paleozoic, and so true polar wander cannot be precluded,

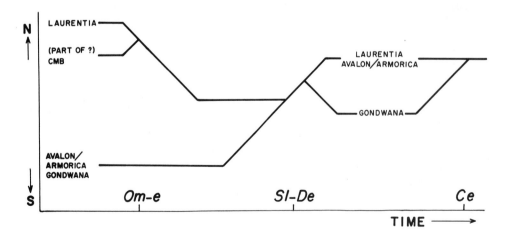

Figure 8. Cladogram, in the style of Young (1986), showing the separations and collisions between the tectonic elements discussed in this paper. The vertical axis generally represents relative north-south positioning, whereas the horizontal axis represents time. CMB = Central Mobile Belt of the northern Appalachians. O = Ordovician, S = Silurian, D = Devonian, C = Carboniferous, e = early, m = middle, l = late.

but at this time, the idea remains highly speculative. Confirmation with more Gondwana data for the Late Ordovician through Early Devonian apparent polar wander path is urgently needed.

The Late Devonian paleogeography of Armorica, Gondwana, and Laurentia also needs corroboration with reliable paleomagnetic results. The climatic setting of northwest Africa in the framework of low- versus high-temperature gradient models (Boucot and Gray, 1983; Stanley, 1984; Copper, 1986) is a topic that certainly merits further study.

Lastly, it remains unclear what the paleogeographic setting of the southern British Isles has been in the middle Paleozoic. Pre-Ordovician paleomagnetic data and geological affinities argue for juxtaposition with Armorica, but the Ordovician and Silurian paleolatitudes for southern England and Wales do not fully support this. The Ordovician paleopoles show high as well as intermediate paleolatitudes; the latter are, on average, lower than those expected if the area were part of Armorica. On the other hand, the Silurian paleopoles (although not very reliable) show latitudes which are higher, on the average, than those expected for Armorica. The paleogeographic position of the southern British Isles is a very important element in the overall framework of the assembly of Pangea and needs further study.

ACKNOWLEDGMENTS

This paper benefitted from valuable remarks made by John Crowell, Rob Hargraves, Michel Robardet, Chris Scotese, Carola Stearns, Ben van der Pluijm, F. B. Van Houten, and Pete Ziegler. Critical reviews and comments were received from A. Boucot and the journal's reviewers, Neil Opdyke and Nick Rast. The paleogeographic reconstructions were made with the Terra Mobilis (trademark) software of C. Denham and C. R. Scotese. My recent research discussed in this paper was supported by the Division of Earth Sciences, the National Science Foundation, Grants 84-07007 and 86-12469.

REFERENCES CITED

André, L., Hertogen, J., and Deutsch, S., 1986, Ordovician-Silurian magmatic provinces in Belgium and the Caledonian orogeny in middle Europe: Geology, v. 14, p. 879–882.

Bachtadse, V., and Van der Voo, R., 1986, Paleomagnetic evidence for crustal and thin-skinned rotations in the European Hercynides: Geophysical Research Letters, v. 13, p. 161–164.

Bachtadse, V., Heller, F., and Kroener, A., 1983, Paleomagnetic investigations in the Hercynian mountain belt of western Europe: Tectonophysics, v. 91, p. 285–299.

Bachtadse, V., Van der Voo, R., and Haelbich, I. W., 1987, Paleomagnetism of the western Cape Fold Belt, South Africa, and its bearing on the Paleozoic apparent polar wander path for Gondwana: Earth and Planetary Science Letters, v. 84, p. 487–499.

Bambach, R. K., Scotese, C. R., and Ziegler, A. M., 1980, Before Pangea: The geographies of the Paleozoic world: American Scientist, v. 68, p. 26–38.

Barron, E. J., 1984, Climatic implications of the variable obliquity explanation of Cretaceous-Paleogene high-latitude floras: Geology, v. 12, p. 595–598.

Beales, F. W., Carracedo, J. C., and Strangway, D. W., 1974, Paleomagnetism and the origin of Mississippi-Valley type ore deposits: Canadian Journal of Earth Sciences, v. 11, p. 211–223.

Beuf, S., Biju-Duval, B., De Charpal, O., Rognon, P., Gabriel, O., and Bennacef, A., 1971, Les grès du Paléozoïque inférieur au Sahara, sédimentation et discontinuité, évolution structurale d'un craton: Publications de l'Institut Français du Pétrole, Editions Technip, v. 18, 464 p.

Bird, J. M., and Dewey, J. F., 1970, Lithosphere plate–continental margin tectonics and the evolution of the Appalachian orogen: Geological Society of America Bulletin, v. 81, p. 1031–1060.

Boucek, B. V., 1972, The paleogeography of Lower Ordovician graptolite faunas: A possible evidence of continental drift: International Geological Congress, 24th, Section 7, Paleontology, p. 266–272.

Boucot, A. J., and Gray, J., 1983, A Paleozoic Pangaea: Science, v. 222, p. 571–580.

Briden, J. C., 1968a, Paleoclimatic evidence of a geocentric axial dipole field, *in* Phinney, R. A., ed., History of the Earth's crust: Princeton, New Jersey, Princeton University Press, p. 178–211.

——1968b, Paleomagnetism of the Ntonya ring structure, Malawi: Journal of Geophysical Research, v. 73, p. 725–733.

Briden, J. C., and Duff, B. A., 1981, Pre-Carboniferous paleomagnetism of Europe north of the Alpine orogenic belt, *in* McElhinny, M. W., and Valencio, D. A., eds., Paleoreconstruction of the continents: American Geophysical Union and Geological Society of America Geodynamics Series, Volume 2, p. 137–149.

Briden, J. C., and Mullan, A. J., 1984, Superimposed Recent, Permo-Carboniferous and Ordovician paleomagnetic remanence in the Builth Volcanic Series, Wales: Earth and Planetary Science Letters, v. 69, p. 413–421.

Briden, J. C., Turnell, H. B., and Watts, D. R., 1984, British paleomagnetism, Iapetus Ocean, and the Great Glen Fault: Geology, v. 12, p. 428–431.

Brock, A., 1967, Paleomagnetic results from the Hook intrusives, Zambia: Nature, v. 216, p. 359–360.

Brown, P. M., and Van der Voo, R., 1983, A paleomagnetic study of Piedmont metamorphic rocks in northern Delaware: Geological Society of America Bulletin, v. 94, p. 815–822.

Burrett, C. F., 1985, Comment on "Paleozoic evolution of the Armorica plate on the basis of paleomagnetic data": Geology, v. 13, p. 380.

Caputo, M. V., and Crowell, J. C., 1985, Migration of glacial centers across Gondwana during the Paleozoic Era: Geological Society of America Bulletin, v. 96, p. 1020–1036.

Chamalaun, F. H., and Creer, K. M., 1964, Thermal demagnetization studies of the Old Red Sandstone of the Anglo-Welsh cuvette: Journal of Geophysical Research, v. 69, p. 1607–1616.

Cocks, L.R.M., 1972, The origin of the Silurian Clarkeia shelly fauna of South America, and its extension to West Africa: Paleontology, v. 15, p. 623–630.

Cocks, L.R.M., and Fortey, R. A., 1982, Faunal evidence for oceanic separation in the Paleozoic of Britain: Geological Society of London Journal, v. 139, p. 465–478.

Coles, K. S., and Snyder, W. S., 1985, Significance of lower and middle Paleozoic phosphatic chert in the Toquima Range, central Nevada: Geology, v. 13, p. 573–576.

Cook, T. D., and Bally, A. W., eds., 1975, Stratigraphic atlas of North and Central America: Princeton, New Jersey, Princeton University Press, 272 p.

Copper, P., 1986, Frasnian/Famennian mass extinction and cold-water oceans: Geology, v. 14, p. 835–839.

Creer, K. M., 1968, Palaeozoic palaeomagnetism: Nature, v. 219, p. 246–250.

Dankers, P., 1982, Implications of Early Devonian poles from the Canadian Arctic archipelago for the North American apparent polar wander path: Canadian Journal of Earth Sciences, v. 19, p. 1802–1809.

Deutsch, E. R., 1984, Mid-Ordovician paleomagnetism and the Proto-Atlantic Ocean in Ireland, *in* Van der Voo, R., Scotese, C. R., and Bonhommet, N., eds., Plate reconstructions from Paleozoic paleomagnetism: American Geophysical Union Geodynamics Series, Volume 12, p. 116–119.

Duff, B. A., 1979, The paleomagnetism of Cambro-Ordovician redbeds, the Erquy Spilite Series, and the Trégastel-Ploumanac'h granite complex, Armorican Massif (France and the Channel Islands): Royal Astronomical Society Geophysical Journal, v. 59, p. 345–365.

Dunlop, D. J., and Stirling, J. M., 1985, Post-tectonic magnetizations from the Cordova gabbro, Ontario and Palaeozoic reactivation in the Grenville province: Royal Astronomical Society Geophysical Journal, v. 91, p. 521–550.

Edel, J. B., Lacaze, M., and Westphal, M., 1981, Paleomagnetism in the North-Eastern Central Massif (France and the Channel Islands): Evidence for Carboniferous rotations of the Hercynian orogenic belt: Earth and Planetary Science Letters, v. 55, p. 48–52.

Embleton, B.J.J., 1972, The palaeomagnetism of some Paleozoic sediments from central Australia: Journal and Proceedings of the Royal Society of New South Wales, v. 105, p. 86–93.

Fahraeus, L. E., 1976, Possible Early Devonian conodontophorid provinces: Palaeogeography, Palaeoclimatology, Palaeoecology, v. 19, p. 201–217.

Fang Wu, Van der Voo, R., and Johnson, R. J., 1986, Eocambrian paleomagnetism of the Boston Basin: Evidence for displaced terrane: Geophysical Research Letters, v. 13, p. 1450–1453.

Fortuin, A. R., 1984, Late Ordovician glaciomarine deposits (Orea Shale) in the Sierra de Albarracin, Spain: Palaeogeography, Palaeoclimatology, Palaeoecology, v. 48, p. 245–261.

French, A. N., and Van der Voo, R., 1979, The magnetization of the Rose Hill Formation at the classical site of Graham's fold test: Journal of Geophysical Research, v. 84, p. 7688–7696.

Hargraves, R. B., Dawson, E. M., and Van Houten, F. B., 1987, Paleomagnetism and age of mid-Paleozoic ring complexes in Niger, West Africa, and tectonic implications: Royal Astronomical Society Geophysical Journal, v. 90, 705–729.

Hartnady, C., 1985, Comment on "Paleozoic evolution of the Armorica plate on the basis of paleomagnetic data": Geology, v. 13, p. 589.

Hurley, N. F., and Van der Voo, R., 1986, Late Devonian magnetostratigraphy from a condensed limestone, Canning Basin, Western Australia [abs.]: EOS (American Geophysical Union Transactions), v. 67, p. 265.

——1987, Paleomagnetism of Upper Devonian reefal limestones, Canning Basin, Western Australia: Geological Society of America Bulletin, v. 98, p. 123–137.

Irving, E., and Strong, D. F., 1984, Paleomagnetism of the early Carboniferous Deer Lake Group, western Newfoundland: No evidence for mid-Carboniferous displacement of "Acadia": Earth and Planetary Science Letters, v. 69, p. 379–390.

Jackson, M. J., and Van der Voo, R., 1985, A Lower Ordovician paleomagnetic pole from the Oneota Dolomite, Upper Mississippi Valley: Journal of Geophysical Research, v. 90, p. 10449–10461.

Johnson, R. J., and Van der Voo, R., 1985, Middle Cambrian paleomagnetism of the Avalon terrane in Cape Breton Island, Nova Scotia: Tectonics, v. 4, p. 629–651.

——1986, Paleomagnetism of the Late Precambrian Fourchu Group, Cape Breton Island, Nova Scotia: Canadian Journal of Earth Sciences, v. 23, p. 1673–1685.

Jones, D. S., McFadden, B. J., Opdyke, N. D., and Smith, D. L., 1983, Paleomagnetism of lower Paleozoic rocks of the Florida basement [abs.]: EOS (American Geophysical Union Transactions), v. 64, p. 690.

Kaljo, D. L., and Klaamann, E. R., 1973, Ordovician and Silurian corals, *in* Hallam, A., ed., Atlas of paleobiogeography: Amsterdam, the Netherlands, Elsevier, p. 37–45.

Kent, D. V., 1986, Separation of pre-folding and secondary magnetizations from the Bloomsburg Formation from the southern limb of the Pennsylvania re-entrant [abs.]: EOS (American Geophysical Union Transactions), v. 67, p. 270.

Kent, D. V., and Opdyke, N. D., 1978, Paleomagnetism of the Devonian Catskill red beds: Evidence for motion of the coastal New England-Canadian Maritime region relative to cratonic North America: Journal of Geophysical Research, v. 83, p. 4441–4450.

——1980, Paleomagnetism of Siluro-Devonian rocks from eastern Maine: Canadian Journal of Earth Sciences, v. 17, p. 1653–1665.

——1985, Multicomponent magnetizations from the Mississippian Mauch Chunk Formation of the central Appalachians and their tectonic implications: Journal of Geophysical Research, v. 90, p. 5371–5383.

Kent, D. V., Dia, O., and Sougy, J.M.A., 1984, Paleomagnetism of lower-Middle Devonian and upper Proterozoic-Cambrian (?) rocks from Mejeria (Mauritania, West Africa), *in* Van der Voo, R., Scotese, C. R., and Bonhommet, N., eds., Plate reconstructions from Paleozoic paleomagnetism: American Geophysical Union Geodynamics Series, Volume 12, p. 99–115.

Lapointe, P., 1979, Paleomagnetism and orogenic history of the Botwood Group and Mount Peyton batholith, Central Mobile Belt, Newfoundland: Canadian Journal of Earth Sciences, v. 16, p. 866–876.

Leo, G. W., 1985, Trondhjemite and metamorphosed quartz keratophyre tuff of the Ammonoosuc Volcanics (Ordovician), western New Hampshire and adjacent Vermont and Massachusetts: Geological Society of America Bulletin, v. 96, p. 1493–1507.

Lewandowski, M., 1987, Results of the preliminary paleomagnetic investigations of some lower Paleozoic rocks from the Holy Cross Mountains (Poland): Kwartalnik Geologicne (in press).

Luck, G. R., 1972, Palaeomagnetic results from Palaeozoic sediments of northern Australia: Royal Astronomical Society Geophysical Journal, v. 28, p. 475–487.

Ludman, A., 1986, Timing of terrane accretion in eastern and east-central Maine: Geology, v. 14, p. 411–414.

Manzoni, M., and Nanni, T., Paleomagnetism of Ordovician lamprophyres from Taylor Valley, Victoria Land, Antarctica: Pageoph, v. 115, p. 961–977.

Matte, P., 1986, Tectonics and plate tectonics model for the Variscan belt of Europe: Tectonophysics, v. 126, p. 329–374.

McCabe, C., Van der Voo, R., Wilkinson, B. H., and Devaney, K., 1985, A Middle/Late Silurian paleomagnetic pole from limestone reefs of the Wabash Formation (Indiana, U.S.A.): Journal of Geophysical Research, v. 90, p. 2959–2965.

McKerrow, W. S., and Cocks, L.R.M., 1976, Progressive faunal migration across the Iapetus Ocean: Nature, v. 263, p. 304–306.

McKerrow, W. S., and Ziegler, A. M., 1972, Palaeozoic oceans: Nature Physical Science, v. 240, p. 92–94.

Merrill, R. T., and McElhinny, M. W., 1983, The Earth's magnetic field: London, England, Academic Press, 401 p.
Miller, J. D., and Kent, D. V., 1986, Paleomagnetism of the Upper Devonian Catskill Formation from the southern limb of the Pennsylvania salient: Geophysical Research Letters, v. 13, p. 1173–1176.
Morel, P., and Irving, E., 1978, Tentative paleocontinental maps for the early Phanerozoic and Proterozoic: Journal of Geology, v. 86, p. 535–561.
Morrison, J., and Ellwood, B. B., 1986, Paleomagnetism of Silurian-Ordovician sediments from the Valley and Ridge province, northwest Georgia: Geophysical Research Letters, v. 13, p. 189–192.
Murthy, G. S., and Rao, K. V., 1976, Paleomagnetism of Steel Mountain and Indian Head anorthosites from western Newfoundland: Canadian Journal of Earth Sciences, v. 13, p. 75–83.
Neuman, R. B., 1984, Geology and paleobiology of islands in the Ordovician Iapetus Ocean: Review and implications: Geological Society of America Bulletin, v. 95, p. 1188–1201.
Oliver, W. A., 1976, Biogeography of Devonian rugose corals: Journal of Paleontology, v. 50, p. 365–373.
Palmer, A. R., and Secor, D. T., Jr., 1983, Do they really mean Africa?: Geological Society of America Bulletin, v. 94, p. 1380.
Perigo, R., Van der Voo, R., Auvray, B., and Bonhommet, N., 1983, Paleomagnetism of late Precambrian–Cambrian volcanics and intrusives of the Armorican Massif, France: Royal Astronomical Society Geophysical Journal, v. 75, p. 235–260.
Perroud, H., 1986, Paleomagnetic evidence for tectonic rotations in the Variscan mountain belt: Tectonics, v. 5, p. 205–214.
Perroud, H., and Bonhommet, N., 1981, Paleomagnetism of the Ibero-Armorican arc and the Hercynian orogeny in western Europe: Nature, v. 292, p. 445–448.
—— 1984, A Devonian pole for Armorica: Royal Astronomical Society Geophysical Journal, v. 77, p. 839–845.
Perroud, H., and Van der Voo, R., 1985, Paleomagnetism of the Late Ordovician Thouars Massif, Vendée province, France: Journal of Geophysical Research, v. 90, p. 4611–4625.
Perroud, H., Van der Voo, R., and Bonhommet, N., 1984, Paleozoic evolution of the Armorica plate on the basis of paleomagnetic data: Geology, v. 12, p. 579–582.
Perroud, H., Bonhommet, N., and Ribeiro, A., 1985, Paleomagnetism of late Paleozoic igneous rocks from southern Portugal: Geophysical Research Letters, v. 12, p. 45–48.
Perroud, H., Bonhommet, N., and Thebault, J. P., 1986, Paleomagnetism of the Ordovician Moulin de Chateaupanne formation, Vendée, western France: Royal Astronomical Society Geophysical Journal, v. 85, p. 573–582.
Piper, J.D.A., 1982, A paleomagnetic investigation of the Malvernian and Old Radnor Precambrian, Welsh Borderlands: Geological Journal, v. 17, p. 69–88.
Piper, J.D.A., and Briden, J. C., 1973, Palaeomagnetic studies in the British Caledonides—Part I, Igneous rocks of the Builth Wells–Llandridnod Wells Ordovician inlier, Radnorshire, Wales: Royal Astronomical Society Geophysical Journal, v. 34, p. 1–12.
Rao, C. P., 1981, Geochemical differences between tropical (Ordovician) and subpolar (Permian) carbonates, Tasmania, Australia: Geology, v. 9, p. 205–209.
Rao, K. V., and Van der Voo, R., 1980, Paleomagnetism of a Paleozoic anorthosite from the Appalachian Piedmont, northern Delaware: Possible tectonic implications: Earth and Planetary Science Letters, v. 47, p. 113–120.
Robardet, M., and Doré, F., 1988, The Late Ordovician diamictic formations from southwestern Europe: North Gondwana glaciomarine deposits: Palaeogeography, Palaeoclimatology, Palaeoecology (in press).
Robinson, P. L., 1973, Palaeoclimatology and continental drift, in Tarling, D. H., and Runcorn, S. K., eds., Implications of continental drift to the earth sciences: New York, Academic Press, p. 451–476.
Rodgers, J., and Sougy, J., 1984, The West African connection—Evolution of the central Atlantic Ocean and its continental margins: Geology, v. 12, p. 635–636.
Roy, J. L., 1982, Paleomagnetism of Siluro-Devonian rocks from eastern Maine: Discussion: Canadian Journal of Earth Sciences, v. 19, p. 225–232.
Roy, J. L., and Anderson, P., 1981, An investigation of the remanence characteristics of three sedimentary units of the Silurian Mascarene Group of New Brunswick: Journal of Geophysical Research, v. 86, p. 6351–6368.
Roy, J. L., Opdyke, N. D., and Irving, E., 1967, Further paleomagnetic results from the Bloomsburg Formation: Journal of Geophysical Research, v. 72, p. 5075–5086.
Salmon, E., Montigny, R., Edel, J. B., Pique, A., Thuizat, R., and Westphal, M., 1986, The Msissi Norite revisited: K/Ar dating, petrography, and paleomagnetism: Geophysical Research Letters, v. 13, p. 741–743.
Schmidt, P. W., 1987, New palaeomagnetic poles from Australia: International Union of Geophysics and Geodesy, 19th General Assembly, Vancouver, Canada, Abstracts, v. 2, p. 501.
Schmidt, P. W., Embleton, B.J.J., and Palmer, H. C., 1987, Pre- and post-folding magnetizations from the Early Devonian Snowy River Volcanics and Buchan Caves limestone, Victoria: Royal Astronomical Society Geophysical Journal, v. 91, p. 155–170.
Schmidt, P. W., Embleton, B.J.J., Cudahy, T. J., and Powell, C. McA., 1986, Prefolding and premegakinking magnetizations from the Devonian Comerong Volcanics, New South Wales, Australia and their bearing on the Gondwana pole path: Tectonics, v. 5, p. 135–150.
Schutts, L. D., Brecher, A., Hurley, P. M., Montgomery, C. W., and Krueger, H. W., 1976, A case study of the time and nature of paleomagnetic resetting in a mafic complex in New England: Canadian Journal of Earth Sciences, v. 13, p. 898–907.
Scotese, C. R., 1984, Paleozoic paleomagnetism and the assembly of Pangea, in Van der Voo, R., and others, eds., Plate reconstruction from Paleozoic paleomagnetism: American Geophysical Union Geodynamics Series, Volume 12, p. 1–10.
Scotese, C. R., Van der Voo, R., and Barrett, S. F., 1985, Silurian and Devonian base maps: Royal Society of London Philosophical Transactions, v. B 309, p. 57–77.
Scotese, C. R., Bambach, R. K., Barton, C., Van der Voo, R., and Ziegler, A. M., 1979, Paleozoic base maps: Journal of Geology, v. 87, p. 217–277.
Secor, D. T., Jr., Samson, S. L., Snoke, A. W., and Palmer, A. R., 1983, Confirmation of the Carolina slate belt as an exotic terrane: Science, v. 221, p. 649–651.
Seguin, M. K., Rao, K. V., and Pineault, R., 1982, Paleomagnetic study of Devonian rocks from Ste. Cecile-St. Sébastien region, Quebec Appalachians: Journal of Geophysical Research, v. 87, p. 7853–7864.
Spariosu, D. J., and Kent, D. V., 1983, Paleomagnetism of the Lower Devonian Traveler Felsite and the Acadian orogeny in the New England Appalachians: Geological Society of America Bulletin, v. 94, p. 1319–1328.
Spjeldnaes, N., 1961, Ordovician climatic zones: Norsk Geologisk Tidesskrift, v. 21, p. 45–77.
Stanley, S. M., 1984, Temperature and biotic crises in the marine realm: Geology, v. 12, p. 205–208.
Tarling, D. H., 1985, Problems in Palaeozoic palaeomagnetism: Journal of Geodynamics, v. 3, p. 87–103.
Thomas, C., and Briden, J. C., 1976, Anomalous geomagnetic field during the Late Ordovician: Nature, v. 259, p. 380–382.
Turner, S. W., and Tarling, D. H., 1982, Thelodont and other agnathan distribution as tests of lower Palaeozoic continental reconstructions: Palaeogeography, Palaeoclimatology, Palaeoecology, v. 39, p. 295–311.
van der Pluijm, B. A., 1987, Timing and spatial distribution of deformation in the Newfoundland Appalachians: A "multi-stage collision" history: Tectonophysics, v. 135, p. 15–24.
Van der Voo, R., 1982, Pre-Mesozoic paleomagnetism and plate tectonics: Annual Reviews of the Earth and Planetary Sciences, v. 10, p. 191–220.
—— 1988, Paleomagnetism of North America: The craton, its margins and the Appalachians, in Pakiser, L. C., and Mooney, W. D., eds., Geophysical framework of the continental United States: Geological Society of America Memoir (in press).
Van der Voo, R., and French, R. B., 1977, Paleomagnetism of the Late Ordovician Juniata Formation and the remagnetization hypothesis: Journal of Geophysical Research, v. 82, p. 5796–5802.
Van der Voo, R., and Johnson, R. J., 1985, Paleomagnetism of the Dunn Point Formation (Nova Scotia): High paleolatitudes for the Avalon terrane in the Late Ordovician: Geophysical Research Letters, v. 12, p. 337–340.
Van der Voo, R., Perroud, H., and Bonhommet, N., 1985a, Reply (to C. Burrett): Geology, v. 13, p. 380–381.
—— 1985b, Reply (to C. Hartnady): Geology, v. 13, p. 589–590.
Van Houten, F. B., 1985, Oolitic ironstones and contrasting Ordovician and Jurassic paleogeography: Geology, v. 13, p. 722–724.
Van Houten, F. B., and Hargraves, R. B., 1987, Early-Middle Paleozoic drift of Gondwana, in Bowden, P., and Kinnaird, J. A., eds., African geology reviews: New York, Wiley and Sons (in press).
Veevers, J. J., and Powell, C. McA., 1987, Late Paleozoic glacial episodes in Gondwanaland reflected in transgressive-regressive depositional sequences in Euramerica: Geological Society of America Bulletin, v. 98, p. 475–487.
Watts, D. R., and Van der Voo, R., 1979, Paleomagnetic results from the Ordovician Moccasin, Bays, and Chapman Ridge Formations of the Valley and Ridge province, eastern Tennessee: Journal of Geophysical Research, v. 84, p. 645–655.
Weisse, P. A., Haggerty, S. E., and Brown, L. L., 1985, Paleomagnetism and magnetic mineralogy of the Nahant Gabbro and Tonalite, eastern Massachusetts: Canadian Journal of Earth Sciences, v. 22, p. 1425–1435.
Wendt, J., 1985, Disintegration of the continental margin of northwestern Gondwana: Late Devonian of the eastern Anti-Atlas (Morocco): Geology, v. 13, p. 815–818.
Whittington, H. B., and Hughes, C. P., 1972, Ordovician geography and faunal provinces deduced from trilobite distribution: Royal Society of London Philosophical Transactions, v. 263, p. 235–278.
Williams, A., 1973, Distribution of brachiopod assemblages in relation to Ordovician palaeogeography, in Hughes, N. F., ed., Organisms and continents through time: Palaeontological Association Special Paper, no. 12, p. 241–269.
Williams, H., and Hatcher, R. D., Jr., 1983, Appalachian suspect terranes, in Hatcher, R. D., Jr., Williams, H., and Zietz, I., eds., Contributions to the tectonics and geophysics of mountain chains: Geological Society of America Memoir 158, p. 33–53.
Young, G. C., 1986, Cladistic methods in Paleozoic continental reconstructions: Journal of Geology, v. 94, p. 523–537.
Ziegler, A. M., Hansen, K. S., Johnson, M. E., Kelly, M. A., Scotese, C. R., and Van der Voo, R., 1977, Silurian continental distributions, paleogeography, climatology and biogeography: Tectonophysics, v. 40, p. 18–51.
Ziegler, A. M., Scotese, C. R., McKerrow, W. S., Johnson, M. E., and Bambach, R. K., 1979, Paleozoic paleogeography: Annual Reviews of Earth and Planetary Science, v. 7, p. 473–502.
Ziegler, P. A., 1984, Caledonian and Hercynian crustal consolidation of western and central Europe—A working hypothesis: Geologie en Mijnbouw, v. 63, p. 93–108.
Zijderveld, J.D.A., 1968, Natural remanent magnetizations of some intrusive rocks from the Sør Rondane Mountains, Queen Maud Land, Antarctica: Journal of Geophysical Research, v. 73, p. 3773–3785.

Manuscript Received by the Society July 31, 1987
Revised Manuscript Received November 13, 1987
Manuscript Accepted November 16, 1987

CENTENNIAL ARTICLE

Paleoceanography: A review for the GSA Centennial

WILLIAM W. HAY *Museum, Department of Geology* and *Cooperative Institute for Research in Environmental Sciences, University of Colorado, Boulder, Colorado 80309*

ABSTRACT

The central problem of paleoceanography is the history of the circulation of the ocean. Although speculation about ancient oceanic circulation goes back to the past century, the field of paleoceanography was founded in the 1950s as oxygen-isotope studies suggested that oceanic deep waters were warmer in the past than they are today. Extensive coring of deep-sea sediments by numerous expeditions after World War II was followed by the ocean drilling programs, providing a rich data base. Paleoceanographic interpretations have tried to explain the most obvious changes in sea-floor sediments and their contained fossils: changing paleotemperatures indicated by oxygen isotopes, fluctuations in the calcium carbonate compensation depth, accumulations of organic carbon-rich sediments, and the unexpected abundance of hiatuses in a setting which had been thought to be the ultimate sedimentary sink. The result has been the intriguing discovery that although the positions and circulation of the major surface gyres is generally stable, the deep circulation of the ocean may reverse on a variety of time scales. It has been suggested that formation of North Atlantic Deep Water, which causes the uneven distribution of nutrients, alkalinity, and oxygen in the deep sea today, may have been replaced by formation of North Pacific Deep Water during the last deglaciation, reversing the concentration gradients of nutrients, alkalinity, and oxygen. On a longer time scale, the present general circulation, which is dominated by production of oxygen-rich cold deep water in the subpolar regions today, may have replaced a pre-Oligocene general circulation in which warm, saline, oxygen-poor deep waters were formed in warm seas in the arid zones. Paleoceanography is still in its infancy; many new clues to the history of the ocean are being discovered, and many new ideas about conditions in the past are being developed. The beginning of the next century should see continuing rapid growth and maturation in this exciting new field.

INTRODUCTION

Although systematic geologic investigation of the Earth has been under way since the 18th century, until 25 years ago it was essentially restricted to the study of sediments and rocks exposed on land or penetrated by mining or drilling activity. The 70% of the Earth's surface covered by water was virtually inaccessible to geologic investigation and contributed very little to our understanding of the development of the planet. It now seems strange that geologists of the 19th and the first half of the 20th century should have been so confident that they knew the Earth in spite of this obvious gap in knowledge. It has generally been assumed by those dealing primarily with rocks exposed on land that the history of the oceans can be deduced from observation of the deposits of shallow seas and geosynclines. Thus, Schopf's *Paleoceanography* (1980) discusses what is known of Precambrian, Paleozoic, and Mesozoic oceans from indirect knowledge gleaned from the continents. Paleooceanography is the study of the oceans of the past, and ancient oceans have been at least as different from marginal seas as are the modern oceans. As Berger (1979, 1981) has observed, the central problem of paleoceanography is the history of circulation of the ocean. Although marginal seas may offer important clues, real information that can be used to deduce the history of ocean circulation must come from the ocean basins proper.

EARLY IDEAS ABOUT THE DEEP SEA

Deep-sea sediments had been sampled in the 19th century first as part of the effort to lay the trans-Atlantic telegraph cables, then as a routine function of the great voyages of scientific exploration of the sea. As recently as 1970, the deep sea was regarded by the majority of geologists as an unchanging primordial environment which had been the most constant feature on the surface of the Earth throughout most of geologic time. Paleontologists considered the cold lightless deep sea to be the final refuge for such forms of life as the hyalosponges and coelacanths which had been unsuccessful in the competition for Lebensraum in shallower waters. The climate of the continents and the shallow epeiric seas had obviously changed with time, but the open ocean was regarded as the great stabilizing feature of the Earth's surface. So ingrained was this idea in geologic thinking that the pioneer micropaleontologists J. J. Galloway and J. A. Cushman studiously avoided examining the biostratigraphic potential of the planktonic foraminifera and radiolaria because it was well known that oceanic plankton did not evolve. C. G. Wallich (1861) had given all coccoliths a single name, *Coccolithus oceanicus*, for the same reason. Only T. C. Chamberlin (1906) bothered to consider the possibility of a radically different ocean, suggesting that salinity rather than temperature differences may have driven oceanic circulation in the past.

Geology is unique among the sciences in having a historical aspect. Probably because of the vast amount of historical geological evidence which had been accumulated and investigated, geologists tended to be very conservative and slow to accept radical new ideas. The classical maxim

Geological Society of America Bulletin, v. 100, p. 1934–1956, 5 figs., December 1988.

attributed to James Hutton, that "the present is the key to the past," was so thoroughly ingrained in the reasoning of geologists that very little attention was given to the consideration of a world that would have been very different from that observed today. Recall that Alfred Wegener and the idea of "continental drift" were rejected by the great majority of the world's geologists for more than 50 years.

DEVELOPMENT OF SAMPLING TECHNIQUES AND EARLY RESULTS

No unequivocal ancient oceanic deposits were known until the invention of the piston corer by Kullenberg (1947) and its subsequent extensive use by the Swedish Deep-Sea Expedition. The long piston cores recovered by the Swedish Deep-Sea Expedition in 1947–1948 added historical perspective to the sedimentary record of the ocean basins, albeit only the Pleistocene and late Pliocene were penetrated. Arrhenius (1952) noted that the cores displayed significant variations in carbonate content, which could be correlated with glacial and interglacial ages.

Although sedimentological studies were important in demonstrating that the oceanic environment has in fact changed with time, the new branch of science, "Paleooceanography," can be said to have been founded by Cesare Emiliani, who in three classic papers presented an outline of ocean thermal history and demolished the notion of constancy of the oceanic environment. Emiliani and Epstein (1953) presented the results of analysis of $^{18}O/^{16}O$ ratios (now commonly expressed as enrichment or depletion of ^{18}O, $\delta^{18}O$, relative to the PDB standard, a belemnite rostrum from the PeeDee Formation in South Carolina) in planktonic foraminifera in samples from the lower Pleistocene of southern California indicating substantial glacial to interglacial temperature fluctuations. Emiliani and Edwards (1953) and Emiliani (1954) reported the results of oxygen isotopic analysis of benthonic foraminifera from pre-Pleistocene samples from the Pacific Ocean basin. These suggested bottom-water temperatures of 2.2 °C for the late Pliocene, 7.0 °C for the early middle Miocene, and 10.5 °C for the middle Oligocene (the supposed Oligocene sample has subsequently been determined to be Eocene; K. G. Miller, 1988, personal commun.). In 1955 Emiliani published the record of Pleistocene variation of the $^{18}O/^{16}O$ ratio in surface-dwelling planktonic foraminifera; the possibility of phyletic influence on isotopic fractionation was excluded by using a single species. He stated that the record of change in the isotopic composition of the tests must be due to (1) change in the over-all isotopic composition of the ocean, which would be a function of the mass of ice existing as glaciers, and (2) the temperature of the surface waters at the time of shell formation. These papers opened a new perspective on the history of the oceans and introduced the prospect of quantification of ocean temperatures of the past, a major descriptor of global climate. In 1958 Emiliani discussed the results of these studies of ancient ocean temperatures in what may be considered the first summary article on paleoceanography.

The Kullenberg piston corer is limited in its depth of penetration by the strength of the materials involved and by the mechanical properties of the sediment being cored. In effect, these limitations restricted the coring efforts in the deep sea to Pleistocene sediments except in those few areas where unlithified older sediments are exposed or are very close to the sea floor; consequently, knowledge of the older history of the ocean basins was virtually non-existent, and discussions from the middle of this century now seem quite fanciful. Kuenen (1950) acknowledged the remote possibility of continental drift, but preferred to assume that the ocean basins were ancient features, and that deposition in them had been going on continuously since the early Precambrian; he estimated that the average sediment thickness should be about 3 km. He mentioned the adaptation of echo sounding to measurement of sediment thickness, described by Weibull (1947), but very few data were available at the time. Poldervaart (1955) noted that Oliver and others (1953) had found that average sediment thicknesses in the ocean were about 600 m and remarked that such thicknesses could be accounted for by deposition during the past 200 m.y. He suggested that "with perhaps lower relief of the remaining land, absence of *Globigerina* and virtual absence of turbidity currents, practically all deposition could have occurred in shallow waters, and most of the ocean floor could have remained barren of a sedimentary cover until modern conditions prevailed."

The 1950s and 1960s were the golden age of piston coring. Ericson and others (1961) reported on studies of 221 Atlantic and Caribbean cores, 41 of which contained pre-Pleistocene sediments, the oldest being Upper Cretaceous. It was evident that unconformities are common in deep-sea sediments, that turbidity-current deposits are widespread, that rates of accumulation in the deep sea are highly variable, and that "a drastic reorganization of . . . the Earth's crust now covered by the oceans took place at some time during the latter part of the Mesozoic era" (p. 282). It had become evident that the ocean had a complex history, but it was not at all clear what that history was.

THE OCEAN DRILLING PROGRAMS

Because of the great interest aroused by the Pleistocene sequences recovered from the deep sea, it was not surprising that Emiliani championed the cause of deeper coring of the sediment on the ocean floor. It was obvious that the only way in which cores could be recovered from the more deeply buried part of the sedimentary section was through the application of industrial drilling and coring techniques being developed for offshore exploration for oil and gas. In the late 1950s, it had been proposed that there be a national program to drill through the Mohorovičić Discontinuity in the Pacific Ocean basin where it was thought that samples of the mantle could be recovered by penetration to a depth of only about 12 km beneath the sea surface (Bascom, 1958, 1961; Hess, 1959), but these plans called for bypassing the sediment column. Emiliani (1981) has presented a personal recounting of the events of this period, which eventually led to establishment of the National Sediment Coring Program and its operational offspring, the Deep Sea Drilling Project (DSDP), and its successor, the Ocean Drilling Program (ODP).

Scientific drilling operations in the deep sea began on March 6, 1961, using the drilling barge *Cuss I* of Global Marine Exploration Company of Los Angeles in a water depth of 945 m off La Jolla, California; it ended 9 days later after a penetration of 1,315 m into the sea floor. Immediately following this, a second site was drilled 40 nautical miles east of Guadalupe Island, off Baja California. Known as the Experimental Mohole, this effort was conducted in 3,558 m of water and penetrated 183 m of sediment and 13 m of basalt. It demonstrated the feasibility of recovering scientifically valuable cores from depths well below the limit of penetration of the Kullenberg corer. The Mohole effort was conceived in a fixist context; the consensus among geologists was that the ocean basins were permanent features. Project Mohole was abandoned for a variety of reasons, but interest in scientific ocean drilling to recover the sedimentary record continued to develop.

The next effort to recover older oceanic materials was called "Project LOCO" (LOng COres). With funding from the National Science Foundation (NSF), the Global Marine vessel *Submarex* carried out drilling and coring operations on the Nicaragua Rise, from November 27 to December 17, 1963. Weather conditions were unfavorable, but drilling and coring operations penetrated the sediments to a depth of 56.4 m in a water depth of 610 m, with better than 30% recovery (Bolli and others, 1968).

In 1964, four United States oceanographic institutions, The Institute of Marine Sciences (now the Rosenstiel School of Marine and Atmospheric Sciences) of the University of Miami, Lamont (now Lamont-Doherty) Geological Observatory of Columbia University, the Woods Hole Oceanographic Institution, and the Scripps Institution of Oceanography of the University of California, joined together as a consortium which took the name JOIDES (Joint Oceanographic Institutions for Deep Earth Sampling). They proposed and NSF supported a six-hole drilling effort off Jacksonville, Florida, using Global Marine's vessel *Caldrill* in April and May of 1965; these later became known as the JOIDES holes. The six sites were continuously cored to sub-bottom depths of more than 1 km. Examination of the cores revealed that significant changes in the oceanographic conditions on the east Florida shelf and Blake Plateau had occurred since the Late Cretaceous. The samples provided very well preserved assemblages of planktonic and benthonic microfossils; this material was of major importance in developing the biostratigraphic schemes in use today.

The Experimental Mohole, *Submarex* hole and JOIDES drilling off Florida had convinced the scientists involved of two things. (1) Deep-sea sediments contained long sequences of extraordinarily well preserved microfossil assemblages which could be used for global stratigraphic correlation. Before samples from these sites were available, the basis for global stratigraphic correlation had been a fragmentary record preserved on land, often in structurally complex areas such as Trinidad (Bolli, 1957; Bramlette and Wilcoxon, 1967; Hay and others, 1967) where structural complexity made determination of stratigraphic relations difficult. For refinement of biostratigraphy alone, a campaign of further drilling was justified. (2) Although Emiliani's pioneering work of the 1950s had suggested that oceans of the past might have been significantly different from those of today, the stratigraphic sections recovered by these early efforts (less than 5 months of operations at sea over a 4-yr period) confirmed that the history of the ocean basins was anything but simple and unchanging.

JOIDES next proposed an 18-month campaign of drilling in the Atlantic and Pacific Oceans, to be known as the Deep Sea Drilling Project, selecting Scripps to be the operator. The NSF approved and funded the project. Global Marine was able to modify a vessel already under construction to become a research tool specifically designed for scientific ocean drilling, core recovery, and analysis. The new vessel, christened *Glomar Challenger*, set sea from Orange, Texas, on July 20, 1968. This was the culmination of a decade of dreams of marine geologists, and van Andel (1968) summarized the expectations at the time. No one anticipated the full impact of the discoveries which would be made, nor that the at-sea operations of the *Glomar Challenger* would continue for 15 years. By the early 1970s, the project had been so successful (Hammond, 1970) that other United States and foreign institutions joined JOIDES in support of the program (the Department of Oceanography of the University of Washington; the Academy of Sciences of the USSR; the Bundesanstalt für Geowissenschaften und Rohstoffe of the Federal Republic of Germany; the Centre National pour l'Exploitation des Océans (now the Institut français de Recherche pour l'Exploitation de la Mer) of France; the Natural Environment Research Council of the United Kingdom; the Ocean Research Institute of the University of Tokyo, Japan; the Hawaii Institute of Geophysics of the University of Hawaii; the School of Oceanography of Oregon State University; the Graduate School of Oceanography of the University of Rhode Island; and the Department of Oceanography of Texas A&M University). On November 8, 1983, the *Glomar Challenger* returned to port in Mobile, Alabama, for the last time as JOIDES scientific drilling vessel. JOIDES has continued to grow, adding the Institute of Geophysics of the University of Texas at Austin; the Department of Energy, Mines and Resources of Canada; and the European Science Foundation in the 1980s. A larger and more sophisticated drillship was modified to continue the program of scientific drilling in the oceans. Operated by Texas A&M University, the drillship, officially registered as the *Sedco*/BP 471, was named the *JOIDES Resolution* by the scientific community. She set sail on the first of a 10-yr series of cruises on March 20, 1985; this new program carries the experience and ideas developed since 1967 to a new level of sophistication (Hay, 1987).

The impetus for deep-sea drilling was that the recovery of sediment and rocks from the ocean basins was expected to provide a new perspective on the history of the planet. The drive for recovering materials from the ocean floor, however, happened to coincide with the revolution in geological thinking introduced by the theory of sea-floor spreading and plate tectonics. It is one of the most fortuitous events in the history of science that the revolutionary theory of plate tectonics and the geological tools and methods essential for its investigation developed simultaneously but almost entirely independently, and that an unforeseen offspring, the burgeoning field of paleoceanography, would result from this lucky marriage of theory and technology.

THE AT-SEA AND ON-SHORE ANALYSES

To understand the development of ideas in paleoceanography, it is important to know about the kinds of information which have become available, largely as a result of the Deep Sea Drilling Project and its successor, the Ocean Drilling Program. As each core is recovered, a shipboard scientific party performs certain routine tasks. Before the core is cut longitudinally, bulk density and gamma-ray attenuation are measured to determine the volume of pore space and mass of the solid phase; this is essential for later determinations of sediment mass accumulation rates. The core is then cut and described; smear slides are made at frequent intervals to aid in the descriptions. Although many investigators have made highly detailed descriptions of smear slides, including visual estimates of the abundances of different sorts of mineral grains, little of this detail appears in the Initial Reports of the DSDP or the ODP because the information is presented in a synthesized form in the core descriptions. [All of the original observations have been preserved, however, and are incorporated into the data banks for the projects. Copies of the complete data files for the DSDP will soon be available from the National Geophysical Solar-Terrestrial Data Center (NGSDC) of the National Oceanographic and Atmospheric Administration (NOAA) in Boulder, Colorado, in CD ROM format.] The split core is then sampled for a variety of other studies to be undertaken by the shipboard scientific party and by shore-based laboratories. The shipboard party includes micropaleontologists who study the planktonic foraminifera, calcareous nannofossils, radiolaria, diatoms, and silicoflagellates to provide a biostratigraphic and paleoenvironmental framework for the lithologic studies. Shore-based laboratories provide additional biostratigraphic determinations, grain-size distribution in terms of three fractions, $CaCO_3$ content, organic carbon content, and X-ray mineralogy. In the later phases of the DSDP and in the ODP, paleomagnetic determinations, Rockeval analysis, and more chemical analyses have become routine. The descriptive data and results of analyses of the cores have been published in the *Initial Reports of the DSDP* and *Proceedings Part A: Initial Reports of the ODP*, with volume numbers corresponding to the cruise legs, which were ordinarily two months long. It is in large part due to the homogeneity of this vast body of data that regional and global analyses and paleoceanographic interpretation are possible.

DEVELOPMENT OF THE STRATIGRAPHIC BASE

Paleoceanographic interpretation requires that stratigraphic resolution be as fine as possible. Because of the inertia inherent in the system, the ocean's response to short-term climatic change is on the order of hundreds to thousands of years; unfortunately, this is the same order of magnitude as the blurring of stratigraphic resolution resulting from bioturbation of the sediment. Hence it requires unusual preservation of the sedimentary record in order to resolve the short-term changes in the ocean which would relate directly to the observations of physical oceanographers. The record of longer-term climate change, such as that from glacial to interglacial and on the order of thousands to tens of thousands of years, is preserved in most oceanic sediments and can be determined from study of stable isotopes. At the time the ocean-drilling studies began, it was expected that the study of materials from the deep sea would result in significant improvements in biostratigraphy, magnetic stratigraphy, and isotope stratigraphy and would improve the level of stratigraphic resolution to that required to relate paleoceanography directly to modern physical oceanography.

For the most part, oceanic oozes are made up of microfossils, and it was widely assumed that it would be possible to greatly refine existing biostratigraphic schemes and achieve much finer stratigraphic resolution. The biostratigraphic zonation schemes for the major calcareous oceanic microfossils, the planktonic foraminifera, and calcareous nannoplankton fossils had been developed just prior to the DSDP using mostly samples from land and the JOIDES holes. The stratigraphic zonation schemes for the calcareous groups were little affected by the new information from the deep-sea samples because it was found that differential dissolution of species confused their apparent ranges and because hiatuses were much more common than had been expected. For the development of biostratigraphic zonations for the siliceous microfossil groups, the diatoms, silicoflagellates, and particularly the radiolarians, the deep-sea material was critically important. The use of fish debris ("ichthyoliths") for stratigraphic work was a direct outgrowth of the need of scientists involved with the DSDP to develop a technique for correlating red-clay sequences. Instead of documenting and refining a global biostratigraphic zonation scheme, the new materials have documented differing species distributions in different areas. The contrasts between tropical, temperate, subpolar, and polar assemblages have become well known, but much remains to be done in correlating species distributions in different environments. The enormous abundance of the microfossils and the global coverage of sampling has permitted statistical studies of oceanic assemblages which have been of great importance to the development of paleoceanographic ideas.

Measurement of natural remanent magnetism in samples from deep-sea cores can determine whether the Earth's magnetic field had normal or reversed polarity at the time of deposition of the sediments. The sequence of reversals observed in the sediments should match that known from areal geophysical surveys of the magnetic-anomaly patterns on the sea floor. The stratigraphic reversal sequence had been worked out during the 1960s and 1970s using piston cores from the deep sea and sections exposed on land, but some parts of the record were uncertain or incomplete. Unfortunately, during the earlier part of the DSDP, disturbance of soft sediment by the drilling and coring process rendered cores unsuitable for paleomagnetic investigation. The DSDP cores became much more useful for paleomagnetic studies after it was decided that all sections should be continuously cored and after the development of the hydraulic piston corer (HPC). The HPC is a hydraulic ram coring device which is used with the drill pipe and which can recover relatively undisturbed cores at sub-bottom depths to as much as 300 m. Since Leg 64 of the DSDP in 1978, it has been standard practice to continuously core the upper part of the section at each drill site with the HPC and to continuously core the deeper parts of the section with the standard coring device. During Legs 73 and 74 of the DSDP, a special effort was made to correlate magnetic stratigraphy with the calcareous plankton zonations and to document the magnetostratigraphic position of important paleontologic datum levels (Moore, Rabinowitz, and others, 1984; Hsü, LaBrecque, and others, 1984; Hsü and Weissert, 1985). The horizons of magnetic reversal are thought to be essentially instantaneous and global, and they can serve as precise references for paleoceanographic studies, but correlation in the intervening intervals is often limited by the uncertainties inherent in biostratigraphy.

Emiliani had envisioned the use of varying ratios of stable isotopes as a routine means of stratigraphic correlation, and although this was proposed at the inception of the DSDP, it required such large numbers of mass spectrometric measurements that it was considered impractical. By the middle 1970s, improvements in mass spectrometry reduced the sample size and time required for such measurements, and more recently stable-isotope stratigraphy has become almost routine. Studies of stable isotopic variation in pre-Pleistocene sediments have shown that there are significant signals which should be useful for stratigraphic correlation; although this technique has to date rarely been utilized, it has in some instances achieved spectacular results, permitting resolutions of 10,000–20,000 yr in the late Pleistocene, 40,000–50,000 yr in the early Pleistocene, and 100,000 yr in the Pliocene and Miocene (Williams and Trainor, 1986; Trainor and Williams, 1987). Integrated biostratigraphic, magnetostratigraphic, and isotopic stratigraphies for part of the Cenozoic have been prepared by Miller and others (1985a, 1988) and Keigwin (1987). At present a number of studies of both the stable isotopes and the variations in $CaCO_3$ (for example, Dean and Gardner, 1985) are underway, seeking a breakthrough that will permit Milankovich cycles to be recognized throughout the Mesozoic and Cenozoic, at last providing the temporal resolution required to relate paleoceanographic changes to modern processes.

THE DEVELOPMENT OF PALEOCEANOGRAPHY

Emiliani's pioneering work with stable isotopes of oxygen opened the field of paleoceanography in the 1950s. He had discovered that the deep waters of the Pacific had been much warmer during the Paleogene than they are at present. The DSDP confirmed the changes in the thermal structure of the ocean and added a number of new facets to paleoceanography, including clues to the circulation of the past oceans, such as fluctuations of the carbonate compensation depth; deposition of organic carbon-rich sediments over extensive areas; changes in biogeographic provinces; changes in oceanic productivity; and changes in ocean chemistry; as well as direct evidence bearing on the surface, mid-depth, and deep circulation of the ocean and the modes of vertical mixing.

CHANGES IN THE THERMAL STRUCTURE OF THE OCEAN

The ocean today is characterized by a thin (200–500 m), warm layer of tropical-subtropical water lying between the Antarctic Convergence in the south and the Arctic Convergence in the north. This warm-water "sphere," characterized by waters of relatively low density (<1.0270), with temperatures ranging from 12 to 30 °C and salinities from 34 to 37.5 per mil, overlies denser (>1.0270) intermediate and deep waters, most of which have a much narrower range of potential temperature (−1 to 5 °C) and salinity (34.4 to 35.0 per mil). Numerical paleotemperatures can be interpreted from ratios of the stable isotopes of oxygen (Emiliani, 1971) and indicate that the present thermal structure of the oceans is characteristic only of the late Cenozoic and Pleistocene-Holocene.

The long-term changes in the thermal structure of the ocean since the middle Cretaceous have been documented in a number of works and summarized by Savin and others (1975, 1985), Shackleton and Kennett (1975), Hecht (1976), Savin (1977, 1983), Keigwin (1979, 1980, 1982a, 1982b), Woodruff and others (1981), Savin and Yeh (1981), Shackleton and Boersma (1981), Saltzman and Barron (1982), Hodell and others (1983, 1985, 1986), Muza and others (1983), Shackleton (1983), Haq (1984), Hsü and others (1984), Loutit and others (1984), Shackleton and others (1984), Johnson (1985), Kennett (1985), McKenzie and Oberhänsli (1985), Murphy and Kennett (1985), Oberhänsli and Toumarkine (1985), Savin and Douglas (1985), Weissert and Oberhänsli (1985), Williams and others (1985), Frakes (1986), Keigwin and Corliss (1986), Miller and others (1987a), Barrera and others (1987), and Mix (1987). The temperature of oceanic surface and bottom waters is interpreted from the $^{18}O/^{16}O$ ratios in the calcite tests of selected species of planktonic and benthonic foraminifera and in the prismatic layer of shells of the Mesozoic pelecypod *Inoceramus* (Barron and others, 1982). General trends for the surface and deep ocean are illustrated in Figure 1. The heavy line shows the long-term average trend; for the Cenozoic, this was established by averaging a number of measurements from different sites over intervals of sub-epoch length; but for the Cretaceous, it represents a small number of single data points from deep-sea benthonic foraminifera, *Inoceramus* (triangles; data from Saltzman and Barron, 1982), and Antarctic shelf benthonic foraminifera (Barrera and others, 1987). More-detailed records from surface-dwelling planktonic foraminifera and abyssal benthonic foraminifera from the Angola Basin are also shown (after Shackleton, 1983). Three major states of thermal structure can be recognized: a state which prevailed during Cretaceous-Eocene time, with Cretaceous tropical surface water temperatures about the same as or warmer than today (27–32 °C), as suggested by Barron (1983a, 1983b, 1984, 1986), and slightly cooler tropical surface-water temperatures in the Eocene (23 °C), as indicated from the studies of Shackleton and Boersma (1981). High-latitude surface-water temperatures may have been as high as 17 °C or as low as 0 °C in the Cretaceous (Barron, 1986), but bottom-water temperatures ranged from as high as 23 °C in the Angola Basin during the Coniacian to 13 °C in the later Cretaceous (Saltzman and Barron, 1982; Barron and others, 1982) and were 10–12 °C in the Eocene (Shackleton and Boersma, 1981).

According to Barrera and others (1987), high-latitude Antarctic (Seymour Island) shelf waters had temperatures between 4 and 10.5 °C during late Campanian through Paleocene time.

A very different state began to develop in the early Oligocene. Frakes and Kemp (1972) suggested that ocean waters were warmer during the Eocene and cooler during the Oligocene. Kennett and Shackleton (1976) found oxygen isotopic evidence that bottom-water temperatures dropped sharply 38 Ma. Miller and Fairbanks (1983, 1985) confirmed that bottom-water temperatures decreased markedly near the Eocene-Oligocene boundary and noted that there were shorter periods when bottom-water temperatures became as cold as or colder than they are today (2 °C). They reported that the five coldest episodes were 36–35 Ma (early Oligocene), 31–28 Ma (middle Oligocene), 25–24 Ma (Oligocene/Miocene boundary), and starting at 15 Ma (middle Miocene), corresponding to episodes of glaciation in Antarctica. They suggested that these cold excursions coincide with 30- to 50-m glacio-isostatic falls of sea level. Keigwin and Corliss (1986) have analyzed $\delta^{18}O$ data from a global suite of tropical and mid-latitude locations. They found that the deep-sea benthonic foraminifera record major cooling steps at the middle/late Eocene boundary and near the Eocene/Oligocene boundary, but the oxygen isotopic record of planktonic foraminifera does not indicate an increase in planetary temperature gradient from the Eocene into the Oligocene. They concluded that the deep sea cooled in the early Oligocene by 2° at some localities and by 1° overall; they suggested that the Eocene/Oligocene cooling was associated with the development of continental glaciation. During most of Oligocene–early Miocene time, tropical surface-water temperatures were

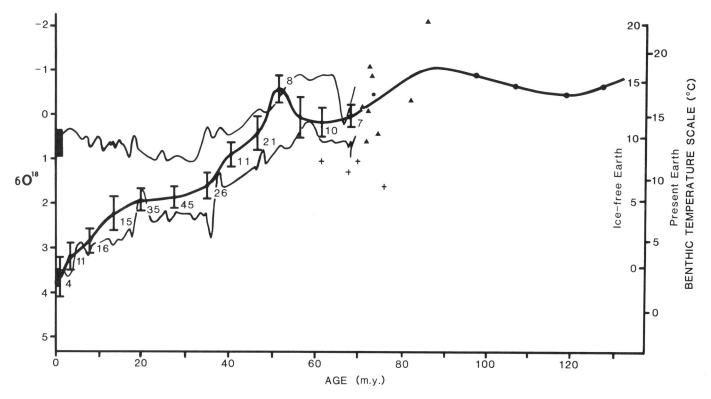

Figure 1. Cenozoic and Cretaceous ocean-surface and bottom-water temperatures indicated by oxygen-isotope ratios. Heavy line is an average of multiple measurements of benthonic foraminifera over subepoch or stage time intervals with error bars (two standard deviations) shown. Upper thin line is for surface planktonic foraminifera from the Angola Basin; lower thin line is from benthonic foraminifera from the Angola Basin (both after Shackleton, 1983). Filled circles are single data points based on benthonic foraminifera. Triangles are single data points based on *Inoceramus* (Saltzman and others, 1982; Barron and others, 1982). Crosses are data points for Antarctic (Seymour Island) shelf benthonic foraminifera (Barrera and others, 1987).

about 20 °C and deep-water temperatures about 3-7 °C. A third state has existed since the middle Miocene, with tropical surface-water temperatures about 25-29 °C and deep-water temperatures of 0-3 °C. The thermal contrast between tropical surface and deep waters during the Cretaceous was 10°-15 ° (Hay, 1983); in the Eocene, the thermal contrast was similar, about 13° (Shackleton and Boersma, 1981); in the Oligocene, the thermal contrast remained 13°-15° as both surface and bottom waters cooled (Brass and others, 1982); but from the middle Miocene to Holocene, the thermal contrast between tropical surface and deep waters has increased from about 20 °C to its present value of 25 °C as the deep waters remain cold and the surface waters warm. The significant drop in bottom-water temperatures that began near the end of the Eocene, between 40 and 35 Ma, marked the origin of the psychrosphere, the cold, constant deep-sea environment (Benson, 1975; Kennett and Shackleton, 1976; Benson and others, 1985).

Paleotemperatures have been interpreted from stable oxygen isotopes, but the conversion from the isotope ratio to temperature is not straightforward; an excellent review of the state of the art of interpretation has been presented by Mix (1987). Fractionation by vital effects between different species is eliminated insofar as possible by making the measurements on material from a single or closely related species. The interpretation of temperature, however, must also make assumptions about the isotopic composition of ocean water. At present, the ice on Greenland and Antarctica has a composition very different from that of sea water, being highly depleted in ^{18}O; Greenland ice ranging from -25 to -35 per mil, and Antarctic ice ranging to -60 per mil, whereas the range in the ocean is only from +1 to -0.5 per mil (Garlick, 1972). Even if the isotopic composition of the ice caps were exactly known, it would still be impossible to determine the exact isotopic composition of the ocean on an ice-free Earth because the isotopes are also fractionated in the carbonate-rock output from the ocean. Nevertheless, it is possible to estimate the over-all isotopic composition of glacial ice and establish an approximate temperature scale for an ice-free world. There is, however, no general agreement as to when (if ever) the world was ice free. It is widely assumed that the Earth was ice free during the Cretaceous. Savin and Douglas (1985) have interpreted the oxygen isotopic data as indicating that the Earth was essentially ice free during the Paleocene, Eocene, parts of the Oligocene, and during the early Miocene. Shackleton and others (1984), Savin and Douglas (1985), and Miller and others (1987) wrote that there have been significant ice caps on the Earth during parts of the Oligocene and from middle Miocene to present. Matthews and Poore (1980) have presented an alternative interpretation of the oxygen isotopic data, suggesting that the Earth has usually had significant polar ice.

The intermediate and deep waters of the ocean are formed by modification of surface waters to increase their density and cause them to sink as downwelling plumes. The formation of denser water can be the result of chilling by cold air, or of increase in salinity as a result of evaporation or sea-ice formation. The ites of formation of such dense water are geographical accidents; the surface water must be quasi-isolated from the main body of ocean water so that it can be strongly modified by local air-sea interaction. The differing thermal structure of the oceans prior to the middle Miocene suggests that intermediate- and bottom-water sources then may have been different from those of today.

Deeper water temperatures have generally been assumed to correspond to high-latitude surface temperatures, following the early ideas of Emiliani (1954, 1961) and assuming that the sites of deep-water formation have always been polar or subpolar. Peterson (1979), however, noted that the deep water of the ocean will be produced by the dense-water source or downwelling plume which has the highest buoyancy flux. The buoyancy flux is the product of the density difference and the volume flow of the plume per unit time. The downwelling plume having the highest buoyancy flux need not be polar or subpolar, driven by the density increase from chilling of the water or increased salinity from sea-ice formation, but may form in the subtropics from density increases as a result of increased salinity caused by evaporation. This is likely when mediterranean seas or shallow shelf seas are developed in the arid subtropics and are able to feed the world ocean. The polar or subpolar origin of deep water in the ocean is unequivocal only for the Oligocene-Holocene. Prior to the Oligocene, the sources of deep water may have been at lower latitudes, and the deep-water temperatures cannot be equated with high-latitude surface temperatures. Furthermore, for the pre-middle Miocene, it is questionable how much of the difference in oxygen-isotope ratios between near-surface-dwelling planktonic microfossils and benthonic microfossils and *Inoceramus* is due to temperature and how much is due to salinity differences between the surface and deep waters. Water evaporating from the sea surface is depleted in ^{18}O, so that as the salinity increases by 1 per mil, the oxygen isotopes shift by 0.1 to 0.6 per mil, depending on whether the isotopic composition of the evaporated vapor and the precipitation are the same, as in the tropics, or significantly different, as at higher latitudes. Unless taken into account, the shift in oxgyen isotopic composition as the salinity increases by 1 per mil would be interpreted as though the temperature had decreased by .5° to 2.7 °C, depending on the assumptions made about the isotopic composition of evaporated vapor and precipitation during the warm geological times. The question as to whether bottom waters are formed at high latitudes even during the times when the polar regions were much warmer than today, as suggested by Saltzman and Barron (1982) and by Barrera and others (1987), or were formed in the subtropics, as suggested by Brass and others (1982), remains to be resolved.

Shorter-term fluctuations of the oxygen-isotope ratios, in the order of thousands to hundreds of thousands of years, have been known since Emiliani's pioneering work of 1955. Lidz (1966) demonstrated that the variations in the oxygen-isotope record could be precisely matched by variations in assemblages of planktonic foraminifera, but the interpretation of these variations became a matter of controversy. Emiliani (1955, 1971) concluded that in the Caribbean and tropical Atlantic the change in oxygen-isotope ratio from a glacial to an interglacial was about 1.8 per mil, that the average composition of sea water during glacial ages was +0.5 per mil, and that the composition of glacial ice was -15 per mil. This would mean that about 30% of the isotopic signal is due to ice-volume changes from glacial to interglacial, and 70% due to changes in the water temperature. Dansgaard and Tauber (1969), on the basis of study of a Greenland ice core, suggested that the average composition of Pleistocene ice was probably -30 per mil; this would mean a glacial-interglacial change of 1.2 per mil and that 70% of the isotopic signal would be due to ice-volume changes. Shackleton (1967), assuming that bottom-water temperatures remained constant between glacials and interglacials, estimated the isotopic change to be 1.4 to 1.6 per mil; in this case, virtually all of the isotopic signal in tropical planktonic foraminifera is due to ice-volume changes. Shackleton and Opdyke (1973) noted that the isotopic variations in *Globigerinoides sacculifer*, the shallowest-dwelling planktonic foraminifer, in the equatorial Pacific are about 1.2 per mil between glacials and interglacials. They concluded that this is entirely a glacial ice-volume effect because of the similarity of their isotopic record to those established by Emiliani in the tropical Atlantic and Caribbean, because of the similarity of the curve with that for benthonic foraminifera and because of the good correlation of the curve with known sea-level variations. This estimate of the ice-volume effect (0.1 per mil for 10 m of sea-level change) was independent of the assumption that interglacial-glacial bottom-water temperatures did not change. Fairbanks and Matthews

(1978), from study of the isotopic composition of coral specimens from levels with known vertical distances from uplifted interglacial terraces on Barbados, were able to make an independent estimate that the oxygen isotopic composition of sea water varies between 0.10 and 0.12 per mil for 10 m of sea-level change. Shackleton (1986) has presented a recent review of interpretation of the history of the ocean from Pliocene-Pleistocene stable-isotope data.

Birchfield (1987) has made new determinations of the deep-water oxygen isotopic values for the Atlantic, Indian, and Pacific Oceans; he has concluded that the water isotopic change related to ice volume during the last glacial maximum is 74% of the total signal in the Atlantic, 78% in the Indian Ocean, and 84% in the Pacific Ocean. He also concluded that the temperature of deep water in the Atlantic was about 2° cooler during the glacial maximum than it is today; in the Pacific, the deep water was about 1.1° cooler; in the Indian Ocean, it was about 1.4° cooler. Labeyrie and others (1987), assuming that the temperature of bottom waters in the deep Pacific probably remained constant during glaciation, derived a mean global $\delta^{18}O$ record by piecing together the interglacial $\delta^{18}O$ record from a core from the Norwegian Sea with the glacial record of core V19-30 from the Pacific (corrected by the known difference in $\delta^{18}O$ for the part of the cores that overlap). The result was that they estimated that Atlantic deep waters were about +4 °C at 6 Ka and −1.8 °C at 25 Ka; the southern Indian Ocean deep waters were about +1.6 °C at 6 Ka and −1.8 °C at 25 Ka; the equatorial Pacific has been about +2 °C during the Holocene but was −1 °C at 8.4 Ka. They found that the interglacial/glacial shift is smaller ($1.1^0/_{00}$) than earlier estimates ($1.6^0/_{00}$).

FLUCTUATION OF THE CARBONATE COMPENSATION DEPTH

The CCD is the boundary between carbonate-rich sediments (for example, calcareous oozes) and carbonate-free sediments (for example, red clay) and reflects an excess of dissolution demand over supply of pelagic carbonate to the deep sea. The CCD in the modern ocean was first described in detail by Schott (1935), on the basis of studies of transects across the South Atlantic by the Meteor Expedition of 1925-1927. The CCD is such an obvious boundary on the sea floor that it is inevitable that it should be considered one of the fundamental clues to the manner in which the ocean processes material.

Bramlette (1958) had noted that in the central Pacific calcareous oozes underlie red clays in depths well below the present CCD. Riedel and Funnell (1964) documented additional cores in which carbonate oozes lie beneath red clay. Heath (1969) suggested that the CCD had descended to depths greater than 5,000 m during the Oligocene and early Miocene, because cores with calcareous oozes of that age had been recovered from abyssal depths in the equatorial Pacific. Both Bramlette and Heath cited the higher abyssal temperatures of the middle Cenozoic as a possible cause for the apparently deeper CCD.

It was discovered on DSDP Leg 2 (Peterson and others, 1970) that there had been significant fluctuations of the calcite compensation depth in the ocean. It was expected that red clays might overlie calcareous oozes as was known from the Pacific, but on Leg 2 it was found that in the North Atlantic Pleistocene and Pliocene calcareous foraminifer-nannoplankton oozes overlie Miocene red clay, which in turn overlies older calcareous oozes, documenting that the CCD had moved both up and down. In the Initial Report for DSDP Leg 3, Hsü and Andrews (1970) followed the common knowledge of the day in assuming that the dissolution of carbonate and the level of compensation depth were determined by pressure. They concluded that the >1-km fluctuations of the CCD documented in the South Atlantic by DSDP Leg 3 were a direct reflection of changes in the depth of the South Atlantic ocean. Because these apparent depth changes did not correlate with known sea-level changes, they suggested that there must have been widespread vertical motions of the ocean floor.

The first compilations of the fluctuations of the CCD through time were by Peterson and others (1970) and Hay (1970), but their work antedated the publication of age-depth curves for the subsidence of ocean crust (Sclater and others, 1971), so that they were unable to determine the true magnitude of fluctuations of the CCD. More sophisticated compilations, taking sea-floor subsidence into account, were published by Berger (1972, 1973), van Andel and Moore (1974), Berger and Winterer (1974), van Andel and others (1975, 1977), Berger and Roth (1975), van Andel (1975), Heath and others (1977b), Le Pichon and others (1978), Melguen (1978), Thierstein (1979), Tucholke and Vogt (1979), Hsü and others (1984), Dean and others (1984), Moore and others (1985), and Hsü and Wright (1985). The effects of the fluctuations on sedimentary lithofacies are shown by the maps of McCoy and Zimmermann (1977). Figure 2 is a recent compilation of data on fluctuations of the CCD in different parts of the ocean. It shows that the CCD has fluctuated by more than 2 km in some parts of the ocean. It is also clear that the CCD behaved differently in different ocean basins during the Cretaceous but began to change in the same way during the Eocene, although the magnitude of the fluctuations differs from basin to basin. Individual basins, notably the Angola Basin, shown in Figure 2, and the semi-isolated Venezuelan Basin (Hay, 1985b) may show CCD fluctuations significantly different from the major ocean basins proper. Although fluctuation of the CCD has become a well-documented phenomenon, its causes are still not well understood.

The term "lysocline" was originally coined by Berger (1970) to indicate the depth zone through which the tests of planktonic foraminifera become dissolved. Broecker and Peng (1982) have noted that this is the depth at which the effects of dissolution become obvious in the sediment. According to Berger and others (1982), the lysocline occurs where about 20% of the carbonate has been dissolved; it is only slightly less obvious than the CCD where dissolution reaches 100%. The lysocline usually marks a transition from *Globigerina* ooze to nannofossil ("coccolith") ooze. Although most writers continue to use the term lysocline to refer to the zone of dissolution of planktonic foraminifera (for example, Berger, 1985), Kennett (1982) applied it in a general sense for any zone in which dissolution of microfossils occurs, that is, foraminiferal lysocline, pteropod lysocline, and coccolith lysocline. Because coccoliths and some other calcareous nannofossils are more resistant to dissolution than most other calcareous bioliths in the sea, the coccolith lysocline immediately overlies the CCD. Although the foraminiferal lysocline tracks the temporal fluctuations of the CCD, the vertical distance between the two does not remain constant, indicating that the slope of the dissolution gradient also changes with time, as was noted by van Andel and others (1975).

Peterson (1966) presented results of an experiment on dissolution of calcite spheres suspended in the Pacific, noting that the rate of dissolution increased rapidly below the top of the cold, CO_2-rich Antarctic Bottom Water mass (AABW). Berger (1967) related this to his studies of dissolution of planktonic foraminifera on the sea floor, demonstrating that the level of their rapid dissolution (the lysocline) also coincides with the top of the AABW. Considering the implications for the circulation of the ocean in the past, Berger (1972) attributed the high CCD prior to the late Miocene to more intense production of Antarctic bottom water. During the late Miocene and Pliocene, there was increasingly more significant production of North Atlantic Deep Water (NADW). He speculated that NADW replaced and mixed with the AABW to lower the CCD to its present level. Although this is attractive as an explanation for the Miocene-Holocene behavior of the CCD, it cannot be extrapolated into the Paleogene because the drop in the CCD from the Eocene into the Oligocene is

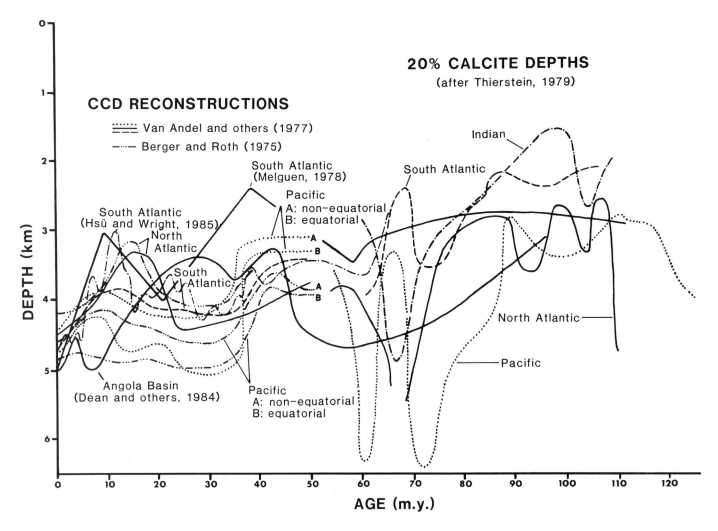

Figure 2. Fluctuations of the calcite compensation depth (CCD; 0% carbonate) in different ocean basins, showing complications by several investigators. For the Cretaceous and Paleocene, the lines represent 20% carbonate content.

accompanied by a marked lowering of bottom-water temperatures, as interpreted from the stable-oxygen-isotope ratios in benthonic foraminifera (Kennett and Shackleton, 1976). Re-evaluating the evidence, Edmond (1974) suggested that the lysocline in the Pacific may be related to increased deep-water current velocities. Adding to the uncertainty of the underlying cause of the development of the lysocline, recent studies of calcite dissolution in the Atlantic by Thunell (1982) have shown that in the western Atlantic the NADW/AABW boundary, as defined by a deep thermocline, coincides with the transition from saturation to undersaturation and with the lysocline; in the eastern basins of the Atlantic, however, no such deep thermocline occurs. Although there is no well-defined boundary between NADW and AABW, the lysocline is sharply developed but is significantly deeper than the saturation/undersaturation transition.

Milliman (1974), Berger and Winterer (1974), and Hay and Southam (1977) discussed why fluctuations of the CCD should be expected to follow sea level: higher sea levels would flood a larger shelf area and increase rates of sedimentation of shallow-water carbonates on the shelves; this would decrease the supply of carbonate to the deep sea, and without any concomitant decrease in dissolution demand, the CCD must rise. Verification of the sea-level control hypothesis was possible after the publication of the sea-level curves of Vail and others (1977). The first- and second-order changes in sea level described by Vail and others do seem to be well correlated with the over-all trends for the CCD, at least for most of the Cenozoic (Davies and Worsley, 1981). Hay and others (1982, 1984) have even suggested that Eocene and younger third-order cycles of Vail and others (1977) may be recorded as CCD fluctuations in the sediments of the Angola Basin.

The study of the distribution of carbonates in the Mesozoic and early Cenozoic has demonstrated that, until the Eocene, fluctuations of the CCD in different ocean basins were not synchronous and equal (see Fig. 2). The major ocean basins behaved differently, suggesting that there were local competing deep-water sources with different chemical characteristics; in particular, at the end of the Mesozoic, the CCD sank in the Atlantic and rose in the Pacific (see Fig. 2). This has been attributed to fractionation of carbonate between ocean basins, in much the same way that a balance between equatorial and extra-equatorial compensation has been postulated for the Pacific.

It is now clear that dissolution of different forms of calcite and aragonite may occur throughout the entire water column. Between the lysocline and the CCD, the sediments consist of calcareous nannoplankton which are relatively more resistant to dissolution than are the tests of planktonic foraminifera. As Broecker and Peng (1982) have noted, the percentage of $CaCO_3$ in the sediment is a deceptive index of dissolution because the rain rate of noncarbonate sediment is usually a small fraction of the carbonate rain rate. As a result, the proportion of carbonate in the sediment does not change appreciably until most of the carbonate has been dissolved.

In retrospect, it has become evident that calcium-carbonate deposition in the ocean is a highly complex phenomenon which is influenced by a number of variable factors (Broecker and Peng, 1982). The supply of calcium carbonate to the ocean may vary with time (Budyko and Ronov, 1979; Southam and Hay, 1977, 1981; Hay, 1985a; Budyko and others, 1985); the supply from subaerial erosion, the supply of Ca^{++} from submarine weathering of mid-ocean–ridge basalts (Honnorez, 1983; Mottl, 1983), and the supply from submarine dissolution of older sediments (Moore and Heath, 1977; Moore and others, 1978; Thierstein, 1979; Thiede, 1979, 1981; Thiede and Ehrmann, 1986) are all subject to change with time and may even be interrelated in complex ways. If the model of Berner and others (1983) is correct, then an increase in the rate of sea-floor spreading would result in increased submarine weathering of mid-ocean–ridge basalts, releasing more Ca^{++}; it would also result in more volcanic activity and increased input of CO_2 into the atmosphere and ocean, as had been suggested by Budyko and Ronov (1979) and discussed more thoroughly by Budyko and others (1985). Because the CO_2 content of soils is already high with respect to that of the atmosphere, the increased CO_2 content of the atmosphere would not itself directly affect the rate of weathering of rocks on land, as suggested by Budyko and Ronov (1979) and Budyko and others (1985), but would result in general warming of the Earth which would in turn increase the rates of chemical reactions and cause subaerial weathering rates to increase (Berner and others, 1983; Lasaga and others, 1985). Evidence that the CO_2 content of the atmosphere has indeed changed has become available in recent years, first from ice cores which demonstrated that during the last glacial atmospheric CO_2 content was one-half to one-third that of the present (Delmas and others, 1980; Neftel and others, 1982), and subsequently from deep-sea cores (Shackleton and others, 1983). It has also been suggested that the changing CO_2 content of the atmosphere affects dissolution directly by increasing the total CO_2 in the ocean and by causing changes in oceanic circulation and productivity (Vincent and Berger, 1985; Berger, 1985; Boyle, 1986b; Berger and Mayer, 1987).

The partitioning of carbonate deposition between the continental shelves, marginal seas, and the ocean basins may be highly significant and is probably the first-order control on the CCD (Milliman, 1974; Berger and Winterer, 1974; Hay and Southam, 1977). The biological fixation processes for $CaCO_3$ may vary with time; fixation of calcium carbonate as calcite by marine plankton appears to have been significant only during the past 100–150 m.y. (Poldervaart, 1955; Hay and Southam, 1975; Sibley and Vogel, 1976). The apportionment between aragonite and calcite, two phases of $CaCO_3$ with different solubilities (Sandberg, 1983, 1985) may have varied with time. The apportionment between benthonic and planktonic fixers and even between biologic and abiologic fixation may also have varied with time. Calcite and aragonite are secreted by marine organisms in different ways, and as a result, the mineral phases produced by different taxa may have significantly different solubilities. The calcite spicules secreted by soft corals (gorgonians, plexaurids, and so on) dissolve in waters which are supersaturated with respect to calcite; at the other extreme, minute calcite coccoliths are more resistant to dissolution than are the much larger tests of planktonic foraminifera. The rate of fixation by any of these processes may vary with changes in the composition of sea water and because biologic agents are involved with the availability of nutrients. Changes in the rate of biological fixation in the surface waters has been invoked to explain the rapid fluctuations of the CCD between the last glacial and the Holocene. The mechanism of transport of the calcareous particles to the deep-sea floor (Honjo, 1980), whether incorporated in fecal pellets or as discrete particles, may vary with time. The flux of carbonate produced in shallow water and carried to the deep sea after winnowing and transport by turbidity currents may be significant at some times. The apportionment of C between the organic carbon and $CaCO_3$ reservoirs may be a significant variable. At present about one-fifth of the carbon leaves the ocean as organic carbon and four-fifths as $CaCO_3$ (Broecker and Peng, 1982), but variations have occurred in the past. Hay's (1985) analysis of Ronov's (1980) and DSDP data indicates that the ratio of C_{org} to $CaCO_3$ is one-seventh for the Cenozoic and one-sixth for the Mesozoic; for the Paleozoic rocks still preserved on the continents, however, the ratio is less than one-tenth C_{org} to nine-tenths $CaCO_3$. The preservation of $CaCO_3$ depends on its not being dissolved or subsequently eroded. Factors which affect dissolution of carbonate significantly as it descends through the water column or at the sediment surface have been discussed by Li and others (1969), Heath and Culberson (1970), Takahashi and Broecker (1977), Sclater and others (1979), and Broecker and Peng (1982); these factors include alkalinity and CO_2 content of the waters, temperature, pressure, and the velocity and turbulence of the water over the sediment-water interface (Edmond, 1974). Dissolution may occur within the sediment due to internal production of CO_2 as a result of oxidation of organic matter (Emerson and Bender, 1981), and even the rate at which burrowing organisms rework the sediment may be significant (Schink and Guinasso, 1977). Recently Delaney and Boyle (1988) have shown that information on the fluctuations of the CCD can be combined with information on the mean carbon isotopic ratio, the strontium isotopic ratio, the lithium-to-calcium ratio and the strontium-to-calcium ratio to assess possible changes in continental weathering rates and fluxes to the ocean, changes in the isotopic composition of materials being weathered on the continents, amount of carbonate deposited on shelves as opposed to in the deep sea, changes in the deposition or weathering rates of sediments, and changes in hydrothermal circulation. They concluded that, during the Tertiary, hydrothermal fluxes have been constant or have decreased, continental weathering and continent-to-ocean fluxes may have increased, and the proportion of carbonate deposited on the shelves has decreased. In spite of this great complex of factors which may affect $CaCO_3$ deposition in the deep sea, I believe the primary long-term control to be sea-level change and that the secondary controls are related to water-mass formation.

SEDIMENTS WITH HIGH ORGANIC CARBON CONTENT

Prior to the DSDP, young deep-sea sediments with high organic carbon (C_{org}) content were known from the Black Sea (Ross and Degens, 1974) and eastern Mediterranean, and ancient sediments with high C_{org} were known from the western North Atlantic (Windisch and others, 1968; Habib, 1968). Although more sediments with high C_{org} content were recovered in the western Atlantic during Leg 1 of the DSDP (Ewing and others, 1969), the notion that any significant area of the deep sea could become anaerobic was rejected by ocean geochemists (Broecker, 1969). It was not until extensive drilling had been carried out that a pattern of widespread deposition of organic carbon in oceanic basins began to emerge (Fig. 3). "Black shales" have been found in the Cretaceous of all ocean basins; however, the term is used in a manner somewhat different from the term "black shales" on land. In the deep sea, the black shales are thin, C_{org}-enriched layers (millimeters to several tens of centimeters) incorporated in C_{org}-poor sequences of green, brown, or red clays or chalks. The black shales of the deep sea never exceed tens of centimeters in thickness, whereas the Chattanooga Shale, the black shales of the Pennsylvanian cyclothems, and the contemporaneous black shales of the

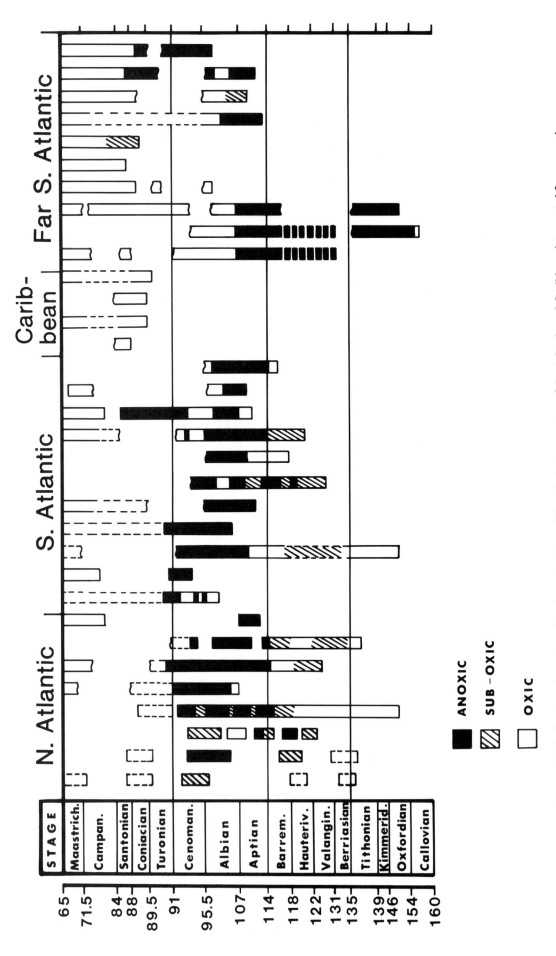

Figure 3. Distribution of anoxic, sub-oxic, and oxic conditions in bottom waters of the Atlantic and Caribbean interpreted from organic carbon content and sedimentary features of the sediments (after Krashenninikov and Basov, 1985).

Cretaceous western interior seaway of North America are usually meters to tens of meters thick.

As shown in Figure 3, the most extensive and intensive episode of deposition of C_{org}-rich sediments in the ocean basins occurred during the Aptian and early Albian (110–100 Ma) in the South Atlantic, which at that time was a long narrow sea connected with the world ocean only at its southern end, through the Cape and Argentine Basins. High organic carbon contents (up to 21%) have been found in sediments recovered from the Angola and Cape Basins (Bolli and others, 1978; Hay and others, 1984), and black shale layers extend south to the Falkland Plateau (Barker, Dalziel, and others, 1977; Ludwig, Krasheninnikov, and others, 1983). In the Angola Basin, the C_{org}-enriched sediments overlying an evaporite sequence of early Aptian age that was deposited as ocean crust appeared during the opening of the South Atlantic. The paleodepth range through which this Aptian–early Albian organic carbon enrichment occurs is 2.2 km, from the basin bottom at 3 km to about 0.8 km where the highest C_{org} content occurred. It is thought that this episode involved development of an extensive oxygen minimum throughout the entire Angola-Brazil basin complex up to the level of the sills (Walvis Ridge and Rio Grande Rise) connecting it with the Cape-Argentine Basins (Arthur and Natland, 1979). On the Maurice Ewing Bank of the Falkland Plateau, the black shales are more massive than are those from the basins proper and are of Late Jurassic, possibly Neocomian and Aptian–early Albian age. They were formed in waters of outer-shelf, upper-rise depths (Basov and Krasheninnikov, 1983). In the later Cretaceous, another extensive episode of deposition of C_{org}-rich sediments occurred during the Turonian–early Santonian in mid-depth waters of the Angola Basin, extending west to the Sao Paulo Plateau between the Brazil and Argentine Basins (Perch-Nielsen and others, 1977). Again, sediments from the deeper parts of the Angola Basin have a lower C_{org} content than do those at shallower depths, suggesting mid-depth intensification of the oxygen-minimum zone. Detailed examination of the middle Cretaceous organic-enriched sequence on DSDP Leg 75 (Hay and others, 1984) showed that the age range for all of the black-shale layers extends from late Albian to early Santonian, with the greatest abundance of C_{org}-enriched layers straddling the Cenomanian-Turonian boundary. At Site 530, where the stratigraphic section and occurrence of black shales is typical of both the South and North Atlantic, there are 262 beds of black shale in the middle Cretaceous, with C_{org} content up to 19%. The black-shale layers range from 1 to 62 cm in thickness, averaging 4.3 cm. They are separated by green or red shales which are virtually devoid of organic carbon. The C_{org} content of the black shales was found to be inversely related to the thickness of the enriched layer. Calculations show that the C_{org} content of each layer is approximately that which would be expected if (1) the deep basin waters of that time had had a nutrient concentration comparable to that of the modern deep sea, (2) the entire water column with its nutrients had been cycled through the photic zone, and (3) all of the nutrients had been fixed with organic carbon in the Redfield ratio characteristic of modern plankton. The average C_{org} content of the entire sequence is less than 0.2%, so that there is no over-all enrichment of C_{org} over average marine sediment, but there is instead a sharp separation into enriched and depleted layers. The black layers, with high C_{org} contents, are often sandwiched between green shales, and green-shale layers are often sandwiched between red-shale layers. Dean and others (1984) concluded that it was impossible to be certain that the bottom waters were anoxic at the time of deposition of all of the black-shale beds. About 80% of the black-shale layers were laminated, but about 40% of the layers were also bioturbated to some extent (Stow and Dean, 1984). Although most of the organic matter in the black shales was of marine origin (Deroo and others, 1984), the entire sequence containing the black shales included a number of fine-grained turbidites (Degens and others, 1986). It was not possible, therefore, to determine an exclusive mode of origin for the organic matter, nor was it possible to exclude the possibility that anaerobic conditions existed only in the bottom sediments themselves and not in the overlying water column (Dean and others, 1984).

The eastern and western North Atlantic behaved differently. In the North Atlantic, black shales occur in Aptian and early Albian strata throughout the depth of the basins, but even the layers most enriched in C_{org} have relatively low contents (up to 4%). The greatest C_{org} enrichments occur in late Albian–Turonian sediments and are restricted to the deepest part of the basins. In the western North Atlantic, moderate C_{org} contents occur in sediments of Hauterivian to Aptian age and C_{org} contents are greatest in mid-depths. Samples with up to 11% have been recovered from Cenomanian-Turonian strata. In the North Atlantic, however, organic matter in sediments from the western basins is mostly terrigenous, with woody fragments a common component, whereas organic matter in the black shales of the eastern basins is dominated by material of marine origin.

Organic carbon–rich sediments also occur in the Venezuelan Basin of the Caribbean, although except at Site 146/149 the C_{org}-enriched layers were overlooked by the DSDP shipboard scientific party (Hay, 1985b). At Sites 146/149, 150, and 153, high organic carbon content occurs in beds of Coniacian/Santonian age presently at depths of about 4,700 m; The C_{org} contents range up to 11.2%. The location and paleodepth of the floor of the Venezuelan Basin in the Cretaceous are uncertain, but because sites 146/149, 150, and 153 are presently in the deeper part of the basin, it seems likely that they are due to brief pulses of warm saline bottom water which produced anoxia beneath a deep pycnocline. Organic carbon–rich layers also occur at Site 151, which is shallower, but are late Santonian; here they may be due to intensification of a mid-depth oxygen minimum.

No deposits containing significantly more than 1% C_{org} have been found in the Indian Ocean except for Aptian-Albian samples which contain contents up to nearly 4%.

In the Pacific, significant C_{org} contents have been found in isolated samples from middle to shallow depths in Barremian strata (9%, 1.8-km paleodepth, Site 306) and in Barremian-Aptian strata (28%, near surface paleodepth, Site 317); these also appear to represent intensification of the oxygen minimum.

Although the correlation of individual layers of black shale is almost always questionable, there were times at which the deposition of C_{org}-rich layers was favored on a regional or global basis. The Aptian–early Albian was a time of global tendency toward black-shale deposition, and the interval of the Cenomanian-Turonian boundary is exceptional both because it was a time when there was again a global tendency to black-shale deposition and because the boundary itself records a moment when black-shale deposition occurred simultaneously over a very wide area (Schlanger and Jenkyns, 1976; Graciansky and others, 1984; Arthur and others, 1985; Herbin and others, 1986). Known as the "Bonarelli bed" in the southern Alps, the Cenomanian-Turonian–boundary bed is a black-shale layer which can be specifically identified by a carbon-isotope signature (Scholle and Arthur, 1980; Schlanger and others, 1986), suggesting a global episode of carbon burial (Arthur and others, 1985).

Interpretation of the meaning of C_{org} enrichment of sediments is more complex than merely the result of salinity stratification (by analogy with the modern-day Black Sea) or expansion and intensification of the oxygen minimum, as had been assumed (Schlanger and Jenkyns, 1976; Ryan and Cita, 1977; Fischer and Arthur, 1977; Thiede and van Andel, 1977; Arthur and Schlanger, 1979). In the simplest scenario, it was assumed that the rate of fixation of carbon in the surface waters, and hence the supply of carbon to the depths, remained constant, but the supply of oxygen is retarded by either salinity stratification or reduced rates of vertical mixing in the ocean, resulting in anoxic conditions. Alternatively, the supply of oxygen to the deeper waters could remain constant and the productivity of the surface waters might increase, again resulting in the development of anoxic conditions. The dissolved-oxygen content of intermediate and deep waters of the ocean reflects a dynamic balance between supply of oxygen to the surface waters by diffusion from the atmosphere, oxygen production within the surface waters as a result of photosynthesis, oxygen demand in the decomposition of settling organic matter, and advection of cold and/or saline former surface waters. Excellent discussions of these factors have been presented by Wyrtki (1962) and Broecker and Peng (1982).

Tissot and others (1979), Herbin and Deroo (1982), and Meyers and others (1986), however, have noted that much of the organic material in Cretaceous strata of the Atlantic, especially in the western basins, is of terrigenous origin and consists of forms of kerogen and charcoal, which are less readily oxidized than organic matter of marine origin and upon maturation is likely to produce gaseous hydrocarbons. The organic matter of the eastern Atlantic basins is mostly from marine plankton, and upon maturation, it is more likely to produce liquid hydrocarbons. Because the sources of the C_{org} which becomes incorporated into the sediment are so different, it is difficult to give credence to a scenario which requires a constant supply of C_{org}. Rather, the accumulation of C_{org} in sediments on the sea floor is a function of many of the same factors which affect the accumulation of $CaCO_3$: (1) variations in the supply of C from older rocks or volcanoes; (2) partitioning of C between the C_{org} and $CaCO_3$ reservoirs; (3) variations in the rate of fixation of C_{org}, which is complex because fixation in significant quantities may occur on land as well as in shallow seas, marginal basins, and the deep sea; (4) the oxidation profile in the water column; (5) the oxidation capacity of the bottom and interstitial waters; and (6) the degree to which the sediments are worked by burrowing organisms. In spite of this complex of factors which may affect burial of C_{org}, its ultimate preservation in the sediment appears to be a nonlinear function of the sedimentation rate (Heath and others, 1977a; Mueller and Suess, 1979; Southam and Hay, 1981) such that as the sedimentation rate increases, the rate of C_{org} burial and preservation increases rapidly. Although it had been considered that the overriding control on preservation might simply be the rapidity of burial, more recent studies (Emerson, 1985) suggest that the increase in amount of C_{org} buried with increasing sedimentation rate is a result of a concomitant increase in the rain rates of both the C_{org} and other sedimentary material. Lyle and others (1988) consider the rain rate of C_{org} to be the most significant factor controlling the C_{org} content of the sediments, and they noted that fluctuations are related to productivity on a regional scale.

It had been thought that anaerobic conditions in the bottom waters were indicated by the finely stratified or laminated sediments which are evidence of the absence of benthic burrowing organisms, but recently it has been discovered that in the basins off southern California mats of sulfide-oxidizing bacteria may occur at the sediment water interface, so that anaerobic sediments can be deposited beneath moving aerobic waters (Williams and Reimers, 1982).

Evidence that anaerobic conditions are due to stagnation induced by hypersaline conditions has been suggested, but it is less easy to document. Fischer and Arthur (1977) proposed that the accumulations of C_{org} were the result of general oceanic stagnation resulting from sluggish oceanic circulation during "polytaxic" times, that is, episodes when the diversity of oceanic species was high. Arthur and Natland (1979) favored stable stratification as a result of sinking of dense saline waters produced in shelf seas or restricted low-latitude ocean basins. Drawing analogies with studies of saline lakes by Hay (1966), Natland (1978) had argued that the presence of phillipsite with authigenic kaolinite and illite at Site 361 in the Cape Basin and Site 364 in the Angola Basin strongly suggested elevated salinities. Brass and others (1982a, 1982b) suggested that during the Cretaceous the bottom waters of the ocean may have been warm saline waters, rather than cold waters as at present. Warm saline bottom (or intermediate) waters would be formed in marginal seas in the arid zones through evaporation. Because the temperature of the water at the site of formation is likely to be very warm, and because the solubility of oxygen decreases both as the temperature and as the salinity of the water increases (Weiss, 1970), dense warm saline waters would carry only about half as much oxygen to intermediate or deep levels in the ocean. Further, the basins of young opening oceans are often relatively isolated by shallow intervening ridges, and during the Cretaceous, there may have been many local sources of intermediate and deep waters.

C_{org}-rich sediments are not confined only to the Cretaceous, but are also especially common in the Pleistocene-Holocene deposits of several marginal seas, notably the Black Sea (Ryan, Hsü, and others, 1973; Degens and Ross, 1974), the eastern Mediterranean (Thunell, 1969; Thunell and others, 1977, 1983, 1984; Cita and Grignani, 1982; Thunell and Williams, 1983; Ganssen and Troelstra, 1987; van Hinte and others, 1987), Sulu Sea (Linsley and others, 1986), Sea of Japan (to be investigated on a future Leg of the Ocean Drilling Program), Cariaco Trench (Edgar, Saunders, and others, 1973), in the basins off southern California and in small basins on the northern slope of the Gulf of Mexico (Shokes and others, 1977; Trabant and Presley, 1978; Leventer and others, 1983). In these cases, anoxic conditions in the bottom waters have been caused by stratification of the water column, most often associated with the effects of lowered sea level, large volumes of glacial runoff during the deglaciations, and with strengthened monsoons supplying large quantities of fresh water to the region.

PRODUCTIVITY

A fundamental question which has not been satisfactorily answered is whether organic productivity in the ocean waters during the Cretaceous was significantly different from that of today. Bralower and Thierstein (1984), making use of the relationships between productivity, sedimentation rate, and C_{org} preservation determined for young sediments by Mueller and Suess (1979) and other data on Holocene sediments, have estimated that primary productivity in the ocean during the middle Cretaceous was an order of magnitude lower than it is today. De Boer (1986) also concluded that black-shale deposition was a result of decreased circulation but emphasized the importance of nutrients from land. In contrast, modeling exercises by Southam and others (1982) have indicated that because the oxygen content of the intermediate and deep waters and the supply of nutrients to the surface ocean are both linked to the vertical circulation of the ocean, development of extensive anoxia should occur when the thermohaline (or halothermal) circulation is rapid. This is not in conflict with the conclusion of Bralower and Thierstein that Cretaceous productivity was much lower than that at present, but it would indicate that turnover in the Cretaceous ocean may have been more rapid than at present.

The most obvious indicator of productivity in the ocean is the occurrence of biogenic silica (Lisitzin, 1972, 1974; Berger, 1976; Molina-Cruz and Price, 1977; Barron and Whitman, 1981; Broecker and Peng, 1982; Monin and Lisitzin, 1983; Leinen and others, 1986). At present biogenic silica accumulates in a belt around the Antarctic continent, in the equatorial Pacific, in the western Bering Sea, and in the Gulf of California (Calverg, 1966, 1968) and Sea of Okhotsk (Bezrukov, 1955), as well as in other upwelling regions. In each case, upward mixing of nutrient-rich waters containing dissolved silica results in high productivity in the surface waters, and a high proportion of silica-secreting plankton, such as diatoms and radiolarians, in the upwelling communities. The Gulf of California is a site of very high output of biogenic silica because of the year-around upwelling which shifts from one side of the Gulf to the other with the seasonal changes in wind direction. The Gulf of California and Sea of Okhotsk draw on the silica supply of Pacific intermediate waters and are, effectively, global sinks, with output rates per unit area two orders of magnitude greater than in the circum-Antarctic (Heath, 1974). The analysis of opal deposition by Miskell and others (1985) suggests that accumulation rates were much lower in the past than DeMaster's (1981) estimate of present-day inputs and output. Even taking the loss of sea floor through subduction into account, the opal accumulation on ocean floor investigated by the DSDP is much less than expected. It seems likely that in the geologic past a significant part of the opal output of the ocean was sequestered in small regions with intense upwelling, and that these remain unknown.

Recently (for example, Pokras, 1987) the recognition of particular species of diatom, such as *Ethmodiscus rex*, as productivity indicators has shown great promise toward development of an index of productivity, but a means of quantifying productivity with diatoms remains to be defined.

Ramsay (1974) had suggested that fluctuations of the $CaCO_3$ compensation depth might reflect changes in oceanic productivity, but it now seems more likely that the overriding control on calcite dissolution is changes in sea level; in any case, there are so many factors which can influence carbonate dissolution in the deep sea that the possibility of using the accumulation rates of carbonate sediments as indices of productivity has not been further explored as yet.

A recent study by Lyle and others (1988), comparing sedimentation of different biogenic components, confirmed the idea that the distribution of $CaCO_3$ is controlled by dissolution processes but found that the distribution of C_{org} and opal reflect changes in productivity. They noted, however, that the distribution of fluctuations in C_{org} can be traced over wide areas, whereas the fluctuations in opal-accumulation rates are highly variable.

Moody and others (1981, 1988) have pioneered the study of output of phosphorus into deep-sea sediments as an index for productivity. Although there are complications with partitioning of the output of phosphorus on shelves versus in the deep sea and with the addition of phosphorus from eolian dust transport, their studies indicate that significant increases in the intensity of upwelling and productivity are associated with the build-up of the Antarctic ice sheet and with northern hemisphere glaciation. Upwelling increased as a result of higher wind stress on the sea surface as the latitudinal thermal gradient developed.

The use of shifts in $\delta^{13}C$ as an index of global productivity is being developed, but it suffers from the fact that there is no unique cause for such shifts (Berger and Vincent, 1986; Berger and Mayer, 1987). In specific instances, such as at the Cenomanian/Turonian boundary (Scholle and Arthur, 1980; Schlanger and others, 1986) and in the middle Miocene (Vincent and Berger, 1985), the shift of $\delta^{13}C$ toward heavier values is clearly related to significantly increased organic carbon burial in the ocean in response to episodes of high productivity.

Recently Sarnthein and others (1987) have proposed a formula for estimating paleoproductivity based on the relationships between carbon-accumulation rate, water depth, and carbon-free bulk sedimentation rates calibrated with $\delta^{13}C$ in a benthic foraminifer. Using this method, they concluded that glacial productivity off northwestern Africa was three times greater than interglacial productivity.

The occurrence of barium in deep-sea sediments has been cited as having great potential as an index of paleoproductivity (Bostrom and others, 1979). It is thought that it may be a better measure of surface-water productivity than is silica, because barite is relatively insoluble, and the amount buried in the sediment is a close approximation of the supply to the sea floor. There are complications because of the introduction of particulate barite into the marine environment from terrigenous sources. Its occurrence in sediments in the Indian Ocean has been used by Schmitz (1987) to locate the belt of equatorial high productivity and to determine rates of plate motion, but a quantitative formulation of barium content as a measure of surface-water productivity has not yet been established.

A new line of investigation of ocean productivity has opened with the combined use of Cd/Ca and $\delta^{13}C$ ratios (Boyle, 1986a, 1986b) which may lead to applications in evaluating productivity; this technique can be used to evaluate (1) the relative amount of phosphorus entering the deep sea as organic particulates as a result of fixation in the surface waters and (2) the amount of phosphorus entering as preformed nutrient, that is, as dissolved phosphorus present in a water mass at the time it sinks to the deep ocean. This work has particular implications for ocean circulation and atmospheric climate and CO_2 content during the Pleistocene.

Recently, models of atmospheric circulation have been applied to ancient ocean configurations to attempt to identify ancient upwelling sites. Parrish (1982) and Parrish and Curtis (1982) assumed ancient atmospheric circulation systems to be analogous to that of today, so that predicted sites of upwelling in the past are located mostly on the western margins of landmasses in the subtropics. Barron (1985) has found that numerical climate models produce significantly different results for regions outside the subtropics; his "most realistic" Cretaceous simulation suggested upwelling around the margins of the Arctic Ocean. Parrish and Barron (1986) have recently summarized the state of knowledge on paleoclimates and economic deposits, including an extensive discussion of ancient upwelling systems. Unfortunately, many of the continental margin regions which would be critically important to understanding coastal upwelling are poorly known at present.

HIATUSES

Saito and others (1974) reported that the Lamont-Doherty core library contained more than 900 pre-Quaternary cores and dredgings, attesting to the fact that hiatuses are widespread in the deep sea. After the DSDP had been operational for a number of years and a large number of stratigraphic sections sampled, it became apparent that there was paleoceanographic information content in the distribution of hiatuses in the deep sea. Moore and Heath (1977) noted a tendency for a greater proportion of older sections to be represented by hiatuses, so that for sediments older than the Cenozoic, only about 30% of the section is still present. Moore and others (1978) interpreted the increasing abundance of hiatuses in older sediments to be the result of dissolution of biogenic carbonate and opaline silica and erosion by bottom currents. It is also apparent that at certain times (middle and late Miocene, late Eocene–early Eocene, early Paleocene), hiatus generation was much more common than at other times. During the late Oligocene–early Miocene and late Paleocene–middle Eocene, hiatus abundance is less than a simple decay curve predicts. Subsequent studies have focused on the interpretation of hiatuses in terms of rates of production of bottom waters. The interpretation of hiatuses is not straightforward because (1) it is difficult to date the time of their formation, (2) dissolution alone may create hiatuses in carbonate and opaline sediments, and (3) the effect of formation and movement of ocean bottom waters on the sediments is controlled to a large extent by the local topography. Nevertheless, occurrence of hiatuses in particular locations have offered important clues to sites of bottom-water formation and flow paths in the past (Kennett and others, 1975; Kennett and Watkins, 1976;

Kennett, 1977; Ledbetter and others, 1978; Kaneps, 1979; Ledbetter, 1979, 1981; Ledbetter and Ciesielski, 1982; Johnson, 1982, 1983; Barron and Keller, 1983; Keller and Barron, 1983, 1987; Miller and Tucholke, 1983; Miller and others, 1985b, 1987a). Recently it has become possible to interpret seismic stratigraphic sections on continental margins and in the deep sea in terms of paleoceanographic events; Schlager and others (1984) suggested that hiatuses in the Late Cretaceous of the southeastern Gulf of Mexico may be the result of contour currents which entered the Gulf from the Atlantic before the Cuban arc arrived south of Florida. Pinet and Popenoe (1985), interpreting seismic profiles of the Blake Plateau, suggested that the initiation of the flow of the Gulf Stream over the Blake Plateau occurred in the early Tertiary. They were able to recognize two flow paths, one across the inner plateau along the Florida-Hatteras slope and the other across the central plateau; they attributed the shift from one path to the other to the effects of sea-level changes. Mullins and others (1987), from examination of a seismic stratigraphic succession off western Florida, suggested that an oceanographic event, that is, intensification of the Loope Current/Gulf Stream, occurred in the middle Miocene (12–15 Ma); they attributed this event to closure of the Isthmus of Panama to major ocean currents, intensifying the northward water transport of the western Atlantic.

CIRCULATION OF THE SURFACE OCEAN

The first attempt to reconstruct ancient oceanic circulation patterns was the study of Berggren and Hollister (1974) who used the theory of wind-driven surface currents and paleobiogeographic data to postulate current flow directions in the Atlantic, western Tethys, and Caribbean from middle Triassic to present. A major flaw in their paleogeographic reconstructions was that they used modern shorelines, even for the Cretaceous when the Gulf of Mexico and Arctic Ocean were connected through the western interior of North America. Biostratigraphic schemes for the zonation of deep-sea sediments had become standardized by the early 1970s, and a number of micropaleontologists took up the new approach of using biogeographic distributions of oceanic microfossils to outline ocean-surface currents (Berggren and Hollister, 1977; Haq, 1980, 1981, 1984; Reyment, 1980; Thunell and Belyea, 1982; Boersma and Premoli-Silva, 1983; Poag and Miller, 1986; Berggren and Olsson, 1986; Boersma and others, 1987; Morley and others, 1987). Starting from the premise that assemblages of planktonic microfossils record surface-water temperatures (Schott, 1935), Imbrie and Kipp (1971) developed a quantitative method to relate a particular assemblage of planktonic foraminifera to oceanographic parameters such as sea-surface temperature through a "transfer function." A refinement of this technique (Kipp, 1976) has been used extensively to develop maps of distribution of temperature and salinity at the sea surface and at 100-m depth for summer and winter year at 18,000 Ma (McIntyre and others, 1976; CLIMAP Project Members, 1981). Ruddiman and McIntyre (1976) have shown that in the North Atlantic, the polar front migrates south to about 45°N during glacials, retreating very rapidly to the far north during interglacials.

THE DEVELOPING PALEOCEANOGRAPHIC SYNTHESIS: THREE-DIMENSIONAL CIRCULATION OF THE OCEAN

The information from oxygen isotopes, distribution of carbonate sediments reflecting fluctuations of the CCD, occurrence of C_{org}-rich sediments, occurrence of productivity indicators, location of hiatuses, and the paleobiogeography of oceanic plankton and benthos is currently being coupled with theoretical and model studies to provide a synthesis of the past behavior of the ocean. Rather than location of surface currents per se, it is the location of oceanic fronts and sites of intermediate- and deep-water production which are critical to the geologic history of the surface of the planet. Fronts mark significant climatic boundaries, and oceanic fronts, where waters of different temperature and salinity converge, separate major circulation systems which are, in effect, biogeographic provinces. Oceanic fronts are also marked by major changes in sediment type on the sea floor. The Antarctic Divergence, which closely surrounds the Antarctic continent, mixes nutrient-rich water upward, resulting in the high productivity belt of the Southern Ocean. South of the divergence, the sea floor is covered with ice-rafted debris; north of the divergence lies the circum-Antarctic band of diatomaceous sediment. Convergences mark absolute limits of transport of floating material. The Antarctic Convergence, which is about 50°S except off South America, where it lies more than 60°S, marks the northerly limit of drift ice and hence of the occurrence of ice-rafted debris on the ocean floor. Because of the complex distribution of land and sea in the Northern Hemisphere, the Antarctic Divergence and Convergence have no well-defined counterparts in the north. Using the occurrence of ice-rafted debris in Quaternary sediments, however, Lisitzin (1974) was able to estimate that during the glacials of the Pleistocene, the southern polar front extended to 40°S, and the northern polar front reached 40°N in the North Atlantic and Pacific. Subsequent studies by CLIMAP (1981), based on the analysis of fossil plankton, have confirmed Lisitzin's work. The subtropical convergences, which lie at about 30°N and 30°S, separate the westerly drifts of the higher latitudes from the easterly currents of the equatorial region, and mark the regions of lowest productivity in the ocean. The Equatorial Convergence, which is well developed in the Pacific and less well in the Atlantic, lies close to the thermal equator of the ocean; the Equatorial divergences on either side of it delineate the zone of open-ocean equatorial upwelling, with high productivity.

One of the most effective methods of learning about changes in atmospheric and oceanic circulation in the past is to trace the position of oceanic fronts back in time. This has been done specifically by Ciesielski and Grinstead (1986) and Hodell and Kennett (1985). Wind-driven upwelling systems can offer valuable insights into the strength and direction of winds in the past (Sarnthein and others, 1981, 1982; Rea and others, 1985; Pisias and Rea, 1988).

Studies of production of intermediate and deep waters in the present-day ocean have shown that they result from a complex interaction of factors. For water to sink and become intermediate or deep water in the ocean, it must be modified by interaction with the atmosphere so that its density is increased; this can be either (1) through evaporation, which increases the salinity of the surface water and hence its density; (2) through the formation of sea ice, which, because sea ice is less saline than the sea water from which it forms, expels salt into the surrounding surface water, increasing its density; and (3) through chilling of the water surface by conduction of heat to the atmosphere or by loss of heat due to evaporation. The relations between salinity, temperature, pressure, and density are defined by the equation of state of sea water and are illustrated in Figure 4. As dense water forms at the surface, it sinks, and in doing so, it entrains the surrounding water, thereby decreasing its density contrast. Peterson (1979) found that the source of dense water which produces a downwelling plume having the highest buoyancy flux will overcome all other sources and fill the depths of the ocean. The buoyancy flux is the product of the volume flux of the plume and its density excess over the surrounding water. As it descends, the density of the water is increased by the pressure of the overlying water; cold water, especially colder than 7 °C, is more compressible than warm water, and thus gains density faster. Figure 4 shows density at the surface and at a pressure of 400 bars, corresponding

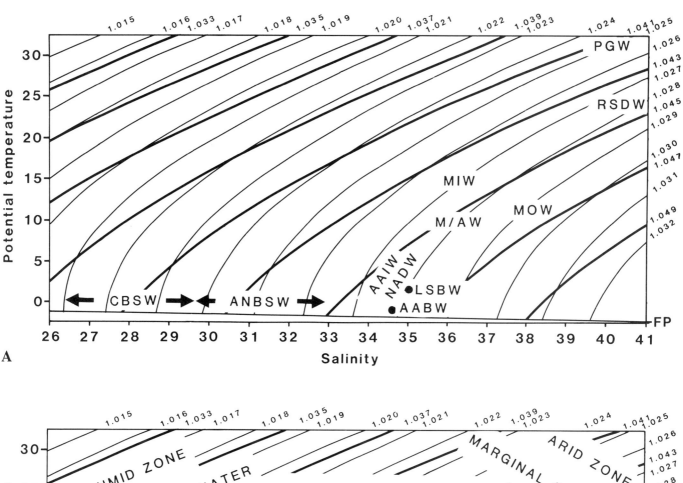

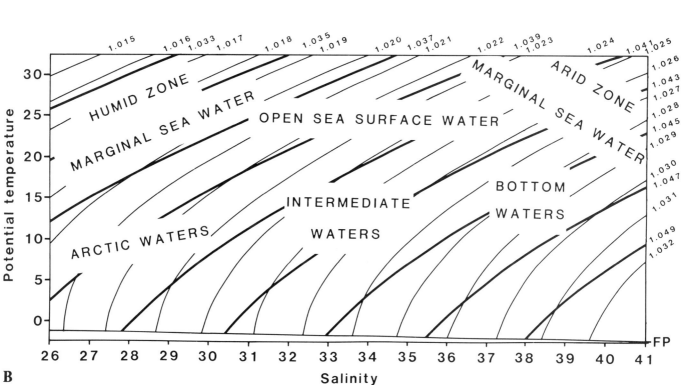

Figure 4. A. Relation of potential temperature, salinity, and density, with fields occupied by some important modern water masses. Potential temperature is the temperature the water would have if raised adiabatically to the surface, that is, it would eliminate the increase in temperature of the water due to compression. Thin lines are surface densities (1.105–1.032) calculated for the sea surface beneath one atmosphere of pressure; thick lines are ocean-bottom densities (1.033–1.049) calculated for a pressure of 400 bars (approximately equal to 4,000 m depth. CBSW = Canadian Basin (Arctic) Surface Water; ANBSW = Amundsen-Nansen Basin (Arctic) Surface Water; AAIW = Antarctic Intermediate Water; NADW = North Atlantic Deep Water; AABW = Antarctic Bottom Water; LSBW = Labrador Sea Bottom Water; M/AW = Mediterranean/Atlantic Mixed Water; MIW = Mediterranean Inflow; MOW = Mediterranean Outflow; RSDW = Red Sea Deep Water; PGW = Persian Gulf Water.

B. Relation of potential temperature, salinity, and density, with approximate fields for some Cretaceous water masses.

approximately to a depth of 4,000 m, near the average depth of the ocean (3,730 m); the densities were calculated from the recent high-pressure equation of state of sea water and the new International Equation of State of Sea Water (Millero and others, 1980; Millero and Poisson, 1981). The deep circulation of today's ocean is driven by thermohaline circulation, that is, the major sources of bottom water obtain their excess density by virtue of their coldness. As T. C. Chamberlin (1906) speculated, however, the ocean may in the past have had a halothermal circulation in which the deep-water sources obtained their excess density by increase of salinity.

The formation of deep waters in the present ocean illustrates the complex interplay of factors. The salinity, temperature, and density of many of the world's major water masses are indicated in Figure 4. The major water masses of the Atlantic are shown in Figure 5A. The Mediterranean Sea plays a special role in global circulation today; almost isolated from the world ocean and located in a region where the annual evaporation exceeds precipitation by about 1 m, the eastern Mediterranean is the site of formation of the densest water in the ocean. The salinity of water flowing into the Mediterranean through Gibraltar is increased to over 39.2 by the high evaporation in the eastern Mediterranean (Miller and others, 1970). As a result, the density increases to as much as 1.0292 g/cc (Worthington and Wright, 1970) in the winter. This dense water sinks in the eastern Mediterranean to flow back into the western basins and then into the Atlantic over the Gibraltar sill. Enroute it mixes with a larger volume flux of less saline but cooler deep water formed in the winter in the northwestern Mediterranean (Medoc Group, 1970; Bryden and Stommel, 1982, 1984). Although the Mediterranean outflow, with a density as high as 1.0291 g/cc, is still one of the densest waters found in the ocean (see Fig. 4), the volume of outflow is not sufficient to give it a buoyancy flux adequate to make it become the bottom water of the North Atlantic. Instead, the Mediterranean outflow becomes less dense through entrainment of the surrounding waters so that at a depth of about 1 km it begins to spread on the 1.02751 g/cc (calculated from Fuglister, 1960), isopycnal (equal-density) surface. Advection of water in the ocean takes place predominantly on the isopycnal surfaces, and the 1.02751 isopycnal reaches the surface of the North Atlantic near the entrance to the Norwegian-Greenland Sea (Reid, 1979, 1981; see also Gorshkov, 1977). As the North Atlantic water, which is anomalously warm and saline because of its admixture of Mediterranean outflow water, enters the Norwegian-Greenland Sea near the pycnocline depth between Greenland and Scotland, it mixes with water of similar density and sinks to become the major source of North Atlantic Deep Water (NADW) (Peterson and Rooth, 1976; Swift and others, 1980). As can be seen in Figure 4, when two oceanic waters of equal density but different salinity and temperature mix, the resulting mixture is more dense and sinks. Other sources, notably the Labrador Sea, supplement the Norwegian-Greenland sea source, and the NADW fills the intermediate levels of the Atlantic with water which is nutrient-poor but oxygen-rich (Bainbridge, no date). NADW, diluted by entrainment of other waters it has encountered on the way south, returns to the surface at the Antarctic Divergence, where it is again chilled. The water which flows south from the divergence becomes involved in sea-ice formation in the Weddell Sea; its salinity is increased and it sinks as nutrient-rich, silica-rich, but oxygen-poor Antarctic Bottom Water, which underflows NADW to the north. The Coriolis effect, deflecting the flow to the left in the Southern Hemisphere, and the barrier of Walvis Ridge in the eastern South Atlantic cause AABW to flow northward through the western basins of the South Atlantic. Water which flows north from the Antarctic divergence escapes the sea-ice area, but it is both chilled and diluted by mixing. It sinks beneath the surface but is relatively light, and it overrides the NADW as it flows north as nutrient-rich but silica- and oxygen-poor Antarctic Intermediate Water (AAIW). Thus, each of the

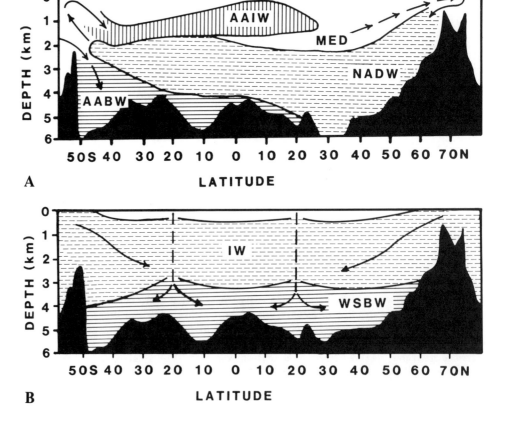

Figure 5. A. Schematic diagram of the relation of some important water masses in the modern Atlantic. AABW = Antarctic Bottom Water; AAIW = Antarctic Intermediate Water; MED = Mediterranean Outflow mixing into Atlantic water at 1.5 km depth; NADW = North Atlantic Deep Water.

B. Schematic diagram of possible water masses in the Cretaceous. IW = Intermediate Water; WSBW = Warm Saline Bottom Water.

major water masses has a different chemistry. The upwelling systems of the South Atlantic margin draw mostly from AAIW. AABW with its incorporated component of NADW also flows into the basins of the Indian and Pacific Oceans (Reid and Lynn, 1971; see also Gorshkov, 1974). Concentrations of nutrients, bicarbonate ion, and dissolved oxygen all indicate that the flow of bottom waters must be from south to north in the Indian and Pacific Oceans. There is no possibility of a northerly deep-water source for the Indian Ocean, but the warm saline outflows from the Red Sea–Gulf of Oman and Persian Gulf form a major intermediate water mass in the Arabian Sea, as is evident in Gorshkov (1977). The Bay of Bengal is diluted by the large fresh-water outflow from the Ganges-Brahmaputra river complex, and under these conditions, it could never become a deep- or intermediate-water source. There is presently no source of deep water in the North Pacific, but the potential of a source area certainly exists in the complex geography of the Bering Sea and Sea of Okhotsk. Sea-level changes could readily increase the isolation of the marginal seas and make the waters in them more susceptible to modification by the atmosphere.

Streeter (1973) and Schnitker (1974), in the pioneering studies of the differences in interglacial and glacial benthonic foraminifera, had noted that the distribution of certain modern deep-sea benthonic foraminifera in the North Atlantic coincides with the present distribution of bottom waters from different sources. Documentation of the changes in glacial-interglacial distribution of the benthonic foraminifera initially led to the conclusion that deep waters were warmer in glacial times and cooler in the interglacials, although the absolute magnitude of the changes could only be estimated. Schnitker (1974) assumed that this was the result of more limited distribution of Antarctic bottom water. The situation was clarified when Duplessy and others (1975) compared the oxygen-isotope record from a core from the Norwegian Sea with the record from Shackleton and Opdyke's (1973) west equatorial Pacific core. The amplitude of glacial-interglacial changes in the west equatorial Pacific core was 1.1 per mil greater than in the core from the Norwegian Sea. It seemed unreasonable to suggest that the Pacific bottom waters would have been 4.5° cooler during the glacial than they are today (1°), and more reasonable to assume that Norwegian Sea bottom waters were 4.5° warmer during the glacial than they are today. Streeter and Shackleton (1979) suggested that this effect was a result of cessation of North Atlantic Deep Water (NADW) production as a result of ice cover of the Norwegian-Greenland sea. Duplessy and others (1980) then compared the oxygen-isotope record from a core in the northeast Atlantic with that from a core taken beneath Circumpolar Deep Water in the southern Indian Ocean, where they assumed that the bottom-water temperatures had remained constant. The difference in glacial-interglacial amplitudes was 0.3 per mil, and from this they concluded that the bottom waters of the North Atlantic were 1.3° cooler during the glacial than they are at present; they also concluded that the North Atlantic has been a source of deep water for the past 75,000 yr.

In 1982 Hester and Boyle announced the discovery that the shells of benthic foraminifera deposited in oxygen-depleted water are enriched in cadmium. Boyle and Keigwin (1982) applied this discovery to paleoceanographic interpretation; they examined the Cd/Ca ratio, $\delta^{18}O$, and $\delta^{13}C$ in benthic foraminifera from a Pleistocene core taken northwest of the Azores at 42°N, 32°W at a depth of 3.2 km, the level of the NADW core. They found that the Cd/Ca ratio is always lower than it is in the Pacific, indicating that at no time had the flux of low-nutrient water ceased. The highest Cd/Ca ratios occur during the extreme glacials, and the lowest ratios, in the interglacials. The Cd concentration is inversely proportional to the flow rate; they concluded that during glacials the flux from the nutrient-depleted NADW sources was diminished in importance relative to that from southern deep and intermediate sources, but that the flux of NADW never ceased. Boyle and Keigwin (1987) interpreted Cd/Ca and $^{13}C/^{12}C$ data from shells of benthic foraminifera in the North Atlantic as indicating that there was a marked depletion of nutrients in intermediate water masses in association with the reduced NADW flux during the last glacial. They suggested that cold high-latitude sea-surface temperatures enhanced intermediate-water formation at the expense of deep-water formation. The complex history of changing surface conditions and bottom-water sources during the Pleistocene has been recently reviewed by Duplessy and Shackleton (1985), Corliss and others (1986), Williams and Fillon (1986), Ruddiman and others (1986), Ruddiman (1987), and Kellogg (1987). An important driving factor in altering thermohaline circulation in the Atlantic may be the nature of the Mediterranean outflow water (MOW); Zahn and Sarnthein (1987) concluded from a study of stable isotopes in benthic foraminifera near the Strait of Gibraltar that the MOW outflow decreased significantly during the deglaciations, and that its temperature was decreased by as much as 5° during glacial time. They suggest that this indicates that the importance of saline MOW to the formation of NADW may be less than has been assumed, but the time of reduced MOW flux cited by Zahn and Sarnthein corresponds to a time glacial meltwater flux to the North Atlantic was a maximum and when production of NADW was also reduced (Berger and Vincent, 1986; Broecker and others, 1988). The longer-term possible change in Mediterranean/Atlantic relations has been discussed by Thunell and others (1987).

Berger (1968), Moore and others (1977), and Broecker and others (1985) have suggested that oceanic circulation during glacials may be fundamentally different from that of today. It has been speculated that the Norwegian-Greenland Sea may become covered by permanent sea ice during the glacials (Broecker, 1975; Hughes and others, 1977; Lindstrom and MacAyeal, 1986), and the formation of NADW may be reduced or cease. Greatly increased sea-ice formation in the north Pacific may transform that region into a source area for oceanic deep waters. This would result in a complete reversal of gradients of nutrients, dissolved oxygen, and alkalinity in the deep sea. At present nutrients and bicarbonate have their lowest concentrations in the North Atlantic and highest concentrations in the North Pacific, but with a transfer of the Northern Hemisphere sites of deep-water formation to the Pacific and cessation of deep-water formation in the North Atlantic, the highest concentrations of nutrients and bicarbonate would be in the North Atlantic, and the lowest concentrations, in the North Pacific. Similarly, the related oxygen concentration would reverse, with the North Atlantic waters becoming depleted and the North Pacific waters oxygen-rich. Berger and Vincent (1986) have suggested that production of NADW may have been shut off during the glacial-Holocene transition (Dryas), which would again cause redistribution of biolimited materials in the oceanic deep waters, and Berger (1987) suggested that there was indeed deep-water production in the North Pacific during the glacial and deglaciation.

Cenozoic paleoceanography of the North Atlantic has been reviewed by Tucholke and Mountain (1986), of the South Atlantic by Hsü and others (1984) and Williams and others (1985), and of the Southern Ocean by Wise and others (1985).

The most significant changes in surface circulation during the Pliocene occurred with the final closure of the Central American Isthmus and resulting complete separation of tropical and subtropical circulation between the Atlantic and Pacific (Emiliani and others, 1972; Malfait and Dinkleman, 1972; van Andel, 1976; Keigwin, 1982a; Gartner and others, 1983; Haq, 1984). The final separation of Atlantic and Pacific appears to have occurred about 2.5 Ma, and the last interchange of water was probably between the Pacific and the Gulf of Mexico, across Isthmus of Tehuantepec, bypassing the Caribbean (Gartner and others, 1987). The salinity contrasts between the Arctic Ocean, where the surface waters of the Canadian Basin are diluted below a salinity of 30 to as little as 27.5 by the plume of the Mackenzie River (Gorshkov, 1980), and the salty (34.92) Atlantic and less salty (34.60) Pacific (Worthington, 1981) resulting from large-scale atmospheric water transport across the Central American Isthmus (Weyl, 1968) are important underlying factors promoting the development of the modern circulation system with a strong NADW source. The Pliocene ocean clearly differed from the modern ocean in one very significant way: upwelling was greatly enhanced in the eastern South Atlantic (Dean and others, 1984), about the Antarctic (Brewster, 1980), and probably elsewhere. The cause of this intensification of upwelling has yet to be determined. Changes in the thermohaline circulation during the Pliocene have been discussed by Weissert and Oberhänsli (1985) and Murray and others (1986).

During the late Miocene (Messinian), a special event occurred—the "salinity crisis" which occurred as the Mediterranean became desiccated and extracted about 1×10^6 km^3 of salts from the world ocean (Ryan, Hsü, and others, 1973; Cita, 1982; Cita and McKenzie, 1986; Mueller and Hsü, 1987). The fall of sea level associated with build-up of the Antarctic ice cap was a contributing factor in causing the isolation of the Mediterranean. The Red Sea had become a salt-extraction basin in the earlier Miocene, and these two salt extractions may have lowered the salinity of the world ocean by as much as 7% (Southam and Hay, 1981).

Paleoceanography of the Miocene ocean on a global scale was the subject of GSA Memoir 163 (Kennett, 1985). The interpretations are based largely on analysis of oceanic microfossils and stable isotopes. It was during the Miocene that the deep-water passage from the Pacific into the Indian Ocean became restricted as Australia–New Guinea collided with the Indonesian Arc. Kennett and others (1985) have determined that the Pacific Equatorial Undercurrent developed during the middle Miocene, and they concluded that this was related to the closing of the Indonesian Seaway, which also increased the intensity of gyral circulation in the Pacific. Intensification of the Countercurrent is associated with increased productivity and output of biogenic silica. The middle Miocene was also the time of re-establishment of a strong North Atlantic Deep Water source after it has weakened in the early Miocene (Schnitker, 1980, 1986). The Miocene was a time of increasing latitudinal temperature gradients and decrease in bottom-water temperature. These both reflect the increasing isolation of the Antarctic continent with the development of the West Wind Drift and the expansion of the ice cap. The sharp middle Miocene cooling of oceanic deep waters is assumed to be in response to an increasing buoyancy flux from a source in the Southern Ocean or along the Antarctic margin. Woodruff (1985) has shown how benthic foraminifera can be used to recognize Miocene water masses in the Pacific and to estimate their relative oxygen concentrations.

The Oligocene was the age during which the ocean took on its modern aspect, when the deep waters became significantly cooler than the surface waters (Corliss and Keigwin, 1986). During some short episodes, the bottom waters may have cooled to temperatures comparable to those of today (Miller and Fairbanks, 1985). Ice may have been present on the Antarctic continent by the early Oligocene (Wise and others, 1987), and deep-water production in the surrounding seas may have been restricted by fresh-water runoff. It is not clear what other deep-water sources may have existed during the Oligocene.

In the Eocene, Paleocene, and Cretaceous, the ocean was fundamentally different from that of today (Kraus and others, 1978; Barron and others, 1981; Hay, 1983; Oberhänsli and Hsü, 1986; Roth, 1986; Barron, 1987); the circulation appears to have been halothermal rather than thermohaline. Brass and others (1982a, 1982b) have suggested that warm saline waters that formed in marginal seas in the arid zones filled the ocean basins. Barrera and others (1987) have presented evidence for Antarctic shallow-water temperatures and concluded that the polar regions were sites of deep-water formation for at least the Pacific Basin during the Late Cretaceous and Paleocene, although they acknowledged that the Atlantic basins may have had local sources. Miller and others (1987b) found that the Antarctic region was the source of bottom water for the Cape Basin during the late Paleocene, but that the supply was reduced or eliminated by the early Eocene. They also noted that the nutrient-depleted Antarctic waters failed to enter the western basins of the South Atlantic, probably being blocked by the Islas Orcadas Rise.

As noted above, warm saline waters contain only about one-half the dissolved oxygen of young modern deep waters; hence, intermediate and deep waters that formed in arid marginal or shelf seas would have been much more likely to become anoxic. The great variability in the CCD between different ocean basins during the Cretaceous (Fig. 2) is readily understood if the ocean depths were being filled from numerous, relatively small and weak, locally warm, saline-water sources. Based on the oxygen-isotopic evidence from inoceramids reported by Barron and others (1982), Hay (1983) suggested that in the Late Cretaceous South Atlantic, surface-water temperatures were similar to those of today, but bottom waters were about 10–12 °C, and intermediate waters were about 8 °C (see Fig. 4B). This implies that the intermediate waters were produced at higher latitudes than were the deep waters (Figure 5B shows a general scheme for halothermal circulation). Bottom waters were formed in the mid-latitude arid regions; the source areas would be in shallow seas or nearly isolated basins on the eastern margins of the oceans. In each hemisphere, these waters would flow poleward, following the eastern basins or as contour currents along the eastern margin of each ocean basin. These warm saline waters would return to the surface in the polar regions. On reaching the surface, the waters would be chilled by conductive cooling and, more importantly, by evaporation in the relatively warm polar regions. They could then sink and flow equatorward as intermediate waters; during their equatorward flow, they would be concentrated as contour currents on the western margins of the oceans. It is interesting to note that in the Late Cretaceous the over-all salinity of the ocean was about the same as it is today; but in the middle Cretaceous, after the very large salt extraction in the opening South Atlantic, over-all salinities may have been 20% higher (Southam and Hay, 1981), so that the average salinity would have been about 41.6‰. This means that the major water masses would fall into a temperature/salinity field near and beyond the right side of Figures 4A and 4B.

THE FUTURE

The most exciting prospects for the future of paleoceanographic studies lie in attempting to understand how plate tectonics, atmospheric and oceanic circulation, and biologic evolution of marine animals and plants are interrelated. Do plate-tectonic cycles drive the climate and ocean systems on long-term time scales as proposed by Worsley and others (1986)? It is particularly important to understand why the Earth has different climatic states and how the ocean has responded in the past. What was the nature of oceanic circulation in warm geological periods when the deep ocean was filled with relatively warm water? Because most of the older ocean floor has been subducted, the data base for the oceans proper is very incomplete. To understand the oceans of the Eocene, Paleocene, Mesozoic, and Paleozoic, it will be necessary to understand the relation of the record preserved in marginal seas and on the continents. The separation between those working in the deep sea and those working on epeiric seas must end. To understand how the Earth has operated in the past, we make use of all of the available data in a global context. Modeling oceanographic circulation for the earlier Paleogene and Mesozoic will undoubtedly generate many new ideas. Before modeling can be realistic enough to provide output that can be tested against geological observations, however, the shape of the surface of the Earth in the past must be reconstructed as accurately as possible. Plate-tectonic reconstructions show the positions of the major continental blocks and bathymetry of the ocean floor based on simple subsidence curves. The present reconstructions are, unfortunately, gross oversimplifications because they do not take into account the many large terranes which have moved independently of the major continental blocks and because they do not include the residual depth anomalies (areas of the ocean floor which are more than 500 m shallower or deeper than predicted by the thermal decay curve) which form a significant part of the sea floor. Preliminary reconstructions by Wilson (1987) and Wilson and others (in press) reveal that drifting terranes obstruct many passages which appear to be clear on maps that show only the major continental blocks. Coupled atmospheric and oceanic models run on such a realistic paleogeographic base are necessary to achieve the better resolution needed before the output can be used for site-specific geologic predictions.

Much of paleoceanographic interpretation by geologists in the past has reflected a lack of familiarity of the principles of physical oceanography as they are understood today, although many recent works have begun to bridge the gap and integrate the two fields. The languages of the two fields are sufficiently different that most physical oceanographic literature is not readily accessible to geologists. Fortunately several general works have appeared in recent years which summarize modern concepts in physical oceanography and provide the background necessary to understand developments in that field: Harvey (1976), which provides a general introduction; Warren and Wunsch (1981), which contains a series of summary papers; Pickard and Emery (1982); Pond and Pickard (1983), which includes a glossary of commonly used symbols and a review of the mathematical notation commonly used; and Tolmazin (1985), which has a good discussion of flow resulting from pressure gradients especially applicable to studies of marginal seas.

The new field of paleoceanography was born three decades ago, but it is still very much in its infancy. Only recently have we come to realize that the oceans of the past may have operated in modes very different from that of today. The new proxy indicators being used to investigate distribution of properties and materials in the ancient oceans will make it possible to understand major processes at a level which could not have been anticipated when the field was born.

SUMMARY AND CONCLUSIONS

Paleoceanography, a field born little more than 30 yr ago, has become a major branch of geology. It is an interdisciplinary field making use of stratigraphy, geochemistry (particularly stable isotopes), geophysics (particularly reflection seismics and paleomagnetics), paleontology, sedimentology, and other aspects of geology and relating these to progress in physical, chemical, and biological oceanography. Coring of deep-sea sediments began after World War II and developed into the ocean drilling and coring programs, notably the Deep Sea Drilling Project and the Ocean Drilling Program. These have provided a rich sample library and data base, permitting the development of a global science. Among the most significant advances in the field are the following.

1. Development of a refined stratigraphic base incorporating micropaleontology, magnetostratigraphy, and isotope stratigraphy so that resolution of 100,000 yr or less is now possible for parts of the Cenozoic.

2. Documentation and interpretation of the thermal history of the ocean, from the equable climates of the Mesozoic through the development of the pole-to-equator temperature gradient during the course of the Cenozoic to the present condition of alternating glacials and interglacials.

3. Documentation of fluctuations of the calcite compensation depth and development of an understanding of the complex interrelation of factors which control the oceans' buffering capacity and how these are related to changes in the concentration of CO_2 in the atmosphere.

4. Investigation of the accumulation of organic carbon-rich sediments in the deep sea, and their implications for changes in surface productivity and the nature of deep-ocean circulation.

5. The discovery that the deep sea is not the site of continuous sedimentation as had been assumed by earlier geologists, but that the deep-sea record is riddled by hiatuses. The hiatuses testify to the development of strong bottom currents and corrosive deep waters, and they offer valuable clues to ancient deep circulation.

6. Documentation of changes in paleobiogeography of ocean plankton which indicate changes in surface currents as the Earth changed from a planet with latitudinal oceanic circulation in the subtropics (Tethys) to one with meridional oceans connected in the subpolar region.

7. Insight into alternative modes of circulation of the ocean in which the deep circulation was not thermohaline, that is, driven by cold waters sinking in the subpolar or polar regions, but may have been halothermal, driven by dense plumes of warm saline waters formed in marginal seas and shelves of the subtropics.

8. The field of paleoceanography is burgeoning with new ideas and offering a view of the history of the Earth's surface in a global context. It offers a bright future for documentation and modeling of global oceanic processes and the opportunity to relate marine geologic observations to atmospheric and oceanic climate and circulation models, and relation of atmospheric and ocean climate studies of the deep circulation.

ACKNOWLEDGMENTS

The writer has benefited from conversations with a number of colleagues, particularly John R. Southam, Michael J. Rosol, and Christopher G. A. Harrison. This work has been supported by Grants OCE 8409369 and OCE 8716408 from the National Science Foundation and 19274-AC2 from the Petroleum Research Fund of the American Chemical Society and by gifts from Texaco, Inc.

REFERENCES CITED

Arrhenius, G., 1952, Sediment cores from the East Pacific: Swedish Deep-Sea Expedition, 1947–1948, Report 5, p. 1–228.

Arthur, M. A., and Natland, J. H., 1979, Carbonaceous sediments in the North and South Atlantic: The role of salinity in stable stratification of early Cretaceous basins, in Talwani, M., Hay, W., and Ryan, W.B.F., Deep drilling results in the Atlantic Ocean: Continental margins and paleoenvironment: Washington, D.C., American Geophysical Union, Maurice Ewing Series, Volume 3, p. 375–401.

Arthur, M. A., and Schlanger, S., 1979, Cretaceous "oceanic anoxic events" as causal factors in the development of reef-reservoired giant oil fields: American Association of Petroleum Geologists Bulletin, v. 63, p. 870–885.

Arthur, M. A., Schlanger, S. O., and Jenkyns, H. C., 1985, The Cenomanian-Turonian anoxic event, II. Paleoceanographic controls on organic matter production and preservation, in Brooks, J., and Fleet, A., eds., Marine petroleum source rocks: Geological Society of London Special Publication 26, p. 401–420.

Bainbridge, A. E., no date, GEOSECS Atlantic Expedition, Volume 2, Sections and Profiles: Washington, D.C., National Science Foundation, xiv + 198 p.

Barker, P., Dalziel, I.W.D., and others, 1977, Initial reports of the Deep Sea Drilling Project, Volume 36: Washington, D.C., U.S. Government Printing Office, xxii + 1,079 p.

Barrera, E., Hubner, B. T., Savin, S. M., and Webb, P.-N., 1987, Antarctic marine temperatures: Late Campanian through early Paleocene: Paleoceanography, v. 2, p. 21–47.

Barron, E. J., 1983a, A warm, equable Cretaceous: The nature of the problem: Earth Science Reviews, v. 19, p. 305–338.

——— 1983b, The ocean and atmosphere during warm geologic periods: Joint Oceanographic Assembly 1982, General Symposia, Proceedings, p. 64–71.

——— 1984, Ancient climates: Investigation with climate models: Reports of Progress in Physics, v. 47, p. 1563–1599.

——— 1985, Numerical climate modeling, a frontier in petroleum source rock prediction: Results based on Cretaceous simulations: American Association of Petroleum Geologists Bulletin, v. 69, p. 448–459.

——— 1986, Physical paleoceanography: A status report, in Hsü, K. J., ed., Mesozoic and Cenozoic oceans: American Geophysical Union Geodynamics Series, Volume 15, p. 1–9.

——— 1987, Eocene equator-to-pole surface ocean temperatures: A significant climate problem?: Paleoceanography, v. 2, p. 729–740.

Barron, J. A., and Keller, G., 1983, Widespread Miocene deep-sea hiatuses: Coincidence with periods of global cooling: Geology, v. 10, p. 577–581.

Barron, E. J., and Whitman, J. M., 1981, Oceanic sediments in space and time, in Emiliani, C., The sea, Volume 7, The oceanic lithosphere: New York, Wiley-Interscience, p. 689–731.

Barron, E. J., Thompson, S. L., and Schneider, S. H., 1981, An ice-free Cretaceous? Results from climate model simulations: Science, v. 212, p. 501–508.

Barron, E. J., Saltzman, E., and Price, D., 1982, Inoceramus: Occurrence in South Atlantic and oxygen isotope paleotemperatures at Hole 530A, in Hay, W., Sibuet, J. C., and others, Initial reports of the Deep Sea Drilling Project, Volume 75: Washington, D.C., U.S. Government Printing Office, p. 893–904.

Bascom, W., 1958, The Mohole: Scientific American, v. 200, p. 41–49.

Basov, I. A., and Krasheninnikov, V. A., 1983, Benthic foraminifers of the Mesozoic and Cenozoic sediments of the southwestern Atlantic as an indicator of paleoenvironment, in Ludwig, W. J., Krasheninnikov, V. A., and others, Initial reports of the Deep Sea Drilling Project, Volume 71: Washington, D.C., U.S. Government Printing Office, p. 739–787.

——— 1961, A hole in the bottom of the sea: Garden City, New York, Doubleday, 352 p.

Benson, R. H., 1975, The origin of the psychrosphere as recorded in changes of deep-sea ostracode assemblages: Lethaia, v. 8, p. 69–83.

Benson, R. H., Chapman, R. E., and Deck, L. T., 1985, Evidence from the Ostracoda of major events in the South Atlantic and worldwide over the past 80 million years, *in* Hsü, K. J., and Weissert, H. J., eds., South Atlantic paleoceanography: Cambridge, England, Cambridge University Press, p. 325–350.

Berger, W. H., 1967, Foraminiferal ooze: Solution at depths: Science, v. 156, p. 383–385.

——— 1968, Planktonic foraminifera: Shell production and preservation [Ph.D. thesis]: San Diego, California, University of California at San Diego, 241 p.

——— 1970, Planktonic foraminifera: Selective solution and the lysocline: Marine Geology, v. 8, p. 111–138.

——— 1972, Deep sea carbonates: Dissolution facies and age-depth constancy: Nature, v. 236, p. 392–395.

——— 1973, Cenozoic sedimentation in the eastern tropical Pacific: Geological Society of America Bulletin, v. 84, p. 1941–1954.

——— 1976, Biogenous deep-sea sediments: Production, preservation and interpretation, *in* Riley, J. P., and Chester, R., eds., Chemical oceanography, Volume 5: New York, Academic Press, p. 265–374.

——— 1979, Impact of deep-sea drilling on paleooceanography, *in* Talwani, M., Hay, W., and Ryan, W.B.F., Deep drilling results in the Atlantic Ocean: Continental margins and paleoenvironment: Washington, D.C., American Geophysical Union, Maurice Ewing Series, Volume 3, p. 297–314.

——— 1981, Paleoceanography: The deep-sea record, *in* Emiliani, C., The sea, Volume 7, The oceanic lithosphere: New York, Wiley-Interscience, p. 1437–1519.

——— 1985, CO_2 increase and climate prediction: Clues from deep-sea carbonates: Episodes, v. 6, p. 163–168.

——— 1987, Ocean ventilation during the last 12,000 years: Hypothesis of counterpoint deep water production: Marine Geology, v. 78, p. 1–10.

Berger, W. H., and Mayer, L. A., 1987, Cenozoic paleoceanography 1986, An introduction: Paleoceanography, v. 2, p. 613–624.

Berger, W. H., and Roth, P. H., 1975, Oceanic micropaleontology: progress and prospects: Reviews of Geophysics and Space Physics, v. 13, p. 561–585 and 624–635.

Berger, W. H., and Vincent, E., 1986a, Sporadic shutdown of North Atlantic deep water production during the Glacial-Holocene transition?: Nature, v. 324, p. 53–55.

——— 1986b, Deep-sea carbonates: Reading the carbon-isotope signal: Geologische Rundschau, v. 75, p. 249–269.

Berger, W. H., and Winterer, E. L., 1974, Plate stratigraphy and the fluctuating carbonate line, *in* Hsü, K. J., and Jenkyns, H. C., eds., Pelagic sediments on land and under the sea: International Association of Sedimentologists Special Publication 1, p. 11–48.

Berger, W. H., Bonneau, M.-C., and Parker, F. L., 1982, Foraminifera on the deep-sea floor: Lysocline and dissolution rate: Oceanologica Acta, v. 5, p. 249–258.

Berggren, W. A., and Hollister, C. D., 1974, Paleogeography, paleobiogeography and the history of circulation in the Atlantic Ocean, *in* Hay, W. W., ed., Studies in paleo-oceanography: Society of Economic Paleontologists and Mineralogists Special Publication No. 20, p. 126–186.

——— 1977, Plate tectonics and paleocirculation—Commotion in the ocean: Tectonophysics, v. 38, p. 11–48.

Berggren, W. A., and Olsson, R. K., 1986, North Atlantic Mesozoic and Cenozoic paleobiogeography, *in* Vogt, P. R., and Tucholke, B. E., The western North Atlantic region: The geology of North America, Volume M: Boulder, Colorado, Geological Society of America, Decade of North American Geology, p. 565–588.

Berner, R. A., Lasaga, A. C., and Garrels, R. M., 1983, The carbonate-silicate geochemical cycle and its effect on atmospheric carbon dioxide over the past 100 million years: American Journal of Science, v. 283, p. 641–683.

Bezrukov, P. L., 1955, Distribution and rate of sedimentation of silica silts in the Sea of Okhotsk: Akademiya Nauk SSSR Doklady, v. 103, p. 473–476.

Birchfield, G. E., 1987, Changes in deep-ocean water $\delta^{18}O$ and temperature from the last glacial maximum to the present: Paleoceanography, v. 2, p. 431–442.

Boersma, A., and Premoli-Silva, I., 1983, Paleocene planktonic foraminiferal biogeography and the paleoceanography of the Atlantic Ocean: Micropaleontology, v. 29, p. 355–381.

Boersma, A., Premoli-Silva, I., and Shackleton, N. J., 1987, Atlantic Eocene planktonic foraminiferal paleohydrographic indicators and stable isotope paleoceanography: Paleoceanography, v. 2, p. 287–331.

Bolli, H. M., 1957, Planktonic foraminifera from the Oligocene-Miocene Cipero and Lengua Formations of Trinidad: U.S. National Museum Bulletin 215, p. 97–123.

Bolli, H. M., Boudreaux, J. E., Emiliani, C., Hay, W. W., Hurley, R. J., and Jones, J. I., 1968, Biostratigraphy and paleotemperatures of a section core on the Nicaragua Rise, Caribbean Sea: Geological Society of America Bulletin, v. 79, p. 459–470.

Bolli, H. M., Ryan, W.B.F., and others, 1978, Initial reports of the Deep Sea Drilling Project, Volume 40: Washington, D.C., U.S. Government Printing Office, xxi + 1,079 p.

Bostrom, K., Moore, C., and Joensuu, O., 1979, Biological matter as a source for Cenozoic deep-sea sediments in the Equatorial Pacific: Ambio Special Reports, v. 6, p. 11–17.

Boyle, E. A., 1986a, Deep ocean circulation, preformed nutrients, and atmospheric carbon dioxide: Theories and evidence from oceanic sediments, *in* Hsü, K. J., ed., Mesozoic and Cenozoic oceans: Washington, D.C., American Geophysical Union, Geodynamics Series, Volume 15, p. 49–59.

——— 1986b, Paired carbon isotope and cadmium data from benthic foraminifera: Implications for changes in oceanic phosphorus, oceanic circulation, and atmospheric carbon dioxide: Geochimica et Cosmochimica Acta, v. 50, p. 265–276.

Boyle, E. A., and Keigwin, L. D., 1982, Deep circulation of the North Atlantic over the last 200,000 years: Geological evidence: Science, v. 218, p. 784–787.

——— 1987, North Atlantic thermohaline circulation during the past 20,000 years linked to high-latitude surface temperature: Nature, v. 330, p. 35–40.

Bralower, T. J., and Thierstein, H. R., 1984, Low productivity and slow deep-water circulation in mid-Cretaceous oceans: Geology, v. 12, p. 614–618.

Bramlette, M. N., 1958, Significance of coccolithophorids in calcium-carbonate deposition: Geological Society of America Bulletin, v. 69, p. 121–126.

Bramlette, M. N., and Wilcoxon, J. A., 1967, Middle Tertiary calcareous nannoplankton of the Cipero Section, Trinidad, W. I.: Tulane Studies in Geology, v. 5, p. 93–131.

Brass, G. W., Saltzman, E., Sloan, J. L., II, Southam, J. R., Hay, W. W., Holser, W. T., and Peterson, W. H., 1982a, Ocean circulation, plate tectonics and climate, *in* Berger, W. H., and Crowell, J. C., eds., Climate in Earth history: Washington, D.C., National Academy Press, p. 83–89.

Brass, G. W., Southam, J. R., and Peterson, W. H., 1982b, Warm saline bottom water in the ancient ocean: Nature, v. 296, p. 620–623.

Brewster, N. A., 1980, Cenozoic biogenic silica sedimentation in the Antarctic Ocean: Geological Society of America Bulletin, v. 91, p. 337–347.

Broecker, W. S., 1969, Why the deep sea remains aerobic: Geological Society of America Abstracts with Programs for 1969, pt. 7, p. 20–21.

——— 1975, Floating glacial caps in the Arctic Ocean: Science, v. 188, p. 116–118.

Broecker, W. S., and Peng, T.-H., 1982, Tracers in the sea: Palisades, New York, Lamont-Doherty Geological Observatory, Eldigio Press, 690 p.

Broecker, W. S., Peteet, D. M., and Rind, D., 1985, Does the ocean-atmosphere system have more than one stable mode of operation?: Nature, v. 315, p. 21–26.

Broecker, W. S., Andree, M., Wolfli, W., Oeschger, H., Bonani, G., Kennett, J., and Peteet, D., 1988, The chronology of the last deglaciation: Implications to the cause of the Younger Dryas event: Paleoceanography, v. 3, p. 1–19.

Brumsack, H. J., 1986, The inorganic geochemistry of Cretaceous black shales (DSDP Leg 41) in comparison to modern upwelling sediments from the Gulf of California, *in* Summerhayes, C. P., and Shackleton, N. J., eds., North Atlantic paleoceanography: London, Geological Society Special Publication No. 21, p. 447–462.

Bryden, H. L., and Stommel, H. M., 1982, Origin of the Mediterranean outflow: Journal of Marine Research, supp., v. 40, p. 55–71.

——— 1984, Limiting processes that determine basic features of the circulation in the Mediterranean: Oceanologica Acta, v. 7, p. 289–296.

Budyko, M. I., and Ronov, A. B., 1979, Chemical evolution of the atmosphere in the Phanerozoic [in Russian]: Geokhimiya, no. 5, p. 643–653; [English translation] Geochemistry International, v. 15, p. 1–9.

Budyko, M. I., Ronov, A. B., and Yanshin, A. L., 1985, History of the Earth's atmosphere: New York, Springer-Verlag, vi + 139 p.

Calvert, S. E., 1966, Accumulation of diatomaceous silica in the sediments of the Gulf of California: Geological Society of America Bulletin, v. 77, p. 569–596.

——— 1968, Silica balance in the ocean and diagenesis: Nature, v. 219, p. 919–920.

Chamberlin, T. C., 1906, On a possible reversal of deep-sea circulation and its influence on geologic climates: Journal of Geology, v. 14, p. 363–373.

Ciesielski, P. F., and Grinstead, G. P., 1986, Pliocene variations in the position of the Antarctic Convergence in the southwest Atlantic: Paleoceanography, v. 1, p. 197–232.

Cita, M. B., 1982, The Messinian salinity crisis in the Mediterranean: A review, *in* Berckhemer, H., and Hsü, K. J., eds., Alpine-Mediterranean geodynamics: Washington, D.C., American Geophysical Union, Geodynamics Series, Volume 7, p. 113–140.

Cita, M. B., and Grignani, D., 1982, Nature and origin of late Neogene Mediterranean sapropels, *in* Schlanger, S. O., and Cita, M. B., eds., Nature and origin of Cretaceous carbon-rich facies: New York, Academic Press, p. 165–195.

Cita, M. B., and McKenzie, J. A., 1986, The terminal Miocene event, *in* Hsü, K. J., ed., Mesozoic and Cenozoic oceans: Washington, D.C., American Geophysical Union, Geodynamics Series, Volume 15, p. 123–140.

CLIMAP Project Members, 1981, Seasonal reconstructions of the Earth's surface at the Last Glacial Maximum: Geological Society of America Map and Chart Series MC-36.

Corliss, B. C., and Keigwin, L. D., Jr., 1986, Eocene-Oligocene paleoceanography, *in* Hsü, K. J., ed., Mesozoic and Cenozoic oceans: Washington, D.C., American Geophysical Union, Geodynamics Series, Volume 15, p. 101–118.

Corliss, B. H., Martinson, D. G., and Keffer, T., 1986, Late Quaternary deep-ocean circulation: Geological Society of America Bulletin, v. 97, p. 1106–1121.

Dansgaard, W., and Tauber, H., 1969, Glacier oxygen-18 content and Pleistocene ocean temperatures: Science, v. 166, p. 499–502.

Davies, T. A., and Worsley, T. R., 1981, Paleoenvironmental implications of oceanic carbonate sedimentation rates, *in* Warme, J. E., Douglas, R. G., and Winterer, E. L., eds., The Deep Sea Drilling Project: A decade of progress: Society of Economic Paleontologists and Mineralogists Special Publication 32, p. 169–179.

Dean, W. E., and Gardner, J., 1985, Cyclic variations in calcium carbonate and organic carbon in Miocene to Holocene sediments, Walvis Ridge, South Atlantic Ocean, *in* Hsü, K. J., and Weissert, H. J., eds., South Atlantic Paleoceanography: Cambridge, England, Cambridge University Press, p. 61–78.

Dean, W. E., Arthur, M. A., and Stow, D.A.V., 1984a, Origin and geochemistry of Cretaceous deep-sea black shales and multicolored claystones, with emphasis on Deep Sea Drilling Project Site 530, southern Angola Basin, *in* Hay, W. W., Sibuet, J.-C., and others, Initial reports of the Deep Sea Drilling Project, v. 75: Washington, D.C., U.S. Government Printing Office, p. 819–844.

Dean, W. E., Hay, W. W., and Sibuet, J.-C., 1984b, Geologic evolution, sedimentation and paleoenvironments of the Angola Basin and adjacent Walvis Ridge: Synthesis of results of Deep Sea Drilling Project Leg 75, *in* Hay, W. W., Sibuet, J.-C., and others, Initial reports of the Deep Sea Drilling Project, Volume 75: Washington, D.C., U.S. Government Printing Office, p. 509–544.

De Boer, P. L., 1986, Changes in the organic carbon burial during the Early Cretaceous, *in* Summerhayes, C. P., and Shackleton, N. J., eds., North Atlantic paleoceanography: London, Geological Society Special Publication No. 21, p. 321–331.

Degens, E. T., and Ross, D. A., 1974, The Black Sea: Geology, chemistry, and biology: Tulsa, Oklahoma, American Association of Petroleum Geologists Memoir 20, ix + 633 p.

Degens, E. T., Emeis, K.-C., Mycke, B., and Wiesner, M. G., 1986, Turbidites, the principal mechanism yielding black shales in the early deep Atlantic ocean, *in* Summerhayes, C. P., and Shackleton, N. J., eds., North Atlantic paleoceanography: London, England, Geological Society Special Publication No. 21, p. 361–376.

Delaney, M. L., and Boyle, E. A., 1988, Tertiary paleoceanic chemical variability: Unintended consequences of simple geochemical models: Paleoceanography, v. 3, p. 137–156.

Delmas, R. J., Ascencio, J. M., and Legrand, M., 1980, Polar ice evidence that atmospheric CO_2 20,000 B.P. was 50% of present: Nature, v. 284, p. 155–157.

DeMaster, D. J., 1981, The supply and accumulation of silica in the marine environment: Geochimica et Cosmochimica Acta, v. 45, p. 1715–1732.

Deroo, G., Herbin, J. P., and Huc, A. Y., 1984, Organic geochemistry of Cretaceous black shales from Deep Sea Drilling Project Site 530, Leg 75, eastern South Atlantic, *in* Hay, W. W., Sibuet, J.-C., and others, Initial reports of the Deep Sea Drilling Project, Volume 75: Washington, D.C., U.S. Government Printing Office, p. 983–999.

Duplessy, J. C., and Shackleton, N. J., 1985, Response of global deep water circulation to Earth's climatic change 135,000–107,000 years ago: Nature, v. 316, p. 500–507.

Duplessy, J. C., Chenouard, L., and Vila, F., 1975, Weyl's theory of glaciation supported by isotopic study of Norwegian Core K11: Science, v. 188, p. 1208–1209.

Duplessy, J. C., Moyes, J., and Pujol, C., 1980, Deep water formation in the North Atlantic Ocean during the last ice age: Nature, v. 286, p. 479–482.

Edgar, N. T., Saunders, J. B., and others, 1973, Initial reports of the Deep Sea Drilling Project, Volume 15: Washington, D.C., U.S. Government Printing Office, xii + 1,137 p.

Edmond, J. M., 1974, On the dissolution of carbonate and silicate in the deep ocean: Deep-Sea Research, v. 21, p. 455–480.

Ehrmann, W. U., and Thiede, J., 1985, History of Mesozoic and Cenozoic sediment fluxes to the North Atlantic Ocean: Contributions to Sedimentology, Volume 15: Stuttgart, West Germany, E. Schweizerbart'sche Verlagsbuchhandlung, 109 p.

Emerson, S., 1985, Organic carbon preservation in marine sediments, *in* Sundquist, E., and Broecker, W. S., eds., The carbon cycle and atmospheric CO_2: Natural variations, Archaean to present: American Geophysical Union Geophysical Monograph Series, Volume 32, p. 78–87.

Emerson, S., and Bender, M., 1981, Carbon fluxes at the sediment-water interface of the deep sea: Dissolution of calcium carbonate rich sediments: American Journal of Science, v. 278, p. 344–353.

Emiliani, C., 1954, Temperatures of Pacific bottom waters and polar superficial waters during the Tertiary: Science, v. 119, p. 853–855.

——— 1955, Pleistocene temperatures: Journal of Geology, v. 63, p. 538–578.

——— 1958, Ancient temperatures: Scientific American, v. 198, p. 154–163.

——— 1961, The temperature decrease of surface sea water in high latitudes and of abyssal-hadal water in open oceanic basins during the past 75 million years: Deep-Sea Research, v. 8, p. 144–147.

——— 1971, The amplitude of Pleistocene climatic cycles at low latitudes and the isotopic composition of glacial ice, *in* Turekian, K. K., ed., The late Cenozoic glacial ages: New Haven, Connecticut, Yale University Press, p. 183–197.

——— 1981, A new global geology, *in* Emiliani, C., ed., The sea, Volume 7, The oceanic lithosphere: New York, Wiley-Interscience, p. 1687–1728.

Emiliani, C., and Edwards, G., 1953, Tertiary ocean bottom temperatures: Nature, v. 171, p. 887–889.

Emiliani, C., and Epstein, S., 1953, Temperature variations in the lower Pleistocene of southern California: Journal of Geology, v. 61, p. 171–181.

Emiliani, C., Gartner, S., and Lidz, B., 1972, Neogene sedimentation on the Blake Plateau and the emergence of the Central American land bridge: Palaeogeography, Palaeoclimatology, Palaeoecology, v. 11, p. 1–10.

Ericson, D. B., Ewing, M., Wollin, G., and Heezen, B. C., 1961, Atlantic deep-sea sediment cores: Geological Society of America Bulletin, v. 72, p. 193–286.

Ewing, M., Worzel, J. L., Beall, A. O., Berggren, W. A., Bukry, D., Burk, C. A., Fischer, A. G., and Pessagno, E. A., Jr., 1969, Site 5, *in* Ewing, M., Worzel, J. L., and others, Initial reports of the Deep Sea Drilling Project, Volume 1: Washington, D.C., U.S. Government Printing Office, p. 214–242.

Fairbanks, R. G., and Matthews, R. K., 1978, The marine oxygen isotope record in Pleistocene coral, Barbados, West Indies: Quaternary Research, v. 10, p. 181–196.

Fischer, A. G., and Arthur, M. A., 1977, Secular variations in the pelagic realm, *in* Cook, H. E., and Enos, P., eds., Deep-water carbonate environments: Society of Economic Paleontologists and Mineralogists Special Publication 25, p. 19–50.

Frakes, L. A., 1986, Mesozoic-Cenozoic climatic history and causes of the glaciation, *in* Hsü, K. J., ed., Mesozoic and Cenozoic oceans: Washington, D.C., American Geophysical Union, Geodynamics Series, Volume 15, p. 33–48.

Frakes, L. A., and Kemp, E. M., 1972, Influence of continental positions on early Tertiary climates: Nature, v. 240, p. 97–100.

Fuglister, F. C., 1960, Atlantic Ocean Atlas: Temperature and salinity profiles and data from the International Geophysical Year of 1957–1958: Woods Hole, Massachusetts, Woods Hole Oceanographic Institution, 209 p.

Ganssen, G., and Troelstra, S. R., 1987, Paleoenvironmental changes from stable isotopes in planktonic foraminifera from eastern Mediterranean sapropels: Marine Geology, v. 75, p. 221–230.

Garlick, G. D., 1972, Oxygen isotope geochemistry, *in* Fairbridge, R. W., ed., The encyclopedia of geochemistry and environmental sciences: New York, Van Nostrand Reinhold Company, p. 864–874.

Gartner, S., Chen, M. P., and Stanton, R. J., Jr., 1983, Late Neogene nannofossil stratigraphy and paleooceanography of the northeastern Gulf of Mexico and adjacent areas: Marine Micropaleontology, v. 8, p. 17–50.

Gartner, S., Chow, J., and Stanton, R. J., Jr., 1987, Late Neogene paleooceanography of the eastern Caribbean, Gulf of

Mexico and the eastern equatorial Pacific: Marine Micropaleontology, v. 12, p. 255–304.
——— 1974, Atlas okeanov, Tichiy Okean: Ministerstvo Oboronyi SSSR, Voenno-Morskoe Flot, 302 + 20 p.
——— 1977, Atlas okeanov, Atlanticheskiy i Indiyskiy Okean'i: Ministerstvo Oboronyi SSSR, Voenno-Morskoe Flot, 306 + 27 p.
Gorshkov, S. G., 1980, Atlas okeanov, Severn'i Ledovit'i Okean: Ministerstvo Oboronyi SSSR, Voenno-Morskoe Flot, 184 + 4 p.
Graciansky, P. C., Deroo, G., Herbin, J. P., Montadert, L., Muller, C., Schaaf, A., and Sigal, J., 1984, Ocean wide stagnation episode in the Late Cretaceous: Nature, v. 308, p. 346–349.
Habib, D., 1968, Spores, pollen and microplankton from the Horizon Beta outcrop: Science, v. 162, p. 1480–1482.
Hammond, A. L., 1970, Deep sea drilling: A giant step in geological research: Science, v. 170, p. 520–521.
Haq, B. U., 1980, Biogeographic history of Miocene calcareous nannoplankton and paleoceanography of the Atlantic Ocean: Micropaleontology, v. 26, p. 414–443.
——— 1981, Paleogene paleoceanography: Early Cenozoic oceans revisited: Oceanologica Acta, no. Sp., p. 71–82.
——— 1984, Paleoceanography: A synoptic overview of 200 million years of ocean history, *in* Haq, B. U., and Milliman, J. D., Marine geology and oceanography of Arabian Sea and coastal Pakistan: New York, Van Nostrand Reinhold Co., p. 201–231.
Hart, M. B., and Ball, K. C., 1986, Late Cretaceous anoxic events, sea level changes and the evolution of the planktonic foraminifera, *in* Summerhayes, C. P., and Shackleton, N. J., eds., North Atlantic paleoceanography: Geological Society of London Special Publication No. 21, p. 67–78.
Harvey, J. G., 1976, Atmosphere and ocean: Our fluid environments: London, England, Artemis Press, 143 p.
Hay, R. L., 1966, Zeolites and zeolitic reactions in sedimentary rocks: Geological Society of America Special Paper 85, 129 p.
Hay, W. W., 1970, Calcium carbonate compensation, *in* Bader, R. A., Gerard, R. D., and others, Initial reports of the Deep Sea Drilling Project, Volume 3: Washington, D.C., U.S. Government Printing Office, p. 669, 672.
——— 1983, The global significance of regional Mediterranean Neogene paleoenvironmental studies: Utrecht Micropaleontological Bulletin, v. 30, p. 1–23.
——— 1985a, Potential errors in estimates of carbonate rock accumulating through geologic time, *in* Sundquist, E. T., and Broecker, W. S., eds., The carbon cycle and atmospheric CO_2: Natural variations, Archaean to present: American Geophysical Union Geophysical Monograph 32, p. 573–583.
——— 1985b, Paleoceanography of the Venezuelan Basin: First Geological Conference of the Geological Society of Trinidad and Tobago, Proceedings, p. 302–307.
——— 1987, The past and future of scientific ocean drilling: International Geological Congress, 27th, Moscow, 4–14 August, 1984, General Proceedings, p. 27–41.
Hay, W. W., and Southam, J. R., 1975, Calcareous plankton and loss of CaO from the continents: Geological Society of America Abstracts with Programs, v. 7, p. 1105.
——— 1977, Modulation of marine sedimentation by the continental shelves, *in* Anderson, N. R., and Malahoff, A., eds., The role of fossil fuel CO_2 in the oceans: New York, Plenum Press, p. 569–605.
Hay, W. W., Mohler, H. P., Roth, P. H., Boudreaux, J. E., and Schmidt, R. R., 1967, Calcareous nannoplankton zonation of the Cenozoic of the Gulf Coast and Caribbean-Antillean area, and transoceanic correlation: Gulf Coast Association of Geological Societies Transactions, v. 17, p. 428–480.
Hay, W. W., Sibuet, J.-C., Barron, E. J., Boyce, R. E., Brassell, S., Dean, W. E., Huc, A. Y., Keating, B. H., McNulty, C. L., Meyers, P. A., Nohara, M., Schallreuter, R. E., Steinmetz, J. C., Stow, D., and Stradner, H., 1982, Sedimentation and accumulation of organic carbon in the Angola Basin and on Walvis Ridge: Preliminary results of Deep Leg Drilling Project Leg 75: Geological Society of America Bulletin, v. 93, p. 1038–1050.
——— 1984, Site 530, *in* Hay, W. W., Sibuet, J.-C., and others, Initial reports of the Deep Sea Drilling Project, Volume 75: Washington, D.C., U.S. Government Printing Office, p. 29–285.
Heath, G. R., 1969, Carbonate sedimentation in the abyssal equatorial Pacific during the past 50 million years: Geological Society of America Bulletin, v. 80, p. 689–694.
——— 1974, Dissolved silica and deep-sea sediments, *in* Hay, W. W., ed., Studies in paleo-oceanography: Society of Economic Paleontologists and Mineralogists Special Publication No. 20, p. 77–93.
Heath, G. R., and Culberson, 1970, Calcite: Degree of saturation, rate of dissolution, and the compensation depth in the deep oceans: Geological Society of America Bulletin, v. 81, p. 3157–3160.
Heath, G. R., Moore, T. C., Jr., and Dauphin, J. P., 1977a, Organic carbon in deep-sea sediments, *in* Anderson, N. R., and Malahoff, A., eds., The role of fossil fuel CO_2 in the oceans: New York, Plenum Press, p. 605–625.
Heath, G. R., Moore, T. C., Jr., and van Andel, Tj. H., 1977b, Carbonate accumulation and dissolution in the equatorial Pacific during the past 45 million years, *in* Anderson, N. R., and Malahoff, A., eds., The role of fossil fuel CO_2 in the oceans: New York, Plenum Press, p. 627–639.
Hecht, A. D.,1976, The oxygen isotopic record of foraminifera in deep-sea sediment, *in* Hedley, R. H., and Adams, C. G., eds., Foraminifera Volume 2: London, England, Academic Press, p. 1–43.
Herbin, J. P., and Deroo, G., 1982, Sedimentologie de la matiere organique dans les formations du Mesozoique de l'Atlantique Nord: Bulletin de la Societe geologique de France, ser. 7, v. 24, p. 497–510.
Herbin, J. P., Montadert, L., Muller, C., Gomez, R., Thurow, J., and Wiedmann, J., 1986, Organic-rich sedimentation at the Cenomanian-Turonian boundary in oceanic and coastal basins in the North Atlantic and Tethys, *in* Summerhayes, C. P., and Shackleton, N. J., eds., North Atlantic paleoceanography: Geological Society of London Special Publication No. 21, p. 389–422.
Hess, H. H., 1959, The AMSOC hole to the Earth's mantle: American Geophysical Union Transactions, v. 40, p. 340–345.
Hester, K., and Boyle, E. A., 1982, Water chemistry control of cadmium content in recent benthic foraminifera: Nature, v. 298, p. 260–262.
Hodell, D. A., and Kennett, J. P., 1985, Miocene paleoceanography of the South Atlantic Ocean at 22, 16, and 8 Ma, *in* Kennett, J. P., ed., The Miocene ocean—Paleoceanography and biogeography: Geological Society of America Memoir 163, p. 317–337.
Hodell, D. A., Kennett, J. P., and Leonard, K. A., 1983, Climatically induced changes in vertical water mass structure of the Vema Channel during the Pliocene: Evidence from Deep Sea Drilling Project Holes 516A, 517, and 518, *in* Barker, P. F., Carlson, R. L., Johnson, D. A., and others, Initial reports of the Deep Sea Drilling Project, Volume 72: Washington, D.C., U.S. Government Printing Office, p. 907–919.
Hodell, D. A., Elmstrom, K. M., and Kennett, J. P., 1986, Latest Miocene benthic $d^{18}O$ changes, global ice volume, sea level and the 'Messinian salinity crisis': Nature, v. 320, p. 411–414.
Honjo, S., 1980, Material fluxes and modes of sedimentation in the mesopelagic and bathypelagic zones: Journal of Marine Research, v. 38, p. 53–97.
Honnorez, J., 1983, Basalt-seawater exchange: A perspective from an experimental viewpoint, *in* Rona, P. A., Bostrom, K., Laubier, L., and Smith, K. L., Jr., Hydrothermal processes at seafloor spreading centers: NATO Conference Series IV, Marine Sciences, v. 12: New York, Plenum Press, p. 169–176.
Hsü, K. J., and Andrews, J. E., 1970, Mid-Atlantic ridge sequence, lithology, *in* Maxwell, A. E., von Herzen, R. P., and others, Initial reports of the Deep Sea Drilling Project, Volume 3: Washington, D.C., U.S. Government Printing Office, p. 445–453.
Hsü, K. J., and Weissert, H. J., 1985, Introduction, *in* Hsü, K. J., and Weissert, H. J., eds., South Atlantic paleoceanography: Cambridge, England, Cambridge University Press, p. 1–9.
Hsü, K. J., and Wright, R., 1985, History of calcite dissolution of the South Atlantic Ocean, *in* Hsü, K. J., and Weissert, H. J., eds., South Atlantic paleoceanography: Cambridge, England, Cambridge University Press, p. 149–187.
Hsü, K. J., LaBrecque, J. L., and others, 1984, Initial reports of the Deep Sea Drilling Project, Volume 73: Washington, D.C., U.S. Government Printing Office, xxv + 798 p.
Hsü, K. J., McKenzie, J. A., and Oberhänsli, H., 1984, South Atlantic Cenozoic paleoceanography, *in* Hsü, K. J., LaBrecque, J. L., and others, Initial reports of the Deep Sea Drilling Project, Volume 73: Washington, D.C., U.S. Government Printing Office, p. 771–785.
Hughes, T. J., Denton, G. H., and Grosswald, M. G., 1977, Was there a late-Wurm Arctic ice sheet?: Nature, v. 266, p. 596–602.
Imbrie, J., and Kipp, N. G., 1971, A new micropaleontological method for quantitative paleoclimatology: Application to a late Pleistocene core, *in* Turekian, K. K., ed., The late Cenozoic glacial ages: New Haven, Yale University Press, p. 71–181.
Johnson, D. A., 1982, Abyssal teleconnections: Interactive dynamics of the deep ocean circulation: Palaeogeography, Palaeoclimatology, Palaeoecology, v. 38, p. 93–128.
——— 1983, Paleocirculation of the South Atlantic, *in* Barker, P. F., Johnson, D. A., and others, Initial reports of the Deep Sea Drilling Project, Volume 72: Washington, D.C., U.S. Government Printing Office, p. 977–994.
Johnson, D. A., 1985, Abyssal teleconnections II. Initiation of Antarctic Bottom Water flow in the southwestern Atlantic, *in* Hsü, K. J., and Weissert, H. J., eds., South Atlantic paleoceanography: Cambridge, England, Cambridge University Press, p. 243–281.
Kaneps, A. G., 1979, Gulf Stream: Velocity fluctuations during the late Cenozoic: Science, v. 204, p. 297–301.
Keigwin, L. D., Jr., 1979, Late Cenozoic stable isotopic stratigraphy and paleoceanography of DSDP Sites from the east equatorial and central north Pacific Ocean: Earth and Planetary Science Letters, v. 45, p. 361–382.
——— 1980, Paleoceanographic change in the Pacific at the Eocene-Oligocene boundary: Nature, v. 287, p. 722–725.
——— 1982a, Isotopic paleoceanography of the Caribbean and East Pacific: Role of Panama uplift in late Neogene time: Science, v. 217, p. 350–353.
——— 1982b, Stable isotope stratigraphy and paleoceanography of Sites 502 and 503, *in* Prell, W. L., Gardner, J. V., and others, Initial reports of the Deep Sea Drilling Project, Volume 68: Washington, D.C., U.S. Government Printing Office, p. 445–453.
——— 1987, Toward a high-resolution chronology for latest Miocene paleoceanographic events: Paleoceanography, v. 2, p. 639–660.
Keigwin, L. D., and Corliss, B. H., 1986, Stable isotopes in late middle Eocene to Oligocene foraminifera: Geological Society of America Bulletin, v. 97, p. 335–345.
Keller, G., and Barron, J. A., 1983, Paleoceanographic implications of Miocene deep sea hiatuses: Geological Society of America Bulletin, v. 94, p. 590–613.
——— 1987, Paleodepth distribution of Neogene deep-sea hiatuses: Paleoceanography, v. 2, p. 697–714.
Kellogg, T. B., 1987, Glacial-interglacial changes in global deepwater circulation: Paleoceanography, v. 2, p. 259–271.
Kennett, J. P., 1977, Cenozoic evolution of Antarctic glaciation, the circum-Antarctic current and their impact on global paleooceanography: Journal of Geophysical Research, v. 82, p. 3843–3860.
——— 1982, Marine geology: Englewood Cliffs, New Jersey, Prentice-Hall, Inc., xv + 752 p.
——— editor, 1985, The Miocene ocean—Paleoceanography and biogeography: Geological Society of America Memoir 163, vi + 337 p.
——— 1986, Miocene to early Pliocene oxygen and carbon isotope stratigraphy in the southwest Pacific; Deep Sea Drilling Project Leg 90, *in* Kennett, J. P., von der Borch, C. C., and others, Initial reports of the Deep Sea Drilling Project, Volume 90: Washington, D.C., U.S. Government Printing Office, p. 1383–1411.
Kennett, J. P., and Shackleton, N. J., 1976, Oxygen isotopic evidence for the development of the psychrosphere 38 m.y. ago: Nature, v. 260, p. 513–515.
Kennett, J. P., and Watkins, N. D., 1976, Regional deep-sea dynamic processes recorded by late Cenozoic sediments of the southeastern Indian Ocean: Geological Society of America Bulletin, v. 87, p. 321–339.
Kennett, J. P., Houtz, R. E., Andrews, P. B., Edwards, A. R., Gostin, V. A., Hajos, M., Hampton, M., Jenkins, D. G., Margolis, S. V., Ovenshine, A. T., and Perch-Nielsen, K., 1975, Cenozoic paleoceanography in the southwest Pacific Ocean, Antarctic glaciation and the development of the Circum-Antarctic Current, *in* Kennett, J. P., Houtz, R. E., and others, Initial reports of the Deep Sea Drilling Project, Volume 29: Washington, D.C., U.S. Government Printing Office, p. 1155–1169.
Kennett, J. P., Keller, G., and Srinivasan, M. S., 1985, Miocene planktonic foraminiferal biogeography and paleoceanographic development of the Indo-Pacific region, *in* Kennett, J. P., ed., The Miocene ocean: Paleoceanography and biogeography: Geological Society of America Memoir 163, p. 197–236.
Kipp, N. G., 1976, New transfer function for estimating past sea-surface conditions from sea-bed distribution of planktonic foraminiferal assemblages in the North Atlantic, *in* Cline, R. M., and Hays, J. D., Investigation of late Quaternary paleoceanography and paleoclimatology: Geological Society of America Memoir 145, p. 3–41.
Krasheninnikov, V. A., and Basov, I. A., 1985, Stratigrafiya mela Yuzhnogo Okeana: Izdatelstvo "Nauka," Moscow, 174 p.
Kraus, E. B., Petersen, W. H., and Rooth, C. G., 1978, The thermal evolution of the ocean: International Conference, Evolution of Planetary Atmospheres and Climatology of the Earth: Centre national d'etudes spatiales (France), p. 201–211.
Kuenen, Ph. H., 1950, Marine geology: New York, John Wiley & Sons, x + 568 p.
Kullenberg, B., 1947, The piston core sampler: Svenska Hydro-Biol. Komm. Skrifter, S. 3, Bd. 1, Hf. 2, 46 p.
Labeyrie, L. D., Duplessy, J. C., and Blanc, P. L., 1987, Variations in mode of formation and temperature of oceanic deep waters over the past 125,000 years: Nature, v. 327, p. 477–482.
Lasaga, A. C., Berner, R. A., and Garrels, R. M., 1985, An improved geochemical model of atmospheric CO_2 fluctuations over the past 100 million years, *in* Sundquist, E. T., and Broecker, W. S., eds., The carbon cycle and atmospheric CO_2: Natural variations, Archaean to present: American Geophysical Union Geophysical Monograph 32, p. 397–411.
Le Pichon, X., Melguen, M., and Sibuet, J.-C., 1978, A schematic model of the evolution of the South Atlantic, *in* Charnock, H., and Deacon, G., eds., Advances in oceanography: New York, Plenum Press, p. 1–48.
Ledbetter, M. T., 1979, Fluctuations of the Antarctic Bottom Water velocity in the Vema Channel during the last 160,000 years: Marine Geology, v. 33, p. 71–89.
——— 1981, Paleoceanographic significance of bottom-current fluctuations in the southern ocean: Nature, v. 294, p. 554–556.
Ledbetter, M. T., and Ciesielski, P. F., 1982, Bottom-current erosion along a traverse in the South Atlantic sector of the Southern Ocean: Marine Geology, v. 46, p. 329–341.
Ledbetter, M. T., Williams, D. F., and Ellwood, B. B., 1978, Late Pliocene climate and southwest Atlantic abyssal circulation: Nature, v. 272, p. 237–239.
Leinen, M., Cwienk, D., Heath, G. R., Biscaye, P. E., Kolla, V., Thiede, J., and Dauphin, J. P., 1986, Distribution of biogenic silica and quartz in recent deep-sea sediments: Geology, v. 14, p. 199–203.
Leventer, A., Williams, D. F., and Kennett, J. P., 1983, Relationships between anoxia, glacial meltwater and microfossil preservation in the Orca Basin, Gulf of Mexico: Marine Geology, v. 53, p. 23–40.
Li, Y. H., Takahashi, T., and Broecker, W. S., 1969, Degree of saturation of $CaCO_3$ in the oceans: Journal of Geophysical Research, v. 74, p. 5507–5525.
Lidz, L., 1966, Deep-sea Pleistocene biostratigraphy: Science, v. 154, p. 1448–1452.
Lindstrom, D. R., and MacAyeal, D. R., 1986, Paleoclimatic constraints on the maintenance of possible ice shelf cover in the Norwegian and Greenland Seas: Paleoceanography, v. 1, p. 313–338.
Linsley, B. K., Thunell, R. C., Morgan, C., and Williams, D. F., 1985, Oxygen minimum expansion in the Sulu Sea, western equatorial Pacific, during the last glacial low stand of sea level: Marine Micropaleontology, v. 9, p. 395–418.
Lisitzin, A. P., 1972, Sedimentation in the world ocean: Society of Economic Paleontologists and Mineralogists Special Publication 17, 218 p.
——— 1974, Osadkoobrazovanie v Okeanach: Moscow, Izdatel'stvo "Nauka," 438 p.
Loutit, T. S., Kennett, J. P., and Savin, S. M., 1984, Miocene equatorial and southwest Pacific paleoceanography from stable isotope evidence: Marine Micropaleontology, v. 8, p. 215–233.
Ludwig, W. J., Krasheninikov, V. A., Basov, I. A., Bayer, U., Bloemendal, J., Bornhold, B., Ciesielski, P. F., Goldstein, E. H., Robert, C., Salloway, J., Usher, J. L., Von der Dick, H., Weaver, F. M., and Wise, S. W., Jr., 1983, Site 511, *in* Ludwig, W. J., Krasheninikov, V. A., and others, Initial reports of the Deep Sea Drilling Project, Volume 71: Washington, D.C., U.S. Government Printing Office, p. 21–109.
Lyle, M., Murray, D. W., Finney, B. P., Dymond, J., Robbins, J. M., and Brooksforce, K., 1988, The record of late Pleistocene biogenic sedimentation in the eastern tropical Pacific ocean: Paleoceanography, v. 3, p. 39–59.
Malfait, B. T., and Dinkleman, M. G., 1972, Circum-Caribbean tectonic and igneous activity and the evolution of the Caribbean plate: Geological Society of America Bulletin, v. 83, p. 251–272.
Matthews, R. K., and Poore, R. Z., 1980, Tertiary $\delta^{18}O$ record and glacio-eustatic sea-level fluctuations: Geology, v. 8, p. 501–504.
McCoy, F. W., and Zimmermann, H. B., 1977, A history of sediment lithofacies in the South Atlantic Ocean, *in* Supko, P. R., Perch-Nielsen, K., and others, Initial reports of the Deep Sea Drilling Project, Volume 39: Washington, D.C., U.S. Government Printing Office, p. 1047–1079.
McIntyre, A., Kipp, N. G., Be, A.W.H., Crowley, T., Kellogg, T., Gardner, J. V., Prell, W., and Ruddiman, W. F., 1976, Glacial North Atlantic 18,000 years ago: A CLIMAP reconstruction, *in* Cline, R. M., and Hays, J. D., Investigation of late Quaternary paleoceanography and paleoclimatology: Geological Society of America Memoir 145, p. 43–76.
McKenzie, J. A., and Oberhänsli, H., 1985, Paleoceanographic expressions of the Messinian salinity crisis, *in* Hsü, K. J., and Weissert, H. J., eds., South Atlantic paleoceanography: Cambridge, England, Cambridge University Press, p. 99–123.
Medoc Group, 1970, Observation of formation of deep water in the Mediterranean Sea: Nature, v. 227, p. 1037–1040.
Melguen, M., 1978, Facies evolution, carbonate dissolution cycles in sediments from the eastern South Atlantic (DSDP Leg 40) since the Early Cretaceous, *in* Bolli, H. M., Ryan, W.B.F., and others, Initial reports of the Deep Sea Drilling Project, Volume 40: Washington, D.C., U.S. Government Printing Office, p. 981–1003.
Meyers, P. A., Dunham, K. W., and Dunham, P. L., 1986, Organic geochemistry of Cretaceous organic-carbon–rich

shales and limestones from the western North Atlantic ocean, *in* Summerhayes, C. P., and Shackleton, N. J., eds., North Atlantic paleoceanography: London Geological Society Special Publication No. 21, p. 333–345.
Miller, A. R., Tchernia, P., and Charnock, R., 1970, Mediterranean Sea atlas of temperature-salinity-oxygen; profiles and data from cruises of R. V. Atlantis and R. V. Chain: Woods Hole, Massachusetts, Woods Hole Oceanographic Institution, 190 p.
Miller, K. G., and Fairbanks, R. G., 1983, Evidence for Oligocene–middle Miocene abyssal circulation changes in the western North Atlantic: Nature, v. 306, p. 250–253.
——— 1985, Oligocene-Miocene global carbon and abyssal circulation changes, *in* Sundquist, E., and Broecker, W. S., eds., The carbon cycle and atmospheric CO_2: Natural variations, Archaean to present: American Geophysical Union Geophysical Monograph Series, Volume 32, p. 469–486.
Miller, K. G., Aubry, M. P., Khan, M. J., Melillo, M. J., Kent, D. V., and Berggren, W. A., 1985, Oligocene to Miocene biostratigraphy, magnetostratigraphy and isotopic stratigraphy of the western North Atlantic: Geology, v. 13, p. 257–261.
Miller, K. G., Fairbanks, R. G., and Mountain, G. S., 1987a, Tertiary oxygen isotope synthesis, sea level history, and continental margin erosion: Paleoceanography, v. 2, p. 1–19.
Miller, K. G., Janacek, T. R., Katz, M. E., and Keil, D. J., 1987b, Abyssal circulation and benthic foraminiferal changes near the Paleocene/Eocene boundary: Paleoceanography, v. 2, p. 741–761.
Miller, K. G., Feigenson, M. D., Kent, D. V., and Olsson, R. K., 1988, Upper Eocene to Oligocene isotope ($^{87}Sr/^{86}Sr$, $\delta^{18}O$, $\delta^{13}C$) standard section, Deep Sea Drilling Project Site 522: Paleoceanography, v. 3, p. 223–233.
Miller, K. J., and Tucholke, B. E., 1983, Development of Cenozoic abyssal circulation south of the Greenland-Scotland Ridge, *in* Bott, M., Saxov, S., Talwani, M., and Thiede, J., eds., Structure and development of the Greenland-Scotland Ridge: New York, Plenum Press, p. 549–589.
Miller, K. J., Mountain, G. S. and Tucholke, B. E., 1985, Oligocene glacio-eustasy and erosion on the margins of the North Atlantic: Geology, v. 13, p. 10–13.
Millero, F. J., and Poisson, A., 1981, International one-atmosphere equation of state of seawater: Deep-Sea Research, v. 28A, p. 625–629.
Millero, F. J., Chen, C.-T., Bradshaw, A., and Schleicher, K., 1980, A new high-pressure equation of state for sea water: Deep-Sea Research, v. 27A, p. 255–264.
Milliman, J. D., 1974, Marine carbonates, *in* Milliman, J. D., Mueller, G., and Foerster, U., eds., Recent sedimentary carbonates: Heidelberg, West Germany, Springer-Verlag, pt. 1, xv + 375 p.
Miskell, K. J., Brass, G. W., and Harrison, C.G.A., 1985, Global patterns in opal deposition from Late Cretaceous to late Miocene: American Association of Petroleum Geologists Bulletin, v. 69, p. 996–1012.
Mix, A., 1987, The oxygen isotope record of glaciation, *in* Ruddiman, W. F., and Wright, H. E., Jr., eds., North America and adjacent oceans during the last deglaciation: The Geology of North America, Volume K-3, p. 111–136.
Molina-Cruz, A., and Price, P., 1977, Distribution of opal and quartz on the ocean floor of the subtropical southeastern Pacific: Geology, v. 5, p. 81–84.
Monin, A. S., and Lisitzin, A. P., editors, 1983, Biogeochimiya Okeana: Moscow, Izdatel'stvo "Nauka," 368 p.
Moody, J. B., Worsley, T. R., and Mangoonian, P. R., 1981, Long term phosphorus flux to deep sea sediments: Journal of Sedimentary Petrology, v. 51, p. 307–312.
Moody, J. B., Chaboudy, L. R., Jr., and Worsley, T. R., 1988, Pacific pelagic phosphorus accumulation during the last 10 m.y.: Paleoceanography, v. 3, p. 113–136.
Moore, T. C., Jr., and Heath, G. R., 1977, Survival of deep sea sedimentary sections: Earth and Planetary Science Letters, v. 37, p. 71–80.
Moore, T. C., Jr., Pisias, N. G., and Heath, G. R., 1977, Climate changes and lags in Pacific carbonate preservation, surface temperatures and global ice volume, *in* Anderson, N. R., and Malahoff, A., eds., The fate of fossil fuel CO_2 in the oceans: New York, Plenum Press, p. 145–165.
Moore, T. C., van Andel, Tj. H., Sancetta, C., and Pisias, N., 1978, Cenozoic hiatuses in pelagic sediments: Micropaleontology, v. 24, p. 113–138.
Moore, T. C., Jr., Rabinowitz, P. D., and others, 1984, Initial reports of the Deep Sea Drilling Project, Volume 74: Washington, D.C., U.S. Government Printing Office, xxii + 894 p.
Moore, T. C., Jr., Rabinowitz, P. D., Borella, P. E., Shackleton, N. J., and Boersma, A., 1985, History of the Walvis Ridge. A precis of the results of DSDP Leg 74, *in* Hsü, K. J., and Weissert, H. J., eds., South Atlantic paleoceanography: Cambridge, England, Cambridge University Press, p. 57–60.
Morley, J. J., Pisias, N. G., and Leinen, M., 1987, Late Pleistocene time series of atmospheric and oceanic variables recorded in sediments from the Subarctic Pacific: Paleoceanography, v. 2, p. 21–48.
Mottl, M. J., 1983, Hydrothermal processes at seafloor spreading centers: Application of basalt-seawater experimental results, *in* Rona, P. A., Bostrom, K., Laubier, L., and Smith, K. L., Jr., Hydrothermal processes at seafloor spreading centers: NATO Conference Series IV, Marine Sciences, Volume 12: New York, Plenum Press, p. 199–223.
Mueller, D. W., and Hsü, K. J., 1987, Event stratigraphy and paleoceanography in the Fortuna Basin (southeast Spain): A scenario for the Messinian salinity crisis: Paleoceanography, v. 2, p. 679–696.
Mueller, P. J., and Suess, E., 1979, Productivity, sedimentation rate and sedimentary organic matter in the oceans; I. Organic carbon preservation: Deep-Sea Research, v. 26, p. 1347–1367.
Mullins, H. T., Gardulski, A. F., Wise, S. W., Jr., and Applegate, J., 1987, Middle Miocene oceanographic event in the eastern Gulf of Mexico: Implications for seismic stratigraphic succession and Loop Current/Gulf Stream circulation: Geological Society of America Bulletin, v. 98, p. 702–713.
Murphy, M. G., and Kennett, J. P., 1985, Development of latitudinal thermal gradients during the Oligocene: Oxygen-isotope evidence from the southwest Pacific, *in* Kennett, J. P., von der Borch, C. C., and others, Initial reports of the Deep Sea Drilling Project, Volume 90: Washington, D.C., U.S. Government Printing Office, p. 1347–1360.
Murray, J. W., Weston, J. F., Haddon, C. A., and Powell, A.D.J., 1986, Miocene to Recent bottom water masses of the north-east Atlantic: An analysis of benthic foraminifera, *in* Summerhayes, C. P., and Shackleton, N. J., eds., North Atlantic palaeoceanography: Geological Society of London Special Publication 21, p. 219–230.
Muza, J. P., Williams, D. F., and Wise, S. W., Jr., 1983, Paleogene oxygen isotope records for the subantarctic South Atlantic Ocean: DSDP Sites 511 and 512: Antarctic Journal of the United States, v. 18, p. 146–147.
Natland, J. H., 1978, Composition, provenance and diagenesis of Cretaceous clastic sediments drilled on the Atlantic continental rise off southern Africa, DSDP Site 361—Implications for the early circulation of the South Atlantic, *in* Bolli, H. M., Ryan, W.B.F., and others, Initial reports of the Deep Sea Drilling Project, Volume 40: Washington, D.C., U.S. Government Printing Office, p. 1025–1061.
Neftel, A., Oeschger, H., Swander, J., Stauffer, B., and Zumbrunn, R., 1982, Ice core sample measurements give atmospheric CO_2 content during the past 40,000 years: Nature, v. 295, p. 220–223.
Oberhänsli, H., and Hsü, K. J., 1986, Paleocene-Eocene paleoceanography, *in* Hsü, K. J., ed., Mesozoic and Cenozoic oceans: American Geophysical Union Geodynamics Series, Volume 15, p. 85–100.
Oberhänsli, H., and Toumarkine, M., 1985, The Paleogene oxygen and carbon isotope history of Sites 522, 523, and 524 from the central South Atlantic, *in* Hsü, K. J., and Weissert, H. J., eds., South Atlantic paleoceanography: Cambridge, England, Cambridge University Press, p. 125–147.
Oliver, J., Ewing, M., and Press, F., 1953, Crustal structure and surface wave dispersion; part 4. The Atlantic and Pacific Ocean basins: Palisades, New York, Lamont Geological Observatory Technical Report 26, 39 p.
Parrish, J. T., 1982, Upwelling and petroleum source beds, with reference to the Paleozoic: American Association of Petroleum Geologists Bulletin, v. 66, p. 750–774.
Parrish, J. T., and Barron, E. J., 1986, Paleoclimates and economic geology: Society of Economic Paleontologists and Mineralogists Short Course No. 18, 162 p.
Parrish, J. T., and Curtis, R. L., 1982, Atmospheric circulation, upwelling, and organic-rich rocks in the Mesozoic and Cenozoic Eras: Palaeogeography, Palaeoclimatology, Palaeoecology, v. 40, p. 31–66.
Perch-Nielsen, K., Supko, P. R., and others, 1977, Initial reports of the Deep Sea Drilling Project, Volume 39: Washington, D.C., U.S. Government Printing Office, xxv + 1,079 p.
Peterson, M.N.A., 1966, Calcite: Rates of dissolution in a vertical profile in the central Pacific: Science, v. 154, p. 1542–1544.
Peterson, M.N.A., Edgar, N. T., von der Borch, C. C., and Rex, R. W., 1970, Cruise leg summary and discussion, *in* Peterson, M.N.A., Edgar, N. T., and others, Initial reports of the Deep Sea Drilling Project, Volume 2: Washington, D.C., U.S. Government Printing Office, p. 413–427.
Peterson, W. H., 1979, A steady state thermohaline convection model [Ph.D. thesis]: Miami, Florida, Rosenstiel School of Marine and Atmospheric Sciences, University of Miami, 160 p.
Peterson, W. H., and Rooth, C.G.H., 1976, Formation and exchange of deep water in the Greenland and Norwegian Seas: Deep-Sea Research, v. 23, p. 273–283.
Pickard, G. L., and Emery, W. J., 1982, Descriptive physical oceanography: Oxford, England, Pergamon Press, xiv + 249 p.
Pinet, P. R., and Popenoe, P., 1985, A scenario of Mesozoic-Cenozoic ocean circulation over the Blake Plateau and its environs: Geological Society of America Bulletin, v. 96, p. 618–626.
Pisias, N. G., and Rea, D. K., 1988, Late Pleistocene paleoclimatology of the central equatorial Pacific: Sea surface response to the southeast trade winds: Paleoceanography, v. 3, p. 21–38.
Poag, C. W., and Miller, K. G., 1986, Neogene marine microfossil biofacies of the western North Atlantic, *in* Vogt, P. R., and Tucholke, B. E., The western North Atlantic region: The geology of North America, Volume M: Boulder, Colorado, Decade of North American Geology, p. 547–564.
Pokras, E. M., 1987, Diatom record of late Quaternary climatic change in the eastern equatorial Atlantic and tropical Africa: Paleoceanography, v. 2, p. 273–286.
Poldervaart, A., 1955, Chemistry of the Earth's crust, *in* Poldervaart, A., ed., Crust of the Earth: Geological Society of America Special Paper 62, p. 119–144.
Pond, S., and Pickard, G. L., 1983, Introductory dynamical oceanography: Oxford, England, Pergamon Press, xx + 329 p.
Ramsay, A.T.S., 1974, The distribution of calcium carbonate in deep sea sediments, *in* Hay, W. W., ed., Studies in paleo-oceanography: Society of Economic Paleontologists and Mineralogists Special Publication No. 20, p. 58–76.
Rea, D. K., Leinen, M., and Janecek, T. R., 1985, Geologic approach to the long-term history of atmospheric circulation: Science, v. 227, p. 721–725.
Reid, J. L., 1979, On the contribution of the Mediterranean Sea outflow to the Norwegian-Greenland Sea: Deep-Sea Research, v. 26, p. 1199–1223.
——— 1981, On the mid-depth circulation of the world ocean, *in* Warren, B. A., and Wunsch, C., Evolution of physical oceanography, scientific surveys in honor of Henry Stommel: Cambridge, Massachusetts, MIT Press, p. 70–111.
Reid, J. L., and Lynn, R. J., 1971, On the influence of the Norwegian-Greenland and Weddell Seas upon the bottom waters of the Indian and Pacific Oceans: Deep-Sea Research, v. 18, p. 1063–1088.
Reyment, R. A., 1980, Paleo-oceanology and paleobiogeography of the Cretaceous south Atlantic Ocean: Oceanologica Acta, v. 3, p. 127–133.
Riedel, W. R., and Funnell, B. M., 1964, Tertiary sediment cores and microfossils from the Pacific ocean floor: Geological Society of London Quarterly Journal, v. 120, p. 305–368.
Ronov, A. B., 1980, The Earth's sedimentary shell (quantitative patterns of its structures, compositions and evolution). The 20th V. I. Vernadsky Lecture, March 12, 1978 [in Russian], *in* Yaroshevskii, A. A., ed., The Earth's sedimentary shell (quantitative patterns of its structures, compositions and evolution): Moscow, Nauka, 80 p. [English translation *in* International Geology Review, v. 24, p. 1313–1363, 1365–1388 (1982); also, American Geological Institute Reprint Series, no. 5, 73 p.].
Ross, D. A., and Degens, E. T., 1974, Recent sediments of Black Sea, *in* Degens, E. T., and Ross, D. A., eds., The Black Sea—Geology, chemistry, and biology: American Association of Petroleum Geologists Memoir 20, p. 183–199.
Roth, P. H., 1986, Mesozoic paleoceanography of the North Atlantic and Tethys Oceans, *in* Summerhayes, C. P., and Shackleton, N. J., eds., North Atlantic paleoceanography: Geological Society of London Special Publication No. 21, p. 299–320.
Ruddiman, W. F., 1987, Northern oceans, *in* Ruddiman, W. F., and Wright, H. E., Jr., eds., North America and adjacent oceans during the last deglaciation: The geology of North America, Volume K-3: Boulder, Colorado, Decade of North American Geology, p. 137–154.
Ruddiman, W. F., and McIntyre, A., 1976, Northeast Atlantic paleoclimatic changes over the past 600,000 years, *in* Cline, R. M., and Hays, J. D., Investigation of late Quaternary paleoceanography and paleoclimatology: Geological Society of America Memoir 145, p. 111–146.
Ruddiman, W. F., Shackleton, N. J., and McIntyre, A., 1986, North Atlantic sea-surface temperatures for the last 1.1 million years, *in* Summerhayes, C. P., and Shackleton, N. J., eds., North Atlantic palaeoceanography: Geological Society of London Special Publication 21, p. 155–174.
Ryan, W.B.F., and Cita, M. B., 1977, Ignorance concerning episodes of oceanwide stagnation: Marine Geology, v. 23, p. 197–215.
Ryan, W.B.F., Hsü, K. J., and others, 1973, Initial reports of the Deep Sea Drilling Project, Volume 13: Washington, D.C., U.S. Government Printing Office, xxiii + 1,447 p.
Saito, T., Burckle, L. H., and Hays, J. D., 1974, Implications of some pre-Quaternary sediment cores and dredgings, *in* Hay, W. W., ed., Studies in paleo-oceanography: Society of Economic Paleontologists and Mineralogists Special Publication No. 20, p. 6–36.
Saltzman, E., and Barron, E. J., 1982, Deep circulation in the Late Cretaceous, oxygen isotope paleotemperatures from *Inoceramus* remains in D.S.D.P. cores: Palaeogeography, Palaeoclimatology, Palaeoecology, v. 40, p. 167–181.
Sandberg, P. A., 1983, An oscillating trend in Phanerozoic nonskeletal carbonate mineralogy: Nature, v. 305, p. 19–22.
——— 1985, Nonskeletal aragonite and pCO_2 in the Phanerozoic and Proterozoic, *in* Sundquist, E., and Broecker, W. S., eds., The carbon cycle and atmospheric CO_2: Natural variations, Archaean to present: American Geophysical Union Geophysical Monograph Series, Volume 32, p. 585–594.
Sarnthein, M., Tetzlaff, G., Koopmann, B., Wolter, K., and Pflaumann, U., 1981, Glacial and interglacial wind regimes over the eastern subtropical Atlantic and NW Africa: Nature, v. 293, p. 193–196.
Sarnthein, M., Thiede, J., Pflaumann, U., Erlenkeuser, K., Fuetterer, D., Koopmann, B., Lange, H., and Seibold, E., 1982, Atmospheric and oceanic circulation patterns off northwest Africa during the past 25 million years, *in* von Rad, U., Hinz, K., and others, eds., Geology of the northwest African continental margin: Berlin, West Germany, Springer-Verlag, p. 545–604.
Sarnthein, M., Winn, K., and Zahn, R., 1987, Paleoproductivity of oceanic upwelling and the effect on atmospheric CO_2 and climatic change during deglaciation times, *in* Berger, W. H., and Lebeyrie, L. D., eds., Abrupt climatic change: Dordrecht, Netherlands, D. Riedel Publishing Co., p. 311–337.
Savin, S. M., 1977, The history of the Earth's surface temperature during the past 100 million years: Annual Reviews of Earth and Planetary Science, v. 5, p. 318–355.
——— 1983, Stable isotopes in climatic reconstructions, *in* Berger, W. H., and Crowell, J. C., eds., Climate in Earth history: Washington, D.C., National Academy Press, p. 164–171.
Savin, S. M., and Douglas, R. G., 1985, Sea level, climate and the Central American land bridge, *in* Stehli, F. G., and Webb, S. D., eds., The great American biotic interchange: New York, Plenum Press, p. 303–324.
Savin, S. M., and Yeh, H.-W., 1981, Stable isotopes in ocean sediments, *in* Emiliani, C., ed., The sea, Volume 7, The oceanic lithosphere: New York, Wiley-Interscience, p. 1521–1554.
Savin, S. M., Douglas, R. G., and Stehli, F. G., 1975, Tertiary marine paleotemperatures: Geological Society of America Bulletin, v. 86, p. 1499–1510.
Savin, S. M., Abel, L., Barrera, E., Hodell, D., Keller, G., Kennett, J. P., Killingley, J., Murphy, M., and Vincent, E., 1985, The evolution of Miocene surface and near surface marine temperatures: Oxygen isotopic evidence, *in* Kennett, J. P., ed., The Miocene ocean—Paleoceanography and biogeography: Geological Society of America Memoir 163, p. 49–82.
Schink, D. R., and Guinasso, N. L., Jr., 1977, Modelling the influences of bioturbation and other processes on calcium carbonate dissolution at the sea floor, *in* Anderson, N. R., and Malahoff, A., eds., The role of fossil fuel CO_2 in the oceans: New York, Plenum Press, p. 375–400.
Schlager, W., Buffler, R. T., Angstadt, D., Bowdler, J. L., Cotillon, P. H., Dallmeyer, R. D., Halley, R. B., Kinoshita, H., Magoon, L. B., III, McNulty, C. L., Patton, J. W., Pischiotto, K. A., Premoli-Silva, I., Suarez, O. A., Testarmata, M. M., Tyson, R. V., and Watkins, D. K., 1984, Deep Sea Drilling Project, Leg 77, southeastern Gulf of Mexico: Geological Society of America Bulletin, v. 95, p. 226–236.
Schlanger, S. O., and Jenkyns, H. C., 1976, Cretaceous ocean anoxic events, causes and consequences: Geologie en Mijnbouw, v. 55, p. 179–184.
Schlanger, S. O., Arthur, M. A., Jenkyns, H. C., and Scholle, P. A., 1986, The Cenomanian Turonian oceanic anoxic event. I. Stratigraphy and distribution of organic carbon rich beds and the marine $\delta^{13}C$ excursion, *in* Brooks, J., and Fleet, A., eds., Marine petroleum source rocks: Geological Society of London Special Publication 26, p. 371–399.
Schmitz, B., 1987, Barium, equatorial high productivity, and the northward wandering of the Indian continent: Paleoceanography, v. 2, p. 63–77.
Schnitker, D., 1974, West Atlantic abyssal circulation during the past 120,000 years: Nature, v. 248, p. 385–387.
——— 1980, North Atlantic oceanography as possible cause of Antarctic glaciation and eutrophication: Nature, v. 284, p. 615–616.
Schnitker, D., 1986, North-east Atlantic benthic foraminiferal faunas: tracers of deep water palaeoceanography, *in* Summerhayes, C. P., and Shackleton, N. J., eds., North Atlantic palaeoceanography: Geological Society of London Special Publication 21, p. 191–204.
Scholle, P. A., and Arthur, M. A., 1980, Carbon isotopic fluctuations in Cretaceous pelagic limestones: Potential stratigraphic and petroleum exploration tool: American Association of Petroleum Geologists Bulletin, v. 64, p. 67–87.
Schopf, T.J.M., 1980, Paleoceanography: Cambridge, Massachusetts, Harvard University Press, 341 p.

Schott, W., 1935, Die Foraminiferen in dem aequatorialen Teil des Atlantischen Ozeans: Deutscher atlantischer Expedition, Meteor 1925–1927, v. 3, p. 43–134.
Sclater, J. G., Anderson, R. N., and Bell, M. L., 1971, Elevation of ridges and evolution of the central eastern Pacific: Journal of Geophysical Research, v. 76, p. 7888–7915.
Sclater, J. G., Boyle, E., and Edmond, J. M., 1979, A quantitative analysis of some factors affecting carbonate sedimentation in the oceans, *in* Talwani, M., Hay, W., and Ryan, W.B.F., Deep drilling results in the Atlantic Ocean: Continental margins and paleoenvironment: American Geophysical Union, Maurice Ewing Series, Volume 3, p. 235–248.
Shackleton, N. J., 1967, Oxygen isotope analyses and Pleistocene temperatures re-assessed: Nature, v. 215, p. 15–17.
——— 1983, Climatic crises in the Cenozoic: Proceedings of the Joint Oceanographic Assembly 1982, General Symposia, p. 131–132.
——— 1986, The Plio-Pleistocene ocean: Stable isotope history, *in* Hsü, K. J., Mesozoic and Cenozoic oceans: American Geophysical Union Geodynamics Series, v. 15, p. 141–153.
Schackleton, N., and Boersma, A., 1981, The climate of the Eocene ocean: Geological Society of London Journal, v. 138, p. 153–157.
Shackleton, N. J., and Kennett, J. P., 1975, Paleotemperature history of the Cenozoic and the initiation of Antarctic glaciation: Oxygen and carbon isotope analyses in DSDP Sites 277, 279, and 281, *in* Kennett, J. P., Houtz, R. E., and others, Initial reports of the Deep Sea Drilling Project, Volume 29: Washington, D.C., U.S. Government Printing Office, p. 743–755.
Shackleton, N. J., and Opdyke, N. D., 1973, Oxygen isotope and paleomagnetic stratigraphy of equatorial Pacific core V28-238: Oxygen isotope temperatures and ice volumes on a 10^5 year and 10^6 year scale: Quaternary Research, v. 3, p. 39–55.
Shackleton, N. J., Hall, M. A., Line, J., and Shuxi, C., 1983, Carbon isotope data in core V19-30 confirm reduced carbon dioxide concentration in the ice age atmosphere: Nature, v. 306, p. 319–322.
Shackleton, N. J., Backman, J., Zimmermann, H., Kent, D. V., Hall, M. A., Roberts, D. G., Schnitker, D., Baldauf, J. G., Desprairies, A., Homrighausen, R., Huddlestun, P., Keene, J. B., Kaltenback, A. J., Krumsiek, K.A.O., Morton, A. C., Murray, J. W., and Westberg-Smith, J., 1984, Oxygen isotope calibration of the onset of ice-rafting and history of glaciation in the North Atlantic region: Nature, v. 307, p. 620–623.
Shokes, R. F., Trabant, P. K., Presley, B. J., and Reid, D. F., 1977, Anoxic hypersaline basin in the northern Gulf of Mexico: Science, v. 196, p. 1443–1446.
Sibley, D. F., and Vogel, T. A., 1976, Chemical mass balance of the earth's crust: The calcium dilemma and the role of pelagic sediments: Science, v. 192, p. 551–553.
Southam, J. R., and Hay, W. W., 1977, Time scales and dynamic models of deep-sea sedimentation: Journal of Geophysical Research, v. 82, p. 3825–3842.
——— 1981, Global sedimentary mass balance and sea level changes, *in* Emiliani, C., The sea, Volume 7, The oceanic lithosphere: New York, Wiley-Interscience, p. 1617–1684.
Southam, J. R., Peterson, W. H., and Brass, G. W., 1982, Dynamics of anoxia: Palaeogeography, Palaeoclimatology, Palaeoecology, v. 40, p. 183–198.
Stow, D.A.V., and Dean, W. E., 1984, Middle Cretaceous black shales at Site 530 in the southeastern Angola Basin, *in* Hay, W. W., Sibuet, J.-C., and others, Initial reports of the Deep Sea Drilling Project, Volume 75: Washington, D.C., U.S. Government Printing Office, p. 809–817.
Streeter, S. S., 1973, Bottom waters and benthonic foraminifera in the North Atlantic—Glacial-interglacial contrasts: Quaternary Research, v. 3, p. 131–141.
Streeter, S. S., and Shackleton, N. J., 1979, Paleocirculation of the deep North Atlantic: A 150,000-year record of benthic foraminifera and ^{18}O: Science, v. 203, p. 168–171.
Swift, J. H., Aagaard, K., and Malmberg, S.-A., 1980, The contribution of the Denmark Strait overflow to the deep North Atlantic: Deep-Sea Research, v. 27A, p. 29–42.
Takahashi, T., and Broecker, W. S., 1977, Mechanisms for calcite dissolution on the sea floor, *in* Anderson, N. R., and Malahoff, A., eds., The role of fossil fuel CO_2 in the oceans: New York, Plenum Press, p. 455–478.
Thiede, J., 1979, History of the North Atlantic Ocean: Evolution of an asymmetric zonal paleoenvironment in a latitudinal basin, *in* Talwani, M., Hay, W., and Ryan, W.B.F., Deep drilling results in the Atlantic Ocean: Continental margins and paleoenvironment: American Geophysical Union, Maurice Ewing Series, Volume 3, p. 275–296.
——— 1981, Reworking in upper Mesozoic and Cenozoic central Pacific deep-sea sediments: Nature, v. 289, p. 667–670.
Thiede, J., and Ehrmann, W. U., 1986, Late Mesozoic and Cenozoic sediment flux to the central North Atlantic Ocean, *in* Summerhayes, C. P., and Shackleton, N. J., eds., North Atlantic paleoceanography: Geological Society of London Special Publication No. 21, p. 3–15.
Thiede, J., and van Andel, Tj. H., 1977, The paleoenvironment of anaerobic sediments of the late Mesozoic south Atlantic Ocean: Earth and Planetary Science Letters, v. 33, p. 301–309.
Thierstein, H. R., 1979, Paleoceanographic implications of organic carbon and carbonate distribution in Mesozoic deep sea sediments, *in* Talwani, M., Hay, W., and Ryan, W.B.F., Deep drilling results in the Atlantic Ocean: Continental margins and paleoenvironment: American Geophysical Union, Maurice Ewing Series, Volume 3, p. 249–274.
Thunell, R. C., 1979, Pliocene-Pleistocene paleotemperatures and paleosalinity history of the Mediterranean Sea; Results from Deep Sea Drilling Project Sites 125 and 132: Marine Micropaleontology, v. 4, p. 173–187.
——— 1982, Carbonate dissolution and abyssal hydrography in the Atlantic Ocean: Marine Geology, v. 47, p. 165–180.
Thunell, R. C., and Belyea, P., 1982, Neogene planktonic foraminiferal biogeography of the Atlantic Ocean: Micropaleontology, v. 28, p. 381–398.
Thunell, R. C., and Williams, D. F., 1983, Paleotemperature and paleosalinity history of the eastern Mediterranean during the late Quaternary: Palaeogeography, Palaeoclimatology, Palaeoecology, v. 44, p. 23–29.
Thunell, R. C., Williams, D. F., and Kennett, J. P., 1977, Late Quaternary paleoclimatology, stratigraphy and sapropel history in eastern Mediterranean deep-sea sediments: Marine Micropaleontology, v. 2, p. 371–388.
Thunell, R. C., Williams, D. F., and Cita, M. B., 1983, Glacial anoxia in the eastern Mediterranean: Journal of Foraminiferal Research, v. 13, p. 283–290.
Thunell, R. C., Williams, D. F., and Belyea, P. R., 1984, Anoxic events in the Mediterranean Sea in relation to the evolution of late Neogene climates: Marine Geology, v. 59, p. 105–134.
Thunell, R. C., Williams, D. F., and Howell, M., 1987, Atlantic-Mediterranean water exchange during the late Neogene: Paleoceanography, v. 2, p. 661–678.
Tissot, B., Deroo, G., and Herbin, J. P., 1979, Organic matter in Cretaceous sediments of the North Atlantic: Contribution to sedimentology and paleogeography, *in* Talwani, M., Hay, W., and Ryan, W.B.F., Deep drilling results in the Atlantic Ocean: Continental margins and paleoenvironment: American Geophysical Union, Maurice Ewing Series, Volume 3, p. 362–374.
Tolmazin, D., 1985, Elements of dynamic oceanography: Winchester, Massachusetts, Allen and Unwin, Inc., 181 p.
Trabant, P. K., and Presley, B. J., 1978, Orca Basin, an anoxic depression on the continental slope, northwest Gulf of Mexico, *in* Bouma, A. H., Moore, G. T., and Coleman, J. M., eds., Frame facies and oil trapping characteristics of the upper continental margin: American Association of Petroleum Geologists, Studies in Geology, Volume 7, p. 289–303.
Trainor, D., and Williams, D. F., 1987, Isotope chronostratigraphy: High resolution stratigraphic correlations in deep-water exploration tracts of the northern Gulf of Mexico: Gulf Coast Association of Geological Societies Transactions, v. 37, p. 247–254.
Tucholke, B. E., and Mountain, G. S., 1986, Tertiary paleoceanography of the western North Atlantic Ocean, *in* Vogt, P. R., and Tucholke, B. E., The western North Atlantic region: The Geology of North America, Volume M, p. 631–650.
Tucholke, B. E., and Vogt, P. R., 1979, Western North Atlantic: Sedimentary evolution and aspects of tectonic history, *in* Tucholke, B. E., Vogt, P. R., and others, Initial reports of the Deep Sea Drilling Project, Volume 43: Washington, D.C., U.S. Government Printing Office, p. 791–825.
Vail, P. R., Mitchum, R. M., Jr., and Thompson, S., III, 1977, Global cycles of relative changes of sea level, *in* Payton, C. E., ed., Seismic stratigraphy—Applications to hydrocarbon exploration: American Association of Petroleum Geologists Memoir 26, p. 83–97.
van Andel, Tj. H., 1968, Deep-sea drilling for scientific purposes: A decade of dreams: Science, v. 160, p. 1419–1424.
——— 1975, Mesozoic/Cenozoic calcite compensation depth and the global distribution of calcareous sediments: Earth and Planetary Science Letters, v. 26, p. 187–194.
van Andel, Tj. H., 1976, An eclectic overview of plate tectonics, paleogeography and paleoceanography, *in* Gray, J., and Boucot, A. J., eds., Historical biogeography, plate tectonics, & the changing environment, Oregon State University Press, Corvallis, Oregon, p. 9–25.
van Andel, Tj. H., and Moore, T. C., 1974, Cenozoic calcium carbonate distribution and calcite compensation depth in the central equatorial Pacific: Geology, v. 2, p. 87–92.
van Andel, Tj. H., Heath, G. R., and Moore, T. C., 1975, Cenozoic tectonics, sedimentation and paleo-oceanography of the central equatorial Pacific: Geological Society of America Memoir 143, p. 1–134.
van Andel, Tj. H., Thiede, J., Sclater, J. G., and Hay, W. W., 1977, Depositional history of the south Atlantic Ocean during the last 125 million years: Journal of Geology, v. 85, p. 651–698.
van Hinte, J. E., Cita, M. B., and van der Weijden, C. H., eds., 1987, Extant and ancient anoxic basin conditions in the eastern Mediterranean: Marine Geology, v. 75 (Special Issue), 281 p.
Vincent, E., and Berger, W. H., 1985, Carbon dioxide and polar cooling in the Miocene: The Monterey hypothesis, *in* Sundquist, E. T., and Broecker, W. S., eds., The carbon cycle and atmospheric CO_2: Natural variations, Archaean to present: American Geophysical Union Geophysical Monograph 32, p. 455–468.
Wallich, C. G., 1861, Remarks on some novel phases of organic life, and on the boring powers of minute annelids, at great depths in the sea: Annals and Magazine of Natural History, ser. 3, v. 43, p. 52–58.
Warren, B. A., and Wunsch, C., 1981, Evolution of physical oceanography; Scientific surveys in honor of Henry Stommel: Cambridge, Massachusetts, MIT Press, xxxiii + 623 p.
Weibull, W., 1947, The thickness of ocean sediments measured by a reflection method: Meddelelser Ocean. Inst. Gøteborg, v. 12, p. 2–17.
Weiss, R. F., 1970, The solubility of nitrogen, oxygen and argon in water and seawater: Deep-Sea Research, v. 17, p. 721–735.
Weissert, H. J., and Oberhänsli, H., 1985, Pliocene oceanography and climate: An isotope record from the southwestern Angola Basin, *in* Hsü, K. J., and Weissert, H. J., eds., South Atlantic paleoceanography: Cambridge, England, Cambridge University Press, p. 79–97.
Weyl, P. K., 1968, The role of the oceans in climatic change: a theory of the icea ages: Meteorological Monographs, v. 8, p. 37–62.
Williams, D. F., and Fillon, R. H., 1986, Meltwater influences and paleocirculation changes in the North Atlantic during the last glacial termination, *in* Summerhayes, C. P., and Shackleton, N. J., eds., North Atlantic palaeoceanography: Geological Society of London Special Publication 21, p. 175–180.
Williams, D. F., and Trainor, D., 1986, Application of isotope chronostratigraphy in the northern Gulf of Mexico: Gulf Coast Association of Geological Societies Transactions, v. 36, p. 589–600.
Williams, D. F., Thunell, R. C., Hodell, D. A., and Vergnaud-Grazzini, C., 1985, Synthesis of Late Cretaceous, Tertiary and Quaternary stable isotope records of the South Atlantic based on Leg 72 DSDP core material, *in* Hsü, K. J., and Weissert, H. J., eds., South Atlantic paleoceanography: Cambridge, England, Cambridge University Press, p. 203–241.
Williams, L. A., and Reimers, C., 1982, Recognizing organic mats in deep water environments: Geological Society of America Abstracts with Programs, v. 14, p. 647.
Wilson, K. M., 1987, Circum-Pacific suspect terranes and lost microcontinents: Chips off the old blocks: Pacific Rim Congress, 1987, AUS. IMM, Proceedings: Victoria, New South Wales, Geological Society of Australia, p. 927–930.
Wilson, K. M., Rosol, M. J., and Hay, W. W., in press, Global Mesozoic reconstructions using revised continental data and terrane histories: A progress report, *in* Hillhouse, J. W., ed., Deep structure and past kinematics of accreted terranes: American Geophysical Union Geophysical Monographs.
Windisch, C. C., Leyden, R. J., Worzel, J. L., Saito, T., and Ewing, J., 1968, Investigation of Horizon Beta: Science, v. 162, p. 1473–1475.
Wise, S. W., Gombos, A. M., and Muza, J. P., 1985, Cenozoic evolution of polar water masses, southwest Atlantic Ocean, *in* Hsü, K. J., and Weissert, H. J., eds., South Atlantic paleoceanography: Cambridge, England, Cambridge University Press, p. 283–324.
Wise, S. W., Hay, W. W., O'Connell, S., Barker, P. F., Kennett, J. P., Burckle, L. H., Egeberg, P. K., Futterer, D. K., Gersonde, R. E., Golovchenko, X., Hamilton, N., Lazarus, D., Mohr, B., Nagao, T., Pereira, C.P.G., Pudsey, C. J., Robert, C. M., Shandl, E., Speiss, V., Stott, L. D., Thomas, E., and Thompson, F.K.M., 1987, Early Oligocene ice on the Antarctic continent: Geological Society of America Abstracts with Programs, v. 19, p. 893.
Woodruff, F., 1985, Changes in Miocene deep-sea benthic foraminiferal distribution in the Pacific Ocean: Relationship to paleoceanography, *in* Kennett, J. P., ed., The Miocene ocean—Paleoceanography and biogeography: Geological Society of America Memoir 163, p. 131–175.
Woodruff, F., Savin, S. M., and Douglas, R. G., 1981, Miocene stable isotope record: A detailed deep Pacific Ocean study and its paleoclimatic implications: Science, v. 212, p. 665–668.
Worsley, T. R., Nance, R. D., and Moody, J. B., 1986, Tectonic cycles and the history of the Earth's biogeochemical and paleoceanographic record: Paleoceanography, v. 1, p. 233–263.
Worthington, L. V., 1981, The water masses of the world ocean: Some results of a fine scale census, *in* Warren, B. A., and Wunsch, C., Evolution of physical oceanography; Scientific surveys in honor of Henry Stommel: Cambridge, Massachusetts, MIT Press, p. 42–69.
Worthington, L. V., and Wright, W. R., 1970, North Atlantic Ocean atlas of potential temperature and salinity in the deep water, including temperature, salinity and oxygen profiles from the Erika Dan Cruise of 1962: Woods Hole, Massachusetts, Woods Hole Oceanographic Institution, 77 p.
Wyrtki, K., 1962, The oxygen minima in relation to ocean circulation: Deep-Sea Research, v. 9, p. 11–23.
Zahn, R., and Sarnthein, M., 1987, Benthic isotope evidence for changes of the Mediterranean outflow during the late Quaternary: Paleoceanography, v. 2, p. 543–559.

Manuscript Received by the Society March 8, 1988
Revised Manuscript Received July 27, 1988
Manuscript Accepted July 29, 1988

Printed in U.S.A.

CENTENNIAL ARTICLE

The mechanical paradox of large overthrusts

RAYMOND A. PRICE* *Geological Survey of Canada, Department of Energy, Mines and Resources, Ottawa K1A 0E4 Canada*

ABSTRACT

The mechanical paradox of large overthrusts originates with a fallacious assumption in the conceptual model upon which mechanical analyses of overthrust faulting have been based. The notion that the maximum width of a sheet of rock that may be displaced along an overthrust fault can be determined by using the Mohr-Coulomb failure criterion to compare the strength of the rocks with the stress required to overcome the *total* frictional resistance to slip (and the total cohesion or adhesion?) over the *entire* fault surface tacitly assumes that slip is initiated, and occurs, simultaneously over the entire fault surface. This unstated and untenable assumption, which is incompatible with what is known about real faults, is the real explanation for the "mechanical paradox."

A realistic model for overthrust faulting must be compatible with the nature of real overthrust faults, and particularly with the way in which displacements, including earthquake displacements, occur on real faults. Most large overthrusts are discrete shear surfaces within a coherent mass of rock that is physically continuous around their ends. Total displacement clearly varies from place to place over these discrete shear surfaces and, moreover, must be the cumulative result of many relatively small incremental displacements that also varied from place to place over the finite part of the fault surface on which they occurred. Each increment began as a local shear failure and propagated along the fault as a "smeared-out" (Somigliana) dislocation at a velocity that was small considering the time required for propagation over the total area of the fault. During any individual displacement "event," whether a megathrust earthquake or a creep event, only a small part of the fault surface was in relative motion at any one time. Fault displacements may appear to occur simultaneously across laboratory specimens that are very small relative to the velocity of propagation of displacement because the transit time is so short. When the linear scale is expanded over six orders of magnitude to encompass an entire surface of a large overthrust fault, however, the transit time is much longer because the velocity of propagation is very small relative to the area of the fault surface, and the displacement clearly involves only a very small part of the fault surface at any one time.

The shape and size of overthrust faults result from the propagation of dislocations, and this process is controlled by the strength heterogeneity and anisotropy of the rock mass as well as by variations in the regional stress field, but not by the frictional resistance to sliding integrated over the entire fault surface.

INTRODUCTION

Far better an approximate answer to the right question, which is often vague, than an exact answer to the wrong question, which can always be made precise.

—John W. Tukey, as quoted by Krumbein and Greybill, 1965, p. 1

Our thinking in tectonics is done with models (Carey, 1962). We use models to represent the geometric, kinematic, and rheologic attributes of real tectonic phenomena because our knowledge of the real thing invariably is quite incomplete, and also because what we do know generally involves details of such complexity that simplifying assumptions are necessary in order to make the problems in tectonic analysis fit the capabilities of the available analytical tools. Tectonic models span a vast spectrum, ranging from simple, and sometimes rather vague, concepts of structural geometry, of the kinematics of tectonic processes, and of the rheology of rocks; through very precise mathematical expressions of empirical "laws" concerning the mechanical behavior of rocks; to highly sophisticated representations of the mechanical, geometrical, and other physical attributes of geologic structures. Nevertheless, all of these models are subject to the same fundamental constraint—because they are models, the results of any analysis in which they are used are only realistic to the degree to which they approach the real thing. An inappropriate choice of models can lead to erroneous conclusions, and nothing better illustrates the significance of this point than the "mechanical paradox of large overthrusts" (Hubbert and Rubey, 1959). My basic objective in this paper is to critically review the origin, evolution, and widespread application of a mechanical model for overthrust faulting that is intuitively appealing, but fallacious, and consequently has impeded progress in understanding the mechanics of overthrust faulting.

The mechanical paradox of large overthrust faults originates with a fallacious assumption in the conceptual model upon which the mechanical analyses of overthrust faulting have been based. The model assumes, tacitly, that sliding is initiated, and occurs, simultaneously over the entire fault surface. Widespread use of this fallacious assumption ignores a large body of data from structural geology and from earthquake seismology that shows that slip on thrust faults involves the propagation of dislocations, each of which affects only a small portion of the fault at any one time

---

*Present address: Department of Geological Sciences, Queen's University, Kingston, Ontario K7L 3N6 Canada.

Geological Society of America Bulletin, v. 100, p. 1898–1908, 11 figs., December 1988.

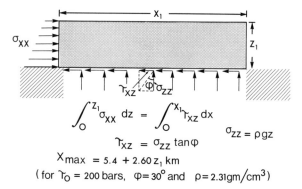

Figure 1. Thrust block on a horizontal surface (after Hubbert and Rubey, 1959, Fig. 5 and eqs. 10, 11, and 15). The maximum length of a rigid rectangular thrust block is calculated on the basis of a static equilibrium of forces in the x direction, as limited by the shear strength of the rock within the block, which is assumed to be $\tau = \tau_0 + \sigma \tan \phi$. The thrust fault, which is represented by the base of the block, is assumed to have propagated as a dislocation, with the result that along the base of the block, $\tau_0 = 0$.

15 miles, while the breaking stress of granite corresponds to a height of only about 2 miles. Thus we may press the block with whatever force we like; we may eventually crush it, but we cannot succeed in moving it. (Smoluchowski, 1909, p. 204–205)

With this statement, Smoluchowski had enunciated the mechanical paradox of large overthrusts—that the maximum possible width of an overthrust mass, as calculated in terms of the limiting condition for static frictional equilibrium of a rigid rectangular block on a horizontal plane surface, is substantially less than the widths of many actual overthrust masses, as determined on the basis of field relationships. Smoluchowski (1909, p. 205) offered two possible solutions to the paradox:

First, it may be remarked, the bed may not be horizontal but inclined; in this case the component of gravity is sufficient, at an inclination of 1:6.5, to put the block in sliding motion, and we need not apply any external pressure at all. And what seems still more important, nobody ever will explain Alpine overthrusts in any other way than as a phenomenon of rock-plasticity. Suppose a layer of plastic material, say pitch, interposed between the block and the underlying bed; or suppose the bed to be composed of such material: then the law of viscous liquid friction will come into play, instead of the friction of solids; therefore any force however small will succeed in moving the block.

The first of these, the gravitational sliding model, has had brief periods of widespread popularity (de Jong and Scholten, 1973), but only very local long-term acceptance, because it requires a significant slope and also because it requires that the compression and/or overlap at the front of the detached mass be balanced by an equivalent stretching and/or gap at the back, whereas in general, the slope of the overthrust faults is in the wrong direction for gravitational sliding and there is no tectonic gap or stretching at the back of the overthrust masses that compensates for the compression and tectonic overlap at the front (Price, 1971). The second model, which requires that the overthrust mass be underlain by a material which has been sheared, as a viscous fluid, by an amount equivalent to the displacement of the overthrust mass, also has had only very limited acceptance because large overthrusts normally involve the abrupt juxtaposition of rocks from a lower crustal level over rocks from a much higher crustal level, and although some large overthrusts follow layers of "weak" rock over part of their course, or have zones of distributed shear strain marked by mylonite along their hanging walls, most involve the sharply defined juxtaposition of rocks from a deep structural level over rocks from a shallow structural level, and many are discrete, sharply defined shear surfaces, as will be documented below.

Oldham's elegantly simple solution to the problem (Oldham, 1921) apparently was too far ahead of its time. It appeared before dislocations became widely known as the answer to the analogous mechanical paradox between the observed strengths of crystalline materials and the theoretical strengths of the ideal crystalline structures and, consequently, was virtually ignored. There were other attempts, some involving rather unrealistic hypotheses concerning slopes for gravitational sliding and/or coefficients of sliding friction (for example, Lawson, 1922; Longwell, 1945; Reeves, 1946), but the notion of a mechanical paradox continued to dominate thinking about overthrust faulting, until Hubbert and Rubey (1959) used Terzaghi's (1943, 1950) concept of effective stress, and the Mohr-Coulomb failure criterion, to argue that high pore-fluid pressures can reduce the frictional resistance to sliding required by Smoluchowski's model and thereby resolve the paradox.

Hubbert and Rubey used the same basic model as did Smoluchowski. They assumed that the limiting condition for static frictional equilibrium of a rigid rectangular block on a plane surface could be used to calculate both the maximum possible width of an overthrust mass that can be pushed from the rear on a horizontal surface, and the minimum slope angle necessary for gravitational sliding (Fig. 1). On the basis of empirical rela-

(Price, 1973a; see also De Braemaecker, 1987). This answer to the "mechanical paradox" is not entirely new. Oldham pointed out in 1921 that "the hypothesis of origin needs correction, not the facts of observation," but his explanation "that the thrusts did not move simultaneously over the whole of their extent, but partially, first in one part then in another, each separate movement involving an area limited by the strength of the rocks and their power to transmit or resist the effects of pressure . . . the movement would not be like that of a sledge, pushed bodily forward over the ground, but more akin to the crawl of a caterpillar which advances one part of its body at a time, and all parts in succession," (Oldham, 1921, p. lxxvii–xcii) has been ignored by virtually everyone who has subsequently written about the phenomenon. Accordingly, it is useful to review the evolution of thought on the subject.

HISTORICAL OVERVIEW

Large overthrust faults are an intuitively implausible concept when viewed from the perspective of human experience with the mechanical behavior of rocks. Almost everyone has had some experience with the frictional resistance that must be overcome in sliding large blocks of rock at the Earth's surface. Thus, it is not surprising that at the beginning of the twentieth century, after most of the initial skepticism about the very existence of large overthrusts had been dispelled by the overwhelming weight of unequivocal geologic field evidence, the intuitively vexing question of the mechanical basis of large overthrust faults emerged as one of the most perplexing enigmas in geology. The problem was outlined succinctly in 1909 by Smoluchowski, and subsequent thinking about the matter has been constrained by his conceptual model of the mechanics of large overthrust faults.

It is easy enough to calculate the force required to put a block of stone into sliding motion on a plane bed, even if its length and breadth be 100 miles . . . Let us indicate the length, breadth, and height of the block by a, b, c, its weight per unit volume by w, the coefficient of sliding friction by e; then, according to well known physical laws, a force a, b, c, w, e will be necessary to overcome the friction and to put the block into motion. Now, the pressure exerted by this force would be distributed over the cross-section a, c; hence the pressure on unit area will be equal to the weight of a column of a height b, e. Putting $b = 100$ miles, we get a height of

Figure 2. The Lewis thrust fault on the north side of the valley of Spionkop Creek, Clark Range, Rocky Mountains, Alberta. Light gray dolomite of the middle Proterozoic Siyeh Formation is juxtaposed, along the Lewis thrust, over dark gray shales of the Upper Cretaceous Alberta Group. Stratigraphic separation is approximately 7 km, and net slip, about 100 km (Price, 1962, 1981). The pick handle is 30 cm long (Geological Survey of Canada photograph 136844).

tions derived from the results of triaxial experiments on the brittle shear failure of small cores of apparently homogeneous, isotropic rocks, they concluded that the limiting condition leading to thrust faulting within the rigid rectangular block that they used to represent the overthrust mass is given by a linear Mohr-Coulomb–type shear failure criterion,

$$\tau = \tau_0 + \sigma \tan \phi,$$

where τ and σ are the critical shear stress and critical normal stress, respectively, on the surface of failure; $\tan \phi$, the "coefficient of internal friction," is the constant of proportionality between the critical shear stress and the critical normal stress, has a representative value of 0.577 ($\phi = 30°$), and is assumed to be equal to the coefficient of static friction for sliding on the fault surface; and τ_0 is the critical shear stress when the critical normal stress is zero and has a representative value of 200 bars (20 MPa). Using these values and the Mohr-Coulomb criterion for shear failure of the material within the rectangular block, they specified the limiting condition of static equilibrium for sliding on the fault in terms of the balance of forces acting on the block in the direction of ultimate displacement (Fig. 1). These forces consist of the total force applied to the rear edge of the block (left-hand term) and the frictional resistance to sliding summed over the entire base of the block (right-hand term). They assumed, *explicitly*, that the resistance due to τ_0 could be neglected because the initial rupture would propagate as a dislocation and, therefore, that the area over which the stress τ_0 would be involved at any one time would comprise only a minute fraction of the total area of the fault, but they also assumed, *implicitly*, and presumably without realizing the implications, that the actual slip on the fault would be initiated, and would occur, simultaneously over the entire fault surface.

Hubbert and Rubey (1959) showed that on the basis of these relationships, the maximum width, x_{max}, of an "overthrust" block of any given thickness, z, that can be pushed from the rear should be

$$x_{max} = 5.4 + 2.6 \, z \text{ km},$$

and they noted that x_{max} is much less than the observed width of many actual overthrust masses. Using the same basic criteria, they also concluded that the minimum slope (θ) for gravitational sliding of a rigid rectangular block is given by

$$\tan \theta = \tan \phi,$$

and that because ϕ is about 30°, this slope is, in general, much too steep to account for real overthrust faults. On the basis of this restatement of Smoluchowski's (1909) model, they concluded that the problem was a matter of reducing the frictional resistance to sliding and that the concept of effective stress and the known occurrence of high pore pressures in rocks provided the solution to the problem. They showed that if the ratio between the pore pressure, p, and the weight per unit area, S_{zz}, of a vertical column of rock at some depth, z, is given by

$$p = \lambda \, S_{zz},$$

the expression for the maximum width, x_{max}, of a block that can be pushed from the rear becomes

$$x_{max} = \frac{1}{1 - \lambda} 5.4 + (2.6 - 1.73 \, \lambda) z \text{ km},$$

and the expression for the minimum slope angle for gravitational sliding becomes

$$\tan \theta = (1 - \lambda) \tan \phi,$$

and, therefore, that as the pore pressure becomes very high, the maximum width of a block that can be pushed from the rear may become very large, and the minimum slope angle required for gravitational sliding may become very small.

The Hubbert and Rubey solution to the mechanical paradox of large overthrust faults and of gravitational sliding on shallow slopes has been widely acclaimed for more than a quarter century, in spite of some substantive criticism that was focussed on one vexing point. Birch (1961) and Hsü (1969a) both challenged the assumption that τ_0 could be neglected in calculating the maximum width of a thrust fault or of the minimum slope angle required for gravitational sliding, and Hsü went to great lengths to illustrate the effect of including τ_0 or an adhesive strength (τ_a) in the calculations. The criticisms by both Birch and Hsü seem to have been prompted by the inconsistency inherent between the explicit assertion that the rupture propagates along the fault as a dislocation and the implicit assumption that displacement of the fault is initiated and occurs simultaneously over the entire fault surface. Although neither Birch nor Hsü seems to have recognized that the latter is an unstated (and untenable) assumption, both challenged the justification for the former assumption. These criticisms notwithstanding, the Hubbert and Rubey model, which relies on the balance of forces that drive and resist the motion of a thrust sheet, has become part of the dogma of structural geology. Equations for the static equilibrium limit for frictional sliding on thrust faults, which

Figure 3. Mount Crandall thrust fault on the south slope of Mount Crandall, Waterton Lakes National Park, Alberta (Douglas, 1952). Altyn Formation dolomite in the hanging wall is juxtaposed over Appekunny Formation argillite. A splay branches upward from the thrust into the Altyn Formation. Stratigraphic separation is about 250 m, and net displacement, about 6 km (Geological Survey of Canada photograph 113780-E).

unwittingly and incorrectly rely on the assumption that slip is initiated and occurs simultaneously over the entire fault surface, have been used extensively in mechanical analysis of various aspects of overthrust faulting (for example, Raleigh and Griggs, 1963; Hsü, 1969a, 1969b; Forristall, 1972; Wiltschko, 1979a, 1979b, 1981; Chapman, 1979). This fundamental flaw in the Hubbert and Rubey model, and in these subsequent modifications of it, can be elucidated by considering some of the characteristics of real thrust faults, as well as contemporary observations and ideas on earthquake mechanics and active faults.

SOME CHARACTERISTICS OF THRUST FAULTS

Real thrust faults differ fundamentally from the hypothetical faults in the mathematical models. In the models, the thrust sheets undergo rigid-body translation, and the magnitude of the displacement is the same over the entire fault surface, but on real faults, the magnitude of the displacement changes conspicuously, both along the direction of displacement and along the direction perpendicular to the direction of displacement, and the rock mass is not rigid but deforms penetratively during displacement.

Figure 4. Thrust fault cutting shale and cherty limestone of the lower part of the Mississippian Banff Formation within the Lewis thrust sheet in a roadcut in Crowsnest Pass, Alberta. Stratigraphic separation is about 10 m, and net displacement is less than 1 km. Pick handle is 30 cm long (Geological Survey of Canada photograph 136869).

Figure 5. Small-contraction (thrust) faults offsetting Lower Cretaceous (Blairmore Group) sandstone beds have displacements of 10–20 cm and form 30° ramps connecting bedding glide zones in shales, Livingstone thrust sheet, east of Hailstone Butte, Rocky Mountain foothills, Alberta. Pick handle is 30 cm long (Geological Survey of Canada photograph 154555).

Thrust faulting involves smaller pervasively distributed displacements within the rock mass in addition to the large "rigid-body" displacements that occur along discrete shear surfaces. Our perceptions of the nature of thrust faults change with the scale at which we observe them. This critical relationship, which has been either overlooked or else ignored without justification in the models, has been well known and thoroughly documented for many years by geologic mapping in many thrust belts, including the southern Canadian Rockies (Douglas, 1958; Price, 1965, 1967, 1973b; Dahlstrom, 1969, 1970), from which some examples will be drawn for this discussion.

At the scale of the individual outcrop, even the largest thrust faults characteristically appear as "knife-sharp" surfaces juxtaposing rocks that were far apart before the thrust faulting (Price, 1965, 1973b; Gretener, 1972, 1977). This relationship is typical for the Lewis thrust fault in southernmost Alberta, where the net displacement is about 100 km and the stratigraphic separation is about 7 km (Fig. 2), as well as for the Mount Crandall thrust fault (Douglas, 1952), the roof thrust of Dahlstrom's (1970) prototype duplex structure, at Mount Crandall (Fig. 3), where it has a net displacement of about 6 km and a stratigraphic separation of about 0.25 km (see Boyer and Elliott, 1982, Fig. 17), and also for a small unnamed thrust fault that cuts the Mississippian strata within the Lewis

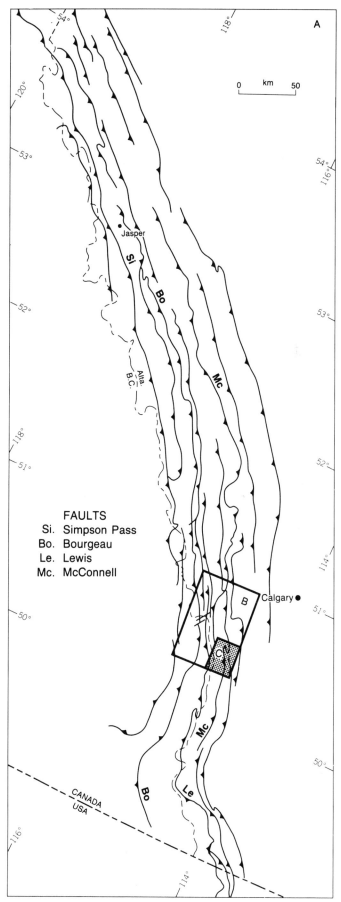

Figure 6. Lateral variation in displacement along thrust faults in the Canadian Rockies. (A) Regional relationships among major thrust faults (after Wheeler, 1987). Locations of Figures 6B and 6C as shown.

FAULTS

Bo. Bourgeau
Sm. Sulphur Mountain
Le. Lewis
Et. Etherington
Ru. Rundle
Mi. Misty
Ba. Baril

Co. Coleman
Hi. Highwood
Mc. McConnell
Se. Sentinel Peak
Be. Bear Creek
Li. Livingstone
Dm. Dyson Mountain

Paleozoic rocks

Mesozoic rocks

Figure 6. (B) Relationships among the larger thrust faults in the Kananaskis area. Location of Figure 6C as shown.

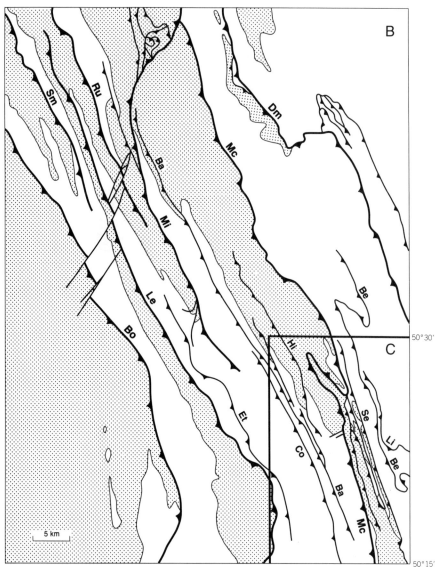

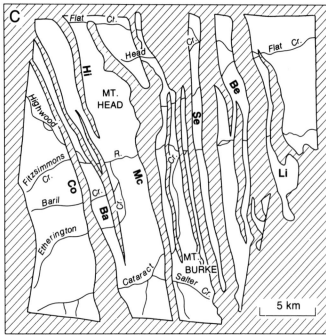

thrust sheet at Crowsnest Pass (Fig. 4) and has a net displacement of less than 1 km and a stratigraphic separation of about 10 m. Of course, the same relationship obtains for thrust faults on which the displacement is measured in tens of centimeters and the stratigraphic separation in centimeters or millimeters (Fig. 5). At the outcrop scale, the most conspicuous attribute of the thrust faults is that they provide evidence of rigid-body translation or rotation following brittle fracture. Evidence of penetrative ductile deformation may be discernible, as in the case of crystalline thrust sheets or ophiolite sheets that are detached along zones of high ductility in the middle or lower crust or upper mantle (Oxburgh, 1972; Armstrong and Dick, 1974). The ductile deformation, however, involves much smaller relative displacements than those which have occurred along the thrust faults themselves and is much less conspicuous. Moreover, in the

Figure 6. (C) Palinspastic reconstruction of the Mount Head map area (after Douglas, 1958). Restored fault traces are identified by the same labelling as in Figure 6B.

Figure 7. Small conjugate contraction (thrust) faults offsetting the interface between limestone and shale beds in the Upper Mississippian Mount Head Formation, west flank of Livingstone Range anticlinorium, Livingstone thrust sheet, at Oldman River Gap, Alberta (Douglas, 1950). Slicken striae are perpendicular to rough stylolitic joints and parallel calcite-coated joints (after Price, 1967, Fig. 9). Pick handle is 30 cm long (Geological Survey of Canada photograph 180255).

examples illustrated above, the absence of extensive penetrative strain in the rocks adjacent to the thrusts, and the "knife-sharp" juxtaposition of rocks that were very far apart before thrusting, is incompatible with the notion that viscous or plastic deformation of a basal layer provides a *general* explanation for large thrust faults (Smoluchowski, 1909; Kehle, 1970; Elliott, 1976a, 1976b; Chapple, 1978).

At the other end of the spectrum, at a scale of observation that encompasses the whole of the southern Canadian Rockies (Fig. 6A), even the largest of the thrust faults, such as the Lewis, McConnell, Bourgeau, and Simpson Pass, on which net displacements exceed 50 km, die out along strike within the deformed rock mass, and the rock mass is physically continuous around the edges of each and every one of them. At this scale, we see a northeasterly tapering wedge of supracrustal rocks, about 200 km wide and as much as 20 km thick, that has been compressed by more than 50% and thickened proportionately as a result of shear displacements that are distributed over a myriad array of overlapping and interfingering discrete discontinuous shear surfaces. The rock mass has undergone profound changes in shape without over-all loss of cohesion, and the deformation can be described as a type of ductile flow (Price, 1973b).

At a scale of observation that is an order of magnitude larger and encompasses a segment of the thrust belt about 100 km long (Fig. 6B), some of the larger faults appear as throughgoing discontinuities that divide the belt into discrete "thrust sheets" or nappes (Dyson Mountain, McConnell, Lewis, Bourgeau), but others (Sulphur Mountain, Rundle, Misty) clearly die out along strike; they, along with many other smaller discontinuous faults and the folds that are related to them, show that significant inhomogeneous ductile strain occurs "within the thrust sheets." Real "thrust sheets," unlike those in most of the models, deform internally as they are displaced.

Increasing the scale of observation by almost another order of magnitude, to focus on an area of about 500 km^2 that was carefully mapped and palinspastically reconstructed some 30 yr ago by Douglas (1958), shows (Fig. 6C) that the internal deformation within the thrust sheets involves displacements on even smaller thrust faults. These faults also form arrays of discrete, discontinuous, interfingering, overlapping, and branching dis-

locations, around the ends of which the deformed rock mass is physically continuous and has retained its cohesion.

A further change in scale of observation by more than three orders of magnitude, to a field of view that encompasses a surface area of less than 10 m^2 in an individual outcrop, shows (Fig. 7) that folds and other "pervasive" inhomogeneous strains within thrust sheets involve displacements on arrays of still smaller faults. At this scale of observation, mesoscopic contraction faults and extension faults occur in association with stylolitic joints, fissure veins, and cleavage, and this ensemble of small-scale structures commonly shows complex crosscutting relations that record a complex polyphase history of deformation, involving the superposition of episodes of bedding-parallel shortening and bedding-parallel stretching (Price, 1967).

Sharply defined strain discontinuities over which the displacement is non-uniform, and is characterized by conspicuous gradients, are a typical feature of overthrust belts at various scales, ranging from thrust faults that are many hundreds of kilometers long and have net displacements of as much as 100 km to contraction and extension faults, fissure veins, and stylolites that are centimeters long and have displacements ranging to as small as millimeters. Displacement discontinuities of this type have been described as "smeared dislocations" or "Somigliana dislocations" to distinguish them from the "Volterra dislocations" in crystalline materials over which the displacement discontinuity is constant (Eshelby, 1973). Realistic models of overthrust faulting must accommodate the role of Somigliana dislocations. They also must accommodate contemporary knowledge and observations on active thrust faults and the earthquakes that are associated with them.

ACTIVE THRUSTS AND EARTHQUAKES

The megathrust faults associated with subduction and collision zones produce many of the most powerful earthquakes known. Accordingly, they have received a great deal of attention during the past few decades and have generated powerful new insights on the thrust-faulting process (Kanamori, 1986). Perhaps the best studied megathrust earthquake is the great Alaskan Earthquake of March 1964 (Plafker, 1965, 1967). It involved overthrust displacements ranging to at least 20 m on a shallow (5°–10°) north-dipping thrust fault over an area as much as 900 km long and 290 km wide (260,000 km^2) extending westward along the Aleutian arc-trench gap from near the intersection of the Queen Charlotte–Fairweather transform fault and the Aleutian trench. The hypocenter, at which the rupture began, was at a depth of between 20 and 50 km in the northeastern part of the area that ruptured. Although the rupture propagated initially in various directions, after an elapsed time of 44 s, it continued only in a southwesterly direction (Wyss and Brune, 1967), mainly as a screw dislocation. The average velocity of propagation was 3.5 km/s, which means that only a very small part of the fault surface was in motion at any one time. The amount of displacement varied conspicuously and systematically over the portion of the fault that was affected by this particular earthquake and its associated aftershocks. The focus of maximum displacement was more than 100 km south of the epicenter. The overthrust mass underwent substantial distortion in conjunction with the displacement on the overthrust, and this deformation continued after the main earthquake. During the succeeding 9 months, there were more than 600 aftershocks of Richter magnitude 4.4 or greater, including six greater than magnitude 6, and these were distributed over the same area as that of the main rupture. A large area of the overthrust mass, updip from a line that passes through the hypocenter and is perpendicular to the direction of horizontal displacement of the overthrust mass, was uplifted 3 to 4 m and locally as much as 10 m. A corresponding area over the downdip part of the overthrust subsided 1–2 m. There was substantial internal deformation within the overthrust mass, including horizontal shortening of 3–6 m across the Patton Bay fault, which is located approximately midway across the width of the area of the rupture and more than 150 km south of the epicenter. The Patton Bay fault, together with other subsidiary, northwest-dipping reverse faults across which there was a total relative southeast displacement of 10 m of the hanging-wall rocks, would, in the currently fashionable jargon for thrust faulting, be called "out-of-sequence thrust faults."

The overthrust displacement associated with the great Alaskan Earthquake of March 1964 also differs fundamentally from the hypothetical faults in the mathematical models of Smoluchowski (1909), Hubbert and Rubey (1959), Hsü (1969a), and others. The overthrust mass was not displaced as a rigid block; it underwent substantial internal deformation in the course of a single episode of displacement on the overthrust. The displacement associated with the earthquake was not uniform and did not occur over the entire fault surface. It involved only part of the total fault surface, and within that part, it varied significantly in magnitude. It was a type of "smeared" or Somigliana dislocation, and it represents one increment in the development of the total net slip over the whole overthrust fault, which is itself a very much larger Somigliana dislocation resulting from the superposition of many overlapping smaller Somigliana dislocations like that associated with the great Alaskan Earthquake of March 1964.

The mechanism of incremental displacement on the Aleutian arc-trench megathrust during the great Alaskan Earthquake of March 1964 provides a prototype for an actualistic process model for displacements on overthrusts in general. Displacement on overthrusts occurs as relatively small overlapping increments, each of which involves the propagation of a Somigliana dislocation over a discrete area that generally comprises only part of the fault surface. During each increment of slip, the displacement propagates outward from the initial point(s) of rupture at velocity that may range to as much as 4 or 5 km/s.

The average velocity of propagation of fault displacement during the great Alaskan Earthquake of March 1964, as estimated by Wyss and Brune (1967) using the radiation of P waves, was 3.5 km/s. Weertman (1969, 1975), from theoretical considerations, concluded that the maximum velocity of propagation of an earthquake dislocation should be equal to the shear wave velocity for screw dislocations and the Raleigh wave velocity for edge dislocations or else should be supersonic. Johnson and Scholtz (1976), using a biaxial loading frame, with a 20-cm-long sliding surface, and arrays of strain gauges, measured rupture velocities during stick-slip that ranged to as much as 4.6 km/s, but generally were lower than the shear wave velocity of the rocks involved, and increased with increasing stress drop. The velocity of propagation of fault displacement may be very much lower than these maximum values. Although it is common to draw a sharp distinction between stick-slip and creep as mechanisms for slip on faults, fault creep also involves the propagation of dislocations along the fault surface (Frank, 1973; Nason and Weertman, 1973). King and others (1973) have monitored creep events propagating along the San Andreas fault at velocities of 10 km/day or less, which is roughly 5 orders of magnitude smaller than the corresponding rate of seismic faulting. The largest creep event recorded by them had a rupture length of 6 km and a maximum offset of 9 mm, which is comparable to the length and offset parameters of a shallow earthquake of magnitude 4.7.

DUCTILE FAULTS

The Somigliana dislocation model is also appropriate for the description of ductile shear zones; moreover, it may be helpful in describing the propagation of "episodes" of deformation along ductile faults, within which processes such as pressure solution pressure-shadow precipitation and crystalline plasticity and grain size reduction are involved.

The concept of a ductile bead along the tip line of a propagating thrust fault was advanced by Elliott (1976b) on the basis of observations by Douglas (1958), Price (1965, 1967), Dahlstrom (1969, 1970), and others concerning the way in which some thrust faults in the Canadian Rockies and elsewhere end in the cores of folds. The concept has gained widespread popularity (see Hossack, 1983; Williams and Chapman, 1983; Farrell, 1984; Chapman and Williams, 1984; Higgs and Williams, 1987; among others) but only in the context of the propagation of the initial

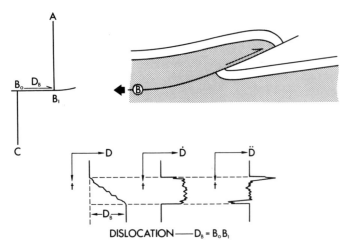

Figure 8. Passage of a dislocation (D_B) at a point (B) on a fault surface (after Aki, 1967); at the point B, the dislocation offsets the reference line AC from B_0 to B_1; D = displacement of one side of the fault relative to the other side, t = time, $\dot{D}$ = velocity of displacement, $\ddot{D}$ = acceleration of displacement.

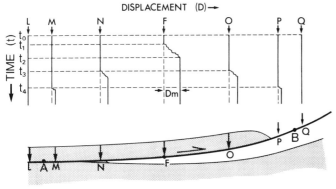

Figure 9. Transit time of a propagating thrust-fault dislocation; t = successive instants of time, F = focus, LMNOPQ = other points on the fault, D = displacement of hanging wall relative to footwall at individual points on the fault; displacement is arbitrarily shown as symmetrical about the focus.

fracture. The same kind of ductile deformation, however, also can be expected to occur along the tip line of a slipped patch that has reactivated part of an already-established fault. Moreover, insofar as fault displacements are smeared-out dislocations, they must be characterized by displacement gradients over most or all of their areal extent, and therefore, brittle and/or ductile strain gradients must occur almost everywhere on both sides of the fault, not just near the tip line.

THE DISLOCATION MODEL

Aki (1967) has provided an instructive conceptual model for the passage of a dislocation at a point on a fault (Fig. 8). If displacement is plotted against time, the passage of a dislocation is represented by a smeared-out ramp (D_B). The height of the ramp is the magnitude of the dislocation, and the slope is the average velocity of propagation ($\dot{D}$). The passage of the dislocation can be viewed as comprising a brief interval of acceleration, which precedes a longer period of slip at a stable velocity, that is, in turn, followed by a brief interval of deceleration. Extending this model to the propagation of a smeared-out displacement *along* a thrust fault (Fig. 9) provides a clear illustration of the way the magnitude of the displacement and the time at which the displacement occurs can vary from one location to another along the fault. Although displacement in the model has arbitrarily been made symmetrical about the focus, the great Alaskan Earthquake of March 1964 shows that the maximum displacement need not occur at the focus and that the displacement need not vary symmetrically about the maximum.

It is instructive to compare the pattern of variation in the amount and the time of displacement in the Somigliana dislocation model of thrust faulting (Fig. 9) with the pattern of variation in amount and the time of displacement in the Smoluckowski (1909)–Hubbert and Rubey (1959) model of thrust faulting (Fig. 10). The idea that maximum length of the overthrust block is controlled by the total frictional resistance to slip, summed over the entire length of the fault surface, clearly implies that displacement is initiated, occurs simultaneously, and is equal in magnitude over the whole length of the fault surface. This is tantamount to assuming that the overthrust block is perfectly rigid. These tacit assumptions obviously lack any physical basis in reality and clearly are contradicted by

what is known about real faults. Nevertheless, these assumptions have survived without serious challenge for a long time. It is appropriate to consider why.

If we recognize that our thinking in tectonics is done with models, that tectonic processes range through many orders of magnitude in length and in time, and that phenomena which are quite insignificant on one scale may be the predominant ones on other scales, our models must be continuously selected to fit the particular scale of our deliberations at the moment (Carey, 1962). The transit time of a propagating shear displacement across a laboratory specimen, on one hand, and across a large thrust fault, on the other hand, is a good example. It can be ignored conveniently when dealing with models for shear failure in laboratory specimens, but it cannot be ignored when dealing with models for shear failure over large thrust faults.

Brittle shear displacements propagating at 3 km/s take about 10 microseconds to transit a laboratory specimen a few centimeters across, but about 10 s to transit a segment of thrust fault that is about 25 km long (Fig. 11). Empirical relationships based on the experimental deformation of cylindrical specimens a few centimeters long cannot be assumed to apply without modification when extrapolated over a linear scale of six orders of magnitude to encompass a thrust sheet tens of kilometers long. The rate of propagation of the shear displacement has been ignored in the experiments, but cannot be ignored in the case of major thrust faults.

Experimental rock deformation has played a fundamental role in the development of the conceptual models that form the basis for the mechanical analysis of thrust faulting by Hubbert and Rubey (1959) and by the many others who have emulated them. These experiments have been concerned with the determination of the limiting state of stress leading to a shear failure that is marked by an abrupt loss of strength, generally associated with displacement on a shear fracture that extends right through the specimen. They have provided a sound empirical basis for the Mohr-Coulomb criterion for brittle shear failure, and verification of the role of pore-fluid pressure in reducing the effective normal stress on the surface of shear failure, but they have ignored the question of the rate of propagation of the shear displacement, tacitly assuming that the displacement is instantaneous or commences and occurs simultaneously over the entire shear fracture. It may be justifiable to assume that displacement is instantaneous over a linear scale of a few centimeters when formulating an empirical criterion for the limiting state of stress leading to shear failure at a point in

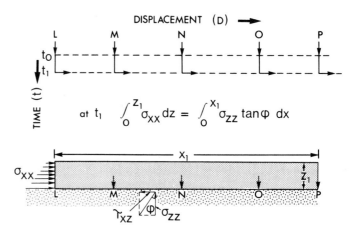

Figure 10. Transit time of displacement on a thrust fault according to the Smoluchowski (1909) and Hubbert and Rubey (1959) model. Symbols as in Figures 1 and 9.

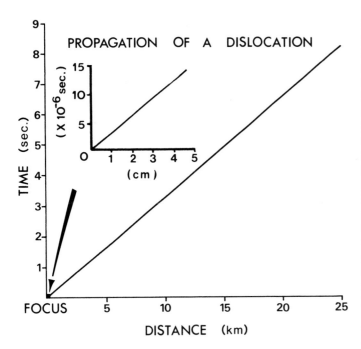

Figure 11. Total transit time for a dislocation of average velocity 3 km/s along a thrust fault 25 km long (8.3 s) and across a laboratory test specimen 2 cm long (6.6×10^{-6} s).

a rock mass, but it clearly is not justifiable to assume that displacement is instantaneous over a linear scale of tens of kilometers when estimating the maximum possible length of a thrust sheet. The time required for the propagation of a shear displacement may be relatively inconspicuous when observing shear failure in laboratory specimens, but it is a predominant consideration when observing shear failure over large thrust faults.

Our perceptions of the fundamental nature of thrust faulting change dramatically with changes in the scale at which we observe the thrust faulting (Price, 1973). The most obvious example is the apparent contradiction between the over-all ductility of an orogenic wedge (such as the Canadian Rockies or a submarine accretionary complex) on one hand, and the obvious brittle deformation within the orogenic wedge on the other hand. The ductility is implicit in the fact that the wedge retains its continuity and mechanical integrity while undergoing shortening, thickening, and lateral spreading in response to the gravitational potential associated with its surface slope. The brittle deformation is manifest in "knife-sharp" thrust faults within the wedge that juxtapose relatively unstrained rocks from widely separated stratigraphic levels. The change in appearance from discontinuous brittle deformation to pervasive ductile flow occurs when the scale of observation becomes substantially greater than the linear scale of the largest Somigliana dislocations (thrust faults). It is analogous to the situation in crystalline plasticity when the scale of observation is substantially greater than the linear scale of the Volterra dislocations, and the deformation of the crystal appears to involve homogeneous continuous flow.

CRITICAL-TAPER WEDGES

The recognition and reconciliation of this apparent contradiction between the brittle and ductile attributes of deformation in orogenic wedges have provided important new insights on their mechanics and regional tectonic significance. Bucher's (1956) concept of lateral gravitational spreading, and the analogy with the lateral gravitational spreading of glaciers and ice sheets, provided the basis for a new model by Price and Mountjoy (1970) for the evolution of the Canadian Rockies as well as a new concept for the origin of foreland basins, involving lithospheric flexure in response to thrust loading (Price, 1973b). This, in turn, provided the basis for Elliott's (1976a, 1976b) mathematical models of thrust faulting driven by gravitational spreading and for Beaumont's (1981) mathematical models of foreland basin evolution in response to thrust loading. The ensuing models for orogenic wedge deformation by Chapple (1978) and Stockmal (1983) that were based on a perfectly plastic rheology have been superceded by the critical Coulomb wedge theory of Davis and others (1983), Dahlen and others (1984), and Dahlen (1984), which appears to have completely reconciled the putative contradiction between the application of the Mohr-Coulomb brittle-shear failure criterion to the deformation within the orogenic wedge and the over-all ductile behavior of the wedge. They have argued that an actively accreting orogenic wedge maintains a critical taper that corresponds to an internal state of stress that is everywhere on the verge of brittle shear (Coulomb) failure and to a basal shearing stress that is on the verge of frictional sliding failure according to Byerlee's (1978) law.

The critically tapered wedge theory provides a convincing explanation for the over-all shape of actively accreting orogenic wedges, and for the sliding on the basal décollement. It is, however, an over-simplified model, according to which the cohesion, coefficient of internal friction, and coefficient of basal sliding friction each can be represented by single over-all values similar to those established by laboratory measurements (Dahlen and others, 1984). Woodward (1987) has discussed some of the problems in reconciling this simplified model with what is known about the internal structural evolution of real foreland thrust-and-fold belts. These considerations notwithstanding, to find an explanation for the size and configuration of individual thrust faults within an orogenic wedge, it is necessary to look beyond the generalized over-all strength parameters of the rocks, to the detailed internal variations in cohesion, coefficient of internal friction, coefficient of sliding friction, and pore-fluid pressure within the rocks. Because thrust faults originate and grow as Somigliana dislocations, their configuration and size must be controlled, at each locality in the rocks, by the strength anisotropy and heterogeneity within the rocks, as well as by the local state of stress arising from the over-all configuration and mass distribution of the wedge and from the external boundary stresses acting on the wedge.

CONCLUSIONS

The shape and size of overthrust faults result from the propagation of Somigliana dislocations, and this process is controlled by the strength heterogeneity and anisotropy of the orogenic wedge, as well as by variations in the external stress field and the gravitational stresses arising from the thrust displacements themselves, but not by the frictional resistance to sliding, integrated over the entire fault surface.

ACKNOWLEDGMENTS

It is a pleasure to acknowledge the assistance of C. Patenaude and J. Smalldridge in the preparation of the manuscript. The basic ideas which were presented 15 yr ago at a Geological Society of America Annual Meeting benefited from discussions with David Elliott and Bill Chapple.

The manuscript has been revised to incorporate helpful comments from each of the separate GSA reviewers: S. Edelman, R. D. Hatcher, Jr., D. Wiltschko, and N. B. Woodward.

J. Cl. De Braemaecker kindly provided me with a preprint of his 1987 paper on thrust sheet motion and earthquakes.

REFERENCES CITED

Aki, K., 1967, Scaling law of seismic spectrum: Journal of Geophysical Research, v. 72, p. 1217–1231.
Armstrong, R. L., and Dick, H.J.B., 1974, A model for the development of thin overthrust sheets of crystalline rock: Geology, v. 1, p. 35–40.
Beaumont, C., 1981, Foreland basins: Royal Astronomical Society Geophysical Journal, v. 65, p. 291–329.
Birch, F., 1961, Role of fluid pressure in mechanics of overthrust faulting: Discussion: Geological Society of America Bulletin, v. 72, p. 1441–1444.
Boyer, S. M., and Elliott, D., 1982, Thrust systems: American Association of Petroleum Geologists Bulletin, v. 66, p. 1196–1230.
Bucher, W. H., 1956, The role of gravity in orogenesis: Geological Society of America Bulletin, v. 67, p. 1295–1318.
Byerlee, J., 1978, Friction of rocks: Pure Applied Geophysics, v. 116, p. 615–626.
Carey, S. W., 1962, Scale of geotectonic phenomena: Geological Society of India Journal, v. 3, p. 97–105.
Chapman, R. E., 1979, Mechanics of unlubricated sliding: Geological Society of America Bulletin, v. 90, p. 19–28.
Chapman, T. J., and Williams, G. D., 1984, Displacement-distance methods in the analysis of fold-thrust structures and linked-fault systems: Geological Society of London Journal, v. 141, p. 121–128.
Chapple, W. M., 1978, Mechanics of thin-skinned fold-and-thrust belts: Geological Society of America Bulletin, v. 89, p. 1189–1198.
Dahlen, F. A., 1984, Noncohesive critical Coulomb wedges—An exact solution: Journal of Geophysical Research, v. 89, p. 10125–10133.
Dahlen, F. A., Suppe, J., and Davis, D., 1984, Mechanics of fold-and-thrust belts and accretionary wedges—Cohesive Coulomb theory: Journal of Geophysical Research, v. 89, p. 10087–10101.
Dahlstrom, C.D.A., 1969, Balanced cross sections: Canadian Journal of Earth Sciences, v. 6, p. 743–757.
———— 1970, Structural geology in the eastern margin of the Canadian Rocky Mountains: Bulletin of Canadian Petroleum Geology, v. 18, p. 332–406.
Davis, D., Suppe, J., and Dahlen, F. A., 1983, Mechanics of fold-and-thrust belts and accretionary wedges: Journal of Geophysical Research, v. 88, p. 1153–1172.
De Braemaecker, J. Cl., 1987, Thrust sheet motion and earthquake mechanisms: Earth and Planetary Science Letters, v. 83, p. 159–166.
de Jong, K. A., and Scholten, R., eds., 1973, Gravity and tectonics: New York, John Wiley and Sons, 502 p.
Douglas, R.J.W., 1950, Callum Creek, Langford Creek and Gap map-areas, Alberta: Geological Survey of Canada Memoir 255, 124 p.
———— 1952, Waterton, Alberta: Geological Survey of Canada Paper 52-10 with preliminary map.
———— 1958, Mount-Head map-area, Alberta: Geological Survey of Canada Memoir 291, 241 p.
Elliott, D., 1976a, The motion of thrust sheets: Journal of Geophysical Research, v. 81, p. 949–963.
———— 1976b, The energy balance and deformation mechanisms of thrust sheets: Royal Society of London Philosophical Transactions, Series A, v. 283, p. 289–312.
Eshelby, J. D., 1973, Dislocation theory for geophysical applications: Royal Society of London Philosophical Transactions, Series A, v. 274, p. 331–338.
Farrell, S. G., 1984, A dislocation model applied to slump structures, Ainsa Basin, south central Pyrenees: Journal of Structural Geology, v. 6, p. 727–736.
Forristall, G. Z., 1972, Stress distribution and overthrust faulting: Geological Society of America Bulletin, v. 83, p. 3073–3082.
Frank, F. C., 1973, Dislocation models for fault creep processes: Royal Society of London Philosophical Transactions, Series A, v. 273, p. 331–338.
Gretener, P. E., 1972, Thoughts on overthrust faulting in a layered sequence: Bulletin of Canadian Petroleum Geology, v. 20, p. 583–607.
———— 1977, On the character of thrust faults with particular reference to the basal tongues: Bulletin of Canadian Petroleum Geology, v. 25, p. 110–122.
Higgs, W. G., and Williams, G. D., 1987, Short notes—Displacement efficiency of faults and fractures: Journal of Structural Geology, v. 9, p. 371–374.
Hossack, J. R., 1983, A cross-section through the Scandinavian Caledonides constructed with the aid of branch-line maps: Journal of Structural Geology, v. 5, p. 103–111.
Hsü, K. J., 1969a, Role of cohesive strength in the mechanics of overthrust faulting and of landsliding: Geological Society of America Bulletin, v. 80, p. 927–952.
———— 1969b, A preliminary analysis of the statics and kinematics of the Glarus overthrust: Eclogae Geologicae Helvetiae, v. 62, p. 143–154.
Hubbert, M. K., and Rubey, W. W., 1959, Role of fluid pressure in mechanics of overthrust faulting—I. Mechanics of fluid-filled porous solids and its application to overthrust faulting: Geological Society of America Bulletin, v. 70, p. 115–166.
Johnson, T. L., and Scholtz, C. H., 1976, Dynamic properties of stick-slip friction of rock: Journal of Geophysical Research, v. 81, p. 881–888.
Kanamori, H., 1986, Rupture process of subduction-zone earthquakes: Annual Review of Earth and Planetary Sciences, v. 14, p. 293–322.
Kehle, R. O., 1970, Analysis of gravity sliding and orogenic translation: Geological Society of America Bulletin, v. 81, p. 1641–1644.
King, C.-Y., Nason, R. D., and Tocher, D., 1973, Kinematics of fault creep: Royal Society of London Philosophical Transactions, Series A, v. 274, p. 355–360.
Krumbein, W. C., and Greybill, F. A., 1965, An introduction to statistical models in geology: New York, McGraw-Hill, 475 p.
Lawson, A. C., 1922, Isostatic compensation considered as a cause of thrusting: Geological Society of America Bulletin, v. 33, p. 337–352.
Longwell, C. R., 1945, The mechanics of orogeny: American Journal of Science, v. 243, p. 417–447.
Nason, R., and Weertman, J., 1973, A dislocation theory analysis of fault creep events: Journal of Geophysical Research, v. 78, p. 7745–7757.
Oldham, R. D., 1921, Know your faults: Geological Society of London Quarterly Journal, v. 77, p. 77–92.
Oxburgh, E. R., 1972, Flake tectonics and continental collision: Nature, v. 239, p. 202–209.
Plafker, G., 1965, Tectonic deformation associated with the 1964 Alaska earthquake: Science, v. 148, p. 1675–1687.
———— 1967, The Alaska earthquake, March 27, 1964: Regional effects: U.S. Geological Survey Professional Paper 543I, p. I1–I74.
Price, R. A., 1962, Fernie map-area E/2, Alberta and British Columbia: Geological Survey of Canada Paper 61-24, 65 p.
———— 1965, Flathead map-area, British Columbia and Alberta: Geological Survey of Canada Memoir 336, 221 p.
———— 1967, The significance of mesoscopic subfabrics on the Southern Rocky Mountains of Alberta and British Columbia: Canadian Journal of Earth Sciences, v. 4, p. 39–70.
———— 1971, Gravitational sliding and the foreland thrust and fold belt of the North American Cordillera—Discussion: Geological Society of America Bulletin, v. 82, p. 1133–1138.
———— 1973a, The mechanical paradox of large overthrusts: Geological Society of America Abstracts with Programs, v. 5, p. 772.
———— 1973b, Large scale gravitational flow of supracrustal rocks, Southern Canadian Rockies, in de Jong, K. A., and Scholten, R., eds., Gravity and tectonics: New York, John Wiley and Sons, p. 491–502.
———— 1981, The Cordilleran foreland thrust-and-fold belt in the Southern Canadian Rocky Mountains, in McClay, K. J., and Price, N. J., eds., Thrust and nappe tectonics: Geological Society of London Special Publication 9, p. 427–448.
Price, R. A., and Mountjoy, E. W., 1970, The geological structure of the Southern Canadian Rockies between Bow and Athabasca Rivers—A progress report, in Wheeler, J. O., ed., A structural cross-section of the Southern Canadian Cordillera: Geological Association of Canada Special Paper 6, p. 7–25.
Raleigh, C. B., and Griggs, D. T., 1963, Effect of the toe in the mechanics of overthrust faulting: Geological Society of America Bulletin, v. 74, p. 819–830.
Reeves, F., 1946, Origin and mechanics of the thrust faults adjacent to the Bearpaw Mountains, Montana: Geological Society of America Bulletin, v. 57, p. 1033–1048.
Smoluchowski, M. S., 1909, Some remarks on the mechanics of overthrusts: Geological Magazine, v. 6, p. 204–205.
Stockmal, G. S., 1983, Modeling of large-scale accretionary wedge deformation: Journal of Geophysical Research, v. 88, p. 8271–8287.
Terzaghi, K. C., 1943, Theoretical soil mechanics: New York, John Wiley and Sons, 510 p.
———— 1950, Mechanics of landslides, in Application of geology to engineering practice (Berkey Volume): Geological Society of America, p. 83–123.
Weertman, J., 1969, Dislocation motion on an interface with friction that is dependent on sliding velocity: Journal of Geophysical Research, v. 74, p. 6617–6622.
———— 1975, Theory of velocity of earthquake dislocations: Geological Society of America Memoir 142, p. 175–183.
Wheeler, J. O., 1987, Tectonic assemblage map of the Canadian Cordillera and adjacent parts of the United States of America: Geological Survey of Canada Open-File 1565.
Williams, G., and Chapman, T., 1983, Strains developed in the hanging walls of thrusts due to their slip/propagation rate—A dislocation model: Journal of Structural Geology, v. 5, p. 563–571.
Wiltschko, D. V., 1979a, A mechanical model for thrust sheet deformation at a ramp: Journal of Geophysical Research, v. 84, p. 1091–1104.
———— 1979b, Partitioning of energy in a thrust sheet and implications concerning driving forces: Journal of Geophysical Research, v. 84, p. 6050–6058.
———— 1981, Thrust sheet deformation at a ramp: Summary and extensions of an earlier model, in McClay, T. J., and Price, N. J., eds., Thrust and nappe tectonics: Geological Society of London Special Publication 9, p. 55–63.
Woodward, N. B., 1987, Geological applicability of critical-wedge thrust-belt models: Geological Society of America Bulletin, v. 99, p. 827–832.
Wyss, M., and Brune, J. N., 1967, The Alaska earthquake of 28 March 1964—A complex multiple rupture: Seismological Society of America Bulletin, v. 57, p. 1017–1023.

MANUSCRIPT RECEIVED BY THE SOCIETY APRIL 18, 1988
REVISED MANUSCRIPT RECEIVED JUNE 17, 1988
MANUSCRIPT ACCEPTED JUNE 24, 1988
GEOLOGICAL SURVEY OF CANADA CONTRIBUTION NO. 11888

Material balance in Alpine orogeny

CENTENNIAL ARTICLE

HANS LAUBSCHER *Geologisch-paläontologisches Institut der Universität Basel, Bernoullistrasse 32, CH-4056 Basel, Switzerland*

ABSTRACT

Recent geophysical data are combined with older information for an updated picture of Alpine kinematics, using semiquantitative considerations of material balance. In the Alps, a conventional pile of thin peel nappes is disrupted by a central longitudinal system of steep faults and steeply limbed folds, both affecting disharmoniously the entire edifice of otherwise flat-lying nappes. It is particularly from late Tertiary phases of deformation that structures of the second type, the Insubric system with the Insubric line and the belt of late Alpine windows, acquired prominence for the present aspect of the Alps. These late structures formed as a dextrally transpressive intracontinental branch of the Africa-Europa plate boundary, mostly between areas of extension. Early motions in the Late Cretaceous to the late Eocene were probably responsible for at least half of the 300-km dextral displacement required for the palinspastic restoration of the inner West Carpathians and the Dinarides. During Oligocene extension, the late Alpine batholiths intruded in a belt roughly along the Insubric line while deep crustal and lithospheric roots were destroyed. In the latest Oligocene to middle Miocene, the Adriatic plate moved dextrally to the west along the Insubric transfer fault. From its frontal Insubric indenter, a lower crust–upper mantle flake, probably first obducted in the Cretaceous, was detached and wedged into the Penninic nappes of the western Alps ("bird's head" of the Ivrea body). Both the northern and the southern transfer faults of the Insubric indenter were disrupted and inactivated by late Miocene to later events, the Giudicarie and Neo–North Apennine events. Instead of the inactivated Insubric fault, the Windows belt of *en echelon* folds assumed the role of dextral transpression. All of these successive zones of motion are entangled in the "Ligurian knot," largely hidden under the sea or young sediments. A large part of the pre-Alpine crust and practically the whole mantle were subducted; only in some cases were high-pressure rocks of shallow continental origin re-obducted. The lithospheric root of the Neo-Alps is largely contained in a vertical slab under the Alps, but some parts seem to have been removed and others disharmonically displaced with respect to the surface structures.

INTRODUCTION

Classical Alpine kinematics was concerned foremost with cross-sectional nappism. The nappes, where not deformed subsequently, are subhorizontal thin peels. In sharp contrast to these peels are the steep, discordant faults, such as the Insubric line, and the steeply limbed folds affecting the entire pile of nappes, such as the external massifs or the Tauern window (Figs. 1 and 2). Attempts at integrating them into cross-sectional nappe development were rather unconvincing (compare, for example, Cornelius and Furlani-Cornelius, 1930). Their geometry suggested three-dimensional kinematics with strike-slip components; it became abundantly clear that the Alps are not just a pile of nappes, although it proved difficult to harmonize the kinematics of overthrust and strike-slip tectonics.

A number of articles in the Geological Society of America *Bulletin* have been important or even seminal for my own attempts at clarifying Alpine kinematics. I should like to single out the work of Harry Hess, a man with vision, and his students in the southern Caribbean (for example, Hess and Maxwell, 1953; Dengo, 1953). I met them in the field more than 30 years ago while employed in Venezuela. Although well aware of transpressive tectonics from work in the Jura, I tended to compare what I saw with what I had learned about the Alps, which was conditioned by Argandian cylindrism (Argand, 1916). I thought that I recognized nappes, some of them similar in facies and metamorphism to the Penninic nappes of the Alps. The Hess team, on the other hand, emphasized the maze of steep young faults that dissect the compressional units of the Caribbean Coast Range and that belong to what now would be called the "southern dextral transform boundary," in some cases transpressive, in some transtensive, of the Caribbean plate. It was obvious that both types of structures played a role and that the composite was definitely non-cylindrical. Moreover, an intriguing analogy turned up. Consider the arc of the western Alps as a reduced replica of the Antillean arc, and turn it around by 180°. The central Alps with their steep faults now become a reduced equivalent of the Caribbean Coast Range, a dextral transfer (which I prefer to "transform") boundary of the Adriatic or Insubric block, which in turn becomes the equivalent of the Caribbean block (plate tectonics had not yet been born at that time).

Back in Switzerland, the Alps therefore appeared in a new perspective. In a series of papers, I have attempted, for the past 20 years, to integrate the known facts of Alpine geology into crustal sections and map-view kinematical schemes. Based as they were on very sketchy information in important aspects, they had to be quite schematical. I tried to outline the main kinematic problems and to suggest ways of solving them.

The European Geotraverse (EGT; Galson and Müller, 1985, 1986) with its crucial seismic lines across the Alps is in the final stage of its execution. Already, important new data have been produced and await integration with old and new surface geological and geophysical information. In this article, an effort is made to this end. The meaning of the combined data for kinematics will be investigated, using material-balance considerations.

Material balance has played a fundamental role in Alpine tectonics since its beginning. Mostly, however, its application was essentially qualitative, involving rather vague scenarios, and the concept of material balance was not explicitly stated. Before the turn of the century, after the discovery of large-scale thrusting, nappism faced the necessity to reconsti-

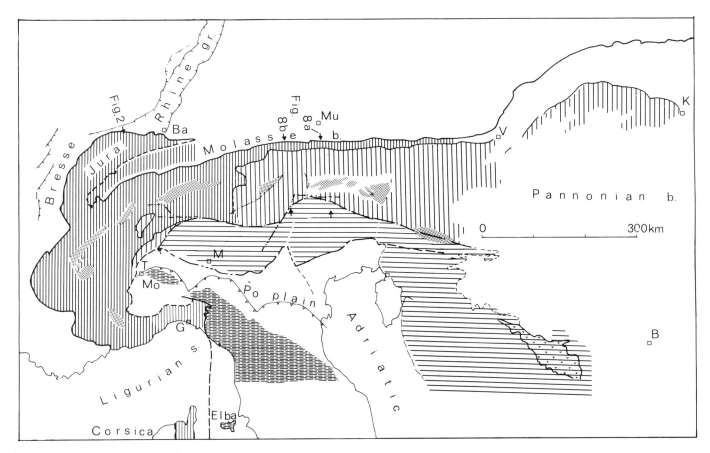

Figure 1. The tectonic elements of the Alps and their surroundings. Narrow vertical ruling: units derived from the European continental margin of Tethys, including Ligurian Alps and associated units such as Sestri-Voltaggio zone, Antola nappe, Alpine Corsica. Wide ruling: units derived from the African continental margin of Tethys. Wide vertical: Austro-Alpine nappes and internal units of the western Carpathians. Wide horizonal: South Alpine units and Dinarides. With V-pattern added: innermost Dinarides carrying the ophiolites. Horizontal dashes: northern Apennines. Stippled: late Alpine dextral *en echelon* belt of brachyanticlines (windows, external massifs) = Windows belt (a part of the Tauern window is left blank in order to indicate its polyphase origin). B = Belgrade, Ba = Basel, G = Genoa, K = Košice, M = Milan, Mo = Monferrato, Mu = Munich, T = Turin, V = Vienna.

tute paleogeographic domains now piled on top of each other. These palinspastics were, however, essentially cross sectional and based on the ambiguous notion of the "geosyncline."

My own views on Alpine tectonics have been conditioned, in addition to the Caribbean experience, by the continuous pursuit of material-balance problems on different scales and with various degrees of quantitative tolerance. From the outset, it was obvious that palinspastic restoration, even if perfectly feasible, would be only a first step for constructing balanced sections. Only the bed length of the most surficial layers would be preserved. Mapping had revealed long ago that only the top few kilometers of the crystalline crust are involved in the Alpine nappes (compare Fig. 2 and Laubscher, 1970a, 1983a). If several hundred kilometers of shortening had occurred as palinspastic reconstructions seemed to imply, where had the underpinnings of these surficial peels gone? Ampferer (1906), in prescient cross-sectional material-balance considerations, concluded that underneath the surface structures there must be an equivalent shortening by flow in the plastically deformable hot rocks ("Unterströmung"). Verification and refinement of these ideas had to wait for the results of modern geophysical techniques and insight into the rheology of rocks and minerals. Thin-skinned subsystems such as the Jura were mod-

eled early on with roughly balanced sections (Buxtorf, 1907, 1916). Much later, for the Alps as a whole, crustal refraction seismic surveys combined with considerations of isostasy permitted the crude reconstruction of palinspastic sections involving the entire crust. Estimates of water depths at the time of deposition, necessary for guessing the palinspastic position of the Moho, were very speculative, but the earliest palinspastic crustal sections through the Alps (for example, Laubscher, 1967) have not been dramatically improved till now. There is still no precise method for deriving depth of deposition, particularly for pelagic and turbidite sequences, from sedimentology or paleontology.

From these crude reconstructions, it was obvious that even for a continental crust close to 30 km thick, only the topmost few kilometers escaped subduction. It was soon evident that contrary to a commonly held belief, large parts of the continental crust, although less dense than the mantle, did not remain afloat. This conclusion was supported by another line of evidence; occasionally, subducted continental crust has re-emerged at the surface. Examples in the western Alps, recognized long ago, are the eclogitic gneisses and pyrope-bearing quartzites of the Sesia and Dora Maira units (Dal Piaz and others, 1972; Vialon, 1966; Borghi and others, 1985), the second even containing coesite relics (Chopin, 1987). But al-

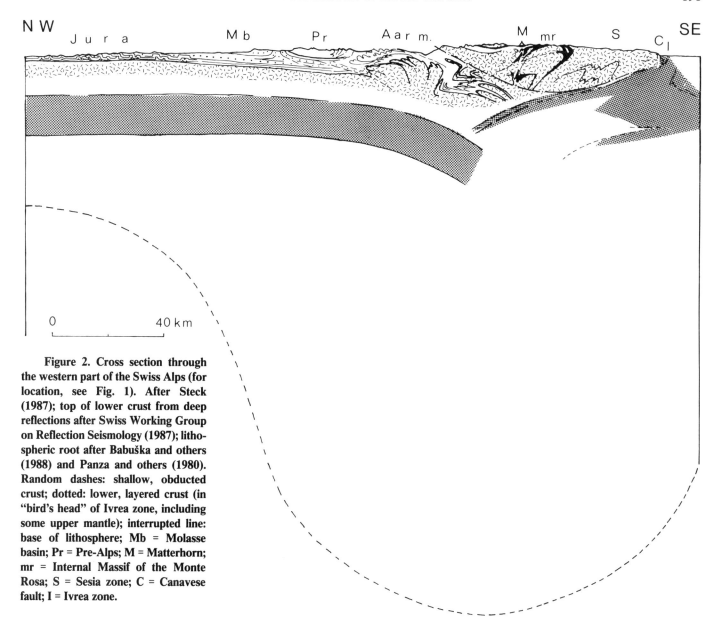

Figure 2. Cross section through the western part of the Swiss Alps (for location, see Fig. 1). After Steck (1987); top of lower crust from deep reflections after Swiss Working Group on Reflection Seismology (1987); lithospheric root after Babuška and others (1988) and Panza and others (1980). Random dashes: shallow, obducted crust; dotted: lower, layered crust (in "bird's head" of Ivrea zone, including some upper mantle); interrupted line: base of lithosphere; Mb = Molasse basin; Pr = Pre-Alps; M = Matterhorn; mr = Internal Massif of the Monte Rosa; S = Sesia zone; C = Canavese fault; I = Ivrea zone.

though more originally subducted continental crust is being found all the time, also in areas outside the Alps (Griffin, 1987), its volume does not nearly account for the "disappeared" crustal masses. Consequently, large portions of continental crust supposedly still float somewhere in the mantle or have been absorbed in it; jadeite and coesite instead of feldspar and quartz would increase density as well as velocity to mantle values.

The new information on the deep structure now available may be used for updating the crustal kinematics. Although the recent deep crustal reflection surveys have only partly been processed and interpreted as of the time of writing (Bayer and others, 1987; Swiss Working Group on Reflection Seismology, 1988), their results warrant new thinking. Some of the results are shown in figures in this paper, together with teleseismic information on the base of the lithosphere (Babuška and others, 1988). The fate of the lithosphere is still only sketchily documented. Lithospheric roots or subduction zones are sometimes found where expected from surface data and sometimes mysteriously fail to show up; perhaps they broke off and sank into the mantle. New techniques such as seismic tomography may one day provide the necessary information (compare Spakman, 1986).

In addition to cross-sectional aspects, material-balance considerations of a general and three-dimensional nature will be applied, such as combined strike-slip and thrust kinematics, source-sink relations (the fundamental, most general notion of any material transport field), and also palinspastic restoration, to what seem to me the most outstanding and enigmatic features of the Alps. Quantitative estimates of shortening and strike-slip will be attempted, albeit with considerable tolerance as present knowledge dictates.

One important feature of three-dimensional material balance, particularly stressed in this article, is what may be called "kinematic continuity"; motion in a certain time interval, measured or estimated as shortening, extension, strike-slip, or a combination of them, cannot simply disappear laterally. It has to be integrated in a closed source-sink system, similar to that of plate tectonics in a global frame. Elementary as this may seem, it receives scant attention in many kinematic models of the Alps.

In the following discussion, a number of problems are singled out and taken up one at a time, although they are all interdependent. This entails repeated viewing of the same phenomena from different angles. The se-

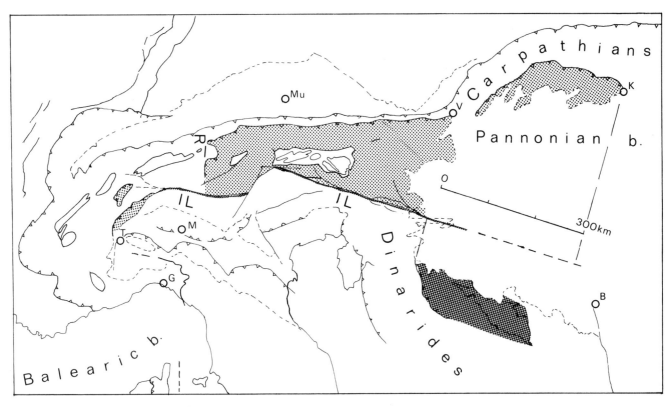

Figure 3. The Alps-Dinarides problem. For background information, see Figure 1. Fine dots: Austro-Alpine nappes and the internal West Carpathians (triangles indicate vergence); IL: Insubric fault system; heavy dots: internal Dinarides; R = Rhine line.

quence of presentation is generally one from the older to the younger events, although geographical association of phenomena of different ages precludes strict adherence to this principle. In the conclusions, the time sequence is briefly restated.

THE ALPS-DINARIDES PROBLEM

The north-vergent Austro-Alpine nappes, interrupted at the surface by the late Neogene extensional basins of the Pannonian system (Royden and others, 1982, 1983), continue into the internal West Carpathians till Košice in Slovakia. In the west, the southern boundary of the Austro-Alpine nappes is the steep Insubric fault system, with important Miocene displacements, cutting discordantly through a number of nappes. South of it follow the south-vergent thrusts and nappes, first of the southern Alps and then of the Dinarides (Figs. 1 and 3). All of these units, the Austro-Alpine nappes as well as the southern Alps and the Dinaric nappes, had essentially the same facies development in the Mesozoic, with a gradient toward an ocean basin in the north (the "central Tethys" with its ophiolites), placing them palinspastically at the southern continental margin of Tethys. The ophiolite-bearing nappes are at the base of the Austro-Alpine units but on top of the Dinaric units. This fact, in addition to the opposite vergence, makes cross-sectional palinspastic restoration impossible. In order to reconstitute the southern margin of Tethys, it is necessary to have recourse to the third dimension. Inversion of dextral strike-slip motion on the order of 300 km permits the lining up of the eastern end of the internal West Carpathians with the western end of the internal Dinarides, after which operation cross-sectional palinspastics is roughly feasible.

This is briefly the problem as stated by Laubscher (1970b), and it is the solution suggested. In the meantime, much argument has raged, particularly as to the postulated strike-slip, its amount, and its timing (Schmid and others, 1987), but to my knowledge, the basic problem has never been removed nor has another solution been proposed. Qualitatively, dextral strike-slip has now been reported from practically every part of the Insubric line, although timing and quantification of motion remain a problem. Laubscher (1971, 1973), meanwhile, pursued the consequences of his postulate, especially for the western part of the Insubric fault system, the western Alps, and their relation to the Apennines.

THE INSUBRIC (PERI-ADRIATIC) FAULT SYSTEM AND THE BELT OF LATE ALPINE WINDOWS ("WINDOWS BELT")

The Insubric fault system, fundamental not only for the Alps-Dinarides problem, consists of two straight segments, the Jorio-Tonale line in the west and the Pustertal-Gailtal line in the east (Fig. 4). They are offset sinistrally about 70 km by the Giudicarie line, a fact which complicates the kinematics of dextral strike-slip postulated for the solution of the Alps-Dinarides problem and which has been one of the arguments against dextral strike-slip. In the east, the Insubric line is lost in the late Tertiary maze of normal faults of the Pannonian basins (Royden and others, 1982) but is usually thought to continue into the Vardar zone, as farther north, sinistral shear seems to dominate (Royden and others, 1982; Fig. 5). Before late Tertiary extension in the Pannonian basins, however, the situation was different, and branches farther north seem plausible.

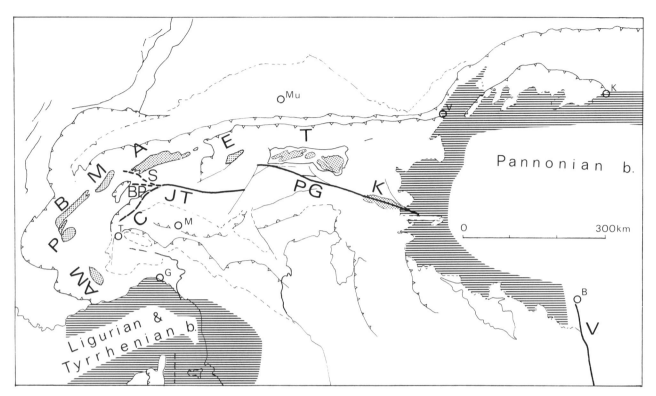

Figure 4. The Insubric system. Heavy line: the straight segments along which the main dextral strike-slip is concentrated; JT = Jorio-Tonale segment, PG = Pustertal-Gailtal segment, V = Vardar zone, C = Canavese fault, BP = Bognanco-Portjengrat fault, S = Simplon fault. Shaded: Windows belt with the External Massifs in the west (AM = Argentera-Mercantour, P = Pelvoux, B = Belledonne, M = Montblanc, A = Aar) and the windows in the east (E = Engadin window, T = Tauern window complex); K = Karawanks flower structure; horizontal ruling: Tertiary extensional basins.

For the Jorio-Tonale line, post-Oligocene dextral displacement has been estimated at between 50 and 80 km, based on a correlation of the eastern end of the Oligocene Bergell granite north of the line and the upper Oligocene to Miocene conglomerates containing large boulders of this granite south of the line (Fumasoli, 1974; Lardelli, 1981; Heitzmann, 1987). The correlation, however, is ill defined and subject to modification by other data, as will be discussed further on in the paper. At the western end, the fault zone disintegrates into several branches that are important for the tectonics of the western Alps (Schmid and others, 1987; Steck, 1987). Of these, the Canavese fault (C) is the most important as it continues to separate the main body of the Alpine nappe edifice from the southern Alps. The complex tectonics of the Canavese fault has a shallow, surface aspect and a deep, geophysical aspect which are not easy to fit into one harmonic kinematical model. The shallow part (Schmid and others, 1987) in most cases consists of two subparallel fault bands, the northern one containing shear-sense indicators of southeastward thrusting, the southern one of continuing dextral strike-slip. Furthermore, a thin sliver of Permian-Mesozoic sediments and Oligocene volcanics is found between the two in most cases, suggestive of a relative drop of the northwestern (Sesia nappe) flank in post-Oligocene time (Fig. 2, and section "The arc of the western Alps and the Insubric-Ivrea phase"), which is at odds with the southeastward thrusting inferred from shear-sense indicators. Considerable southeastward tilting of units both northwest and southeast of the line in post-Oligocene time is directly observable in places and is also suggested by paleomagnetic data (for example, Heller and Schmid, 1974; Heller, 1980; Handy, 1986; Schmid and others, 1987). Finally, the Canavese line is joined by young faults issuing from the southern Alps, such as the Cremosina fault, which again modify its appearance. All this, unfortunately, is of little help in explaining the deep geophysical structure of the Ivrea zone, the famous "bird's head" of the refraction seismologists (for example, Berckhemer, 1968; Kaminski and Menzel, 1968; compare Fig. 2 and further discussion in the section "The arc of the western Alps and the Insubric-Ivrea phase").

The more northerly branches, the Bognanco-Portjengrat (BP) and the Simplon (S) fault zones, penetrate discordantly the Alpine nappe edifice. Originally described by Bearth (1956a, 1956b), they have received much attention recently (for example, Mancktelow, 1985; Steck, 1987). Although interpretation is still controversial, the following facts seem incontestable. The fault zones begin with a ductile phase (cooling ages between 19 and 12 m.y.; Hunziker, 1970; compare Mancktelow, 1985; Laubscher and Bernoulli, 1982). The Simplon ductile shear zone, apparently in the late Miocene (biotite K-Ar cooling ages, Ar-Ar ages, of 10 m.y.; Frank, 1979; Frank and Stettler, 1979), passes into a cataclastic phase (Steck, 1987; Mancel and Merle, 1987). The Bognanco fault is dextral strike-slip, whereas the Simplon fault is mostly extensional dip-slip; the covering Penninic nappes on the southwest side moved southwest on the hot Lepontine nappes, thereby causing their cooling by tectonic denudation. At its northern end in the Rhone valley, the Simplon fault swings again into a westerly direction with almost pure dextral strike-slip, joining the dextral Windows belt. It also joins the "Penninic front" (see section "The arc of

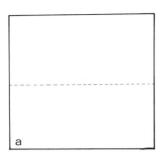 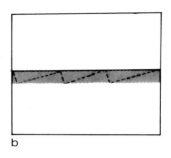 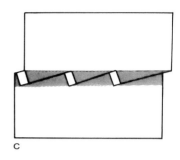

Figure 5. A simple model for separating the compressional and the strike-slip parts of an *en echelon* belt of brachyanticlines. a. Before transpression; b. shortening by the amount indicated by the shaded stripe; c. staggered dextral translation within the belt of shortening with the development of pull-apart depressions. In reality, diffuse distribution of strike-slip and extension and the addition of rotation make accurate quantification difficult.

the western Alps and the Insubric-Ivrea phase"), the surface expression of a band of strong discordant reflections recorded by the recent deep reflection seismic surveys discussed later.

As both the Bognanco-Portjengrat and Simplon branches cut into the Alpine edifice, they should displace dextrally already existing Alpine nappes. This is most easily established where those nappes had a north-south boundary. The western boundary of the Cretaceous Austro-Alpine nappes at the Rhine line (Fig. 3) may provide an example.

Laubscher (1988b), after an analysis of the tectonics of the Rätikon, concluded that this margin of the Austro-Alpine nappes is largely original, the western border of the basement part of a Cretaceous arc convex to the north. When, in the Eocene, the Austro-Alpine nappes with their sole of South Penninic units, including Tethyan ophiolites, were thrust as "traîneau écraseur" (or orogenic lid) on top of the Central and North Penninic domains, this margin exerted a profound influence, a constraint at the upper boundary, on the décollement and deformation of these domains, there provoking a new transverse boundary which is also a part of the Rhine line. Similar conditioning of the Helvetic nappes took place at an even later phase. The Rhine line heads almost perpendicularly toward the Insubric line, although the actual contact is obscured by deformations along the Insubric and Engadin lines and the intrusion of the Bergell granite.

A dextral displacement of the Cretaceous Rhine line by the Insubric line is indeed suggested by the reappearance of the Austro-Alpine Sesia–Dent Blanche basement complex in the western Alps, south of the Simplon and possibly a part of the Bognanco branches. Conventional interpretation ascribes this reappearance of Austro-Alpine nappes to erosion on the intervening Ticino culmination. This is the natural explanation from a cylindristic point of view. Moreover, there must have been considerable denudation as implied by the high metamorphic grade of the Ticino nappes. On the other hand, a scenario involving a non-cylindric cover (there are pronounced transverse tectonics in the Ticino units, as will be discussed) and tectonic denudation attending dextral strike-slip is at least possible (compare tectonic denudation at the Simplon and Brenner lines, section "The Oligocene intrusions along the Insubric line"). No attempt is made herein to model Eocene ("Meso-Alpine") motions which were primarily responsible for burial of the Lepontine domain. Apparent displacement of the western margin of the Austro-Alpine nappes is about 150 km; if taken at face value, it would include motion along pre-Miocene forerunners of the Insubric line. On the other hand, late motion has evidently penetrated the pile of Alpine nappes from its southern boundary, the Insubric line.

The same inference may be made if the sinistral displacement of the Insubric line by the Giudicarie line is accepted, provisionally, as the disruption of an originally straight, continuous fault. Here, kinematic continuity demands a further conclusion; late motions from the southern front of the southern Alps, at the south end of the Giudicarie line, have penetrated across the Insubric line into the eastern Alps (see discussion, section "The Giudicarie system and the Adige embayment"). It should therefore come as no surprise that a very late dextral belt, the *en echelon* belt of late Alpine windows, or the "Windows belt" for short, is located in the middle of the nappe edifice, crossing it discordantly in an arcuate zone which is subparallel to the Insubric line. Unlike this, it is not displaced by the Giudicarie fault and is therefore younger. Apparently, the Insubric line was inactivated by the sinistral offset, and dextral motion was initiated along a new system (Laubscher, 1988a).

This view is at variance with conventional interpretations which have been based on cross-sectional rather than three-dimensional aspects. The late Alpine windows are traditionally considered as belonging to two groups that are entirely unrelated. In the west, the external massifs, where the "autochthonous" units (displaced by comparatively small amounts) of the northern foreland appear in windows of the Helvetic nappes, are thought to end with the Rhine line (R in Fig. 3), east of which the Austro-Alpine nappes become the predominant feature. East of the Rhine line, the Lower Engadin and the Tauern windows, although lined up in the same belt with the external massifs and, at least partly, of late Tertiary age, pierce the Austro-Alpine nappes with the exposure of Penninic units. This difference, however, is to be expected for a late Alpine, post-nappe belt of crustal folds that crosses the Rhine line.

The belt has an internal consistency inasmuch as its units are arranged as brachyanticlines in a dextral *en echelon* pattern. I should like to associate to the eastern end of this belt, before it disappears on entering the Pannonian basin, the late Miocene flower structure of the Karawanks (Schönlaub, 1980; Prey, 1980), which so ill fits the picture of the Alps as a pile of nappes. It straddles the Insubric line and thus is located even more internally than the windows of the eastern Alps. Formerly, the Karawanks had been assumed by Laubscher (1983b) to date stratigraphically movements along the Insubric line as late Miocene, which is at odds with radiometric data farther west (latest Oligocene to middle Miocene). The inclusion of the Karawanks in the Windows belt, albeit with possible forerunners, does away with this inconsistency.

There are other, older folds affecting the nappe system, which have resulted in the creation of windows (particularly the "Internal Massifs"). They seem to be related to rapid denudation along the late Oligocene–early Miocene Insubric line and are discussed in the section "The Oligocene intrusions along the Insubric line."

In cross section, the late Alpine windows are upright anticlines typically some 20 km wide with steep to overturned limbs (Fig. 2), thus resembling crustal "flower structures" (Harding and Lowell, 1979; Meier and others, 1987). Their style contrasts sharply with that of the basement

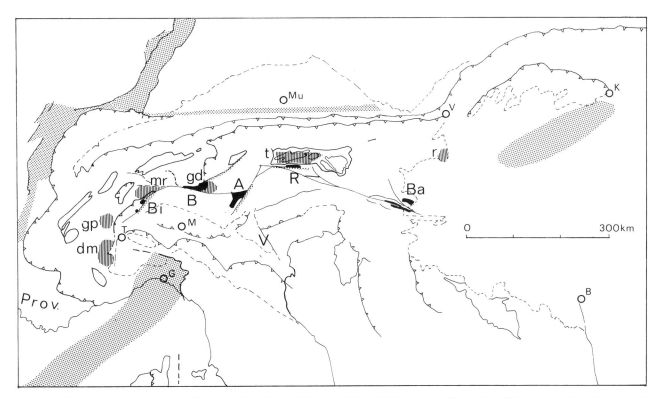

Figure 6. The late Alpine intrusions (black) and the belt of "Internal Massifs" (vertical ruling). Bi = Biella-Traversella, B = Bergell, A = Adamello, R = Rieserferner, Ba = Bacher, V = Vicenza volcanics; dm = Dora Maira, gp = Gran Paradiso, mr = Monte Rosa, gd = Gruf-Disgrazia, t = Tauern, r = Rechnitz; shaded: late Eocene to early Miocene extensional belts (schematic); Prov = Provence.

lobes which they fold: thin units of uppermost crystalline crust that constitute the basement part of the northern Helvetic nappes (Fig. 2; compare Laubscher, 1983a; Steck, 1987). In this system, the Tauern window (Fig. 4 and section on "The Giudicarie system and the Adige embayment") requires special treatment. It is by far the largest and appears to have been active during various Oligocene-Miocene phases, playing a fourfold role. The Windows belt seems to have been the very last gasp of the Alpine orogeny perhaps 7–5 m.y. ago (compare the apatite fission-track ages in the Tauern window of Grundmann and Morteani, 1985; exact age brackets are not known at this time), as a kind of substitute for transpression along the Insubric line when this was inactivated by the Giudicarie displacement.

As to material balance, transpressional *en echelon* systems require a three-dimensional analysis. They are made up of features of axial extension in addition to the compression manifest in the brachyanticlines, and such relative extension or reduction of compression appears in depressions of the nappe edifice, for example, in the Rawil depression between the Mont Blanc and the Aar massifs. Material-balance estimates are possible for a very simple model which replaces the actual diffuse transpression by a compressional translation perpendicular to the belt of the brachyanticlines and a superposed staggered strike-slip with pull-aparts (Fig. 5). The amount of stretching in the pull-aparts, measured in longitudinal section as negative area with respect to the brachyanticlines, gives the strike-slip component of total translation. A rough estimate results in a dextral strike-slip of several tens of kilometers, probably more than 20 km. Taking the western margin of the Austro-Alpine nappes at the Rhine line as a cue, a distributed displacement of that amount seems possible.

Kinematic continuation demands that this amount again be accounted for at the southern margin of the Insubric plate as defined by the Windows belt; it should appear as a sinistral component along the Pelvoux–Argentera–Gulf of Genoa zone (Fig. 4). No marker for a verification of this postulate, however, is readily visible. The belt, being the latest addition to the Alpine edifice, deforms and disrupts all of the earlier members, in cross section as well as in map view. Late Oligocene and Miocene motions along the Insubric line and the Windows belt together account for possibly about half of the total 300 km of dextral slip postulated for the solution of the Alps-Dinarides problem. The remaining 150 km would have to be attributed to some forerunners. The latest of these is the belt of Oligocene intrusions.

THE OLIGOCENE INTRUSIONS ALONG THE INSUBRIC LINE

The only batholiths of Alpine age are small intrusive bodies strung out along the Insubric line (Fig. 6). Most of them are late Oligocene (30 m.y.), although the southern part of the Adamello is late Eocene and the Bacher pluton at the eastern end is early Miocene (19 m.y., compare Exner, 1976). Most of them are centered on side branches of the Insubric line rather than on the line itself and are faulted by it. The general geographic association suggests some relation, a localization along a belt which was a forerunner of the Miocene Insubric line as mapped today. The mere fact of the intrusions suggests an extensional scenario at that time, in continuation of the vast extensional episode in the western Mediterranean (opening of the Balearic sea) and its northeastern foreland (late Eocene to early Miocene Rhine-Bresse-Limagne graben complex).

There is also a belt of normal faults of that age extending all the way from the Rhine graben through the Austrian molasse basin (schematically shown in Fig. 7). Within the Miocene Pannonian basin, there is also an

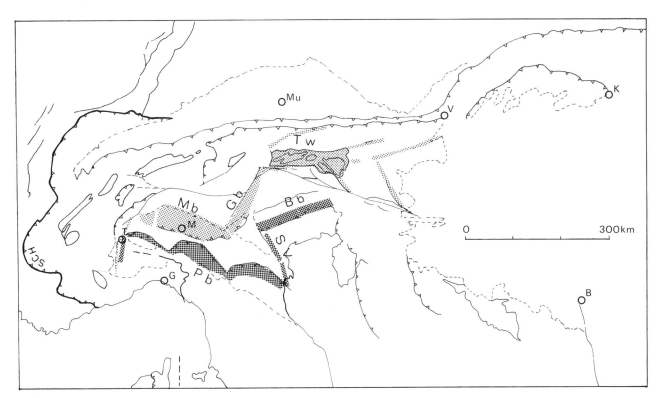

Figure 7. The Giudicarie and the Schio-Vicenza systems. Giudicarie system: Mb = Milan belt, Gb = Giudicarie belt, Tw = Tauern window. SCH = Sub-Alpine chains. Schio-Vicenza system: Pb = Po belt, SV = Schio-Vicenza line, Bb = Belluno belt. Width and quantities (shortening, strike-slip) are poorly known for several branches, and the width of the shading is very schematical. This is particularly true for the amount of Pliocene-Pleistocene translation along the Schio-Vicenza line.

Eocene-Oligocene forerunner (Royden and others, 1982; shown schematically in Fig. 7). The impression is that most if not all of the Alpine domain was extensional at least during a part of that time ("Oligocene lull" before "Miocene revolution," Laubscher, 1983b). Some of the intrusions are located on northeasterly striking branches with, among other deformations, indications of sinistral strike-slip. This may be attributed largely to Miocene transpression as along the Giudicarie, but Oligocene sinistral extensional gashes are not entirely ruled out; they may even suggest an inversion of the shear sense during part of the Oligocene lull.

Associated with this Oligocene lull scenario are some important phenomena usually tied to the kinematics of the Insubric line. One is the rapid cooling in late Oligocene–early Miocene time of the Bergell granite and the Lepontine area to the west of it (Wagner and others, 1977). It is usually attributed exclusively to very rapid uplift and erosion (the "Himalayan model," Dokka and others, 1986), but participation of thermal uplift combined with tectonic denudation in an extensional domain (the "Basin and Range" model) would fit the scenario better. On the other hand, the "Meso-Alpine" (Eocene) nappes were folded by the belt of the "Internal Massifs" and other possibly coeval windows in the Pennine and Austro-Alpine nappes (Figs. 2 and 6; see discussion in text). This belt has not yet been analyzed systematically, and information at present is somewhat ambiguous. Nevertheless, a provisional assessment of its kinematic significance is possible. Cooling ages for several minerals (Wagner and others, 1977; Hurford, 1986; Hurford and Hunziker, 1988) suggest an interval of formation between about 31 and 23 Ma. The apparent ages are difficult to interpret in terms of three-dimensional kinematics because they are affected by Meso-Alpine nappe formation, inhomogeneous Oligocene extension and intrusion, inhomogeneous folding of the nappes, and Miocene compression. Apparent sequences in time (Wagner and others, 1977) may, at least in part, be attributed to contemporaneous inhomogeneous folding. The cooling history along an *en echelon* belt, for example, is different for the brachyanticlines and the depressions between them. The folds of the Internal Massifs belt and the corresponding uplifts seem largely to postdate the late Alpine batholiths (compare Bucher-Nurminen and Droop, 1983) and the associated dikes (30 m.y., except the Novate granite with an ill-defined age of 25 m.y.; Gulson, 1973) but predate late motions on the Insubric line. From these data, it would appear that the Oligocene extensional phase in this part of the Alps was short lived, ceasing already about 2 m.y. after the intrusions. Mylonites of ages between 30 and 28 m.y. in the Lepontine area, particularly where steeply tilted along the Insubric line, have been attributed to southward thrusting, "backthrusting," as the shear sense indicates (Heitzmann, 1987). Present information does not rule out their association with the Internal Massifs belt. An alternative interpretation as tilted extensional mylonites should also be considered, however, as tilting and folding are common along the Insubric line.

A special position in this late Oligocene–early Miocene scenario is occupied by the Lepontine culmination of the Meso-Alpine nappe complex (Fig. 6, between mr and gd). Although not yet satisfactorily integrated into over-all kinematics, its synmetamorphic Maggia transverse structure, for example, is reminiscent of other branches of the Insubric line, for example, the Simplon branch, albeit in the ductile domain and somewhat distorted by dextral motion (compare the tectonic map of Switzerland, scale 1:500,000, or the map by Bigi and others, 1986; Wenk, 1955).

What is the kinematic significance of the Internal Massifs belt? Its

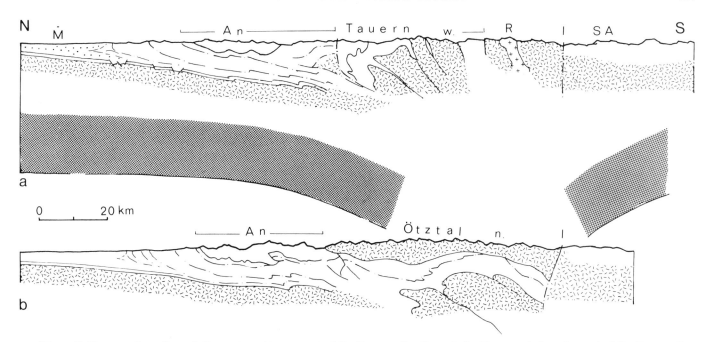

Figure 8. Cross sections through the eastern Alps; a. east of the Brenner line through the Tauern window, b. west of the Brenner line through the Ötztal nappe (after Prey, 1980; for location, see Fig. 1). Random dashes: uppermost, obducted crust; dotted: lower crust (schematic); M = Molasse basin; An = Austro-Alpine cover nappes; R = Oligocene Rieserferner intrusion; I = Insubric line; SA = southern Alps.

similarity to the External Massifs or Windows belt suggests inception of dextral transpression, after Oligocene extension, along the Insubric line. When and how the transition from a belt of brachyanticlines to sharply defined strike-slip faulting took place is not yet clear. Local as well as regional data favor division of total late Oligocene–Miocene motion along the Insubric fault into two phases, the first being the Internal Massifs phase, the second the "Insubric-Ivrea phase" (see discussion, section "The arc of the western Alps and the Insubric-Ivrea phase").

THE GIUDICARIE SYSTEM AND THE ADIGE EMBAYMENT

Perhaps the most controversial transverse feature of the Alps is the Giudicarie belt and the Adige embayment (between Gb and SV in Fig. 7) that dissect the southern Alps into a western and an eastern half with radically different structures (for example, Castellarin, 1984). This transverse embayment had a long history going back at least as far as the Early Jurassic (for example, Castellarin and Vai, 1982; Doglioni and Bosellini, 1987), although its present prominent tectonic and orographic expression is due to late Miocene motions. Speculations have repeatedly centered on a possible connection with the Sestri-Voltaggio line in the Ligurian Alps, but this line is sealed by the Oligocene transgression (for example, Elter and Pertusati, 1973; Gelati and Gnaccolini, 1980; Bigi and others, 1986).

The late Miocene Giudicarie motions in turn have been assigned various kinematic roles. For Trevisan (1939), it was a sinistral belt, with sinistral strike-slip faults connecting east-west–striking folds and thrusts. Castellarin (for example, Castellarin and Sartori, 1986), on the other hand, impressed by the many east-vergent thrusts, considered it an east-vergent segment of a late Miocene arcuate thrust belt. Both elements are evident, and it would appear that the Giudicarie belt is the sinistrally transpressive eastern boundary (or transform or transfer boundary) of the late Miocene frontal Milan belt of the southern Alps (Pieri and Groppi, 1981). In particular, the straight eastern boundary fault of the Adamello massif is suggestive of strike-slip, and the abrupt change of direction at its southern end is hard to harmonize with a purely compressional convex arc; such divergence would require, for material balance, a large amount of axial stretching which is not obvious. Moreover, subduction of the rather rigid Adriatic subplate under the Milan belt would not produce a purely compressional, divergent arc. A translation of the Adriatic subplate slightly east from north would fit better the over-all picture. In a transpressional Giudicarie belt, a roughly estimated 30 km east-southeast compressional component, based on the profiles in Castellarin and Sartori (1986), is vectorially consonant with about 70 km sinistral strike-slip, that is, the amount by which the Insubric line is displaced by the Giudicarie. It is also consistent with about 80 km of late Miocene shortening in the Milan belt, mostly hidden under the Po plain. There are reasons to assume that this may be the right order of magnitude (Roeder, 1985), although direct evidence is lacking. Laubscher (1973) mentioned this and illustrated it, as a possibility, although practically nothing was known at the time about the subsurface of the Po plain. He pointed out that the consequence would be a transfer of 70 km of shortening to a structure north of the Insubric line of late Miocene age and that this role could be played only by the Tauern window and associated structures.

Recently, Selverstone (1988) has mapped and evaluated pervasive extensional lineation in the western part of the Tauern window and across the Brenner line, similar to that across the Simplon line. The results of this investigation seem to suggest that east-west stretching has been the predominant deformation in the window and its western margin. Such a mechanism, however, cannot explain the conspicuous north-south shortening required by all cross sections through the Tauern window (for example, Fig. 8). As its flanks are steep to overturned and faulted, a mere vertical (for example, isostatic) uplift is out of the question. This is one of those examples wherein fabric analysis and large-scale material-balance requirements do not yet seem to harmonize. Perhaps a part of the solution for the problem lies in the fact that the northern part of the Brenner line strikes north-northwest and consequently was a transtensional rather than a transpressional boundary for north-northeast shortening of its eastern flank. In the ductile domain, this would call for axial stretching. The southwestern part of the window, which strikes north-northeast, may have been dextrally rotated during the subsequent phase of the Windows belt. South of the window, east-west stretching is also manifested by a number of young dextral and sinistral strike-slip faults (Senarclens-Grancy, 1972; compare Bigi and others, 1986).

Folding of the pile of nappes to produce the Tauern window (Fig. 8) east of the Brenner line occurred between 30 and 10 m.y. B.P. (for example, Frank and others, 1987; compare Selverstone, 1988), with apatite fission-track ages (Grundmann and Morteani, 1985) indicating even later stages in parts of the window. As discussed above, the Oligocene to early Miocene ages probably document the extensional and Internal Massifs phases, and the very young apatite fission-track ages, the Windows belt phase of the Tauern window. Again, however, it should be emphasized that the correlation of radiometric ages and three-dimensional kinematics in such complex areas is a delicate matter.

The Tauern window seems to have about the right order of cumulative crustal shortening, as estimated on the basis of folding and thrusting of the Ötztal nappe in Figure 8a. Particularly when adding the Inntal fault (Fig. 7, north of Tw), with an additional post-Oligocene shortening of perhaps 20 km (compare Ampferer, 1921; Tscheuschner, 1985), shortening on the order of 100 km is suggested. Conversely, this considerable shortening in the Tauern window requires some late Miocene kinematic continuity at both ends, and at the western end, the only one visible, and obvious at that, is the connection across the severely tectonized Mauls-Penserjoch area (late Miocene ages, Hammerschmidt, 1986; Frank and others, 1987) into the Giudicarie and the Milan belt. Approaching the material-balance problem from both the Giudicarie and the Tauern thus leads to the same conclusion. This would be the third phase of the structuration of the Tauern window, preceding immediately that of the superposed Windows belt and succeeding those of Oligocene extension with the forming of a "metamorphic core complex" (Selverstone and Hodges, 1987; Selverstone, 1988) and the folding of the Internal Massifs. The post-nappe kinematics of the Tauern window, although studied in considerable detail, for example, by Selverstone (1985, 1988), is obviously not yet sufficiently understood to go beyond the suggestions as outlined above.

At this point, it may be useful to sum up the benefits of a late Miocene 70-km sinistral translation along the Giudicarie-Brenner zone for a coherent kinematics of otherwise unrelated elements. (1) The belt of Oligocene intrusions is straightened out. (2) The belt of Internal Massifs is straightened out. (3) The Insubric line is straightened out and is well suited for strike-slip. (4) Inactivation of the Insubric line by the Giudicarie phase led to the branching off, in the Karawanks, of the Windows belt which elegantly curves around the kink in the Insubric line, passing through the Tauern window where it superposes the latest deformation. (5) The Milan belt, the Giudicarie belt, the Tauern shortening, and the distributed compression of Miocene basins in the eastern Alps (see below) all find the kinematic continuity they require.

Along the northern part of the Giudicarie, there is no thrust belt for estimating the compressional part of transpression, the Dolomites block, uplifted in post-latest Oligocene (Cros, 1966), extending westward all the way to the Giudicarie. Here, shortening by deep-seated thrusting is postulated to be at least partly responsible for the uplift of the westernmost Dolomites and the adjacent eastern Alps, but confirmation has to wait for a detailed geophysical survey. A minor part of the Giudicarie belt may even branch off into the Valsugana structure (compare Doglioni and Bosellini, 1987).

The late Miocene Giudicarie system continues westward into the frontal Milan thrust belt (Castellarin and Vai, 1982; Castellarin and Sartori, 1986). Its western end, although unknown, would have to be postulated somewhere east of the internal border of the western Alps northeast of Turin. It would appear plausible that the crust under the thrust belt and its foreland was depressed by several kilometers, with a north-dipping slope (compare Roeder, 1985). In the Pliocene-Pleistocene, this slope was inverted to a certain degree, the young strata now dipping south into the foreland trough of the Apennines (Pieri and Groppi, 1981). The Miocene position of the Milan belt basement at its western end is difficult to estimate on present information, but it may be assumed that by its depression, it deformed the Ivrea zone. Perhaps the remarkable eastward drop of the Ivrea body is partly due to this influence.

The Giudicarie system cannot end with the western termination of the Milan belt. Kinematic continuity demands its swinging to the south, thereby again assuming the role of a sinistral transfer zone ("Turin line"). In this area, between Turin and the Mediterranean, are concentrated several loose ends of important zones of movement between the southern Alps, the western Alps, the Ligurian Alps, the northern Apennines, and the Ligurian and Tyrrhenian seas ("Ligurian knot"). The problem of unraveling this knot would be formidable even under favorable circumstances and is made more difficult because much is hidden under late Tertiary and Quaternary sediments. Nevertheless, an attempt at sketching a coordinated scenario will be made after discussion of the main constituents.

One candidate for a continuation of the Milan belt might be the arc of Nice, the southwestern end of the Sub-Alpine chains. Their northeastern end, the Jura, seems to have roughly the same age (Laubscher, 1988a). The Sub-Alpine chains form the external margin of the arc of the western Alps and may be assumed to reflect the complex polyphase kinematics of the internal margin discussed in this article. Kinematic models for parts of the Sub-Alpine chains have been proposed, but an integrated picture that satisfies three-dimensional material-balance requirements has yet to be worked out. One element that seems to link the kinematics of the external with those of the internal arc is the Kandertal zone (Laubscher and Bernoulli, 1982).

Whatever the kinematics of the Miocene external arc, however, the source for the Milan belt motions, that is, the southern margin of the Adriatic plate active at the time, ought to be looked for at least as far south as the Gulf of Genoa. The source must have been either a push from the African plate or a pull due to back-arc spreading (hardly sufficient at this time; compare Rehault and others, 1986; Fanucci and Nicolich, 1984; Burrus, 1984) in the Ligurian-Tyrrhenian domain. Structures in this area, and particularly their deep crustal and lithospheric roots, apparently have been obliterated by the subsequent extension (Giese, 1985; Giese and others, 1982). At any rate, the southern margin of the Adriatic plate was south of the Ligurian Alps and the northern Apennines and their dividing fault, the Villalvernia-Varzi-Levanto line (VVL). This line is the only discontinuity separating the pre-Oligocene Ligurian Alps from the post–early Miocene northern Apennines to the north and east. Its western end is an almost straight east-west fault cutting lower and middle Miocene formations, whereas its southeastern part is not so obvious, as it seems to have been folded along with the Apennine nappes (Gelati and Gnaccolini, 1980; Bigi and others, 1986). The position of the southern margin of the Adriatic plate during the Milan belt phase implies a late Miocene northward translation of the Ligurian Alps, the VVL, and the northern Apennines by 80 km along the postulated Turin line.

A problem similar to that in the Ligurian domain is encountered east of the Tauern window, where extensional basins marginal to the Pannonian "back-arc" basin enter the picture and complicate material-balance estimates. The quotation marks indicate that these basins in many cases cut

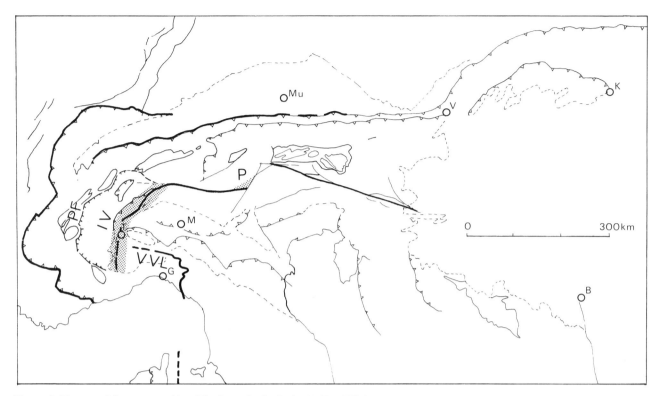

Figure 9. The arc of the western Alps. IV = Ivrea body; faults: VVL = Villalvernia-Varzi-Levanto, PF = Penninic front; P = Pejo zone; heavy line with triangles: external front, with the Sub-Alpine chains and the Jura; whether a belt of this age continues eastward to the south of Munich is doubtful.

across arcs and even penetrate into the foreland. Laubscher (1983b, 1988a) therefore proposed the term "pores" (in a convergent plate boundary). Miocene extensional basins east of the Tauern window and inside the eastern Alps were compressed in late Miocene times (Fuchs, 1980), this compression somehow continuing late Miocene Tauern folding eastward, but a quantitative estimate for the different branches schematically shown in Figure 7 is not available at this time. The lithospheric root shown here by Babuška and others (1988) looks oddly athwart the Alpine structures, suggesting a profound disharmony and possibly displacement by Pannonian extension.

THE ARC OF THE WESTERN ALPS AND THE INSUBRIC-IVREA PHASE

The very tight arc of the western Alps (Figs. 9–12) poses many problems of material balance. As it is situated at the southwestern end of the Insubric line, at the front of the Adriatic indenter, it appears as the frontal transverse range of an east-west–directed transform boundary. This, however, is not obvious from current plate-tectonic models of Africa-Europa convergence in the Tertiary, which is nearly perpendicular to the strike of the central Alps (Dewey and others, 1973). Apparently, the Adriatic plate has a kinematics of its own, to some extent independent from that of Africa and conditioned by such secondary plate-boundary processes as back-arc spreading. At a first glance, all of the Tertiary motions along the Insubric line shown in Figures 3 through 9 are oblique intracontinental links between the extensional areas of the western Mediterranean and those of the Pannonian basin and the Aegean (via the Vardar zone). Unfortunately for this simple picture, however, no or little extension has been reported so far from the Balearic and Tyrrhenian basins between about 19 and 6 Ma—the time of major motion along the Insubric and Giudicarie systems.

For cross-sectional kinematics in the western Alps, Laubscher (1970a, 1974, 1985a) proposed a lithospheric profile and sought to incorporate geophysical as well as geological information. Taking the "bird's head" model of the refraction seismologists (Berckhemer, 1968) as an indication for the westward obduction of lower crust and some upper mantle of South Alpine or Adriatic provenance, he arrived at a profile similar to that in Figure 2 as the most plausible solution. The alternative, letting the west-vergent overthrust at the base of the Ivrea zone continue eastward under the Po plain, would require the corresponding subduction zone to be located somewhere to the east, presumably under the central Po plain. There is no indication anywhere, geophysical or other. Indeed, the most reasonable place for this subduction zone would be right under the central low-gravity belt of the Alps (Fig. 11), and this in turn would require the Ivrea zone to be a lithospheric delta structure. Crustal flakes or delta structures coupled with bivergent subvertical lithospheric roots have also been proposed for the central and eastern Alps, on similar grounds (Laubscher, 1970a; Oxburgh, 1972). Recently, the application of seismic tomography to southern California has revealed a similar vertical lithospheric root under the Transverse Ranges (Hager and others, 1987).

Many geologists balk at this notion, but no reasoned alternative has ever been offered, to my knowledge (compare Zingg, 1983). As descending lithospheric slabs may break off under yet unknown conditions, the depth of the lithospheric root may vary irregularly. Recently, this model has found some resonance in the geophysical community because it agrees with newer results, first from surface wave dispersion analysis (Panza and

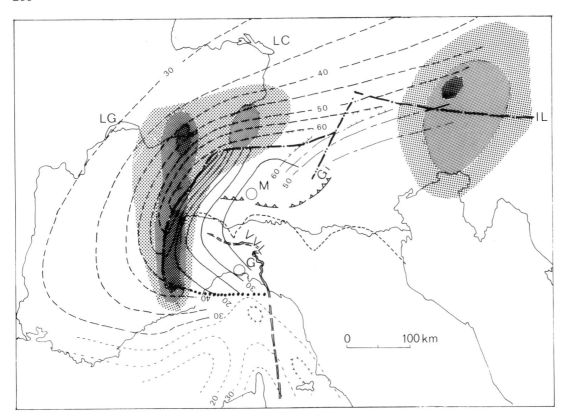

Figure 10. The deep structure of the Alps. Moho contours (in kilometers) after Giese (1982), modified in the Ligurian sea; solid lines: front of Insubric indenter, Ivrea body. Base lithosphere after Babuška and others (1988); dark shading: more than 200 km, intermediate shading: 180–200 km, light shading: 160–180 km. IL = Insubric line, Gi = Giudicarie line, VVL = Villalvernia line; dotted line = position of VVL before northward transport, particularly by the Neo–North Apennine event. LG = Lake Geneva, LC = Lake Constance.

others, 1980), then from teleseismic traveltime residuals (Babuška and others, 1988). The geometry changes, however, where the Ligurian sea in the southwest and the Pannonian basin in the east are approached (Fig. 10).

The new reflection data (Bayer and others, 1987) suggest that the Ivrea bird's head may be composed of several lower crust and upper mantle slices which were pushed as tectonic wedges into the middle crust of the Alpine edifice as it existed then (Fig. 12c). This picture is reminiscent of recent reflection results from another intracontinental branch of the Africa-Europa plate boundary, the Pyrenees, where lower crust and upper mantle of the European side were apparently squeezed into the middle crust of the Pyrenean-Iberian side (Fig. 12d). These interpretations of the crustal reflection surveys are preliminary, however, and will probably change somewhat in future revisions.

The wedging of the Ivrea body into the western Alps poses some problems of timing. Inasmuch as it is tied to the Miocene Insubric line, the first conjecture would be to assign it a Miocene age, too, as argued by Laubscher (1971). On the other hand, association with Canavese sediments and Sesia basement, both metamorphosed in the Cretaceous, suggests a possible connection with Cretaceous nappe-forming processes (Zingg, 1983; Martinotti and Hunziker, 1984). I think that the arguments for a Miocene age are more cogent and are becoming more so, given the latest information. The Cretaceous nappes of the western Alps, the Sesia–Dent Blanche and the Ophiolite nappe, were thrust on top of the Penninic realm in the late Eocene; the Ivrea body, on the other hand, lies below at least some of the Penninic nappes. Even though it may well have participated in Cretaceous orogeny, its wedging into the western Alps would still

Figure 11. The Bouguer gravity anomalies of the western Alps. After Vecchia (1968). Narrow ruling: maxima in the center of the Ivrea body; wide ruling: minima of the crustal root of the Alps; depth of the lithospheric root added (after Babuška and others, 1988). External massifs (stippled): A = Argentera, P = Pelvoux, Be = Belledonne. Added is the approximate location of the recent reflection seismic surveys and of the refraction profile of Figure 12b.

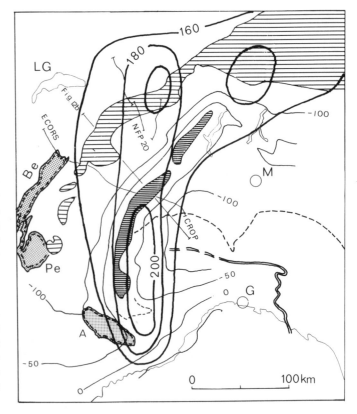

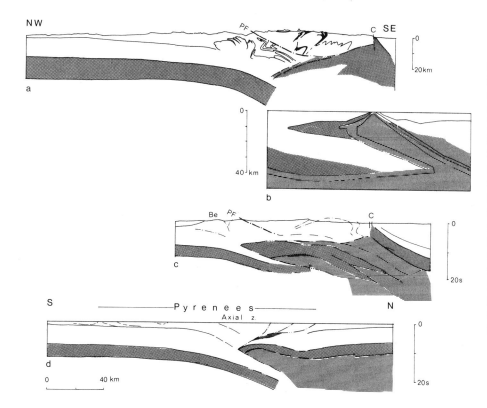

Figure 12. A comparison of the deep structure of the western Alps and the Pyrenees. a. The upper part of Figure 2, with data from Swiss Working Group on Reflection Seismology (1988). b. After Kaminski and Menzel (1968), shows lines of equal density; stippled: more than about 2.8 g/cm^3; ruled: more than 3 g/cm^3; this density model is only crudely interpretable in terms of lower crust and upper mantle. The horizontal scales are approximately the same, whereas the vertical scales for c (after Bayer and others, 1987) and d (after ECORS Pyrenees team, 1988) are in two-way traveltimes. Dotted = lower crust (including upper mantle in a, ruled = upper mantle. The interpretations shown are provisional; other, although similar, models are possible. Faults: PF = Penninic front with strong discordant reflections (dash-dot), C = Canavese line.

have had to occur after the Eocene. The strong crustal reflections from the "Penninic front" (Bayer and others, 1987; Swiss Working Group on Reflection Seismology, 1988; Figs. 2, 9, and 12) indicate a very young, discordant event; they merge with the top of the Ivrea body. The deep lithospheric root under the western Alps, obviously due to the latest phases of Alpine compression, fits well the scenario of Miocene westward translation of the Insubric indenter, with the Ivrea flake at its front. A serious material-balance problem, unsolved as yet, then arises, however, concerning the fate of the upper crust originally on top of the Ivrea body (compare Fig. 12). It may have been eroded from the Cretaceous on as a result of the original obduction of the Ivrea body and related lower crustal masses in the Sesia–Dent Blanche nappe complex.

Figures 9 and 12 suggest the existence of a late Miocene partial orogenic lid, bounded frontally by the Penninic front, in the rear by the Canavese fault, in the north by the Simplon fault, and in the south by the Larche shear zone between the Argentera massif and the southern end of the Ivrea body ("Simplon lid"). The seismic data indicate a basal truncation of the Internal Massifs folds. Surface and radiometric information in the Simplon area show cooling of the hanging wall between 30 and 23 Ma (about the range of Internal Massifs folding) and cooling of the footwall between 19 and 10 Ma (about the range of the Insubric phase). The Simplon lid was underthrust from the east by the Ivrea bird's head and simultaneously or subsequently pushed to the west at the Penninic front, causing the pull-apart at the Simplon boundary. It was deformed frontally by the subsequent Windows belt and in the rear by the addition of the Milano belt and the Neo–North Apennines.

Laubscher (1971, 1973) developed the scenario of an Insubric (or Adriatic) indenter, sharply bounded, pushed into a yielding mass, thereby producing a diffusion of deformation in ever-widening arcs, which is characteristic of the western Alps. The widest of these arcs is the external western front of the Alps in the Sub-Alpine chains and the Jura (compare with discussion in section "The Giudicarie system and the Adige embayment"). They are young, largely of late Tortonian age (Laubscher, 1988a), and thus approximately coeval with the Giudicarie-Milan belt; however, they may well have started with the Insubric-Ivrea event or even the Internal Massifs event.

As to material balance at transpressional systems such as the Insubric line, it is clear that after 100 km or so of slip, the original spatial relations between the two sides of the fault were disrupted, some fault slivers were probably excised and subjected to erosion or depressed into the subsurface (compare Meier and others, 1987), and rotation of pre-existing fabrics may be severe. One important question is what happened to the southern portions of the Alpine nappes now cut off by the Insubric line. Are they buried below the southern Alps as cross-sectional kinematics would seem to demand? This appears to be precluded at least in some places by the presence of the Moho at a depth not much exceeding 30 km south of the Insubric line (Müller and others, 1980). The continuation of Alpine structures cut off by the Insubric line is one of the important three-dimensional material-balance problems which still wait for a solution. One conjecture would be that they were wedged back into the main body of the Alps, another that they were subducted during late Miocene compression. Some wedging is suggested by the gravity calculations of Kissling (1980).

If one accepts a total post-Eocene westward motion of the Adriatic indenter of more than 130 km as intimated by the Alps-Apennines relation at its southern margin (see below), a remarkable correlation emerges. The northeast end of the Ivrea body, moved back to the east by 130 km, joins the southwest end of the Pejo zone (P in Fig. 9). This zone is known for its highly metamorphic Kinzigite gneisses (Andreatta, 1951, 1954), comparable to those of the Ivrea zone.

The southern end of the western Alps in the Ligurian knot is enigmatic, as stated in this paper. Still, some highly unusual and significant features are known and may be used, in connection with material-balance considerations, for conjecture concerning some kinematic relations.

The Insubric indenter here calls for a left-lateral southern transfer boundary (Laubscher, 1971), but instead, the following situation is observed (Figs. 7, 9, and 10–12). The Ligurian Alps as bounded by the Villalvernia-Varzi-Levanto line (VVL, already discussed) extend to the north of the expected transfer boundary; this suggests that they were emplaced there by later events, and particularly by the Milan belt–Giudicarie phase. Indeed, this distortion is comparable to that of the northern transfer boundary of the indenter, the Insubric line, by the Giudicarie line. Even farther north, near Turin, is the front of the north-vergent Apennines, with the puzzling hills of the Monferrato immediately behind. Their Cretaceous to Tertiary rocks are practically identical with those found much farther east in the allochthonous northern Apennines (Elter and others, 1966), but they are separated from them by late Tertiary depressions. Their allochthony so far has not been proved, as the pertinent contacts are nowhere exposed. They can hardly be autochthonous, however, as their eastern continuation, the northern Apennines, have been pushed on their northern foreland by hundreds of kilometers in post–early Miocene times (Elter and Pertusati, 1973). Geophysical data, although not conclusive (Stein and others, 1978; Losecke and Scheelke, 1978) also point toward allochthony.

How did the Monferrato get behind the Ligurian Alps? The straightforward model would be, if the timing as perceived now remains valid, (1) an original sinistral southern transfer boundary of the Adriatic indenter during the Internal Massifs and Insubric-Ivrea events, taking off at the southern end of the Ivrea body and swinging eastward into the Gulf of Genoa, the original VVL. Total sinistral translation between the western tip of the Monferrato and the eastern border of the Ligurian Alps exceeds 130 km, and this would be the sum of the Internal Massifs and the Insubric-Ivrea phases. Where the VVL resumes a southerly direction, it developed a compressional boundary between the Oligocene to early Miocene Paleo–North Apennines (Kligfield and others, 1986; coeval with the Internal Massifs phase) and the rather rigid block of the Ligurian Alps. This block may have undergone some sinistral rotation during the Paleo-Apennines–Internal Massifs phase (for example, Laubscher, 1988a). (2) A late Miocene to Quaternary northward translation of the northern Apennines, now including the still comparatively rigid Ligurian Alps, by roughly 80 km along the Turin line. Most of it may be attributed to the Milan belt–Giudicarie phase, although a small amount was added subsequently (Pieri and Groppi, 1981; Castellarin and others, 1985).

The deep structure for a region of such large young motions is of singularly low relief. It is apparently quite chaotic, and the interpretation of the refraction data is rather uncertain, possible remnants of abandoned subduction zones appearing on some interpretations and disappearing again on others (Giese and others, 1982; Giese, 1985; compare Fig. 10 and Laubscher, 1988a). One may suspect that Tyrrhenian extension has obliterated much of the deep structure here as it did in the Tyrrhenian itself.

Considerations so far have concerned translations of the Insubric plate. The over-all curvature of the various segments of the Insubric line, including the Vardar line, when taken together, however, suggests an anticlockwise rotation as well, and this has probably some bearing, as yet not quantifiable, on the Ligurian knot.

To sum up: What are the benefits of the Internal Massifs phase and the Insubric-Ivrea phase as outlined above for unifying Alpine kinematics? (1) Separation of the westward motion of the Adriatic indenter into an earlier Internal Massifs and a subsequent Insubric-Ivrea phase helps clarify local as well as regional correlations. Locally, at the Simplon fault, cooling ages of the hanging wall may be attributed to the Internal Massifs phase, whereas cooling ages of the footwall correspond to the Insubric-Ivrea phase. The Penninic front reflections as a continuation of the Simplon fault cut discordantly through the base of the Internal Massifs. Regionally, the Internal Massifs phase is coeval with the compression in the Paleo–North Apennines at a time of prevailing extension in the Alps and their surroundings. The Insubric-Ivrea phase begins with the resumption of generalized compression in that area (Miocene revolution). (2) At the Insubric line, the Internal Massifs phase helped in unroofing the Bergell granite north of the line and transporting boulders of the granite into the Molasse basin south of the line after a dextral motion of 60–80 km. The Insubric-Ivrea phase separated the boulders and the granite dextrally by another 60–80 km. The two displacements together account for the total dextral separation of the Ivrea zone with respect to the Pejo zone in the north (P in Fig. 9) and of the Monferrato with respect to the North Apennines in the south.

CONCLUSIONS

When three-dimensional material-balance considerations are used to connect the bewildering observations reported for more than 100 years of Alpine geology, a relatively coherent picture of some of the most prominent Alpine structures and kinematics emerges. The Cretaceous (Eo-Alpine) Austro-Alpine nappes formed a continuous north- to west (Europe)-vergent arc from the Rhine line in eastern Switzerland to Košice in Slovakia. There, it was bounded by a transfer zone along which shortening of the southern continental margin of Tethys was transferred to the west- to south(Africa)-vergent Dinarides. In the subsequent approximately 100 m.y., this transform zone was dextrally displaced by at least 300 km along the Insubric fault system and its forerunners. As this component of motion is to be added to considerable north-south compression, the Alps turn out to be an oblique, dextrally transpressive branch of the Africa-Europa plate boundary zone.

From the Eocene on, when the Piemontese ocean basin of Tethys had been totally consumed, the Alps became the narrow intracontinental deformation belt they are now. In the Oligocene, but in places beginning in the late Eocene and continuing into the early Miocene, extension or "back-arc" spreading invaded much of the Alpine domain. The late Alpine batholiths intruded into a belt of gashes that approximately coincides with the Miocene Insubric fault zone. The intrusions were succeeded almost immediately by the dextrally transpressive (60–80 km) phase of the Internal Massifs belt with uplift and erosion in brachyanticlines and the deposition of Bergell granite boulders at the end of the phase, which is coeval with the Paleo–North Apennines. The first half of a total left-lateral displacement of at least 130 km along the Villalvernia-Varzi-Levanto line may be attributed to the Internal Massifs phase.

In the Miocene (beginning at 19 Ma?), some of the more conspicuous and eye-catching orographic and tectonic features of the Alps were either added or given their final form. The sequence of events that best seems to satisfy the data is as follows.

(1) Another 60–80 km of dextral displacement along the Insubric line, separating the Bergell granite boulders south of the line from the intrusion north of it. At the front of the combined Oligocene-Miocene westward translation of the Adriatic plate (Insubric indenter), some striking structures developed. The lower crust–upper mantle masses of the Ivrea zone, perhaps originally obducted in the Cretaceous, were wedged into the western Alps, under the Simplon lid, which is bounded everywhere by young shear zones that cut discordantly through all previous structures. A part of the external zone of the western Alps, the Sub-Alpine chains, probably developed at this time. The frontal part of the Adriatic lithosphere was subducted to the west almost vertically by about 100 km. No equivalent lithospheric root is known under the northern Apennines. The southern continuation of this kinematic system passes through the Alps-Apennines suture between Elba and Corsica, and particularly the deep structure was largely obliterated by post-Miocene extension in the Tyrrhenian.

(2) Pre-Messinian sinistral dissection and displacement of the Insubric fault by the Giudicarie system (70 km). This comprises, from west to east, the Torino line, the Milan belt, the Giudicarie belt, a part of the Tauern window with the frontal Inntal fault, and the compressed Miocene basins east of the Tauern window. In the western Alps, another part of the Sub-Alpine chains may be linked to this event. It continues, connecting the Tyrrhenian and Pannonian extensional domains. No definite lithospheric root can be attributed to this phase. The Ligurian Alps, the Villalvernia-Varzi-Levanto line, and the Monferrato Apennines are displaced sinistrally with respect to the western Alps (80 km).

(3) Dextral transpression resumes but cannot use the disrupted Insubric fault. It is transferred to the *en echelon* belt of the crustal brachyanticlines, the Windows belt, known in the west as "External Massifs" and east of the Rhine line as "windows" in the Austro-Alpine nappes. Dextral translation may be estimated at some tens of kilometers.

(4) Additional post-Messinian to Quaternary transport of the Neo-Apennines, probably minor at their western end.

These kinematics illustrate the complexity of real plate interactions. In the Alps, every few million years, the direction of convergence seems to have changed. Typically, the regional structures are composed of dip-slip and strike-slip segments, the latter often transpressional but occasionally transtensional. Subsequent phases in most cases produced structures that cut disharmoniously across the existing structures, disrupting them in various ways. Compressional phases alternated with extensional ones in space-time patterns that are poorly understood at present. The role of the various crustal-lithospheric layers is complex also. As a rule, the topmost few kilometers of the crust exclusively participate in observable thrusting. The rest, comprising the middle crust and, particularly, as recent reflection surveys show, the lower crust, are subducted. Rarely, lower crust and upper mantle slivers are obducted as in the zone of Ivrea, and frequently upper crust is subducted to great depths and is only in some cases re-obducted. The geometry of subduction seems to deviate considerably from that usually assumed. Wedging of crustal layers of one plate into those of the other plate and subvertical lithospheric roots are intimated by seismic data.

ACKNOWLEDGMENTS

Besides to Harry Hess and his work in the Caribbean, I am indebted to Peter Bearth, who beginning more than 40 years ago, introduced me to the intricacies of the Monte Rosa and the Simplon and Bognanco areas. Martin Frey and Hannes Hunziker provided valuable literature. Albert Bally and Rick Groshong helped clarify the paper by their thoughtful and constructive critique. Ongoing discussions with a number of colleagues, and particularly with Daniel Bernoulli, Alberto Castellarin, Stefan Schmid, and André Zingg, kept and keep ideas in flux. The printing of this article was financed by the Freiwillige Akademische Gesellschaft, Basel.

REFERENCES CITED

Ampferer, O., 1906, Ueber das Bewegungsbild von Faltengebirgen: Jahrbuch der Kaiserlich Königlichen Geologischen Reichsanstalt, v. 56, p. 539–622.

——— 1921, Ueber die regionale Stellung des Kaisergebirges: Jahrbuch der Geologischen Staatsanstalt, v. 71, p. 159–172.

Andreatta, C., 1951, Carta Geologica delle tre Venezie, 1:100,000, Monte Cevedale F. 9: Venice, Italy, Ministero dei lavori pubblici.

——— 1954, La Val di Peio e la catena Vioz-Cevedale, studio geo-petrotettonico e minerario di una parte del massiccio dell'Ortles: Acta Geologica Alpina, v. 5.

Argand, E., 1916, Sur l'arc des Alpes occidentales: Eclogae Geologicae Helvetiae, v. 14, p. 145–191.

Babuška, V., Plomerová, J., and Silen'y, J., 1988, Structural model of the subcrustal lithosphere in central Europe, *in* Fuchs, K., and Froidevaux, C., eds., Composition, structure and dynamics of the lithosphere-asthenosphere system: American Geophysical Union Geodynamics Series, v. 16, p. 239–251.

Bayer, R., Cazes, M., Dal Piaz, G. V., Damotte, B., Elter, G., Gosso, G., Hirn, A., Lanza, R., Lombardo, B., Mugnier, J.-L., Nicolas, A., Nicolich, R., Polino, R., Roure, F., Sacchi, R., Scarascia, S., Tabacco, I., Tapponnier, P., Tardy, M., Taylor, M., Thouvenot, F., Torreilles, G., and Villien, A., 1987, Premiers résultats de la traversée des Alpes occidentales par sismique réflexion verticale (Programme ECORS-CROP): Comptes rendus des séances de l'Académie des Sciences Série II, v. 305, p. 1461–1470.

Bearth, P., 1956a, Zur Geologie der Wurzelzone östlich des Ossolatales: Eclogae Geologicae Helvetiae, v. 49, p. 267–278.

——— 1956b, Geologische Beobachtungen im Grenzgebiet der lepontinischen und penninischen Alpen: Eclogae Geologicae Helvetiae, v. 49, p. 279–290.

Berckhemer, H. (German Research Group for Explosion Seismology), 1968, Topographie des Ivrea Körpers, abgeleitet aus seismischen und gravimetrischen Daten: Schweizerische Mineralogische und Petrographische Mitteilungen, v. 48, p. 235–246.

Bigi, G., Cosentino, D., Parotto, M., Sartori, R., and Scandone, P., eds., 1986, Structural model of Italy, Consiglio nazionale delle Ricerche, progetto finalizzato geodinamica: Rome, Italy, Centro Nazionale Ricerche, sheet 1, scale 1:500,000.

Borghi, A., Cadoppi, P., Porro, A., and Sacchi, R., 1985, Metamorphism in the north part of the Dora-Maira Massif (Cottian Alps): Bollettino Museo regionale di scienze naturali, v. 3, no. 2, p. 369–380.

Bucher-Nurminen, K., and Droop, G., 1983, The metamorphic evolution of garnet-cordierite-sillimanite gneisses of the Gruf-Complex, eastern Pennine Alps: Contributions to Mineralogy and Petrology, v. 84, p. 215–227.

Burrus, J., 1984, Contribution to geodynamic synthesis of the Provençal basin (northwestern Mediterranean): Marine Geology, v. 55, p. 247–270.

Buxtorf, A., 1907, Zur Tektonik des Kettenjura: Berichte über die Versammlungen des oberrheinischen geologischen Vereins, v. 40, p. 29–38.

——— 1916, Prognosen und Befunde beim Hauensteinbasis und Grenchenbergtunnel und die Bedeutung der letzteren für die Geologie des Juragebirges: Verhandlungen der Natforschenden Gesellschaft in Basel, v. 27, p. 184–254.

Castellarin, A., 1984, Schema delle deformazioni tettoniche sudalpine: Bolletino Oceanologia Teorica Applicata, v. 2, p. 105–114.

Castellarin, A., and Sartori, R., 1979, Struttura e significato della linea delle Giudicarie Sud: Rendiconti della Societa geologica italiana, v. 2, p. 29–32.

——— 1986, Il sistema tettonico delle Giudicarie, della Val Trompia e del sottosuolo dell'alta pianura lombarda: Memorie della Societa geologica italiana, v. 26, p. 31–37.

Castellarin, A., and Vai, G. B., 1982, Introduzione alla geologia strutturale del Sudalpino, *in* Guida alla geologia del Sudalpino centro-orientale: Guide geologia regionale, Societa geologica italiana, p. 1–22.

Castellarin, A., Eva, C., Giglia, G., and Vai, G. B., 1985, Analisi strutturale del Fronte Appenninico Padano: Giornale di Geologia, Sez. 3a, v. 4771-2, p. 47–76.

Chopin, Ch., 1987, Very-high-pressure metamorphism in the western Alps: New petrologic and field data: Terra Cognita 7, p. 94.

Cornelius, H.-P., and Furlani-Cornelius, M., 1930, Die Insubrische Linie vom Tessin bis zum Tonalepass: Denkschriften der Akademie der Wissenschaft in Wien, Mathematisch-naturwissenschaftliche Klasse, Bd. 102, p. 29–101.

Cros, P., 1966, Age oligocène d'un poudingue (du Monte Parei) dans les Dolomites centrales italiennes: Compte rendu sommaire des séances de la société géologique de France, p. 205–252.

Dal Piaz, G. V., Hunziker, J. C., and Martinotti, G., 1972, La zona Sesia-Lanzo e l'evoluzione tettonico-metamorfica delle Alpi nordoccidentali interne: Memorie della Societa geologica italiana, v. 11, p. 433–466.

Dengo, G., 1953, Geology of the Caracas region, Venezuela: Geological Society of America Bulletin, v. 64, p. 7–40.

Dewey, J. F., Pitman, W. C., Ryan, W.B.F., and Bonin, J., 1973, Plate tectonics and the evolution of the Alpine system: Geological Society of America Bulletin, v. 84, p. 3137–3184.

Doglioni, C., and Bosellini, A., 1987, Eoalpine and mesoalpine tectonics in the Southern Alps: Geologische Rundschau, v. 76, p. 735–754.

Dokka, R. K., Mahaffie, M. J., and Snoke, W., 1986, Thermochronologic evidence of major tectonic denudation: Tectonics, v. 5, p. 995–1006.

ECORS Pyrenees Team, 1988, The ECORS deep reflexion seismic survey across the Pyrenees: Nature, v. 331, p. 508–510.

Elter, G., Elter, P., Sturani, C., and Weidmann, C., 1966, Sur la prolongation du domaine ligure de l'Appennin dans le Monferrat et les Alpes et sur l'origine de la nappe de la Simme s.l. des Préalpes Romandes et chablaisiennes: Archives des Sciences, Genève, v. 19.

Elter, P., and Pertusati, P., 1973, Considerazioni sul limite Alpi-Appennino e sulle sue relazioni con l'arco delle Alpi occidentali: Memorie della Societa geologica italiana, v. 12, p. 359–375.

Exner, Ch., 1976, Die geologische Position der Magmatite des periadriatischen Lineaments: Verhandlungen der Geologischen Bundesanstalt, 1976, p. 3–64.

Fanucci, F., Nicholich, R., 1984, Il Mar Ligure: nuove acquisizioni sulla natura, genesi ed evoluzione di un bacino marginale: Memorie della Societa geologica italiana, v. 27, 97 p.

Frank, E., 1979, Metamorphose mesozoischer Gesteine im Querprofil Brig-Verampio: mineralogisch-petrographische und isotopengeologische Untersuchungen [Ph.D. thesis]: Berne, Switzerland, University of Berne, 204 p.

Frank, E., and Stettler, A., 1979, K-Ar and ^{39}Ar-^{40}Ar systematics of white K-mica from an Alpine metamorphic profile in the Swiss Alps: Schweizerische mineralogische und petrographische Mitteilungen, v. 59, p. 375–394.

Frank, W., Kralik, M., Scharbert, S., and Thöni, M., 1987, Geochronological data from the eastern Alps, *in* Flügel, H. W., and Faupl, P., eds., Geodynamics of the eastern Alps: Vienna, Austria, Franz Deuticke, p. 272–281.

Fuchs, W., 1980, Das Inneralpine Tertiär, *in* Oberhauser, R., ed., Der geologische Aufbau Österreichs: Wien, Austria, Springer-Verlag, p. 452–483.

Fumasoli, M. W., 1974, Geologie des Gebietes nördlich und südlich der Jorio-Tonale Linie im Westen von Gravedona: Mitteilungen des geologischen Instituts der Universität u. ETH Zürich NF 194.

Galson, D., and Müller, St., eds., 1985, Proceedings of the Second Workshop on the European Geotraverse (EGT) Project 2: EGT Workshop, 2nd, The Southern Segment, Venice, Italy, 7–9 February, 143 p.

——— eds., 1986, The European Geotraverse, Part 2: Tectonophysics, v. 128, p. 163–396.

Gelati, R., and Gnaccolini, M., 1980, Significato dei corpi arenacei di conoide sottomarina (Oligocene–Miocene inferiore) nell'evoluzione tettonico-sedimentaria del Bacino terziario ligure-piemontese: Rivista italiana di paleontologia e stratigrafia, v. 86, p. 167–186.

Giese, P., 1985, The structure of the upper lithosphere between the Ligurian Sea and the southern Alps. Part B: The consolidated crust and the uppermost mantle, *in* Galson, D., and Müller, St., eds., 2nd, The southern segment: European Geotraverse Workshop, Venice, Italy, 7–9 February, Proceedings, 143 p.

Giese, P., Nicolich, R., and Reutter, K. J., 1982, Explosion seismic crustal studies in the Alpine-Mediterranean region and their implications to tectonic processes, *in* Berckhemer, H., and Hsü, K., eds., Alpine-Mediterranean geodynamics: Geodynamics Series, v. 7, p. 39–74.

Griffin, W. L., 1987, On the eclogites of Norway—65 years later: Mineralogical Magazine, v. 51, p. 333–343.

Grundmann, G., and Morteani, G., 1985, The Young uplift and thermal history of the central eastern Alps (Austria/Italy) evidence from apatite fission track ages: Jahrbuch der Geologischen Bundesanstalt, v. 128, p. 197–216.

Gulson, B. L., 1973, Age relations in the Bergell region of the southeast Swiss Alps: With some geochemical comparisons: Eclogae Geologicae Helvetiae, v. 66, p. 293–313.

Hager, B. H., Humphreys, E., and Clayton, R. W., 1987, Upper mantle structure beneath southern California: Observations using seismic tomography and interpretation of resulting flow and stress, *in* Noller, J. S., Kirby, St. H., and Nielson-Pike, J. E., eds., Geophysics and petrology of the deep crust and upper mantle: Workshop sponsored by the U.S. Geological Survey and Stanford University, U.S. Geological Survey Circular 956.

Hammerschmidt, K., 1982, K/Ar and $^{40}Ar/^{39}Ar$ resolution from illites of the Trias of Mauls; Mesozoic cover of the Austroalpine basement, eastern Alps (South Tyrol): Schweizerische mineralogische und petrographische Mitteilungen, v. 62, p. 113–133.

Handy, M., 1986, The structure and rheological evolution of the Pogallo fault zone, deep crustal dislocation in the Southern Alps of northwestern Italy (Prov. Novara) [Ph.D. thesis]: Basel, Switzerland, University of Basel, 327 p.

Harding, T. P., and Lowell, J. D., 1979, Structure styles, their plate-tectonics habitats, and hydrocarbon traps in petroleum provinces: American Association of Petroleum Geologists Bulletin, v. 63, p. 1016–1058.

Heitzmann, P., 1987, Evidence of late Oligocene/early Miocene backthrusting in the central Alpine root zone: Geodynamica Acta, v. 1, p. 183–192.

Heller, F., 1980, Palaeomagnetic evidence for late Alpine rotation of the Lepontine area: Eclogae geologicae Helvetiae, v. 73, p. 607–618.

Heller, F., and Schmid, R., 1974, Paläomagnetische Untersuchungen in der Zone Ivrea-Verbano (Prov. Novara, Norditalien): Vorläufige Ergebnisse: Schweizerische mineralogische und petrographische Mitteilungen, v. 54, p. 229–242.

Hess, H. H., and Maxwell, J. C., 1953, Caribbean research project: Geological Society of America Bulletin, v. 64, p. 1–6.

Hunziker, J. C., 1970, Polymetamorphism in the Monte Rosa, Western Alps: Eclogae geologicae Helvetiae, v. 63, p. 151–161.

Hunziker, J. C., and Martinotti, G., 1984, Geochronology and evolution of the Western Alps: A review: Memorie della Società Geologica Italiana, v. 29, p. 43-56.

Hurford, A. J., 1986, Cooling and uplift patterns in the Lepontine Alps south central Switzerland and an age of vertical movement on the Insubric fault line: Contributions to Mineralogy and Petrology, v. 92, p. 413–427.

Hurford, A. J., and Hunziker, J. C., 1988, A revised thermal history for the Gran Paradiso Massif: Contributions to Mineralogy and Petrology (in press).

Kaminski, W., and Menzel, H., 1968, Zur Deutung der Schwereanomalie des Ivrea-Körpers: Schweizerische mineralogische und petrographische Mitteilungen, v. 48, p. 255–260.

Kissling, E., 1980, Krustenaufbau und Isostasie in der Schweiz [Ph.D. thesis]: Zurich, Switzerland, Eidgenössische Technische Hochschule, no. 6655, 165 p.

Kligfield, R., Hunziker, J., Dallmeyer, R. D., and Schamel, St., 1986, Dating of deformation phases using K-Ar and $^{40}Ar/^{39}Ar$ techniques: Results from the Northern Apennines: Journal of Structural Geology, v. 8, p. 781-798.

Lardelli, T., 1981, Die Tonale Linie im unteren Veltlin: il. Fakultät II der Univ. Zürich, Juris Druck & Verlag Zürich.

Laubscher, H. P., 1967, Geologie und Paläontologie: Tektonik. Verhandlungen der Naturforschenden Gesellschaft in Basel, v. 78, p. 24–34.

—— 1970a, Bewegung und Wärme in der alpinen Orogenese: Schweizerische mineralogische und petrographische Mitteilungen, v. 5, p. 565–596.

—— 1970b, Das Alpen-Dinariden-Problem und die Palinspastik der südlichen Tethys: Geologische Rundschau, v. 60, p. 813–833.

—— 1971, The large-scale kinematics of the western Alps and the northern Apennines and its palinspastic implications: American Journal of Science, v. 271, p. 193–226.

—— 1973, Alpen und Plattentektonik. Das Problem der Bewegungsdiffusion an kompressiven Plattengrenzen: Zeitschrift der Deutschen geologischen Gesellschaft, v. 124, p. 295–308.

—— 1974, Evoluzione e struttura delle Alpi: Le Scienze (edizione italiana di Scientific American), v. 72, p. 48–59.

—— 1983a, Detachment, shear and compression in the central Alps: Geological Society of America Memoir, v. 158, p. 191–211.

—— 1983b, The late Alpine (Periadriatic) intrusions and the Insubric line: Memorie della Società geologica italiana, v. 26, p. 21–30.

—— 1985a, The tectonics of the western and southern Alps: Correlation between surface observations and deep structure, in Galson, D., and Müller, St., eds., The southern segment: European Geotraverse Workshop, 2nd, Venice, Italy, 7-9 February, Proceedings, p. 93–101.

—— 1988a, The arcs of the Western Alps and the Northern Apennines: An updated view: Tectonophysics, v. 146, p. 67–78.

—— 1988b, The tectonics of the southern Alps and the Austroalpine nappes: A comparison, in Coward, M. P., Dietrich, D., and Park, R. G., eds., Alpine tectonics: Geological Society of London Special Publication (in press).

Laubscher, H. P., and Bernoulli, D., 1982, History and deformation of the Alps, in Hsü, K. H., ed., Mountain building processes: London, United Kingdom, Academic Press, p. 170–180.

Losecke, W., and Scheelke, I., 1978, Results of magnetotelluric measurements in the south-western Po plain, Italy, in Closs, H., Roeder, D., and Schmidt, K., eds., Alps Apennines Hellenides geodynamic investigations along geotraverses by an international group of geoscientists: Stuttgart, Germany, E. Schweizerbart'sche Verlagsbuchhandlung, p. 228–230.

Mancel, P., and Merle, O., 1987, Kinematics of the northern part of the Simplon line (Central Alps): Tectonophysics, v. 135, p. 265–275.

Mancktelow, N., 1985, The Simplon line: Major displacement zone in the western Lepontine Alps: Eclogae geologicae Helvetiae, v. 78, p. 73–96.

Meier, B., Schwander, M., and Laubscher, H., 1987, The tectonics of Táchira—key to North Andean tectonics, in Schaer, J.-P., and Rodgers, J., eds., The anatomy of mountain ranges: Princeton, New Jersey, Princeton University Press, p. 229–237.

Müller, St., Ansorge, J., Egloff, R., and Kissling, E., 1980, A crustal cross section along the Swiss Geotraverse from the Rhinegraben to the Po plain: Eclogae geologicae Helvetiae, v. 73, p. 463–483.

Oxburgh, E. R., 1972, Flake tectonics and continental collision: Nature, v. 239, p. 202–204.

Panza, G. F., Mueller, St., Calcagnile, G., 1980, The gross features of the lithosphere-asthenosphere system in Europe from seismic surface waves and body waves: Pure and Applied Geophysics, v. 118, p. 1209–1213.

Pieri, M., and Groppi, G., 1981, Subsurface geological structure of the Po plain, Italy: Consiglio Nazionale Ricerche Progetto Finalizzato Geodinamica, v. 414, p. 1–13.

Prey, S., 1980, Die Geologie Österreichs in ihrem heutigen geodynamischen Entwicklungsstand sowie die geologischen Bauteile und ihre Zusammenhänge, in Oberhauser, R., ed., Der geologische Aufbau Österreichs: Wien, Austria, Springer-Verlag, p. 79–117.

Rehault, J. P., Mascle, J., and Boillot, G., 1986, Evolution géodynamique de la Méditerranée depuis l'Oligocène: Memorie della Società geologica italiana, v. 27, p. 85.

Roeder, D., 1985, Geodynamics of southern Alps: Seminar paper given at University of Milan, November 28, Anschutz Corporation Exploration Research Division, Denver, Colorado, 22 p.

Royden, L., Horváth, F., and Burchfiel, B. C., 1982, Transform faulting, extension, and subduction in the Carpathian Pannonian region: Geological Society of America Bulletin, v. 93, p. 717–725.

Royden, L., Horváth, F., and Rumpler, J., 1983, Evolution of the Pannonian basin system, 1. Tectonics: Tectonics, v. 2, p. 63–90.

Schmid, S. M., Zingg, A., and Handy, M. R., 1987, The kinematics of movements along the Insubric line and the emplacement of the Ivrea zone: Tectonophysics, v. 135, p. 47–66.

Schönlaub, H. P., 1980, Die Südalpen (Karnische Alpen - Südkarawanken), in Oberhauser, R., ed., Der geologische Aufbau Österreichs: Wien, Austria, Springer-Verlag, p. 427–451.

Schweizerische Geologische Kommission, ed., 1972, Tektonische Karte der Schweiz, 1:500,000, Wabern-Bern, Eidgenössische Landestopographie.

Selverstone, J., 1985, Petrologic constraints on imbrication, metamorphism, and uplift in the SW Tauern window, eastern Alps: Tectonics, v. 4, p. 687–704.

—— 1988, Evidence for east-west crustal extension in the eastern Alps: Implications for the unroofing history of the Tauern window: Tectonics, v. 7, p. 87–105.

Selverstone, J., and Hodges, K. V., 1987, Unroofing history of the western Tauern window: Evidence for west-directed removal of the austroalpine nappe sequence: Terra Cognita, v. 7, p. 89.

Semenza, E., 1974, La fase giudicariense, nel quadro di una nuova iptesi sull'orogenesi alpina nell'area italo-dinarica: Memorie della Società geologica italiana, v. 13, p. 187–226.

Senarclens-Grancy, W., 1972, Geologische Karte der westlichen Defereger Alpen, Osttirol, scale 1:25,000, Geologische Bundesanstalt, Vienna.

Spakman, W., 1986, Subduction beneath Eurasia in connection with the Mesozoic Tethys: Geologie en Mijnbouw, v. 65, p. 145–153.

Steck, A., 1984, Structures de déformations tertiaires dans les Alpes centrales (transversale Aar-Simplon-Ossola): Eclogae geologicae Helvetiae, v. 77, p. 55–100.

—— 1987, Le massif du Simplon-Réflexions sur la cinématique des nappes de gneiss: Schweizerische mineralogische und petrographische Mitteilungen, v. 67, p. 27–45.

Stein, A., Vecchia, O., and Froelich, R., 1978, Seismic model of refraction profile across the western Po valley, in Closs, H., Roeder, D., and Schmidt, K., eds., Alps Apennines Hellenides geodynamic investigations along geotraverses by an international group of geoscientists: Stuttgart, Germany, E. Schweizerbart'sche Verlagsbuchhandlung, p. 180–189.

Swiss Working Group on Reflection Seismology, 1988, Preliminary results of two seismic profiles across the northern part of the Swiss Alps by NFP-20: Vereinigung Schweizerischer Petroleumgeologen und Ingenieure (in press).

Trevisan, L., 1939, Il Gruppo di Brenta: Memorie degli Instituti di geologia e mineralogia dell'Universita di Padova, v. 13, p. 1–128.

Tscheuschner, E., 1985, Die Geologie am Nordrand des Zahmen Kaisers östlich Ebbs (Nördliche Kalkalpen, Tirol) [Diploma thesis]: Technische Universität-München, 188 p.

Vecchia, O., 1968, La zone Cuneo-Ivrea-Locarno, element fondamental des Alpes. Geophysique et geologie: Schweizerische mineralogische und petrographische Mitteilungen, v. 48, p. 215–226.

Vialon, P., 1966, Etude géologique du Massif cristallin Dora-Maira (Alpes Cottiennes): Travaux du Laboratoire de géologie de la Faculté des sciences de l'Université de Grenoble, v. 4, 293 p.

Wagner, G., Reimer, G. M., and Jäger, E., 1977, Cooling ages derived by apatite fission-track, mica Rb-Sr and K-Ar dating: The uplift and cooling history of the Central Alps: Memorie degli Instituti di Geologia e Mineralogie dell'Universita di Padova, v. 30, 27 p.

Wenk, E., 1955, Eine Strukturkarte der Tessineralpen: Schweizerische mineralogische und petrographische Mitteilungen, v. 35, p. 311–319.

Zingg, A., 1983, The Ivrea and Strona-Ceneri zones (southern Alps, Ticino and N Italy)—Review: Schweizerische mineralogische und petrographische Mitteilungen, v. 63, p. 361–392.

MANUSCRIPT RECEIVED BY THE SOCIETY JANUARY 19, 1988
REVISED MANUSCRIPT RECEIVED APRIL 18, 1988
MANUSCRIPT ACCEPTED APRIL 19, 1988

CENTENNIAL ARTICLE

Strike-slip faults

ARTHUR G. SYLVESTER *Department of Geological Sciences, University of California, Santa Barbara, California 93106*

ABSTRACT

The importance of strike-slip faulting was recognized near the turn of the century, chiefly from investigations of surficial offsets associated with major earthquakes in New Zealand, Japan, and California. Extrapolation from observed horizontal displacements during single earthquakes to more abstract concepts of long-term, slow accumulation of hundreds of kilometers of horizontal translation over geologic time, however, came almost simultaneously from several parts of the world, but only after much regional geologic mapping and synthesis.

Strike-slip faults are classified either as transform faults which cut the lithosphere as plate boundaries, or as transcurrent faults which are confined to the crust. Each class of faults may be subdivided further according to their plate or intraplate tectonic function. A mechanical understanding of strike-slip faults has grown out of laboratory model studies which give a theoretical basis to relate faulting to concepts of pure shear or simple shear. Conjugate sets of strike-slip faults form in pure shear, typically across the strike of a convergent orogenic belt. Fault lengths are generally less than 100 km, and displacements along them are measurable in a few to tens of kilometers. Major strike-slip faults form in regional belts of simple shear, typically parallel to orogenic belts; indeed, recognition of the role strike-slip faults play in ancient orogenic belts is becoming increasingly commonplace as regional mapping becomes more detailed and complete. The lengths and displacements of the great strike-slip faults range in the hundreds of kilometers.

The position and orientation of associated folds, local domains of extension and shortening, and related fractures and faults depend on the bending or stepping geometry of the strike-slip fault or fault zone, and thus the degree of convergent or divergent strike slip. Elongate basins, ranging from sag ponds to rhombochasms, form as result of extension in domains of divergent strike slip such as releasing bends; pull-apart basins evolve between overstepping strike-slip faults. The arrangement of strike-slip faults which bound basins is tulip-shaped in profiles normal to strike. Elongate uplifts, ranging from pressure ridges to long, low hills or small mountain ranges, form as a result of crustal shortening in zones of convergent strike slip; they are bounded by an arrangement of strike-slip faults having the profile of a palm tree.

Paleoseismic investigations imply that earthquakes occur more frequently on strike-slip faults than on intraplate normal and reverse faults. Active strike-slip faults also differ from other types of faults in that they evince fault creep, which is largely a surficial phenomenon driven by elastic loading of the crust at seismogenic depths. Creep may be steady state or episodic, pre-seismic, co-seismic, or post-seismic, depending on the constitutive properties of the fault zone and the nature of the static strain field, among a number of other factors which are incompletely understood. Recent studies have identified relations between strike-slip faults and crustal delamination at or near the seismogenic zone, giving a mechanism for regional rotation and translation of crustal slabs and flakes, but how general and widespread are these phenomena, and how the mechanisms operate that drive these detachment tectonics are questions that require additional observations, data, and modeling.

Several fundamental problems remain poorly understood, including the nature of formation of en echelon folds and their relation to strike-slip faulting; the effect of mechanical stratigraphy on strike-slip–fault structural styles; the thermal and stress states along transform plate boundaries; and the discrepancy between recent geological and historical fault-slip rates relative to more rapid rates of slip determined from analyses of sea-floor magnetic anomalies. Many of the concepts and problems concerning strike-slip faults are derived from nearly a century of study of the San Andreas fault and have added much information, but solutions to several remaining and new fundamental problems will come when more attention is focused on other, less well studied strike-slip faults.

INTRODUCTION

The earliest scientific record of strike slip was of a New Zealand earthquake in 1888, the same year the Geological Society of America was founded. Many other strike-slip faults have been discovered since that time, also as a result of observed surface ruptures. The great San Francisco earthquake of 1906 on the San Andreas fault is particularly noteworthy in this regard. Strike-slip faults, however, are regional structures and require regional studies to document their existence and history. Several decades elapsed after that New Zealand earthquake before the geological community had sufficient data to extrapolate from empirical, instantaneous-strike-slip in earthquakes to more abstract interpretations that permitted tens or even thousands of kilometers of crustal translation by strike slip over geologic time. Then plate tectonics, with its startling concepts of crustal mobility, allowed geologists to overcome the limitations of fixist tectonics which prevailed before the 1960s and to understand the mechanical complexities and tectonics of strike-slip.

Extensive field investigations, innovative experimental studies, images of the third dimension by seismic reflection and drilling, refined dating techniques, painstaking interpretations of paleoseismicity, and analyses of modern earthquakes show that crustal slabs and plates have indeed slipped horizontally great distances during geologic time, as Wegener's hypotheses presupposed. Now we have new ideas of how elongate uplifts and basins rise and fall along active strike-slip faults; we see increasingly com-

Geological Society of America Bulletin, v. 100, p. 1666–1703, 31 figs., 1 table, November 1988.

pelling evidence that crustal "miniplates" have rotated about vertical axes in broad zones of simple shear; we have good theoretical and experimental models supported with well-documented geological examples to explain the once thorny question of how major strike-slip faults terminate. Our understanding of how they behave mechanically has improved, but we still know little about these faults in the third dimension or about their thermo-mechanical behavior. Seismology and paleoseismology have been especially important in providing mechanical understanding, because in continental areas such as southern California, New Zealand, and Asia, active strike-slip faults can be studied to give ideas about how they have behaved during geologic time.

Perhaps because it is so well exposed along much of its length, because it is an active fault so close to a highly populated part of the United States having several major academic and governmental scientific groups living almost directly upon it, and because of the presence of petroleum in structural traps along it, the San Andreas fault is the most extensively studied strike-slip fault in the world (see summary papers by Crowell, 1979; Hill, 1981; Allen, 1981). For that reason, the San Andreas fault has been the source of many ideas pertaining to strike-slip tectonics (Hill, 1981); I therefore give the San Andreas fault what may seem to some readers an undue provincial emphasis in this review.

The purposes of this paper are to collect and review a number of principles, concepts, and notions about strike-slip faults; to summarize what is known about their geometric, kinematic, and tectonic implications; to discuss some of the outstanding problems; and to call attention to some challenging problems that must receive attention over the next decade or so to gain better understanding of these structures. A plethora of literature citations is given to show the sources of those principles, concepts, and notions, and to guide the reader toward more specific information. Many references to the older literature demonstrate that many of our beliefs about strike-slip faults were perceived, even if only dimly, by remarkably insightful geologists well before the plate-tectonic revolution.

TERMINOLOGY AND CLASSIFICATION

Strike-slip faults and dip-slip faults are the end members of the spectrum in a kinematic classification of faults (Reid and others, 1913; Perry, 1935). A strike-slip fault is "a fault on which most of the movement is parallel to the fault's strike" (Bates and Jackson, 1987).

The term "wrench fault" was popularized by Moody and Hill (1956), who borrowed the term from Kennedy (1946). Kennedy, in turn, was influenced by E. M. Anderson, who used the term in 1905, because "wrench planes" was used for a long time by the Scottish Geological Survey in the Highlands (Anderson, written discussion in Kennedy, 1946). All of these writers used the term for a deep-seated, regional, nearly vertical strike-slip fault which involves igneous and metamorphic basement rocks as well as supracrustal sedimentary rocks (Moody and Hill, 1956; Wilcox and others, 1973; Biddle and Christie-Blick, 1985). Other writers referred to such faults as "transcurrent faults" (Geikie, 1905); many still do, and it is a good term for any major strike-slip fault whose relation to a genetic classification (for example, Woodcock, 1986) is not clear.

Many writers have used "wrench fault" with increasing frequency for any and all strike-slip faults, whether or not they are regional, vertical, or involve crystalline basement. Many major strike-slip faults are not vertical, however, nor can they be shown to involve basement. Moreover, "wrench fault" carries the kinematic implication of torsion, which adds to the confusion. For these reasons, I recant my previous infatuation with "wrench faults" (Sylvester, 1984) and recommend that *strike-slip fault* be used for a fault of any scale along which movement is parallel to the strike of the fault, that *transform fault* be retained for a plate-bounding

TABLE 1. CLASSIFICATION OF STRIKE-SLIP FAULTS

INTERPLATE (deep-seated)	INTRAPLATE (thin-skinned)
TRANSFORM faults (delimit plates, cut lithosphere, fully accommodate motion between plates)	TRANSCURRENT faults (confined to the crust)
Ridge transform faults* • Displace segments of oceanic crust having similar spreading vectors • Present examples: Owen, Romanche, and Charlie Gibbs fracture zones	Indent-linked strike-slip faults* • Separate continent-continent blocks which move with respect to one another because of plate convergence • Present examples: North Anatolian fault (Turkey); Karakorum, Altyn Tagh, and Kunlun fault (Tibet)
Boundary transform faults* • Join unlike plates which move parallel to the boundary between the plates • Present examples: San Andreas fault (California), Chaman fault (Pakistan), Alpine fault (New Zealand)	Tear faults • Accommodate differential displacement within a given allochthon, or between the allochthon and adjacent structural units (Biddle and Christie-Blick, 1985) • Present examples: northwest- and northeast-striking faults in Asiak fold-thrust belt (Canada)
Trench-linked strike-slip faults* • Accommodate horizontal component of oblique subduction; cut and may localize arc intrusions and volcanic rocks; located about 100 km inboard of trench • Present examples: Semanko fault (Burma), Atacama fault (Chile), Median Tectonic Line (Japan)	Transfer faults • Transfer horizontal slip from one segment of a major strike-slip fault to its overstepping or en echelon neighbor • Present examples: Lower Hope Valley and Upper Hurunui Valley faults between the Hope and Kakapo faults (New Zealand), Southern and Northern Diagonal faults (eastern Sinai)
	Intracontinental transform faults • Separate allochthons of different tectonic styles • Present example: Garlock fault (California)

*See Woodcock (1986, p. 20) for additional examples, both ancient and modern, and for their geometric and kinematic characteristics.

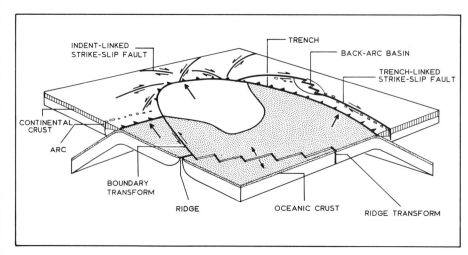

Figure 1. Plate-tectonic settings of major classes of transform faults (redrawn from Woodcock, 1986). Reproduced with permission of The Royal Society (London).

strike-slip fault, and that *transcurrent fault* be revived as the general term for the variety of strike-slip faults which do not cut the lithosphere (Table 1).

A transform fault is a kind of plate-bounding strike-slip fault, regional in scale, that cuts through the lithosphere and fully accommodates the motion between plates. Woodcock (1986) has presented a useful, although genetic, classification of transform faults according to their plate-tectonic setting (Fig. 1). The primary attributes of ridge transforms, the first of Woodcock's four types, are that they link spreading oceanic ridges and cut oceanic lithosphere (DeLong and others, 1979; Woodcock, 1986). They have a short active life as a strike-slip strand, but a long later history as an oceanic fracture zone with dip-slip displacements, and a sense of strike slip that is opposite to the sense of ridge offset (Wilson, 1965). Moreover, the apparent offset of ridge crests may be very much greater than the actual displacement along the ridge transform itself. Ancient ridge transforms are presently restricted to ophiolite complexes, posing difficulties for their recognition, but fairly compelling examples have been described in ophiolite complexes in California (Saleeby, 1977; Cannat, 1985), Washington (Miller, 1985), Newfoundland (Karson and Dewey, 1978), New Caledonia (Prinzhofer and Nicolas, 1980), the Apennines (Abbate and others, 1972), and Turkey (Gianelli and others, 1972). They are recognized by the presence of one or more of the following features: a zone of intensely deformed peridotite having a vertical foliation and horizontal lineation, syntectonic dikes parallel to the shear zone, hydrothermal alteration, and less than normal crustal thickness owing to reduced partial melting. In addition, the penetrative shear strain decreases away from the ancient transform zone, and the sense of shear through the zone may be determined by the bending of the foliation or from the mineral fabrics.

Trench-linked strike-slip faults (Fig. 1) are parallel to the trench and located within, or immediately bordering, the magmatic arc. Woodcock (1986) does not regard them as true transforms, but they are, because they cut through the lithosphere and delimit a fore-arc plate against a hinterland plate. Although displacements measurable in hundreds of kilometers are typical of trench-linked strike-slip faults, these faults accommodate only part of the total displacement behind the trench at a convergent plate boundary (Woodcock, 1986, p. 22). The Semangko fault zone (Fitch, 1972; Page and others, 1979; Karig, 1980; Hla Muang, 1987) may be regarded as the prototype of this kind of fault. The Atacama fault inboard of the Chile trench in South America (Allen, 1965) may be another example. Ancient trench-linked strike-slip faults may be partially or completely obliterated or buried by the very plutons and volcanic ejecta that they channel to the surface along the volcanic arc. Thus, major structural and lithologic discontinuities in the country rocks between batholiths in the Peruvian Andes may be vestiges of a trench-linked strike-slip fault which localized a long narrow line of subarc intrusions in an "Andinotype" orogenic environment (Pitcher and Bussell, 1977; Pitcher, 1979); similarly, a Mesozoic trench-linked strike-slip fault may have been responsible for localization of plutons in the Sierra Nevada batholith of California to the extent that they eventually obliterated the fault. Tobisch and others (1986) give evidence that the batholith formed in a long-lived extensile tectonic regime related to tumescence, arising from subduction-zone heating and magmatism, and they implied that the extension may have been related to orogen-parallel strike slip. The Foothills fault system, active in Late Jurassic time in the western Sierra Nevada, may have been part of such a system of faults (Clark, 1960; Cebull, 1972); so also may have been the proto–Kern Canyon fault which was active in the southern Sierra Nevada when the batholith was emplaced in Mesozoic time (Saleeby and Busby-Spera, 1986). In addition, an increasing body of field evidence shows that intrusions of granitic plutons are related spatially, temporally, and genetically to strike-slip faults. Shear-heating at depth along strike-slip faults is an added factor in enhancing crustal anatexis (Michard-Vitrac and others, 1980), and several workers have demonstrated the synkinematic intrusion of granitic plutons into pull-apart structures along major transform fault zones which, of course, provide a domain of extension for the intrusions (Davies, 1982; Hutton, 1982; Castro, 1985; Gapais and Barbarin, 1986; Guineberteau and others, 1987).

The San Andreas (California), El Pilar (Venezuela), Chaman (Pakistan), and Chugach-Fairweather–Queen Charlotte (Alaska and Canada) faults are examples of active and dormant boundary transform faults in Woodcock's classification (Fig. 1). They accommodate the horizontal displacement between continental or, rarely, oceanic plates which move horizontally with respect to each other. Boundary transforms have long lives and large displacements, comparable to those of trench-linked strike-slip faults (Woodcock, 1986, p. 23). They may evolve naturally from indent-linked strike-slip faults, or they may reactivate old, steep faults having a variety of orientations and mechanisms, including ancient subduction zones, as Hill (1971) proposed for the Newport-Inglewood zone in southern California, and as Freeland and Dietz (1972) postulated for the El Pilar fault in northern Venezuela. Conversely, arc and trench structures may owe their position to the former existence of regional strike-slip faults (Sarewitz and Karig, 1986).

Indent-linked strike-slip faults (Fig. 1) are not true transform faults, because they do not cut the lithosphere. They juxtapose pieces of continental lithosphere, especially in zones of plate convergence and tectonic escape. Displacements on single faults may range from tens to hundreds of kilometers, and they may reactivate any kind of available pre-existing, steep type of fault. Present examples include several strike-slip faults in Tibet and southern China that formed in response to collision with India, in the central part of southern Japan where the Pacific plate converges with the Asian plate (Sugimura and Matsuda, 1965), and in Iran where the northeast edge of the Arabian plate converges with Eurasia (Berberian, 1981; Tirrul and others, 1983). Indent-linked faults may be the main cause of the pervasive lineament networks in Precambrian continental crust (Watterson, 1978; Burtman, 1980; Woodcock, 1986, p. 23).

Intraplate, or intracontinental, transform faults are regional strike-slip faults which are similar to indent-linked strike-slip faults in that they are restricted to the crust (Lemiszki and Brown, 1988), but they need not be genetically related to "indentor tectonics," although they typically separate regional domains of extension, shortening, or shear. The Garlock fault, southern California, separates the southern end of the extended Basin and Range province from the Mojave Desert characterized by dextral shear and regional rotation about a vertical axis (Davis and Burchfiel, 1973). A series of strike-slip faults similarly terminates the northern end of the Basin and Range province from a domain of plate convergence and associated arc tectonics in Oregon (Lawrence, 1976).

A *tear fault* accommodates the differential displacement within a given allochthon, or between the allochthon and adjacent structural units (Biddle and Christie-Blick, 1985). Tear faults generally strike transverse to the strike of the deformed rocks and are sometimes called transverse faults or even transcurrent faults for that reason. Tear faults have been known for a long time in many fold-thrust belts (for example, in the Jura Mountains, Switzerland; Heim, 1919, p. 613–623; Lloyd, 1964), and early model studies reproduced them experimentally (Cloos, 1933; Lee, 1929).

Transfer fault has been used informally but increasingly for strike-slip faults that connect overstepping segments of parallel or en echelon strike-slip faults. Commonly located at the ends of pull-aparts, they "transfer" the displacement across the stepover from one parallel fault segment to the other. They have also been called "oblique faults" (Mann and others, 1983).

DEVELOPMENT OF THE CONCEPT OF STRIKE-SLIP FAULTING

The Book of Zechariah, written in 347 B.C., contains what may be the first reference to strike-slip faulting (Freund, 1971); in fact, one may regard it as a prediction, although the sense of slip is not specified: "and the Mount of Olives shall cleave in the midst thereof toward the east and toward the west, and there shall be a very great valley; and half the mountain shall remove to the north, and half of it to the south" (Zechariah 14:4).

The Swiss geologist, Arnold Escher von der Linth, may have been the first geologist to discover and correctly interpret the geology of a strike-slip fault (Şengör, 1987, written commun.; Şengör and others, 1985). Escher noted horizontal slickensides and the surprising linearity of the 8-km-long trace of the what is now called the "Sax Schwendi fault" which cuts the Säntis folds south of Wildkirchli in the canton of Appenzell, and he showed them to Suess (1885, p. 153, 154) in the 1850s. Escher's mapping clearly showed that the displacement is sinistral and ranges from 500 to 800 m.

The earliest report of strike slip during an earthquake may be the anecdote that a sheep corral was transected by the surface rupture of the great 1857 earthquake on California's San Andreas fault and was thus deformed into a structure shaped like the letter "S" (Wood, 1955). Freund (1971), however, gave credit for the first published record of strike slip to McKay (1890, 1892), one of New Zealand's most distinguished field workers, who documented surface strike slip associated with the earthquake of September 1, 1888, on what is now called the "Hope fault" on the South Island of New Zealand. Kotó (1893) described left-lateral displacement of up to 2 m and vertical displacement up to 6 m at Midori, Japan, associated with the Mino-Owari earthquake of 1891. The Chaman fault in Pakistan was discovered in 1892 when the Quetta-Chaman railroad was sinistrally offset 75 cm during an earthquake on that fault (Griesbach, 1893). Those earthquakes were in fairly isolated places, however, and their descriptions were published in journals with limited circulation.

The phenomenon of strike-slip faulting was roundly evinced to the scientific world when a maximum of 4.7 m of right-lateral slip abruptly occurred on the San Andreas fault in the great San Francisco earthquake of 1906. "Had the San Andreas been classified before 1906, it would probably have been described as a 'normal' fault" (Willis, 1938a, p. 799), because no clear indication of strike-slip faulting was recognized on any of its segments that had been mapped prior to 1906 (Hill, 1981). Many geomorphic features considered today as being characteristic of strike-slip faults, however, were recognized before the earthquake along a linear zone several hundred kilometers long (Fairbanks, 1907, p. 324; Gilbert, 1907, p. 228). It was the nearly 300-km-long surface rupture itself, in fact, which clearly linked the sundry segments together and demonstrated the length and mechanism of the fault. The San Francisco earthquake incontrovertibly taught contemporary geologists that substantial horizontal movements occur and recur along faults, especially in California (Lawson and others, 1908; Wood, 1916; Lawson, 1921). As late as 1950, however, some geologists were not so certain: "to judge from meager reports available, primary displacement of the Californian wholly transcurrent kind is uncommon in other regions on a major scale. Some of the San Andreas fractures seem to be almost unique" (Cotton, 1950, p. 750).

Geologists were very slow to extrapolate the observed movements during earthquakes to repetition of those movements over geologic time, producing displacements of hundreds of kilometers. The early English and Scottish geological literature contains many descriptions of strike-slip faults having from a few tens of meters to a few thousands of meters of horizontal displacement (Cunningham Craig in Horne and Hinxman, 1914, p. 70; Read, 1923), and Argand and his students certainly had a mobilist view of crustal deformation that included room for much horizontal displacement through geologic time (Şengör, in press), but many years of patient geologic mapping were needed in every part of the world to depict geometries and cumulative displacements among correlative rock units over great spans of time and space before acceptable interpretations of many kilometers of strike slip were eventually made.

Noble (1926, 1927) was one of the first investigators to propose tens of kilometers of strike slip on a major strike-slip fault, although Vickery (1925) gave little-noticed geologic documentation for 20 km of right slip on the Sunol-Calaveras fault, a major branch of the San Andreas fault system in central California. Noble correlated some distinctive nonmarine sandstone units across the San Andreas fault in the Mojave Desert in southern California. That correlation required 40 km of strike slip since late Miocene or early Pliocene time, a displacement which now we realize is too small, based on a partially incorrect correlation of a very similar lithofacies (Woodburne, 1975). Some contemporary geologists were receptive, but cautious: "The total horizontal displacement of the southwest mass northwestward past the northeast body [of the San Andreas fault] appears to be very notable, though not definitely determinable" (Willis, 1938a, p. 798), whereas others realized concurrently that much of coastal California is a long, narrow zone of horizontal shear strain with right slip on several faults (Lawson, 1921, p. 580; Buwalda, 1937b), perhaps much of California itself (Locke and others, 1940). Wallace (1949) mapped a 30-km-long segment of the San Andreas and concluded that strike slip on the San Andreas fault must exceed 120 km since mid-Tertiary time, based on his provisional slip rate determined from stream offsets; however, prior to 1950, the opinion of many California geologists coincided with that of the influential N. L. Taliaferro (1941, p. 161), who maintained that the horizontal displacement "has not been greater than one mile and probably less." He had mapped 70 km of the fault in central California, but that was not enough to reveal the magnitude of displacements recognized today. Taliaferro believed that the San Andreas fault was a late Pleistocene structure which coincided with a profound dip-slip fault of Eocene age.

Similar histories of recognition and rejection of great strike slip on faults in other parts of the world could be cited, but one that stands out is Dubertret's (1932) hypothesis for 160 km of left slip in the Dead Sea Rift. It was a remarkable hypothesis for the fact that the 160 km were required by the consequences of *continental drift and rotation* that Dubertret proposed to account for the arrangement of the Sinai and Arabian peninsulas relative to Africa, although the total slip along the Dead Sea transform is presently regarded as 105 km since Miocene time (Bartov and others, 1980). Bailey Willis (1938b), who translated relevant parts of Dubertret's article for the *Bulletin*, used contemporary dogma to reject Dubertret's hypothesis by saying, just as others said to Wegener, that it was an entertaining idea but one which lacked substance because no known force causes continents to drift.

Trümpy (1977, p. 1) related a delightful anecdote regarding the regionality of strike-slip faults and the attendant difficulties in recognizing them:

In the autumn of 1957, I had the pleasure of guiding Professor Biq Chingchang through part of the Swiss Alps. When we arrived in the Engadine, I told him about our difficulties in correlating the geological structures across the valley. In his very modest and cautious way, he suggested that these discrepancies might be due to a sinistral wrench fault. I thought about it for a moment, and then I slapped my forehead: Professor Biq's explanation was the only possible one, and how could we have been so foolish not to have seen it before?

How indeed the Swiss geologists failed to recognize such a conspicuous feature is a curious and in some ways an instructive story. Much of it is due to parochial shortcomings, because the Engadine line cuts across national boundaries. On a wider scale, Alpine geologists, convinced that they were working in the navel of the geological world, took little account of the discovery of large wrench faults in the Circum-Pacific orogens....

Now the existence of strike-slip faults in the Alps is an acceptable hypothesis (Laubscher, 1971), and recent microstructural studies demonstrate an episode of right slip on the Insubric Line in Neogene time (Ratschbacher, 1986; Schmid and others, 1987).

Outside of California, several workers proposed major strike slip on several faults; these included Hess (1938), who postulated that the left separation of Haiti and Cuba could be attributed to post-Miocene lateral movement on a fault in the Bartlett Trough. I believe, however, that the first work to give a strong geologic basis to significant horizontal displacement on a fault, at least in the western world, was Kennedy's (1946) paper on the Great Glen fault in Scotland, a paper which was read to the Royal Geological Society of London in 1939, but which was not published until seven years later. Kennedy correlated rocks and structures produced in several different geologic events, now sinistrally separated 100 km across the Great Glen fault since middle Carboniferous time. Kennedy was quick to acknowledge assistance in his thinking from E. M. Anderson, who had already published the basis of "Andersonian" fault dynamics in 1905.

I believe Kennedy's paper spawned several nearly simultaneous papers in the late 1940s and early 1950s that presented evidence for great displacements on diverse strike-slip faults the world over, including the Alpine fault (450 km, Wellman in Benson, 1952), the San Andreas fault (more than 560 km, Hill and Dibblee, 1953), and the Dead Sea rift (100 km, Quennell, 1958, 1959).

Following these seminal papers by Kennedy, Quennell, Wellman, and Hill and Dibblee, geologists began to recognize strike-slip faults in many places in space and time, largely on the basis of physiography, and they were increasingly bold about publishing their findings during the 1960s, especially in western North America where the evidence is sufficiently well exposed over adequately long distances. Important papers include those about the San Andreas fault system (Crowell, 1952, 1960, 1962; Allen, 1957; Allen and others, 1960); about strike-slip faults in Alaska (Tocher, 1960; St. Amand, 1957); in Canada (Wilson, 1962; Webb, 1969; Roddick, 1967); and in the Basin and Range province (Nielsen, 1965; Shawe, 1965; Albers, 1967; Hill and Troxel, 1966; Stewart, 1967). Recognition of the geologic activity of strike-slip faults followed also in South America (Wilson, 1940, updated in 1968; Bucher, 1952; Rod, 1956; Campbell, 1968; Feininger, 1970); in the circum-Pacific (Biq, 1959; Allen, 1962, 1965; Sugimura and Matsuda, 1965; Kanenko, 1966); in Europe, the Middle East, and Asia (Ketin, 1948; Ketin and Roesli, 1953; Wilson and Ingham, 1958; Pavoni, 1961a; Burtman and others, 1963; Bagnall, 1964; Wellman, 1965; Burton, 1965), and in Africa (Rod, 1962; de Swardt and others, 1965).

As bold as the writers were in the 1950s and early 1960s, it is almost amusing today to read their struggles to explain the mechanics of the great faults. Many geologists, with some notable exceptions (for example, Schofield, 1960), could not visualize, mechanically, how tens or hundreds of kilometers of strike slip could diminish to zero at the ends of the faults, nor could they distinguish between interpretations obtained from fault geometry (separation) or from fault movement (slip) (Hill, 1981). A coherent kinematic explanation was given by Wilson (1965) in his paper on transform faults. Wilson postulated how tens or hundreds of kilometers of horizontal movement on an oceanic transform fault may be transferred into divergent or convergent movements of the Earth's crust. This concept may have been the key revelation in the formulation of plate tectonics and the mechanics of strike-slip faults (Hill, 1981), because it transformed contemporary thinking from fixed continents dominated by vertical movements to crustal slabs and plates having great horizontal movements.

One of the most cited of all *Bulletin* papers in this regard is Atwater's 1970 paper about the implications of plate tectonics for the Cenozoic tectonic evolution of North America. She used Wilson's transform mechanism for ocean-floor tectonics to formulate a tectonic history among several plates in the eastern Pacific and to bring plate tectonics onto the land. Her hypothesis made use not only of the spatial and temporal implications from sea-floor magnetic lineations, but it also explained the origin and history of one of the Earth's major transforms, the San Andreas fault, in a way consistent with the accumulated tangle of geological data.

Now, within the plate-tectonics framework provided by seismotectonics, geology, seismic reflections, geometry, and deep drilling, strike-slip faults are not as mysterious as formerly, but they are no less complicated. In fact, these same investigative tools and new concepts continually reveal the compound mechanical and tectonic roles and complexities of strike-slip faults. With increasing frequency, investigators are appealing to strike-slip faults to translate, rotate, and juxtapose great slabs of the crust within and along orogenic belts over thousands of kilometers throughout geologic time and in ways perhaps even Wegener could not have imagined.

RECOGNITION OF STRIKE-SLIP FAULTS AND THEIR DISPLACEMENTS

Active strike-slip faults are recognized by co-seismic surface displacements (Lawson and others, 1908; Ketin and Roesli, 1953; Ambraseys, 1963), by distinctive physiographic features as outlined below, and by earthquake focal mechanisms (Julian and others, 1982). The modern place and rate of strike slip on active faults are documented by geodetic studies (Reid, 1910; Thatcher, 1979, 1986; Crook and others, 1982; Prescott and Yu, 1986). Their paleoseismic behavior is studied by detailed microstratigraphic studies (Sieh, 1978, 1984).

Physiographic Features

The most distinctive characteristic of active or recently active strike-slip faults is their extreme structural and topographic linearity over very long distances together with an array of distinctive physiographic features which were succinctly described by Noble (1927, p. 37) in reference to the San Andreas fault: "It has a curiously direct course across mountains and plains with little regard for gross physiographic features, yet it influences profoundly the local topographic and geologic features within it." Several of the strike-slip faults of the heavily vegetated parts of Asia and the western Pacific were recognized simply by the great length and linearity of their "rift" topography (Willis, 1937; Biq, 1959; Allen, 1962, 1965). However, recognition of strike-slip faults by geomorphic structures is limited by the durability of small, easily eroded landforms, such as sag ponds and deflected streams whose preservation depends mainly on climate. Even clearly offset cultural features, such as roads, walls, fences, and winery facilities (Steinbrugge and others, 1960; Rogers and Nason, 1971; Zhang and others, 1986, 1987) have a finite life expectancy.

The "rift" or trough may be up to 10 km wide (Gilbert, 1907, p. 234; Lawson and others, 1908, p. 25–52; Noble, 1927, p. 26–27; Davis, 1927) with a variety of fault-formed structures (Fig. 2), including pressure ridges (Wallace, 1949, p. 793), closed depressions called "sag ponds" if presently or once filled with water (Ransome, 1906, p. 286; Cotton, 1950), shutter-ridges (Buwalda, 1937a), and systematically deflected streams (Russell, 1926; Rand, 1931; Cotton, 1952; Wallace, 1968, 1976; Kuchay and Trifonov, 1977; Burtman, 1980; Sieh and Jahns, 1984; Zhang and others, 1987). These characteristic landforms are clearly illustrated in neotectonic strip maps of the Bocono and La Victoria faults in Venezuela (Schubert, 1982a, 1986a), the Agua Blanca fault in Baja California (Allen and others, 1960), and the San Andreas fault (Vedder and Wallace, 1970; Schubert, 1982b; Clark, 1984; Davis and Duebendorfer, 1987). Neither side of the "rift" need be higher than the other. When the opposing blocks of the fault move in strike slip, elongate blocks and slivers subside between parallel or en echelon fault strands, or they warp, sag, or tilt to form

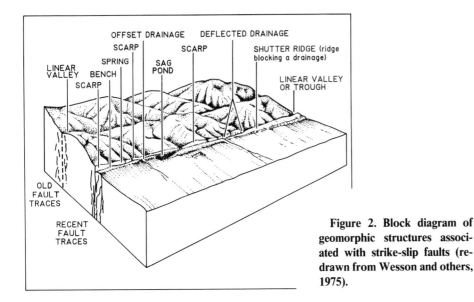

Figure 2. Block diagram of geomorphic structures associated with strike-slip faults (redrawn from Wesson and others, 1975).

closed depressions in the "rift" zone (Gilbert, 1907, p. 228; Fairbanks, 1907, p. 323), and they do so from the smallest to the largest scale. Other blocks may rise, tilt, or slide obliquely to produce pressure ridges. Notched ridges, fault-parallel trenches, or troughs along the fault may reflect increased erosion of the crushed and broken rocks in the fault zone (Fairbanks, 1907, p. 326; Allen and others, 1960; Vedder and Wallace, 1970; Wallace, 1976). Deflected streams must be used with caution to determine direction of lateral displacement. The offsets must be in an uphill direction, or else stream piracy or differential erosion may be invoked to explain them (Higgins, 1961; Allen, 1962, 1965; Wallace, 1976; Patterson, 1979).

The height of a fault scarp generally indicates its minimum vertical displacement, but this may not be true along a strike-slip fault which transects rugged topography, or where the vertical component of slip varies concurrently with the strike of the fault (Peltzer and others, 1988). On a small scale, shutterridges are formed by lateral or oblique displacement on faults transecting a ridge and canyon topography. A displaced part of a ridge shuts off a canyon, hence the term (Buwalda, 1937a; Sharp, 1954). On a larger scale, an entire mountain front may be exposed by lateral displacement of its toes or its other half, as Noble (1932) postulated for the south face of the San Bernardino Mountains in southern California.

Where several scarps are present within a strike-slip fault zone, it is typical that they will have opposing senses of vertical displacements or "scissors" geometry, due to topographic irregularities or because of the variable oblique displacements on the faults themselves. Indeed, a strike-slip fault may display reversals of the direction of throw along its trace, giving rise to descriptive phrases in the older literature as "scissoring" and "propeller faults" (Tomlinson, 1952).

Locally, the relatively uplifted block may slide by gravity or be thrust onto the adjacent, downdropped block. Where "scissoring" has apparently occurred along the strike of the strike-slip fault, a series of thrusts with alternat-

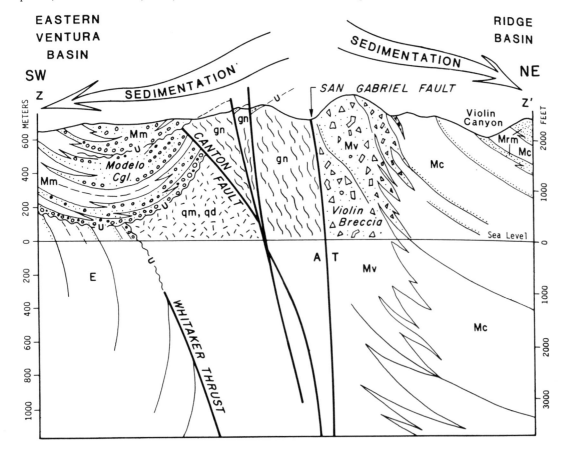

Figure 3. Diagrammatic cross section showing sedimentation mismatch across the San Gabriel fault, southern California (from Crowell, 1982a). Conglomerate of the Modelo Formation, derived from the northeast, now lies faulted against Violin Breccia which was derived from the southwest. The source areas for both stratigraphic units have been displaced by right slip on the fault zone. T = displacement toward observer; A = displacement away from observer. Reproduced by permission of the Pacific Section of Society of Economic Paleontologists and Mineralogists.

ing and opposing throw may be produced, perhaps effectively hiding the deeper presence of the responsible strike-slip fault. This may be the explanation for the alternating and opposed verging folds and thrust directions in the Columbia Plateau of Washington (G. Davis, 1985, personal commun.; R. Bentley, 1987, personal commun.).

Geologic Features

The presence of a strike-slip fault is frequently indicated by en echelon arrays of fractures, faults, and folds in narrow elongate zones. In addition to truncating reference features such as stratification, foliation, folds, dikes, sills, and other faults, strike-slip faults juxtapose rocks of dissimilar lithology, facies, age, origin, and structure. Sedimentary facies may be telescoped or "stretched" across some faults. Locally, great mismatches in sedimentation history may be juxtaposed across a fault (Fig. 3). Slickensides or mullions, where present on fault surfaces, are mainly horizontal. The apparent simultaneous development of extended and shortened structures, together with the variable vertical separation along strike, are typical aspects of strike-slip faults. These features are caused by localized and alternating convergent and divergent components of lateral displacement along the fault in combination with alternating local uplift and subsidence of blocks and slices developed within the fault zone over time. Resultant local uplift and erosion of those blocks and slices yield unconformities of the same age as thick, but not laterally extensive, sedimentary sequences that were deposited very rapidly on adjacent subsided blocks. A great discordance between clast size or lithology of detritus in sedimentary units and possible sources across a fault typically reflects the horizontal translation of the source relative to the deposits.

Structural cross sections reveal variable senses of vertical separation on faults near the surface that coalesce downward into the main fault. Serial cross sections may bear little systematic relation to one another in terms of structure, style, throw, and juxtaposed lithology over distances of 1–2 km (Tomlinson, 1952) in contrast to those at the same scale, for example, across a major normal or thrust fault system.

The criteria for determination of the magnitude of horizontal displacement on a strike-slip fault are those that are least affected by depth of erosion (Gabrielse, 1985). They include offset of geologic lines having considerable horizontal extent that yield piercing points (Fig. 4; Campbell, 1948; Crowell, 1959, 1962), such as strand lines and shelf-to-basin transition zones (Addicott, 1968); sedimentary and metamorphic facies boundaries (Suggate and others, 1961; Roddick, 1967); formation pinch-outs; and isopachous lines (Stewart, 1983); channels and shoestring sands; constructional lines such as "fold axes" and intersections of surfaces (Crowell, 1962), of unique isotopic and geochemical trends (Silver and Mattinson, 1986; James, 1986), and of surfaces of considerable vertical extent such as dikes (Smith, 1962; Speight and Mitchell, 1979), plutons (Sharp, 1967), batholiths (Kennedy, 1946), and even suture zones (Şengör, 1979). Ideally the reference lines and surfaces should be at a high angle to the fault. In many other instances, a rare rock type or unique assemblage of rocks has been shed or erupted from one side of a strike-slip fault and deposited on the other side, so that a minimum horizontal displacement may be determined by the distance the deposit has been removed from its source (Crowell, 1952; Fletcher, 1967; Ross, 1970; Ross and others, 1973; Ehlig and others, 1975; Matthews, 1976; Ramirez, 1983). Juxtaposition of provinces having great dissimilarities in geochemistry or paleomagnetic orientations, together with the slicing of long slivers of oceanic crust, are evidences of strike-slip faulting in the Semangko fault zone (Page and others, 1979).

Syntectonic igneous activity is notably sparse along strike-slip faults except locally in zones of transtension (Şengör, 1979; Şengör and Canitez, 1982; Hempton and Dunne, 1984) and along trench-linked strike-slip faults which are imbedded in the volcanic arc. Major strike-slip faults also lack a conspicuous, peaked heat-flow anomaly which should be present if the average dynamic frictional resistance exceeds a hundred bars in the seismogenic zone (Brune and others, 1969; Lachenbruch and Sass, 1980). The heat flow in the Coast Ranges, within which the San Andreas fault is located, however, is moderately high and is comparable to that in the Basin and Range province (Lachenbruch and Sass, 1980). Where the San Andreas fault is transformed at its south end to a spreading rift, the heat flow is very high and spatially associated with young rhyolitic volcanic rocks and geothermal activity (Muffler and White, 1969; Robinson and others, 1976).

Separation, Slip, and Trace Slip

In any discussion of how to determine the amount and direction of displacement on any fault, it is essential to distinguish carefully between the terms "slip" and "separation" (Hill, 1959; Crowell, 1959, 1962). Separation is the apparent displacement of a plane in two dimensions and has geometric significance only, whereas slip, the actual displacement of lines, has kinematic significance. Separation has clear and long-standing geometric meaning in the context of faulting and should be retained for use in its original context. Regrettably, however, some writers (Rodgers, 1980; Segall and Pollard, 1980; Barka and Cadinsky-Cade, 198) have given the word different meaning in association with strike-slip faults.

Of equal importance is the recognition and understanding of "trace slip" (Beckwith, 1941;

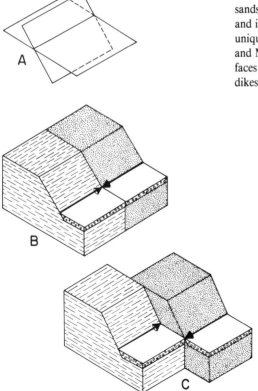

Figure 4. Diagrams illustrating the concept of piercing points of a displaced geological line. (A) Intersection of two surfaces defines a line. Block diagrams show a vertical fault between sandstone (stippled) and shale (dashed). The sloping surface represents a stream bank. (B) The intersection between the stream bank surface and the upper surface of the alluvium layer defines a line. The intersection of that line with the fault defines a piercing point. (C) Amount and direction of slip of the fault blocks can be determined from the displacement of the piercing points.

Crowell, 1962) where the displacement is parallel to the trace of reference markers, such as bedding, on the fault surface (Fig. 5). Trace slip especially hinders recognition of strike-slip faults in seismic sections which are sensitive only to vertical separation. Strike slip out of the plane of the section, parallel to the line or trace of the layering on the fault surface, may displace the layers laterally but not vertically. I suspect that many strike-slip faults have gone unrecognized in many areas of nearly flat-lying strata for this reason and will be eventually "discovered" when sufficient regional data are accumulated. Many orogen-parallel strike-slip faults have gone unnoticed also because of regional trace slip, or because geologic reference lines make low angles with the fault.

A criterion which should force the suspicion of the existence of a strike-slip fault in seismic or structure sections is the presence of unresolved space problems after palinspastic restoration of the structural geometry has been attempted (Sylvester and Smith, 1976; Harding and Lowell, 1979; Harding, 1983b), problems which necessitate movement of rock volume in and out of the plane of the section. It is this factor that makes it difficult or impossible to achieve balance in structural cross sections. Conversely, lack of balance in faulted sequences is one of the keys to recognition that strike slip must be considered in the deformation picture (Sylvester and Smith, 1976). A second criterion is the presence of several adjacent, nearly vertical faults which show apparent, opposing senses of vertical separation (Harding and Tuminas, 1988, their fig. 11). The vertical separations may indeed represent true slip, but it is difficult to construct a plausible tectonic story of normal and reverse faulting which alternates over a short distance in time and space. In my experience, two-dimensional separations on steeply dipping faults should be regarded with considerable suspicion and may be better interpreted as vertical components of three-dimensional strike or oblique slip.

MECHANICS OF STRIKE-SLIP FAULTING

The presence of shortening structures such as folds and thrust faults, of extensile structures including normal faults and dikes, and structures representing horizontal shear on nearly vertical surfaces—all together in a strike-slip regime—is involved in the concept of "wrench tectonics" (Anderson, 1942; Moody and Hill, 1956; Wilcox and others, 1973). The complexity and variety of these structures individually or in combination have three main aspects (Naylor and others, 1986): (1) the en echelon nature of

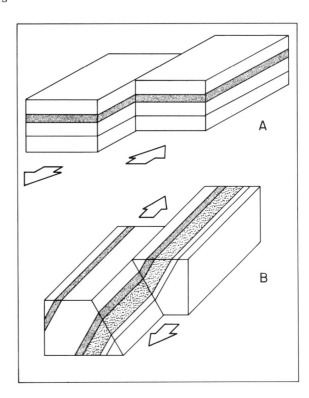

Figure 5. Diagrams illustrating the concept of trace slip. (A) Slip, which is parallel to the trace that the bed makes on the vertical fault surface, is not revealed in a vertical cross section perpendicular to the fault, (B), even if the beds and fault surface are not orthogonal to the ground surface or to each other.

faults and folds; (2) complications due to components of reverse or normal dip-slip on the basement fault; and (3) lateral offsets of basement-involved strike-slip faults which create local extensile or shortening structures. Two principal mechanisms explain the geometric and dynamic relations among these faults and associated structures: pure shear, sometimes called the Coulomb-Anderson model, and simple or "direct" shear (Fig. 6). Pure shear produces relatively short, typically conjugate sets of strike-slip faults which help to accommodate the brittle component of strain in tectonic regimes of crustal shortening, such as overthrust belts. Bulk pure shear is irrotational and has an orthorhombic symmetry. Simple shear has a monoclinic symmetry and rotational component of bulk strain and accounts for the kinematics of strike-slip faults at all dimensions (Tchalenko, 1970).

Pure Shear

This mechanism was originally proposed by Anderson (1905) to explain the orientations of faults relative to a triaxial stress field in a homogeneous medium. In the case of strike-slip faults, it predicts that a conjugate set of complementary sinistral and dextral strike-slip faults will form at an angle of ϕ and $-\phi$ about the shortening direction (Fig. 6a), where ϕ is the angle of internal friction. It predicts that extension fractures or normal faults will form perpendicular to elongation axis, and that folds and thrust faults will form perpendicular to the shortening axis. Notice in Figure 6 that the orientations of structures are shown relative to incremental strain axes rather than to stress axes. In most published diagrams, the principal stress axes are assumed to be parallel to the principal strain axes because of the assumed homogeneity of the medium and of the instantaneous strain, theoretical assumptions which are rarely achieved by heterogeneous rocks in prolonged natural deformation.[1]

The conjugate faults can accommodate irrotational bulk strain as long as they operate simultaneously; otherwise space problems ensue which can be solved only by rotation and alternating differential slip on each of the conjugate faults (Fig. 7). Strike-slip faults in domains of pure shear do not evince offsets measurable in hundreds of kilometers, because of room problems that result from convergence of large crustal masses (Fig. 8). Anderson (1941) himself recognized the space problem caused by the convergence of large crustal masses if the faults have a high junction angle and if they do not operate simultaneously (Fig. 8), but he either did not recognize or acknowledge the role of simple

---

[1]Readers may wonder about the lack of the term "compression" in this paper, as in the context of "compressional structures." Many geological writers use "compression" indiscriminately for both stress and strain. I follow the convention from rock mechanics that "tension" and "compression" are terms that should be used in the context of *stress*, whereas the corresponding *strain* terms are "extension," "elongation," or "lengthening" and "contraction" or "shortening" or even "constriction."

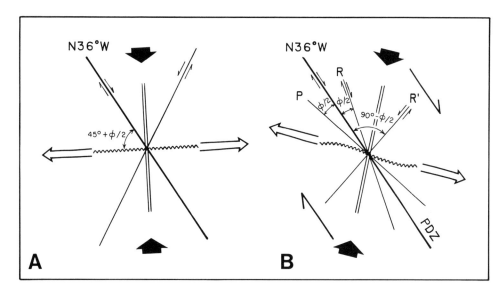

Figure 6. Plan view of geometric relations among structures according to two-dimensional, strike-slip, tectonic models for a vertical fault which strikes N36°W (adapted with modifications from Aydin and Page, 1984). (A) Coulomb-Anderson model of pure shear; (B) Riedel model of right simple shear. Double parallel line represents orientation of extension (T) fractures; wavy line represents orientation of fold axes. P = P fracture, R and R' are synthetic and antithetic shears, respectively; PDZ = principal displacement zone; ϕ = angle of internal friction. Short black arrows = shortening axis; open arrows = axis of lengthening.

shear in strike-slip faulting of the crust. In fact, in discussion of Kennedy's (1946) paper about strike slip on the Great Glen fault in Scotland, Anderson maintained that strike-slip faults occur in complementary X-shaped pairs, 50° apart, that shear on the Great Glen fault is left-lateral, and that equally great right-lateral faults should be found nearby. Complementary right-lateral faults having a magnitude of strike-slip equivalent to that of the Great Glen fault have not been found in Scotland, because the Great Glen fault is a product of simple shear and not pure shear as Anderson assumed.

The requirements of the pure shear mechanism have been missed by many geologists who have misapplied it to field areas of strike-slip faulting, yielding simplistic misinterpretations of the kinematics and dynamics of strike slip. In southern California, for example, some writers assumed that the Garlock and Big Pine faults were the left-lateral counterparts to the right-lateral San Andreas fault in a pure shear system (Fig. 9), even though the apical angle between them across the presumed direction of shortening is about 120° rather than the standard 60° (Hill and Dibblee, 1953). The interpretation was based on the X-shaped pattern and the senses of slip on the three faults. It is an incorrect interpretation, however, when the amount and timing of the fault displacement are considered. The San Andreas fault came into existence about 24 Ma, accumulating 330 km of dextral slip (Crowell, 1979), whereas movement on the Garlock fault commenced 10 m.y. ago and totals about 60 km of left slip (Davis and Burchfiel, 1973; Burbank and Whistler, 1987; Loomis and Burbank, 1988). The Big Pine fault had an Oligocene episode of displacement, mainly dip-slip, and a post–late Miocene episode of left slip (Crowell, 1962).

The notions of McKinistry (1953) and Moody and Hill (1956), which simplistically explained strike-slip faulting on all scales in terms of the Coulomb-Anderson mechanism, did much to confuse understanding of strike-slip tectonics for several decades in my opinion, and judging from critical discussions of their papers (Prucha, 1964; Maxwell and Wise, 1958; Laubscher, 1958). The Coulomb-Anderson mechanism is an easy concept to grasp, and many writers did so, misapplying the concept to domains of simple shear, but finding that all of the structures were not explained, wondering why, and struggling with *ad hoc* arguments to explain the lack of the conjugate set of equal magnitude that the Andersonian theory promises (Rod, 1958; Allen and others, 1960). They did this in spite of the fact that a considerable volume of literature was available, especially from the 1920s, about simple shear experiments and their relation to observed structures in the field (Mead, 1920; Fath, 1920; Hubbert, 1928; Brown, 1928; Cloos, 1928; Riedel, 1929).

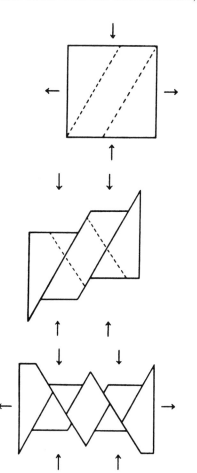

Figure 7. Bulk pure shear strain by alternating slip on conjugate shears with rigid body rotation (from Anderson, 1905).

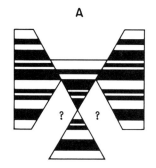
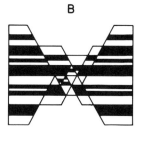

Figure 8. Space problems with conjugate slip in bulk pure shear (adapted with modifications from Ramsay, 1979). (A) Synchronous faulting; (B) alternating fault activity.

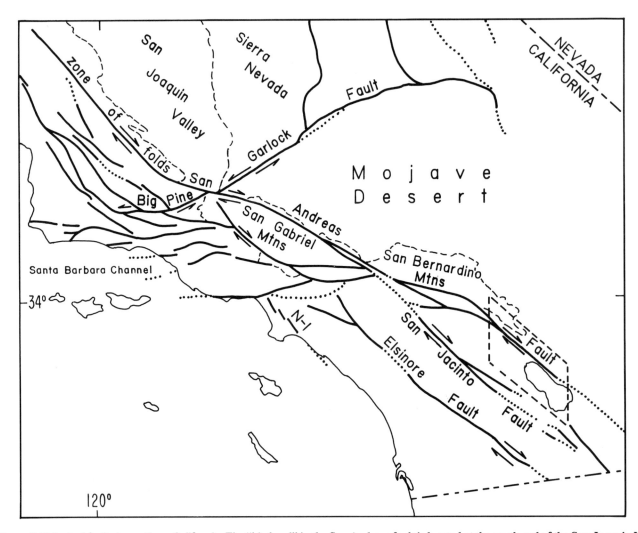

Figure 9. Principal faults in southern California. The "big bend" in the San Andreas fault is located at the south end of the San Joaquin Valley and at the northwest end of the San Gabriel Mountains where the Big Pine and Garlock faults intersect the fault. The dashed parallelogram indicates the area shown in the map in Figure 23.

Domains of regional pure shear with conjugate strike-slip faults having many kilometers of displacement are real, however, and are well documented. Typically these domains are in fold-thrust belts where conjugate strike-slip faults transect the fold trends. A good example is in the Wopmay orogen of northern Canada where the Asiak foreland thrust-fold belt (Hoffman and St-Onge, 1981; Tirrul, 1982, 1984) is offset by a conjugate set of east-northeast-striking, right-slip faults which are typically longer than 50 km, and by west-northwest-striking, left-slip faults (Figs. 10 and 11). The faults are developed at all scales and are in mutually exclusive domains for a given scale (Tirrul, 1984). As much as 15 km of displacement has occurred across some of the faults. Geometric arguments and palinspastic map reconstructions indicate that both fault sets formed initially at 25°–30° to the east-west shortening direction and rotated thereafter about a vertical axis away from it (Tirrul, 1984), as Freund (1970a) postulated for conjugate faults in the Sistan District of Iran (see also Tirrul and others, 1983). The bulk strain approaches the magnitude of the regional pure shear: east-west shortening is up to 25% with north-south extension (Tirrul, 1984). Other notable regional domains of strike slip caused by bulk pure shear in areas of crustal convergence have been identified in the Apennines of central Italy (Lavecchia and Pialli, 1980, 1981), the Carpathian Pannonian basin in southeastern Europe (Royden and others, 1982), the Makran of southwest Pakistan (Platt and others, 1988), the southern Chilean Andes (Katz, 1962), and in the accretionary prism of the Aleutian trench (Lewis and others, 1988).

Simple Shear

The major strike-slip faults of the world are in domains of simple shear which may be thousands of kilometers long and tens of kilometers wide, and they have displacements measured in hundreds of kilometers. Within the domain of simple shear, the most recently active strand may be a zone of active faulting that is only a few meters wide.

Simple shear has a monoclinic symmetry of strain because it is rotational, and a greater variety of structures forms in simple shear than in pure shear (Fig. 6b). The structures typically form en echelon arrangements in relatively narrow zones (Fig. 12). Five sets of fractures form in simple shear in model experiments, in experimental deformation of homogeneous rocks under confining pressure, and in alluvium deformed by surface rupturing during earthquakes (Fig. 6b): (1) Riedel (R) shears (Tchalenko, 1970) or "synthetic" (Cloos, 1928) or "pinnate" (Ma and Deng, 1965) strike-slip faults; (2) conjugate Riedel (R′) shears (Tchalenko, 1970) or "antithetic" (Cloos, 1928) strike-slip faults; (3) secondary synthetic strike-slip faults at an angle of $-\phi/2$ to the direction of applied shear (P shears of Skempton, 1966; Tchalenko, 1970;

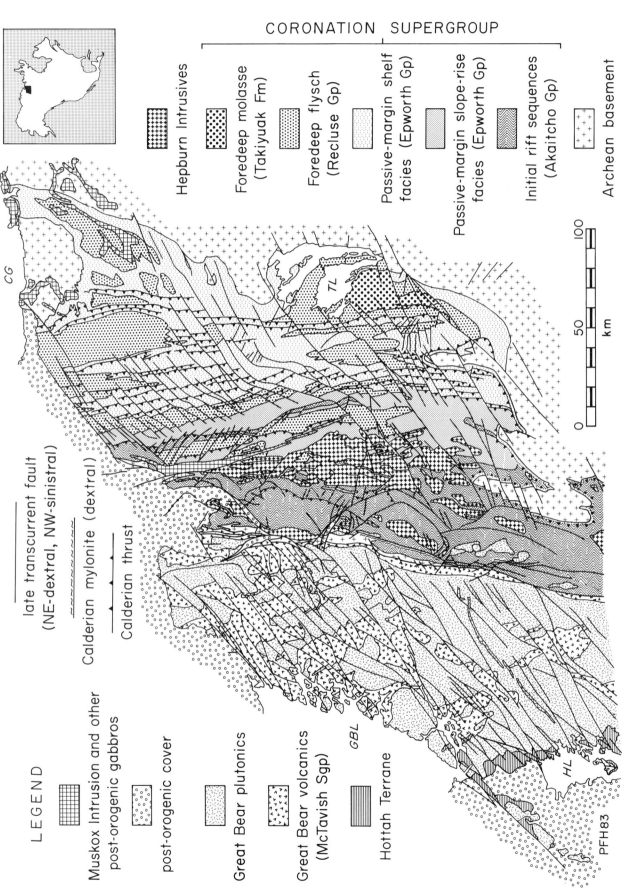

Figure 10. Simplified geologic map of the northern two-thirds of the Wopmay orogen, showing regional setting of conjugate strike-slip faults which transect the Asiak fold-thrust belt (from Hoffman and others, 1984). Northeast-striking faults are right-slip faults; northwest-striking faults are left-slip faults. Inset shows location of map. Prominent water bodies are: CG, Coronation Gulf; GBL, Great Bear Lake; TL, Takijug Lake; HL, Hottah Lake.

Figure 11. Distribution of strike-slip faults in the northern Wopmay orogen. As indicated in the ellipsoidal inset, all northeast-striking faults are dextral, northwest faults are sinistral, and east-west faults are normal. WFZ = Wopmay fault zone. Dark arrows indicate shortening axis; open arrows indicate the extension axis (P. F. Hoffman, M. R. St-Onge, and R. Tirrul, Geological Survey of Canada, unpub.).

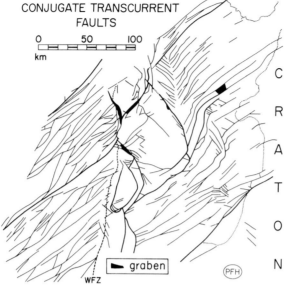

Tchalenko and Ambraseys, 1970); (4) extension fractures (T fractures of Tchalenko and Ambraseys, 1970) or normal faults which develop at about 45° to the principal displacement zone; and (5) faults parallel to the principal displacement zone (Y shears of Morgenstern and Tchalenko, 1967).

The laboratory studies (Cloos, 1928; Riedel, 1929; Tchalenko, 1979; Wilcox and others, 1973) simulated a rigid basement by two stiff boards overlain by a cake of unbroken clay, analogous to a cover of sedimentary rocks. When the boards slipped parallel to one another at depth, the first-formed structures in the overlying clay were en echelon R shears whose overstepping sense is directly related to the sense of slip of the underlying boards: that is, left-stepping in right simple shear, and right-stepping in left simple shear, although reversals locally occurred. The width of the zone of fracturing in plan view is a function of the thickness of the clay cake. These experiments have been repeated numerous times by academic and industry scientists, with results that vary according to the type of material being sheared (Skempton, 1966; Morgenstern and Tchalenko, 1967; Emmons, 1969; Hoeppener and others, 1969; Lowell, 1972; Courtillot and others, 1974; Mandl and others, 1977; Groshong and Rodgers, 1978; Graham, 1978; Bartlett and others, 1981; Gamond, 1983; Deng and Zhang, 1984; Macdonald and others, 1986; Hempton and Neher, 1986; Naylor and others, 1986). The same arrangement of structures also forms in alluvium deformed by surface rupturing during earthquakes (Gianella and Callaghan, 1934; Florensov and Solenenko, 1963; Brown and others, 1967; Clark, 1968; Tchalenko and Ambraseys, 1970; Philip and Megard, 1977; Terres and Sylvester, 1981; Sharp and others, 1972; Deng and Zhang, 1984).

The sense of strike slip along the R, P, and Y shears is the same as that of the basement fault, whereas that of the R' shear is opposite. All of the faults, except the thrust faults, are nearly vertical when they form. The R and R' shears make angles of $\phi/2$ and $90°-\phi/2$, respectively, with the principal displacement zone (Fig. 6b), where ϕ is the angle of internal friction. This means that the R shears strike from 15° to 20° to the principal displacement zone; and the R' shears, from 60° to 75° (Tchalenko and Ambraseys, 1970, p. 56). In sand model experiments, the actual angle depends on the thickness of "overburden" above the "basement" fault, being at a low angle when the "overburden" is thin, and being greater than 15° when the "overburden" is thick (Naylor and others, 1986). The extension fractures bisect the angle between the R and R' shears and are oriented parallel to the incremental axis of shortening and at an angle of 45° to the direction of applied shear. R' shears rarely develop in nature (Keller and others, 1982) except where there is a substantial overlap between adjacent R shears (Tchalenko and Ambraseys, 1970; Naylor and others, 1986).

Soon after the initial formation of the R shears in the model studies, the incremental strain field is locally modified in the deformation zone, giving rise to the development of P shears and short-lived splay faults (Naylor and others, 1986). Splay faults form at the tips of the R shears and curve toward parallelism with the extension fractures so that an R shear will be a strike-slip fault in the central part of the deformation zone, but it will be a normal fault with little or no strike slip at its extremities. The P shears form as a consequence of the reduction of shearing resistance along the R shears, so that all

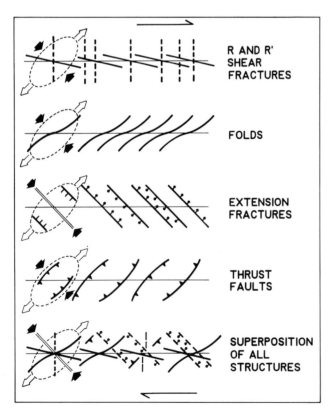

Figure 12. Orientation of folds and faults in bulk right simple shear.

of the basement displacement cannot be taken up on the discontinuous en echelon R shears in the overburden (Tchalenko, 1970). Thus, typically between two overlapping R shears, the shortening axis will be reoriented toward the R shear, producing new faults in this local strain field that strike at an angle of $-\phi/2$ to the principal displacement zone (Naylor and others, 1986).

The most advanced stage of the deformation yields a principal displacement zone of braided or anastomosing, vertical fractures wherein the main surficial shear strain is centered on an irregular, through-going fracture whose path may be composed variously of the R and P fractures (Naylor and others, 1986). Tchalenko (1970) also showed that the assemblage of fractures formed in simple shear is geometrically similar through the microscopic to the macroscopic scales.

Folds and thrust faults form initially perpendicular to the axis of shortening (Fig. 12), and thus, also at an angle of 45° to the principal displacement zone. If deformation continues, then the fold axes will rotate according to the amount of shearing: as much as 19° for a shear strain of unity (Ramsay, 1967, p. 88).

The theory of simple shear in the strict sense (Ramsay, 1967, 1980) is probably not applicable to any large part of the Earth's crust with any mathematical rigor because of uncertainties in mechanical behavior and strain profile and because of the heterogeneous nature of rocks (Aydin and Page, 1984). "Simple shear," however, is a general approximation of the theoretical concept when applied to strike-slip faulting as was done by Tchalenko (1970) and by Wilcox and others (1973). Typically it is applied to any zone between strike-slip faults where crustal strain is the direct result of pervasive horizontal shear in a consistent sense and direction (Aydin and Page, 1984). Even though the magnitude and rate of shear strain may vary greatly from place to place within the zone, the entire domain of strike slip is made up of many subzones, each undergoing simple shear at a particular rate (Aydin and Page, 1984).

Only a few good geological field examples (Keller and others, 1982; Erdlac and Anderson, 1982) show the idealized among R and R' fractures depicted in Figure 6b, because natural structures develop sequentially rather than nearly instantaneously as they do in laboratory models and earthquakes, because rocks are heterogeneous, and because early-formed structures may be internally rotated with protracted shear strain. Thus, many of the early-formed fractures may be cut off and will be inactive during subsequent shear on the principal displacement zone. The resultant plethora of R, R', P, T, and Y fractures in nature is perplexing to unravel sequentially, especially in strike-slip fault zones that have had a long history of movement. The style of natural structures is also affected by convergent or divergent strike slip, as is discussed more fully in a subsequent section. The end result may be that some of the folds and faults will form or be rotated into orientations parallel to the principal displacement zone. The resultant faults will look like dip-slip growth faults, and the folds will have all of the characteristics of drape folds when viewed in two-dimensional seismic or structure sections (Harding and others, 1985).

In some cases, the strike-slip at depth is sufficiently small or deep that a through-going fracture fails to develop at the surface (Naylor and others, 1986). Instead, only a long, narrow zone of en echelon normal faults or R shears forms (Erdlac and Anderson, 1982), perhaps associated with en echelon folds as is illustrated, for example, in the Columbus basin of offshore Trinidad (Leonard, 1983). En echelon arrays composed only of extension fractures, gash fractures, or normal faults are common above a buried strike-slip fault. Thus, most of the individual faults of the Lake Basin and Nye-Bowler fault zones in Montana are normal faults that strike ~45° to the trend of the strike-slip fault zone (Wilson, 1936; Alpha and Fanshawe, 1954; Smith, 1965). R shears and even extension fractures may have significant strike slip on them, as does the Newport-Inglewood zone in southern California (Barrows, 1974; Yeats, 1973; Harding, 1973), where as much as 800 m of right slip has occurred across the fault zone, truncating early-formed en echelon folds into half-anticlines and domes. The simple shear mechanism for the Newport-Inglewood zone was modeled and clearly understood by Ferguson and Willis (1924) and later by Wilcox and others (1973).

R fractures have a helicoidal shape in three dimensions (Fig. 13), which is a consequence of three factors (Naylor and others, 1986): (1) the en echelon nature of the shears at the surface; (2) their concave-upward geometry when formed in shear without components of convergence or divergence; and (3) the need to join a single basement fault at depth. In cross section, such a fault looks like an upthrust at depth and like a landslide near the surface: Crowell

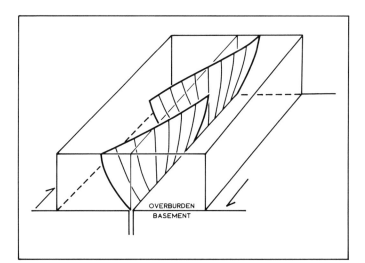

Figure 13. Helicoidal form of individual Riedel shears in right simple shear, reconstructed from horizontal serial sections in sandbox model experiments (redrawn from Naylor and others, 1986). Reproduced with permission of *Journal of Structural Geology*.

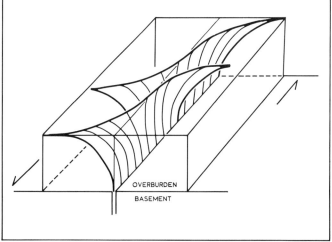

Figure 14. Helicoidal form of axial surfaces of two en echelon folds in left simple shear. Inspired from analogous diagram for R fractures (Naylor and others, 1986) and from descriptions of natural folds (Sylvester and Smith, 1976; Gamond and Odonne, 1983).

(1982a, p. 40) informally refers to this fault form as a "slust" or a "thride." The imprecise and non-euphonic terminology represents the difficulty of characterizing and describing with a single word or phrase an oblique-slip fault, the surface of which may be vertical at depth and convex-upward but which flattens to horizontal at the surface (Allen, 1965; Lowell, 1972; Wilcox and others, 1973; Sylvester and Smith, 1976) or which is nearly vertical at the surface and concave-upward but flattens at depth into the main strike-slip fault (Harding, 1983a; Harding and others, 1985; Naylor and others, 1986).

EN ECHELON FOLDS

Folds associated with strike-slip faults are typically arranged in en echelon pattern oblique to the principal direction of shear, and they have received much attention from the petroleum industry, because they are attractive prospective traps for hydrocarbons (Harding, 1974; Dibblee, 1977b; Harding and Lowell, 1979; Harding and Tuminas, 1988). Typically, en echelon folds are distributed in a relatively narrow and persistent zone above or adjacent to a master strike-slip fault. They may form in a broad zone between two major strike-slip faults as they do in the East Bay Hills in the San Francisco Bay region of California (Aydin and Page, 1984). The presence of en echelon folds or faults parallel to a zone of deformation, however, is not restricted to strike-slip faulting (Sherill, 1929; Campbell, 1958; Christie-Blick and Biddle, 1985). En echelon folds may reflect the complex influence of basement structures at depth, or they may represent the superposition of differently oriented folds in time and space (Harding, 1988).

Ideally, the crestal traces of en echelon folds should make an angle of 45° in plan view to the shear direction (Fig. 6b), representing the shortening component of the bulk strain, but trends of real folds are gently twisted and vary from 10° to 35° to the strike of the fault zone (Harding and Lowell, 1979). In three dimensions (Fig. 14), the axial surfaces of echelon folds in a sequence of strata overlying a rigid basement are nearly vertical and parallel to the fault at basement level, but higher in the overburden, they flatten upward and twist away from the strike of the fault (Gamond and Odonne, 1984; Koral, 1983), and they plunge away from the principal displacement zone (Harding and Tuminas, 1988). The axial surface of an en echelon fold has a helical geometry, similar to that of an R shear (Naylor and others, 1986), except that an R shear steepens upward, whereas a fold axial surface flattens upward. This implies that the observed angular relation between en echelon fold axes and principal displacement zone in plan view may depend locally on the depth of erosion, as well as on the amount of internal rotation within the shear zone.

En echelon folds have been popularly called "drag folds" (Moody, 1973) where they curve into parallelism with a strike-slip fault, but the term and notion are misnomers in the context of strike-slip faulting, because the folds are born in an en echelon orientation as many model studies show (Pavoni, 1961b; Wilcox and others, 1973; Dubey, 1980). Another reason why natural folds do not always fit the optimal orientation predicted by heterogeneous simple shear, therefore, is that they may be rotated or internally sheared by piecemeal slip on R fractures and smeared, thereby, into a "dragged" appearance.

En echelon folds are useful structural indicators, because they tell three things about the associated fault and its related structures: (1) that a strike-slip fault is probably nearby laterally or at depth; (2) the direction of slip on that fault by the overstepping direction of the folds; and (3) the expected orientations of related faults. These concepts are discussed in the next two paragraphs.

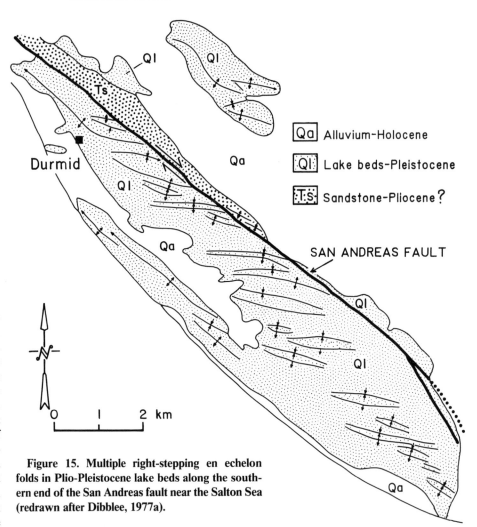

Figure 15. Multiple right-stepping en echelon folds in Plio-Pleistocene lake beds along the southern end of the San Andreas fault near the Salton Sea (redrawn after Dibblee, 1977a).

Clay-model studies of en echelon folds imply that the folds form symmetrically above the principal zone of displacement (Wilcox and others, 1973; Odonne and Vialon, 1983). At the south end of the San Andreas fault (Fig. 15), however, en echelon folds are rare on the northeast side of the fault, relative to their abundance on the southwest side (Babcock, 1974; Dibblee, 1977a). Their apparent lack on the northeast side is partly due to the fact that they are covered by younger deposits; but it is just as reasonable to expect that folds which formed on the northeast side were displaced out of the picture by strike slip; or that the rocks on the northeast side were incapable of folding as readily as the thin-bedded, gypsiferous, lacustrine strata on the southwest side. Rather than project the most likely locus of the strike-slip fault symmetrically beneath the field of folds, as one would tend to do in analogy with clay-model experiments, it is better to infer only that a strike-slip fault is nearby or at depth.

The direction of the horizontal movement on the strike-slip fault is revealed by the stepping direction of the folds (Fig. 16): Right-stepping folds form in right slip; left-stepping folds form in left slip. Then from the geometry of strain in simple shear, one can deduce the expected directions of associated Riedel shears, normal faults, and thrust faults (Fig. 6b).

In divergent strike slip, the crestal traces of folds are typically parallel to the principal displacement zone and have a parallel or relay pattern (Burkart and others, 1987) resembling those having formed in pure shear. The folds have a cross-sectional geometry and evolution similar to those of drape, or forced folds that form above normal fault blocks complete with anticlines or monoclinal knees next to, and parallel with, the relatively higher side of the principal displacement zone; they also have synclines or monoclinal "ankle flexures" adjacent to, and parallel with, the edge of the apparently downdropped fault block (Harding, 1974; Harding, 1983a; Harding and others, 1985). The faults beneath the folds may have the geometry of growth faults in cross section, and may well have played a "growth fault" role during syntectonic sedimentation (Sylvester and Smith, 1976). In convergent strike slip, en echelon folds may have any or all of the profile geometries found in other convergent tectonic styles, even thrust-associated types.

The factors which control the style and development of natural en echelon folds in simple shear have not been clearly defined. Initially they form perpendicular to the shortening axis within a shear couple in clay-model studies (Fig. 6b), but that relation may be an oversimplification, because their formation in model studies requires a ductile material or interlayered member, such as a thin sheet of tin foil, rubber, or plastic, which will deform continuously rather than by shearing (Mead, 1920; Wilcox and others, 1973). Thus, in order to form a neat arrangement of en echelon folds in clay, Wilcox and others (1973) placed a thin sheet of plastic film beneath the surface in some of their clay models. They found that the fold spacing, orientation, size, and rate of growth depended on the cohesion of the clay, the strain rate, and the degree of strike-slip convergence of the boards beneath the clay cake. Therefore, in convergent strike slip, a component of shortening is imposed above the zone of strike-slip deformation, and the folds form readily and in a distinct, consistent en echelon arrangement (Wilcox and others, 1973; Babcock, 1974).

Harding and Lowell (1979) depicted the maturation of a strike-slip fault zone and its associated structures from the evolutionary sequence of structures observed in clay-model studies (Wilcox and others, 1973). Early in the deformation history, an en echelon array of simple folds is formed (Fig. 17). With more defor-

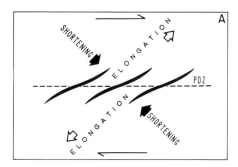

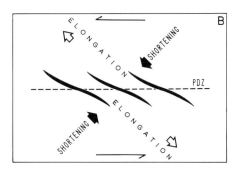

Figure 16. Geometry of en echelon folds in right simple shear (A) and left simple shear (B). PDZ = principal displacement zone. Bulk strain axes are labeled.

mation, R shears break the surface and deform the folds into domes, half anticlines, and synclines. Then the deformation zone broadens, and parts of the early-formed structures may be deeply eroded or faulted out of the picture, leaving vestiges of steep, secondary faults or parts of folds now cut complexly by thrust and normal faults and by R shears. Folding extends progressively farther from the principal displacement zone with increasing displacement over time. The largest amplitude folds are at depth near the fault, and the most recent folds are farthest from the fault at the margins of the deformation zone (Harding, 1976). Some tectonicists object that such a narrow zone of actual faulting can produce a zone of deformation as wide as 30 km. The main zone of modern strike slip is narrow indeed, but the zone of pervasive simple shear is wide and heterogeneous over geological time; at the latitude of central California, it may be from 500 km (Hamilton, 1961; Minster and Jordan, 1987; Ward, 1988) to 1,000 km wide (Atwater, 1970).

The simple shear mechanism and scale of the process involved in the formation of the remarkable series of en echelon folds associated with the San Andreas fault in the San Joaquin Valley was appreciated and understood clearly by Lawson (1921, p. 580): "With regard to the force which is responsible for the lateral movement on the these faults, all the evidence tends to show that there is a northward creep of the mass beneath the ocean with respect to the mainland mass or shield of the Sierras [sic], and that the region of the coast ranges represents the shear zone beneath these two great masses. The mass beneath the ocean, moving northward, presses against the northwesterly trending coast line and slides northwest along it, exerting, of course, great northeasterly pressure against it which is responsible for the parallel northwesterly trending folds in the more sharply folded portions of the coast ranges."

A strong case has been made by Harding (1976) in support of Lawson's (1921) and Hamilton's (1961) suggestion that the wide zone of contemporaneous faults and en echelon folds in San Joaquin Valley of southern California (Fig. 9) was produced by large-scale, convergent strike slip along the San Andreas fault rather than by contraction at a high angle, as was often assumed. The stratigraphy recorded by Harding in the folds, which are some of the most prolific producers of petroleum in North America, shows that they originated in mid-Miocene time when the San Andreas fault became active. Their growth increased and new folds developed during the Pliocene and Pleistocene epochs when the rate of movement increased on the San Andreas fault (Harding, 1976). If the Pliocene and early Pleistocene strata are palinspastically "unfolded," then the number of pre-Pliocene anticlines and synclines that remain in the underlying strata are minimal except on those structures that are closest to the fault. If the exercise is repeated for Miocene formations, then only rarely can a coincident pre-Miocene fold be shown to have existed at the same site, although underlying Paleogene and Cretaceous rocks are generally discordant with respect to the Miocene strata. In addition, the older, more tightly folded cores of anticlines nearest the San Andreas fault are commonly disrupted by young, southwest-dipping thrust faults or northeast-dipping reverse faults (Dibblee, 1973; Fuller and Real, 1983; Stein, 1983, 1984; Stein and King, 1984), faults which Harding (1976) regards as late-stage reactions to prolonged deformation.

An alternate hypothesis for the formation of the en echelon folds outside the narrow zone of strike slip in the San Joaquin Valley maintains that present-day shortening deformation is not controlled by distributed shear associated with drag on a high-friction San Andreas fault. Instead, based on interpretation of structural styles (Namson and Davis, 1988) together with borehole elongations and well breakouts, the present maximum stress is inferred to be oriented nearly perpendicular to the fault and therefore, "transpressive tectonics in central California can be better described as decoupled transcurrent and compressive deformation, operating simultaneously and largely independently" (Mount and Suppe, 1987, p. 1146). Separate and largely in-

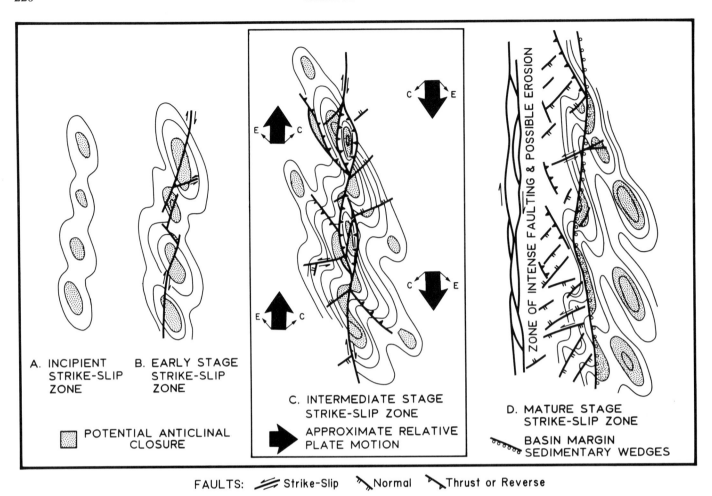

Figure 17. Schematic diagrams of structural assemblage associated with major strike-slip fault and their evolutionary history (redrawn from Harding and Lowell, 1979). Arrows C and E represent inferred principal directions of contraction and extension, respectively, that arise in the right simple shear couple represented by the heavy black arrows. Reproduced with permission of American Association of Petroleum Geologists.

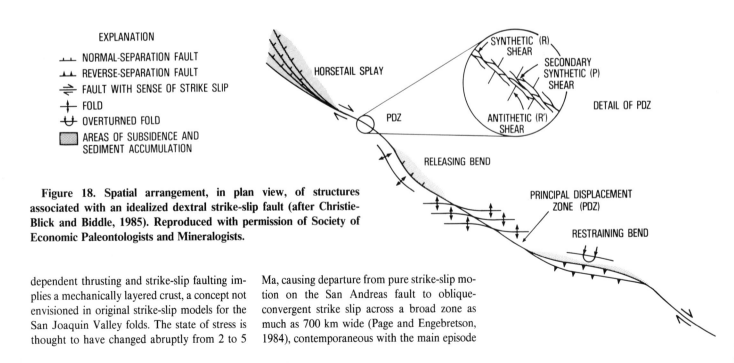

Figure 18. Spatial arrangement, in plan view, of structures associated with an idealized dextral strike-slip fault (after Christie-Blick and Biddle, 1985). Reproduced with permission of Society of Economic Paleontologists and Mineralogists.

dependent thrusting and strike-slip faulting implies a mechanically layered crust, a concept not envisioned in original strike-slip models for the San Joaquin Valley folds. The state of stress is thought to have changed abruptly from 2 to 5 Ma, causing departure from pure strike-slip motion on the San Andreas fault to oblique-convergent strike slip across a broad zone as much as 700 km wide (Page and Engebretson, 1984), contemporaneous with the main episode

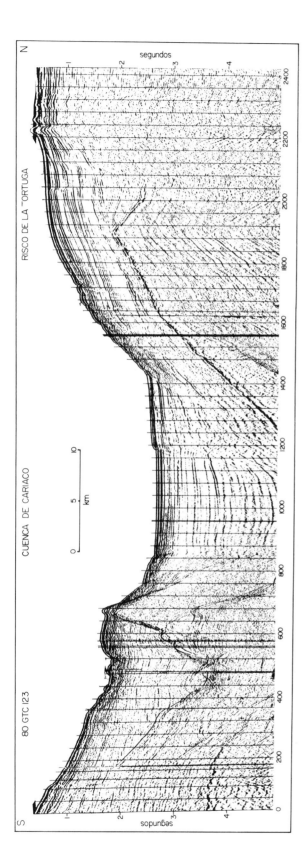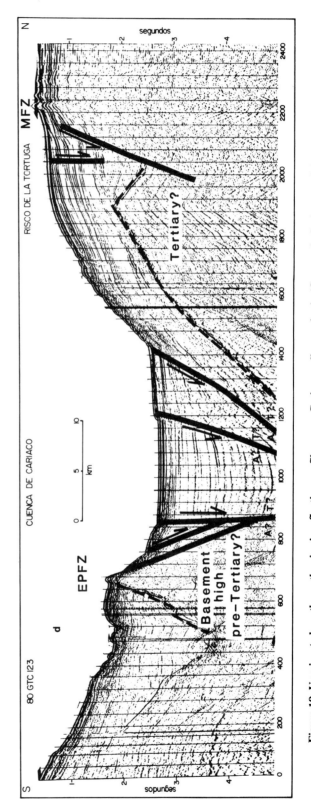

Figure 19. Unmigrated south-north seismic reflection profile across Caricao pull-apart basin (Cuenca de Cariaco) and the Tortuga Rise (Risco de la Tortuga), northern Venezuela (Schubert, 1986b). Vertical exaggeration is approximately 10:1. (A) Uninterpreted; (B) interpreted by C. Schubert, 1988. EPFZ = El Pilar fault zone; MFZ = Morón fault zone; segundos = seconds; A = movement away from observer; T = movement toward observer.

of crustal shortening and folding which created the California Coast Ranges. The young, contractile faults in the cores of the en echelon anticlines are therefore believed to accommodate recent interplate crustal shortening of about 5 mm/yr across the California Coast Ranges (Minster and Jordan, 1984), whereas the San Andreas fault takes up the much greater transform displacement (~35 mm/yr).

Excellent descriptions are in hand to explain the existence, timing, and developmental sequence of en echelon folds, but their mode of coupling to strike-slip faults is not really understood, nor is the influence of mechanical layering in their development, and it is not known why folds rather than fractures should form at one place or another. Natural en echelon folds lack interbedded sheets of tin foil, rubber, or plastic needed for their formation in experimental models, but the vast literature on the genesis of buckle and bending folds, wherein such factors as bedding thickness, strain rate, and bedding-plane slip are of fundamental importance, will certainly have relevance to the problem of generating en echelon folds in strike-slip fault zones.

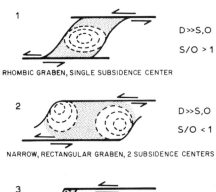

Figure 20. Geometry and locations of basins in overstepping domains of parallel faults (with modifications after Deng and others, 1986). D = depth to basement; S = spacing between parallel or en echelon fault strands; O = overstep. Reproduced with permission of Birkhäuser Verlag AG, Basel, Switzerland.

CONVERGENT AND DIVERGENT STRIKE-SLIP FAULT TECTONICS

The geometry and style of structures associated with strike-slip faulting depend greatly on several factors at different times and different places along and within a strike-slip fault zone, including the nature of the rocks being deformed, the configuration of pre-existing structures, the amount of horizontal slip, the contribution of the vertical component of slip, and the strain rate, which is a major consideration for clay models wherein the strain rate is rapid.

The most important factor governing uplift or subsidence along a strike-slip fault is the bending geometry of the fault surface relative to its slip vector (Fig. 18), because that determines whether local convergence or divergence will occur (Fairbanks, 1907, p. 322; Pakiser, 1960; Clayton, 1966; Crowell, 1974b). Where strike-slip movement is inhibited by restraining bends (Crowell, 1974b), convergent strike slip or transpression (Harland, 1971) occurs associated with crowding, crustal shortening, and uplift. Releasing bends (Crowell, 1974b) provide for transtension (Harland, 1971) or divergent strike slip accompanied by stretching, crustal extension, subsidence, and formation of pull-apart basins.[2]

An appreciation of crustal mobility and flexibility, both vertical and horizontal, is of particular importance for arriving at tectonic understanding of deformation in strike-slip fault zones (Crowell and Sylvester, 1979). Even while crustal blocks move laterally over time, they may alternately rise and sink (Kingma, 1958; Clayton, 1966). In experimental studies, elongate blocks in the zone of strike-slip faulting are bounded by various arrangements of R and P fractures (Woodcock and Fisher, 1986; Bartlett and others, 1981; Tchalenko, 1970; Wilcox and others, 1973; Naylor and others, 1986). Similarly, detailed mapping of strike-slip fault zones (Grose, 1959; Bridwell, 1975; Moore, 1979) shows that strike-slip faults, which are typically belts of braided, nearly vertical shears, bound elongate blocks which are squeezed upward along those shears to make high-standing source areas for sediments (Crowell, 1974b). Where strike-slip movement is divergent, crustal blocks may sag, subside, or tilt between or adjacent to

---

[2]This kind of basin has also been called a tectonic depression (Clayton, 1966), strike-slip graben (Belt, 1968), rhomb graben (Freund, 1971), negative flower structure (Harding, 1983a), tulip structure (Naylor and others, 1986), or rhombochasm (Carey, 1958) if it has a volcanic floor like the Cayman trough (Rosenkrantz and others, 1988) and the Black and southern Caspian Seas (Apol'skiy, 1974).

bounding faults, making local sites for deposition of sediments whose stratigraphic characteristics reveal much about the related tectonic activity (Ballance and Reading, 1980). Crowell and Sylvester (1979) drew an analogy between this structural process and a porpoise swimming parallel to the fault strike, alternately arching above the sea surface and diving below it.

Bailey Willis (1938b, p. 664) was well aware of the uplift and subsidence of elongate ridges of rocks, bounded by steep, inward-dipping minor faults in strike-slip fault zones: "Furthermore, it is a common mechanical result of continued pressure in a zone traversed by vertical shears that the crushed rock in the zone is both squeezed up and down, with the result that the stresses are carried back into the adjacent masses. Thrust faults are thus developed in the latter, and they curve upward in the direction of least resistance. They thus become up-curving thrusts or ramps," and (Willis, 1938a, p. 795) "blocks within such an area (fault zone) are completely isolated by minor faults. They are pushed about, mayhap up or down, mayhap over or under, perhaps lengthwise along strike." Willis termed this process "wedge-block faulting," and he regarded the uplifts as wedges both in horizontal and vertical views similar to the notion that Wallace (1949) had for the origin of "center-trough ridges" or pressure ridges, as they are now called.

Basins Related to Strike-Slip Faults

Form and Shape. Becker (1934) and Lotze (1936) were among the first to realize that strike-slip faulting may generate large and complex basins (Şengör and others, 1985). Between curved or releasing overstepped fault segments, depressions develop as sharp, rhomb-shaped basins (Crowell, 1974b; Garfunkel, 1981) or as lazy S- or Z-shaped basins (Schubert, 1980; Mann and others, 1983) due to local crustal extension. The basins range in size from small sag ponds along a strike-slip fault to rhomb-shaped basins up to 500 km long and 100 km wide, such as the Cayman trench located between the releasing overstepped Oriente and Swan Islands transform faults along the north edge of the Caribbean plate (Mann and others, 1983; Rosenkrantz and others, 1988).

Considerable literature concerns the geometry of basins related to strike slip (Dibblee, 1977a; Aydin and Nur, 1982; Mann and others, 1983; Harding and others, 1985), their three-dimensional structure (Ben-Avraham and others, 1979; Ben-Avraham, 1983; Howell and others, 1980; Ginzburg and Kashai, 1981; Fuis and others, 1982; Bally, 1983; Schubert, 1982b), sedimentation (Crowell, 1974a; Ballance and

Reading, 1980; Hempton, 1983; Hempton and Dunne, 1984), origin (Crowell, 1974b, 1981, 1987; Christie-Blick and Biddle, 1985), evolution (Bahat, 1983; Şengör and others, 1985), tectonic setting (Crowell, 1974b, 1981), and thermal history (Royden, 1985; Karner and Dewey, 1986). The close attention focused on strike-slip basins reveals that they are much more varied and complex than originally envisioned in the early, frequently cited paper about the Death Valley "pull-apart" by Burchfiel and Stewart (1966).

Crowell (1974a) depicted pull-apart basins as deep, rhomb-shaped depressions bounded on their sides by two, subparallel, overlapping strike-slip faults, and at their ends by perpendicular or diagonal dip-slip faults, termed "transfer faults," which link the ends of the strike-slip faults. Early writers inferred that the bounding strike-slip faults merge at depth to a single master fault (Kingma, 1958; Clayton, 1966; Sharp and Clark, 1972), and whereas some do merge at depth (D'Onfro and Glagola, 1983), others just as clearly must be vertical at depth and bound a down-dropped block in between; for example, the Dead Sea basin (Manspeizer, 1985; Eyal and others, 1986) and the Caricao basin of Venezuela (Fig. 19).

Typical pull-apart basins have an aspect ratio of 3:1 in plan view (Aydin and Nur, 1982), although that value may vary widely, depending on whether the structural, physiographic, or active dimensions of the basin are measured. Crowell (1974b) implied, and Quennell (1959) and Freund and Garfunkel (1976) suggested, that the dip-slip faults at each end of a rhombic pull-apart were a single fault prior to extension between the parallel segments of the main strike-slip fault. In such a model, the length of the graben would therefore reflect the amount of horizontal displacement (Eyal and others, 1986), and the graben would be characterized by great depth relative to its areal dimensions. This may be true for some small or young basins such as those between closely spaced strike-slip faults; for example, Mesquite basin in Imperial Valley, southern California, is only 5 km wide and 5 km long, but it is filled by at least 5 km of sedimentary rocks (Fuis and Kohler, 1984). Other pull-apart basins are much more complex than the simple pull-apart concept implies; a range of possibilities is depicted by Harding and others (1985).

High-resolution seismic data from the Gulf of Elat illustrate some of the main structural features of pull-apart basins (Ben-Avraham and others, 1979; Ben-Avraham, 1985). There major rift faults having a component of left slip are arranged en echelon, and between them are three sharp, north-trending, pull-apart basins, separated by low sills, wherein the structural relief may exceed 5 km. The sedimentary fill, composed of turbidites and pelagic sediments, is more than 7 km in the deepest of the three basins. The basin floors tilt eastward in the form of a half-graben so that the stratal thicknesses are asymmetric, being thickest on the east sides of the basins. Locally the basins are evidently being shortened in an east-west direction, because the strata are arched upward in a series of large anticlines; however, the seismic lines were not spaced sufficiently close to determine the continuity of the folds from one profile to another. Ben Avraham and his colleagues concluded that the folds trend north-south, parallel to the main basin-bounding faults, but it is also possible that they have an en echelon arrangement, related to syntectonic strike slip associated with the opening of the basins.

The structural geometry of the sedimentary cover in the overstep between strike-slip fault segments depends on the length of overlap, the width of the gap between the fault segments, and the depth to the main fault in the basement (Fig. 20). Recent analyses of the patterns of basin formation and faulting related to overstepped strike-slip faults were published by Mann and others (1983), Rodgers (1980), Segall and Pollard (1980), and Hempton and Dunne (1984). They analyzed a discontinuous fault composed of interacting segments. Rodgers based his analysis on infinitesimal strain theory and assumed constant slip along the entire length of the fault. Segall and Pollard, however, maximized the displacement near the middle of the fault segments and allowed it to go to zero at the ends of the faults, comparable to what is observed in nature. When taken together, both analyses provide clues to the displacement geometry of pull-apart basins, the orientation and kinds of faults which may be found in these basins, as well as the state of stress in and around those basins.

The extended domain within a releasing bend has been depicted as having a meshlike arrangement of extension fractures and strike-slip faults between segments of the bounding strike-slip faults (Sibson, 1986, 1987), an arrangement which was confirmed for the North China Basin by analysis of the complex 1976 Tangshan earthquake sequence (Nábělek and others, 1987). The basin is a large, hydrocarbon-producing basin that began to form in early to middle Eocene time. It is located, and evidently evolved, between two, right-stepping, master, right-slip faults of a larger dextral-slip fault system, and the dominant focal mechanism of the earthquake sequence was right slip. Lesser right-slip, normal, and thrust faults are present within the basin in a geometrical arrangement that mimics the pattern of master faults, and they produced lesser shocks in the earthquake sequence that revealed the structural nature and tectonic mechanism of the basin. The interplay among the intrabasin faults outlines domains of uplift and subsidence within the basin. Secondary pull-aparts in the basin form unconnected domains of local subsidence which, taken in combination, impart an intense, wholesale subsidence to the entire basin. Unfortunately, the earthquake data are insufficient to determine how the basin-bounding faults project to depth or to determine the nature of the basin floor (Nábělek and others, 1987).

With regard to the nature of their floors, pull-apart basins seem to lie between two end members.

At one end are true rifts that extend at depth into hot rocks of the upper mantle, such as those expected above an oceanic spreading ridge. Older rocks, largely continental, are ripped asunder, first by attenuation of the upper crust, and then by actually breaking apart as the mantle material wells up into the widening gash. Under these circumstances, older parts of the basin floor on which sediments are laid down may be missing completely. These basins lack a true basement, and a well drilled to depth would go through sediments into a sill and dike complex of volcanic rocks. Because they intrude the oldest sediments in the basin, the volcanic rocks are younger than sediments at the base of this type of pull-apart. If the well were drilled vastly deeper, it would presumably reach hot and even molten rocks of the lower crust and upper mantle. The Salton basin is an example of this kind of basin.

At the other end of the spectrum of pull-apart basin types are those that bottom-out on a detachment or decollement surface within the upper crust, and they may be grouped into two subtypes: 1) those that bottom against flat tectonic surfaces, or the detachments or flat faults themselves, and 2) those that bottom unconformably against older basement where the detachment is deeper still. Seismic profiling by CO-CORP and CALCRUST and other geophysical studies, along with down-dip extrapolation of surface observations, show that much of southern California is underlain by subhorizontal tectonic surfaces (Cheadle et al., 1986; Frost and Okaya, unpub. ms.). It is quite likely that some of the through-going reflections are from structural discontinuities on which crustal blocks pull apart and rotate. The profiles disclose several suspect detachments, however, so it is not yet clear on which, if any, the rotations and pull-aparts occur. Perhaps the reflections are stacked decollements, and intermediate blocks between them rotate differently from their underlying and overlying neighbors (J. C. Crowell, unpub. data).

Examples of basins that terminate in a detachment at a relatively shallow level include the Vienna basin (Royden, 1985) and the Dead Sea basin (Manspeizer, 1985).

SEDIMENTATION RELATED TO STRIKE-SLIP FAULTS

Much literature has been written about sedimentation in basins along strike-slip faults, the most recent and comprehensive of which has been edited by Ballance and Reading (1980), Crowell and Link (1982), and Biddle and Christie-Blick (1985). Many basins are typified by high sedimentation rates, scarce igneous and metamorphic activity, abrupt facies changes, abrupt thickening of sedimentary sequences over short distances, numerous unconformities which reflect syntectonic sedimentation, and the presence of a locally derived, skewed fan-body of fault-margin breccia facies representing talus detritus or alluvial fans (Crowell, 1974a, 1974b; Mitchell and Reading, 1978; Hempton and others, 1983; Dunne and Hempton, 1984; Nilsen and McLaughlin, 1985). The coarse, basin-margin facies forms a narrow band along the fault at the edge of the basin. It is volumetrically subordinate to, and contrasts strongly with, the main sequence of flood basin and lacustrine strata with which it interfingers and mixes in the basin, and which is much finer grained, farther traveled, and commonly deposited by turbidity currents (Hempton and others, 1983; Sadler and Demirer, 1986).

The most distinctive stratigraphic feature of basins that form in association with strike slip is the extreme thickness of onlapping sedimentary sequences in pull-apart basins relative to their area (Fig. 21). This happens because of migration of the depocenter by means of syndepositional strike slip (Crowell, 1974b, 1982a). The center of deposition migrates in the direction opposite to that of strike-slip movement of the basin, so that the basin lengthens over time, and the sediments are deposited in an overlapping "venetian blind" arrangement or "stratal shingling" which youngs toward the depocenter (Crowell, 1982a, 1982b; Hempton and Dunne, 1984). The areal extent of Hornelen Basin of western Norway is less than 1,250 km^2, but the stratigraphic thickness of the Devonian sedimentary sequence therein approaches 25 km in a basin 60–70 km long, 15–25 km wide (Steel and others, 1977; Steel and Gloppen, 1980). The true maximum vertical thickness of the succession at any point, however, is probably less than 8 km (Steel and Gloppen, 1980). Ridge basin in southern California is 30–40 km long, 6–15 km wide, about 400 km^2 in areal extent, with a cumulative fill of about 13 km (Crowell and Link, 1982), but the thickness of strata in any single drill hole would be considerably less (Fig. 21). Such thick, asymmetric, sedimentary fillings are characteristic of other basins of various sizes, ages, and tectonic settings (Aspler and Donaldson, 1985; Guiraud and Seguret, 1985; Manspeizer, 1985; Nilsen and McLaughlin, 1985).

Uplifts Related to Strike-Slip Faults

Convergent strike slip or transpression (Harland, 1971; Sanderson and Marchini, 1984) provides a component of horizontal shortening across the strike-slip fault zone which is necessarily accompanied by compensatory uplift of rocks in the fault zone. This is clearly demonstrated in laboratory-model studies where an elongate, fault-bounded welt forms above the zone of principal displacement (Fig. 22) because of the accommodation of the component of shortening strain by uplift (Lowell, 1972; Wil-

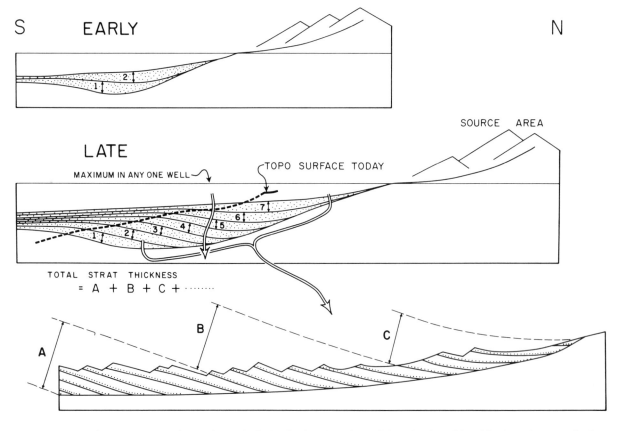

Figure 21. Diagrammatic arrangement of stratal units in Ridge Basin, viewed parallel to the depositional basin and perpendicular to a right slip fault which lies between the source area and the basin (from Crowell, 1982a). The center of deposition migrates toward the principal source area which lies north of the basin and across the fault behind the basin. The upper and center cross sections show the arrangement at early and late states, and the lower cross section shows the way the region is exposed today after uplift and erosion. The total stratigraphic thickness is obtained by adding thicknesses measured along the topographic surface today, and is reproducible by any stratigrapher. Reproduced with permission of Society of Economic Paleontologists and Mineralogists.

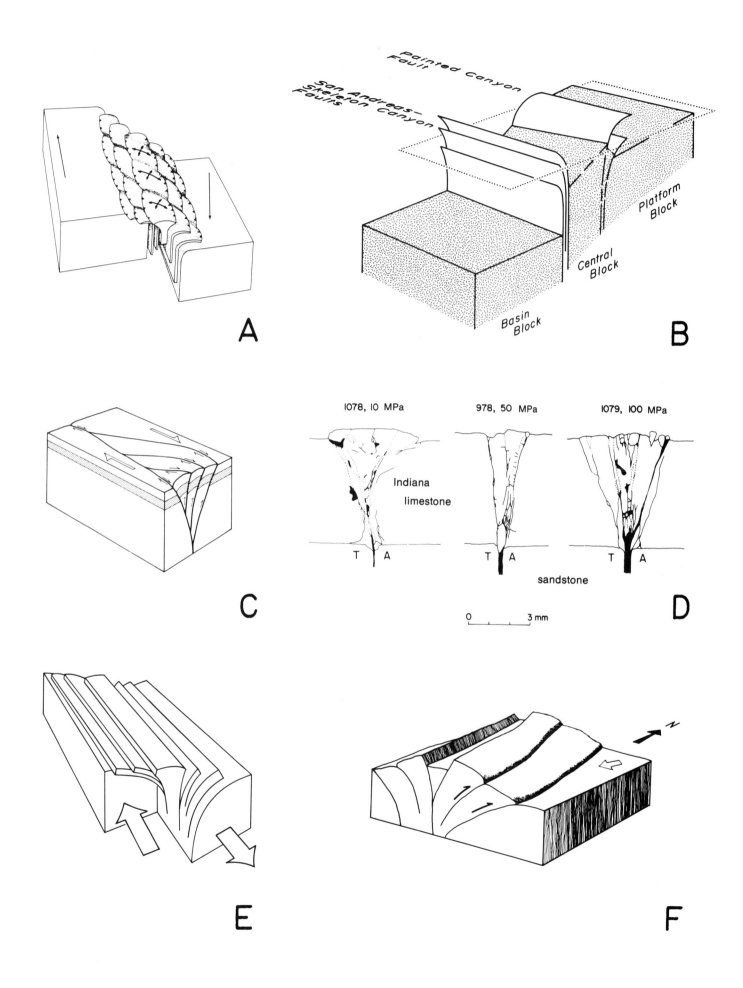

Figure 22. Conceptual diagrams of palm tree structures in right simple shear. (A) After Lowell (1972, p. 3099), reproduced with permission of Geological Society of America and of Lowell, 1988; (B) after Sylvester and Smith (1976), reproduced with permission of American Association of Petroleum Geologists and of Sylvester, 1988; (C) after Woodcock and Fisher (1986), reproduced with permission of Journal of Structural Geology; (D) after Bartlett and others (1981), reproduced with permission of Elsevier Science Publishers and of Bartlett, 1988; (E) adapted with modifications from Ramsay and Huber (1987, p. 529); (F) with axial graben after Steel and others (1985), reproduced with permission of Society of Economic Paleontologists and Mineralogists.

---

cox and others, 1973; Bartlett and others, 1981). The welt is bounded by sinuous faults which, in cross section, are nearly vertical at depth and flatten upward, carrying parts of the welt short distances outward upon the adjacent, stable blocks. This upward-branching arrangement of faults in strike-slip fault zones is mentioned by Willis (1938a), Kingma (1958), and Clayton (1966); was mapped by several writers (Wallace, 1949; Sylvester and Smith, 1976; Burke, 1979; Davis and Duebendorfer, 1982); and was produced especially clearly in experimental studies in layered media (Emmons, 1968; Bartlett and others, 1981; Naylor and others, 1986).

Lowell's (1972) conceptual block diagram of transpression (Fig. 22A) shows a central zone of vertical slabs which rise upward and outward on convex-up faults over the adjacent blocks like a stack of imbricate thrusts. Wilcox and others (1973) termed the upward and outward branching arrangement of faults in Figure 22A "positive flower structure."[3] Because the surficial traces of the faults are sinuous in plan view, most geologists would interpret field outcrop and map patterns of the faults as evincing thrusts rather than strike-slip faults, because, almost by definition, strike-slip faults are "transcurrent"; that is, they cut straight across rocks, structures, and topography as vertical, throughgoing faults.

---

[3]Descriptions of the two-dimensional arrangement of the faults in interpretations of seismic sections and structural cross sections have a variety of botanical appellations, including "positive flower structure" (Wilcox and others, 1973) or "palm tree structure" (Sylvester and Smith, 1976), as well as "pop-up," "squeeze-up," and "tectonic wedge" in convergent strike slip, and "negative flower structure" (Harding and Lowell, 1979) or "tulip structure" (Naylor and others, 1986) in divergent strike slip. Because real flowers have a variety of shapes as well as appellations, I prefer the term "palm tree structure" to describe the convex-upward geometry of faults in profile that bounds an uplifted block in a strike-slip fault zone, even though flower structure has a few years' precedence (Biddle and Christie-Blick, 1985). I believe the term "tulip structure" evokes a clear image of the concave-upward geometry of faults in profile that form in divergent strike-slip, and I propose that "tulip structure" be used in place of "negative flower structure," even though these structures are not circular in plan view.

The near-surface, low-angle segments of strike-slip faults are common along many major convergent strike-slip faults, particularly where the faults mark the base of a steep mountain front. In the Mecca Hills (Fig. 22B), the San Andreas, Skeleton Canyon, and Painted Canyon faults dip 60°–70° toward the central block in the deepest exposures. The faults flatten upward into short, oblique-slip thrust faults beneath rocks of the central block that have been thrust from 50 to 200 m upon the footwall of the adjacent block.

The upward flattening of strike-slip faults in zones of transpression has been documented also on the Alpine fault of New Zealand (Wellman, 1955), and along the Banning and San Jacinto faults of southern California (Allen, 1957; Sharp, 1967). Similar, more extensive nappes traveled up to 1 km in the "big bend" segment of the San Andreas fault (Davis and Duebendorfer, 1982), and as much as 5 km in West Spitsbergen (Lowell, 1972; Kellogg, 1975; Craddock and others, 1985).

Wellman (1955) considered that the surficial thrusting results from downslope creep under gravity which bends the fault in mass movement, but according to Allen (1965, p. 84): "Another important causal factor here and elsewhere may be related to the origin of the steep mountain front itself: if, as seems likely, the presence of the mountain front is caused by a local vertical component of displacement along the predominantly transcurrent fault, then vertical motion constrained at depth to a vertical plane must necessarily result in localized low-angle thrusting at the surface, as has been demonstrated analytically and in models by Sanford (1959)" and by Hafner (1951). Allen pointed out that these surficial thrusts may conceal the major underlying strike-slip faults and may have delayed recognition of the dominance of horizontal displacements on faults in many parts of the world.

Allen's hypothesis is a good explanation for the local surficial flattening of strike-slip faults, but I believe that the larger-scale thrusting results from a mechanical delamination of an uplifted block coupled with shortening that permits thin structural flakes to move obliquely across the adjacent block and appear as though they have been thrust up and out of the deformation zone. This must be the explanation for those thrust segments which seemingly have come out of a strike-slip fault zone which is very much narrower than the thrust segment itself.

Sylvester and Smith (1976) were influenced by Lowell's block diagram to the extent that they also visualized an uplifted central block nearly 2 km wide underlain by a zone of crushed metamorphic rocks between the two, subparallel faults (Fig. 22B). Analyses of fractured and faulted gneiss layering show that the crystalline basement responded to convergent strike slip by cataclastic flow and piecemeal slip along fractures and faults at all scales, just as Willis (1938b) realized for strike-slip faults elsewhere. The overlying sedimentary sequence between the faults, however, deformed passively in response to cataclastic flow and differential uplift of the basement (Sylvester and Smith, 1976). Outside of the main fault zone, the Painted Canyon and lesser strike-slip faults branch away from the San Andreas fault to form an array of splay faults (extensile fan in the terminology of Woodcock and Fischer, 1986; Fig. 22C) in plan view (Fig. 23), just as they do in experimental studies (Naylor and others, 1986). The lesser faults are characterized by normal separation where they strike at high angles to the San Andreas fault (Harding and others, 1985; Fig. 23).

The well-exposed structure of the footwall at the edges of the relatively down-dropped blocks and beneath the thrust in the Mecca Hills (Fig. 24) shows that the uplift necessitated by the convergent strike slip is accommodated by an accordion style of folding and by a variety of faults, some of which project into the main fault that bounds the uplifted central block, and others which dip away from the central block as back-thrusts and die out in bedding surfaces (Sylvester and Smith, 1976). Together, all of the faults bound a triangular-shaped domain of folded strata which resembles the triangle or delta zone (Jones, 1982; Butler, 1982; Lowell, 1985, p. 282) seen at the toe of some overthrust faults, including the Prospect thrust near Jackson, Wyoming (Dorr and others, 1977; Dixon, 1982), the Findley structure in the northern

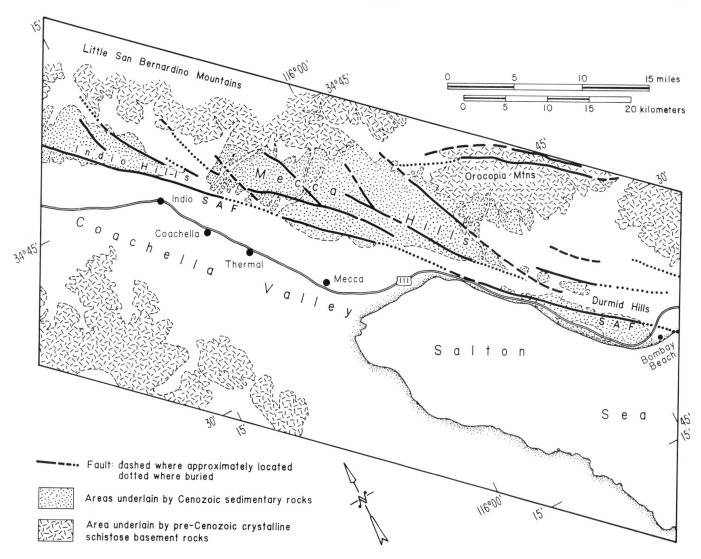

Figure 23. Simplified geologic map of fault splays and of tectonic culminations along the fault zone, represented by the Indio Hills, Mecca Hills, and Durmid Hills at the southern end of the San Andreas fault (SAF) in Salton Trough.

foothills of Alberta, Canada (Jones, 1982), that beneath the Nushman anticline in the western foothills belt of south-central Taiwan (Suppe, 1980), and the toe of the Alpine front and Molasse Basin, Bavaria (Bachman and Koch, 1983). The axes of the folds in the triangle zone along strike-slip faults are oriented at low angles to the main faults and constitute a transected series of en echelon folds. Similar, larger triangle zones along convergent strike-slip faults are alleged to have been documented in the subsurface of producing offshore oil fields of southern California and in the Taranaki basin of New Zealand.

Seismic profiles have been obtained over palm tree structures (for example, Fig. 25) and tulip structures (Harding and Lowell, 1979; Bally, 1983; Harding, 1985) and are typified by having reverse- and dip-separation faults side by side across the crest of the structure (Figs. 22D, 22F). To be sure, some of the faults are indeed thrusts that reflect the over-all uplift of the central block and shortening across it, but others are just as clearly normal faults reflecting the extension across the top of the uplifted and laterally spreading block as documented by Sylvester and Smith (1976) and shown diagrammatically in Figure 22F. To a greater degree, however, the variable apparent vertical displacement is due to oblique slip or pure strike slip in and out of the plane of the vertical section.

SEISMOTECTONICS

Segmentation

Different segments of active strike-slip faults behave differently in terms of (1) the maximum magnitude of earthquakes they have generated or are capable of generating; (2) the maximum amount of surface displacement they display with those earthquakes; (3) the return frequency of earthquakes; and (4) the rate of aseismic fault creep (Allen, 1968; Wallace, 1970). For example, the strike of the San Andreas fault is within 5° of the plate slip vector, with three main exceptions where the fault trace departs from this trend: one north of San Francisco and located largely offshore (Thatcher and Lisowski, 1987), the second on the north side of the San Gabriel Mountains (Fig. 9), and the third along the south side of the San Bernardino Mountains (Fig. 9). Each is a restraining bend and a center of great earthquakes where the fault is considered to be locked (Allen, 1968) relative to the straight, creeping segment of the fault in central California which is typified by rather frequent minor to major earthquakes (Brown and Wallace, 1968; Wallace, 1970).

Fractal analyses reveal that the San Andreas fault has slight, but statistically significant, varia-

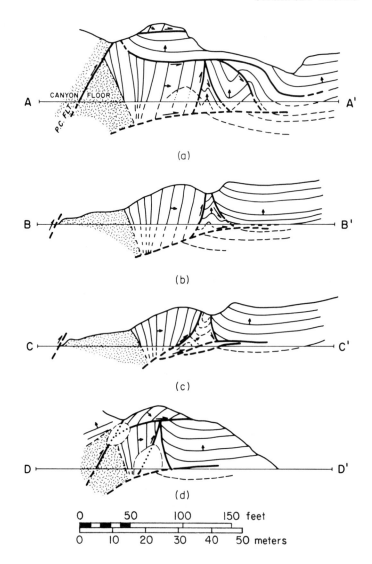

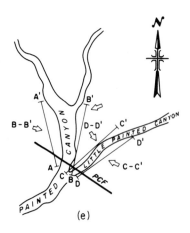

Figure 24. Field sketches of triangle zone exposed on canyon walls beneath Painted Canyon fault (P.C. fault; PCF), Mecca Hills, southern California. (a) View west-southwest; (b) view east-southeast (reversed); (c) view northwest; (d) view southeast (reversed); (e) locations of cross sections relative to each other.

tions in fractal dimensions from segment to segment, variations that correlate with Allen's (1968) subdivisions (Okubo and Aki, 1987; Aviles and Scholz, 1987). Although fault jaggedness increases southeastward along the strike of the fault, fault segments having variations of creep, seismic slip, or microearthquake activity cannot be distinguished from one another on the basis of their fractal dimension (Aviles and Scholz, 1987).

The jagged surface trace of the San Andreas fault may be subdivided into 12- to 13-km segments which have abrupt changes in trend of 6° ± 2° (Clark, 1984). Tectonic depressions are present where the segments are parallel to the plate slip vector (N40°W), and uplift occurs where the fault segments are oblique to the plate slip vector (Bilham and Williams, 1985; Fig. 26). Across the San Andreas opposite the uplifts are some of the deepest parts of the Salton Trough; judging from gravity data (Biehler and others, 1964), the depth to basement in the basin block is 4,000 m only a few kilometers from the southwest edge of the Mecca Hills (Fig. 22B).

Many authors have recently addressed the short-term implications of segmentation insofar as earthquake mechanics and hazards are concerned, but the long-term geological implications have yet to be elucidated. Presently defined segment lengths are determined by the depth to the seismogenic zone, by the presence and position of releasing or restraining bends, by fault bends, and by "asperities" (Tang and others, 1984; Barka and Kadinsky-Cade, 1988). Within major zones of strike slip, geologists have recognized and mapped numerous shears which predate the presently active trace, and which may be reactivated in future earthquakes. How and why should slip transfer from segment to segment across strike, and how does this slip-swapping affect the length of the fault segments? When a strike-slip fault is bent beyond some angle that accommodates easy slip, does a new fault take a shorter path, thereby short-cutting the bend and abandoning it? The curved San Gabriel fault (Fig. 9), active in Miocene and Pliocene time, is regarded as an old segment of the San Andreas fault which was abandoned when the San Andreas fault shifted to its younger, straighter locus on the northeast side of the San Gabriel Mountains (Crowell, 1979). How do strike-slip faults become bent in the first place?

The average strike of the San Andreas fault is N40°W north of the Transverse Ranges, it strikes exactly east-west for a distance of 10 km in the "big bend" segment where the Big Pine and Garlock faults intersect the San Andreas fault (Fig. 9), and then it strikes about N60°W for ~300 km into the Salton Trough. Several writers have speculated that the "big bend" is a result of left slip on the Garlock fault as it plays its role as an *intraplate transform* (Fig. 27). In each of the models, Basin and Range crust north of the Garlock fault extends westward, relative to a stationary Mojave Desert, carrying also westward the Sierra Nevada off the Kingston Peak core complex (G. A. Davis, 1988, personal commun.), and the Great Valley, together with the San Andreas fault (Eaton, 1932; Hamilton and Myers, 1966; Davis and Burchfiel, 1973; Wright, 1976; Hill, 1982; Bohannon and Howell, 1982). At the east end of the Garlock fault at the Kingston Range, therefore, the slip is evidently zero; at the center part of the fault, sinis-

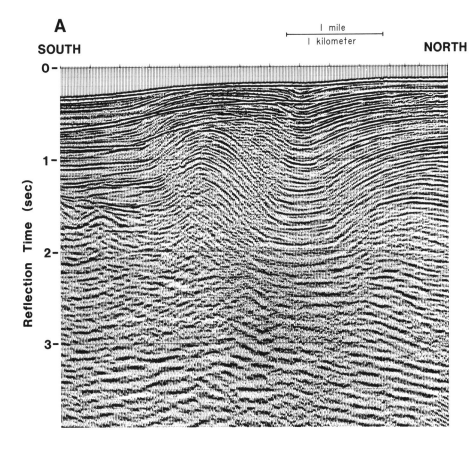

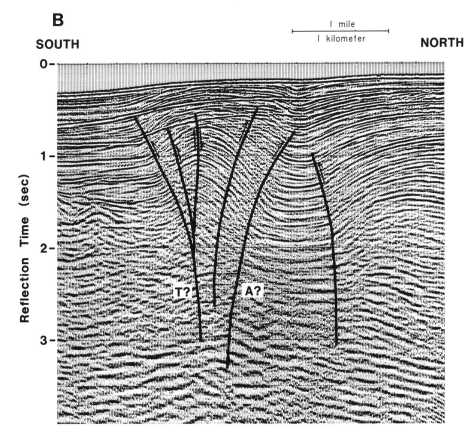

Figure 25. Seismic expression of a palm tree structure along the subsea extension of the Oak Ridge fault in the Santa Barbara Channel (Fig. 9), southern California. (A) Migrated reflection profile; (B) interpreted, migrated reflection profile. A = movement away from observer; T = movement toward observer. The near-surface ends of the faults cannot be projected to the sea floor because of interference of reflectors by sea-floor multiples. The east-west length of the structure is nearly 6.5 km.

tral displacement is about 60 km, and it increases westward to the San Andreas fault (Davis and Burchfiel, 1973). Because of its low angle of intersection, one might guess that sinistral displacement during a large earthquake on the Garlock fault will transfer into the dextral San Andreas fault for a short distance. Then when the San Andreas fault slips right laterally in a subsequent earthquake, the displacement will bypass the end of the Garlock fault and cancel the bit of sinistral slip imparted to it by the earlier event on the Garlock fault (Bohannon and Howell, 1982).

The net effect of the westward transport of blocks north of the Garlock fault is to place a restraining bend in the edge of the North American plate around which the Pacific plate must pass. The difficulty in that passage is reflected by the profound structural complexities in southern California around the "big bend" manifested by the Transverse Ranges, only one of two east-west–trending mountain ranges in North America.

Paleoseismicity

One of the most far-reaching papers concerning frequency of earthquakes and slip rate on any kind of fault was published by Kerry Sieh in 1978. He defined the timing of Holocene movements on the San Andreas fault and, thereby, quantified the earthquake hazard for the major population centers of southern California. By means of microstratigraphic analyses of peat beds interlayered with strata of fluvial sand and gravel in a drained sag pond, Sieh (1978) recognized vertical separation, stratigraphic disruption, and liquefaction related to nine events that deformed the strata similar to the way that the M 8 earthquake of 1857 disturbed the same strata. He showed that nine earthquakes, probably like that of 1857, occurred in the past 2,000 yr, yielding a slip rate of 30 mm/yr and an average recurrence interval of

140 ± 40 yr, but with a range of from 70 to 300 yr. Dendrochronologic studies, however, show that major earthquakes on this part of the San Andreas fault do not cluster closely about the 131-yr average, nurturing "doubts about hypotheses of uniform fault-strain accumulation and relief" (Jacoby and others, 1988, p. 196). Recent ultra-precise ^{14}C ages have revised the average of the intervals to 131 ± 20 yr, and a range from 60 to 400 yr (Sieh and others, unpub. ms.). That is quite frequent, given the fact that similar studies on faults, mainly normal faults, in other tectonic domains of western North America yield return frequencies on many time scales, ranging in general from several thousands of years (Swan and others, 1980; Bucknam and others, 1980; Scott and others, 1985; Malde, 1987; Pearthree and Calvo, 1987; Lubetkin and Clark, 1988) to several hundreds of thousands of years (Bull and Pearthree, 1988; Wallace, 1987), as do major strike-slip faults in other parts of the world (Japan: Okada, 1983; China: Allen and others, 1984; Peru: Schwartz, 1988; north Africa: Meghraoui and others, 1988, Peltzer and others, 1988). Field sites having the favorable combinations of well-preserved, readily datable material, intimately interlayered with a distinctive stratigraphic sequence on the active segment of a fault are rare along the San Andreas fault. Other major strike-slip faults therefore need to be searched and given the kind of systematic paleoseismic attention that the San Andreas fault has received to provide more information about how these faults work.

Earthquakes are temporally clustered in the Pallett Creek segment of the San Andreas fault: the earthquakes in each cluster are separated by decades, but the clusters are separated by dormant periods from two to three centuries (Sieh and others, in press). The occurrence of large historic earthquakes on the North Anatolian fault in Turkey and on predominantly normal faults in the Basin and Range province are also clustered in time (Ambraseys, 1970, 1971; Wallace, 1987). On a longer time scale, Clifton (1968) found sedimentary cycles suggestive of tectonic "events (or closely spaced flurries of events) with a periodicity of tens of thousands of years" possibly related to movements on the San Andreas fault.

King (1987), following Savage's (1971) hypothesis that migratory pulses of earthquake activity along the southern half of the San Andreas fault are driven by creep waves induced by episodic magma injections at the East Pacific Rise, has offered a comprehensible method of loading episodic horizontal strain energy into the fault system. The injections propagate along the broad transform boundary between the Pacific

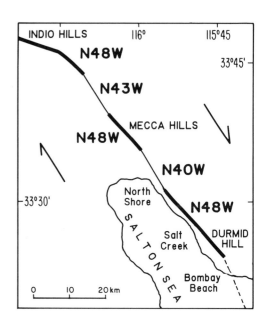

Figure 26. Segments of parallel slip and convergent slip along the southern end of the San Andreas fault (redrawn from Bilham and Williams, 1985). The plate slip vector is parallel to the N40°W segment. Folding and uplift occur on the northeast sides of the N48°W fault segments. Compare with Figure 23.

and North American plates at subseismogenic depths.

Recently Sibson (1987) proposed an intriguing idea of wide interest that links strike-slip earthquakes, ore deposits, rock mechanics, segmentation, and fluid flow. "Paleoseismic studies show that segments of some faults tend to rupture at fairly regular intervals in characteristic earthquakes of about the same size" (Sibson, 1985, p. 248; Schwartz and Coppersmith, 1984). The regularity of the ruptures has been questioned recently (Wallace, 1987; Sieh and others, in press), but the characteristic size and rupture dimensions are evidence that segment terminations, which are typically releasing or restraining bends or oversteps, arrest or perturb fault rupture propagation (Sibson, 1985, 1987; Barka and Kadinsky-Cade, 1988). At releasing bends, abrupt extension locally reduces fluid pressure, leading to brecciation by hydraulic im-

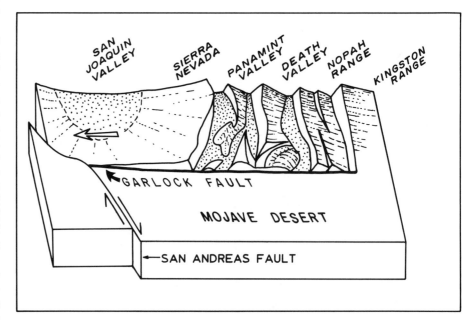

Figure 27. Northward diagrammatic view of the Garlock fault, southern California, as an intracontinental transform between the northern, westward-extended Basin and Range province, and the southern, nondistended Mojave Desert (redrawn with modifications from Davis and Burchfiel, 1973, p. 1417, and reprinted with permission of Geological Society of America and of Davis, 1988).

plosion and to sudden, concentrated influx of fluid into the bend. Arrest of the rupture may be followed by delayed slip transfer through the bend during fluid pressure re-equilibration by diffusion (Sibson, 1986). Where high boiling events are triggered by the arrest mechanism, Sibson (1987) postulated that episodic mineral deposition is induced in the top 1-2 km of the bends by the dynamic effects of rupturing on the flanking strike-slip faults. Other writers have already borrowed these ideas to explain how strike-slip faults controlled the structure of some hydrothermal mineral deposits (Willis and others, 1987), and I suspect this is only the beginning of a renaissance in structural economic geology.

Aseismic Slip

One of the most curious aspects of strike-slip faults not readily demonstrated by other kinds of faults is that of aseismic slip, or creep, which is a surficial phenomenon restricted to the upper 100 m or so of the fault (Goulty and Gilman, 1978; Sharp and others, 1986a). The phenomenon has been documented on only two faults to date: the North Anatolian fault in Turkey (Ambraseys, 1970; Aytun, 1980) and the San Andreas and related faults in central and southern California (Steinbrugge and others, 1960; Brown and Wallace, 1968; Schulz and others, 1982; Louie and others, 1985; Galehouse, 1987). The maximum rate of creep on the central segment of the San Andreas fault is 35 mm/yr (Burford and Harsh, 1980) where the fault zone is narrow and straight and is contained in oceanic basement rocks of the Franciscan Formation (Irwin and Barnes, 1975). That creep rate is equal to the historic slip rate for the entire fault zone as determined from geodetic evidence (Thatcher, 1979), and for Holocene time as inferred from paleoseismic studies (Sieh, 1978, 1984; Sieh and Jahns, 1984). The creep rate is only about 3.4 mm/yr at the southern end of the San Andreas fault (Louie and others, 1985).

Creep is believed to be driven by elastic loading of the crust at seismogenic depths. Its occurrence at the surface represents either deep aseismic motion which is expressed as seismicity in one area and to aseismic slip in another, or the accumulation of seismic dislocations away from slipping faults (Louie and others, 1985).

The tectonic significance of creep is a topic of debate (Wallace, 1970; Sylvester, 1986; Williams and others, 1988): some investigators believe that creep represents steady-state slip which relieves buildup of stress on strike-slip faults so that large earthquakes are precluded in a creeping fault segment (Brown and Wallace, 1968; Prescott and Lisowski, 1983), and that notion seems to have gained support, at least for the San Andreas fault, during the past 15 yr by a massive amount of horizontal strain data (Langbein, 1981). Alternatively, creep is postulated to be the first step in progressive failure leading to a large earthquake (Nason, 1973, 1977), although it is clear that several southern California strike-slip earthquakes were not preceded by surficial pre-seismic creep (Cohn and others, 1982).

Strike-slip faults also exhibit a greater degree of *afterslip* than do other kinds of faults so far as is known. Afterslip is fault slip that occurs on the fault in the days, weeks, or even months following the main earthquake. It may be gradual and continuous or episodic, but its principal character is that the slip rate decreases logarithmically over time (Wallace and Roth, 1967; Sylvester, 1986; Wesson, 1987). The amount of displacement of afterslip following strike-slip earthquakes may equal or exceed the coseismic slip (Smith and Wyss, 1968; Ambraseys, 1970; Burford, 1972; Bucknam and others, 1978), whereas it is very much less in thrust and normal fault earthquakes: For example, Sylvester and Pollard (1975) found that afterslip totaled only 1% of the coseismic slip during the year following a thrust earthquake which produced a maximum of 2 m of coseismic net slip at the surface. Whether afterslip is truly aseismic has not been clearly established, although the afterslip following the 1979 Homestead Valley, California, earthquake (M_L = 5.8) was much greater than the summed moment of the aftershocks, leading Stein and Lisowski (1983) to conclude that the afterslip, which constituted about 10% of the seismic slip, was aseismic. Wesson (1987) modeled afterslip with a "stuck" or locked patch at depth on the fault surface, surrounded by an area that creeps in response to the applied stress.

Still another unique seismotectonic phenomenon exhibited to date only by strike-slip faults is that of *triggered slip*, which is coseismic slip on a fault or faults other than the causative fault outside the epicentral area of the main shock (Sylvester, 1986). The phenomenon has been observed repeatedly in moderate earthquakes in the Salton Trough of southern California where up to 30 mm slip has occurred on faults as far as 40 km from the causative fault and its epicenter (Allen and others, 1972; Fuis, 1982; Sieh, 1982; Williams and others, 1988). The mechanism for the triggered slip is problematic, although geodetic and seismologic evidence in the California earthquakes suggests that parts of the affected faults may be variably prestressed or have different shear strengths. Thus those faults which are near failure will slip small amounts either because of shaking dynamically induced by the main earthquake (Fuis, 1982; Sieh, 1982), or because the regional static strain field is perturbed so that stress is concentrated on other faults which then yield by creep. Allen and others (1972) preferred the mechanism of dynamic strain to explain slip triggered on faults in the Salton Trough by the 1968 Borrego Mountain earthquake (M = 6.8), because they found that the change in the static strain field was an order of magnitude less than the dynamic strain due to ground shaking.

Resolution of the various problems of creep will require more and continued monitoring of slip by geodetic and instrumental methods, careful searches for minor slip on associated faults near the seismogenic faults, and better determinations of components of pre-seismic slip, co-seismic slip, and afterslip in historical and young prehistoric offset data, not only for the San Andreas fault, but also on other active strike-slip faults.

TECTONIC ROTATION IN SIMPLE SHEAR

Tectonic rotation of slabs of the Earth's crust about a vertical axis in simple shear was suspected or postulated for parts of the American Pacific coast and the Dead Sea regions by several writers (Hamilton and Myers, 1966; Freund, 1970a, 1970b; Teissere and Beck, 1973; Garfunkel, 1974; Beck, 1976; Jones and others, 1976; Simpson and Cox, 1977; Hamilton, 1978) before the notion took root in southern California and flowered.

Hamilton and Myers (1966) postulated that the Transverse Ranges rotated clockwise about a vertical axis at their east end from a position alongside Peninsular Ranges. Jones and others (1976), noting the 90° difference in structural trends in Mesozoic (?) rocks on Catalina and Santa Cruz Islands, relative to the north-south trend in the southern Sierra Nevada foothills, postulated that the islands had rotated 90° without specifying the sense.

Luyendyk and others (1980, 1985), following these and other earlier indications of rotation (Teissere and Beck, 1973; Beck, 1976; Simpson and Cox, 1977; Greenhaus and Cox, 1979), and building on theoretical considerations of crustal rotation (Freund, 1974), thought that the direction and amount of rotation could be determined paleomagnetically. Accordingly, Luyendyk and his students systematically sampled early Neogene rocks, principally volcanic rocks in the western Transverse Ranges, which should have been rotated clockwise by distributed late Neogene shear within the San Andreas fault system (Kamerling and Luyendyk, 1979, 1985; Terres and Luyendyk, 1985). Their results supported the clockwise rotation inferred by Hamilton and Myers (1966), but not the mechanism. More importantly, they were able to determine dimensions of the domains of rotation (Fig. 28), postulating that several blocks or slabs in south-

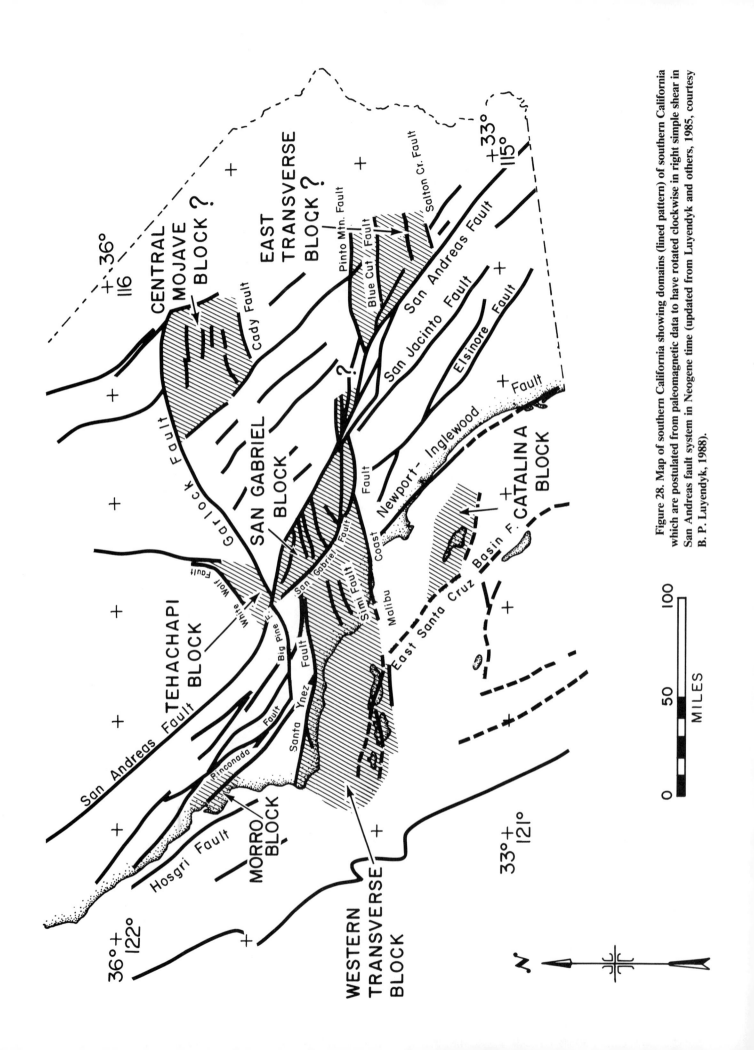

Figure 28. Map of southern California showing domains (lined pattern) of southern California which are postulated from paleomagnetic data to have rotated clockwise in right simple shear in San Andreas fault system in Neogene time (updated from Luyendyk and others, 1985, courtesy B. P. Luyendyk, 1988).

ern California had rotated at least 90° clockwise in a broad zone of simple shear, and perhaps as much as 120° in Neogene time. Initially, they predicted that blocks bounded by east-west-striking, left-slip faults and by northwest-striking, right-slip faults would also be found to have rotated clockwise across a broad zone of simple shear between the two great crustal plates. By and large, subsequent paleomagnetic studies by Luyendyk and his colleagues have substantiated this model. Hornafius and others (1986) extended the hypothesis into the fourth dimension by showing that the magnetic declination vectors having the greatest degree of clockwise rotation are in the older rocks, and the rotations decrease progressively in progressively younger rocks. The western Transverse Ranges (Fig. 28) therefore rotated from 20 Ma to 4 Ma.

Whereas several writers embraced the rotation concepts and found paleomagnetic evidence for rotations in other major strike-slip fault zones (Rotstein, 1984; Ron and Eyal, 1985; Ron and others, 1986; Kissel and others, 1987), other workers have had difficulties understanding how such relatively long, narrow crustal slabs could behave so rigidly during rotation (Nelson and Jones, 1987), and why supporting geologic evidence of the paleomagnetic rotation is not especially evident. This is because existing models of the proposed rotations are geometrically simplified or theoretical (Fig. 29), but few of those models are defined by compelling field data. We lack sufficient information about the three-dimensional geometry of natural strike- and oblique-slip faults, the mechanism of translation and rotation of fault blocks around strike-slip fault bends, or about the sequential development of structures in the complex zone of heterogeneous strain. Geologists have searched for geologic evidence, such as associated deformation, or rotation of paleocurrent directions of Paleogene rocks away from their provenances with variable success (Karner and Dewey, 1986) or inconclusive results (Crowell, 1987, p. 227; Howard, 1987). In southern California, the paleomagnetic evidence seems to point strongly to kinematics of clockwise rotation, but the present models need to be tested by more detailed mapping and structural studies of well-exposed areas.

It seems clear that the rotated blocks must in fact be flakes (Oxburgh, 1972), slabs, or crustal panels (Dickinson, 1983), which detach on a shallow horizontal shear surface as Brown (1928) observed in model studies that rotation of upper layers occurred where horizontal shear took place in a weak, underlying layer. In a 1-m-wide zone of right slip associated with the 1979 Imperial earthquake, Terres and Sylvester (1981) found that elongate blocks of soil, which

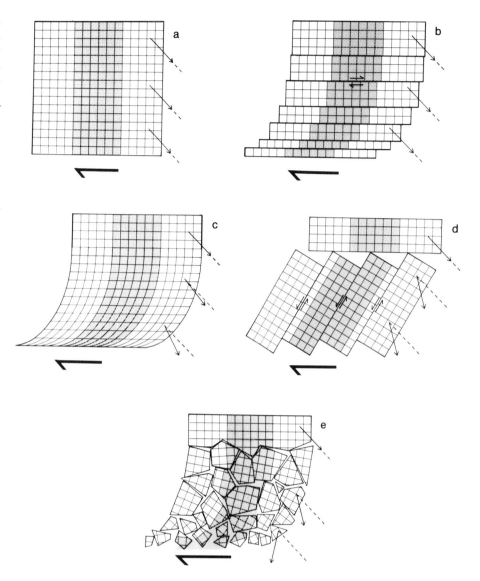

Figure 29. Mechanisms of oroflexural bending and rotation in simple shear zone (after Nelson and Jones, 1987). (a) Undeformed domain; (b) shearing on faults parallel to the main shear zone without rotation; (c) pervasive, continuous bulk simple shear; (d) block rotation with internal antithetic shear; (e) small block model with variable, internal rotation.

were defined and bounded by pre-earthquake plow cuts, rotated clockwise from 20° to 40°, forming the deltoid "basins" and slipping internally on the left-slip faults predicted by kinematic models (Freund, 1974; Luyendyk and others, 1980). The rotated blocks had detached at a mechanical anisotropy 15 cm below the surface: the dry-soil–wetted-soil interface.

On a larger scale, Wilson (1960) described how slabs of ice rotated in a simple shear zone between the Filchner Ice Shelf and the Antarctic mainland. He saw that the slabs were "pinned" on either side of the chasm and rotated rigidly in response to the oceanic current which moves the ice shelf. The ice slabs were bounded by exten-

sion fractures or R shears which, during prolonged shear across the ice chasm, twisted and extended in response to rigid body rotation of the slabs (Fig. 30).

Evidence that a mechanical discontinuity or detachment (or several) must underlie the Transverse Ranges in order for them to rotate comes from analyses of P-wave delays from quarry blasts and earthquakes in southern California (Hadley and Kanamori, 1977; Nicholson and others, 1986). The presence of horizontal zones of mechanical detachment is also supported by deep reflection seismic profiling in the Mojave Desert and Transverse Ranges which shows a series of nearly flat reflectors in the

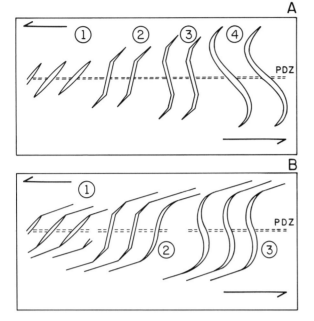

Figure 30. Rotation and progressive propagation of fractures in a left-shear couple (redrawn after Wilson, 1960). Extension (T) fractures form (1), then in progressive simple shear, they rotate counter-clockwise but continue to propagate either parallel to the direction of the initial fracture (A) or parallel to the direction favorable for the formation of R fractures (B), that is, at an angle $\phi/2$ to the principal displacement zone (PDZ). See also Figure 6.

upper crust, the highest one of which shoals eastward and rises to the surface at the traces of one or two major faults in the Mojave Desert (Cheadle and others, 1986), although several of the COCORP reflectors have been reinterpreted recently as "side-swipe" effects (Serpa, 1988).

Additional P-wave analyses revealed that the distribution of relatively "fast" and "slow" crust is moderately well outlined by the positions of the major faults in the upper crust, but not in the lower crust, suggesting that the major strike-slip faults do not extend beneath the seismogenic zone into the lower crust (Hearn and Clayton, 1986a, 1986b). When the focal mechanisms of a very large sample of earthquakes are analyzed, the only ones with thrust mechanisms correct for a detachment below rotated slabs are not only the deepest in southern California, but are also just those deep earthquakes located beneath the central Transverse Ranges (Hearn and Clayton, 1986a, 1986b). Not coincidentally, the central Transverse Ranges are not isostatically compensated; they lack a root.

These studies, together with the thrust-fault interpretation of the Whittier Narrows earthquake of 1 October 1987 (Hauksson and others, 1988), have revived the notion of "flake tectonics" (Oxburgh, 1972), that is, of "thin-skinned tectonics" in southern California (Yeats, 1968, 1981), wherein much or most of the Transverse Ranges is detached on deep and blind, north-dipping thrust faults (Webb and Kanamori, 1985). Above them, high-strength, flake-like slabs up to 15 km thick have detached from the lower crust to offset strike-slip faults at depth (Hearn and Clayton, 1986a, 1986b; Dewey and others, 1986; Lemiszki and Brown, 1988). In this model, the slabs may slide away from one another and are free to rotate externally about a vertical axis in the prevailing regime of simple shear. Webb and Kanamori (1985) postulated that perhaps even the San Andreas fault is offset at depth, but interpretations of seismic reflection data across several segments of the fault seem to negate that hypothesis (Lemiszki and Brown, 1988).

These are profound hypotheses and revelations in southern California, where notions of a simple transform plate boundary have prevailed since Hamilton (1961) and Wilson (1965), following Carey (1958), gave plate-tectonic bases to explain how horizontal displacements, measurable in hundreds of kilometers on the San Andreas fault, terminated abruptly at its southern and northern ends. The earthquake hazards posed by the blind thrust faults beneath the Transverse Ranges give southern California considerable reason for concern.

STRIKE-SLIP FAULTS IN CONVERGENT PLATE MARGINS

Orogen-Parallel Strike Slip

Suture zones make ideal strike-slip boundaries to accommodate much of the horizontal plate motion, as proposed for the Indo-Eurasia collision (Fitch, 1972; Karig, 1980; Tapponier and others, 1986). These motions result in part from the conversion of the horizontal component of oblique convergence onto a discrete strike-slip fault behind or within the magmatic arc, resulting in strike-parallel shuffling of terranes along major, trench-linked, strike-slip faults (Sarewitz and Karig, 1986). Orogen-parallel strike-slip faults are particularly prevalent within present-day, subduction-arc complexes characterized by oblique plate convergence (Fitch, 1972; Oxburgh, 1972; Saleeby, 1978), the most popularly cited example of this phenomenon being the Semangko fault system (Page and others, 1979; Karig, 1980; Hla Muang, 1987). Now that we have a better idea of the various roles and settings of strike-slip faults, a revival is occurring in the recognition of strike slip along the length of many ancient orogenic belts (Wilson, 1962; Reed and Bryant, 1964; Webb, 1969; Tapponier, 1977; Bradley, 1982; Gates and others, 1986; Sarewitz and Karig, 1986; Costain and others, 1987; Ferrill and Thomas, 1988; Şengör, in press).

Strike-slip faulting is a ubiquitous process in volcanic arcs of most subduction zones that have a continental overriding plate (Jarrard, 1986), and arc volcanoes such as Mount St. Helens are located in extensile zones above a deep shear zone which itself is preferentially oriented to accommodate horizontal strike slip within the volcanic arc (Weaver and others, 1987). It is noteworthy that the majority of volcanic eruptions are preceded by earthquakes having strike-slip focal mechanisms (Zobin, 1972; Weaver and others, 1981).

Geologic and paleomagnetic evidence is building an interesting and complex story of orogen-parallel lithospheric translation in mid-Jurassic to early Tertiary time in the great Cordilleran orogenic belt from the west coasts of Mexico, United States, and Canada on into Alaska as predicted by Atwater (1970), and which may be a precursor of the kinds of motions to be expected in orogenic belts elsewhere. In the Cordilleran orogen, the zone of transport lay within the trench or bordered the volcanic arc, perhaps as a series of trench-linked, strike-slip faults or boundary transforms at various places in time and space. Strands of the fault system had 175 km of dextral displacement in Late Cretaceous or early Tertiary time in what is now the trans-Mexican volcanic belt (Gastil and Jensky, 1973), as much as 1,000 km of dextral displacement along western Mexico (Karig and others, 1978), at least 1,200 km of dextral displacement west of what is now Baja California (Hagstrum and others, 1985), and at least 900 km of dextral displacement off the present coasts of northern California and Oregon (Champion and others, 1984; Bourgeois and Dott, 1985). The orogen is bent in Washington and northeast Oregon (Carey, 1958; Wise, 1963); then it regains its straight northwest trend north of the Shuswap terrane and emerges again in the northern Rocky Mountain–Tintina Trench system in western Canada, where at least 750 km, probably more than 900 km, of dextral displacement occurred in Middle Jurassic to early Cenozoic time (Gabrielse, 1985).

Opposed to these dextral motions are mid-Cretaceous sinistral movements of from 700 to 800 km that have been postulated for other major faults in the orogen, including the Sur-Nacimiento fault in central California (Seiders, 1983; Dickinson, 1983), and the Mojave-Sonora megashear in northern Mexico in mid-Mesozoic time, which is postulated to extend into eastern California (Silver and Anderson, 1974; Anderson and Schmidt, 1983). The direction of fault slip, whether dextral or sinistral, on the various strike-slip faults in the trench-linked, strike-slip fault system in the Cordilleran orogen would certainly have depended on the angle of oblique convergence and the relative motion of one plate relative to the other at any given time. Displacement along faults which bound translating terranes may also have opposing senses of movement at any given time.

Much of the discussion about the origin, movement, and eventual resting place of suspect terranes along the west margin of the Cordilleran orogen in Mesozoic time (Jones and others, 1983; Beck, 1986) is involved in this history, simply because of the requirement of horizontal translation for terrane movement as is seen today along active, oblique convergent plate boundaries (Sarewitz and Karig, 1986).

Tectonic Escape

Along irregular convergent or collision fronts, nodes of constriction cause discrete pieces of crustal fragments and splinters termed "scholles"[4] to be expelled sideways on strike-slip faults having opposed senses of slip toward zones of overthrusting or free faces formed by subduction zones. This process of "tectonic escape" (Burke and Şengör, 1982) is a consequence of the inability of trenches to consume continental crust (McKenzie, 1972) and is possible because of the extreme heterogeneity and low shear strength of continental rocks (Şengör and others, 1985). The boundary force is produced by the buoyancy of the continental crust,

---

[4]The term "Scholle" is a German word for a floe (of ice), a flake, or a clod (of soil), and it is an excellent descriptive word for a tectonic "flake," "slab," "block," "crustal panel," or even "terrane," where the tectonic domain in question is detached from the lithosphere and therefore cannot be termed a "plate" (Dewey and Şengör, 1979). Burke and Şengör (1982) seem to have extended scholle to include small lithospheric domains, but I believe that "miniplate" or "microplate" are good terms for small plates, and I recommend that scholle be used as Dewey and Şengör originally intended.

not its strength. "The surface expression of this buoyancy is a mountain belt, and it is therefore possible to regard the gravitational forces caused by surface elevations as the driving forces for the motion of small plates" (McKenzie, 1972, p. 181).

One of the best examples of tectonic escape is illustrated by the Anatolian scholle in Turkey (Dewey and Şengör, 1979); it is driven westward between the dextral North Anatolian fault and the sinistral East Anatolian fault in response to the northward collision of the Arabian plate into the southern part of the Eurasian plate (Fig. 31). Similarly and on a larger scale, a great wedge of south China is being expelled eastward and southeastward between enormous intracontinental transform faults, the sinistral Altyn Tagh and Kunlung faults and the dextral Red River fault, onto the Philippine and Pacific plates (Tapponier and others, 1982, 1986). There, escape tectonics has progressed from south to north, causing reversals of movement on the participating strike-slip faults, and shunting ever larger scholles of the Asian lithosphere eastward as India continues its northward penetration into southern Asia (Tapponier and others, 1986). On a smaller, intracontinental scale, the wedge-shaped Mojave Desert in southern California (Fig. 9) is a scholle being driven eastward between the sinistral Garlock fault and the dextral San Andreas fault in response to the localized upper-crustal shortening between irregular edges of the Pacific and North American plates (McKenzie, 1972; Cummings, 1976).

An interesting case of the tectonics of a block mosaic (Hill, 1982) is in southern China where north-south shortening and east-west elongation of a tectonic domain composed of a number of blocks cause horizontal slip of several blocks relative to one another, thus opening large basins between blocks that are wedged away from one another along strike-slip faults (Wu and Wang, 1988).

FOUR PROBLEMS

Much of this paper has focused on the peculiarities of the San Andreas fault, but just how representative are the kinematics, dynamics, and seismic behavior of the San Andreas fault, of strike-slip faults in general, and for boundary transform faults in particular? The answer is important for earthquake-hazard assessment and for understanding strike-slip mechanics and tectonics. The lessons learned from the boundary transform in California will be watched with interest by the tectonic community worldwide, and just as many strike-slip concepts have evolved in southern California and spread elsewhere, so also will these latest revelations.

Four fundamental problems bearing on the geophysical nature of strike-slip faults are exemplified by lessons learned from the San Andreas fault. (1) What are the implications of mechanical stratigraphy for determination of strike-slip structural styles? (2) Why does the San Andreas fault lack a heat flow anomaly (Lachenbruch and Sass, 1980)? (3) Why are measurements of present surficial stress discordant with predictions made from kinematic models (Mount and Suppe, 1987)? (4) Why do paleoseismic determinations of relative plate velocity across the fault fall short of those determined from paleomagnetic data from the sea floor (Weldon and

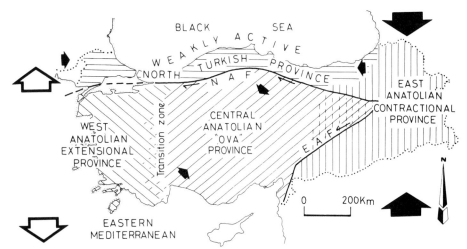

Figure 31. Westward escape of Anatolian scholle between North Anatolian (NAF and East Anatolian (EAF) faults due to collision and continued convergence of the Arabian and Eurasian plates (from Şengör and others, 1985). Reproduced with permission of Society of Economic Paleontologists and Mineralogists and of Şengör, 1988.

Humphreys, 1986; Minster and Jordan, 1984, 1987)?

Mechanical Stratigraphy

Only recently have the implications of mechanical stratigraphy become apparent in considerations of strike-slip structural styles. Although Sylvester and Smith (1976) certainly acknowledged the role of detachments in the transpressional deformation that molded the Mecca Hills, it has been the work by Yeats (1981), Terres and Sylvester (1981), Webb and Kanamori (1985), Nicholson and others (1986), Namson and Davis (1988), and others in central and southern California that has raised many fundamental questions about the widely held assumption that strike-slip faults are vertical to great depth. That dogma grew chiefly out of clay-model laboratory experiments in which the clay lacked mechanical layering. Just as the revelations of detachment have altered our concepts of extensional deformation, so also will they alter our views of strike-slip deformation.

Heat Flow

At this writing, a deep hole is being drilled near the San Andreas fault where it transects the Transverse Ranges to answer the second two questions posed above. Near-surface stress measurements (0 to 1 km) show an increase of stress with depth consistent with estimates based on laboratory frictional experiments (about 90 bars/km). Extensive conductive heat-flow measurements made near the fault, however, show no discernible effects of frictional heating, suggesting that there is little frictional heating on the fault, and that the upper limit for the average shear stress on the fault is less than 200 bars. The near-surface stress measurements therefore cannot be extrapolated to depths greater than 3 km without violating existing heat-flow constraints. Several hypotheses have been advanced to explain the lack of a heat-flow anomaly. Among them is the idea that movement along the fault is in response to very low stress, or that some unknown "cooling mechanism" is at depth, such as subsurface water flow, which reduces temperature (O'Neil and Hanks, 1980). Geologic mapping, seismic studies, and regional gravity data suggest that the shallow low-angle thrusts or detachment faults believed to underlie the eastern Transverse Ranges would disperse the heat and support a thermo-mechanical model that explains the low heat flow on the main fault trace.

Recent geophysical studies suggest that the surface trace of the San Andreas fault is offset from 50 to 200 km westward on a subhorizontal detachment from its continuation at depth down to the top of the seismogenic zone (Webb and Kanamori, 1985; Meisling and Weldon, in press). Furlong (1987, p. 1509) has outlined the thermo-mechanical controls on the three-dimensional geometry and lithospheric evolution in the San Francisco Bay region as follows: "Because of the lithospheric geometry inherited from the period of active subduction, the thermal regime evolves to produce a mechanical environment where shear deformation in the lower crust and upper mantle is concentrated in a near vertical region offset approximately 50 km east of the surface San Andreas. The offset produces a complex geometry for the plate boundary in which a sub-horizontal detachment (shear zone?) must connect the surface and deeper segments of the plate boundary." Furlong continued: "the near surface fault zone formed in this way will have an orientation controlled by the orientation of the deeper shear boundary rather than that associated with the regional stress regime. In addition, through this process slivers of the North American plate are 'acquired' by the Pacific plate, producing the complex pattern of juxtaposed terranes outboard of the San Andreas fault in central California."

The presence and structure of deeper shear boundaries and detachments will be difficult to elucidate, because the seismogenic zone is confined to the upper continental crust in regions of active strike-slip faulting (Chen and Molnar, 1983; Sibson, 1983). This means that seismicity cannot give direct information on the location and nature of aseismic shear zones and detachments that may exist in the lower crust and mantle.

The Stress Paradox

Certainly it is an oversimplification to state that the oblique resolution of stress across a strike-slip fault yields folds or thrusts perpendicular to the shortening axis, especially where the evidence at hand says that the present maximum compression, which is ostensibly parallel to the shortening axis, is normal to the San Andreas fault both at Cajon Pass (Zoback and others, 1987) and in central California (Mount and Suppe, 1987). Such stresses would have to be maintained by the plate motion; otherwise, they would be relaxed quickly by earthquakes. Mackenzie (1972, p. 176) maintained that "it is therefore simpler at present to neglect the stress field, which must be exceedingly complicated, and account for the features produced by major deformation directly in terms of plate motion."

The stresses, however, have to be accounted for eventually, and so these kinds of models must be refined and studied more extensively to resolve the paradox that observed fault-parallel well-breakouts and bore-hole elongations imply that the compression axis is normal close to the fault, whereas simple shear theory says that the regional compression axis ought to be 45° to the fault. Stress measurements at various levels in the Cajon Pass deep hole suggest provisionally that *fault-normal compression results from a generally high level of deviatoric stress in the crust but which, in the vicinity of a weak fault, must be oriented in such a manner that the level of shear stress resolved on the fault is low* (Zoback and others, 1987); that is, either the least or the greatest principal stress must be nearly perpendicular to the fault.

Plate Velocity Discrepancies

Paleomagnetic reversals provide the principal control on the velocity of the Pacific relative to the North American plate and yield a rate of 55 mm/yr (Weldon and Humphreys, 1986; Minster and Jordan, 1984, 1987), whereas historic geodetic evidence yields a rate of about 35 mm/yr (Thatcher, 1979). This discrepancy has been known for quite some time, and the practice has been to explain it by apportioning the residual on faults east of the San Andreas fault well into the Great Basin, or on faults west of the San Andreas fault in the Pacific Ocean. Another rational explanation is that plate movements are episodic on a scale of thousands of years, so that the historic geodetic movements give too short a sample of geologic time. Other writers say that the discrepant rates prove that the plate motion has decreased from 55 mm/yr to 35 mm/yr. Still others have reappraised the interpretations of the paleomagnetic data and conclude that the rate may be only about 45 mm/yr. The lower figure, then, yields a smaller discrepancy which might be explained by combinations of all of the other hypotheses. I believe that the solution will come from acquisition of more geodetic data over longer time periods more representative of geologic time, together with more geologic work focused on the timing of local tectonic events to gain a better understanding of Neogene plate movements. The question may be resolved in the next decade with the Global Positioning Satellite system providing data on the scale of the plates themselves at distances, sampling frequencies, and precision undreamed of only a decade ago.

CONCLUDING STATEMENT

The recent revelations pertaining to concepts of strike-slip faults have dramatically broadened our views about the tectonic roles of strike-slip faults in time and space. For example, the terrane concept has changed notions about how deformation in mobile belts is accomplished by bringing a third dimension to views of microplate tectonics along continental margins. Strike-slip faults are the key structures in tectonic migration along plate boundaries of terranes. Now intensive efforts should focus on reviewing the tectonic nature and history of plate boundaries to place limits on their reconstructed geometry.

Of at least equal importance to understanding strike-slip tectonics are recent interpretations of seismic reflection studies of the intermediate and deep crust which imply the presence of detachments, perhaps at several levels within the crust, and which indicate that some strike-slip faults are cut at depths by those detachments. These interpretations have shaken the widely held dogmas about all strike-slip faults being vertical to great depth, about the orientation and partitioning of stress at different crustal levels near strike-slip faults, and about the strength of strike-slip faults and their capacity to store seismic strain energy. The implications of these concepts are far-reaching and concern such diverse disciplines as mineral and hydrocarbon exploration, paleotectonics, paleogeography, and seismic hazards. These revelations provide enormously stimulating ideas to challenge those who study tectonics in general and strike-slip faults in particular, and they certainly portend exciting times of discovery and understanding in the future.

ACKNOWLEDGMENTS

I am grateful to Kris Meisling, Rick Sibson, John Crowell, Mason Hill, Tod Harding, Carlos Schubert, Bob Wallace, Stacey Zeck, Emery Goodman, Wendy Bartlett, and Celal Şengör for sharing with me some of their ideas in preprints and discussions at various times and for their critique on an earlier version of this paper. Each of them would have emphasized some things more, some things less, and they should not be held responsible for my errors of fact or omission, particularly because this version bears little resemblance to the one they read. Craig Nicholson, Bob Nason, and Tom Blenkinsop straightened out my thoughts on several topics for the final manuscript. Paul Hoffman, Bruce Luyendyk, Wendy Bartlett, and Carlos Schubert provided updated figures for inclusion in this paper.

REFERENCES CITED

Abbate, E., Bortolotti, V., and Passerini, P., 1972, Paleogeographic and tectonic considerations on the ultramafic belts in the Mediterranean area: Bollettino della Societa Geologica Italiana, v. 91, p. 239–282.

Addicott, W. O., 1968, Mid-Tertiary zoogeographic and paleogeographic discontinuities across the San Andreas fault, California, *in* Dickinson, W. R., and Grantz, A., eds., Proceedings of Conference on Geologic Problems of the San Andreas Fault System: Stanford University Publications in the Geological Sciences, v. 11, p. 144–165.

Albers, J. P., 1967, Belt of sigmoidal bending and right-lateral faulting in the western Great Basin: Geological Society of America Bulletin, v. 78, p. 143–156.

Allen, C. R., 1957, San Andreas fault zone in San Gorgonio Pass, southern California: Geological Society of America Bulletin, v. 68, p. 315–350.

——— 1962, Circum-Pacific faulting in the Philippines-Taiwan region: Journal of Geophysical Research, v. 67, p. 4795–4812.

——— 1965, Transcurrent faults in continental areas: Royal Society of London Philosophical Transactions, v. 258, p. 82–89.

——— 1968, The tectonic environments of seismically active and inactive areas along the San Andreas fault system, *in* Dickinson, W. R., and Grantz, A., eds., Proceedings of Conference on Geologic Problems of the San Andreas Fault System: Stanford University Publications in the Geological Sciences, v. 11, p. 70–82.

——— 1981, The modern San Andreas fault, *in* Ernst, W. G., ed., The geotectonic development of California: Englewood Cliffs, New Jersey, Prentice-Hall Inc., p. 511–534.

Allen, C. R., Silver, L. T., and Stehli, F. G., 1960, Agua Blanca fault—A major transverse structure of northern Baja California: Geological Society of America Bulletin, v. 71, p. 457–482.

Allen, C. R., Wyss, M., Brune, J. N., Grantz, A., and Wallace, R. E., 1972, Displacements on the Imperial, Superstition Hills and the San Andreas faults triggered by the Borrego Mountain earthquake: U.S. Geological Survey Professional Paper 787, p. 87–104.

Allen, C. R., Gillespie, A. R., Han, Y., Sieh, K. E., Zhang, B., and Zhu, C., 1984, Red River and associated faults, Yunnan Province, China: Quaternary geology, slip rates, and seismic hazard: Geological Society of America Bulletin, v. 95, p. 686–700.

Alpha, A. G., and Fanshawe, J. R., 1954, Tectonics of northern Bighorn basin area and adjacent south-central Montana: Billings Geological Society, 5th annual field conference, Pryor Mountains–Northern Bighorn Basin, Montana, Guidebook, p. 72–79.

Ambraseys, N. N., 1963, The Buyin-Zara (Iran) earthquake of September, 1962: A field report: Seismological Society of America Bulletin, v. 53, p. 705–740.

——— 1970, Some characteristic features of the Anatolian fault zone: Tectonophysics, v. 9, p. 143–165.

——— 1971, Value of historical records of earthquakes: Nature, v. 232, p. 375–380.

Anderson, E. M., 1905, The dynamics of faulting: Edinburgh Geological Society Transactions, v. 8, pt. 3, p. 387–402.

——— 1942, The dynamics of faulting and dyke formation with application to Britain: Edinburgh, Scotland, Oliver and Boyd, 206 p.

Anderson, T. H., and Schmidt, V. A., 1983, The evolution of Middle America and the Gulf of Mexico–Caribbean Sea region during Mesozoic time: Geological Society of America Bulletin, v. 94, p. 941–966.

Apol'skiy, O. P., 1974, Origin of the Black and south Caspian Sea troughs: Geotectonics, v. 4, p. 310–311.

Aspler, L. B., and Donaldson, J. A., 1985, The Nonacho basin (early Proterozoic), Northwest Territories, Canada: Sedimentation and deformation in a strike-slip setting, *in* Biddle, K. T., and Christie-Blick, N., eds., Strike-slip deformation, basin formation, and sedimentation: Society of Economic Paleontologists and Mineralogists Special Publication 37, p. 193–209.

Atwater, T., 1970, Implications of plate tectonics for the Cenozoic tectonics of western North America: Geological Society of America Bulletin, v. 81, p. 3513–3536.

Aviles, C. A., and Scholz, C. H., 1987, Fractal analysis applied to characteristic segments of the San Andreas fault: Journal of Geophysical Research, v. 92 (B1), p. 331–344.

Aydin, A., and Nur, A., 1982, Evolution of pull-apart basins and their scale independence: Tectonics, v. 1, p. 91–105.

Aydin, A., and Page, B. M., 1984, Diverse Pliocene-Quaternary tectonics in a transform environment, San Francisco Bay region, California: Geological Society of America Bulletin, v. 95, p. 1303–1317.

Aytun, A., 1980, Creep measurements in the Ismetpasa region of the North Anatolian fault zone, *in* Isakara, A. M., and Vogel, A., eds., Multidisciplinary approach to earthquake prediction 2: Braunschweig/Wiesbaden, Friedrich Vieweg & Sohn, p. 279–292.

Babcock, E. A., 1974, Geology of the northeast margin of the Salton Trough, Salton Sea, California: Geological Society of America Bulletin, v. 85, p. 321–332.

Bachman, G. H., and Koch, K., 1983, Alpine front and Molasse Basin, Bavaria, *in* Bally, A. W., ed., Seismic expression of structural styles, Volume 3: American Association of Petroleum Geologists Studies in Geology, ser. 15, p. 3.41-27-3.4.1-32.

Bagnall, P. S., 1964, Wrench faulting in Cyprus: Journal of Geology, v. 72, p. 327–345.

Bahat, D., 1983, New aspects of rhomb structures: Journal of Structural Geology, v. 5, p. 591–601.

Ballance, P. F., and Reading, H. G., eds., 1980, Sedimentation in oblique-slip mobile zones: International Association of Sedimentologists Special Publication 4, 265 p.

Bally, A. W., ed., 1983, Seismic expression of structural styles: American Association of Petroleum Geologists Studies in Geology, ser. 15, 3 volumes.

Barka, A. A., and Kadinsky-Cade, K., 1988, Strike-slip fault geometry in Turkey and its influence on earthquake activity: Tectonics, v. 7, p. 663–684.

Barrows, A. A., 1974, A review of the geology and earthquake history of the Newport-Inglewood structural zone, southern California: California Division of Mines and Geology Special Report 114, 115 p.

Bartlett, W. L., Friedman, M., and Logan, J. M., 1981, Experimental folding and faulting of rocks under confining pressure, Part IX. Wrench faults in limestone layers: Tectonophysics, v. 79, p. 255–277.

Bartov, Y., Steinitz, G., Eyal, M., and Eyal, Y., 1980, Sinistral movement along the Gulf of Aqaba—Its age and relation to opening of the Red Sea: Nature, v. 285, p. 220–221.

Bates, R. L., and Jackson, J. A., 1987, Glossary of geology (3rd edition): Alexandria, Virginia, American Geological Institute, 788 p.

Beck, M. E., Jr., 1976, Discordant paleomagnetic pole positions as evidence of regional shear in the western Cordillera of North America: American Journal of Science, v. 276, p. 694–712.

——— 1986, Model for late Mesozoic–early Tertiary tectonics of coastal California and western Mexico and speculations on the origin of the San Andreas fault: Tectonics, v. 5, p. 49–64.

Becker, H., 1934, Die Beziehungen zwischen Felsengebirge und Grossen Becken im westlichen Nordamerica: Zeitschrift der deutschen geologischen Gesellschaft, v. 86, p. 115–120.

Beckwith, R. H., 1941, Trace-slip faults: American Association of Petroleum Geologists Bulletin, v. 25, p. 2181–2193.

Belt, E. S., 1968, Carboniferous continental sedimentation, Atlantic Provinces, Canada, *in* Klein, G. deV., ed., Symposium on continental sedimentation, northeastern North America: Geological Society of America Special Paper 106, p. 127–176.

Ben-Avraham, Z., 1985, Structural framework of the Gulf of Elat (Aqaba), northern Red Sea: Journal of Geophysical Research, v. 90, p. 703–726.

Ben-Avraham, Z., Almagor, G., and Garfunkel, Z., 1979, Sediments and structure of the Gulf of Elat (Aqaba)–northern Red Sea: Sedimentary Geology, v. 23, p. 239–267.

Benson, W. N., 1952, Meeting of the Geological Division of the Pacific Science Congress in New Zealand, February 1949: International Proceedings: Geological Society of America Bulletin, v. 63, p. 11–13.

Berberian, M., 1981, Active faulting and tectonics of Iran, *in* Gupta, H. K., and Delany, F. M., eds., Zagros, Hindu Kush, Himalaya geodynamic evolution: Geodynamics Series 3: Washington, D.C., and Boulder, Colorado, American Geophysical Union and Geological Society of America, p. 33–49.

Biddle, K. T., and Christie-Blick, N., 1985, Glossary—Strike-slip deformation, basin formation, and sedimentation, *in* Biddle, K. T., and Christie-Blick, N., eds., Strike-slip deformation, basin formation, and sedimentation: Society of Economic Paleontologists and Mineralogists Special Publication 37, p. 375–385.

Biehler, S., Kovach, R. L., and Allen, C. R., 1964, Geophysical framework of northern end of the Gulf of California structural province: American Association of Petroleum Geologists Memoir 3, p. 126–143.

Bilham, R., and Williams, P., 1985, Sawtooth segmentation and deformation processes on the southern San Andreas fault, California: Geophysical Research Letters, v. 12 (9), p. 557–560.

Biq Chingchang, 1959, The structural pattern of Taiwan compared with the eta type of shear form: Geological Society of China Proceedings, v. 2, p. 33–46.

Bohannon, R. G., and Howell, D. G., 1982, Kinematic evolution of the junction of the San Andreas, Big Pine, and Garlock faults, California: Geology, v. 10, p. 358–363.

Bourgeois, J., and Dott, R. H., Jr., 1985, Stratigraphy and sedimentology of Upper Cretaceous rocks in coastal southwest Oregon: Evidence for wrench fault tectonism in a postulated accretionary terrane: Geological Society of America Bulletin, v. 96, p. 1007–1019.

Bradley, D. C., 1982, Subsidence in late Paleozoic basins in the northern Appalachians: Tectonics, v. 1, p. 107–123.

Bridwell, R. J., 1975, Sinuosity of strike-slip fault traces: Geology, v. 3, p. 630–632.

Brown, R. D., and Wallace, R. E., 1968, Current and historic fault movement along the San Andreas fault between Paicines and Camp Dix, California, *in* Dickinson, W. R., and Grantz, A., eds., Proceedings of Conference on Geologic Problems of the San Andreas Fault System: Stanford University Publications in the Geological Sciences, v. 11, p. 22–39.

Brown, R. D., Jr., Vedder, J. G., Wallace, R. E., Roth, E. F., Yerkes, R. F., Castle, R. O., Waananen, A. O., Page, R. W., and Eaton, J. P., 1967, The Parkfield-Cholame, California, earthquakes of June-August 1966: Surface geologic effects, water-resources aspects, and preliminary seismic data: U.S. Geological Survey Professional Paper 579, 66 p.

Brown, R. W., 1928, Experiments relating to the results of horizontal shearings: American Association of Petroleum Geologists Bulletin, v. 12, p. 715–720.

Brune, J. N., Henyey, T. L., and Roy, R. F., 1969, Heat flow, stress, and rate of slip along the San Andreas fault, California: Journal of Geophysical Research, v. 74, p. 3821–3827.

Bucher, W. H., 1952, Geologic structure and orogenic history of Venezuela: Geological Society of America Memoir 49.

Bucknam, R. C., Plafker, G., and Sharp, R. V., 1978, Fault movement (after-slip) following the Guatemala earthquake of February 4, 1976: Geology, v. 6, p. 170–173.

Bucknam, R. C., Algermissen, S. T., and Anderson, R. E., 1980, Patterns of late Quaternary faulting in western Utah and an application in earthquake hazard evaluation, *in* Evernden, J. F., convener, Proceedings of Conference X—Earthquake Hazards along the Wasatch/Sierra Nevada Frontal Fault Zones: U.S. Geological Survey Open-File Report 80-801, 679 p.

Bull, W. B., and Pearthree, P. A., 1988, Frequency and size of Quaternary surface ruptures of the Pitaycachi fault, northeastern Sonora, Mexico: Seismological Society of America Bulletin, v. 78 (2), p. 956–978.

Burbank, D. W., and Whistler, D. P., 1987, Temporally constrained tectonic rotations derived from magnetostratigraphic data: Implications for the initiation of the Garlock fault, California: Geology, v. 15, p. 1172–1175.

Burchfiel, B. C., and Stewart, J. H., 1966, "Pull-apart" origin of the central segment of Death Valley, California: Geological Society of America Bulletin, v. 77, p. 439–442.

Burford, R. O., 1972, Continued slip on the Coyote Creek fault after the Borrego Mountain earthquake, *in* Sharp, R. V., ed., The Borrego Mountain earthquake of April 9, 1968: U.S. Geological Survey Professional Paper 787, p. 105–111.

Burford, R. O., and Harsh, P. W., 1980, Slip on the San Andreas fault in central California from alinement array surveys: Seismological Society of America Bulletin, v. 70, p. 1233–1262.

Burkart, B., Deaton, B. C., Dengo, C., and Moreno, G., 1987, Tectonic wedges and offset Laramide structures along the Polochic fault of Guatemala and Chiapas, Mexico: Reaffirmation of large Neogene displacement: Tectonics, v. 6, p. 411–422.

Burke, D. B., 1979, Log of a trench in the Garlock fault zone, Fremont Valley, California: U.S. Geological Survey Map MF-1028, scale 1:20.

Burke, K., and Şengör, A.M.C., 1986, Tectonic escape in the evolution of the continental crust, *in* Barazangi, M., and Brown, L. D., eds., Reflection seismology: The continental crust: American Geophysical Union Geodynamics Series, Volume 14, p. 41–53.

Burtman, V. S., 1980, Faults of middle Asia: American Journal of Science, v. 280, p. 725–744.

Burtman, V. S., Peive, A. V., and Ruzhentsev, S. V., 1963, The main strike-slip faults of the Tien Shan and Pamir, *in* Faults and horizontal movements of the Earth's crust [in Russian]: Academy of Sciences of the USSR, Institute of Geology Transactions, Issue 80, p. 152–171.

Burton, C. K., 1965, Wrench faulting in Malaya: Journal of Geology, v. 73, p. 781–798.

Butler, R.W.H., 1982, The termination of structures in thrust belts: Journal of Structural Geology, v. 4, p. 239–245.

Buwalda, J. P., 1937a, Shutterridges, characteristic physiographic features of active faults [abs.]: Geological Society of America Proceedings for 1936, p. 307.

——— 1937b, Recent horizontal shearing in the coastal mountains of California [abs.]: Geological Society of America Proceedings for 1936, p. 341.

Campbell, C. J., 1968, The Santa Marta wrench fault of Colombia and its regional setting: 4th Caribbean Geological Conference (Trinidad), Transactions, p. 247–262.

Campbell, J. D., 1958, En echelon folding: Economic Geology, v. 53, p. 448–472.

Campbell, N., 1948, West Bay fault, *in* Structural geology of Canadian ore deposits: Canadian Institute of Mining and Metallurgy, Jubilee Volume, p. 244–259.

Cannat, M., 1985, Tectonics of the Seiad massif, northern Klamath Mountains, California: Geological Society of America Bulletin, v. 96, p. 15–26.

Carey, S. W., 1958, The tectonic approach to continental drift, *in* Continental drift, a symposium: Hobart, Tasmania, Geology Department, University of Tasmania.

Castro, A., 1985, The Central Extremadua Batholith: Geotectonic implications (European Hercynian belt): Tectonophysics, v. 120, p. 57–68.

Cebull, S. E., 1972, Sense of displacement along Foothills fault system: New evidence from the Melones fault zone, western Sierra Nevada, California: Geological Society of America Bulletin, v. 83, p. 1185–1190.

Champion, D. E., Howell, D. G., and Grommé, C. S., 1984, Paleomagnetic and geologic data indicating 2,500 km of northward displacement for the Salinian and related terranes: Journal of Geophysical Research, v. 89, p. 7736–7752.

Cheadle, M. J., Czuchra, B. L., Byrne, T., Ando, C. J., Oliver, J. E., Brown, L. D., Kaufman, S., Malin, P. E., and Phinney, R. A., 1986, The deep crustal structure of the Mojave Desert, California, from COCORP seismic reflection data: Tectonics, v. 5, p. 293–320.

Chen, W-P., and Molnar, P., 1983, Focal depths of intracontinental and intraplate earthquakes and their implications for the thermal and mechanical properties of the lithosphere: Journal of Geophysical Research, v. 88, p. 4183–4214.

Christie-Blick, N., and Biddle, K. T., 1985, Deformation and basin formation along strike-slip faults, *in* Biddle, K. T., and Christie-Blick, N., eds., Strike-slip deformation, basin formation, and sedimentation: Society of Economic Paleontologists and Mineralogists Special Publication 37, p. 1–34.

Clark, L. D., 1960, Foothills fault system, western Sierra Nevada, California: Geological Society of America Bulletin, v. 71, p. 483–496.

Clark, M. M., 1968, Surface rupture along the Coyote Creek fault, the Borrego Mountain earthquake of April 9, 1968: U.S. Geological Survey Professional Paper 787, p. 55–86.

——— 1984, Map showing recently active breaks of the San Andreas fault and associated faults between Salton Sea and Whitewater River, Mission Creek, California: U.S. Geological Survey Miscellaneous Geologic Investigations Map I-1483, scale 1:24,000.

Clayton, L., 1966, Tectonic depressions along the Hope fault, a transcurrent fault in North Canterbury, New Zealand: New Zealand Journal of Geology and Geophysics, v. 9, p. 95–104.

Clifton, H. E., 1968, Possible influence of the San Andreas fault on middle and probable late Miocene sedimentation, southeastern Caliente Range, *in* Dickinson, W. R., and Grantz, A., eds., Proceedings of Conference on Geologic Problems of the San Andreas Fault System: Stanford University Publications in the Geological Sciences, v. 11, p. 183–190.

Cloos, H., 1928, B. Experimente zur innern Tektonik: Zentralblatt für Mineralogie und Palaeontologie, v. 1928, p. 609–621.

——— 1933, Zur tektonischen Stellung des Saargebietes: Zeitschrift der deutschen Geologischen Gesellschaft, v. 85, p. 307–315.

Cohn, S. N., Allen, C. R., Gilman, R., and Goulty, N. R., 1982, Pre-earthquake and post-earthquake creep on the Imperial fault and the Brawley fault zone, *in* The Imperial Valley, California, earthquake of October 15, 1979: U.S. Geological Survey Professional Paper 1254, p. 161–168.

Cotton, C. A., 1950, Tectonic scarps and fault valleys: Geological Society of America Bulletin, v. 61, p. 717–758.

——— 1952, Geomorphology: New York, John Wiley and Sons, 505 p.

Costain, J. K., Coruh, C., and Hatcher, R. D., 1987, Geophysical signature of an inclined strike-slip duplex in the southeastern United States: Geological Society of America Abstracts with Programs, v. 19, p. 628.

Courtillot, V., Tapponier, P., and Varet, J., 1974, Surface features associated with transform faults—A comparison between observed examples and an experimental model: Tectonophysics, v. 24, p. 317–329.

Craddock, C., Hauser, E. C., Maher, H. D., Sun, A. Y., and Zhu, G.-Q., 1985, Tectonic evolution of the West Spitsbergen fold belt: Tectonophysics, v. 114, p. 193–211.

Crook, C. N., Mason, R. G., and Wood, P. R., 1982, Geodetic measurements of horizontal deformation on the Imperial fault, *in* The Imperial Valley, California, earthquake of October 15, 1979: U.S. Geological Survey Professional Paper 1254, p. 183–191.

Crowell, J. C., 1952, Probable large lateral displacement on the San Gabriel fault, southern California: American Association of Petroleum Geologists Bulletin, v. 36, p. 2026–2035.

——— 1959, Problems of fault nomenclature: American Association of Petroleum Geologists Bulletin, v. 43, p. 2653–2674.

——— 1960, The San Andreas fault in southern California: International Geological Congress, 21st Session, Norden, Part 17, p. 45–52.

——— 1962, Displacement along the San Andreas fault, California: Geological Society of America Special Paper 71, 61 p.

——— 1974a, Sedimentation along the San Andreas fault, California, *in* Dott, R. H., Jr., and Shaver, R. H., eds., Modern and ancient geosynclinal sedimentation: Society of Economic Paleontologists and Mineralogists Special Publication No. 19, p. 292–303.

——— 1974b, Origin of Late Cenozoic basins in southern California, *in* Dickinson, W. R., ed., Tectonics and sedimentation: Society of Economic Paleontologists and Mineralogists Special Publication No. 22, p. 190–204.

——— 1976, Implications of crustal stretching and shortening of coastal Ventura basin, California, *in* Howell, D. G., ed., Aspects of the geologic history of the California continental borderland: American Association of Petroleum Geologists, Pacific Section, Miscellaneous Publication 24, 561 p.

——— 1979, The San Andreas fault system through time: Geological Society of London Quarterly Journal, v. 136, p. 293–302.

——— 1981, Juncture of the San Andreas transform system and the Gulf of California rift: Oceanologica Acta, n. SP, p. 137–141.

——— 1982a, The tectonics of Ridge basin, southern California, *in* Crowell, J. C., and Link, M. H., eds., Geologic history of Ridge Basin, Southern California: Society of Economic Paleontologists and Mineralogists, Pacific Section, Guidebook, p. 25–41.

——— 1982b, The Violin Breccia, Ridge basin, southern California, *in* Crowell, J. C., and Link, M. H., eds., Geologic history of Ridge Basin, southern California: Society of Economic Paleontologists and Mineralogists, Pacific Section, Guidebook, p. 89–97.

——— 1987, Late Cenozoic basins of onshore southern California: complexity is the hallmark of their tectonic history, *in* Ingersoll, R. V., and Ernst, W. G., eds., Cenozoic development of coast California (Rubey Volume VI): Englewood Cliffs, New Jersey, Prentice-Hall, Inc., chap. 9, p. 207–241.

Crowell, J. C., and Link, M. H., eds., 1982, Geologic history of Ridge basin, southern California: Society of Economic Paleontologists and Mineralogists, Pacific Section, Guidebook, 304 p.

Crowell, J. C., and Sylvester, A. G., 1979, Introduction to the San Andreas–Salton Trough juncture, *in* Crowell, J. C., and Sylvester, A. G., eds., Tectonics of the juncture between the San Andreas fault system and the Salton Trough, southeastern California—A guidebook: Department of Geological Sciences, University of California, p. 1–14.

Cummings, D., 1976, Theory of plasticity applied to faulting, Mojave Desert, southern California: Geological Society of America Bulletin, v. 87, p. 720–724.

Davies, F. B., 1982, Pan-African granite intrusion in response to tectonic volume changes in a ductile shear zone from northern Saudi Arabia: Journal of Geology, v. 90, p. 467–483.

Davis, G. A., and Burchfiel, B. C., 1973, Garlock fault—An intracontinental transform structure, southern California: Geological Society of America Bulletin, v. 84, p. 1407–1422.

Davis, T. L., and Duebendorfer, E. M., 1987, Strip map of the western big bend segment of the San Andreas fault: Geological Society of America Map and Chart Series, Map 60, scale 1:31,682.

Davis, W. M., 1927, The rifts of southern California: American Journal of Science, v. 13, p. 57–72.

DeLong, S. E., Dewey, J. F., and Fox, P. J., 1979, Topographic and geologic evolution of fracture zones: Geological Society of London Journal, v. 136, p. 303–310.

Deng, Q., and Zhang, P., 1984, Research on the geometry of shear fracture zones: Journal of Geophysical Research, v. 8 (B7), p. 5699–5710.

Deng, Q., Wu, D., Zhang, P., and Chen, S., 1986, Structure and deformation character of strike-slip fault zones: Pure and Applied Geophysics, v. 124, p. 204–223.

de Swardt, A.M.J., Garrard, P., and Simpson, J. G., 1965, Major zones of transcurrent dislocation and superposition of orogenic belts in part of central Africa: Geological Society of America Bulletin, v. 76, p. 89–102.

Dewey, J. F., and Şengör, A.M.C., 1979, Aegean and surrounding regions: Complex multi-plate and continuum tectonics in a convergent zone: Geological Society of America Bulletin, v. 90, p. 84–92.

Dewey, J. F., Hempton, M. R., Kidd, W.S.F., Şaroğlu, F., and Şengör, A.M.C., 1986, Shortening of continental lithosphere: The neotectonics of eastern Anatolia—a young collision zone, *in* Coward, M. P., and Reis, A. C., eds., Collision tectonics: Geological Society of London Special Publication 19, p. 3–36.

Dibblee, T. W., Jr., 1973, Regional geologic map of San Andreas and related faults in Carrizo Plain, Temblor, Caliente, and La Panza ranges and vicinity, California: U.S. Geological Survey Map I-757, scale 1:125,000.

——— 1977a, Strike-slip tectonics of the San Andreas fault and its role in Cenozoic basin evolvement, *in* Nilsen, T. H., ed., Late Mesozoic and Cenozoic sedimentation and tectonics in California: Bakersfield, California, San Joaquin Geological Society, p. 26–38.

——— 1977b, Relations of hydrocarbon accumulations to strike-slip tectonics of the San Andreas fault system, *in* Nilsen, T. H., ed., Late Mesozoic and Cenozoic sedimentation and tectonics in California: Bakersfield, California, San Joaquin Geological Society, p. 135–143.

Dickinson, W. R., 1983, Cretaceous sinistral strike-slip along Nacimiento fault in central California: American Association of Petroleum Geologists Bulletin, v. 67, p. 624–645.

Dixon, J. S., 1982, Regional structural synthesis, Wyoming salient of western overthrust belt: American Association of Petroleum Geologists Bulletin, v. 66, p. 1560–1580.

D'Onfro, P., and Glagola, P., 1983, Wrench fault, southeast Asia, *in* Bally, A. W., ed., Seismic expression of structural styles, Volume 3: American Association of Petroleum Geologists Studies in Geology, ser. 15, p. 4.2-9–4.2-12.

Dorr, J. A., Spearing, D. R., and Steidtman, J. R., 1977, The tectonic and synorogenic depositional history of the Hoback basin and adjacent area, *in* Heisey, E. L., Lawson, D. E., Norwood, E. R., Wach, P. H., and Hale, L. A., eds., Rocky Mountain thrust belt geology and resources: 1977 Wyoming Geological Association Guidebook, p. 549–562.

Dubertret, L., 1932, Les formes structurales de la Syrie et de la Palestine; leur origine: Paris, France, Académie des Sciences, Comptes Rendus, t. 195.

Dubey, A. K., 1980, Model experiments showing simultaneous development of folds and transcurrent faults: Tectonophysics, v. 65, p. 69–84.

Dunne, L. A., and Hempton, M. R., 1984, Strike-slip basin sedimentation at Lake Hazar (eastern Taurus Mountains): International Symposium of the Geology of the Taurus Belt, MTA Ankara, Turkey, p. 229–235.

Eaton, J. E., 1932, Decline of Great Basin, southwestern United States: American Association of Petroleum Geologists Bulletin, v. 16, p. 1–49.

Ehlig, P. L., Ehlert, K. W., and Crowe, B. M., 1975, Offset of the upper Miocene Caliente and Mint Canyon Formations along the San Gabriel and San Andreas faults, *in* Crowell, J. C., ed., San Andreas fault in southern California: California Division of Mines and Geology Special Report 118, p. 83–92.

Emmons, R. C., 1969, Strike-slip rupture patterns in sand models: Tectonophysics, v. 7, p. 71–87.

Erdlac, R. J., Jr., and Anderson, T. H., 1982, The Chixoy-Polochic fault and its associated fractures in western Guatemala: Geological Society of America Bulletin, v. 93, p. 57–67.

Eyal, Y., Eyal, M., Bartov, Y., Steinitz, G., and Folkman, Y., 1986, The origin of the Bir Zreir rhomb-shaped graben, eastern Sinai: Tectonics, v. 5 (2), p. 267–277.

Fairbanks, H. W., 1907, The great earthquake rift of California, *in* Jordan, D. S., ed., The California earthquake of 1906: San Francisco, A. M. Robertson, 371 p.

Fath, E. A., 1920, The origin of faults, anticlines and buried "granite ridges" of the north part of the mid-continental oil and gas field: U.S. Geological Survey Professional Paper 128 C, p. 75–89.

Feininger, T., 1970, The Palestina fault, Colombia: Geological Society of America Bulletin, v. 81, p. 1201–1216.

Ferguson, R. N., and Willis, C. G., 1924, Dynamics of oil-field structure in southern California: American Association of Petroleum Geologists Bulletin, v. 8, p. 576–583.

Ferrill, B. A., and Thomas, W. A., 1988, Acadian dextral transpression and synorogenic sedimentary successions in the Appalachians: Geology, v. 16, p. 604–608.

Fitch, T. J., 1972, Plate convergence, transcurrent faults and internal deformation adjacent to southeast Asia and the western Pacific: Journal of Geophysical Research, v. 77, p. 4432–4460.

Fletcher, G. L., 1967, Post late Miocene displacement along the San Andreas fault zone, central California, *in* Marks, J. G., Chairman, Gabilan Range and adjacent San Andreas fault: American Association of Petroleum Geologists, Pacific Section, and Society of Economic Paleontologists and Mineralogists, Pacific Section, Guidebook, 110 p.

Florensov, N. A., and Solonenko, V. P., 1963, The Gobi-Altai earthquake: Jerusalem, Israel Program for Scientific Translations, 424 p.

Freeland, G. L., and Dietz, R. S., 1972, Plate tectonic evolution of Caribbean–Gulf of Mexico region: Nature, v. 232, p. 20–23.

Freund, R., 1970a, Rotation of strike-slip faults in Sistan, southeastern Iran: Journal of Geology, v. 78, p. 188–200.

——— 1970b, The geometry of faulting in Galilee: Israel Journal of Earth Sciences, v. 19, p. 117–140.

——— 1971, The Hope fault—A strike slip fault in New Zealand: New Zealand Geological Survey Bulletin, v. 86, p. 1–47.

——— 1974, Kinematics of transform and transcurrent faults: Tectonophysics, v. 21, p. 93–134.

Freund, R., and Garfunkel, Z., 1976, Guidebook to the Dead Sea Rift: Jerusalem, Israel, Department of Geology, Hebrew University, 27 p.

Fuis, G. S., 1982, Displacement on the Superstition Hills fault triggered by the earthquake, *in* The Imperial Valley, California, earthquake of October 15, 1979: U.S. Geological Survey Professional Paper 1254, p. 145–154.

Fuis, G. S., and Kohler, W. M., 1984, Crustal structure and tectonics of the Imperial Valley region, California, *in* Rigsby, C. S., ed., The Imperial Basin—Tectonics, sedimentation, and thermal aspects: Los Angeles, Society of Economic Paleontologists and Mineralogists, Pacific Section, p. 1–13.

Fuis, G. S., Mooney, W. D., Healey, J. H., McMechan, G. A., and Lutter, W. J., 1982, Crustal structure of the Imperial Valley region, *in* The Imperial Valley, California, earthquake of October 15, 1979: U.S. Geological Survey Professional Paper 1254, p. 25–50.

——— 1984, A seismic refraction survey of the Imperial Valley region, California: Journal of Geophysical Research, v. 89 (B2), p. 1165–1189.

Fuller, D. R., and Real, C. R., 1983, High-angle reverse faulting, a model for the 2 May 1983 Coalinga earthquake, *in* Bennett, J. H., and Sherburne, R. W., eds., The 1983 Coalinga, California earthquakes: California Division of Mines and Geology Special Publication 66, p. 177–184.

Furlong, K. P., 1987, Thermo-mechanical controls on the 3-D structure and evolution of the San Andreas fault zone: Implications for geodetic, geologic and geophysical observations [abs.]: EOS (American Geophysical Union Transactions), v. 68, p. 1507.

Gabrielse, H., 1985, Major dextral transcurrent displacements along the northern Rocky Mountain Trench and related lineaments in north-central British Columbia: Geological Society of America Bulletin, v. 96, p. 1–14.

Galehouse, J. S., 1987, Theodolite measurements of creep rates on San Francisco Bay region faults, *in* Jacobsen, M. L., and Rodriguez, T. R., compilers, National Earthquake Hazards Reduction Program, Summaries of technical reports, Volume 25: U.S. Geological Survey Open-File Report 88-16, p. 339–344.

Gamond, J.-F., 1983, Displacement features associated with fault zones: A comparison between observed examples and experimental models: Journal of Structural Geology, v. 5, p. 33–45.

Gamond, J.-F., and Odonne, F., 1984, Critères d'identification des plis induits par un décroachment profond: modélisation analogique et données de terrain: Bulletin de la Sociètè Gèologique de France, v. 7, p. 115–128.

Gapais, D., and Barbarin, B., 1986, Quartz fabric transition in a cooling syntectonic granite (Hermitage Massif, France): Tectonophysics, v. 125, p. 357–370.

Garfunkel, Z., 1974, Model for the late Cenozoic tectonic history of the Mojave Desert, California, and for its relation to adjacent regions: Geological Society of America Bulletin, v. 85, p. 1931–1944.

——— 1981, Internal structure of the Dead Sea leaky transform (rift) in relation to plate kinematics: Tectonophysics, v. 80, p. 81–108.

Gastil, R. G., and Jensky, W., 1973, Evidence for strike-slip displacement beneath the trans-Mexico volcanic belt, *in* Kovach, R. L., and Nur, A., eds., Conference on Tectonic Problems of the San Andreas Fault System, Proceedings: Stanford University Publications in the Geological Sciences, v. 13, p. 171–180.

Gates, A. E., Simpson, C., and Glover, L., III, 1986, Appalachian Carboniferous dextral strike-slip faults: An example from Brookneal, Virginia: Tectonics, v. 5, p. 119–133.

Geikie, J., 1905, Structural and field geology: Edinburgh, Scotland, Oliver and Boyd, 435 p.

Gianella, V. P., and Callaghan, E., 1934, The earthquake of December 20, 1932, at Cedar Mountain, Nevada, and its bearing on the genesis of Basin Range structure: Journal of Geology, v. 42, p. 1–22.

Gianelli, G., Passerini, P., and Squazzoni, A., 1972, Some observations on mafic and ultramafic complexes north of the Bolkardag (Taurus, Turkey): Bollettino della Societá Geologica Italiana v. 91, p. 439–488.

Gilbert, G. K., 1907, The investigation of the California earthquake of 1906, *in* Jordan, D. S., ed., The California earthquake of 1906: San Francisco, A. M. Robertson, 371 p.

Ginzburg, A., and Kashai, E., 1981, Seismic measurements in the southern Dead Sea: Tectonophysics, v. 80, p. 67–80.

Goulty, N. R., and Gilman, R., 1978, Repeated creep events on the San Andreas fault near Parkfield, California, recorded by a strainmeter array: Journal of Geophysical Research, v. 83, p. 5415–5419.

Graham, R. H., 1978, Wrench faults, arcuate fold patterns and deformation in the southern French Alps: Geological Association Proceedings, v. 89, p. 125–143.

Greenhaus, M. R., and Cox, A., 1979, Paleomagnetism of the Morro Rock–Islay Hill complex as evidence for crustal block rotation in central coastal California: Journal of Geophysical Research, v. 84, p. 2392–2400.

Griesbach, C. L., 1893, Notes on the earthquake in Baluchistan on the 20th December 1892: Records of the Geological Survey of India, v. 26, pt. 2, p. 57–61.

Grose, L. T., 1959, Structure and petrology of the northeast part of the Soda Mountains, San Bernardino County, California: Geological Society of America Bulletin, v. 70, p. 1509–1548.

Groshong, R. H., and Rodgers, D. A., 1978, Left-lateral strike-slip fault model, *in* Structural style of the Arbuckle region: Geological Society of America, South-Central Section Field Trip Guide no. 3, p. 1–7.

Guineberteau, B., Bouchez, J.-L., and Vigneresse, J.-L., 1987, The Mortagne granite pluton (France) emplaced by pull-apart along a shear zone: Structural and gravimetric arguments and regional implication: Geological Society of America Bulletin, v. 99, p. 763–770.

Guiraud, M., and Seguret, M., 1985, A releasing solitary overstep model for the Late Jurassic–Early Cretaceous (Wealdian) Soria strike-slip basin (northern Spain), *in* Biddle, K. T., and Christie-Blick, N., eds., Strike-slip deformation, basin formation, and sedimentation: Society of Economic Paleontologists and Mineralogists Special Publication 37, p. 159–175.

Hadley, D., and Kanamori, H., 1977, Seismic structure of the Transverse Ranges, California: Geological Society of America Bulletin, v. 88, p. 1469–1478.

Hafner, W., 1951, Stress distributions and faulting: Geological Society of America Bulletin, v. 62, p. 373–398.

Hagstrum, J. T., McWilliams, M. O., Howell, D. G., and Grommé, C. S., 1985, Mesozoic paleomagnetism and northward translation of the Baja California Peninsula: Geological Society of America Bulletin, v. 96, p. 1077–1090.

Hamilton, W., 1961, Origin of the Gulf of California: Geological Society of America Bulletin, v. 72, p. 1307–1318.

——— 1978, Mesozoic tectonics of the western United States, *in* Howell, D. G., and McDougall, K. A., eds., Mesozoic paleogeography of the western United States, Pacific Coast Paleogeography Symposium 2: Society of Economic Paleontologists and Mineralogists, Pacific Section, p. 33–70.

Hamilton, W., and Myers, W. B., 1966, Cenozoic tectonics of the western United States: Reviews of Geophysics, v. 4, p. 509–549.

Harding, T. P., 1973, Newport-Inglewood trend, California—An example of wrenching style of deformation: American Association of Petroleum Geologists Bulletin, v. 57 (1), p. 97–116.

——— 1974, Petroleum traps associated with wrench faults: American Association of Petroleum Geologists Bulletin, v. 58, p. 1290–1304.

——— 1976, Tectonic significance and hydrocarbon trapping consequence of sequential folding synchronous with San Andreas faulting, San Joaquin Valley, California: American Association of Petroleum Geologists Bulletin, v. 58 (7), p. 356–378.

——— 1983a, Divergent wrench fault and negative flower structure, Andaman Sea, *in* Bally, A. W., ed., Seismic expression of structural styles, Volume 3: American Association of Petroleum Geologists Studies in Geology, ser. 15, p. 4.2-1–4.2-8.

——— 1983b, Convergent wrench fault and positive flower structure, Ardmore basin, Oklahoma, *in* Bally, A. W., ed., Seismic expression of structural styles, Volume 3: American Association of Petroleum Geologists Studies in Geology, ser. 15, p. 4.2-13–4.2-16.

——— 1985, Seismic characteristics and identification of negative flower structures, positive flower structures, and positive structural inversion: American Association of Petroleum Geologists Bulletin, v. 69, p. 582–600.

——— 1988, Criteria and pitfalls in the identification of wrench faults with exploration data [abs.]: American Association of Petroleum Geologists Bulletin, v. 72, p. 193.

Harding, T. P., and Lowell, J. D., 1979, Structural styles, their plate-tectonic habitats, and hydrocarbon traps in petroleum provinces: American Association of Petroleum Geologists Bulletin, v. 63, p. 1016–1058.

Harding, T. P., and Tuminas, A. C., 1988, Interpretation of footwall (lowside) fault traps sealed by reverse faults and convergent wrench faults: American Association of Petroleum Geologists Bulletin, v. 72 (6), p. 738–757.

Harding, T. P., Vierbuchen, R. C., and Christie-Blick, N., 1985, Structural styles, plate-tectonic settings, and hydrocarbon traps of divergent (transtensional) wrench faults, *in* Biddle, K. T., and Christie-Blick, N., eds., Strike-slip deformation, basin formation, and sedimentation: Society of Economic Paleontologists and Mineralogists Special Publication 37, p. 51–77.

Harland, W. B., 1971, Tectonic transpression in Caledonian Spitsbergen: Geological Magazine, v. 108, p. 27–42.

Hauksson, E., and others, 1988, The 1987 Whittier Narrows earthquake in the Los Angeles metropolitan area, California: Science, v. 239, p. 1409–1412.

Hearn, T. M., and Clayton, R. W., 1986a, Lateral velocity variations in southern California. I. Results for the upper crust from Pg waves: Seismological Society of America Bulletin, v. 76, p. 495–509.

——— 1986b, Lateral velocity variations in southern California. II. Results for the lower crust from Pn waves: Seismological Society of America Bulletin, v. 76, p. 511–520.

Heim, A., 1919, Geologie der Schweiz, Volume I, Leipzig, Germany, Tauchnitz.

Hempton, M. R., 1983, Evolution of thought concerning sedimentation in pull-apart basins, *in* Boardman, S. J., ed., Revolution in the earth sciences: Dubuque, Iowa, Kendal/Hunt, p. 167–180.

Hempton, M. R., and Dunne, L. A., 1984, Sedimentation of pull-apart basins: Active examples in eastern Turkey: Journal of Geology, v. 92, p. 513–530.

Hempton, M. R., and Neher, K., 1986, Experimental fracture, strain and subsidence patterns over en échelon strike-slip faults: Implications for the structural evolution of pull-apart basins: Journal of Structural Geology, v. 8, p. 597–605.

Hempton, M. R., Dunne, L. A., and Dewey, J. F., 1983, Sedimentation in an active strike-slip basin, southeastern Turkey: Journal of Geology, v. 91, p. 401–412.

Hess, H. H., 1938, Gravity anomalies and island arc structure with particular reference to the West Indies: American Philosophical Society Proceedings, v. 79, p. 71–96.

Higgins, C. G., 1961, San Andreas fault north of San Francisco, California: Geological Society of America Bulletin, v. 72, p. 51–68.

Hill, D. P., 1982, Contemporary block tectonics, California and Nevada: Journal of Geophysical Research, v. 87, B7, p. 5433–5450.

Hill, M. L., 1959, Dual classification of faults: American Association of Petroleum Geologists Bulletin, v. 43, p. 217–237.

——— 1971, Newport-Inglewood zone and Mesozoic subduction, California: Geological Society of America Bulletin, v. 82, p. 2957–2962.

——— 1981, San Andreas fault: History of concepts: Geological Society of America Bulletin, Part I, v. 92, p. 112–131.

Hill, M. L., and Dibblee, T. W., Jr., 1953, San Andreas, Garlock, and Big Pine faults, California—A study of the character, history, and tectonic significance of their displacements: Geological Society of America Bulletin, v. 64, p. 443–458.

Hill, M. L., and Troxel, B. W., 1966, Tectonics of Death Valley region, California: Geological Society of America Bulletin, v. 77, p. 435–438.

Hla Maung, 1987, Transcurrent movements in the Burma-Andaman Sea region: Geology, v. 15, p. 911–912.

Hoeppener, R., Kalthoff, E., and Schrader, P., 1969, Zur physikalischen Tektonik: Bruchbildung bei verschiedenen Deformationen im Experiment: Geologische Rundschau, v. 59, p. 179–193.

Hoffman, P. F., and St-Onge, M. R., 1981, Contemporaneous thrusting and conjugate transcurrent faulting during the second collision in Wopmay Orogen: Current Research, Part A, Geological Survey of Canada, Paper 81-1A, p. 251–257.

Hoffman, P. F., Tirrul, R., Grotzinger, J. P., Lucas, S. B., and Eriksson, K. A., 1984, The externides of Wopmay Orogen, Takijuq Lake and Kikerk Lake map areas, District of Mackenzie: Geological Survey of Canada, Current Research, Part A, p. 383–395.

Hornafius, J. S., Luyendyk, B. P., Terres, R. A., and Kamerling, M. J., 1986, Timing and extent of Neogene tectonic rotation in the western Transverse Ranges, California: Geological Society of America Bulletin, v. 97, p. 1476–1487.

Horne, J., and Hinxman, L. W., 1914, *in* The geology of the country round Beauly and Inverness: Scotland Geological Survey Memoir, Sheet 83.

Howard, J. L., 1987, Paleoenvironments, provenance, and tectonic implications of the Sespe Formation, southern California [Ph.D. dissert.]: Santa Barbara, University of California, 306 p.

Howell, D. G., Crouch, J. K., Greene, H. G., McCulloch, D. S., and Vedder, J. G., 1980, Basin development along the late Mesozoic and Cainozoic California margin: A plate tectonic margin of subduction, oblique subduction and transform tectonics, *in* Ballance, P. T., and Reading, H. G., eds., Sedimentation in oblique-slip mobile zones: International Association of Sedimentologists Special Publication 4, p. 43–62.

Hubbert, M. K., 1928, Direction of stresses producing given geologic strains: Journal of Geology, v. 36, p. 75–84.

Hutton, D.H.W., 1982, A tectonic model for the emplacement of the Main Donegal Granite, NW Ireland: Geological Society of London Journal, v. 139, p. 615–631.

Irwin, W. P., and Barnes, I., 1975, Effect of geologic structure and metamorphic fluids on seismic behavior of the San Andreas fault system in central and northern California: Geology, v. 3 (12), p. 713–716.

Jacoby, G. C., Shepard, P. R., and Sieh, K. E., 1988, Irregular recurrence of large earthquakes along the San Andreas fault: Evidence from trees: Science, v. 241, p. 196–199.

James, E. W., 1986, Pre-Tertiary paleogeography along the northern San Andreas fault [abs.]: EOS (American Geophysical Union Transactions), v. 67, p. 1215.

Jarrard, R. D., 1986, Relations among subduction parameters: Reviews of Geophysics, v. 24, p. 217–284.

Jones, D. L., Blake, M. C., Jr., and Rangin, C., 1976, The four Jurassic belts of northern California and their significance to the geology of the southern California borderland, *in* Howell, D. G., ed., Aspects of the geologic history of the California continental borderland: American Association of Petroleum Geologists Miscellaneous Publication 24, p. 343–376.

Jones, D. L., Howell, D. G., Coney, P. J., and Monger, J., 1983, Recognition, character, and analysis of tectonostratigraphic terranes in western North America, *in* Hashimoto, M., and Uyeda, S., eds., Accretion tectonics in the circum-Pacific region: Tokyo, Terra Scientific, p. 21–35.

Jones, P. B., 1982, Oil and gas beneath east-dipping underthrust faults in the Alberta foothills, Canada, *in* Power, R. G., ed., Geologic studies of the Cordilleran thrust belt: Rocky Mountain Association of Geologists, p. 61–74.

Julian, B. R., Zirbes, M., and Needham, R., 1982, The focal mechanism from the Global Digital Seismograph Network, *in* The Imperial Valley, California, earthquake of October 15, 1979: U.S. Geological Survey Professional Paper 1254, p. 77–81.

Kamerling, M. J., and Luyendyk, B. P., 1979, Tectonic rotations of the Santa Monica Mountains region, western Transverse Ranges, California, suggested by paleomagnetic vectors: Geological Society of America Bulletin, Part I, v. 90, p. 331–337.

——— 1985, Paleomagnetism and Neogene tectonics of the northern Channel Islands, California: Journal of Geophysical Research, v. 90, B14, p. 12,485–12,502.

Kanenko, S., 1966, Transcurrent displacement along the Median Line, southwestern Japan: New Zealand Journal of Geology and Geophysics, v. 9, p. 45–49.

Karig, D. E., 1980, Material transport within accretionary prisms and the "knocker" problem: Journal of Geology, v. 88, p. 27–39.

Karig, D. E., Cardwell, R. K., Moore, G. F., and Moore, D. G., 1978, Late Cenozoic subduction and continental margin truncation along the northern Middle America trench: Geological Society of America Bulletin, v. 89, p. 265–276.

Karner, G. D., and Dewey, J. F., 1986, Rifting: Lithospheric versus crustal extension as applied to the Ridge Basin of southern California, *in* Halbouty, M.T., ed., Future petroleum provinces of the world: American Association of Petroleum Geologists Memoir 40, p. 317–337.

Karson, J., and Dewey, J. F., 1978, Coastal complex, western Newfoundland: An Early Ordovician oceanic fracture zone: Geological Society of America Bulletin, v. 89, pp. 1037–1049.

Katz, H. R., 1962, Fracture patterns and structural history of the sub-Andean belt of southernmost Chile: Journal of Geology, v. 70, p. 595–603.

Keller, E. A., Bonkowski, M. S., Korsch, R. J., and Shlemon, R. J., 1982, Tectonic geomorphology of the San Andreas fault zone in the southern Indio Hills, Coachella Valley, California: Geological Society of America Bulletin, v. 93, p. 46–56.

Kellogg, H. E., 1975, Tertiary stratigraphy and tectonism in Svalbard and continental drift: American Association of Petroleum Geologists Bulletin, v. 59, p. 465–485.

Kennedy, W. Q., 1946, The Great Glen fault: Geological Society of London Quarterly Journal, v. 102, p. 47–76.

Ketin, I., 1948, Über die tektonisch-mechanischen Folgerungen aus den grossen anatolischen Erdbeben des letzten Dezenniumns: Geologische Rundschau, v. 36, p. 77–83.

Ketin, I., and Roesli, F., 1953, Makroseismische Untersuchungen über das nordwestanatolische Beben vom 18. März 1953: Eclogae Geologicae Helvetiae, v. 346 (2), p. 187–208.

King, C.-Y., 1987, Migration of historical earthquakes in California [abs]: EOS (American Geophysical Union Transactions), v. 68, p. 1369.

Kingma, J. T., 1958, Possible origin of piercement structures, local unconformities, and secondary basins in the Eastern Geosyncline, New Zealand: New Zealand Journal of Geology and Geophysics, v. 1, p. 269–274.

Kissel, C., Laj, C., Şengör, A.M.C., and Poisson, A., 1987, Paleomagnetic evidence for rotation in opposite senses of adjacent blocks in northeastern Aegea and western Anatolia: Geophysical Research Letters, v. 14, p. 907–910.

Koral, H., 1983, Folding of strata within shear zones: Inferences from the azimuths of en echelon folds along the San Andreas fault [M.S. thesis]: Troy, New York, Rensselaer Polytechnic Institute, 100 p.

Kotó, B., 1893, On the cause of the great earthquake in central Japan, 1891: Journal of the College of Science, Imperial University of Japan, v. 5, pt. 4, p. 296–353.

Kuchay, V. K., and Trifonov, V. G., 1977, A young left-lateral displacement in the Darvaz-Karakul fault zone: Geotectonics, v. 11 (3), p. 218–226.

Lachenbruch, A. H., and Sass, J., 1980, Heat flow and energetics of the San Andreas fault zone: Journal of Geophysical Research,v. 85 (B11), p. 6185–6222.

Langbein, J. O., 1981, An interpretation of episodic slip on the Calaveras fault near Hollister, California: Journal of Geophysical Research, v. 86, p. 4941–4948.

Laubscher, H. P., 1958, Critical examination of the Moody and Hill principle of wrench fault tectonics: Boletino Informativo, Association Venezolana de Geologia, Mineria y Petroleo, v. 1, p. 14–26.

——— 1971, The large scale kinematics of the Western Alps and the northern Apennines and its palinspastic implications: American Journal of Science, v. 271, p. 193–226.

Lavecchia, G., and Pialli, G., 1980, Appunti per uno schema strutturale dell'-Appennino Umbro-Marchigiano, 2) La copertura: Studi Geologici Camerti VI, p. 23–30.

——— 1981, Appunti per uno schema strutturale dell'Appennino Umbro-

Marchigiano, 1) Il basamento: Geologica Roma, v. 20, p. 183–195.
Lawrence, R. D., 1976, Strike-slip faulting terminates the Basin and Range province in Oregon: Geological Society of America Bulletin, v. 87, p. 846–850.
Lawson, A. C., 1921, The mobility of the Coast Ranges in California: University of California Publications in Geology, v. 12 (7), p. 431–473.
Lawson, A. C., and others, 1908, The California earthquake of April 18, 1906; Report of the State Earthquake Investigation Commission: Carnegie Institution of Washington Publication 87, v. 1.
Lee, J. S., 1929, Some characteristic structural types in east Asia: Geological Magazine, v. 66, p. 413–431.
Lemiszki, P. J., and Brown, L. D., 1988, Variable crustal structure of strike-slip fault zones as observed on deep seismic reflection profiles: Geological Society of America Bulletin, v. 100, p. 665–676.
Leonard, R., 1983, Geology and hydrocarbon accumulations, Columbus basin, offshore Trinidad: American Association of Petroleum Geologists Bulletin, v. 67, p. 1081–1093.
Lewis, S. D., Ladd, J. W., and Bruns, T. R., 1988, Structural development of an accretionary prism by thrust and strike-slip faulting: Shumagin region, Aleutian Trench: Geological Society of America Bulletin, v. 100, p. 767–782.
Lloyd, A. J., 1964, Cover folding in the Sonmartel Chain (Jura Neuchatelois): Geologische Rundschau, Bd. 53, p. 551–580.
Locke, A., Billingsley, P., and Mayo, E. B., 1940, Sierra Nevada tectonic pattern: Geological Society of America Bulletin, v. 51, p. 513–540.
Loomis, D. P., and Burbank, D. W., 1988, The stratigraphic evolution of the El Paso basin, southern California: Implications for the Miocene development of the Garlock fault and uplift of the Sierra Nevada: Geological Society of America Bulletin, v. 100, p. 12–28.
Lotze, F., 1936, Zur Methodik der Forschungen über saxonische Tektonik: Geotektonische Forschungen, v. 1, p. 6–27.
Louie, J. N., Allen, C. R., Johnson, D. C., Haase, P. C., and Cohn, S. N., 1985, Fault slip in southern California: Seismological Society of America Bulletin, v. 75, p. 811–834.
Lowell, J. D., 1972, Spitsbergen Tertiary orogenic belt and the Spitsbergen fracture zone: Geological Society of America Bulletin, v. 83, p. 3091–3102.
——— 1985, Structural styles in petroleum geology: Tulsa, Oklahoma, Oil and Gas Consultants Inc., 477 p.
Lubetkin, L.K.C., and Clark, M. M., 1988, Late Quaternary activity along the Lone Pine fault, eastern California: Geological Society of America Bulletin, v. 100, p. 755–766.
Luyendyk, B. P., Kamerling, M. J., and Terres, R. A., 1980, Geometric model for Neogene crustal rotations in southern California: Geological Society of America Bulletin, v. 91, p. 211–217.
Luyendyk, B. P., Kamerling, M. J., Terres, R. R., and Hornafius, J. S., 1985, Simple shear of southern California during Neogene time suggested by paleomagnetic declinations: Journal of Geophysical Research, v. 90 (B14), p. 12,454–12,466.
Ma, Z., and Deng, Q., 1965, On the determination of mechanical properties and formation stages of the joints and its grouping [in Chinese], *in* Zhang, W., ed., Problems on geostructures: Beijing, China, Science Publishing House.
Macdonald, K. C., Castillo, D. A., Miller, S. P., Fox, P. J., Kastens, K. A., and Bonatti, E., 1986, Deep-tow studies of the Vema fracture zone 1. Tectonics of a major slow slipping transform fault and its intersection with the mid-Atlantic ridge: Journal of Geophysical Research, v. 91, p. 3334–3354.
Malde, H. E., 1987, Quaternary faulting near Arco and Howe, Idaho: Seismological Society of America Bulletin, v. 77, p. 847–867.
Mandl, G., De Jong, L.N.J., and Maltha, A., 1977, Shear zones in granular material: Rock Mechanics, v. 9, p. 95–144.
Mann, P., Hempton, M. R., Bradley, D. C., and Burke, K., 1983, Development of pull-apart basins: Journal of Geology, v. 91, p. 529–554.
Manspeizer, W., 1985, The Dead Sea rift: Impact of climate and tectonism on Pleistocene and Holocene sedimentation, *in* Biddle, K. T., and Christie-Blick, N., eds., Strike-slip deformation, basin formation, and sedimentation: Society of Economic Paleontologists and Mineralogists Special Publication 37, p. 143–158.
Matthews, V., 1976, Correlation of Pinnacles and Neenach volcanic formations and their bearing on San Andreas fault problem: American Association of Petroleum Geologists Bulletin, v. 60, p. 2128–2141.
Maxwell, J. C., and Wise, D. U., 1958, Wrench-fault tectonics: A discussion: Geological Society of America Bulletin, v. 69, p. 927–928.
McKay, A., 1890, On the earthquakes of September 1888 in the Amuri and Marlborough Districts of the South Island: New Zealand Geological Survey Report of Geological Explorations 1888–1889, v. 20, p. 1–16.
——— 1892, On the geology of Marlborough and southeast Nelson: New Zealand Geological Survey Report of Geological Explorations 1890–1891, v. 21, p. 1–28.
McKenzie, D. P., 1972, Active tectonics of the Mediterranean region: Royal Astronomical Society Geophysical Journal, v. 30, p. 109–185.
McKinstry, H. E., 1953, Shears of the second order: American Journal of Science, v. 251, p. 401–414.
Mead, W. J., 1920, Notes on the mechanics of geologic structures: Journal of Geology, v. 28, p. 505–523.
Meghraoui, M., Philip, H., Albarede, F., and Cisternas, A., 1988, Trench investigations through the trace of the 1980 El Asnam thrust fault: Evidence for paleoseismicity: Seismological Society of America Bulletin, v. 78 (2), p. 979–999.
Meisling, K. E., and Weldon, R. J., 1988, The late Cenozoic tectonics of the northwestern San Bernardino Mountains, southern California: Geological Society of America Bulletin (in press).
Michard-Vitrac, A., Albarede, F., Dupuis, C., and Taylor, H. P., 1980, The genesis of Variscan plutonic rocks—Inferences from Sr, Pb, and U studies on the Maladeta Igneous Complex, central Pyrenees, Spain: Contributions to Mineralogy and Petrology, v. 72, p. 57–72.
Miller, R. B., 1985, The ophiolitic Ingalls complex, north-central Cascade Mountains, Washington: Geological Society of America Bulletin, v. 96, p. 27–42.
Minster, J. B., and Jordan, T. H., 1984, Vector constraints on Quaternary deformation of the western United States east and west of the San Andreas fault, *in* Crouch, J. K., and Bachman, S. B., eds., Tectonics and sedimentation along the California margin: Society of Economic Paleontologists and Mineralogists, Pacific Section, Volume 38, p. 1–16.
——— 1987, Vector constraints on western U.S. deformation from space geodesy, neotectonics, and plate motion: Journal of Geophysical Research, v. 92 (B6), p. 4798–4804.
Mitchell, A.H.G., and Reading, H. G., 1978, Sedimentation and tectonics, *in* Reading, H. G., ed., Sedimentary environments and facies: Oxford, England, Blackwell Scientific Publications, p. 439–476.
Moody, J. D., 1973, Petroleum exploration aspects of wrench-fault tectonics: American Association of Petroleum Geologists Bulletin, v. 57, p. 449–476.
Moody, J. D., and Hill, M. J., 1956, Wrench-fault tectonics: Geological Society of America Bulletin, v. 67, p. 1207–1246.
Moore, J. McM., 1979, Tectonics of the Najd transcurrent fault system, Saudi Arabia: Geological Society of London Journal, v. 136, p. 441–454.
Morgenstern, N. R., and Tchalenko, J. S., 1967, Microscopic structures in kaolin subjected to direct shear: Geotechnique, v. 17, p. 309–328.
Mount, V. S., and Suppe, J., 1987, State of stress near the San Andreas fault: Implications for wrench tectonics, Geology, v. 15, p. 1143–1146.
Muffler, L.P.J., and White, P. E., 1969, Active metamorphism of upper Cenozoic sediments in the Salton Sea geothermal field and the Salton Trough, southeastern California: Geological Society of America Bulletin, v. 80, p. 157–182.
Nábělek, J., Chen, W-P., and Ye, H., 1987, The Tangshan earthquake sequence and its implications for the evolution of the North China Basin: Journal of Geophysical Research, v. 92, p. 12,615–12,628.
Namson, J. S., and Davis, T. L., 1988, Seismically active fold and thrust belt in the San Joaquin Valley, central California: Geological Society of America Bulletin, v. 100, p. 257–273.
Nason, R. D., 1973, Fault creep and earthquakes on the San Andreas fault, *in* Kovach, R. L., and Nur, A., eds., Conference on Tectonic Problems of the San Andreas Fault System, Proceedings: Stanford University Publications in the Geological Sciences, v. 13, p. 275–285.
——— 1977, Observations of premonitory creep before earthquakes on the San Andreas fault, *in* Evernden, J. F., ed., Proceedings of Conference II—Experimental studies of rock friction with application to earthquake prediction: Menlo Park, California, U.S. Geological Survey, 701 p.
Naylor, M. A., Mandl, G., and Sijpesteijn, C.H.K., 1986, Fault geometries in basement-induced wrench faulting under different initial stress states: Journal of Structural Geology, v. 8, p. 737–752.
Nelson, M. R., and Jones, C. H., 1987, Paleomagnetism and crustal rotations along a shear zone, Las Vegas Range, southern Nevada: Tectonics, v. 6, p. 13–33.
Nicholson, C., Seeber, L., Williams, P., and Sykes, L. R., 1986, Seismic evidence for conjugate slip and block rotation within the San Andreas fault system, southern California: Tectonics, v. 5, p. 629–648.
Nielsen, R. L., 1965, Right lateral strike-slip faulting in the Walker Lane, west-central Nevada: Geological Society of America Bulletin, v. 76, p. 1301–1307.
Nilsen, T. H., and McLaughlin, R. J., 1985, Comparison of tectonic framework and depositional patterns of the Hornelen strike-slip basin of Norway and the Ridge and Little Sulphur Creek strike-slip basins of California, *in* Biddle, K. T., and Christie-Blick, N., eds., Strike-slip deformation, basin formation, and sedimentation: Society of Economic Paleontologists and Mineralogists Special Publication 37, p. 79–103.
Noble, L. F., 1926, The San Andreas rift and some other active faults in the desert region of southeastern California: Carnegie Institution of Washington Year Book, v. 25, p. 415–428.
——— 1927, San Andreas rift and some other active faults in the desert region of southern California: Seismological Society of America Bulletin, v. 17, p. 25–39.
——— 1932, The San Andreas rift in the desert region of southeastern California: Carnegie Institution of Washington Year Book, v. 31, p. 355–372.
——— 1954, The San Andreas fault zone from Soledad Pass to Cajon Pass, California; Chapter IV—Structural features, *in* Jahns, R. H., ed., Geology of southern California: California Division of Mines Bulletin 170, p. 37–48.
Odonne, F., and Vialon, P., 1983, Analogue models of folds above a wrench fault: Tectonophysics, v. 99, p. 31–46.
Okada, A., 1983, Trenching excavation at the Atotsugawa fault (central Japan): Report of the Coordinating Committee for Earthquake Prediction 30, p. 376–381.
Okubo, P. G., and Aki, K., 1987, Fractal geometry of San Andreas fault system: Journal of Geophysical Research, v. 92 (B1), p. 345–355.
O'Neil, J. R., and Hanks, T. C., 1980, Geochemical evidence for water-rock interaction along the San Andreas and Garlock faults of California: Journal of Geophysical Research, v. 85, p. 6286–6292.
Oxburgh, E. R., 1972, Flake tectonics and continental collision: Nature, v. 239, p. 202–204.
Page, B.G.N., Bennett, J. D., Cameron, N. R., Bridge, D. McM., Jeffery, D. H., Keats, W., and Thaib, J., 1979, A review of the main structural and magmatic features of northern Sumatra: Geological Society of London Journal, v. 136, p. 569–579.
Page, B. M., and Engebretson, D. C., 1984, Correlation between the geologic record and computed plate motions for central California: Tectonics, v. 3, p. 133–155.
Pakiser, L. C., 1960, Transcurrent fault and volcanism in Owens Valley, California: Geological Society of America Bulletin, v. 71, p. 153–160.
Patterson, R. H., 1979, Tectonic geomorphology and neotectonics of the Santa Cruz Island fault, Santa Barbara County, California [M.A. thesis]: Santa Barbara, California, University of California, 141 p.
Pavoni, N., 1961a, Die nordanatolische Horizontalverschiebung: Geologische Rundschau, v. 51, p. 122–139.
——— 1961b, Faltung durch Horizontalverschiebung: Eclogae Geologiae Helvetiae, v. 54, p. 515–534.
Pearthree, P. A., and Calvo, S. S., 1987, The Santa Rita fault zone: Evidence for large magnitude earthquakes with very long recurrence intervals, Basin and Range province of southeastern Arizona: Seismological Society of America Bulletin, v. 77 (1), p. 97–116.
Peltzer, G., Tapponier, P., Gaudemer, Y., Meyer, B., Guo Shumin, Yin Kelun, Chen Zhitai, and Dai Huagang, 1988, Offsets of late Quaternary morphology, rate of slip, and recurrence of large earthquakes on the Chang Ma fault (Gansu, China): Journal of Geophysical Research, v. 93, p. 7793–7812.
Perry, E. L., 1935, Flaws and tear faults: American Journal of Science, v. 29, p. 112–124.
Philip, H., and Megard, F., 1977, Structural analysis of the superficial deformation of the 1969 Pariahuanca earthquakes (central Peru): Tectonophysics, v. 38, p. 259–278.
Pitcher, W. S., 1979, The nature, ascent and emplacement of granitic magmas: Geological Society of London Journal, v. 136, p. 627–662.
Pitcher, W. S., and Bussell, M. A., 1977, Structural control of batholithic emplacement in Peru: A review: Geological Society of London Journal, v. 133, p. 239–246.
Platt, J. P., Leggett, J. K., and Alam, S., 1988, Slip vectors and fault mechanisms in the Makram accretionary wedge, southwest Pakistan: Journal of Geophysical Research, v. 93, p. 7716–7728.
Prescott, W. H., and Lisowski, M., 1983, Strain accumulation along the San Andreas fault system east of San Francisco Bay, California: Tectonophysics, v. 97, p. 41–56.
Prescott, W. H., and Yu, S-B., 1986, Geodetic measurement of horizontal deformation in the northern San Francisco Bay region, California: Journal of Geophysical Research, v. 91, B7, p. 7475–7484.
Prinzhofer, A., and Nicolas, A., 1980, The Bogata Peninsula, New Caledonia: A possible oceanic transform fault: Journal of Geology, v. 88, p. 387–398.
Prucha, J. J., 1964, Moody and Hill system of wrench fault tectonics: Discussion: American Association of Petroleum Geologists Bulletin, v. 48, p. 106–111.
Quennell, A. M., 1958, The structural and geomorphic evolution of the Dead Sea rift: Geological Society of London Quarterly Journal, v. 114, p. 1–24.
——— 1959, Tectonics of the Dead Sea rift, *in* International Geological Congress, 20th, Mexico City: Association of the Geological Service of Africa, p. 385–405.
Ramirez, V. R., 1983, Hungry Valley Formation: Evidence for 220 kilometers of post-Miocene offset on the San Andreas fault, *in* Andersen, D. W., and Rymer, M. J., eds., Tectonics and sedimentation along faults of the San Andreas system: Society of Economic Paleontologists and Mineralogists, Pacific Section, p. 33–44.
Ramsay, J. G., 1967, Folding and fracturing of rocks: New York, McGraw-Hill, 568 p.
——— 1979, Shear zones, *in* Evernden, J. F., convener, Proceedings of Conference VIII—Analysis of actual fault zones in bedrock: U.S. Geological Survey Open-File Report 79-1239, p. 1–35.
——— 1980, Shear zone geometry: A review: Journal of Structural Geology, v. 2, p. 83–89.
Ramsay, J. G., and Huber, M. I., 1987, The techniques of modern structural geology: Orlando, Florida, Academic Press, 700 p.
Rand, W. W., 1931, Geology of Santa Cruz Island: California Division of Mines Special Report 27, p. 214–219.
Ransome, F. L., 1906, The probable cause of the San Francisco earthquake: National Geographic Magazine, v. 17, p. 280–296.
Ratschbacher, L., 1986, Kinematics of Austro-Alpine cover nappes: Changing translation path due to transpression: Tectonophysics, v. 125, p. 335–356.
Read, H. H., 1923, *in* The geology of Corrour and the Moor of Rannoch: Scotland Geological Survey Memoir, p. 62–64.
Reed, J. C., Jr., and Bryant, B., 1964, Evidence for strike-slip faulting along the Brevard zone in North Carolina: Geological Society of America Bulletin, v. 75, p. 1177–1196.
Reid, H. F., 1910, Permanent displacements of the ground, *in* The California earthquake of April 18, 1906: Report of the State Earthquake Investigation Commission: Washington, D.C., Carnegie Institution of Washington, Volume 2, p. 16–28.
Reid, H. F., Davis, W. M., Lawson, A. C., and Ransome, F. L., 1913, Report of the Committee on the nomenclature of faults: Geological Society of America Bulletin, v. 24, p. 163–186.
Riedel, W., 1929, Zur Mechanik geologischer Brucherscheinungen: Zentralblatt für Mineralogie, Geologie und Palaeontologie, Abhandlung B, p. 354–368.
Robinson, P. T., Elders, W. A., and Muffler, L.P.J., 1976, Quaternary volcanism in the Salton Sea geothermal field, Imperial Valley, California: Geological Society of America Bulletin, v. 87, p. 347–360.
Rod, E., 1956, Strike-slip faults of northern Venezuela: American Association of Petroleum Geologists Bulletin, v. 40, p. 457–476.
——— 1958, Application of principles of wrench-fault tectonics of Moody and Hill to northern South America: Geological Society of America Bulletin, v. 69, p. 933–936.

——— 1962, Fault pattern, northwest corner of Sahara shield: American Association of Petroleum Geologists Bulletin, v. 46 (4), p. 529–534.
Roddick, J. A., 1967, Tintina Trench: Journal of Geology, v. 75, p. 23–33.
Rodgers, D. A., 1980, Analysis of basin development produced by en echelon strike slip faults, *in* Ballance, P. F., and Reading, H. G., eds., Sedimentation in oblique-slip mobile zones: International Association of Sedimentologists Special Publication 4, p. 27–41.
Rogers, T. H., and Nason, R. D., 1971, Active fault displacement on the Calaveras fault zone at Hollister, California: Seismological Society of America Bulletin, v. 61, p. 399–416.
Ron, H., and Eyal, Y., 1985, Interplate deformation by block rotation and mesostructures along the Dead Sea transform, northern Israel: Tectonics, v. 4 (1), p. 85–105.
Ron, H., Aydin, A., and Nur, A., 1986, Strike-slip faulting and block rotation in the Lake Mead fault system: Geology, v. 14, p. 1020–1023.
Rosenkrantz, E., Ross, M. I., and Sclater, J. G., 1988, Age and spreading history of the Cayman trough as determined from depth, heat flow, and magnetic anomalies: Journal of Geophysical Research, v. 93 (B3), p. 2141–2157.
Ross, D. C., 1970, Quartz gabbro and anorthositic gabbro: Markers of offset along the San Andreas fault in the California Coast Ranges: Geological Society of America Bulletin, v. 81, p. 3647–3662.
Ross, D. C., Wentworth, C. M., and McKee, E. H., 1973, Cretaceous mafic conglomerate near Gualala offset 350 miles by San Andreas fault from oceanic crustal source near Eagle Rest Peak, California: U.S. Geological Survey Journal of Research, v. 1, p. 45–52.
Rotstein, Y., 1984, Counterclockwise rotation of the Anatolian block: Tectonophysics, v. 108, p. 71–91.
Royden, L. H., 1985, The Vienna basin: A thin-skinned pull-apart basin, *in* Biddle, K. T., and Christie-Blick, N., eds., Strike-slip deformation, basin formation, and sedimentation: Society of Economic Paleontologists and Mineralogists Special Publication 37, p. 319–338.
Royden, L. H., Horváth, F., and Burchfiel, B. C., 1982, Transform faulting, extension, and subduction in the Carpathian Pannonian region: Geological Society of America Bulletin, v. 93, p. 717–725.
Russell, R. J., 1926, Recent horizontal offsets on the Haywards fault: Journal of Geology, v. 34, p. 507–511.
Sadler, P. M., and Demirer, A., 1986, Pelona schist clasts in the Cenozoic of the San Bernardino Mountains, southern California, *in* Ehlig, P. L., compiler, Neotectonics and faulting in southern California: Geological Society of America, Cordilleran Section Meeting, Los Angeles, Guidebook, p. 129–146.
Saleeby, J. B., 1977, Fracture zone tectonics, continental margin fragmentation, and emplacement of the Kings-Kaweah ophiolite belt, *in* Coleman, R. G., and Irwin, W. P., eds., North American ophiolites: Oregon Department of Geology and Mineral Industries Bulletin, v. 95, p. 141–160.
——— 1978, Kings River ophiolite, southwest Sierra Nevada foothills, California: Geological Society of America Bulletin, v. 89, p. 617–636.
Saleeby, J. B., and Busby-Spera, C., 1986, Fieldtrip guide to the metamorphic framework rocks of the Lake Isabella area, southern Sierra Nevada, California, *in* Dunne, G. C., ed., Mesozoic and Cenozoic structural evolution of selected areas, east-central California: Geological Society of America, Cordilleran Section Meeting, Guidebook, p. 81–94.
Sanderson, D. J., and Marchini, W.R.D., 1984, Transpression: Journal of Structural Geology, v. 6, p. 449–458.
Sanford, A. R., 1959, Analytical and experimental study of simple geologic structures: Geological Society of America Bulletin, v. 70, p. 19–62.
Sarewitz, D. R., and Karig, D. E., 1986, Processes of allochthonous terrane evolution, Mindoro Island, Philippines: Tectonics, v. 5, p. 525–552.
Savage, J. C., 1971, Theory of creep waves propagating along a strike-slip fault: Journal of Geophysical Research, v. 76 (8), p. 1954–1966.
Schmid, S. M., Zingg, A., and Handy, M., 1987, The kinematics of movements along the Insubric Line and the emplacement of the Ivrea zone: Tectonophysics, v. 135, p. 47–66.
Schofield, J. C., 1960, Some theoretical structures associated with transcurrent faulting applied to the Alpine fault: New Zealand Journal of Geology and Geophysics, v. 3, p. 461–466.
Schubert, C., 1980, Late-Cenozoic pull-apart basins, Boconó fault zone, Venezuelan Andes: Journal of Structural Geology, v. 2, p. 463–468.
——— 1982a, Neotectonics of Boconó fault, western Venezuela: Tectonophysics, v. 85, p. 205–220.
——— 1982b, Neotectonics of a segment of the San Andreas fault, southern California (USA): Eiszeitalter und Gegenwart, v. 32, p. 13–22.
——— 1986a, Aspectos neotectonicos de la zona de falla de La Victoria y origen de la cuenca de Santa Lucia–Ocumare del Tuy, Venezuela: Acta Cientifica Venezolana, v. 37, p. 278–286.
——— 1986b, Neotectonic aspects of the southern Caribbean plate boundary: Transactions of the First Geological Conference of the Geological Society of Trinidad and Tobago, p. 265–269.
Schulz, S. S., Mavko, J. M., Burford, R. O., and Stuart, W. D., 1982, Long-term fault creep observations in central California: Journal of Geophysical Research, v. 87, p. 6977–6982.
Schwartz, D. P., 1988, Paleoseismicity and neotectonics of the Cordillera Blanca fault zone, northern Peruvian Andes: Journal of Geophysical Research, v. 93, p. 4712–4730.
Schwartz, D. P., and Coppersmith, K. J., 1984, Fault behavior and characteristic earthquakes: Examples from the Wasatch and San Andreas faults: Journal of Geophysical Research, v. 89, p. 5681–5698.
Scott, W. E., Pierce, K. L., and Hait, M. H., Jr., 1985, Quaternary tectonic setting of the 1983 Borah Peak earthquake, central Idaho: Seismological Society of America Bulletin, v. 75, p. 1053–1066.
Segall, P., and Pollard, D. D., 1980, Mechanics of discontinuous faults: Journal of Geophysical Research, v. 85, p. 4337–4350.
Seiders, V. M., 1983, Correlation and provenance of upper Mesozoic chert-rich conglomerate of California: Geological Society of America Bulletin, v. 94, p. 875–888.
Şengör, A.M.C., 1979, The North Anatolian transform fault: Its age, offset and tectonic significance: Geological Society of London Journal, v. 136, p. 269–282.
——— in press, Plate tectonics and orogenic research after 25 years: A Tethyan perspective, *in* Hilde, T.W.C., and Carlson, R., eds., Silver Anniversary of plate tectonics: Tectonophysics, Special Issue.
Şengör, A.M.C., and Canitez, N., 1982, The North Anatolian fault: Alpine-Mediterranean geodynamics: American Geophysical Union Geodynamics Series, Volume 7, p. 205–216.
Şengör, A.M.C., Görür, N., Şaroğlu, F., 1985, Strike-slip faulting and related basin formation in zones of tectonic escape: Turkey as a case study: Society of Economic Paleontologists and Mineralogists Special Publication No. 37, p. 227–264.
Serpa, L., 1988, Reinterpretation of Mojave COCORP data: Implications for the structure of the Mojave rift: Geological Society of America Abstracts with Programs, v. 20, p. 230.
Sharp, R. P., 1954, Physiographic features of faulting in southern California; Chapter V—Geomorphology, *in* Jahns, R. H., ed., Geology of southern California: California Division of Mines Bulletin 170, p. 21–28.
Sharp, R. V., 1967, San Jacinto fault zone in the Peninsular Ranges of southern California: Geological Society of America Bulletin, v. 78, p. 705–730.
Sharp, R. V., and Clark, M. M., 1972, Geologic evidence of previous faulting near the 1968 rupture on the Coyote Creek fault: U.S. Geological Survey Professional Paper 787, p. 131–140.
Sharp, R. V., Rymer, M. J., and Lienkaemper, 1986, Surface displacements on the Imperial and Superstition Hill faults triggered by the Westmorland, California, earthquake of 26 April, 1981: Seismological Society of America Bulletin, v. 76, p. 949–965.
Shawe, D. R., 1965, Strike-slip control of basin and range structure indicated by historical faults in western Nevada: Geological Society of America Bulletin, v. 76, p. 1361–1378.
Sherill, R. E., 1929, Origin of the en echelon faults in north-central Oklahoma: American Association of Petroleum Geologists Bulletin, v. 13, p. 31–37.
Sibson, R. H., 1983, Continental fault structure and the shallow earthquake source: Geological Society of London Journal, v. 140, p. 741–767.
——— 1985, Stopping of earthquake ruptures at dilational fault jogs: Nature, v. 316, p. 248–251.
——— 1986, Rupture interaction with fault jogs: American Geophysical Union Geophysical Monograph 37 (Maurice Ewing Volume 6), p. 157–167.
——— 1987, Earthquake rupturing as a mineralizing agent in hydrothermal systems: Geology, v. 15, p. 701–704.
Sieh, K. E., 1978, Prehistoric large earthquakes produced by slip on the San Andreas fault at Pallett Creek, California: Journal of Geophysical Research, v. 83 (B8), p. 3907–3939.
——— 1982, Slip along the San Andreas fault associated with the earthquake, *in* The Imperial Valley, California, earthquake of October 15, 1979: U.S. Geological Survey Professional Paper 1254, p. 155–160.
——— 1984, Lateral offsets and revised dates of large earthquakes at Pallett Creek, California: Journal of Geophysical Research, v. 89, p. 7641–7670.
Sieh, K. E., and Jahns, R. H., 1984, Holocene activity on the San Andreas fault at Wallace Creek, California: Geological Society of America Bulletin, v. 95 (8), p. 883–896.
Silver, L. T., and Anderson, T. H., 1974, Possible left-lateral early to middle Mesozoic disruption of the southwestern North American craton margin: Geological Society of America Abstracts with Programs, v. 6, p. 955.
Silver, L. T., and Mattinson, J. M., 1986, "Orphan Salinia" has a home [abs.]: EOS (American Geophysical Union Trnasactions), v. 67, p. 1215.
Simpson, R. W., and Cox, A., 1977, Paleomagnetic evidence for tectonic rotation of the Oregon Coast Range: Geology, v. 5, p. 585–589.
Skempton, A. W., 1966, Some observations on tectonic shear zones: First International Congress on Rock Mechanics, Proceedings, v. 1, p. 329–335.
Smith, G. I., 1962, Large lateral displacement on Garlock fault, California, as measured from offset dike swarm: American Association of Petroleum Geologists Bulletin, v. 46, p. 85–104.
Smith, J. G., 1965, Fundamental transcurrent faulting in northern Rocky Mountains: American Association of Petroleum Geologists Bulletin, v. 49, p. 1398–1409.
Smith, S. W., and Wyss, M., 1968, Displacement on the San Andreas fault subsequent to the 1966 Parkfield earthquake: Seismological Society of America Bulletin, v. 58, p. 1955–1973.
Speight, J. M., and Mitchell, J. G., 1979, The Permo-Carboniferous dyke-swarm of northern Argyll and its bearing on dextral displacement on the Great Glen fault: Geological Society of London Journal, v. 136, p. 3–11.
St. Amand, P., 1957, Geological and geophysical synthesis of the tectonics of portions of British Columbia, Yukon Territory and Alaska: Geological Society of America Bulletin, v. 68, p. 1343–1370.
Steel, R. J., and Gloppen, T. G., 1980, Late Caledonian (Devonian) basin formation, western Norway: Signs of strike-slip tectonics during infilling, *in* Ballance, P. F., and Reading, H. G., eds., Sedimentation in oblique-slip mobile zones: International Association of Sedimentologists Special Publication 4, p. 79–103.
Steel, R. J., Mæhle, S., Nilsen, H., Røe, S. L., and Spinnangr, Å, 1977, Coarsening-upward cycles in the alluvium of Hornelen Basin (Devonian), Norway: Sedimentary response to tectonic events: Geological Society of America Bulletin, v. 88, p. 1124–1134.
Steel, R. J., Gjelberg, J., Helland-Hansen, W., Kleinspehn, K., Nøttvedt, A., and Rye-Larsen, M., 1985, The Tertiary strike-slip basins and orogenic belt of Spitsbergen, *in* Biddle, K. T., and Christie-Blick, N., eds., Strike-slip deformation, basin formation, and sedimentation: Society of Economic Paleontologists and Mineralogists Special Publication 37, 386 p.
Stein, R. S., 1983, Reverse slip on a buried fault during the 2 May 1983 Coalinga earthquake: Evidence from geodetic elevation changes, *in* Bennett, J. H., and Sherburne, R. W., eds., The 1983 Coalinga, California earthquakes: California Division of Mines and Geology Special Publication 66, p. 151–163.
——— 1984, Coalinga's caveat: EOS (American Geophysical Union Transactions), v. 64 (45), p. 794–795.
Stein, R. S., and King, G.C.P., 1984, Seismic potential revealed by surface folding: 1983 Coalinga, California, earthquake: Science, v. 224, p. 869–872.
Stein, R. S., and Lisowski, M., 1983, The 1979 Homestead Valley earthquake sequence, California: Control of aftershocks and post-seismic deformation: Journal of Geophysical Research, v. 88, p. 6477–6490.
Steinbrugge, K. V., Zacher, E. G., Tocher, D., Whitten, C. A., and Clair, C. N., 1960, Creep on the San Andreas fault: Fault creep and property damage: Seismological Society of America Bulletin, v. 50, p. 389–404.
Stewart, J. H., 1967, Possible large right-lateral displacement along fault and shear zones in the Death Valley–Las Vegas area, California and Nevada: Geological Society of America Bulletin, v. 78, p. 131–142.
——— 1983, Extensional tectonics in the Death Valley area, California: Transport of the Panamint Range structural block 80 km northwestward: Geology, v. 11, p. 153–157.
Suess, E., 1885, Das Anlitz der Erde: Vienna, Tempsky, 778 p.
Suggate, R. P., Gair, H. S., and Gregg, D. R., 1961, The southwest extension of the Atwatere fault: New Zealand Journal of Geology and Geophysics, v. 4, p. 264–269.
Sugimura, A., and Matsuda, T., 1965, Atera fault and its displacement vectors: Geological Society of America Bulletin, v. 76, p. 509–522.
Suppe, J., 1980, Imbricated structure of western foothills belt, south-central Taiwan: Petroleum Geology of Taiwan, v. 17, p. 1–16.
Swan, F. H., III, Schwartz, D. P., and Cluff, L. S., 1980, Recurrence of surface faulting and moderate to large magnitude earthquakes on the Wasatch fault zone at the Kaysville and Hobble Creek sites, Utah, *in* Evernden, J. F., convener, Proceedings of Conference X—Earthquake Hazards along the Wasatch/Sierra Nevada Frontal Fault Zones: U.S. Geological Survey Open-File Report 80-801, 679 p.
Sylvester, A. G., compiler, 1984, Wrench fault tectonics: American Association of Petroleum Geologists Reprint Series No. 28, 374 p.
——— 1986, Near-field tectonic geodesy, *in* Wallace, R. E., ed., Active tectonics: Washington, D.C., National Academy Press, p. 164–180.
Sylvester, A. G., and Pollard, D. D., 1975, Afterslip on the Sylmar fault segment, *in* Oakeshott, G. B., ed., San Fernando, California, earthquake of 9 February 1971: California Division of Mines and Geology Bulletin, v. 196, p. 227–233.
Sylvester, A. G., and Smith, R. R., 1976, Tectonic transpression and basement-controlled deformation in the San Andreas fault zone, Salton Trough, California: American Association of Petroleum Geologists Bulletin, v. 60 (12), p. 2081–2102.
Taliaferro, N. L., 1941, Geologic history and structure of the central Coast Ranges of California, *in* Jenkins, O. P., ed., Geologic formations and economic development of the oil and gas fields of California, Part Two, Geology of California and the occurrence of oil and gas: California Division of Mines Bulletin no. 118, p. 119–163.
Tang, R., and others, 1984, On the recent tectonic activity and earthquake of the Xianshuihe fault zone, *in* A collection of papers of International Symposium on Continental Seismicity and Earthquake Prediction (ISCSEP): Beijing, China, Seismological Press, p. 347–363.
Tapponier, P., 1977, Èvolution tectonique du système alpin en Méditerranée: poinçonnement et écrasement rigide-plastique: Bulletin de la Sociètè Gèologique de France 7 (XIX), p. 437–460.
Tapponier, P., Peltzer, G., Le Dain, A. Y., Armijo, R., and Cobbold, P., 1982, Propagating extrusion tectonics in Asia: New insights from simple experiments in plasticene: Geology, v. 10, p. 611–616.
Tapponier, P., Peltzer, G., and Armijo, R., 1986, On the mechanics of the collision between India and Asia, *in* Coward, M. P., and Reis, A. C., eds., Collision tectonics: Geological Society of London Special Publication 19, p. 115–157.
Tchalenko, J. S., 1970, Similarities between shear zones of different magnitudes: Geological Society of America Bulletin, v. 81, p. 1625–1640.
Tchalenko, J. S., and Ambraseys, N. N., 1970, Structural analysis of the Dasht-e Bay̌az (Iran) earthquake fractures: Geological Society of America Bulletin, v. 81, p. 41–66.
Teissere, R., and Beck, M., 1973, Divergent Cretaceous paleomagnetic pole position for the southern California batholith, U.S.A.: Earth and Planetary Science Letters, v. 18, p. 269–300.
Terres, R. R., and Luyendyk, B. P., 1985, Neogene tectonic rotation of the San Gabriel region, California, suggested by paleomagnetic vectors: Journal of Geophysical Research, v. 90, B14, p. 12,467–12,484.
Terres, R. R., and Sylvester, A. G., 1981, Kinematic analysis of rotated fractures and blocks in simple shear: Seismological Society of America Bulletin, v. 71, p. 1593–1605.
Thatcher, W., 1979, Systematic inversion of geodetic data in central California: Journal of Geophysical Research, v. 84, p. 2283–2295.
——— 1986, Geodetic measurement of active-tectonic processes, *in* Wallace, R. E., ed., active tectonics: Washington, D.C., National Academy Press, p. 155–163.
Thatcher, W., and Lisowski, M., 1987, 1906 earthquake slip on the San Andreas fault in offshore northwestern California: EOS (American Geophysical Union Transactions), v. 68 (44), p. 1507.
Tirrul, R., 1982, Frontal thrust zone of Wopmay Orogen, Takijuq Lake map area, District of Mackenzie: Current Research, Part A: Canada Geologi-

cal Survey, Paper 82-1A, p. 119–122.
—— 1984, Regional pure shear deformation by conjugate transcurrent faulting, externides of Wopmay Orogen, N.W.T.: Geological Association of Canada, Programs with Abstracts, v. 9, p. 111.
Tirrul, R., Bell, I. R., Griffis, R. J., and Camp, V. E., 1983, The Sistan suture zone of eastern Iran: Geological Society of America Bulletin, v. 94, p. 134–150.
Tobisch, O. T., Saleeby, J. B., and Fiske, R. S., 1986, Structural history of continental volcanic arc rocks, eastern Sierra Nevada, California: A case for extensional tectonics: Tectonics, v. 5, p. 65–94.
Tocher, D., 1960, The Alaska earthquake of July 10, 1958: Movement on the Fairweather fault and field investigation of southern epicentral region: Geological Society of America Bulletin, v. 71, p. 267–292.
Tomlinson, C. W., 1952, Odd geologic structures of southern Oklahoma: American Association of Petroleum Geologists Bulletin, v. 36, p. 1820–1840.
Trümpy, R., 1977, The Engadine line: A sinistral wrench fault in the central Alps: Geological Society of China Memoir 2, p. 1–12.
Vedder, J. G., and Wallace, R. E., 1970, Map showing recently active breaks along the San Andreas and related faults between Cholame Valley and Tejon Pass, California: U.S. Geological Survey Map I-574, scale 1:24,000.
Vickery, F. P., 1925, Structural dynamics of the Livermore region: Journal of Geology, v. 33, p. 608–628.
Wallace, R. E., 1949, Structure of a portion of the San Andreas fault in southern California: Geological Society of America Bulletin, v. 60, p. 781–806.
—— 1968, Notes on stream channels offset by the San Andreas fault, southern Coast Ranges, California, in Dickinson, W. R., and Grantz, A., eds., Proceedings of Conference on Geologic Problems of the San Andreas Fault System: Stanford University Publications in the Geological Sciences, v. 11, p. 374.
—— 1970, Earthquake recurrence intervals on the San Andreas fault: Geological Society of America Bulletin, v. 81, p. 2875–2890.
—— 1976, The Talas-Fergana fault, Kirghiz and Kazakh, U.S.S.R.: Earthquake Information Bulletin, v. 8, p. 4–13.
—— 1987, Grouping and migration of surface faulting and variation in slip rates on faults in the Great Basin province: Seismological Society of America Bulletin, v. 77 (3), p. 868–876.
Wallace, R. E., and Roth, E. F., 1967, Rates and patterns of progressive deformation, in Brown, R., Jr., ed., Parkfield-Cholame, California, earthquakes of June–August 1966—Surface geologic effects, water resources aspects, and preliminary seismic data: U.S. Geological Survey Professional Paper 579, p. 23–40.
Ward, S. N., 1988, North America–Pacific plate boundary, an elastic-plastic megashear: Evidence from very long baseline interferometry: Journal of Geophysical Research, v. 93, p. 7716–7728.
Watterson, J., 1978, Proterozoic intraplate deformation in the light of southeast Asian neotectonics: Nature, v. 273, p. 636–640.
Weaver, C. S., Grant, W. C., Malone, S. D., and Endo, E. T., 1981, Post–May 18 seismicity: Volcanic and tectonic implications, in Lipman, P. W., and Mullineaux, D. R., eds., The 1980 eruptions of Mount St. Helens, Washington: U.S. Geological Survey Professional Paper 1250, p. 109–121.

Weaver, C. S., Grant, W. C., and Shemeta, J. E., 1987, Local crustal extension at Mount St. Helens, Washington: Journal of Geophysical Research, v. 92 (B10), p. 10,170–10,178.
Webb, G.W., 1969, Paleozoic wrench faults in the Canadian Appalachians, in North Atlantic geology and continental drift: American Association of Petroleum Geologists Memoir 12, p. 754–786.
Webb, T. H., and Kanamori, H., 1985, Earthquake focal mechanisms in the eastern Transverse Ranges and San Emigdio Mountains, southern California, and evidence for a regional decollement: Seismological Society of America Bulletin, v. 75, p. 737–757.
Wegener, A., 1929, Die Enstehung der Kontinente und Ozeane: Braunschweig, Germany, Friedrich Vieweg und Sohn.
Weldon, R., and Humphreys, E., 1986, A kinematic model of southern California: Tectonics, v. 5, p. 33–48.
Wellman, H. S., 1965, Active wrench faults in Iran, Afghanistan and Pakistan: Geologische Rundschau, v. 55, p. 716–735.
Wellman, H. W., 1955, The geology between Bruce Bay and Haast River, South Westland: New Zealand Geological Survey Bulletin 48 (n.s.), (2nd edition), 46 p.
Wesson, R. L., 1987, Modelling aftershock migration and afterslip of the San Juan Bautista, California, earthquake of October 3, 1972: Tectonophysics, v. 144, p. 215–229.
Wesson, R. L., Helley, E. J., Lajoie, K. R., and Wentworth, C. M., 1975, Faults and future earthquakes, in Borchardt, R. D., ed., Studies for seismic zonation of the San Francisco Bay region: U.S. Geological Survey Professional Paper 941A, p. 5–30.
Wilcox, R. E., Harding, T. P., and Seely, D. R., 1973, Basic wrench tectonics: American Association of Petroleum Geologists Bulletin, v. 57, p. 74–96.
Williams, P. L., Fagerson-McGill, S., Sieh, K. E., Allen, C. R., and Louie, J. N., 1988, Triggered slip along the San Andreas fault after the 8 July 1986 North Palm Springs earthquake: Seismological Society of America Bulletin, v. 78, p. 1112–1122.
Willis, B., 1937, Geologic observations in the Philippines archipelago: National Research Council of the Philippines Bulletin, v. 13.
—— 1938a, San Andreas rift, California: Journal of Geology, v. 46, p. 793–827.
—— 1938b, Wellings' observations of Dead Sea structure: Geological Society of America Bulletin, v. 49, p. 659–668.
Willis, G. F., Tosdal, R. M., and Manske, S. L., 1987, The Mesquite Mine, southeastern California: epithermal gold mineralization in a strike-slip fault system: Geological Society of America Abstracts with Programs, v. 19, p. 892.
Wilson, C. C., 1940, The Los Bajos fault of south Trinidad, B.W.I.: American Association of Petroleum Geologists Bulletin, v. 24, p. 2102–2125.
—— 1968, The Los Bajos fault: 4th Caribbean Geological Conference, Trinidad, Transactions, p. 87–89.
Wilson, C. W., Jr., 1936, Geology of the Nye-Bowler lineament, Stillwater and Carbon Counties, Montana: American Association of Petroleum Geologists Bulletin, v. 20, p. 1161–1168.
Wilson, G., 1960, The tectonics of the 'Great Ice Chasm,' Filchner Ice Shelf, Antarctica: Geologists Association Proceedings, v. 71, p. 130–138.
Wilson, J. T., 1962, Cabot fault, an Appalachian equivalent of the San Andreas and Great Glen faults and some implications for continental displacement: Nature, v. 195, p. 135–138.

—— 1965, A new class of faults and their bearing on continental drift: Nature, v. 207, p. 343–347.
—— 1966, Some rules for continental drift: Royal Society of Canada Special Publication 9, p. 3–17.
Wilson, R.A.M., and Ingham, F. T., 1958, The geology of the Xeros-Troodos area with an account of the mineral resources: Cyprus Geological Survey Memoir No. 1.
Wise, D. U., 1963, An outrageous hypothesis for the tectonic pattern of the North American Cordillera: Geological Society of America Bulletin, v. 74, p. 357–362.
Wood, H. O., 1916, California earthquakes: Seismological Society of America Bulletin, v. 6 (2–3), p. 55–180.
—— 1955, The 1857 earthquake in California: Seismological Society of America Bulletin, v. 45 (1), p. 47–67.
Woodburne, M. O., 1975, Cenozoic stratigraphy of the Transverse Ranges and adjacent areas, southern California: Geological Society of America Special Paper 162, 91 p.
Woodcock, N. H., 1986, The role of strike-slip fault systems at plate boundaries: Royal Society of London Philosophical Transactions, ser. A. v. 317, p. 13–29.
Woodcock, N. H., and Fischer, M., 1986, Strike-slip duplexes: Journal of Structural Geology, v. 8 (7), p. 725–735.
Wright, L., 1976, Late Cenozoic fault pattern and stress fields in the Great Basin and westward displacement of the Sierra Nevada block: Geology, v. 4, p. 489–494.
Wu, F. T., and Wang, P., 1988, Tectonics of western Yunnan Province, China: Geology, v. 16 (2), p. 153–157.
Yeats, R. S., 1968, Rifting and rafting in the southern California borderland, in Dickinson, W. R., and Grantz, A., eds., Proceedings of Conference on Geologic Problems of the San Andreas Fault System: Stanford University Publications in the Geological Sciences, v. 11, p. 307–322.
—— 1973, Newport-Inglewood fault zone, Los Angeles Basin, California: American Association of Petroleum Geologists Bulletin, v. 57 (1), p. 117–135.
—— 1981, Quaternary flake tectonics of the California Transverse Ranges: Geology, v. 9, p. 16–20.
Zhang Buchun, Liao Yuhua, Guo Shunmin, Wallace, R. E., Bucknam, R. C., and Hanks, T. C., 1986, Fault scarps related to the 1739 earthquake and seismicity of the Yinchuan graben, Ningxia Huizu Zizhiqu, China: Seismological Society of America Bulletin, v. 76 (5), p. 1253–1287.
Zhang Weiqi, Jiao Decheng, Zhang Peichen, Molnar, P., Burchfiel, B. C., Deng Qidong, Wang Yipeng, and Song Fangmin, 1987, Displacement along the Haiuan fault associated with the great 1920 Haiyuan, China, earthquake: Seismological Society of America Bulletin, v. 77 (1), p. 117–131.
Zoback, M. D., and others, 1987, New evidence on the state of stress of the San Andreas fault system: Science, v. 238, p. 1105–1111.
Zobin, V. M., 1972, Focal mechanisms of volcanic earthquakes: Bulletin Volcanologique, v. 36, p. 561–571.

MANUSCRIPT RECEIVED BY THE SOCIETY FEBRUARY 8, 1988
REVISED MANUSCRIPT RECEIVED JUNE 13, 1988
MANUSCRIPT ACCEPTED JUNE 14, 1988

CENTENNIAL ARTICLE

Basin and Range extensional tectonics at the latitude of Las Vegas, Nevada

BRIAN WERNICKE
GARY J. AXEN } *Department of Earth and Planetary Sciences, Harvard University, Cambridge, Massachusetts 02138*
J. KENT SNOW

ABSTRACT

The Basin and Range province at the latitude of Las Vegas, Nevada (approximately 36°N), is ideally suited for reconstructing Neogene extension owing to an abundance of structural markers, primarily Mesozoic thrust faults, developed within the generally conformable Cordilleran miogeocline. In map view, extension is heterogeneous and is divisible into two major extensional domains, the Las Vegas and Death Valley normal fault systems, that lie east and west (respectively) of a relatively unextended median block. We determined horizontal relative-motion vectors between pairs of reference points across the province, chosen so as to best allow geologic markers to constrain the relative motion of the pair during extension. We recognize three sequences of pairs, two in the Las Vegas system and one in the Death Valley system, that define an unbroken path across the entire province. The vectors along these paths sum to give 247 ± 56 km of net extension oriented N73° ± 12°W.

Timing considerations indicate that extension occurred principally during the past 15 m.y. Westward motion of the Sierra Nevada away from the Colorado Plateau occurred at a rate of 20–30 mm/yr in the interval 10–15 m.y. ago, but was no greater than 10 mm/yr over the past 5 m.y. Strike-slip faulting was an important component in the extending system and absorbed perhaps 40–50 km of north-south shortening of the region during extension, indicating a constrictional strain field for the crust as a whole. If one assumes no major rotations of the Sierra Nevada during Cenozoic extension, and about 100 km of pre–15-m.y.-ago extension in the central portion of the northern Great Basin, the crust in the Las Vegas region extended by a factor of 3–4, whereas the wider Great Basin region extended by only a factor of 2. This difference may explain the contrast in regional elevation between the two areas (the northern Great Basin is on average about 1,000 m higher) and the constrictional strain in the Las Vegas region. The more widely distributed extension to the north may not have kept pace with the larger extension to the south, such that the south lost gravitational potential more rapidly. Thus, comparatively buoyant northern Great Basin lithosphere was (and continues to be) forced down the potential gradient into the Las Vegas region. Resolved parallel to the northern San Andreas fault, our reconstruction accounts for 214 ± 48 km of right-lateral shear along the Pacific–North America transform plate boundary.

INTRODUCTION

In the two decades since R. Ernest Anderson's first studies of large-magnitude Cenozoic extensional tectonism in the Lake Mead area of the Basin and Range, the significance of the structures described in his initial report on the area, published in the Geological Society of America *Bulletin* (Anderson, 1971), has grown from that of a freak occurrence of local importance to a benchmark in the recognition of how the crust extends. Simultaneous work in the late 1960s and early 1970s of Lauren Wright and Bennie Troxel (1973) in the Death Valley region, and of John Proffett (1977) in the Yerington mining district in west-central Nevada, led them independently to the same conclusion as that of Anderson, that in at least some regions of the Basin and Range, shallowly dipping normal faults separating steeply tilted fault blocks had accommodated large-magnitude extension of the continental crust in Neogene time. In addition, Armstrong's (1972) perceptive synthesis of low-angle faults in east-central Nevada showed that Cenozoic low-angle faulting was important over a wide region of the Basin and Range.

These studies were not the first to recognize structures now widely believed to accommodate large-magnitude extensional tectonism. Ransome and others (1910) recognized the structural style of imbricate normal faulting of Tertiary volcanic strata in the Bullfrog mining district near Death Valley and the fact that the entire faulted package lay tectonically upon a metamorphic complex. They developed the hypothesis that the basal fault was normal and of significant displacement but favored the interpretation that the fault was an overthrust. They considered it unlikely that the force of gravity alone could have moved the volcanic strata on such a shallowly dipping fault and suggested that the undulose geometry of the basal fault facilitated extensional faulting in the hanging wall of the overthrust. Subsequent works, notably Noble (1941), Longwell (1945), Curry (1954), Young (1960), Misch (1960), Pashley (1966), and Hunt and Mabey (1966), all recognized similar features in the Basin and Range, yet as in the case of Ransome and others (1910), none of them concluded that the deformation resulted from large-magnitude Cenozoic extension (although Hunt and Mabey argued for Mesozoic extension). In reading these older works today, one simultaneously feels a sense of loss over how long it took to begin to understand extensional tectonism, and elation over the fact that there is still so much exciting work to be done.

In the wake of Anderson's and other studies that compellingly demonstrated large-magnitude extension, strong sentiment against low-angle normal faulting as a major mechanism of extension in the Basin and Range province and elsewhere remained, with a tendency to consider areas of shallowly dipping normal faults as exceptional to an over-all style of steeply dipping, widely spaced normal faults that accommodated relatively modest crustal extension (10%–30% increase over original width). Explanations excluding crustal extension, including surficial gravity sliding, special circumstances during thrust faulting, or a combination of the two, were still often invoked to explain the enigmatic low-angle faults. These explanations defended the classical view (for example, of G. K. Gilbert) of a Basin and Range that was folded and thrust faulted in Mesozoic time, blanketed by ignimbrite in early to middle Tertiary time, and block faulted on steep faults in the late Tertiary. The observations of large faults that place high crustal levels on low, and the involvement of steeply dipping Tertiary strata along them, flew in the face of Gilbert's Basin and Range. The lukewarm reception of a decidedly non-Gilbertian Basin and Range is exemplified in the citations of Anderson (1971), Armstrong (1972), Wright and Troxel (1973), and Proffett (1977, but submitted to the *Bulletin* in 1972) shown in Figure 1. Most of this work had been completed and reported on at Geological Society of America meetings by 1972; yet, it was nearly a decade before its importance was widely realized in the geological community. Stewart (1978) best summarized the thinking on the province in the late 1970s, when it was thought that locally large-magnitude extension had been accommodated in areas such as the Yerington district but that most of the province probably had not extended more than about 10%–35%. In contrast to the prevalent view, the mobilistic syntheses of Hamilton and Myers (1966) and Hamilton (1969) argued for a doubling in width of the province, based on crustal structure and thickness, the possible distortion of pre-Cenozoic tectonic belts by extension, and the possibility that steep range-bounding faults flatten downward.

The sudden appreciation of the significance of these early studies was catalyzed by the 1977 Penrose Conference on Cordilleran Metamorphic Core Complexes, at which a number of workers argued that widespread regions of metamorphic tectonite in the Basin and Range were Tertiary in age and related to low-angle faulting and crustal extension (Davis and Coney, 1979; articles in Crittenden and others, 1980), first hypothesized for east-central Nevada by Armstrong (1963, 1972). The provocative reflection profiles from the starved passive margin of the Bay of Biscay (for example, de Charpal and others, 1978) and the arguments of McKenzie (1978) for major crustal extension in rifts and on passive margins also contributed to this appreciation. These insightful field studies had ushered in a new era of Basin and Range *observation*, unencumbered with the need to explain away, case by case, the first-order field relations of the province as flukes.

This paper is a progress report summarizing the results and implications of field studies in the southern Great Basin region by the Program in Extensional Tectonics at Harvard. We build on Anderson, Wright, and Troxel's studies in the region to present for the first time measurement of Cenozoic extensional strain across the entire Basin and Range, accurate to two significant figures, and demonstration of the slowing of extension between 10 m.y. ago and the present. In an earlier report on the region, Wernicke and others (1982) used offsets on selected strike-slip faults to reconstruct the extension, concluding that at least 140 km of extension had occurred. In this report, we incorporate new data into reconstructing the province, in particular an improved understanding of the northern Death Valley–Furnace Creek fault zone, which was not considered in Wernicke and others (1982). Using a similar but more detailed approach, we present our reconstruction as a series of vectors that describe the motion between fiducial points such that we can quantify (that is, bound uncertainties on) the magnitude, direction, and rate of extension of a number of subregions of the province, then sum the vectors and their uncertainties to place bounds on the westward motion of the Sierra Nevada block relative to the Colorado Plateau in Neogene time (compare with Minster and Jordan, 1984, 1987). The vectors are based on palinspastic reconstructions of areas mapped by us and many other workers. Documentation of the field relations of each reconstruction is well beyond the scope of a single paper and is presented elsewhere (Axen and Wernicke, 1987; Wernicke and others, 1988a, 1988b; J. K. Snow and B. Wernicke, unpub. data; numerous reports by other workers cited below), but the key structural markers in each are outlined below. Our results indicate substantially more extension than the 140-km minimum determined by Wernicke and others (1982) and have important implications for the nature of extensional tectonism.

GEOLOGIC SETTING

The Basin and Range province near the latitude of Las Vegas has an ideal regional tectonic setting for a province-wide reconstruction of Cenozoic extension (Fig. 2). The pre-extension geology is more straightforward than at other latitudes because the regionally conformable Cordilleran miogeocline is exposed across the entire width of the province (Figs. 2 and 3). The miogeocline is disrupted by east-vergent Mesozoic thrust faults that make local reconstructions more complicated than they might be in the absence of faults. The thrusts, however, are distinctive enough and the extensional separation of crustal blocks great enough that they provide the markers necessary to tightly constrain large-scale reconstructions. The thrusts are thus more an aid than a complication, for discrete markers within the miogeoclinal section are few, and in most cases, the determination of fault offsets based purely on isopachs and facies trends is limited by the uncertainty in their precise location and with the assumption of their initial geometry (for example, Stewart, 1983; Prave and Wright, 1986). The great thickness of the thrust-faulted miogeocline gives excellent depth control on cross-sectional reconstructions, locally in excess of 15 km (Fig. 3). Exposure is generally excellent in the region because it lies at low elevation and in the rain shadow of the Sierra Nevada and more southerly

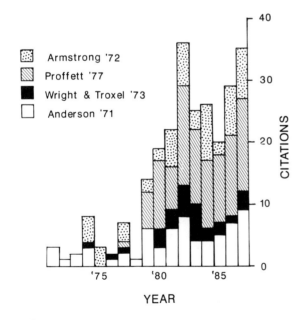

Figure 1. Number of citations in refereed journals of articles discussed in text. Note tenfold increase from 1979 to 1982, roughly a decade after studies were completed.

ranges. In addition, much of the geology is developed within carbonate rocks, which crop out well in desert regions.

The regionally averaged topographic pattern of the Basin and Range at the latitude of Las Vegas is one of high flanks, comprising the Sierra Nevada on the west and the Colorado Plateau on the east, and two broad, low-lying areas on either side of a median high (Eaton and others, 1978). This pattern resembles that of the northern Basin and Range, but at smaller scale because the province here is half the width (Fig. 2). The median high is centered on the Spring Mountains, Sheep Range, and Las Vegas Range, whereas the lows include the Colorado River trough/Lake Mead area on the east and the Death Valley region on the west (Fig. 4). As discussed below, the two low-lying areas are highly extended, whereas the median high is less extended.

Basement, Proterozoic Basin, and Miogeoclinal Wedge

Precambrian Y (mostly ca. 1.7–1.4 Ga) crystalline basement in the region lies nonconformably beneath unmetamorphosed sediments of Precambrian Y (?), Precambrian Z, or Cambrian age (Fig. 3). Precambrian Y (?) and Z strata of the Pahrump Group (Fig. 3) are locally present in ranges west and southwest of the Spring Mountains between basement and regionally persistent Precambrian Z to Cambrian strata that form the base of the Cordilleran miogeocline in the region (Stewart, 1970, 1972). Although the lower portion of the Pahrump is probably Precambrian Y in age, the upper part appears to be in gradational contact with the Cordilleran miogeocline, and thus is probably Precambrian Z in age (Miller, 1987). The west-thickening Precambrian Z and Paleozoic miogeocline (Figs. 2 and 3) is overlain disconformably or with mild angular unconformity by locally thick accumulations of Mesozoic strata (Fig. 3).

The most significant stratigraphic feature beneath the miogeoclinal strata is the northward pinchout of the Pahrump Group in the southern Death Valley region (Wright and others, 1974, 1981). South of the pinchout, as much as 3,000 m of Pahrump strata is present below the basal units of the miogeocline in the southern Black Mountains, Kingston Range, and Panamint Range (Fig. 4). Over a distance of less than 10 km, the basal miogeoclinal unconformity cuts downsection through the Pahrump Group and onto crystalline rocks.

Lithologically, the miogeocline is divisible into two main parts, including a Middle Cambrian and older clastic wedge and a Middle Cambrian and younger carbonate succession (Fig. 3). The clastic wedge thickens from less than 100 m on the craton to the east, where basal strata are Lower Cambrian, to more than 5,000 m in western areas, where most of the sequence lies below basal Cambrian beds. The Paleozoic sequence is entirely marine, except for some Permian strata that are partly nonmarine (Wright and others, 1981; Stone and Stevens, 1987). Westward thickening of the carbonate succession occurs in part by thickening of individual units and in part by the pinching in of Ordovician, Silurian, and Devonian strata beneath a major sub–Upper Devonian disconformity (Fig. 3). On the craton, Upper Devonian strata lie disconformably on Upper Cambrian. To the west, they lie on progressively younger strata until a fully developed Ordovician, Silurian, and Devonian section is present. The youngest marine strata in the region are Triassic and are overlain in eastern areas by nonmarine clastics locally as young as Cretaceous and in western areas by lower Mesozoic volcanics (for example, Wright and others, 1981). In sections in the transition zone between craton and miogeocline, the highest Paleozoic strata present on the craton, including the Kaibab and Toroweap Formations, pinch out westward beneath the basal Mesozoic unconform-

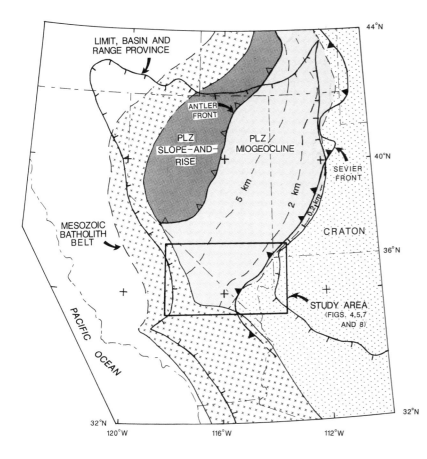

Figure 2. Regional tectonic setting of the Las Vegas area Basin and Range, showing isopach trends of the Precambrian Z–Cambrian clastic wedge of the Cordilleran miogeocline, Paleozoic Antler and Mesozoic Sevier orogenic fronts, and the position of the Mesozoic batholith belt (crosses). Note that the position of the study area of this report resides largely in the miogeoclinal prism and craton.

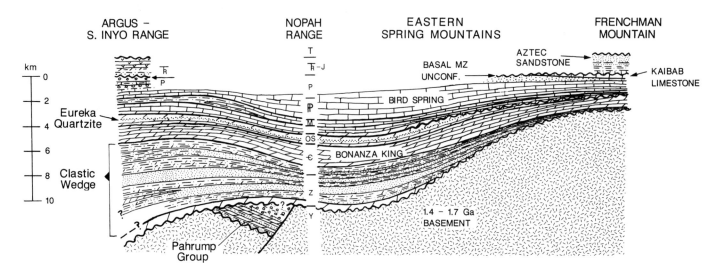

Figure 3. Highly simplified stratigraphic cross section of Mesozoic and older rocks exposed in the region. Modified from Stewart (1970) and Wright and others (1981). See Figure 4 for locations.

ity (Fig. 3; for example, Tschanz and Pampeyan, 1970; Burchfiel and others, 1974). In westernmost sections of the miogeocline, complex unconformities and facies changes in Carboniferous and Permian strata indicate Permian tectonism (Stone and Stevens, 1984, 1988a).

The primary lateral facies changes within the miogeocline include a transition from quartzite and siltstone in eastern exposures of the Precambrian Z–Cambrian clastic wedge to predominantly shale and carbonate in the west (Stewart, 1970), and a transition from mostly shallow marine limestone in Carboniferous and Permian strata in eastern areas to locally deep-marine shale, sandstone, and limestone in the west (Dunne and others, 1981; Stone and Stevens, 1988b; Fig. 3). The increase in fine clastics and carbonate in the clastic wedge indicates a transition from shelf to slope-and-rise facies (Stewart, 1972), but early Paleozoic slope-and-rise deposits are not present east of the Sierran batholith at this latitude (Fig. 2). The westward increase in clastics in Carboniferous strata probably represents the distal effects of the mid-Paleozoic Antler orogeny (for example, Dunne and others, 1981), which farther north in central Nevada is expressed by the eastward thrusting of early Paleozoic slope-and-rise facies strata onto the shelf facies rocks, forming a broad, asymmetrical foreland basin (for example, Poole and Sandberg, 1977). Structural effects of the Antler and Permian-Triassic Sonoma orogenic events may be present in western portions of the region (for example, Nelson, 1981), including possible truncation of the continental margin in Permian and Triassic time (Burchfiel and Davis, 1981; Stone and Stevens, 1988a, 1988b).

Mesozoic Thrust-and-Fold Belt

Mesozoic structures in the region are predominantly east-vergent folds and thrusts. The region lies at a major transition zone in the Mesozoic Cordilleran thrust belt. Within the region and to the north, the thrusts are developed within the miogeoclinal wedge such that many of the thrusts cut slowly through it, in many cases with ramp-flat geometries. To the south, however, the thrust belt changes trend from north-northeast, parallel to isopach and facies trends in the miogeoclinal wedge, to southwest, parallel to the east margin of the Mesozoic batholiths (Figs. 2 and 5). In so doing, the thrust belt leaves the miogeoclinal wedge, and the thrusts lose their ramp-flat geometry, tending instead to cut more steeply across the layers of cratonic sediments and involve crystalline basement (for example, Burchfiel and Davis, 1975, 1981). Thus, within any given thrust plate, the stratigraphic section thins from north to south as the thrusts curve southward out of the miogeocline. In the Spring and Clark Mountains blocks, the thrusts tend to converge upon one another to the south (Burchfiel and Davis, 1975, 1981).

Throughout most of the region, exposures of pre-Tertiary rock are separated by alluviated valleys, hampering correlations of thrusts between ranges (Fig. 5). In the median, relatively unextended block, and within a number of highly extended blocks, the nature of the thrust belt can be deduced across three or four major thrust plates (Figs. 5 and 6). Thrusts that now lie in widely separated blocks may be correlated by examining each of these fragments. As such, the thrusts represent useful markers for the reconstruction of Cenozoic tectonism. Below, we describe the principal characteristics of each of the Mesozoic thrust plates from east to west across the province (Figs. 5 and 6). Although somewhat tedious, these details are of great importance in measuring the directions and magnitudes of Cenozoic displacements discussed in the following section.

Keystone System. The lowest major thrust is the Keystone thrust and correlatives, which can be traced for more than 200 km along strike (for example, Burchfiel and others, 1982). Its hanging wall is characteristically a décollement in Middle Cambrian dolostones of the Bonanza King Formation (Fig. 3). The footwall geology of the thrust is complex, locally including blocks of a lower, older thrust similar to the Keystone. The thrust has a regionally persistent ramp and hanging-wall ramp syncline or synclinorium that is cored with Mesozoic strata. Along its entire trace, the hanging wall contains a portion of the miogeocline characterized by (1) rapid westward pinchin of Ordovician strata beneath the sub–Upper Devonian disconformity and (2) the westward pinchout of the Kaibab and Toroweap Formation beneath the basal Mesozoic unconformity (Figs. 3 and 6).

The stratigraphic and structural uniformity of the Keystone system, and the fact that it is continuous for large distances within range blocks, suggests strong stratigraphic control on the thrust. Subtle details of the miogeoclinal stratigraphy change little in character and trend along strike. The only major break in the trend of the Keystone system occurs along the Las Vegas Valley shear zone (Figs. 4 and 5), where geologic lines defined by the intersection of the ramp with footwall Mesozoic strata are appar-

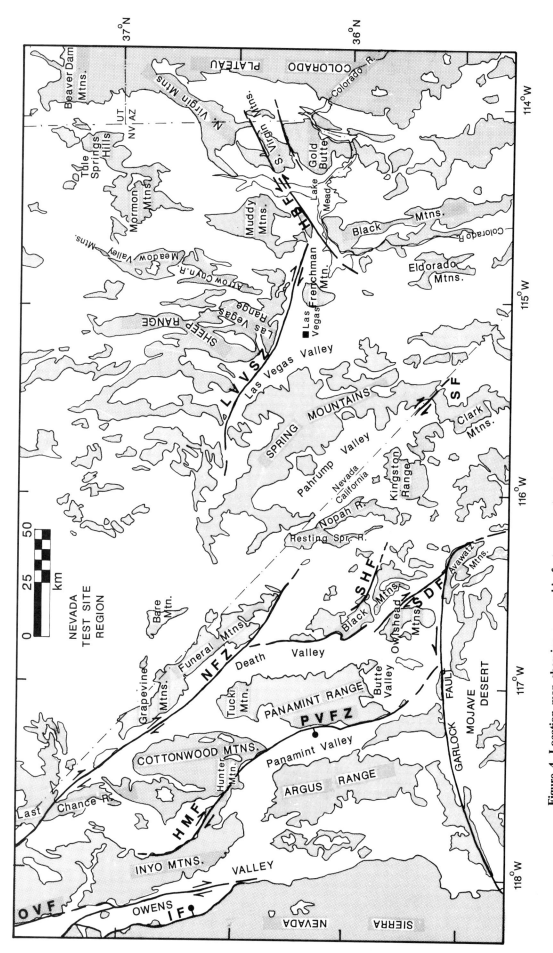

Figure 4. Location map showing geographic features mentioned in text, outcrop area of pre-Cenozoic rocks (shaded), and selected Cenozoic fault zones discussed in text (bold lines). HBF, Hamblin Bay fault; HMF, Hunter Mountain fault; IF, Independence fault; NFZ, northern Death Valley–Furnace Creek fault zone; OVF, Owens Valley fault; PVFZ, Panamint Valley fault zone; SDF, southern Death Valley fault zone; SF, State Line fault; SHF, Sheephead fault.

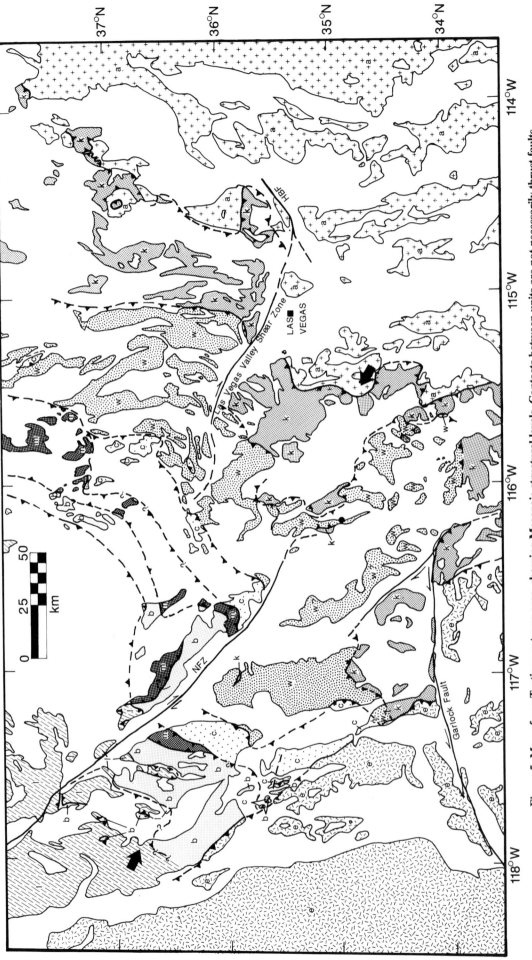

Figure 5. Map of pre-Tertiary outcrop area, showing Mesozoic structural levels. Contacts between units are not necessarily thrust faults and include later faults that juxtapose thrust plates and the axial traces of anticlines in asymmetric fold pairs; some areas in westernmost region are intruded post-tectonically. a = sub-Keystone system, autochthonous and parautochthonous rocks, k = rocks above Keystone thrust system and below Wheeler Pass system, w = rocks above the Wheeler Pass system and below the Clery thrust, c = rocks above the Clery thrust and below the Marble Canyon thrust, m = rocks above the Marble Canyon thrust and the west-vergent White Top Mountain backfold/thrust system, b = rocks below the White Top Mountain structure and the Last Chance system, l = rocks above the Last Chance system, e = rocks above the East Sierran thrust system. Bold arrows show approximate line of restored section in Figure 6. Arrow on west-vergent structure in Grapevine Mountains shows sense of rotation required to realign it with other exposures of the structure. HBF, Hamblin Bay fault; NFZ, northern Death Valley–Furnace Creek fault zone.

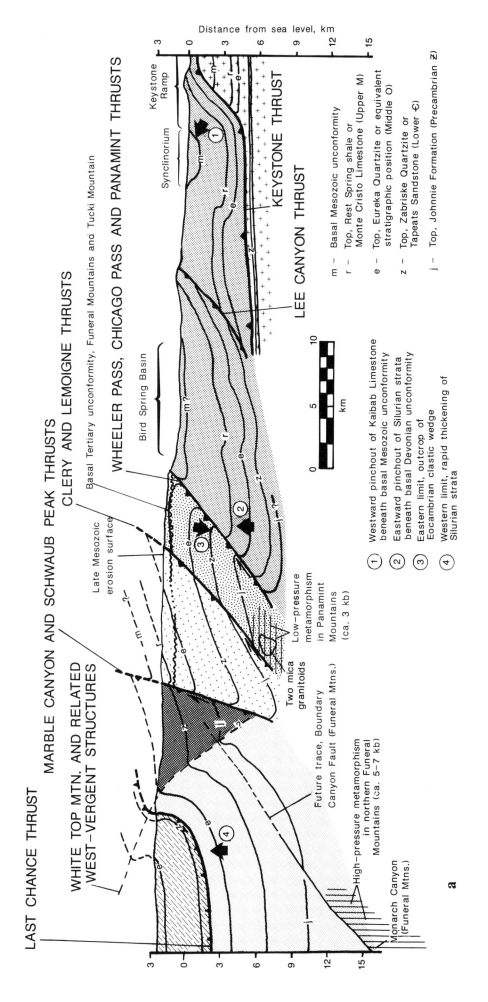

Figure 6. Restored cross section between bold arrows in Figure 5, showing geometry of thrust belt prior to Cenozoic extension. a. Patterned after Figure 5; pressure estimates of metamorphic terrains from Labotka and Albee (1988).

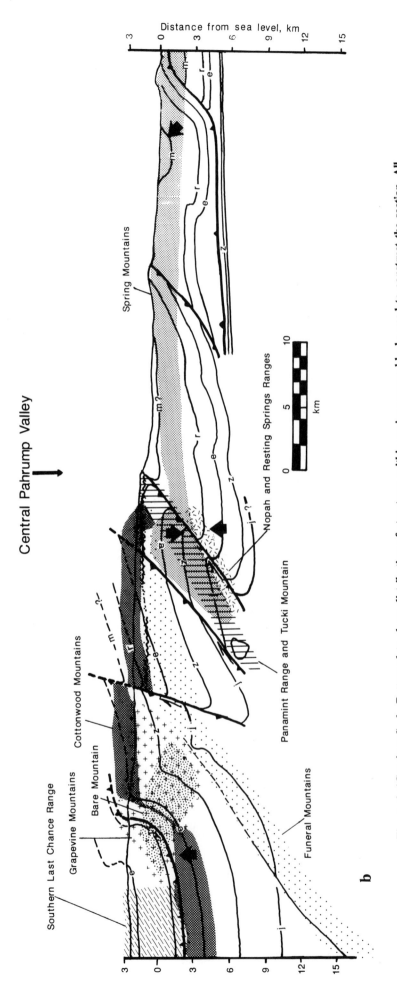

Figure 6. (*Continued*). b. Patterned to show distribution of structures within various range blocks used to construct the section. All geology is taken from within 20 km of the line between the bold arrows. Patterned areas depict structural levels exposed in ranges at present erosion levels. Overlapping ranges occur on either side of the NFZ, providing independent constraints on the section (see Fig. 4 for geographic locations).

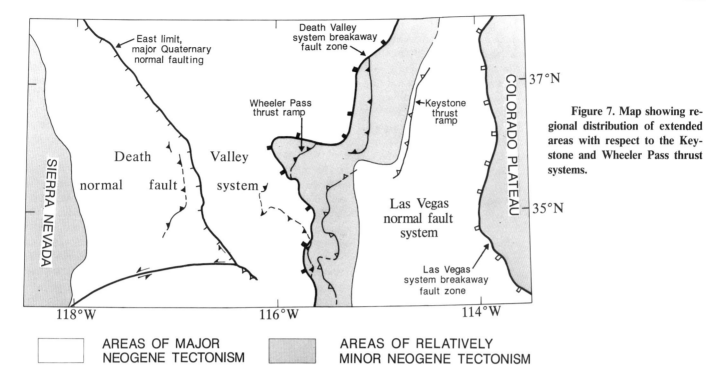

Figure 7. Map showing regional distribution of extended areas with respect to the Keystone and Wheeler Pass thrust systems.

ently offset right laterally about 50 km (Fig. 5). The likelihood that this offset has an origin as a tear structure in the Keystone thrust plate (for example, Royse, 1983) or was controlled in some way by an abrupt change in trend of isopachs in the miogeocline seems remote in view of the similarities of the thrust system north and south of the shear zone. The strong control of facies and isopach trends on thrust geometry observed in most thrust belts suggests that major paleogeographic anomalies are usually associated with major transverse structures, with a change in both the character of the faulted sediments and in the number and spacing of thrusts across them (for example, Price, 1981).

West of the synclinorium in the hanging wall of the Keystone, the Paleozoic section thickens by thickening within individual units and also by the appearance of the Middle Ordovician Eureka Quartzite, Upper Ordovician Ely Springs Dolomite, and Silurian strata below the sub-Devonian unconformity and of Middle and Lower Devonian strata above the unconformity (Figs. 3 and 6; Burchfiel and others, 1974). Within this region, there are a number of relatively small east-vergent folds and thrust faults that carry units as old as the Middle Cambrian Bonanza King Formation over rocks as young as the Carboniferous-Permian Bird Spring Formation, such as the Lee Canyon thrust (Fig. 6a). These structures tend not to be as laterally persistent as the Keystone system. West of the synclinorium, the Bird Spring Formation thickens from about 600 to >2,000 m (Figs. 3 and 6; Burchfiel and others, 1974).

Wheeler Pass System. The next highest major thrust system is the Wheeler Pass system. It can be traced for nearly 120 km along strike in the Sheep Range, Las Vegas Range, and Spring Mountains, interrupted only by the Las Vegas Valley shear zone (Fig. 5; Longwell and others, 1965; Burchfiel, 1965; Guth, 1981). Unlike the Keystone system, the Wheeler Pass system is not continuously exposed within the Spring Mountains-Clark Mountains block (Fig. 5). In its northern exposures in the Spring Mountains, the thrust strikes at high angle to the boundary between the Spring Mountains block and the highly extended area to the west (Fig. 7), projecting into the alluvium of Pahrump Valley (Figs. 4 and 5). Southwest of the Spring Mountains, in the highly extended Death Valley region, the thrust is present in a number of range blocks and is found as far west as the Panamint Range (Wernicke and others, 1988a, 1988b; Figs. 4 and 5). To the south, the thrust system reappears in the Clark Mountains block (Fig. 5).

At present erosion levels, the thrust usually carries Precambrian Z clastics over the Bird Spring Formation (Figs. 3 and 6). The thrust is probably in part a décollement in northern areas (for example, Guth, 1981; Burchfiel and others, 1974), but south of the Spring Mountains, it typically is not a hanging-wall décollement within miogeoclinal strata, as it cuts rapidly through the miogeoclinal section (for example, Burchfiel and Davis, 1971; Burchfiel and others, 1983; Wernicke and others, 1988b). At structurally deep levels, where hanging-wall basement overrides Precambrian Z clastics, the thrust has a décollement geometry (Burchfiel and Davis, 1971).

The Wheeler Pass system is the lowest thrust plate to contain exposures of the Precambrian Z clastic section and underlying basement and Pahrump strata (Fig. 6). It carries the thinnest sections of these strata, which thicken rapidly westward from about 2,000–2,500 m immediately above the thrust to more than 5,000 m to the west (Figs. 3 and 6). Silurian strata pinch in just beneath the thrust in the Nopah Range area (Figs. 4 and 6), but farther north, the Silurian is present well to the east of the thrust plane (for example, Langenheim and others, 1962). Maximum thicknesses of Silurian strata are found only in higher thrust plates (Fig. 6a).

The features that distinguish the Wheeler Pass system include its structural style, position in the miogeocline, and the fact that it is the only thrust in the region that emplaces rocks as old as Precambrian Z on top of post-Mississippian strata, with the exception of portions of the Last Chance system (described below), which is clearly a structurally higher system. The stratigraphic throw of the Wheeler Pass system is consistently about 5,000 m.

Higher Thrusts and Backfold. Above the Wheeler Pass system, we recognize a belt of two east-vergent structures (Clery-Lemoigne and Marble Canyon–Schwaub Peak thrusts) and a third, structurally higher west-vergent structure (White Top Mountain and related west-vergent structures, Figs. 5 and 6a; Reynolds, 1974; J. K. Snow and B. Wernicke, unpub. data). Correlations of these structures are difficult because they all lie west of the unextended, median topographic high and are obscured by extensive Tertiary volcanic cover in the Nevada Test Site region (Figs. 4 and 5). Nonetheless, the sequence is found continuously exposed in each of two range blocks in the Death Valley region, the Funeral-Grapevine Mountains block and the Cottonwood Mountains block (Figs. 4 and 5; J. K. Snow and B. Wernicke, unpub. data). Despite the large distance between the two blocks, the three structures have similar stratigraphic throw, position in the miogeocline, and relative spacing. As shown in Figure 6b, the geology of the Funeral-Grapevine Mountains block and the Cottonwood Mountains block fits together well on the same pre-extensional cross section. The presence of the west-vergent structure is particularly diagnostic of their correlation, as it is the only major west-vergent structure in the Death Valley region (J. K. Snow and B. Wernicke, unpub. data). All three structures characteristically cut upsection rapidly in both hanging wall and footwall and have a stratigraphic throw of 2,000 to 3,000 m. They occupy a position in the miogeocline characterized by rapid increase in thickness of the Silurian section as it pinches in beneath the sub-Devonian unconformity, and the transition from carbonate facies to shale, sandstone, and limestone facies in Carboniferous strata (Figs. 3 and 6a).

Last Chance System. Structurally above the west-vergent system, a major thrust system carries the thickest sections of the Precambrian Z and Cambrian clastic wedge over Carboniferous shale, sandstone, and limestone (Stewart and others, 1966; Reynolds, 1974). The Last Chance system differs from structurally underlying thrusts in the Death Valley region in that it commonly is a décollement in both hanging wall and footwall and has nearly twice the stratigraphic throw (generally 5,000–6,000 m). There are numerous windows into Carboniferous strata throughout the Last Chance Range–Inyo Mountains area that show that the thrust cuts gradually downsection in Precambrian Z strata to the west (Stewart and others, 1966; Nelson, 1981). The transition from quartzite and siltstone facies to shale and carbonate facies of the Precambrian Z clastic wedge (Fig. 3) occurs within the hanging wall of the thrust, although the onset of the change may occur in the footwall (Stewart, 1970).

East Sierran Thrust System and Sierran Batholith. The eastern margin of the Sierra Nevada batholith and a coincident zone of thrust faults trend about N30°W across the western part of the region, cutting obliquely across the northeast-trending isopachs, facies lines, and thrust faults developed in the miogeoclinal wedge (Fig. 5; Dunne, 1986). The East Sierran system was apparently localized by the thermal contrast between the batholith and cooler lithosphere to the east (for example, Burchfiel and Davis, 1975). It is younger than the higher thrusts developed in the miogeocline, as a suite of Early Jurassic alkalic plutons cuts the miogeoclinal thrusts, whereas younger plutons of the batholith are cut by strands of the East Sierran system (Dunne, 1986). The hanging wall of the system seems to override progressively lower thrust plates southward, but the large proportion of plutonic rock in the hanging wall of the thrust system precludes identification of offset traces of the older thrusts. For a discussion of relative and absolute timing constraints on the thrust systems in the region, the reader is referred to Dunne and others (1978), Burchfiel and Davis (1981), and Dunne (1986).

CENOZOIC EXTENSION

The first-order pattern of extensional tectonism is that of two topographically low regions pervaded by down-to-the-west normal faults, separated by a central unextended block (Fig. 7). The system to the east of the unextended block is herein referred to as the "Las Vegas normal fault system" and the system to the west as the "Death Valley normal fault system" (Fig. 7). The Mesozoic structure of the region is in part reflected in the position of these extended domains. The Las Vegas system is developed almost entirely below the Keystone thrust system and involves crystalline basement (Anderson, 1971); major normal faults involve the upper plate of the thrust only at highest structural levels in the northern part of the region (Wernicke and others, 1984, 1985; Smith and others, 1987). The east limit of major extension, or breakaway zone, for the Las Vegas system is developed within cratonic strata but, as is evident from extension magnitudes discussed below, initially lay no more than a few tens of kilometers east of the Keystone system.

The breakaway zone for the Death Valley system closely follows the trace of the Wheeler Pass thrust system, in some places leaving the thrust behind on the stable block, in others carrying it within extensional allochthons more than 100 km to the west (Figs. 5, 6, and 7). It is this fact in particular that affords considerable precision in reconstructing the Death Valley extensional terrane. The localization of the two principal breakaway zones near the Keystone and Wheeler Pass systems leaves a stable terrane between them composed of the Keystone and higher thrust plates that lie below the Wheeler Pass system (Fig. 7).

Our strategy for constraining both local and province-wide extension is to determine horizontal relative motion vectors V_i between the ith pair of reference points, which are chosen so as to best allow geologic markers to constrain the pair's relative motion. Any sequence of n pairs that defines an unbroken path between endpoints then defines the relative motion V_e between the endpoints (compare with Minster and Jordan, 1984, 1987).

$$\sum_{i=1}^{n} V_i = V_e$$

We ignore curvature of the Earth and vertical motions of the reference points, as they are negligible in comparison with the horizontal offsets.

The reference points and key geologic markers are shown in Figure 8. Below, we discuss constraints on the relative motion between pairs of points that define paths suitable for both local and province-wide reconstruction, summarized in Table 1. We emphasize determination of the uncertainties in each of the vectors, which in most cases are based on simple strain compatibility arguments, principally the condition that geologic markers do not overlap in the reconstruction.

Las Vegas System

The Las Vegas system is composed predominantly of southwest-directed normal faults. Geologic data allow constraint of the motion between the Spring Mountains block and the Colorado Plateau using two independent paths, including point pairs A1A2, A2A3, and A3A4, in the Lake Mead area, and C1C2, C2B1, and B1B2, in the Mormon Mountains–Las Vegas Valley area (Fig. 8).

Reconstruction via Lake Mead Fault System. Anderson (1971) and Anderson and others (1972) first recognized the large-magnitude extension in the Lake Mead area and concluded that most of the deformation occurred between 15 and 11 Ma, although the precise magnitude was not determined. Anderson (1973) and Bohannon (1979, 1984) also suggested that large-magnitude strike-slip faulting was present in the region, indicating about 65 km of translation of the Frenchman Mountain block southwestward away from the Virgin Mountains area (V_{A2A3}; Figs. 4 and 8). The evidence for offset includes (1) a distinctive stratigraphic sequence in basal Tertiary (Miocene) sedimentary rocks (Bohannon, 1979, 1984); (2) the geometry of the basal Tertiary unconformity, which at Frenchman Mountain and in the Virgin Mountains, gradually cuts out Mesozoic section when followed from north to south (Bohannon, 1979, 1984); and (3) the presence of numerous landslide breccias within the Miocene section in the Frenchman Mountain block, composed of at least 14 differ-

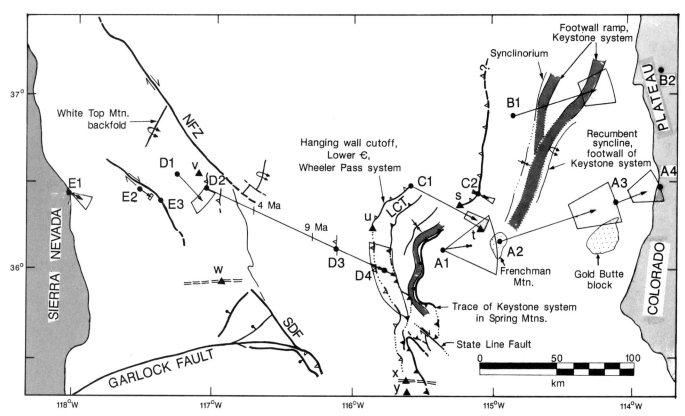

Figure 8. Map showing location of points used for reconstruction paths, key structural features used to constrain offsets, and "best-fit" restoration vectors and their individual uncertainties. Small-lettered points (triangles) are reference points for strain compatibility arguments discussed in text. Capital-lettered points (filled circles) each belong to one or more point pairs used for reconstruction paths. Dotted lines, structural features projected beneath Quaternary alluvium; light shading, uncertainty regions of vectors in Table 1; note irregular geometry of V_{D2D4}; double dashed lines, northern limits of abundant Mesozoic plutons in southern Panamint Range and Clark Mountains area; NFZ, northern Death Valley–Furnace Creek fault zone; SDF, southern Death Valley fault zone; LCT, Lee Canyon thrust. Doubling of northernmost exposures of footwall ramp in Keystone system is due to offset on Mormon Peak detachment. Vertical lines on segment D2D3 show western limit of Tertiary strata that overlap major extensional structures at times indicated (from Wright and others, 1984). See Table 1 for numerical values of vectors, and text for discussion of their derivation.

ent rock types that match those seen in the Gold Butte block, many of which are not common in other crystalline blocks exposed in the region (Figs. 3 and 8; Anderson, 1973; Longwell, 1974; Parolini and others, 1982; Parolini and Rowland, 1988).

The Frenchman Mountain block must have been in a position close enough to the Gold Butte block to receive megabreccia deposits for which the transport direction was to the north (Anderson, 1973; Longwell, 1974; Parolini and Rowland, 1988). Restoration of Frenchman Mountain to a position north of the Gold Butte block is also supported by the southward pinchout of the Jurassic Aztec Sandstone and Triassic Chinle Formation beneath the basal Tertiary unconformity in the Frenchman Mountain block and in the fault blocks in the South Virgin Mountains just to the north of Gold Butte (Bohannon, 1979, 1984). On the basis of the proximal-channel facies of the megabreccias (Parolini and Rowland, 1988), a distance of no more than 10 km between the Frenchman Mountain and the Gold Butte blocks prior to extension is likely, giving a minimum of 50 km of west-southwest relative motion between the two (lower bound of displacement for V_{A2A3}; Fig. 8). Palinspastic reconstruction of the Gold Butte and other blocks in the South Virgin Mountains (for example, Wernicke and Axen, 1988), however, requires that they restore at least 10 km east toward the Colorado Plateau, but no farther than the edge of the plateau itself, giving a minimum westward translation of the Frenchman Mountain block relative to the plateau of 60 km (minimum length of $V_{A2A3} + V_{A3A4}$; Fig. 8, Table 1). The maximum possible translation is 90 km, the current distance from Frenchman Mountain to the plateau.

The azimuth of displacement suggested by matching the southward pinchouts of the Jurassic Aztec sandstone beneath the basal Tertiary unconformity (Bohannon, 1979) is S70°W, consistent with other kinematic studies of extension direction in the region (for example, Anderson, 1971; Angelier and others, 1985; Wernicke and others, 1985; Smith and others, 1987). The pinchout, however, is so gradual that its pre-extension trend is poorly constrained and is valid as a piercing point only if Frenchman Mountain restores directly atop the blocks north of Gold Butte (see analysis of Proffett, 1977, Fig. 11). An uncertainty in the azimuth of displacement of 10° for V_{A2A3} from this direction, however, places Frenchman Mountain too far north to receive the proximal megabreccias on the northern extreme, and too far south to accept crystalline detritus from the south on the southern extreme. We assume the same direction for V_{A3A4}, but assign a 20° azimuthal uncertainty, constrained on the north by the overlap of Phanerozoic strata at A3 with those in the North Virgin Mountains (Fig. 4) and on the south by the improbability that the blocks had a northward component of motion relative to the plateau (Fig. 8, Table 1).

TABLE 1. RELATIVE MOTION VECTORS BETWEEN SELECTED POINTS IN THE BASIN AND RANGE PROVINCE NEAR THE LATITUDE OF LAS VEGAS, NEVADA

	Vector	Displacement (km)	Azimuth
I. Las Vegas system			
A. Lake Mead path	V_{A1A2}	8 ± 8	N84° ± 20°E
	V_{A2A3}	65 ± 15	N70° ± 10°E
	V_{A3A4}	20 ± 10	N70° ± 20°E
B. Mormon Mountains/ Las Vegas Valley path	V_{B1B2}	54 ± 10	N75° ± 10°E
	V_{B1C2}	0 ± 10	N68° ± 0°W
	V_{C1C2}	48 ± 7	N69° ± 3°W
II. Spring Mountains rotation	V_{D4C1}	0 ± 10	N65° ± 0°W
III. Death Valley system	V_{D2D4}	125 ± 7	N65° ± 7°W
	V_{D1D2}	22 ± 3	N45° ± 20°W
	V_{E1E2}	9 ± 6	N60° ± 20°W
	V_{E2E3}	9 ± 1	N35° ± 10°W

Anderson (1973) and Bohannon (1979, 1984) proposed that the motion of Frenchman Mountain relative to the Virgin Mountains was accommodated by left-lateral strike-slip faulting and recognized that it may be kinematically associated with normal faulting (for example, Hamilton and Myers, 1966; Davis and Burchfiel, 1973). The extent to which the large translations are a product of crustal thinning versus strike slip, however, is not clear from the field relations. Although there are clearly large left-lateral strike-slip offsets present in the region (a stratovolcano centered on the fault system is offset about 20 km by the Hamblin Bay fault, Fig. 4; Anderson, 1973), it is not clear to what extent the translation of the Frenchman Mountain block away from the plateau is a product of *crustal* strike slip (deep-seated relative translation without crustal thinning) versus crustal extension. As emphasized by Anderson (1984), plane strain by sets of strike-slip faults may combine with normal faulting that is otherwise not kinematically coordinated with (or even coeval with) the strike-slip faulting to produce crustal extension. Thus, the left-lateral faults in the Lake Mead region may combine with the right-lateral Las Vegas Valley shear zone to accommodate east-west extension and coeval north-south shortening. The ambiguity of how much extension is absorbed by north-south shortening versus crustal extension is a problem throughout in Death Valley region as well. There is clearly a major component of strike slip in the extending system (for example, Wright and others, 1981; Anderson, 1984; see faults in Fig. 4), and we will attempt to quantify its contribution to the over-all strain pattern below.

In order to obtain the displacement of the Spring Mountains block relative to the plateau, the displacement between the Spring Mountains and Frenchman Mountain must be determined (V_{A1A2}; Fig. 8, Table 1). This is not well known as there are no geologic markers between the two that can be used as a basis for reconstruction. The relative stability of the Spring Mountains contrasts with the highly extended fault blocks on Frenchman Mountain (Longwell and others, 1965), and presumably there has been significant separation of the two, as suggested by the presence of Las Vegas Valley between them. The limits on magnitude of relative motion for V_{A1A2} are 0 and 36 km, assuming no extension and complete closure of Las Vegas Valley (that is, no overlap of autochthonous Phanerozoic section), respectively. The uncertainty in azimuth is difficult to evaluate; we chose an uncertainty of ±30° as a conservative limit (Fig. 8, Table 1), which assumes that it is not exceptional to known regional extension directions in either the Death Valley or Las Vegas systems.

Reconstruction via Las Vegas Valley Shear Zone and Mormon Mountains-Tule Springs Hills Area. An alternative to the path between A1 and A4 is combining a cross-section palinspastic reconstruction drawn between the Colorado Plateau and the Meadow Valley Mountains (V_{B1B2}; Axen and Wernicke, 1987, unpub. data) with the offset along the Las Vegas Valley shear zone (V_{C1C2}; for example, Burchfiel, 1965). The Mormon Mountains and Tule Springs Hills are the principal ranges along the transect and have been mapped at scales of 1:12,000 to 1:24,000 along the entire length of the palinspastic reconstruction. On the basis of varying the geometry of faults at depth that are significant to the reconstruction, Axen and Wernicke (1987) determined that there has been 54 ± 10 km of extension (Fig. 8, Table 1). The azimuth of the extension direction is constrained by mesoscopic studies of fault striae and fault mullions in the Beaver Dam Mountains, fault dips and striae in the Tule Springs Hills, and trends of tear faults in the Tule Springs Hills (Smith and others, 1987). Farther west, the orientation of an elongate dome in the highest major detachment in the system (Mormon Peak detachment); the net slip determined on one of the normal faults in the Mormon Mountains; and the bisectrix of a system of small, conjugate normal faults exposed throughout the western half of the Mormon Mountains constrain the extension direction (Wernicke and others, 1985). These indicators yield an extension direction between S60°W and S80°W for V_{B1B2} (Fig. 8, Table 1).

The Meadow Valley Mountains are structurally contiguous with the Las Vegas Range and are thus part of the central stable block (Figs. 4 and 7). Although the block comprises a number of ranges, including the Meadow Valley Mountains and the Las Vegas, Sheep, and Arrow Canyon Ranges (Fig. 4), their bounding faults are steep and discontinuous along strike, and Mesozoic structures within the ranges are not cut by major detachments (Langenheim, 1988). In particular, the Keystone and Gass Peak systems are continuous and maintain their relative spacing from north to south between B1 and C2 (Fig. 8; although note offset of the Keystone system ramp by the Mormon Peak detachment). A small amount of extension (2–5 km), however, is probable. In addition, the block may have rotated about a vertical axis during extension, although rotation of more than 10° in either direction seems unlikely in that it would misalign structural elements of the thrust belt from their regional north-northeast to north-south trends. This is supported by paleomagnetic studies at three sites located 7, 20, and 25 km due north of C2, indicating little rotation of the block following thrusting (Nelson and Jones, 1987). Thus, we assign a value of 5 ± 10 km of motion S65 ± 10°W to V_{B1C2} to account for the extension and possible small rotations of the entire block (Fig. 8, Table 1).

Offset on the Las Vegas Valley shear zone is the last vector needed to complete the closing of the Las Vegas system (V_{C1C2}). A displacement of 48 km best aligns the Gass Peak thrust with the Wheeler Pass thrust, also closely aligning the positions of the Keystone system ramp and its hanging-wall syncline. Displacement of more or less than about 7 km either way results in significant misalignment of these features. The displacement includes oroflexure in the Las Vegas Range area (for example, Burchfiel, 1965), which has been confirmed by paleomagnetic studies (Nelson and Jones, 1987). A northern limit for the azimuth of reconstruction is firmly set by the requirement that the thrust traces do not overlap (C1 cannot restore north of reference point s, Fig. 8). The southern limit is constrained by the maximum amount of north-south compressive strain accommodated in range blocks and in Las Vegas Valley. Although there may be minor north-south shortening in the ranges north of the shear zone, the oroflexural bending in these ranges seems best considered as a response of the brittle crust to deep-seated simple shear on vertical planes (for example, Shawe, 1965). As concluded by Nelson and Jones (1987), a model of block rotation above a smoothly shearing medium below (Jackson and McKenzie, 1983) fits paleomagnetic and structural data well. Along the portion of the shear zone between C1 and C2, a wide alluviated valley is present (Las Vegas Valley, Fig. 4), favoring net transtension over transpression. Minimal north-south distance between exposures of the thrusts prior to motion on the shear zone places C1 no farther south than reference point t in Figure 8; hence, we assign a 3° uncertainty to the azimuth of V_{C1C2} (Fig. 8, Table 1).

Death Valley System

Down-to-the-west normal faulting of the Death Valley normal fault system is superimposed on the Wheeler Pass system and higher thrusts (Fig. 7). By reconstructing the Mesozoic orogen in the Spring Mountains and in ranges to the west, we have established firm correlations between individual thrust faults discontinuously exposed in the range blocks and determined their relative spacing. An important factor in the precision of the reconstruction is the correlation of the Panamint thrust fault, exposed in the Panamint Range (point D2, Fig. 8), with the Chicago Pass thrust, exposed in the Nopah Range (point D3, Fig. 8; Wernicke and others, 1988a, 1988b). Correlations of structurally higher thrusts confirm this, because they tie together the thrust stack in the Tucki Mountain–Cottonwood Mountains area with that in the Funeral and Grapevine Mountains areas (J. K. Snow and B. Wernicke, unpub. data), showing that the entire system, now exposed across an area more than 150 km wide, was initially slightly less than 30 km wide prior to extension (compare scales of Figs. 5 and 6). The principal marker constraining the reconstruction is the White Top Mountain backfold in the Cottonwood Mountains and a correlative fold system in the Funeral Mountains (Figs. 5, 6, and 8; J. K. Snow and B. Wernicke, unpub. data).

Tucki Mountain–Nopah Range Reconstruction. The Panamint and Chicago Pass thrusts exposed in these two areas share the distinguishing features of the Wheeler Pass system described above. In addition, the normal fault blocks of the two ranges reconstruct such that the ranges structurally overlap one another in map view (Fig. 6b; Wernicke and others, 1988b). The geology of both ranges is characterized by steep dips of Tertiary strata that lie with mild angular unconformity on Paleozoic strata (Burchfiel and others, 1983; Wernicke and others, 1986, 1988a). Proximity of the two ranges prior to extension was proposed by Stewart (1970, 1983), based on similarities in stratigraphy of the miogeoclinal clastic wedge between them. Stewart's (1983) reconstruction restores the strong anomaly in isopach trends across the northern Death Valley–Furnace Creek fault zone (NFZ, Fig. 4), indicating about 80 km of displacement N55°W of the Panamint Range block relative to the Nopah–Resting Springs Range block. The reconstruction was questioned by Prave and Wright (1986), who argued that the isopach trends could be reasonably restored with only 50 km of displacement.

Identification of the Panamint thrust at Tucki Mountain confirms Stewart's (1983) placement of the Panamint Range adjacent to the Nopah–Resting Springs Range block because it provides a structural marker that can be used to precisely determine the relative offset. As discussed above, the Panamint and Chicago Pass thrusts have about 5,000 m of stratigraphic throw, cut steeply (40°–60°) across miogeoclinal layering, and occur in identical positions in the miogeocline. In addition, the structural details of the exposure of the two thrusts permit cross-section reconstruction of the two range blocks *directly adjacent* to one another without holes or overlap (Fig. 6b). In the Nopah Range, the thrust places lowest Cambrian on Devonian-Mississippian strata (at point D3), whereas at Tucki Mountain, the thrust places Middle Cambrian strata on Permian (at point D2). In both blocks, now tilted owing to extension, a normal fault system dips slightly more shallowly than the thrust, such that it cuts downward to the west from the footwall into the hanging wall of the thrust, moving the footwall westward over the hanging wall. In both ranges, the normal fault crosses the thrust at very shallow angle where the thrust emplaces high Lower or low Middle Cambrian on Pennsylvanian or Permian strata (at points D3 and D2, respectively). Figure 6b shows that the two ranges fit directly against one another, restoring into a crustal sliver only 2–3 km wide that contains the trace of the thrust.

If it is assumed that the thrusts do correlate and require juxtaposition of the two ranges, they do not specify azimuthal control on V_{D2D3}. We suspect that points D2 and D3 fit directly against one another; otherwise, the structural details of the normal fault system relative to the thrust in the two ranges would have to persist for significant distances along strike. As we show below, however, an azimuth similar to the one suggested by Stewart (1983) is indicated from independent strain-compatibility arguments.

Closing Pahrump Valley. The placement of the combined Nopah–Resting Springs–Panamint crustal sliver in its position with respect to the Spring Mountains (V_{D2D4}, Table 1) can be done with precision by considering the trace of the Wheeler Pass relative to other thrusts in the Spring Mountains block. When followed from north to south, (1) the thrusts curve from northeast strikes to due north or north-northwest, (2) individual thrust plates carry progressively thinner sections of the miogeoclinal prism, and (3) the relative spacing between major thrusts decreases. At large scale, the Death Valley system breakaway fault zone makes a concave-west scoop across the west side of the Spring Mountains such that the Wheeler Pass system projects beneath the alluvium of Pahrump Valley for a distance of 60 km and reappears in the Clark Mountains area as the Winters Pass thrust (Figs. 4, 7, and 8; Burchfiel and Davis, 1971, 1981). In the Clark Mountains area, spacing between the three major thrusts, the Keystone, Mesquite Pass, and Wheeler Pass thrusts, is only about a kilometer or two, whereas to the north in the Spring Mountains, the spacing is about 10–15 km (Figs. 6 and 8). Thus, a geologic line associated with the Wheeler Pass thrust (for example, the intersection of the thrust plane with footwall Mississippian strata) projected southward into Pahrump Valley would lie within 30 km of the trace of the Keystone system and within 10 km of the Lee Canyon thrust, the major thrust between the Keystone and Wheeler Pass (Figs. 6 and 8). The choice of geologic line is not critical, as the thrust cuts steeply through the miogeoclinal section where exposed, and thus, any fault-bed intersection originally lay within a few kilometers of any other along the thrust. These arguments suggest that D2 and D3 restore to a position at least as far east as that line, *regardless of their initial position relative to one another* (Fig. 8).

The Panamint Range block, although tilted and extended, is a relatively coherent homocline of miogeoclinal strata and underlying basement that contains the trace of the thrust. It has experienced little, if any, north-south internal strain (for example, Wernicke and others, 1988a; Albee and others, 1981). At the southern end of the block, exposures of upper Paleozoic and Mesozoic strata in Butte Valley (Fig. 4) may represent the footwall of the thrust, as they are juxtaposed against basement and Pahrump strata along the steeply dipping Butte Valley fault zone (Johnson, 1957). We favor the interpretation that the Butte Valley fault juxtaposes the hanging wall of the Wheeler Pass system with its footwall (Figs. 5 and 7), such that the Precambrian is downthrown along the fault relative to the younger rocks. If so, then the Panamint Range homocline is everywhere within a kilometer or two of the thrust plane, tightly constraining the position of the Panamints with respect to the Spring Mountains block, in light of the western limit on the original position of the Wheeler Pass system in Pahrump Valley discussed above. Even if the upper Paleozoic and Mesozoic strata in Butte Valley are not part of the footwall of the thrust, the shallow dip of the thrust where exposed at Tucki Mountain (Wernicke and others, 1988b) suggests that it cannot have strayed too far beneath the Panamint homocline south of Tucki Mountain.

An independent argument supporting correlation of the Panamint thrust with the Wheeler Pass system is the apparent structural continuity of rocks in the Panamints with those exposed in the hanging wall of the Winters Pass thrust in the Clark Mountains area. As indicated by Burchfiel and others (1983), the region between the Nopah–Resting Springs Range block and the Panamints is devoid of exposures of older-over-younger faults. It is composed of a number of steeply east-dipping, north-striking homoclines repeated on numerous low-angle normal faults (Wright and Troxel, 1973, 1984). These blocks are apparently fragments of a once-contiguous homocline exposed from the Winters Pass thrust to the Panamints, including the southern Nopah Range, the Kingston Range, the southern Black Mountains, and other smaller blocks (Fig. 4). Northward pinchouts of the Pahrump beneath Precambrian Z strata are preserved within each of the blocks and fall on a single west-northwest–trending line between the Kingston and southern Panamint Ranges (Wright and others, 1974), and the basal Tertiary unconformity in most places rests on Lower or Middle Cambrian strata, which throughout the area are fairly uniform in thickness (for example, Stewart, 1970). These relations indicate that it is unlikely that a thrust fault with 5 km of stratigraphic throw disrupts the blocks between the Winters Pass area and the Panamints. All of the blocks likely belong to the same Mesozoic thrust plate, and their lower bounds (Chicago Pass, Winters Pass, and Panamint thrusts) are thus parts of the same thrust system (Fig. 7).

Azimuthal limits on V_{D2D4} are prescribed on the northern extreme by the condition that Cambrian strata intersecting the thrust plane at Tucki Mountain (reference point v, Fig. 8) do not overlap those at the southern limit of exposure of the Wheeler Pass thrust (reference point u, Fig. 8). On the south, a limit is set by the presence of a roughly east-west–trending boundary between both hanging-wall and footwall strata intruded by Mesozoic plutons to the south and a pluton-free area to the north (Fig. 8). If the upper Paleozoic and Mesozoic strata in Butte Valley are not in the footwall of the thrust system, the reconstruction is less tightly constrained, but not seriously compromised. The Panamint block must restore to a position such that unintruded hanging-wall rocks do not overlap intruded hanging wall in the Clark Mountains area. The precise location of the thrust plane in the Panamints is not critical to this constraint, as in both areas, the northern limit of the plutonic belt is laterally persistent for tens of kilometers. In Figure 8, this constraint means that point w in the Butte Valley area cannot restore to a position south of point x in the Clark Mountains area. The possibility exists, however, that there has been north-south shortening in the southern part of the Spring Mountains block, in which case the southern limit of the plutonic belt would shift farther south. The shortening may be accommodated by as much as 20 km of right-lateral movement along the State Line fault (Fig. 8; Hewitt, 1956), although the geology on both sides of the fault does not require major displacement. Because significant displacement is possible, however, we consider a reference point y 10 km to the south of x as a southern limit for the azimuth of the reconstruction. These constraints give an azimuth of N65 ± 7°W for V_{D2D4}. This corresponds closely with the azimuth inferred for restoration of Tucki Mountain and Chicago Pass discussed above, but is based on independent constraints.

Distance limits on V_{D2D4} are given by the need to restore D2 at least as far east as the projection of the Wheeler Pass system into Pahrump Valley, but not so far east that it would overlap the interpolated trajectory of the next-lower thrusts (Lee Canyon and Green Monster thrusts) into Pahrump Valley (Fig. 8). These limits place D2 at D4, 125 ± 7 km S65° ± 7°E of its present location, but with the distance uncertainty skewed such that the minimum follows the western limit for the position of the Wheeler Pass system (Fig. 8).

The final vector needed to close Pahrump Valley is that for the restoration of the Cottonwood Mountains relative to Tucki Mountain (V_{D1D2}, Table 1). Palinspastic reconstruction of the Cottonwood Mountains relative to D2 on Tucki Mountain requires restoration of the Emigrant fault system on the east side of the Panamint block. Detailed mapping and structural analysis of this area (see Wernicke and others, 1986, 1988b; Hodges and others, 1987) show that 20–25 km of extension has occurred between these points, oriented N45° ± 20°W, assuming liberal uncertainty in the extension direction from several hundred measurements of fault striae and mylonitic stretching lineations in the extended blocks at Tucki Mountain (Walker and others, 1986).

Cottonwood Mountains to the Sierra Nevada. Extension between the Cottonwood Mountains and the Sierra Nevada is modest and is best constrained by closing the northern part of Panamint Valley along the Hunter Mountain fault (Burchfiel and others, 1987). Piercing points across this structure indicate 9 ± 1 km of motion oriented N55° ± 10°W (Burchfiel and others, 1987) for V_{E2E3} (Table 1). The area between D1 and E3 is occupied by the Hunter Mountain batholith terrain and does not appear to be highly extended or rotated about a vertical axis.

The last vector to be considered is one that connects the northern Argus–southern Inyo Range area with the Sierra Nevada, which takes into account extension related to the opening of Owens Valley (V_{E1E2}, Table 1). There are no major detachments or major normal faults other than those bordering Owens Valley between E1 and E2 (see Dunne, 1986, for a review of extensional structures in this region). The geometry of faulting at depth in Owens Valley is not known but is probably of the type that is steep and fairly deeply penetrating (for example, Anderson and others, 1983). Two major known faults, the Owens Valley and Independence faults (Fig. 4), are steeply dipping at the surface and show evidence of oblique slip, being primarily right-lateral strike slip for the Owens Valley fault and dip slip for the Independence fault (Zoback and Beanland, 1986; Gillespie, 1982). We assume that the region is generally pervaded by high-angle faults with modest offset; an estimate of 15% ± 10% extension reasonably bounds the extension, giving a displacement of 9 ± 6 km. An over-all extension direction of N60° ± 20°W, parallel to the geodetically determined direction (Savage, 1983; see also review of pertinent data in Jones, 1987), bounds the azimuth.

Rotation of the Spring Mountains Block. The two vector paths within the Las Vegas system and the one in the Death Valley system have different endpoints in the Spring Mountains block that are separated by about 50 km. Thus, although the Spring Mountains block is negligibly extended internally (for example, Burchfiel and others, 1974; Axen, 1984), rotation about a vertical axis of the entire range block could significantly affect the relative position of two widely separated points. In particular, rotation of segment C1D4 introduces significant error into determination of east-west extension. The amount of rotation has yet to be constrained paleomagnetically, but the over-all north to north-northeast trend of the thrust faults in the area suggests that major rotation (in excess of 10°) has probably not occurred. This uncertainty will be considered in the discussion of putting the other vectors together into a whole-province reconstruction.

DISCUSSION

Vector Addition

A whole-province reconstruction may be obtained by adding the displacements of the Las Vegas system to those of the Death Valley system. Because we have two paths in the Las Vegas system based on independent constraints, we can narrow the uncertainty limits on the Las Vegas system reconstruction to that region of uncertainty common to both paths. We can then define a new vector for the Las Vegas system and add it to the Death Valley system to obtain a whole-province reconstruction, taking into account possible rotation about a vertical axis of the Spring Mountains block.

In adding vectors with uncertainties, we assume constant probability distribution within each uncertainty domain. We generate the combined uncertainty region by sweeping one uncertainty region around the other, placing the "best-fit" vector of one on the perimeter of the other. The new area is that in which any combination of the summed vectors may lie. This area is a conservative estimate of the uncertainty, as two randomly chosen vectors from the original uncertainty fields are less likely to sum to a point on the perimeter of the combined uncertainty field than to a point near the center (see Monte Carlo simulation, below). The uncertainty is also considered conservative to the extent that the probability distribution within each of the uncertainty regions to be summed is not everywhere equal, but generally concentrated in the center near the "best-fit" vector. Because it is difficult to quantify the probability distribution for each vector, we have chosen an even distribution in order to provide an upper limit on the uncertainty.

Adding the vector paths between the Spring Mountains and the Colorado Plateau shows that the uncertainty for the Lake Mead path is considerably larger than that for the Mormon Mountains area path (Figs 9a and 9b, respectively); however, there is a relatively small area of overlap between the two (Fig. 10). The Lake Mead path suggests a more northerly over-all extension direction; most of the uncertainty lies in the positioning of the Frenchman Mountain block (Fig. 8). The more easterly trend of the Mormon Mountains path is due primarily to the southeasterly motion on the Las Vegas Valley shear zone. The two paths are thus best reconciled by southeasterly motion between the Frenchman Mountain block and the Spring Mountains, parallel to the Las Vegas shear zone. If

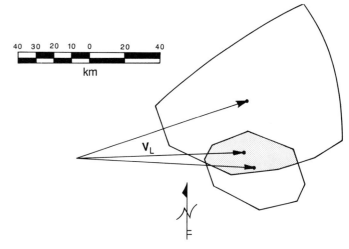

Figure 10. Derivation of Las Vegas system vector V_L using overlap of combined uncertainty fields of two independent paths (Figs. 9a and 9b).

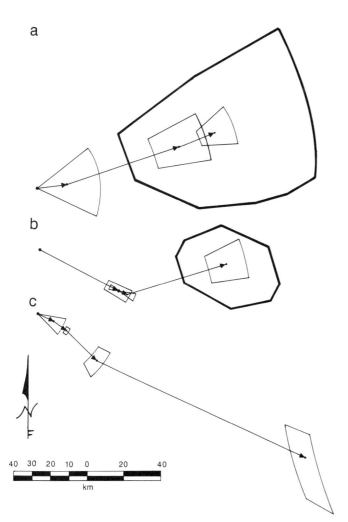

Figure 9. Summation of vectors in reconstruction paths. a. Las Vegas normal fault system via Lake Mead area path; b. Las Vegas system via Mormon Mountains–Las Vegas Valley area path; c. Death Valley system. Bold outlines in a and b show limits of combined uncertainties. See text for discussion.

so, it appears likely that a strand of the shear zone passes to the south of Frenchman Mountain (for example, Anderson, 1973). The area of overlap defines a new vector V_L with an uncertainty region that comprises all the combinations of vectors from the two paths that are consistent between the two sets. We chose a "best fit" that lies in the center of this combined uncertainty field (Fig. 10).

Combining the vector derived from the two paths in the Las Vegas system with those in the Death Valley system must account for possible rotation of the Spring Mountains block about a vertical axis, because the paths for the Las Vegas and Death Valley systems have distant endpoints within the Spring Mountains (Fig. 8). Although such a rotation of the whole block is probably small, a significant differential displacement of points D4 and C1 with respect to the plateau is possible. To account for as much as 10° of rotation of either sense, we let $V_{C1D4} = 0 \pm 10$ km N65°W (Fig. 8, Table 1).

The total displacement of the Sierra Nevada relative to the Colorado Plateau V_t is

$$V_t = V_{E1E2} + V_{E2E3} + V_{D1D2} + V_{D2D4} + V_{D4C1} + V_L \quad (1)$$

We depict the result of this addition in the form of an error cloud produced by a Monte Carlo simulation of 10,000 runs of equation 1, where vectors were chosen randomly from the uncertainty regions of each vector (Fig. 11). A curve parallel to the density distribution of points which excludes 5% of them is taken as an estimate of the uncertainty in the province-wide reconstruction. From this, we obtain $V_t = 247 \pm 56$ km S73° $\pm$ 12°E, using the "best-fit" vectors and considering the extremes of the error curve in Figure 11.

Strain Rate

The timing of extensional tectonism in the region is constrained to have occurred principally between 20 m.y. ago and the present. Deposition of orogenic conglomerates in the oldest preserved Tertiary strata began in Oligocene time (for example, Reynolds, 1974), but major extension, as indicated by angular unconformities within the Tertiary section and the depositional overlap of extensional structures, appears not to have begun until after 20 m.y. ago (Bohannon, 1984; Anderson and others, 1972; Cemen and others, 1985; Wright and others, 1983). On the basis of the age of strata cut by major extensional features, the bulk of the exten-

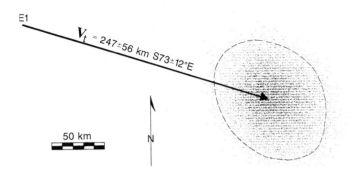

Figure 11. Neogene displacement vector and uncertainty region of point E1 (Fig. 8) with respect to the Colorado Plateau. See text for discussion.

sion appears to be post–15 m.y. ago. The peak period of extension on the Las Vegas system occurred between 15 and 11 m.y. ago (Anderson and others, 1972; Bohannon, 1984; Smith and others, 1987). In the Death Valley system, most of the deformation in the eastern part of the path occurred between about 14 and 4 m.y. ago (Cemen and others, 1985), but in western areas, tens of kilometers of displacement have probably occurred since 4 m.y. ago (for example, Wernicke and others, 1986; Burchfiel and others, 1987; Butler and others, 1988), including motion on major low-angle normal faults. If it is assumed that most of the translation of the Sierra Nevada away from the plateau occurred after 15 m.y. ago, the average displacement rate for the past 15 m.y. is 16.7 ± 4.5 mm/yr.

Given that the province has at least doubled in width (lower limit of displacement of about 190 km compared with a current width of 360 km) and may have experienced a sixfold increase in width (about 300 km of extension or 500% increase over original width), the time-averaged strain rate of the lithosphere as a whole is in the range 2.1×10^{-15} to 1.9×10^{-14} s^{-1}. The lower bound of our average displacement rate is greater than the upper bound of Minster and Jordan's (1987) Holocene opening rate of 9.7 ± 2.1 mm/yr derived from considerations of geodetic data, the RM2 plate model, and strain west of the San Andreas fault. The azimuth of opening from this study and that derived by Minster and Jordan (1984, 1987) are similar. Combined, these data indicate that Basin and Range extension has slowed significantly over the past 15 m.y. Such slowing would require faster displacement rates during earlier parts of the Neogene. For example, if the opening rate of the Basin and Range has been 8 mm/yr for the past 5 m.y., then the average rate between 5 and 15 m.y. ago would have been 21 ± 7 mm/yr.

The concept of slowing of Basin and Range extension with time has been hypothesized based on the observation that widely spaced, steep normal faults commonly overprint younger, more closely spaced normal-fault systems (for example, Zoback and others, 1981), but measurements supporting such a hypothesis have heretofore been lacking. The difference in timing between extension in the Las Vegas system and the Death Valley system places bounds on how the displacement rate of the Sierra relative to the plateau varied in time. Figure 8 shows points on V_{D2D4} that constrain the timing of movement of the Panamint Range block northwestward relative to the Spring Mountains block. Tertiary overlap of major extensional structures occurred by 9 m.y. ago in the Resting Springs Range area and by 4 m.y. ago in the Furnace Creek Wash area (Fig. 8; Wright and others, 1983, 1984; Cemen and others, 1985; McAllister, 1973). Thus, of the 125 km of motion represented by D2D4, at least 50 km had occurred by 10 m.y. ago, at least 90 km had occurred by 5 m.y. ago, and no more than about 35 km occurred between 5 m.y. ago and the present. If it is assumed that all motion on vectors west of D2 occurred in the past 5 m.y., the Death Valley system has accommodated no more than 50 km of extension in the past 5 m.y., giving a rate of 10 mm/yr, in good agreement with the Holocene rate of Minster and Jordan (1987). Because it is clear that extension in the Las Vegas system was complete by 5 m.y. ago (although it was probably mostly complete by 10 m.y. ago; Anderson and others, 1972; Bohannon, 1984), slowing of extension with time is required.

The most likely displacement history, neglecting the effects of locally variable extension direction, includes about 150 km of extension accommodated on both systems between 10 and 15 m.y. ago (100 km on the Las Vegas system, 50 km on the Death Valley system, for a total of 30 mm/yr) and an additional 100 km accommodated on the Death Valley system in the past 10 m.y. (10 mm/yr), with over-all slowing occurring between 5 and 10 m.y. ago (Fig. 12, curve 1). Alternatively, if extension in the Las Vegas system were evenly distributed across the time interval 15–5 m.y. ago, then the displacement rate would be 20 mm/yr for that interval, slowing to 10 mm/yr for the past 5 m.y. (Fig. 12, curve 2). The actual displacement rate probably slowed from a value in excess of 20 mm/yr to one near 10 mm/yr over the interval 5–15 m.y. ago (Fig. 12, curve 3). Timing data are not yet precise enough to meaningfully bound displacement rates at greater precision than over 5-m.y. intervals.

Although the slowing documented in this study is in accord with that previously suggested on the basis of changing structural style with time in the Basin and Range, we stress that our results are independent of assumptions of structural style. In fact, we note that the Holocene opening rate of 9.7 ± 2.1 mm/yr (Minster and Jordan, 1987; Fig. 12) must be accommodated principally in the western part of the Death Valley system at the latitude of Las Vegas (Fig. 7), which has a structural style similar to that of earlier extensional regimes (Wernicke, 1981; Wernicke and others, 1986; Hamilton, 1987; Burchfiel and others, 1987). According to the analysis above, neither the Death Valley system nor the Las Vegas system individually needs to have spread at a rate in excess of 10 mm/yr at any time in their histories. Thus, our results do not necessarily support the concept that slowing of Basin and Range extension is associated with a change in structural style or that such change, if any, constrains the displacement rate.

The magnitudes of strain rate and the changes of strain rate with time are all broadly consistent with the physical model presented by Sonder and others (1987) and Wernicke and others (1987), suggesting that extension is controlled by the gravitational collapse of crust overthickened during Mesozoic time (for example, Coney and Harms, 1984). According to the calculations of Sonder and others (1987), the peak magnitude of strain rate

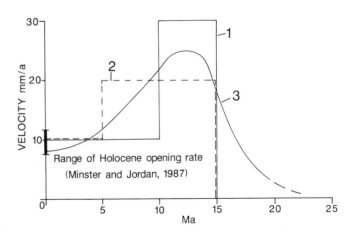

Figure 12. Opening rates of Las Vegas area Basin and Range averaged over 5-m.y. intervals. Curve 1, assuming Las Vegas system ceased moving by 10 m.y. ago; curve 2, assuming Las Vegas system ceased moving by 5 m.y. ago; curve 3, possible smoothed path. See text for discussion.

is approximately $1.0-3.0 \times 10^{-15}$ s^{-1} with a time scale of slowing on the order of a factor of 2–3 per 10 m.y. following the peak of extensional strain rate. The calculations therefore broadly agree with these observations, but of course do not rule out forces other than gravitational ones for the origin of extension.

An important difference between the results of Sonder and others (1987) and of this report is the magnitude of strain. Figure 5 shows that the total shortening across the thrust belt is about 100 km or roughly a factor of 2. Given an initially cold Moho temperature (<600 °C) for the Las Vegas region (Sonder and others, 1987) and shortening on the order of a factor of 2 during thrusting, the gravitational-collapse model is difficult to reconcile with the "best-fit" value of $\beta = 3.5$, although it is consistent with the lower bound (extension factor $\beta > 2$). The results of Sonder and others (1987) show that in general, it is possible to have a greater magnitude of extension than of compression because of the excess potential added to the lithosphere by upward advection of heat during extension. For a broad range of assumptions as to how such advection occurs and the mechanical properties of the lithosphere, however, the extension in the Las Vegas area, if significantly greater than the lower bound of $\beta = 2$, may not be entirely explained by the gravitational collapse mechanism. As we discuss below, two factors may contribute to the possible discrepancy between our measurements and the gravitational collapse model: (1) a large component of constrictional strain in the evolution of the Las Vegas area Basin and Range and (2) the availability of a driving force for extension other than gravitational collapse, possibly the tangential shear traction exerted on the west margin of the North American plate by the Pacific plate in Neogene time.

Constrictional Strain Component

A substantial percentage of the east-west extensional strain in the region may be absorbed by north-south crustal shortening rather than crustal thinning, resulting in an over-all constriction of the crust during extension. The effect of constriction on the gravitational collapse model, which is one dimensional and assumes that all crustal extension is accommodated by plane-strain crustal thinning, is to allow the crust to extend more than the driving force of gravity alone might permit. For example, if 30% of the total extensional strain is balanced by north-south shortening, then sufficient energy may be available from gravitational collapse to account for the remaining component of extension. Such a solution to the problem is not entirely satisfactory, however, because it does not specify the driving force for the constriction itself. A detailed assessment of this problem is beyond the scope of this paper, but we can make some statements about the likely importance of constriction in the region and propose a model for its origin.

As pointed out by Wright (1976), the presence of numerous, large strike-slip faults in the Las Vegas area relative to other parts of the Basin and Range may indicate that the region has been extended more than have areas to the north and south, responding to the difference in extension by accommodating much of it along conjugate strike-slip zones in addition to

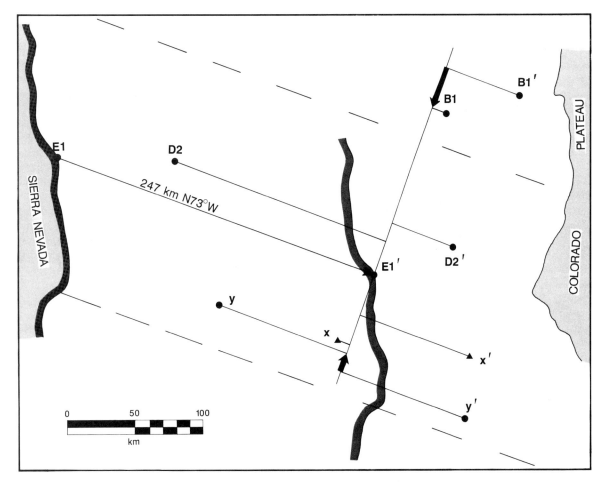

Figure 13. Map showing selected points from Figure 8 (unprimed) restored to their pre-extension configurations (primed) relative to the Colorado Plateau, using "best-fit" vector V_t from Figure 11. See text for discussion.

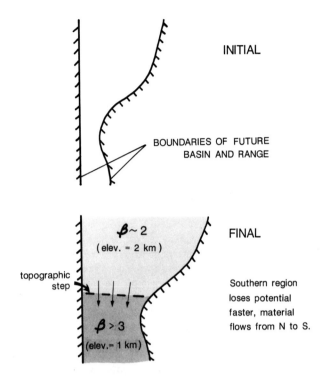

Figure 14. Model for origin of constrictional strain in the Las Vegas area Basin and Range. See text for discussion.

normal faulting. Hamilton and Myers (1966), Davis and Burchfiel (1973), and Lawrence (1976) viewed strike-slip faulting at the northern and southern extremes of the Great Basin as large tear or transform faults bounding regions of relatively large extension from those of little or no extension. The major strike-slip faults in the Great Basin region characteristically have apparent left-lateral offset where northeast striking and right-lateral offset where northwest striking (Shawe, 1965). This is particularly clear in the Las Vegas region, where the Lake Mead and Garlock fault systems are left lateral and the Las Vegas Valley shear zone and Death Valley fault zone and several other northwest-striking faults are right lateral (Fig. 4). Shawe's (1965) theory of Basin and Range extension ascribes a fundamental role to deep-seated conjugate strike-slip faults in the development of Basin and Range structure. Such an origin implies a predominantly map-view plane-strain pattern in which the crust need not thin appreciably in the accommodation of extension, thus representing an end-member case in which conjugate shears accommodate most or all of the extension. At the other extreme, the "intracontinental transform" (Davis and Burchfiel, 1973) or "transfer" (Gibbs, 1984) fault model of these strike-slip faults implies that none of the strike slip be attributed to north-south shortening. These end members are herein referred to as "transfer" and "conjugate" faulting.

The relative importance of conjugate faulting can be assessed by determining the component of north-south motion of crustal blocks relative to the net extension direction of N73° ± 12°W. Consider two points B1 and y in the extending system (Figs. 8 and 13), which initially lie at the northern and southern extremes of the region characterized by strike-slip faulting. Point y lies in the northern Mojave Desert and thus accounts for approximately 65 km of left slip on the Garlock fault in the reconstruction (vector not listed in Table 1). The initial positions of selected points from Figure 8 are shown in Figure 13 as primed points. Measured relative to the net extension direction, points B1 and y converge 43 km, using the "best-fit" vectors. Because of the uncertainty in the net extension direction and in the individual displacements, there is a great deal of uncertainty in this estimate. In addition, if extension directions were different at different times or in different areas (for example, Zoback and others, 1981), it is possible to produce apparent shortening perpendicular to the net extension direction in the absence of true conjugate faulting. For example, a combination of southwest extension of points B1 and y followed by northwest extension involving only y would produce apparent convergence of B1 and y without conjugate faulting or shortening perpendicular to the net extension direction. Considering these difficulties, we can estimate, but not reasonably bound, the constrictional component. It does not seem likely, however, that the shortening is substantially more than 43 km, which represents about a 20% decrease in original north-south distance between B1 and y. Thus, the component of crustal extension accommodated by horizontal shortening may be significant but is probably relatively small. For an extension factor $\beta = 3.3$ and north-south shortening factor of 0.8, the crustal thinning factor is 0.38, as opposed to a thinning factor of 0.30 in the absence of horizontal shortening.

The modest amount of constriction is probably related to the narrowness of the province at the latitude of Las Vegas (for example, Wright, 1976). If we consider the motion of two rigid blocks relative to one another with unequal widths of material accommodating the strain between them, crustal thinning will be greater in the narrow areas than in the wide ones (Fig. 14). The thin-sheet calculations of Sonder and others (1987) indicate that significant differential strain between two regions (say, thinning by a factor $\beta = 3.5$ versus $\beta = 2$) may lead to a significant gradient in gravitational potential of the lithosphere between them. We suggest that such a gradient may have led to flow of material parallel to the gradient (Fig. 4). Such a contrast in buoyancy of the lithosphere may still be affecting the region, where a strong north-south gradient in regional topography is present all across the province (for example, Eaton and others, 1978) and north-south shortening is active, as in the Death Valley region (Fig. 7). In addition, the Intermountain Seismic Belt (for example, Smith, 1978) is developed along the topographic discontinuity and is perhaps driven by the current buoyancy contrast. Such a contrast is not present to the south of the region, but the contrast in width of the province to the south is negligible compared with that to the north.

The mode of strain accommodation for the constrictional component seems to be similar to that proposed by Hill (1982), whereby a system of conjugate faults bounds rigid, crudely hexagonal blocks. In this model, the strike-slip faults are both conjugate and transform, because they transfer extensional displacement from one zone of pull-apart to another, while simultaneously accommodating extensional strain in a manner similar to that proposed by Shawe (1965). The join between the Las Vegas Valley shear zone and the Hamblin Bay fault (Fig. 4) may represent the southern apex of a relatively stable block that moved southward into the region of severe pull-apart between the Spring Mountains and the Colorado Plateau, as predicted by Hill's (1982) block model (see also Anderson, 1984).

CONCLUSIONS

We conclude that the Sierra Nevada moved 247 ± 56 km N73° ± 12°W relative to the Colorado Plateau in Neogene time. The rate of motion appears to be on average more easterly and of greater magnitude (20–30 mm/yr) in the early phases of extension between 10 and 15 m.y. ago, slowing to its current rate of less than 10 mm/yr. Although the implications of these measurements for the Cenozoic tectonics of western North America and for processes of extension in general are significant, we emphasize their contribution to the model of Atwater (1970), who proposed that Basin and Range extension was in part the result of diffuse extensional shearing of the continent in response to the growing right-lateral San Andreas transform. Since her study, global plate reconstructions and improved knowledge of the San Andreas fault have confirmed that Pacific–North America relative plate motion during the past 20 m.y.

exceeds the offset on the San Andreas fault by more than a factor of 2 (for example, Stock and Molnar, 1988).

Resolved parallel to the northern San Andreas, our reconstruction adds at least 166 km and as much as 262 km of right-lateral motion along the plate boundary at the latitude of the central Colorado Plateau, coeval with growth of the transform plate boundary. Within the uncertainties of all the data bearing on the probem, the extension magnitude proposed herein supports the hypothesis that all of the right-lateral displacement between the growing San Andreas transform and North America was accommodated as relatively diffuse deformation within the North American plate. Much work remains to be done, however, on reconciling the details and timing of the deformation with plate-tectonic constraints. For example, the west-southwest direction of extension within portions of the Las Vegas normal fault system may be the last phases of extension prior to a province-wide reorientation of extension direction to west-northwest at about 10 m.y. ago, thought to signal the influence of the San Andreas transform on Basin and Range extension (for example, Zoback and others, 1981). Major extension (probably on the order of 100 km or more) in the northern Great Basin region occurred in Eocene to mid-Miocene time (for example, Coney and Harms, 1984; Wernicke and others, 1987), and as such, the 15- to 10-m.y.-ago extension in the Las Vegas region may have an origin more closely related to pure gravitational collapse than to forces applied to the edge of North America by the San Andreas transform system. Continuum models of the San Andreas transform as a finite-width deformation zone attached firmly to the edge of North America have yielded promising comparisons with the actual zone of plate boundary deformation (Sonder and others, 1986).

Our results are one of a long series of attempts to constrain timing and magnitude of Basin and Range extension. We feel that the strength of our approach is to introduce quantitative rigor to the problem, including the major challenge of developing criteria for bounding uncertainties in geologic reconstructions. Although construction of balanced and restored cross sections yields possible strain fields (and spares us the embarrassment of proposing an impossible one), we have not yet developed a general set of techniques for determining all possible strain fields from a given body of geologic data. This study represents a crude attempt at doing so, but there is considerable room for debating, point by point, our methods for determining uncertainties, and we hope that from such debate a set of more general principles on how to handle these problems might emerge. Because the principal basis for assigning uncertainties has been the simple compatibility condition that no two points occupy the same place upon reconstruction, we think that we have not grossly underestimated the uncertainties. New geologic, geochronologic, and paleomagnetic data may expand, but one hopes, mostly contract, the uncertainty regions presented in Figure 8 and Table 1. In addition, new data to the north and south may provide more independent paths across all or parts of the province that will serve to reduce uncertainties. We envision an advanced stage of research in the Basin and Range where individual displacement vectors form a tight grid over the province, ultimately allowing inverse modeling of the strain field to obtain the stress field.

Our results fully confirm earlier suggestions that extension in the Basin and Range province is quite large. Hamilton and Myers (1966) proposed that the province may have doubled in width, based in part on the possibility that range-bounding faults may flatten with depth. Since then, a major geologic revolution has fundamentally changed how geologists view extension of the Earth's lithosphere, based principally on Basin and Range field studies. The lower limit of our measurement of extension (which is likely too small) exceeds those made by a handful of workers in the 1960s and 1970s (with the exception of Hamilton and Myers), who at that time were regarded by most other workers as the liberal fringe.

The magnitude of the shift in thinking shows the importance not only of these early field studies on the problem, but also of field-oriented geologic research in general. The revolution in thinking about how the crust extends is likely only a minor sampling of what geologic field relations have yet to teach us about the continental lithosphere. We are probably in a period of development of research into the nature of the continental lithosphere where we know a great deal about large areas of the continents in reconnaissance but have yet to ask the right questions of the rocks to shed proper light on major processes; this is certainly the feeling one gets from comparing the voluminous pre-Anderson literature on the Basin and Range with that of today. Far from the nineteenth-century descriptive science that field geology is perceived to be by the growing number of earth scientists comfortably insulated from terrestrial reality by machines and elegant calculations, it seems to us that major advances in understanding the continental lithosphere will be unlikely in the absence of well-posed, field-oriented research.

ACKNOWLEDGMENTS

We thank R. E. Anderson, J. M. Bartley, R. G. Bohannon, B. C. Burchfiel, M. D. Carr, J. C. Crowell, G. A. Davis, G. C. Dunne, P. L. Guth, W. B. Hamilton, K. V. Hodges, M. W. Reynolds, J. M. Stock, B. W. Troxel, J. D. Walker, and L. A. Wright for discussions that contributed substantially to the content of this report. We thank Yahn Bernier for developing the code for the Monte Carlo simulation. Thoughtful reviews by Richard W. Allmendinger and Peter J. Coney contributed greatly to the clarification of ideas presented herein. This research was supported by grants from Texaco, Incorporated; Exxon Production Research Company; the Shell Foundation; and National Science Grants EAR-84-51181 and EAR-86-17869 awarded to B. P. Wernicke.

REFERENCES CITED

Albee, A. L., Labotka, T. C., Lanphere, M. A., and McDowell, S. D., 1981, Geologic map of the Telescope Peak quadrangle, California: U.S. Geological Survey Map GQ-1532.
Anderson, R. E., 1971, Thin-skin distension in Tertiary rocks of southwestern Nevada: Geological Society of America Bulletin, v. 82, p. 43–58.
——— 1973, Large-magnitude late Tertiary strike-slip faulting north of Lake Mead, Nevada: U.S. Geological Survey Professional Paper 794, 18 p.
——— 1984, Strike-slip faults associated with extension in and adjacent to the Great Basin: Geological Society of America Abstracts with Programs, v. 16, no. 6, p. 429.
Anderson, R. E., Longwell, C. R., Armstrong, R. L., and Marvin, R. F., 1972, Significance of K-Ar ages of Tertiary rocks from the Lake Mead region, Nevada-Arizona: Geological Society of America Bulletin, v. 83, p. 273–288.
Anderson, R. E., Zoback, M. L., and Thompson, G. A., 1983, Implications of selected subsurface data on the structural form and evolution of some basins in the northern Basin and Range province, Nevada and Utah: Geological Society of America Bulletin, v. 94, p. 1055–1072.
Angelier, J., Coletta, B., and Anderson, R. E., 1985, Neogene paleostress changes in the Basin and Range; a case study at Hoover Dam, Nevada-Arizona: Geological Society of America Bulletin, v. 96, p. 347–361.
Armstrong, R. L., 1963, Geochronology and geology of the eastern Great Basin in Nevada and geology of the eastern Great Basin in Nevada and Utah [Ph.D. thesis]: New Haven, Connecticut, Yale University, 202 p.
——— 1972, Low-angle (denudation) faults, hinterland of the Sevier orogenic belt, eastern Nevada and western Utah: Geological Society of America Bulletin, v. 83, p. 1729–1754.
Atwater, T., 1970, Implications of plate tectonics for the Cenozoic tectonic evolution of western North America: Geological Society of America Bulletin, v. 81, p. 3513–3536.
Axen, G. J., 1984, Thrusts in the eastern Spring Mountains, Nevada: Geometry and mechanical implications: Geological Society of America Bulletin, v. 95, p. 1202–1207.
Axen, G. J., and Wernicke, B. P., 1987, Magnitude and style of Miocene upper-crustal extension in the southern Nevada area: Geological Society of America Abstracts with Programs, v. 19, p. 576.
Bohannon, R. G., 1979, Strike-slip faults of the Lake Mead region of southern Nevada, in Armentrout, J. M., and others, eds., Cenozoic paleogeography of the western United States: Society of Economic Paleontologists and Mineralogists, Pacific Section, Pacific Coast Paleogeography Symposium, 3rd, p. 129–139.
——— 1984, Nonmarine sedimentary rocks of Tertiary age in the Lake Mead region, southeastern Nevada and northwestern Arizona: U.S. Geological Survey Professional Paper 1259, 69 p.
Burchfiel, B. C., 1965, Structural geology of the Specter Range quadrangle, Nevada and its regional significance: Geological Society of America Bulletin, v. 76, p. 175–192.
Burchfiel, B. C., and Davis, G. A., 1971, Clark Mountain thrust complex in the Cordillera of southeastern California: Geologic summary and field trip guide: University of California, Riverside, Museum Contribution, v. 1, p. 1–28.
——— 1975, Nature and controls of Cordilleran orogenesis in the western United States: Extensions of an earlier synthesis: American Journal of Science, v. 275A, p. 363–396.
——— 1981, Mojave desert and environs, in Ernst, W. G., The geotectonic development of California: Englewood Cliffs, New Jersey, Prentice-Hall, p. 218–252.
Burchfiel, B. C., Fleck, R. J., Secor, D. I., Vincelette, R. R., and Davis, G. A., 1974, Geology of the Spring Mountains, Nevada: Geological Society of America Bulletin, v. 85, p. 1013–1022.
Burchfiel, B. C., Hamill, G. S., and Wilhelms, D. I., 1982, Geologic map with discussion of stratigraphy of the Montgomery Mountains and the northern half of the Nopah and Resting Springs Ranges, Nevada and California: Geological Society of America Map and Chart Series MC 44.
——— 1983, Structural geology of the Montgomery Mountains and the northern half of the Nopah and Resting Springs Ranges, Nevada and California: Geological Society of America Bulletin, v. 94, p. 1359–1376.
Burchfiel, B. C., Hodges, K. V., Walker, J. D., Klepacki, D. W., Tilke, P. G., Crowley, P. D., Jones, C. H., and Davis, G. A., 1985, The Kingston Range detachment system: Structures at the eastern edge of the Death Valley extensional zone, southeastern California: Geological Society of America Abstracts with Programs, v. 17, no. 6, p. 345.
Burchfiel, B. C., Hodges, K. V., and Royden, L. H., 1987, Geology of Panamint Valley-Saline Valley pull-apart system, California: Palinspastic evidence for low-angle geometry of a Neogene range-bounding fault: Journal of Geophysical Research, v. 92, p. 10422–10426.
Butler, P. R., Troxel, B. W., and Verosub, K. L., 1988, Late Cenozoic history and styles of deformation along the southern Death Valley fault zone, California: Geological Society of America Bulletin, v. 100, p. 402–410.
Cemen, I., Wright, L. A., Drake, R. E., and Johnson, F. C., 1985, Cenozoic sedimentation and sequence of deformational

events at the southeastern end of the Furnace Creek strike-slip fault zone, Death Valley region, California, *in* Biddle, K. T., and Christie-Blick, N., eds., Strike-slip deformation, basin formation, and sedimentation: Society of Economic Paleontologists and Mineralogists Special Publication 37, p. 127–141.

Coney, P. J., and Harms, T. A., 1984, Cordilleran metamorphic core complexes: Cenozoic extensional relics of Mesozoic compression: Geology, v. 12, p. 550–554.

Crittenden, M. D., Jr., Coney, P. J., and Davis, G. H., eds., 1980, Cordilleran metamorphic core complexes: Geological Society of America Memoir 153, 490 p.

Curry, H. D., 1954, Turtlebacks in the central Black Mountains, Death Valley, California, *in* Jahns, R. H., ed., Geology of southern California: California Division of Mines and Geology Bulletin, v. 170, p. 53–59.

Davis, G. A., and Burchfiel, B. C., 1973, Garlock fault: An intra-continental transform structure, southern California: Geological Society of America Bulletin, v. 84, p. 1407–1422.

Davis, G. H., and Coney, P. J., 1979, Geologic development of Cordilleran metamorphic core complexes: Geology, v. 7, p. 120–124.

de Charpal, D., Montadert, L., Guennoc, P., and Roberts, D. G., 1978, Rifting, crustal attenuation and subsidence in the Bay of Biscay: Nature, v. 275, p. 706–710.

Dunne, G. C., 1986, Mesozoic evolution of the southern Inyo Mountains, Darwin Plateau, and Argus and Slate Ranges, *in* Dunne, G. C., ed., Mesozoic and Cenozoic structural evolution of selected areas, east-central California: Geological Society of America, Cordilleran Section Field Trip 2, Guidebook, p. 3–21.

Dunne, G. C., Gulliver, R. M., and Sylvester, A. G., 1978, Mesozoic evolution of rocks of the White, Inyo, Argus, and Slate Ranges, eastern California, *in* Howell, D., and McDougall, K., eds., Mesozoic paleogeography of the western United States: Society of Economic Paleontologists and Mineralogists, Pacific Section, Los Angeles, California, v. 2, p. 189–206.

Dunne, G. C., Gulliver, R. M., and Stevens, C. H., 1981, Correlation of Mississippian shelf-to-basin strata, eastern California: Geological Society of America Bulletin, v. 92, Part II, p. 1–38.

Eaton, G. P., Wahl, R. R., Prostka, E. J., Mabey, D. R., and Kleinkopf, M. D., 1978, Regional gravity and tectonic patterns: Their relation to late Cenozoic epeirogeny and lateral spreading in the western Cordillera, *in* Smith, R. B., and Eaton, G. P., eds., Cenozoic tectonics and regional geophysics of the western Cordillera: Geological Society of America Memoir 152, p. 51–92.

Gibbs, A. D., 1984, Structural evolution of extensional basin margins: Geological Society of London Journal, v. 141, p. 609–620.

Gillespie, A. R., 1982, Quaternary glaciation and tectonism in the southern Sierra Nevada, Inyo County, California [Ph.D. thesis]: Pasadena, California, California Institute of Technology, 695 p.

Guth, P. L., 1981, Tertiary extension north of the Las Vegas Valley shear zone, Sheep and Desert Ranges, Clark County, Nevada: Geological Society of America Bulletin, v. 92, p. 763–771.

Hamilton, W., 1969, Mesozoic California and the underflow of Pacific mantle: Geological Society of America Bulletin, v. 80, p. 2409–2430.

——— 1987, Crustal extension in the Basin and Range province, southwestern United States, *in* Coward, M. P., and others, eds., Continental extensional tectonics: Geological Society of London Special Publication 28, p. 155–176.

Hamilton, W., and Myers, W. B., 1966, Cenozoic tectonics of the western United States: Reviews of Geophysics, v. 4, p. 509–549.

Hewitt, D. F., 1956, Geology and mineral resources of the Ivanpah quadrangle, California and Nevada: U.S. Geological Survey Professional Paper 275, 172 p.

Hill, D. P., 1982, Contemporary block tectonics: California and Nevada: Journal of Geophysical Research, v. 87, p. 5433–5450.

Hodges, K. V., Walker, J. D., and Wernicke, B. P., 1987, Footwall structural evolution of the Tucki Mountain detachment system, Death Valley region, southeastern California, *in* Coward, M. P., and others, eds., Continental extensional tectonics: Geological Society of London Special Publication 28, p. 393–408.

Hunt, C. B., and Mabey, D. R., 1966, Stratigraphy and structure, Death Valley, California: U.S. Geological Survey Professional Paper 494-A, 162 p.

Jackson, J. A., and McKenzie, D., 1983, The relationship between strain rates, crustal thickening, paleomagnetism, finite strain and fault movements within a deforming zone: Earth and Planetary Science Letters, v. 65, p. 182–202.

Johnson, B. K., 1957, Geology of a part of the Manly Peak quadrangle, southern Panamint Range, California: University of California Publications in Geological Sciences, v. 30, p. 353–423.

Jones, C. H., 1987, Is extension in Death Valley accommodated by thinning of the mantle lithosphere beneath the Sierra Nevada, California?: Tectonics, v. 6, p. 449–474.

Labotka, T. C., and Albee, A. L., 1988, Metamorphism and tectonics of the Death Valley region, California and Nevada, *in* Ernst, W. G., ed., Metamorphism and crustal evolution of the western United States: Englewood Cliffs, New Jersey, Prentice-Hall, p. 714–736.

Langenheim, R. L., 1988, Extensional and other structures in the Arrow Canyon and Las Vegas Ranges, Clark County, Nevada: Geological Society of America Abstracts with Programs, v. 20, no. 3, p. 175.

Langenheim, R. L., Carss, B. W., Kennerly, J. B., McCuctheon, V. A., and Waines, R. H., 1962, Paleozoic section in Arrow Canyon Range, Clark County, Nevada: American Association of Petroleum Geologists Bulletin, v. 46, p. 592–609.

Lawrence, R. D., 1976, Strike-slip faulting terminates the Basin and Range province in Oregon: Geological Society of America Bulletin, v. 87, p. 846–850.

Longwell, C. R., 1945, Low-angle normal faults in the Basin and Range province: American Geophysical Union Transactions, v. 26, p. 107–118.

——— 1974, Measure and rate of movement on Las Vegas Valley shear zone, Clark County, Nevada: Geological Society of America Bulletin, v. 85, p. 985–990.

Longwell, C. R., Pampeyan, E. H., Bowyer, B., and Roberts, R. J., 1965, Geology and mineral deposits of Clark County, Nevada: Nevada Bureau of Mines and Geology Bulletin 62, 218 p.

McAllister, J. F., 1973, Geologic map and sections of the Amargosa Valley borate area, southeast continuation of the Furnace Creek area, Inyo County, California: U.S. Geological Survey Miscellaneous Geological Investigations Map I-782.

McKenzie, D. P., 1978, Some remarks on the development of sedimentary basins: Earth and Planetary Science Letters, v. 40, p. 25–32.

Miller, J.M.G., 1987, Paleotectonic and stratigraphic implications of the Kingston Peak–Noonday contact in the Panamint Range, eastern California: Journal of Geology, v. 95, p. 75–85.

Minster, J. B., and Jordon, T. H., 1984, Vector constraints on Quaternary deformation of the western United States east and west of the San Andreas fault, *in* Crouch, J. K., and Bachman, S. B., eds., Tectonics and sedimentation along the California margin: Society of Economic Paleontologists and Mineralogists, Pacific Section, v. 38, p. 1–16.

——— 1987, Vector constraints on western U.S. deformation from space geodesy, neotectonics, and plate motions: Journal of Geophysical Research, v. 92, p. 4798–4804.

Misch, P., 1960, Regional structural reconnaissance in central-northeast Nevada and some adjacent areas: Observation and interpretations: Intermountain Association of Petroleum Geologists, Annual Field Conference, 11th, Guidebook, p. 17–42.

Nelson, C. A., 1981, Basin and Range province, *in* Ernst, W. G., ed., The geotectonic development of California: Englewood Cliffs, New Jersey, Prentice-Hall, p. 203–216.

Nelson, M. R., and Jones, C. H., 1987, Paleomagnetism and crustal rotations along a shear zone, Las Vegas, southern Nevada: Tectonics, v. 6, p. 13–33.

Noble, L. F., 1941, Structural features of the Virgin Spring area, Death Valley, California: Geological Society of America Bulletin, v. 52, p. 941–1000.

Parolini, J. R., and Rowland, S. M., 1988, Proximal Miocene debris flows in the Frenchman Mountain structural block: Implications for extension in the Lake Mead region: Geological Society of America Abstracts with Programs, v. 20, no. 3, p. 220.

Parolini, J. R., Smith, E. I., and Wilbanks, J. R., 1982, Landslide masses in the Rainbow Gardens, Clark County, Nevada; lithology, emplacement, and significance: Geological Society of America Abstracts with Programs, v. 14, p. 223.

Pashley, E. F., 1966, Structure and stratigraphy of the central, northern, and eastern parts of the Tucson basin, Pima County, Arizona [Ph.D. thesis]: Tucson, Arizona, University of Arizona, 273 p.

Poole, F. G., and Sandberg, C., 1977, Mississippian paleogeography and tectonics of the western United States, *in* Stewart, J. H., and others, eds., Paleozoic paleogeography of the western United States: Society of Economic Paleontologists and Mineralogists, Pacific Coast Paleogeography Symposium, 1st, p. 67–86.

Prave, A. R., and Wright, L. A., 1986, Isopach pattern of the Lower Cambrian Zabriske Quartzite, Death Valley region, California-Nevada: how useful in tectonic reconstructions?: Geology, v. 14, p. 251–254.

Price, R. A., 1981, The Cordilleran foreland thrust and fold belt in the southern Canadian Rocky Mountains, *in* McClay, K. R., and Price, N. J., eds., Thrust and nappe tectonics: Geological Society of London, p. 427–448.

Proffett, J. M., 1977, Cenozoic geology of the Yerington district, Nevada, and implications for the nature and origin of Basin and Range faulting: Geological Society of America Bulletin, v. 88, p. 247–266.

Ransome, F. L., Emmons, W. H., and Garrey, G. H., 1910, Geology and ore deposits of the Bullfrog district, Nevada: U.S. Geological Survey Bulletin 407, 130 p.

Reynolds, M. W., 1974, Geology of the Grapevine Mountains, Death Valley, California, a summary: Guidebook: Death Valley region, California and Nevada: Shoshone, California, Death Valley Publishing Company, p. 92–97.

Royse, F., 1983, Comment on "Magnitude of crustal extension in the southern Great Basin": Geology, v. 11, p. 495–496.

Savage, J., 1983, Strain accumulation in western United States: Annual Reviews of Earth and Planetary Sciences, v. 11, p. 11–43.

Shawe, D. R., 1965, Strike-slip control of Basin-Range structure indicated by historical faults in western Nevada: Geological Society of America Bulletin, v. 76, p. 1361–1378.

Smith, E. I., Anderson, R. E., Bohannon, R. G., and Axen, G. J., 1987, Miocene extension, volcanism, and sedimentation in the eastern Basin and Range province, southern Nevada, *in* Davis, G. H, and Van der Dolder, E. M., eds., Geologic diversity of Arizona and its margins: Excursions to choice areas: Arizona Bureau of Geology and Mineral Technology Special Paper 5, p. 383–397.

Smith, R. B., 1978, Seismicity, crustal structure, and intraplate tectonics of the western Cordillera, *in* Smith, R. B., and Eaton, G. P., eds., Cenozoic tectonics and regional geophysics of the western Cordillera: Geological Society of America Memoir 152, p. 111–144.

Sonder, L. J., England, P. C., and Houseman, G. A., 1986, Continuum calculations of continental deformation in transcurrent environments: Journal of Geophysical Research, v. 91, p. 4797–4810.

Sonder, L. J., England, P. C., Wernicke, B. P., and Christiansen, R. L., 1987, A physical model for Cenozoic extension of western North America, *in* Coward, M. P., and others, eds., Continental extensional tectonics: Geological Society of London Special Publication 28, p. 187–201.

Stewart, J. H., 1970, Upper Precambrian and Lower Cambrian strata in the southern Great Basin, California and Nevada: U.S. Geological Survey Professional Paper 620, 206 p.

——— 1972, Initial deposits in the Cordilleran geosyncline: Evidence of a late Precambrian (<850 m.y.) continental separation: Geological Society of America Bulletin, v. 83, p. 1345–1360.

——— 1978, Basin-range structure in western North America: A review, *in* Smith, R. B., and Eaton, G. P., Cenozoic tectonics and regional geophysics of the western Cordillera: Geological Society of America Memoir 152, p. 1–32.

——— 1983, Extensional tectonics in the Death Valley area, California: Transport of the Panamint Range structural block 80 km northwestward: Geology, v. 11, p. 153–157.

Stewart, J. H., Ross, D. C., Nelson, C. A., and Burchfiel, B. C., 1966, Last Chance thrust, a major fault in the eastern part of Inyo County, California: U.S. Geological Survey Professional Paper 550D, p. D23–D34.

Stock, J., and Molnar, P., 1988, Uncertainties and implications of the Late Cretaceous and Tertiary position of North America relative to the Farallon, Kula, and Pacific plates: Tectonics (in press).

Stone, P., and Stevens, C., 1984, Stratigraphy and deposition history of Pennsylvanian and Permian rocks in the Owens Valley–Death Valley region, eastern California, *in* Lintz, J. P., ed., Western geological excursions, Volume 4: Reno, Nevada, Mackay School of Mines, p. 94–119.

——— 1987, Stratigraphy of the Owens Valley Group (Permian), southern Inyo Mountains, California: U.S. Geological Survey Bulletin 1962, 19 p.

——— 1988a, An angular unconformity in the Permian section of east-central California: Geological Society of America Bulletin, v. 100, p. 547–551.

——— 1988b, Pennsylvanian and Early Permian paleogeography of east-central California: Implication for the shape of the continental margin and the timing of continental truncation: Geology, v. 16, p. 330–333.

Tschanz, C. M., and Pampeyan, E. H., 1970, Geology and mineral deposits of Lincoln County, Nevada: Nevada Bureau of Mines Bulletin 73, 188 p.

Walker, J. D., Hodges, K. V., and Wernicke, B. P., 1986, The relation of tilt geometry to extension direction: Geological Society of America Abstracts with Programs, v. 18, no. 2, p. 194.

Wernicke, B., 1981, Low-angle normal faults in the Basin and Range province: Nappe tectonics in an extending orogen: Nature, v. 291, p. 645–648.

Wernicke, B., and Axen, G. J., 1988, On the role of isostasy in the evolution of normal fault systems: Geology, v. 16 (in press).

Wernicke, B., Spencer, J. E., Burchfiel, B. C., and Guth, P. L., 1982, Magnitude of crustal extension in the southern Great Basin: Geology, v. 10, p. 499–502.

Wernicke, B., Guth, P. L., and Axen, G. J., 1984, Tertiary extension in the Sevier orogenic belt, southern Nevada, *in* Lintz, J. P., ed., Western geological excursions, Volume 4: Reno, Nevada, Mackay School of Mines, p. 473–510.

Wernicke, B., Walker, J. D., and Beaufait, M. S., 1985, Structural discordance between Neogene detachments and frontal Sevier thrusts, central Mormon Mountains, southern Nevada: Tectonics, v. 4, p. 213–246.

Wernicke, B., Hodges, K. V., and Walker, J. D., 1986, Geological setting of the Tucki Mountain area, Death Valley National Monument, California, *in* Dunne, G. C., ed., Mesozoic and Cenozoic structural evolution of selected areas, east-central California: Geological Society of America, Cordilleran Section, Los Angeles, p. 67–80.

Wernicke, B., Christiansen, R. L., England, P. C., and Sonder, L. J., 1987, Tectonomagmatic evolution of Cenozoic extension in the North American Cordillera, *in* Coward, M. P., and others, eds., Continental extensional tectonics: Geological Society Special Publication 28, p. 203–221.

Wernicke, B. P., Snow, J. K., and Walker, J. D., 1988a, Correlation of early Mesozoic thrusts in the southern Great Basin and their possible indication of 250–300 km of Neogene crustal extension, *in* Weide, D. L., and Faber, M. L., eds., This extended land, geological journeys in the southern Basin and Range: Geological Society of America, Cordilleran Section, Field Trip Guide, p. 255–269.

Wernicke, B., Walker, J. D., and Hodges, K. V., 1988b, Field guide to the northern part of Tucki Mountain fault system, Death Valley region, California, *in* Weide, D. L., and Faber, M. L., eds., This extended land, geological journeys in the southern Basin and Range: Geological Society of America, Cordilleran Section, Field Trip Guide, p. 58–65.

Wright, L. A., 1976, Late Cenozoic fault patterns and stress fields in the Great Basin and westward displacement of the Sierra Nevada block: Geology, v. 4, p. 489–494.

Wright, L. A., and Troxel, B. W., 1973, Shallow-fault interpretation of Basin and Range structure, southwestern Great Basin, *in* de Jong, K. A., and Scholten, R., eds., Gravity and tectonics: New York, John Wiley and Sons, p. 397–407.

——— 1984, Geology of the northern half of the Confidence Hills 15′ quadrangle, Death Valley region, eastern California: The area of the Amargosa chaos: California Division of Mines and Geology Map Sheet 34.

Wright, L. A., Otton, J. K., and Troxel, B. W., 1974, Turtleback surfaces of Death Valley viewed as phenomena of extension: Geology, v. 2, p. 53–54.

Wright, L. A., Troxel, B. W., Burchfiel, B. C., Chapman, R., and Labotka, T., 1981, Geologic cross section from the Sierra Nevada to the Las Vegas Valley, eastern California to southern Nevada: Geological Society of America Map and Chart Series MC-28M.

Wright, L. A., Troxel, B. W., and Drake, R. E., 1983, Contrasting space-time patterns of extension-related, late Cenozoic faulting, southwestern Great Basin: Geological Society of America Abstracts with Programs, v. 15, p. 287.

Wright, L. A., Drake, R. E., and Troxel, B. W., 1984, Evidence for the westward migration of severe Cenozoic extension, southeastern Great Basin, California: Geological Society of America Abstracts with Programs, v. 16, no. 6, p. 701.

Young, J. C., 1960, Structure and stratigraphy in north-central Schell Creek Range: Intermountain Association of Petroleum Geologists, Annual Field Conference, 11th, Guidebook, p. 158–172.

Zoback, M. L., and Beanland, S., 1986, Temporal variations in stress magnitude and style of faulting along the Sierran frontal fault system: Geological Society of America Abstracts with Programs, v. 18, no. 6, p. 801.

Zoback, M. L., Anderson, R. E., and Thompson, G. A., 1981, Cainozoic evolution of the state of stress and style of tectonism in the Basin and Range province of the western United States: Royal Society of London Philosophical Transactions, v. A300, p. 407–434.

MANUSCRIPT RECEIVED BY THE SOCIETY MARCH 10, 1988
REVISED MANUSCRIPT RECEIVED JULY 12, 1988
MANUSCRIPT ACCEPTED JULY 12, 1988

Seismic imaging of extended crust with emphasis on the western United States

CENTENNIAL ARTICLE

JILL MCCARTHY   *U.S. Geological Survey, M.S. 977, 345 Middlefield Road, Menlo Park, California 94025*
GEORGE A. THOMPSON   *Geophysics Department, Stanford University, Stanford, California 94305*

ABSTRACT

Understanding of the crust has improved dramatically following the application of seismic reflection and refraction techniques to studies of the deep crust. This is particularly true in areas where the last tectonic event was extensional, such as the Basin and Range province of the western United States and much of western Europe. In these regions, a characteristic reflective pattern has emerged, whereby the lower crust is highly reflective and the upper crust and upper mantle are either poorly reflective or strikingly nonreflective. In the metamorphic-core-complex belt in the western United States, where extension can be as much as an order of magnitude greater than in the more classic continental rift zones, the lower crustal reflectivity thickens and rises, yielding a picture of a crust that is reflective throughout. Synthetic seismic studies have documented that the reflectivity in these regions can be modeled by numerous laminae tens of meters thick and hundreds of meters across, characterized by interlayered high and low velocities. Two geologic factors are interpreted as contributing to this layered character: ductile strain, responding to stress in the thermally weakened middle and lower crust, and intrusive layering, corresponding to injection of subhorizontal sheets of mantle-derived magmas. These two processes yield a variety of geologic structures, including transposed compositional layering, mylonitic ductile shear zones, and intrusive mafic sheets, all of which occur at the proper scales to cause the prominent reflectivity observed. If metamorphic core complexes are representative of extended continental crust world-wide, then these results suggest that magmatism and ductile flow have also contributed to the evolution of the middle and lower crust in many other areas around the world.

INTRODUCTION AND SCOPE

On this occasion of the hundredth anniversary of the Geological Society of America (GSA), our objective is to review the current status of crustal imaging and the role the *Bulletin* and its daughter journal *Geology* have played in illuminating this progress. In this paper, we focus on the reflection character of extended continental crust, building on a variety of exciting new results that have reshaped our thinking as to the composition of the crust and how it forms. We illustrate some of the more significant progress and problems by means of selected examples. A variety of more comprehensive reviews of seismic imaging that include other types of geologic terranes are already available (for example, Barazangi and Brown, 1986a, 1986b; Matthews and Smith, 1987). We begin by highlighting major milestones in the seismic study of the Earth's crust.

Historical Background of Deep Crustal Studies and the Role of the GSA

From a bold interpretation of scattered earthquake arrival times in Europe, Mohorovičić discovered a lower boundary to the low-velocity, upper layer of the Earth, which is now known as the *crust* (Fig. 1; Mohorovičić, 1909; Byerly, 1955). As generally defined, the crust is divided from the underlying mantle by a seismic velocity contrast, the Mohorovičić discontinuity or Moho, without any implication of a rheological contrast that might be implied by the word "crust." On the other hand, the major rheological transition, between *lithosphere* and *asthenosphere*, generally falls at greater depth than the Moho and forms the lower boundary of the tectonic plates.

Geologists realized early that the continental crust, with its abundant granitic rocks, differs profoundly from oceanic crust, with its predominant basalts. Encircling the Pacific Ocean, the andesite line marks this compositional difference between the two crusts (Suess, 1888; Born, 1933; Macdonald, 1949), but it does not necessarily indicate a difference in thickness. Pratt isostasy could be easily satisfied by a contrast in composition alone (Fig. 2).

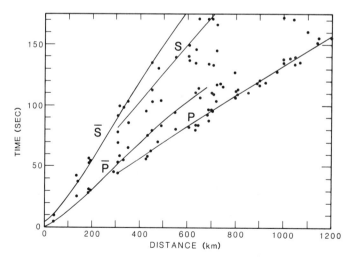

Figure 1. Earthquake traveltime curves used by Mohorovičić to infer a low-velocity upper layer of the Earth, the crust. $\bar{P}$ and $\bar{S}$ are direct longitudinal and shear waves through the crust. P and S, now generally designated P_n and S_n, are refracted within the upper mantle. From Byerly, 1955.

Geological Society of America Bulletin, v. 100, p. 1361–1374, 15 figs., September 1988.

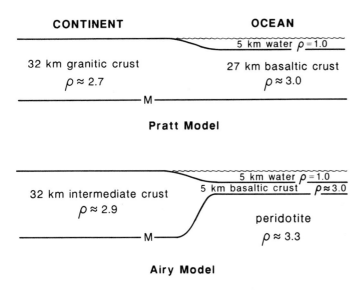

Figure 2. Conceptual Pratt and Airy isostatic models of continental and oceanic crust. Refraction seismology verified the Airy model.

The enormous contrast in thickness (a factor of 5 to 10) between oceanic and continental crust was discovered by explosion seismic experiments in the late 1940s and early 1950s and constitutes the most dramatic demonstration of Airy isostasy. Pioneering refraction experiments, supported in part by Penrose grants from the Geological Society of America, were begun on the Atlantic coastal plain (Ewing and others, 1937) and gradually extended onto the submerged continental shelf. By 1949, Raitt reported definitive evidence of a 5-km-thick crust in the Pacific Ocean west of San Diego, California, and soon Ewing and his colleagues reported similar evidence in the Atlantic Ocean near Bermuda (Ewing and others, 1950; see also Ewing's reinterpretation in Gutenberg, 1951, footnote on p. 434). The general pattern of a thin crust in ocean basins unfolded quickly in the following years. The crust under the mid-ocean ridges, however, was a surprise. By 1965, seismic refraction results had demonstrated that the broadly elevated ridges are not supported by a thicker crust; the crust under the ridges is as thin as under ocean basins. A combined analysis of seismic and gravity data across the Mid-Atlantic Ridge produced a model that anticipated the thickened mantle asthenosphere and thinned lithosphere that we now associate with spreading ridges (Talwani and others, 1965). Clearly, the compensation for the ridges is within the mantle.

Meanwhile, in continental areas, seismic refraction investigations were finding broad variations in crustal thickness (Gutenberg, 1955): generally thicker crust under high mountains and plateaus and thinner crust under lowlands (Fig. 3). Shield areas tend to have higher mean crustal velocities and greater crustal thickness. Upper crustal velocities of 5.9 to 6.2 km/s for compressional waves in most cases are replaced in the lower crust by velocities of 6.5 to 6.9 km/s, similar to the range of velocities in the oceanic crust. This boundary between upper and lower crust, where it is well defined, is sometimes called the "Conrad discontinuity." In the early investigations, exceptions to the broad pattern of crustal thickness varying with elevation were also appearing (Tatel and Tuve, 1955). For example, young, elevated, and tectonically active regions with a high heat flow commonly have an abnormally thin crust (about 30 km) and an upper-mantle velocity less than the usual 8.0–8.3 km/s (that is, the Basin and Range province, Tatel and Tuve, 1955; Thompson and Burke, 1974).

A new wave of insight on crustal structure emerged when the seismic reflection techniques of the petroleum industry were adapted to look into the deep crust (Oliver and others, 1976). Prior to this time, the limited resolution of crustal-scale refraction experiments led to averaged models consisting of a simple, layered crust (Fig. 3). This image was so unlike the varied lithologies and structure seen in deep geologic exposures that little correlation between the refraction and the visual images could be made. The reflection method, on the other hand, is capable of resolving gently dipping ($<40°$) geologic boundaries within tens to hundreds of meters. Currently, modifications of both refraction and reflection techniques allow exploitation of the advantages of each and bridge the gap between them (Mooney and Brocher, 1987).

The Geological Society of America has played an active role throughout this period of advancement and discovery. The pioneering experiments on the Atlantic coast of North America were both sponsored by the Society and reported in the *Bulletin* (Ewing and others, 1937), and

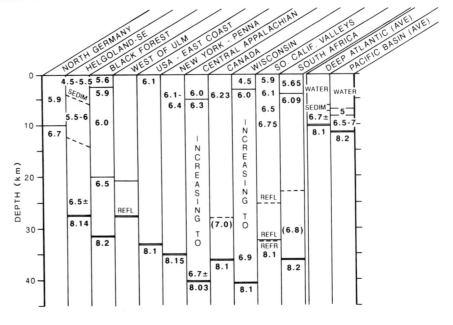

Figure 3. Early explosion seismic models of continental and oceanic crust. Note the variation in thickness of continental crust from less than 30 to more than 40 km and the variable intracrustal boundary between upper and lower crust. Numbers are P-wave velocities. From Gutenberg, 1955.

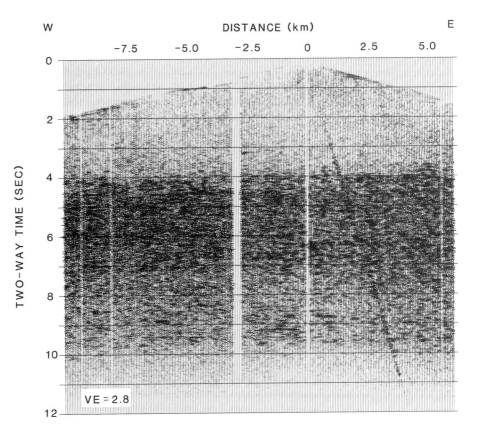

Figure 4. Shot gather showing transparent upper crust near Stillwater Range in northern Nevada. Data collected by C. M. Jarchow and others (unpub. data) as a piggyback experiment on the 1986 PASSCAL refraction experiment. The source is a 500-lb explosive shot located on the western side of Buena Vista Valley. Amplitudes have been corrected for spherical divergence only. Horizontal axis is distance in meters from the shotpoint. Moho is at about 10 seconds.

discovery of the thin ocean crust was also reported there (Raitt, 1949; Gutenberg, 1951). Much of the modern deep-reflection work in the United States has also been published in the *Bulletin* and *Geology*. The results are especially interesting in regions of extended crust.

REFLECTIVE CHARACTER OF EXTENDED CRUST

Major advancement in our understanding of the crust has come from reflection profiles in regions where the last tectonic event was extensional. This includes vast areas of Europe, Great Britain, Australia, the United States, and Canada. Although the reflective character is variable from region to region, as well as across individual profiles, data collected across crystalline terranes nevertheless define a consistent pattern of strong reflectivity in the lower crust sandwiched between a poorly reflective upper crust and a generally transparent upper mantle (Fig. 4).

The regional significance of this lower-crustal reflectivity was not initially apparent, despite its early identification more than two decades ago in a number of areas (Junger, 1951; Clowes and others, 1968; Fuchs, 1969; and Clowes and Kanasewich, 1970). The dramatic discovery by the Consortium for Continental Reflection Profiling (COCORP) of a subhorizontal package of reflections that sweeps across the lower crust of Nevada (Klemperer and others, 1986; Allmendinger and others, 1987) and appears to be largely unaffected by the varying geologic and tectonic boundaries identified at the surface was the first indication of the magnitude and apparent youth of this deep reflectivity. Furthermore, the subhorizontal, knife-sharp nature of the reflection Moho was viewed as strong evidence that the present crust-mantle boundary has no relation to its earlier, Mesozoic counterpart.

Subsequent studies from other extended regions around the world (Great Britain: Matthews and Cheadle, 1986; France: Bois and others, 1986; Germany: Lueschen and others, 1987; Australia: Moss and Mathur, 1986; the eastern United States: Hutchinson and others, 1987; and so on) confirmed and expanded these initial COCORP observations, resulting in an evolving picture of a rheologically weak and seismically layered lower crust that responds to stress by ductile flow (Meissner and Strehlau, 1982; Chen and Molnar, 1983; Smith and Bruhn, 1984). Although occasional dipping reflections can be traced from the upper crust down into the reflective lower crust, rarely, if ever, can they be shown to offset this deep reflective packet. Furthermore, no root zones associated with the Paleozoic orogeny in Europe or the Mesozoic orogeny in western North America have been identified in these areas, supporting a post-compressional age for the Moho (Klemperer and others, 1986) and, by inference, for the reflective lower crust as well.

We do not mean to imply that only extensional deformation can yield reflective crust. Clearly, the numerous regions listed above as being characterized by extension have all experienced a complex evolution, encompassing a wide range of deformational events through time. A reflective crust has also been imaged by Lithoprobe off Vancouver Island (Clowes and others, 1986), by the U.S. Geological Survey in Alaska (Fisher and others, 1988), and by ECORS across the Alps and Pyrenees mountain belts (Roure and Banda, 1987). Where this reflectivity is subhorizontal, where the crust is thin, and where the reflection Moho is typically well defined, however, we conclude that extension has been the primary factor affecting this reflectivity. Finally, it is important to note that extension in the lower crust can occur over a much broader zone than is observed in the upper crust (Matthews and Cheadle, 1986), suggesting that

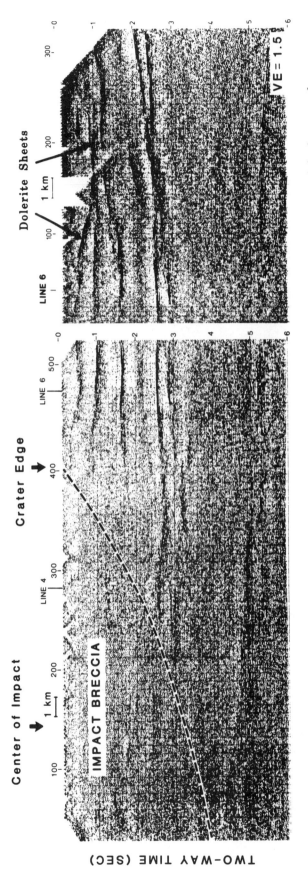
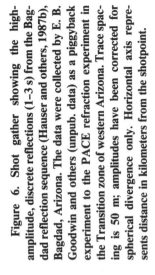

Figure 5. Multichannel seismic reflection section across Siljan meteor impact structure in Sweden. The rocks are Precambrian igneous and metamorphic, and the impact age is 340 m.y. The uppermost three reflectors were identified as dolerite sheets in an exploratory borehole, and there are Precambrian dolerite dikes at the surface. Dashed line shows generalized interpretation of the boundary of the impact breccia, based on termination or disruption of the reflectors. From Dahl-Jensen and others, 1987; Juhlin and Pedersen, 1987.

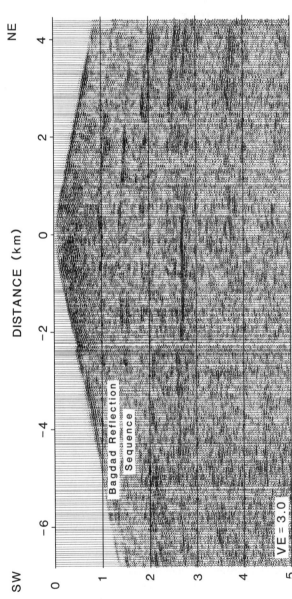

Figure 6. Shot gather showing the high-amplitude, discrete reflections (1–3 s) from the Bagdad reflection sequence (Hauser and others, 1987b), Bagdad, Arizona. The data were collected by E. B. Goodwin and others (unpub. data) as a piggyback experiment to the PACE refraction experiment in the Transition zone of western Arizona. Trace spacing is 50 m; amplitudes have been corrected for spherical divergence only. Horizontal axis represents distance in kilometers from the shotpoint.

Figure 7. Recumbent ductile isoclinal fold in West Greenland. Height of cliff, 860 m. White marble in the core of the fold is Proterozoic, and the other rocks are Archean gneisses. Photograph courtesy of T.C.R. Pulvertaft and M. C. Andersen, Geological Survey of Greenland.

extensional faulting exposed at the surface is not necessarily a prerequisite for extension at depth (see also Gans, 1987).

ORIGIN OF REFLECTIONS IN THE CRYSTALLINE CRUST

It is interfaces between rocks of different mineralogic (not necessarily chemical) compositions or different textures that produce reflections. For example, chemically identical gabbro and eclogite have very different velocities and densities, and a boundary between them, if sharp compared to seismic wavelength, is highly reflective. The textural contrast between rocks of identical mineralogic composition (for example, a granite and granite gneiss) may also produce reflections because oriented minerals make rocks seismically anisotropic (Christensen, 1966). This has been demonstrated for aligned micaceous minerals from ductile fault zones (Fountain and others, 1984; Jones and Nur, 1984), where the velocity perpendicular to foliation is slower than the velocity parallel to foliation.

Reflections from the Deep Crust

Against this general background, we list four of the major candidates for the origin of subhorizontal reflectors in the deep crust.

(1) Tabular mafic or other igneous intrusions ("sills"). Multiple dolerite sheets in granitic rocks have been confirmed by drilling at Siljan, Sweden (Dahl-Jensen and others, 1987), and can be correlated to laterally continuous reflections imaged to a depth of about 17 km (Fig. 5). A similar intrusive origin may account for the gently dipping events that comprise the Bagdad, Arizona, reflection sequence (Hauser and others, 1987a; Fig. 6). We will discuss the evidence for these sheets and the stress relations that favor their formation.

In the lower crust, where temperatures are higher and tectonic (deviatoric) stress tends to be relaxed by creep, gravitational stratification becomes more important. Trapping of mafic beneath silicic magma (for example, Hildreth, 1981) and cumulate layering in mafic magma may contribute to subhorizontal reflections (Lynn and others, 1981). We believe that intrusion of mafic sheets may be particularly common at the base of the crust, where underplating of mantle-derived melts has been postulated (Lachenbruch and Sass, 1978; Furlong and Fountain, 1986). We will discuss underplating further in connection with stress requirements.

(2) Transposed compositional layering. Sandwiched between the brittle upper crust and the more refractory mantle, the hot, plastic lower crust deforms by flowage, transposing any originally dipping layers into parallelism with the subhorizontal regional foliation (Phinney and Jurdy, 1979; Ramsay, 1982). Subhorizontal, isoclinal folds, banded layers, and nappe structures are produced and in many cases contain compositional layers of the proper scale to generate coherent seismic reflectors (Fig. 7; Smithson and others, 1980).

(3) Aligned anisotropic minerals. Zones of mylonites in which anisotropic minerals (especially micas) are oriented parallel to the regional foliation and juxtaposed next to nonmylonitized rock of equivalent composition are also a proven source of reflectivity (Jones and Nur, 1982, 1984). These zones have been well documented in the ductile shear zones of metamorphic core complexes (Fountain and others, 1984) as well as the mylonitic zones bounding thrust sheets in compressional orogens (Christensen and Szymanski, 1988).

(4) Microcrack porosity. Fluid-filled cracks or pores drastically lower seismic velocities, especially if the fluids are overpressured (Jones and Nur, 1984; Feng and McEvilly, 1983). Pods of fluid-bearing rock (including partial melt) in the lower crust could therefore result in strong impedance contrasts suitable for the generation of reflections (Matthews and Cheadle, 1986; Klemperer, 1987). Microcracking and water have been reported in the deep Kola well in Russia (Borevsky and others, 1984), and free water has also been invoked to explain the widespread observation of surprisingly high electromagnetic conductivity in the lower crust (Shankland and Ander, 1983; Jones, 1987).

Finally, each of the four models presented above can be expanded to include multiple layers of high and low velocity, regardless of whether the source is due to variations in composition, texture, or fluids. If the layering is on a scale of tens to hundreds of meters, the seismic waves will interfere, resulting in a tuning effect that can greatly enhance the observed reflectivity (Fuchs, 1969). Velocity variations less regular than layering may also scatter seismic waves and produce an appearance of discontinuous layered reflectors that is quite different from the actual geologic variations (Gibson and Levander, 1988).

Nonreflective Upper Crust

One feature that is perhaps more enigmatic than the abundant reflectivity in the lower crust is the *lack* of reflectivity in the upper crust (Mooney and Brocher, 1987). This is clearly demonstrated in the PASSCAL reflection data (Jarchow and others, 1987; Fig. 4), where the upper crust is unusually transparent down to 4 s two-way traveltime (TWTT). Ground roll and other forms of source-generated noise are clearly not the cause of the transparent upper crust in this example, requiring that geologic factors such as composition and/or physical properties are instead involved. Fountain (1986) postulated that much of the transparent appearance may be due to the lack of significant impedance contrasts among rocks typical of the upper crust (for example, granite, gneiss, schist). This explanation, however, does not account for the transparent character observed on the BIRPS' profile WINCH (Fig. 8), where Lewisian granulites, exposed at the surface and presumed to be uplifted equivalents of the highly reflective crust imaged at depth, are strikingly devoid of reflectivity (Matthews and Cheadle, 1986). Although it is possible that the Lewisian granulites are in fact *not* representative of the lower crust (exposures of Kapuskasing granulites have proven to be reflective; R. Clowes, 1988, personal commun.), it is more likely that their transparent character reflects a change that has occurred during, or following, uplift (Matthews and Cheadle, 1986; Klemperer and others, 1987).

One possible explanation for the lack of reflectivity of the Lewisian granulites, as well as the upper crust in general, can be tied to the rheologic character of the crust. Detailed studies of the velocity structure and reflection character of fault zones indicate that cataclastic deformation experienced at shallow crustal levels results in a reduction in velocity (Mooney and Ginzburg, 1986) and a degradation in the quality of the reflected signal over a broad (~10 km wide) zone. Reflection profiles across major fault zones such as the San Andreas fault in California (Feng and McEvilly, 1983), the Bray fault in France (Bois and others, 1986), and a splay of the Alpine fault in New Zealand (Stern and Davey, 1988) all yield a transparent image bracketing the fault zone, across which major reflections cannot be traced. Although some of this transparency may be due to

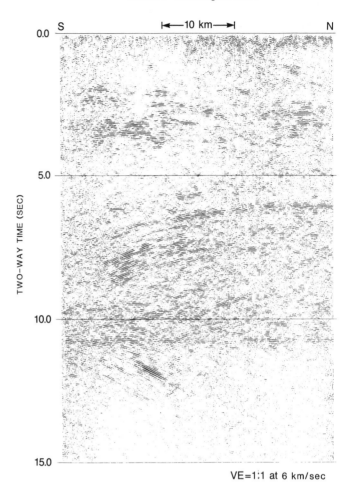

Figure 9. COCORP Washington line 8 stacked reflection section (unmigrated) from northern Washington across the Okanogan dome. Southward-dipping reflections in the lower crust pinch out against the subhorizontal reflection Moho at approximately 11 s TWTT. In addition, deeper, and more steeply dipping reflections between 11.0 and 12.5 s are prominent; these reflections, however, shift upward above the reflection Moho following migration. No vertical exaggeration at 6.0 km/s. After Potter and others, 1986.

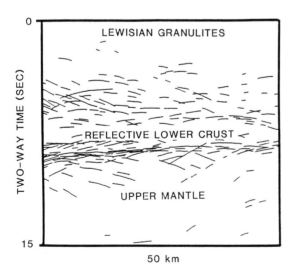

Figure 8. Line drawing of part of BIRPS line WINCH 2 west of Scotland, showing unreflective upper crust of Lewisian granulites above a highly reflective lower crust. From Klemperer and the BIRPS group (Klemperer and others, 1987).

complex, steeply dipping structures in the vicinity of the fault zone that cannot be resolved using conventional seismic methods (Louie and others, 1988), a similar result is also observed on the reflection profile recorded across the Siljan impact crater in Sweden. As can be seen in Figure 5, the quality of this reflection profile is significantly degraded where it crosses from the outer rim into the area affected by the impact (Juhlin and Pedersen, 1987). Thus, the brecciation and microcracking experienced during impact has disrupted reflector continuity and results in a comparatively transparent image for the upper crustal region underlying this structure.

In a broad sense, the entire crystalline upper crust behaves in a fashion analogous to the fault zones and impact regions described above. Cataclasis, microcracking, and brecciation are well-documented phenomena in the uppermost crust, and results from the Kola superdeep well in the northwestern Soviet Union document a wide distribution of fissuring and

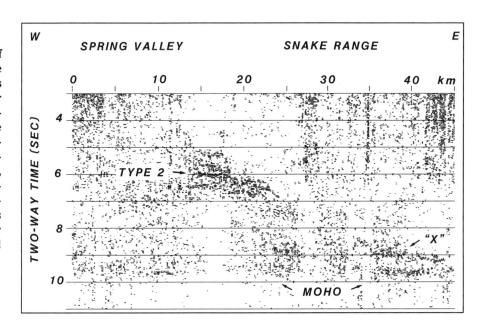

Figure 10. 100%-stacked profile of COCORP line 5 across the northern Snake Range, Nevada. The data have been bandpass filtered, trace balanced, and corrected for spherical divergence; only reflections continuous across a minimum of seven traces are plotted. No coherency filtering has been performed. Lower crustal reflectivity is particularly well developed beneath Spring Valley, where it consists of both discontinuous reflecting segments as well as a few longer, continuous reflections, such as the one at 6 s TWTT. The X and M events of Klemperer and others (1986) can be clearly seen at 9 and 10 s TWTT, respectively.

fracturing at deeper depths. Furthermore, when this behavior is complicated in extended provinces by abundant normal faulting, it is not surprising that only thick, laterally continuous units such as those associated with sedimentary basins (Behrendt and others, 1988) or master detachment faults (Allmendinger and others, 1983) are clearly imaged on seismic reflection profiles. Other layers, particularly those within the crystalline crust, are in many cases too discontinuous laterally to produce coherent reflections. Only at deeper depths, where cracks are closed and brittle failure is replaced by aseismic creep, will reflector continuity be preserved.

MAGMATISM AND EXTENSION IN THE CRUST

The nearly universal correlation of lower crustal reflectivity with tectonically extended provinces raises the question: What roles have magmatism and ductile extension played in the formation of this reflectivity? The metamorphic core complexes of the western United States are an ideal area to study this problem; not only do they record significantly larger amounts of extension (on the order of hundreds of percent) than observed in the more standard intracontinental rift zones like the Rhinegraben (Illies, 1975) and the East African rift (Baker and others, 1978), but also they are characterized by a more strongly reflective upper and middle crust than is observed in most extensional provinces. Furthermore, the core complexes expose rocks from mid-crustal depths, thereby facilitating studies of the lithologies and processes common in the middle crust.

Reflection transects by COCORP across three metamorphic core complexes, the Okanogan dome in Washington (Potter and others, 1986; Sanford and others, 1988), the northern Snake Range in Nevada (Klemperer and others, 1986; Hauser and others, 1987b), and the Buckskin/Rawhide Mountains in Arizona (Hauser and others, 1987a), serve as the foundation for a growing data base of geophysical studies across highly extended terranes. In addition, two refraction studies conducted by the U.S. Geological Survey (USGS) across the Whipple and Buckskin/Rawhide metamorphic core complexes in southeastern California and western Arizona provide the velocity information necessary to analyze the origin of the major units identified on the reflection profiles. This is particularly true for the refraction study across the Buckskin/Rawhide Mountains (McCarthy and others, 1987a), which was largely coincident with the COCORP transect. A variety of smaller-scale studies in the vicinity of the core complexes have also been carried out by the University of Wyoming (Hurich and others, 1985; Smithson and others, 1986; Valasek and others, 1987), CALCRUST (Okaya and Frost, 1986), and various industry groups (for example, Reif and Robinson, 1981; McCarthy, 1986; Goodwin and Thompson, 1988). Although the latter studies do not have the necessary coverage to address the lateral variation in the properties of the crust across the extended regions, they are in general of higher resolution and provide more detailed information about the specific character of the crust from region to region. For this reason, we concentrate our summary on the COCORP and USGS seismic results, but supplement our discussion with these more detailed studies when appropriate.

The Moho is the most prominent event on all of the COCORP profiles and is located at the base of a high-amplitude band of discontinuous reflections that extend many tens of kilometers in a subhorizontal fashion across the core complex terrane, as well as across the Basin and Range in general. Because the Moho crosscuts dipping events in the lower crust (Figs. 9 and 10), it has been interpreted as a relatively young feature (Klemperer and others, 1986), formed during Tertiary Basin and Range extension. The subhorizontal (±1 km) geometry of the Moho has been confirmed by seismic refraction studies (McCarthy and others, 1987b), which show that despite variable amounts of extension observed at the surface, the thickness of the crust remains uniform regionally.

Mid- and lower-crustal reflectivity is also a prominent feature in broad regions surrounding the core complexes and, in the case of the Okanogan dome and Snake Range (Figs. 9 and 10), can be shown to rise locally beneath these centers of extension. Thus, in areas where much of the upper crust has been tectonically removed, the reflectivity extends well up into the middle crust, in many cases producing an image of a rising welt of reflectivity. Beneath both the Okanogan dome and the northern Snake Range, the upper boundary of this reflective zone is characterized by a sharp, convex-upward reflection located at approximately 6 s TWTT (Fig. 9). The reflective packet between this upper surface and the Moho at the base of the crust is characterized by two types of seismic reflectors: (1) hundreds of short, discontinuous events (≤2 km across) that combine to yield a laminated character to the lower crust and (2) a few discrete reflecting horizons which are quite strong in amplitude and are laterally continuous for several kilometers (see Fig. 9 at 6 s and Fig. 10 at 6 and 9 s TWTT).

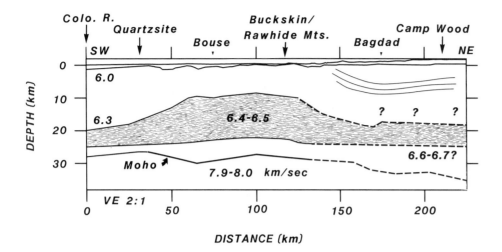

Figure 11. Preliminary reflection-refraction model across the Buckskin/Rawhide metamorphic core complex in western Arizona. A thickening package of mid-crustal material is observed to rise beneath, and slightly west of, the core complex and is characterized by a velocity step to 6.4–6.5 km/s. This packet is highly reflective, particularly on the flanks of the thickening body, and coincides with a change to more mafic compositions in the crust.

Although interpretations are still preliminary, the wide-angle reflection data collected by the USGS (J. McCarthy, unpub. data) appear to define a similar thickening of the reflective middle crust approaching the Buckskin/Rawhide metamorphic core complex (Fig. 11). A strong, discrete reflection from a depth of 20 km at the Colorado River can be observed to rise sharply toward the core complexes up to a depth of only 8 km below sea level. This event corresponds to the onset in reflectivity observed on the COCORP-Arizona profiles and is particularly well defined on the flanks of the culmination (Hauser and others, 1987a). The velocity of the thickened mid-crustal layer is constrained from reversed refracted arrivals to be 6.4 ± 0.1 km/s. Although the rise in the reflective crust in the Buckskin/Rawhide region results in a domal structure that is remarkably similar in shape to the reflective zone imaged by COCORP across the Snake and Okanogan metamorphic core complexes, the depths to the latter are as much as 10 km greater than in Arizona. Nevertheless, we suspect that the source of these dome-shaped mid-crustal layers may be related.

The reflectivity in the upper crust in the vicinity of the core complexes is typically not as strong as it is in the lower crust, although prominent reflective packages are apparent (Fig. 9), in contrast to the transparent upper crust in adjacent regions in the Basin and Range and in extended provinces in general (McCarthy, 1986). This upper crustal reflectivity is similar in character to the laminated, discontinuous reflections (type 1) described above for the deeper crust and has been correlated to the compositionally variable mylonitic rocks exposed at the surface or sampled in deep drill holes (for example, Hurich and others, 1985; Smithson and others, 1986; Goodwin and Thompson, 1988). Where there are differences between the upper and lower crustal reflectivity, these differences lie largely in the absence of the coherent, high-amplitude (type 2) reflectors in the upper crust and may reflect somewhat more silicic compositions (Fountain, 1986). In addition, the added effects of faulting, open cracks, and high fluid pressures at shallow levels may degrade the strength and continuity of upper-crustal events.

From the Okanogan, Snake Range, and Buckskin/Rawhide COCORP surveys, it is apparent that there are two characteristic reflections that must be accounted for in any model of the evolution of these highly extended terranes: the type 1 reflections, which are generally restricted to the middle and lower crust in the Basin and Range province but can be observed in the upper crust as well in the vicinity of the metamorphic core complexes, and the more occasional type 2 reflections, which are laterally more continuous and are higher in amplitude. Examples of the latter are the deep reflections located typically 1 s above the reflection Moho (labeled "X" by Klemperer and others, 1986) as well as the 6 to 6.5 s reflections marking the top of the reflective crust beneath the northern Snake Range (Fig. 10).

Goodwin and Thompson (1988), following the pioneering work of Fuchs (1969) and more recent studies by Fountain and others (1984), Jones and Nur (1984), and Hurich and Smithson (1987), observed that the type 1 reflections can be produced by interference effects through many thin layers of high and low velocity. Their modeling of reflection data from the Picacho Mountains metamorphic core complex in southern Arizona constrains these layers to be tens of meters thick and hundreds of meters long with gradual variations in velocity laterally. Furthermore, they deduced that the velocity variations within these layers are necessarily small (5%–10%); otherwise, a significantly larger component of diffracted energy would be observed.

Although the specific geologic source of these thin high- and low-velocity bodies can be interpreted in a variety of ways (that is, compositional or textural differences, ductile fault zones, or fluids), their predominance in the vicinity of the core complexes and their correlation to ductile fabrics exposed at the surface suggest that the layer geometries are a by-product of extensional strain superimposed on the entire crust. In an eloquent discussion of magmatism and strain in the eastern Basin and Range, Gans (1987) noted not only that it was problematic that the crust *maintained* its thickness beneath regions of large extension, but also that it was not observed to be *unusually thick* beneath areas of little to no extension. He concluded that the most reasonable way to maintain a nearly uniform crustal thickness across a region of such diverse supracrustal extension is to allow lower crustal material from adjacent regions to flow laterally into the more highly extended zones. Whether this smoothing is accommodated by homogeneous, coaxial flow or a network of anastomosing ductile shear zones (Hamilton, 1982; Kligford and others, 1984; Frost and Okaya, 1988), the process would predict the formation of a range of possible geologic structures and fabrics suitable for the generation of the subhorizontal, discontinuous reflections (for example, transposed layering, aligned mineral fabrics, compositional variations, and so on). It thus seems clear that compositional and textural variations on the scale of tens to hundreds of meters are responsible for much of the type 1 reflectivity observed most commonly in the lower crust on the seismic profiles. We believe that the more continuous, type 2 reflections, however, may owe their existence to a somewhat different cause.

Lachenbruch (1988) and Thompson and McCarthy (1986, 1988), on the basis of a regional review of elevations and gravity, emphasized the importance of magmatic replenishment of the crust in the regions of large

Figure 12. Diabase sheets in sandstone in the Dry Valley region of Antarctica. The two 200-m concordant sills near the center of the picture are connected by an equally thick inclined discordant sheet at the right. Another crosscutting sheet and a feeder dike are shown at left. From Hamilton, 1965.

extension. Because crustal thinning reduces the buoyancy of the lithosphere, the surface elevation of the highly extended region should be *lower* than in less extended areas, except in the special case of the extension coinciding with an area originally so high, relative to its surroundings, that it remains high. Clearly, the Mesozoic orogenic events in the western United States resulted in regional crustal thickening (Armstrong, 1968; Coney and Harms, 1984); nevertheless, it is unlikely that there were local variations in crustal thickness sufficient to account for the unusually high elevations of the core complexes today (Gans, 1987). Furthermore, even if one considers only the average amount of extension across the eastern Great Basin, it is difficult to account for the observed 1- to 2-km elevations without assuming an unusually thick crust (55–60 km) prior to Tertiary extension (Gans, 1987). A more viable alternative for these high elevations is that the mass removed by tectonic denudation is replaced at depth by intrusions of similar density.

The seismic data support the latter model of a magmatically rejuvenated crust. The velocity increase in the middle crust to 6.4 ± 0.1 km/s indicates that the onset of reflectivity in western Arizona is associated with a transition to somewhat more mafic compositions at depth. This transition, although not extreme, corresponds to material intermediate between granite and gabbro and may reflect a mixing between basaltic magma and derivatives of crustal melting. Deeper layers corresponding to even higher velocities within the lowest 5–10 km of the crust have also been identified in the Buckskin/Rawhide region (Fig. 11) and are similar to wide-angle reflections correlated to Klemperer's X-event in western Nevada (Catchings, 1987). These more continuous, type 2 reflections thus may mark major compositional changes or interlayers of mafic material ranging from hundreds of meters to as much as several kilometers thick in the middle and lower crust (Louie and Clayton, 1987; Thompson and others, 1988).

In summary, we believe that both magmatic additions to the crust and regional ductile flow are responsible for the highly reflective character of the crust in the vicinity of metamorphic core complexes, as well as in extended regions in general. If igneous layers contribute to the subhorizontal character of the reflections, however, they must be either intruded as subhorizontal sheets or later transposed into such a geometry. Furthermore, the identification of subhorizontal diabase sheets intruded into the crust in the Mohave Mountains in Arizona (Howard and others, 1982; Nakata, 1982), in the Siljan region of Sweden (Dahl-Jensen and others, 1987), and by inference, in the Transition zone of Arizona, confirms that subhorizontal intrusions can occur as primary structures. Although it is uncertain what the stress regime was during the time of emplacement of these sheets, we nevertheless speculate below, in a corollary to our discussion of magmatism, on the mechanisms required to produce these geometries in an extensional stress regime.

Stress Requirements for Tabular Subhorizontal Intrusions

As mentioned earlier, dense magma tends to pond beneath material of lower density, particularly in the absence of local tectonic (deviatoric) stress, and this may explain mafic underplating of the crust and mafic intrusions below caldera granites. The mushrooming of salt domes below low-density sediments is analogous. We also note that the high viscosity of both granitic magma and salt favor buoyant rise in large, comparatively equidimensional bodies, wherein the ratio of volume (buoyant body forces) to surface area (viscous resistive forces) is maximized. Mafic magma, on the other hand, has a low viscosity and may readily form extensive dikes or sills. At or near the base of the crust, we expect tectonic stress to be comparatively relaxed relative to that of the elastic, seismogenic upper crust or the more refractory mantle (Meissner and Strehlau, 1982; Chen and Molnar, 1983; Smith and Bruhn, 1984). Under these conditions, a sufficiently dense mafic magma, possibly loaded with a mush of ultramafic crystals, will be trapped beneath or within lower crustal rocks. Crystal settling and solidification, followed by repeated episodes of intrusion, will produce a layered lower continental crust somewhat like that of lower oceanic crust and ophiolites (Fig. 15, McCarthy and others, 1988). In the cooler upper crust and the refractory upper mantle, tectonic stresses are not so readily relaxed and may play an important role in determining the shape of intrusions.

Tectonic stresses are the key to understanding subhorizontal intrusive sheets like those identified in the upper crust at Siljan, Sweden, and those inferred near Bagdad, Arizona (Figs. 5 and 6). Although sills concordantly emplaced in stratified sedimentary rocks are commonplace, these discordant subhorizontal sheets in deformed crystalline rocks are more difficult to understand. That dikes tend to intrude along planes perpendicular to the least horizontal principal stress is firmly established (Anderson. 1951; Nakamura, 1977; Delaney and others, 1986) and is confirmed on a broad scale by hydrofracturing experience (Zoback and Zoback, 1980). Horizontal intrusive sheets, however, imply a vertical least principal stress. How, then, can dikes and dike-fed sheets coexist, as is observed (for example, Hamilton, 1965; note that some of the subhorizontal sheets shown in Fig. 12 are discordant and that some are emplaced in granite)?

We propose the following sequential mechanism for the generation of subhorizontal intrusions; for simplicity, we neglect edge effects of dikes and sills on the stress field, a reasonable assumption for thin extensive sheets such as those at Siljan (Fig. 5). In a tectonically extending volcanic region, dikes intrude perpendicular to the least principal stress direction, the direction of extension (Fig. 13A). If the magma supply is plentiful during an intrusive episode, dilation of the dike walls relaxes the extensional strain and increases the least compressional stress in the wall rocks.

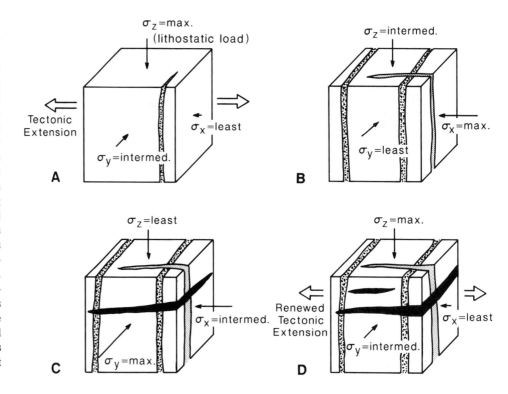

Figure 13. Schematic explanation of intrusion of horizontal sheets fed by dikes in an extending terrane. The sequential diagrams represent a small element at depth. A. Dike intruding perpendicular to σ_x, the least principal stress; σ_z, the lithostatic load, is constant in all the diagrams. B. Inflation of dikes increases σ_x, and to a lesser extent σ_y, controlled by Poisson's ratio, until σ_y or σ_z becomes least; in the former case, a new dike intrudes perpendicular to σ_y. C. Dike intruding perpendicular to σ_y increases σ_y until σ_z becomes least and horizontal sheets are emplaced; inflation of the sheets does not increase σ_z (lithostatic) appreciably, and intrusion can continue. D. Continued or renewed far-field tectonic extension restores original stress relations. Note that emplacement of the secondary dike set (B) is not required in every case to make the vertical stress least, because intrusion of the first set (A) increases both horizontal stresses.

This initial dike emplacement not only increases the horizontal principal stress normal to the dike, but to a lesser degree, also increases the other horizontal principal stress parallel to the dike. This is true because of the tendency of material to bulge laterally when compressed longitudinally (the ratio of the two strains is Poisson's ratio). At this stage, if the principal stresses are originally of approximately the same magnitude (as is probable near the Colorado Plateau; Zoback and Zoback, 1980), their relative order changes, and either the horizontal stress parallel to the dike or the vertical principal stress becomes least. If the vertical stress is least, intrusion of horizontal sheets is immediately favored (Fig. 13C). If not, a new dike orthogonal to the first is favored (Fig. 13B; for field examples of dikes with a predominant set in the direction normal to tectonic extension and an occasional dike in nearly orthogonal directions, see maps in Hamilton, 1965; Anderson and others, 1955; and Reynolds, 1985). Inflation of the new dike increases the stress normal to it and makes the vertical (lithostatic) stress least (Fig. 13C). As a result, horizontal sheets are emplaced, fed by dikes (see examples in Reynolds, 1985, and Hamilton, 1965). In this latter case, both horizontal stresses are greater than the vertical lithostatic load.

One may ask how the magma pressure can exceed the lithostatic load. We visualize equal magma and rock pressure at the depth of melt generation (Fig. 14). At any shallower depth, the pressure in a static magma column will be less by an amount equal to the weight of the melt column, whereas the pressure in the adjacent rock column will be less by an amount equal to the weight of the rock column. Thus, magma pressure will exceed lithostatic load, and magma can be erupted even from high volcanoes. Eaton and Murata (1960) used this reasoning to calculate the origin depth of Hawaiian eruptions.

In the more realistic case of upward flow in a dike, the magma pressure will be somewhat lower because of viscous losses (Fig. 14), but for the dike to remain open, the pressure must exceed the smaller of the two horizontal principal stresses. Because the magma pressure exceeds the vertical (least) principal stress, horizontal sheets can potentially propagate out from the dike. The depth where sheets form will probably depend on a critical overpressure to produce the initial horizontal fracture. This overpressure will be smallest in horizontally stratified rocks with a low tensile

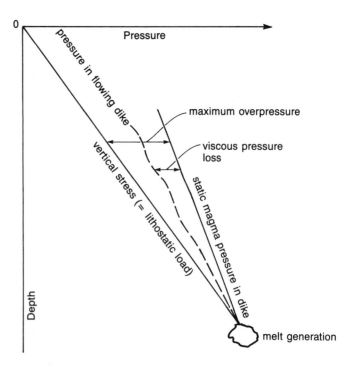

Figure 14. Simplified conceptual diagram relating magma pressure in a dike to lithostatic load.

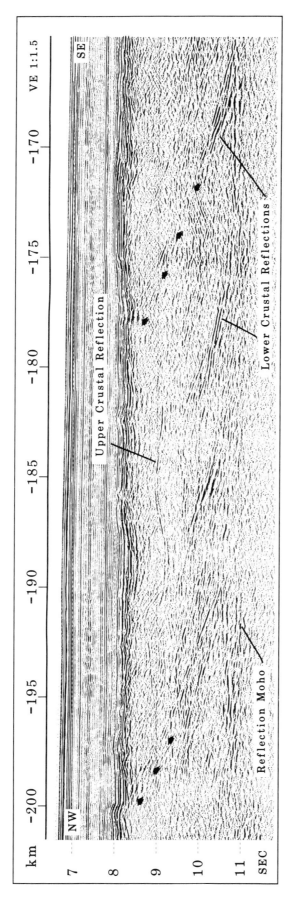

Figure 15. Reflection section from the North Atlantic Transect (NAT) two-ship experiment in the western North Atlantic (McCarthy and others, 1988). Dipping reflections (20°–40°) are imaged in lower crust (9–11 s) of the western North Atlantic. In some cases, the reflections can be traced upward toward the basement surface, where they can be correlated with faults that offset the basement. In the lower crust, however, these events are more laminated and more reflective and terminate abruptly against the Moho, leading some to interpret them as reflections from mafic-ultramafic layering formed along the inclined walls of the paleo–Mid-Atlantic Ridge (Mutter and others, 1985; McCarthy and others, 1988).

strength perpendicular to stratification. Note that the maximum overpressure (difference between static pressure in the dike conduit and the lithostatic load) increases with decreasing depth (Fig. 14). Sheet emplacement will also depend on the dynamics of flow, heat loss, and pressure in the dike, the varying tectonic stress in the wall rocks, and perhaps (as in the case of sills) on inhomogeneities that favor incipient sheet intrusion.

The emplacement of sheets, unlike the case of the dikes, has little effect on the vertical stress, because inflation of a sheet simply lifts a constant load with a free surface at the top. Thus, sheets can grow or repeat as long as the magma supply lasts or until enough far-field tectonic extension accumulates (Fig. 13D) to re-establish the least principal stress in the horizontal direction.

DISCUSSION AND CONCLUSIONS

Although we have based our discussion of the origin of lower crustal reflectivity on geophysical and geologic studies of the highly extended regions of the western United States, similar results have been obtained in other areas, particularly in the Black Forest of southwest Germany. P- and S-wave refraction studies, together with the more conventional vertical incidence reflection data, have yielded a similar picture of a transparent upper crust, a laminated lower crust (for P-waves), and a strong, subhorizontal Moho overlying a transparent upper mantle (Fuchs and others, 1987; Gajewski and others, 1987a, 1987b; Holbrook and others, 1987; Lueschen and others, 1987; Sandmeier and Wenzel, 1986; Sandmeier and others, 1987). Moreover, the onset of reflectivity observed in the middle crust can, in some areas, be correlated to an increase in seismic P-wave velocity, corresponding to more mafic compositions with depth (Fuchs and others, 1987; Gajewski and others, 1987a, 1987b; Holbrook and others, 1987).

The acquisition of coincident S-wave data in the Black Forest, however, provides additional constraints as to the possible origin of the lower-crustal reflections. Holbrook and others (1987) noted that the amplitudes and over-all reflectivity of the S-wave arrivals from the lower crust were significantly lower than those of their equivalent P-wave arrivals, implying abrupt, high and low variations in Poisson's ratio in the lower crust. One proposed scenario which could produce this contrasting P- and S-wave reflectivity and the associated variations in Poisson's ratio consists of thin layers of alternating mafic and silicic compositions (Holbrook and others, 1987). Because Poisson's ratio is sensitive to quartz content, it will be high in the mafic layers and low in the silicic layers. The interlayered mafic-silicic rocks will thus produce strong P-wave reflectivity but weak S-wave reflectivity. Conversely, if the laminations observed from the P-wave reflections are interpreted as solid-fluid interlayers, the S-waves would be strongly reflective, owing to their high sensitivity to the presence of pore fluids (Holbrook and others, 1988). Significantly, no such reflectivity is observed.

Largely on the basis of these results from Germany, and on the predominance of dehydrated mineral assemblages in granulite terranes from the lower crust, we find it difficult to believe that fluids are the major contributor to the reflectivity of the deep crust in extended areas. Studies of the exposed mid-crustal rocks of the metamorphic core complexes indicate that interlayering of dry high- and low-velocity layers are sufficient to generate the reflectivity observed (Hurich and Smithson, 1987; Goodwin and Thompson, 1988). Furthermore, the presence of high velocities ($V_p > 6.6$ km/s) in the lowermost crust (for example, Mooney and Brocher, 1987) of most extended regions is difficult to reconcile with numerous overpressured zones caused by fluid-filled cracks at deep crustal depths.

In a few respects, our hypothesis of how the crust extends and the origin of the reflectivity is similar to that of Meissner and Wever (1986) and Wever and others (1987), although instead of predicting an over-all decrease in the thickness of the reflective crust, the observations from metamorphic core complexes require that reflectivity actually *increases* in these regions of high extension. We therefore favor a model in which the reflective crust evolves during extension not only by differentiation but also by ductile flow of lower-crustal material from adjacent, less extended regions and by injection and/or underplating of mantle-derived magmas (of crustal densities) into the lower crust.

Throughout this discussion, we have assumed that the processes controlling the evolution of the middle and lower crust of the metamorphic core complexes are equally applicable to regions of more moderate continental extension. This, however, tends to be a major source of debate, particularly because the classic metamorphic core complexes, characterized by exposures of mid-crustal rocks and hundreds of percent extension, have been identified in few areas outside of the western North American Cordillera. Several explanations, such as a correlation with the passing of the Mendocino triple junction (Glazner and Bartley, 1984) or extension following a thermally weakened and over-thickened crust (Coney and Harms, 1984; Glazner and Bartley, 1985; Holt and others, 1986), have thus been proposed to account for their anomalous existence. Certainly, more work is necessary before the principal factors leading to their formation can be understood. Nevertheless, the systematic progression of reflectivity from the Basin and Range province into the core-complex terrane suggests an interpretation whereby the metamorphic core complexes are simply one end member in a range of possible structures developed during different degrees of crustal extension. Furthermore, it is difficult to imagine that core-complex evolution differs significantly from that of other areas characterized by extension of a previously thickened crust, such as the Caledonides and Variscides of western Europe. Thus, we believe that the geophysical studies of metamorphic core complexes described herein are applicable to numerous other areas characterized by thinned crust and well-developed reflectivity. This working hypothesis, however, must ultimately await confirmation from future studies.

Magmatism and Faulting in Oceanic Crust

The discussion of crustal extension would be incomplete without a brief mention of the mid-ocean ridges. Although the exact relationship between magmatism and extension in the continents is far more controversial, clearly magmatism is the predominant process controlling crustal evolution in the ocean basins, generating two-thirds of the Earth's crust. As in the case of the continents, reflection studies from ocean basins reveal the close interplay between brittle faulting in the upper crust and magmatism and intrusion at depth. This is perhaps best documented in the North Atlantic Transect seismic reflection data collected in the western North Atlantic (Mutter and others, 1985; McCarthy and others, 1988). This two-ship reflection profile reveals basement-cutting faults that in some cases can be traced downdip into anomalous packages of laminated, high-amplitude reflections confined to within 1 s of the reflection Moho (Fig. 15). These deeper reflection packages dip as much as 20°–40° back toward the paleoridge and thicken downward, before abruptly terminating at the subhorizontal reflection Moho (~11 s). Although the lower crustal events can be interpreted as the downdip equivalents of the basement-cutting faults, a more intriguing possibility is that their high amplitudes and laminated character are the result of constructive interference through a package of interlayered mafic-ultramafic cumulates crystallized along the walls of the Mid-Atlantic Ridge magma chamber at the time of crustal accretion (Mutter and others, 1985). Furthermore, the apparent continuation of the upper crustal faults downward into these layered sets of mafic-ultramafic material is suggestive of a spatial correlation between faulting and magmatism. Episodic intrusions into the lower crust (on a time scale of every 250,000 to 500,000 yr) serve to inflate and extend this deeper region, forcing faulting and extension to occur in the brittle, upper crust as well. These faults sole into the sides of the magma chamber, following the thermally weakened borders of the magma body.

In the scenario outlined above, the magmatic and extensional processes thus are linked, with magmatism building the crust from below and extension controlling the faulting and deformational patterns observed at shallow levels. Although the faulting and magmatic histories of the continental regions are much more difficult to decipher (particularly as most of these regions have experienced multiple periods of deformation), a similar pattern may prevail. Numerous geologic and geophysical studies have shown that extension of the upper crust is controlled by normal faulting along listric or planar surfaces that penetrate down to mid-crustal depths. In addition, there is strong evidence for magmatism contemporaneous with extension; high heat flow (Lachenbruch and Sass, 1978), elevated rates of ^{3}He degassing (Oxburgh and others, 1986), and regionally high elevations (Lachenbruch, 1988; Thompson and McCarthy, 1988), together with the abundant volcanic rocks exposed at the surface (Christiansen and Lipman, 1972), have all been used to infer a magmatic source at depth. The volume of this mantle-derived melt and how it is distributed in the crust, however, are both poorly understood. One possibility is that the magma ponds as sills and/or laccoliths in the middle and lower crust, creating a hot and soupy substrate into which the upper crustal faults sole. This basic model is supported by the seismic data, wherein the faulted and brecciated upper crust is characterized by a weakly reflective, transparent image, whereas the hot, ductile lower continental crust is characterized by discontinuous and laminated reflections that are rarely if ever offset by normal faults from above. Whether the magmatic material is intruded into the crust or underplated at its base is presently not well constrained, but continued analysis of the PACE wide-angle reflection and refraction data will shed more light on this question in the future.

We conclude with the observation that others (see, for example, Varga and Moores, 1985; Karson and Winters, 1987; Norrell and Harper, 1987) have made before us, namely that despite the differences in composition and crustal thickness, the fundamental processes controlling continental extension in the metamorphic core complexes appear to be remarkably similar to those active at slow-spreading mid-ocean ridges. They share the same styles of faulting, rates of extension, and apparent interplay between faulting and magmatism. What is yet unclear, however, is what causes rifting to stop in the metamorphic core complexes before reaching completion. Perhaps the answer to this question and others can be found by integrating studies of highly extended crust with those of passive margins and mid-ocean ridges. Many new studies are being planned in each of these end-member terranes and will undoubtedly be reported here, in the *Bulletin*, in the years to come.

ACKNOWLEDGMENTS

We gratefully acknowledge C. M. Jarchow and E. B. Goodwin for the donation of previously unpublished data shown in Figures 4 and 6, respectively. S. Klemperer and the members of COCORP generously donated data tapes of the 100% stacked-record section of the 40°N transect, a portion of which is displayed in Figure 10. T. Dahl-Jensen, T.C.R. Pulvertaft, M. C. Anderson, and W. Hamilton provided the seismic section and geologic photographs shown in Figures 5, 7, and 12, respectively. D. Okaya provided the line-drawing algorithm used to generate Figure 10. This manuscript has also benefited from the critical reviews of T. Brocher, D. Fountain, E. Goodwin, W. Hamilton, W. S. Holbrook, W. Mooney, D. Pollard, and A. Snoke. The work was supported in part by National Science Foundation Grant EAR84-17309.

REFERENCES CITED

Allmendinger, R. W., Sharp, J. W., Von Tish, D., Serpa, L., Brown, L., Kaufman, S., Oliver, J., and Smith, R. B., 1983, Cenozoic and Mesozoic structure of the eastern Basin and Range from COCORP seismic reflection data: Geology, v. 11, p. 532–536.
Allmendinger, R. W., Hauge, T. A., Hauser, E. C., Potter, C. J., Klemperer, S. L., Nelson, K. D., Knuepfer, P., and Oliver, J., 1987, Overview of the COCORP 40°N Transect, western United States: The fabric of an orogenic belt: Geological Society of America Bulletin, v. 98, p. 308–319.
Anderson, C. A., Scholz, E. A., and Strobell, J. D., Jr., 1955, Geology and ore deposits of the Bagdad area, Yavapai County, Arizona: U.S. Geological Survey Professional Paper 278, 99 p.
Anderson, E. M., 1951, The dynamics of faulting and dyke formation with applications to Britain (2nd edition): Edinburgh, United Kingdom, Oliver and Boyd.
Armstrong, R. L., 1968, Sevier orogenic belt in Nevada and Utah: Geological Society of America Bulletin, v. 79, p. 429–458.
Baker, B. H., Crossley, R., and Goles, G. G., 1978, Tectonic and magmatic evolution of the southern part of the Kenya rift valley, *in* Neumann, E. R., and Ramberg, I. B., eds., Petrology and geochemistry of continental rifts: Dordrecht, Holland, D. Reidel, p. 29–50.
Barazangi, M., and Brown, L., eds., 1986a, Reflection seismology: A global perspective, Volume 13: American Geophysical Union Geodynamics Series, v. 13, 311 p.
—————1986b, Reflection seismology: The continental crust, Volume 14: American Geophysical Union Geodynamic Series, v. 14, 399 p.
Behrendt, J. C., Green, A. G., Cannon, W. F., Hutchinson, D. R., Lee, M. W., Milkereit, B., Agena, W. F., and Spencer, C., 1988, Crustal structure of the Midcontinent rift system: Results from GLIMPCE deep seismic reflection profiles: Geology, v. 16, p. 81–85.
Bois, C., Cazes, M., Damotte, B., Galdeano, A., Hirn, A., Mascle, A., Matte, P., Raoult, J. F., and Torreilles, G., 1986, Deep seismic profiling of the crust in northern France: The ECORS project, *in* Barazangi, M., and Brown, L., eds., Reflection seismology: A global perspective, Volume 13: Washington, D.C., American Geophysical Union Geodynamics Series, v. 13, p. 21–30.
Borevsky, L. V., Vartanyan, G. S., and Kulikov, T. B., 1984, Hydrogeological essay, *in* Kozlovsky, Ye.-A., ed., The superdeep well of the Kola Peninsula: Moscow, Soviet Union, Springer-Verlag, p. 271–287.
Born, A., 1933, Der geologische Aufbau der Erde, Handbuch der Geophysik, Volume 2: Berlin, Germany, v. 2, p. 565–867 (see p. 759 and Fig. 306).
Byerly, P., 1955, Subcontinental structure in the light of seismological evidence, advances in geophysics, Volume 3: New York, Academic Press, p. 105–152.
Catchings, R. D., 1987, Crustal structure of the northwestern United States [Ph.D. thesis]: Stanford, California, Stanford University, 183 p.
Chen, W.-P., and Molnar, P., 1983, Focal depths of intracontinental and intraplate earthquakes and their implications for the thermal and mechanical properties of the lithosphere: Journal of Geophysical Research, v. 88, p. 4183–4214.
Christensen, N. I., 1966, Elasticity of ultrabasic rocks: Journal of Geophysical Research, v. 71, p. 5921–5931.
Christensen, N. I., and Szymanski, D. L., 1988, Origin of reflections from the Brevard fault zone: Journal of Geophysical Research, v. 93, p. 1087–1102.
Christiansen, R. L., and Lipman, P. W., 1972, Cenozoic volcanism and plate-tectonic evolution of the western United States. II. Late Cenozoic: Royal Society of London Philosophical Transactions, v. 271, p. 249–284.
Clowes, R. M., and Kanasewich, E. R., 1970, Seismic attenuation and the nature of reflecting horizons within the crust: Journal of Geophysical Research, v. 75, p. 6693–6705.
Clowes, R. M., Kanasewich, E. R., and Cumming, G. L., 1968, Deep crustal seismic reflections at near-vertical incidence: Geophysics, v. 33, p. 441–451.
Clowes, R. M., Spence, G. D., Ellis, R. M., and Waldron, D. A., 1986, Structure of the lithosphere in a young subduction zone: Results from reflection and refraction studies, *in* Barazangi, M., and Brown, L., eds., Reflection seismology: The continental crust, Volume 14: Washington, D.C., American Geophysical Union Geodynamics Series, v. 14, p. 313–322.
Coney, P. J., and Harms, T. A., 1984, Cordilleran metamorphic core complexes: Cenozoic extensional relics of Mesozoic compression: Geology, v. 12, p. 550–554.
Dahl-Jensen, T., Dyrelius, D., Juhlin, C., Palm, H., and Pedersen, L. B., 1987, Deep reflection seismics in the Precambrian of Sweden: Royal Astronomical Society Geophysical Journal, v. 89, p. 371–378.
Delaney, P. T., Pollard, D. D., Ziony, J. I., and McKee, E. H., 1986, Field relations between dikes and joints: Emplacement processes and paleostress analysis: Journal of Geophysical Research, v. 91, p. 4920–4938.
Eaton, J. P., and Murata, K. J., 1960, How volcanoes grow: Science, v. 132, p. 925–938.
Ewing, M., Crary, A. P., and Rutherford, H. M., 1937, Geophysical investigations in the emerged and submerged Atlantic coastal plain, Part I: Methods and results: Geological Society of America Bulletin, v. 48, p. 753–802.
Ewing, M., Worzel, J. L., Hersey, J. B., Press, F., and Hamilton, G. R., 1950, Seismic refraction measurements in the Atlantic Ocean basin: Part I: Seismological Society of America Bulletin, v. 40, p. 233–242.
Feng, R., and McEvilly, T. V., 1983, Interpretation of seismic reflection profiling data for the structure of the San Andreas fault zone: Seismological Society of America Bulletin, v. 73, p. 1701–1720.
Fisher, M. A., Brocher, T. M., Smith, G. L., and Nokleberg, W. J., 1988, Seismic reflection images of the crust of the northern part of the Chugach terrane, Alaska: Preliminary results of a survey for the Trans Alaska Crustal Transect (TACT): Journal of Geophysical Research (in press).
Fountain, D. M., 1986, Implications of deep crustal evolution for seismic reflection interpretation, *in* Barazangi, M., and Brown, L., eds., Reflection seismology: The continental crust, Volume 14: Washington, D.C., American Geophysical Union Geodynamics Series, v. 14, p. 1–7.
Fountain, D. M., Hurich, C. A., and Smithson, S. B., 1984, Seismic reflectivity of mylonite zones in the crust: Geology, v. 12, p. 195–198.
Frost, E. G., and Okaya, D. A., 1988, Continuity of exposed mylonitic rocks to middle- and lower-crustal depths within the Whipple Mountains detachment terrane, SE California, and its implications for continental extensional tectonics: Geological Society of America Bulletin (in press).
Fuchs, K., 1969, On the properties of deep crustal reflectors: Journal of Geophysics, v. 35, p. 133–149.
Fuchs, K., Bonjer, K.-P., Gajewski, D., Lueschen, E., Prodehl, C., Sandmeier, K.-J., Wenzel, F., and Wilhelm, H., 1987, Crustal evolution of the Rheingraben area: I. Exploring the lower crust in the Rhinegraben Rift by unified geophysical experiments: Tectonophysics, v. 141, p. 261–275.
Furlong, K. P., and Fountain, D. M., 1986, Continental crustal underplating: Thermal considerations and seismic-petrologic consequences: Journal of Geophysical Research, v. 91, p. 8285–8294.
Gajewski, D., Holbrook, W. S., and Prodehl, C., 1987a, A three-dimensional crustal model of southwest Germany derived from seismic-refraction data: Tectonophysics, v. 142, p. 149–170.
—————1987b, Combined seismic reflection and refraction profiling in southwest Germany—Detailed velocity mapping by the refraction survey: Royal Astronomical Society Geophysical Journal, v. 89, p. 333–338.
Gans, P. G., 1987, An open-system, two-layer crustal stretching model for the eastern Great Basin: Tectonics, v. 6, p. 1–12.
Gibson, B. S., and Levander, A. R., 1988, Modeling and processing of scattered waves in seismic reflection surveys: Geophysics, v. 53, p. 466–478.
Glazner, A. F., and Bartley, J. M., 1984, Timing and tectonic setting of Tertiary low-angle normal faulting and associated magmatism in the southwestern United States: Tectonics, v. 3, p. 385–396.
—————1985, Evolution of lithospheric strength after thrusting: Geology, v. 13, p. 42–45.
Goodwin, E. B., and Thompson, G. A., 1988, The seismically reflective crust beneath highly extended terranes: Evidence for its origin in extension: Geological Society of America Bulletin (in press).
Gutenberg, B., 1951, Crustal layers of continents and oceans: Geological Society of America Bulletin, v. 62, p. 427–440.
—————1955, Wave velocities in the Earth's crust, in Poldervaart, A., ed., Crust of the Earth: Geological Society of America Special Paper 62, p. 19–34.
Hamilton, W., 1965, Diabase sheets of the Taylor glacier region, Victoria Land, Antarctica: U.S. Geological Survey Professional Paper 456-B, p. B1–B71.
—————1982, Structural evolution of the Big Maria Mountains, northeastern Riverside County, southeastern California, *in* Frost, E. G., and Martin, D. L., eds., Mesozoic-Cenozoic tectonic evolution of the Colorado River region, California, Arizona, and Nevada: San Diego, California, Cordilleran Publishers, p. 1–28.
Hauser, E. C., Gephart, J., Latham, T., Brown, L., Kaufman, S., Oliver, J., and Lucchitta, I., 1987a, COCORP Arizona

transect: Strong crustal reflections and offset Moho beneath the transition zone: Geology, v. 15, p. 1103–1106.

Hauser, E. C., Potter, C. J., Hauge, T. A., Burgess, S., Burtch, S., Mutschler, J., Allmendinger, R. W., Brown, L., Kaufman, S., and Oliver, J., 1987b, Crustal structure of eastern Nevada from COCORP deep seismic reflection data: Geological Society of America Bulletin, v. 99, p. 833–844.

Hildreth, W., 1981, Gradients in silicic magma chambers: Implications for lithospheric magmatism: Journal of Geophysical Research, v. 86, p. 10153–10192.

Holbrook, W. S., Gajewski, D., and Prodehl, C., 1987, Shear-wave velocity and Poisson's ratio structure of the upper lithosphere in southwest Germany: Geophysical Research Letters, v. 14, p. 231–234.

Holbrook, W. S., Gajewski, D., Krammer, A., and Prodehl, C., 1988, An interpretation of wide-angle compression and shear wave data in southwest Germany: Poisson's ratio and petrological implications: Journal of Geophysical Research (in press).

Holt, W. E., Chase, C. G., and Wallace, T. C., 1986, Crustal structure from three-dimensional modeling of a metamorphic core complex: A model for uplift, Santa Catalina–Rincon Mountains, Arizona: Geology, v. 14, p. 927–930.

Howard, K. A., Goodge, J. W., and John, B. E., 1982, Detached crystalline rocks of the Mohave, Buck, and Bill Williams Mountains, western Arizona, in Frost, E. G., and Martin, D. L., eds., Mesozoic-Cenozoic tectonic evolution of the Colorado River region, California, Arizona, and Nevada: San Diego, California, Cordilleran Publishers, p. 377–390.

Hurich, C. A., and Smithson, S. B., 1987, Compositional variation and the origin of deep crustal reflections: Earth and Planetary Science Letters, v. 85, p. 416–426.

Hurich, C. A., Smithson, S. B., Fountain, D. M., and Humphreys, M. C., 1985, Seismic evidence of mylonite reflectivity and deep structure in the Kettle Dome metamorphic core complex, Washington: Geology, v. 13, p. 577–580.

Hutchinson, D. R., Klitgord, K. D., and Trehu, A. M., 1987, Structure of the lower crust beneath the Gulf of Maine: Royal Astronomical Society Geophysical Journal, v. 89, p. 189–194.

Illies, J. H., 1975, Recent and paleo-intraplate tectonics in stable Europe and the Rhinegraben rift system: Tectonophysics, v. 29, p. 251–264.

Jarchow, C. M., Walker, D., Smithson, S. B., Karl, J., and PASSCAL Basin and Range seismic experiment participants, 1987, Analysis of the PASSCAL Basin and Range lithospheric seismic experiment reflection data [abs.]: EOS (American Geophysical Union Transactions), v. 68, p. 1360.

Jones, A. G., 1987, MT and reflection: An essential combination: Royal Astronomical Society Geophysical Journal, v. 89, p. 7–18.

Jones, T., and Nur, A., 1982, Seismic velocity and anisotropy in mylonites and the reflectivity of deep crustal fault zones: Geology, v. 10, p. 260–263.

——— 1984, The nature of seismic reflections from deep crustal fault zones: Journal of Geophysical Research, v. 89, p. 3153–3171.

Juhlin, C., and Pedersen, L. B., 1987, Reflection seismic investigations of the Siljan impact structure, Sweden: Journal of Geophysical Research, v. 92, p. 14113–14122.

Junger, A., 1951, Deep reflections in Big Horn County, Montana: Geophysics, v. 16, p. 499–505.

Karson, J. A., and Winters, A. T., 1987, Tectonic extension on the Mid-Atlantic Ridge [abs.]: EOS (American Geophysical Union Transactions), v. 68, p. 1508.

Klemperer, S. L., 1987, A relation between continental heat flow and the seismic reflectivity of the lower crust: Journal of Geophysics, v. 61, p. 1–11.

Klemperer, S. L., Hauge, T. A., Hauser, E. C., Oliver, J. E., and Potter, C. J., 1986, The Moho in the northern Basin and Range province, Nevada, along the COCORP 40°N seismic reflection transect: Geological Society of America Bulletin, v. 97, p. 603–618.

Klemperer, S. L., and the BIRPS group, 1987, Reflectivity of the crystalline crust: Hypotheses and tests: Royal Astronomical Society Geophysical Journal, v. 89, p. 217–222.

Kligfield, R., Crespi, J., Naruk, S., and Davis, G. H., 1984, Displacement and strain patterns of extensional orogens: Tectonics, v. 3, p. 577–609.

Lachenbruch, A. H., 1988, Continental extension, magmatism, and elevation; formal relations and rules of thumb, in Lucchitta, I., and Morgan, P., eds., Heat and detachment in crustal extension on continents and planets: Tectonophysics special volume (in press).

Lachenbruch, A. H., and Sass, J. H., 1978, Models of an extending lithosphere and heat flow in the Basin and Range province, in Smith, R. B., and Eaton, G. P., eds., Cenozoic tectonics and regional geophysics of the western Cordillera: Geological Society of America Memoir 152, p. 209–250.

Louie, J. N., and Clayton, R. W., 1987, The nature of deep crustal structures in the Mojave Desert, California: Royal Astronomical Society Geophysical Journal, v. 89, p. 125–132.

Louie, J. N., Clayton, R. W., and LeBras, R. J., 1988, Three-dimensional imaging of steeply dipping structure near the San Andreas fault, Parkfield, California: Geophysics, v. 53, p. 176–185.

Lueschen, E., Wenzel, F., Sandmeier, K.-J., Menges, D., Ruehl, Th., Stiller, M., Janoth, W., Keller, F., Soellner, W., Thomas, R., Krohe, A., Stenger, R., Fuchs, K., Wilhelm, H., and Eisbacher, G., 1987, Near-vertical and wide-angle seismic surveys in the Black Forest, SW Germany: Journal of Geophysics, v. 62, p. 1–30.

Lynn, H. B., Hale, L. D., and Thompson, G. A., 1981, Seismic reflections from the basal contacts of batholiths: Journal of Geophysical Research, v. 86, p. 10633–10638.

Macdonald, G. A., 1949, Hawaiian petrographic province: Geological Society of America Bulletin, v. 60, p. 1541–1596.

Matthews, D. H., and Cheadle, M. J., 1986, Deep reflections from the Caledonides and Variscides west of Britain and comparison with the Himalayas, in Barazangi, M., and Brown, L., eds., Reflection seismology: A global perspective, Volume 13: Washington, D.C., American Geophysical Union Geodynamics Series, v. 13, p. 5–20.

Matthews, D., and Smith, C., eds., 1987, Deep seismic profiling of the continental lithosphere, Volume 89: Royal Astronomical Society Geophysical Journal special issue, v. 89, p. 1–447.

McCarthy, J., 1986, Reflection profiles from the Snake Range metamorphic core complex: A window into the mid-crust, in Barazangi, M., and Brown, L., eds., Reflection seismology: The continental crust, Volume 14: Washington, D.C., American Geophysical Union Geodynamics Series, v. 14, p. 281–292.

McCarthy, J., Fuis, G. S., Wilson, J., McKissick, C., and Larkin, S., 1987a, PACE seismic refraction and wide-angle reflection data recorded across the central Arizona transition zone [abs.]: EOS (American Geophysical Union Transactions), v. 68, p. 1359.

McCarthy, J., Fuis, G., and Wilson, J., 1987b, Refraction profiling across the Whipple Mountains, southeastern California: Crustal structure across a region of large continental extension: Royal Astronomical Society Geophysical Journal, v. 89, p. 119–124.

McCarthy, J., Mutter, J. C., Morton, J. L., Sleep, N. H., and Thompson, G. A., 1988, Relict magma chamber structures preserved within the Mesozoic North Atlantic crust?: Geological Society of America Bulletin (in press).

Meissner, R., and Strehlau, J., 1982, Limits of stresses in continental crusts and their relation to the depth-frequency distribution of shallow earthquakes: Tectonics, v. 1, p. 73–89.

Meissner, R., and Wever, T., 1986, Nature and development of the crust according to deep reflection data from the German Variscides, in Barazangi, M., and Brown, L., eds., Reflection seismology: A global perspective, Volume 13: Washington, D.C., American Geophysical Union Geodynamics Series, v. 13, p. 31–42.

Mohorovičić, A., 1909, Das beben vom 8.X. 1909: Jahrb. Meteorol. Obs. Zagreb, v. 9, Teil 4, p. 1–63.

Mooney, W. D., and Brocher, T. M., 1987, Coincident seismic reflection/refraction studies of the continental lithosphere: A global review: Reviews of Geophysics, v. 25, p. 723–742.

Mooney, W. D., and Ginzburg, A., 1986, Seismic measurements of the internal properties of fault zones: Pageoph, v. 124, p. 141–157.

Moss, F. J., and Mathur, S. P., 1986, A review of continental reflection profiling in Australia, in Barazangi, M., and Brown, L., eds., Reflection seismology: A global perspective, Volume 13: Washington, D.C., American Geophysical Union Geodynamics Series, v. 13, p. 67–76.

Mutter, J. C., and North Atlantic Transect (NAT) Study Group, 1985, Multichannel seismic images of the oceanic crust's internal structure: Evidence for a magma chamber beneath the Mesozoic Mid-Atlantic Ridge: Geology, v. 13, p. 629–632.

Nakamura, K., 1977, Volcanoes as possible indicators of tectonic stress orientation—Principle and proposal: Journal of Volcanology and Geothermal Research, v. 2, p. 1–16.

Nakata, J. K., 1982, Preliminary report on diking events in the Mohave Mountains, Arizona, in Frost, E. G., and Martin, D. L., eds., Mesozoic-Cenozoic tectonic evolution of the Colorado River region, California, Arizona, and Nevada: San Diego, California, Cordilleran Publishers, p. 85–90.

Norrell, G. T., and Harper, G. D., 1987, Upper mantle detachment faulting in the Josephine ophiolite: Implications for slow-spreading mid-ocean ridges [abs.]: EOS (American Geophysical Union Transactions), v. 68, p. 1509.

Okaya, D. A., and Frost, E. G., 1986, Crustal structure of the Whipple–Turtle–Old Woman Mtns. region based on CALCRUST and industry seismic profiles: Geological Society of America Abstracts with Programs, v. 18, p. 166.

Oliver, J., Dobrin, M., Kaufman, S., Meyer, R., and Phinney, R., 1976, Continuous seismic reflection profiling of the deep basement, Hardeman County, Texas: Geological Society of America Bulletin, v. 87, p. 1537–1546.

Oxburgh, E. R., O'Nions, R. K., and Hill, R. I., 1986, Helium isotopes in sedimentary basins: Nature, v. 324, p. 632–635.

Phinney, R. A., and Jurdy, D. M., 1979, Seismic imaging of deep crust: Geophysics, v. 44, p. 1637–1660.

Potter, C. J., Sanford, W. E., Yoos, T. R., Prussen, E. I., Keach, W., II, Oliver, J. E., Kaufman, S., and Brown, L. D., 1986, COCORP deep seismic reflection traverse of the interior of the North American Cordillera, Washington and Idaho: Implications for orogenic evolution: Tectonics, v. 5, p. 1007–1026.

Raitt, R. W., 1949, Studies of ocean-bottom structure off southern California with explosive waves: Geological Society of America Abstracts with Programs, v. 60, p. 1915.

Ramsay, J. G., 1982, Rock ductility and its influence on the development of tectonic structures in mountain belts, in Hsü, K. J., ed., Mountain building processes: London and New York, Academic Press, p. 111–127.

Reif, D. M., and Robinson, J. P., 1981, Geophysical, geochemical, and petrographic data and regional correlation from the Arizona state A-1 well, Pinal County, Arizona: Arizona Geological Society Digest, v. 13, p. 99–109.

Reynolds, S. J., 1985, Geology of the South Mountains, central Arizona: Arizona Bureau of Geology and Mineral Technology Bulletin 195, 61 p.

Roure, F., and Banda, E., 1987, Crustal geometry of an orogenic belt: The deep structure of the Pyrenees along the ECORS seismic profile [abs.]: EOS (American Geophysical Union Transactions), v. 67, p. 1480.

Sandmeier, K.-J., and Wenzel, F., 1986, Synthetic seismograms for a complex crustal model: Geophysical Research Letters, v. 13, p. 22–25.

Sandmeier, K.-J., Walde, W., and Wenzel, F., 1987, Physical properties and structure of the lower crust revealed by one- and two-dimensional modeling: Royal Astronomical Society Geophysical Journal, v. 89, p. 339–344.

Sanford, W. E., Potter, C. J., and Oliver, J. E., 1988, Detailed three-dimensional structure of the deep crust based on COCORP data in the Cordilleran interior, north-central Washington: Geological Society of America Bulletin, v. 100, p. 60–71.

Shankland, T. J., and Ander, M. E., 1983, Electrical conductivity, temperatures, and fluids in the lower crust: Journal of Geophysical Research, v. 88, p. 9475–9484.

Smith, R. B., and Bruhn, R. B., 1984, Intraplate extensional tectonics of the eastern Basin and Range: Inferences on structural style from seismic reflection data, regional tectonics, and thermal-mechanical models of brittle-ductile deformation: Journal of Geophysical Research, v. 89, p. 5733–5762.

Smithson, S. B., Brewer, J. A., Kaufman, S., Oliver, J. E., and Zawislak, R. L., 1980, Complex Archean lower crustal structure revealed by COCORP crustal reflection profiling in the Wind River Range, Wyoming: Earth and Planetary Science Letters, v. 46, p. 295–305.

Smithson, S. B., Johnson, R. A., and Hurich, C. A., 1986, Crustal reflections and crustal structure, in Barazangi, M., and Brown, L., eds., Reflection seismology: The continental crust, Volume 14: Washington, D.C., American Geophysical Union Geodynamics Series, v. 14, p. 21–32.

Stern, T. A., and Davey, F. J., 1988, Crustal structure associated with basins formed behind the Hikurangi subduction zone, North Island, New Zealand, in Price, R. A., ed., Sedimentary basins, Volume 20: American Geophysical Union Geodynamics Series, v. 20 (in press).

Suess, E., 1888, Das Antlitz der Erde, Volume 2: Vienna, Austria, v. 2, p. 256–264.

Talwani, M., Le Pichon, X., and Ewing, M., 1965, Crustal structure of the mid-ocean ridges, 2. Computed model from gravity and seismic refraction data: Journal of Geophysical Research, v. 70, p. 341–352.

Tatel, H. E., and Tuve, M. A., 1955, Seismic exploration of a continental crust, in Poldervaart, A., ed., The crust of the Earth: Geological Society of America Special Paper 62, p. 35–60.

Thompson, G. A., and Burke, D. B., 1974, Regional geophysics of the Basin and Range province: Annual Review of Earth and Planetary Science, v. 2, p. 213–237.

Thompson, G. A., and McCarthy, J., 1986, Geophysical evidence for igneous inflation of the crust in highly extended terranes [abs.]: EOS (American Geophysical Union Transactions), v. 67, p. 1184.

——— 1988, A gravity constraint on the origin of highly extended terranes, in Lucchitta, I., and Morgan, P., eds., Heat and detachment in crustal extension on continents and planets: Tectonophysics special volume (in press).

Thompson, G. A., Catchings, R. C., Goodwin, E. B., Holbrook, W. S., Jarchow, C. M., McCarthy, J., Mann, C. E., and Okaya, D. A., 1988, Geophysics of the western Basin and Range province, in Pakiser, L., and Mooney, W. D., eds., Geophysical framework of the continental U.S.: Geological Society of America Memoir Series (in press).

Valasek, P. A., Hawman, R. B., Johnson, R. A., and Smithson, S. B., 1987, Nature of the lower crust and Moho in eastern Nevada from "wide-angle" reflection measurements: Geophysical Research Letters, v. 14, p. 1111–1114.

Varga, R. J., and Moores, E. M., 1985, Spreading structure of the Troodos ophiolite, Cyprus: Geology, v. 13, p. 846–850.

Wever, Th., Trappe, H., and Meissner, R., 1987, Possible relations between crustal reflectivity, crustal age, heat flow, and viscosity of the continents: Annales Geophysicae, v. 5B, p. 255–266.

Zoback, M. L., and Zoback, M., 1980, State of stress in the conterminous United States: Journal of Geophysical Research, v. 85, p. 6113–6156.

MANUSCRIPT RECEIVED BY THE SOCIETY FEBRUARY 13, 1988
REVISED MANUSCRIPT RECEIVED MAY 4, 1988
MANUSCRIPT ACCEPTED MAY 9, 1988

CENTENNIAL ARTICLE

Significance of past and recent heat-flow and radioactivity studies in the Southern Rocky Mountains region

EDWARD R. DECKER Department of Geological Sciences, University of Maine, Orono, Maine 04469

HENRY P. HEASLER
KENNETH L. BUELOW*
KEITH H. BAKER*
JAMES S. HALLIN*
Department of Geology and Geophysics, University of Wyoming, Laramie, Wyoming 82071

ABSTRACT

As reported in *The Geological Society of America Bulletin* of 1950, Francis Birch's innovative heat-flow research in the Colorado Front Range introduced frequently used terrain correction and thermal-conductivity measurement methods. That report also presented the first empirical evidence for a positive correlation between above-normal flux and radioactive roots in an isostatically compensated mountain area with a thick crust, an observation that strongly influenced continental heat-flow studies for the next 18 years. Birch's Front Range study clearly showed that reliable continental heat-flow research requires knowledge of bedrock radioactivity. This concept reached an unprecedented level of acceptance in 1968, when linear relations between heat flow and radiogenic heat production were discovered for three contrasting provinces in the United States. Subsequently, such lines have been central to virtually all heat-flow and radioactivity research on land.

Heat-flow (Q) data for 139 boreholes provide new detail on thermal regimes in the Southern Rocky Mountains and bordering areas, as do radiogenic heat-production (A) data for 60 locales. Interpretations also emphasize reduced and residual heat-flow values; reduced values correspond to the intercepts of regional Q-A lines, whereas a residual value is the flux at a locale after the probable effect of near-surface heat production has been subtracted. The observed Q-A lines demonstrate that the Front Range and other easterly frontal ranges of the Southern Rocky Mountains in Colorado and northern New Mexico are characterized by a reduced flux (54–58 mWm^{-2}) that is dramatically higher than that (~27 mWm^{-2}) in the Wyoming Basin–Southern Rocky Mountains area in southeastern Wyoming. Because the transitions between these provinces are narrow ($\leqslant$50–60 km), sources in the upper crust must explain some of the contrasting reduced heat-flow values. In southeastern Wyoming, normal heat flow in Archean and early Proterozoic basement terranes probably reflects deep erosion that produced a thin (~7 km) near-surface granitic layer that overlies a low-radioactivity lower crust. In the Colorado Front Range, Proterozoic, Mesozoic, and Cenozoic silicic rocks with relatively enriched radiogenic heat could comprise a 20- to 25-km-thick granitic layer in the upper crust that produces a large part of the above-normal reduced and residual flux. Here, partial melting of deep protolithic rocks in late Mesozoic and Cenozoic times could have produced a lower crust with low radiogenic heat production. By these views, Birch's "high heat flow–radioactive mountain root" model of the Colorado Front Range is confirmed if a large part of the topography is isostatically compensated by low-density pre-Miocene crystalline masses in the upper crust.

Background reduced heat flow in Colorado parts of the Southern Rocky Mountains and the eastern Colorado Plateau is high (54–68 mWm^{-2}). Zones of unusually high residual flux (88–118 mWm^{-2}) occur in the Rio Grande rift zone in the environs of the Colorado mineral belt in the Leadville–northern Sawatch Range region, eastern parts of the San Juan Mountains in southern Colorado, and in the Park Range–Mountain Parks area near the Colorado-Wyoming border. The flux in these areas implies unrealistically high equilibrium temperatures near the crust-mantle boundary, and the 50- to 60-km-wide borders of the Leadville–northern Sawatch Range residual heat-flow anomaly must be caused by sources in the upper crust. Therefore, young (10- to 1-m.y.-old) intrusions in a late Cenozoic rhyolitic complex in the upper crust are preferred to explain gravity lows, late Cenozoic uplift and igneous activity, and the high residual flux in the Leadville–northern Sawatch Range area. Similar models may apply elsewhere in the northern rift zone. If this interpretation is correct, magmatic thickening of the crust, not extensional-subsidence mechanisms, probably explains late Cenozoic uplift and extension of the northern Rio Grande rift–Southern Rocky Mountains system. Because very high regional flux in the northern rift zone and the San Juan Mountains areas implies above-liquidus equilibrium temperatures in the lower crust and upper mantle, serious inconsistencies arise when steady-state thermal models are examined relative to volumes of late Tertiary volcanism, uniform crustal thickness, and the absence of unusually low P-wave velocities in the mantle in the Southern Rocky Mountains.

---

*Present addresses: (Buelow) 5051 Eagle Heights, Madison, Wisconsin 53705; (Baker) 2610 Hunters Locke, Sugar Land, Texas 77479; (Hallin) Conoco, Inc., P.O. Box 2197, Houston, Texas 77252.

Geological Society of America Bulletin, v. 100, p. 1851–1885, 18 figs., 5 tables, December 1988.

INTRODUCTION

Francis Birch, internationally acclaimed geophysicist and Sturgis Hooper Professor of Geological Sciences at Harvard University, published nine papers in *The Geological Society of America Bulletin* (Birch, 1943, 1950, 1953, 1956, 1958, 1961, 1965; Birch and Dow, 1936; Birch and Law, 1935). Three focused on thermal research: "Flow of heat in the Front Range, Colorado" (Birch, 1950); "Heat flow at Eniwetok Atoll" (Birch, 1956); and, as his Presidential Address to the Society in 1965, "Speculations on the earth's thermal history" (Birch, 1965). The paper on heat flow in the Colorado Front Range, like most of Birch's work, was an outstanding, insightful contribution in the geological sciences that was extremely important to subsequent development of geothermal research on land. The introductory portions of this report briefly trace the impact of that paper, an influence that lasted until at least 1968 when Birch and others (1968) published a paper on heat flow and thermal history in New England and New York that inspired much of the terrestrial heat-flow and radioactivity research that has been done since that time. New heat-flow and radioactivity data for the Southern Rocky Mountains region are then presented and discussed.

IMPACT OF THE FRONT RANGE PAPER BY FRANCIS BIRCH

There are at least three "classic" parts of Birch's (1950) Front Range research. First, new procedures for topographic corrections of near-surface Earth temperatures were developed and tested. Second, new methodology for divided-bar measurements of the thermal conductivities of rocks were presented and used. Finally, observational and theoretical bases were established for increased thermal output in mountains with radioactive, isostatically thickened crusts. Brief details on the impact of these elements of Birch's paper are given below.

Corrections for Steady-State and Time-Dependent Topography

Observations of underground temperatures in mountainous terrane acquire their full significance in heat-flow studies only after corrections are made for topographic relief. Birch (1950, p. 569–590) clearly recognized this while discussing the potential value of a large amount of inadequately reduced temperature data for tunnels in the Alps and in his highly original development of methods for correcting temperatures in the Adams and Moffat Tunnels in the Colorado Front Range. Although numerical and approximate in application, Birch's new corrections were not restricted to relief that is small compared to temperature depth, and two- and three-dimensional topography could be treated. In contrast to previous corrections (Lees, 1910; Bullard, 1938; Jeffreys, 1938), evolution of topography (uplift and erosion) in a study area could be treated. For the Adams Tunnel in the Front Range, the correction for indefinitely persisting (steady-state) three-dimensional topography led to a gradient of 24.1 °C/km (Birch, 1950). After corrections for uniform uplift (~2.1 km) and erosion (0–2 km) for the past 1 to 4 m.y., the reduced gradients for this tunnel are in the range 20.5–22.5 °C/km, and the "steady-state" and "best values" of corrected heat flow are 81 ± 4 and 71 ± 8 mWm^{-2}, respectively (Birch, 1950). This seemingly small difference of the corrected regional heat flows could be extremely important in thermal modeling in an area such as the Front Range. For example, calculated equilibrium temperatures near the Moho (48–52 km) increase by 180–220 °C, if an extra 10 mWm^{-2} of flux is assumed to cross the crust-mantle boundary and be in equilibrium with the heat flow at the surface.

Post-1950 heat-flow research on land was substantially enhanced by the availability of the topographic corrections developed by Birch (1950). From comparable reduction of mine adit temperatures at Grass Valley in California, Clark (1957) demonstrated the presence of unusually low heat flow in the Sierra Nevada. Clark and Niblett (1956) and Clark (1961) also showed that the heat flow in four out of five alpine railroad tunnels was greater than the world-wide average that could be calculated in the early 1960s. Decker (1969) similarly reduced temperature data for the Roberts Tunnel in Colorado and showed that the flux in the central Front Range was high. A considerable body of heat-flow data that was published in 1968 and 1969 essentially outlined the thermal regimes of the conterminous United States. As described by Birch and others (1968), Roy and others (1968a), Decker (1969), and Blackwell (1969), most of these heat-flow values were corrected for the effects of steady-state topography using the methods of Birch (1950). A few of the more recent uses of these corrections in heat-flow studies include the works of Sass and others (1971b), Costain and Wright (1973), Decker and Smithson (1975), Chapman and others (1981), Bodell and Chapman (1982), Decker and others (1980, 1984), and Decker (1987).

Birch (1950) noted that his correction can lead to inaccurate gradients at stations at shallow depths under sharp irregularities because it neglects second-order effects arising from lateral changes in the vertical gradient in rugged relief. This realization clearly motivated Lachenbruch (1968a, 1969) in his development of a new method for properly correcting shallow temperature data below rugged terrain on land, or along the ocean bottom. Similarly, the inadequacies of Birch's correction motivated Blackwell (1969) and Blackwell and others (1980) during their developments of a method that corrects for many of the second-order effects (for example, exposure to sun, irregular ground-surface temperatures) in rugged mountainous terrane.

From modeling constrained by metamorphic mineral closure temperatures, reasonable crustal radioactivity distributions, and regional heat flows corrected after methods like those in Birch (1950), Clark and Jaeger (1969) and Clark (1979) predicted denudation rates (0.4–1.0 mm/yr) near central Alpine tunnels that closely agree with those estimated from regional geochronological data. Perhaps such modeling would have been less definitive or never accomplished if Birch's transient and steady-state terrain corrections had not been developed.

Procedures and Equipment for Thermal Conductivity Measurements

Less than two pages of Birch's 1950 paper were devoted to new divided-bar methods that were created for thermal conductivity measurements of disks of rocks. Yet it is difficult to visualize any things that had greater impact on subsequent laboratory determinations of conductivity in heat-flow studies in the United States and, to a lesser extent, elsewhere. Many of the earliest applications were by Birch and/or his students or post-doctoral associates. For example, the original divided-bar and procedures, or close variants thereof, were used for heat-flow or basic thermal-conductivity research by Birch (1956), Herrin and Clark (1956), Clark (1957, 1961), Clark and Niblett (1956), Diment and Robertson (1963), Diment (1964), Diment and Weaver, 1964; Diment and Werre (1964), Diment and others (1965a, 1965b), Walsh and Decker (1966), and Decker (1969).

Later uses of divided-bars, also often by Birch, his students, and co-researchers, mainly streamlined the earlier theory and equipment design in the 1950 paper. This is evident from Figure 1, where the essential elements of Birch's divided-bar are depicted with portrayals of single bars of four-column systems that were used later by Roy and others (1968b) and Sass and others (1971b).

The system designed by Birch used resistance heaters at the ends of the divided-bar and an external shield to reduce lateral heat transfer. Samples, standards, and reference disks were carefully machined to the same thicknesses and diameters and then placed in the shielded bar; measurements were made after a long equilibration interval.

The bars employed by Roy, Sass, and their

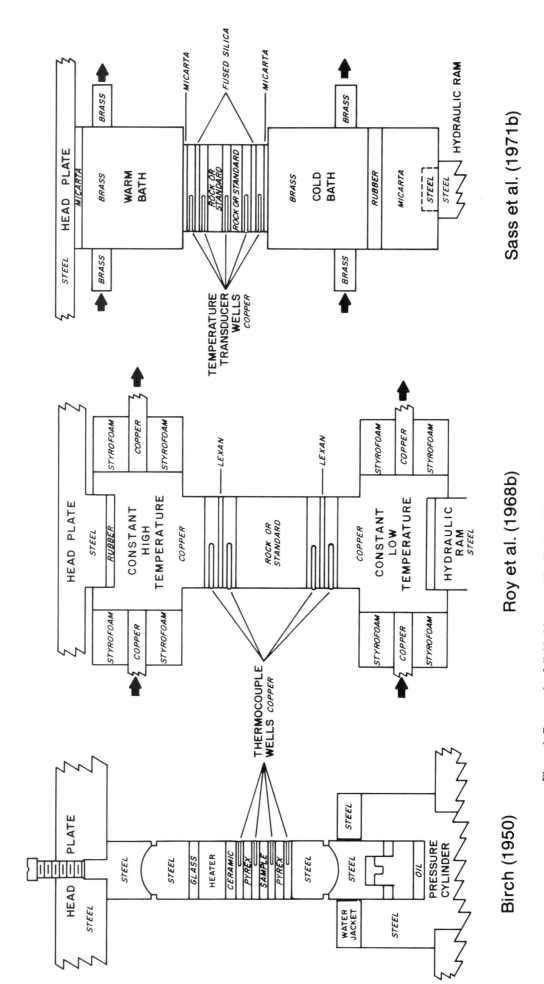

Figure 1. Portrayals of divided-bars used by Birch (1950), Roy and others (1968b), and Sass and others (1971b).

colleagues permitted quicker (30 minutes or so) measurements of conductivity by using bath-controlled constant-temperature fluids at upper and lower bar ends to produce temperature gradients across the composite columns. Reference disks, standards, and samples of different thicknesses and diameters also could be used, and carefully machined styrofoam insulators were placed around the bars to minimize lateral heat transfer. Otherwise, all depicted bars were used in essentially identical ways: the flat faces of disks of unknowns or standards were treated with low-viscosity greases or fluids and measured under uniaxial stress to reduce contact resistances; measurements of temperature differences in a bar were required for determinations; and rock or sample conductivities were determined relative to references that were calibrated with materials with "known" conductivity (for example, pyrex, silica glass, or quartz (see Birch, 1950; Roy and others, 1968b; Sass and others, 1971b)). Obviously, the divided-bar systems and procedures of Birch (1950) played a large role in measurements and equipment design some 20 years into the future, when realistic government funding for geothermal research in the United States required "modern" design due to the need for increased measurement throughput.

The thermal conductivity sections in Roy and others (1968b) and Sass and others (1971b) have been cited in numerous reports (Blackwell, 1969; Decker, 1969; Warren and others, 1969; Decker and Smithson, 1975; Blackwell and Baag, 1973; Blackwell and others, 1982a, 1982b; Brott and others, 1978, 1981; Chapman and others, 1981; Bodell and Chapman, 1982; Lachenbruch and others, 1976; Sass and others, 1985; Steele and Blackwell, 1982; Ziagos and others, 1985). These works are cited here because they illustrate the continuing value of the concepts, theory, and equipment that were initially developed by Birch (1950). The methodology of Birch also led to comparable systems that were used in geothermal research in England (for example, Richardson and Oxburgh, 1978), and similar equipment and procedures are cited in very recent heat-flow studies (Ballard and others, 1987).

Vertical Distribution of Crustal Radioactivity and Heat Flow

In a theoretical study of isostatically balanced crustal sections for mountains, high plains, and lowlands, Birch (1947a) demonstrated that formation of radioactive roots could produce augmented heat flow in the Colorado Front Range and other mountain areas. Pre-1950 heat-flow data on land, however, included no determinations for regions at elevations above about 1,890 m (Birch, 1947a); many existing values were estimates or uncertain (Birch, 1947a, 1947b; Birch and Clark, 1945); and, as a result, it was difficult to perceive any difference of thermal output between mountains and lower areas. This problem of inadequate data was resolved by Birch's paper in the 1950 volume of the *Bulletin* (v. 61). There it was shown that all topographically corrected heat-flow values (69–84 mWm^{-2}) for the Adams and Moffat Tunnels, at elevations >2,500 m in the Colorado Front Range, were significantly above normal, and unequivocal empirical evidence for increased thermal output in a mountain zone was presented (Birch, 1950). With reasonable assumptions as to thickness, density, and radioactivity of crustal units, Birch (1950) then demonstrated for the first time that granitic or intermediate mountain roots would explain isostatic equilibrium and the increased flux (~25 mWm^{-2}) in the Front Range, whereas the observed heat flow was consistent with an approximately uniform distribution of radioactivity in a 20- to 30-km-thick granitic layer that made up the upper crust.

Birch's hypothesis for a correlation between high heat flow, increased elevation, and radioactive roots in the Front Range clearly influenced a number of subsequent geothermal projects. From heat-flow determinations in tunnels in Switzerland and Austria, as examples, Clark and Niblett (1956) and Clark (1961) concluded that different, above-normal flux values in the central Alps were most likely due to laterally and vertically variable radioactivity in the upper parts of an isostatically thickened crust. Passages from Clark's (1957, p. 239) paper on heat flow at Grass Valley, California, in the Sierra Nevada also vividly reflect the impact of Birch's earlier results: "The mines are located on the flanks of a high mountain range, and Bouguer anomalies in their vicinity reach about –60 mgal. . . . Hence there is reason to expect that the heat flow here should be higher rather than lower than 'normal.' . . . This expectation is based on the high rate of heat production in the acidic rocks that are usually thought to make up a mountain root (Birch, 1950)." Clark's (1957) best estimates for the flux in this area, however, were in the range 31–28 mWm^{-2}, surprisingly low values that were difficult to explain in terms of nonradioactive mountain roots.

The actual causes of low heat flow in the Sierra Nevada remained obscure until at least 1968, even though the ideas of plate tectonics and associated transient thermal phenomena near continent margins were clearly formulated or postulated in the years 1963–1967 (Cox, 1973). During the same time period, Birch's (1950) "high heat flow–mountain root" model remained a plausible if not accepted hypothesis. This is evident in the classic paper by A. H. Lachenbruch on the Sierra Nevada, where, while discussing part of the rationale for the U.S. Geological Survey's heat-flow work in the years 1963–1968, it is stated that (Lachenbruch, 1968b, p. 6977) "it was generally expected that heat flow (q) would be high in the Sierra (in spite of the low value measured at Grass Valley (Clark, 1957)) because of the abundance of granitic rocks and the great crustal thickness."

From a broader perspective, an extremely important result of Birch's Front Range work was the demonstration that combined heat flow and radioactivity studies are required for reliable studies of the lateral and vertical distributions of heat sources in the outer parts of the Earth. This was made clear by Birch's comprehensive compilation of contrasting heat-production data for mafic, intermediate, and silicic rocks, and his recognition that the above-normal heat flow in the Front Range could be partly due to high radioactivity in near-surface rocks (Birch, 1950, p. 612–614). Modeling of heat-flow data for the Alps further illustrate the validity of this point because Clark and Niblett (1956) were compelled to "design" (after data in Birch, 1950) nonuniform distributions of crustal radioactivity to explain 67 to 92 mWm^{-2} values in southern Switzerland. Later, Clark (1961) presented other "artificial" crustal models which suggested that high flux in most of the central Alps was best explained by radioactive heat from a granitic root in a thickened crust.

The obvious need for combined measurements of flux and bedrock radioactivity on land was not fully realized until 1967. Then, Birch discovered a linear relation between heat flow and radioactive heat production of plutonic rocks in New England and New York. As reported by Birch and others (1968), the 5- to 7-km slope of that line was consistent with extreme upward differentiation of radioactive elements in bodies having comparable thicknesses, and the 35–39 mWm^{-2} intercept could be a measure of the flux from the underlying crust and upper mantle.

Discovery of the remarkable New York–New England heat-flow–heat-production line strongly influenced two other landmark papers that also appeared in 1968. One by Roy and others (1968a) presented lines for contrasting heat-flow provinces in the eastern United States, the Basin and Range province, and the Sierra Nevada, and focused on crustal layers with vertically uniform radioactive heat to explain the slopes of the lines. The other by Lachenbruch (1968b) obtained an identical relation for the Sierra Nevada and proposed that the linear relation and areal differential erosion there implied exponentially decreasing radiogenic heat production in the crust. Later, Roy and others (1972) discussed probable world-wide application of linear relations between heat flow and heat production and proposed that the contrasting intercepts of known lines were the characteristic thermal parameters of a province because they were measures of the flux from the lower crust and/or upper mantle. This point was illustrated with portrayals of reduced heat flow for all provinces in the United States where map and profile control were calculated reduced values that were valid for vertically uniform or exponentially decreasing distributions of crustal radiogenic heat production.

From a synthesis of data in the *Science Citation Index*, the just-mentioned works by Birch, Roy, Lachenbruch, and their colleagues have been cited at least 700 times in the period 1968 to mid-1988, a remarkable testimony to the fact that these papers inspired a major new era of terrestrial heat flow, bedrock radioactivity, and other geothermal research. Moreover, 19 papers in a special section entitled "Heat Production in the Continental Lithosphere" in the March 1987 issue of *Geophysical Research Letters* show that verification and understanding of continental heat-flow–heat-production lines remain high priorities of modern geophysical and geological research. Detailed discussion of these and other post-1972 geothermal research papers is not attempted in this paper. Rather, we now proceed to the results of relatively recent heat-flow and radioactivity research in the Southern Rocky Mountains region, and refer to relevant works when appropriate and needed.

INTRODUCTORY REMARKS ON SOUTHERN ROCKY MOUNTAIN REGION HEAT FLOW

Nearly two decades followed Birch's Front Range work before Decker (1969) presented five new heat-flow values for Colorado and northern New Mexico and suggested that high flux was characteristic of all of the Southern Rocky Mountains. In the same year, Blackwell (1969) published three estimates for the southern Wyoming Basin and implied that the regional flux there was normal. Several more values of heat flow have been presented for these and bordering areas in the past 20 years (Bodell and Chapman, 1982; Costain and Wright, 1973; Decker and Birch, 1974; Decker and others, 1980; Edwards and others, 1978; Eggleston and Reiter, 1984; Reiter and Mansure, 1983; Reiter and others, 1975, 1979; Roy and others, 1968b; Sass and others, 1971b), and, as a result, more detail on regional thermal regimes is emerging. One obvious feature of the present data is that most of the values in the Colorado mountains are high. In contrast, most of the values in southern Wyoming are low or normal. There is limited evidence also for a linear relation between heat flow and radiogenic heat production in portions of the Colorado Front Range (Roy and others, 1972; Decker and others, 1984), and parts of the northern Rio Grande rift zone in Colorado appear to be unusually high heat-flow zones with narrow borders (Decker and others, 1984). The basic heat-flow and radioactive heat-production data that were used to formulate the latter conjectures, however, have not been formally published. In the following sections, we tabulate a large body of unpublished basic heat-flow and radioactivity data for Colorado, northern New Mexico, eastern Utah, and southern Wyoming; we also elaborate on previously proposed and new interpretations of the thermal regimes in these areas.

One concern of most geothermal research on land is the issue of whether deep boreholes are required for reliable conductive heat-flow studies of young, tectonically active areas (Reiter and Clarkson, 1983; Reiter and Mansure, 1983). In our research, we have emphasized diamond-cored holes in crystalline rocks with relatively low porosity and permeability, and for which most of the conductivity and radioactivity control is based on down-hole samples. Additionally, we place appreciable faith in combined heat-flow and radiogenic heat-production data that agree with regional geology and geophysics. Thus lengthy speculation on "possible" effects of environmental factors (precipitation, paleoclimate, hydrologic recharge, geologic structure, and so on) in heat-flow studies is avoided herein, except to state that such things were considered in all of our research and to discuss them as needed. Likewise, we do not speculate on the possible merits of deeper drilling at our localities; very deep measurements of flux and radioactivity are always desirable, even near our 1-km-deep and deeper holes in low-porosity crystalline rocks.

We also avoid lengthy discussion of our techniques for collection and reduction of basic data because all procedures directly followed well-established philosophies and methods such as those described by Roy and others (1968b), Sass and others (1971a, 1971b), Decker (1973), and Birch (1950). Throughout the remaining text, heat flow (Q) is expressed in mWm^{-2}, radioactive heat production (A) is in μWm^{-3} and thermal conductivity (C or COND) is in $Wm^{-1}K^{-1}$. Standard errors ($\pm$) are used to indicate measures of the internal consistencies of data.

PHYSIOGRAPHY, GEOGRAPHY, AND GENERALIZED GEOLOGY

Generalized geology of the northern Rio Grande rift–Southern Rocky Mountains system and bordering areas is shown in Figure 2 (see folded insert accompanying this issue), as are physiographic, geologic, and geographic terms that are frequently used below.

The Southern Rocky Mountains occupy a small area in southeastern Wyoming. The Wyoming Basin is essentially a group of intermontane basins and small uplifts or arches that merges into the Southern Rocky Mountains to the south and east, and into the Great Plains to the northeast. Precambrian rocks are exposed in the cores of the mountains, and the Wyoming Basin is filled with Paleozoic, Mesozoic, and Cenozoic sediments. The Cheyenne Belt (CHYB in Fig. 2), a strongly deformed series of lithotectonic blocks in southeastern Wyoming, separates northerly Archean basement from southerly Proterozoic basement; it probably developed during early to middle Proterozoic accretionary-continent collision events (Duebendorfer and Houston, 1987; Karlstrom and Houston, 1984; Karlstrom and others, 1983). The present distribution of mountains and basins in Wyoming is essentially the result of the Laramide orogeny. Laramide deformation occurred from Late Cretaceous through early to middle Eocene times, and is characterized by compression, uplift, and thrust faulting (Blackstone, 1971, 1975; Houston, 1963, 1969).

The tectonic style in Wyoming after the Miocene is characterized by extension and associated normal faulting in the neighborhood of existing arches and uplifts. Most of Wyoming was uplifted in the late Pliocene, and subsequent erosion has produced the present topography (Blackstone, 1971, 1975).

Seismic data indicate a 37- to 40-km crustal thickness in the Rocky Mountains in southern Wyoming (Allmendinger and others, 1982; Jackson and Pakiser, 1965; Prodehl and Pakiser, 1980; Prodehl and Lipman, 1988). The crust thickens to the south, and Johnson and others (1984) suggested that broad gravity gradients near the Colorado-Wyoming border partly reflect crustal-thickness variations across the Cheyenne Belt that have persisted since the Proterozoic. Transient magnetic studies in southern Wyoming are consistent with low electrical conductivity in the upper mantle (Camfield and others, 1971; Gough, 1974; Porath and others, 1971). From geochemical studies of Devonian kimberlites, Tertiary igneous rocks and mantle-derived xenoliths in the Wyoming Basin–Southern Rocky Mountains area, the lithosphere below southern Wyoming is cold and perhaps 170–200 km thick (Eggler, 1987).

Several of the mountain ranges in Colorado are elements of late Paleozoic uplifts that experienced rapid uplift and erosion in middle to late Cretaceous times (Tweto, 1975; Bryant and Naeser, 1980). Associated upwarping and erosion appears to have continued into the Eocene (Tweto, 1975; Bryant and Naeser, 1980; Trimble, 1980). The Front Range, Park, Wet, and Sangre de Cristo Mountains are early Laramide uplifts that are cored by Precambrian rocks. The Sawatch and Mosquito Ranges and the Elk Mountains may be part of a large anticline that formed in Laramide time; Precambrian rocks make up most of the Sawatch and Mosquito Ranges, whereas the Elk Mountains consist of folded and thrust-faulted Paleozoic rocks that are intruded by Oligocene and Miocene igneous rocks (Tweto, 1975). The San Juan volcanic field rests on earlier uplifts that formed wholly or partially in the Laramide (Steven, 1975; Tweto, 1975; Lipman and others, 1970, 1978). As described by Tweto (1975), North, Middle, and South Parks are Laramide structural and sedimentary basins that contain significant thicknesses of Laramide and other sediments. The Uncompahgre Highland was uplifted in late Tertiary as well as Laramide time (Curtis, 1960; Tweto, 1975), and the La Plata Mountains were the site of Laramide intrusions into Paleozoic sediments (Tweto, 1975).

Parts of the northern Rio Grande rift–Southern Rocky Mountains system experienced significant uplift in Oligocene and Miocene times (Axelrod and Bailey, 1976; Bryant and Naeser, 1980; Kelley and Duncan, 1984, 1986; Lindsley and others, 1983; Lipman and others, 1986; Scott, 1975; Taylor, 1975; Trimble, 1980). Chapin (1979) suggested that rapid uplift of the system occurred in the Pliocene, and Larson and others (1975) found evidence for rapid uplifts of the Elk Mountains–Sawatch Range area in north-central Colorado in the late Miocene and in the Quaternary. Trimble (1980) summarized geologic evidence for about 1.5 km of uplift of the Colorado Front Range since the Miocene.

Sizable masses of Laramide igneous rocks crop out in the environs of the Colorado mineral belt (Fig. 2). In the Oligocene, extrusive activity created a major volcanic field that covered most of the mountains in southern Colorado and northern New Mexico, and much Park area to the north (Steven, 1975). The youngest Oligocene volcanism created a large ash-flow field in the San Juan Mountains and perhaps smaller fields near source areas in the Sawatch Range and the Never Summer Range near North Park (Steven, 1975). Miocene and Pliocene extrusive activity in the San Juan Mountains consisted of less voluminous bimodal volcanism (Lipman and others, 1978).

The Rio Grande rift in Colorado developed since latest Oligocene and/or earliest Miocene time (Lipman and Mehnert, 1975; Lipman and others, 1978; Chapin, 1979; Tweto, 1979). Although the main graben zone ends about 20 km north of Leadville (Chapin, 1979), Tweto (1977, 1979) indicated that rift-related extension, faulting, and igneous activity extend northward to areas near the Wyoming border. The rift between Alamosa and Leadville is notable for a near absence of Neogene volcanism in the axial basins (Chapin, 1979). In contrast, Neogene basalts with alkali affinities crop out in the rift system north of Leadville (Larson and others, 1975). Rift-related Neogene intrusive rocks occur in the Spanish Peaks–Raton Basin area, the Leadville–Sawatch Range–Elk Mountains region, and perhaps in the Park Range near the Wyoming border (Tweto, 1979).

From gravity lows and epizonal plutons that trend northeastward across Colorado, a batholith of Oligocene rocks is inferred to exist beneath the San Juan Mountains, and a batholith composed of Laramide through Oligocene and perhaps early Miocene rocks is considered to exist below the mineral belt in the Elk Mountains–Sawatch Range–western Front Range region (Tweto and Case, 1972; Isaacson and Smithson, 1976; Plouff and Pakiser, 1972; Tweto, 1975; Lipman and others, 1978; White and others, 1981). Gravity and elevation data imply that the Colorado mountains are in isostatic equilibrium (Birch, 1950; Qureshy, 1960; Plouff and Pakiser, 1972). Seismic data indicate that the Front Range and western mountain ranges have slightly different crustal thicknesses (52 versus 48 km) and comparable P-wave velocities (7.9–8.0 km/sec) in the upper mantle (Archambeau and others, 1969; Jackson and Pakiser, 1965; Prodehl and Pakiser, 1980; Prodehl and Lipman, 1988). Seismic interpretations suggest distinct crustal layering in the Front Range (Prodehl and Pakiser, 1980; Prodehl and Lipman, 1988). Transient magnetic data for the Southern Rocky Mountains have been interpreted by Porath (1971) with models of shallow (45 km) or deep (100 km) upwarpings of high electrical conductivity in the mantle.

HEAT-FLOW DATA

Basic heat-flow data for 139 boreholes are summarized in Tables 1 and 2. Figures 3 and 4 show relative temperature-depth data for all relevant holes to illustrate the reliability of the observed gradient data that were used for flux calculations. For example, the irregular temperature profiles for the deepest parts of both holes at Poison Ridge, Colorado (see PR1 and PR2 in Fig. 3), suggest major hydrologic flow at this locale, a disturbance that is not evident from the statistically precise mean conductivities, least-squares gradients, and observed heat flows that are shown in Table 1 for these stations. Decker and Birch (1974) tabulated basic data for 27 of the holes listed in Table 1 (also see Decker, 1969; Roy and others, 1968b); updated results for many of these stations are listed herein for regional completeness and because new flux and radioactivity data are now available for several of these locales (see Canon City, Dove Creek, Fulford, Golden, Hesperus, and Ouray, Colorado; and Red River, New Mexico, in Tables 1 and 3). The basic data for the other 112 holes have not been formally published elsewhere, although interested readers may find detail on several of the locales in the masters theses that are indicated in Table 1 and the references below.

The boreholes near Cisco, Utah, and Albany, Wyoming, were drilled for heat-flow and radioactivity research. The hole near Canon City, Colorado, was deepened for geothermal studies. All other holes were drilled by companies or government agencies for non-geothermal resource evaluation purposes. For all research, temperatures were measured at 5-, 10-, or 20-m depth intervals in the holes using thermistor probes in combination with 3 or 4 lead cables and electronic null-detectors or digital multimeters. Water-saturated samples of core or rotary drill chips with known or estimated porosities were used for thermal-conductivity measurements; diameters and thicknesses of the disks of core were in the ranges 2.22–4.76 cm and 1.27–1.91 cm, respectively, and the 40- to 70-gm samples of fragments were measured in thin-walled (.15–.16 cm) plexiglass containers with 2.54-cm diameters and 1.91-cm heights. Conductivities were measured with divided-bar systems that were calibrated with silica glass (GE 101, GE 102, or Dynasil) and quartz (heat flow normal to the optic-axis). Ratcliffe's (1959) curves for the conductivities of silica glass and quartz were used for calibrations.

Observed temperatures and conductivities were used to calculate the uncorrected heat-flow values, Q_{unc}, in both tables. The heat-flow values in Table 1 that are corrected for steady-state topography are designated Q_{cor}. The correction followed procedures in Birch (1950) and was made for two- or three-dimensional topography out to lateral distances equal to 10–20 times the depth of each hole; other corrections (for example, Lachenbruch, 1968a, 1969; Blackwell and others, 1980) were not attempted because most of the locales are characterized by three-dimensional terrain, and we had no control on possible second-order effects near each station. Corrections for evolution of terrain were not applied at any of the sites listed in Table 1 because detailed physiographic histories could not be obtained for all of the stations. Terrain corrections were not applied at the locales listed in Table 2 because the topographic relief near many of these stations is small, and because larger random errors could exist in the approximate conductivity values that were used to calculate these heat-flow values.

Table 1 contains our most reliable heat-flow data because most of the indicated holes were cored, most penetrated known rock units, most of the holes are in low-porosity igneous and metamorphic rocks, and because a large num-

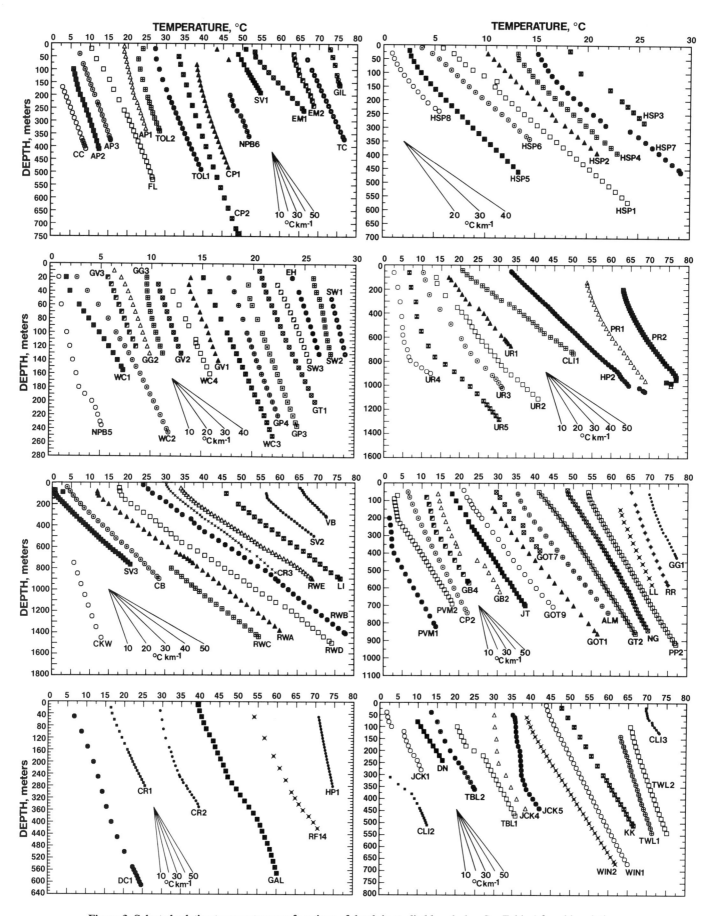

Figure 3. Selected relative temperatures as functions of depth in studied boreholes. See Table 1 for abbreviations.

TABLE 1. BASIC HEAT-FLOW DATA FOR MOST-RELIABLE BOREHOLES

Area Locality, state[1]	Plot[1,2] code	N. lat. (deg-min)	W. long. (deg-min)	Elev. (m)	Depth range (m)	COND[3] ($W\ m^{-1}K^{-1}$)	GRAD (°C km^{-1})	Heat flow[4,5] (mW m^{-2}) Qunc	Qcor
Frontal Ranges									
Apex, CO	AP1(b)	39-52.4	105-33.4	3171	100–375	2.46(26)	24.8	61 [1]	70
	AP2	39-52.1	105-33.1	3270	230–330	2.72(12)	23.1	63 [2]	75
	AP3	39-52.2	105-33.3	3262	280–410	2.47(15)	24.0	59 [5]	66
							Mean values:	61	71
							Geologically best value:	79 ± 8	
Canon City, CO (b)	CC	38-29.8	105-19.6	1937	380–410	3.17(33)	25.3	80 [3]	77
Elk Horn, CO	EH	40-44.5	105-31.8	2362	50–130	3.68(16)	24.1	89 [4]	82
Empire, CO	EM1	39-45.7	105-41.7	2902	160–230	3.78(9)	67.5	255 [16]	192
	EM2	39-46.0	105-41.9	2637	40–250	2.99(11)	28.4	85 [3]	85
							Mean values:	91	89
							Best value:	85	
Gem Park, CO (b)	GP3	38-16.0	105-32.2	2446	60–235	3.23(30)	25.5	82 [3]	86
	GP4	38-16.3	105-32.2	2438	100–220	2.68(23)	26.2	70 [3]	71
							Mean values:	77	79
Geneva Basin, CO (h)	GB2	39-34.4	105-48.7	3575	490–610	4.77(11)	28.6	136 [9]	130
							Geologically best value:	105 ± 8	
	GB4	39-34.5	105-48.9	3493	300–560	3.55(19)	30.2	107 [6]	96
							Best mean value:	99	
Georgetown, CO (h)	GT1	39-29.0	105-44.4	3440	100–200	3.56(19)	34.1	121 [4]	109
	GT2	39-29.0	105-44.4	3427	120–860	3.68(35)	30.3	112 [3]	105
							Mean values:	114	106
							Geologically best value:	92 ± 10	
Golden area, CO									
Golden Gate Canyon	GG1	39-46.3	105-15.9	2049	60–425	3.15(13)	20.5	65 [5]	63
	GG2(ba)	39-46.4	105-17.8	2238	40–130	3.01(32)	24.5	74 [2]	74
	GG3(ba)	39-46.5	105-17.8	2294	60–130	3.01(32)	19.8	60 [3]	69
	DDH-1(a,b)			1905	100–530	3.01(32)	22.2	70 [2]	64
							Mean values:	66	65
Grapevine (ba)	GV1	39-40.9	105-14.4	2073	90–140	3.12(9)	23.9	75 [4]	67
	GV2	39-40.5	105-14.4	2052	50–130	2.81(11)	23.1	65 [5]	57
	GV3	39-40.8	105-14.4	2052	50–120	2.93(6)	28.2	83 [7]	72
							Mean values:	73	64
Swartzwalder Mine (ba)	SW1	39-50.3	105-17.3	2198	50–130	3.91(8)	18.1	71 [3]	74
	SW2	39-51.3	105-17.3	2189	60–130	2.33(13)	13.6	32 [2]	31
	SW3	39-50.6	105-17.1	2259	50–140	4.05(15)	27.5	111 [7]	120
							Best value:	74 ± 4	
Jamestown, CO (h)	JT	40-07.7	105-23.3	2478	280–680	2.47(17)	32.1	80 [2]	88
							Geologically best value:	100 ± 8	
Northgate, CO (bu)	NG	40-55.9	106-16.6	2609	60–840	3.59(64)	27.0	97 [2]	95
Tolland, CO (h)	TOL1	39-54.7	105-35.9	2869	210–490	2.38(13)	30.1	72 [9]	72
	TOL2	39-54.8	105-36.2	2984	200–340	2.42(6)	24.5	59 [3]	63
							Mean values:	61	64
							Geologically best value:	84–105	
Urad, CO	UR1(b)	39-46.3	105-50.4	3156	100–670	3.37(17)	29.3	99 [3]	83
	UR2(b)	39-45.6	105-50.6	3439.6	930–1110	3.59(27)	34.7	125 [3]	126
	UR3[6]	39-45.7	105-50.2	3109	701–1021	3.84(11)	32.6	125 [4]	125
	UR4[6]	39-45.5	105-49.6	3109	793–899	3.84(11)	53.6	206 [8]	206
	UR5[6]	39-45.4	105-50.4	3109	854–1281	3.84(11)	34.6	133 [4]	133
							Best mean values:	140	140
							Best value:	125 ± 8	
Westcliffe, CO (b)	WC1	38-08.4	105-27.0	2417.5	100–155	1.71(8)	42.9	73 [.4]	69
	WC2	38-08.7	105-27.3	2433	100–245	1.90(28)	35.6	67 [.4]	66
	WC3	38-08.7	105-25.9	2467.4	100–250	2.30(22)	30.2	69 [3]	68
	WC4	38-08.7	105-26.2	2460.8	120–160	2.15(8)	31.3	67 [2]	67
							Mean values:	70	68
Red River, NM (b)	RR	36-43.0	105-24.0	2695	280–580	2.84(42)	29.7	85 [2]	79
Western Ranges–southern Sawatch Range									
Cumberland Pass, CO	CP1(b)	38-41.1	106-29.5	3605.4	270–480	2.79(71)	25.2	70 [1]	78
	CP2	38-40.8	106-30.0	3536	300–740	3.05(6)	24.7	75 [3]	79
							Mean values:	74	79
Fulford, CO (b)	FL	39-30.9	106-39.1	3038	440–530	4.15(11)	24.1	103 [1][7]	100
							Geologically best value:	88 ± 8	
Gilman, CO (b)	GIL	39-33.0	106-24.0	3119	1010–1070	3.52(17)	20.4	89 [1][7]	94
Gothic, CO	GOT1(ba)	38-59.3	106-56.7	3250	220–860	3.29(16)	39.2	129 [6]	112
	GOT7	38-59.3	106-56.7	3232	300–380	3.59(24)	36.5	131 [6]	108
	GOT9	38-59.3	106-56.7	3262	160–680	3.41(29)	39.0	133 [7]	105
							Mean values:	131	108
Lake Irwin, CO	LI	38-54.0	107-6.7	3290	300–900	3.30(29)	39.5	130 [7][7]	129
North Pole Basin, CO	NPB5	39-01.3	107-05.5	3665	210–235	4.27(6)	23.1	99 [8]	100
	NPB6	39-01.4	107-05.4	3649	300–350	3.54(6)	26.5	94 [2]	96
							Mean values:	94	96
Paradise Pass, CO	PP2	38-59.7	107-03.5	3445	280–90	3.13(21)	27.6	86 [2]	90
	DH-1 (a,b)			3380	100–210	2.94(33)	26.5	78 [1]	65
							Weighted mean values:	85	86
							Best values:	86	90
Redwell Basin, CO	RWA(ba)	38-53.5	107-03.4	3462	710–1360	3.69(39)	36.7	135 [6]	150
	RWB(ba)	38-53.4	107-03.3	3359	860–1410	3.70(10)	36.3	135 [6]	148
	RWC(ba)	38-53.5	107-03.2	3327	840–1440	3.69(29)	36.8	136 [7]	149
	RWD(ba)	38-53.4	107-03.6	3439	800–1100	3.29(21)	42.0	138 [6]	155
						1200–1500	3.69(39)	35.8	132 [5]	143
	RWE	38-53.6	107-03.1	3350	200–820	3.18(53)	45.0	143 [4]	160
						820–920	3.53(11)	33.5	118 [3]	132
							Mean values:	136	150
Tincup, CO	TC	38-46.2	106-28.9	3051	110–370	2.93(21)	31.7	92 [7]	85

TABLE 1. (Continued)

Area Locality, state[1]	Plot[1,2] code	N. lat. (deg-min)	W. long. (deg-min)	Elev. (m)	Depth range (m)	COND[3] (W m^{-1}K^{-1})	GRAD (°C km^{-1})	Heat flow[4,5] (mW m^{-2}) Qunc	Qcor
Leadville-Rift area, eastern San Juan Mountains									
Alma, CO	ALM	39-19.5	106-7.7	3610	400-740	3.21(22)	37.4	121 [8]	113
Climax, CO	CI11	39-22.1	106-11.0	3463	442-473	3.62(2)	40.7	147 [6]	140
					473-564	4.05(13)	32.5	132 [4]	125
					564-732	3.57(12)	44.6	160 [8]	152
	CI12	39-22.1	106-11.0	3180	80-170	4.27(14)	32.3	138 [9]	149
	CI13	39-22.3	106-10.1	3440	90-125	4.63(13)	51.0	236 [12]	214
							Mean values:	166	163
							Best value:	150 ± 10	
Del Norte, CO	DN	37-47.0	106-22.0	2578	80-240	2.80(19)	40.1	112 [8]	112
Hahns Peak, CO (bu)	HP1	40-50.3	106-55.4	2987	50-280	2.61(66)	21.8	57 [.8]	102
	HP2	40-50.3	106-55.4	2851	50-330	2.52(18)	37.1	94 [1]	107
					330-850	2.65(45)	35.4	94 [1]	106
					1020-1050	2.11(6)	50.4	106 [6]	107
							Best value:	103 ± 3	
Jack Creek, CO	JCK1	40-24.3	105-56.4	3241	240-279	3.82(35)	26.0	100	90
	JCK4	40-23.9	105-58.4	3055	360-440	3.67(9)	42.6	156 [5]	144
	JCK5	40-24.4	105-57.5	3213	250-440	4.07(12)	36.7	149 [16]	138
							Best value:	140 ± 20	
Kokomo, CO (b)	KK	39-26.6	106-08.3	3201.2	480-510	3.49(7)	38.5	134 [8]	117
Parkview Mtn., CO	PVM1	40-20.1	106-07.5	3354	420-800	2.72(28)	29.9	81 [2]	93
	PVM2(bu)	40-19.9	106-08.0	3475	380-680	2.63(26)	30.3	80 [4]	97
							Mean values:	81	94
Poison Ridge, CO (bu)	PR1	40-19.2	106-14.9	2994	440-980	2.47(67)	30.3	75 [2]	74
	PR2	40-19.1	106-15.1	2937	610-920	2.47(67)	27.3	67 [2]	65
							Mean values:	71	70
Summitville, CO	SV1(b)	37-25.6	106-35.7	3489	100-190	2.63(15)	41.0	108 [2]	103
	SV2	37-25.4	106-36.0	3530	300-500	2.75(19)	37.2	102 [5]	107
							Mean values:	105	105
Alum Creek	SV3	37-23.0	106-35.7	3384	116-769	3.32-3.70(15)	31.0	[8]103-115 [4]	106-118
					657-769	3.32-3.70(15)	32.0	[8]106-118 [4]	109-122
							Mean values:	104-116	108-120
						Summitville area best mean value:		107	
Timberline, CO (ba)	TBL1	39-17.5	106-28.6	3269	260-470	3.33(28)	35.6	118 [3]	107
	TBL2	39-17.6	106-28.8	3315	270-360	4.07(10)	35.8	146 [4]	138
							Mean values:	131	122
Twin Lakes, CO	TWL1	39-0.1	106-30.3	3948	200-540	3.51(12)	21.0	74 [7]	88
	TWL2	38-59.0	106-31.5	3963	240-540	3.39(10)	24.5	83 [7]	100
							Mean values:	78	94
Winfield, CO	WIN1	38-58.4	106-26.1	3536	400-660	3.39(29)	35.4	120 [6]	119
	WIN2	38-59.0	106-27.4	3255	300-670	3.01(27)	41.2	124 [3]	114
							Mean values:	123	116
San Juan Mountains									
Creede, CO	CR1	37-55	106-57	3550	80-280	2.24(6)	38.1	85 [3]	98
	CR2	37-55	106-57	3550	183-250	2.41(4)	31.1	75 [4]	83
					290-350	2.90(5)	50.1	145 [12]	159
	CR3	37-56	106-57	3580	430-560	2.47(8)	38.1	94 [4]	100
					620-671	2.80(6)	43.3	121 [1]	121
					671-838	2.95(15)	32.3	95 [2]	94
							Mean values:	99	100
							Best value:	100 ± 5	
Ouray, CO (b)	OU	37-54.2	107-40.3	2962.6	80-375	3.86(25)	55.5	213 [8]	155
Rico, CO	RIC1	37-42.3	108-0.7	2745	300-475	2.84(7)-2.92(6)	73.8-86.2	218-247 [33]	
	RIC2	37-42.3	108-2.0	2730	140-165	2.72(8)	87.4	238 [29]	
							Geologically best value:	141-150 (md)	
Colorado Plateau									
Dove Creek, CO (b)	DC1	37-47.2	108-51.4	2102	560-610	3.15(14)	37.6	128 [3][7]	125
	DC8	37-47.1	108-46.2	1918	130-170	2.62(6)	69.8	177 [8][7]	99
	DC9	37-46.5	108-45.6	1925	170-190	3.02(10)	44.1	121 [.4][7]	82
							Mean values:	135	112
							Best value:	125	
Glade Park, CO (b)	GLP1	38-56.7	108-36.8	2089	140-175	2.47(4)	27.0	67 [2]	67
	GLP2	38-56.1	108-37.4	2165	170-225	2.24(9)	25.5	57 [3]	59
	GLP3	38-56.7	108-37.0	2089	150-375	2.39(34)	23.3	56 [1]	56
							Mean values:	60	60
Hesperus, CO (h)	HSP1	37-24.5	108-05.3	3112	180-560	2.26(15)	31.9	72 [4]	65
	HSP2	37-24.5	108-05.4	3171	130-390	2.34(22)	30.9	72 [3]	65
	HSP4	37-24.5	108-05.4	3191	340-390	2.13(12)	30.5	65 [2]	61
	HSP5	37-24.3	108-05.6	3310	210-450	2.26(15)	27.6	62 [3]	64
	HSP7	37-24.4	108-05.2	3077	340-450	2.38(6)	32.7	78 [4]	70
							Mean values:	68	64
	HSP3[6]	37-24.4	108-05.5	3252	160-570	2.22(71)	28.3	63 [2]	59
	HSP6[6]	37-24.4	108-05.4	3170	80-340	2.22(71)	31.2	69 [2]	57
	HSP8[6]	37-24.4	108-05.5	3307	110-240	2.22(71)	25.6	57 [3]	57
							Grand mean values:	66	60
	DH-1(a,b)			2770	270-540	2.90(24)	37.6	109 [4]	87
							Best Hesperus value:	75 ± 8	
Rifle, CO (b)	RF14	39-58.2	108-21.5	1914	280-480	.87(14)	56.0	46 [.8][7]	44
	RF28	39-56.3	108-24.0	1951	100-220	1.62(13)	37.2	61 [.8][7]	60
							Mean values:	54	52
Gallup, NM (b)	GAL	35-39.0	108-31.0	2169	500-570	2.94(8)	22.8	67 [7]	67
Blanding, UT (h)	BL53	37-43.9	109-21.8	2108	210-230	3.89(7)	17.0	66 [5]	68
	BL55[6]	37-43.9	109-21.8	2109	220-240	3.89(7)	17.3	67 [2]	69
							Mean values:	67	69

TABLE 1. (Continued)

Area Locality, state[1]	Plot[1,2] code	N. lat. (deg-min)	W. long. (deg-min)	Elev. (m)	Depth range (m)	COND[3] ($W\ m^{-1}K^{-1}$)	GRAD (°C km^{-1})	Heat flow[4,5] (mW m^{-2}) Qunc	Qcor
Cisco, UT (h)	CIS	39-03.0	109-33.0	1597	60-270	1.80(11)	51.9	92 [2]	92
Needles Mountains									
Chicago Basin, CO	CB	37-36.2	107-36.8	3438	200-900	3.68(38)	30.6	112 [4][7]	102
Lilly Lake, CO	LL	37-34.8	107-35.6	3840	300-580	3.63(34)	21.9	80 [.8]	88
Vallecito Basin, CO	VB	37-35.1	107-35.7	3720	130-320	3.59(17)	28.2	101 [3]	96
Wyoming Southern Rocky Mountains-Wyoming Basin-Powder River Basin									
Albany, WY	AB	41-12.4	106-10.8	2835	90-170	3.30(12)	20.2	67 [2]	68
Eden, WY	EDN1	41-51.0	109-32.5	2049.1	100-280	2.05(9)	32.6	67 [1]	67
	EDN2	41-51.0	109-32.5	2049.3	100-330	2.05(9)	33.9	71 [2]	71
							Mean values:	68	68
Meadow Creek, WY (bu)	MC1	43-55.6	109-17.2	3056	210-340	2.65(33)	35.4	95 [3]	75
Medicine Bow Mts., WY	MB5(bu)	41-28.5	106-13.9	3069	100-330	5.54(51)	7.2	40 [1]	45
	MB13(bu)	41-27.6	106-20.9	2843	60-170	5.66(68)	7.6	43 [1]	41
	SL	41-29.8	106-13.3	3048	100-300	5.78(16)	9.7	56 [3]	58
							Mean values:	41	44
Powder River Basin, WY	PRB1	42-49.5	105-18.1	1558	340-500	2.05(15)	43.8	89 [12]	89
	PRB2	43-12.5	105-04.7	1421	100-540	2.03(15)	33.8	69 [9]	69
Rawlins, WY	RAW	41-44.3	107-26.9	2076	400-760	2.80(11)	16.4	46 [4]	46
Rock Springs, WY	RKS	41-35.6	109-22.4	2259	60-145	[9]1.63-1.75	24.7-28.4	40-50	43-54
Sierra Madre, WY	SM11(bu)	41-13.5	107-08.2	2838	64-256	2.64(20)	15.2	40 [3]	40
	SM8(bu)	41-10.6	106-53.8	2682	60-200	3.61(12)	8.3	30 [3]	33
							Mean values:	36	36
	CKW	41-13.8	107-14.4	2346	650-1450	3.65(66)	9.8	36 [2]	36
							Mean values:	36	36
Squaw Springs, WY	SR	42-27.4	106-03.9	2338	180-250	4.03(12)	18.1	73 [4]	74

[1]References to basic data wholly or partially reported elsewhere: (a) Decker (1969), and (b) Decker and Birch (1974). References to M.S. theses: (ba) Baker (1976), (bu) Buelow (1980), (h) Hallin (1973), and (md) Medlin (1983).
[2]Abbreviation shown for hole in temperature-depth figures.
[3]Number of conductivity samples shown in parentheses.
[4]Uncorrected heat flow calculated as product of least-squares gradient multiplied by mean conductivity, unless indicated otherwise.
[5]Standard errors of uncorrected heat-flow values shown in brackets after values.
[6]Mean conductivity based on measurements of disks of core from nearby holes in the same area and/or pluton.
[7]Uncorrected heat flow calculated by interval or resistivity integral method.
[8]Conductivity based on 5%-10% porosity range determined from large fragments or core samples of penetrated volcanics.
[9]Conductivity estimated from measured values for shales/oil shales elsewhere in drilling area.

ber of the holes are deep (>200 m). In contrast, the individual values in Table 2 are considered less reliable because all of the holes are in porous sediments, not one of the holes was cored, and many of the holes are shallow (<200 m). Additionally, estimated porosities were used for the fragment conductivity measurements that are summarized in this table: an average porosity of 25% was assumed when detailed lithology was not known; otherwise, we followed lithologic logs, representative porosity data (D. L. Blackstone, Jr., 1985, personal commun.), and

TABLE 2. BASIC HEAT-FLOW DATA FOR BOREHOLES IN SEDIMENTS IN BASINS NEAR THE SOUTHERN ROCKY MOUNTAINS IN WYOMING

Locality, state	Plot[1] code	N. lat. (deg-min)	W. long. (deg-min)	Elev. (m)	Depth range (m)	COND[2] ($W\ m^{-1}K^{-1}$)	GRAD (°C km^{-1})	Heat flow[3] (mW m^{-2}) Qunc
Baggs, WY	BGS	41-3.8	107-45.9	2018	130-217	2.51(24)	26.4	66 [12]
Walck Ranch, WY	WR1	41-26.0	106-55.1	2172	70-170	6.0 (18)	21.1	53 [9]
	WR2	41-26.2	106-53.8	2166	30-171	6.0 (18)	20.2	50 [8]
								Mean value: 52[4]
Finley Reservoir, WY	FR1	41-27.3	106-44.1	2116	70-204	1.96(7)	33.0	65 [7]
	FR2	41-27.5	106-40.0	2201	20-83	2.47(8)	34.5	85 [18]
								Best value: 65
Rattlesnake Pass, WY	RLP	41-43.6	106-37.1	2220	60-227	2.04(5)	20.8	42 [7]
Hanna, WY	HAA	41-46.8	106-40.3	2142	30-268	2.05(5)	21.2	44 [5]
Rock River area, WY	RC1	42-4.1	105-47.4	2196	30-190	2.02(15)	18.2	37 [14]
(Boot Heel quad.)	RC2	42-8.4	105-47.8	2140	10-100	2.38(11)	30.0	71 [25]
	RC3	42-9.1	105-46.1	2140	90-150	2.43(18)	21.0	51 [10]
	RC4	42-9.1	105-46.1	2146	40-137	2.43(15)	23.3	57 [16]
	RC5	42-9.3	105-45.8	2140	40-83	2.80(10)	14.1	40 [15]
	RC6	42-9.9	105-58.0	2122	20-90	2.38(10)	22.4	53 [15]
	RC7	42-10.5	105-49.4	2121	20-90	2.43(10)	23.5	57 [12]
	RC8	42-10.5	105-49.7	2121	20-87	2.72(10)	25.8	70 [19]
								Mean values: 52[4]-53[5]

[1]Abbreviation shown for hole in temperature-depth figures.
[2]Number of conductivity samples shown in parentheses.
[3]Standard error of heat flow shown in brackets.
[4]Mean weighted by inverse squares of standard errors.
[5]Mean weighted by borehole depth ranges.

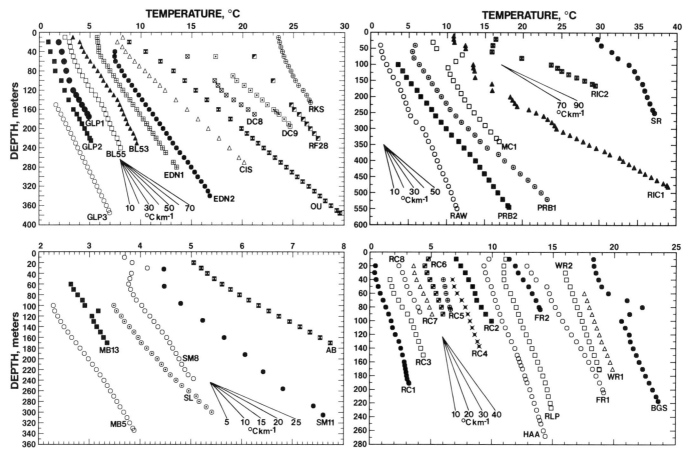

Figure 4. Selected relative temperatures as functions of depth in studied boreholes. See Tables 1 and 2 for abbreviations.

porosity-depth curves for shallow sediments (Chapman and others, 1984, Fig. 8, p. 462) and assumed 30% porosity for sandstones, 20% for siltstones and claystones, and 10% for shales. Thus, a large uncertainty in these heat-flow values could lie in the use of "approximate" mean conductivities in each calculation. We cannot be more specific about the heat-flow values in Table 2, except to note that most of the results agree with previously published determinations for the Wyoming Basin. For example, Blackwell's (1969) 52 mWm^{-2} flux for Rock River, Reiter and others' (1979) 62 mWm^{-2} value for the Quealy Dome, and our 52–53 mWm^{-2} for the Boot Heel quadrangle (Table 2) all suggest that the regional flux in the Shirley Basin–Laramie Basin area in southern Wyoming is normal.

Some of the localities in Tables 1 and 2 show different values of observed or corrected heat flow. Thus, "Best values" have been entered if some of the data are considered to be of superior quality. For other locales, "Weighted mean values" were calculated using two different rationales: (1) by weighting with inverse squares of the standard errors of the individual values if data quality is similar and all holes are of comparable depth or (2) by weighting by borehole depths or depth-ranges if some data are from much greater depth or significantly longer hole intervals. Finally, "Geologically best values" are listed for areas where the geology is sufficiently well known to permit corrections to be estimated for complexities in local thermal, radioactivity, and/or geologic structure.

COMBINED HEAT-FLOW AND RADIOACTIVITY DATA

Basic radioisotope (U, Th, and K) data for 60 locales are compiled in Table 3, with corresponding heat-production and heat-flow values, and relevant comments on generalized geology, radioactivity sampling, and reliability of results. Data for six locales discussed by Decker and others (1980) are also listed for completeness, and because we have new regional radioactivity data for the Keystone, Wyoming, area. None of the other *combined* heat-flow, heat-production, and generalized geology information has been published elsewhere.

The radioactivity data for the Adams Tunnel, the Moffat Tunnel, and the Roberts Tunnel in the Colorado Front Range are not based on sampling that was done for geothermal research, as is the case for the data that are summarized for the Sierra Madre and Medicine Bow Mountains in southern Wyoming, and portions of the data shown for Questa, New Mexico, and Canon City and Elk Horn, Colorado (see Table 3). All other data are based on gamma-ray analyses of samples that we collected and prepared for our research.

Gamma-ray analyses were done at the University of California at Berkeley (Wollenberg and Smith, 1964), Rice University (Adams, 1964), the University of Wyoming (Decker, 1973), and the University of Maine (after Decker, 1973) (see Table 3). Analyses at Rice University were done using 300-gm samples of crushed core or outcrop; all other analyses were done using crushed samples with weights in the range 550–850 gm. From close agreement of calculations using analyses of the same or comparable samples at Berkeley, Rice, Wyoming, Maine, and the U.S. Geological Survey in Denver, Colorado, the heat-production values that are based on gamma-ray methods are considered accurate to within ±10% (Table 3; also Decker and others, 1980).

Our radioactivity research with a wide variety

TABLE 3. URANIUM (U), THORIUM (Th), AND POTASSIUM (K) DATA WITH HEAT PRODUCTION VALUES (A), GENERALIZED GEOLOGY, AND COMMENTS

Area Locality, state[1]	n[2]	U (ppm)	Th (ppm)	K (%)	A[3] ($\mu W\ m^{-3}$)	Q ($mW\ m^{-2}$)	Plot[4] code	Geology/comments
Frontal Ranges								
Adams Tunnel, CO (B,p)	79	4.0	24.5	4.2	3.2	80	AT	Precambrian schist, gneiss, granite, lesser diorite.
Apex, CO								
Core, holes AP2 & AP3 (w)	6	4.3	18.1	4.7	3.0 [.6]			Tertiary quartz monzonite/quartz latite in tabular body in
Core, other regional holes (w)	16	5.5	19.7	5.5	3.3 [.2]			Precambrian metamorphic rocks. Best flux value after model for
				Best A value, mean Q:	3.2	71	APm	possible refraction.
				Best A value, best Q:	3.2	79	APb	
Canon City, CO								
Core, surface samples (r,m)	13	1.1	14.4	3.9	1.7 [.1]			Core and surface samples mainly Precambrian granite. Large
Pikes Peak Granite (PG)		4.0	13.0	4.1	2.3			mass Precambrian Pikes Peak granite to the north-northeast.
				Observed A value, best Q:	1.7	77	CCo	
				Mean A value, best Q:	2.0	77	CCb	
Elk Horn, CO								
Core samples, granite (w)	6	15.6	46.3	5.8	7.6 [.6]			Core of U-enriched fractured vein. Surface samples and data
Surface samples, granite (w)	8	8.1	28.5	4.0	4.4 [.9]			for Log Cabin batholith of relatively fresh Precambrian
Log Cabin batholith (PG)	11	6.7	24.5	4.0	3.9			Silver Plume granite.
				Best A value, observed Q:	4.2	82	EH	
Empire, CO (w)	9	7.2	26.2	4.6	4.1 [.8]			Tertiary monzonite porphyry in Precambrian granite/metamorphics.
				Observed A value, mean Q:	4.1	89	EMm	A, samples of deep core (>1,127 m) in monzonite.
				Observed A value, best Q:	4.1	85	EMb	
Gem Park, CO								
Core, ultramafic (w)	3	0.4	3.2	0.3	0.4 [.2]			Holes in ultramafic dike in small (3–4 km radius) funnel-shaped
Surface granitic rocks (m)	6	3.2	25.9	4.7	3.1 [.2]			Cambrian pyroxenite intrusion in Proterozoic rocks. Surface
Metamorphic rocks (m)	11	3.1	15.3	2.1	2.2 [.02]			samples are of Proterozoic country rocks within 1–3 km
				Mean A value, observed Q:	2.2	79	GPm	of Cambrian intrusion.
			Best A value (after surface samples), observed Q:	2.5	79	GPb	
Geneva Basin, CO (w)	15	5.0	19.9	3.8	3.1 [.3]			Tertiary porphyries of high conductivity in Precambrian granite
				Observed A value, best Q:	3.1	99	GB	and gneiss. Flux after model for possible refraction.
Georgetown, CO (w)	11	7.1	27.9	3.7	4.2 [.6]			Small (0.3 km radius) Tertiary alaskite plug in Precambrian
				Observed A value, mean Q:	4.2	106	GTm	metamorphic rocks. Best flux after model for possible
				Observed A value, best Q:	4.2	92	GTb	refraction.
Golden area, CO								
Golden Gate Canyon (w)	10	2.0	5.5	1.4	1.1 [.04]			Precambrian hornblende gneiss from deep hole in area that was
Golden Gate Canyon (r)	13	1.8	5.0	1.4	1.0 [.1]			studied by Decker (1969).
				Mean A value, best Q:	1.0	65	GG	
Grapevine (w)	15	2.6	8.3	2.0	1.6 [.1]	64	GV	Precambrian feldspathic gneiss with granite pegmatites.
Swartzwalder Mine (w)	8	2.5	7.9	2.5	1.6 [.2]	74	SW	Precambrian hornblende-microcline gneisses, some mica schist.
				Golden area means:	1.4–1.5	67	GO	
Jamestown, CO (w)	14	12.3	18.6	5.0	4.9 [.4]			Tertiary granite/monzonite that joins Tertiary granodiorite,
				Observed A value, observed Q:	4.9	88	JTo	all dip 45°–75° in Precambrian igneous and metamorphic units.
				Observed A value, best Q:	4.9	100	JTb	Best flux after model for refraction.
Moffat Tunnel, CO (B)					3.3	84	MT	Mostly Precambrian schist, gneiss, Tertiary monzonite. Value of A after data for comparable rocks along Roberts Tunnel.
Northgate, CO (w)	14	9.9	28.0	4.2	4.8 [.2]	95	NG	Cupola(?) of Precambrian quartz monzonite, faulted locally.
Roberts Tunnel, CO (D,p)	105	6.0	22.0	4.2	3.5	100	RT	Eastern portion–mostly Precambrian schist, gneiss, granite.
Tolland, CO								
All core (w)	24	19.9	53.1	4.4	9.1 [.1]			Laramide monzonite stock in Precambrian rocks. Observed Q = 67
Intermediate A phase(?) (w)	4	6.6	22.4	4.1	3.6 [.6]		n/u	$mW\ m^{-2}$; refraction model yields Q = 84–105 $mW\ m^{-2}$. Unknown
Low A phase(?) (w)	2	2.1	5.9	2.4	1.1 [.1]			complex radioactivity distribution.
Urad, CO (w)	15	14.1	60.4	4.6	8.1 [.8]			Tertiary granite, age 28–23 m.y. (White and others, 1981). A, samples
				Observed A value, mean Q:	8.1	140	URm	of core >610 m deep.
				Observed A value, best Q:	8.1	125	URb	
Westcliffe, CO (r)	4	1.1	6.6	2.3	1.1 [.1]	68	WC	Precambrian gneiss below altered Tertiary volcanics.
Questa, NM (D,RE)								
Composite of core (w)	1	3.0	25.9	6.1	3.1			Upper Oligocene–lower Miocene granite and quartz monzonite. Core of granite; other of granite or granite outcrop. Flux
Areal plutons (L)	8	6.5	24.2	3.5	3.6 [.5]			is mean of areal values of 64, 72, 75, 79, and 85 $mW\ m^{-2}$.
				Weighted A value, mean Q:	3.5	74	QU	
Red River, NM (w)	5	3.4	11.8	3.4	2.0 [.1]	79	RR	Tertiary granite.
Rift zone area								
Alma, CO (w)	5	9.6	35.2	5.2	4.9 [1.3]			Hole mainly Precambrian orthofeldspathic gneiss and granite,
				Best A value, observed Q	3.6	113	ALM	with Tertiary granodiorite sills and dikes. From geology, best A based on local borehole value and values for Kokomo to north.
Climax, CO (w,m)	7	14.3	30.9	4.2	6.2 [.6]			Late Tertiary granites/granite porphyries. Age 33–18 m.y.
				Mean A value, mean Q:	6.2	163	CLlm	(White and others, 1981). A, samples of deep core >450 m below
				Mean A value, best Q:	6.2	150	CLlb	274-m level in mine.
Del Norte, CO (w)	4	0.9	4.5	1.9	0.8 [.1]	112	DN	Tertiary(?) granodiorite intrusion.
Hahns Peak, CO (w)	91	3.0	7.3	6.3	1.7 [.04]	103	HP	Core of Upper Miocene quartz latite stock in Mesozoic sediments.
Jack Creek, CO (cb,us)	8	3.8	16.5	3.9	2.4 [.4]	140	JCK	Tertiary granites below hornfels or Precambrian gneiss, schist.
Kokomo, CO								
Precambrian gneiss (w)	4	2.2	15.3	4.1	2.0 [.1]			Core of granite gneiss. Other samples of large exposures of
Tertiary quartz monzonite (w)	9	3.6	16.3	3.3	2.3 [.1]			Silver Plume granite and Lincoln and Humbug quartz monzonite
Precambrian granite (w)	4	4.6	87.9	5.3	7.7 [.1]			porphyries near flux locale.
				Mean A value, observed Q:	3.5	117	KK	
Parkview Mtn., CO (w)	28	3.9	13.0	3.6	2.2 [.1]	94	PVM	Tertiary(?) quartz latite stock, core.
Poison Ridge, CO (w)	14	4.4	16.8	3.6	2.6 [.1]	70	PR	Tertiary monzonite/quartz latite. Deep-water flow evident.
Summitville area, CO								
Summitville Mine								
Core of volcanics (w)	4	4.6	6.4	0.7	1.6 [.1]			Core of porphyritic rhyolite and quartz latite. Surface
Alamosa River stock (w)	4	4.7	17.4	3.6	2.7 [.2]			samples of Tertiary monzonite in Alamosa stock. Gosnold (1987)
				Mean A value, observed Q:	2.2	105	SV	suggested A = 2.2 $\mu W\ m^{-3}$ for less-altered parts of Alamosa stock.

TABLE 3. (Continued)

Area Locality, state[1]	n[2]	U (ppm)	Th (ppm)	K (%)	A[3] ($\mu W\ m^{-3}$)	Q ($mW\ m^{-2}$)	Plot[4] code	Geology/comments
Alum Creek area								Hole mainly in andesite volcanics. Areal surface and core
Alamosa River stock (m)	2	1.9	5.9	2.3	1.1 [.1]			samples of monzonite in Alamosa River stock, Alum Creek
Lookout Mtn. Porphyry (m)	3	4.6	14.6	3.7	2.6 [.1]			monzonite porphyry, and Lookout Mountain quartz latite porphyry.
Alum Creek Porphyry (m)	3	3.4	11.6	3.0	2.0 [.1]			All units have mid-Tertiary ages.
				Best A value, best Q:	2.3	115	ALC	
Summitville area means				Best A value, mean Q:	2.4	107	SVm	
Timberline, CO (w)	28	3.3	13.0	3.9	2.1 [.1]	122	TBL	Tertiary breccias above feldspathic porphyry/granodiorite. Precambrian gneiss and granites border.
Twin Lakes, CO (w)	4	2.8	7.7	2.3	1.6 [.1]	94	TWL	Outcrop of stock of Tertiary Twin Lakes quartz monzonite/ granodiorite in/near boreholes.
Winfield, CO (w)	7	5.7	14.3	5.0	2.9 [.5]	116	WIN	Eocene monzonite/diorite, lesser rhyolite/granite porphyries.
Western Ranges								
Cumberland Pass, CO								
Quartz monzonite (w)	4	1.8	7.2	2.5	1.2 [.03]			Precambrian granite gneiss in boreholes. Other samples of
Quartz diorite (w)	4	1.6	12.5	2.9	1.5 [.01]			large Tertiary quartz monzonite pluton and small Tertiary
Gneiss (w)	12	3.5	4.7	3.8	1.6 [.05]			quartz diorite stock <1 km from boreholes.
				Mean A value, observed Q:	1.5	79	CP	
Fulford, CO (w)	5	1.6	14.8	3.4	1.7 [.1]			Hole in shales, limestones, and dolomites. Best flux after model
				Observed A value, best Q:	1.7	88	FLb	for possible refraction. A samples are of large stock of Tertiary quartz monzonite <1.5 km from flux locale.
Gilman, CO								
Metamorphics, migmatite (r,w)	11	2.6	17.9	4.0	2.4 [.1]			Core of Precambrian gneiss off 914-m level of mine. Other A
Granodiorite (w)	4	1.8	12.3	2.7	1.6 [.02]			samples of large exposures of Precambrian schist, gneiss,
				Weighted mean A value, observed Q:	2.2	94	GIL	migmatite, and Precambrian granodiorite in the Gilman area.
Gothic area, CO								
Granodiorite (w)	6	2.8	9.9	1.2	1.6 [.4]		n/u	Tertiary granodiorite/quartz monzonite stock (White Rock
Quartz monzonite (w)	7	2.6	9.8	.9	1.5 [.4]			pluton) with significant granite in boreholes and in areal
Granitic dikes (w)	5	29.5	48.1	4.5	11.2 [.5]			surface exposures.
North Pole Basin, CO (w)	5	13.6	36.5	4.3	6.4 [.1]		n/u	Holes mainly in limestones and dolomites, parts in Tertiary granite porphyry. A, samples of granite. Pluton age is 12.5 m.y. (Obradovich and others, 1969). Unknown complex structure.
Paradise Basin, CO (w)	12	2.9	6.6	3.8	1.6 [.2]	86	PBm	A, samples of Tertiary granodiorite stock in indurated shales.
				Observed A value, best Q:	1.6	90	PBb	
Redwell Basin, CO								
Shale (w)	11	4.2	13.2	2.7	2.3 [.2]			Felsite breccia above upper Tertiary granite porphyry stock. Age of
Felsite (w)	4	9.5	30.5	4.1	4.9 [.4]			stock is 18–16 m.y. (Thomas and Galey, 1982; White and others,
Deep granite (w)	41	13.7	47.0	4.2	7.1 [.3]			1981). Shales border shallow igneous rocks.
				Best A value, mean Q:	7.1	150	RW	
Tincup, CO (w)	5	1.7	6.8	5.0	1.3 [.1]	85	TC	Tertiary(?) quartz monzonite intrusion, core.
Needles Mts., San Juan Mts.								
Creede, CO (m)	16	4.9	15.5	5.4	2.6 [.1]		n/u	Late Oligocene volcanics.
Chicago Basin, CO (w)	3	14.3	43.4	4.7	7.0 [.3]	102	CB	Miocene [10 Ma (Bookstrom, 1981)] rhyolite in mostly Precambrian granite and quartz monzonite.
Ouray, CO (r)	24	4.1	11.6	1.5	2.0 [.4]		n/u	Tertiary tuff above inclined siliceous shale and dolomite.
Rico, CO (w)	4	2.6	10.4	4.0	1.8 [.1]		n/u	Lower Tertiary(?) monzonite in area of Rico and flux stations.
Vallecito Basin, CO (w)	4	17.5	36.6	4.7	7.8 [.6]	96	VB	Altered Precambrian granite and small bodies of Precambrian
& Lilly Lake, CO (w)					7.8 [.6]	88	LL	granites/quartz monzonites.
Eastern Colorado Plateau								
Glade Park, CO (r)	11	0.1	0.8	0.5	0.1 [.04]	60	GLP	Core of altered Precambrian gabbroic rocks below sediments.
Hesperus, CO (w)	54	4.7	19.4	7.7	3.1 [.2]	75	HSP	Cretaceous(?) or Tertiary(?) syenite stock in shales, sands, mudstones. Core of syenite from five holes.
Wyoming Mts., Wyoming Basin								
Albany, WY (w)	5	3.3	25.9	4.1	3.1 [.5]	68	AB	Proterozoic Sherman granite from sites on 5- to 6-km traverse in pluton. A = 2.8 $\mu W\ m^{-3}$ of core of sheared granite in hole.
Bosler, WY (DE)	3				0.0	25	BS	Precambrian anorthosite.
Copper King, WY (DE)	13				3.3	52	CK	Proterozoic(?) quartz monzonite(?), granodiorite, Sherman granodiorite.
Jeffrey City, WY (DE)	132				5.1	63	JC	Archean granite.
Keystone, WY								
Quartz diorite (DE)	3				0.9			Proterozoic quartz diorite pluton with numerous "older"
Older granite, (m)	4	1.5	9.6	3.1	1.4 [.02]			Proterozoic granites/granite bodies distributed throughout its
Older granite, (m)	4	1.1	4.2	3.3	0.8 [.01]			outcrop area.
				Observed A values, observed Q:	.9–1.1	50	KS	
Medicine Bow and Sierra Madre Mts., WY								
Regional rock groups (K)								
[5]Libby Creek Group (118,83,24)		6.4	22.0	1.2	3.5 [.1]			Proterozoic quartzite, conglomerate, phyllite, schist, slate, metadolomite, amphibolite.
[5]Deep Lake Group (162,116,110)		2.0	8.0	1.2	1.2 [.1]			Proterozoic quartzite, phyllite, marble.
[5]Phantom Lake Suite (276,249,238)		4.3	14.0	1.9	2.3 [.3]			Archean(?) quartzite, conglomerate, metavolcanics.
[5]Archean granitics (33,26,27)		2.0	11.0	1.8	1.5 [.2]			Archean quartzofeldspathic gneiss.
[5]Southern basement (110,71,107)		2.4	8.0	1.6	1.4 [.1]			Proterozoic basement, except granite, south of Cheyenne Belt.
Medicine Bow Mts. regional combined A and Q values								Two holes in Deep Lake units, one in Phantom Lake units.
				Minimum A value, best Q:	1.8	44	MBm	50-50 mix of Deep Lake and Phantom Lake Groups.
				Maximum A value, best Q:	2.0	44	MBm	5 times more Phantom Lake units than Deep Lake units.

TABLE 3. (Continued)

Area Locality, state[1]	n[2]	U (ppm)	Th (ppm)	K (%)	A[3] ($\mu W\,m^{-3}$)	Q ($mW\,m^{-2}$)	Plot[4] code	Geology/comments
Sierra Madre regional combined A and Q values								Holes in Deep Lake units, <0.5 km from Phantom Lake units.
				Maximum A value, best Q:	1.8	36	SMm	50-50 mix of Deep Lake and Phantom Lake Groups.
				Minimum A value, best Q:	1.4	36	SMm	4–5 times more Archean gneisses than Deep Lake Group.
C&K well, WY Sierra Madre (m)	5	1.6	5.3	1.3	1.0 [.2]	36	CKW	Proterozoic metagabbro, phyllite, quartzite, conglomerate.
Pedro Mountain (DE)	152				4.1	54	PM	Archean granite.
Shirley Rim, WY (m)	7	10.2	36.8	4.4	5.7 [.7]	74	SR	Archean(?) granite-granitic basement, rotary drill chips.
Vee Dauwoo, WY (DE)	44				4.9	46	VW	Proterozoic Sherman granite.

[1]Sources for data not new or directly from Tables 1 and 2: (B) Birch (1950), (D) Decker (1969), (DE) Decker and others (1980), (K) Karlstrom and others (1981a), (PG) Phair and Gottfried (1964), (RE) Reiter and others (1975), Edwards and others (1978), and (L) Lipman (1983). Sources/laboratories for radioelement data for heat production determinations: (p) G. Phair (1965, written commun.), (r) Rice University, (cb) University of California at Berkeley, (us) U.S. Geological Survey, (w) University of Wyoming, (m) University of Maine.
[2]n refers to number of radioactivity samples.
[3]Standard errors shown in brackets after heat generation, when possible.
[4]Plot code is the abbreviation used in heat flow (Q)–heat generation (A) figures. n/u means not used.
[5]Successive numbers in parentheses are numbers of samples used for U, Th, and K determinations, respectively, of indicated rock sequences, groups, or suites.

of fresh and altered rocks raises the question of whether some of the our heat-production values contain nonexperimental uncertainties related to variable geology, anisotropy of metamorphic rocks, and mobilization of uranium and thorium near some areas. Some of these "effects" were investigated through comprehensive areal sampling, investigations of local geology and geophysics, and study of isotope values or ratios. For example, the "best" A values that are shown for Canon City, Cumberland Pass, Gem Park, Gilman, Gothic, Kokomo, and Redwell Basin, Colorado; and Albany and Keystone, Wyoming, follow measurements of borehole and/or surface samples of representative rock units in the environs of these locales (Table 3). The heat generations for metamorphic rocks in Colorado are considered reliable regional values for at least three reasons: (1) radioelement concentrations for different rock units near several stations are similar or uniform (for example, Golden, Cumberland Pass, Gilman), (2) gravimetric modeling suggests that near-surface density contrasts between igneous and metamorphic rocks extend to appreciable depths (≥9–12 km) near some stations (Brinkworth and others, 1968; Tweto and Case, 1972; Isaacson and Smithson, 1976; Plouff and Pakiser, 1972), and (3) combined data for metamorphic terranes in New England and England agree with heat-flow–heat-production lines obtained using values for areal igneous units (Jaupart and others, 1982; Richardson and Oxburgh, 1978). Finally, the low U values for core and surface samples for Canon City, Colorado, suggest near-surface leaching of uranium near this site, and the data for samples of core from the Elk Horn, Colorado, and Albany, Wyoming, holes may be consistent with hydrothermal enrichment and shear-zone depletion, respectively, at these localities (Table 3). The "best" heat-generation value for the Canon City locale, therefore, is partly based on published U values for nearby Pikes Peak granite, and those for Elk Horn and Albany are based on analyses of surface samples of less-altered representative parts of large Precambrian granite bodies in both regions (Table 3).

We could not sample core for radioactivity research at our heat-flow stations in the Sierra Madre and northern Medicine Bow Mountains in Wyoming; therefore, mean heat-flow values were calculated for these areas and combined with regional heat-production values that were determined from extensive radioactivity data in Karlstrom and others (1981a). Three groups of Precambrian metamorphic rocks were considered for heat-production calculations: the Proterozoic Deep Lake Group, the Archean Phantom Lake Metamorphic Suite, and regional Archean Archean granitic gneisses. Rocks in the Libby Creek Group were not considered because geologic maps and gravity modeling indicate that this metamorphic sequence does not occur below or near any of our heat-flow stations in these mountains (after Karlstrom and others, 1981a, 1981b; Johnson and others, 1984).

The average flux in the northern Medicine Bow Mountains is 44 mWm^{-2} (Table 1). Values for three holes determine this mean; two are in Deep Lake Group rocks, whereas the other is in

TABLE 4. PARAMETERS OF Q-A LINES

	n[1]	Intercept[2] ($mW\,m^{-2}$)	Slope[2] (km)	r[3]
Frontal Ranges, Colorado and northern New Mexico				
Terrain-corrected Q values, observed A values	20	55.3 [4.4]	9.2 [1.2]	.88
Best Q and A values, terrain-corrected Q elsewhere	20	58.2 [3.4]	8.0 [1.0]	.89
Best regional values:		55–58	9	
Colorado Front Range				
Terrain-corrected Q values, observed A values	16	54.0 [5.3]	9.6 [1.4]	.88
Best Q and A values, terrain-corrected Q elsewhere	16	57.6 [4.0]	8.2 [1.0]	.90
Best regional values:		54–58	9	
Wyoming Rocky Mountains–Basin				
All non-transition points (no KS, AB)	11	27.5 [3.4]	6.7 [1.0]	.91
All non-transition points except Vee Dauwoo (no KS, AB, VW)	10	26.2 [1.8]	7.7 [.6]	.98
Best regional values:		27	7	
Western Ranges				
Best Q values	6	68.0 [2.5]	11.6 [.8]	.99
Best regional values:		68	12	
Rift zone–eastern San Juan Mountains				
All values	11	88.6 [6.4]	10.1 [1.9]	.88
Best Q values	10	93.3 [6.5]	7.9 [2.1]	.80
Best regional values:		91	9	
Colorado Plateau–Needles Mountains				
All values	5	60.7 [6.3]	4.6 [1.1]	.93

[1]n is number of points.
[2]Standard errors shown in brackets.
[3]r is correlation coefficient.

parts of the Phantom Lake Metamorphic Suite that are near (<1 km) Deep Lake Group units (Karlstrom and others, 1981b). The Deep Lake Group is stratigraphically higher than the Phantom Lake Metamorphic Suite, and gravity models suggest that about 5 km of Phantom Lake Suite rocks could underlie as little as 1-2 km of Deep Lake Group units in these mountains (Karlstrom, 1981a, 1981b; Johnson and others, 1984). Therefore, two slightly different heat-production values may be reasonable for basement in the environs of our heat-flow stations: 1.8 μWm^{-3} may be assumed if underlying units are composed of equal amounts of Deep Lake Group and Phantom Lake Suite rocks; or 2.0 μWm^{-2} is reasonable if Phantom Lake Suite units are 5 times more abundant than Deep Lake Group units below the boreholes (Table 3).

The mean flux for the Sierra Madre is 36 mWm^{-2}, based on values of 33 and 40 mWm^{-2} for two holes (Table 1). Both holes penetrated Deep Lake Group rocks that are near (<0.5 km) stratigraphically lower Phantom Lake Metamorphic Suite units, and these metasedimentary sequences determine a 3- to 11-km wide, east-west–trending area of basement between our heat-flow stations. This band of Deep Lake Group and Phantom Lake Metamorphic Suite rocks thickens to the west and may be underlain by early Archean granitic rocks or granitic gneisses that border northern parts of the Phantom Lake Suite (Flurkey and others, 1981). For one reasonable estimate of average heat production near our boreholes, 1.8 μWm^{-3}, the mean of Deep Lake Group and Phantom Lake Suite heat-production values is listed (Table 3). Alternatively, 1.4 μWm^{-3} is reasonable if we follow regional gravity modeling (Johnson and others, 1984) and assume that large masses of Archean granitic gneiss occur below a relatively thin (<2 km) section of Phantom Lake Suite rocks near our heat-flow stations (Table 3).

The hole near Ouray, Colorado, penetrated a variable sequence of inclined volcanic rocks and sediments, and we conclude that the tabulated radioactivity data are not reliable estimates for the true bedrock/basement heat production near this locality. Similarly, we could not resolve obvious uncertainties due to complex geology (inclined sediments, sills, dikes, laccoliths, or composite plutons) in the environs of North Pole Basin, the Gothic area, and Tolland, Colorado (Table 3); therefore, we avoid quantitative use of the combined data for these locales because they probably do not provide accurate measures of the true areal values. For all remaining localities in Table 3, we have no unequivocal evidence for significant alterations of the areal radioelement distributions. Hence, qualitative assessments of the reliabilities of the tabulated heat productions for these locales are avoided herein.

DISCUSSION AND INTERPRETATIONS

Recent modeling of possible conductivity and radioactivity contrasts in continental crust raises questions about the validity of simple one-dimensional interpretations of the parameters of heat-flow–heat-production lines (England and others, 1980; Furlong and Chapman, 1987; Jaupart, 1983; Nielsen, 1987; Vasseur and Singh, 1986). It should be recalled, however, that possible edge effects of geologically and geophysically likely conductivity and radioactivity contrasts to depths equal to line slopes were treated during the original establishment of relations for provinces in the United States (see Birch and others, 1968; Roy and others, 1968a; Lachenbruch, 1968b), and the essentially identical corrected and observed results suggest that the linear relation remains fundamental to understandings of continental flux. Moreover, one limitation of the recent numeral modeling is that it focuses on the impact of crustal heterogeneity in one heat-flow province, whereas the principles of potential theory show that analyses of juxtaposed contrasting provinces are likely to provide otherwise unobtainable information on the meanings of line parameters, and the distributions and kinds of heat sources in the crust and upper mantle. An unprecedented opportunity for tests of the latter points is provided by at least two well-studied surface and deep heat-flow transitions that are now evident in the Southern Rocky Mountain region in Colorado and nearby areas. Before proceeding with interpretations of these and other indicated transitions, we first examine regional heat-flow data, attempt to obtain average corrections for regional uplift and erosion, and discuss the correlations between heat flow and local bedrock radioactivity.

In Figure 5 (see folded insert included in this issue), all new and previously published heat-flow values for the areas discussed herein are located on the map shown in Figure 2. The combined flux and radiogenic heat-production values that we emphasize are summarized in Table 3, where the symbols that are shown in Q-A figures are listed under *Plot Code*. Least-squares parameters of representative Q-A lines for emphasized regions are listed in Table 4. Many other sets of terrain corrected, best, or most reliable and best heat-flow data were linearly regressed with observed or best heat-generation values to bracket line parameters for each study area, and the nature of the Q-A correlations for each region was essentially the same as of those that are summarized in Table 4.

To emphasize regional contrasts, all of our Q-A plots show a zone for "normal" parts of the Basin and Range high heat-flow province, and one shows estimated upper and lower bounds for the Battle Mountain, Nevada, unusually high heat-flow zone. Normal parts of the Basin and Range were assumed to be within ±15% of calculated values of flux on the Q-A line that Roy and others (1968a) determined for this province. The ±10% bounds for the Battle Mountain, Nevada, anomaly follow data in Lachenbruch and Sass (1977, 1978).

In text below, the term "reduced heat flow" refers to the intercept flux on a regional Q-A plot. We also emphasize "residual heat flow" values. Here, a residual heat flow is the flux at a site after the probable effect of local near-surface heat production has been subtracted according to Q-bA, where Q is near-surface flux, A is heat production, and b is the slope of the regional Q-A line. Thus, residual values are defined locally at a station, whereas a reduced value may be regarded as a regional or average parameter that applies to an entire province or zone. We avoid further discussion of the definitions of reduced and residual heat-flow values, except to state that both may provide estimates of the flux from the lower crust and/or upper mantle in a region, and to note that our residual heat-flow determinations directly follow previously described reduced heat-flow calculations by Roy and others (1972).

Regional Heat Flow in the Southern Rocky Mountains and Southern Wyoming Basin, and Average Corrections for Evolution of Terrain

Heat flows at 7 localities in the Southern Rocky Mountains in Colorado and northern New Mexico are in the range 62–71 mWm^{-2} (Fig. 5). All other measurements in these mountains are in the range 71–152 mWm^{-2}. The large number of values >71 mWm^{-2} confirm earlier suggestions (Decker, 1969; Roy and others, 1972; Reiter and others, 1975, 1979; Edwards and others, 1978) that most of the Southern Rocky Mountains in Colorado and northern New Mexico are characterized by high flux.

The Colorado Front Range, the Wet Mountains, and the Sangre de Cristo Mountains are eastern frontal ranges of the Southern Rocky Mountains that are adjacent to the Great Plains (Figs. 2 and 5). Hence, we refer to this 500-km-long zone of mountains as the "Frontal Ranges." The range of the average flux values in this zone is 85–88 ± 5 mWm^{-2}, based on 21–24 sepa-

TABLE 5. UPLIFT, EROSION, AND ESTIMATED CORRECTIONS TO HEAT FLOW IN THE RIO GRANDE RIFT-SOUTHERN ROCKY MOUNTAIN SYSTEM

Area	Time[1] (m.y.)	Uplift, rate (km, m/m.y.)	Uplift effect[2] (-%)	Erosion[3] (km)	Erosion effect[2] (-%)	Total effect[2] (-%)	Source
Colorado Front Range	75–65[g]	6–7, 100	<2	3.5(?)	<6.5	<<<8.5[4]	Tweto (1975)
Colorado Front Range	130–90, 65[a]	6–7, 411	<2	0.7–3.2	<6	<<<8[4]	Bryant and Naeser (1980)
Colorado Front Range	67.5[g]	6.1, 90	≤1.6	2.6	<5	<<<7[4]	Trimble (1980)
Sawatch Range, Colorado	50[a]	?		?			Bryant and Naeser (1980)
Frontal Ranges, High Plains	35–20[g]	1.0, 28–67	<<1.5	0.5	<<3	<<4–5[5]	Trimble (1980)
Frontal Ranges, High Plains	28–5[g]	1.0, 36–43	<<1	0.5	<<2	<<3.4[5]	Trimble (1980)
Sawatch Range, Colorado	0–20, 30[a]	1.5(?), ?	<1	1.5	<5	<6	Bryant and Naeser (1980)
Sangre de Cristo Mtns., Colorado	15–24[g]	0.7, 70	<<1	0.35–0.7	<<4	<<<6[5]	Lindsay and others (1983)
Sangre de Cristo Mtns., Colorado	0–7[g]	1.2, 170	<1	0.6	<4	<5	Scott (1975), Taylor (1975)
Frontal Ranges, High Plains	5[g]	1.5, 300	<1.5	0.8	<6	<8	Trimble (1980)
Sawatch Range–Elk Mtns.–Roaring Fork River, Colorado	10–8[g]	>0.6 [1.2(?)], rapid	<1	0.6	<4	<<6[5]	Larson and others (1975)
Sawatch Range–Elk Mtns.–Roaring Fork River, Colorado	1.5–0[g]	>0.3 [0.6(?)], rapid	<1.5	0.3	<4	<<6[5]	Larson and others (1975)
Colorado Front Range	4, 1[g]	2.1, 525–2100	..	0.6–1.1	..	10–15	Birch (1950)
Colorado Front Range–Southern Rocky Mtns.	4[g]	1.0–2.0, 250–500	..	0.3–1.0	..	5–10	Decker (1969)
Southern Rocky Mountain region	7, 4[g]	?, ?		?		?	Chapin (1979)
Espanola-Albuquerque Basins, New Mexico	0–14[f]	1.2, 8.5	<1	0.6	<3	<4–5	Axelrod and Bailey (1976)
Sandia Mtns., New Mexico	15–30[a]	1.2, 81	<1	0.7	≤4.3	<<<6[5]	Kelley and Duncan (1984, 1986)
	0–15[a]	3.5, 230	<3	1.8	≤6	<10[6]	Kelley and Duncan (1984, 1986)
	0–30[a]	4.7, 157	<2.5	2.0	≤5	<8[6]	Kelley and Duncan (1984, 1986)
Santa Fe Mtns., New Mexico	55–70[a]	1.9, 125	<1.5	1.0	<7	<<<8[5]	Kelley and Duncan (1984, 1986)
	0–55[a]	3.0, 55	<1.5	1.5	<3	<5–6[6]	Kelley and Duncan (1984, 1986)
	0–70[a]	4.9, 70	<1.5	2.5	≤4.1	<6–7[6]	Kelley and Duncan (1984, 1986)
Taos Range, New Mexico	0–14[a]	2.9, 210	<2	1.5	≤5.2	<7–8	Kelley and Duncan (1984, 1986)

[1]From sources, times arrived at in different ways: a, largely by fission-track ages of apatites; f, ages based on temperature dependence of flora during uplift; and g, combined geologic control and age-date data.
[2]All calculations assume linear uplift and erosion (same rate) for time period, a thermal diffusivity of 63 km^2/m.y., and a ground-surface temperature change with elevation of -4.5 °C/km (Birch, 1950; Decker, 1969). Calculations directly after expressions in Birch (1950, p. 590). Negative percentage effects (-%) show that corrections would produce heat-flow values that are less than those obtained after correction for steady-state topography.
[3]Erosion assumed to be one-half (50%) of total uplift for most areas.
[4]All evidence suggests that Laramide uplift and erosion ended sometime in the Eocene, when a mature Front Range landscape had developed. Therefore, the <<< symbol means that effects of Laramide uplift and erosion would be dramatically less than the tabulated values that were calculated assuming linear uplift and erosion over the indicated times.
[5]Actual effect on present gradients or flux likely much less than indicated percentages because uplift and/or erosion would have ceased by youngest indicated time.
[6]Calculations for linear uplift and erosion over time intervals shown, starting with 0 m.y. Source actually shows two time intervals with different rates, for which calculations are also tabulated.

rated (>10 km) values of terrain-corrected, best and/or geologically corrected heat flow. Most of the heat-flow values for the eastern frontal ranges are in the Colorado Front Range, and one is in the Medicine Bow Mountains in northern Colorado. For these areas, the means of 18 and 19 separated values are in the range 89–90 ± 5 mWm^{-2}.

The average of 13–15 values in the northern Rio Grande rift zone in Colorado (eastern San Juan Mountains, Rio Grande rift graben, the Leadville area, northern and central Sawatch Range, and Middle and North Parks) is 114 ± 7 mWm^{-2}. That for the 9 "best" values in the San Juan Mountains is 115 ± 8 mWm^{-2}.

We place data for the Elk Mountains, Tincup, Cumberland Pass, Gilman, and Fulford areas in one group, and we separate it from other parts of the Southern Rocky Mountains in Colorado. For brevity, we refer to these stations as being in "Western Ranges" (see Table 1), but note that Tincup and Cumberland Pass are in the southern part of the Sawatch Range (Figs. 2 and 5). The range of the averages that are based on 9 observed and/or best heat-flow values in these Western Ranges is 100–102 ± 8 mWm^{-2}.

Thirty-one heat-flow values are shown for southern Wyoming. The 29 values for the southern Wyoming Basin–Southern Rocky Mountains range from 25–74 mWm^{-2} and average 53 ± 2 mWm^{-2}. The average flux for the Laramie Mountains, the northern Medicine Bow Mountains, and the northern Sierra Madre is about 44 ± 6 mWm^{-2}, and the mean of 9 values in crystalline rocks north of the Cheyenne Belt in southern Wyoming is about 49 ± 5 mWm^{-2}. These calculations did not use the values for the Albany and Keystone, Wyoming, areas south of the Cheyenne Belt because these stations may be part of a transition to high-heat flux in northern Colorado.

These mean heat-flow values may represent regional maximums because the calculations did not involve data that are corrected for areal uplift and erosion. To obtain values for regional minimums, areal uplift and erosion data were analyzed, and average corrections were estimated and applied.

The data that we considered for the Rio Grande rift–Southern Rocky Mountains system are summarized in Table 5. Regions south of our study areas were considered to illustrate possible regional effects, many of the estimated areal erosions are halves of the tabulated regional uplifts (after data in Birch, 1950; Decker, 1969), and linear uplift and erosion were assumed to estimate maximum effects (see Table 5; also Birch, 1950). From this tabulation, present effects of major uplift (6–7 km) and erosion (3.2–3.5 km) of the Southern Rocky Mountains since the Late Cretaceous (130–65 Ma) would be much less than 5%–6%, as would be the effects of inferred uplift and erosion that started in the late Paleocene (55 Ma) and the early Eocene (50 Ma). Similarly, 1–1.5 km of uplift during Oligocene through early Miocene times (35–20 Ma) would have small effects (<5%) on present flux values, as would uplift of 0.6–2.8 km of these mountains at different times (24 Ma, 20 Ma, 14–15 Ma, 7 Ma) in the Miocene, or uplift of about 1 km in the Pliocene or the early Pleistocene. The principal questions pertain to the likely effects of late Miocene and later erosion of this region. In one extreme view, Neogene and later erosion of 0.5–1.6 km is assumed to have affected all parts of an area by the same amount and, as a result, present flux and gradients could be artificially high by as much as 7%–10% (Table 5). Conversely, as Scott (1975) has suggested, most of the current relief in the Colorado and New Mexico Rocky Mountains could have existed during much of Miocene time, with later elevation of the higher ranges leading to localized river-canyon erosion between resistant topographic highs. By this view, average erosion of the entire rift-mountain system in late Miocene through Holocene times would be much less than the local rates shown in Table 5, effects of likely amounts of uplift would be small, and the combined effect on current regional heat flow would be much less than 5%–6%.

We are thus led to conjecture that Birch (1950) and Decker (1969) actually obtained maximum corrections for the effects of late Cenozoic evolution of topography in the rift–Southern Rocky Mountains system in Colorado and northern New Mexico. In those papers, very short evolution times of 4 m.y. and/or 1 m.y. were assumed; regional uplift and erosion were estimated to be in the ranges 1.0–2.1 km and 0.5–1.0 km, respectively; and resulting corrected heat-flow values for 7 widely separated stations were 10%–15% less than values obtained after corrections for steady-state topography (Table 5). Birch's corrections for 2.1 km of regional uplift and as much as 1 km of regional erosion during the past 1 m.y., however, are not realistic for the Southern Rocky Mountains; less regional uplift appears to have occurred since the Pliocene, and late Pleistocene and younger erosion is largely restricted to rapid downcutting of river canyons between resistant topographic highs that could have been uplifted to their present elevations in much earlier times (Table 5). Thus, a maximum average correction of about –10% for topographic evolution may be applied to present data for the mountains in Colorado and northern New Mexico and the following minimum regional heat flows are implied: 77–79 mWm^{-2} for the Frontal Ranges area; 80–81 mWm^{-2} for the Colorado Front Range; 103 mWm^{-2} for the rift zone in Colorado; 104 mWm^{-2} for the San Juan Mountains; and 90–92 mWm^{-2} for the Western Ranges.

Blackstone (1975) argued that early Paleocene and older (?) mountain uplifts in southern Wyoming stood well above bordering basins by the end of the Eocene, and that a mature, thoroughly eroded landscape existed throughout this region at that time. Ensuing deposition started in the Oligocene and continued until late Miocene or early Pliocene time, when only higher parts of the mountain masses occurred above the plain of aggradation (Blackstone, 1975). During late Pliocene through present time, regional denudation of the basins in southern Wyoming has occurred, enhanced by periods of climatic cooling, glacial erosion, and perhaps epeirogenic uplift (Blackstone, 1975; D. L. Blackstone, Jr., 1987, personal commun.).

From our calculations for the effects of pre-Oligocene evolution of topography in Colorado mountains, we may conclude that the heat-flow values in southern Wyoming probably are not significantly affected by Paleocene through Eocene development of local terrain, unless amounts of erosion and uplift during that time interval exceeded 6–7 km (see Table 5). The effects of Pliocene and younger uplift also could be small; for example, uniform uplift of 1 km of the Medicine Bow Mountains in the past 4 m.y. would increase gradients in these mountains by only 3%–6%. Moreover, lack of extensive Pliocene sediments in southern Wyoming basins (Blackstone, 1975) implies that the Sierra Madre, Medicine Bow, and Laramie Mountains were not extensively eroded in late Neogene and later time. Apparently, the thermal effects of late Cenozoic evolution of topography in the mountains in southern Wyoming are not large, and a maximum regional correction of 5%–10% may be suggested. Application of a 5%–10% correction, in turn, implies 49–51 mWm^{-2} for the minimum regional flux in the southern Wyoming Basin-Rocky Mountain region, a 40–42 mWm^{-2} value for northern parts of the Sierra Madre and Medicine Bow Mountains and a 45–47 mWm^{-2} value for crystalline terranes north of the Cheyenne Belt.

Q and A Relations in the Eastern Frontal Ranges in Colorado and New Mexico, the Western Ranges in Colorado, and in Southern Wyoming

The preceding discussion clearly shows that the average heat flow in the mountains in Colorado and northern New Mexico dramatically exceeds the areal values in southeastern Wyoming. Regional heat-flow calculations, however, smooth out lateral flux changes; hence they provide little quantitative understanding of the heat-source distributions in these areas. Consequently, we explain the indicated regional differences using (1) combined flux and radiogenic heat-production values and (2) excellent control on the lateral changes of these parameters in critical areas.

Q-A Lines and Near-Surface Radioactive Heat. One important result of our research is the close correlation between heat flow and radiogenic heat of igneous and metamorphic bedrocks in the eastern frontal ranges in Colorado and northern New Mexico, the Western Ranges in Colorado, and the mountain-basin area in southern Wyoming. This is shown by the Q-A plots in Figures 6 and 7, and the least-squares straight-line calculations that are summarized in Table 4.

The calculations suggest a slope and intercept of 9 km and 55–58 mWm^{-2}, respectively, for the Frontal Ranges, 9 km and 54–58 mWm^{-2} values for the Colorado Front Range, and all combined data for the eastern frontal ranges plot in or near the Q-A zone for the Basin and Range province (Fig. 6). The Q and A data for the Western Ranges also plot in or just above the Basin and Range zone (Fig. 7), with a 12-km slope and a 68-mWm^{-2} intercept for these areas being suggested. In marked contrast, lines for the southern Wyoming sites indicate a regional slope of about 7 km and an average intercept of about 27 mWm^{-2}, and, except for possible transition points near Albany (AB) and Keystone (KS) in the Medicine Bow Mountains, all of the data for this region plot unambiguously below the values for mountain areas to the south (Fig. 6). The eastern frontal ranges of the Southern Rocky Mountains in Colorado and northern New Mexico, and the Colorado Front Range thus have heat-flow–heat-production characteristics that are comparable to those of the Basin and Range province heat-flow high, as may the Elk Mountains and other parts of the Western Ranges (Tincup, Cumberland Pass, Gilman) in west-central Colorado. These mountain areas with high-intercept flux, in turn, are major heat-flow highs compared to the Wyoming Basin–Southern Rocky Mountains area in Wyoming, where the intercept heat flow is unequivocally low to normal.

One interpretation of the Front Range lines is

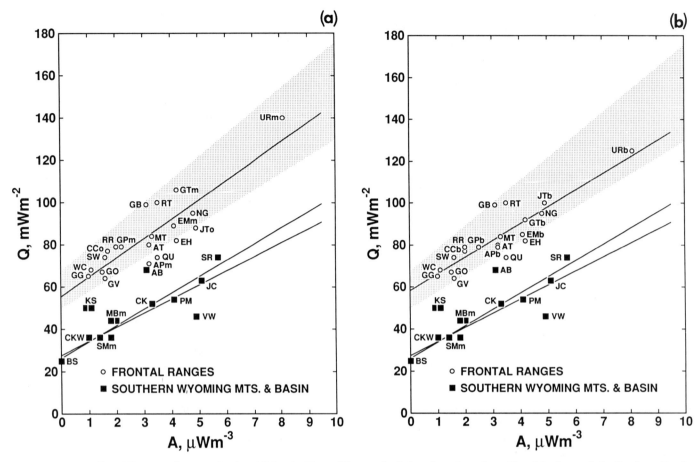

Figure 6. Heat flow (Q) versus heat production (A) for the Frontal Ranges in Colorado and northern New Mexico, and the Southern Rocky Mountains–Wyoming Basin region in southeastern Wyoming. Figure 6a shows terrain-corrected and mean terrain-corrected Q values and observed A values. Figure 6b shows best Q and A values when possible, and terrain-corrected Q data and observed A data for other locales. See Table 3 for locale abbreviations and Table 4 for parameters of the least-squares Q-A lines. The shaded Q-A zone is within ±15% bounds of heat-flow values on the least-squares line that Roy and others (1968a) determined for the Basin and Range high heat-flow province.

that an 8- to 9-km thickness of granite with the radioactivity (A = 8.1) of that measured at Urad (UR), Colorado, generates most of the above-intercept flux (70–85 mWm^{-2}) near this locality. Similarly, an 11- to 12-km-thick layer of the granite (A = 7.1) below Redwell Basin (RW) in the Elk Mountains could account for most of the excess flux (80–85 mWm^{-2}) there. Such thicknesses are in reasonable accord with areal gravity modeling; Brinkworth and others (1968) and Tweto and Case (1972) explained gravity lows along the Colorado mineral belt not far from Urad using 9- to 10-km-thick granite plutons in the subsurface; Isaacson and Smithson (1976) explained a major gravity low in the Redwell Basin–Crested Butte area using a 12- to 25-km-thick Tertiary granite batholith in the upper crust. Additionally, gravimetric evidence for an ~10-km-thick pluton of Sherman granite in the Medicine Bow Mountains east of Northgate (NG) in northern Colorado (Johnson and others, 1984) agrees with the 8- to 9-km slopes of the Front Range Q-A lines. These observations suggest that the slopes of the lines for the Front Range and Western Ranges in Colorado may correspond to gravimetric or other geophysical estimates for near-surface pluton thicknesses, a view that contrasts with those of Jaupart (1983) and Vasseur and Singh (1986) in their discussions of the relations between pluton thicknesses and line slopes in heat-flow provinces elsewhere. Similar agreements between the slopes of observed Q-A lines and the depths of density contrasts inferred from gravity modeling have been discussed by Birch and others (1968) for New York and New England and by Roy and others (1968a) for the western United States.

A near-surface origin for much of the heat flow in the mountains in Colorado is in any case required by the rates of horizontal variation. One good example occurs in the Front Range where the changes from Q = 64–74 mWm^{-2} and A = 1–2 μWm^{-3} near Golden (GO) to Q = 125–140 and A = 8.1 near Urad (UR) occur over a lateral distance <50 km. Large concomitant changes of flux and radioactivity also occur in the Elk Mountains; from Paradise Basin (PB), with Q = 86–90 and A = 1.6, to Redwell Basin (RW), with Q = 150 and A = 7.1, the horizontal distance is 10–12 km. Finally, from Apex (AP) in the Front Range, with Q = 71–79 and A = 3.2, to Empire (EM), Q = 85–89 and A = 4.0, to Urad (UR), Q = 125–140 and A = 8.1, the lateral distances are about 25 km and 15 km, respectively. Clearly, shallow sources are needed to explain such rapid heat-flow changes, and substantial contributions from near-surface radioactivity is strongly implied. This conclusion shows that very high heat flows of 125–150 mWm^{-2} at places such as Urad and Redwell Basin in the Colorado mountains should be examined with care before nonradiogenic sources (magmatic heat, hydrologic convection) are invoked to explain them; high Q in these mountains may be associated with high A of local bedrock and, as a result, the actual deep heat

Figure 7. Heat flow (Q) versus heat production (A) for stations in the Western Ranges. See Table 3 for locale abbreviations and Table 4 for parameters of least-squares Q-A lines. The shaded Q-A zone is within ±15% bounds of the heat-flow values on the least-squares straight line that Roy and others (1968a) determined for the Basin and Range high heat-flow province.

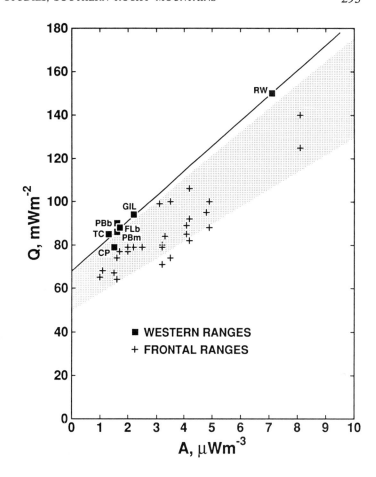

flows may be "typical" conductive values for this high heat-flow province. Normal Q of 60–70 mWm^{-2} at locales such as the Golden area, Canon City (CC), and Westcliffe (WC) may, by themselves, be misleading because local A could be low (1–3 μWm^{-3}), and, as a result, the reduced and residual flux in such areas could be high.

Source of the Reduced Heat Flow in the Front Range–Medicine Bow Mountains Area in Northern Colorado. What explains the intercept or reduced heat flow in "normal" parts of the mountains in northern Colorado? Is part of the high reduced flux due to sources in the upper crust, or is all of the reduced value due to sources in the lower crust and mantle? Near the Colorado-Wyoming border, excellent control on lateral changes of residual heat flow permits resolution of these questions.

Figure 8 illustrates the surface and residual heat-flow variations that occur in this area. Residual-flux calculations were made using slope values as follows (see Table 4): 7- and 9-km values were used for indicated transition points near Albany and Keystone, Wyoming, in the Medicine Bow Mountains; otherwise, 9- and 7-km values were used for locales in Colorado and Wyoming, respectively. From the data plotted, there are remarkable northerly changes from high values in Colorado to lower values in the Wyoming mountains along the entire width of the Southern Rocky Mountains, and all surface and residual flux transitions are ≤50–60 km wide (Fig. 8). For Keystone (KS) and Albany (AB), Wyoming, the residual values are in the range 40–46 mWm^{-2}, and those for Northgate (NG), Colorado, to the south are 50–51 mWm^{-2}. Those for the Medicine Bow Mountains to the north are 30–34 mWm^{-2}. Because the Keystone and Albany locales are centrally located between Northgate and the northern Medicine Bow Mountains sites (Fig. 8), the half-width of this residual heat-flow transition is 25–30 km.

From data along profiles B–B' and D–D' in Figure 8, the residual heat flow near Northgate (NG), Colorado, is 20–23 mWm^{-2} higher than values to the north-northwest. Near Elk Horn (EH), Colorado, in the Front Range, the residual flux is 15–17 mWm^{-2} more than values in the Laramie Mountains of Wyoming, if data along profile A–A' are used for calculations. Thus, compared to southern Wyoming, the residual heat-flow anomaly in northernmost Colorado is +15–23 mWm^{-2}. The anomaly is +13–20 mWm^{-2}, if 5%–10% corrections for late Cenozoic evolution of terrain are applied to the plotted residual heat-flow values. As seen in Figure 9, 25- to 30-km half-widths of the 50- to 60-km-wide transitions to higher residual flux values in northern Colorado cannot be explained by steady-state or transient heat sources that are deeper than 25–30 km in the crust below the high heat-flow areas.

There is no evidence for extensive middle to late Tertiary plutonism and tectonism in the mountains in southern Wyoming, or in the Front Range in northern Colorado (Blackstone, 1975). In the Medicine Bow Mountains near Northgate, Colorado, the youngest "rift-like" mineralization (fluorspar) occurred in post-Oligocene (?) time (Steven, 1960), and movement along the Independence Mountain thrust fault in this area is "inferred" to have occurred in the Neogene (Tweto, 1979). To the west, extensive late Neogene faulting and folding is well documented on the west flanks of the Park Range, 12- to 7-m.y.-old intrusive and extrusive igneous activity occurred in the Elkhead Mountains, and 9- to 10-m.y.-old plutonic rocks occur at Hahns Peak in the western Park Range (McDowell, 1971; Izett, 1975; Tweto, 1979; see Figs. 2 and 5). The late Neogene igneous and tectonic activity in the western Park Range–Elkhead Mountains areas implies that transient, nonradioactive heat sources in the upper crust account for some of the unusually high residual flux at Hahns Peak (HP), Colorado (see Fig. 8; also text below). To the east, lack of unequivocal evidence for late Neogene plutonism and tectonism suggests that nonmagmatic heat sources in the upper crust must explain sizable fractions of the higher residual flux near Northgate and Elk Horn, Colorado.

Considering locales listed in Tables 1 and 3, a 3.0–3.3 Wm^{-1} K^{-1} average thermal conductivity is calculated for near-surface bedrocks and basement in the mountains and Park areas in northern Colorado. For basement in southeastern Wyoming, an average of 3.0–3.9 Wm^{-1} K^{-1} is obtained. Hence, we find *no* convincing evidence that the indicated residual heat-flow transitions are caused by thermal conductivity contrasts in the upper crust. There is also no evidence for pervasive hydrologic disturbances everywhere in the normal heat-flow area of

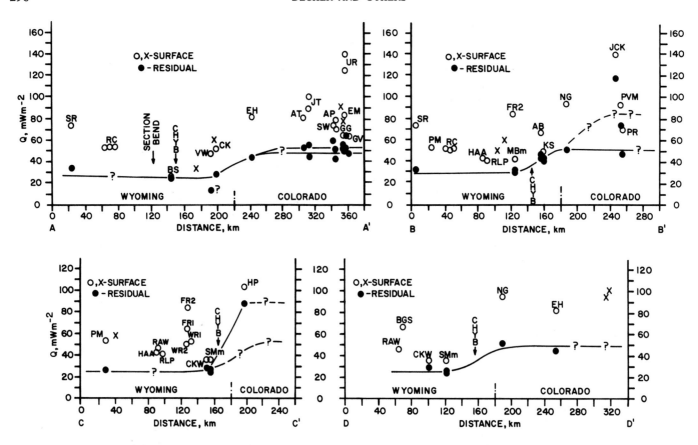

Figure 8. Profiles of surface and residual heat flow in northern Colorado and southern Wyoming. Open circles and solid dots represent surface and residual heat-flow values, respectively, from this report; X's represent surface heat flows from other papers (see references for Fig. 5). Locations of profiles are shown in Figure 5. Locale abbreviations directly follow symbols in Tables 1–3.

about 50,000 km² in southeastern Wyoming. We thus conclude that crustal radioactivity contrasts must explain a significant part of the high residual flux in northern Colorado and the narrow transitions to lower values to the north.

We may speculate on the amount of radiogenic heat that is needed to explain the residual heat-flow anomaly in the Northgate and Elk Horn areas, assuming that the sources must occur above 25 km, or so in the crust. Suppose, for example, that the 10- to 25-km-depth interval of the crust contains higher radiogenic heat (ΔA) sufficient to produce an extra 13–23 mWm^{-2} of flux; a $\Delta A = 0.9$–2.2 μWm^{-3} would then be implied for this 10- to 15-km-thick section. Alternatively, an exponentially decreasing radioactivity contrast in the 10- to 25-km-deep part of the crust could explain 80%–90% of the anomaly, if the ΔA at 10 km is assumed to be 1.1–2.5 μWm^{-3} and the radiogenic heat below decreases from those values according to a logarithmic decrement of 10–15 km.

As to actual radioactivity data (Table 3), we obtain near-surface heat-production values of 4.2 and 4.8 μWm^{-3} for Northgate and Elk Horn, respectively, and the average of 19 values for the Frontal Ranges in Colorado and northern New Mexico is 3.4 ± 0.4 μWm^{-3}. In contrast, data for the thickest basement terranes in the Sierra Madre and Medicine Bow Mountains north of the Cheyenne Belt in Wyoming are consistent with supracrustal heat production of 1.4–2.0 μWm^{-3}. Moreover, Archean granites and granitic gneisses with low heat production of about 1.4 μWm^{-3} could compose most of the basement north of the Cheyenne Belt in the Laramie Mountains to the northeast (after maps in Johnson and others, 1984; and Duebendorfer and Houston, 1987). We may conjecture, therefore, that the average upper crust in the Southern Rocky Mountains in northern Colorado is more radioactive than the average upper crust in the Sierra Madre, the northern Medicine Bow Mountains, and the northern Laramie Mountains, with heat production differences of at least 1.4–1.6 μWm^{-3} being implied. Perhaps more radioactive upper-crustal rocks extend to greater depths in the mountains in northern Colorado, and, as observed, the reduced and residual heat flow there exceeds the values in the mountains in southern Wyoming.

Our estimates of average radiogenic heats and equilibrium temperatures below profiles from the Front Range–Medicine Bow Mountains areas in northern Colorado into the mountains in southern Wyoming are shown in Figure 10. Near-surface flux and heat-production values follow data herein. Otherwise, thermal properties and estimated heat productions for the lower crust are after reasonable values in Roy and others (1968a, 1972), Blackwell (1971), Lachenbruch (1970), and Smithson and Decker (1974). Crustal-thickness data are after Prodehl and Lipman (1988), and modeling was done using finite-difference techniques.

In all depicted models, 15–16 mWm^{-2} of the residual heat-flow transitions in the environs of the Colorado-Wyoming border is explained by radioactivity differences in the upper 25 km of the crust. Thus, the average heat production between 9- and 25-km depths in the Colorado Front Range is about 1.3 μWm^{-3}, a value that is 1.0 μWm^{-3} greater than value in the same depth interval below Wyoming. Other reasonable radiogenic heat contrasts could be selected for this part of the Colorado crust; for example, selection of a contrast of 1.3–1.5 μWm^{-3} (A = 1.6–1.8 μWm^{-3}) would produce the same sur-

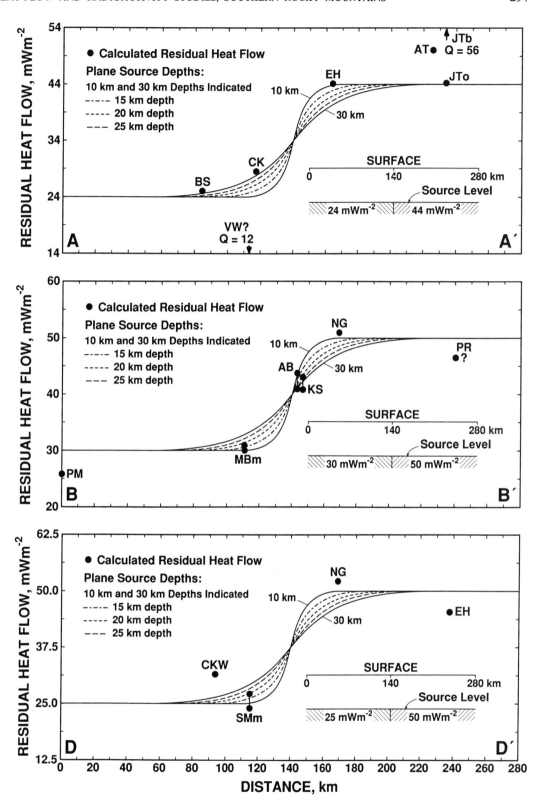

Figure 9. Heat flow produced at the surface above steady plane heat sources at various depths compared to residual heat-flow values for locales in northern Colorado and southern Wyoming. See Figure 5 for profile locations; see Figure 8 and Table 3 for locale abbreviations.

face heat-flow profiles, but the mantle flux below Colorado would be lowered to 25 mWm^{-2}, and the geotherms would be shifted to greater depths. If some of the surface flux in all areas is from the mantle, one inescapable implication of our modeling is that the radiogenic heat production of the lower crust in all areas is low (0.2–0.3 μWm^{-3}). Like earlier studies of other areas by Birch and others (1968), Roy and others (1968a, 1972), and Lachenbruch (1968b), therefore, our combined data for the mountains in northern Colorado and southern Wyoming imply that continental crustal radioactivity is dramatically upward differentiated and virtually independent of *total* crustal thickness.

Rio Grande Rift in Central and Southern Colorado

A strong case can be made that parts of the Rio Grande rift in Colorado are major heat-flow highs compared to mountain areas to the east. One excellent example is the Leadville–northern Sawatch Range region because the combined data for Alma (ALM), Climax (CLI), Kokomo (KK), Timberline (TBL), Twin Lakes (TWL), and Winfield (WIN) plot conspicuously above all Frontal Ranges points, and the Q-A band shown for the Basin and Range province (Fig. 11). Likewise, data for the Summitville (SV, ALC) and Del Norte (DN) areas in the eastern San Juan Mountains plot unambiguously above the Frontal Ranges data, and the Basin and Range zone. Except for the slightly low values at Alma (ALM) and Twin Lakes (TWL), all data in the Leadville–northern Sawatch Range area and the eastern San Juan Mountains also plot within or on the ±10% bounds that are shown for the unusually high heat-flow area in the vicinity of Battle Mountain, Nevada. This provides additional evidence for very high flux in the rift zone between northern New Mexico and central Colorado.

Two least-squares lines with contrasting correlation coefficients of 0.9 and 0.8 suggest an average slope of about 9 km and an average intercept of about 91 mWm^{-2} for the rift zone in the Leadville, northern Sawatch Range, and eastern San Juan Mountains areas (Table 4). This intercept exceeds those for the Frontal Ranges and the Western Ranges by 23–36 mWm^{-2}. The reduced flux anomaly in these parts of the rift zone is +20–32 mWm^{-2}, if a –10% correction for late Cenozoic evolution of topography is applied to the intercepts of the lines for these areas.

The anomaly in the eastern San Juan Mountains is poorly controlled because the nearest heat-flow values outside the Summitville and Del Norte areas are at least 100 km away in all directions. In contrast, we have several stations with combined data in the Sawatch Range–Leadville area, and profiles of surface and residual flux were constructed to further investigate this anomaly.

Residual flux calculations for the Front Range and the rift zone were made using an average slope of 9 km (Table 4). The parameters of the Q-A line for the Western Ranges may be questioned because they are based on only 6 points. Therefore, computations for these locales were made using an assumed slope of 9 km and

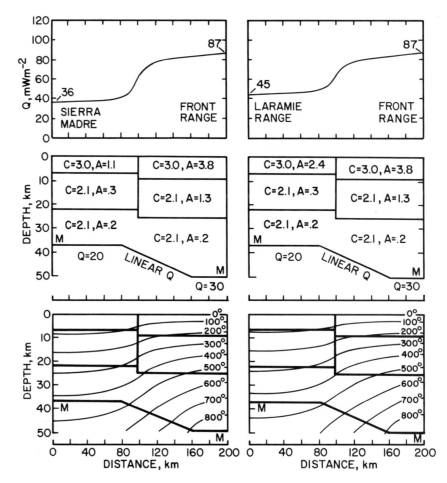

Figure 10. Crustal thickness, thermal parameters, calculated surface heat flow, and subsurface geotherms along generalized north-south profiles from the Front Range–Medicine Bow Mountains area (87–90 mWm^{-2}) in northern Colorado into the Sierra Madre (36 mWm^{-2}) and Medicine Bow Mountains–Laramie Mountains area (44–45 mWm^{-2}) in Wyoming. C is thermal conductivity in Wm^{-1} K^{-1}, A is radiogenic heat production in μWm^{-3}, and Q is heat flow in mWm^{-2}. M locates the Moho.

the calculated line slope of about 12 km; use of a 9-km slope probably leads to maximum residual values for each site, whereas calculations with 12 km may yield minimum values of residual flux. These calculations do not affect the widths of indicated transitions. Instead, they raise the question of whether the residual flux is unusually high in parts of the Elk Mountains. Unfortunately, data elsewhere in the Elk Mountains do not resolve this question because published Q and A data for Crested Butte (Edwards and others, 1978) may be for a thin laccolith in that area, an unquantified distribution of markedly different A values (1.5–12 μWm^{-3}) exists below our sites near Gothic (Table 3), and a complicated interlayering of granite (A = 6.4 μWm^{-3}), dolomites and other sediments may occur beneath our boreholes in North Pole Basin (Table 3). Without more combined data, use of two slope values seems a reasonable way of determining the probable range of residual heat-flow values in these mountains.

In Figure 12, the resulting profile along the mineral belt (A"–A"') unambiguously shows that the residual flux in the Front Range changes from 50–60 mWm^{-2} to 86–103 mWm^{-2} in the northern Sawatch Range–Leadville area, about 30–60 km to the west. Similarly, the profile across the mineral belt (E–E') indicates a large residual heat flow high with narrow borders in the eastern part of the northern Sawatch Range, regardless of whether a 9- or 12-km slope is used to calculate residual values for Fulford (FL) and Gilman (GIL) to the north, and Tincup (TC) and Cumberland Pass (CP) to the south.

Using a 12-km slope, the residual heat-flow values for the Elk Mountains (Paradise Basin (PB), Redwell Basin (RW)), Tincup, and Cum-

Figure 11. Heat flow (Q) versus heat production (A) for the Leadville–northern Sawatch Range area and the eastern San Juan Mountains. See Table 3 for locale abbreviations and Table 4 for parameters of least-squares Q-A lines. The shaded Q-A zone is within ±15% bounds of the heat-flow values on the least-squares line that Roy and others (1968a) determined for the Basin and Range high-heat province; Frontal Ranges and Western Ranges points are indicated in and near this area. The dashed lines represent ±10% bounds for the Battle Mountain, Nevada, unusually high heat-flow zone (after Lachenbruch and Sass, 1977, 1978).

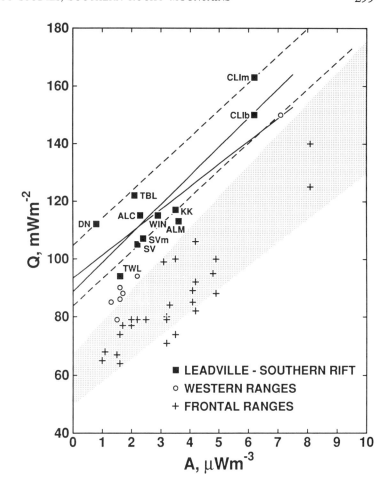

berland Pass are in the range 61–71 mWm^{-2}, and an 50- to 60-km-wide transition to much higher values of 86–103 mWm^{-2} in the northern Sawatch Range–Leadville area to the northeast is implied (Fig. 12). For a 9-km slope, residual values range from 66 to 73 mWm^{-2} for Tincup, Cumberland Pass, and Paradise Basin, and an 86-mWm^{-2} value is obtained for Redwell Basin in the Elk Mountains farther to the west. From both sets of calculations, therefore, the south-southwestern border of the residual heat-flow anomaly in the northern Sawatch Range–Leadville area anomaly is ≤50–60 km wide, and the calculations for a 9-km slope imply a smaller residual anomaly in the western part of the rift zone and suggest that the residual flux in western parts of the Elk Mountains could be unusually high.

The half-widths of the transitions to unusually high residual flux in the Leadville–northern Sawatch Range parts of the rift zone do not exceed 25–30 km. Hence, deep sources cannot account for this anomaly, and a crustal origin is suggested. Moreover, unrealistic temperatures are implied for the lower crust and upper mantle, if the average flux in the anomalous area is explained in terms of reasonable crustal distributions of radioactivity and conductivity, and a large part of the residual flux is assumed to originate in the mantle. Briefly, equilibrium thermal calculations based on the very high surface and residual heat flow suggest above-liquidus temperatures for basalt near the crust-mantle boundary, and massive amounts of basaltic and other melting in the lower crust and upper mantle are implied (Fig. 13). Large amounts of basaltic melting at depth here do not agree with the near-absence of volcanics within the rift north of Alamosa, Colorado (Chapin, 1979), or the lack of widespread exposures of Neogene and younger mafic intrusions in the thermally anomalous region (Tweto, 1979).

Pre-rifting plutonism in the northeastern mineral belt is characterized by episodic emplacement of calc-alkaline intrusions along northeast-trending Precambrian shear zones (Tweto and Sims, 1963; Tweto, 1975). Such plutonism occurred between about 70 and 35 m.y. ago within and east of the Sawatch Range, and between about 29 and 27 m.y. ago in the Elk Mountains (Pearson and others, 1962; Tweto, 1968, 1975; Cunningham and others, 1977; Bookstrom, 1981; White and others, 1981). Subsequently, rift-related (?), fluorine-rich granite porphyries [often called "rhyolite porphyries" (White and others, 1981; Bookstrom, 1981)] intruded parts of the area; such "Climax-type" bodies were emplaced between 33 and 18 m.y. ago at Climax (CLI) and Urad (UR) east of the Sawatch Range, and at about 17 m.y. ago in the Mount Emmons–Redwell Basin area in the Elk Mountains (White and others, 1981; Thomas and Galey, 1982). The Elk Mountains also were intruded by F-rich granites at Treasure Mountain and Round Mountain about 12 and 14 m.y. ago, respectively (Obradovich and others, 1969; Cunningham and others, 1977; Christiansen and others, 1986). As discussed by Tweto (1979, p. 39), there are 22- to 25-m.y.-old intrusive rhyolites on the lower slope of the Sawatch Range, and thin felsite and vitrophyre dikes are present in faults in the rift system in the Arkansas Valley and at Leadville. Tweto (1979, p. 39) emphatically stated that geologic relations suggest a post-Miocene age for several rhyolitic explosion breccias in the Leadville area. Although a K-Ar analysis yields an Oligocene age for these intrusions, large quantities of Precambrian and Laramide rocks in them suggest that this date may reflect contamination and could be too old (Tweto, 1979).

We have not found evidence for late Cenozoic basaltic igneous activity in the Leadville–northern Sawatch Range area. Instead, the nearest Neogene basalts occur to the north in the Gore Range, to the southeast in the South Park and Canon City areas, and 15–20 km south of the Redwell Basin area in the Elk Mountains (Tweto, 1979; see also Figs. 2 and 5). Neogene and younger basalts also crop out at least 25–30 km to the north-northwest along the Colorado River and in other areas considerably north of the Elk Mountains (Tweto, 1979; Figs. 2 and 5). Tweto (1979, p. 40) noted that all "known" basalts in the rift system in Colorado north of the San Luis Valley are alkalic. They therefore probably originated below a thick crust.

The calc-alkaline plutonic rocks along the mineral belt have been explained in terms of partial melting of the lower crust during low-

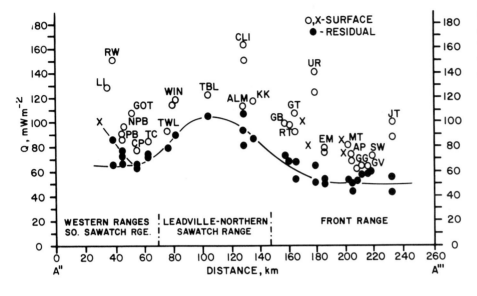

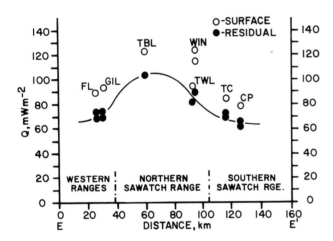

Figure 12. Profiles of surface and residual heat flow in the environs of rift zone in the Leadville–Sawatch Range area. Open circles and solid dots represent surface and residual heat-flow values, respectively, from this report; X's represent surface heat flows from other papers (see references for Fig. 5). Locations of profiles are shown in Figure 5. Locale abbreviations directly follow symbols in Tables 1–3.

angle subduction and associated compression in the Southern Rocky Mountains between about 80 and 40 m.y. ago (Lipman and others, 1971; White and others, 1981). In contrast, the younger Climax-type granitic porphyries are considered to be highly fractionated end products of anorogenic rhyolitic magmas that formed during atectonic and/or extensional intervals that followed the subduction-related calc-alkaline magmatism (White and others, 1981; Bookstrom, 1981). Presumably, these parent magmas formed when rising mafic magmas partially melted high-grade Precambrian metamorphic rocks in the lower and middle crust, and an extensional or atectonic setting allowed batches of collected magma to rise quickly into the upper crust and fractionate into bodies with high concentrations of uranium and other lithophile elements (White and others, 1981; Christiansen and others, 1983, 1986).

From the preceding discussion, we strongly suggest that the Leadville–Sawatch Range–Elk Mountains area is underlain by a late Cenozoic rhyolitic complex; basalts typically erupt around the peripheries of shallow rhyolitic systems while failing to erupt in their silicic centers (Hildreth, 1981, p. 10183), and young plutons in such a complex could produce heat-flow highs. We therefore believe that a large part of the unusually high residual heat-flow anomaly in the rift zone here is due to the transient cooling effects of unexposed granitic intrusions in the upper crust. We further suggest that some of these intrusions have relatively high radioactivity. Our new heat-production data for F-rich plutons in Colorado support this suggestion; for example, granites from Climax (33–18 Ma), Urad (28–23 Ma), Redwell Basin (18–16 Ma), the North Pole Basin–Treasure Mountain area (12.5 Ma), and Chicago Basin (10 Ma) have A values in the range 6.1–8.1 μWm^{-3}, and average 7.1 ± .4 μWm^{-3} (Table 3). The uranium contents of other F-rich rhyolites and "rhyolite intrusions" in the northern rift system in Colorado also are high (5–13 ppm, after Christiansen and others, 1986, Table 2, p. 29–37). We do not suggest, however, that the highly fractionated Climax-type porphyries provide direct measures of the radiogenic heats in the unexposed intrusions. Instead, the parent magmas could have lower A values of 3–5 μWm^{-3} (after data supplied by E. H. Christiansen, 1988, personal commun.), and partial melting of the lower crust during late Oligocene, Miocene, and later times could have produced source rocks and resulting magmas with concentrations of lithophile elements that decreased with time.

A major gravity low coincides with the mineral belt in the Leadville–Sawatch Range–Elk Mountains region (Fig. 14). Parts of this anomaly have been explained in terms of low-density plutons in the upper crust (Fig. 14), and one interpretation of the regional low is that it reflects a batholith of Late Cretaceous and/or early to middle Tertiary granitic rocks under most of the thermally anomalous zone (Tweto and Case, 1972; Isaacson and Smithson, 1976). From data for 94 thermal-conductivity samples of core from Urad, Climax, and Redwell Basin, we calculate a 2.61-Mgm^{-3} bulk density for 33- to 17-m.y.-old Climax-type intrusions in this part of the mineral belt. Because this value is remarkably close to the 2.62- to 2.63-Mgm^{-3} densities that Tweto and Case (1972) and Isaacson and Smithson (1976) determined for late Mesozoic and middle Tertiary plutonic rocks in this region, we may infer that similar young granitic rocks comprise the lower parts of the batholith without modifying their interpretations of the gravity field.

We therefore avoid detailed refinements of the gravity models of Tweto and Case (1972) and Isaacson and Smithson (1976), except to state that a 20- to 25-km-thick batholith of late Mesozoic through late Cenozoic rocks is assumed. From the 9-km slopes of the Q-A lines for the rift zone and Frontal Ranges areas, the upper 9–10 km of the batholith is considered to be "cool" Laramide, Oligocene, and perhaps early Miocene rocks which contribute nothing

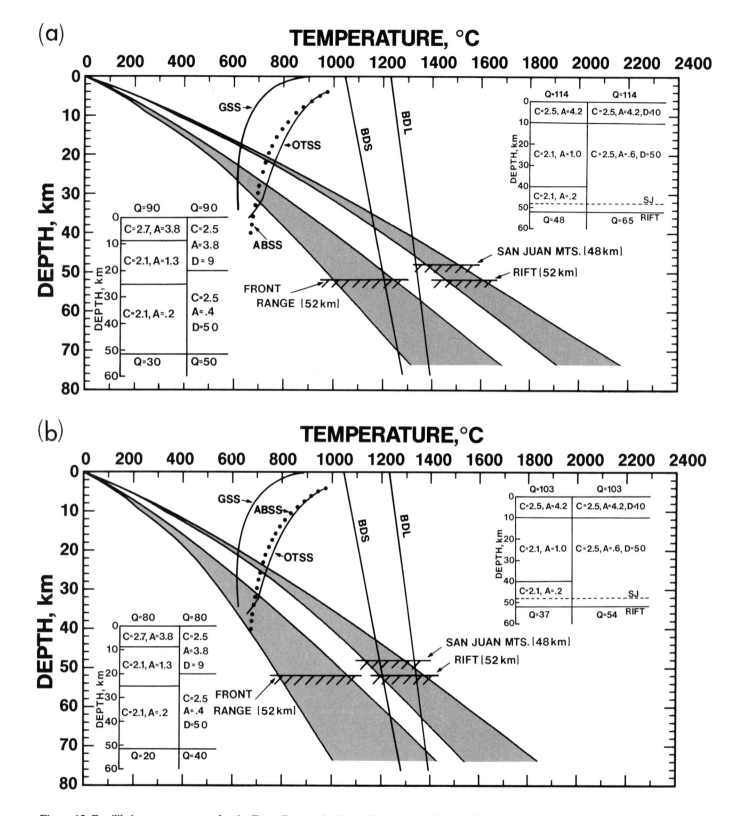

Figure 13. Equilibrium temperatures for the Front Range, the Leadville–northern Sawatch Range area, and the San Juan Mountains. Figure 13a is based on mean surface heat flows that were calculated using values that are corrected for steady-state terrain. Figure 13b is based on mean surface heat flows that are reduced for late Cenozoic evolution of topography. Bases of crusts are indicated by hachures. The upper bound of each temperature band is based on exponentially decreasing radiogenic heat in crustal layers; the lower bound of each band is based on vertically uniform radiogenic heat in each crustal layer. C is thermal conductivity in $Wm^{-1} K^{-1}$, A is radiogenic heat production in μWm^{-3}, Q is heat flow in mWm^{-2}, and D is logarithmic decrement in kilometers for exponentially decreasing heat production according to A exp (-x/D) where x is depth. Melting curves after Wyllie (1971), Lambert and Wyllie (1972), and Yoder and Tilley (1962): GSS, granodiorite saturated solidus; ABSS, alkali basalt saturated solidus; OTSS, olivine tholeiite saturated solidus; BDS, basalt dry solidus; and BDL, basalt dry liquidus.

to the thermal anomaly. In the lower 10–15 km of the batholith, however, late Miocene and/or Quaternary granitic intrusions are placed and used to explain the excess residual flux. Times since emplacement of the intrusions follow evidence for late Cenozoic uplift of the Leadville–Sawatch Range–Elk Mountains region (Larson and others, 1975) and geologic evidence for post-Miocene rhyolitic breccia pipes in the Leadville area (Tweto, 1979). From the 25- to 30-km half-widths of the residual heat-flow anomalies, the 90- to 100-km elongate nature of the regional gravity anomaly, and the north-south widening of the gravity low across the Sawatch Range (Figs. 12 and 14), the composite batholith is assumed to have lateral dimensions of 50 and 60 km (25- and 30-km half-widths) across and parallel to the trend of the mineral belt, respectively.

Along the mineral belt and for a slope of 12 km, the residual flux values in the Elk Mountains and southern Sawatch Range areas are about 10 mWm^{-2} higher than the values in the Front Range about 80 km east of the midpoint of the Leadville–northern Sawatch Range anomaly (Fig. 12). The 2- to 6-mWm^{-2} errors of the intercepts of the Western Ranges, Front Range, and rift zone Q-A lines (Table 4) imply that only 5 mWm^{-2} of this difference may be considered significant, and realistic use of models that exactly match the indicated shape of this anomaly would require that we know the actual crustal structure below this area. Therefore, we use pluton models which produce a simple, symmetrical residual flux profile along the mineral belt that is about 5 mWm^{-2} less than the indicated average residual value (~67

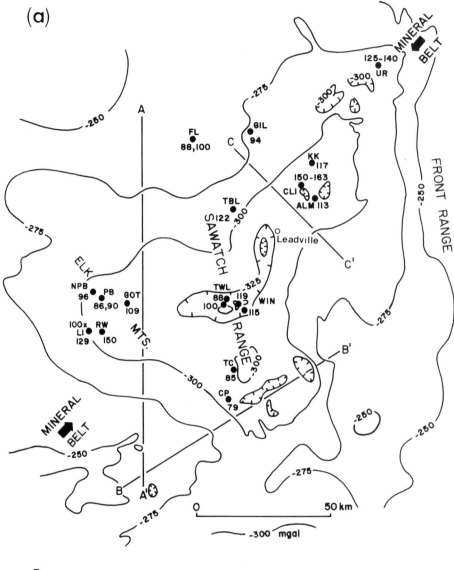

Figure 14. Bouguer gravity (a) and gravity interpretations (b) in the environs of the mineral belt in the Elk Mountains–Leadville–Sawatch Range area. Gravity contours from Behrendt and Bjawa (1974). Two-dimensional gravity models along profiles A–A' and B–B' after models in Isaacson and Smithson (1976). Two-dimensional gravity model along profile C–C' after model in Tweto and Case (1972). Density (ρ) values and density contrasts ($\Delta\rho$) are in Mgm^{-3}. Heat-flow values in mWm^{-2} are from Table 1 and Figure 5.

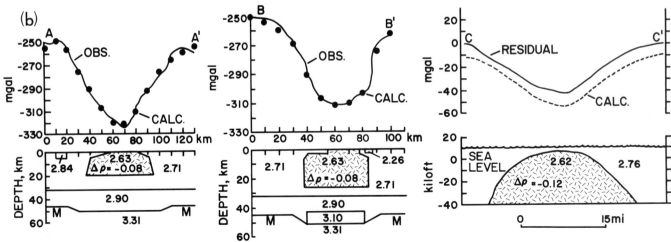

mWm^{-2}) in the Western Ranges, and 5 mWm^{-2} higher than the values (55–58 mWm^{-2}) in the Front Range at comparable distances to the northeast. This approach yields a useful first approximation of this anomaly, and shows that more complex models that would match details of the plotted anomaly could be designed if needed. The profile across the mineral belt shows a rift-zone anomaly that is symmetric and of nearly uniform magnitude, regardless of whether a 9- or 12-km slope is used to calculate residual values at Western Ranges sites to the north and south (Fig. 12). Hence, pluton models that produce a symmetrical distribution of flux along this profile are reasonable. For both profiles, rift-zone anomalies that are based on estimated corrections for steady-state and time-dependent topography are shown and modeled.

Conductive cooling calculations directly followed analytical procedures and crustal thermal parameters outlined by Birch and others (1968) and Jaeger (1968). Two-dimensional rectangular plutonic bodies were placed under each profile, and the thermal diffusivity and conductivity of all parts of the upper crust were taken to be 50 km^2/m.y. and 2.7 Wm^{-1} K^{-1}, respectively. This diffusivity for the silicic upper crust exceeds the 32 km^2/m.y. value that is often used for modeling of entire crustal sections (see Birch and others, 1968; Jaeger, 1968); hence, our models may reflect minimum heating times that are not likely to be more than 50% low. For all modeling, the younger intrusions were emplaced at initial temperatures (1000–1100 °C) equal to rhyolite melting temperatures (800–900 °C) plus a value (200 °C) that permitted consideration of latent heats of fusion of silicic melts (after Jaeger, 1968; Lachenbruch and others, 1976; Wyllie, 1981; Christiansen and others, 1983, 1986). The granitic plutons were emplaced in crusts with equilibrium temperatures like those calculated for the Front Range; therefore, the intrusions below the rift zone initially were at temperatures about 700 °C above those (300–500 °C) of bordering units in 9- to 25-km depth intervals (see Fig. 13).

The young F-rich plutons are visualized as having formed during limited (10%) partial melting of a granulitic protolith in the lower crust (after Christiansen and others, 1983, 1986). For one extreme case, we assume that the heat productions of the late Miocene (10–8 m.y.) intrusions are about half of the values of the 12- to 10-m.y.-old, Climax-type intrusions in the Elk Mountains and the Needles Mountains. The values are then 3.0 to 3.5 μWm^{-3}, and the contrasts with bordering rocks outside the thermally anomalous zone are in the range 2.0–3.0 μWm^{-3}. Because partial melting of the lower crust since Oligocene time could have produced progressively less radioactive source rocks and partial melts, 2.0 to 3.0 μWm^{-3} values were assumed for the youngest and deepest plutons, and contrasts of 1.0–2.0 μWm^{-3} with surrounding rocks are implied. Steady-state effects of the radioactivity contrasts in the prismatic intrusions were estimated using methods outlined by Simmons (1967). Then, emplacement times, thicknesses, and depth intervals were considered, and first-order corrections were made for non-equilibrium heat flow at the surface (see caption for Fig. 15).

Selected results of modeling are shown in Figure 15. For the indicated parameters and conditions, 10 ± 2 m.y.-old late Miocene intrusions in the lower 10–25 km of a composite batholith would contribute only about 5–7 mWm^{-2} to present flux and thus cannot explain the residual heat-flow anomalies, unless their average radiogenic heat production is unusually high (>6–7 μWm^{-3}). In contrast, 5- to 7-km-thick, 1.5- to 1-m.y.-old early Quaternary granitic intrusions in the 13- to 20-km interval of the inferred batholith could explain most or all of the residual anomalies and match the shapes of the transitions. Relatively higher radioactivity in the late Miocene and Quaternary plutons in the lower part of the batholith has a small, significant effect. In particular, downward-decreasing contrasts of 3.0–1.0 μWm^{-3} in the 9- to 20-km depth interval could produce 5–10 mWm^{-2} of the present residual flux, increments sufficient to produce near-matches of the anomalies near the centers of the profiles (Fig. 15).

Larson and others (1975) show that the Roaring Fork River Canyon in the Elk Mountains–Sawatch Range region was rapidly eroded between 10 and 8 m.y. ago and in the past 1.5 m.y. For the density contrasts in Figure 15, emplacement of 5- to 7-km-thick granite plutons in the late Miocene and early Pleistocene could have produced 180–250 m of isostatic uplift and thus increased erosion during these times. Additional upwarping and ensuing erosion could have occurred then if the intrusions evolved from rhyolitic magmas that were created when material from the mantle rose, thickened the lower crust, and produced partial melting of bordering materials. For example, 5 km of new basaltic material in the lower crust could have led to an additional .6–.7 km of isostatic uplift if the crust and mantle had mean densities of 2.8 and 3.2 Mgm^{-3}, respectively (after Morgan, 1983; Morgan and Golombek, 1984). Hence we suggest a close correlation between present residual heat-flow anomalies, late Cenozoic plutonism, and late Cenozoic uplift of the Leadville–Sawatch Range part of the northern rift zone.

Sources in the upper crust must explain the 25- to 30-km half-widths of the Leadville–northern Sawatch Range residual heat-flow anomaly. Otherwise, most parts of the models depicted in Figure 15 may be considered artificial and others involving effects of different plutons, unusually high crustal radioactivity, hydrologic convection, or extension of the upper crust could be designed. The main argument against radioactivity as the only cause of the excess residual flux in this part of the rift is that unrealistically high heat production that greatly exceeds 6–7 μWm^{-3} in the 10- to 25-km part of the upper crust would be needed to account for the magnitude and narrow borders of the indicated residual anomaly (see Fig. 15a). Also, we have no evidence for pervasive regional hydrologic disturbances in the crystalline terrane of about 3,500 km^2 that comprises the Leadville–northern Sawatch Range region, and, in simple form, young extension of the crust could imply crustal thinning and subsidence (McKenzie, 1978; Morgan, 1983; Lachenbruch, 1978; Lachenbruch and Sass, 1978), views that are not consistent with a 48- to 52-km crust, late Cenozoic uplift, and gravity lows in these parts of the rift zone. Thus we favor late Cenozoic granitic intrusions into the upper crust to explain part of the gravity low, late Cenozoic upwarpings, and most of the residual heat-flow anomaly in this area.

An important topic for future geothermal research in Colorado is the determination of a statistically reliable Q-A relation for the Elk Mountains. If a 12- km or larger slope is confirmed, reduced flux values in these mountains would be about 65 mWm^{-2}, and arguments against large, rift-like heat-flow zones there could be presented. On the other hand, confirmation of a 9-km or lower slope would imply unusually high residual flux for parts of the Elk Mountains. Regional geology does not preclude rift-like heat-flow anomalies in these mountains: Tweto (1979) suggested that a rift-related igneous body exists in the basement along a north-northwest–trending zone of plutons that passes through Redwell Basin, Obradovich and others (1969) and Cunningham and others (1977) documented 12.5- and 14-m.y.-old plutonism at Treasure Mountain and Round Mountain, respectively, and Larson and others (1975) showed that alkalic basalts were erupted about 1.3 and 2.0 m.y. ago in areas immediately to the north. The Elk Mountains probably experienced uplift in late Miocene and early Pleistocene times, based on rapid erosion of the Roaring Fork River Canyon between 10 and 8 m.y. ago and in the past 1.5 m.y. (Larson and others, 1975). Evidently, a cooling-intrusion model like that for the Leadville–northern Sawatch Range region could be designed for portions of these mountains; late Miocene and Plio-Pleistocene intrusions in the upper and lower crust could explain regional uplift during these times, and late Pliocene and younger granitic intrusions in

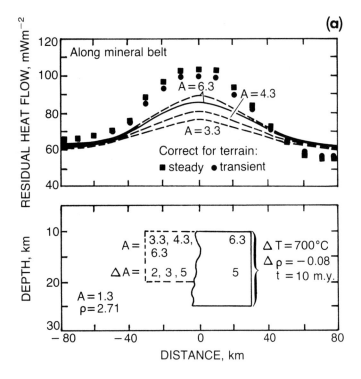

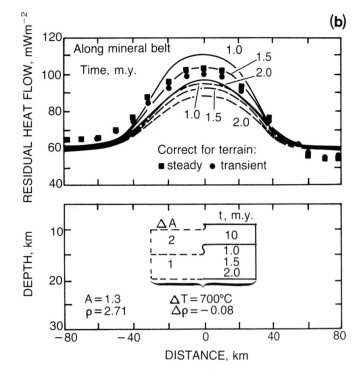
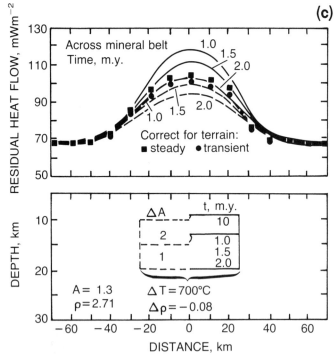

Figure 15. Cooling-intrusion models of the northern rift zone in the Elk Mountains–Leadville–Sawatch Range area. Solid squares represent residual heat-flow values corrected for steady-state terrain that were estimated from profiles in Figure 12. Solid circles represent residual heat-flow values that include a 10% reduction for late Cenozoic evolution of terrain, and adjustment by a constant so that all residual values in the same areas outside the anomalous rift-zone area are essentially identical. Half-width dimensions of modeled tabular plutonic bodies are indicated in the figures; total widths of all bodies in Figures 15a and 15b are 60 km, and total widths of all bodies in Figure 15c are 50 km. Dashed curves are residual heat-flow profiles produced by composite tabular plutonic bodies that are outlined by dashed borders; solid curves are residual heat-flow profiles produced by composite tabular plutonic bodies that are outlined by solid borders. Vertical thicknesses and times since emplacement (t) are indicated for each plutonic body. Radiogenic heat production (A, ΔA), temperature contrasts (ΔT), and density (ρ, $\Delta \rho$) are in μWm^{-3}, °C, and Mgm^{-3}, respectively; other parameters and units are given in the text. The effects of transient radioactive heat were calculated from $Q(0, t) = \Delta A \sqrt{4kt}\,[ierfc(H_t/\sqrt{4kt}) - ierfc(H_b/\sqrt{4kt})]$, where t is time since emplacement; H_t and H_b are the top and bottom depths of the layer, respectively; k is thermal diffusivity; and ΔA is the heat production contrast; $Q(0, t)$ is flux at the surface after time t; and the function *ierfc* is the integral of the complementary error function.

the upper crust could produce a present-day residual heat-flow anomaly. We cannot be more specific without more reliable residual heat-flow values for the Elk Mountains area.

Rift Zone in Northern Colorado: Park Range, North and Middle Parks

On the basis of regional geology and a preliminary heat-flow value of 120 mWm^{-2} for one borehole in Middle Park, Decker and others (1980) speculated that very high flux like that in the Rio Grande rift in New Mexico occurs in parts of the Park areas in the mountains in northern Colorado. Our new Q and A data for Hahns Peak (HP) in the Park Range and Jack Creek (JCK) in Middle Park support this suggestion because both points plot well above all of our Frontal Ranges values and the upper bound of the Q-A zone that we show for the Basin and Range high heat-flow province (Fig. 16). The thermal output near Parkview Mountain (PVM) in Middle Park also may be unusually high. Here, combined data plot above values for Frontal Ranges stations with comparable heat production, and above the depicted Basin and Range zone (Fig. 16).

Residual flux in the Hahns Peak and Jack Creek areas must be unusually high. Suppose, for extreme calculations, that a hypothetical Q-A line with a slope of 15 km is appropriate for the Parks–Park Range area; the residual values at Hahns Peak and Jack Creek would then be 78 and 104 mWm^{-2}, respectively. Higher residual values are obtained if the 9- to 12-km slopes of lines for other Southern Rocky Mountain terrains also apply here; 83–88 mWm^{-2} is then calculated for Hahns Peak, and 111–118 mWm^{-2} is determined for Jack Creek. Corresponding values in the Parkview Mountain area are in the range 68–74 mWm^{-2}.

Buelow (1980) has shown that Pliocene and younger granitic intrusions in the upper 10–15 km of the crust could explain 30–40 mWm^{-2} of the unusually high residual heat flow in the Middle Park–Park Range region. This conjecture agrees with much evidence for late Cenozoic development of the Rio Grande rift system in northern Colorado: Oligocene and Miocene (?) volcanics, intrusives, and faults abound in the Middle Park area (Steven, 1975; Izett, 1975; Tweto, 1977, 1979); Miocene to Pliocene (?) volcanism at Specimen Mountain (Blackstone, 1975) is consistent with late Neogene extension or rifting in the Never Summer Range near North Park; and 12- to 7-m.y.-old intrusive and extrusive rocks in the Park Range–Elkhead Mountains region are chemically similar to rift-related alkalic rocks that occur in the Raton Basin in the south (Tweto, 1979). We have not, however, been able to find unequivocal evidence for late Pliocene and/or early Pleistocene intrusions and associated crustal upwarpings at our study areas in the Park Range or in the Middle and North Parks areas. Perhaps such masses exist in the subsurface in these parts of northern Colorado, but their possible effects were slight or not yet delineated in surface investigations.

Eastern Colorado Plateau–Needles Mountains–Western San Juan Mountains

New heat-flow (60 mWm^{-2}) and heat-production (0.3 μWm^{-3}) data for Precambrian gabbroic rocks in Glade Park (GLP) in the Colorado Plateau about 20 km southwest of Grand Junction, Colorado, plot within the Q-A band that we show for the Basin and Range province, and near our data for the Frontal Ranges in Colorado and northern New Mexico (Fig. 17). The new surface flux near Cisco, Utah, about 75 km to the west is 92 mWm^{-2}, and those for the Dove Creek and Rico, Colorado, areas in the Plateau about 125–150 km to the south are 125 mWm^{-2} and 141–150 mWm^{-2}, respectively (Fig. 5; Table 1). The residual heat flow at Glade Park, therefore, is high (~60 mWm^{-2}; see Fig. 17), as may be residual values at Dove Creek, Rico, and Cisco.

New Q and A data have been obtained for three closely spaced drill holes in the Chicago Basin (CB)–Vallecito Basin (VB)–Lilly Lake (LL) area in the Needles Mountains in southwestern Colorado. Because the combined values for these locales plot below our Frontal Ranges data and the Basin and Range Q-A zone (Fig. 17), one view is that the Needles Mountains are not characterized by above-normal residual flux. The Q and A data for Glade Park (GLP), the

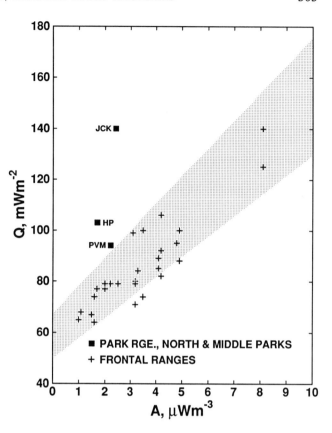

Figure 16. Heat flow (Q) versus heat production (A) for the Park Range–Mountain Parks region in northern Colorado. See Table 3 for locale abbreviations. The shaded Q-A zone is within ±15% bounds of heat-flow values on the least-squares line that Roy and others (1968a) determined for the Basin and Range high heat-flow province; Frontal Ranges points are indicated in and near this area.

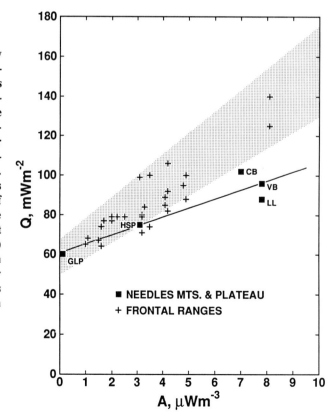

Figure 17. Heat flow (Q) versus heat production (A) for the Needles Mountains–eastern Colorado Plateau areas. See Table 3 for locale abbreviations; see Table 4 for the parameters of the indicated least-squares line. The shaded Q-A zone is within ±15% bounds of heat-flow values on the least-squares line that Roy and others (1968a) determined for the Basin and Range high heat-flow province; Frontal Ranges points are indicated in and near this area.

Needles Mountains (CB, VB, LL), and Hesperus (HSP) in the La Plata Mountains, however, determine a straight line with a slope and intercept of 4.6 km and 61 mWm^{-2}, respectively (Table 4; Fig. 17). Another hypothesis is that these areas have reduced flux that is comparable to that in Frontal Ranges and the Basin and Range province. This new evidence for high reduced heat flow in the eastern Plateau–Needles Mountains area agrees with the hypothesis by Bodell and Chapman (1982) that parts of the border of the eastern Colorado Plateau could be high heat-flow zones that may be related to lateral warming associated with the Southern Rocky Mountains to the east. Elevated deep flux in the eastern Plateau also agrees with late Tertiary uplift of the Uncompahgre Highland (Curtis, 1960; Tweto, 1975) and the relatively high elevations (1,800–2,400 m) of the Colorado Plateau–Southern Rocky Mountains border (Tweto, 1979).

Our research is the first to show that above-normal reduced and residual heat-flow values occur in the eastern Colorado Plateau and the Needles Mountains. We also present evidence for narrow transitions to unusually high surface flux in these areas; the 141–150 mWm^{-2} values near Rico in the eastern Plateau are only 30–50 km away from the La Plata and Needles Mountains, and 128–155 mWm^{-2} values in the western San Juan Mountains near Ouray, and locales to the north are 30–50 km from the Needles Mountains sites to the south (see Fig. 5). Clearly, deep sources cannot explain such narrow transitions. Moreover, steady-state interpretations of average surface heat flow $\geqslant$103 mWm^{-2} in the San Juan Mountains region imply equilibrium temperatures that would be consistent with unrealistic, massive amounts of basaltic and other melting in the lower crust and upper mantle (Fig. 13). Our view, therefore, is that some of the unusually high heat-flow values in the western San Juan Mountains–eastern Plateau area must be due to convective or magmatic sources in the upper crust. We also speculate that parts of the eastern Colorado Plateau and bordering mountain areas are characterized by high reduced and residual heat-flow values (~60 mWm^{-2}) like those for the Colorado Front Range and the western Great Basin. Hence, a complex distribution of crustal radioactivity and transient heat sources probably is needed to explain juxtaposed values of high and unusually high surface and residual heat flow in the Plateau-mountain region in southwestern Colorado. This view contrasts with interpretations by Reiter and Clarkson (1983) which suggested that steady-state step functions of heat sources and elevated temperatures in the lower crust and upper mantle account for above-normal surface flux in the San Juan Basin–La Plata Mountains area, and a smooth transition to surface flux $\geqslant$113 mWm^{-2} everywhere in the western San Juan Mountains–Needles Mountains region.

CONCLUDING REMARKS

The linear relationships between heat flow and surface heat production presented in this study unequivocally demonstrate that the Front Range and other eastern frontal ranges in Colorado are characterized by a high reduced flux of 54–58 mWm^{-2} that substantially exceeds the reduced heat flow of about 27 mWm^{-2} in the Wyoming Basin–Southern Rocky Mountains area in Wyoming. In southeastern Wyoming, ubiquitous normal surface and residual heat flow strongly suggest that the most radioactive layer of the upper crust is thin ($\leqslant$7–10 km), and steady-state modeling implies that the heat production of the lower 20 km or so of the 37- to 40-km-thick crust is low (0.2–0.3 μWm^{-3}). This normal heat-flow zone coincides with 2,500-m.y.-old and older cratonic basement rocks, and we have not found conclusive evidence for extensive tectonic and/or magmatic reworking of the crust in southern Wyoming during Phanerozoic time. Therefore, a reasonable interpretation is that the normal regional flux reflects long periods of Archean and early Proterozoic erosion that produced a thinned granitic layer and a nearly normal crustal thickness. This structural configuration may have existed for most of the Phanerozoic and perhaps since late Proterozoic times.

From analyses of the narrow transitions ($\leqslant$50–60 km) to lower values in southeastern Wyoming, we conclude that radioactivity in the upper 20–25 km of the crust could account for about 15 mWm^{-2} of the reduced and residual heat-flow excess (20–25 mWm^{-2}) in the Front Range–Medicine Bow Mountains area in northern Colorado. Our evidence that near-surface bedrocks in the Colorado Front Range are more radioactive than early Proterozoic and Archean rocks in Wyoming supports this conclusion, as does other research (Phair and Gottfried, 1964; Phair and Jenkins, 1975) which has shown that many of the crystalline bedrocks in the Colorado mountains are generally enriched in radioactivity. Late Mesozoic and Cenozoic plutonic rocks in the Southern Rocky Mountains in Colorado are comonly explained by partial melting of the lower crust, a process that led to low-density silicic magmas that were emplaced into somewhat radioactive Proterozoic rocks in the upper crust. Hence, supracrustal Proterozoic, late Mesozoic, and Cenozoic silicic igneous units could collectively comprise a 20- to 25-km-thick, radioactive "granitic" layer beneath the Front Range and nearby areas, whereas prolonged partial melting of deep protolithic rocks through late Mesozoic and Cenozoic times could have produced low-heat-production, refractory-like residuum rocks in the lower crust and upper mantle.

Present geothermal evidence for a relatively thick radioactive layer in the upper crust is consistent with negative Bouguer gravity and isostatic equilibrium in the Colorado Front Range area, if compensation is partly achieved in the upper crust. Compensation within the upper crust, in turn, agrees with seismic data which show that these mountains are not balanced locally by a thick root in the lower crust; seismically estimated crustal thicknesses across the Great Plains–Front Range join near Denver, Colorado, are in the small range 48–52 km (Prodehl and Lipman, 1988), despite the dramatic increase to 2- to 3-km higher elevations along the continental divide in the mountains. For the assumption that radiogenic heat in the upper crust is in equilibrium with the present surface flux, this model suggests that much of the shallow crust and elevation of the Front Range was established before the late Cenozoic. For example, it can be shown that a 20- to 25-m.y.-old, 25-km-thick, radioactive layer could be out of equilibrium by as much as 20%, and, as a result, Miocene and younger rocks in deeper parts of the upper crust would have to have unusually high radioactive heat to explain an appreciable part of the increased flux in these mountains. By this view, most of the upper crust and topographic relief in the Front Range were created by Oligocene and earlier times, during periods of uplift, magmatism, and tectonism that preceded later development of the northern Rio Grande rift.

We present *combined* heat-flow and heat-production evidence for enriched or thickened radioactivity in the upper 20–25 km of the crust in Colorado Front Range, an interpretation that agrees with seismic evidence for a 20- to 25-km-thick granitic layer there (Prodehl and Pakiser, 1980; Prodehl and Lipman, 1988). We also present a new, refined interpretation of the background heat-flow field in the eastern frontal ranges of the Southern Rocky Mountains; lateral changes of radiogenic heat in shallow (7–10 km) rocks probably produce most of the surface heat-flow changes in "normal" parts of these mountains in Colorado, Wyoming, and northern New Mexico, whereas relatively enriched radioactivity in lower parts of a 20- to 25-km-thick granitic layer partially explain above-normal reduced and residual heat flow in the Medicine Bow Mountains–Front Range area in northern Colorado. Our data and those of Decker and others (1980) also show, however, that previous conjectures (Blackwell, 1971, 1978; Roy and others, 1972; Prodehl and Pakiser, 1980) for high flux in the Southern Rocky Mountains in southeastern Wyoming are not correct; the average flux there is normal ($\sim$44 mWm^{-2}) as is the reduced heat flow ($\sim$27 mWm^{-2}). Perhaps a thinner ($<$10 km) granitic layer occurs in Wyoming parts of the Southern Rocky Mountains, a new conjecture that could be tested with additional seismic profiling along north-south traverses in the Sierra Madre, Medicine Bow, and Laramie Mountains.

Birch (1950) presented two models of isostatically compensated crust with estimated radioactivities that would be consistent with above-normal heat flow in the Colorado Front Range. One model is composed of a granitic layer and crustal thickness of 28.5 km and 48.5 km, respectively, whereas the other has a 21-km-thick granitic layer and a 56-km-thick crust; the lower crust is "basaltic" in both models, and compensating roots occurred in the upper or lower crust (Birch, 1950, Fig. 15 and Table 26, p. 616 and 619). Clearly, these models could be adjusted *slightly* to agree with modern seismic evidence for crustal layering, a 20- to 25-km-thick granitic layer, and a 48- to 52-km-thick crust in the Front Range (Jackson and Pakiser, 1965; Prodehl and Pakiser, 1980; Prodehl and Lipman, 1988). Additionally, there is geophysical evidence for thick ($\leq$25 km) granitic batholiths in the upper crust in the mountains in Colorado (Tweto and Case, 1972; Plouff and Pakiser, 1972; Isaccson and Smithson, 1976), and a strong observational case can now be made that radioactivity in different levels of a 20- to 25-km-thick granitic layer explains rapid changes of surface and residual flux in the Front Range. Hence, Birch (1950) probably was correct in suggesting that the granitic layer in these mountains is thicker and more radioactive than normal, and present-day confirmation of his "high heat flow–mountain root" model of the Front Range is implied.

The average surface heat flow in Archean basement north of the Cheyenne Belt in southern Wyoming is about 50 mWm^{-2}, whereas that for Proterozoic and Phanerozoic bedrock terranes in the eastern frontal ranges in Colorado and northern New Mexico is in the range 85–90 mWm^{-2}. Different meanings of similar heat-flow contrasts between Archean and younger terranes have been presented; Morgan (1985) has suggested that most of the difference is due to contrasting crustal-heat production, whereas Ballard and Pollack (1986) and Ballard and others (1987) suggested that universally lower flux for cratonic nuclei imply a fundamentally different deep thermal structure that extends several hundreds of kilometers into the lithosphere. For the Front Range in Colorado, concomitant changes of Q and A over short lateral distances provide convincing evidence that crustal radioactivity accounts for lateral changes of near-surface heat-flow values that are greater than the reduced flux for this region. Radiogenic heat above $\sim$25 km in the crust could produce the magnitude and narrow borders of the residual heat-flow anomaly in northern Colorado. Therefore, convection in the mantle and/or lateral diversion of mantle heat from the north (after Ballard and Pollack, 1986) probably do not control all of the residual heat-flow increases in the Southern Rocky Mountains south of Wyoming. Thermal and geochemical models of the Archean terranes in Wyoming, however, do support other views of Ballard and his co-workers; steady-state models of the normal surface and reduced heat flow there imply unusually cool, submelting temperatures in a uniquely different upper mantle below this cratonic region (Fig. 18), as do isotopic data for mantle xenoliths which suggest a $\geq$1.5- to 1.8-b.y.-old mantle and a 170- to 200-km-thick lithosphere (Eggler, 1987). Perhaps a combination of crustal radioactivity and mantle diversion of heat is needed to explain increased reduced heat flow in the Southern Rocky Mountains in Colorado, where the upper mantle is hot and the lithosphere is thin ($\leq$80–100 km) based on projected intersections of basalt-melting curves and estimated equilibrium temperature profiles (Fig. 18).

One of the most important results of our work is the observation that an unusually high residual heat-flow anomaly with narrow borders occurs in the northern Rio Grande rift zone in the Leadville–northern Sawatch Range area in the environs of the Colorado mineral belt. There is evidence also for very high residual heat-flow zones with narrow borders in the eastern San Juan Mountains, Middle Park, and the Park Range in Colorado. For the Leadville–northern Sawatch Range region, we believe that late Cenozoic fluorine-rich granite porphyries, Neogene intrusive rhyolites, peripheral alkali basalts, and post-Miocene (?) rhyolitic breccias in the Leadville Mining District are consistent with a late Tertiary rhyolitic complex in the lower parts of a composite silicic batholith in the upper crust below this portion of the mineral belt. Therefore, we conclude that Miocene through Quaternary granitic intrusions in the upper crust explain the magnitude and shape of the residual heat-flow anomaly in this region, a significant part of the associated Bouguer gravity low, and uplift of the Sawatch Range–Elk Mountains areas between 10 and 8 m.y. ago and in the past 1.5 m.y. High residual flux in this part of the rift zone also could be explained in terms of convective heat sources in an extending crust. The rift north of Alamosa, Colorado, however, is not characterized by synrift volcanics (Chapin, 1979); Miocene and younger alkali basalts that occur outside the thermal anomaly probably originated below a thick crust (Tweto, 1979); seismic data do not imply crustal thinning below the northern rift zone (Prodehl and Pakiser, 1980; Prodehl and Lipman, 1988); and widespread hydrologic circulation is not suggested by our temperature-depth data for the crystalline bedrock terranes in the Leadville–Sawatch Range region. Hence there is no unequivocal evidence for shallow convection in the lithosphere north of Alamosa, and, from the lack of crustal thinning in the mountains in Colorado, we suggest that the Leadville–Sawatch Range part of the rift zone has not experienced extensive subsidence-producing extension in the late Cenozoic.

From the temperature-depth bands and melting curves shown in Figures 13 and 18, above-normal surface and reduced heat flow in the Colorado Front Range imply that basaltic melts could have been created and/or sustained in 60- to 80-km depths in the mantle throughout the Cenozoic. Below a developing rift to the west, therefore, rising alkali basalts from the mantle could have penetrated the lower crust and led to high-temperature partial melting. Such melting, in turn, could have produced fluorine-rich rhyolitic magmas that subsequently rose into the upper crust (Christiansen and others, 1983, 1986; Eichelberger, 1978; Eichelberger and Gooley, 1977). A two-stage melting process like this may account for extensive bimodal volcanism and topaz rhyolites in rapidly extending parts of the southern Rio Grande rift and the western Great Basin (Christiansen and others, 1983, 1986; Christiansen and Lipman, 1972; Eaton, 1979; Lipman, 1980; Lipman and others, 1971, 1972). For the rift area north of Alamosa, Colorado, however, a relatively tight, less-extended crust would imply slower ascent of such silicic magmas and encourage solidification within the crust. Furthermore, early Miocene, Oligocene, and older intrusions in the upper crust below the Leadville–Sawatch Range area could have hindered ascent of younger magmas, despite the evidence (Tweto and Sims, 1963; Warner, 1978; Tweto, 1979) for pre-existing zones of weakness in the lithosphere. Hence, it does not seem unreasonable that this area could be underlain by low-density plutons in a rhyolitic complex that was trapped in the crust during late Miocene and younger times. Such young plutons, in turn, could compose the lower part of the gravimetrically indicated 20- to 25-km-thick granitic batholith in this part of the mineral belt and explain the unusually high residual heat-flow anomaly in the Leadville–northern Sawatch Range portion of the northern rift.

If our interpretation of the Leadville–northern Sawatch Range region is correct, simple extensional-subsidence models probably cannot explain the late Cenozoic elevation history of the rift–Southern Rocky Mountains

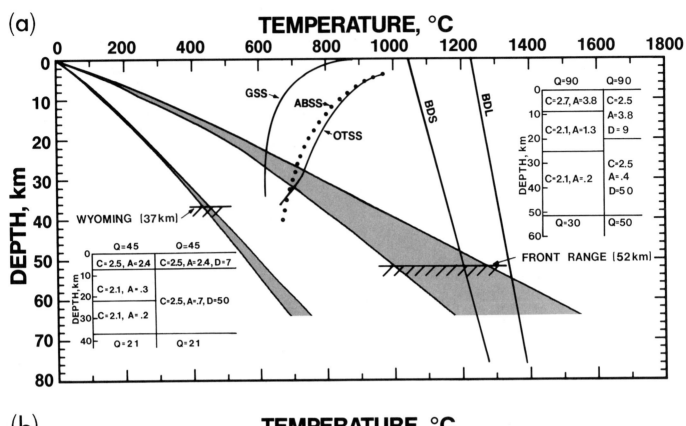

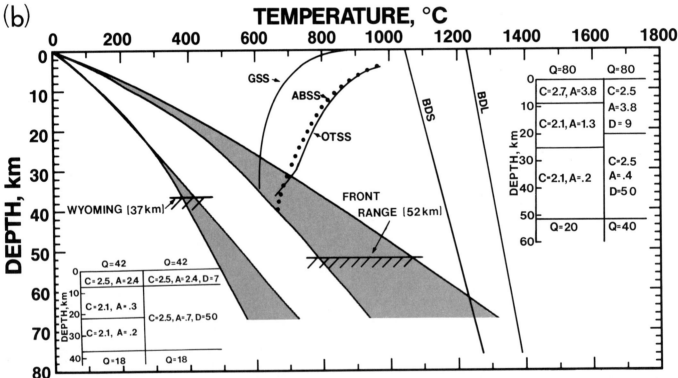

Figure 18. Equilibrium temperatures for the Colorado Front Range and southern Wyoming. Figure 18a is based on mean surface heat flows that were calculated using values that are corrected for steady-state terrain. Figure 18b is based on mean surface heat flows that are reduced for late Cenozoic evolution of topography. Bases of crusts are indicated by hachures. The upper bound of each temperature band is based on exponentially decreasing radiogenic heat in crustal layers; the lower bound of each band is based on vertically uniform radiogenic heat in each crustal layer. C is thermal conductivity in $Wm^{-1} K^{-1}$, A is radiogenic heat production in μWm^{-3}, Q is heat flow in mWm^{-2}, and D is logarithmic decrement in kilometers for exponentially decreasing heat production according to A exp (–x/D) where x is depth. Melting curves after Wyllie (1971), Lambert and Wyllie (1972), and Yoder and Tilley (1962): GSS, granodiorite saturated solidus; ABSS, alkali basalt saturated solidus; OTSS, olivine tholeiite saturated solidus; BDS, basalt dry solidus; and BDL, basalt dry liquidus.

complex in Colorado and northern New Mexico. Instead, magmatic thickening of the crust likely accounts for epeirogenic uplift of this region in the past 10 m.y., and extension in the northern rift zone probably was associated with addition of new material to the crust. Moreover, contemporaneous upwarping of the Southern Rocky Mountains, the Colorado Plateau, and the western Great Plains in the late Cenozoic (Axelrod and Bailey, 1976; Trimble, 1980; see also Table 5) suggests that recent uplift of the northern rift zone may be related to larger regional events that reflect complex interrelations between regional stress fields, areal magmatic events, and the compositions of deep protolithic crustal rocks. Morgan and Golombek (1984) and Morgan and others (1986) have suggested a similar late Cenozoic development of the rift in New Mexico, on the basis of significant uplift since the late Miocene, lack of evidence for large regional subsidence associated with extension, and evidence for magmatic thickening of the crust in the Colorado Plateau and the rift zone.

Eaton (1979, 1980) presented convincing evidence that east-west tension in the western United States has played an important role in late Cenozoic development of the Rio Grande rift. In the Southern Rocky Mountains, the northerly trend of the rift zone is approximately perpendicular to the Colorado-Wyoming border, where folding, faulting, and linear gravity and magnetic anomalies suggest that a fundamental east-west-trending structure exists in the basement (Tweto, 1979; Karlstrom and others, 1983; Johnson and others, 1984). This feature is the remnant of an early Proterozoic Atlantic–type cratonic margin that coincides with the boundary between <1,800-m.y.-old and >2,500-m.y.-old rocks in Precambrian terranes in Colorado and Wyoming, respectively (Karlstrom and others, 1983; Karlstrom and Houston, 1984) and an associated 10- to 12-km change to a thinner crust (37–40 km) in Wyoming (Prodehl and Pakiser, 1980; Prodehl and Lipman, 1988). An ancient crosscutting suture zone here may indirectly account for the narrow widths ($\leqslant$50–60 km) of the heat-flow transitions in the mountains along the Colorado-Wyoming border by marking the mechanical juxtaposition of two fundamentally different crusts with contrasting thicknesses, ages, and, from present thermal data, different average radioactivities. Our studies also suggest that this Proterozoic feature is rheologically impeding additional northerly development of the rift. In particular, the complex distribution of high and unusually high surface and residual flux in the Park Range–Mountain Parks areas in northern Colorado implies that the northernmost part of the rift zone is incompletely developed owing to an east-west mechanical barrier and/or a relatively short period of east-west regional tension since the Miocene.

Finally, modeling of transient magnetic data suggests that anomalously high electrical conductivity could exist between 45- and 100-km depths below the Southern Rocky Mountains in Colorado (Porath, 1971). Estimated Pn velocities for this region are in the range 7.9–8.0 km/sec, and seismic data do not suggest an appreciably thinner crust in the San Juan Mountains–northern rift areas relative to the mountains farther to the east (Prodehl and Lipman, 1988; Prodehl and Pakiser, 1980; Archambeau and others, 1969). To a first approximation, our heat-flow and heat-production data for most of the mountains in Colorado imply near-melting equilibrium temperatures in the upper mantle (Figs. 13 and 18), an interpretation that is consistent with lower Pn velocities (<8.1 km/sec), shallower asthenosphere, and an upwarping of high electrical conductivity in the mantle beneath these areas. Anomalously high electrical conductivity at depths of 45 km, however, implies pervasive partial melting in the lower crust and upper mantle, as do equilibrium temperature profiles that are calculated for $\geqslant$103-mWm^{-2} regional heat flows in the San Juan Mountains and the rift area near Leadville (Fig. 13). Massive amounts of melting in the lower crust and upper mantle, however, would imply (1) Pn velocities much lower than the estimated values of 7.9–8.0 km/sec, (2) unsubstantiated crustal thinning below the rift–Southern Rocky Mountains complex, and (3) appreciably more late Tertiary bimodal volcanism and uplift of the rift–San Juan Mountains region than are presently visualized. Therefore, our steady-state modeling undoubtedly predicts unrealistically high temperatures for the lower-crust and upper-mantle region below the rift–San Juan Mountains region, as may transient magnetic models that imply anomalously high electrical conductivity at depths as shallow as 45 km. This conclusion has important implications with regard to the depths (upper crust or deeper) and kinds of heat sources (steady or transient) that should be used to explain unusually high surface and residual heat-flow zones in the Southern Rocky Mountains, a region that has a thick crust and above-normal background heat flow.

ACKNOWLEDGMENTS

Partial financial support was received from U.S. Department of Energy Agreement DE-FC07-791D12026, DOSECC, INC. contract CMD 86-01, and National Science Foundation Grants GA-18450, DES 75-15119, EAR-8207036, and RII-8011448. The University of Maine provided financial support for drafting of figures. We thank Gerald Bucher, Steve Ridley, and Brad Hahn for careful reduction and plotting of data, and Malcolm McCallum and Paul Graff for collecting radioactivity samples from the Medicine Bow Mountains in Wyoming. Numerous companies and organizations allowed us access to drill holes for geothermal research; the complete, indispensible cooperation of these groups is gratefully acknowledged. David Chapman, Paul Morgan, and Richard Williams provided critical, helpful reviews of the manuscript.

REFERENCES

Adams, J.A.S., 1964 Laboratory γ-ray spectrometer for geochemical studies, in Adams, J.A.S., and Lowder, W. M., eds., The natural radiation environment: Chicago, Illinois, The University of Chicago Press, p. 485–497.
Allmendinger, R. W., Brewer, J. A., Brown, L. D., Kaufman, S., Oliver, J. E., and Houston, R. S., 1982, COCORP profiling across the Rocky Mountain front in southern Wyoming; Part 2, Precambrian, basement structure and its influence on Laramide deformation: Geological Society of America Bulletin, v. 93, p. 1253–1263.
Archambeau, C. E., Flinn, E. A., and Lambert, D. G., 1969, Fine structure of the upper mantle: Journal of Geophysical Research, v. 74, p. 5825–5866.
Axelrod, D. I., and Bailey, H. P., 1976, Tertiary vegetation, climate, and altitude of the Rio Grande depression, New Mexico–Colorado: Paleobiology, v. 2, p. 235–254.
Baker, K. H., 1976, Heat flow studies in Colorado and Wyoming [M.Sc thesis]: Laramie, Wyoming, University of Wyoming, 142 p.
Ballard, S., and Pollack, H. N., 1986, Diversion of heat by Archean cratons: Implications for "reduced" heat flow [abs.]: EOS (American Geophysical Union Transactions), v. 67, p. 388.
Ballard, S., Pollack, H. N., and Skinner, N. J., 1987, Terrestrial heat flow in Botswana and Namibia: Journal of Geophysical Research, v. 92, p. 6291–6300.
Behrendt, J. C., and Bajwa, L. Y., 1974, Bouguer gravity and generalized elevation maps of Colorado: U.S. Geological Survey Geophysical Investigations Map GP-896.
Birch, F., 1943, Elasticity of igneous rocks at high temperatures and pressures: Geological Society of America Bulletin, v. 54, p. 263–286.
——— 1947a, Crustal structure and surface heat flow near the Colorado Front Range: American Geophysical Union Transactions, v. 28, p. 792–797.
——— 1947b, Temperature and heat flow in a well near Colorado Springs: American Journal of Science, v. 245, p. 733–753.
——— 1950, Flow of heat in the Front Range, Colorado: Geological Society of America Bulletin, v. 61, p. 567–630.
——— 1953, Uniformity of the earth's mantle: Geological Society of America Bulletin, v. 64, p. 601–602.
——— 1956, Heat flow at Eniwetok Atoll: Geological Society of America Bulletin, v. 67, p. 941–942.
——— 1958, Differentiation of the mantle: Geological Society of America Bulletin, v. 69, p. 483–485.
——— 1961, Role of fluid pressure in mechanics of overthrust faulting: Discussion: Geological Society of America Bulletin, v. 72, p. 1441–1444.
——— 1965, Speculations on the earth's thermal history: Geological Society of America Bulletin, v. 76, p. 133–154.
Birch, F., and Clark, H., 1945, An estimate of the surface flow of heat in the west Texas Permian Basin: American Journal of Science, v. 243-A, p. 69–74.
Birch, F., and Dow, R. B., 1936, Compressibility of rocks and glasses at high temperatures and pressures: Seismological application: Geological Society of America Bulletin, v. 47, p. 1235–1255.
Birch, F., and Law, R. R., 1935, Measurement of compressibility at high pressures and high temperatures: Geological Society of America Bulletin, v. 46, p. 1219–1250.
Birch, F., Roy, R. F., and Decker, E. R., 1968, Heat flow and thermal history in New England and New York, in Zen, E-an, and others, eds., Studies of Appalachian geology: Northern and maritime: New York, Interscience, p. 437–451.
Blackstone, D. L., Jr., 1971, Travelers guide to the geology of Wyoming: Geological Survey of Wyoming Bulletin 55, 90 p.
——— 1975, Late Cretaceous and Cenozoic history of Laramie Basin region, southeast Wyoming, in Curtis, B. F., ed., Cenozoic history of the Southern Rocky Mountains: Geological Society of America Memoir 144, p. 249–279.
Blackwell, D. D., 1969, Heat flow determinations in the northwestern United States: Journal of Geophysical Research, v. 74, p. 992–1007.
——— 1971, The thermal structure of the continental crust, in Heacock, J. C., ed., The structure and physical properties of the earth's crust: American Geophysical Union Geophysical Monograph 14, p. 169–184.
——— 1978, Heat flow and energy loss of the western United States, in Smith, R. B., and Eaton, G. P., eds., Cenozoic tectonics and regional geophysics of the Western Cordillera: Geological Society of America Memoir 152, p. 175–208.
Blackwell, D. D., and Baag, C., 1973, Heat flow in a "blind" geothermal area near Marysville, Montana: Geophysics, v. 38, p. 941–956.

Blackwell, D. D., Steele, J. L., and Brott, C. A., 1980, The terrain effect on terrestrial heat flow: Journal of Geophysical Research, v. 85, p. 4757–4772.

Blackwell, D. D., Bowen, R. G., Hall, D. A., Riccio, J., and Steele, J. L., 1982a, Heat flow, arc volcanism, and subduction in northern Oregon: Journal of Geophysical Research, v. 87, p. 8735–8754.

Blackwell, D. D., Murphy, C. F., and Steele, J. L., 1982b, Heat flow and geophysical log analysis for OMF-7A geothermal test well, Mount Hood, Oregon, *in* Priest, G. R., and Vogt, B. F., eds., Geology and geothermal resources of the Mount Hood area, Oregon: Oregon Department of Geology and Mineral Resources Special Paper 14, p. 47–56.

Bodell, J. M., and Chapman, D. S., 1982, Heat flow in the north-central Colorado Plateau: Journal of Geophysical Research, v. 87, p. 2869–2884.

Bookstrom, A. A., 1981, Tectonic setting and generation of Rocky Mountain porphyry molybdenum deposits, *in* Dickinson, W. R., and Payne, W. D., eds., Relations of tectonics to ore deposits in the Southern Cordillera: Tucson, Arizona, Arizona Geological Society Digest Volume XIV, p. 215–226.

Brinkworth, G., Behrendt, J. C., and Popenoe, P., 1968, Geophysical investigations in the Colorado mineral belt [abs.]: American Geophysical Union Transactions, v. 49, p. 668.

Brott, C. A., Blackwell, D. D., and Mitchell, J. C., 1978, Tectonic implications of the heat flow of the western Snake River Plain, Idaho: Geological Society of America Bulletin, v. 89, p. 1697–1707.

Brott, C. A., Blackwell, D. D., and Ziagos, J. P., 1981, Thermal and tectonic implications of heat flow in the eastern Snake River Plain, Idaho: Journal of Geophysical Research, v. 86, p. 11709–11734.

Bryant, B., and Naeser, C. W., 1980, The significance of fission-track ages of apatite in relation to the tectonic history of the Front and Sawatch Ranges, Colorado: Geological Society of America Bulletin, v. 91, p. 156–164.

Buelow, K. L., 1980, Geothermal studies in Wyoming and northern Colorado, with a geophysical model of the Southern Rocky Mountains near the Colorado-Wyoming border [M. Sc. thesis]: Laramie, Wyoming, University of Wyoming, 150 p., 1 pl.

Bullard, E. C., 1938, The disturbance of the temperature gradient in the earth's crust by inequalities of height: Royal Astronomical Society Geophysical Supplement Monthly Notices, v. 4, p. 360–362.

Camfield, P. A., Gough, D. I., and Porath, H., 1971, Magnetometer array studies of the northwestern United States and southwestern Canada: Royal Astronomical Society Geophysical Journal, v. 22, p. 201–221.

Chapin, C. E., 1979, Evolution of the Rio Grande rift—A summary, *in* Riecker, R. E., ed., Rio Grande rift: Tectonics and magmatism: Washington, D.C., American Geophysical Union, p. 1–5.

Chapman, D. S., Clement, M. D., and Mase, C. W., 1981, Thermal regime of the Escalante Desert, Utah, with an analysis of the Newcastle geothermal system: Journal of Geophysical Research, v. 86, p. 11375–11746.

Chapman, D. S., Keho, T. H., Bauer, M. S., and Picard, M. D., 1984, Heat flow in the Uinta Basin determined from bottom hole temperature (BHT) data: Geophysics, v. 49, no. 4, p. 453–466.

Christiansen, E. H., Sheridan, M. F., and Burt, D. M., 1986, The geology and geochemistry of Cenozoic topaz rhyolites from the western United States: Geological Society of America Special Paper 205, 82 p.

Christiansen, E. H., Burt, D. M., Sheridan, M. F., and Wilson, R. T., 1983, The petrogenesis of topaz rhyolites from the western United States: Contributions to Mineralogy and Petrology, v. 83, p. 16–30.

Christiansen, R. L., and Lipman, P. W., 1972, Cenozoic volcanism and plate tectonic evolution of the western United States II. Late Cenozoic: Royal Society of London Philosophical Transactions, ser. A, v. 271, p. 249–284.

Clark, S. P., Jr., 1957, Heat flow at Grass Valley, California: American Geophysical Union Transactions, v. 38, p. 239–244.

——1961, Heat flow from the Austrian Alps: Royal Astronomical Society Geophysical Journal, v. 6, p. 54–63.

——1979, Thermal models of the Central Alps, *in* Jaeger, E., and Hunziker, J. C., eds., Lectures in isotope geology: Federal Republic of Germany, Springer-Verlag, p. 225–230.

Clark, S. P., Jr., and Jager, E., 1969, Denudation rate in the Alps from geochronologic and heat flow data: American Journal of Science, v. 267, p. 1143–1160.

Clark, S. P., Jr., and Niblett, E. R., 1956, Terrestrial heat flow in the Swiss Alps: Royal Astronomical Society Geophysical Supplement Monthly Notices, v. 7, p. 176–195.

Costain, J. K., and Wright, P. M., 1973, Heat flow at Spor Mountain, Jordan Valley, Bingham, and LaSal, Utah: Journal of Geophysical Research, v. 78, p. 8687–8698.

Cox, Allan, 1973, Plate tectonics and geomagnetic reversals; readings, selected, edited, and with introductions by Allan Cox: San Francisco, California, W. H. Freeman and Company, 702 p.

Cunningham, C. G., Naeser, C. W., and Marvin, R. F., 1977, New ages for intrusive rocks in the Colorado mineral belt: U.S. Geological Survey Open-File Report 77-573, 7 p.

Curtis, B. F., 1960, Major geologic features of Colorado, *in* Weimer, R. J., and Haun, J. D., eds., Guide to the geology of Colorado: Geological Society of America, Rocky Mountain Association of Geologists, and Colorado Scientific Society, p. 1–8.

Decker, E. R., 1969, Heat flow in Colorado and New Mexico: Journal of Geophysical Research, v. 74, p. 550–559.

——1973, Geothermal measurements by the University of Wyoming: Wyoming University Contributions to Geology, v. 12, p. 21–24.

——1987, Heat flow and basement radioactivity in Maine: First-order results and preliminary interpretations: Geophysical Research Letters, v. 14, p. 256–259.

Decker, E. R., and Birch, F., 1974, Basic heat-flow data from Colorado, Minnesota, New Mexico and Texas, *in* Sass, J. H., and Munroe, R. J., compilers, Basic heat flow data from the United States: U.S. Geological Survey Open-File Report 74-9, p. 5-1–5-60.

Decker, E. R., and Smithson, S. B., 1975, Heat flow and gravity interpretation across the Rio Grande rift in southern New Mexico and west Texas: Journal of Geophysical Research, v. 80, p. 2542–2552.

Decker, E. R., Baker, K. H., Bucher, G. J., and Heasler, H. P., 1980, Preliminary heat flow and radioactivity studies in Wyoming: Journal of Geophysical Research, v. 85, p. 311–321.

Decker, E. R., Bucher, G. J., Buelow, K. L., and Heasler, H. P., 1984, Preliminary interpretation of heat flow and radioactivity in the Rio Grande rift zone in central and northern Colorado, *in* Baldridge, W. S., Dickerson, P. W., Riecker, R. E., and Zidek, J., eds.: New Mexico Geological Society 35th Field Conference, Rio Grande Rift: Northern New Mexico, Guidebook, p. 45–50.

Diment, W. H., 1965, Thermal conductivity of serpentinite from Mayaguez, Puerto Rico, and some other localities, *in* A study of serpentinite: The AMSOC core hole near Mayaguez, Puerto Rico: NAS-NRC Publication 1188, p. 92–100.

Diment, W. H., and Robertson, E. C., 1963, Temperature, thermal conductivity, and heat flow in a drilled hole near Oak Ridge, Tennessee: Journal of Geophysical Research, v. 68, p. 5035–5047.

Diment, W. H., and Weaver, J. D., 1964, Subsurface temperatures and heat flow in the AMSOC core hole near Mayaguez, Puerto Rico, *in* A study of serpentinite: The AMSOC core hole near Mayaguez, Puerto Rico: NAS-NRC Publication 1188, p. 75–91.

Diment, W. H., and Werre, R. W., 1964, Terrestrial heat flow near Washington, D.C.: Journal of Geophysical Research, v. 69, p. 2143–2149.

Diment, W. H., Marine, I. W., Neiheisel, J., and Siple, G. E., 1965a, Subsurface temperature, thermal conductivity and heat flow near Aiken, South Carolina: Journal of Geophysical Research, v. 70, p. 5635–5644.

Diment, W. H., Raspet, R., Mayhew, M. A., and Werre, R. W., 1965b, Terrestrial heat flow near Alberta, Virginia: Journal of Geophysical Research, v. 70, p. 923–929.

Duebendorfer, R. M., and Houston, R. S., 1987, Proterozoic accretionary tectonics at the southern margin of the Archean Wyoming craton: Geological Society of America Bulletin, v. 98, p. 554–568.

Eaton, G. P., 1979, A plate-tectonic model for late Cenozoic crustal spreading in the western United States, *in* Riecker, R. E., ed., Rio Grande rift: Tectonics and magmatism: Washington, D.C., American Geophysical Union, p. 7–32.

——1980, Geophysical and geological characteristics of the crust of the Basin and Range province, *in* Continental tectonics: Washington, D.C., National Academy of Science, p. 96–113.

Edwards, C. L., Reiter, M., Shearer, C., and Young, W., 1978, Terrestrial heat flow and crustal radioactivity in northeastern New Mexico and southeastern Colorado: Geological Society of America Bulletin, v. 89, p. 1341–1350.

Eggler, D. H., 1987, Geochemistry of upper mantle and lower crust beneath Colorado and Wyoming: Geological Society of America Abstracts with Programs, v. 19, p. 272–273.

Eggleston, R. E., and Reiter, M., 1984, Terrestrial heat flow estimates from bottom-hole temperature data in the Colorado Plateau and the eastern Basin and Range province: Geological Society of America Bulletin, v. 95, p. 1027–1034.

Eichelberger, J. C., 1978, Andesitic volcanism and crustal evolution: Nature, v. 275, p. 21–27.

Eichelberger, J. C., and Gooley, R., 1977, Evolution of silicic magma chambers and their relationships to basaltic volcanism, *in* Heacock, J. G., ed., Keller, G. V., Oliver, J. E., and Simmons, G., assoc. eds., The earth's crust: Its nature and physical properties: Washington, D.C., American Geophysical Union Geophysical Monograph 20, p. 57–77.

England, P. C., Oxburgh, E. R., and Richardson, S. W., 1980, Heat refraction and heat production in and around granite plutons in north-east England: Royal Astronomical Society Geophysical Journal, v. 62, p. 439–455.

Fenneman, N. M., 1946, Physical divisions of the United States: map prepared in cooperation with the Physiographic Committee of the U.S. Geological Survey, scale 1:7,000,000.

Flurkey, A. J., Houston, R. S., Karlstrom, K. E., and Kratochvil, A. L., 1981, The geology of Archean and early Proterozoic terranes of the Sierra Madre, Wyoming, *in* Karlstrom, K. E., Houston, R. S. and others, A summary of the geology and uranium potential of Precambrian conglomerates in southeastern Wyoming: Bendix Field Engineering Corporation report DJBX-139-81, Volume 1, Part three, p. 401–512.

Furlong, K. P., and Chapman, D. S., 1987, Crustal heterogeneities and the thermal structure of the continental crust: Geophysical Research Letters, v. 14, p. 314–317.

Gosnold, W. D., Jr., 1987, Redistribution of U and Th in shallow plutonic environments: Geophysical Research Letters, v. 14, p. 291–294.

Gough, D. I., 1974, Electrical conductivity under western North America in relation to heat flow, seismology, and structure: Journal of Geomagnetism and Geoelectricity, v. 26, p. 105–123.

Hallin, J. S., 1973, Heat flow and radioactivity studies in Colorado and Utah, 1971–1972 [M.S. thesis]: Laramie, Wyoming, University of Wyoming, 108 p.

Herrin, E., and Clark, S. P., Jr., 1956, Heat flow in west Texas and eastern New Mexico: Geophysics, v. 21, no. 4, p. 1087–1099.

Hildreth, W., 1981, Gradients in silicic magma chambers: Implications for lithospheric magmatism: Journal of Geophysical Research, v. 86, p. 10153–10192.

Houston, R. S., 1963, Non-paleontological methods of correlation of rocks of Tertiary age in Wyoming, III, The petrographic calendar: Wyoming University Contributions to Geology, v. 3, p. 15–26.

——1969, Aspects of the geologic history of Wyoming related to formation of uranium deposits: Wyoming University Contributions to Geology, v. 8, p. 67–79.

Isaacson, L. B., and Smithson, S. B., 1976, Gravity anomalies and granite emplacement in west-central Colorado: Geological Society of America Bulletin, v. 87, p. 22–28.

Izett, G. A., 1975, Late Cenozoic sedimentation and deformation in northern Colorado and adjoining areas, *in* Curtis, B. F., ed., Cenozoic history of the Southern Rocky Mountains: Geological Society of America Memoir 144, p. 179–209.

Jackson, W. H., and Pakiser, L. C., 1965, Seismic study of crustal structure in the Southern Rocky Mountains: U.S. Geological Survey Professional Paper 525-D, p. D85–D92.

Jaeger, J. C., 1968, Cooling and solidification of igneous rocks, *in* Hess, H. H., and Poldervaart, A., eds., Basalts—The Poldervaart treatise on rocks of basaltic composition, Volume 2: New York, Interscience Publishers, p. 508–536.

Jaupart, C., 1983, Horizontal heat transfer due to radioactivity contrasts: Causes and consequences of the linear heat flow relation: Royal Astronomical Society Geophysical Journal, v. 75, p. 411–435.

Jaupart, C., Mann, J. R., and Simmons, G., 1982, A detailed study of the distribution of heat flow and radioactivity in New Hampshire (U.S.A.): Earth and Planetary Science Letters, v. 59, p. 262–287.

Jeffreys, H., 1938, The disturbance of the temperature gradient in the earth's crust by inequalities of height: Royal Astronomical Society Geophysical Supplement Monthly Notices, v. 4, p. 309–312.

Johnson, R. A., Karlstrom, K. E., Smithson, S. B., and Houston, R. S., 1984, Gravity profiles across the Cheyenne Belt, a Proterozoic suture in southeastern Wyoming: Journal of Geodynamics, v. 1, p. 445–472.

Karlstrom, K. E., and Houston, R. S., 1984, The Cheyenne Belt: Analysis of a Proterozoic suture in southern Wyoming: Precambrian Research, v. 25, p. 415–446.

Karlstrom, K. E., Houston, R. S., Coolidge, C. M., Flurkey, A. J., and Sever, C. K., 1981a, The geology of Archean and early Proterozoic terranes of the Medicine Bow Mountains, Wyoming, *in* Karlstrom, K. E., Houston, R. S., and others, A summary of the geology and uranium potential of Precambrian conglomerates in southeastern Wyoming: Bendix Field Engineering Corporation Report DJBX-139-81, Volume 1, Part two, p. 195–400.

Karlstrom, K. E., Houston, R. S., Schmidt, T. G., Inlow, D., Flurkey, A. J., Kratochvil, A. L., Coolidge, C. M., Sever, C. K., and Quimby, W. F., 1981b, Drill-hole data, drill-site geology, and geochemical data from the study of Precambrian uraniferous conglomerates of the Medicine Bow Mountains and Sierra Madre of southeastern Wyoming: Bendix Field Engineering Corporation Report DJBX-139-81, Volume 2, 682 p.

Karlstrom, K. E., Flurkey, A. J., and Houston, R. S., 1983, Stratigraphy and depositional setting of Proterozoic metasedimentary rocks in southeastern Wyoming: Record of an early Proterozoic Atlantic-type cratonic margin: Geological Society of America Bulletin, v. 94, p. 1257–1294.

Kelley, S. A., and Duncan, I. J., 1984, Tectonic history of the northern Rio Grande rift derived from apatite fission-track geochronology, *in* Baldridge, W. S., Dickerson, P. W., Riecker, R. E., and Zidek, J., eds., New Mexico Geological Society 35th Field Conference, Rio Grande Rift: Northern New Mexico, Guidebook, p. 67–73.

——1986, Late Cretaceous to middle Tertiary tectonic history of the northern Rio Grande rift, New Mexico: Journal of Geophysical Research, v. 91, p. 6246–6262.

King, P. B., and Beikman, H. M., 1974, compilers, Geologic map of the United States (exclusive of Alaska and Hawaii): Reston, Virginia, U.S. Geological Survey, scale 1:2,500,000.

Lachenbruch, A. H., 1968a, Rapid estimation of the topographic disturbances to superficial thermal gradients: Reviews of Geophysics, v. 6, p. 365–400.

——1968b, Preliminary geothermal model of the Sierra Nevada: Journal of Geophysical Research, v. 73, p. 6977–6989.

——1969, The effect of two-dimensional topography on superficial thermal gradients: U.S. Geological Survey Bulletin 1203-E, 86 p.

——1970, Crustal temperatures and heat production: Implications of the linear heat flow relation: Journal of Geophysical Research, v. 75, p. 3291–3300.

——1978, Heat flow in the Basin and Range province and thermal effects of tectonic extension: Pure and Applied Geophysics, v. 117, p. 34–50.

Lachenbruch, A. H., and Sass, J. H., 1977, Heat flow in the United States and the thermal regime of the crust, *in* Heacock, J. G., ed., Keller, G. V., Oliver, J. E., and Simmons, G., assoc. eds., The earth's crust: Its nature and physical properties: Washington, D.C., American Geophysical Union Geophysical Monograph 20, p. 626–675.

——1978, Models of an extending lithosphere and heat flow in the Basin and Range province, *in* Smith, R. B., and Eaton, G. P., eds., Cenozoic tectonics and regional geophysics of the western Cordillera: Geological Society of America Memoir 152, p. 209–250.

Lachenbruch, A. H., Sass, J. H., Munroe, R. J., and Moses, T. H., Jr., 1976, Geothermal setting and simple heat conduction models for the Long Valley Caldera: Journal of Geophysical Research, v. 81, p. 769–784.

Lambert, I. B., and Wyllie, P. J., 1972, Melting of gabbro (quartz eclogite) with excess water to 35 kilobars, with geological implications: Journal of Geology, v. 80, p. 693–708.

Larson, E. E., Ozima, M., and Bradley, W. C., 1975, Late Cenozoic basic volcanism in northwestern Colorado and its implications concerning tectonism and the origin of the Colorado River system, *in* Curtis, B. F., ed., Cenozoic history of the Southern Rocky Mountains: Geological Society of America Memoir 144, p. 155–178.

Lees, C. H., 1910, On the shape of isogeotherms under mountain ranges in radio-active districts: Royal Society of London Proceedings, v. A83, p. 339–346.

Lindsey, D. A., Johnson, B. R., and Andriessen, P.A.M., 1983, Laramide and Neogene structure of the Sangre de Cristo Range, south-central Colorado: Rocky Mountain Association of Geologists, Symposium, p. 219–228.

Lipman, P. W., 1980, Cenozoic volcanism in the western United States: Implications for continental tectonics, *in* Continental tectonics: Washington, D.C., National Academy of Science, p. 161–174.

——— 1983, The Miocene Questa caldera, northern New Mexico: Relation to batholith emplacement and associated molybdenum mineralization, *in* The genesis of Rocky Mountain ore deposits: Changes with time and teconics: Denver, Colorado, Denver Region Exploration Geologists Society Symposium Proceedings, p. 133–148.

Lipman, P. W., and Mehnert, H. H., 1975, Late Cenozoic basaltic volcanism and development of the Rio Grande depresson in the Southern Rocky Mountains, *in* Curtis, B. F., ed., Cenozoic history of the Southern Rocky Mountains: Geological Society of America Memoir 144, p. 119–154.

Lipman, P. W., Steven, T. A., and Mehnert, H. H., 1970, Volcanic history of the San Juan Mountains, Colorado, as indicated by potassium-argon dating: Geological Society of America Bulletin, v. 81, p. 2329–2352.

Lipman, P. W., Prostka, H. J., and Christiansen, R. L., 1971, Evolving subduction zones in the western United States, as interpreted from igneous rocks: Science, v. 174, p. 821–825.

——— 1972, Cenozoic volcanism and plate-tectonic evolution of the western United States I. Early and middle Cenozoic: Royal Society London Philosophical Transactions, ser. A, v. 271, p. 217–248.

Lipman, P. W., Doe, B. R., Hedge, C. E., and Steven, T. A., 1978, Petrologic evolution of the San Juan volcanic field, southwestern Colorado: Pb and Sr isotope evidence: Geological Society of America Bulletin, v. 89, p. 59–82.

Lipman, P. W., Mehnert, H. H., and Naeser, C. W., 1986, Evolution of the Latir volcanic field, northern New Mexico, and its relation to the Rio Grande rift, as indicated by potassium-argon and fission track dating: Journal of Geophysical Research, v. 91, p. 6329–6345.

McDowell, F. W., 1971, K-Ar ages of igneous rocks from the western United States: Isochron/West, no. 2, p. 1–16.

McKenzie, D. P., 1978, Some remarks on the development of sedimentary basins: Earth and Planetary Science Letters, v. 40, p. 25–32.

Medlin, W. E., 1983, Modeling local thermal anomalies: Constraints from conductivity gravity and heat flow [M. Sc. thesis]: Laramie, Wyoming, University of Wyoming, 142 p.

Morgan, P., 1983, Constraints on rift thermal processes from heat flow and uplift: Tectonophysics, v. 94, p. 277–298.

——— 1985, Crustal radiogenic heat production and the selective survival of ancient continental crust: Journal of Geophysical Research, v. 90, supp., p. C561–C570.

Morgan, P., and Golombek, M. P., 1984, Factors controlling the phases and styles of extension in the northern Rio Grande rift, *in* Baldridge, W. S., Dickerson, P. W., Riecker, R. E., and Zidek, J., eds., New Mexico Geological Society 35th Field Conference, Rio Grande Rift: Northern New Mexico, Guidebook, p. 13–19.

Morgan, P., Seager, W. R., and Golombek, M. P., 1986, Cenozoic thermal, mechanical and tectonic evolution of the Rio Grande rift: Journal of Geophysical Research, v. 91, p. 6263–6276.

Nielsen, S. B., 1987, Steady state heat flow in a random medium and the linear heat flow–heat production relationship: Geophysical Research Letters, v. 14, p. 318–321.

Obradovich, J. D., Mutschler, F. E., and Bryant, B., 1969, Potassium-argon ages bearing on the igneous and tectonic history of the Elk Mountains and vicinity, Colorado—A preliminary report: Geological Society of America Bulletin, v. 80, p. 1749–1756.

Pearson, R. C., Tweto, O., Stern, T. W., and Thomas, H. W., 1962, Age of Laramide porphyries near Leadville, Colorado: U.S. Geological Survey Professional Paper 450-C, p. C78–C80.

Phair, G., and Gottfried, D., 1964, The Colorado Front Range, Colorado, U.S.A., as a uranium and thorium province, *in* Adams, J.A.S., and Lowder, W. M., eds., The natural radiation environment: Chicago, Illinois, The University of Chicago Press, p. 7–38.

Phair, G., and Jenkins, L. B., 1975, Tabulation of uranium and thorium data on the Mesozoic-Cenozoic intrusive rocks of known chemical composition in Colorado: U.S. Geological Survey Open-File Report 75-501, 98 p.

Plouff, D., and Pakiser, L. C., 1972, Gravity study of the San Juan Mountains, Colorado: U.S. Geological Survey Professional Paper 800-B, p. B183–B190.

Porath, H., 1971, Magnetic variation and seismic low-velocity zone in the western United States: Journal of Geophysical Research, v. 76, p. 2643–2648.

Porath, H., Gough, D. I., and Camfield, P. A., 1971, Conductive structures in the northwestern United States and southwest Canada: Royal Astronomical Society Geophysical Journal, v. 23, p. 387–398.

Prodehl, C., and Lipman, P. W., 1988, Crustal structure of the Rocky Mountain region, *in* Pakiser, L., and Mooney, W. D., eds., Geophysical framework of the continental United States: Geological Society of America Special Paper (in press).

Prodehl, C., and Pakiser, L. C., 1980, Crustal structure of the southern Rocky Mountains from seismic measurements: Geological Society of America Bulletin, v. 91, p. 147–155.

Qureshy, M. N., 1960, Airy-Heiskanen anomaly map of Colorado, *in* Weimer, R. J., and Haun, J. D., eds., Guide to the geology of Colorado: Geological Society of America, Rocky Mountain Association of Geologists, and Colorado Scientific Society, p. 8–9.

Ratcliffe, E. H., 1959, Thermal conductivities of fused and crystalline quartz: British Journal of Applied Physics, v. 10, p. 22–25.

Reiter, M., and Clarkson, G., 1983, Geothermal studies in the San Juan Basin and the Four Corners area of the Colorado Plateau II. Steady-state models of the thermal source of the San Juan volcanic field: Tectonophysics, v. 93, p. 253–269.

Reiter, M., and Mansure, A. J., 1983, Geothermal studies in the San Juan Basin and the Four Corners area of the Colorado Plateau I. Terrestrial heat flow measurements: Tectonophysics, v. 93, p. 233–251.

Reiter, M., Edwards, C. L., Hartman, H., and Weidman, C., 1975, Terrestrial heat flow along the Rio Grande rift, New Mexico and southern Colorado: Geological Society of America Bulletin, v. 86, p. 811–818.

Reiter, M., Mansure, A. J., and Shearer, C., 1979, Geothermal characteristics of the Rio Grande rift within the southern Rocky Mountain complex, *in* Riecker, R. E., ed., Rio Grande rift: Tectonics and magmatism: Washington, D.C., American Geophysical Union, p. 253–267.

Richardson, S. W., and Oxburgh, E. R., 1978, Heat flow, radiogenic heat production and crustal temperatures in England and Wales: Geological Society of London Journal, v. 135, p. 323–337.

Roy, R. F., Blackwell, D. D., and Birch, F., 1968a, Heat generation of plutonic rocks and continental heat flow provinces: Earth and Planetary Science Letters, v. 5, p. 1–12.

Roy, R. F., Decker, E. R., Blackwell, D. D., and Birch, F., 1968b, Heat flow in the United States: Journal of Geophysical Research, v. 73, p. 5207–5221.

Roy, R. F., Blackwell, D. D., and Decker, E. R., 1972, Continental heat flow, *in* Robertson, E. C., ed., The nature of the solid earth: New York, McGraw-Hill, p. 506–543.

Sass, J. H., Lachenbruch, A. H., and Munroe, R. J., 1971a, Thermal conductivity of rocks from measurements on fragments and its application to heat flow determinations: Journal of Geophysical Research, v. 76, p. 3391–3401.

Sass, J. H., Lachenbruch, A. H., Munroe, R. J., Greene, G. W., and Mosses, T. H., Jr., 1971b, Heat flow in the western United States: Journal of Geophysical Research, v. 76, p. 6376–6412.

Sass, J. H., Lawver, L. A., and Munroe, R. J., 1985, A heat-flow reconnaissance of southeastern Alaska: Canadian Journal of Earth Sciences, v. 22, p. 416–421.

Scott, G. R., 1975, Cenozoic surfaces and deposits in the Southern Rocky Mountains, *in* Curtis, B. F., ed., Cenozoic history of the Southern Rocky Mountains: Geological Society of America Memoir 144, p. 227–248.

Simmons, G., 1967, Interpretation of heat flow anomalies. I. Contrasts in heat production: Reviews of Geophysics, v. 5, p. 43–52.

Smithson, S. B., and Decker, E. R., 1974, A continental crustal model and its geothermal implications: Earth and Planetary Science Letters, v. 22, p. 215–225.

Spicer, H. C., 1964, Geothermal gradients and heat flow in the Salt Valley anticline, Utah: Bollettino di Geofisica Teorica Applicata, v. 6, p. 263–282.

Steele, J. L., and Blackwell, D. D., 1982, Heat flow in the vicinity of the Mount Hood Volcano, Oregon, *in* Priest, G. R., and Vogt, B. F., eds., Geology and geothermal resources of the Mount Hood area, Oregon: Oregon Department of Geology and Mineral Resources Special Paper 14, p. 31–42.

Steven, T. A., 1960, Geology and fluorspar deposits, Northgate district, Colorado: U.S. Geological Survey Bulletin 1082-F, p. 323–422.

——— 1975, Middle Tertiary volcanic field in the Southern Rocky Mountains, *in* Curtis, B. F., ed., Cenozoic history of the Southern Rocky Mountains: Geological Society of America Memoir 144, p. 75–94.

Taylor, R. B., 1975, Neogene tectonisms in south-central Colorado, *in* Curtis, B. F., ed., Cenozoic history of the Southern Rocky Mountains: Geological Society of America Memoir 144, p. 211–226.

Thomas, J. A., and Galey, J. T., Jr., 1982, Exploration of the Mt. Emmons molybdenite deposits, Gunnison County, Colorado: Economic Geology, v. 77, p. 1085–1104.

Trimble, D. E., 1980, Cenozoic tectonic history of the Great Plains contrasted with that of the Southern Rocky Mountains: A synthesis: The Mountain Geologist, v. 17, p. 59–69.

Tweto, O., 1968, Leadville district, Colorado, *in* Ridge, J. D., ed., Ore deposits of the United States, 1933–1967 (Graton-Sales Volume): New York, American Institute of Mining and Metallurgical and Petroleum Engineers, p. 681–705.

——— 1975, Laramide (Late Cretaceous–early Tertiary) orogeny in the southern Rocky Mountains, *in* Curtis, B. F., ed., Cenozoic history of the Southern Rocky Moutnains: Geological Society of America Memoir 144, p. 1–44.

——— 1977, Tectonic map of the Rio Grande rift system in Colorado: U.S. Geological Survey Open-File Report 77-750.

——— 1979, The Rio Grande rift system in Colorado, *in* Riecker, R. E., ed., Rio Grande rift: Tectonics and magmatism: Washington, D.C., American Geophysican Union, p. 33–56.

Tweto, O., and Case, J. E., 1972, Gravity and magnetic features as related to geology in the Leadville 30-minute quadrangle, Colorado: U.S. Geological Survey Professional Paper 726-C, 31 p.

Tweto, O., and Sims, P. K., 1963, Precambrian ancestry of the Colorado mineral belt: Geological Society of America Bulletin, v. 74, p. 991–1014.

Vasseur, G., and Singh, R. N., 1986, The effects of random horizontal variations in radiogenic heat source distribution on its relationship with heat flow: Journal of Geophysical Research, v. 91, p. 10397–10404.

Walsh, J. B., and Decker, E. R., 1966, Effect of pressure and saturating fluid on the thermal conductivity of compact rock: Journal of Geophysical Research, v. 71, p. 3053–3061.

Warner, L. A., 1978, The Colorado Lineament: A middle Precambrian wrench fault system: Geological Society of America Bulletin, v. 89, p. 161–171.

Warren, R. E., Sclater, J. G., Vacquier, V., and Roy, R., 1969, Comparison of terrestrial heat flow and transient magnetic fluctuations in the southwestern United States: Geophysics, v. 34, p. 463–478.

White, W. H., Bookstrom, A. A., Kamilli, R. J., Ganster, M. W., Smith, R. P., Ranta, D. E., and Steininger, R. C., 1981, Character and origin of Climax-type molybdenum deposits, *in* Skinner, B. J., ed., Economic Geology Seventy-Fifth Anniversary Volume 1905–1980: El Paso, Texas, The Economic Geology Publishing Company, p. 270–316.

Wollenberg, H. A., and Smith, A. R., 1964, Studies in terrestrial γ radiation, *in* Adams, J.A.S., and Lowder, W. M., eds., The natural radiation environment: Chicago, Illinois, The University of Chicago Press, p. 513–566.

Wyllie, P. J., 1971, Experimental limits for melting in the earth's crust and upper mantle, *in* Heacock, J. G., ed., The structure and physical properties of the earth's crust: Washington, D.C., American Geophysical Union Geophysical Monograph 14, p. 279–301.

——— 1981, Magma sources in Cordilleran settings, *in* Dickinson, W. R., and Payne, W. D., eds., Relations of tectonics to ore deposits in the Southern Cordillera: Tucson, Arizona, Arizona Geological Society Digest Volume XIV, p. 39–48.

Yoder, H. S., and Tilley, C. E., 1962, Origin of basalt magmas: An experimental study of natural and synthetic rock systems: Journal of Petrology, v. 3, p. 342–539.

Ziagos, J. P., Blackwell, D. D., and Mooser, F., 1985, Heat flow in southern Mexico and the thermal effects of subduction: Journal of Geophysical Research, v. 90, p. 5410–5420.

Manuscript Received by the Society March 28, 1988
Revised Manuscript Received September 13, 1988
Manuscript Accepted September 13, 1988

Progress in understanding jointing over the past century

CENTENNIAL ARTICLE

DAVID D. POLLARD *Departments of Applied Earth Sciences and Geology, Stanford University, Stanford, California 94305*
ATILLA AYDIN *Department of Earth and Atmospheric Sciences, Purdue University, West Lafayette, Indiana 47907*

ABSTRACT

Joints are the most common result of brittle fracture of rock in the Earth's crust. They control the physiography of many spectacular landforms and play an important role in the transport of fluids. In its first century, the *Geological Society of America Bulletin* has published a significant number of papers on joints and jointing. One hundred years ago, there were lively debates in the literature about the origin of joints, and detailed descriptions of joints near the turn of the century catalogued most geometric features that we recognize on joints today. In the 1920s, theories relating joint orientation to the tectonic stress field and to other geologic structures led to a proliferation of data on the strike and dip of joints in different regions. The gathering of orientation data dominated work on joints for the next 50 yr. In the 1960s, key papers re-established the need to document surface textures, determine age relations, and measure relative displacements across joints in order to interpret their origins. At about this time, fundamental relationships from the fields of continuum and fracture mechanics were first used to understand the process of jointing. In the past two decades, we have witnessed an effort to use field data to interpret the kinematics of jointing and to understand the initiation, propagation, interaction, and termination of joints. Theoretical methods have been developed to study the evolution of joint sets and the mechanical response of a jointed rock mass to tectonic loading. Although many interesting problems remain to be explored, a sound conceptual and theoretical framework is now available to guide research into the next century.

INTRODUCTION

Joints are the most ubiquitous structure in the Earth's crust, occurring in a wide variety of rock types and tectonic environments. They profoundly affect the physiography of the Earth's surface by controlling the shapes of coastlines (Hobbs, 1904; Nilson, 1973), drainage systems (Daubree, 1879; Hobbs, 1905), lakes (Tarr, 1894; Plafker, 1964), and continental lineaments (Nur, 1982). Many spectacular natural sites in America (for example, Devils Postpile of California, Devils Tower of Wyoming, Zion and Arches National Parks of Utah, and the Finger Lakes of New York), enjoyed by millions of people every year, owe their unique form to joints. One of the early devotees of joints (King, 1875, p. 616) put it this way:

An admirer of Nature may be excused becoming enraptured when he takes a view from any of these noble terraces (in County Clare, Ireland). Looking north, or south, his eyes are riveted on vast surfaces of gray limestone rocks, split up to an extent, and with a regularity of direction, truly wonderful. . . .The observer becomes so absorbed with the scene that he unconsciously begins to feel as if the rocks under and around him were in process of being illimitably cleft from north to south—as if the earth's crust were in course of splitting up from one pole to the other; and he only rids himself of the feeling to become bewildered with the question, as to what mysterious agent produced the singular phenomenon he is contemplating.

Today there is no doubt that joints reveal rock strain accommodated by brittle fracture. Establishment of reliable relationships between joints and their causes can provide the structural geologist with important tools for inferring the state of stress and the mechanical behavior of rock. By determining the relative ages of joints and other structures, different phases of brittle deformation can be mapped through geologic time, contributing to efforts to understand the structural history of the Earth's crust.

There is also a very practical side to the study of joints. They influence mineral deposition by guiding ore-forming fluids, and they provide fracture permeability for water, magma, geothermal fluids, oil, and gas. Because joints may significantly affect rock deformability and fluid transport, they are carefully considered by engineering geologists in the design of large structures, including highways, bridges, dams, power plants, tunnels, and nuclear-waste repositories. Two of the three sites (Yucca Mountain, Nevada; and Hanford, Washington) selected as candidates for the nation's nuclear-waste repository are extensively jointed. Knowledge of the spacing, orientation, aperture, and connectivity of joints at these sites is crucial for construction of the repository and for prediction of the long-term behavior of ground-water flow.

Partly because of their practical importance and partly because of the mystery behind their knife-edge sharpness and intriguing patterns, joints have been the subject of many scientific papers (10,933 citations were listed under joint, joints, and jointing in GEOREF for the period 1785 to 1987). We were asked to review papers on joint formation published in the *Geological Society of America Bulletin* over its first one hundred years. The literature that we review goes beyond the *Bulletin*, because the contributions of the *Bulletin* papers are best appreciated in a broader context. On the other hand, our scope is narrowed by our own predilections, the availability of other review articles (Friedman, 1975; Kranz, 1983; Engelder, 1987; Pollard and Segall, 1987), and the constraints of time and pages. We focus on the geometric characteristics of joints and the mechanics of jointing.

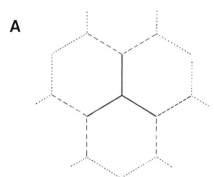
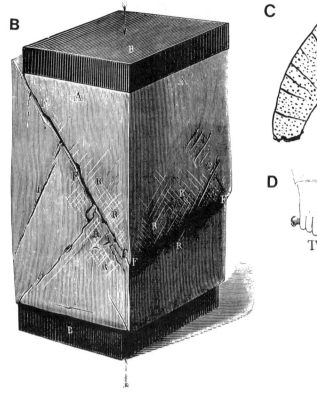

Figure 1. A. Drawing of "primary tension cracks" from Becker (1893). B. Drawing of an experiment demonstrating two sets of fractures in beeswax under uniaxial compression (Daubree, 1879). C. Drawing of "radial cracks due to tension in sharply flexed stratum" from Van Hise (1896). D. Drawing of an experiment demonstrating two sets of tensile fractures in a glass plate under torsion (Daubree, 1879).

Rather than merely recite the accomplishments of the past one hundred years, we attempt to critique this work and to identify the most insightful concepts and fruitful methods of inquiry. Our goal is to offer a beacon for the course of future research on joints and jointing.

HISTORICAL PERSPECTIVE

The Geological Society of America began publishing the *Bulletin* at a time when there was little consensus about the origins of joints. A lively discussion on this topic was started by Gilbert's note (1882a) on a rectangular drainage system, apparently controlled by two joint sets, in post-Glacial sediments of the Great Salt Lake Desert, Utah. The presence of joints in young sediments was surprising news because many geologists believed that joints in sedimentary rocks were of mechanical origin and were associated with tectonic deformation after lithification. Gilbert stirred a controversy by asserting that no "satisfactory explanation has ever been given of the origin of the jointed structure in rocks," (p. 27) and he had none to propose. LeConte (1882) suggested that Gilbert's structures were contraction or shrinkage joints caused by desiccation. Gilbert (1882b) argued against this origin by noting that the joints in the young sediments differed from typical shrinkage cracks by their parallel arrangement and crosscutting geometry. He was unsatisfied with King's (1875) theory of joint origin based on magnetic forces, and he considered a shearing force origin to be an "absolutely baseless hypothesis" (1882b, p. 53). Earthquakes or cooling due to erosion and uplift were invoked by Crosby (1882) as possible causes of parallel joints, and McGee (1883) suggested that incipient slaty cleavage developed into joints upon denudation, cooling, and desiccation. Gilbert (1884) discounted the relationship between slaty cleavage and joints, but he was intrigued by Crosby's earthquake hypothesis. These geologists overlooked Hopkins' (1835, 1841) insightful concept of joint origin. Working about half a century earlier in Great Britain, Hopkins interpreted joints as discontinuities caused by tensile stresses, and he attributed the tension to uplift.

The first major paper on jointing in the *Bulletin* (Becker, 1893) was motivated by the observation that rock exposures in the Sierra Nevada are nearly always broken by joints, fissures, and faults (Becker, 1891). Becker argued that orogeny can never be discussed in a meaningful way until the mechanical significance of joints and faults is understood. To this end, he presented a remarkably complete treatment of finite strain analysis, anticipating the popularity of this technique among structural geologists of the past few decades (Ramsay, 1967). According to Becker, the divisional surfaces in rocks were caused by tensile or compressive stresses. He cited cooling cracks (Fig. 1A) as good examples of tensile fractures because they clearly gape when first formed. Other divisional surfaces, which geologists of the time referred to as "joints," displayed small shear displacements. Becker suggested that two sets of fractures would form in planes of maximum shear strain, oblique to the direction of maximum compression. He supported this thesis by citing Daubree's (1879) experiments in which two sets of fractures formed in prisms of beeswax and resin subjected to uniaxial compression (Fig. 1B). In a later publication, a shear origin of joints was emphasized by Becker (1895), and slickensides were used as an important genetic indicator of jointing.

In a study of North American Precambrian geology, Van Hise (1896) also ascribed the origin of joints to tension or compression. Possible mechanisms for tensile joints, however, were broadened to include tension produced by folding (Fig. 1C). Van Hise asserted that joints intersected bedding at right angles and formed at right angles to the tensile forces in the outer arc of folds. Two orthogonal sets, called "strike and dip joints," formed in doubly plunging folds. Warping in torsion, illustrated by

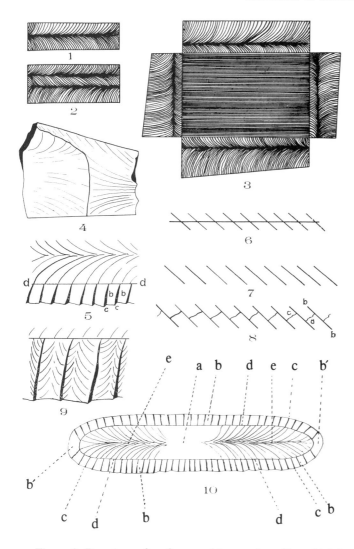

Figure 2. Drawings of surface markings and profiles of joints from pelitic rocks of the Mystic River region, Massachusetts, by Woodworth (1893): hackle and plume structure (Parts 1–4); joint fringe region with echelon joint segments (Part 5); profiles through echelon segments (Parts 6–8); plumose patterns on fringe segments (Part 9); and the entire joint surface (Part 10).

He recognized plumose structures on fractures that opened within the pelitic rocks of the Mystic River near Somerville, Massachusetts. Woodworth differentiated plumose structures from slickensides and, contrary to Becker (1893, 1895), interpreted slickensides as secondary features caused by later shear displacement on some of the joint planes. Most structural details that we recognize today on joint surfaces were carefully illustrated by Woodworth (Fig. 2) and interpreted as inherent characteristics of fractures that open. Woodworth described joint initiation at a point and outward propagation to form plumose structures. In a remarkable leap of scale, he applied the discontinuous nature of single joint surfaces to large-scale structures such as lines of volcanoes in the central Pacific and circum-Pacific. For a modern version of this application, see Pollard and Aydin (1984).

The first female American structural geologist to publish on joints was Sheldon (1912a, 1912b), who worked in the Finger Lakes region of upstate New York. Unlike many of her predecessors, she gathered abundant field data on joint orientations (3,046 measurements) and illustrated her reports with instructive photographs and maps. On the basis of the geometric relations between joints, faults, and regional folds, she suggested that the joints formed during the early period of Appalachian folding. Sheldon reviewed the theories of joint origin and found that not all of her data were consistent with any single theory. Reflecting the uncertainty of her colleagues, she concluded that the conditions under which joints formed probably included translation, rotation, compression, pure shear, torsion, and earthquake shock. Sheldon proposed that more detailed field and experimental work was required to understand the formation of joints.

James (1920) recorded detailed observations on joint surface features in lava flows. He attributed the horizontal bands on column faces to incremental growth of the joints, and suggested that columns were produced by cooling from both the top and bottom surfaces of a lava flow. James explained that columns in the rapidly cooled upper part of a flow are long and thin, whereas columns in the slowly cooled lower part are short and thick.

Balk (1925) introduced readers of the *Bulletin* to the work of H. Cloos, in which rock fabrics (oriented minerals, xenoliths, and schlieren) are used to infer the flow, shear, and stretching directions in igneous plutons. Fractures (veins, joints, and dikes) are related to these fabrics through their relative orientations (Fig. 3). According to Balk (1937), steeply dipping joints that are perpendicular or parallel to flow lines (cross joints or longitudinal joints, respectively) are tensile fractures. Steep joints that strike at angles of about 45° to flow lines (diagonal joints)

another of Daubrée's (1879) experiments (Fig. 1D), was cited by Van Hise as a special case of his folding hypothesis. He recognized that two sets of tensile fractures formed in the experiments, and he related these to orthogonal joint sets in folded terranes. In the noteworthy appendix to Van Hise's paper, Hoskins (1896) defined the relationship between fluid pressure in cavities and the applied stresses, and he recognized the importance of the principal stress difference for determining fracture and flow of rock. He suggested that tension could produce parallel planes of separation (joints) of considerable regularity. Hoskins noted that rock cannot separate along planes of maximum compression, and so what some geologists had referred to as joints formed by compression were shear fractures inclined to the direction of maximum compression.

None of the papers cited above included detailed maps or field descriptions of joints and associated structures. Indeed, the pertinent observations were made in one case by a field assistant (Gilbert, 1882a) and in another by the geologist's son (LeConte, 1882). In contrast, a very careful description of joint surfaces was published by Woodworth (1896).

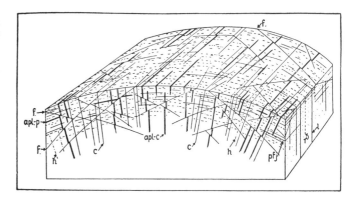

Figure 3. Drawing of "primary structure elements and directions of parting" in the Strehlen massif from H. Cloos as redrawn by Balk (1937). Classes of fractures include cross joints (c), longitudinal joints (l), and primary flat-lying joints (pfl).

are interpreted as shear fractures. Horizontal joints (flat-lying joints) apparently are related to volume change of the pluton. This geometric fracture model (Fig. 3) formed the basis for interpretations of joints in many plutons (for example, Balk and Grout, 1934; Grout and Balk, 1934; Hutchinson, 1956).

Bucher (1920, 1921) observed two sets of conjugate fractures in the crest of a small anticline in Kentucky and, without documenting slickensides or measuring offset, he asserted that "there could be little doubt that these joints represented planes of shearing" (p. 708). To justify this interpretation, Bucher compared the conjugate pattern with that of shear fractures produced in laboratory experiments on rocks and steel. Theories of shear fracture for brittle materials (Hartmann, 1896; Mohr, 1900) provided Bucher with a mathematical rationale for the two conjugate sets and led him to infer that the largest compressive stress at the time of fracture bisected the acute angle, whereas the least compressive stress bisected the obtuse angle. Bucher reinterpreted the experiments of Daubree (1879) and the field observations of Hobbs (1905) and Sheldon (1912a), concluding that his concept of shear fracturing was the appropriate mechanism.

About this time, Griffith (1921, 1924) laid down the experimental and theoretical framework of modern fracture mechanics by developing concepts that explicitly accounted for stress concentration at fracture tips and conservation of energy. Geologists found the approaches of Bucher and Balk more appealing, however, and these formed the basis of most work on jointing for the next half-century. The conjugate pattern of fractures, or the geometric relationship of fractures to other structures, was used to interpret the origin of joints. Fracture surface textures, displacements across fractures, and the relative ages of joint sets as emphasized by Woodworth (1896), Swanson (1927), and Sheldon (1912a), respectively, were de-emphasized or neglected, and many joint studies relied primarily on geometric pattern recognition. Accordingly, the field methods for measuring joint orientations and the presentation and statistical analysis of these data were a major concern (Pincus, 1951; Reches, 1976; Wise, 1982). Pincus (1951, p. 127) quoted Muller (1933) as suggesting that the best method to ensure unbiased sampling was to measure "completely blindly." Over the years, the number of joint-orientation measurements in particular studies increased with unreserved enthusiasm from about 4,000 (Melton, 1929), to 6,798 (Parker, 1942), to an apparent all-time record of 25,000 (Spencer, 1959).

During this period, opinions were divided on the interpretation of joint-orientation data. Spencer (1959), working in the Beartooth Mountains of Montana and Wyoming, lamented that "unique solutions to the problem of the nature of stress fields responsible for the formation of fractures are rarely found" (p. 501). In the same vein, after measuring 4,787 joint orientations, Holst and Foote (1981) were unable to determine a specific mechanism and origin for joints in the Devonian rocks of the Michigan Basin. On the other hand, Stearns (1969) identified extension and shear fractures in the Teton anticlines of Montana and inferred their age relationships based on orientation data. He inferred the principal stress directions by grouping three fracture orientations into an assemblage of two conjugate shear fractures and one bisecting extension fracture and associated all of these fractures with one stress field. Friedman and Stearns (1971) concluded that "this assemblage is too ubiquitous in the field and the laboratory to doubt the geometric interpretation" (p. 3153).

Joints in Paleozoic rocks of the northern Appalachian Plateau of New York and Pennsylvania were featured in a *Bulletin* article by Parker (1942). Differing somewhat with Sheldon (1912a, 1912b), he interpreted the joints as older than the folds. Parker identified two conjugate joint sets intersecting at a small dihedral angle (averaging about 19°), and he interpreted these as shear fractures because of their vertical, clean-cut planes. Unusually small dihedral angles were rationalized by Muehlberger (1961) by using a Mohr fracture envelope that accounted for the pressure dependence of this angle. He realized that plumose structure on the joints reported by Parker might be indicative of extension fractures, and he cited slickensides, offset features, and gouge as field criteria for shear fractures. Muehlberger postulated a continuous transition from a single extension fracture to two sets of extension fractures of small dihedral angle to two sets of conjugate shear fractures. This opened the way for interpretation of paired fracture sets covering the entire range of angular relationships from 0° to 90°. Hoppin (1961) went so far as to interpret one set of fractures as the limiting case of shear fractures of small dihedral angle.

It is not surprising that confusion arose during this period regarding the interpretation of plumose structures, because they were reported on both what were believed to be shear fractures and tension fractures. This contradictory evidence led Roberts (1961) to conclude that plumose structures were "not indicative of either shearing or tensile stress, but demonstrate . . . a high rate of propagation" (p. 489). Hodgson (1961a) clarified the subject by returning to the original interpretations of Woodworth (1896). He described examples of large plumose structures, and he slightly revised Woodworth's classification. In a study of the regional jointing in the Comb Ridge–Navajo Mountain area of Arizona and Utah, Hodgson (1961b) contributed a set of instructive descriptions of joint trace patterns on bedding surfaces and cross sections. Curiously, Hodgson's (1961b) definition of a joint, "a fracture that traverses a rock and is not accompanied by any discernible displacement of one face of the fracture relative to the other" (p. 12) is not quite in line with Woodworth's insight that joints are fractures that gape so as to preserve the plumose structure. Beautiful illustrations of joint surface structures and their implications for joint initiation, propagation direction, and arrest are found in the publications of Bankwitz (1965, 1966, 1984). Techniques for studying and interpreting these structures are described by Lutton (1969), Kulander and others (1979), and Kulander and Dean (1985). Gramberg (1966) and Rummel (1987) have experimentally produced plumose structures in rock and plexiglass.

A new perspective on regional jointing was achieved by Nickelsen and Hough (1967), who worked on the Appalachian Plateau of Pennsylvania. They repudiated application of the conjugate shear fracture concept of jointing by stating that "unless there is proof of equal age of apparently conjugate joint sets, their angles of intersection and relations to other structures cannot be used as a basis of mechanical interpretation" (p. 624). Joints, they argued, are characterized by opening displacements perpendicular to the joint surfaces and by plumose structures. Nickelsen and Hough also introduced the concept of a fundamental joint system consisting of two unequal joint sets—a set of systematic continuous joints, as defined by Hodgson (1961a), and a set of nonsystematic joints approximately perpendicular to the first. They interpreted systematic joints as early fractures, formed independently of folds and faults, and nonsystematic joints as late release-type fractures formed during unloading and erosion. Engelder and Geiser (1980) used regional joint sets in the Appalachian Plateau of New York to infer the trajectories of the paleostress field. On the basis of the plumose structures, calcite filling, and lack of shear displacements, Engelder (1982a, 1982b) interpreted one set of these joints as extension fractures that formed in a principal stress plane associated with the Alleghanian orogeny. This set formed due to high fluid pressure in deeper undercompacted levels of the Catskill Delta, whereas another set is associated with shallow, normally compacted levels and apparently formed during unloading (Engelder, 1985; Engelder and Oertel, 1985).

Price (1959) related jointing to possible changes in the stress state with depth during unloading, and placed a limit of about 3 km on the depth of tensile joints in rock that has undergone only moderate tectonic

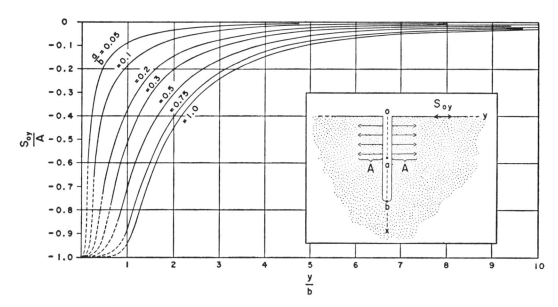

Figure 4. "Theoretical values of elastic stress relief due to a vertical tension crack" from Lachenbruch (1962). Plot of horizontal stress at the surface versus distance from the fracture for different loading distributions. Inset shows idealization of the fracture and defines symbols.

compression. This and other calculations (Anderson, 1951, p. 159; Hubbert, 1951) going back to Hoskins (1896) made geologists reluctant to accept the formation of tensile fractures at great depth. Secor (1965), however, used the concept of effective stress as outlined by Hubbert and Rubey (1959) to show that tensile fractures could form in an environment of compressive stress if the pore-fluid pressure was great enough. Later, these ideas were put on a firm theoretical foundation by Secor (1969), using the elastic solution for a crack loaded by internal fluid pressure (Sneddon, 1946). This removed the obstacle to interpreting joints formed at depth as tensile fractures.

In two remarkable papers, one published as a *Geological Society of America Special Paper*, Lachenbruch (1961, 1962) ushered in a new era of joint studies by interpreting his field observations in light of solutions for stress distributions around cracks (introduced by Inglis, 1913), the physics of crack propagation (developed by Griffith, 1921), and a modern approach to fracture mechanics (established by Irwin, 1957, 1958). He addressed the general problem of the depth and spacing of tension cracks growing down from the surface, and the particular problem of the mechanics of thermal contraction cracks in permafrost. Lachenbruch's analysis (Fig. 4) showed that the release of tensile stresses is confined mostly to a region whose horizontal dimension scales with joint depth, thus providing a rationale for the observed correspondence between joint depth and spacing.

During the past two decades, joint studies reported in the *Bulletin* have combined the principles of experimental and theoretical fracture mechanics with detailed observations of joint geometry. Following the work of Peck and Minakami (1968) on columnar joints in lava, Ryan and Sammis (1978) used cyclic fatigue concepts to infer that the striations on column faces result from incremental joint growth. DeGraff and Aydin (1987) determined the sequence and direction of fracture propagation on column faces from a detailed kinematic analysis of plumose structure. Pollard and others (1982) used predicted stress distributions around the tip of a joint under various loading conditions to explain breakdown into multiple echelon segments and curving patterns of joint traces. Joint sets in plutonic igneous rocks were mapped by Segall and Pollard (1983b), who documented opening displacements across the joints and estimated the stress conditions for their formation. In companion papers, Segall (1984a, 1984b) examined how, and at what rates, a vast array of joints can grow in the Earth's crust, although only a single major fracture is produced in typical laboratory tension experiments.

FUNDAMENTAL CONCEPTS—OLD AND NEW

Much of the confusion, apparent in our historical perspective, about the definition, formation, and interpretation of joints, can be attributed to an incomplete recording of pertinent geologic data and to the lack of a sound conceptual framework. Here we propose a conceptual framework, based on principles of fracture mechanics (Lawn and Wilshaw, 1975; Kanninen and Popelar, 1985), that admits the full range of complexity contained in geologic data. We point out shortcomings in some of the older concepts and suggest that these be abandoned.

Three basic physical characteristics of fractures in rock (Fig. 5A) are that (1) fractures have two parallel surfaces that meet at the fracture front; (2) these surfaces are approximately planar; and (3) the relative displacement of originally adjacent points across the fracture is small compared to fracture length (Pollard and Segall, 1987). Following the traditions of fracture mechanics (Irwin, 1958; Lawn and Wilshaw, 1975), we idealize a small portion of the front as a straight line (Fig. 5B) and identify the modes of fracture. Opening displacements (mode I) are perpendicular to the fracture surface; shearing displacements are parallel to the fracture surface, and are either perpendicular (mode II) or parallel (mode III) to the propagation front. Broadly speaking, joints are associated with the opening

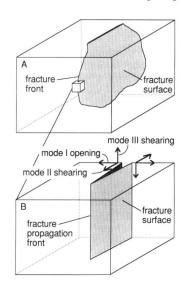

Figure 5. A. Idealization of a fracture in a rock mass illustrating two planar surfaces joined at the fracture front. B. Enlargement of a small region containing the front. Three displacement components are associated with Mode I (opening), Mode II (sliding perpendicular to the front), and Mode III (sliding parallel to the front).

mode, whereas faults are associated with the shearing modes. Because the mode may vary along the fracture front and may involve mixtures of modes I, II, and III, however, one should not be too categorical about these associations. Other geologic structures associated with opening-mode fracture are veins and igneous dikes.

We propose that the word "joint" be restricted to those fractures with field evidence for dominantly opening displacements. Because the displaced markers, surface textures, and fillings that could be diagnostic of relative displacements may be absent or not readily observed, the history of relative displacements and the fracture modes involved may not be determinable. Be that as it may, to attempt a geological interpretation of fractures without this information is a risky endeavor that should not be encouraged. Lacking this information, geologists should avoid speculative interpretations and refer to the structures simply as fractures.

Several popular concepts related to joints and jointing appear to be in conflict with those proposed above. For example, joints have been defined as fractures that show no discernible relative displacements (Hodgson, 1961b, p. 12; Price, 1966, p. 110; Ramsay and Huber, 1987, p. 641). Joints have also been defined as unopened extension fractures (Griggs and Handin, 1960, p. 351) and as discontinuities that show no relative movement between the two surfaces (Bles and Feuga, 1986, p. 73, 79). The small magnitude of relative displacement across many joints has motivated these definitions; however, they tend to obscure two fundamental facts. There must have been some relative displacement, or we are faced with the mechanical enigma of no stress concentration and no energy available for propagation (Pollard and Segall, 1987, p. 289–305). Even if the joint closed somewhat after formation, the existence of the two surfaces demonstrates that some relative displacement remains.

Some early workers (for example, Van Hise, 1896) used the term "tension fracture" for joints because they thought that joints opened in response to remote tensile stresses. By the term "remote," we simply mean at a distance that is large compared to fracture length. Theoretical arguments, however, indicate that internal fluid pressure can promote jointing deep in the Earth without remote tension (Secor, 1969), and laboratory experiments show that opening fractures can form even if the applied stress is compressive (Griggs and Handin, 1960, p. 348–351). Because laboratory samples extended perpendicular to the fracture plane, the term "extension fracture" was adopted for joints. The term "extension," however, is just as problematic as the term "tension." For example, in the case of joints caused by cooling or desiccation, the total strain component perpendicular to the joints is a contraction. Terminology based on the geometric relationship of joints to other structures, as in normal, cross, longitudinal, diagonal, ac, hkl, dip, strike, and so on (Billings, 1972, p. 146; Hancock, 1985; Dennis, 1987, p. 197–199), draws attention away from the fundamental characteristics of the fractures themselves and from direct determination of the fracture mode. We suggest that all of these adjectives should be dropped.

The concept of shear joints requires special attention because it is so well established in the literature. From Bucher (1921) to Scheidegger (1982) to Hancock (1985), geologists have identified joints that formed as shear fractures based on their conjugate pattern or geometric relationship to other geologic structures. In Nickelsen and Hough's (1967, 1969) discussion of Parker's (1942, 1969) data, in Engelder's (1982b) reply to Scheidegger's (1982) assertions, and in Barton's (1983) discussion of Muecke and Charlesworth's (1966) work, however, additional evidence indicates that opening fractures were misinterpreted as shear fractures. In other areas, there is good evidence for shear displacements and for the fact that the fractures formed predominantly by opening and were sheared at a later date (Segall and Pollard, 1983a). Because the shear displacements now dominate, these fractures are called faults. There are cases where all

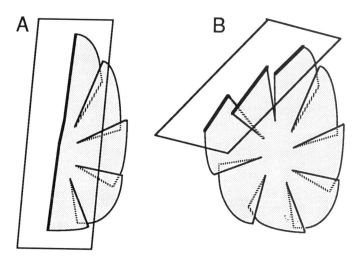

Figure 6. Joint trace geometries as a function of orientation of exposure surface. A. A nearly straight and continuous trace. B. A segmented, discontinuous trace.

fractures with small relative displacements are called joints, as on Kelley and Clinton's (1960) map of the Colorado Plateau. Some of these fractures formed with a small amount of shear displacement and thus originated as faults (Aydin, 1978), whereas others formed as joints and were sheared later (Dyer, 1983). As we have defined them, joints and faults are mechanically different because they have unique stress, strain, and displacement fields (Pollard and Segall, 1987). They can have different surface textures and fracture fillings, and they certainly accommodate tectonic strains differently. Because it needlessly combines these distinct fracture types, the concept of shear joints is sheer nonsense.

Failure envelopes on Mohr diagrams have been used by Bucher (1921), Muehlberger (1961), Hancock (1985), and many others to analyze and interpret joints. This tool is exploited in modern structural-geology textbooks to explain the origin of joints (Hobbs and others, 1976, p. 323–325; Davis, 1984, p. 351; Suppe, 1985, p. 190–196; Dennis, 1987, p. 232–236), but the inquiring student should ask: "Where's the joint?" Although a Mohr diagram is useful for representing a homogeneous stress or strain field and for providing an empirical failure criterion, it does not represent the heterogeneous fields associated with a joint. As a tool for studying joints, therefore, a Mohr diagram cannot further our understanding of the process of joint initiation, propagation, and arrest. The methods that explicitly treat the heterogeneous fields of fractures, as pioneered by Inglis (1913), Griffith (1921), and Irwin (1958), should replace this reliance on the Mohr diagram.

GEOMETRY OF JOINTS

In the following sections, we exploit the new conceptual framework as we review the geometry of joints and the kinematics and mechanics of jointing. A joint is a three-dimensional structure composed of two matching surfaces (joint faces), that are commonly idealized as smooth, continuous, and planar. However, virtually all joints have some roughness, minor and major discontinuities, and occasional curves and kinks. An observer commonly sees only the joint trace, whose geometry depends on the orientation and location of the exposure surface relative to the joint surface. For example, traces of the same joint may be nearly straight and continuous or segmented (Fig. 6). We will call the structure a single joint if continuous paths connect the segments and the parent joint surface.

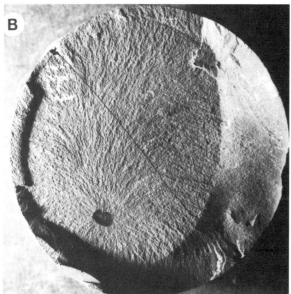
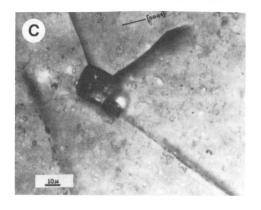

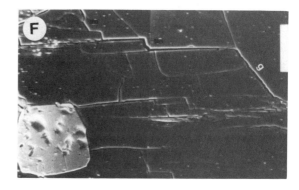

Figure 7. Joint initiations at various inhomogeneities. A. An elliptical hole in siltstone. B. A brachiopod fossil (from Kulander and others, 1979). C. An opaque mineral inclusion. Scale bar is 10 μm. (from Mitra, 1978). D. A synformal cusp at the bottom of a siltstone layer. E. An antiformal cusp nearby. F. Microcracks in experimentally deformed Westerly granite. Scale bar is 50 μm. (From Wong, 1982.)

Surface Morphology of a Single Joint

Joint surfaces are decorated by characteristic textures such as origins, hackles, and rib marks that record joint initiation as well as the direction and relative velocity of joint propagation. Joint origins commonly are marked by geometric or material inhomogeneities (Hodgson, 1961a; Wise, 1964; Kulander and others, 1979; Bahat and Engelder, 1984). Examples shown in Figure 7 are an elliptical cavity, a brachiopod fossil, an opaque mineral inclusion, and sole marks with sharp cusps. Equally common (Fig. 8A) are joint origins expressed by dimples without any obvious heterogeneity. Microcracks (Fig. 7F), which are believed to be common joint initiators, usually are too small to identify on natural joint surfaces.

Some of the most distinctive ornaments on joint surfaces are hackle marks (Figs. 7A, 7B, 7D, 7E, and 8), which are curvilinear boundaries with differential relief between adjacent surfaces. Hackles either radiate from the origin (Figs. 7A, 7B, and 8A) or fan away from a curvilinear axis (Figs. 8B–8D). The chevronlike pattern was named "feather structure" by Woodworth (1896), "plume" by Parker (1942), and "plumose" by Hodgson (1961b). We consider the collection of an origin, axis, and hackles to be a plumose structure, which may display a variety of forms. Asymmetric

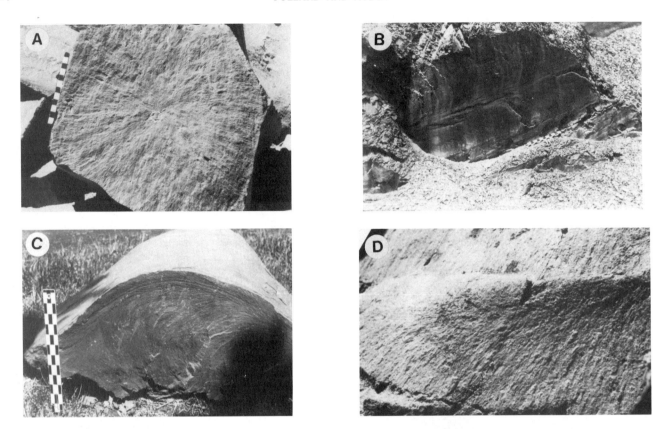

Figure 8. Plumose patterns on joint surfaces. A. A radial pattern on a joint perpendicular to column axis in columnar basalt. B. An axial symmetric pattern in shale. C. An asymmetric pattern about a curved axis in a siltstone concretion. D. Half of a plumose pattern on a columnar joint surface (from DeGraff and Aydin, 1987).

plumose structure is common in nature (Kulander and others, 1979; Bahat and Engelder, 1984; and DeGraff and Aydin, 1987), and Figures 8C and 8D illustrate asymmetric patterns with a curved and straight plume axis, respectively. A half plume (Fig. 8D) is the end member of asymmetric plumose patterns (DeGraff and Aydin, 1987).

Some hackles, especially those near a joint origin, are so fine that they can be noticed only under optimal oblique illumination. At the distal margin or fringe region of the joint (Woodworth, 1896; and Hodgson, 1961b), however, hackles can be identified easily as intersection lines of adjacent joint segments (Fig. 9A) or as oblique to perpendicular cross fractures (fracture lances of Sommer, 1969) that link discrete segments (Fig. 9B). The line drawings of Bankwitz (1965, 1966) clearly show that fine hackles may gradually evolve into large hackles at the joint fringe, or that the transition may be abrupt. The segments are small surfaces that emanate from, and are continuous with, the parent joint surface (Fig. 9). The attitude of the segments diverges from that of the parent surface by a twist of the segment plane about an axis parallel to the hackles. The separation and overlap of adjacent segments increase proportionally from

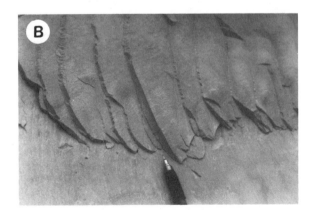

Figure 9. Large hackle or breakdown at joint fringes. A. Gradual transition from minute hackles to large hackles in sandstone. B. Abrupt initiation of breakdown in limestone.

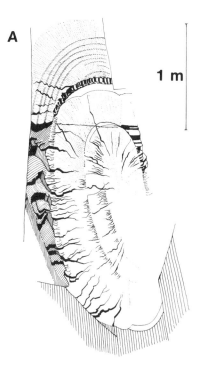

Figure 10. Conchoidal rib marks. A. Line drawing of concentric rib marks and hackles (from Bankwitz, 1965). B. Closely spaced rib marks with approximately rounded profiles in sandstone. C. Rib marks with sharp profiles in sandstone.

the point of breakdown to the distal end (Woodworth, 1896). Each segment may have its own plumose structure (Bankwitz, 1965, 1966), and hackles on each segment curve sharply to become nearly normal to the segment's edge or large hackle (Woodworth, 1896; see part 9 of Fig. 2).

Rib marks, also known as conchoidal structures, are curvilinear ridges or furrows (Fig. 10A) oriented at right angles to hackles and plume axes (Woodworth, 1896; Bankwitz, 1965; Kulander and others, 1979). In a profile that is parallel to hackle, rib marks express themselves as curves and kinks. They may be rounded in profile (Fig. 10B), or have sharp apexes (Fig. 10C). Wallner lines are a textural feature similar to ribs (Kulander and others, 1979; Engelder, 1987), but they occur as one or two sets of oblique to the hackles.

Figure 11. Composite joint surfaces with multiple plumose patterns. A. Cooling joint in a lava flow (from DeGraff and Aydin, 1987). B. Dessication joint in mud. C. Tectonic joint in a sequence of siltstones and shales. Some hackles are highlighted by white chalk (photograph by Daniel Helgeson).

Joints with Multiple Plumose Structure

Most natural joints are composed of multiple joint segments, each of which has its own initiation point, plumose pattern, and terminal boundary. Figure 11 shows composite surfaces of a cooling joint in a lava flow, a desiccation joint in mud, and a tectonic joint in a sequence of siltstone and shale layers. DeGraff and Aydin (1987) demonstrated that the growth of cooling joints in lava flows occurs by successive addition of new segments to previous ones. Each segment has its origin on the leading edge of an older segment, and it spreads primarily laterally (Fig. 12A). A new segment is either coplanar with, or diverges from, the previous one. In profile, a new segment curves to approach the old one, thereby producing a cusp on the joint surface. The old segment, however, is not curved and leaves a blind tip (Fig. 12B). Incremental growth of desiccation joints (Fig. 11B) is kinematically similar to cooling joints, but the nature of joint growth in layered rocks (Fig. 11C) is not well known.

Shape and Dimension of Joints

The shape and size of joints can be controlled by the geometry of the rock mass. For example, most joints within layered sedimentary rocks are oriented perpendicular to the layering and are roughly rectangular. Joint dimension perpendicular to the layering is controlled by the thickness of the jointed unit, which may be composed of many layers, but is rarely more than several tens of meters. The dimensions of joints parallel to layering are not well documented, but these may be hundreds of meters. In massive rock units, joint shape depends on the fracture process, especially mechanical interaction among neighboring joints. Although little is known about joint shape in massive rocks, rib marks (Fig. 10A) suggest that an elliptical geometry may be common. In some cases (Segall and Pollard, 1983b), the long dimension is rarely more than one hundred meters.

Joints produced by cooling of a lava flow or by desiccation of a sediment layer are rectangular in shape and usually are perpendicular to the flow surface or layer. The long (vertical) dimension of the rectangle is limited by the thickness of the jointed unit and usually does not exceed several tens of meters. The short (horizontal) dimension is controlled by the fracture process, as adjacent joints interact and intersect to produce surfaces with widths less than the thickness of the layer.

It is interesting that although joints form in a wide variety of environments, their reported dimensions rarely exceed several hundred meters. We arbitrarily set the lower limit on joint dimension as several times the characteristic grain size of the rock; fractures smaller than this are called "microcracks" (Kranz, 1983).

Spacing and Density of Joints

Curiously, joints never occur alone, but as a series of subparallel fractures defining a joint set. Methods for characterizing the spacing, density, and trace lengths of joints in sets are described by La Pointe and Hudson (1985). The spacing of joints in some sets in intrusive igneous rocks is not uniform. For example, distances between joints mapped in granodiorite of the Sierra Nevada range from about 20 cm to nearly 25 m, and clusters of joints crop out sporadically (Figs. 2 and 3 of Segall and Pollard, 1983b). Joints in sedimentary rocks, like those described by Hodgson (1961b) and Dyer (1983) on the Colorado Plateau, do have a regular distribution, and the spacing of joints can scale with the thickness of the fractured layer (Crosby, 1882; Lachenbruch, 1962; Hobbs, 1967; Ladeira and Price, 1981). Field data suggest that other factors also influence joint spacing. First, two joint sets in the same lithologic unit often have different spacings (Hodgson, 1961a; Barton, 1983). Second, spacings of joints in different lithologic units of comparable thicknesses can be different (Harris and others, 1960; Price, 1966). Third, joint spacing can change as a joint set evolves. For example, columnar joints (Fig. 13A) that initiated at a flow base show an increase in spacing toward the interior (Aydin and DeGraff, 1988), and the number of joints in a sedimentary unit (Fig. 13B) decreases with distance from the initiation surface. The spacing of cooling joints that grow down from the top of a lava flow is smaller than the spacing of those that grow up from the base. This has been attributed to a faster cooling rate at the flow top (James, 1920; Spry, 1962; Saemundsson, 1970; Long and Wood, 1986).

Patterns of Multiple Joint Sets

Joint patterns comprising more than one joint set are common in nature (Spencer, 1959; Harris and others, 1960; Hodgson, 1961a; Nickelsen, 1976; Engelder and Geiser, 1980), and variations in joint patterns across a given region define joint domains. Aerial views (Fig. 14A, 14B) of intersecting joint sets in sandstone from Arches National Park (Dyer, 1983) show two different patterns: one has a high joint intersection angle and the other has a low angle. Each joint pattern has sets that mutually cross and sets that truncate. The age relationship of truncating sets is unambiguous; truncated joints belong to a later episode (for example, Crosby, 1882; Lachenbruch, 1961; Bankwitz, 1984). It is a challenge, however, to determine in the field whether crosscutting joint sets formed during the same or different deformation episodes. Dyer (1983) used offset relationships to infer that each crosscutting set is of a different deformation episode. Barton (1983) distinguished different jointing episodes from

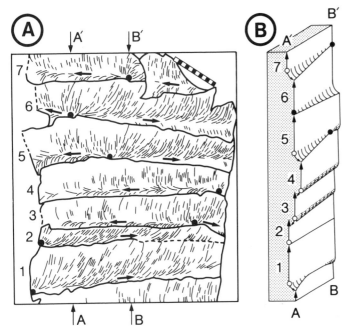

Figure 12. Multiple plumose patterns on a cooling joint near a flow base (from DeGraff and Aydin, 1987). A. A discrete crack event is represented by a single plumose pattern with an origin (dot) on the edge of the previous segment. Short thin lines show hackles, and arrows indicate local propagation direction. Numbers on the left give the sequence of the crack formation. B. Block diagram of a cut-out section A-A′ and B-B′ showing out-of-plane addition of new segments. The vertical component of propagation is indicated by arrows and the sequence by numbers along the profile A-A′.

Figure 13. Joint elimination and change of joint spacing. A. A system of joints initiating from the bottom of a lava flow and selectively growing upward with increasing joint spacing (from Aydin and DeGraff, 1988). B. A system of joints initiating from the top of a shale siltstone sequence and selectively growing downward with increasing joint spacing (Aydin was introduced to this outcrop by Terry Engelder).

Figure 14. Multiple joint patterns in various rocks. A. Two joint sets with a high intersection angle in sandstone. B. Joint sets from the same area with low intersection angle. C. Polygonal joint patterns in mud predominantly with orthogonal intersections. D. Polygonal patterns in a lava flow with non-orthogonal intersections (from Aydin and DeGraff, 1988).

crosscutting relationships of mineral-filled joints in the Canadian Rockies. Nickelson and Hough (1967) and Nickelson (1976) attributed each continuous joint set of the Appalachian Plateau to a single deformation episode. Nur (1982) proposed that the number and orientation of joint sets depend upon the ratio of sealed joint strength to rock strength during different episodes of tension. On the other hand, some orthogonal sets of crosscutting joints are interpreted as forming in a single episode (Ramsay, 1967, p. 112).

Methods for interpreting the time frame of initiation and growth of two or more sets of the same episode are available in some cases. For example, cooling or desiccation joints of many orientations may form during one deformation episode, yet the joint geometry is closely related to the order of formation of individual joints (Aydin and DeGraff, 1988). Figure 14C shows a network of desiccation joints on the surface of a mud layer at the bottom of a dry reservoir in California. All of the joints terminate approximately orthogonally against others. A similar, but more regular polygonal joint pattern with nonorthogonal terminations occurs within lava flows (Fig. 14D). Both patterns form sequentially within one episode. The sequence and time frame for multiple sets of tectonic joints of the same episode, however, have not been clearly explained.

Although there are many varieties of joint patterns in nature, types of joint intersection geometries are few. Following Lachenbruch (1961) joint intersection geometries can be classified as orthogonal and nonorthogonal (Figs. 15A, 15B). Both types can be divided into three groups according to the continuity of the joints at intersections: all continuous joints (Figs. 15A, 15B); some continuous and some discontinuous sets (Figs. 15C, 15D); and all discontinuous (Figs. 15E, 15F, 15G, 15H). The continuous set is called systematic by Hodgson (1961b). Depending on the angles, two continuous sets form either + or X type intersections. The intersections of orthogonal joints with one set continuous and the other discontinuous (Fig. 15C) are known as T or curved T. Nonorthogonal joints with one set continuous and the other discontinuous (Fig. 15D) either have very low intersection angles or the discontinuous set becomes parallel to the continuous set (Dyer, 1983). Orthogonal and nonorthogonal joints with both sets discontinuous (Figs. 15E, 15F) are observed also. Triple intersections known as Y are common in discontinuous cooling joints (Figs. 15G, 15H) whose intersection angles vary a great deal (Aydin and DeGraff, 1988), but they are commonly near 120° (Fig. 15H).

KINEMATICS OF JOINTING

Interpretation of the geometric and textural features of joint surfaces and joint patterns provides valuable information about the kinematics of fracturing. In fact, joints are distinctive among major structures in rock because many of their geometric and textural features can be related unambiguously to the processes of joint propagation, interaction, and termination.

Joint Propagation

Hackles radiate from the joint origin (Fig. 7), indicating the local propagation direction (Woodworth, 1896), and the convergence of hackles can be used to locate the origin even where it is not marked by an obvious flaw. The plume axis marks the leading tip of the joint front (Kulander and others, 1979). The open end of the feather or chevron pattern points in the direction of fracture propagation (Woodworth, 1896).

Composite joints with multiple origins and plumose patterns have an overall propagation direction that differs from the local propagation direction as indicated by individual hackle (Fig. 12). The overall growth direction of a composite joint is the direction in which new segments are added to the previous segments (DeGraff and Aydin, 1987). For example, overall growth direction of cooling joints (Fig. 12) can be determined using three criteria: (1) the origin of a new segment is on the leading edge of an older segment; (2) the larger part of asymmetric plumose structure is on the side of the plume axis in the direction of joint growth; and (3) viewed in profile, the straight blind tip of a segment points in the direction of joint growth.

The overall joint propagation direction can be determined also from a set of joints that initiate at a surface and decrease in number away from the surface by selective elimination. The joint-elimination direction is the same as the propagation direction of the joint system. For example, joint systems near the base of lava flows show upward elimination, indicating an upward overall growth direction (Fig. 13A). A joint system with downward elimination in a sedimentary unit (Fig. 13B), indicates downward propagation. Joint elimination is best observed on exposures that are perpendicular to both the initiating surface and the joint set.

Joint Propagation Velocity

Although propagation velocities of natural joints are difficult to determine from field data alone, relative velocities in various directions on a single joint may be deduced. Conchoidal rib marks indicate the positions of joint fronts at unspecified times. In the absence of rib marks, positions of joint fronts can be constructed by drawing curves that are perpendicular to hackle marks (Kulander and others, 1979; DeGraff and Aydin, 1987). A comparison of distances along hackles between two consecutive rib marks or joint fronts (Fig. 10) gives an estimate of the relative velocities (Kulander and Dean, 1985). This method also provides information about relative velocity variations along a given hackle and along a joint front.

It may be possible to estimate absolute velocities for some special cases. Engineering experiments (Lawn and Wilshaw, 1975, p. 100–105) and theory (Kanninen and Popelar, 1985, p. 192–280) associate branching of a crack into a fork-shaped geometry with dynamic crack propagation velocities, approaching those of elastic waves. Joint branching has been interpreted as indicative of dynamic joint propagation (Lachenbruch, 1962; Bahat, 1979). Breakdown into echelon segments at quasi-static

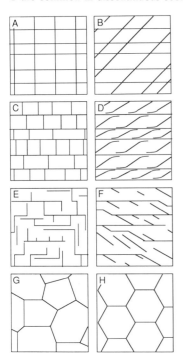

Figure 15. Schematic illustration of major multiple joint patterns. A. Orthogonal and continuous (+ intersections). B. Nonorthogonal and continuous (X intersections). C. Orthogonal, one continuous and the other discontinuous (T intersections). D. Nonorthogonal, one continuous and the other discontinuous. E. Orthogonal, both sets being discontinuous. F. Nonorthogonal, both sets discontinuous. G. Triple intersections with all joints discontinuous at various angles. H. Triple intersections at 120° angles.

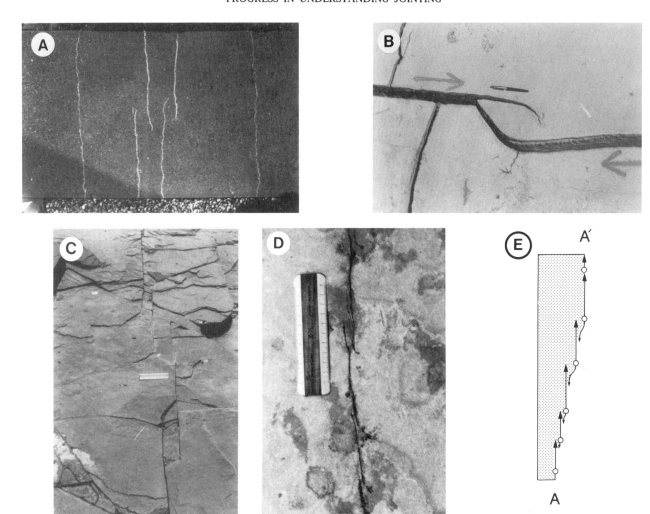

Figure 16. Geometries of interacting joints. A. Cracks in concrete sidewalk at the front of the Geoscience–Civil Engineering Building, Purdue University. Closely spaced cracks at the center interacted and arrested; two outer cracks did not. Sidewalk width about 2 m. B. Characteristic hook-shaped geometry of two interacting cracks in mud (pen for scale). Arrows indicate propagation direction of the joints as determined from plumose patterns. C. A series of echelon joints with right and left steps in siltstone (20-cm ruler for scale). Adjacent joints are linked by cross fractures. D. Interacting echelon joints in sandstone with one joint tip diverging from the neighboring tip (20-cm ruler for scale). E. Drawing of joint profile in lava flow (height is about 1 m). Tips of the younger segments approach, and become parallel to, older segments (from DeGraff and Aydin, 1987).

propagation velocities (Sommer, 1969) and the asymptotic intersection of two joints also can produce fork-like profiles, however. Kulander and others (1979) described how to estimate velocities of joint propagation using Wallner lines (Wallner, 1939), but they commented that these features are rare on natural joint surfaces.

Joint Interaction

Observed geometries of joints indicate the widespread occurrence of mechanical interaction. The stress field associated with one joint can have an important effect on the growth of neighboring joints if the distance between them is small with respect to their dimensions (Pollard and others, 1982). For example, the four closely spaced fractures in a concrete sidewalk (Fig. 16A) have interacted to inhibit each other from crossing the sidewalk, whereas two widely spaced fractures traverse the sidewalk. The two interacting fractures on the right of the group of four have a characteristic hook-shaped geometry. A similar geometry (Fig. 16B) occurs where two desiccation joints in mud are linked. Here plumose structures on the uppermost segments indicate that the joints propagated toward one another. Examples of echelon joints in siltstone (Fig. 16C) are linked by curved joints, by nearly straight cross joints, and even by an oblique joint that extends beyond the overlap area. The type of linking fractures does not seem to depend on the sense of step, as it does for echelon faults (Segall and Pollard, 1980; Aydin and Nur, 1982). Examples of echelon joints in sandstone (Fig. 16D) consistently diverge away from the neighboring joint; whereas in some lava flows, new joint segments approach and then turn sharply to become parallel to older segments without linking (Fig. 16E). A greater overlap typically corresponds to a greater joint spacing for a wide range of scales (Pollard and others, 1982; Pollard and Aydin, 1984).

Joint Termination and Intersection

Surprisingly, the form of joint terminations in a rock mass and the associated deformation are not well known. Experimental data from engineering materials (for example, Kanninen and Popelar, 1985, p. 172–182, 281–391) and rocks (Friedman and others, 1972; Hoagland and others, 1973; Peck and others, 1985a, 1985b; Swanson, 1987) indicate the existence of a plastic zone (Fig. 17A) or a zone of microcracking (Fig. 17B) at fracture terminations. Such inelastic deformation reduces the stress concentration at joint tips and, as the joint tip traverses from its origin to final termination, a relic zone will be left behind that should influence joint surface morphology. For example, joint tip blunting associated with temporary arrest and repropagation may produce a type of symmetric rib mark (Fig. 17C), unlike rib marks shown in Figure 10. This type of surface texture can be used to indicate episodic joint growth (Bahat and Engelder, 1984; Kulander and Dean, 1985).

Joints commonly terminate at a discontinuity such as a lithologic boundary, a fault, or another joint; for example, orthogonal joint terminations (Fig. 18) occur against the convex side of a continuous joint to form "curved-T" intersections (DeGraff and Aydin, 1987). This pattern is attributed to high stresses on the convex side of a curved fracture (Lachenbruch, 1962). The younger abutting joints may propagate either toward or away from the existing joint.

Examples abound of joints that apparently cut across bedding interfaces and other joints. The + or X types of intersections (Figs. 15A, 15B) seem to contradict the notion that older surfaces act as barriers to joint propagation, as implied by T intersections (Fig. 18). Kulander and others (1979) offered several possible explanations for + and X type intersections. It is possible that the older joint was closed, but now is open, or that it was sealed by mineral precipitates which were removed subsequently. Another possibility is that the younger joint traveled around the surfaces of the older joint, leaving the impression that it cut across the older joint without being influenced. Also, two independent joints may intersect an older joint so close together that they are misinterpreted as crosscutting joints when, in fact, they form a double T intersection (Aydin and DeGraff, 1988).

Strain Accommodated by Jointing

The original opening (dilation) across a single barren joint is difficult to determine because the joint reflects the total displacement, including opening and closing, since formation of the joint. A joint filled by mineral precipitate, however, can provide the magnitude and direction of relative displacement at the time of precipitation from cross-cut mineral grains (Segall and Pollard, 1983b) or fibrous minerals (Ramsay and Huber, 1983, 1987). The average extensional strain accommodated by joint dilation can be calculated by first measuring the sum of joint openings along a traverse perpendicular to the set and then dividing the sum by the traverse length. The total strain additionally includes that amount accommodated by deformation of the intervening host rock (Segall and Pollard, 1983b). The distribution of opening displacements along a joint has not been studied. If the form of the displacement distribution were known, however, aspects of the causative stress field, the surrounding strain field, and the rock stiffness could be determined. Interested readers are referred to Pollard (1987) for a review of these methods applied to igneous dikes.

Shear displacements across fractures, inferred to have been opened initially as joints, were recognized a long time ago (Woodworth, 1896; Van Hise, 1896). Evidence of shearing comes from field observations of slickensides overprinting plumose patterns (Barton, 1983, p. 83–85) and from sheared joint fillings (Nickelson and Hough, 1967; Segall and Pollard, 1983a). This evidence indicates at least two episodes of deformation

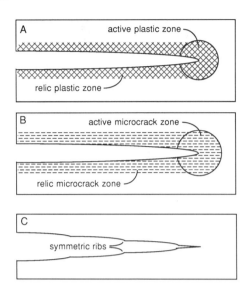

Figure 17. Joint-termination structures. A. Plastic zone. B. Microcrack zone. C. Symmetric rib marks associated with temporary arrest and joint-tip blunting.

with markedly different strain fields. For example, two episodes in the development of fractures in granodioritic rocks of the Sierra Nevada (Segall and Pollard, 1983b) began with extensional strains of about 10^{-4} accommodated by joint dilation (Fig. 19A). Later shear strain (Fig. 19B) was accommodated by slip on some of these fractures to produce small

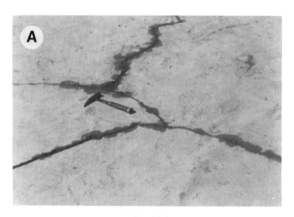

Figure 18. Orthogonal joint terminations (curved T intersections) on the convex side of continuous joints. A. In sandstone (hammer for scale). B. In mud (scale is about 25 cm).

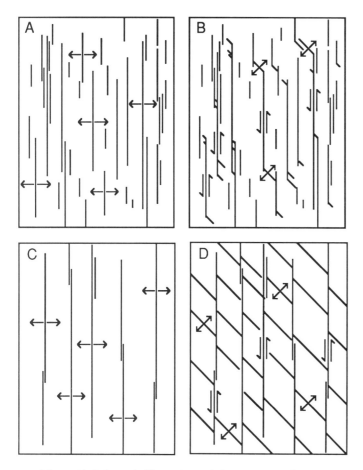

Figure 19. Schematic illustrations of sheared joints after two case studies. A. A set of unevenly spaced joints in granite; after Segall and Pollard (1983b). B. The joints are sheared during a second deformation episode; splay cracks opened in harmony with the sense of shear. C. A set of evenly spaced joints in sandstone; after Dyer (1983). D. Another joint set formed during a second deformation episode; shearing on the older set and opening of the younger set are consistent.

faults. Relic joints having undeformed filling minerals coexist with small faults having highly deformed fillings. Splay cracks opened at the ends of some of the small faults.

In a second example, the older of two sets of nonorthogonal joints accommodates an extensional strain perpendicular to the joint trace when it forms (Fig. 19C). The strain field associated with the younger set, however, will resolve into some shear strain across the older set (Fig. 19D). If the normal and shear components of the resolved stresses on the older set satisfied the Coulomb criterion (Jaeger and Cook, 1969, p. 65–68), there will be frictional sliding along these fractures. The sense of shear displacement across the older fractures should be consistent with the shear strain inferred from opening of the younger joint set. This concept has been used to distinguish deformation episodes by Dyer (1983) at Arches National Monument.

MECHANICS OF JOINTING

The view that natural forces acting in a rock mass cause jointing was expounded by Hopkins (1835) and later echoed by Hoskins (1896). To interpret the initiation, propagation path, surface textures, spacing, and patterns of joints, it is necessary to relate field observations to their causative forces. The science of mechanics provides the principles and tools to establish these relationships, as exemplified in the second half of this century by the work of Lachenbruch (1961), Secor (1969), Sowers (1972), and Segall (1984a), among others. It is not within the scope of this review paper to derive equations governing these relationships, but we will highlight some of the important concepts and results.

Joint Initiation

Field observations (Fig. 7) indicate that joints initiate at flaws. They do so because flaws perturb the stress field in such a way that the magnitude of local tensile stresses at the flaw exceeds the tensile strength of the rock. This may happen under a variety of remotely applied tensile or compressive stresses. We identify two circumstances of importance: concentration of an insufficient remote tension and conversion of remote compression into local tension.

Fossils, grains, clasts, and other objects with different elastic properties than the surrounding rock can concentrate a remote tension. To demonstrate this, we consider a circular inclusion (Fig. 20A) with elastic shear modulus, μ_1, in a rock mass with shear modulus, μ_2; the ratio of moduli is $k = \mu_1/\mu_2$. For simplicity we assume that Poisson's ratio, ν, is 0.25 for both materials. A remote tension, σ_3^r, induces a uniform tension, σ_i, throughout the inclusion and a tangential component, σ_0, at two points on the boundary (Jaeger and Cook, 1969, p. 248–251).

$$\sigma_i = \sigma_3^r [3k/(2k+1)]$$
$$\sigma_0 = \sigma_3^r [3/(2k+1)]$$

For stiffer inclusions (high k ratio), a remote tensile stress is amplified by factors up to 1.5 inside the inclusion, and the tangential stress outside the inclusion is diminished. Near softer inclusions, the tangential stress is amplified, and for an open cavity or pore, this amplification is a factor of 3.0.

Some flaws, such as cavities, microcracks, and grain boundaries, are very eccentric (Kranz, 1982). For example, the cavity in Figure 7A may be idealized as an elliptical hole with long and short axes, a and b (Fig. 20B). The local stress can exceed the remote tension by orders of magnitude (Inglis, 1913) if the axial ratio, a/b is very large.

$$\sigma_e = \sigma_3^r [1 + (2a/b)]$$

Accordingly, it is not surprising that cavities are places of joint initiation. To calculate the stress concentration associated with notches, Inglis derived the relationship

$$\sigma_n \approx \sigma_3^r [2a/r]^{1/2}$$

where r is the radius of curvature and a is the characteristic length (Fig. 20C). The local tensile stress is greatest where the radius of curvature is smallest. The example of joint initiation at a sharp cusp (Fig. 7E) is consistent with this result. Pinch-and-swell structure in a stretched layer (Fig. 20D) caused by a periodic instability can also lead to a tensile stress concentration and jointing (Sowers, 1972).

Because of overburden weight, the vertical stress in the Earth is compressive, and its magnitude increases with depth. Stress measurements (McGarr and Gay, 1978; McGarr, 1982) show that horizontal stresses in the shallow crust also are compressive. Knowing that joints form at great depths, we are faced with the paradox of how joints open in a compressive stress field. The answer is suggested by laboratory experiments (Hoek and Bieniawski, 1965; Gramberg, 1965; Peng and Johnson, 1972; Tapponnier

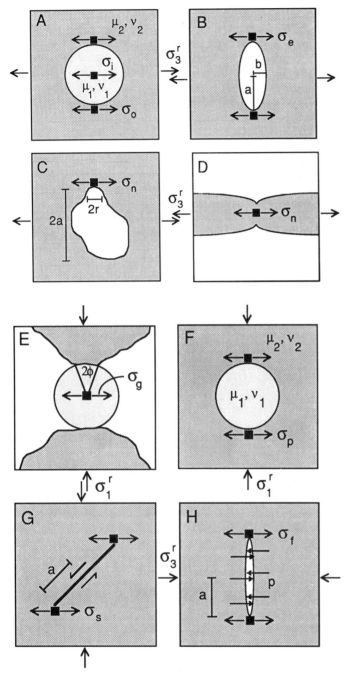

Figure 20. Idealizations of joint-initiation mechanisms based on stress concentration and changing remote compressive stress into local tensile stress. A. Circular inclusion under tension. B. Elliptical hole under tension. C. Irregular cavity under tension. D. Cusp in stretch layer. E. Circular grain compressed between two grains. F. Circular inclusion under compression. G. Inclined elliptical hole under compression. H. Internally pressurized elliptical hole perpendicular to a remote compression. In all cases, a is a length, μ is the elastic shear modulus, ν is Poisson's ratio, σ^r is the remote stress, p is internal fluid pressure, and the various local stresses are referred to in the text.

and Brace, 1976) which have demonstrated that flaws in rock subjected to a compressive stress field can induce local tensile stresses and thus facilitate joint formation. Here we will cite a few theoretical results that enable one to estimate the magnitude of these stresses.

Grain contacts (Fig. 20E) can induce local tensile stresses in a compressive remote field. For a small angle of grain contact, 2ϕ, stress at the grain center is

$$\sigma_g = -\sigma_1^r [2\phi/\pi]$$

(Jaeger and Cook, 1969, p. 245–247). For compressive remote stress, σ_1^r, the resulting stress in the grain is tensile. Experimental (Borg and Maxwell, 1956; Gallagher and others, 1974) and field observations (Aydin, 1978) confirm the initiation of opening cracks at grain contacts.

Inclusions, pores, and microcracks also can change remote compression into a local tension. The tangential stress, σ_p, along the boundary of a circular inclusion (Fig. 20F) at the ends of the diameter parallel to the applied compression is

$$\sigma_p = \sigma_1^r [(1-k)/(2k+1)]$$

For inclusions less stiff than the surrounding rock, the tangential stress is tensile. This principle may apply to situations like Figure 7C; but in that case, we lack information about relative stiffnesses. In the limiting case of a cavity, the applied compressive stress induces tension of the same magnitude at the two points (Fig. 20F), regardless of cavity shape. Because compressive stresses are large in the Earth's crust, and the tensile strength of rock is small relative to the compressive strength, the change of sign provides an attractive mechanism for joint initiation.

Griffith (1924) showed that sliding of the walls of an elliptical crack inclined to the remote compression (Fig. 20G), induces a tensile stress, σ_s, approximated as

$$\sigma_s \approx -\sigma_1^r [a/4b]$$

For typical microcrack geometries in rock (Nur and Simmons, 1970; Sprunt and Brace, 1974; Wong, 1982; Padovani and others, 1982; Kranz, 1983), the term $a/4b$ is much larger than one, and so the tensile stresses necessary to initiate joints should be achieved easily. Several experimental studies (Brace and Bombolakis, 1963; Hoek and Bieniawski, 1965; Ingrafea, 1981; Nemat-Naser and Horii, 1982), however, have shown that cracks propagating from such a sheared microcrack first turn into the direction of maximum compressive stress and then tend to stop. Thus this mechanism alone is not sufficient to explain joints with dimensions much greater than the flaw size.

Perhaps the most effective mechanism of joint initiation is based on cavities and microcracks subjected to internal fluid pressure, p, and aligned with their longest dimension perpendicular to the remote least compressive stress, σ_3^r (Fig. 20H). For an elliptical hole with axial ratio a/b, the local stress at each end is

$$\sigma_f = (p - \sigma_3^r)[2(a/b)] - (p + \sigma_3^r)$$

(Jaeger and Cook, 1969, p. 266–267). It has been known for a long time that fluid pressures can exceed the least compressive stress (Rubey and Hubbert, 1959; Suppe, 1985), in which case σ_f is tensile. Because a/b for microcracks is much greater than one, tensile stresses greater than the rock can withstand can be induced at microcrack tips, thereby initiating joints. Because the stress concentration increases with joint length, joints can propagate as long as adequate fluid pressure is maintained.

The mechanisms described here and illustrated in Figure 20 result in the initial joint surface being approximately perpendicular to the least compressive stress, σ_3^r. For the following discussion, we assume this orientation for the remote principle stresses (Fig. 21A) for the earliest stage of joint propagation and then consider the effects of spatial or temporal changes in the stress field.

Joint Propagation

The propagation of a joint is controlled by the stress field near the joint tip. From linear elastic fracture theory (Lawn and Wilshaw, 1975; Kanninen and Popelar, 1985), we know that this stress field is likely to be very heterogeneous, the region of stress concentration is small, and the stresses decrease as one over the square root of distance from the tip. Fortunately all stress components, σ_{ij}, in this region (radius D in Fig. 21B) are proportional to quantities called the "stress intensity factors," which we will use to establish a joint propagation criterion.

The three stress intensity factors (K_I, K_{II}, K_{III}) each correspond to a fracture mode (mode I, mode II, mode III of Fig. 5), and each is associated with a unique stress distribution near the fracture tip. Equations for, and numerically computed values of, stress intensity factors, which depend on fracture geometry and loading configuration, are available for many interesting cases in engineering handbooks (Sih, 1973; Tada and others, 1973). As a particular example (Fig. 21A), the stress intensity for an opening mode fracture subject to uniform remote stress, σ_3^r, and internal fluid pressure, p, is

$$K_I = (p - \sigma_3^r)[\pi a]^{1/2} \quad (1)$$

The stress intensity is proportional to the driving stress, $(p - \sigma_3^r)$, and the square root of the fracture length, $2a$. When the stress intensity factor reaches a critical value,

$$K_I = K_{IC} \quad (2)$$

the fracture, or joint in our case, will propagate (Lawn and Wilshaw, 1975, p. 65). Unlike a failure criterion that uses the concept of tensile strength of a homogeneously stressed material (Jaeger and Cook, 1969, p. 83–86), this criterion is based on the heterogeneous stress field at a fracture tip. The fracture toughness, K_{IC}, is a property of the material. Measured values of toughness for rock fractures a few centimeters long at low confining pressures and temperatures typically are in the range 0.3 to 3 MPa m$^{1/2}$ (Atkinson and Meredith, 1987b, p. 477–525). Although data are limited, K_{IC} for some rocks increases with increasing confining pressure and decreases with increasing temperature (Barton, 1982).

Expressions for stress intensity factors (such as eq. 1) combined with the propagation criterion (eq. 2) allow us to make important inferences about the behavior of joints. For two joints of unequal lengths subjected to the same increasing driving stress, the longest joint will meet the propagation criterion first. On the other hand, the joint subjected to the greatest driving stress will propagate first among a set of joints with equal lengths in a spatially varying stress field. Variations in fluid pressure, p, and remote stress, σ_3^r, along the face of a single joint will determine the magnitude of the driving stress and consequently the stress intensity factor for nonuni-

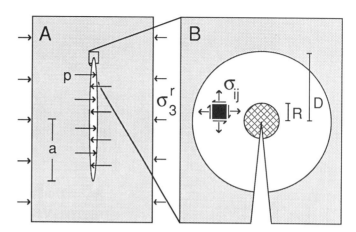

Figure 21. A. Idealization of a mode I fracture of length $2a$ loaded by remote least compressive stress, σ_3^r, and internal fluid pressure, p. B. The tip region where the local stress field, σ_{ij}, is approximated by the elastic solution within radius D and inelastic deformation occurs in the process zone within radius R.

form loading conditions (Lachenbruch, 1961; Weertman, 1971; Secor and Pollard, 1975; Pollard, 1976). Indeed, all of the joint-initiation mechanisms described above involve incipient joint propagation through a nonuniform stress field. Rummel (1987) provided an example of the analysis of such a problem.

Irwin (1957) showed that a propagation criterion based on stress intensities is equivalent to Griffith's (1921) energy-balance criterion

$$G = G_C \quad (3)$$

for crack growth in a brittle elastic material. Here G is the change in energy with respect to a presumed extension of the fracture plane, whereas G_C is the critical value of this change in energy required for propagation actually to occur. We will refer to G as the fracture propagation energy (other names in the literature are energy release rate and crack extension force). For growth in the fracture plane, the propagation energy, G_i, is related to the three stress intensity factors by

$$G_i = [(K_I^2 + K_{II}^2)(1 - \nu) + K_{III}^2]/2\mu \quad (4)$$

For a pure mode I fracture subjected to uniform loading, the propagation energy is found by combining equations 1 and 4 to find

$$G_I = (p - \sigma_3^r)^2 \pi a (1 - \nu)/2\mu \quad (5)$$

If inelastic deformation in a process zone at the fracture terminations (Fig. 17) is restricted to a small region of radius R ($R < D$ in Fig. 21B), the propagation criteria mentioned above are applicable (Irwin, 1957).

Joint propagation is intimately related to joint opening. For example, the opening, Δu_I, at the center of the uniformly loaded, mode I fracture (Fig. 21A) is

$$\Delta u_I = (p - \sigma_3^r) 2a (1 - \nu)/\mu \quad (6)$$

(Pollard and Segall, 1987, eq. 8.35). Negative values of Δu_I are not

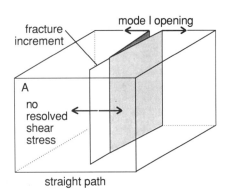 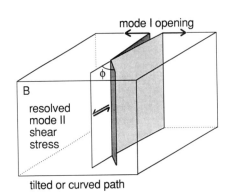 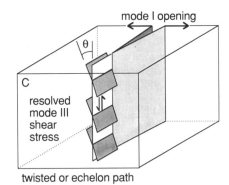

Figure 22. Illustrations of the fracture tip with a small increment of new fracture. A. Planar path associated with pure mode I. B. Tilted path associated with shearing on the increment and mixed mode I + II. C. Twisted path associated with shearing on the increment and mixed mode I + III.

admissible; the joint walls will remain closed, $\Delta u_I = 0$, if internal fluid pressure does not exceed the remote stress. Comparing equations 1, 5, and 6 shows that this condition also implies no stress concentration and no energy for propagation. Without an internal fluid pressure, the remote stress acting across a joint plane must be tensile for opening and propagation.

Joint-Propagation Path

The joint-propagation path depends on the joint tip stress field. We can understand this dependence using a method proposed by Gell and Smith (1967). They determined fracture-propagation paths by considering the stress intensity factors associated with a small increment of new crack surface in an isotropic material (Fig. 22). Equation 4 and expressions for the three stress-intensity factors are used to calculate the propagation energy (Fig. 23) for growth of the increment as a function of its orientation. The orientation that produces the maximum propagation energy is the preferred propagation path. This approximate method is in qualitative agreement with more exact calculations (Cotterell and Rice, 1980; Kariha-loo and others, 1980).

A pure mode I loading produces a maximum propagation energy for an increment oriented in the plane of the joint (Fig. 23A), thus leading to in-plane propagation (Fig. 22A). A joint oriented perpendicular to the remote least compressive stress, σ_3^r, will satisfy this condition. If this principal stress direction does not change along the extension of the joint plane, in-plane propagation will produce a planar joint surface. If the principal stress direction does change, shear stresses are resolved on the extension of the joint plane. The two orientations of these shear stresses correspond to mixed modes I + II (Fig. 22B) and mixed modes I + III (Fig. 22C) loading. In both cases, the preferred joint path is out of plane and tends to be oriented perpendicular to the local maximum tension, thereby reducing the resolved shear. Mixed modes I + II results in a tilt of the joint path about an axis parallel to the joint front (Fig. 22B), and so the joint surface turns along a smoothly curved or sharply kinked path. Mixed modes I + III result in a path that twists about an axis perpendicular to the joint front (Fig. 22C). When this occurs, the entire joint front does not twist and maintain its continuity, but the next increment breaks down into echelon segments. From energetic considerations, breakdown is favored because the surface area of multiple echelon segments is less than that for a single twisted surface (Pollard and others, 1982).

The sense of tilt or twist (clockwise or counterclockwise) depends on the sense of shearing, and the magnitudes of the tilt and twist angles increase with increasing ratios of K_{II}/K_I (Fig. 23B) and K_{III}/K_I (Fig. 23C). Changes in orientation of the joint surface can be sharp or smooth.

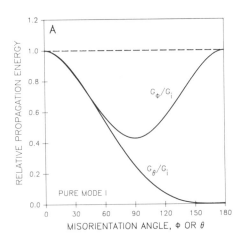 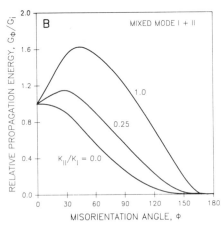 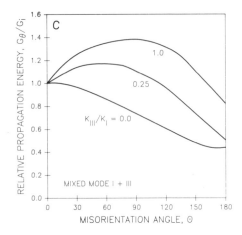

Figure 23. Relative propagation energy as a function of the misorientation (out of the original fracture plane) of the new increment (from Lawn and Wilshaw, 1975). Angles ϕ and θ are defined in Figure 22. A. Pure mode I loading. B. Mixed mode I + II loading with different ratios of the stress intensity factors K_{II} and K_I. C. Mixed mode I + III loading with different ratios of K_{III} to K_I.

A temporal change in the loading conditions, demonstrated by laboratory experiments (Ernsberger, 1960; Ingraffea, 1981), can produce sharp changes in orientation. On the other hand, a stress field varying continuously in space can produce a smoothly curving joint path.

Joint surface morphology is a direct expression of the joint path. Tilted paths produce rib structures (Fig. 10) and large kinks (Figs. 15E, 15F), whereas twisted paths produce hackles (Figs. 7, 8), fringe structures (Fig. 9), and echelon joint segments (Fig. 6). Using these common structures, we can infer the mode of fracture and, hence, the stress state for the entire history of joint propagation. Hackle geometry (plumose patterns) also can be used to infer specific loading conditions. Because the driving stress is greatest across a joint along its plume axis (Kulander and others, 1979), the symmetry of the plumose pattern can be used to infer the symmetry of the loading (Fig. 24, modified from DeGraff and Aydin, 1987).

The complexity of single joint surfaces (Figs. 7 through 12) attests to the fact that the local principal stress directions can vary considerably. The natural heterogeneity of rock also can produce shear stresses at the front of a propagating joint, resulting in tilted and twisted paths. We have emphasized conditions where the stress state controls the propagation path of joints, but for highly anisotropic rock, planes of weakness may be favored for jointing even though they are not perpendicular to the least compressive stress. In this case, joint propagation is of a mixed mode type, and the opening displacement is oblique to the joint surfaces.

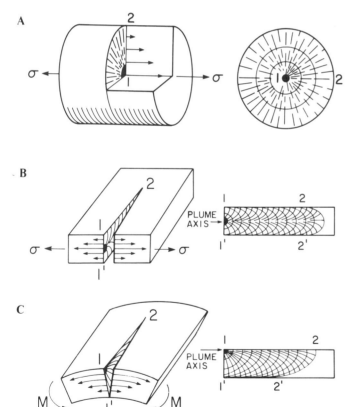

Figure 24. Possible relationships between the symmetry of plumose patterns and loading configurations. A. Axisymmetric loading and radially symmetric plumose pattern. B. Mirror image symmetry of loading configuration and pattern symmetric about the plume axis. C. Asymmetric loading produced by bending and asymmetric half-plume pattern. Modified from DeGraff and Aydin (1987).

In our discussion, we have ignored possible dynamic effects on the propagation criterion and the joint path because so little is known about joint-propagation velocity. For propagation velocities approaching the elastic wave speeds, inertial forces must be considered (Kanninen and Popelar, 1985, p. 192–280). The near-tip stress field is different than that for quasi-static conditions (Freund, 1972), and the crack path may fork into a Y-shape (Yoffe, 1951). This geometry and the loading conditions leading to high velocities, however, are probably uncommon in nature (Segall, 1984b). At the other extreme, Segall (1984b) has suggested that joints may propagate very slowly. Laboratory cracks in rock are known to propagate when the available energy is less than G_C but greater than a threshold value of the propagation energy, G_O. Crack velocities under these subcritical conditions ($G_O < G < G_C$) are very small, typically less than 10^{-2} m/s, and are known to be strongly influenced by chemical reactions at the crack tip (Atkinson and Meredith, 1987a).

Joint Arrest

The mechanics of joint arrest are summarized by examining the two competing terms, G_I and G_C (or G_O), in the joint-propagation criterion (eq. 3). The important factors are those that decrease the energy available for propagation, or that increase the energy required for propagation. For simplicity we consider the propagation energy (eq. 5) of a pure mode I joint subjected to a uniform load. Because G_I increases with joint length (a in eq. 5), a joint should not stop propagating if all other factors remain constant. Arrest can occur if either the fluid pressure, p, decreases or the remote compressive stress, σ_3^r, increases sufficiently. The difference $(p - \sigma_3^r)$ is squared in equation 5, so that small changes in these terms can easily offset the linear increase in G_I with a. A drop in fluid pressure is a natural consequence of the increasing cross-sectional area of a growing joint (Secor, 1969). The effectiveness of this arrest mechanism depends on how readily the pore fluid can recharge the fluid pressure in the joint. Possible mechanisms for increasing the compressive stress, σ_3^r, are described in the following section. Arrest mechanisms solely depending on rock properties may occur when a joint propagates into a stiffer (greater μ in eq. 5) or a more incompressible rock (greater ν in eq. 5), or when a joint intersects another joint or discontinuity that can open or slide (see below).

The other group of arrest mechanisms involves an increase in G_C on the right side of equation 3. Greater critical propagation energy can be related to changes in environmental conditions such as greater confining pressure (Atkinson and Meredith, 1987b). Greater temperature can increase G_C by increasing ductility (Ryan and Sammis, 1978) or by crack-tip blunting (DeGraff, 1987), but it can also decrease G_C if chemical reaction rates are increased at the joint tip (Atkinson and Meredith, 1987a). Ritchie and Yu (1986) reviewed several micromechanical mechanisms for increasing G_C of metals. Two of these, microcracking (Kobayashi and Fourney, 1978) and friction along an interlocking crack surface (Swanson, 1987) have been identified in experiments on rocks.

Mechanical Interaction

In many outcrops (Fig. 16), joint geometry suggests that mechanical interactions between nearby joints or between a joint and a local heterogeneity influence joint growth and arrest and, consequently, joint pattern and spacing. We examine first the characteristic patterns of echelon joints and joint segments (Fig. 25A, inset) in terms of joint interaction. Pollard and others (1982) showed that the propagation energy for significantly underlapped joints is essentially the same as that for an isolated joint (Fig. 25A), but it sharply increases as the tips propagate toward one another. Each joint enhances propagation of its neighbor by inducing a tensile stress in

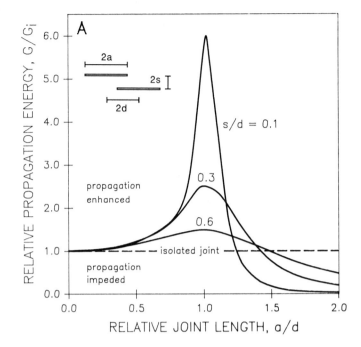

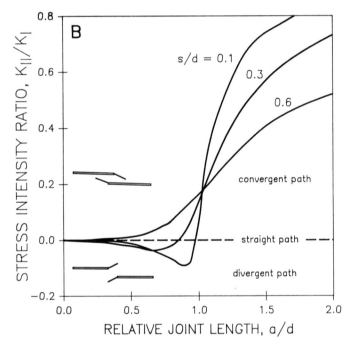

Figure 25. A. Plot of relative propagation energy versus joint length for parallel (echelon) joints with central spacing, $2d$, and orthogonal spacing, $2s$. Energy for an isolated joint is indicated. B. Plot of ratio of mode II to mode I stress intensity versus joint length for echelon geometry. Inferences regarding joint path are illustrated in the insets and discussed in the text.

the vicinity of the neighbor's tip. As the tips overlap, however, the induced stress becomes compressive, and the propagation energy drops precipitously, so that joint growth will tend to stop. This explains why slightly overlapped echelon joints are so common in nature (Fig. 16).

The stress field from one joint tip induces shear stresses on the neighboring echelon joint, and the resulting mode II deformation promotes a curved path (Fig. 23B, inset). The ratio K_{II}/K_I for underlapped joints (Fig. 25B) is negative, especially for closely spaced joints, and so the propagation paths should diverge slightly (Pollard and others, 1982). As the joints overlap, the stress-intensity ratio increases markedly and changes sign, and so the paths should converge sharply. This result rationalizes the common hook-shaped patterns of echelon joints (Fig. 17B) and explains why joints almost always join tip-to-plane rather than tip-to-tip (Swain and Hagan, 1974; Melin, 1982). Exceptions to this geometry may occur because of anisotropic material properties (Lawn and Wilshaw, 1975, p. 71) or because of a compressive stress acting parallel to the joint array (Cotterell and Rice, 1980), resulting in straight or even diverging paths (Fig. 17D).

Following Segall and Pollard (1983b), we consider two parallel joints of different lengths (Fig. 26, inset) and illustrate the effect of one on the other's growth. The normalized propagation energy is plotted against the ratio of joint lengths for different spacings (Fig. 26). The propagation energies are identical for two joints of equal length but are less than that for an isolated joint because the two adjacent joints must share the available energy. As the length ratio increases, the propagation energy for the longer joint approaches that of the isolated joint, whereas the energy for the shorter joint drops toward zero. This result quantifies how longer joints shield nearby shorter joints. As a joint set evolves in a homogeneous rock, more and more joints are shielded, and the resulting population is characterized by many short joints and a few long joints, consistent with measured distributions of joint lengths in some granite outcrops (Segall and Pollard, 1983b).

A similar shielding effect occurs when a series of parallel cracks interact (Nemat-Nasser and others, 1978). Normalized propagation energy (Fig. 27) for each of 20 joints are calculated for different increments in the length of the central joint (DeGraff, 1987). In response to a small increase in length (the upper curve), the propagation energy of the two adjacent joints decreases, suggesting that they will propagate less readily. In contrast, there is little effect on the propagation energy of more distal joints. As

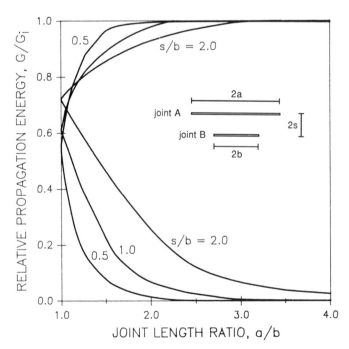

Figure 26. Plot of relative propagation energy versus joint length for two parallel joints of lengths a and b with orthogonal spacing s. Upper three curves refer to joint A, and lower three refer to joint B. See text for interpretation.

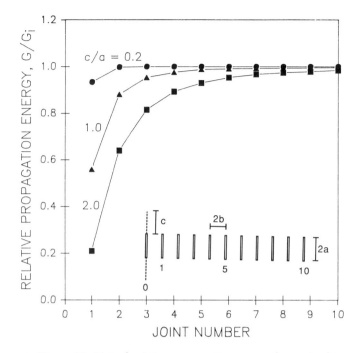

Figure 27. **Plot of relative propagation energy for each of ten joints in a symmetric array of twenty as the length of the central joint (number 0) is incrementally increased by** c. **Joint length is** $2a$ **and central spacing is** $2b$. **Three curves represent three different growth increments. After DeGraff (1987).**

the set of joints grows, the propagation of alternate joints is inhibited by the shielding effect of nearest neighbors. As the length of the central joint increases (lower curves), so does the influence of the central joint on the more distal joints. The elimination of more and more distal joints from those that can propagate results in an increasing spacing between the surviving cracks. This phenomenon, which is observed also in experiments in glass (Geyer and Nemat-Nasser, 1982), is responsible for joint elimination in lava flows and sedimentary rocks (Aydin and DeGraff, 1988).

Interaction also occurs between joints and planar discontinuities, as when a joint approaches a bedding interface or another joint (Fig. 28A, inset). The propagation energy for a joint approaching an open interface steadily increases (Fig. 28A) as the interface attracts the joint tip (Cook and Erdogan, 1972; Pollard and Holzhausen, 1979). Upon intersection, the tip will be blunted as the interface slides open, and the joint will terminate at the interface. If the interface is closed, but can slip, the same qualitative conclusions hold (Weertman, 1980; Keer and Chen, 1981). If the interface cannot slip because of greater cohesion, coefficient of friction, or normal stress, the joint may propagate across the interface.

Material properties can change substantially across interfaces between dissimilar beds or formations. Consider a closed and bonded interface (no slip or opening) with elastic shear modulus μ_1 for the rock containing the joint and μ_2 for rock across the interface (Cook and Erdogan, 1972; Erdogan and Biricikoglu, 1973). If the rock across the interface is stiffer ($\mu_1 < \mu_2$), the propagation energy decreases as the joint approaches the interface (Fig. 28), thereby tending to arrest the joint before such contacts. If the rock across the interface is less stiff ($\mu_1 > \mu_2$), the propagation energy increases as the joint approaches the interface, and one would expect the joint to propagate across the interface. Clearly, joint propagation across bedding planes, formation boundaries, and other structures depends on many factors. It should not be surprising that joints terminate before or at some boundaries, and propagate across others.

SUMMARY AND RECOMMENDATIONS

We have highlighted the bright spots as well as the shortcomings in research on joints and jointing over the past century. Joints are defined as dominantly opening mode fractures, and, as such, they are associated with characteristic stress, strain, and displacement fields. Joints are distinguished from small faults by distinctive surface textures and a lack of shear dis-

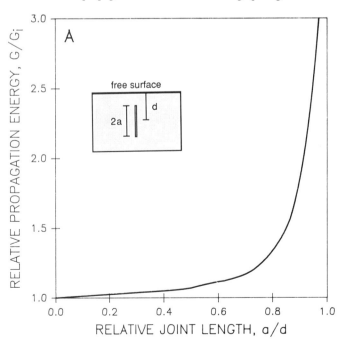
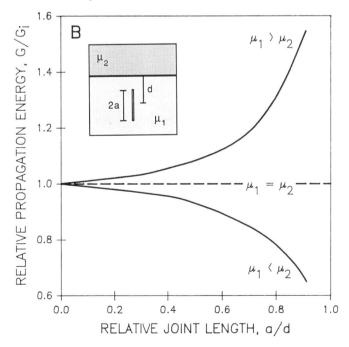

Figure 28. **Plot of relative propagation energy versus length** $2a$ **for a joint that is perpendicular to an interface at a distance of** d. **A. The interface is a free surface or open fracture. Inset illustrates the geometry. B. Joint approaching a closed interface across which the rock is either stiffer,** $\mu_1 < \mu_2$, **or less stiff,** $\mu_1 > \mu_2$. **Inset illustrates geometry. See text for interpretation.**

placements. Surface morphology provides valuable information about the kinematics and mechanics of jointing and can be related unambiguously to joint initiation, propagation, interaction, and termination.

Field observations suggest that joints commonly initiate at material inhomogeneities such as fossils, grains, clasts, sole marks, pores, and microcracks. These inhomogeneities can concentrate local tensile stresses in a rock mass subjected to remote compression. Internal fluid pressure slightly in excess of the remote least compressive stress is acknowledged as an effective driving force for jointing and joints tend to be oriented perpendicular to this remote stress. Changes in the joint-propagation path caused by minor shearing (mixed mode loading) produces hackles, rib marks, and fringe structures. The local direction and relative velocity of joint propagation are recorded by hackles and consecutive rib marks, whereas the overall growth direction of a composite joint is defined by the sequential addition of joint segments.

The mechanical interaction of joints helps to rationalize the ratio of overlap to separation of echelon joints, their hook-shaped geometry, and the length distribution of joints in a set. In addition, interaction can explain why some joints selectively terminate, whereas others propagate, and why joints stop before or at some bedding interfaces but cut across others. Joint intersections (T, curved T, Y, +, and X) are a key element in the interpretation of joint patterns, and so information regarding the continuity, formation order, and propagation direction of joints at intersections is very desirable. Methods for documenting the sequential formation of cooling and desiccation joints are available. The sense of shear displacement across the older of two tectonic joint sets can be a useful criterion for establishing relative age. On the other hand, alternating crosscutting relationships between two joint sets is an indication that both sets belong to the same tectonic episode.

Several areas of research are particularly attractive at this time. These include better knowledge of the origins of high fluid pressure responsible for jointing, and the maintenance of this pressure during joint growth. In isotropic rocks, the joint-propagation path is strongly dependent on the fluid pressure and the stress field, but the exact nature of this dependence is not completely understood, and factors that control joint paths in anisotropic rocks have attracted little attention. Surface textures on joint faces depend on the path and on the mechanical processes at the joint tip. The nature of these inelastic processes can be established by careful field and laboratory observations. The joint process zone may be the site of chemical reactions that are influential, if not the controlling factors in propagation. These reactions should be identified and their importance assessed in the propagation energy balance. The development of diagnostic methods to estimate propagation velocities of joints would be a major contribution. The sequential development of multiple joint segments in sedimentary rocks require further evaluation. The occurrence of joints in some lithologic units but not others, and varied joint orientations and spacings in different lithologic units of the same sequence remain to be exploited. Further work on the arrest of joints at interfaces would greatly help our understanding of the mechanisms controlling joint size and shape. The systematic spacing of joints in sediment sequences depends in part on these arrest mechanisms and also on mechanical interaction, but spacing and clustering of joints remains an enigmatic problem. The interpretation of crosscutting joints in terms of deformation episodes and relative age can be deduced in some cases, but improved field criteria should be formulated. Finally, further consideration of the relationships among joints and major geologic structures, such as faults and folds, should improve our predictive capabilities for problems encountered in petroleum exploration and production, mineral resource recovery, and containment of toxic waste.

We opened this review by recounting some of the questions raised by geologists one hundred years ago concerning the origin and interpretation of joints. Progress has been slow, and several unproductive concepts have diverted the attention of researchers; therefore, some of these questions and many new ones are unanswered. The good news is that field methods and theoretical tools are now available to unravel many of the mysteries of joints and jointing.

ACKNOWLEDGMENTS

This work was supported by National Science Foundation Grant EAR-8707314 to Pollard and Grant EAR-8415113 to Aydin and by a grant from ARCO. The last draft of this paper was written while Aydin was on sabbatical at the Department of Applied Earth Sciences, Stanford University. He appreciates the support from that department and from the Rock Physics Project of A. Nur. We thank D. Secor, T. Engelder, and R. Groshong for their thoughtful reviews for the *Bulletin*. Aydin thanks T. Engelder for introducing him to interesting outcrops in central New York. Pollard thanks E. F. Poncelet for his skeptical attitude regarding shear cracks. Our colleagues and students, including C. Barton, J. DeGraff, P. Delaney, D. Helgeson, M. Jackson, M. Linker, L. Mastin, G. Ohlmacher, J. Olson, A. Rubin, R. Schultz, P. Segall, S. Saltzer, and P. Wallmann have helped with suggestions and constructive criticisms, for which we are very grateful.

REFERENCES CITED

Anderson, 1951, The dynamics of faulting: Edinburgh, Scotland, Oliver and Boyd Ltd., 206 p.
Atkinson, B. K., and Meredith, P. G., 1987a, The theory of subcritical crack growth with applications to minerals and rocks, *in* Atkinson, B. K., ed., Fracture mechanics of rock: London, England, Academic Press, p. 111–166.
—— 1987b, Experimental fracture mechanics data for rocks and minerals, *in* Atkinson, B. K., ed., Fracture mechanics of rock: London, England, Academic Press, p. 477–525.
Aydin, A., 1978, Small faults formed as deformation bands in sandstone: Pure and Applied Geophysics, v. 116, p. 913–930.
Aydin, A., and DeGraff, J. M., 1988, Evolution of polygonal fracture patterns in lava flows: Science, v. 239, p. 471–476.
Bahat, D., 1979, Theoretical considerations on mechanical parameters of joint surfaces based on studies on ceramics: Geology Magazine, v. 116, p. 81–92.
Bahat, D., and Engelder, T., 1984, Surface morphology on cross-fold joints of the Appalachian Plateau, New York and Pennsylvania: Tectonophysics, v. 104, p. 299–313.
Balk, R., 1925, Primary structure of granite massives: Geological Society of America Bulletin, v. 36, p. 679–696.
—— 1937, Structural behavior of igneous rock: Geological Society of America Memoir 5, 177 p.
Balk, R., and Grout, F. F., 1934, Structural study of the Snowbank Stock: Geological Society of America Bulletin, v. 45, p. 621–636.
Bankwitz, P., 1965, Uber klufte I. Beobachtungen im Thuringischen Schiefergebirge: Geologie, v. 14, p. 241–253.
—— 1966, Uber klufte II. Die bildung der kluftlache und eine systematik ihrer strukturen: Geologie, v. 15, p. 896–941.
—— 1984, Die symmetrie von kluftoberflachen und ihre nutzung fur eine palaospannungsanalyse: Zeitschrift fuer geologische Wissenschaften, v. 12, p. 305–334.
Barton, C. C., 1982, Variables in fracture energy and toughness testing of rock, *in* Goodman, R. E., and Heuze, F. E., eds., Proceedings of the 23rd Symposium on Rock Mechanics, Issues in Rock Mechanics: American Institute of Mining, Metallurgical, and Petroleum Engineers, p. 449–462.
—— 1983, Systematic jointing in the Cardium Sandstone along the Bow River, Alberta, Canada [Ph.D. thesis]: New Haven, Connecticut, Yale University, 301 p.
Becker, G. F., 1891, The structure of a portion of the Sierra Nevada of California: Geological Society of America Bulletin, v. 2, p. 49–74.
—— 1893, Finite homogeneous strain, flow and rupture of rocks: Geological Society of America Bulletin, v. 4, p. 13–90.
—— 1895, The torsional theory of joints: American Institute of Mining Engineering Transactions, v. 24, p. 130–138.
Billings, M. P., 1972, Structural geology: Englewood Cliffs, New Jersey, Prentice-Hall, Inc., 606 p.
Bles, J. L., and Feuga, B., 1986, The fracture of rock: New York, Elsevier, 131 p.
Borg, I. Y., and Maxwell, J. C., 1956, Interpretation of fabrics of experimentally deformed sands: American Journal of Science, v. 254, p. 71–81.
Brace, W. F., and Bombolakis, E. G., 1963, A note on brittle crack growth in compression: Journal of Geophysical Research, v. 68, p. 3709–3712.
Bucher, W. H., 1920, The mechanical interpretation of joints, Part I: Journal of Geology, v. 28, p. 707–730.
—— 1921, The mechanical interpretation of joints, Part II: Journal of Geology, v. 29, p. 1–28.
Cook, T. S., and Erdogan, F., 1972, Stresses in bonded materials with a crack perpendicular to the interface: International Journal of Engineering Science, v. 10, p. 677–697.
Cotterell, B., and Rice, J. R., 1980, Slightly curved or kinked cracks: International Journal of Fracture, v. 16, p. 155–169.

Crosby, W. O., 1882, On the classification and origin of joint structures: Proceedings of the Boston Society of Natural History, v. 22, p. 72–85.
Daubree, A., 1879, Etudes synthetiques de geologie experimentale: Paris, France, Dunod, 828 p.
Davis, G. H., 1984, Structural geology of rocks and regions: New York, John Wiley & Sons, 492 p.
DeGraff, J. M., 1987, Mechanics of columnar joint formation in igneous rocks [Ph.D. thesis]: West Lafayette, Indiana, Purdue University, 221 p.
DeGraff, J. M., and Aydin, A., 1987, Surface morphology of columnar joints and its significance to mechanics and directions of joint growth: Geological Society of America Bulletin, v. 99, p. 605–617.
Dennis, J. G., 1987, Structural geology, an introduction: Dubuque, Iowa, Wm. C. Brown, 448 p.
Dyer, J. R., 1983, Jointing in sandstones, Arches National Park, Utah [Ph.D. thesis]: Stanford, California, Stanford University, 202 p.
Engelder, T., 1982a, Is there a genetic relationship between selected regional joints and contemporary stress within the lithosphere of North America: Tectonics, v. 1, p. 161–177.
——— 1982b, Reply: Tectonics, v. 1, p. 465–470.
——— 1985, Loading paths to joint propagation during a tectonic cycle: An example from the Appalachian Plateau, U.S.A.: Journal of Structural Geology, v. 7, p. 459–476.
——— 1987, Joints and shear fractures in rock, *in* Atkinson, B. K., ed., Fracture mechanics of rock: London, England, Academic Press, p. 27–69.
Engelder, T., and Geiser, P., 1980, On the use of regional joint sets as trajectories of paleostress fields during the development of the Appalachian Plateau, New York: Journal of Geophysical Research, v. 85, p. 6319–6341.
Engelder, T., and Oertel, G., 1985, Correlation between abnormal pore pressure and tectonic jointing in the Devonian Catskill Delta: Geology, v. 13, p. 863–866.
Erdogan, F., and Biricikoglu, V., 1973, Two bonded half planes with a crack going through the interface: International Journal of Engineering Science, v. 11, p. 745–766.
Ernsberger, F. M., 1960, Detection of strength-impairing surface flaws in glass: Royal Society of London Proceedings, ser. A, v. 257, p. 213–223.
Friedman, M., 1975, Fracture in rock: Reviews of Geophysics and Space Physics, v. 13, p. 352–358.
Friedman, M., and Stearns, D. W., 1971, Relations between stresses inferred from calcite twin lamellae and macrofractures, Teton Anticline, Montana: Geological Society of America Bulletin, v. 82, p. 3151–3162.
Friedman, M., Handin, J., and Alani, G., 1972, Fracture energy of rocks: International Journal of Rock Mechanics and Mining Science, v. 9, p. 757–766.
Freund, L. B., 1972, Crack propagation in an elastic solid subjected to general loading: Journal of the Mechanics and Physics of Solids, v. 20, p. 129–152.
Gallagher, J. J., Friedman, M., Handin, J., and Sowers, G. M., 1974, Experimental studies relating to microfracture in sandstone: Tectonophysics, v. 21, p. 203–247.
Gell, M., and Smith, E., 1967, The propagation of cracks through grain boundaries in polycrystalline 3% silicon-iron: Acta Metallurgica, v. 15, p. 253–258.
Geyer, J. F., and Nemat-Nasser, S., 1982, Experimental investigation of thermally induced interacting cracks in brittle solids: International Journal of Solids and Structures, v. 18, p. 349–356.
Gilbert, G. K., 1882a, Post-glacial joints: American Journal of Science, v. 123, p. 25–27.
——— 1882b, On the origin of jointed structure: American Journal of Science, v. 124, p. 50–53.
——— 1884, On the origin of jointed structure: American Journal of Science, v. 127, p. 47–49.
Gramberg, J., 1965, Axial cleavage fracturing, a significant process in mining and geology: Engineering Geology, v. 1, p. 31–72.
Griffith, A. A., 1921, The phenomena of rupture and flow in solids: Royal Society of London Transactions, v. 221, p. 163–198.
——— 1924, The theory of rupture, *in* Biezeno, C. B., and Burgers, J. M., eds., First International Congress on Applied Mechanics, Proceedings: Delft, J. Waltman, p. 55–63.
Griggs, D. T., and Handin, J., 1960, Observations on fracture and an hypothesis of earthquakes, *in* Griggs, D. T., and Handin, J., eds., Rock deformation: Geological Society of America Memoir 79, p. 347–364.
Grout, F. F., and Balk, R., 1934, Internal structures in the Boulder Batholith: Geological Society of America Bulletin, v. 45, p. 877–896.
Hancock, P. L., 1985, Brittle microtectonics: Principles and practice: Journal of Structural Geology, v. 7, p. 437–457.
Harris, J. F., Taylor, G. L., and Walper, J. L., 1960, Relation of deformational fractures in sedimentary rocks to regional and local structures: American Association of Petroleum Geologists Bulletin, v. 44, p. 1853–1873.
Hartmann, L., 1896, Distribution des deformations dans les metaux soumis a des efforts: Paris, France, Berger-Levrault.
Hoagland, R. G., Hahn, G. T., and Rosenfield, A. R., 1973, Influence of microstructure on fracture propagation in rock: Rock Mechanics, v. 5, p. 77–106.
Hobbs, B. E., Means, W. D., and Williams, P. F., 1976, An outline of structural geology: New York, John Wiley & Sons, 571 p.
Hobbs, D. W., 1967, The formation of tension joints in sedimentary rocks: An explanation: Geological Magazine, v. 104, p. 550–556.
Hobbs, W. H., 1904, Lineaments of the Atlantic border regions: Geological Society of America Bulletin, v. 15, p. 483–506.
——— 1905, Examples of joint-controlled drainage from Wisconsin and New York: Journal of Geology, v. 13, p. 363–374.
Hodgson, R. A., 1961a, Classification of structures on joint surfaces: American Journal of Science, v. 259, p. 493–502.
——— 1961b, Regional study of jointing in Comb Ridge–Navajo Mountain area, Arizona and Utah: American Association of Petroleum Geologists Bulletin, v. 45, p. 1–38.
Hoek, E., and Bieniawski, Z. T., 1965, Brittle fracture propagation in rock under compression: International Journal of Fracture Mechanics, v. 1, p. 137–155.
Hopkins, W., 1835, Researches in physical geology: Transactions of the Cambridge Philosophical Society, v. 9, p. 1–84.
——— 1841, On the geological structure of the Weldon District and of the Bas Boullonais: Geological Society of London Transactions, 2nd ser., v. 12, p. 1–51.
Hoppin, R. A., 1961, Precambrian rocks and their relationship to Laramide structure along the east flank of the Bighorn Mountains near Buffalo, Wyoming: Geological Society of America Bulletin, v. 72, p. 351–368.
Holst, T. B., and Foote, G. R., 1981, Joint orientation in Devonian rocks in the northern portion of the lower peninsula of Michigan: Geological Society of America Bulletin, v. 92, p. 85–93.
Hoskins, L. M., 1896, Flow and fracture of rocks as related to structure, *in* Van Hise, C. R., Principles of North American Pre-Cambrian Geology: U.S. Geological Survey 16th Annual Report, p. 845–879.
Hubbert, M. K., 1951, Mechanical basis for certain familiar geologic structures: Geological Society of America Bulletin, v. 62, p. 355–372.
Hubbert, M. K., and Rubey, W. W., 1959, Role of fluid pressure in mechanics of overthrust faulting: Geological Society of America Bulletin, v. 70, p. 115–166.
Hutchinson, R. M., 1956, Structure and petrology of Enchanted Rock Batholith, Llano and Gillespie Counties, Texas: Geological Society of America Bulletin, v. 67, p. 763–806.
Inglis, C. E., 1913, Stresses in a plate due to the presence of cracks and sharp corners: Royal Institute of Naval Architects Transactions, v. 55, p. 219–230.
Ingrafea, A. R., 1981, Mixed-mode fracture initiation in Indiana limestone and Westerly granite: U.S. Symposium on Rock Mechanics, 22nd, Proceedings, p. 186–191.
Irwin, G. R., 1957, Analysis of stresses and strains near the end of a crack traversing a plate: Journal of Applied Mechanics, v. 24, p. 361–364.
——— 1958, Fracture, *in* Flugge, S., ed., Encyclopedia of physics: Berlin, Springer-Verlag, p. 551–590.
Jaeger, J. C. and Cook, N.G.W., 1969, Fundamentals of rock mechanics: London, England, Methuen and Co., 513 p.
James, A.V.G., 1920, Factors producing columnar structure in lavas and its occurrence near Melbourne, Australia: Journal of Geology, v. 28, p. 458–469.
Kanninen, M. F., and Popelar, C. H., 1985, Advanced fracture mechanics: New York, Oxford University Press, 563 p.
Karihaloo, B. L., Keer, L. M., and Nemat-Nasser, S., 1980, Crack kinking under nonsymmetric loading: Engineering Fracture Mechanics, v. 13, p. 879–888.
Keer, L. M., and Chen, S. H., 1981, The intersection of a pressurized crack with a joint: Journal of Geophysical Research, v. 86, p. 1032–1038.
Kelley, V. C., and Clinton, N. J., 1960, Fracture systems and tectonic elements of the Colorado Plateau: University of New Mexico Publications in Geology, v. 6, 104 p.
Kobayashi, T., and Fourney, W. L., 1978, Experimental characterization of the development of the micro-crack process zone at a crack tip in rock under load, *in* Kim, Y. S., ed., Symposium on Rock Mechanics, 19th, Proceedings, p. 243–246.
King, W., 1875, Report on the superinduced divisional structure of rocks, called jointing; and its relation to slaty cleavage: Royal Irish Academy Transactions, v. 25, p. 605–662.
Kranz, R. L., 1983, Microcracks in rocks: A review: Tectonophysics, v. 100, p. 449–480.
Kulander, B. R., Barton, C. C., and Dean, S. L., 1979, The application of fractography to core and outcrop fracture investigations: Morgantown Energy Technology Center, METC/SP-79/3, 174 p.
Kulander, B. R., and Dean, S. L., 1985, Hackle plume geometry and joint propagation dynamics, *in* International Symposium on Fundamentals of Rock Joints, Proceedings, p. 85–94.
Lachenbruch, A. H., 1961, Depth and spacing of tension cracks: Journal of Geophysical Research, v. 66, p. 4273–4292.
——— 1962, Mechanics of thermal contraction cracks and ice-wedge polygons in permafrost: Geological Society of America Special Paper 70, 69 p.
Ladeira, F. L., and Price, N. J., 1981, Relationship between fracture spacing and bed thickness: Journal of Structural Geology, v. 3, p. 179–183.
La Pointe, P. R., and Hudson, J. A., 1985, Characterization and interpretation of rock mass joint patterns: Geological Society of America Special Paper 199, 37 p.
Lawn, B. R., and Wilshaw, T. R., 1975, Fracture of brittle solids: Cambridge, England, Cambridge University Press, 204 p.
LeConte, J., 1882, Origin of jointed structure in undisturbed clay and marl deposits: American Journal of Science, v. 123, p. 233–234.
Long, P. E., and Wood, B. J., 1986, Structures, textures, and cooling histories of Columbia River basalt flows: Geological Society of America Bulletin, v. 97, p. 1144–1155.
Lutton, R. J., 1969, Systematic mapping of fracture morphology: Geological Society of America Bulletin, v. 80, p. 2061–2065.
McGarr, A. M., 1982, Analysis of stress between provinces of constant stress: Journal of Geophysical Research, v. 87, p. 9279–9288.
McGarr, A. M., and Gay, N. C., 1978, State of stress in the earth's crust: Annual Reviews of Earth and Planetary Science, v. 6, p. 401–436.
McGee, W. J., 1983, Note on jointed structure: American Journal of Science, v. 125, p. 152–153.
Melin, S., 1982, Why do cracks avoid each other?: International Journal of Fracture, v. 23, p. 37–45.
Melton, F. A., 1929, A reconnaissance of the joint-systems in the Ouachita Mountains and central plains of Oklahoma: Journal of Geology, v. 37, p. 729–746.
Mohr, O., 1900, Welche Umstande bedingen die elastizitatsgrenze und den bruch eines materials: Zeitschrift des Vereines Deutscher Ingenieure, v. 44, p. 1524.
Muecke, G. K., and Charlesworth, H.A.K., 1966, Jointing in folded Cardium sandstones along the Bow River, Alberta: Canadian Journal of Earth Sciences, v. 3, p. 579–596.
Muehlberger, W. R., 1961, Conjugate joint sets of small dihedral angle: Journal of Geology, v. 69, p. 211–218.
Muller, L., 1933, Untersuchengen uber statistische kluftmessung: Geologie und Bauwesen, Jarb. 5, p. 206–234.
Nemat-Nasser, S., and Horii, H., 1982, Compression-induced nonplanar crack extension with application to splitting, exfoliation, and rockburst: Journal of Geophysical Research, v. 87, p. 6805–6821.
Nemat-Nasser, S., Keer, L. M., and Parihar, K. S., 1978, Unstable growth of thermally induced interacting cracks in brittle solids: International Journal of Solids and Structures, v. 14, p. 49–430.
Nickelsen, R. P., 1976, Early jointing and cumulative fracture patterns: International Conference on the New Basement Tectonics, 1st, Proceedings, Utah Geological Association, Publication 5, p. 193–199.
Nickelsen, R. P., and Hough, V.N.D., 1967, Jointing in the Appalachian Plateau of Pennsylvania: Geological Society of America Bulletin, v. 78, p. 609–629.
——— 1969, Jointing in south-central New York: Reply: Geological Society of America Bulletin, v. 80, p. 923–926.
Nilsen, T. H., 1973, The relation of joint patterns to the formation of fjords in western Norway: Norsk Geologisk Tidsskrift, v. 53, p. 183–194.
Nur, A., 1982, The origin of tensile fracture lineaments: Journal of Structural Geology, v. 4, p. 31–40.
Nur, A., and Simmons, G., 1970, The origin of small cracks in igneous rocks: International Journal of Rock Mechanics and Mining Science, v. 7, p. 307–314.
Padovani, E. R., Shirly, S. B., and Simmons, G., 1982, Characteristics of microcracks in amphibolite and granulite facies grade rocks from southeastern Pennsylvania: Journal of Geophysical Research, v. 87, p. 8605–8630.
Parker, J. M., 1942, Regional systematic jointing in slightly deformed sedimentary rocks: Geological Society of America Bulletin, v. 53, p. 381–408.
——— 1969, Jointing in south-central New York: Discussion: Geological Society of America Bulletin, v. 80, p. 19–22.
Peck, D. L., and Minakami, T., 1968, The formation of columnar joints in the upper part of Kilauean lava lakes, Hawaii: Geological Society of America Bulletin, v. 79, p. 1151–1166.
Peck, L., Barton, C. C., and Gordon, R. B., 1985b, Measurement of the resistance of imperfectly elastic rock to the propagation of tensile cracks: Journal of Geophysical Research, v. 90, p. 7827–7836.
Peck, L., Nolen-Hoeksema, R. C., Barton, C. C., and Gordon, R. B., 1985a, Microstructure and the resistance of rock to tensile fracture: Journal of Geophysical Research, v. 90, p. 11,533–11,546.
Peng, S., and Johnson, A. M., 1972, Crack growth and faulting in cylindrical specimens of Chelmsford granite: International Journal of Rock Mechanics and Mining Science, v. 9, p. 37–86.
Pincus, H. J., 1951, Statistical methods applied to the study of rock fractures: Geological Society of America Bulletin, v. 62, p. 81–130.
Pollard, D. D., 1976, On the form and stability of open hydraulic fractures in the earth's crust: Geophysical Research Letters, v. 3, p. 513–516.
——— 1987, Elementary fracture mechanics applied to the structural interpretations of dykes, *in* Halls, H. C., and Fahrig, W. F., eds., Mafic dyke swarms: Geological Association of Canada Special Paper 34, p. 5–24.
Pollard, D. D., and Aydin, A., 1984, Propagation and linkage of oceanic ridge segments: Journal of Geophysical Research, v. 89, p. 10,017–10,028.
Pollard, D. D., and Holzhausen, G., 1979, On the mechanical interaction between a fluid-filled fracture and the earth's surface: Tectonophysics, v. 53, p. 27–57.
Pollard, D. D., and Segall, P., 1987, Theoretical displacements and stresses near fractures in rock, with applications of faults, joints, veins, dikes, and solution surfaces, *in* Atkinson, B. K., ed., Fracture mechanics of rock: London, England, Academic Press, p. 277–349.
Pollard, D. D., Segall, P., and Delaney, P. T., 1982, Formation and interpretation of dilatant echelon cracks: Geological Society of America Bulletin, v. 93, p. 1291–1303.
Plafker, G., 1964, Oriented lakes and lineaments of northeastern Bolivia: Geological Society of America Bulletin, v. 75, p. 503–522.
Price, N., 1959, Mechanics of jointing in rocks: Geology Magazine, v. 96, p. 149–167.
——— 1966, Fault and joint development in brittle and semi-brittle rock: Oxford, England, Pergamon Press, 176 p.
Ramsay, J. G., 1967, Folding and fracturing of rocks: New York, McGraw-Hill Book Co., 568 p.
Ramsay, J. G., and Huber, M. I., 1983, The techniques of modern structural geology, Volume I: London, England,

Academic Press, 307 p.
——1987, The techniques of modern structural geology, Volume II: London, England, Academic Press, 700 p.
Reches, Z., 1976, Analysis of joints in two monoclines in Israel: Geological Society of America Bulletin, v. 87, p. 1654–1662.
Ritchie, R. O., and Yu, W., 1986, Short crack effects in fatigue: A consequence of crack tip shielding, in Ritchie, R. O., and Lankford, J., eds., Small fatigue cracks: Engineering Foundation International Conference 2nd, Proceedings: Warrendale, Pennsylvania, Metallurgical Society of AIME, p. 167–189.
Roberts, J. C., 1961, Feather-fracture, and the mechanics of rock-jointing: American Journal of Science, v. 259, p. 481–492.
Rubey, W. W., and Hubbert, M. K., 1959, Role of fluid pressure in mechanics of overthrust faulting: Geological Society of America Bulletin, v. 70, p. 167–206.
Rummel, F., 1987, Fracture mechanics approach to hydraulic fracturing stress measurements, in Atkinson, B. K., ed., Fracture mechanics of rock: London, England, Academic Press, p. 217–239.
Ryan, M. R., and Sammis, C. G., 1978, Cyclic fracture mechanisms in cooling basalt: Geological Society of America Bulletin, v. 89, p. 1295–1308.
Saemundsson, K., 1970, Interglacial lava flows in the lowlands of southern Iceland and the problem of two-tiered columnar jointing: Jökull, v. 20, p. 62–77.
Scheidegger, A. E., 1982, Comment on 'Is there a genetic relationship between selected regional joints and contemporary stress within the lithosphere of North America?': Tectonics, v. 1, p. 465.
Secor, D. T., 1965, Role of fluid pressure in jointing: American Journal of Science, v. 263, p. 633–646.
——1969, Mechanics of natural extension fracturing at depth in the earth's crust, in Research in tectonics: Geology Survey of Canada Paper 68-52, p. 3–47.
Secor, D. T., and Pollard, D. D., 1975, On the stability of open hydraulic fractures in the earth's crust: Geophysical Research Letters, v. 2, p. 510–513.
Segall, P., 1984a, Formation and growth of extensional fracture sets: Geological Society of America Bulletin, v. 95, p. 454–462.
——1984b, Rate-dependent extensional deformation resulting from crack growth in rock: Journal of Geophysical Research, v. 89, p. 4185–4195.
Segall, P., and Pollard, D. D., 1980, Mechanics of discontinuous faults: Journal of Geophysical Research, v. 85, p. 4337–4350.
——1983a, Nucleation and growth of strike-slip faults in granite: Journal of Geophysical Research, v. 88, p. 555–568.
——1983b, Joint formation in granitic rock of the Sierra Nevada: Geological Society of America Bulletin, v. 94, p. 563–575.
Sheldon, P., 1912a, Some observations and experiments on joint planes, I: Journal of Geology, v. 20, p. 53–79.
——1912b, Some observations and experiments on joint planes, II: Journal of Geology, v. 20, p. 164–183.
Sih, G. C., 1973, Handbook of stress intensity factors: Bethlehem, Pennsylvania, Institute of Fracture and Solid Mechanics, Lehigh University.
Sneddon, I. N., 1946, The distribution of stress in the neighborhood of a crack in an elastic solid: Royal Society of London Proceedings, v. 187, p. 229–260.
Sommer, E., 1969, Formation of fracture lances in glass: Engineering Fracture Mechanics, v. 1, p. 539–546.
Sowers, G. M., 1972, Theory of spacing of extension fracture, in Pincus, H., ed., Geological factors in rapid excavation: Engineering Geology Case History No. 9: Boulder, Colorado, Geological Society of America, p. 27–53.
Spencer, E. W., 1959, Geologic evolution of the Beartooth Mountains, Montana and Wyoming, Part 2. Fracture patterns: Geological Society of America Bulletin, v. 70, p. 467–508.

Sprunt, E. S., and Brace, W. E., 1974, Direct observation of microcavities in crystalline rocks: International Journal of Rock Mechanics and Mining Science, v. 11, p. 139–150.
Spry, A. H., 1962, The origin of columnar jointing, particularly in basalt flows: Geological Society of Australia Journal, v. 8, p. 191–216.
Stearns, D. W., 1969, Certain aspects of fracture in naturally deformed rocks, in Riecker, R.E., ed., Rock mechanics seminar: Bedford, Massachusetts, Air Force Cambridge Research Laboratory, p. 97–118.
Suppe, J., 1985, Principles of structural geology: Englewood Cliffs, New Jersey, Prentice-Hall, Inc., 537 p.
Swain, M. V., and Hagan, J. T., 1978, Some observations of overlapping interacting cracks: Engineering Fracture Mechanics, v. 10, p. 299–304.
Swanson, C. O., 1927, Notes on stress, strain, and joints: Journal of Geology, v. 35, p. 193–223.
Swanson, P. L., 1987, Tensile fracture resistance mechanisms in brittle polycrystals: An ultrasonics and in situ microscopy investigation: Journal of Geophysical Research, v. 92, p. 8015–8036.
Tada, H., Paris, P. C., and Irwin, G. R., 1973, The stress analysis of cracks handbook: Hellertown, Pennsylvania, Del Research Corporation.
Tapponnier, P., and Brace, W. F., 1976, Development of stress-induced microcracks in Westerly granite: International Journal of Rock Mechanics and Mining Science, v. 13, p. 103–112.
Tarr, R. S., 1894, Lake Cayuga, a rock basin: Geological Society of America Bulletin, v. 5, p. 339–356.
Van Hise, C. R., 1896, Principles of North American Pre-Cambrian geology: U.S. Geological Survey 16th Annual Report, p. 581–874.
Wallner, H., 1939, Linienstruckturen an Bruchflachen: Zeitschrift für Physik, v. 114, p. 368–378.
Weertmann, J., 1971, Theory of water-filled crevasses in glaciers applied to vertical magma transport beneath oceanic ridges: Journal of Geophysical Research, v. 76, p. 1171–1183.
——1980, The stopping of a rising, liquid-filled crack in the earth's crust by a freely slipping horizontal joint: Journal of Geophysical Research, v. 83, p. 967–976.
Wise, D. U., 1964, Microjointing in basement, middle Rocky Mountains of Montana and Wyoming: Geological Society of America Bulletin, v. 75, p. 287–306.
——1982, A new method of fracture analysis: Azimuth versus traverse distance plots: Geological Society of America Bulletin, v. 93, p. 889–897.
Wong, T-f., 1982, Micromechanics of faulting in Westerly granite: International Journal of Rock Mechanics and Mining Science, v. 19, p. 49–64.
Woodworth, J. B., 1896, On the fracture system of joints, with remarks on certain great fractures: Boston Society of Natural Historical Proceedings, v. 27, p. 163–183.
Yoffe, E. H., 1951, The moving Griffith crack: Philosophical Magazine, v. 42, p. 739–750.

MANUSCRIPT RECEIVED BY THE SOCIETY OCTOBER 20, 1987
REVISED MANUSCRIPT RECEIVED APRIL 5, 1988
MANUSCRIPT ACCEPTED APRIL 5, 1988

Low-temperature deformation mechanisms and their interpretation

CENTENNIAL ARTICLE

RICHARD H. GROSHONG, JR. *Department of Geology, The University of Alabama, Tuscaloosa, Alabama 35487-1945*

ABSTRACT

Low-temperature deformation is characterized by heterogeneous strain in which the bulk of the material clearly retains its primary texture. Deformation is by grain-scale crystal plasticity, rotation, fracture, and pressure solution, and by transgranular mechanisms that crosscut numerous grains. The important low-temperature crystal-plastic features are twin lamellae, deformation bands, and undulatory extinction. Subgrain formation by recrystallization or crystal-plastic strain of more than 15% marks the upper limit of the low-temperature regime. Grain rotation may produce foliations in soft sediments or rocks. Microscopic to mesoscopic kinks and crenulations of bedding occur in soft clay and shale. Transgranular features include Lüders' bands, cooling and desiccation cracks, joints, extension-fracture cleavage, clastic dikes, mineral-filled veins of several types, recrystallization/replacement veins, vein arrays, boudins, faults, stylolites, slickolites, solution cleavages that range from widely spaced to slaty and pencil cleavage. Pressure fringes form adjacent to relatively rigid grains and have fabrics analogous to those in veins. Faults include conjugate fault pairs (Andersonian faults) multiple simultaneous conjugates (Oertel faults), and Riedel shear-zone configurations. The sense of fault displacement is determined from bends, steps, trails, tails, and feather fractures. Superplasticity, especially if aided by diffusion in grain-boundary water, might be important at low temperatures. Fault textures are diagnostic of the environment of deformation but have yet to be uniquely correlated with the presence or absence of earthquakes. Riedel shears and pseudotachylite may form in earthquake source regions, although pseudotachylite is evidently rare in brittle fault zones. The best indicators of stress magnitudes are the critical resolved shear stress for deformation twinning and the presence of tensile fractures. Strain magnitudes and stress and strain tensor orientations can be determined with a variety of methods that are based on mechanical twins, platy grain orientation, grain center distribution, and fault geometry and slip directions. Different deformation mechanism associations, expressed by the partitioning of the total strain into different mechanisms, are related to the ductility and environment of deformation. Deformation fronts separating different mechanism associations are defined on the basis of changes in the crystal-plastic component of strain.

Rocks do not suffer deformation; they enjoy it.
Rob Knipe, 1982

INTRODUCTION

This Centennial article reviews the identification and interpretation of low-temperature deformation mechanisms as seen in outcrop and thin section. Interpretations are given in terms of the stress or strain that causes the features to develop and the stress or strain that may be inferred from the features. Low-temperature deformation generally means deformation at one-third or less of the melting temperature of the framework minerals. Under these conditions, individual crystals are strained less than about 15%–20%, a magnitude that leaves the rock looking relatively undeformed (Fig. 1). There is minimal recrystallization except in fine-grained, water-wet materials which may recrystallize as a result of either diagenesis or deformation. Framework grains do not react to form new minerals. Phyllosilicates may change species but retain their constituent sheets of silica tetrahedra (Oertel, 1983). Uncemented grains may be free to slide past one another under shallow-burial or high–pore-pressure conditions. Large strains usually occur by the transgranular mechanisms of extension fractures, faults, stylolites, or solution cleavage. It is the material between these transgranular features that remains relatively undeformed. Strain usually reduces porosity and permeability in high-porosity materials but may increase it in low-porosity materials.

Low-temperature deformation is often characterized as brittle, where brittle phenomena include a number of processes that are dominated by fracture as a mechanism or are controlled by the presence of pre-existing fractures (Logan, 1979). A fracture is defined as a surface along which a loss of cohesion has taken place (Griggs and Handin, 1960). Brittle defor-

Figure 1. Low-temperature deformation of the Upper Cretaceous Schrattenkalk, eastern Switzerland (sample 11, Groshong and others, 1984). Based on the twin strain, the rock has enjoyed a shortening of 6% at a temperature on the order of 150 °C (from the vitrinite reflectance of about 3.0% in nearby rocks, using the well-standard curve of Bostick, 1974). The fossil was discovered and photographed by Laurel Pringle-Goodell.

Geological Society of America Bulletin, v. 100, p. 1329–1360, 2 figs., September 1988.

mation is usually contrasted with ductile deformation; however, the term "ductile" is used in several different senses (Rutter, 1986): (1) plastic deformation of crystals, (2) homogeneous deformation by any mechanism (the uniform flow of Griggs and Handin, 1960), and (3) more than a specified amount of permanent strain (1%–2%; Logan, 1979) prior to faulting. A problem in using the terms in the field is whether soft-sediment faults and deformation by transgranular pressure solution should be considered brittle or ductile. Current usage favors the former term, and so "brittle" is here redefined as deformation with less than 1%–2% crystal-plastic strain prior to transgranular failure. "Semi-brittle" (suggested by C. Simpson, 1987) is deformation with 2%–20% crystal-plastic strain prior to transgranular failure, and "ductile" is deformation with >20% crystal-plastic strain with or without transgranular strain. Low-temperature strain is brittle to semi-brittle. Large homogeneous deformation may occur by soft-sediment flow, cataclastic flow (Stearns, 1969), or pressure solution, each being brittle or semi-brittle if the crystal-plastic strain is small. Thus "brittle" is the opposite of "crystal-plastic" (Rutter, 1986), rather than the opposite of "flow."

The review begins with crystal-plastic deformation features and moves on to the mainly transgranular mechanisms of extension fracture, pressure solution, and faulting. The stress, strain, and sense-of-shear interpretations of the mechanisms are discussed along with their descriptions. Techniques for inferring stress or strain from the bulk deformation are reviewed next. The final discussions are of strain partitioning, flow laws, deformation fronts, and deformation-mechanism associations. Many field observations have not been clearly understood until reproduced in the laboratory. An attempt is made here to bring together data from the field and the laboratory. Numerous illustrations would be desirable but are precluded by length constraints, and so the references cited in the text include page numbers that refer to key photographs, diagrams, or original definitions.

CRYSTAL PLASTICITY

Crystal plasticity is the permanent deformation without loss of cohesion of single crystals by internal deformation mechanisms. Mechanisms quantitatively important at low temperatures are twin glide and translation glide. Higher-temperature dislocation climb and diffusion mechanisms result in polygonization and recrystallization, mentioned here because they define the upper limit of low-temperature deformation. Grain-boundary diffusion may be important at both low and high temperatures. The term "plastic" is also widely used for the permanent deformation without loss of cohesion of the bulk rock, approximately synonymous with "flow" (Griggs, 1942). Stress-strain curves typically show a yield stress that marks the change from recoverable to permanent strain (Nádai, 1931), and so "plastic" is often used in "those cases where the presence of a yield value can be assumed" (Burgers and Burgers, 1935). The first experimental proof that normally brittle rocks could be made to flow was by Adams and Nicolson (1901), and the importance of both temperature and pressure was demonstrated by Adams and Coker (1910) and von Kármán (1911). The modern era of experimentation began with the high-pressure experiments of Griggs (1936). Analysis of plastic deformation usually requires oriented samples; see Prior and others (1987) for collecting procedures.

Twin Lamellae

Twin lamellae are thin bands of twinned material usually found in sets of many parallel lamellae. First recognized in calcite by Huygens in the late 1600s, they were shown by Reusch in the mid-1800s to be the result of deformation according to Klassen-Neklyudova (1964). Sorby (1908, Pl. 18) recognized them as a deformation feature in limestone. A thick twin is an optically visible domain that goes to extinction under crossed polarizers at a different position than the host grain. Thin twins do not contain distinguishable twinned material under the optical microscope but appear as thin dark lines (Knopf, 1949) that were called "untwinned" lamellae by Turner (1953). Conel (1962) showed that "untwinned" lamellae exhibit the interference fringes predicted to be associated with twins, and under the scanning electron microscope they have been seen to be twins (Spang and others, 1974). Twins formed at low temperatures in calcite are thin (1–10 μ) and plane sided, whereas twins formed at high temperatures are thicker and tend to be convex sided (Schmid and others, 1980).

Twin gliding occurs on a specific glide plane and in a specific direction along the glide line. The direction and sense of shear can always be uniquely determined for a given twin set (Bell, 1941; Turner and others, 1954a), which leads to the use of twins in dynamic and strain analysis, discussed later. The twinning geometry in dolomite was initially proposed from field samples by Fairbairn and Hawkes (1941) and later confirmed in laboratory experiments by Handin and Fairbairn (1955) and Turner and others (1954b, p. 483). Mechanical Dauphiné twinning in quartz produces strong preferred orientations but causes no permanent strain (Tullis, 1970; Tullis and Tullis, 1972). According to Vance (1961), deformation twinning was first recognized in plagioclase by van Werveke in 1883 and first produced experimentally by Mügge and Heide in 1931. Deformation twinning occurs according to (a) the albite law, for which the twin plane is (010) and the glide line is irrational and (b) the pericline law, for which the twin plane is a rhombic section and the glide line is <010>, both with positive sense of shear (Borg and Heard, 1970). The two directions of twinning are about 94° apart, and both develop during experimental deformation (Borg and Handin, 1966; Borg and Heard, 1970, p. 387). Mechanical twinning has been suggested for microcline (Marmo and Binns, in Spry, 1969) but has not yet been produced in the laboratory. The twin laws for 69 different crystal species are listed by Handin (1966, p. 241–243).

Deformation versus growth as the cause of twinning has long been debated. Growth twins tend to be single twins that divide a crystal into two units; deformation twins are lamellar or polysynthetic (Spry, 1969). In feldspar, lamellar twins are attributed to both growth and deformation. Vance (1961) suggested that growth twins in plagioclase are wide and highly variable in thickness, occur in small numbers, and often change width or terminate abruptly and independently of other twins, whereas deformation twins are thin, numerous, and, if they die out within a crystal, tend to do so along a kink boundary or parallel to the trend of the undulatory extinction. On the other hand, this difference closely resembles the difference between high- and low-temperature deformation twins produced experimentally in calcite (Schmid and others, 1980). Spry (1969) concluded that most polysynthetic twins in metamorphic plagioclase, microcline, calcite, and dolomite are the result of deformation. Calcite and dolomite twins have been very useful in the study of deformation; more attention should be given to the feldspars.

Deformation Lamellae and Undulatory Extinction

Translation glide indirectly results in deformation lamellae, certain types of undulatory extinction, and kinks in single crystals. Handin (1966, p. 238–240) gave the translation glide systems for 81 different crystal species. Deformation lamellae are parallel sets of very thin, optically distinct bands within a crystal that are not mineral cleavage, open fractures, or twins. In the optical microscope, one type of deformation lamellae in quartz consists of narrow, closely spaced subplanar features having slightly different refractive indices from the host quartz (Fairbairn, 1941, Pl. 2; Ingerson and Tuttle, 1945; Carter and others, 1964, Pl. 2; Christie and others, 1964), causing them to appear as parallel bands of brighter and darker material that terminate at crystal boundaries. They are subbasal or basal in quartz (Carter, 1971). Certain inclusion-rich lamellae in quartz have been called "Böhm" or "Boehme" lamellae (Böhm, 1883); however, this usage is vague and includes healed fractures (Griggs and Bell, 1938) and so should not be used (Fairbairn, 1941). I suggest the term "Fairbairn lamellae" for deformation lamellae resulting from a difference in refractive index relative to the host. In quartz they are usually associated with, and at a high angle to, parallel bands of undulatory extinction. Irrational Fairbairn lamellae have been produced in plagioclase (Borg and Heard, 1970). Quartz contains rational Fairbairn lamellae that are usually parallel to the basal plane or, less commonly, parallel to a rhomb. TEM reveals that the basal lamellae may be thin zones of glass with associated dislocation loops (S. White, 1973c; Christie and Ardell, 1974) or paired stacking faults interpreted as Brazil twins (S. White, 1973a, p. 27) and that the prismatic lamellae are zones of high dislocation density (Twiss, 1974). Subbasal Fairbairn lamellae in quartz are not parallel to any crystallographic plane but usually make an angle of ~30° to the basal plane. Ingerson and Tuttle (1945) recognized on the basis of field data that these lamellae formed in planes of high shear stress, a fact later demonstrated experimentally by Hansen and Borg (1962). In TEM, irrational lamellae are seen to have several different origins (S. White, 1973c, p. 27; Christie and Ardell, 1974; Ardell and others, 1976; Twiss, 1974, 1976): (1) walls of dislocations that form narrow, elongated basal or prismatic subgrains and are decorated with bubbles; (2) subgrains that are too thin to be optically resolvable; (3) dislocation tangles; and (4) alternating slabs of high- and low-dislocation density. Early experiments (Griggs, 1936; Griggs and others, 1960b) required unusually high stresses to produce this type of deformation in quartz. Griggs and Blacic (1965) accidentally discovered that small amounts of water drastically weaken the crystals. This hydrolitic weakening results in more realistic experimental stresses. The relationship of the lamellae orientation to the crystal structure is a function of pressure, temperature, and strain rate (Carter, 1971) with the rational lamellae being the result of shock deformation and irrational lamellae being formed under normal tectonic strain rates.

Sander (1930) made the first quantitative measurements of undulatory extinction attributed to deformation in quartz. The first experimental study was by Griggs and Bell (1938), who produced undulatory extinction but no other deformation features. Lattice bending results in a sweeping extinction, and the formation of new boundaries within the crystal leads to blocky and patchy extinction.

Kinking in a single crystal results from a slip instability on a weak direction within the crystal structure and results in extinction bands. Conjugate kink bands may form in crystals with very strong planar anisotropy, such as brucite and gypsum (Turner and Weiss, 1965, p. 359–364) or mica (Borg and Handin, 1966, p. 331). A single direction of kinking is characteristic of crystals with a weaker anisotropy (Carter, 1968), such as dolomite (Higgs and Handin, 1959), basal glide in quartz (Christie and others, 1964), r glide in calcite (Turner and Heard, 1965; Carter, 1976, p. 311) f glide in calcite (Wenk and others, 1973; Nicolas and Poirier, 1976), or slip on (010) in plagioclase (Seifert, 1965, p. 1470–1471). Gentle curvature at the kink boundaries gives a linear undulatory extinction, whereas sharp curvature produces well-defined kink bands (Carter and others, 1964, Pls. 4 and 5). Thin, well-defined bands that have a different crystallographic orientation from the host may be called "deformation bands" (Carter and others, 1964; Christie and others, 1964). Strongly developed deformation bands may cause a single crystal to look like several very elongate individual crystals (Carter and others, 1964, Pl. 5). The deformation origin of this type of undulatory extinction was first shown by lattice distortions seen in X-ray patterns (Bailey and others, 1958).

Submicroscopic microcracking and crushing between relatively undeformed zones can produce a patchy undulatory extinction that resembles poorly defined subgrains (Tullis and Yund, 1987). Patchy extinction involves more continuous and irregular changes in extinction position as compared to the sharp boundaries of subgrains (Tullis and Yund, 1987, p. 607). Patchy extinction has been produced experimentally in the low-temperature cataclastic flow of albite (Tullis and Yund, 1987). The fabric seems to penetrate the entire grain, rather than forming first at grain boundaries. This texture should form during low-temperature deformation.

Subgrains and Recrystallized Grains

Subgrains are regions within a crystal that are misoriented from the host by an angle of greater than one degree (Nicolas and Poirier, 1976, p. 139). Under the microscope, they are small, clear areas separated from one another by distinct, low-angle boundaries (Tullis and Yund, 1987). Polygonization texture was defined by Nicolas and Poirier (1976, p. 137) as having subgrains misoriented from the host by angles of 1°–15° and a recrystallization texture as having grains that are misoriented from the host by more than 15°. The typical polygon size is 50–100 μm but may be up to 1 mm (Nicolas and Poirier, 1976, p. 139). White (1976, p. 74) accepted a 10° misorientation as sufficient to define a new grain boundary that represents recrystallization. Subgrains produce a blocky undulatory extinction. In the early stages of development, subgrains are likely to be localized along older deformation bands or grain boundaries (White, 1973b, 1976), forming serrated boundaries (Hobbs and others, 1976, p. 115). This should be a relict texture in rocks deformed at low temperature.

Subgrain formation occurs by dislocation-glide-related processes that are thermally activated: dynamic recovery and dynamic recrystallization (Tullis and Yund, 1985). Dynamic recovery (polygonization of Poirier, 1985, p. 174) results in the formation of subgrains with less than 10° misorientation (Tullis and Yund, 1985). Progressive subgrain rotation can produce higher angle mismatches within this regime (Poirier, 1985; Tullis and Yund, 1985). Dynamic recrystallization occurs as either rotation recrystallization or migration recrystallization (Poirier, 1985). In rotation recrystallization, low-angle-boundary subgrains rotate to high-angle mismatches (Poirier, 1985, p. 181). In migration recrystallization, grain boundaries migrate between undeformed subgrains and deformed larger grains (Poirier, 1985, p. 182). Syntectonic migration recrystallization tends to produce jigsaw-puzzle-like grain boundaries (Hobbs, 1968, p. 387), whereas static migration recrystallization produces a foam texture. Dynamic recovery and recrystallization textures are generally relict in the low-temperature environment, but a foam texture might develop at low temperatures in very fine-grained rocks.

In a mortar texture (Spry, 1969, Pl. XXVIII) or core and mantle structure (White, 1976), large original grains are surrounded by and invaded by small subgrains or recrystallized grains (Carter and others, 1964, Pls. 8–10; Neumann, 1969, Pl. 1; Tullis and others, 1973, p. 305). A similar texture may also occur within large host grains which recrystallize along high-energy sites at kink-band boundaries or twin lamellae (Hobbs, 1968, p. 356–380). The orientation of the new grains tends to be related to that of the host, such as by a rotation about an a-axis in quartz (White, 1976). New grain formation at grain boundaries indicates that the recrystallization occurred *in situ* and was not inherited from the source terrane.

The size of dynamically recrystallized grains is inversely related to differential stress in single-phase materials (Twiss, 1977; Christie and Ord, 1980; Kohlstedt and Weathers, 1980; Etheridge and Wilkie, 1981; Ord and Christie, 1984; Ranalli, 1984; Poirier, 1985), although the exact relationship remains in doubt. If grain sizes are stabilized by a second phase, such as mica, then there may be no relationship to stress, and the relationship may be a function of temperature and strain rate (White, 1979). Subgrain size may reflect the maximum stress in a history of variable stress magnitudes (Poirier, 1985). In a rotational strain experiment, Friedman and Higgs (1981, p. 20) found that the larger recrystallized grains showed the most consistent relationship between grain size and magnitude of stress.

The Ambiguous Foam Texture

In a foam texture (Stanton, 1972, p. 233), the grains are plane-sided polygons having triple-point boundaries, as seen in a closely packed foam. The triple-point junctions between 3 grains are usually at 120° angles (Voll, 1960) when viewed normal to the boundary. The angle may vary from 120°, depending on the relative surface energies of the materials in contact (Stanton, 1972). This texture is characteristic of foam, compressed lead shot, and a variety of other organic and inorganic materials (Folk, 1965), certain undeformed and deformed sedimentary rocks, and is well known as the end result of the high-temperature static recrystallization of deformed rocks. Griggs and others (1960a, Pl. 1) published the first example of annealing in quartz. Annealing recrystallization does not produce a crystallographic orientation fabric, although traces of the pre-existing fabric remain because the orientation of the new crystals is related to that of the old (Griggs and others, 1960a; Hobbs, 1968, p. 357).

In unmetamorphosed limestones, foam texture is characteristic of micrite (2- to 3-μ grain size) and microspar (5- to 10-μ grain size) illustrated by Folk (1965, p. 34, 38). Microspar tends to be loaf shaped (Folk, 1965). Many examples of micrite and microspar are interpreted to be the result of diagenetic recrystallization of unstrained magnesian calcite or aragonite needles in an environment with a low Mg/Ca ratio or where Mg is removed from the rock (Folk, 1974). Diagenetic recrystallization of fine-grained calcite has been accomplished experimentally by Baker and others (1980) who found that the presence of clay minerals retarded the process. This suggests that relatively pure fine-grained materials will readily recrystallize at low temperatures and may likewise anneal or deform at low temperatures by mechanisms usually considered to be restricted to high temperatures. Keller and others (1985, p. 1354–1356) showed foam-textured chert (in the Ouachita thrust belt) in which the grain size systematically increases with degree of metamorphism. The thermal data of Guthrie and others (1986) show that the initial recrystallization began at a vitrinite reflectance of about 1.5% with a 1-μm crystal size and that crystal size increased up to 4 μm or more at a reflectance of 3%. This represents temperatures of 150–200 °C according to the "well-standard" correlation of Bostick (1974, p. 9). All of the rocks are folded or thrusted. Syntaxial cement may also produce a foam texture.

Recently a foam-like texture has been recognized as characteristic of superplastic deformation in which the dominant deformation mechanism is diffusion-aided, grain-boundary sliding (see discussion below of superplastic fault zones). In superplastic deformation, grains are relatively undeformed internally, the crystallographic fabric is weak, and the grains tend to be square or loaf shaped.

GRAIN ROTATION

Parallel Fabric

Primary ellipsoidal or platy grains may show a high degree of parallelism within a rock. This is commonly observed in shales (Ingram, 1953; Moon, 1972) where the fabric is contrasted to that of mudstones in which the grain alignment is absent. Parallelism of platy minerals is characteristic of slates (Sorby, 1853; flow cleavage of Leith, 1905), and parallelism of pebbles and cobbles is seen in deformed conglomerates (Leith, 1905). In fine-grained, phyllosilicate-rich rocks, the fabric is observed in thin section as a tendency for the whole section to go to extinction at the same time (Sorby, 1880; Dale and others, 1914). The mineral fabric can be observed directly using the SEM (Davies, 1982, p. 65) and indirectly as point concentrations in X-ray diffraction patterns (Odom, 1967, p. 612; Oertel, 1970; Oertel and Curtis, 1972, p. 2601). Sorby (1853) correlated the degree of preferred phyllosilicate orientation in slate to the quality of the cleavage, and Ingram (1953) appears to have been the first to quantitatively correlate the degree of preferred orientation in shale to the perfection of the fissility.

The parallel orientation of platy minerals has been attributed to deposition, deformation, compaction, and oriented grain growth. Clay may be deposited with a horizontal preferred orientation from fresh water but is usually deposited with an open house-of-cards framework of small packets of clay flakes (Moon, 1972; Collins and McGowan, 1974; Oertel, 1983). It has been demonstrated that constant-volume strain can cause randomly oriented clay flakes to become aligned perpendicular to the direction of shortening experimentally by Sorby (1853, 1908), Clark (1970), and Fernandez (1987). Shortening with dewatering produces only a slightly greater degree of preferred orientation than does shortening alone (Clark, 1970). Oertel and Curtis (1972) demonstrated that compaction was the cause of the preferred orientation seen in one shale, and it is a likely cause for the fabric in many others. The fabrics obtained by syndeformational synthesis or recrystallization experiments on mica produce the same degree of preferred orientation as strain without recrystallization (Means and Paterson, 1966; Tullis, 1976), leading these authors to conclude that rotation was the predominant orienting mechanism.

Slumping or tectonic deformation of soft sediments can produce planar and/or linear fabrics in the unconsolidated materials. In this case, the experiments of Sorby (1853, 1908) and Clark (1970) apply directly, rather than as rock analogs. A well-developed phyllosilicate foliation is seen in clay-rich rocks from a subduction zone (Lundberg and Moore, 1986, p. 19) where it is associated with a slickensided scaley fabric. Soft-sediment slump and buckle folds develop foliations that are caused by grain rotation (Moore and Geigle, 1974, p. 509; Brodzikowski and others, 1987). Williams and others (1969, p. 421–422) described both a single axial-plane parallel-mineral foliation in folded fine-grained shale (Pl. 2) and double planar-mineral foliations, separated by up to about 20° and bisected by the axial plane. Slump folds in a shale-siltstone sequence may have an axial-plane preferred orientation of platy minerals that is penetrative in thin section (Woodcock, 1976). In a soft-sediment fold in sandstone described by Yagishita and Morris (1979, p. 109), the quartz grains have a long-axis lineation parallel to the fold axis and weak axial-plane foliation defined by the intermediate dimensions of ellipsoidal grains. On the other hand, soft-sediment folds are usually characterized as not having *cleavage* (Helwig, 1970; Pickering, 1987). My personal experience is that axial-plane mineral foliations are reasonably common in folded soft sediments and that sediments or poorly consolidated rocks will usually break into rough slabs parallel to the foliation. This foliation does not resemble the near-penetrative smooth planes of splitting that characterize slaty cleavage or the well-defined compositional banding of spaced cleavage, nor are the rocks lithified to the same degree as slate.

Rotation of pebbles in a conglomerate having a fine-grained matrix produces a tectonic polish and aligned microstriations (Judson and Barks, 1961, p. 373). Clifton (1965, p. 871) documented an area in which the striations from all clasts were oriented parallel to dip and concluded that the striations were the result of structural deformation, although striations also occur in conglomerates that are not deformed tectonically (Judson and Barks, 1961; Wiltschko and Sutton, 1982). External rotation of a pebble due to slip along internal shear planes was found by Tyler (1975, p. 507) in a small fault zone. Mosher (1987) attributed a major component of the preferred clast orientation in a deformed conglomerate to rotation during deformation.

Interweaving Fabric

A mixture of equant grains with platy grains can result in an anastomosing fabric that might be interpreted as being the result of multiple deformation. The fabric is, however, common in natural soils (Brewer, 1964, p. 311–333; Collins and McGowan, 1974, p. 238–241) and is shown to occur in certain clays as well as in silt-clay mixtures. Brewer (1964) attributed the fabric to rotation of platy minerals caused by movements resulting from expansion and contraction caused by wetting and drying. Large phyllosilicate grains in a matrix of finer grains may also show oblique orientations (Weaver, 1984, p. 13). Lennox (1987) produced a similar fabric experimentally by deforming a mica–quartz-sand mixture under conditions favoring recrystallization, showing that, just as with parallel orientations, similar fabrics can be found in soils and metamorphic rocks and that both are probably caused primarily by deformation of a mixture of platy and equant grains.

Microlamination

The fissility or property of splitting along approximately parallel surfaces in a number of shales and slates correlates with the presence of microlaminations rather than grain fabric. The fissility of shale is significantly enhanced by fine-scale interbedding of clay with bands of organic material (Ingram, 1953; Gipson, 1965; Odom, 1967; Curtis and others, 1980, p. 336) or the presence or organic-rich and organic-poor clay layers (Spears, 1976). The fissility of slate may be caused by closely spaced laminae of different compositions (Sorby, 1880, p. 73; Plessman, 1965, p. 77). Fissility along such spaced cleavage (see discussion later) is more the result of compositional difference than grain orientations. Microlamination fabrics are common, and so fissility should not be assumed to be the result of grain orientation only.

Kinks and Crenulations

Asymmetrical kink bands and symmetrical crenulations occur as microscopic- to hand-specimen–scale deformation mechanisms in well-foliated but unlithified clay and sand. The first report of kinked clay appears to be the field examples and laboratory experiments of Tchalenko (1968, p. 169) in which kinks occur in and adjacent to fault zones. His experimental kinks were produced in a shear box in which the clay fabric began parallel to the shear plane. During shear, the foliation rotated in the direction of the shear couple and then kinked with the short limbs rotating in a sense opposite to that of the shear couple associated with faulting. This is evidently the same type of kink band that has been produced experimentally in card decks, slate (Gay and Weiss, 1974, p. 291, 294), and rubber strips (Honea and Johnson, 1976, p. 207; Reches and Johnson, 1976, p. 305, 307) where it is the result of unstable bedding-plane slip. In the experiments of Tchalenko (1968), the kink bands grew in width until all of the clay foliation had the kink orientation, about normal to the maximum compressive stress within the fault zone. At an intermediate stage, the structure looks like kinks with the opposite sense of asymmetry. Kinks have also been produced experimentally in clay by Foster and De (1971, Fig. 13) and Maltman (1977, p. 430) and prove to be relatively common in the deformed accretionary prism sediments of modern subduction zones (Lundberg and Moore, 1986, p. 22), as well as in other areas (Reches and Johnson, 1976, p. 298). Van Loon and others (1984) illustrated natural centimeter-scale folds from unconsolidated but well-bedded sands that have the geometry of chevron folds (p. 355), kinks (p. 357, 359, 368), and kinks with highly attenuated steep limbs (p. 358) resembling faults. The latter geometry is characteristic of the early stages of displacement on faults in experiments with unconsolidated sand (Hubbert, 1951, Pl. 2) and indicates the close connection between faults and certain folds having a kink geometry.

Maltman (1977, p. 422), using a pure-shear experimental configuration with clay, produced symmetric crenulations (p. 426, 430) and very tight hinged folds that he called creases (p. 432). Crenulations are found naturally in clay-rich accretionary prism sediments (Lundberg and Moore, 1986, p. 23) and in sedimentary rocks (Nickelsen, 1979, p. 258–259; Geiser and Engelder, 1983, p. 163; Nickelsen, 1986, p. 365). The symmetric geometry of the crenulations is favored when the maximum compressive stress is parallel to layering (Reches and Johnson, 1976).

EXTENSION FRACTURE

An extension fracture (Griggs and Handin, 1960) forms perpendicular to the least principal stress (σ_3). A tensile fracture (mode I of Lawn and Wilshaw, 1975) is formed if the least principal stress is tensile and the crack opens perpendicular to the fracture plane. Tensile stresses may occur in a compressive environment as a result of stress concentrations (Griffith, 1924) or by the increase of pore-fluid pressure to produce a tensile effective stress (Secor, 1965). Some fractures may form as mixed modes with both opening and shear displacement across the surface. A mode-II fracture (sliding mode, Lawn and Wilshaw, 1975) results from displacement of the crack walls parallel to the surface in a direction normal to the crack front, the slip being analogous to the movement of an edge dislocation. A mode-III fracture (tearing mode, Lawn and Wilshaw, 1975) results from displacement of the crack walls parallel to the surface and to the crack front, the slip being analogous to the movement of a screw dislocation.

Microfractures

Microfractures are microscopic cracks that are confined to a single crystal, grain, or grain boundary or to only a few crystals, grains, or grain boundaries. Cracks that are restricted to single grains imply a strong contrast in material properties between grain and matrix or grain and pore space. Such cracks usually connect grain-to-grain contacts and thus have a wide range of orientations within a single thin section (Gresley, 1895; Gallagher and others, 1974, p. 234–236; Aydin, 1978, p. 920; Batzle and others, 1980, p. 7074–7085). Through-going cracks that continue across several grains without change in orientation (Reik and Currie, 1974, p. 1257; Aydin, 1978, p. 926) imply low contrasts in physical properties between grains and matrix or grains and cement and require some degree of prior cementation in porous rocks (Gallagher and others, 1974). Microfractures that have healed without filling remain visible as planes of solid and fluid inclusions called "Tuttle lamellae" (Spry, 1969, citing N. Rast) after Tuttle (1949), who thoroughly described this feature. Experimental work shows that high temperature (Lemmleyn and Kliya, 1960, p. 127) and the presence of a reactive pore fluid greatly speed the healing process and reduce the time required for crack closure in quartz (Smith and Evans, 1984, p. 4128), implying that open microfractures are young and have remained at relatively low temperature (Kowallis and others, 1987). Field studies show that well-oriented microfracture sets are parallel to joint sets (Tuttle, 1949; Bonham, 1957; Roberts, 1965; Friedman, 1969; Dula, 1981). Filled microfractures that represent microveins may be invisible in plane light or crossed polars. Microveins may stand out in cathodoluminescence (Sippel, 1968; Sibley and Blatt, 1976; Sprunt and Nur, 1979; Kanaori, 1986, p. 137–140; Narahara and Wiltschko, 1986, p. 161) revealing the presence of many more veins than anticipated. These microveins do not resemble Tuttle lamellae. Elongate grains that formed in a low-temperature environment may owe their shape to invisible filled microfractures rather than to plastic strain (Narahara and Wiltschko, 1986, p. 161).

Grain fracturing has been the most conspicuous result of compaction experiments on quartz sands (Maxwell and Verrall, 1954; Borg and Maxwell, 1956; Maxwell, 1960, Pl. 1; Borg and others, 1960, Pl. 2). There has been some discussion about whether microfractures, especially those confined to single grains, solitary fractures, or single fracture sets were initially caused by extension or shear (Borg and others, 1960). A wide range of orientations is typical (Borg and Maxwell, 1956). It is now generally accepted that the origin is extensile (Gallagher and others, 1974; Kranz,

1983) and that shear offsets result from later movement, such as the sliding of a fragment into an adjacent pore, or from noncoaxial strain, as in a fault zone. Brown and Macaudiere (1984, p. 580–581), however, show examples of conjugate microfractures within single crystals of plagioclase having the proper directions and offsets to establish that they are conjugate faults. This pattern is quite different from the microfractures seen in a typical sandstone. The compressibility of porous materials increases with grain size as a result of the greater stress concentrations at the smaller number of grain contacts present in coarse-grained materials (Borg and others, 1960; Friedman, 1967).

Open microfractures have attracted much attention in recent years because of their association with brittle faulting (Kranz, 1983) and as possible porosity- and permeability-enhancing mechanisms. In experiments, microfractures are produced by either mechanical or thermal stresses. Brittle macroscopic shear failure of low-porosity rocks is usually preceded by distributed microfracturing that causes dilatancy, a volume increase (Brace and others, 1966). Significant increases in microfracture density along the trend of the fault occur just before faulting (Gramberg, 1965, p. 40–41). The microfractures nucleate at a variety of stress concentrations (Brace and others, 1966; Olsson and Peng, 1976, p. 55; Beeré, 1978, Pls. 1 and 2; Abdel-Gawad and others, 1987, p. 12914). Thermal stresses cause microfracturing because of anisotropic thermal expansion of adjacent grains, a process that is enhanced by the presence of quartz because of its large and anisotropic coefficients of thermal expansion (Kranz, 1983). Some microfractures form upon stress release when a rock is removed from deep burial and can be closed by subjecting the sample to a confining pressure equivalent to the released stress (Wang and Simmons, 1978). Stress-release microfractures may form parallel to existing joint sets (Carlson and Wang, 1986). Grain boundary microcracks seem to be a common result of surficial weathering and disappear in fresh rock (R. J. Kuryvial, 1976, personal commun.).

Measured microfracture porosities are in the range of 0.01%–2.4% (Brace and others, 1966; Carlson and Wang, 1986; Abdel-Gawad and others, 1987). Abundant weathering-related grain-boundary microfractures in a sandstone resulted in a porosity increase of zero to 1% and a permeability increase that was usually negligible but ranged upward to 5 millidarcys (measured under atmospheric conditions; R. J. Kuryvial, 1976, personal commun.). Microfracture porosity and permeability are much less under elevated confining pressure because of crack closure, unless the cracks are held open by surface irregularities (Batzle and others, 1980; Nelson, 1985).

Joints

A joint is a fracture surface with no visible displacement parallel to the surface (Bates and Jackson, 1980). It is a feature seen on the scale of an outcrop or greater. Wise (1964) defined microjoints as macroscopic subparallel fractures spaced closer than 3 mm apart. It seems likely that all joints form as cracks; that is, they are cohesionless at their inception. Later events might cause slip parallel to the surface or cementation to form veins.

Cooling and Desiccation Cracks. Contraction due to drying causes mudcracks in shaly or clay-rich sediments (Neal and others, 1968; Kahle and Floyd, 1971), and contraction due to cooling causes columnar joints in basalts (Iddings, 1886; Peck and Minakami, 1968; DeGraff and Aydin, 1987) and in permafrost (Lachenbruch, 1962). When fractures propagate from a planar surface, two-dimensional polygonal arrays (Neal and others, 1968) to rectangular arrays (Lachenbruch, 1962) are developed. Fracture intersection angles and, where visible, fracture surface markings indicate a tensile fracture origin for these features. Typical fracture intersection angles are 60°, 90°, and 120° (Lachenbruch, 1962; Peck and Minakami, 1968). A 90° intersection occurs where a propagating tension fracture meets the free surface of another open fracture. Because the free surface is a principal plane, a tension crack must intersect it at right angles (Anderson, 1942; Lachenbruch, 1962). A tension fracture propagating at some other angle will curve to intersect a free face at 90° (Lachenbruch, 1962). Branching of propagating tensile fractures has been observed in time-lapse photography of developing mudcracks to cause a 120°-angle intersection (Anderson and Everett, 1965), an observation that demonstrates that large-angle branching need not result from high-speed fracture propagation as suggested by Lachenbruch (1962). Surface striations on columnar basalt indicate tensile fracture propagation. According to Ryan and Sammis (1978), the striations were first noticed by Iddings (1886) and first related to the fracture process by James (1920). A striation consists of a smooth and rough band resulting from the propagation and cessation of propagation of the growing tensile fracture (Ryan and Sammis, 1978, p. 1296–1299). The stress concentration at the pre-existing fracture segment controls the position of the next fracture increment (DeGraff and Aydin, 1987). A small component of shear parallel to the fracture (combined fracture modes I and II) causes relief on the fracture surfaces called "fracture lances" (Ryan and Sammis, 1978, p. 1300) which trend parallel to the fracture propagation direction. Contraction cracks in soils may form as parallel sets or as orthogonal systems (Lachenbruch, 1962; Brewer, 1964, p. 139) mimicking systematic joints in rocks.

Syneresis fractures are extension fractures resulting from bulk volume reduction due to desiccation and are randomly oriented in three dimensions (Nelson, 1979, p. 2216). The fractures have a scale and spacing that causes them to resemble chicken wire on a plane surface, and so they are commonly called "chicken-wire fractures" (Nelson, 1979).

Systematic Joints. These joints occur as a subparallel set (Hodgson, 1961b). Nearly ubiquitous in surface exposures, they are readily visible on the scale of the outcrop, and a single joint may extend more than 400 ft (Hodgson, 1961b). In outcrop and thin section, uncemented joints are thin cracks that cut straight through most grains (Nickelsen and Hough, 1967, Pl. 6; Jamison and Stearns, 1982, p. 2594; Segall and Pollard, 1983a, p. 568) or may, in part, follow grain boundaries (Nelson, 1985, p. 32). Microscopic shear offsets of grain boundaries are not reported from thin sections of joints, indicating little if any shear displacement (Narr and Burruss, 1984, p. 1091; Ramsay and Huber, 1987, p. 642).

Systematic joints are characterized by their geometric relationships to one another and general surface morphology. Cross joints (Hodgson, 1961b, p. 18) extend between and are approximately normal to a systematic joint set, have irregular surfaces, and commonly terminate on bedding surfaces and against other joints. Nickelsen and Hough (1967, Pl. 4) pointed out that this term had been used previously by Balk (1937) for joints perpendicular to lineation in igneous rocks and renamed them "truncated joints." I prefer a more general terminology, calling the very planar joints "planar systematic" and the rougher-surface–textured joints "rough systematic," because both types occur in sets, and all rough systematic sets are not necessarily perpendicular to a planar set. The standard deviation of a planar set in a single outcrop is small, on the order of 2.4°, so that 95% of the joints should fall within a range of 9° (Groshong, 1965; Engelder and Geiser, 1980). Rough systematic joints may have rough surfaces or be rippled on a scale of centimeters; the orientation variability is usually twice or more that of a planar set. Hodgson (1961b) defined nonsystematic joints as joints that display random rather than oriented patterns in plan and section. He considered cross-joints to be a type of nonsystematic joint, an interpretation incompatible with their being perpendicular to a systematic set. I consider that the best use of the terminology is to define nonsystematic joints as random, not occurring in sets, and as excluding truncated joints.

The small-scale surface morphology of joints has played a major role in the interpretation of joint origin since the review by Hodgson (1961a). Forms are concentric ridges or ribs (Price, 1966, Pl. 2; Syme Gash, 1971, p. 351) and radial ridges and valleys called "hackle" or "hackle marks"

(Price, 1966; Syme Gash, 1971, p. 351; DeGraff and Aydin, 1987, p. 607) or "striations" (Bahat, 1979). Bahat (1979, p. 82; 1986a, p. 201) termed the portion of a joint between concentric ridges that has a rough surface a "hackle," a different use of the term. This will be here termed a "grainy surface" to avoid confusion. Plumose markings (Woodworth, 1896, Pls. 1-6; Hodgson, 1961a, p. 494) are hackle marks that consist of small ridges and valleys curving away from a central axis like the geometry of a feather. The edges of joints with plumose markings, especially at bedding planes, may show what Hodgson (1961a) called a "fringe of small en echelon fractures" that curve obliquely away from the main joint face and consequently are called "en echelons" by Bahat (1986a, p. 199-200; 1986b, p. 185-186). The initiation point of the fracture may be a very smooth "mirror" (Bahat, 1979, 1986a, p. 202). Hodgson (1961b, p. 23) pointed out that "the interlocking nature of the plumose patterns on the opposed faces also precludes transcurrent movement; no such movement can occur without obliterating the plumose pattern." Hodgson (1961b) believed that the plumose pattern arose from an extension fracture propagating along the direction of the plume toward the tip of the feather. Gramberg (1965) and Price (1966) cited experimental evidence that documents this interpretation. The en echelon fractures in the fringe region may be the result of a consistent slight rotation of the tensile stress direction (Pollard and others, 1982) or may be mixed-mode fractures analogous to the fracture lances seen on basalt columns (Ryan and Sammis, 1978; DeGraff and Aydin, 1987). Conchoidal ridges or ribs (Bahat, 1979, Pls. 3 and 4; Bahat and Engelder, 1984, p. 302) have been termed "arrest marks" by Bahat and Engelder (1984) because the same surface morphology is characteristic of the start-stop propagation of fatigue fractures in the laboratory. Fracture propagation is in the direction of convexity of the conchoidal ridges. A grainy surface represents small-scale branching of a rapidly propagating extension fracture (Bahat, 1986a, p. 201-203). Bahat and Engelder (1984, p. 302) and Bahat (1987a) pointed out that different joint sets in the same area may be characterized by different plume types. Inferences have been made about fracture velocity based on plume type, but all of the features evidently occur on both fast- and slow-moving fractures. Plumes with fringes are observed in the laboratory on rapidly moving fractures (Bahat, 1979) but have been observed on mudcracks by Bahat (1979, Pl. 16) and are interpreted by Pollard and others (1982) as occurring on slowly moving fractures. Conchoidal arrest marks may occur on rapidly moving fractures in association with seismic energy release (Ryan and Sammis, 1978) but represent a start-stop movement which implies that the growth of the complete fracture is slow.

The characteristics of joint intersections provide additional information about the origin of the fractures (Hancock, 1985, p. 447; Bahat, 1987b, p. 308-316). Hodgson (1961b, p. 14) pointed out that joints of the same set intersect by curving to end at right angles against each other. This is a common, although not universal, phenomenon and indicates the intersection of a tensile fracture with a free surface (principal plane) as in the propagation of mudcracks or basalt columns. Low-angle intersections pose a more difficult interpretation problem.

Bucher (1920) proposed a shear origin for pairs of joint sets that have a conjugate shear-fracture orientation. The Appalachian Plateau of New York and Pennsylvania provides a classic example of the alternatives. In an influential paper, Parker (1942) interpreted the New York Plateau as having three sets of joints. Parker's Set I (better termed "system I") contains two conjugate directions separated by a 16° angle. Parker's descriptions show the joints of system I to be planar systematic with abundant plumose markings. Parker interpreted these joints as shear fractures formed with a tensile σ_3 and concluded that the plumose markings must indicate a shear origin (see also Roberts, 1961). Discussing the adjacent Pennsylvania Plateau, Nickelsen and Hough (1967) reached the contrary conclusion that the planar systematic joints are extension fractures because of their openness, lack of tangential movements, and the interlocking character of the relief on plumose markings. They interpreted Parker's system I joints to be two independent sets, one overprinting the other, the major alternative to the conjugate hypothesis. Parker (1969) replied that individual joints of system I could be found changing direction along strike from one trend to the other, indicating contemporaneity. In the latest interpretation of system I by Nickelsen (1979), Engelder and Geiser (1980), Engelder (1985), and Engelder and Oertel (1985), there are two independent tensile joint directions formed in three different episodes. Deep in the undercompacted Devonian deltaic sequence (Engelder and Oertel, 1985), the more northwesterly set (Ib) is cut by a spaced cleavage that is contemporaneous with the more northeasterly set (Ia). Higher in the sequence, set Ib joints terminate against Ia joints, indicating that set Ia is here older, and thus indicating a third episode of jointing controlled by residual stress (Evans and Engelder, 1986) that produced joints parallel to those of the first episode.

In summary, it appears that joints are extensional in origin and may form at a variety of different times (Engelder, 1985, 1987) with orientations controlled by the modern stress field (Engelder, 1982), residual elastic strain (Friedman, 1972; Reik and Currie, 1974; Evans and Engelder, 1986), original rock fabrics (Brewer, 1964, p. 329; Nelson and Stearns, 1977), older deformation fabrics (Engelder and Geiser, 1980), or topography (Bradley, 1963; Nelson, 1979). Apparent conjugate relationships of joint (not fault) sets are best explained by overprinting; shear interpretations based on pattern analysis (Parker, 1942; Hancock, 1985) need additional documentation of the origin of the features before being accepted.

Joints may have a significant effect on bulk porosity and permeability, depending on their degree of opening and spacing. Joint widths tend to be in the range of 0.001-0.05 cm, although reduced by a factor of 10 to 1,000 by confining pressures appropriate to hydrocarbon reservoir depths (Nelson, 1985), but may be wider by a factor of 10 to 100 if held open by partial vein filling (Lucas and Drexler, 1976, p. 129-130; Nelson, 1985, p. 51, 55). Maximum fracture porosity is usually less than 2% (Nelson, 1985) but is 6% in the Monterey chert in California (Weber and Bakker, 1981). Not all fracture porosity or permeability is related to joints; faults may be an important factor. Joint permeabilities range from nearly zero to 400 millidarcys at the surface (Shuaib, 1973; Nelson, 1985) but tend to be less than 300 millidarcys under reservoir confining pressures (Nelson, 1985), a value large enough to have a very significant effect on the fluid flow in low-porosity rocks. Permeability increases approximately as the cube of the width of the opening (Engelder and Scholz, 1981).

VEINS AND RELATED FEATURES

A vein is a relatively thin, normally tabular, rock mass of distinctive lithologic character, usually crosscutting the structure of the host rock (Dennis, 1967). This definition does not specify the origin of the vein material; the filling could be from an external source such as an intrusive igneous vein or an epigenetic ore vein (Bates and Jackson, 1980), or could be the result of *in situ* deformation, as in a fault zone. The distinction between externally sourced quartz veins and fault zones containing *in situ* cataclastic filling can be difficult in the field and might require thin-section analysis.

Dilation veins represent openings that are filled by external materials. Active intrusions result from the forceful injection of material into extension fractures. Passive intrusions are pulled into low-pressure zones created by the formation of extension fractures. Nondilational veins (Hobbs and others, 1976, p. 292) result from the alteration or replacement of the wall rock.

Clastic Dikes

A clastic dike is a vein filled with sedimentary debris (Newson, 1903). Neptunian dikes (Hancock, 1985) are surface features such as mudcracks, filled with sediments from above (Heron and others, 1971). Some Neptunian dikes may fill pre-existing fractures that have structural significance but are not related to the filling episode (Smith, 1952), whereas other surface fractures form during an event that controls their orientation, such as faulting (Vintanage, 1954; Harms, 1965), landsliding, or glacial movement (Dionne and Shilts, 1974). Active or passive subsurface sediment intrusions may extend upward or downward from the source bed and are commonly oriented by the contemporaneous stress field in folds or adjacent to faults (Diller, 1890; Harms, 1965; Peterson, 1966; Plessman and Spaeth, 1971; Winslow, 1983). Clastic sills are also known (Bielenstein and Charlesworth, 1965; Truswell, 1972). Injected dikes may show compositional layering, graded layering, and alignment of grains parallel to the vein walls (Diller, 1890; Harms, 1965; Peterson, 1968, p. 180–187; Winslow, 1983, p. 1078).

Mineral-Filled Veins

Dilation veins filled with minerals precipitated from aqueous solutions are of primary interest here. Mosaic-filled and fiber-filled types are recognized. Mosaic-filled veins are characterized by interlocking anhedra of quartz (Adams, 1920) or calcite spar cement (Mišík, 1971) either covering an early fiber-type filling (druse) or perhaps completely filling the vein (Groshong, 1975b, p. 1368; Spang and Groshong, 1981, p. 333). This is called a "rapid opening fabric" by Mišík (1971) because the fracture obviously opened faster than it could be filled. This fabric is characteristic of the lower temperatures of deformation. Fiber-veins are filled with single crystals that do not show prominent crystal-face boundaries and that are elongate oblique to the vein wall. Not uncommonly the outer margin of a vein consists of fibers and a mosaic-filled center, suggesting an increase in opening rate (Mišík, 1971) or a decrease in the production of the filling (Ramsay and Huber, 1987). An opening rate increase may coincide with a change from extension normal to the vein wall to oblique opening (Groshong, 1975b, p. 1368; Beach, 1975, p. 259).

Macroscopic stylolites normal to veins are common (Nelson, 1981, p. 2421), and microstylolites may be found on fiber boundaries within veins (Beach, 1977, p. 217). Both relationships indicate shortening perpendicular to the length of the vein. Many examples have curved fibers. The optic axis of a crystal comprising a fiber is typically straight, not bent, and lacks evidence for large internal plastic deformation, and so the curvature is interpreted to be the result of growth, not the later deformation of originally straight fibers.

In a paper that triggered a great expansion of research in this area, Durney and Ramsay (1973) subdivided fiber veins into three types based upon the mode of growth: syntaxial, antitaxial, and composite (Ramsay and Huber, 1983, p. 241). In a syntaxial vein (Durney and Ramsay, 1973, p. 71) the fibers show a break in optical continuity near the center of the vein, interpreted to be the result of mid-point refracturing and crystal growth from the wall to the vein center. This fabric, however, is also characteristic of some diagenetic pore-filling cements (Sandberg, 1985, p. 35, 50; Pierson and Shinn, 1985, p. 160), where clearly a void is being filled, and so syntaxial veins may be of the void-filling type. Slightly more drusy cement would cause the cement texture in Figure 1 to resemble that of a syntaxial vein. Where the mineralogy of the vein filling and wall rock is the same, the fibers are optically continuous with the wall-rock grains (Mišík, 1971; Durney and Ramsay, 1973) and are usually free of wall-rock inclusions. In an antitaxial vein (Durney and Ramsay, 1973, p. 73; Ramsay and Huber, 1983, p. 241), the fibers are interpreted to have grown by refracturing at one or both of the vein edges. The fibers are optically continuous across the vein but not optically continuous with the wall-rock grains and are likely to contain inclusions of the wall rock. Cox and Etheridge (1983, p. 154–155) show that such veins might contain fibers grown syntaxially over the wall rock and might grow from one side only (p. 155). The vein crystals may become wider in the direction of growth due to the elimination of narrow fibers (Cox, 1987, p. 781).

A special type of antitaxial vein has been called the "product of vein coalescence" by Mišík (1971) and a "crack-seal vein" by Ramsay (1980b, p. 136–138). This vein type contains multiple screens of wall rock called "inclusion bands" (Ramsay and Huber, 1987, p. 576) marking successive openings. The "stretched crystal" veins of Durney and Ramsay (1973, p. 77) are interpreted by Ramsay and Huber (1983) as a type of crack-seal vein. Inclusion trails are isolated crystals in the vein that are generally of the same species and in optical continuity (Ramsay and Huber, 1987, p. 576). Inclusion trails track the direction of extension and may be oblique to the fiber direction (Cox, 1987, p. 781, 783), indicating that the fibers do not necessarily parallel the opening direction.

A composite vein (Durney and Ramsay, 1973, p. 76) has one mineral species at the vein wall and another at the center (but no evidence of void-filling textures). Both mineral phases of the vein are interpreted to have grown simultaneously at the phase boundaries within the vein (Durney and Ramsay, 1973), a notably different interpretation from the common assumption that the two phases represent two different episodes of infilling. The chemistry of this has not been explained.

Bedding-parallel veins are found in undeformed sedimentary rocks in which the fibers are approximately perpendicular to bedding. The variety known as "beef" has parallel fibers (Marshall, 1982). The unusual morphology called "cone-in-cone structure" (Sorby, 1860; Marshall, 1982, p. 618–619) usually contains crystals that diverge at low angles to produce a cone-shaped cross section. The interpreted origin of both types is similar to that of tectonic fiber veins (slow opening of the vein with simultaneous filling), but no explanation is given for the conical morphology which is absent in tectonic veins.

The description of cone-in-cone structure superficially resembles that of a shatter cone. A shatter cone is a distinctively striated conical fragment of rock (or fracture trace), ranging in length from less than a centimeter to several meters, along which fracturing has occurred; it is generally found in nested or composite groups (after Dietz, 1959, Pls. 1–8). Shatter cones are associated with high-velocity impact; the cones point toward the center of impact.

Recrystallization/Replacement Veins

Recrystallization/replacement veins show no evidence for separation of the wall rock (such as offsets of bedding) and appear to be the result of recrystallization or replacement of the host. Mišík (1971) described two types: (1) veins with ghost fabrics in which the grain size of the host is changed and (2) decolorizing veins. A decolorizing vein contains unaltered host fabric that has been lightened in color, usually being nearly clear in plane transmitted light (Mišík, 1971, p. 452). In a vein with ghost fabric, the vein crystals no longer conform to the original host-grain fabric and perhaps not even to the host-grain mineralogy, but undeformed ghosts of the original fabric remain (Mišík, 1971, p. 455). Tectonically oriented ghost-fabric recrystallized veins have been recognized by Spang and Groshong (1981, p. 323) in the same rock with dilation veins; they have ill-defined boundaries in thin section yet are clearly recognizable as veins and are normal to the extension direction inferred from calcite twinning. An analogous structure has been reported in plagioclase crystals by Hanmer (1981, p. T55–T57) where the vein orientation was at least partly controlled by structure within the crystal.

Vein Arrays

Veins quite commonly occur in en echelon arrays, and two end-member configurations are common: veins at nearly 45° to the array boundary and veins at 10°–20° to the array boundary (Hancock, 1972) or at even lower angles (Pollard and others, 1982, p. 1292) with a rather continuous range of examples in between. There are at least four documented origins for single arrays. (1) Small-angle–change kink bands require stretching parallel to bedding or dilation normal to bedding within the kink band and consequently may be the sites of bedding-parallel arrays within the kink band (Anderson, 1968, p. 204–205) or approximately bedding-normal arrays (termed "feather fractures" by Cloos, 1932, 1947b, p. 899; see also Garnett, 1974, p. 132). (2) Veins at 45° to the array boundary are at exactly the angle expected for tension fractures if the maximum shear stress is parallel to the array boundary, such as in a shear zone (Cloos, 1955, Pl. 3). (3) Veins at a low angle to the array boundary may be the result of the breakdown of a propagating planar vein that is parallel to the array boundary where the vein encounters a slight change in the direction of the maximum tensile stress (Pollard and others, 1982). (4) Veins oblique to a pre-existing surface may be produced as a result of slip on the surface (Conrad and Friedman, 1976; Nemat-Nasser and Horii, 1982).

Conjugate vein arrays are common. The conjugate array boundaries make angles of about 40° to more than 90° to one another, usually having the configuration of conjugate faults (Shainin, 1950, Pl. 1). Roering (1968) recognized and Beach (1975) defined two categories of conjugate array: (1) conjugate zones in which the undistorted portions of the veins in one zone are not parallel to those in the other zone and (2) conjugate zones in which the undistorted portions of the veins in both zones are parallel. In a type 2 zone, the veins are all parallel to the bisector of the acute angle between the zones (Beach, 1975, p. 248; Ramsay and Huber, 1987, p. 629). At the intersection of type 1 conjugate arrays, there is commonly a vein or veins parallel to the conjugate angle bisector (Choukroune and Séguret, 1968, p. 243), but veins of opposite trends may cross in this area (Shainin, 1950, Pl. 1). It is generally agreed that the far-field σ_1 axis bisects the acute angle between the vein arrays (Hancock, 1985), but the exact cause of the vein geometries and especially the evolution of the geometry are matters of controversy.

Vein arrays have the geometry of shear zones. Displacement parallel to the array boundary should rotate the veins (Ramsay and Graham, 1970) leading to a sigmoidal vein shape and implying large strains in the wall rock between the veins (Nicholson and Pollard, 1985). Large wall-rock strains are usually indicated in the field by pressure solution within the array in excess of that found outside the array (Beach, 1974; Rickard and Rixon, 1983). Later veins that are parallel to the original trend and that obliquely crosscut the earlier wider veins have been observed (Shainin, 1950, p. 516; Durney and Ramsay, 1973; Rickard and Rixon, 1983, p. 574), supporting the concept of vein rotation. Type 2 array veins may open perpendicular to the vein wall without significant rotation. The sigmoidal vein shapes can be caused by tip interactions between growing en echelon cracks (Pollard and others, 1982, p. 1299) or by the formation of curved country-rock bridges between dilating planar veins (Gorlov, 1971, p. 149; Beach, 1975, p. 256; Nicholson and Pollard, 1985, p. 584–585). The angle between conjugate type 2 arrays is explained by Ramsay and Huber (1987, p. 629) as being related to the dilation; a small conjugate angle implies volume gain, and a large conjugate angle implies volume loss associated with pressure solution. If the differential stress is small, it is possible that some veins will form as conjugate shears of very low dihedral angle (Muehlberger, 1961) and open obliquely (Hancock, 1972). Veins that show evidence for shear offset of external markers and irrotational oblique opening based on fibers might have this origin (Hancock, 1972, Pl. 4; Beach, 1977, p. 204–205; Burg and Harris, 1982, p. 353; Cox, 1987). Methods for computing the strain associated with en echelon veins are given in Ramsay and Huber (1983, 1987) and Collins and DePaor (1986).

Lüders' Bands

Lüders' bands (or lines) were originally described from deformed soft steel by Lüders in 1854 according to Nádai (1950). They are seen on the surface of a sample as conjugate sets of prominent parallel lines in planes of high shear stress. They were first described from experimentally deformed rocks by Heard (1960, Pl. 1). Heard found that Lüders' bands in the Solenhofen Limestone occurred in the brittle-ductile transition region, which is also characteristic of their occurrence in mild steel (Nádai, 1950). Experiments by Friedman and Logan (1973, p. 1466) resulted in two conjugate sets with a dihedral angle of 75°–103°. The dihedral angle is bisected by the σ_1 axis, and the angle increases with confining pressure. In both limestone and sandstone, the effect occurs in the brittle-ductile transition regime but continues into the ductile regime in sandstone. The features are not faults, as they have zero offset, no gouge, and a dihedral angle that is at least 20° greater than the faults that form late in the same experiment (Friedman and Logan, 1973, p. 1466). Optical and SEM analyses by Friedman and Logan reveal the bands to be zones of grain-size reduction (1973, p. 1470) or intense grain microfracturing (1973, p. 1473) with a width of about two original grain diameters. A natural example from sandstone (1973, p. 1475) shows prominent bands on outcrop that in thin section are zones of grain fracture without offset. Lüders' bands in rock thus appear to be arrays of extensional microfractures.

Pressure Shadows and Pressure Fringes

Pressure shadows and pressure fringes are veins, fillings, and relict materials found at the opposite ends of relatively rigid grains or fossils. In a pressure shadow (Spry, 1969), the foliation of the host rock wraps around the boundary of the shadow, and the grains in the shadow are unoriented or may be relics of the host rock. Since the work of Spry (1969), Choukroune (1971), and Durney and Ramsay (1973), a great deal of attention has been paid to pressure fringes, in which the foliation of the host rock abuts the boundary of the fringe, and the fringe contains oriented mineral growths (fibers) according to the definition of Spry (1969). These are well-known fabrics in low-grade metamorphic rocks, but pressure shadows have been recognized in deformed soft sediments (Knipe, 1986, p. 80), and short pressure fringes have been recognized in deformed sedimentary rocks (Fellows, 1943, Pl. 7; Geiser, 1974, p. 1404; Spang and others, 1979, p. 1111; Ramsay and Huber, 1983, p. 267). Such pressure fringes may be much more common than would be inferred from the paucity of published descriptions. Rutter (1983, p. T31) appears to have been the first to produce a pressure fringe experimentally.

Durney and Ramsay (1973) subdivided pressure fringes in a fashion analogous to their classification of veins. A syntaxial or pyrite-type (Ramsay and Huber, 1983) fringe is interpreted to have grown from the fiber/host-rock boundary. Syntaxial fringes are characteristically seen on pyrite and are found where there is a strong contrast in mineral species between the object being overgrown and the host rock, such as between a pyrite grain and argillite host. A more complex form of pyrite-type fringe is called "face controlled" by Ramsay and Huber (1983, p. 269), in which the fibers are oriented perpendicular to the pyrite face and for which the extension direction is oblique to the fiber trend. Inclusion trails and the boundaries between fibers growing from adjacent faces mark the true extension direction. An antitaxial or crinoid-type (Ramsay and Huber, 1983) fringe grows from the rigid-grain/fiber boundary. This form is common on crinoids and is seen where the object and the host material have the same mineralogy. A composite fringe has more than one mineral present, and growth occurs at the phase boundary.

Boudins

The original boudins (Lohest, 1909) are in a quartzite between schist beds and are barrel shaped in cross section with veins forming the top and bottom of the barrel. Bedding-normal veins, regularly spaced at about one or two times the thickness of the bed in which they occur and restricted to the bed (Cloos, 1947a, p. 626), provide a mechanism for the formation of boudins in the low-temperature deformation of rocks. In his review, Cloos (1947a) pointed out that investigators generally agreed that boudins were caused by extension of a brittle layer between more ductile layers but noted that the deformation implied by the barrel shape was an unresolved problem in brittle rocks. The barrel shape results from strain in the stiffer layer (Ramberg, 1955) and has been produced in rock experiments of a more brittle layer by cataclastic thinning (Gay and Jaeger, 1975b). Richter (1963, p. 245) and more recently Mullenax and Gray (1984, p. 65–69) have shown examples in limestone in which the barrel shape is due to differential removal of material by pressure solution.

The opening of vein sets of more than one orientation can lead to the chocolate tablet form of boudinage (Wegmann, 1932). The veins may open at different times, in different directions, and at different rates leading to complex fiber patterns (Casey and others, 1983; Ramsay and Huber, 1983, p. 256–257). Burg and Harris (1982) demonstrated that in a number of examples the vein arrays developed simultaneously, oblique to the extension axis. They interpreted the deformation to begin with necking due to the formation of Lüders' bands having the appropriate orientations, followed by through-going extension fracture parallel to the bands and vein formation.

STYLOLITES AND SLICKOLITES

Pressure solution is defined here descriptively as the process by which material is removed by solution or diffusion from along a discrete surface, the sides of which remain in close contact. This contrasts with vuggy or cavernous solution in which a void is created. Small voids may occur along pressure-solution surfaces (Spang and others, 1979, p. 1113; Tada and Siever, 1986), but large voids represent a later opening event (Wong and Oldershaw, 1981, p. 518). A stylolite is a thin seam or contact surface that is interlocking by mutual interpenetration of the two sides, or irregular or smooth (Logan and Semeniuk, 1976). Sorby (1863) was the first to recognize stylolites as being formed by the mechanism of pressure solution. Stylolites can be subdivided on the basis of their cross sections into columnar, peaked, irregular, hummocky, and smooth (Logan and Semeniuk, 1976, p. 16). A stylolitic surface is nearly always marked by a seam of the relatively insoluble or slowly diffusing components of the adjacent rock. Stylolites are common in carbonates (Stockdale, 1922, p. 8–12; Logan and Semeniuk, 1976, p. 17–20) and siliciclastics (Heald, 1955; 1956, Pls. 1–4) but have also been found in rhyolite (Golding and Conolly, 1962, p. 535–537), pegmatite (Bailly, 1954), chert (Glover, 1969; Cox and Whitford-Stark, 1987), and granite (Burg and Ponce de Leon, 1985, p. 434). Stylolite formation occurs under conditions ranging from diagenesis to at least the lower greenschist grade of metamorphism (Stockdale, 1922; Park and Schott, 1968; Mimran, 1975; Kerrich, 1978; Wong and Oldershaw, 1981; Groshong and others, 1984a; Engelder and Marshak, 1985).

Stylolites may either occur at grain-to-grain contacts or be transgranular, having lengths of tens of meters or more. Stylolites that are parallel to bedding may be the result of burial diagenesis (for example, Wong and Oldershaw, 1981) or vertical tectonics (Dunnington, 1967, p. 347; Johnson and Budd, 1975). Transgranular stylolites oblique to bedding were first recognized as being the result of structural deformation by Blake and Roy (1949, p. 784; see also Rigby, 1953, p. 268; Choukroune, 1969, p. 66; Arthaud and Mattauer, 1969; Plessman, 1972, p. 336) and may be referred to as tectonic stylolites (Jaroszewski, 1969). Smooth or gently curved to anastomosing stylolites oblique to bedding are the major cause of spaced cleavage and may be referred to as solution cleavage (Alvarez and others, 1978, p. 364).

The surface shape of a stylolitic contact is affected by the relative solubilities of the minerals on opposite sides of the contact and by the amount of insoluble residue present. Relative pressure solubilities have been determined from the minerals found within the insoluble residue and from differential grain indentation, summarized by Trurnit (1968) from

most to least soluble as (1) halite and potassium salts; (2) calcite; (3) dolomite; (4) anhydrite; (5) gypsum; (6) amphibolite and pyroxene; (7) chert; (8) quartzite; (9) quartz, glauconite, rutile, and hematite; (10) feldspars and cassiterite; (11) mica and clay minerals; (12) arsenopyrite; (13) tourmaline and sphene; (14) pyrite; (15) zircon; and (16) chromite. It seems possible that local pore-fluid chemistry could alter the relative solubilities, and so this order might not always be followed. It should be noted that an unreactive pore fluid such as liquid hydrocarbon prevents pressure solution in the laboratory (Griggs, 1940) and in the field (Dunnington, 1967) and inhibits precipitation (Hawkins, 1978) as well. Trurnit (1968) found that sutured contacts are characteristic of stylolites with materials of equal solubility on both sides (p. 97–99) and smooth contacts (p. 104–105) are characteristic of unequal solubility materials where only the more soluble material dissolves (Morawietz, quoted by Trurnit, 1968). A thick zone of insoluble residue acts like a more insoluble solution partner and favors smooth solution contacts (Trurnit, 1968). Heald (1955) observed that the amplitude of, and spacing between, sutured stylolites was greater in quartzose sandstones than in calcareous and argillaceous sandstones. Solution cleavages are much more abundant in argillaceous sandstones or carbonates than in purer lithologies. The rate of pressure solution is enhanced by the presence of clay (Heald, 1956; Oldershaw and Scoffin, 1967; Thomson, 1959; Whisonant, 1970; Mossop, 1972), and by the presence of good solvents (Gratier and Guiguet, 1986).

Mechanical indentation of a stiff material into a softer material was proposed as a cause for pebble indentation by Gresley (1895) and revived by Deelman (1975, p. 23) as a major alternative to pressure solution. Mechanical indentation implies large strain in the deformed material, resulting in internal shape perturbations, visible as fractures or crystal-plastic deformation features (Gresley, 1895; Fruth and others, 1966; Gay and Jaeger, 1975a, p. 315; McEwen, 1981, p. 33). Stockdale (1922, p. 56–57) pointed out that the preservation of undistorted bedding adjacent to the stylolite precluded this interpretation, and McEwen (1977, p. 249–250) provided additional examples.

Stylolite and solution cleavage surfaces are believed to form perpendicular to σ_1 (Sorby, 1908; Weyl, 1959; Robin, 1978; Fletcher and Pollard, 1981; Green, 1984) and are regularly observed to be at least approximately perpendicular to the axis of maximum shortening strain (Nickelsen, 1972; Geiser, 1974; Groshong, 1976; Engelder, 1979; Onasch, 1983a, 1983b).

The direction of shortening across a stylolite is parallel to the axis of the columns where this feature is present. Stockdale (1922, p. 39) observed that in square-wave columnar stylolites the caps of the columns contained thick seams of residue, whereas the sides of the columns contained much less material and were slickensided, indicating pressure solution across the caps and slip parallel to the column axis. The plane of a stylolite may curve 20°–30°, especially to follow some compositional heterogeneity of the rock, but the columns remain parallel (Stockdale, 1922, p. 44–54; Arthaud and Mattauer, 1969; 1972, p. 13).

Stylolites that are parallel to bedding usually nucleate along contacts of contrasting lithology. Stylolites oblique to bedding may also nucleate at lithologic contrasts, such as the edges of worm tubes and burrows (Nickelsen, 1972, p. 109). Cleavage patterns observed by Geiser and Sansone (1981, p. 282) led them to propose that cleavage followed pre-existing joints. Residue in stylolites that does not have the composition of insoluble residue of the host rock caused Laubscher (1980, personal commun.) to suggest that some stylolites in the Jura Mountains formed on older, filled-joint surfaces. Orientation control by pre-existing discontinuities should be obvious if the stylolite columns are oblique to the stylolitic surface but may not be obvious for irregular and smooth stylolites. Stylolites at a low angle to pre-existing discontinuities occur (Barrett, 1964) but seem rare.

The response of a stylolite to a noncoaxial strain path has not been firmly established. Irregular to smooth stylolites or solution cleavages might allow pressure solution to occur on planes oblique to σ_1, as suggested by the existence of oblique columnar stylolites. Solution cleavages not perpendicular to the axis of shortening have been observed by Borradaile (1977) and Spang and others (1979) and can be attributed to a noncoaxial strain path. Stylolitic laminae may grow by the asymmetric accretion of material (Oertel, 1983, p. 435). The final stylolite surface reflects the strain history but is not necessarily perpendicular to the axis of finite shortening.

The currently most satisfactory macroscopic description of stylolite formation and evolution is the anticrack model, first stated by Durney (1974) and named and developed by Fletcher and Pollard (1981). According to this concept, a stylolite is mechanically identical to a crack that shortens perpendicular to its surface by the removal of material. This leads to a stress and strain distribution around the stylolite identical to that of an opening crack but with opposite sign. The model suggests that stylolites should propagate by growing at the tips. Natural examples show volume strains along the stylolite and plastic strain adjacent to the stylolite that fit the model prediction (Tapp and Cook, 1988). The rotation of stress on fold limbs causes curvature of the cleavage in perfect conformity to the anticrack model (Groshong, 1975a, p. 411; Tapp and Wickham, 1987). In the kinematic model of stylolite formation by Guzzetta (1984), a sutured stylolite geometry is obtained if the side from which material is removed changes along the surface. If the side from which material is removed remains the same through time, a square-wave stylolite evolves; but if the side changes, then the peaked form evolves. The regular spacing of stylolites and solution cleavage was addressed by Merino and others (1983), who proposed a theory for monomineralic rock in which initial porosity variations localize pressure solution and set up an instability that controls the spacing between stylolites or cleavages; spacing is controlled by porosity, grain size, temperature, stress, mineral compliance, molar volume, and the dissolution rate constant.

The exact mechanism(s) by which pressure solution takes place is a subject of some controversy as shown by the reviews of Elliott (1973), Paterson (1973), de Boer (1977a), McClay (1977), Durney (1978), Kerrich (1978), Rutter (1983), and Green (1984). It seems to be generally agreed that solid-state grain-boundary diffusion is too slow a process to allow significant strain to occur at geological strain rates (Fletcher and Hofmann, 1974; Rutter, 1983; Green, 1984). Pressure solution has been inferred in water-wet experiments where large bulk strain has been produced without cataclastic or plastic strain of the crystals (Griggs, 1940; de Boer, 1977b; de Boer and others, 1977; Sprunt and Nur, 1977; Urai, 1985). The two most likely models are diffusion in a water film adsorbed along the grain boundary (Weyl, 1959; Robin, 1978; Green, 1984) and undercutting by free surface dissolution at grain margins followed by grain crushing (Bathurst, 1958; Weyl, 1959). The experiments by Tada and Siever (1986) demonstrate the first model for halite; the experiments of Gratier and Guiguet (1986) on quartz sand are interpreted by them in terms of the second model. Movement of the pore fluid, even in very low porosity and permeability rocks, will substantially speed up the process (Fletcher and Hofmann, 1974). The term "solution transfer" (Durney, 1972) might be used where movement of pore fluid is important and the term "diffusion transfer" where the pore fluid does not move.

Grain-to-Grain Stylolites

Grain-to-grain stylolites have been observed in a variety of clastic rock types of various grain sizes and imply that the rock is grain supported (Dunham, 1962) or that the contrast in physical properties between grains and matrix (if present) is large. Grain contacts may be sutured (Heald, 1956, Pl. 1; Thomson, 1959, p. 97) or smooth (Trurnit, 1968, p. 104). The first experimental deformation of a granular aggregate to show a grain-to-grain stylolite was by de Boer (1977b, Fig. 5); Shinn and Robbin (1983, p. 613) and Gratier and Guiguet (1986, p. 852) have produced excellent examples. Stylolitic-like contacts in a rock containing grains and cement might represent the result of pressure solution or might be irregular boundaries between cement overgrowths (Land and Dutton, 1978). Evidence of pressure solution is provided by truncation of original contacts of fossils of known original shape, such as ooids (Ramsay, 1967, p. 196; Logan and Semeniuk, 1976, p. 24) or by truncation of dust rings around sand grains (Heald, 1955, Pl. 3; 1956, Pl. 2; Rittenhouse, 1973, Fig. 4). Differential cathodoluminescence of grains and cement facilitates the distinction between stylolitic and overgrowth contacts (Sibley and Blatt, 1976, p. 885–886; Houseknecht, 1987, p. 636). Where the radii of curvature of the grains in contact differ, all other things being equal, small grains indent large grains (Trurnit, 1968, p. 102–105). Stylolitic contacts between grains tend to be approximately parallel to one another (Heald, 1956, Pl. I; Ramsay, 1967, p. 196, Houseknecht, 1987, p. 636) but with a significant variability in orientation, as expected from the fact that not all grain contacts are normal to the bulk shortening direction (analogous to the control on microfracture orientations; Gallagher and others, 1974). A large amount of grain-to-grain pressure solution results in what Wanless (1979) called a fitted fabric in which nearly all grain contacts are stylolitic (Buxton and Sibley, 1981, p. 25).

Grain-to-grain pressure solution reduces pore space by compaction as material is removed and also by cementation if the material is deposited in the nearby pore space (Rittenhouse, 1971b). This process, which may be called "chemical compaction," is now recognized as being very important in the field (Choquette and James, 1987). Experiments on sands have resulted in porosity reductions of 45%–70%, with the greatest reduction in the finest grained material (Renton and others, 1969; Sprunt and Nur, 1976, 1977), a relationship also observed in the field (Houseknecht, 1984). Mimran (1977) demonstrated tectonic control of the process by showing that in a naturally deformed chalk the bulk density increase due to pressure-solution–related porosity loss is directly related to dip.

Transgranular Stylolites

Stylolites that traverse many grains are here termed "transgranular stylolites." Park and Schott (1968) used the term "aggregate stylolites" for stylolites having an amplitude larger than the grain size in which they occur. In a rock lacking matrix, this type of stylolite appears to require the prior existence of cement, otherwise there would be only grain-to-grain stylolites. Tectonic stylolites were first used as a tool for structural analysis by Lindström (1962). He was soon followed by Price (1967) and Plessman (1972).

Much of our quantitative understanding of transgranular sutured stylolites traces to the exceptionally thorough work of Stockdale (1922, 1926) on limestones from the United States craton. He showed that the material in the stylolite seam had the same mineralogical composition, oxide ratios, and percentage of organic matter as the insoluble residue component of the adjacent limestone, thereby establishing that the seam material is insoluble residue. Schwander and others (1981) have verified this conclusion but have also identified host material in the stylolite. In addition, Stockdale showed that the thickness of the seam varies directly with the amplitude of the columns and inversely with the purity of the limestone. These relationships imply that a stylolite begins as a planar surface and that the amplitude of the suturing increases as more material is removed (Stockdale, 1926, p. 402). The first experimental example of this process has been illustrated by Gratier and Guiguet (1986, p. 852). In my experience, where stylolites are found along lithologic contacts, the separation of lithologies to opposite sides of the stylolite is always perfect. If a stylolite began at random centers of solution that joined along small faults to produce the columns, there should be examples where the offset of the lithologic contact is less than the amplitude of the stylolite, something I have yet to see. In some homogeneous units, pressure solution may be the cause of the bedding (Simpson, 1985). Stylolites may also be initiated along irregular boundaries formed during compaction (Shinn and Robbin, 1983). Stylolite amplitude provides a minimum estimate of the material removed because there may be a component of pressure solution that is equal on both sides of the stylolite that does not produce relief on the contact. Stylolite width provides a maximum estimate because an unknown amount of residual-composition material may have been present previously, for example, clay on a bedding plane. Another technique for estimating material removed by pressure solution uses the offset of planar markers such as beds, veins, or large fossils. Perhaps first noted by Conybeare (1949, p. 84), pressure solution oblique to a linear marker produces an offset of the marker without slip parallel to the stylolite. The amount of offset is proportional to the angle between the marker and the stylolite and the amount of material removed.

The total amount of material removed along stylolites may be large. Stockdale (1922, p. 39) illustrated a square-wave stylolite with an amplitude of 13 inches. Stockdale (1926) reported stylolitic bed thinning of 13%–34%, Dunnington (1967) reported 20%–25%, and Johnson and Budd (1975) provided data that indicate a 15% bed-thickness reduction. Hortenbach (1977) determined an increase in percent pressure solution with depth of burial. Closed systems, from which solutions cannot escape, quickly reach thinning values of 4%–6% and remain constant, whereas open systems show a linear thinning up to 30%. Hortenbach suggested that the pressure solution thinning can be used to estimate maximum depth of burial. Cretaceous chalk folded over a salt dome has been shown by Langheinrich and Plessman (1968) to have been thinned along stylolites in an amount up to 28% and to have a porosity that correlates inversely with the amount of stylolitic bed thinning and the dip of bedding. Restoration of a fold that has deformed primarily by pressure solution has demonstrated a minimum pressure-solution area loss in the hinge of 18% (Groshong, 1975a). In a fold examined by Droxler and Schaer (1979), 10%–20% of the original volume was pressure-solved and in part reprecipitated in nearby veins. Large amounts of pressure solution on irregular to smooth bedding-parallel stylolites can produce nodular bedding in which relatively unaffected nodules are surrounded by solution residue (Richter, 1963; Logan and Semeniuk, 1976; Garrison and Kennedy, 1977, p. 115–123; Wanless, 1979, p. 442, 448). In relatively heterogeneous lithologies, the result of extensive pressure solution may be a stylobreccia (Logan and Semeniuk, 1976, p. 44, 47).

In otherwise high-porosity rocks, stylolites are typically associated with zones of low porosity. Very low porosity stylolitic zones in high-porosity reservoirs segment the reservoirs. A single stylolite affects a zone from 0.5 ft (Wong and Oldershaw, 1981, p. 513) to about 5 ft in width to either side (Dunnington, 1967, p. 347; Johnson and Budd, 1975, p. 16, 22). It is commonly assumed that the lack of porosity is the result of local deposition of pressure-solved material, although Nelson (1983) proposed that zones of low porosity control the location of the stylolites. Buxton and Sibley (1981, p. 25) found greater porosity in a zone of fitted texture adjacent to a transgranular stylolite than in the adjacent relatively unstylolitic rock, showing that the pressure-solved material is not necessarily deposited in the zone of pressure solution. Volume balance and chemical changes contemporaneous with pressure solution suggest that nearby redeposition is a general if not universal rule. Wanless (1979, 1982) found dolomite growing adjacent to stylolites where magnesian calcite was being pressure-solved, a fact he attributed to pressure solution of the impure calcite and precipitation as dolomite. This is called "incongruent pressure solution" by Beach (1979). Cements contemporaneous with pressure solution have been observed experimentally (de Boer and others, 1977). On the other hand, changes in oxygen isotopes associated with the pressure solution compaction of a chalk (Mimran, 1977) suggest that some of the precipitated calcite came from outside the chalk bed.

Slickolites

A slickolite, a term coined by combining slickenside and stylolite, was originally discovered in association with shallow substratal solution in limestone (Bretz, 1940, p. 354–355). According to Bretz (1950), a slickolite consists of alternating grooves and ridges or half columns which fit into ridges and grooves opposite them, the series on one face being the exact reverse of the marking on the opposite face. Ends of columns are engaged by sockets in the termini of grooves, and the column-socket relationship on one face is the opposite of that in the other. Bretz (1940) pointed out that no clay films are associated with slickolites, an observation that appears to me to be generally valid, although small amounts of residual material may be found. The slickolites observed by Bretz (1940, 1950) were mainly vertical, downthrown toward the center of zones of solution collapse, but occasionally found on bedding planes and were interpreted to have formed under about 55 ft of overburden. Jaroszewski (1969, Pl. II), Carannante and Guzzetta (1972), and Laubscher (1979, p. 472) called attention to the role of slickolites in structural deformation, although Laubscher did not use the term.

Slickolites evidently form parallel to the displacement direction. This is consistent with the vertical orientation and vertical displacement on the slickolites observed by Bretz (1940, 1950), with the general lack of residual material on the stylolite and with the right-angle relationship between conventional stylolites and the "dextral stylolites" illustrated by Laubscher (1979, p. 472). Blake and Roy (1949) and Jaroszewski (1969, p. 21) reported a continuous gradation with changing orientation from normal stylolites to slickolites, and the same gradation is implied in the illustration of stylolites and oblique stylolites by Arthaud and Mattauer (1969, p. 739).

CLEAVAGE

Cleavage includes all types of secondary planar parallel fabric elements other than coarse schistosity which impart mechanical anisotropy to the rock without apparent loss of cohesion (Dennis, 1967). Spaced cleavage, defined as having cleavage surfaces spaced at finite intervals, however small (Dennis, 1967), is the typical fabric. A rock having spaced cleavage is divided into cleavage laminae and narrow slices of the original rock called "microlithons" (de Sitter, 1956).

Solution Cleavage

The quantitative importance of pressure solution in the formation of slaty cleavage was first demonstrated by Plessman (1965) in the slate of the Rheinisches Schiefergebirge. Plessman showed that the cleavage laminae were zones of pressure-solution residue with grains and fossils in the microlithons truncated against the cleavage by pressure-solution removal (1965, p. 75). Offsets of bedding along cleavage were shown by Plessman (1965) to be the result of the amount of material removed. These observations have been duplicated by Roy (1978, p. 1777) and repeated in other slates by Williams (1972, p. 10), Groshong (1976, p. 1141), and Bell (1978, p. 186–187). The same origin for more widely spaced cleavages has been demonstrated for sandstone and mudstone (Nickelsen, 1972, p. 110; Geiser, 1974, p. 1404; Gray, 1978, p. 579), limestone (Nickelsen, 1972; Groshong, 1975b, p. 1367; Alvarez and others, 1976, p. 699; Spang and others, 1979, p. 1111), and dolomite (Schweitzer and Simpson, 1986, p. 782–783). A large number of examples are given in the compilation edited by Borradaile and others (1982).

Many fabrics that had previously been termed "fracture cleavage" are now seen to be solution cleavages that break easily along the cleavage laminae. In many fabrics previously termed "slip cleavage," the offsets along cleavage can be interpreted in terms of pressure solution normal to axial plane cleavage (Plessman, 1965, p. 81), normal to fanning cleavage (Groshong, 1975a, p. 413) or pressure-solution removal of fold limbs in certain crenulation cleavages (Williams, 1972, p. 19; Gray, 1977b, p. 234). Pressure solution oblique to the solution laminae is also possible as a cause of offsets parallel to cleavage (Groshong, 1975a; Borradaile, 1977). Relatively widely spaced cleavages appear to be transitional to closely spaced slaty cleavages with increasing strain, depth of burial (Engelder and Marshak, 1985), and increasing clay content of the host rock (Marshak and Engelder, 1985).

The descriptive classification of solution cleavages parallels that of stylolites in having stylolitic (sutured), smooth, wavy, and anastomosing geometries (Powell, 1979; Borradaile and others, 1982; Engelder and Marshak, 1985; Schweitzer and Simpson, 1986). The term "rough cleavage" (Gray, 1978, p. 579) is used for the irregular subplanar spaced cleavage typical of psammitic rocks. The same cleavage plane may be both stylolitic and smooth as it passes through beds of different lithology (Helmstaedt and Greggs, 1980, p. 105, 107). Cleavage domains that cannot be resolved with an optical microscope are called "continuous," as is cleavage resulting from platy minerals evenly distributed throughout the rock (Powell, 1979).

Grain-orientation fabrics associated with solution cleavage are explained by pressure solution removal and by recrystallization. Solution cleavages may show large grains in the microlithons elongated parallel to cleavage. The grains are usually dimensionally oriented but not necessarily crystallographically oriented. This is explained by removal of the grain margins adjacent to the cleavage laminae by pressure solution; the width of the grain normal to cleavage is reduced, but its length parallel to cleavage remains unchanged (Williams, 1972, p. 10; Lisle, 1977; Beutner, 1978, p. 11). In many examples, the rock within the microlithons is strained, as shown by the presence of pressure fringes on rigid grains within the microlithons (Geiser, 1974, p. 1404; Roy, 1978, p. 1779; Gray and Durney, 1979, p. 55). Mica can rotate into parallelism with the cleavage in the cleavage laminae as the supporting framework is removed by pressure solution (Williams, 1972, p. 14–15, 39; Knipe and White, 1977, p. 363, 366). In many examples, the lack of bent micas at the border of cleavage laminae indicates that micas within the cleavage laminae have recrystallized by pressure solution/diffusion to achieve their preferred orientation (Oertel and Phakey, 1972, p. 6–7; Holeywell and Tullis, 1975, p. 1299; White and Knipe, 1978, p. 168–169; Wintsch, 1978; Woodland, 1982, p. 108; Lee and others, 1986, p. 773). Soft-sediment rotation of micas during compaction has been proposed as a mechanism to form slate by Maxwell (1962) and Powell (1972) because of a near parallelism between soft-sediment sandstone dikes and cleavage, leading to the inference of contemporaneity. This concept has been largely discredited by the discovery that pressure solution is a viable orienting mechanism and that the sandstone dikes, ostensibly parallel to cleavage, are in fact, not parallel but owe their orientation to a later strain (Geiser, 1975, p. 718–719; Groshong, 1976, p. 1136; Beutner and others, 1977; Beutner, 1980, p. 172–173). None of the true soft-sediment foliations described previously has resulted in the degree of lithification characteristic of slate.

Crenulation cleavage is defined by cleavage planes, whether micaceous layers or sharp breaks, which are separated by thin slices of rock containing a crenulated cross lamination (Rickard, 1961). The cleavage laminae in many crenulation cleavages are explained by pressure solution (Williams, 1972, p. 19; Gray, 1977a, p. 99; 1977b, p. 234; 1979, p. 98–99). Crenulation cleavage may be found in fissile, clay-rich sedimentary rocks (Geiser and Engelder, 1983, p. 126; Nickelsen, 1986, p. 365). My own observations lead me to believe that the cleavage laminae may be relatively independent of the crenulation geometry because the laminae do not have a consistent relationship to the fold geometry. The cleavage may be found on limbs or axial surfaces of the crenulations and may change positions along the axial surface.

Large strains may be associated with solution-cleavage formation. In some examples, the presence of syntectonic veins and pressure fringes suggests that some or all of the material is deposited nearby (Geiser, 1974; Mitra, 1976; Clendenen and others, 1988), but other examples show no evidence of extension of any type, and so the strain must be associated with a volume loss (Wright and Platt, 1982; Beutner and Charles, 1985). Shortening strains of up to 50% have been reported by Plessman (1965), Alvarez and others (1978), and Wright and Platt (1982); 59% was reported by Beutner and Charles (1985), much or all of which can represent volume loss.

Pencil Cleavage

The term "pencil structure" was used as early as 1858 by Naumann for a lineation in gneiss and later for the cleavage-bedding intersection lineation in slate producing pencil slate (Cloos, 1946, p. 8). Graham (1978) and Engelder and Geiser (1979) introduced the term "pencil cleavage" to replace what Crook (1964) called "reticulate cleavage." Crook (1964, p. 527) described the fabric as quasi-planar fractures that are penetrative on the scale of hand specimen and that result (especially after weathering) in a characteristic mass of acutely terminated elongate polygonal fragments. Pencil cleavage is found in shale, mudstone, and siltstone (Crook, 1964, p. 526; Engelder and Geiser, 1979, p. 463; Reks and Gray, 1982, p. 165–168).

Pencil cleavage on the Appalachian Plateau (Engelder and Geiser, 1979) shows almost no evidence of the insoluble residue expected for solution cleavage but can be traced into solution cleavage in interbedded limestones. The pencil cleavage in the inner Valley and Ridge province of the Appalachians is caused by very thin, disconnected solution cleavage laminae that become better connected as the cleavage becomes more perfect (Reks and Gray, 1982, p. 167–168). The cleavage (Reks and Gray, 1982) developed over a range of cleavage-normal shortening strains of 9%–26%, being invisible at lower strains and grading into solution cleavage at larger strains.

Extension-Fracture Cleavage

Fracture cleavage is a parting defined by closely spaced discrete parallel fractures, ideally independent of any planar preferred orientation of grain boundaries that may exist in the rock (Turner and Weiss, 1963, p. 98). Confusion over the use of this term goes back to its first use by Leith (1905) who applied it to wide veins, joints, and pressure-solution cleavage. In recent years, most field examples have been shown to be solution cleavage. A few published examples appear to fit the Turner and Weiss definition of fracture cleavage (Knill, 1960; Hancock, 1965; Price and Hancock, 1972) and notably the example of Foster and Huddleston (1986, p. 88–92). The fractures described by Foster and Huddleston are spaced 0.1–1.0 cm apart and are several centimeters long. They appear parallel in outcrop but are anastomosing in thin section and are often filled with alteration products of the host rock. Refraction of the fractures is noticeable in layers of different composition. Good evidence from offsets shows the extensional nature of the fractures (Foster and Huddleston, 1986, p. 92).

FAULT-ZONE FABRICS

Differences of opinion currently exist about the proper use of such common terms as "fault," "fault zone," "cataclasite," and "mylonite" (Tullis, and others, 1982; White, 1982; Wise and others, 1984, 1985a, 1985b; Mawer, 1985; Raymond, 1985). A *fault zone* is a tabular region across which the displacement parallel to the zone is appreciably greater than the width of the zone (modified after Wise and others, 1984) and in which the deformation is greater than outside the zone. This means that even if bedding or foliation is continuous across the zone, it may be described as a fault zone, a reasonable approach for the discussion of rock-deformation fabric, even if not always appropriate for the description of map patterns (Mawer, 1985). A *fault* is a surface along which displacement has taken place.

A fault surface is typically grooved (Dzulynski and Kotlarczyk, 1965, p. 151) on scales ranging from meters (Nevin, 1949, p. 130; corrugations of Hancock and Barka, 1987, p. 579) to hand sample (molded grooves of Willis, 1923, p. 61) to hand lens (Means, 1987, p. 588) with the groove axes being parallel to the displacement direction. Willis (1923) thought the grooves were the result of asperity plowing, but Means (1987) pointed out that parallel grooves are parallel sided, fit perfectly into ridges on the opposite side, and appear to be unrelated to any asperity. A smoothed and grooved fault surface is a planar feature named a "slickenside" by Conybeare and Phillips (1822) according to Dennis (1967). The grooves or striations on the slickenside are termed "slickenlines" by Fleuty (1975) and give the line of slip of the fault. Slip lineations on fault surfaces are also produced by crystal growth fibers. Although the original definition of the term "slickenside" is for polished or smoothed surfaces, not crystal fibers (Fleuty, 1975), current usage includes both, and this practice will be followed here.

Fault-zone materials include mylonite, cataclasite, gouge, and breccia. Mylonite was originally defined by Lapworth (1885, p. 559) for rocks in fault zones that were "crushed, dragged and ground out into a finely laminated schist composed of shattered fragments of the original crystals of the rock . . ."; however, the type example (Higgins, 1971, p. 18) shows clear evidence of crystal-plastic deformation textures, resulting in a persistent ambiguity in the use of the term. The definition also included a statement that the rock must be foliated, leaving unfoliated fault-zone rocks without a name. Bell and Etheridge (1973) pointed out that "brittle deformation is unnecessary for the formation of a typical mylonite" (p. 337) and defined the term as "a foliated rock, commonly lineated and containing megacrysts, which occurs in narrow, planar zones of intense deformation. It is often finer grained than the surrounding rocks into which it grades" (p. 347). According to Waters and Campbell (1935), the term "cataclasite" was introduced by Grubenmann and Niggli (1924) for an aphanitic, structureless rock that differs from mylonite (in the sense of Lapworth) by an absence of foliation. Recent classifications (Sibson, 1977; Wise and others, 1984; Tullis and others, 1982) use cataclasite for a nonfoliated rock formed by brittle mechanisms, and "mylonite" as a foliated or generally foliated rock formed by crystal-plastic mechanisms. Both soft-sediment fault zones (Mandl and others, 1977) and brittle fault zones (Proctor and others, 1970, Fig. 4; Gay and Ortlepp, 1979, p. 53; Chester and others, 1985) may be foliated, however. *Cataclasite* is here defined as a cohesive rock formed mainly by brittle fracturing and usually showing evidence of grain rotations and grain-size reduction. A cataclasite might be formed by local crushing without significant fault offset (cataclastic flow in the sense of Stearns, 1969) and so is not necessarily restricted to fault zones. *Mylonite* is a rock formed mainly by crystal-plastic deformation mechanisms and shows evidence of internal rotation and grain-size reduction. Cataclasite or mylonite may be foliated or unfoliated, although on the thin-section scale, many cataclasites are unfoliated and most mylonites are foliated.

The noncohesive equivalents of cataclasite are gouge and fault breccia (Higgins, 1971). A fault breccia is composed of angular or rounded fragments formed in the fault zone and consists of more than 30% fragments large enough to be seen by the naked eye. Gouge is a paste-like rock material formed in the fault zone in which less than 30% of the fragments are large enough to be seen by the naked eye. Cataclasite series (Sibson, 1977) represents all cohesive and noncohesive rocks fitting the definitions of cataclasite, gouge, and breccia.

A hydroplastic fault zone is one in which the dominant deformation mechanisms are grain rotation and grain-boundary sliding of unlithified materials (after Petit and Laville, 1987). This type of fault zone is found in unlithified sediments under low effective confining pressures, conditions which allow the grains to move past one another without breaking.

Conjugate Fault and Riedel Shear Geometry

The geometry of conjugate faults, Riedel shears, and related features has proved to be a very important unifying concept in understanding fault zones from the microscopic scale to the map scale. First the conjugate fault geometry is described, then the Riedel geometry.

According to Handin (1969), the recognition that faults occur on planes of high shear stress was due to Coulomb (1776) who determined that faults do not occur on the planes of maximum shear stress (45° to σ_1) but rather occur at an angle on the order of 30° to σ_1. A fault initiated at this angle is often called a "Coulomb fault." Daubrée's (1879) uniaxial compressive tests on wax may have been the first analog experiments to show the conjugate geometry in a geological context. In his classic 1905 paper, Anderson used the concept of a pair of conjugate Coulomb faults at 30° to σ_1 that intersect parallel to σ_2 to explain normal, thrust, and strike-slip fault orientations. The conjugate (Andersonian) fault geometry is seen, although not regularly, in pure shear experiments on rocks (Adams, 1910; Paterson, 1958, Pl. 1; Griggs and Handin, 1960, Pls. 7–8; Hadizadeh and Rutter, 1983, p. 502) and sand (Hubbert, 1951, Pl. 1; Horsfield, 1980, p. 307); is common in clay (H. Cloos, 1930; E. Cloos, 1955, Pl. 1) and is observed on the small scale in the field (Cloos, 1947b, p. 900; Stearns, 1972, p. 163; Lockwood and Moore, 1979, p. 6045; Ramsay, 1980a, p. 85).

Perhaps the first published simple-shear fault-zone experiment was the clay-model study of Cloos (1928). Riedel (1929, p. 361–362) reproduced the experiment, obtaining fault zones defined by en echelon tensile cracks and en echelon faults and correctly interpreting σ_1 as being 45° to the trend of the fault zone. Although Riedel obtained only the conjugate direction having the same sense of displacement as the zone as a whole, similar experiments commonly result in en echelon sets of both conjugate directions (Cloos, 1955, Pl. 3). Hills (1963) stated that the faults in this experiment are known as "Riedel shears." Experimentally, this geometry is characteristic of a fault zone that is forced by the boundary conditions to have an orientation parallel to the displacement direction, as in the Cloos-Riedel experiment or in a closed shear box (Morganstern and Tchalenko, 1967a).

The terminology of Riedel shear zones (Fig. 2) is due mainly to Skempton (1966, p. 330). The fault zone trend is the D direction. Faults parallel to this direction form after the Riedel (R and R') shears. Displacement may be concentrated on one or two D shears that are then called "principal displacement shears." The D shear is equivalent to the C direction in the terminology for crystal-plastic fault zones of Berthé and others (1979) and the Y direction of Bartlett and others (1981). The Riedel shears are conjugate to the σ_1 axis (45° to D): the R shear is at ~15° to the trend of D and has the same sense of displacement, the R' shear is its conjugate at ~75° to the trend of D and having the opposite sense of displacement. A tensile fracture 45° to D and parallel to the σ_1 axis is termed "T." "Thrust-shears," labeled P, have the same sense of slip as R shears and form at 10°–30° to D. In the normal fault zones illustrated by Skempton (1966), the P shears had a thrust orientation, hence the name, but the designation P is from his interpretation that the material was in the passive Rankine state (incipient failure with horizontal stress greater than vertical stress: Jaeger and Cook, 1979, p. 415). In their rock-model experiments, Bartlett and others (1981) observed the aforementioned features and a new direction, termed "X," that is perpendicular to R and has the same sense of displacement as R'. A Riedel shear zone thus contains more than just R and R' shears. A Riedel shear zone has the same over-all geometry as the first- and second-order shears of Moody and Hill (1956) but the opposite sequence of formation; the smaller faults form first.

Foliations are produced by inequant minerals within the fault zone (notably clays) that rotate to positions nearly parallel to the individual faults (Weymouth and Williamson, 1953; Morganstern and Tchalenko, 1967b, p. 151; Tchalenko, 1968, p. 168; Platt and Vissers, 1980; Maltman, 1987), or rotate to approximate the finite strain orientation (Morganstern and Tchalenko, 1967a, p. 316), or kink (Morganstern and Tchalenko, 1967a, Figs. 6–17; Tchalenko, 1968, p. 169). Kinking may be related to the anisotropy of the mineral foliation (Tchalenko, 1968, p. 163) or may represent kinking of a set of parallel faults (Morganstern and Tchalenko, 1967a, p. 319) with passive rotation of the mineral fabric. Solution cleavage may also produce a foliation within the fault zone (Alvarez and others, 1978; Ghisetti, 1987, p. 691–692). A mineral shape foliation oblique to the trend of the fault zone has been termed the "S plane" in crystal-plastic fault zones (Berthé and others, 1979). Rutter and others (1986) believe the foliation in a Riedel shear zone to be parallel to the P shears and call it a "P foliation." Fault zones might show foliations parallel to cleavage; rotated bedding; and R, R', P, and D shears, all being approximately contemporaneous, and none of which necessarily exists in the rock adjacent to the fault zone. Fault zones are seen to grow as Riedel shear zones from the microscopic to the map scale (Morganstern and Tchalenko, 1967a; Tchalenko, 1970), and so multiple Riedel foliations are possible within a single fault zone.

The origin of the P shear, which is not directly related to the stress field causing R, R', and T, has been explained by Gamond (1983) as the result of stress reorientation between en echelon R shears. Using the theory of Segall and Pollard (1980), Gamond showed that the P shear has the expected 30° angle of a Coulomb fault to the reoriented maximum compressive stress axis between en echelon R shears. Gamond (1983, p. 37) also showed that sliding on the P shears can cause the R shears to open. This leads to the important observations that all open or filled cracks in fault zones are not necessarily parallel to T (Gamond, 1983, p. 34–35), and that dilation across the fault zone may favor the formation and opening of P shears rather than the formation of T shears. The X-direction fractures in the experiments of Bartlett and others (1981, p. 266–270) appear to form as connections between R shears and have the orientation and sense of displacement appropriate for a slightly rotated conjugate to P that would form at the ends of the R shears (compare Gamond, 1983, Fig. 15d).

R shears tend to remain active or form anew during the displacement of a fault zone and may cut D shears (Fig. 2; Skempton, 1966, p. 330). Slip on the R shears causes extension of the D shears. Late extensional crenulation cleavage (Platt and Vissers, 1980, p. 402–403) and small brittle faults (Platt and Leggett, 1986, p. 192, 200) may form in the R-shear orientation.

Shear lenses (Skempton, 1966, p. 331) are small regions bounded by slip surfaces. Rhombic zones are bordered by R and D shears; trapezoidal zones, by R, D, and P shears. Slickenlines on shear lenses may be oblique to the direction of displacement of the fault zone as determined from the slickenlines on the principal displacement shears. Shear lenses might also be caused by the intersection of R-foliations with a bedding foliation (Platt and Vissers, 1980, p. 402–403) or by intersecting R and R' foliations (Platt and Vissers, 1980, p. 404). Substantial extension by this mechanism may result in boudinage of relatively stiff layers (Moore and Allwardt, 1980, p. 4746).

The Anderson (1905) and Riedel (1929) models of faulting are two dimensional, applying to plane strain. As a result of very careful analysis of the results of experimental clay deformation, Oertel (1962, p. 29–30; 1965, p. 355) recognized that four conjugate fault directions had formed simultaneously. The faults formed with a geometry that can be visualized as two conjugate pairs for which the dihedral angle of each pair is bisected by σ_1 and the line of intersection of each pair makes a small but non-zero angle with σ_2 such that the resulting pattern is symmetric across the σ_1–σ_2 plane. Three or more conjugate faults formed simultaneously can be termed an "Oertel conjugate geometry" (suggested by Z. Reches, 1977, personal commun.) to distinguish it from the Andersonian pair of conjugates. Aydin (1977) and Reches (1978, p. 113–114) recognized the Oertel geometry in the field and interpreted it to be the result of a three-dimensional strain accomplished by faulting. Field examples of the contemporaneous formation of three or four sets of conjugate faults (as opposed to slip on pre-existing faults) have been interpreted using this concept (Bruhn and Pavlis, 1981, p. 288–289; Aydin and Reches, 1982, p. 109–110; Underhill and Woodcock, 1987) and have been produced in true triaxial rock-deformation experiments (Reches and Dieterich, 1983, p. 114–116). The interpretation of Reches (1978, 1983) is based on the analysis of the number of slip systems required to accommodate a general three-dimensional strain, analogous to the theory for the number of slip systems required in crystal-plastic deformation by Taylor (1938) and Bishop (1953). Oertel conjugate faults may explain the shear lens geometry and slip directions noted by Skempton (1966).

Hydroplastic Fabric

In unlithified materials having a major component of platy minerals, a Riedel shear zone forms the typical fault-zone fabric (compare Petit and Laville, 1987, p. 111). The platy minerals are aligned parallel or nearly parallel to the R, R', P, D, or X shear directions (Morganstern and Tchalenko, 1967b, p. 148; Maltman, 1977, p. 424; Carson and Bergland, 1986, p. 144–145). Maltman (1987, p. 80) experimentally produced R shears that curved into the D direction. Slickensided, bi-pyramidal cone-shaped faults, perhaps analogous to fault cones sometimes produced in the triaxial

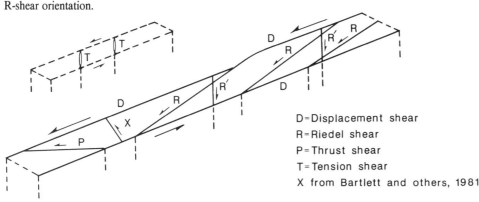

Figure 2. Nomenclature of a Riedel shear zone, after Skempton (1966) and Bartlett and others (1981).

compression of cylindrical samples have been found to form during shale compaction (Guiraud and Séguret, 1987, p. 127). Kinks are also common and may be transitional into fault zones (Tchalenko, 1968, p. 163, 169; Lundberg and Moore, 1986, p. 22–23). Dismembered folds contribute significantly to the texture of accretionary prism fault-zone fabrics (Nelson, 1982, p. 627–628, 632; Lucas and Moore, 1986, p. 95); some are sheath-like folds with axes parallel to the transport direction (Hibbard and Karig, 1987, p. 852). Also characteristic of accretionary prisms is scaley foliation defined by anastomosing polished and slickensided fracture surfaces, pervasive on the scale of millimeters (Lundberg and Moore, 1986, p. 18–19; Moore and others, 1986). This fabric seems to be the same as the shear lenses of Skempton (1966, p. 321) where extension on R shears may be very prominent (Nelson, 1982, p. 628; Cowan, 1985, p. 453) but may represent multiple shear foliations (Maltman, 1988, p. 173). Raymond (1984) defined mélange as a body of rock mappable at a scale of 1:24,000 or smaller and characterized both by the lack of internal continuity of contacts or strata and by the inclusion of fragments and blocks of all sizes, both exotic and native, embedded in a fragmented matrix of finer-grained material. The mélange fabric appears to be typically that of a large-scale, soft-sediment (or cataclastic) Riedel shear zone having large displacements. The same textures may be seen in thin sections of shale-matrix fault zones.

The Riedel R and D shear-zone geometry is also seen in hydroplastic fault zones in equigranular materials such as sand (Petit and Laville, 1987, p. 111). A weak grain-size segregation foliation occurs parallel to the slip direction of individual faults, and elongate particles are rotated to near parallelism with the slip direction (Thomson, 1973, p. 528; Mandl and others, 1977, p. 101, 111, 113–118; Maltman, 1988, p. 172). Very minor cataclasis produces rounding of the more angular grains and leads to an increase in fines within the fault zone (Mandl and others, 1977). The boundaries between the country rock and the fault zone are not sharp on the thin-section scale, and original bedding may be obliterated in the fault zone. Fold and kink-band geometries may occur that are transitional into fault zones (Hubbert, 1951, Pl. 2; Sieh, 1984, p. 7652; Van Loon and others, 1984, p. 358). Dilation is required within the fault zone to allow the grains to move past one another without fracturing (Mandl and others, 1977). The depth to which this can occur is controlled by the fluid pressure; high fluid pressures will permit soft-sediment deformation to considerable depth.

The porosity and permeability of soft-sediment fault zones generally reflect that of the material entrained in the zone. Based on the analysis of normal faults in the Gulf of Mexico that are sealing or nonsealing to hydrocarbon migration (Smith, 1980), the fault zone is a barrier to hydrocarbon migration if it contains mainly shale or is a conduit if it contains mainly sand (Weber and others, 1978, Fig. 11). Where a single sand body is in contact across the fault, the fault zone is not a barrier. Weber and others (1978, Fig. 6) found a zone of dilation on the hanging wall of a small normal fault and a parallel zone of compaction on the footwall, implying that fluids might have been better able to migrate along the hanging-wall side of the fault zone. Mimran (1985, p. 373) described a normal fault in chalk that was a conduit for fluid flow as indicated by the presence of calcite veins having the isotopic composition of meteoric water. Because of the different abilities of oil and water to migrate through water-wet rocks, a fault might be a barrier to oil migration, but not to water migration (Smith, 1980).

Cataclastic Fabric

Cataclastic textures are similar whether or not the material has primary cohesion. Jaeger (1959) produced cohesionless powder (gouge) in dry friction experiments and a compacted dense material (cataclasite) in otherwise identical water-wet experiments, suggesting that cohesion is controlled by the presence of water in the fault zone. Foliation parallel to, or at a low angle to, the fault-zone boundaries is reasonably common, marked by compositional layering (Proctor, 1970, Fig. 4; Bonilla and others, 1978, p. 354; Mitra, 1984, p. 58; Chester and others, 1985, p. 141), elongate grain alignment (House and Gray, 1982, p. 259; Segall and Pollard, 1983b, p. 559; Chester and others, 1985, p. 141), and by variations in grain size (Wojtal and Mitra, 1986, p. 682, 684). Foliations of all three types occur in the fault zone described by Gay and Ortlepp (1979, p. 54).

The most distinctive textural features are grain fracture, fragment separation, and rotation. Grain sizes within the matrix are reduced from that of the host to sizes of 5–25 μm as seen in thin section (Engelder, 1974a; Aydin, 1978), but fragments down to the size range of 0.06–0.1 μm are seen in SEM. The smaller particles tend to be very angular, and the quartz fragments may be basal cleavage flakes, produced by high strain rate (Gay and Ortlepp, 1979, p. 55; Moody and Hundley-Goff, 1980, p. 306; Olgaard and Brace, 1983, p. 14). Larger fragments, on the order of 30 μm or more, may be rounded (Engelder, 1974a, p. 1515; Moody and Hundley-Goff, 1980, p. 306; House and Gray, 1982, p. 359; Mitra, 1984, p. 58; Wojtal and Mitra, 1986, p. 682) or angular (Anderson and others, 1980, p. 229; House and Gray, 1982, p. 259–261; Blenkinsop and Rutter, 1986, p. 671). Bonilla and others (1978) reported rounded and polished pebbles and rock fragments in the San Andreas fault zone. Rounding is due to abrasion during rotation in the fault matrix (Pittman, 1981, p. 2382). Moody and Hundley-Goff (1980, p. 306) found a greater degree of rounding in wet experiments than in dry experiments. Some rounded grains may be relatively unaltered host-rock grains in sandstones (Dunn and others, 1973, p. 2411; Pittman, 1981, p. 2382). Cathodoluminescence of a natural fault zone showed some of the matrix to be undispersed fragments of the adjacent larger grains (Lucas and Moore, 1986, p. 1012). Cataclasite clasts are found within cataclasites (Brock and Engelder, 1977, p. 1669; Anderson and others, 1980; House and Gray, 1982, p. 262; Mitra, 1984, p. 58; Wojtal and Mitra, 1986, p. 683). Gouge injections are reported by Engelder (1974a, p. 1516) and Brock and Engelder (1977, p. 1669). Cataclasite veins (Engelder, 1974a; Gretener, 1977; Labaume, 1987) show that the material is easily deformable.

The grain-scale microfractures are virtually always extensional (Aydin, 1978, p. 920) but may slide after formation, although rotation appears to be more common than sliding. The well-ordered deck-of-cards sliding and rotation characteristic of crystal-plastic shear zones (Simpson and Schmid, 1983, p. 1286; Tullis and Yund, 1987, p. 607) appears to be extremely rare or absent in cataclastic fault zones. Vein fillings in microfractures have been observed by Aydin (1978, p. 920, 926) and Blenkinsop and Rutter (1986, p. 671). Larger-scale transgranular veins occur in the cataclasite (Stel, 1981, p. 587, 590; House and Gray, 1982, p. 262) and may contain fibers (Mitra, 1987, p. 574), dogtooth crystals, or voids (Stel, 1981, p. 590–596; Chester and Logan, 1986, p. 87). Stylolites may occur in the matrix as either the grain-to-grain type (Brock and Engelder, 1977, p. 1669) or the wavy transgranular variety (Moore and Allwardt, 1980, p. 4747; House and Gray, 1982, p. 259–262; Mitra, 1984, p. 57–58; Chester and Logan, 1986, p. 87; Wojtal and Mitra, 1986, p. 682).

Pressure shadows around large grains are rare and where present are filled with fragments of the adjacent large grain (House and Gray, 1982, p. 361; Lucas and Moore, 1986, p. 96; Rutter and others, 1986, p. 6). In contrast, pressure shadows in mylonitic fault zones are relatively common and are elongate and filled with recrystallized grains (Berthé and others, 1979, p. 34; Watts and Williams, 1980, p. 327; Simpson and Schmid, 1983, p. 1282–1283).

First described by Heald (1956, Pl. 2), very thin fault zones having widths of a few millimeters or less but lengths of meters or more contain gouge or cataclasite like that previously described. They are usually but not always lighter in color than the host rock and show offsets in the range of less than a grain diameter to centimeters (Aydin, 1978, p. 917; Aydin and Johnson, 1978, p. 934; Pittman, 1981, p. 2382; Jamison and Stearns, 1982; Byrne, 1984, p. 36). Shiny, slickensided surfaces in rock that appear to have no associated gouge may be called "fault mirrors" (for example, Petit and others, 1983) and are probably fault zones of this type that have just enough gouge adhering to the surface to smooth it (Nelson, 1985, p. 43–47).

The Riedel shear geometry also occurs *within* cataclastic fault zones. Jackson and Dunn (1974, p. 243), Engelder and others (1975), and Byerlee and others (1978, p. 168) produced R shears within experimental gouge zones. R and R' shears (Friedman and Higgs, 1981, p. 14; Logan and others, 1981, p. 131) as well as D shears (Logan and others, 1981) have been produced. Chester and Logan (1986) interpreted the faults in the damage zone adjacent to the Punchbowl fault zone as R and R' shears. In a very important contribution, Gay and Ortlepp (1979) found en echelon conjugate shears that were clearly produced by a seismic event (p. 53) in a mining-induced fault zone, with the R shears being larger and the R' shears forming mainly in the areas of R shear overlap. The resulting shear zone appears to be made up of incompletely connected R, R', and D shears.

Increasing fragmentation toward a large-displacement fault zone has been documented by an increase in the number of small faults and an increase in the number of fractured grains (Engelder, 1974a; Brock and Engelder, 1977, p. 1669; House and Gray, 1982, p. 258; Chester and Logan, 1986, p. 84, 86; Wojtal and Mitra, 1986). The principal displacement fault is not necessarily in the center of the damage zone (Wallace and Morris, 1986, p. 120).

Fault breccias appear to be of two types: vein derived and fault derived. In vein-derived breccias, the host-rock fragments are bounded by veins (Carson and others, 1982, p. 286; Chester and Logan, 1986, p. 90; Gaviglio, 1986, p. 249; Mitra, 1987, p. 583, 585; Ramsay and Huber, 1987). The vein filling may be derived locally by pressure solution, may be introduced during deformation by high-pressure fluids (Rye and Bradbury, 1988), or may be the result of later hydraulic fracturing or opening of the damage zones adjacent to the fault zone. Chester and Logan (1986) found that up to 5%–10% of the clasts in the Punchbowl fault zone are vein fragments, indicating many fracturing and filling events. Veins appear to be common in crystal plastic fault zones as well (White and White, 1983, p. 583; Sibson, 1986a, 1987, Pl. 2; Stel, 1986, p. 295) and usually show evidence of multiple events of vein formation and deformation. Fault-derived breccia blocks are bounded by thin shear zones (Heald, 1956, Pl. 2; Brock and Engelder, 1977, p. 1669; Carson and others, 1982, p. 283; House and Gray, 1982, p. 260; Dokka, 1986, p. 81; Lucas and Moore, 1986, p. 99–100) or injected cataclasite (Labaume, 1987, p. 154). Farmin (1941, p. 148) stated that in the western United States breccia is more common on normal faults than on reverse faults. Phillips (1972) reported significant vein-fill breccia on normal faults, whereas Gretener (1977), in contrast, cited a number of large-displacement thrust faults that are only centimeters thick. The origin and occurrence of fault breccia merits further research.

The thermal alteration of organic components of the rock provides data on the thermal effects in fault zones. Wilson (1971) reported that fault zones with displacements of a few feet in Oklahoma oil basins had no effect on the color of palynomorphs, whereas fault zones with displacements of more than 100 ft caused palynomorphs to change from brown to black within 600–800 ft of the fault plane (equivalent to a temperature change from about 100 to 175 °C; Bostick, 1974). For fault zones in the frontal ranges of the Canadian Rocky Mountains, Bustin (1983) reported vitrinite reflectances of 0.72%–3.09% in the fault zone that drop to regional values of 0.78%–1.21% a meter or less away from the fault zone. For heating times as short as 1 hour, this could correspond to temperatures of 430–650 °C locally compared to the regional background of 250–350 °C (Bustin, 1983).

Glass has been produced in low-temperature friction and faulting experiments (Friedman and others, 1974, p. 938; Jackson and Dunn, 1974, p. 248; Teufel and Logan, 1978; Moody and Hundley-Goff, 1980, p. 309; Spray, 1987). The maximum temperatures are about 1150 °C over the background of 100–125 °C (Teufel and Logan, 1978). Moody and Hundley-Goff (1980) produced glass only in their dry experiments, not in water-wet experiments. Pseudotachylite is a glassy appearing rock, dark colored and of veinlike or pseudointrusive character which bears a strong resemblance to tachylite (after Shand, 1916; and Waters and Campbell, 1935, but removing reference to fusion). It usually contains angular clasts of the country rock (Shand, 1916, Pl. 19). Pseudotachylite is generally glass or devitrified glass (Shand, 1916; Sibson, 1980, p. 168–169; Maddock, 1983, p. 106), but some examples may be ultrafine-grained cataclasite (Waters and Campbell, 1935; Wenk, 1978, p. 508–509) or both (Philpotts, 1964, p. 1015, 1017, 1021).

Porosity is decreased by cataclasis in high-porosity rocks but increased in low-porosity rocks. Aydin (1978) found that a host-rock porosity averaging 24%–25% was reduced to 6%–10% in natural fault zones. Pittman (1981) found the host-rock porosity of 30% and permeability of 51,394 md reduced to 3% and 0.05 md in cores consisting mainly of small natural fault zones. In experiments on clay-rich and clay-absent gouges, Morrow and others (1984) found a wide range of small permeabilities from 10^{-22} to 10^{-18} m^2 (where 1 darcy = 0.987×10^{-12} m^2). Nelson (1985, p. 432) found that polished-gouge slickensides may show significant permeability due to small mismatches in the surface. He reported permeabilities from a core containing a slickensided surface as 18–211 md oblique to the slickenside and 1,532 md parallel to the slickenside. Underground mining experience reported by Wallace and Morris (1986, p. 120–123) is that most fault zones are wet, some produce water, some act as barriers to fluid migration, and that the greatest water production is from the moderately fractured damage zone adjacent to the largest fault zones. Dilatancy prior to (Brace and others, 1966) and during (Teufel, 1981) faulting of low-porosity rocks may increase the porosity by up to 8% (Teufel, 1981, p. 143).

Pressure-Solution–Dominated Fabric

A pressure-solution–dominated fault zone, as defined here, must contain abundant evidence of removal of material along stylolites, slickolites, or solution cleavage as seen in examples by Alvarez and others (1978), Rispoli (1981, p. T31), and Marshak and others (1982, p. 1018). Fault zones dominated by pressure solution may also contain crystal growth-fibers along slickensides (Elliott, 1976, Pl. 12), fiber-filled veins (Spang and Groshong, 1981, p. 330), mosaic-filled veins (Marshak and others, 1982, p. 1018), Riedel shears (Casas and Sabat, 1987, p. 649), and cataclastic material (Ghisetti, 1987, p. 962). Slickenside fibers and vein fillings might be sourced from outside the fault zone and so are not sufficient evidence to define deformation by pressure solution.

The fiber orientations and country-rock screens characteristic of crystal-fiber slickensides were first explained by Durney and Ramsay (1973, p. 87). The initial slip surface contains offsets that open as slip occurs (the offsets may have initiated as an en echelon array of T fractures, Ramsay and Huber, 1983, p. 258). As the displacement on the slip surfaces becomes large, the fiber bundles growing from nearby offsets overlap to bring fibers of different ages into contact. Screens of the wall rock may be preserved between the fiber bundles producing the typical laminated appearance (Ramsay and Huber, 1983, p. 258–261). If the slip direction changes through time, because the fiber bundles are of different ages where they are superimposed, they will have different fiber directions resulting in superimposed slickensides with different slickenline orientations (Durney and Ramsay, 1973, p. 89).

One type of fault zone develops from closely spaced solution cleavage. Alvarez and others (1978, p. 265) documented an increasing intensity of solution cleavage to spacings representing greater than 35% material loss as a fault zone is approached, concomitant with a curving of cleavage to a low angle to the (thrust) fault zone. In the fault zone, there is a chaotic mixture of fragments that are derived from detached cleavage microlithons. Fault zones may consist of closely spaced solution cleavage (Bogacz, 1981) which may be cut by Riedel shears (Koopman, 1983; Lavecchia, 1985). The early development of such a fault zone may be represented by the cleavage duplex recognized by Nickelsen (1986, p. 362, 367, 369). The cleavage within the duplex curves (Nickelsen, 1986, p. 364) to approximate the expected trajectories of the finite elongation direction (Ramsay and Graham, 1970) but does not actually join the fault surface. One of the associated cleavage-duplex thrusts is a zone of sheared shale containing euhedral crystals and cataclastic fragments that were originally precipitated in veins or fault-step pressure shadows and were broken by continuing fault movements (Nickelsen, 1986, p. 367).

A closed-system fault zone consisting of faults, stylolites, and veins (Rispoli, 1981, p. T31) evidently began with the formation of parallel small faults as the slip surfaces in a slightly rotated kink band. The displacements on the small faults terminate in veins and stylolites that make low angles to the kink-band boundary. Additional rotation results in connection along the kink band by the formation of additional small faults, veins, and stylolites to form a through-going fault zone. An orthogonal array of veins and stylolites may have a flow law similar to grain-boundary diffusion (Fletcher and Pollard, 1981). A block size of 1–10 cm and a differential stress of 30 bars gives a reasonable geological strain rate.

Superplastic Fabric

Superplasticity is defined for metals as a quasi-viscous behavior whereby elongation can exceed 1,000% without necking and subsequent failure (Boullier and Gueguen, 1975, p. 93); no specific deformation mechanism is implied (Schmid and others, 1977, p. 258). The interpreted deformation mechanism is diffusion- or dislocation-assisted grain-boundary sliding (Schmid and others, 1977, p. 277). The texture in rocks is characterized by square to loaf-shaped grains (Boullier and Gueguen, 1975; Schmid and others, 1977, p. 271; Behrmann, 1985, p. 104) in which triple-point grain intersections do not usually make 120° angles. The grains are relatively undeformed internally and have no more than a weak crystallographic fabric (Schmid and others, 1977; Behrmann, 1985). Grain sizes on the order of 1–10 μm are required for the processes to be important at the lower experimental temperature ranges (Schmid and others, 1977), and this is the grain size found in field examples (Boullier and Gueguen, 1975; Schmid and others, 1981, p. 156; Behrmann, 1985, p. 104). Superplastic deformation in limestone and quartzite is normally associated with crystal-plastic strain at temperatures above 400 °C for which the small grain size is achieved by syntectonic recrystallization in a mylonite (Schmid and others, 1977; Etheridge and Wilkie, 1979), but temperatures may be as low as 250–400 °C (Behrmann, 1985).

Small (1-6 μm), nearly equant, interlocking grains are relatively common as the matrix in low-temperature fault zones that otherwise appear to be cataclastic (Phillips, 1982, p. 112; personal observations). This texture could be the result of either the recrystallization of gouge (Jaeger, 1959) or superplastic flow; criteria for making the distinction have yet to be developed. The presence of reactive fluids should enhance either process (Etheridge and Wilkie, 1979). The interlocking 4- to 6-μm grain-size texture in a quartz-matrix fault zone formed at 250–350 °C is interpreted by Phillips (1982) as being the result of superplastic flow. The Lochseiten mylonite on the Glarus thrust of eastern Switzerland may be another example. The Lochseiten is a 1- to 2-m-thick foliated limestone consisting of equant, 1- to 3-μm diameter grains (Schmid, 1975, p. 262). The fault zone is characterized by either mylonite or by extreme cataclasis and brecciation in a zone a few tens of meters thick (Schmid, 1975, p. 261). Foliation-parallel stylolites and veins indicate pressure solution in the mylonites (Schmid, 1975, p. 266). In the same area, the hanging wall of the Glarus thrust is characterized by 6%–9% crystal-plastic strain in the limestone, no recrystallization, and vitrinite reflectance of 2.5%–3.5% (Groshong and others, 1984a) equal to perhaps 160–190 °C (well-standard curve, Bostick, 1974, p. 9). Thus superplastic flow may have occurred in the fault zone while the remainder of the rock was only moderately deformed without recrystallization.

Inference of Seismic Activity from Fabric

It has been accepted since about 1900 that most earthquakes occur along pre-existing faults (Howell, 1986) and are caused by the release of stored elastic energy (Reid, 1910). It has long been known that the experimental formation of a fault in intact rock results in the release of elastic energy, but not until the work of Jaeger (1959) was it shown that sliding on an existing fault (or sawcut) in rock would also result in elastic energy release. This behavior is termed "stick-slip" because the energy release occurs during periodic slip episodes characterized by stress drops that are only a fraction of the total differential stress. Brace and Byerlee (1966) and Byerlee and Brace (1969) first proposed that stick-slip sliding could be the cause of earthquakes. Factors that influence the occurrence of stick-slip behavior include temperature, pressure, and strain rate (Byerlee and Brace, 1969), thickness of gouge zone (Summers and Byerlee, 1977), and composition of the gouge (Summers and Byerlee, 1977; Logan, 1979). Nearly all common crustal minerals show stick-slip behavior under suitable conditions, including many phyllosilicates but with the exception of vermiculite and montmorillonite (Summers and Byerlee, 1977). In mixtures the frictional behavior is controlled by the softer mineral (Logan, 1979; Logan and Rauenzahn, 1987).

The formation of gouge and slickensides during stick-slip experiments (Jaeger, 1959) suggests that fault-zone textures might correlate with seismic behavior. Short carrot-shaped grooves on slickenside surfaces were shown experimentally by Engelder (1974b, p. 4388) to correlate with stick-slip events because their lengths are comparable to the slip in a single event and they fail to occur during stable sliding. The features are also seen on natural slickensides (Engelder, 1974b, p. 4390–4391). Application of this criterion to felt earthquakes is problematical because the grooves are only 0.4–2 mm long, much shorter than the slip in most earthquakes. McKenzie and Brune (1972) proposed that pseudotachylite forms by melting during seismic events. Pseudotachylite is rather rare (Sibson, 1986b; Spray, 1987), however, and is most commonly reported in association with mylonite (Sibson, 1980; Hobbs and others, 1986) or in impact craters (Reimold and others, 1987). No texture in cataclastic fault zones has yet been shown to have a unique correlation to felt earthquakes.

Riedel shears form within the gouge in stick-slip experiments (Friedman and Higgs, 1981; Logan and others, 1981; Hiraga and Shimamoto, 1987). Riedel shears also formed in intact rock during an earthquake in what might be the only known earthquake source region ever mapped (Gay and Ortlepp, 1979). Significantly, the double-couple nature of earthquake source mechanisms implies slip along more than one plane during an earthquake (Howell, 1986). Perhaps the presence of Riedel shears within the fault zone will prove to be an index of seismic activity.

Sense-of-Shear Criteria

Sense-of-shear criteria are divided here on the basis of scale into fault-zone features and fault-surface features. Fault-zone features of value include "drag" folds, gash or feather fractures, cleavage, and foliation. A "drag" fold (Nevin, 1949, p. 136) is a fold adjacent to a fault and is convex toward the direction of motion of the block in which it occurs. The drag probably occurs before faulting, not as a result of slip along the fault (Nevin, 1949, p. 142; Dennis, 1987). Extension fractures at 45° to the fault zone (T direction, Fig. 2) were first named "feather joints" by Cloos (1922; also Cloos, 1932), according to Dennis (1967). The same features were termed "gash fractures" by Gilbert (1928) and "pinnate tension joints" by Hills (1953). The sense of slip on the fault is as given by Figure 2. Solution cleavage, stylolites, and mineral-shape foliations within a fault zone are usually approximately parallel to the extension strain direction; the acute angle between foliation and the fault trend points in the direction of slip of the adjacent block.

The first comprehensive review of fault-surface sense-of-shear criteria seems to be in the textbook of Willis (1923, p. 54–62). Willis pointed out that asymmetrical steps commonly occur on slickensided fault surfaces such that a finger run over the surface in the smooth direction is moving in the direction of displacement of that side of the fault. This became the standard sense-of-displacement criterion in later textbooks. The interpretation was challenged by Paterson (1958, Pl. 2), who illustrated experimentally produced fault steps with the opposite asymmetry, and soon afterward Tjia (1964, p. 684) provided field evidence of such reversed-symmetry steps. Both types of steps occur in nature and, in fact, both types were described by Willis (1923).

A general terminology for steps on a fault surface devised by Norris and Barron (1969, p. 153) defines incongruous steps as those oriented to oppose the displacement and congruous steps as the opposite. Accretion steps consist of gouge or cataclasite plastered to the fault plane, and fracture steps are small fractures that cut the fault surface (Norris and Barron, 1969, p. 136). A double bend in the surface of a fault in which a straight fault curves and returns to its original trend is a "restraining bend" if it causes restraint as the fault slips and is a "releasing bend" if it causes extension or void formation as the fault slips (Crowell, 1974, p. 191). The terms were defined for the map-scale geometry of strike-slip faults, but the concept is useful for any type of fault down to the thin-section scale (Spang and Groshong, 1981; Gamond, 1987). A releasing bend causes dilation as the fault slips. The dilation may be manifested by tensile fractures at a high angle to the displacement direction or as a zone of many conjugate faults (Sibson, 1987, p. 701–702). Releasing bends form a major site for vein-filling ore deposits, even if the main fault trend is barren (Sibson, 1987). This type of ore shoot is perpendicular to the fault slip vector.

Pressure solution is common on the resisting face of a restraining bend (Laubscher, 1979, p. 472; Spang and Groshong, 1981, p. 329; Marshak and others, 1982, p. 1018). Pressure solution at restraining bends is frequently contemporaneous with fiber growth at releasing bends (Marshak and others, 1982, p. 1018). The collapse and fragmentation of a restraining bend by vein-filled radiating tensile fractures is illustrated by Laubscher (1979, p. 477–478). Recognition that the feature is a restraining bend allows it to be used as a sense-of-shear criterion. Low-angle restraining bends may be polished, whereas low-angle releasing bends on the same fault are rough (Angelier and others, 1985, p. 352).

Fault-surface features produced by secondary fractures and faults are conveniently classified according to the Riedel system (Petit and others, 1983; Petit, 1987, p. 598). T criteria are in the T direction (Fig. 2). Microscopic T fractures, called "microscopic feather fractures," have been produced experimentally by Friedman and Logan (1970, p. 3418), Dunn and others (1973, p. 2412), Conrad and Friedman (1976, p. 189), and Friedman and Higgs (1981). Crescentic T fractures, concave in the direction of slip of the opposing face (Wegmann and Schaer, 1957, p. 492) have the same geometry as the chatter marks defined by Willis (1923, p. 57) and Tjia (1972, p. 55). Chatter marks are produced by a rounded gouging tool on the opposite face (Chamberlin, 1888; Lawn and Wilshaw, 1975, p. 22), and they may be known as Hertzian fractures after Hertz (1896), who derived the stress distribution (Lawn and Wilshaw, 1975, p. 19; Engelder and Scholz, 1976, p. 157). Wegmann and Schaer (1957, p. 494) and Laubscher (1979, p. 476) illustrated tool impressions bordered on one side by similar markings. Occasionally the tool will remain in place (Angelier and others, 1985, p. 352). Prod marks (Tjia, 1972, p. 54) are the displacement-parallel grooves produced by a tool.

A number of displacement criteria can be related to Riedel shears. An R shear offsetting a D shear causes a low-angle congruous step for which the sense of offset on the R shear is the same as for the D shear (Angelier and others, 1985; Platt and Leggett, 1986, p. 192). The typical incongruous fracture step appears to be caused by a fracture (X?) between an R shear and the D surface (Currie, 1969, p. 169–171; Angelier and others, 1985, p. 352). This is perhaps the most likely explanation of the steps in the experiment of Paterson (1958, Pl. 2). Crescentic gouge marks and lunate fractures (Wegmann and Schaer, 1957, p. 492) are similar in cross section, but the former is convex in plan view in the direction of displacement of the opposing block, and the latter is just the opposite. Pinnate shears (Hills, 1963, p. 174–176) are slickensided incongruous steps making a high angle to the fault surface and having the orientation of R' shears. This is the geometry of the incongruous slickensided steps seen in the field by Tjia (1964, p. 684) as explained by Rod (1966, p. 1164). Shear bands of this orientation, if mistaken for features in the R direction, give the wrong sense of displacement (Behrmann, 1987, p. 661).

Not all sense-of-shear criteria are directly linked to secondary fractures. Any step in the fault surface, whether congruous or incongruous, if it creates a releasing bend, will cause a void or become the site of an accretion step (Norris and Barron, 1969, p. 150–153). When a fault surface is separated, accretion steps typically break at a high angle to the fault surface, leading to congruous fault steps. Accretion steps not found adjacent to releasing bends may result in either congruous or incongruous steps (Norris and Barron, 1969). The apex of a triangular accretion step points in the direction of movement of the fault surface to which it is attached (Norris and Barron, 1969, p. 159). Experiments by Engelder (1974b) and Jackson and Dunn (1974, p. 246) showed that accretion steps on one side of the fault correspond to grooves on the other side. Blocks plucked from the surface leave depressions in the fault surface usually having a steep trailing edge and a low-angle leading edge that may be smoothed or polished by subsequent fault movement. This is the type example of the smooth-direction test for sense of displacement (Willis, 1923, p. 60–61). A trail was first defined by Willis (1923, p. 57, by analogy to glacial spurs: Chamberlain, 1888, p. 245) as a ridge formed in the lee of a projection on the fault surface. The ridge may consist of either wall rock or gouge; as the ridge extends away from the projection, it decreases in height and may diminish in width. Rutter and others (1986, p. 6) showed a good example of a gouge trail behind a resistant grain where the trail is formed of fragments of the grain. Shear tails are elongate, asymmetric accumulations of fragments adjacent to large grains (Lucas and Moore, 1986, p. 100–101). The asymmetry of the accumulations, similar to gouge trails extending from diagonally opposite sides of the grain, give the sense of shear. The geometry is exactly the same as the asymmetric augen that are common in crystal-plastic fault zones (Simpson and Schmid, 1983, p. 1281, 1283). Asymmetric gouge lips across voids intersected by the fault plane provide a good displacement criterion (Angelier and others, 1985, p. 353); the gouge overhangs the void in the direction that the opposite wall slipped.

STRESS AND STRAIN ANALYSIS OF THE AGGREGATE

The first quantitative approach to the interpretation of rock fabrics was macroscopic strain analysis (Sharpe, 1847; Sorby, 1853). Becker (1893) derived many of the properties of the strain ellipsoid but unfortunately interpreted faults as occurring on the planes of zero elongation. This interpretation was refuted by Griggs (1935), but strain analysis did not recover as a respectable subject until the classic study of Cloos (1947b). On the microscopic scale, the symmetry analysis of Sander (1911, 1930; Knopf and Ingerson, 1938) was the first influential interpretive framework. According to Sander (1930, in Turner and Weiss, 1963, p. 385), "symmetry principles form the soundest basis for correlating the tectonite fabric with the physical factors—stress, strain, movement picture, and so on—concerned in their evolution." This approach has been termed "kinematic analysis" (Friedman, 1964; Friedman and Sowers, 1970, p. 480).

It is important to consider what results are theoretically possible given the type of data available. All methods are based on observations of permanent deformation features in the rocks, in other words, strain. It is quite obvious in an experimental context that forces and displacements are measured, and stresses and strains are computed. It is not possible to observe stress directly. Stress may be computed from strain if the relevant physical properties of the rock are known, for example, the elastic constants or the viscosity and strain rate. The strains form the primary measurement, and the stress directions and magnitudes are inferred. The same is true of field data. In situations where the stress and strain tensors do not coincide, either because of rock anisotropy or rotational strain, some methods may be better at predicting stress axes than strain axes, or the reverse.

Strain Measurement

A variety of techniques have been devised for measuring strain. Compaction of porous rocks is considered first. The packing density (Kahn, 1956) is the summation of the total framework-grain lengths along a straight traverse divided by the length of the traverse, given as a percent. As a measure of compaction, it is about equivalent to point counting grain area (Coogan and Manus, 1975). To determine compaction, the original framework grain-packing density or grain area must be known. This in turn depends upon the original packing density (a topic first addressed quantitatively by Sorby in 1908) which ranges from a grain volume of 52.36% to 74.05% for cubic or rhombohedral packing of uniform spheres (Graton and Fraser, 1935). Random packing has an average grain volume of 60%–63% (Graton and Fraser, 1935; Coogan and Manus, 1975), which is a good average for more complex mixtures of equant grains (Fraser, 1935); in very poorly sorted sediments, however, the grain volume may be as high as 72% (Beard and Weyl, 1973). The number of grain contacts per grain is a measure of compaction (Taylor, 1950). Random sections through randomly packed spheres produce 0.63 contacts/grain and 47% floating grains (no contacts visible) according to Chillingarian and Wolf (1975). Experimentally deposited sand ranges from 1.6 contacts/grain and 17% floating grains (Taylor, 1950) to 0.85 contacts/grain and 46% floating grains (Gaither, 1953). In a field study of Wyoming sandstones, Taylor (1950) found that the contacts per grain went from 1.6 to 5.2 as the depth of burial increased from 2,885 ft to 8,343 ft with a negligible number of floating grains at all depths.

The theoretical framework for quantitatively evaluating the importance of compaction by pressure solution together with cementation was developed by Rittenhouse (1971a, 1971b) and extended by Mitra and Beard (1980). Houseknecht (1987) presented a practical method for making the evaluation from point-count data. Mechanical compaction causes more porosity loss at the early stage of compaction by pressure solution even if the dissolved material is deposited nearby. Complete loss of porosity by this process has been interpreted for some rocks (Lowry, 1956; Heald, 1956; Thomson, 1959). The compaction rather than cementation component was shown to be more important by Sibley and Blatt (1976) and Wilson and Sibley (1978).

The amount of rotation of grains free from mutual interactions depends directly upon the magnitude of the strain. March (1932; see Oertel, 1983) developed a model for random markers deformed homogeneously with their matrix that predicts the relationship between strain and fabric orientation. The March method has been applied successfully to experimental phyllosilicate fabrics (Means and Paterson, 1966) and natural slates (Oertel, 1970; reviewed in Oertel, 1983). If the phyllosilicates interfere with one another during rotation, the model might not be appropriate (Siddans, 1976), but the interactions are evidently reduced by pressure solution or some other means, and so the model works well even in low-porosity rocks (Etheridge and Oertel, 1979; Oertel, 1983). Talbot's (1970) strain-measurement technique using the orientations of deformed veins is a similar method and might be suitable for coarse, platy materials.

The technique of Mitra (1976) to determine the strain due to translation glide in quartz is based on the change in normal strain with direction in the strain ellipsoid. The quartz grains contain rutile needles that are buckled or boudinaged depending on their orientation. A plot of final length divided by original length of fiber versus current orientation produces the strain ellipse, a method confirmed experimentally by Mitra and Tullis (1979). The change in shear strain with orientation is the basis of Wellman's (1962) graphical method which is especially useful for deformed fossils in which the original angular relationships are known. The method has proved suitable for strains as small as 7% (Groshong, 1972) as well as for larger magnitudes.

With deformation, grain centers become more closely spaced in the shortening direction, more widely spaced in the extension direction, and the angles between grain centers change. Ramsay (1967, p. 195–197) first proposed the nearest-neighbor strain-measurement technique based on this change. The method developed by Fry (1979) and implemented by Hanna and Fry (1979) has proved effective in mildly deformed sedimentary rocks (Narahara and Wiltschko, 1986). If the initial distribution of grains were truly random (Poisson), then the Fry method might indicate zero strain even if the strain is large (Fry, 1979); but for most reasonable natural distributions, there is no problem (Fry, 1979; Onasch, 1986a, 1986b, 1987) because the size of the grains precludes spacings closer than one diameter. The Fry method finds the deviatoric strain which can be corrected for volume change if necessary (Onasch, 1986a).

The calculation of displacement path and strain from fiber directions is treated by Durney and Ramsay (1973), Wickham (1973), Wickham and Anthony (1977), Ramsay and Huber (1983), Casey and others (1983), Dieterich and Song (1984), and Beutner and Diegel (1985). Fibers or inclusion trails in antitaxial veins and pressure fringes provide the best available data for determining displacement history. Mitra (1976) used the fractional area of pressure fringe to measure the amount of pressure diffusion, a useful technique in a closed system.

Stress Measurement

Stresses cannot be measured directly but may be inferred from a material response that is directly controlled by the stress magnitude. Transgranular extension fracture is an example (Etheridge, 1983, 1984). Experiments indicate that transgranular extension fracture occurs at a unique, tensile effective stress that is a property of the material (Paterson, 1978, p. 23), although failure is highly influenced by stress concentrations, and so a range of tensile strengths is to be expected. The tensile strengths of 7 rocks tabulated by Jaeger and Cook (1979, p. 190) are in the range of 3.5 to 400 bars, very small numbers compared to the compressive strengths. Indirect tensile tests (that is, bending, point load) tend to give larger values (Jaeger and Cook, 1979, p. 191). The hydraulic fracture strength of a number of rocks ranges from 30 to 266 bars (Rummel, 1987, p. 220).

The angle between conjugate faults is a function of the differential stress at failure. Mohr (1900) proposed that fracture occurs when the shear stress on the fracture plane reaches a value that is a function of the normal stress, the Mohr envelope on a Mohr diagram. Two assumptions are associated with the Mohr envelope: fault initiation occurs as the Mohr's circle touches the envelope, and the point of tangency represents the stress on the fault plane. The latter assumption means that the dihedral angle between conjugate faults might be used to infer the differential stress and effective confining pressure for a curved envelope. When all of the stresses are compressive and above about 1 kbar (100 MPa), the experimental Mohr envelope is approximately linear for many rocks (Handin and others, 1963, p. 731, 743; Brace, 1964, p. 163). The typical dihedral angle is 55°–60° and is close to that predicted from the Mohr envelope in experiments on Solnhofen limestone (Heard, 1960), Repetto siltstone, and Muddy shale (Handin and others, 1963) but differs by an average of +6° in Berea sandstone and −6° in Marianna limestone (Handin and others, 1963). The average differences are +8° to 12° on various rock types in the low confining pressure tests of Borg and Handin (1966, p. 262) but only +1° in their high confining pressure tests.

The Mohr envelope for rocks is parabolic in the vicinity of the origin where the confining pressure is less than 1 kbar (Handin and others, 1963; Brace, 1964; Corbett and others, 1987). Muehlberger (1961, p. 214) used the concept of a parabolic Mohr envelope to explain fractures with low dihedral angles as having formed with σ_3 tensile. The later triaxial experiments of Brace (1964, p. 163) on dry rock, however, demonstrated that the dihedral angle is zero when σ_3 is tensile. There seems to be no experimental evidence that the Mohr envelope is continuous into the tensile field. Low-dihedral-angle conjugate faults (~50° or less) evidently represent small compressive effective stresses. The absolute magnitude depends on the rock type. The sequence of events in a Riedel shear zone may be a better micromechanical model for some faults. A very important feature of the Riedel model is the fact that the ultimate fault zone (D shear) is at an angle of 45° to the far-field σ_1, not 30°, and there need be no conjugate.

Deformation by crystal-plastic glide is related to the differential stress. Handin and Griggs (1951, p. 866) introduced into structural geology the metallurgical concept of critical resolved shear stress. Slip occurs on a glide plane when the shear-stress component in the glide direction reaches the critical resolved shear stress. Called "Schmid's law" (Schmid and Boas, 1950), Handin and Griggs (1951) used it to predict fabric changes due to large strain. It has been established experimentally that the critical resolved shear stress for glide is independent of the normal stress across the glide plane (Turner and others, 1954a; Friedman, 1967) but is, in general, a function of temperature and strain rate (Turner and others, 1954a; Higgs and Handin, 1959, p. 276; Carter and Raleigh, 1969; Heard and others, 1978; Spang and Friedman, 1978). Data on critical resolved shear stress are available for calcite (Griggs and others, 1951, 1953; Turner and others, 1954a; Griggs and others, 1960a; Friedman, 1967), dolomite (Higgs and Handin, 1959; Griggs and others, 1960b; Heard and others, 1978; Spang and Friedman, 1978), and clinopyroxene (Tullis, 1980). Tullis (1980) found that the average single-crystal critical resolved shear stress for calcite is 100 bars; for dolomite, 1 kbar; and for clinopyroxene, 1.4 kbar. The values are subject to variation depending on the starting material but, clearly, the presence of twins provides a good measure of the minimum differential stress.

Determining the stress in a grain within a multicrystalline aggregate is a more complex problem. Early in the experimental deformation of rocks, the question arose as to whether the fabric was the result of (nearly) homogeneous strain and consequently stress that changed from grain to grain or (nearly) homogeneous stress with strain that changed from grain to grain. By comparing predicted and observed fabrics, Griggs and Miller (1951, p. 860) and Handin and Griggs (1951, p. 864) determined that the strain was homogeneous, and so the stress must vary in the differently oriented grains. The deformed oolite experiments of Donath and Wood (1976), lower temperature experiments of Schmid and Paterson (500 °C, 1977), and Wood and Holm (1980) confirmed the homogeneity of the strain. Whether the critical resolved shear stress is reached on a glide system in any given crystal depends on the orientation of the glide system with respect to the principal stresses and upon the constraints provided by neighboring crystals (Paterson and Turner, 1970).

A theory based on resolved shear stress (Jamison and Spang, 1976, p. 870) uses the relative numbers of calcite grains containing one, two, and three sets of twin lamellae to measure differential stress. A small differential stress on the aggregate will cause only the most favorably oriented crystals to twin, and mainly single twin sets will form, but a large shear stress will cause twinning in less favorably oriented directions, resulting in two or three sets in some grains. Tested against experimental data, the percentage of grains having three twin sets gave the best stress estimate, agreeing to 21% or better with the experimental values of differential stress. The later approach of Laurent and others (1981, p. 658), based on finding the tensor that maximizes the shear stress on every twin set, underestimates the differential stress in their experimental test by a factor of 3–4 or, alternatively, requires a critical resolved shear stress for calcite of 3 or 4 times the experimental single-crystal value.

The minimum differential stress required to cause pressure solution has never been established but must be very small. Solution-pitted pebbles have been found as shallow as 30–40 m (Trurnit, 1968), and sutured stylolites were found at 90 m burial (Schlanger, 1964). Alvarez and others (1976) found abundant solution cleavage in folded limestones from northern Italy in which the coarse-grained calcite was untwinned.

Dynamic Analysis and Related Strain-Analysis Techniques

Dynamic analysis, a term introduced by Turner (1953), is defined by Friedman and Sowers (1970, p. 488) as "the use of certain fabric elements as criteria for inferring the orientations and relative magnitudes of principal stress axes in the rock at the time these fabric elements develop." The fabric elements are the result of strain, and so dynamic analysis necessarily represents inferences based on strain phenomena. In practice, the term is used for methods in which the direction and sense of glide or slip are known but in which the amount of glide or slip is unknown or not used. In other words, dynamic analysis represents the inference of stress from strain without using strain-magnitude data. Particularly for anisotropic materials, the stress axes so derived may be more accurate than stress axes that are assumed to be parallel to the strain axes, but this has yet to be demonstrated. The current techniques have arisen rather independently within different interpretive traditions, yet fundamentally they are very similar in that all are designed to estimate some or all of the properties of a tensor. The deformed-crystal techniques use the crystal lattice to define an original configuration, and crystal glide systems to determine the direction, sense, and amount of glide. The fault-slip techniques use slickenlines and sense-of-shear indicators to define direction and sense of slip. Twin gliding is constrained by the crystal lattice to occur along a specific glide line with a specific sense of shear. Quartz deformation lamellae are less constrained because the lamellae are irrational planes and the amount of translation glide is not fixed. Fault slip is the least constrained because slip may occur in any direction on any plane and in any amount; for some faults, the slip line may be known but not the slip direction.

Dynamic analysis based on crystal glide was developed by Turner (1953), who first recognized that it was possible to invert glide data to infer average stress orientation. In the Turner technique, the principal stress directions (C and T axes) most favorably oriented to cause twinning in each twin set are plotted on a stereonet for a large number of grains, and the average C and T positions are interpreted as the stress axes most favorably oriented to produce the observed twinning. Turner (1953) demonstrated that this interpretation was consistent with the experimental results, and Friedman (1963) first demonstrated by direct experiment that the technique works. It should be remembered that in these experiments the stress axes and strain axes are coaxial, and so it is not possible to determine which tensor is being estimated. The method was extended to dolomite by Christie (1958) and Crampton (1958) and to plagioclase by Lawrence (1970). Not until the experiments of Heard and Carter (1968) was the proper technique for dynamic analysis of quartz deformation lamellae understood. Carter and Raleigh (1969) pointed out that dynamic analysis techniques could be extended to kinking.

Later work has produced refinements in the original Turner (1953) method. Using only twin sets with high spacing indices produces better results (Turner and Weiss, 1963, p. 414). Nissen (1964a, p. 901) showed that C and T axes computed as most favorable for glide on two planes in doubly twinned crystals produced sharper average axis concentrations than did C and T axes for singly twinned grains. Venkitasubramanyan (1971) and Dietrich and Song (1984) have developed other versions of this approach also with improved results.

The dynamic analysis of faults has developed from several different and independent directions. For Andersonian conjugate fault pairs, slip on the fault planes should be parallel to the maximum shear stress direction at the time of formation of the faults. Wallace (1951) appears to have been the first to relate the slip direction to slip parallel to the maximum shear stress on arbitrarily oriented faults in a true triaxial stress field. Bott (1959) presented a similar analysis for determining the orientation of slip on a plane of weakness in an anisotropic body.

A practical dynamic analysis for slip on multiple (non-Andersonian) faults was developed by Compton (1966) independently of these theoretical treatments. Referring to Hubbert (1951), Compton noted that slip is most likely on lines at 30° to σ_1. In a stress state for which $\sigma_2 = \sigma_3$, a plot of the slip linears on a stereonet is a small circle of points arranged around σ_1 (Compton, 1966, p. 1371), confirmed by Aleksandrowski (1985, p. 77). Compton suggested that for $\sigma_2 \neq \sigma_3$ the distribution would be elliptical, and the examples of Aleksandrowski (1985) show that, although this is correct, it is nearly impossible to recognize in practice. Compton (1966, p. 1373) also performed an analysis very similar to a Turner dynamic analysis but with the compression axis at 30° to the fault. Use of the sense-of-slip data clearly indicates the σ_1 direction. Arthaud (1969) developed a different graphical technique based on the stereonet intersections of the planes containing the slip line and the pole to the fault. This relates to Compton's method as the beta pole fold axis relates to the pi pole. Both methods seem most suitable for biaxial stress tensors. Aleksandrowski (1985) modified Arthaud's method to improve the resolution of the three principal axes and computed the stress ratio from the result.

Spang (1972) introduced the first numerical technique for performing a dynamic analysis based on twin glide in calcite. The Spang numerical dynamic analysis technique (NDA) is produced by transforming with the tensor transformation equation the Turner C and T axes for each twin set, taken as axes of unit length, into the same coordinate system, averaging the components, and finding the principal values of the resulting 3 × 3 matrix. The principal directions correspond to the centers of the C and T distributions produced graphically (Spang, 1972). In experimental results, the magnitudes of the principal NDA values correlate with the tightness of the C-T axis distributions (Groshong, 1974). Spang and Van Der Lee (1975) extended the NDA method to quartz deformation lamellae and dolomite twin lamellae. A new method based on maximizing the shear stress on twinned planes while minimizing the shear stress on untwinned planes (Laurent and others, 1981) is as successful as the Spang technique in finding the principal axis directions.

Groshong (1972) introduced the least-squares strain gauge technique for computing strain in a calcite aggregate from measurements on twin lamellae. The method traces to the calculation by Handin and Griggs (1951, p. 867–868) of the strain produced by complete twinning in calcite from which Conel (1962) derived the equation for the strain in a partially twinned grain. The strain in each grain is related to that in the other grains by the strain-transformation equation (Jaeger and Cook, 1979), which is solved by least squares for the strain in the aggregate (Groshong, 1972, p. 2029). The least-squares technique is a multiple linear regression that simultaneously minimizes the difference between grain-strain components. The result is the magnitude and orientation of the strain tensor. The technique has been demonstrated to be accurate to within about 1% strain in room temperature experiments favoring glide (Groshong, 1974; Groshong and others, 1984b). It is possible to recognize a multiple deformation and, in favorable circumstances, accurately compute the strains involved (Teufel, 1980). On the other hand, Spiers (1979) in experiments at 400° found the strain in individual twinned grains to be consistently twice that predicted from the imposed bulk strain. Spiers (1979, p. 287) suggested that the grains suitably oriented for twinning deformed preferentially, although his photomicrographs show all the grains to be twinned. If the difference is not an algebraic error (the engineering shear strain used to define the twin strain is twice the tensor shear strain used in the transformation equation; for example, Jaeger and Cook, 1979, p. 38), then this result suggests a change in the strain mechanism between 25 and 400 °C that should be further investigated.

Inspired by Bott's (1959) equations for slip on an arbitrary fault plane, the first least-squares dynamic analysis for faults was published by Carey and Brunier (1974). Since then a number of other methods have been published, the differences being mainly in the parameters minimized by least squares. The Carey and Brunier (1974) method minimized the angle between the maximum shear strain on the fault and the slickenline direction, an approach also used by Etchecopar and others (1981) in an iterative solution. Another approach is to minimize the components of the calculated shear stress perpendicular to the slip line (Angelier and Goguel, 1979; Angelier, 1979; Angelier and others, 1982). The method of Michael (1984) uses the idea that the magnitude of the shear stress in the active slip direction should be the same on each fault and so minimizes the difference between the computed shear stress on each slip line and a constant reference value. Because the maximum shear stress directions are controlled by the ratios of the three principal stresses (Wallace, 1951; Bott, 1959), all of the techniques allow calculation of the principal stress ratios.

Dynamic analysis techniques for fault slip directions are very similar to earthquake focal-mechanism solutions. Pavoni (1980) used a focal-mechanism-solution method to determine the compressional and extensional quadrants and best compression and extension axes for fault slip data from outcrop. Pfiffner and Burkhard (1987) have extended this approach to petrofabric analysis. None of the fault dynamic analysis methods has been verified experimentally, but all produce reasonable results. The Angelier (1979) Crete data set gave nearly the same answer when recomputed by the later methods of Angelier and others (1982, p. 615), Michael (1984, p. 11520), and Pfiffner and Burkhard (1987). The method of Pfiffner and Burkhard (1987) has also been compared to twin-strain data computed by the experimentally verified strain-gauge method and found to give about the same axis directions. It seems safe to conclude that all the published tensor-estimation techniques give similar principal axis directions. It has yet to be demonstrated whether the differences reflect merely the different approaches to accommodating natural variation or reflect real differences between the stress and strain tensors.

TEXTURAL INTERPRETATION OF THE AGGREGATE

The aggregate fabric in low-temperature deformation is characterized by simultaneously active deformation mechanisms and heterogeneous strain. The total strain may be the result of steady-state deformation by more than one mechanism or may represent a time sequence of mechanisms in pre-failure strain.

Strain Partitioning

Strain partitioning is the subdivision of the total strain at some scale into components produced by different mechanisms (Groshong and others, 1984a). Perhaps the first quantitative field study of low-temperature strain partitioning into different mechanisms at the same scale was that of Nissen

(1964b), who found that the plastic twinning strain in elliptical crinoid columnals in a graywacke matrix was insufficient to account for the shape change. Nissen attributed the discrepancy to an initial ellipticity of the columnals, but he mentioned that pressure solution was active, and his photos show stylolitic boundaries on the long sides of the ellipses. Engelder and Engelder (1977) and Engelder (1979) obtained the same result in another area and demonstrated that pressure solution at the crinoid boundaries explained the discrepancy. Nickelsen (1966) discovered deformed fossils in an apparently undeformed matrix, a relationship he attributed to strain partitioning between microfolding and extension microfracture in the fossils and lateral compaction in the matrix. The partitioning of strain between grain rotation and grain strain is important because the grain strain will be less than the total strain, and for compaction the grain strain could be zero (Borradaile, 1981).

The strain partitioning changes as strain increases. Donath and others (1971) and Tobin and Donath (1971) have provided the only quantitative experimental analysis of the deformation mechanism transitions that occur with increasing strain before faulting. They defined the mesoscopic deformation modes as extension fracture, brittle fault, "ductile" fault, and uniform flow (Donath and others, 1971, p. 1447). In their terminology, a brittle fault is cohesionless and a "ductile" fault cohesive. They showed that at 25 °C the modes were controlled by confining pressure (1–2 kbar range) and total strain (2%–20% range). The microscopic fabric elements that correlated with the macroscopic modes were (Tobin and Donath, 1971, p. 1468) undeformed grains, grains with twin lamellae, grains with fractures and twin lamellae, and transgranular fractures. Tobin and Donath (1971, p. 1472) found that as the strain increased, the relative proportion of these elements changed in conjunction with the change in macroscopic mode. Uniform flow is characterized by undeformed and twinned grains, with the proportion being a function of the total strain. Grains with twins and fractures become important as microfaulting begins and transgranular fractures characterize the region of faulting. Within the regime of brittle faulting, a moderate percentage of undeformed grains persist but are nearly absent in the low-temperature cohesive faulting regime. Primary undeformed grains must be distinguished from recrystallized undeformed grains in higher-temperature deformation.

Another way to express the difference between mesoscopic faulting and flow is in terms of the strain heterogeneity, first shown by Donath and Wood (1976, Pl. 1) as contours of strain magnitudes on experimentally deformed oolitic limestone and later by Wood and Holm (1980) as the correlation coefficient on a plot of the long axis versus the short axis of oolite strain ellipses. The homogeneity of strain increases with increasing temperature and confining pressure, and little effect of strain rate is seen. In all experiments, the tendency to form fault zones increased with total strain. Gray (1981, p. 231–232) found a somewhat analogous result in the field for deformation by pressure solution. The early stage of a single deformation was by grain-scale stylolites and produced a weak slaty fabric that is truncated by zones of more intense pressure solution that form spaced cleavage.

The change in mesoscopic mode from flow to faulting with increasing strain is to be expected in a work-hardening material. The grain-scale strain may reflect either the total strain when deformation ended or the ductility of the rock, that is, the strain at which deformation shifts to a mesoscopic mode. It is clear in experiments, at least, that when the mesoscopic failure mode is faulting, the grain-scale flow mechanisms are supplanted by fault slip. If the mesoscopic failure mode is solution cleavage, it is not yet certain whether grain-scale deformation continues. The average glide strains in three small folds from one outcrop (Groshong, 1975b, p. 1367) are nearly constant in magnitude (2.5% ± 2.0%) and direction (about layer parallel), even though the interlimb limb angles of 150°, 90°, and 0° imply large differences in total strain. This indicates that the crystal-plastic strain in this example is a measure of the ductility because it does not increase appreciably with total strain. The folding strain is predominantly by pressure solution. A natural suite of rocks (Friedman, 1967, p. 191) showed greater fracture strengths in presumably more deformed samples from within folds than in presumably less deformed samples from horizontal beds away from the folds. Thus, as a result of work hardening, the current strength of the rock may be a function of the total strain.

Flow Laws

Within the work-hardening regime, the aggregate flow law must be a combination of flow laws for the individual mechanisms (Groshong, 1975b; Mitra, 1976). Flow laws for individual mechanisms might be interpreted from experimentally derived deformation-mechanism maps (Stocker and Ashby, 1973; Frost and Ashby, 1982) and can be combined by using standard strength-of-materials concepts as given by, for example, Jaeger and Cook (1979, p. 314–325).

Crystal-plastic dislocation glide mechanisms follow an exponential flow law (Cottrell, 1964; Schmid, 1982). Dislocation glide is resisted by impurities and the presence of grain boundaries, and so small grains deforming by this mechanism are stronger than big grains (Robertson, 1955; Handin and Hager, 1957; Paterson, 1958; Brace, 1961; Friedman, 1967; Hugman and Friedman, 1979). The Hall-Petch law in metallurgy (Nicolas and Poirier, 1976) states that the excess hardening is related to the inverse square root of grain size. Pressure enhances this effect by the suppression of microfractures, whereas temperature reduces the effect (Olsson, 1974). Increasing temperature reduces work hardening due to dislocation tangles, marked by the reduction of yield strength in materials deforming by glide (Griggs and others, 1951; Handin and Hager, 1958; Heard, 1963). Reduction in strain rate also significantly reduces the yield strength (Heard, 1963; Donath and Fruth, 1971).

Grain-boundary diffusion mechanisms follow a linear flow law (Coble, 1963). Elliott (1973) proposed that pressure solution is analogous to grain-boundary diffusion (Coble creep) for which the strain rate should be directly proportional to the differential stress and inversely proportional to the temperature and the square of grain size. The effect of temperature is quite small (Rutter, 1983), however. Kerrich and others (1977, p. 248) found in a field example that pressure solution had a greater effect on small grains than on large grains, substantiating the grain-size effect.

The fault-initiation stress increases with confining pressures but decreases with temperature (Griggs, 1936; Griggs and others, 1951; Handin and Hager, 1958; Paterson, 1958, Pl. 1; Griggs and Handin, 1960; Handin and others, 1963). A completely satisfactory fracture criterion has yet to be developed. The fracture stress should not depend upon the choice of coordinate system and therefore should be a function of the stress invariants (Jaeger and Cook, 1979, p. 23–24). Handin and others (1967) attempted to find a single stress-invariant (Jaeger and Cook, 1979, p. 23–24) fracture criterion that would apply to compression, extension, and torsion tests by plotting the octahedral shear stress (second deviatoric invariant) against the mean stress (first invariant) but did not obtain consistent results. Cherry and others (1968) defined a more complex function of octahedral shear stress, the third stress invariant and the third deviatoric invariant that gives more consistent results; Mogi (1971) showed that a curve relating the octahedral shear stress to the octahedral normal stress provided an acceptable criterion; and White (1973) determined a function of the three invariants. Kirby (1980, p. 6355–6356) found that a plot of differential stress versus σ_3 caused data from different types of tests to fall on a single curve.

Many of the experiments and field examples previously cited show tensile microcracks in the vicinity of faults, and experiments often show that microfracturing and dilation occur just prior to faulting (Gramberg, 1965; Brace and others, 1966; Scholz, 1968a, p. 4794; 1968b; Wawersik and Brace, 1971, p. 74–77; Edmond and Paterson, 1972; Paterson, 1978,

p. 112–137). This is a work-softening process that causes deformation to be localized in the fault zone. Griffith (1921) demonstrated that cracks form flaws that greatly reduce the stress needed for tensile fracture propagation and extended the concept in 1924 to plane-strain shear failure related to obliquely oriented cracks. The Griffith criterion has been modified for closed cracks by Brace (1960) and McClintock and Walsh (1962) and extended to three dimensions by Murrell (1963) as discussed by Jaeger and Cook (1979, p. 101–106). The extended theory predicts a parabolic Mohr envelope for which the uniaxial compressive strength is 12 times the tensile strength. The extended theory plots as a straight line when all stresses are compressive and predicts that σ_2 will be important, all in reasonable accord with observations (Jaeger and Cook, 1979, p. 190–192; Paterson, 1978, p. 52–70).

A problem with theories of failure based on tensile microcracks is how the cracks link together to form a fault. Numerous experiments have shown that tensile cracks form parallel to σ_1 or extend parallel to σ_1 from oblique cracks (Brace and Bombolakis, 1963; Gramberg, 1965, p. 48; Hoek and Bieniawski, 1965, p. 148, 150; Lajtai, 1971, p. 142, 146–147). Peng and Johnson (1972) proposed that a fault develops from the tensile cracks by buckling of the thin beams between the cracks after the cracks have become long enough. The concept is appealing because brittle buckling is an unstable phenomenon, and the granulation of the beams would form gouge almost instantly.

Faulting in porous materials leads to porosity collapse and work hardening in the fault zone. This was first suggested by Aydin and Johnson (1978) based on their observation that increased displacement across fault zones in porous sandstone was accompanied by the formation of new, small-displacement faults in intact rock. The hardening occurs because of cataclastic porosity loss (Aydin and Johnson, 1978; Underhill and Woodcock, 1987). Work hardening due to porosity loss by grain rotation and dewatering in clay-rich sediments is interpreted by Moore and Byrne (1987) to be the cause of fault-zone widening during subduction and to lead to the formation of mélange.

After a fault has formed, further displacements may be by frictional sliding. Bedding-plane slip is also a mechanism of major importance for which slip on a pre-existing surface is significant. Frictional sliding is a function of the normal and shear stress and can be represented by a linear Mohr envelope. The coefficient of friction is the slope of the line. Byerlee (1978) divided frictional sliding into three regions: low pressure having normal stresses of up to 50 bars, intermediate pressure up to 1000 bars, and high pressure up to 17 kbar or more. In the low-pressure region, friction is dominated by the roughness of the sliding surfaces, and consequently the lithology is important, with the coefficient of friction ranging from 0.3 to 10 (Logan and others, 1972). In the intermediate- and high-pressure range, the maximum friction coefficient is relatively independent of rock type. In the intermediate range, the value is 0.85 and in the high-pressure range is 0.6 plus a constant of 0.5 divided by the normal stress. In the high-pressure regime, much lower coefficients of friction are found for vermiculite, montmorillonite, and illite (Byerlee, 1978, p. 624).

Based on the results of Donath (1970), Donath and others (1971), and Tobin and Donath (1971), it might be possible to quantitatively infer stress and strain as given on ductility-depth curves (Handin and others, 1963) from the strain partitioning, but this has not yet been seriously attempted. When the ductile limit has been reached, the deformation is usually dominated by a single mechanism, either pressure solution or fault slip for which a steady-state flow law might be appropriate.

Deformation Fronts and Deformation Mechanism Associations

It has long been known that the character of rock deformation changes from sedimentary to metamorphic from the external to the internal parts of many mountain belts, and from the shallow levels to the deeper levels (Heim, 1878, as discussed by Milnes, 1979). Van Hise (1898) defined an upper zone of fracture, an intermediate zone of fracture and flow, and a lower zone of flow. This type of categorization remains important in tectonic analysis and in the interpretation of structural provenance.

A deformation front was first explicitly defined for quartzite in the central Appalachians by Fellows (1943, p. 1415) as occurring where the quartzites change from having unoriented grains to having grains that show a preferred orientation in shape and crystal lattice orientation. Mitra (1987, p. 588) showed that in the same area the tectonite front for carbonates is more external in the mountain belt than the tectonite front for siliciclastic rocks. Also in the same area, Cloos (1947b, p. 911) defined a front for cleavage in limestone as occurring where the oolite extension first reaches 20% and the cleavage front for shale as visible at a more external position. (In this area, the cleavage in limestone is the result of the grain shape fabric, not pressure solution.) A deformation front could be defined as the first occurrence of any deformation mechanism.

Blake and others (1967, p. C7) introduced the concept of naming the rocks between fronts in a classification currently used for subduction-zone rocks. They defined "textural zones of progressive metamorphism" (including deformation) for graywacke: zone 1 appears unmetamorphosed and shows no evidence of cataclasis (or mylonitization), either in outcrop or hand lens; zone 2 shows well-developed platy cleavage in outcrop and is clearly mylonitic (they used the term "cataclastic" in the sense of Lapworth) under the hand lens; zone 3 is completely recrystallized to quartz-mica schist.

The term "deformation regime" appears to have been introduced by Kerrich and Allison (1978) without explicit definition but in the context of ranges of environmental parameters in which deformation occurs by different mechanisms. Schmid (1982, p. 97) stated that deformation regimes are defined by empirical flow laws and diagnostic microstructural imprints. The regimes identified by Schmid (1982) are cataclastic flow, low-temperature plasticity, power-law creep, and grain-size sensitive creep. Kerrich and Allison (1978) treated pressure solution as a separate regime. Mechanisms in the regimes just identified can, however, occur simultaneously in the same rock. For example, low-temperature plasticity is intimately related to cataclasis under a range of laboratory conditions (Donath and others, 1971), and both types of features are found in the same samples in the field (Spang and Brown, 1981; Mitra, 1987, p. 583). Pressure solution and low-temperature plasticity also occur together (Groshong, 1975b; Kerrich and others, 1977).

A deformation-mechanism association is here defined as the deformation mechanisms formed under the same environmental conditions in the same rock type. Environmental conditions are temperature, confining pressure, differential stress, strain rate, pore fluid type, and pore pressure. A rock deformation experiment produces a single deformation-mechanism association in the sample. Five associations are defined, based on the nature of the deformation of the framework minerals. In Association I, hydroplastic deformation, the framework minerals are undeformed; the mechanism of deformation is cohesionless particulate flow in the terminology of Borradaile (1981). Large strain in soft grains, such as unlithified pellets in limestone or glauconite in sandstone, is allowed: the crystals within the grains are not deformed by the strain of the grain. The Association I-II boundary is the crystal deformation front. In Association II, brittle-framework deformation, crystal deformation occurs by truncation due to fracture or faulting (IIF) or pressure solution (IIS) or both (IIFS). Strain due to crystal-plastic glide is in the range of 0%–2%. The crystal-plastic component is characterized by slight undulatory extinction and rare deformation lamellae. The Association II-III boundary is the crystal-plastic front. Association III, semi–brittle-framework deformation, is defined by obvious crystal-plastic glide that is insufficient to produce a grain-orientation fabric, a range of about 2%–15% strain (for example, Fig. 1). The crystal plastic component may be characterized by obvious undulatory extinction and/or abundant deformation lamellae. Mineral truncation may occur by faulting and possibly by tension or extension fracture (IIIF), pressure solution (IIIS), or both (IIIFS). The Association III-IV boundary is the tectonite front. In Association IV, tectonites, a deformation-induced crystallographic fabric is present. Association IV may be subdivided into (IVG), glide tectonites, in which the crystallographic fabric is caused by glide mechanisms and the original mineral boundaries are preserved; and (IVR), recrystallization tectonites, in which a crystallographic fabric is present but the grain boundaries are no longer original. Association V is annealed, having no more than a weak crystallographic fabric and the original grain boundaries obliterated. Associations I–III fall in the low-temperature regime in which a sedimentary rock continues to maintain its sedimentary texture. Deformation-mechanism associations are related to rock type, and so in any given environment of deformation, the boundary between two zones will not in general occur at the same location for different lithologies.

The primary control on the location of the tectonite front appears to be temperature. Groshong and others (1984a) found this transition to be rather sharp and to occur at the vitrinite reflectance (R_0) of 3.5% in a relatively clean, coarse-grained limestone in the eastern Helvetic Alps. This corresponds to a temperature of about 175 °C based on the well-standard reflectance curve of Bostick (1974, p. 9). Low-temperature metamorphic stages are subdivided by Tissot and Welte (1978) into diagenesis, having an upper limit of $R_0 \simeq 0.5$; catagenesis, upper limit $\simeq 2.0$; and metagenesis, upper limit $\simeq 4.0$, above which greenschist metamorphism begins. This scale can be correlated to coal rank (Teichmüller and Teichmüller, 1981), spore and pollen coloration (Gray and Boucot, 1975), conodont color (Epstein and others, 1977; Rejebian and others, 1987), chitinozoan, graptolite, and scolecodont reflectance (Bertrand and Heroux, 1987), illite crystallinity (Kübler, 1968), and certain mineral reactions (Kübler, 1968). Temperature and time both clearly affect these indicators of metamorphism (Bostick, 1974; Dow, 1978). Confining pressure alone does not have a significant effect (Bostick, 1974; Epstein and others, 1977), although the presence of water affects conodont coloration (Rejebian and others, 1987), and strain alone can alter the illite crystallinity (Flehmig and Langheinrich, 1974). Considerably more work should be done to determine the relationship between deformation-mechanism associations and low-temperature metamorphism.

ACKNOWLEDGMENTS

I would like to thank the GSA editors William A. Thomas and Robert D. Hatcher, Jr., for inviting me to write this paper and Richard P. Nickelsen for providing the original inspiration. The invaluable International Tectonic Dictionary by John Dennis and John himself were of great assistance. I greatly appreciate the hard work and valuable suggestions from the reviewers Terry Engelder, Carol Simpson, Dave Wiltschko, Joe Benson, Mike Lesher, and Greg Guthrie. Work supported by National Science Foundation Grant EAR-8402915 has contributed directly to the interpretations.

REFERENCES CITED

Abdel-Gawad, M., Bulau, J., and Tittmann, B., 1987, Quantitative characterization of microcracks at elevated pressure: Journal of Geophysical Research, v. 92, p. 12,911–12,916.
Adams, F. D., 1910, An experimental investigation into the action of differential pressure on certain minerals and rocks employing the process suggested by Professor Kick: Journal of Geology, v. 18, p. 489–525.
Adams, F. D., and Coker, E. G., 1910, An experimental investigation into the flow of rocks: American Journal of Science, Fourth Series, v. 29, p. 465–487.
Adams, F. D., and Nicolson, J. T., 1901, An experimental investigation into the flow of marble: Royal Society of London Philosophical Transactions, ser. A, v. 195, p. 363–401.
Adams, S. F., 1920, A microscopic study of vein quartz: Economic Geology, v. 15, p. 623–664.
Aleksandrowski, P., 1985, Graphic determination of principal stress directions for slickenside populations: An attempt to modify Arthaud's method: Journal of Structural Geology, v. 7, p. 73–82.
Alvarez, W., Engelder, T., and Lowrie, W., 1976, Formation of spaced cleavage and folds in brittle limestone by dissolution: Geology, v. 4, p. 698–701.
Alvarez, W., Engelder, T., and Geiser, P. A., 1978, Classification of solution cleavage in pelagic limestones: Geology, v. 6, p. 263–266.
Anderson, E. M., 1905, The dynamics of faulting: Edinburgh Geological Society, v. 8, p. 387–402.
—— 1942, The dynamics of faulting and dyke formation with applications to Britain (1st edition): London, England, Oliver and Boyd, 191 p.; revised 2nd edition, 1951, 206 p.
Anderson, J. J., and Everett, J. R., 1965, Mudcrack formation studied by time-lapse photography [abs.]: Geological Society of America Special Paper 82, p. 4–5.
Anderson, J. L., Osborne, R. H., and Palmer, D. R., 1980, Petrogenesis of cataclastic rocks within the San Andreas fault zone of southern California, U.S.A.: Tectonophysics, v. 67, p. 221–249.
Anderson, T. B., 1968, The geometry of a natural orthorhombic system of kink bands: Geological Survey of Canada Paper 68-52, p. 200–220.
Angelier, J., 1979, Determination of the mean principal directions of stresses for a given fault population: Tectonophysics, v. 56, p. T17–T26.
Angelier, J., and Goguel, J., 1979, Sur une methode simple de determination des axes principaux des contraintes pour une population de failles: Comptes Rendus Académie de Science Paris, sér. D, v. 288, p. 307–310.
Angelier, J., Tarantola, A., Valette, B., and Manoussis, S., 1982, Inversion of field data in fault tectonics to obtain the regional stress—I. Single phase fault populations: A new method of computing the stress tensor: Royal Astronomical Society Geophysical Journal, v. 69, p. 607–621.
Angelier, J., Colletta, B., and Anderson, R. E., 1985, Neogene paleostress changes in the Basin and Range: A case study at Hoover Dam, Nevada-Arizona: Geological Society of America Bulletin, v. 96, p. 347–361.
Ardell, A. J., Christie, J. M., Kirby, S. H., and McCormick, J. W., 1976, Electron microscopy of deformation structures in quartz, in Proceedings of the Electron Microscopy and Analysis Group: Bristol, England, Institute of Physics, University of Bristol.
Arthaud, F., 1969, Détermination graphique des directions de raccourcissement, d'allongement et intermédiaire d'une population de failles: Société Géologique de France Bulletin, v. 11, p. 729–737.
Arthaud, F., and Mattauer, M., 1969, Examples de stylolites d'origine tectonique dans le Languedoc, leur relation avec la tectonique cassante: Société Géologique de France Bulletin, v. 11, p. 738–744.
—— 1972, Sur l'origine tectonique de certains joints stylolitiques paralleles a la stratification; leur relation avec une phase de distension (example du Languedoc): Société Géologique de France Bulletin, v. 14, p. 12–17.
Aydin, A., 1977, Faulting in sandstone, Utah [Ph.D. dissert.]: Stanford, California, Stanford University, 246 p.
—— 1978, Small faults formed as deformation bands in sandstone: Pure and Applied Geophysics, v. 116, p. 913–930.
Aydin, A., and Johnson, A. M., 1978, Developments of faults as zones of deformation bands and as slip surfaces in sandstone: Pure and Applied Geophysics, v. 116, p. 932–942.
Aydin, A., and Reches, Z., 1982, Number and orientation of fault sets in the field and in experiments: Geology, v. 10, p. 107–112.
Bahat, D., 1979, Theoretical considerations on mechanical parameters of joint surfaces based on studies on ceramics: Geological Magazine, v. 166, p. 81–92.
—— 1986a, Criteria for the differentiation of en echelons and hackles in fractured rocks: Tectonophysics, v. 121, p. 197–206.
—— 1986b, Joints and en echelon cracks in middle Eocene chalks near Beer Sheva, Israel: Journal of Structural Geology, v. 8, p. 181–190.
—— 1987a, Correlation between fracture surface morphology and orientation of cross-fold joints in Eocene chalks around Beer Sheva: Tectonophysics, v. 136, p. 323–333.
—— 1987b, Jointing and fracture interactions in middle Eocene chalks near Beer Sheva, Israel: Tectonophysics, v. 136, p. 299–321.
Bahat, D., and Engelder, T., 1984, Surface morphology on cross-fold joints of the Appalachian Plateau, New York and Pennsylvania: Tectonophysics, v. 104, p. 299–313.
Bailey, S. W., Bell, R. A., and Peng, C. J., 1958, Plastic deformation of quartz in nature: Geological Society of America Bulletin, v. 69, p. 1443–1466.
Bailly, P. A., 1954, Presence de microstylolites dans des pegmatites et des lentilles de quartz: Société Géologique de France Bulletin, v. 3, p. 299–301.
Baker, P. A., Kastner, M., Byerlee, J. D., and Lockner, D. A., 1980, Pressure solution and hydrothermal recrystallization of carbonate sediments—An experimental study: Marine Geology, v. 38, p. 185–203.
Balk, R., 1937, Structural behavior of igneous rocks: Geological Society of America Memoir 5, 177 p.
Barrett, P. J., 1964, Residual seams and cementation in Oligocene shell calcarenites, Te Kuiti Group: Journal of Sedimentary Petrology, v. 34, p. 523–531.
Bartlett, W. L., Friedman, M., and Logan, J. M., 1981, Experimental folding of rocks under confining pressure, Part IX. Wrench faults in limestone layers: Tectonophysics, v. 79, p. 255–277.
Bates, R. L., and Jackson, J. A., 1980, Glossary of geology, (2nd edition): Falls Church, Virginia, American Geological Institute, 751 p.
Bathurst, R.G.C., 1958, Diagenetic fabrics in some British Dinantian limestones: Liverpool and Manchester Geological Journal, v. 2, p. 11–36.
Batzle, M. L., Simmons, G., and Siegfried, R. W., 1980, Microcrack closure in rocks under stress: Direct observation: Journal of Geophysical Research, v. 85, p. 7072–7090.
Beach, A., 1974, A geochemical investigation of pressure solution and the formation of veins in a deformed greywacke: Contributions to Mineralogy and Petrology, v. 46, p. 61–68.
—— 1975, The geometry of en-echelon vein arrays: Tectonophysics, v. 28, p. 245–263.

——— 1977, Vein arrays, hydraulic fractures and pressure-solution structures in a deformed flysch sequence, S.W. England: Tectonophysics, v. 40, p. 201–225.
——— 1979, Pressure solution as a metamorphic process in deformed terrigenous sedimentary rocks: Lithos, v. 12, p. 51–58.
Beard, D. C., and Weyl, P. K., 1973, Influence of texture on porosity and permeability of unconsolidated sand: American Association of Petroleum Geologists Bulletin, v. 57, p. 349–369.
Becker, G. F., 1893, Finite homogeneous strain, flow and rupture of rocks: Geological Society of America Bulletin, v. 4, p. 13–90.
Beeré, W., 1978, Stresses and deformation at grain boundaries: Royal Society of London Philosophical Transactions, ser. A, v. 288, p. 177–196.
Behrmann, J. H., 1985, Crystal plasticity and superplasticity in quartzite: A natural example: Tectonophysics, v. 115, p. 101–129.
——— 1987, A precautionary note on shear bands as kinematic indicators: Journal of Structural Geology, v. 9, p. 659–666.
Bell, J. F., 1941, Morphology of mechanical twinning in crystals: American Mineralogist, v. 26, p. 247–261.
Bell, T. H., 1978, The development of slaty cleavage across the Nackara Arc of the Adelaide geosyncline: Tectonophysics, v. 51, p. 171–201.
Bell, T. H., and Etheridge, M. A., 1973, Microstructure of mylonites and their descriptive terminology: Lithos, v. 6, p. 337–348.
Berthé, D., Choukroune, P., and Jeqouzo, P., 1979, Orthogneiss, mylonite, and non-coaxial deformation of granites: The example of the South Armorican shear zone: Journal of Structural Geology, v. 1, p. 31–42.
Bertrand, R., and Heroux, Y., 1987, Chitinozoan, graptolite, and scolecodont reflectance as an alternative to vitrinite and pyrobitumen reflectance in Ordovician and Silurian strata, Anticosti Island, Quebec, Canada: American Association of Petroleum Geologists Bulletin, v. 71, p. 951–957.
Beutner, E. C., 1978, Slaty cleavage and related strain in Martinsburg Slate, Delaware Water Gap, New Jersey: American Journal of Science, v. 278, p. 1–23.
——— 1980, Slaty cleavage unrelated to tectonic dewatering: The Siamo and Michigamme slates revisited: Geological Society of America Bulletin, Part I, v. 91, p. 171–178.
Beutner, E. C., and Charles, E. G., 1985, Large volume loss during cleavage formation, Hamburg sequence, Pennsylvania: Geology, v. 13, p. 803–805.
Beutner, E. C., and Diegel, F. A., 1985, Determination of fold kinematics from syntectonic fibers in pressure shadows, Martinsburg Slate, New Jersey: American Journal of Science, v. 285, p. 16–50.
Beutner, E. C., Jancin, M. D., and Simon, R. W., 1977, Dewatering origin of cleavage in light of deformed calcite veins and clastic dikes in Martinsburg slate, Delaware Water Gap, New Jersey: Geology, v. 5, p. 118–122.
Bielenstein, H. U., and Charlesworth, H.A.K., 1965, Precambrian sandstone sills near Jasper, Alberta: Bulletin of Canadian Petroleum Geology, v. 13, p. 405–408.
Bishop, J.F.W., 1953, A theoretical examination of the plastic deformation of crystals by glide: The Philosophical Magazine, v. 44, p. 51–64.
Blake, D. B., and Roy, C. J., 1949, Unusual stylolites: American Journal of Science, v. 247, p. 779–790.
Blake, M. C., Jr., Irwin, W. P., and Coleman, R. C., 1967, Upside-down metamorphic zonation, blueschist facies, along a regional thrust in California and Oregon: U.S. Geological Survey Professional Paper 575-C, p. C1–C9.
Blenkinsop, T. G., and Rutter, E. H., 1986, Cataclastic deformation of quartzite in the Moine thrust zone: Journal of Structural Geology, v. 8, p. 669–681.
Bogacz, W., 1981, The role of cleavage in the shear folding and faulting processes in the vicinity of the Boguszowice thrust (Upper Silesian Coal Basin): Bulletin de l'Academie Polonaise des Sciences, Serie des Sciences de la Terre, v. 29, p. 251–259.
Böhm, A., 1883, Über Gesteine des Wechsels: Tschermaks Mineralogische und Petrographische Mitteilungen, new series, v. 5, p. 197–214.
Bonham, L. C., 1957, Structural petrology of the Pico anticline, Los Angeles County, California: Journal of Sedimentary Petrology, v. 27, p. 251–264.
Bonilla, M. G., Alt, J. N., and Hodgen, L. D., 1978, Trenches across the 1906 trace of the San Andreas fault in northern San Mateo County, California: U.S. Geological Survey Journal of Research, p. 347–358.
Borg, I. Y., and Handin, J., 1966, Experimental deformation of crystalline rocks: Tectonophysics, v. 3, p. 251–367.
Borg, I. Y., and Heard, H. C., 1970, Experimental deformation of plagioclases, *in* Paulitsch, P., ed., Experimental and natural rock deformation: New York, Springer-Verlag, p. 375–403.
Borg, I. Y., and Maxwell, J. C., 1956, Interpretation of fabrics of experimentally deformed sands: American Journal of Science, v. 254, p. 71–81.
Borg, I. Y., Friedman, M., Handin, J., and Higgs, D. V., 1960, Experimental deformation of St. Peter Sand: A study of cataclastic flow, *in* Griggs, D., and Handin, J., eds., Rock deformation (A symposium): Geological Society of America Memoir 79, p. 133–191.
Borradaile, G. J., 1977, On cleavage and strain: Results of a study in West Germany using tectonically deformed sand dykes: Geological Society of London Journal, v. 133, p. 146–164.
——— 1981, Particulate flow of rock and the formation of cleavage: Tectonophysics, v. 72, p. 305–321.
Borradaile, G. J., Bayly, M. B., and Powell, C. McA., eds., 1982, Atlas of deformational and metamorphic rock fabrics: Berlin, Springer-Verlag, 551 p.
Bostick, N. H., 1974, Phytoclasts as indicators of thermal metamorphism, Franciscan assemblage and Great Valley Sequence (upper Mesozoic), California: Geological Society of America Special Paper 153, p. 1–17.
Bott, M.H.P., 1959, The mechanics of oblique slip faulting: Geological Magazine, v. 46, p. 109–117.
Boullier, A. M., and Gueguen, Y., 1975, SP-mylonites: Origin of some mylonites by superplastic flow: Contributions to Mineralogy and Petrology, v. 50, p. 93–104.
Brace, W. F., 1960, An extension of the Griffith theory of fracture of rocks: Journal of Geophysical Research, v. 65, p. 3477–3480.
——— 1961, Dependence of fracture strength of rocks on grain size: Pennsylvania State University Mineral Industries Experiment Station Bulletin, v. 76, p. 99–103.
——— 1964, Brittle fracture of rocks, *in* Judd, W. R., ed., State of stress in the Earth's crust: New York, American Elsevier Publishing Company, Inc., p. 111–174.
Brace, W. F., and Bombolakis, E. G., 1963, A note on brittle crack growth in compression: Journal of Geophysical Research, v. 68, p. 3709–3713.
Brace, W. F., and Byerlee, J. D., 1966, Stick-slip as a mechanism for earthquakes: Science, v. 153, p. 990–992.
Brace, W. F., Paulding, B. W., Jr., and Scholz, C., 1966, Dilatancy in the fracture of the crystalline rocks: Journal of Geophysical Research, v. 71, p. 3939–3953.
Bradley, W. C., 1963, Large-scale exfoliation in massive sandstones of the Colorado plateau: Geological Society of America Bulletin, v. 74, p. 519–528.
Bretz, J. H., 1940, Solution cavities in the Joliet Limestone of northeastern Illinois: Journal of Geology, v. 48, p. 337–384.
——— 1950, Origin of the filled sink-structures and circle deposits of Missouri: Geological Society of America Bulletin, v. 61, p. 789–834.
Brewer, R., 1964, Fabric and mineral analysis of soils: New York, John Wiley & Sons, Inc., 470 p.
Brock, W. G., and Engelder, T., 1977, Deformation associated with the movement of the Muddy Mountain overthrust in the Buffington window, southeastern Nevada: Geological Society of America Bulletin, v. 88, p. 1667–1677.
Brodzikowski, K., Gotowala, R., Haluszczak, A., Krzyszkowski, D., and Van Loon, A. J., 1987, Soft-sediment deformations from glaciodeltaic glaciolacustrine, and fluviolacustrine sediments in the Kleszczów Graben, *in* Jones, M. E., and Preston, R.M.F., eds., Deformation of sediments and sedimentary rocks: Geological Society Special Publication No. 29: Oxford, England, Blackwell Scientific Publications, p. 255–265.
Brown, W. L., and Macaudiere, J., 1984, Microfracturing in relation to atomic structure of plagioclase from a deformed meta-anorthosite: Journal of Structural Geology, v. 6, p. 579–586.
Bruhn, R. L., and Pavlis, T. L., 1981, Late Cenozoic deformation in the Matanuska Valley, Alaska: Three-dimensional strain in a forearc region: Geological Society of America Bulletin, v. 92, p. 282–293.
Bucher, W. H., 1920, The mechanical interpretation of joints: Journal of Geology, v. 28, p. 707–730.
Burg, J. P., and Harris, L. B., 1982, Tension fractures and boudinage oblique to the maximum extension direction: An analogy with Lüders' bands: Tectonophysics, v. 83, p. 347–363.
Burg, J. P., and Inglesias Ponce de Leon, M., 1985, Pressure-solution structures in a granite: Journal of Structural Geology, v. 7, p. 431–436.
Burgers, W. G., and Burgers, J. M., 1935, First report on viscosity and plasticity: Koninklijk Nederlandse Akademie van Wetenschappen Verhandelingen Series 1, v. 15, p. 1–256.
Bustin, R. M., 1983, Heating during thrust faulting in the Rocky Mountains: friction or fiction?: Tectonophysics, v. 95, p. 309–328.
Buxton, T. M., and Sibley, D. F., 1981, Pressure solution features in a shallow buried limestone: Journal of Sedimentary Petrology, v. 51, p. 19–26.
Byerlee, J. D., 1978, Friction of rocks: Pure and Applied Geophysics, v. 116, p. 615–626.
Byerlee, J. D., and Brace, W. F., 1969, High-pressure mechanical instability in rocks: Science, v. 164, p. 713–715.
Byerlee, J., Mjachkin, V., Summers, R., and Voevoda, O., 1978, Structures developed in fault gouge during stable sliding and stick-slip: Tectonophysics, v. 44, p. 161–171.
Byrne, T., 1984, Structural geology of mélange terranes in the Ghost Rocks Formation, Kodiak Islands, Alaska, *in* Raymond, L. A., ed., Mélanges, their nature, origin, and significance: Geological Society of America Special Paper 198, p. 21–52.
Carannante, G., and Guzzetta, G., 1972, Stiloliti e sliccoliti come meccanismo di deformayione delle masse rocciose: Bollettino della Societa dei Naturalisti in Napoli, v. 81, p. 157–170.
Carey, E., and Brunier, B., 1974, Analyse théorique et numérique d'un Modèle mécanique élémentaire appliqué à l'étude d'une population de failles: Comptes Rendus Académie de Science Paris, sér. D, v. 279, p. 891–894.
Carlson, S. R., and Wang, H. F., 1986, Microcrack porosity and *in situ* stress in Illinois borehold UPH 3: Journal of Geophysical Research, v. 91, p. 10421–10428.
Carson, B., and Berglund, P. L., 1986, Sediment deformation and dewatering under horizontal compression: Experimental results: Geological Society of America Memoir 166, p. 135–150.
Carson, B., von Huene, R., and Arthur, M., 1982, Small-scale deformation structures and physical properties related to convergence in Japan Trench slope sediments: Tectonics, v. 1, p. 277–302.
Carter, N. L., 1968, Flow of rock-forming crystals and aggregates, *in* Riecker, R. E., ed., Rock mechanics seminar, Volume 2: U.S. Air Force Cambridge Research Laboratories, Clearinghouse for Federal Science and Technology Information AD669 376, p. 509–594.
——— 1971, Static deformation of silica and silicates: Journal of Geophysical Research, v. 76, p. 5514–5540.
——— 1976. Steady-state flow of rocks: Reviews of Geophysics and Space Science, v. 14, p. 301–360.
Carter, N. L., and Raleigh, C. B., 1969, Principal stress directions from plastic flow in crystals: Geological Society of America Bulletin, v. 80, p. 1231–1264.
Carter, N. L., Christie, J. M., and Griggs, D. T., 1964, Experimental deformation and recrystallization of quartz: Journal of Geology, v. 72, p. 687–733.
Casas, J. M., and Sabat, F., 1987, An example of three-dimensional analysis of thrust-related tectonites: Journal of Structural Geology, v. 9, p. 647–657.
Casey, M., Dietrich, D., and Ramsay, J. G., 1983, Methods for determining history for chocolate tablet boudinage with fibrous crystals: Tectonophysics, v. 92, p. 211–239.
Chamberlin, T. C., 1888, Rock scorings of the great ice invasions: U.S. Geological Survey Seventh Annual Report, p. 147–248.
Cherry, J. T., Larson, D. B., and Rapp, E. G., 1968, A unique description of the failure of a brittle material: International Journal of Rock Mechanics and Mining Science, v. 5, p. 455–463.
Chester, F. M., and Logan, J. M., 1986, Implications for mechanical properties of brittle faults from observations of the Punchbowl fault zone, California: Pure and Applied Geophysics, v. 124, p. 79–106.
Chester, F. M., Friedman, M., and Logan, J. M., 1985, Foliated cataclasites: Tectonophysics, v. 111, p. 139–146.
Chilingarian, I.G.V., and Wolf, K. H., 1975, Compaction of coarse-grained sediments, I. Developments in sedimentology—18A: New York, Elsevier, 552 p.
Choquette, P. W., and James, N. P., 1987, Diagenesis in limestones—3. The deep burial environment: Geoscience Canada, v. 14, p. 3–35.
Choukroune, P., 1969, Un exemple d'analyse microtectonique d'une série calcaire affectée de plis isopaques ("concentriques"): Tectonophysics, v. 7, p. 57–70.
——— 1971, Contribution à l'étude des mechanismes de la déformtion avec schistosité grâce aux cristallisations syncinématiques dan les "zones abriteés" ("pressure shadows"): Société Géologique de France Bulletin, v. 13, p. 257–271.
Choukroune, P., and Séguret, M., 1968, Exemple de relations entre joints de cisaillement, fentes de tension, plis et schistosité: Revue Géography Physique et de Géologie Dynamique, v. 10, p. 239–246.
Christie, J. M., 1958, Dynamic interpretation of the fabric of dolomite from the Moine thrust zone in northwest Scotland: American Journal of Science, v. 256, p. 159–170.
Christie, J. M., and Ardell, A. J., 1974, Substructures of deformation lamellae in quartz: Geology, v. 2, p. 405–408.
Christie, J. M., and Ord, A., 1980, Flow stresses from microstructures of mylonites: Examples and current assessment: Journal of Geophysical Research, v. 85, p. 6253–6262.
Christie, J. M., Griggs, D. T., and Carter, N. L., 1964, Experimental evidence of basal slip in quartz: Journal of Geology, v. 72, p. 734–756.
Clark, B. R., 1970, Mechanical formation of preferred orientation in clays: American Journal of Science, v. 269, p. 250–266.
Clendenen, W. S., Kligfield, R., Hirt, A. M., and Lowrie, W., 1988, Strain studies of cleavage development of the Chelmsford Formation, Sudbury Basin, Ontario: Tectonophysics, v. 145, p. 191–211.
Clifton, H. E., 1965, Tectonic polish of pebbles: Journal of Sedimentary Petrology, v. 35, p. 867–873.
Cloos, E., 1932, Feather joints as indicators of the direction of movements on faults, thrusts, joints and magmatic contacts: Proceedings of the National Academy of Sciences of the United States of America, v. 18, p. 387–395.
——— 1946, Lineation, a critical review and annotated bibliography; Reprinted with supplement, 1953, 1957: Geological Society of America Memoir 18, 122 p.
——— 1947a, Boudinage: American Geophysical Union Transactions, v. 28, p. 626–632.
——— 1947b, Oolite deformation in the South Mountain fold, Maryland: Geological Society of America Bulletin, v. 58, p. 843–918.
——— 1955, Experimental analysis of fracture patterns: Geological Society of America Bulletin, v. 66, p. 241–256.
Cloos, H., 1922, Über Ausbau und Anwendung der granit-tektonischen Methode: Preussische Geologische Landesanstalt Abhandlungen, v. 89, p. 1–18.
——— 1928, Experimente zur Inneren Tektonik: Centralblatt fur Mineralogie, Geologie und Paläontologie Abteilung B, Geologie und Paläontologie, p. 609–621.
——— 1930, Zur Experimentellen Tektonik: Die Naturwissenschaften, Jhq. 18, Hft. 34, p. 741–747.
Coble, R. L., 1963, Model for boundary diffusion controlled creep in polycrystalline materials: Journal of Applied Physics, v. 34, p. 1679–1682.
Collins, D. A., and de Paor, D. G., 1986, A determination of bulk rotational deformation resulting from displacements in discrete shear zones in the Hercynian fold belt of South Ireland: Journal of Structural Geology, v. 8, p. 101–109.
Collins, K., and McGowan, A., 1974, The form and function of microfabric features in a variety of natural soils: Geotechnique, v. 24, p. 223–254.
Compton, R. R., 1966, Analyses of Pliocene-Pleistocene deformation and stresses in northern Santa Lucia Range, California: Geological Society of America Bulletin, v. 77, p. 1361–1380.
Conel, J. E., 1962, Studies of the development of fabrics in some naturally deformed limestones [Ph.D. dissert.]: California Institute of Technology, Pasadena, California, 257 p.

Conrad, R. E., II, and Friedman, M., 1976, Microscopic feather fractures in the faulting process: Tectonophysics, v. 33, p. 187–198.
Conybeare, C.E.B., 1949, Stylolites in Precambrian quartzite: Journal of Geology, v. 57, p. 83–85.
Conybeare, W. D., and Phillips, W., 1822, Outlines of the geology of England and Wales: London, England, William Phillips, 470 p.
Coogan, A. H., and Manus, R. W., 1975, Compaction and diagenesis of carbonate sands, *in* Chilingarian, G. V., and Wolf, K. H., eds., Compaction of coarse-grained sediments, I. Developments in sedimentology—18A: New York, Elsevier, p. 79–166.
Corbett, K., Friedman, M., and Spang, J., 1987, Fracture development and mechanical stratigraphy of Austin Chalk, Texas: American Association of Petroleum Geologists Bulletin, v. 71, p. 17–28.
Cottrell, A. H., 1964, The mechanical properties of matter: New York, John Wiley and Sons, Inc., 430 p.
Coulomb, C. A., 1776, Sur une application des règles de maximis et minimis a quelques problèmes de statique relatifs à l'architecture: Academic des Sciences Memoires de Math et de Physique, v. 7, p. 343–382.
Cowan, D. S., 1985, Structural styles in Mesozoic and Cenozoic melanges in the western Cordillera of North America: Geological Society of America Bulletin, v. 96, p. 451–462.
Cox, M. A., and Whitford-Stark, J. L., 1987, Stylolites in the Caballos Novaculite, west Texas: Geology, v. 15, p. 439–442.
Cox, S. F., 1987, Antitaxial crack-seal vein microstructures and their relationship to displacement paths: Journal of Structural Geology, v. 9, p. 779–787.
Cox, S. F., and Etheridge, M. A., 1983, Crack-seal fibre growth mechanisms and their significance in the development of oriented layer silicate microstructures: Tectonophysics, v. 92, p. 147–170.
Crampton, C. B., 1958, Structural petrology of Cambro-Ordovician limestones of northwest Highlands of Scotland: American Journal of Science, v. 256, p. 145–158.
Crook, K. A., 1964, Cleavage in weakly deformed mudstones: American Journal of Science, v. 262, p. 523–531.
Crowell, J. C., 1974, Origin of Late Cenozoic basins in southern California: Society of Economic Paleontologists and Mineralogists Special Publication No. 22, p. 190–204.
Currie, J. B., 1969, Written comments on "Structural analysis of features on natural and artificial faults" by D. K. Norris and K. Barron: Canada Geological Survey, Paper 68-52, p. 168–172.
Curtis, C. D., Lipshie, S. R., Oertel, G., and Pearson, M. J., 1980, Clay orientation in some Upper Carboniferous mudrocks, its relationship to quartz content and some inferences about fissility, porosity, and compactional history: Sedimentology, v. 27, p. 333–339.
Dale, T. N., and others, 1914, Slate in the United States: U.S. Geological Survey Bulletin 586, 220 p.
Daubrée, A., 1879, Études synthétiques de géologie expérimentale: Paris, Dunod, 828 p.
Davies, W., 1982, Fine structure of slate, *in* Borradaile, G. J., Bayly, M. B., and Powell, C. McA., eds., Atlas of deformational and metamorphic rock fabrics: New York, Springer-Verlag, p. 64–65.
de Boer, R. B., 1977a, On the thermodynamics of pressure solution—interaction between chemical and mechanical forces: Geochimica et Cosmochimica Acta, v. 41, p. 249–256.
——— 1977b, Pressure solution: Theory and experiments: Tectonophysics, v. 39, p. 287–301.
de Boer, R. B., Nagtegaal, P.J.C., and Duyvis, E. M., 1977, Pressure solution experiments on quartz sand: Geochimica et Cosmochimica Acta, v. 41, p. 257–264.
Deelman, J. C., 1975, "Pressure solution" or indentation?: Geology, v. 3, p. 23–24.
DeGraff, M. J., and Aydin, A., 1987, Surface morphology of columnar joints and its significance to mechanics and direction of joint growth: Geological Society of America Bulletin, v. 99, p. 605–617.
Dennis, J. G., 1967, International tectonic dictionary: American Association of Petroleum Geologists Memoir 7, 196 p.
——— 1987, Structural geology, an introduction: Dubuque, Iowa, Wm. C. Brown Publishers, 448 p.
de Sitter, L. U., 1956, Structural geology: New York, McGraw-Hill Book Company, Inc., 552 p.
Dietrich, D., and Song, H., 1984, Calcite fabrics in a natural shear environment, the Helvetic Nappes of western Switzerland: Journal of Structural Geology, v. 6, p. 19–32.
Dietz, R. S., 1959, Shatter cones in cryptoexplosion structures (meteorite impact?): Journal of Geology, v. 67, p. 496–505.
Diller, J. S., 1890, Sandstone dikes: Geological Society of America Bulletin, v. 1, p. 411–442.
Dionne, J.-C., and Shilts, W. W., 1974, A Pleistocene clastic dike, Upper Chaudiere Valley, Quebec: Canadian Journal of Earth Sciences, v. 11, p. 1594–1605.
Dokka, R. K., 1986, Patterns and modes of early Miocene crustal extensions, central Mojave Desert, California: Geological Society of America Special Paper 208, p. 75–95.
Donath, F. A., 1970, Some information squeezed out of rock: American Scientist, v. 58, p. 54–72.
Donath, F. A., and Fruth, L. S., Jr., 1971, Dependence of strain-rate effects on deformation mechanisms and rock types: Journal of Geology, v. 79, p. 347–371.
Donath, F. A., and Wood, D. S., 1976, Experimental evaluation of the deformation path concept: Royal Society of London Philosophical Transactions, ser. A, v. 283, p. 187–201.
Donath, F. A., Faill, R. T., and Tobin, D. G., 1971, Deformation mode fields in experimentally deformed rock: Geological Society of America Bulletin, v. 82, p. 1441–1462.
Dow, W. G., 1978, Petroleum source beds on continental slope and rises: American Association of Petroleum Geologists Bulletin, v. 62, p. 1584–1606.
Droxler, A., and Schaer, J.-P., 1979, Déformation cataclastique plastique lors du plissement sous faible couverture, de strates calcaires: Eclogae Geologicae Helvetiae, v. 72, p. 551–570.
Dula, W. F., Jr., 1981, Correlation between deformation lamellae, microfractures, macrofractures, and *in situ* stress measurements, White River Uplift, Colorado: Geological Society of America Bulletin, Part I, v. 92, p. 37–46.
Dunham, R. J., 1962, Classification of carbonate rocks according to depositional texture, *in* Ham, W. E., ed., Classification of carbonate rocks: American Association of Petroleum Geologists Memoir 1, p. 108–121.
Dunn, D. E., LaFountain, L. J., and Jackson, R. E., 1973, Porosity dependence and mechanism of brittle fracture in sandstone: Journal of Geophysical Research, v. 78, p. 2403–2417.
Dunnington, H. V., 1967, Aspects of diagenesis and shape change in stylolitic limestone reservoirs: World Petroleum Congress, 7th, Mexico, Proceedings, v. 2, p. 339–352.
Durney, D. W., 1972, Solution-transfer, an important geological deformation mechanism: Nature, v. 235, p. 315–317.
——— 1974, The influence of stress concentrations on the lateral propagation of pressure solution zones and surfaces [abs.]: Geological Society of Australia, Tectonics and Structural Newsletter, no. 3, p. 19.
——— 1978, Early theories and hypotheses on pressure-solution-redeposition: Geology, v. 6, p. 369–372.
Durney, D. W., and Ramsay, J. G., 1973, Incremental strains measured by syntectonic crystal growths, *in* DeJong, K. A., and Scholten, R., eds., Gravity and tectonics: New York, Wiley-Interscience, p. 67–96.
Dzulynski, S., and Kotlarczyk, J., 1965, Tectoglyphs on slickensided surfaces: Bulletin de l'Académie Polonaise des Sciences, Serie des sciences géologie et géographie, v. 13, p. 149–154.
Edmond, J. M., and Paterson, M. S., 1972, Volume changes during the deformation of rocks at high pressures: International Journal of Rock Mechanics and Mining Science, v. 9, p. 161–182.
Elliott, D., 1973, Diffusion flow laws in metamorphic rocks: Geological Society of America Bulletin, v. 84, p. 2645–2664.
——— 1976, The energy balance and deformation mechanisms of thrust sheets: Royal Society of London Philosophical Transactions, ser. A, v. 283, p. 289–312.
Engelder, J. T., 1974a, Cataclasis and the generation of fault gouge: Geological Society of America Bulletin, v. 85, p. 1515–1522.
——— 1974b, Microscopic wear grooves on slickensides: Indicators of paleoseismicity: Journal of Geophysical Research, v. 79, p. 4387–4392.
——— 1979, Mechanisms for strain within the Upper Devonian clastic sequence of the Appalachian Plateau, western New York: American Journal of Science, v. 279, p. 527–542.
——— 1982, Is there a genetic relationship between selected regional joints and contemporary stress within the lithosphere of North America?: Tectonics, v. 1, p. 161–177.
——— 1985, Loading paths to joint propagation during a tectonic cycle: An example from the Appalachian Plateau, U.S.A.: Journal of Structural Geology, v. 7, p. 459–476.
——— 1987, Joints and shear fractures in rock, *in* Atkinson, B. K., ed., Fracture mechanics of rock: London, England, Academic Press, p. 27–69.
Engelder, J. T., and Engelder, R., 1977, Fossil distortion and décollement tectonics of the Appalachian Plateau: Geology, v. 5, p. 457–460.
Engelder, J. T., and Geiser, P., 1979, The relationship between pencil cleavage and lateral shortening within the Devonian section of the Appalachian Plateau, New York: Geology, v. 7, p. 460–464.
——— 1980, On the use of regional joint sets as trajectories of paleostress fields during the development of the Appalachian Plateau, New York: Journal of Geophysical Research, v. 85, p. 6319–6341.
Engelder, J. T., and Marshak, S., 1985, Disjunctive cleavage formed at shallow depths in sedimentary rocks: Journal of Structural Geology, v. 7, p. 327–343.
Engelder, J. T., and Oertel, G., 1985, Correlation between abnormal pore pressure and tectonic jointing in the Devonian Catskill Delta: Geology, v. 13, p. 863–866.
Engelder, J. T., and Scholz, C. H., 1976, The role of asperity indentation and ploughing in rock friction—II: International Journal of Rock Mechanics and Mining Science, v. 13, p. 155–163.
——— 1981, Fluid flow along very smooth joints at effective pressures up to 200 Megapascals: American Geophysical Union Monograph 24, p. 147–152.
Engelder, J. T., Logan, J. M., and Handin, J., 1975, The sliding characteristics of sandstone on quartz fault gouge: Pure and Applied Geophysics, v. 113, p. 69–86.
Epstein, A. G., Epstein, J. B., and Harris, L. D., 1977, Conodont color alteration—An index to organic metamorphism: U.S. Geological Survey Professional Paper 995, 27 p.
Etchecopar, A., Vasseur, G., and Daignieres, M., 1981, An inverse problem in microtectonics for the determination of stress tensors from fault striation analysis: Journal of Structural Geology, v. 3, p. 51–65.
Etheridge, M. A., 1983, Differential stress magnitudes during regional deformation and metamorphism: Upper bound imposed by tensile fracturing: Geology, v. 11, p. 231–234.
——— 1984, Reply to comment on "Differential stress magnitudes during regional deformation and metamorphism: Upper bound imposed by tensile fracturing": Geology, v. 12, p. 56–57.
Etheridge, M. A., and Oertel, G., 1979, Strain measurements from phyllosilicate preferred orientation—A precautionary note: Tectonophysics, v. 60, p. 107–120.
Etheridge, M. A., and Wilkie, J. C., 1979, Grain size reduction, grain boundary sliding and the flow strength of mylonites: Tectonophysics, v. 58, p. 159–178.
——— 1981, An assessment of dynamically recrystallized grain size as a palaeopiezometer in quartz-bearing mylonite zones: Tectonophysics, v. 78, p. 475–508.
Evans, K. F., and Engelder, J. T., 1986, A study of stress in Devonian shale: Part 1—3D stress mapping using a wireline microfrac system: Annual Technical Conference and Exhibition of the Society of Petroleum Engineers, 61st, New Orleans, 1986, SPE 15609, 12 p.
Fairbairn, H. W., 1941, Deformation lamellae in quartz from the Ajibik Formation, Michigan: Geological Society of America Bulletin, v. 52, p. 1265–1278.
Fairbairn, H. W., and Hawkes, H. E., Jr., 1941, Dolomite orientation in deformed rocks: American Journal of Science, v. 239, p. 617–632.
Farmin, R., 1941, Host-rock inflation by veins and dikes at Grass Valley, California: Economic Geology, v. 36, p. 143–174.
Fellows, R. E., 1943, Recrystallization and flowage in Appalachian quartzite: Geological Society of America Bulletin, v. 54, p. 1399–1432.
Fernandez, A., 1987, Preferred orientation developed by rigid markers in two-dimensional simple shear strain: A theoretical and experimental study: Tectonophysics, v. 136, p. 151–158.
Flehmig, W., and Langheinrich, G., 1974, Relation between tectonic deformation and illite crystallinity: Neues Jahrbuch für Geologie und Paläontologie Abhandlungen, v. 146, p. 325–346.
Fletcher, R. C., and Hofmann, A. W., 1974, Simple models of diffusion and combined diffusion-infiltration metasomatism: Carnegie Institution of Washington Publication 634, p. 243–259.
Fletcher, R. C., and Pollard, D. D., 1981, Anticrack model for pressure solution surfaces: Geology, v. 9, p. 419–424.
Fleuty, M. J., 1975, Slickensides and slickenlines: Geological Magazine, v. 112, p. 319–322.
Folk, R. L., 1965, Some aspects of recrystallization in ancient limestones, *in* Pray, L. C., and Murray, R. C., eds., Dolomitization and limestone diagenesis, a symposium: Society of Economic Paleontologists and Mineralogists Special Publication 13, p. 14–48.
——— 1974, The natural history of crystalline calcium carbonate: Effect of magnesium content and salinity: Journal of Sedimentary Petrology, v. 44, p. 40–53.
Foster, M. E., and Hudleston, P. J., 1986, "Fracture cleavage" in the Duluth Complex, northeastern Minnesota: Geological Society of America Bulletin, v. 97, p. 85–96.
Foster, R. H., and De, P. K., 1971, Optical and electron microscopic investigation of shear induced structures in lightly consolidated (soft) and heavily consolidated (hard) kaolinite: Clays and Clay Minerals, v. 19, p. 31–47.
Fraser, H. J., 1935, Experimental study of porosity and permeability of clastic sediments: Journal of Geology, v. 43, p. 910–1010.
Friedman, M., 1963, Petrofabric analysis of experimentally deformed calcite-cemented sandstones: Journal of Geology, v. 71, p. 12–37.
——— 1964, Petrofabric techniques for determination of principal stress direction in rocks, *in* Judd, W. R., ed., State of stress in the Earth's crust: New York, American Elsevier Publishing Company, Inc., p. 451–550.
——— 1967, Description of rocks and rock masses with a view toward their physical and mechanical behavior—A general report: International Congress on Rock Mechanics, 1st, Proceedings, v. 3, p. 182–197.
——— 1969, Structural analysis of fractures in cores from Saticoy Field, Ventura County, California: American Association of Petroleum Geologists Bulletin, v. 53, p. 367–389.
——— 1972, Residual elastic strain in rocks: Tectonophysics, v. 15, p. 297–330.
Friedman, M., and Higgs, N. G., 1981, Calcite fabrics in experimental shear zones: American Geophysical Union Monograph 24, p. 11–27.
Friedman, M., and Logan, J. M., 1970, Microscopic feather fractures: Geological Society of America Bulletin, v. 81, p. 3417–3420.
——— 1973, Lüders' bands in experimentally deformed sandstone and limestone: Geological Society of America Bulletin, v. 84, p. 1465–1476.
Friedman, M., and Sowers, G. M., 1970, Petrofabrics: A critical review: Canadian Journal of Earth Sciences, v. 7, p. 447–497.
Friedman, M., Logan, J. M., and Rigert, J. A., 1974, Glass-indurated quartz gouge in sliding-friction experiments on sandstone: Geological Society of America Bulletin, v. 85, p. 937–942.
Frost, H. J., and Ashby, M. F., 1982, Deformation mechanism maps: The plasticity and creep of metals and ceramics: New York, Pergamon, 184 p.
Fruth, L. S., Jr., Orme, G. R., and Donath, F. A., 1966, Experimental compaction effects in carbonate sediments: Journal of Sedimentary Petrology, v. 36, p. 747–754.
Fry, N., 1979, Random point distributions and strain measurement in rocks: Tectonophysics, v. 60, p. 89–105.
Gaither, A., 1953, A study of porosity and grain relationships in experimental sands: Journal of Sedimentary Petrology, v. 23, p. 180–195.
Gallagher, J. J., Jr., Friedman, M., Handin, J., and Sowers, G. M., 1974, Experimental studies relating to microfracture in sandstone: Tectonophysics, v. 21, p. 203–248.
Gamond, J. F., 1983, Displacement features associated with fault zones: A comparison between observed examples and experimental models: Journal of Structural Geology, v. 5, p. 33–45.
——— 1987, Bridge structures as sense of displacement criteria in brittle fault zones: Journal of Structural Geology, v. 9, p. 609–620.

Garnett, J. A., 1974, A mechanism for the development of en-echelon gashes in kink zones: Tectonophysics, v. 23, p. 129–138.
Garrison, R. E., and Kennedy, W. J., 1977, Origin of solution seams and flaser structure in Upper Cretaceous chalks of southern England: Sedimentary Geology, v. 19, p. 107–137.
Gaviglio, P., 1986, Crack-seal mechanism in a limestone: A factor of deformation in strike-slip faulting: Tectonophysics, v. 131, p. 247–255.
Gay, N. C., and Jaeger, J. C., 1975a, Cataclastic deformation of geological materials in matrices of differing composition: I. Pebbles and conglomerates: Tectonophysics, v. 27, p. 303–322.
——— 1975b, Cataclastic deformation of geological materials in matrices of differing composition: II. Boudinage: Tectonophysics, v. 27, p. 323–331.
Gay, N. C., and Ortlepp, W. D., 1979, Anatomy of a mining-induced fault zone: Geological Society of America Bulletin, Part 1, v. 90, p. 47–58.
Gay, N. C., and Weiss, L. E., 1974, The relationship between principal stress directions and the geometry of kinks in foliated rocks: Tectonophysics, v. 21, p. 287–300.
Geiser, P. A., 1974, Cleavage in some sedimentary rocks of the central Valley and Ridge province, Maryland: Geological Society of America Bulletin, v. 85, p. 1399–1412.
——— 1975, Slaty cleavage and the dewatering hypothesis—An examination of some critical evidence: Geology, v. 3, p. 717–720.
Geiser, P. A., and Engelder, T., 1983, The distribution of layer parallel shortening fabrics in the Appalachian foreland of New York and Pennsylvania: Evidence for two non-coaxial phases of the Alleghanian orogeny: Geological Society of America Memoir 158, p. 161–175.
Geiser, P. A., and Sansone, S., 1981, Joints, microfractures, and the formation of solution cleavage in limestone: Geology, v. 9, p. 280–285.
Ghisetti, F., 1987, Mechanisms of thrust faulting in the Gran Sasso chain, central Apennines, Italy: Journal of Structural Geology, v. 9, p. 955–967.
Gilbert, G. K., 1928, Studies in basin-range structure: U.S. Geological Survey Professional Paper 153, 92 p.
Gipson, M., 1965, Application of the electron microscope to the study of particle orientation and fissility in shale: Journal of Sedimentary Petrology, v. 35, p. 408–414.
Glover, J. E., 1969, Observations on stylolites in Western Australian rocks: Royal Society of Western Australia Journal, v. 52, p. 12–17.
Golding, H. G., and Conolly, J. R., 1962, Stylolites in volcanic rocks: Journal of Sedimentary Petrology, v. 32, p. 534–538.
Gorlov, N. V., 1971, The mechanism of the opening of extension fractures (in the instance of pegmatite veins, northwestern White Sea region): Geotectonics, no. 3, p. 148–152.
Graham, R. H., 1978, Quantitative deformation studies in the Permian rocks of the Alpes-Maritimes: Goguel Symposium, Bureau de Recherches Géologiques et Minières, Mémoire 91, p. 219–238.
Gramberg, J., 1965, The axial cleavage fracture 1. Axial cleavage fracturing, a significant process in mining and geology: Engineering Geology, v. 1, p. 31–72.
Gratier, J. P., and Guiguet, R., 1986, Experimental pressure solution–deposition on quartz grains: The crucial effect of the nature of the fluid: Journal of Structural Geology, v. 8, p. 845–856.
Graton, L. C., and Fraser, H. J., 1935, Systematic packing of spheres—with particular relation to porosity and permeability: Journal of Geology, v. 43, p. 785–909.
Gray, D. R., 1977a, Differentiation associated with discrete crenulation cleavages: Lithos, v. 10, p. 89–101.
——— 1977b, Morphologic classification of crenulation cleavage: Journal of Geology, v. 85, p. 229–235.
——— 1978, Cleavages in deformed psammitic rocks from southeastern Australia: Their nature and origin: Geological Society of America Bulletin, v. 89, p. 577–590.
——— 1979, Microstructure of crenulation cleavages: An indicator of cleavage origin: American Journal of Science, v. 279, p. 97–128.
——— 1981, Compound tectonic fabrics in singly folded rocks from southwest Virginia, U.S.A.: Tectonophysics, v. 78, p. 229–248.
Gray, D. R., and Durney, D. W., 1979, Investigations on the mechanical significance of crenulation cleavage: Tectonophysics, v. 58, p. 35–79.
Gray, J., and Boucot, A. J., 1975, Color changes in pollen and spores: A review: Geological Society of America Bulletin, v. 86, p. 1019–1033.
Green, H. W., II, 1984, "Pressure solution" creep: Some causes and mechanisms: Journal of Geophysical Research, v. 89, p. 4313–4318.
Gresley, W. S., 1895, The indentation of the Bunter pebbles: Geological Magazine, v. 32, p. 239.
Gretener, P. E., 1977, On the character of thrust faults with particular reference to the basal tongues: Bulletin of Canadian Petroleum Geology, v. 25, p. 110–122.
Griffith, A. A., 1921, The phenomena of rupture and flow in solids: Royal Society of London Philosophical Transactions, ser. A, v. 221, p. 163–198.
——— 1924, The theory of rupture: International Congress for Applied Mechanics, 1st, Delft, Proceedings, p. 55–63.
Griggs, D. T., 1935, The strain ellipsoid as a theory of rupture: American Journal of Science, 5th ser., v. 30, p. 121–137.
——— 1936, Deformation of rocks under high confining pressure I. Experiments at room temperature: Journal of Geology, v. 44, p. 541–577.
——— 1939, Creep of rocks: Journal of Geology, v. 47, p. 225–251.
——— 1940, Experimental flow of rocks under conditions favoring recrystallization: Geological Society of America Bulletin, v. 51, p. 1001–1022.
——— 1942, Strength and plasticity: Geological Society of America Special Paper 36, p. 107–130.
Griggs, D. T., and Bell, J. F., 1938, Experiments bearing on the orientation of quartz in deformed rocks: Geological Society of America Bulletin, v. 49, p. 1723–1746.
Griggs, D. T., and Blacic, J. D., 1965, Quartz: Anomalous weakness of synthetic crystals: Science, v. 147, p. 292–295.
Griggs, D. T., and Handin, J., 1960, Observation on fractures and a hypothesis of earthquakes: Geological Society of America Memoir 79, p. 347–364.
Griggs, D. T., and Miller, W. B., 1951, Deformation of Yule marble: Part I—Compression and extension experiments on dry Yule marble at 10,000 atmospheres confining pressure, room temperature: Geological Society of America Bulletin, v. 62, p. 853–862.
Griggs, D. T., Turner, F., Borg, I., and Sosoka, J., 1951, Deformation of Yule marble: Part IV: Effects at 150 °C: Geological Society of America Bulletin, v. 62, p. 1385–1406.
——— 1953, Deformation of Yule marble: Part V: Effects at 300 °C: Geological Society of America Bulletin, v. 64, p. 1327–1342.
Griggs, D. T., Paterson, M. S., Heard, H. C., and Turner, F. J., 1960a, Annealing recrystallization in calcite crystals and aggregates: Geological Society of America Memoir 79, p. 21–37.
Griggs, D. T., Turner, F. J., and Heard, H. C., 1960b, Deformation of rocks at 500 to 800 °C: Geological Society of America Memoir 79, p. 39–104.
Groshong, R. H., Jr., 1965, Systematic joints across a syncline in the Valley and Ridge province [abs.]: Pennsylvania Academy of Science Proceedings, v. 39, p. 23.
——— 1972, Strain calculated from twinning in calcite: Geological Society of America Bulletin, v. 83, p. 2025–2038.
——— 1974, Experimental test of least-squares strain gage calculation using twinned calcite: Geological Society of America Bulletin, v. 85, p. 1855–1864.
——— 1975a, "Slip" cleavage caused by pressure solution in a buckle fold: Geology, v. 3, p. 411–413.
——— 1975b, Strain, fractures, and pressure solution in natural single-layer folds: Geological Society of America Bulletin, v. 86, p. 1363–1376.
——— 1976, Strain and pressure solution in the Martinsburg Slate, Delaware Water Gap, New Jersey: American Journal of Science, v. 276, p. 1131–1146.
Groshong, R. H., Jr., Pfiffner, O. A., and Pringle, L. R., 1984a, Strain partitioning in the Helvetic thrust belt of eastern Switzerland from the leading edge to the internal zone: Journal of Structural Geology, v. 6, p. 5–18.
Groshong, R. H., Jr., Teufel, L. W., and Gasteiger, C., 1984b, Precision and accuracy of the calcite strain-gage technique: Geological Society of America Bulletin, v. 95, p. 357–363.
Grubenmann, U., and Niggli, P., 1924, Die Gesteinmetamorphose (Volume 1): Berlin, Borntraeger, 539 p.
Guiraud, M., and Séguret, M., 1987, Soft-sediment microfaulting related to compaction within the fluvio-deltaic infill of the Soria strike-slip basin (northern Spain), *in* Jones, M. E., and Preston, R.M.F., eds., Deformation of sediments and sedimentary rocks; Geological Society Special Publication No. 29: Oxford, England, Blackwell Scientific Publications, p. 123–136.
Guthrie, J. M., Houseknecht, D. W., and Johns, W. D., 1986, Relationships among vitrinite reflectance, illite crystallinity, and organic geochemistry in Carboniferous strata, Ouachita Mountains, Oklahoma and Arkansas: American Association of Petroleum Geologists Bulletin, v. 70, p. 26–33.
Guzzetta, G., 1984, Kinematics of stylolite formation and physics of the pressure-solution process: Tectonophysics, v. 101, p. 383–394.
Hadizadeh, J., and Rutter, E. H., 1983, The low temperature brittle-ductile transition in quartzite and the occurrence of cataclastic flow in nature: Geologische Rundschau, v. 7, p. 493–509.
Hancock, P. L., 1965, Axial-trace-fractures and deformed concretionary rods in south Pembrokeshire: Geological Magazine, v. 102, p. 143–163.
——— 1972, The analysis of en-echelon veins: Geological Magazine, v. 109, p. 269–276.
——— 1985, Brittle microtectonics: Principles and practice: Journal of Structural Geology, v. 7, p. 437–457.
Hancock, P. L., and Barka, A. A., 1987, Kinematic indicators on active normal faults in western Turkey: Journal of Structural Geology, v. 9, p. 573–584.
Handin, J., 1966, Strength and ductility: Geological Society of America Memoir 97, p. 223–289.
——— 1969, On the Coulomb-Mohr failure criterion: Journal of Geophysical Research, v. 74, p. 5343–5348.
Handin, J., and Fairbairn, H. W., 1955, Experimental deformation of Hasmark dolomite: Geological Society of America Bulletin, v. 66, p. 1257–1273.
Handin, J. W., and Griggs, D., 1951, Deformation of Yule Marble: Part II—Predicted fabric changes: Geological Society of America Bulletin, v. 62, p. 863–886.
Handin, J., and Hager, R. V., 1957, Experimental deformation of sedimentary rocks under confining pressure: Tests at room temperature on dry samples: American Association of Petroleum Geologists Bulletin, v. 41, p. 1–50.
——— 1958, Experimental deformation of sedimentary rocks under confining pressure: Tests at high temperature: American Association of Petroleum Geologists Bulletin, v. 42, p. 2892–2934.
Handin, J., Hager, R. V., Friedman, M., and Feather, J. N., 1963, Experimental deformation of sedimentary rocks under confining pressure: Pore pressure tests: American Association of Petroleum Geologists Bulletin, v. 47, p. 717–755.
Handin, J., Heard, H. C., and Magouirk, J. N., 1967, Effects of the intermediate principal stress on the failure of limestone, dolomite, and glass at different temperatures and strain rates: Journal of Geophysical Research, v. 72, p. 611–640.
Hanmer, S., 1981, Segregation bands in plagioclase: Non-dilational quartz veins formed by strain-enhanced diffusion: Tectonophysics, v. 79, p. T53–T61.
Hanna, S. S., and Fry, N., 1979, A comparison of methods of strain determination in rocks from Southwest Dyfed (Pembrokeshire) and adjacent areas: Journal of Structural Geology, v. 1, p. 155–162.
Hansen, E. C., and Borg, I. Y., 1962, The dynamic significance of deformation lamellae in quartz of a calcite-cemented sandstone: American Journal of Science, v. 260, p. 321–336.
Harms, J. C., 1965, Sandstone dikes in relation to Laramide faults and stress distribution in the southern Front Range, Colorado: Geological Society of America Bulletin, v. 76, p. 981–1002.
Hawkins, P. J., 1978, Relationship between diagenesis, porosity reduction, and oil emplacement in late Carboniferous sandstone reservoirs, Bothamsall Oilfield, E Midlands: Journal of Geological Society of London, v. 135, p. 7–24.
Heald, M. T., 1955, Stylolites in sandstones: Journal of Geology, v. 63, p. 101–114.
——— 1956, Cementation of Simpson and St. Peter sandstones in parts of Oklahoma, Arkansas, and Missouri: Journal of Geology, v. 64, p. 16–30.
Heard, H. C., 1960, Transition from brittle to ductile flow in Solenhofen Limestone as a function of temperature, confining pressure, and interstitial fluid pressure: Geological Society of America Memoir 79, p. 193–226.
——— 1963, The effect of large changes in strain rate in the experimental deformation of the Yule marble: Journal of Geology, v. 71, p. 162–195.
Heard, H. C., and Carter, N. L., 1968, Experimentally induced "natural" intragranular flow in quartz and quartzite: American Journal of Science, v. 266, p. 1–42.
Heard, H. C., Wenk, H.-R., and Barber, D. J., 1978, Experimental deformation of dolomite single crystals to 800 °C [abs.]: EOS (American Geophysical Union Transactions), v. 59, p. 249.
Heim, A., 1878, Untersuchungen über den Mechanismus der Gebirgsbildung im Anschluss an die geologische Monographie der Tödi-Wingällen-Gruppe: Basal, Switzerland, Benno Schwabe, 3 volumes (I, II, Atlas).
Helmstaedt, H., and Greggs, R. G., 1980, Stylolitic cleavage and cleavage refraction in lower Paleozoic rocks of the Great Valley, Maryland: Tectonophysics, v. 66, p. 99–114.
Helwig, J., 1970, Slump folds and early structures, northeastern Newfoundland Appalachians: Journal of Geology, v. 78, p. 172–187.
Heron, S. D., Judd, J. B., and Johnson, H. S., 1971, Clastic dikes associated with soil horizons in the North and South Carolina Coastal Plain: Geological Society of America Bulletin, v. 82, p. 1801–1810.
Hertz, H., 1896, Hertz's Miscellaneous Papers: London, England, Macmillan, p. 146–162.
Hibbard, J., and Karig, D. E., 1987, Sheath-like folds and progressive fold deformation in Tertiary sedimentary rocks of the Shimanto accretionary complex, Japan: Journal of Structural Geology, v. 9, p. 845–857.
Higgins, M. W., 1971, Cataclastic rocks: U.S. Geological Survey Professional Paper 687, 97 p.
Higgs, D. V., and Handin, J., 1959, Experimental deformation of dolomite single crystals: Geological Society of America Bulletin, v. 70, p. 245–278.
Hills, E. S., 1953, Outlines of structural geology (3rd ed.): London, England, Methuen, 182 p.
——— 1963, Elements of structural geology: New York, John Wiley & Sons, Inc., 483 p.
Hiraga, H., and Shimamoto, T., 1987, Textures of sheared halite and their implications for the seismogenic slip of deep faults: Tectonophysics, v. 144, p. 69–86.
Hobbs, B. E., 1968, Recrystallization of single crystals of quartz: Tectonophysics, v. 6, p. 353–401.
Hobbs, B. E., Means, W. D., and Williams, P. F., 1976, An outline of structural geology: New York, John Wiley & Sons, Inc., 571 p.
Hobbs, B. E., Ord, A., and Teyssier, C., 1986, Earthquakes in the ductile regime?: Pure and Applied Geophysics, v. 124, p. 309–336.
Hodgson, R. A., 1961a, Classification of structures on joint surfaces: American Journal of Science, v. 259, p. 493–502.
——— 1961b, Regional study of jointing in Comb Ridge–Navajo Mountain area, Arizona and Utah: American Association of Petroleum Geologists Bulletin, v. 45, p. 1–38.
Hoek, E., and Bieniawski, Z. T., 1965, Brittle fracture propagation in rock under compression: International Journal of Fracture Mechanics, v. 1, p. 137–155.
Holeywell, R. C., and Tullis, T. E., 1975, Mineral reorientation and slaty cleavage in the Martinsburg Formation, Lehigh Gap, Pennsylvania: Geological Society of America Bulletin, v. 86, p. 1296–1304.
Honea, E., and Johnson, A. M., 1976, A theory of concentric, kink, and sinusoidal folding and of monoclinal flexuring of compressible, elastic multilayers. IV. Development of sinusoidal and kink folds in multilayers confined by rigid boundaries: Tectonophysics, v. 30, p. 197–239.
Horsfield, W. T., 1980, Contemporaneous movement along crossing conjugate normal faults: Journal of Structural Geology, v. 2, p. 305–310.
Hortenbach, R., 1977, Pressure solution processes in carbonates and their importance: Zeitschrift fur Geologie Wissenschaften, v. 5, p. 617–621.
House, W. M., and Gray, D. R., 1982, Cataclasites along the Saltville thrust, U.S.A., and their implications for thrust-sheet

emplacement: Journal of Structural Geology, v. 4, p. 257–269.
Houseknecht, D. W., 1984, Influence of grain size and temperature on intergranular pressure solution, quartz cementation, and porosity in a quartzose sandstone: Journal of Sedimentary Petrology, v. 54, p. 348–361.
——— 1987, Assessing the relative importance of compaction processes and cementation to reduction of porosity in sandstones: American Association of Petroleum Geologists Bulletin, v. 71, p. 633–642.
Howell, B. F., Jr., 1986, History of ideas on the cause of earthquakes: EOS (American Geophysical Union Transactions), v. 67, p. 1323–1326.
Hubbert, M. K., 1951, Mechanical basis for certain familiar geologic structures: Geological Society of America Bulletin, v. 62, p. 355–372.
Hugman, R.H.H., III, and Friedman, M., 1979, Effects of texture and composition on mechanical behavior of experimentally deformed carbonate rocks: American Association of Petroleum Geologists Bulletin, v. 63, p. 1478–1489.
Iddings, J. P., 1886, The columnar structure in the igneous rocks on Orange Mountain, New Jersey: American Journal of Science, 3rd ser., v. 31, p. 321–331.
Ingerson, E., and Tuttle, O. F., 1945, Relations of lamellae and crystallography of quartz and fabric directions in some deformed rocks: American Geophysical Union Transactions, v. 26, p. 95–105.
Ingram, R. L., 1953, Fissility of mudrocks: Geological Society of America Bulletin, v. 64, p. 869–878.
Jackson, R. E., and Dunn, D. E., 1974, Experimental sliding friction and cataclasis of foliated rocks: International Journal of Rock Mechanics and Mining Science, v. 11, p. 235–249.
Jaeger, J. C., 1959, The frictional properties of joints in rocks: Pure and Applied Geophysics, v. 43, p. 148–158.
Jaeger, J. C., and Cook, N.G.W., 1979, Fundamentals of rock mechanics (3rd edition): London, England, Chapman and Hall, 593 p.
James, A.V.G., 1920, Factors producing columnar structure in lavas and its occurrence near Melbourne, Australia: Journal of Geology, v. 28, p. 458–469.
Jamison, W. R., and Spang, J. H., 1976, Use of calcite twin lamellae to infer differential stress: Geological Society of America Bulletin, v. 87, p. 868–872.
Jamison, W. R., and Stearns, D. W., 1982, Tectonic deformation of Wingate sandstone, Colorado National Monument: American Association of Petroleum Geologists Bulletin, v. 66, p. 2584–2608.
Jaroszweski, E. T., 1969, New site of tectonic stylolites: Bulletin de l'Academie Polonaise des Sciences, Series des sciences géologie et géographie, v. 17, p. 17–23.
Johnson, J.A.D., and Budd, S. R., 1975, The geology of the Zone B and Zone C Lower Cretaceous limestone reservoirs of Asab Field, Abu Dhabi: 9th Arab Petroleum Congress, Paper No. 109 (B-3), p. 1–24.
Judson, S., and Barks, R. E., 1961, Microstriations on polished pebbles: American Journal of Science, v. 259, p. 371–381.
Kahle, C. F., and Floyd, J. C., 1971, Stratigraphic and environmental significance of sedimentary structures in Cayugan (Silurian) tidal flat carbonates, northwestern Ohio: Geological Society of America Bulletin, v. 82, p. 2071–2098.
Kahn, J. S., 1956, The analysis and distribution of packing in sand-size sediments. 1. On the measurement of packing in sandstones: Journal of Geology, v. 64, p. 385–395.
Kanaori, Y., 1986, A SEM cathodoluminescence study of quartz in mildly deformed granite from the region of the Atotsugawa fault, central Japan: Tectonophysics, v. 131, p. 133–146.
Keller, W. D., Stone, C. G., and Hoersch, A. L., 1985, Textures of Paleozoic chert and novaculite in the Ouachita Mountains of Arkansas and Oklahoma and their geological significance: Geological Society of America Bulletin, v. 96, p. 1353–1363.
Kerrich, R., 1978, An historical review and synthesis of research on pressure solution: Zentralblatt fur Geologie und Palaontologie, Teil I, p. 512–550.
Kerrich, R., and Allison, I., 1978, Flow mechanisms in rocks: Geoscience Canada, v. 5, p. 109–118.
Kerrich, R., Beckinsale, R. D., and Durham, J. J., 1977, The transition between deformation regimes dominated by intercrystalline diffusion and intracrystalline creep evaluated by oxygen isotope thermometry: Tectonophysics, v. 38, p. 241–257.
Kirby, S. H., 1980, Tectonic stresses in the lithosphere: Constraints provided by the experimental deformation of rocks: Journal of Geophysical Research, v. 85, p. 6353–6363.
Klassen-Neklyudova, M. V., 1964, Mechanical twinning of crystals: New York, Consultants Bureau, 213 p.
Knill, J. L., 1960, A classification of cleavages, with special references to the Craignish district of the Scottish Highlands: International Geological Congress, 21st, Proceedings, pt. 18, p. 317–325.
Knipe, R. J., 1986, Microstructural evolution of vein arrays preserved in Deep Sea Drilling Project cores from the Japan Trench, Leg 57: Geological Society of America Memoir 166, p. 75–87.
Knipe, R. J., and White, S. H., 1977, Microstructural variation of an axial plane cleavage around a fold—A H.V.E.M. study: Tectonophysics, v. 39, p. 355–380.
Knopf, E. B., 1949, Fabric changes in Yule marble after deformation in compression: Part II, experimental deformation of Yule marble: American Journal of Science, v. 247, p. 537–569.
Knopf, E. B., and Ingerson, E., 1938, Structural petrology: Geological Society of America Memoir 6, 270 p.
Kohlstedt, D. L., and Weathers, M. S., 1980, Deformation-induced microstructures, palaeopiezometers, and differential stresses in deeply eroded fault zones: Journal of Geophysical Research, v. 85, p. 6269–6285.
Koopman, A., 1983, Detachment tectonics in the central Apennines, Italy: Geologica Ultraiectina, v. 30, p. 1–155.
Kowallis, B. J., Wang, H. F., and Jang, B.-A., 1987, Healed microcrack orientations in granite from Illinois borehole UPH-3 and their relationship to the rock's stress history: Tectonophysics, v. 135, p. 297–306.
Kranz, R. L., 1983, Microcracks in rocks: A review: Tectonophysics, v. 100, p. 449–480.
Kübler, B., 1968, Evaluation quantitative du métamorphisme par la cristallinité de l'illite. Etat des progrès realisés ces dernières années: Centre Recherche Pau Bulletin, v. 2, p. 385–397.
Labaume, P., 1987, Syn-diagenetic deformation of a turbidite succession related to submarine gravity nappe emplacement, Autapie Nappe, French Alps, *in* Jones, M. E., and Preston, R.M.F., eds., Deformation of sediments and sedimentary rocks: Geological Society Special Publication No. 29: Oxford, England, Blackwell Scientific Publications, p. 147–163.
Lachenbruch, A. H., 1962, Mechanics of thermal contraction cracks in ice-wedge polygons in permafrost: Geological Society of America Special Paper 70, 69 p.
Lajtai, E. Z., 1971, A theoretical and experimental evaluation of the Griffith theory of brittle fracture: Tectonophysics, v. 11, p. 129–156.
Land, L. S., and Dutton, S. P., 1978, Cementation of Pennsylvanian deltaic sandstone: Isotopic data: Journal of Sedimentary Petrology, v. 48, p. 1167–1176.
Langheinrich, G., and Plessman, W., 1968, Zue Entstehungsweise von Schieferungs-Flächen in Kalksteinen (Turon-Kalke eines Salzauftriebs-Sattels im Harz-Vorland): Geologische Mitteilungen, v. 8, p. 111–142.
Lapworth, C., 1885, The Highland controversy in British history: Its causes, course and consequences: Nature, v. 32, p. 558–559.
Laubscher, H. P., 1979, Elements of Jura kinematics and dynamics: Eclogae Geologicae Helvetiae, v. 72, p. 467–483.
Laurent, Ph., Bernard, Ph., Vasseur, G., and Etchecopar, A., 1981, Stress tensor determination from the study of e twins in calcite: A linear programming method: Tectonophysics, v. 78, p. 651–660.
Lavecchia, G., 1985, Il sovrascorrimento dei Monti Sibillini: Analisi cinematica e strutturale: Bolletino Societa Geologica Italiana, v. 104, p. 161–194.
Lawn, B. R., and Wilshaw, T. R., 1975, Fracture of brittle solids: Cambridge, England, Cambridge University Press, 204 p.
Lawrence, R. D., 1970, Stress analysis based on albite twinning of plagioclase feldspars: Geological Society of America Bulletin, v. 81, p. 2507–2512.
Lee, J. H., Peacor, D. R., Lewis, D. D., and Wintsch, R. P., 1986, Evidence for syntectonic crystallization for the mudstone to slate transition at Lehigh Gap, Pennsylvania, U.S.A.: Journal of Structural Geology, v. 8, p. 767–780.
Leith, C. K., 1905, Rock cleavage: U.S. Geological Survey Bulletin 239, 216 p.
Lemmleyn, G. G., and Kliya, M. O., 1960, Distinctive features of the healing of a crack in a crystal under conditions of declining temperature: International Geology Review, v. 2, p. 125–128.
Lennox, P. G., 1987, Conjugate oblique planar fabrics in rock analogues—by kinking or not by kinking?: Geological Society of Australia, International Conference on Deformation of Crustal Rocks, Abstracts, no. 19, p. 89–90.
Lindström, M., 1962, A structural study of the southern end of the French Jura: Geological Magazine, v. 99, p. 193–207.
Lisle, R. J., 1977, Clastic grain shape and orientation in relation to cleavage from the Aberystwyth grits, Wales: Tectonophysics, v. 39, p. 381–395.
Lockwood, J. P., and Moore, J. G., 1979, Regional deformation of the Sierra Nevada, California, on conjugate microfault sets: Journal of Geophysical Research, v. 84, p. 6041–6049.
Logan, B. W., and Semeniuk, V., 1976, Dynamic metamorphism; processes and products in Devonian carbonate rocks, Canning Basin, Western Australia: Geological Society of Australia Special Publication No. 6, 138 p.
Logan, J. M., 1979, Brittle phenomena: Reviews of Geophysics and Space Physics, v. 17, p. 1121–1132.
Logan, J. M., Higgs, N. G., and Friedman, M., 1981, Laboratory studies on natural gouge from the U.S. Geological Survey Dry Lake Valley No. 1 well, San Andreas fault zone: American Geophysical Union Monograph 24, p. 121–134.
Logan, J. M., Iwasaki, T., Friedman, M., and Kling, S. A., 1972, Experimental investigations of sliding friction in multilithologic specimens, *in* Pincus, H., ed., Geological factors in rapid excavation: Geological Society of America Engineering Geology Case History No. 9, p. 55–67.
Logan, J. M., and Rauenzahn, K. A., 1987, Frictional dependence of gouge mixtures of quartz and montmorillonite on velocity, composition and fabric: Tectonophysics, v. 144, p. 87–108.
Lohest, M., 1909, De l'origine des veines et des géodes des terrains primaires de Belgique: Annales Société Géologique de Belgique, v. 36, B, p. 275–282.
Lowry, W. D., 1956, Factors in loss of porosity by quartzose sandstones of Virginia: American Association of Petroleum Geologists Bulletin, v. 40, p. 489–500.
Lucas, P. T., and Drexler, J. M., 1976, Altamont-Bluebell—A major naturally fractured stratigraphic trap, Uinta Basin, Utah: American Association of Petroleum Geologists Memoir No. 24, p. 121–135.
Lucas, S. E., and Moore, J. C., 1986, Cataclastic deformation in accretionary wedges: Deep Sea Drilling Project Leg 66, southern Mexico, and on-land examples from Barbados and Kodiak Islands: Geological Society of America Memoir 166, p. 89–103.
Lundberg, N., and Moore, J. C., 1986, Macroscopic structural features in Deep Sea Drilling Project cores from forearc regions: Geological Society of America Memoir 166, p. 13–44.
Maddock, R. H., 1983, Melt origin of fault-generated pseudotachylytes demonstrated by textures: Geology, v. 11, p. 105–108.
Maltman, A. J., 1977, Some microstructures of experimentally deformed argillaceous sediments: Tectonophysics, v. 39, p. 417–436.
——— 1987, Shear zones in argillaceous sediments—An experimental study, *in* Jones, M. E., and Preston, R.M.F., eds., Deformation of sediments and sedimentary rocks: Geological Society Special Publication No. 29: Oxford, England, Blackwell Scientific Publications, p. 77–87.
——— 1988, The importance of shear zones in naturally deformed wet sediments: Tectonophysics, v. 145, p. 163–175.
Mandl, G., de Jong, L.N.J., and Maltha, A., 1977, Shear zones in granular material. An experimental study of their structure and mechanical genesis: Rock Mechanics, v. 9, p. 95–144.
March, A., 1932, Mathematische Theorie der Regelung nach der Korngestalt bei affiner Deformation: Zeitschrift für Kristallographie, v. 81, p. 285–297.
Marshak, S., and Engelder, T., 1985, Development of cleavage in limestones of a fold-thrust belt in eastern New York: Journal of Structural Geology, v. 7, p. 345–359.
Marshak, S., Geiser, P. A., Alvarez, W., and Engelder, T., 1982, Mesoscopic fault array of the northern Umbrian Apennine fold belt, Italy: Geometry of conjugate shear by pressure-solution slip: Geological Society of America Bulletin, v. 93, p. 1013–1022.
Marshall, J. D., 1982, Isotopic composition of displacive fibrous calcite veins: Reversals in pore-water composition trends during burial diagenesis: Journal of Sedimentary Petrology, v. 52, p. 615–630.
Mawer, C. K., 1985, Comment on "Fault-related rocks: Suggestions for terminology": Geology, v. 13, p. 378.
Maxwell, J. C., 1960, Experiments on compaction and cementation of sand: Geological Society of America Memoir 79, p. 105–132.
——— 1962, Origin of slaty and fracture cleavage in the Delaware Water Gap area, New Jersey and Pennsylvania, *in* Engel, A.E.J., James, H. L., and Leonard, B. F., eds., Petrologic studies (Buddington volume): Boulder, Colorado, Geological Society of America, p. 281–311.
Maxwell, J. C., and Verrall, P., 1954, Low porosity may limit oil in deep sands: World Oil, v. 138, no. 5, p. 106–113, and no. 6, p. 102–104.
McClay, K. R., 1977, Pressure solution and Coble creep in rocks and minerals: A review: Geological Society of London Journal, v. 134, p. 57–70.
McClintock, F. A., and Walsh, J. B., 1962, Friction on Griffith cracks under pressure: United States National Congress of Applied Mechanics, 4th, Proceedings, p. 1015–1021.
McEwen, T. J., 1977, Pressure solution or indentation: Comment: Geology, v. 5, p. 249–251.
——— 1981, Brittle deformation in pitted pebble conglomerates: Journal of Structural Geology, v. 3, p. 25–37.
McKenzie, D., and Brune, J. B., 1972, Melting on fault planes during large earthquakes: Royal Astronomical Society Geophysical Journal, v. 29, p. 65–78.
Means, W. D., 1987, A newly recognized type of slickenside striation: Journal of Structural Geology, v. 9, p. 585–590.
Means, W. D., and Paterson, M. S., 1966, Experiments on preferred orientation of platy minerals: Contributions to Mineralogy and Petrology, v. 13, p. 108–133.
Merino, E., Ortoleva, P., and Strickholm, P., 1983, Generation of evenly spaced pressure-solution seams during (late) diagenesis: A kinetic theory: Contributions to Mineralogy and Petrology, v. 82, p. 360–370.
Michael, A. J., 1984, Determination of stress from slip data: Faults and folds: Journal of Geophysical Research, v. 89, p. 11517–11526.
Milnes, A. G., 1979, Albert Heim's general theory of natural rock deformation (1878): Geology, v. 7, p. 99–103.
Mimran, Y., 1975, Fabric deformation induced in Cretaceous chalks by tectonic stresses: Tectonophysics, v. 26, p. 309–316.
——— 1977, Chalk deformation and large-scale migration of calcium carbonate: Sedimentology, v. 24, p. 333–360.
——— 1985, Tectonically controlled freshwater carbonate cementation in chalk: Society of Economic Paleontologists and Mineralogists Special Publication 36, p. 371–379.
Mišík, M., 1971, Observations concerning calcite veinlets in carbonate rocks: Journal of Sedimentary Petrology, v. 41, p. 450–460.
Mitra, G., 1984, Brittle to ductile transition due to large strains along the White Rock thrust, Wind River Mountains, Wyoming: Journal of Structural Geology, v. 6, p. 51–62.
Mitra, S., 1976, A quantitative study of deformation mechanisms and finite strain in quartzite: Contributions to Mineralogy and Petrology, v. 59, p. 203–226.
——— 1987, Regional variations in deformation mechanisms and structural styles in the central Appalachian orogenic belt: Geological Society of America Bulletin, v. 98, p. 569–590.
Mitra, S., and Beard, W. C., 1980, Theoretical models of porosity reduction by pressure solution for well-sorted sandstones: Journal of Sedimentary Petrology, v. 50, p. 1347–1360.
Mitra, S., and Tullis, J., 1979, A comparison of intracrystalline deformation in naturally and experimentally deformed quartzites: Tectonophysics, v. 53, p. T21–T27.
Mogi, K., 1971, Effect of the triaxial stress system on the failure of dolomite and limestone: Tectonophysics, v. 11, p. 111–127.
Mohr, O., 1900, Welche Umstände bedingen die Elastizitätsgrenze und den Bruch eines Materials?: Zeitschrift Vereins Deutsches Ingenieure, v. 44, p. 1524-30; 1572-77.
Moody, J. B., and Hundley-Goff, E. M., 1980, Microscopic characteristics of orthoquartzite from sliding friction experiments. II. Gouge: Tectonophysics, v. 62, p. 301–319.
Moody, J. D., and Hill, M. J., 1956, Wrench-fault tectonics: Geological Society of America Bulletin, v. 67, p. 1207–1246.

Moon, C. F., 1972, The microstructure of clay sediments: Earth-Science Reviews, v. 8, p. 303–321.
Moore, J. C., and Allwardt, A., 1980, Progressive deformation of a Tertiary trench slope, Kodiak Islands, Alaska: Journal of Geophysical Research, v. 85, p. 4741–4756.
Moore, J. C., and Byrne, T., 1987, Thickening of fault zones: A mechanism of mélange formation in accreting sediments: Geology, v. 15, p. 1040–1043.
Moore, J. C., and Geigle, J. E., 1974, Slaty cleavage: Incipient occurrences in the deep sea: Science, v. 183, p. 509–510.
Moore, J. C., Roeske, S., Lundberg, N., Schoonmaker, J., Cowan, D. S., Gonzales, E., and Lucas, S. E., 1986, Scaly fabrics from Deep Sea Drilling Project cores from forearcs: Geological Society of America Memoir 166, p. 55–73.
Morganstern, N. R., and Tchalenko, J. S., 1967a, Microscopic structures in kaolin subjected to direct shear: Geotechnique, v. 17, p. 309–328.
——— 1967b, Microstructural observations on shear zones from slips in natural clays: Geotechnical Conference, Oslo, Proceedings, v. 1, p. 147–152.
Morrow, C. A., Shi, L. Q., and Byerlee, J. D., 1984, Permeability of fault gouge under confining pressure and shear stress: Journal of Geophysical Research, v. 89, p. 3193–3200.
Mosher, S., 1987, Pressure-solution deformation of the Purgatory Conglomerate, Rhode Island (U.S.A.): Quantification of volume change, real strains and sedimentary shape factor: Journal of Structural Geology, v. 9, p. 221–232.
Mossop, G. D., 1972, Origin of the peripheral rim, Redwater reef, Alberta: Bulletin of Canadian Petroleum Geology, v. 20, p. 238–280.
Muehlberger, W. R., 1961, Conjugate joint sets of small dihedral angle: Journal of Geology, v. 69, p. 211–219.
Mügge, O., and Heide, F., 1931, Einfache Schiebung am Anorthit: Neues Jahrbuch für Mineralogie, Geologie und Paläontologie Abhandlungen, A., v. 64, p. 161–170.
Mullenax, A. C., Gray, D. R., 1984, Interaction of bed-parallel stylolites and extension veins in boudinage: Journal of Structural Geology, v. 6, p. 63–72.
Murrell, S.A.F., 1963, A criterion for brittle fracture of rocks and concrete under triaxial stress and effect of pore pressure on the criterion, *in* Fairhurst, C., ed., Rock mechanics: Rock Mechanics Symposium, 5th, University of Minnesota, Proceedings: Oxford, England, Pergamon, p. 563–577.
Nádai, A., 1931, Plasticity: New York, McGraw-Hill, 349 p.
——— 1950. Theory of flow and fracture of solids: New York, McGraw-Hill Book Company, 571 p.
Narahara, D. K., and Wiltschko, D. V., 1986, Deformation in the hinge region of a chevron fold, Valley and Ridge province, central Pennsylvania: Journal of Structural Geology, v. 8, p. 157–168.
Narr, W., and Burruss, R. C., 1984, Origin of reservoir fractures in Little Knife Field, North Dakota: American Association of Petroleum Geologists Bulletin, v. 68, p. 1087–1100.
Naumann, C. F., 1858, Lehrbuch der Geonosie (2nd edition): Leipzig, Engelmann, 3 volumes, 2,053 p.
Neal, J. T., Langer, A. M., and Kerr, P. F., 1968, Giant desiccation polygons of Great Basin playas: Geological Society of America Bulletin, v. 79, p. 69–90.
Nelson, K. D., 1982, A suggestion for the origin of mesoscopic fabric in accretionary mélange, based on features observed in the Chrystalls Beach Complex, South Island, New Zealand: Geological Society of America Bulletin, v. 93, p. 625–634.
Nelson, R. A., 1979, Natural fracture systems: Description and classification: American Association of Petroleum Geologists Bulletin, v. 63, p. 2214–2232.
——— 1981, Significance of fracture sets associated with stylolite zones: American Association of Petroleum Geologists Bulletin, v. 65, p. 2417–2425.
——— 1983, Localization of aggregate stylolites by rock properties: American Association of Petroleum Geologists Bulletin, v. 67, p. 313–319.
——— 1985, Geologic analysis of naturally fractured reservoirs: Houston, Texas, Gulf Publishing Company, 320 p.
Nelson, R. A., and Stearns, D. W., 1977, Interformational control of regional fracture orientations: Rocky Mountain Association of Geologists Symposium, p. 95–101.
Nemat-Nasser, S., and Horii, H., 1982, Compression-induced nonplanar crack extension with application to splitting, exfoliation and rockburst: Journal of Geophysical Research, v. 87, p. 6805–6821.
Neumann, E.-R., 1969, Experimental recrystallization of dolomite and comparison of preferred orientations of calcite and dolomite in deformed rocks: Journal of Geology, v. 77, p. 426–438.
Nevin, C. M., 1949, Principles of structural geology (4th edition): New York, John Wiley & Sons, Inc., 410 p.
Newson, J. F., 1903, Clastic dikes: Geological Society of America Bulletin, v. 14, p. 227–268.
Nicholson, R., and Pollard, D. D., 1985, Dilation and linkage of echelon cracks: Journal of Structural Geology, v. 7, p. 583–590.
Nickelsen, R. P., 1966, Fossil distortion and penetrative rock deformation in the Appalachian Plateau, Pennsylvania: Journal of Geology, v. 74, p. 924–931.
——— 1972, Attributes of rock cleavage in some mudstones and limestones of the Valley and Ridge province, Pennsylvania: Pennsylvania Academy of Science, v. 46, p. 107–112.
——— 1979, Sequence of structural stages of the Alleghany orogeny at the Bear Valley strip mine, Shamokin, Pennsylvania: American Journal of Science, v. 279, p. 225–271.
——— 1986, Cleavage duplexes in the Marcellus Shale of the Appalachian foreland: Journal of Structural Geology, v. 8, p. 361–371.
Nickelsen, R. P., and Hough, V.N.D., 1967, Jointing in the Appalachian Plateau of Pennsylvania: Geological Society of America Bulletin, v. 78, p. 609–630.
Nicolas, A., and Poirier, J. P., 1976, Crystalline plasticity and solid state flow in metamorphic rocks: New York, John Wiley & Sons, 444 p.
Nissen, H. U., 1964a, Calcite fabric analysis of deformed oölites from the South Mountain fold, Maryland: American Journal of Science, v. 262, p. 892–903.
——— 1964b, Dynamic and kinematic analysis of deformed crinoid stems in a quartz graywacke: Journal of Geology, v. 72, p. 346–360.
Norris, D. K., and Barron, K., 1969, Structural analysis of features on natural and artificial faults: Geological Survey of Canada Paper 68-52, p. 136–167.
Odom, I. E., 1967, Clay fabric and its relation to structural properties in mid-continent Pennsylvanian sediments: Journal of Sedimentary Petrology, v. 37, p. 610–623.
Oertel, G., 1962, Stress, strain and fracture in clay models of geologic deformation: Geotimes, v. 6, p. 26–31.
——— 1965, The mechanism of faulting in clay experiments: Tectonophysics, v. 2, p. 343–393.
——— 1970, Deformation of a slaty, lapillar tuff in the Lake District, England: Geological Society of America Bulletin, v. 81, p. 1173–1188.
——— 1983, The relationship of strain and preferred orientation of phyllosilicate grains in rocks—A review: Tectonophysics, v. 100, p. 413–447.
Oertel, G., and Curtis, C. D., 1972, Clay-ironstone concretion preserving fabrics due to progressive compaction: Geological Society of America Bulletin, v. 83, p. 2597–2606.
Oertel, G., and Phakey, P. P., 1972, The texture of a slate from Nantlle, Caernarvon, North Wales: Texture, v. 1, p. 1–8.
Oldershaw, A. E., and Scoffin, T. P., 1967, The source of ferroan and non-ferroan calcite cements in the Halkin and Wenlock limestones: Geological Journal, v. 5, p. 309–320.
Olgaard, D. L., and Brace, W. F., 1983, The microstructure of gouge from a mining-induced seismic shear zone: International Journal of Rock Mechanics and Mining Science, v. 20, p. 11–19.
Olsson, W. A., 1974, Grain size dependence of yield stress in marble: Journal of Geophysical Research, v. 79, p. 4859–4862.
Olsson, W. A., and Peng, S. S., 1976, Microcrack nucleation in marble: International Journal of Rock Mechanics and Mining Science, v. 13, p. 53–59.
Onasch, C. M., 1983a, Dynamic analysis of rough cleavage in the Martinsburg Formation, Maryland: Journal of Structural Geology, v. 5, p. 73–81.
——— 1983b, Origin and significance of microstructures in sandstones of the Martinsburg Formation, Maryland: American Journal of Science, v. 283, p. 936–966.
——— 1986a, Ability of the Fry method to characterize pressure-solution deformation: Tectonophysics, v. 122, p. 187–193.
——— 1986b, Ability of the Fry method to characterize pressure-solution deformation—Reply: Tectonophysics, v. 131, p. 201–203.
——— 1987, Ability of the Fry method to characterize pressure-solution deformation—Reply: Tectonophysics, v. 138, p. 326.
Ord, A., and Christie, J. M., 1984, Flow stresses from microstructures in mylonitic quartzites of the Moine thrust zone, Assynt area, Scotland: Journal of Structural Geology, v. 6, p. 639–654.
Park, W. C., and Schott, E. H., 1968, Stylolites: Their nature and origin: Journal of Sedimentary Petrology, v. 38, p. 175–191.
Parker, J. M., III, 1942, Regional systematic jointing in slightly deformed sedimentary rocks: Geological Society of America Bulletin, v. 53, p. 381–408.
——— 1969, Jointing in south-central New York: Discussion: Geological Society of America Bulletin, v. 80, p. 919–922.
Paterson, M. S., 1958, Experimental deformation and faulting in Wombeyan marble: Geological Society of America Bulletin, v. 69, p. 465–476.
——— 1973, Nonhydrostatic thermodynamics and its geologic applications: Reviews of Geophysics, v. 11, p. 355–389.
——— 1978, Experimental rock deformation: Berlin, Springer-Verlag, 254 p.
Paterson, M. S., and Turner, F. J., 1970, Experimental deformation of constrained crystals of calcite in extension, *in* Paulitsch, P., ed., Experimental and natural rock deformation: New York, Springer-Verlag, p. 109–141.
Pavoni, N., 1980, Comparison of focal mechanisms of earthquakes and faulting in the Helvetic zone of the central Valais, Swiss Alps: Eclogae Geologicae Helvetiae, v. 73, p. 551–558.
Peck, D. L., and Minakami, T., 1968, The formation of columnar joints in the upper part of Kilauean lava lakes, Hawaii: Geological Society of America Bulletin, v. 79, p. 1151–1166.
Peng, S., and Johnson, A. M., 1972, Crack growth and faulting in cylindrical specimens of Chelmsford granite: International Journal of Rock Mechanics and Mining Science, v. 9, p. 37–86.
Peterson, G. L., 1966, Structural interpretation of sandstone dikes, northwest Sacramento Valley, California: Geological Society of America Bulletin, v. 77, p. 833–842.
——— 1968, Flow structures in sandstone dikes: Sedimentary Geology, v. 2, p. 177–190.
Petit, J. P., 1987, Criteria for the sense of movement on fault surfaces in brittle rocks: Journal of Structural Geology, v. 9, p. 597–608.
Petit, J. P., and Laville, E., 1987, Morphology and microstructures of hydroplastic slickensides in sandstone, *in* Jones, M. E., and Preston, R.M.F., eds., Deformation of sediments and sedimentary rocks: Geological Society Special Publication No. 29: Oxford, England, Blackwell Scientific Publications, p. 107–121.
Petit, J. P., Proust, F., and Tapponnier, P., 1983, Criteres de sens de mouvement sur les miroirs de faille en roches non calcaires: Bulletin Société Géologique de France, v. 25, p. 589–608.
Pfiffner, O. A., and Burkhard, M., 1987, Determination of paleo-stress axes orientations from fault, twin and earthquake data: Annales Tectonicae, v. 1, p. 48–57.
Phillips, J.-C., 1982, Character and origin of cataclasite developed along the low-angle Whipple detachment fault, Whipple Mountains, California, *in* Frost, E. G., and Martin, D. L., eds., Mesozoic-Cenozoic tectonic evolution of the Colorado River region, California, Arizona, and Nevada: San Diego, California, Cordilleran Publishers, p. 109–116.
Phillips, W. J., 1972, Hydraulic fracturing and mineralization: Geological Society of London Journal, v. 128, p. 337–359.
Philpotts, A. R., 1964, Origin of pseudotachylites: American Journal of Science, v. 262, p. 1008–1035.
Pickering, K. T., 1987, Wet-sediment deformation in the Upper Ordovician Point Learnington Formation: An active thrust-imbricate system during sedimentation, Notre Dame Bay, north-central Newfoundland, *in* Jones, M. E., and Preston, R.M.F., eds., Deformation of sediments and sedimentary rocks: Geological Society Special Publication No. 29: Oxford, England, Blackwell Scientific Publications, p. 213–239.
Pierson, B. J., and Shinn, E. A., 1985, Cement distribution and carbonate mineral stabilization in Pleistocene limestones of Hogsty Reef, Bahamas, *in* Schneidermann, N., and Harris, P. M., eds., Carbonate cements: Society of Economic Paleontologists and Mineralogists Special Publication 36, p. 153–168.
Pittman, E. D., 1981, Effect of fault-related granulation on porosity and permeability of quartz sandstones, Simpson Group (Ordovician), Oklahoma: American Association of Petroleum Geologists Bulletin, v. 65, p. 2381–2387.
Platt, J. P., and Leggett, J. K., 1986, Stratal extension in thrust footwalls, Makran accretionary prism: Implications for thrust tectonics: American Association of Petroleum Geologists Bulletin, v. 70, p. 191–203.
Platt, J. P., and Vissers, R.L.M., 1980, Extensional structures in anisotropic rocks: Journal of Structural Geology, v. 2, p. 397–410.
Plessmann, W., 1965, Gesteinslösung, ein Hauptfaktor beim Schieferungsprozess: Geologische Mitteilungen, v. 4 (1963), p. 69–82.
——— 1972, Horizontal-stylolithen im französisch-schweizerischen Tafel—und Faltenjura und ihre Einpassung in den regionalen Rahmen: Geologische Rundschau, v. 61, p. 332–347.
Plessmann, W., and Spaeth, G., 1971, Sedimentgänge und tektonisches Schichtfliessen (Biegungsfliessen) im Rechtsrheinischen Schiefergebirge: Geologische Mitteilungen, v. 11, p. 137–164.
Poirier, J.-P., 1985, Creep of crystals; high-temperature deformation processes in metals, ceramics and minerals: New York, Cambridge University Press, 260 p.
Pollard, D. D., Segall, P., and Delaney, P. T., 1982, Formation and interpretation of dilatent echelon cracks: Geological Society of America Bulletin, v. 93, p. 1291–1303.
Powell, C. McA., 1972, Tectonic dewatering strain in the Michigamme slate, Michigan: Geological Society of America Bulletin, v. 83, p. 2149–2158.
——— 1979, A morphological classification of rock cleavage: Tectonophysics, v. 58, p. 21–34.
Price, N. J., 1966, Fault and joint development in brittle and semi-brittle rock: London, England, Pergamon Press, 176 p.
Price, N. J., and Hancock, P. L., 1972, Development of fracture cleavage and kindred structures: International Geological Congress, 24th, Montreal, Section 3, p. 584–592.
Price, R. A., 1967, The tectonic significance of mesoscopic subfabrics in the southern Rocky Mountains of Alberta and British Columbia: Canadian Journal of Earth Sciences, v. 4, p. 39–70.
Prior, D. J., Knipe, R. J., Bates, M. P., Grant, N. T., Law, R. D., Lloyd, G. E., Welbon, A., Agar, S. M., Brodie, K. H., Maddock, R. H., Rutter, E. H., White, S. H., Bell, T. H., Ferguson, C. C., and Wheeler, J., 1987, Orientation of specimens: Essential data for all fields of geology: Geology, v. 15, p. 829–831.
Proctor, R. J., Payne, C. M., and Kalin, D. C., 1970, Crossing the Sierra Madre fault zone in the Glendora tunnel, San Gabriel Mountains, California: Engineering Geology, v. 4, p. 5–63.
Ramberg, H., 1955, Natural and experimental boudinage and pinch-and-swell structures: Journal of Geology, v. 63, p. 512–526.
Ramsay, J. G., 1967, Folding and fracturing of rocks: New York, McGraw-Hill, Inc., 568 p.
——— 1980a, Shear zone geometry: A review: Journal of Structural Geology, v. 2, p. 83–99.
——— 1980b, The crack-seal mechanism of rock deformation: Nature, v. 284, p. 135–139.
Ramsay, J. G., and Graham, R. H., 1970, Strain variation in shear belts: Canadian Journal of Earth Sciences, v. 7, p. 786–813.
Ramsay, J. G., and Huber, M. I., 1983, The techniques of modern structural geology, Volume 1: Strain analysis: London, England, Academic Press, 307 p.
——— 1987, The techniques of modern structural geology, Volume 2: Folds and fractures: London, England, Academic Press, p. 309–700.
Ranalli, G., 1984, Grain size distribution and flow stress in tectonites: Journal of Structural Geology, v. 6, p. 443–447.
Raymond, L. A., 1984, Classification of mélanges: Geological Society of America Special Paper 198, p. 7–20.
——— 1985, Comment on "Fault-related rocks: Suggestions for terminology": Geology, v. 13, p. 218.
Reches, Z., 1978, Analysis of faulting in three-dimensional strain field: Tectonophysics, v. 47, p. 109–129.

——— 1983, Faulting in rocks in three-dimensional strain fields. II. Theoretical analysis: Tectonophysics, v. 95, p. 133–156.
Reches, Z., and Dieterich, J. H., 1983, Faulting of rocks in three-dimensional strain fields. I. Failure of rocks in polyaxial, servo-control experiments: Tectonophysics, v. 95, p. 111–132.
Reches, Z., and Johnson, A. M., 1976, A theory of concentric, kink and sinusoidal folding and of monoclinal flexuring of compressible, elastic multilayers: Tectonophysics, v. 35, p. 295–334.
Reid, H. F., 1910, Mechanics of the earthquake, *in* Lawson, A. C., ed., The California earthquake of 1906, Volume II: Washington, D.C., Carnegie Institution of Washington, 192 p.
Reik, G. A., and Currie, J. B., 1974, A study of relations between rock fabric and joints in sandstone: Canadian Journal of Earth Sciences, v. 11, p. 1253–1268.
Reimold, W. U., Oskierski, W., and Huth, J., 1987, The pseudotachylite from Champagnac in the Rochecouart Meteorite Crater, France: Journal of Geophysical Research, v. 92, p. E737–E748.
Rejebian, V. A., Harris, A. G., and Huebner, J. S., 1987, Conodont color and textural alteration: An index to regional metamorphism, contact metamorphism, and hydrothermal alteration: Geological Society of America Bulletin, v. 99, p. 471–479.
Reks, I. J., and Gray, D. R., 1982, Pencil structure and strain in weakly deformed mudstone and siltstone: Journal of Structural Geology, v. 4, p. 161–176.
Renton, J. J., Heald, M. T., and Cecil, C. B., 1969, Experimental investigation of pressure solution of quartz: Journal of Sedimentary Petrology, v. 39, p. 1107–1117.
Richter, D., 1963, Verkürzung von Fossilien und Entstehung von Flaser—und Knollenkalken durch Lösungsvorgänge in geschieferten kalkigen Gesteinen: Geologische Mitteilungen, v. 4, p. 235–248.
Rickard, M. J., 1961, A note on cleavages in crenulated rocks: Geological Magazine, v. 98, p. 324–332.
Rickard, M. J., and Rixon, L. K., 1983, Stress configurations in conjugate quartz-vein arrays: Journal of Structural Geology, v. 5, p. 573–578.
Riedel, W., 1929, Zur Mechanik geologischer Brucherscheinungen: Centralblatt für Mineralogie, Geologie und Paläontologie, Abteilung B, Geologie und Paläontologie, p. 354–368.
Rigby, J. K., 1953, Some transverse stylolites: Journal of Sedimentary Petrology, v. 23, p. 265–271.
Rispoli, R., 1981, Stress fields about strike-slip faults inferred from stylolites and tension gashes: Tectonophysics, v. 75, p. T29–T36.
Rittenhouse, G., 1971a, Mechanical compaction of sands containing different percentages of ductile grains: A theoretical approach: American Association of Petroleum Geologists Bulletin, v. 55, p. 92–96.
——— 1971b, Pore-space reduction by solution and cementation: American Association of Petroleum Geologists Bulletin, v. 55, p. 80–91.
——— 1973, Pore-space reductions in sandstones—controlling factors and some engineering implications: Offshore Technology Conference, 5th Annual, Houston, Texas (preprint), p. I683–I688.
Roberts, J. C., 1961, Feather-fracture, and the mechanics of rock-jointing: American Journal of Science, v. 259, p. 481–492.
——— 1965, Quartz microfracturing in the north crop of the South Wales Coalfield: Geological Magazine, v. 102, p. 59–72.
Robertson, E. C., 1955, Experimental study of the strength of rocks: Geological Society of America Bulletin, v. 66, p. 1275–1314.
Robin, P.-Y. F., 1978, Pressure solution at grain-to-grain contacts: Geochimica et Cosmochimica Acta, v. 42, p. 1383–1389.
Rod, E., 1966, A discussion of the paper: "Fault plane features: An alternative explanation" by R. E. Riecker: Journal of Sedimentary Petrology, v. 36, p. 1163–1165.
Roering, C., 1968, The geometrical significance of natural en-echelon crack-arrays: Tectonophysics, v. 5, p. 107–123.
Roy, A. B., 1978, Evolution of slaty cleavage in relation to diagenesis and metamorphism: a study from the Hunsruckschiefer: Geological Society of America Bulletin, v. 89, p. 1775–1785.
Rummel, F., 1987, Fracture mechanics approach to hydraulic fracturing stress measurements, *in* Atkinson, B. K., ed., Fracture mechanics of rock: London, England, Academic Press, p. 217–239.
Rutter, E. H., 1983, Pressure solution in nature, theory and experiment: Geological Society of London Journal, v. 140, p. 725–740.
——— 1986, On the nomenclature of mode of failure transitions in rocks: Tectonophysics, v. 122, p. 381–387.
Rutter, E. H., Maddock, R. H., Hall, S. H., and White, S. H., 1986, Comparative microstructures of natural and experimentally produced clay-bearing fault gouges: Pure and Applied Geophysics, v. 124, p. 3–30.
Ryan, M. P., and Sammis, C. G., 1978, Cyclic fracture mechanisms in cooling basalt: Geological Society of America Bulletin, v. 89, p. 1295–1308.
Rye, D. M., and Bradbury, H. J., 1988, Fluid flow in the crust: An example from a Pyrenean thrust ramp: American Journal of Science, v. 288, p. 197–235.
Sandberg, P., 1985, Aragonite cements and their occurrence in ancient limestones, *in* Schneidermann, N., and Harris, P. M., eds., Carbonate cements: Society of Economic Paleontologists and Mineralogists Special Publication 36, p. 33–57.
Sander, B., 1911, Über Zusammenhänge zwischen Teilbewegung und Gefüge in Gesteinen: Tschermaks Mineralogische Petrographische Mitteilungen, v. 30, p. 281–314.
——— 1930, Gefügekunde der Gesteine: Vienna, Springer, 352 p.
Schlanger, S. O., 1964, Petrology of the limestones of Guam: U.S. Geological Survey Professional Paper 403-D, 52 p.
Schmid, E., and Boas, W., 1950, Plasticity of crystals: London, England, F. A. Hughes, 353 p.
Schmid, S. M., 1975, The Glarus overthrust: Field evidence and mechanical model: Eclogae Geologicae Helvetiae, v. 68, p. 247–280.
——— 1982, Microfabric studies as indicators of deformation mechanisms and flow laws operative in mountains building, *in* Hsu, J., ed., Mountain building processes: London, England, Academic Press, p. 95–110.
Schmid, S. M., and Paterson, M. S., 1977, Strain analysis in an experimentally deformed oolitic limestone, *in* Saxena, S. K., and Bhattachargi, S., eds., Energetics of geological processes: New York, Springer-Verlag, p. 67–93.
Schmid, S. M., Boland, J. N., and Paterson, M. S., 1977, Superplastic flow in fine grained limestone: Tectonophysics, v. 43, p. 257–292.
Schmid, S. M., Casey, M., and Starkey, J., 1981, The microfabric of calcite tectonites from the Helvetic nappes (Swiss Alps), *in* McClay, K. R., and Price, N. J., eds., Thrust and nappe tectonics: Boston, Massachusetts, Blackwell Scientific Publications, p. 151–158.
Schmid, S. M., Paterson, M. S., and Boland, J. N., 1980, High temperature flow and dynamic recrystallization in Carrara marble: Tectonophysics, v. 65, p. 245–280.
Scholz, C. H., 1968a, Correction to paper by C. H. Scholz "Experimental study of the fracturing process in brittle rock": Journal of Geophysical Research, v. 73, p. 4794.
——— 1968b, Experimental study of the fracturing process in brittle rock: Journal of Geophysical Research, v. 73, p. 1447–1454.
Schwander, H. W., Burgin, A., and Stern, W. B., 1981, Some geochemical data on stylolites and their host rocks: Eclogae Geologicae Helvetiae, v. 74, p. 217–224.
Schweitzer, J., and Simpson, C., 1986, Cleavage development in dolomite of the Elbrook Formation, southwest Virginia: Geological Society of America Bulletin, v. 97, p. 778–786.
Secor, D. T., 1965, Role of fluid pressure in jointing: American Journal of Science, v. 263, p. 633–646.
Segall, P., and Pollard, D. D., 1980, Mechanics of discontinuous faults: Journal of Geophysical Research, v. 85, p. 4337–4350.
——— 1983a, Joint formation in granitic rock of the Sierra Nevada: Geological Society of America Bulletin, v. 94, p. 563–575.
——— 1983b, Nucleation and growth of strike slip faults in granite: Journal of Geophysical Research, v. 88, p. 555–568.
Seifert, K. W., 1965, Deformation bands in albite: American Mineralogist, v. 50, p. 1469–1472.
Shainin, V. E., 1950, Conjugate sets of en-enchelon tension fractures in the Athens Limestone at Riverton, Virginia: Geological Society of America Bulletin, v. 61, p. 509–517.
Shand, S. J., 1916, The pseudotachylyte of Parijs (Orange Free State): Geological Society of London Quarterly Journal, v. 72, p. 198–221.
Sharpe, D., 1847, On slaty cleavage: Geological Society of London Quarterly Journal, v. 3, p. 74–105.
Shinn, E. A., and Robbin, D. M., 1983, Mechanical and chemical compaction in fine-grained shallow-water limestones: Journal of Sedimentary Petrology, v. 53, p. 595–618.
Shuaib, S. M., 1973, Subsurface petrographic study of joints in variegated siltstone-sandstone and Khairbad limestone, Pakistan: American Association of Petroleum Geologists, v. 57, p. 1775–1778.
Sibley, D. F., and Blatt, H., 1976, Intergranular pressure solution and cementation of the Tuscarora orthoquartzite: Journal of Sedimentary Petrology, v. 46, p. 881–896.
Sibson, R. H., 1977, Fault rocks and fault mechanisms: Journal of the Geological Society of London, v. 133, p. 191–213.
——— 1980, Transient discontinuities in ductile shear zones: Journal of Structural Geology, v. 2, p. 165–171.
——— 1986a, Brecciation processes in fault zones: Inferences from earthquake rupturing: Pure and Applied Geophysics, v. 124, p. 159–175.
——— 1986b, Earthquakes and rock deformation in crustal fault zones: Annual Review of Earth and Planetary Sciences, v. 14, p. 149–175.
——— 1987, Earthquake rupturing as a mineralizing agent in hydrothermal systems: Geology, v. 15, p. 701–704.
Siddans, A.W.B., 1976, Deformed rocks and their textures: Royal Society of London Philosophical Transactions, ser. A, v. 283, p. 43–54.
Sieh, K. E., 1984, Lateral offsets and revised dates of large prehistoric earthquakes at Pallett Creek, southern California: Journal of Geophysical Research, v. 89, p. 7641–7670.
Simpson, C., and Schmid, S. M., 1983, An evaluation of criteria to deduce the sense of movement in sheared rocks: Geological Society of America Bulletin, v. 94, p. 1281–1288.
Simpson, J., 1985, Stylolite-controlled layering in an homogeneous limestone: Pseudo-bedding produced by burial diagenesis: Sedimentology, v. 32, p. 495–505.
Sippel, R. F., 1968, Sandstone petrology, evidence from luminescence petrography: Journal of Sedimentary Petrology, v. 38, p. 530–554.
Skempton, A. W., 1966, Some observations on tectonic shear zones: Congress of the International Society for Rock Mechanics, 1st, Proceedings, v. 1, p. 329–335.
Smith, D. A., 1980, Sealing and nonsealing faults in Louisiana Gulf Coast salt basin: American Association of Petroleum Geologists Bulletin, v. 64, p. 145–172.
Smith, D. L., and Evans, B., 1984, Diffusional crack healing in quartz: Journal of Geophysical Research, v. 89, p. 4125–4136.
Smith, K. G., 1952, Structure plan of clastic dikes: American Geophysical Union Transactions, v. 33, p. 889–892.
Sorby, H. C., 1853, On the origin of slaty cleavage: Edinburgh New Philosophical Journal, v. 55, p. 137–149.
——— 1860, On the origin of "cone-in-cone": British Association for the Advancement of Science, report of 29th meeting, 1859, Transactions of Sections, Geology, p. 124.
——— 1863, On the direct correlation of mechanical and chemical forces: Royal Society of London Proceedings, v. 12, p. 538–600.
——— 1880, On the structure and origin of non-calcareous stratified rocks: Geological Society of London Quarterly Journal (Proceedings), v. 36, p. 46–92.
——— 1908, On the application of quantitative methods to the structure and history of rocks: Geological Society of London Quarterly Journal, v. 64, p. 171–233.
Spang, J. H., 1972, Numerical method for dynamic analysis of calcite twin lamellae: Geological Society of America Bulletin, v. 83, p. 467–472.
Spang, J. H., and Brown, S. P., 1981, Dynamic analysis of a small imbricate thrust and related structures, Front Ranges, southern Canadian Rocky Mountains, *in* McClay, K. R., and Price, N. J., eds., Thrust and nappe tectonics: Boston, Massachusetts, Blackwell Scientific Publications, p. 143–149.
Spang, J. H., and Friedman, M., 1978, Analysis of twinning in naturally and experimentally deformed dolomite: EOS (American Geophysical Union Transactions), v. 59, p. 1186.
Spang, J. H., and Groshong, R. H., Jr., 1981, Deformation mechanisms and strain history of a minor fold from the Appalachian Valley and Ridge province: Tectonophysics, v. 72, p. 323–342.
Spang, J. H., and Van Der Lee, J., 1975, Numerical dynamic analysis of quartz deformation lamellae and calcite and dolomite twin lamellae: Geological Society of America Bulletin, v. 86, p. 1266–1272.
Spang, J. H., Oldershaw, A. E., and Groshong, R. H., Jr., 1974, The nature of thin twin lamellae in calcite [abs.]: EOS (American Geophysical Union Transactions), v. 55, p. 419.
Spang, J. H., Oldershaw, A. E., and Stout, M. Z., 1979, Development of cleavage in the Banff Formation at Pigeon Mountain, Front Ranges, Canadian Rocky Mountains: Canadian Journal of Earth Sciences, v. 16, p. 1108–1115.
Spears, D. A., 1976, The fissility of some Carboniferous shales: Sedimentology, v. 23, p. 721–725.
Spiers, C. J., 1979, Fabric development in calcite polycrystals deformed at 400 °C: Bulletin of Mineralogie, v. 102, p. 282–289.
Spray, J. G., 1987, Artificial generation of pseudotachylyte using friction welding apparatus: Simulation of melting on a fault plane: Journal of Structural Geology, v. 9, p. 49–60.
Sprunt, E. S., and Nur, A., 1976, The reduction of porosity by pressure solution: Experimental verification: Geology, v. 4, p. 463–466.
——— 1977, Destruction of porosity through pressure solution: Geophysics, v. 42, p. 726–741.
——— 1979, Microcracking and healing in granites: New evidence from cathodoluminescence: Science, v. 205, p. 495–497.
Spry, A., 1969, Metamorphic textures: Oxford, England, Pergamon Press, 350 p.
Stanton, R. L., 1972, Ore petrology: New York, McGraw-Hill Book Company, 713 p.
Stearns, D. W., 1969, Fracture as a mechanism of flow in naturally deformed layered rocks, *in* Baer, A. J., and Norris, D. K., eds., Conference on Research in Tectonics, Proceedings: Canada Geological Survey Paper 68-52, p. 79–90.
——— 1972, Structural interpretation of fractures associated with the Bonita fault: New Mexico Geological Society Field Conference Guidebook No. 23, p. 161–164.
Stel, H., 1981, Crystal growth in cataclasites: Diagnostic microstructures and implications: Tectonophysics, v. 78, p. 585–600.
——— 1986, The effect of cyclic operation of brittle and ductile deformation on the metamorphic assemblage in cataclasites and mylonites: Pure and Applied Geophysics, v. 124, p. 289–307.
Stockdale, P. B., 1922, Stylolites: Their nature and origin: Indiana University Studies, v. 9, p. 1–97.
——— 1926, The stratigraphic significance of solution in rocks: Journal of Geology, v. 34, p. 399–414.
Stocker, R. L., and Ashby, M. F., 1973, On the rheology of the upper mantle: Reviews of Geophysics and Space Physics, v. 11, p. 391–426.
Summers, R., and Byerlee, J., 1977, A note on the effect of fault gouge composition on the stability of frictional sliding: International Journal of Rock Mechanics and Mining Sciences, v. 14, p. 155–160.
Syme Gash, P. J., 1971, A study of surface features relating to brittle and semi-brittle fracture: Tectonophysics, v. 12, p. 349–391.
Tada, R., and Siever, R., 1986, Experimental knife-edge pressure solution of halite: Geochimica et Cosmochimica Acta, v. 50, p. 29–36.
Talbot, C. J., 1970, The minimum strain ellipsoid using deformed quartz veins: Tectonophysics, v. 9, p. 47–76.
Tapp, B., and Cook, J., 1988, Pressure solution zone propagation in naturally deformed carbonate rocks: Geology, v. 16, p. 182–185.
Tapp, B., and Wickham, J., 1987, Relationships of rock cleavage fabrics to incremental and accumulated strain in the Conococheague Formation, U.S.A.: Journal of Structural Geology, v. 9, p. 457–472.
Taylor, G. I., 1938, Plastic strain in metals: Journal of Institute of Metals, Great Britain, v. 62, p. 307–324.

Taylor, J. M., 1950, Pore-space reduction in sandstones: American Association of Petroleum Geologists Bulletin, v. 34, p. 701–716.
Tchalenko, J. S., 1968, The evolution of kink-bands and the development of compression textures in sheared clays: Tectonophysics, v. 6, p. 159–174.
——— 1970, Similarities between shear zones of different magnitude: Geological Society of America Bulletin, v. 81, p. 1625–1640.
Teichmüller, M., and Teichmüller, R., 1981, The significance of coalification studies to geology—A review: Bulletin Centres Recherches Exploration-Production Elf-Aquitaine, v. 5, p. 491–534.
Teufel, L. W., 1980, Strain analysis of experimental superposed deformation using calcite twin lamellae: Tectonophysics, v. 65, p. 291–309.
——— 1981, Pore volume changes during frictional sliding of simulated faults: American Geophysical Union Monograph 24, p. 135–145.
Teufel, L. W., and Logan, J. M., 1978, Effect of displacement rate on the real area of contact and temperatures generated during frictional sliding of Tennessee Sandstone: Pure and Applied Geophysics, v. 116, p. 840–865.
Thomson, A., 1959, Pressure solution and porosity, *in* Ireland, H. A., ed., Silica in sediments—A symposium with discussions: Society of Economic Paleontologists and Mineralogists Special Publication No. 7, p. 92–109.
——— 1973, Soft-sediment faults in the Tesnus Formation and their relationship to paleoslope: Journal of Sedimentary Petrology, v. 43, p. 525–528.
Tissot, B. P., and Welte, D. H., 1978, Petroleum formation and occurrence: New York, Springer-Verlag, 538 p.
Tjia, H. D., 1964, Slickensides and fault movements: Geological Society of America Bulletin, v. 75, p. 683–686.
——— 1972, Fault movement, reoriented stress field and subsidiary structures: Pacific Geology, v. 5, p. 49–70.
Tobin, D. G., and Donath, F. A., 1971, Microscopic criteria for defining deformational modes in rock: Geological Society of America Bulletin, v. 82, p. 1463–1476.
Trurnit, P., 1968, Pressure solution phenomena in detrital rocks: Sedimentary Geology, v. 2, p. 89–114.
Truswell, J. F., 1972, Sandstone sheets and related intrusions from Coffee Bay, Transkei, South Africa: Journal of Sedimentary Petrology, v. 42, p. 578–583.
Tullis, J. A., 1970, Preferred orientation in rocks produced by Dauphiné twinning: Science, v. 168, p. 1342–1344.
Tullis, J. A., and Tullis, T. E., 1972, Preferred orientation of quartz produced by mechanical Dauphiné twinning: American Geophysical Union Geophysical Monograph Series, v. 16, p. 67–82.
Tullis, J. A., and Yund, R. A., 1985, Dynamic recrystallization of feldspar: A mechanism for ductile shear zone formation: Geology, v. 13, p. 238–241.
——— 1987, Transition from cataclastic flow to dislocation creep of feldspar: Mechanisms and microstructures: Geology, v. 15, p. 606–609.
Tullis, J. A., Christie, J. M., and Griggs, D. T., 1973, Microstructures and preferred orientations of experimentally deformed quartzites: Geological Society of America Bulletin, v. 84, p. 297–314.
Tullis, J. A., Snoke, A. W., and Todd, V. R., 1982, Penrose conference report: Significance and petrogenesis of mylonitic rocks: Geology, v. 10, p. 227–230.
Tullis, T. E., 1976, Experiments on the origin of slaty cleavage and schistosity: Geological Society of America Bulletin, v. 87, p. 745–753.
——— 1980, The use of mechanical twinning in minerals as a measure of shear stress magnitudes: Journal of Geophysical Research, v. 85, p. 6263–6268.
Turner, F. J., 1953, Nature and dynamic interpretation of deformation lamellae in calcite of three marbles: American Journal of Science, v. 251, p. 276–298.
Turner, F. J., and Heard, H. C., 1965, Deformation of calcite crystals at different strain rates: University of California, Los Angeles, Geological Science Publications, v. 46, p. 103–126.
Turner, F. J., and Weiss, L. E., 1963, Structural analysis of metamorphic tectonites: New York, McGraw-Hill Book Co., 545 p.
——— 1965, Deformation kinks in brucite and gypsum: National Academy of Science Proceedings, v. 54, p. 359–364.
Turner, F. J., Griggs, D. T., and Heard, H. C., 1954a, Experimental deformation of calcite crystals: Geological Society of America Bulletin, v. 65, p. 883–934.
Turner, F. J., Griggs, D. T., Heard, H. C., and Weiss, L. E., 1954b, Plastic deformation of dolomite rock at 380 °C: American Journal of Science, v. 252, p. 477–488.
Tuttle, O. F., 1949, Structural petrology of planes of liquid inclusions: Journal of Geology, v. 57, p. 331–356.
Twiss, R. J., 1974, Structure and significance of planar deformation features in synthetic quartz: Geology, v. 2, p. 329–332.
——— 1976, Some planar deformation features, slip systems, and submicroscopic structures in synthetic quartz: Journal of Geology, v. 84, p. 701–724.
——— 1977, Theory and applicability of a recrystallized grain size palaeopiezometer: Pure and Applied Geophysics, v. 115, p. 227–244.
Tyler, J. H., 1975, Fracture and rotation of brittle clasts in a ductile matrix: Journal of Geology, v. 83, p. 501–510.
Underhill, J. R., and Woodcock, N. H., 1987, Faulting mechanisms in high-porosity sandstones; New Red Sandstone, Arran, Scotland, *in* Jones, M. E., and Preston, R.M.F., eds., Deformation of sediments and sedimentary rocks, Geological Society Special Publication No. 29: Oxford, England, Blackwell Scientific Publications, p. 91–105.
Urai, J. L., 1985, Water-enhanced dynamic recrystallization and solution transfer in experimentally deformed carnalite: Tectonophysics, v. 120, p. 285–317.
Van Hise, C. R., 1898, Metamorphism of rocks and rock flowage: Geological Society of America Bulletin, v. 9, p. 269–328.
Van Loon, A. J., Brodzikowski, K., and Gotowala, R., 1984, Structural analysis of kink bands in unconsolidated sands: Tectonophysics, v. 104, p. 351–374.
van Werveke, L., 1883, Eigenthümliche Zwilligsbildung an Feldspat und Diallag: Neues Jahrbuch, v. 2, p. 97–101.
Vance, J. A., 1961, Polysynthetic twinning in plagioclase: American Mineralogist, v. 46, p. 1097–1119.
Venkitasubramanyan, C. S., 1971, Qualitative analysis of three-dimensional strain using twins in calcite and dolomite: Tectonophysics, v. 11, p. 217–231.
Vintanage, P. W., 1954, Sandstone dikes in the South Platte area, Colorado: Journal of Geology, v. 62, p. 493–500.
Voll, G., 1960, New work on petrofabrics: Liverpool and Manchester Geological Journal, v. 2, p. 503–567.
von Kármán, T., 1911, Festigheitversuche unter allseitigem Druck: Zeitschrift des Vereines Deutsche Ingenieure, v. 55, p. 1749–1757.
Wallace, R. E., 1951, Geometry of shearing stress and relation to faulting: Journal of Geology, v. 59, p. 118–130.
Wallace, R. E., and Morris, H. T., 1986, Characteristics of faults and shear zones in deep mines: Pure and Applied Geophysics, v. 124, p. 107–126.
Wang, H. F., and Simmons, G., 1978, Microcracks in crystalline rock from 5.3-km depth in the Michigan Basin: Journal of Geophysical Research, v. 83, p. 5849–5856.
Wanless, H. R., 1979, Limestone response to stress: Pressure solution and dolomitization: Journal of Sedimentary Petrology, v. 49, p. 437–462.
——— 1982, Limestone response to stress: Pressure solution and dolomitization—Reply: Journal of Sedimentary Petrology, v. 52, p. 328–332.
Waters, A. C., and Campbell, C. D., 1935, Mylonites from the San Andreas fault zone: American Journal of Science, v. 29, p. 473–503.
Watts, M. J., and Williams, G. D., 1980, Fault rocks as indicators of progressive shear deformation in the Guingamp region, Brittany: Journal of Structural Geology, v. 1, p. 323–332.
Wawersik, W. R., and Brace, W. F., 1971, Post-failure behavior of a granite and diabase: Rock Mechanics, v. 3, p. 61–85.
Weaver, C. E., 1984, Shale-slate metamorphism in southern Appalachians: Amsterdam, the Netherlands, Elsevier, 239 p.
Weber, K. J., and Bakker, M., 1981, Fracture and vuggy porosity: Society of Petroleum Engineers, 56th Annual Fall Technical Conference and Exhibition, San Antonio, Texas, 11 p.
Weber, K. J., Mandl, G., Pilaar, W. F., Lehner, F., and Precious, R. G., 1978, The role of faults in hydrocarbon migration and trapping in Nigerian growth fault structures: 10th Annual Society of Petroleum Engineers of American Institute of Mechanical Engineers Offshore Technology Conference Preprint No. OTC-3356, p. 2643–2653.
Wegmann, C. E., 1932, Note sur le Boudinage: Bulletin Société Géologique de France, series 5, II, p. 477–489.
Wegmann, C. E., and Schaer, J. P., 1957, Lunules tectoniques et traces de mouvements dans les plis du Jura: Eclogae Geologicae Helvetiae, v. 50, p. 492–496.
Wellman, H. W., 1962, A graphical method for analysing fossil distortion caused by tectonic deformation: Geological Magazine, v. 49, p. 348–353.
Wenk, H. R., 1978, Are pseudotachylites products of fracture or fusion: Geology, v. 6, p. 507–511.
Wenk, H. R., Venkitasubramanyan, C. S., and Baker, D. W., 1973, Preferred orientation in experimentally deformed limestone: Contributions to Mineralogy and Petrology, v. 38, p. 81–114.
Weyl, P. K., 1959, Pressure solution and the force of recrystallization—A phenomenological theory: Journal of Geophysical Research, v. 64, p. 2001–2025.
Weymouth, J. H., and Williamson, W. O., 1953, The effects of extrusion and some other processes on the microstructure of clay: American Journal of Science, v. 251, p. 89–108.
Whisonant, R. C., 1970, Influence of texture upon the response of detrital quartz to deformation of sandstones: Journal of Sedimentary Petrology, v. 40, p. 1018–1025.
White, J. C., and White, S. H., 1983, Semi-brittle deformation within the Alpine fault zone, New Zealand: Journal of Structural Geology, v. 5, p. 579–589.
White, J. W., 1973, An invariant description of failure for an isotropic medium: Journal of Geophysical Research, v. 78, p. 2438–2441.
White, S., 1973a, Deformation lamellae in naturally deformed quartz: Nature Physical Science, v. 245, p. 26–28.
——— 1973b, Syntectonic recrystallization and texture development in quartz: Nature, v. 244, p. 276–278.
——— 1973c, The dislocation structures responsible for the optical effects in some naturally deformed quartzites: Journal of Material Science, v. 8, p. 490–499.
——— 1976, The effects of strain on the microstructures, fabrics, and deformation mechanisms in quartzites: Royal Society of London Philosophical Transactions, ser. A, v. 283, p. 69–86.
——— 1982, Fault rocks of the Moine thrust zone: A guide to their nomenclature: Texture and Microstructures, v. 4, p. 211–221.
White, S. H., 1979, Difficulties associated with paleostress estimates: Bullétin Mineralogique, v. 102, p. 210–215.
White, S. H., and Knipe, R. J., 1978, Microstructure and cleavage development in selected slates: Contributions to Mineralogy and Petrology, v. 66, p. 165–174.
Wickham, J. S., 1973, An estimate of strain increments in a naturally deformed carbonate rock: American Journal of Science, v. 273, p. 23–47.
Wickham, J. S., and Anthony, M., 1977, Strain paths and folding of carbonate rocks near Blue Ridge, central Appalachians: Geological Society of America Bulletin, v. 88, p. 920–924.
Williams, P. F., 1972, Development of metamorphic layering and cleavage in low grade metamorphic rocks at Bermagui, Australia: American Journal of Science, v. 272, p. 1–47.
Williams, P. F., Collins, A. R., and Wiltshire, R. G., 1969, Cleavage and penecontemporaneous deformation structures in sedimentary rocks: Journal of Geology, v. 77, p. 415–425.
Willis, B., 1923, Geologic structures: New York, McGraw-Hill Book Company, Inc., 295 p.
Wilson, L. R., 1971, Palynological techniques in deep-basin stratigraphy: Shale Shaker, v. 21, p. 124–139.
Wilson, T. V., and Sibley, D. F., 1978, Pressure solution and porosity reduction in shallow buried quartz arenite: American Association of Petroleum Geologists Bulletin, v. 62, p. 2329–2334.
Wiltschko, D. V., and Sutton, S. J., 1982, Deformation by overburden of a coarse quartzite conglomerate: Journal of Geology, v. 90, p. 725–733.
Winslow, M. A., 1983, Clastic dike swarms and the structural evolution of the foreland fold and thrust belt of the southern Andes: Geological Society of America Bulletin, v. 94, p. 1073–1080.
Wintsch, R. P., 1978, A chemical approach to the preferred orientation of mica: Geological Society of America Bulletin, v. 89, p. 1715–1718.
Wise, D. U., 1964, Microjointing in basement, middle Rocky Mountains of Montana and Wyoming: Geological Society of America Bulletin, v. 75, p. 287–306.
Wise, D. U., Dunn, D. E., Engelder, J. T., Geiser, P. A., Hatcher, R. D., Kish, S. A., Odom, A. L., and Schamel, S., 1984, Fault-related rocks: Suggestions for terminology: Geology, v. 12, p. 391–394.
——— 1985a, Reply to Mawer on "Fault-related rocks: Suggestions for terminology": Geology, v. 13, p. 379.
——— 1985b, Reply to Raymond on "Fault-related rocks: Suggestions for terminology": Geology, v. 13, p. 218–219.
Wojtal, S., and Mitra, G., 1986, Strain hardening and strain softening in fault zones from foreland thrusts: Geological Society of America Bulletin, v. 97, p. 674–687.
Wong, P. D., and Oldershaw, A., 1981, Burial cementation in the Devonian, Kaybob Reef Complex, Alberta, Canada: Journal of Sedimentary Petrology, v. 51, p. 507–520.
Wood, D. S., and Holm, P. E., 1980, Quantitative analysis of strain heterogeneity as a function of temperature and strain rate: Tectonophysics, v. 66, p. 1–14.
Woodcock, N. H., 1976, Structural style in slump sheets: Ludlow Series, Powys, Wales: Geological Society of London Journal, v. 132, p. 399–415.
Woodland, B. G., 1982, Gradational development of domainal slaty cleavage, its origin and relation to chlorite porphyroblasts in the Martinsburg Formation, eastern Pennsylvania: Tectonophysics, v. 82, p. 89–124.
Woodworth, J. B., 1896, On the fracture system of joints, with remarks on certain great fractures: Boston Society of Natural History Proceedings, v. 27, p. 163–184.
Wright, T. O., and Platt, L. B., 1982, Pressure dissolution and cleavage in the Martinsburg Shale: American Journal of Science, v. 282, p. 122–135.
Yagishita, K., and Morris, R. C., 1979, Microfabrics of a recumbent fold in cross-bedded sandstones: Geological Magazine, v. 116, p. 105–116.

MANUSCRIPT RECEIVED BY THE SOCIETY NOVEMBER 9, 1987
REVISED MANUSCRIPT RECEIVED APRIL 20, 1988
MANUSCRIPT ACCEPTED APRIL 21, 1988

Printed in U.S.A.

The formation of continental crust: Part 1. A review of some principles; Part 2. An application to the Proterozoic evolution of southern North America

CENTENNIAL ARTICLE

M. E. BICKFORD *Department of Geology, University of Kansas, Lawrence, Kansas 66045*

ABSTRACT

Understanding of the evolution of continental crust has been dominated by work in tectonics, experimental petrology and geochemistry, and geochronology and isotope geochemistry. In part 1, these approaches are discussed in the light of keynote articles, most published by the Geological Society of America, in an attempt to show how new data and new thinking have influenced understanding of the formation and modification of continental crust. In part 2, the evolution of continental crust in southern North America from 1.8 Ga until 1.3 Ga is discussed as an application of ideas generated during the past 30 years.

INTRODUCTION

It is a great privilege to be asked to write this article on the evolution of continental crust for the Centennial *Bulletin*. My own contributions are minor compared to those I will discuss, but I hope a 25-year career of research and teaching will afford a useful perspective. Two points need to be made at the outset. First, the evolution of continental crust is the subject of several books, not an article; thus, I will necessarily emphasize certain aspects of the subject at the expense of others. Second, I write from the perspective of one who has spent his professional life studying the Proterozoic of North America; most of my own ideas deal with the Proterozoic. Although I will briefly address the earliest origins of continental crust, the emphasis of this article will be upon modern processes and the patterns of crustal development through time.

In part 1, I have chosen three major research areas which seem to be central to understanding of the evolution of continental crust. These are (1) tectonics, (2) experimental petrology and geochemistry, and (3) geochronology and isotope systematics. My plan is to select two or three important papers in each area and use their impact to develop some ideas about how continental crust has formed and evolved. The papers were not chosen for historical precedence, and in each case, there are earlier papers which also had major influence. In part 2, I will review the Proterozoic history of the southern part of the North American continental crust from the perspective of my own research and offer some suggestions on how the major topics above have influenced our understanding of its evolution.

PART 1. SOME PRINCIPLES

The Earliest Crust

Earth history prior to about 3.8 Ga is not directly accessible to our study and has largely been inferred from theoretical considerations and the study of the Moon. There is general agreement that Earth formed through cold accretion of planetesimals but subsequently underwent total or partial melting because of the release of gravitational potential energy and radioactive heating. Anderson (1984), however, favored hot accretion with first-order differentiation accompanying planetesimal infall. These events caused the formation of a metallic core and the early separation of a crust. The earliest crust was apparently not permanently preserved, because of vigorous convection and the continued infall of planetesimals, until about 3.8 Ga.

The oldest crustal remnants known are the 3.7–3.8 Ga Isua supracrustals of Greenland (for example, Moorbath and others, 1977; Michard-Vitrac and others, 1977; Hamilton and others, 1978); rocks of similar age that may have once been contiguous with the Isua supracrustals occur in Labrador (for example, Bridgwater and others, 1978; Baadsgaard and others, 1979; Collerson and Bridgwater, 1979). Rocks only slightly younger (3.5–3.6 Ga) are known in the Minnesota River valley (Goldich and Wooden, 1980), in the Marenisco-Watersmeet region of northern Michigan (Peterman and others, 1980), and elsewhere in the world. Although the discovery of 4.1–4.2 Ga detrital zircons in quartzite in the Yilgarn Block of western Australia indicates that crustal remnants of this age may be found (Froude and others, 1983), it seems clear that Earth did not develop significant stable continental crust until about 3.8 Ga. This article deals primarily with the evolution of continental crust from this time until the present. I will use the impact of selected papers to facilitate discussion of how the three chosen research areas, tectonics, experimental petrology and geochemistry, and geochronology and isotope systematics, have influenced our understanding of the processes and time-space patterns of continental crustal evolution.

Tectonics

Thirty years ago, tectonics would not have been viewed as a major factor in the evolution of continental crust, at least not in Britain and North America. Today, it is recognized that the tectonics of lithospheric plates is *the* major factor. The development of plate tectonics is well known today, beginning with the work of Wegener (1915), Argand (1924), du Toit (1937), and Holmes (1944), and culminating in the brilliant insights of Dietz (1961, 1963), Wilson (1965a, 1965b, 1966, 1968), Vine and Matthews (1963), Le Pichon (1968), Morgan (1968), and many others. These workers showed that the Earth's lithosphere is mobile and that new crust is generated largely at either divergent

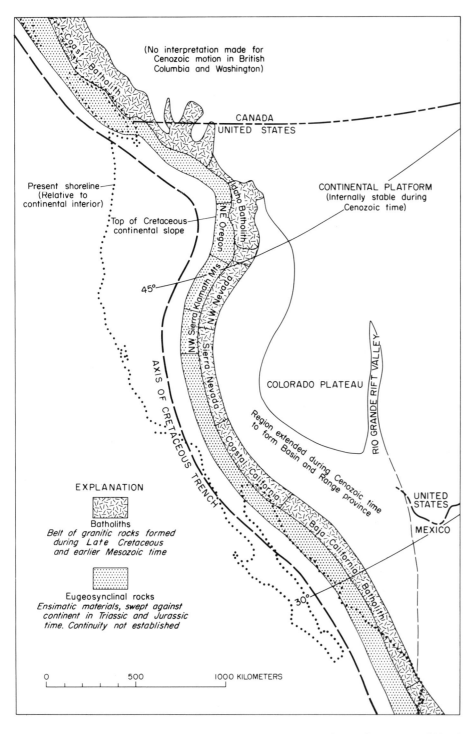

Figure 1. Paleogeographic map of the continental-margin complexes of west-central North America in middle Late Cretaceous time, with Cenozoic extension removed. After Hamilton (1969).

plate boundaries (oceanic crust) or at convergent plate boundaries (first-order continental crust). Many workers (for example, Helwig, 1974, a general collage model; Hoffman, 1980, a model for the accretion of Wopmay orogen) soon recognized that blocks of continental crust, some originating thousands of kilometers distant, have been accreted to pre-existing continental crust through plate motions. These concepts have recently been expanded by those working in western North America (for example, Coney and others, 1980).

I have chosen two papers, both published in the *Bulletin* of the Geological Society of America, as keynotes for this portion of my discussion. The first is "Mesozoic California and the underflow of Pacific mantle" by Warren Hamilton (1969). The second is "Ancient borderland terranes of the North American Cordillera: Correlation and microplate tectonics" by Michael Churkin, Jr., and G. Donald Eberlein (1977). Although I shall not discuss it in as much detail, an equally important early paper was "Implications of plate tectonics for the Cenozoic evolution of western North America," published by Tanya Atwater in the *Bulletin* of the Geological Society of America (Atwater, 1970).

In his paper, Hamilton attempted to synthesize the Mesozoic tectonics of California in terms of the subduction ("underflow") of thousands of kilometers of Pacific lithosphere beneath the western margin of North America. This paper was one of the first to use the new plate-tectonic concepts to synthesize the geology of a major, tectonically active region. Hamilton showed that most of the "eugeosynclinal" rocks of western California are accreted oceanic material, including both abyssal sedimentary rocks and rocks of probable island-arc affinity, and contrasted these with miogeosynclinal rocks farther east. The great Mesozoic batholiths (Baja California, Southern California, Sierra Nevada, Idaho, Coast Ranges) were seen as formed above a great, east-dipping subduction system. In a reconstruction removing later transcurrent faults (Fig. 1), Hamilton compared the Mesozoic system of trench sediments and batholiths with the modern Andean system of western South America. The section called "The Klamaths and ocean-floor sweeping" presented a speculative history for the accumulation of the Klamath terranes involving the accretion of what would later be called "suspect terranes" (Coney and others, 1980). Hamilton showed clearly that plate motions had built North America westward by several hundred kilometers during the Paleozoic and Mesozoic.

Only 1 year later, Atwater published her brilliant paper in which the geometrical principles of plate tectonics and the record of sea-floor spreading preserved in the magnetic anomalies of the northeastern Pacific were used to develop a tectonic history for the western United States, including subduction in the vicinity of the present coast and broad extension in the Basin and Range province. Although Atwater's models were to some extent constrained by the known record of the region, their major value was to predict an igneous and structural history which could be tested by field observation. Testing and the development of refined models continues vigorously today (for example, Eaton, 1979, 1982; Eaton and others, 1979; Wernicke, 1981, 1985; Glazner and Bartley, 1984).

In a paper published 8 years later, Churkin and Eberlein, both of whom had worked for many years in the complex tectonic terranes of

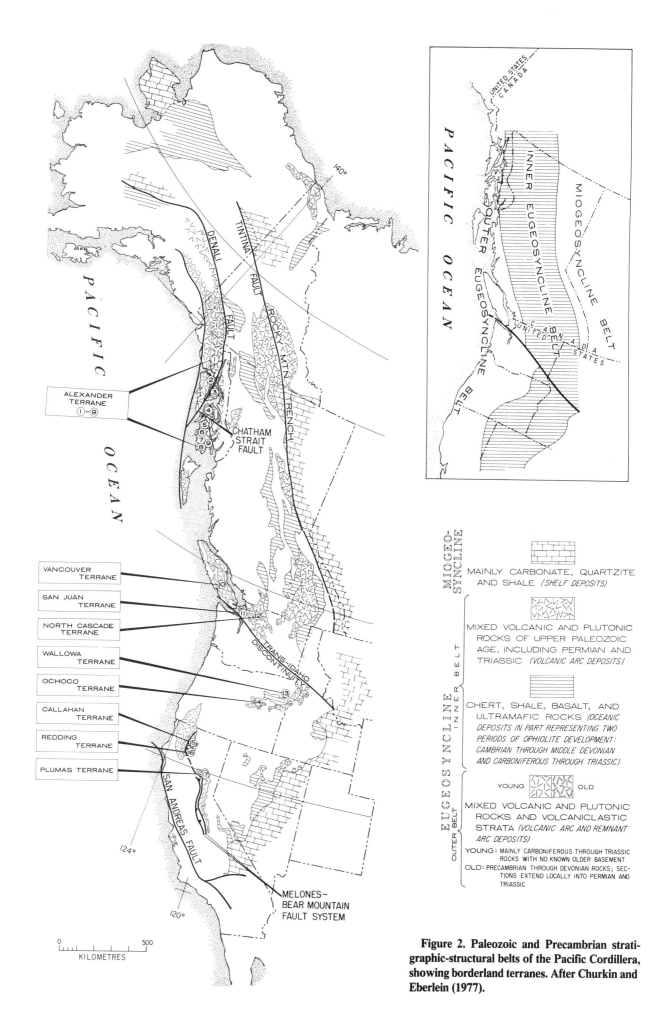

Figure 2. Paleozoic and Precambrian stratigraphic-structural belts of the Pacific Cordillera, showing borderland terranes. After Churkin and Eberlein (1977).

coastal Alaska, showed that western North America, from the coast to the edge of the pre-Mesozoic craton, is a collage of accreted terranes, each of which has a distinctive lithology and stratigraphy (Fig. 2). Through careful stratigraphic analysis, Churkin and Eberlein identified nine terranes which they believed to have formed exotically to North America. Faunal studies within these terranes showed that some of them have Asian affinities and apparently originated thousands of kilometers from their current geographic locations. The occurrence of major thrust faults, commonly with associated ophiolitic sequences and rocks in blueschist facies, between these terranes suggested that they had been accreted against the edge of the pre-existing continental crust.

Subsequent work has added detail to ideas presented in both of these papers. The ensimatic terranes on the west side of the Sierra Nevada include Permian-Triassic and Lower Jurassic arc and ophiolitic rocks as well as older blocks (Saleeby, 1981, 1982; Saleeby and others, 1982; Day and others, 1984) that were evidently thrust against the continent prior to the development of the major Jurassic-Cretaceous subduction system. Some of the exotic terranes of Churkin and Eberlein were considered to be parts of larger terranes by Jones and others (1977) and Coney and others (1980). It is now established, however, that the North American continent has grown through accretionary processes involving plate motions during the Mesozoic; some of the accreted terranes are as old as Paleozoic and, on the basis of paleomagnetic measurements, had their origin far to the south of their current locations.

The papers cited, and those of many other workers, have clearly shown that all major processes of growth of continental crust involve tectonics. In the 1960s, many thought in terms of the formation of subduction zones and outboard island arcs along the margins of continents; these were believed to be subsequently accreted to the continents as convergent plate motion continued. This process is probably locally important, as in parts of the modern Andes; however, it now appears that the predominant process is the formation of intermediate to silicic rocks in island-arc systems and the sweeping of these and oceanic plateaus (some rifted continental fragments and some of oceanic origin) against larger continental masses (Ben Avraham, 1981; Ben Avraham and Nur, 1982).

Processes and patterns identified by workers in western North America for Mesozoic and younger rocks are now being applied to older terranes, with exciting results. Hatcher (1978) and Williams and Hatcher (1982) have proposed a terrane accretion model for the evolution of the Appalachian system of eastern North America during the Paleozoic (Fig. 3). Many workers have proposed tectonic models for the evolution of continental crust during the Proterozoic. For North America, special attention is called to the models of Hoffman (1980) and Hoffman and others (1982) for the Early Proterozoic Wopmay orogen of the Canadian Northwest Territories (Fig. 4); of Lewry (1981) and Lewry and others (1985) for the Early Proterozoic Trans-Hudson orogen (Fig. 4); and of Van Schmus (1976) and Cambray (1978) for the Early Proterozoic Penokean orogen. In a particularly exciting recent paper in *Geology*, Hoffman (1987), amplifying an earlier suggestion by Gibb (1978), has compared the collision of the Archean Slave craton with the Churchill platform and the development of the Great Slave shear zone in the Early Proterozoic (Fig. 5) with the Tertiary collision of India and Eurasia and the development of the Sagaing transform. Condie (1986; Fig. 6), Bickford and Boardman (1984), and Silver (1984), among numerous others, have proposed accretion models for the evolution of lower Proterozoic rocks in the southwestern United States. Writing in *Geology*, Van Schmus, Zietz, and I (Bickford and others, 1986) have proposed that these accreted terranes also extend across the buried midcontinent region (see last section).

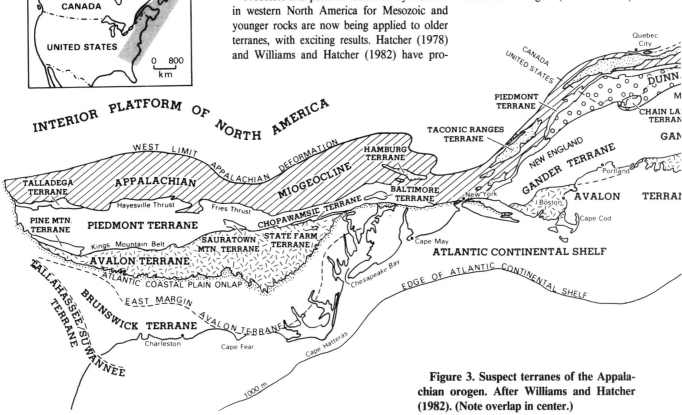

Figure 3. Suspect terranes of the Appalachian orogen. After Williams and Hatcher (1982). (Note overlap in center.)

Whereas the role of tectonics during the Archean is not disputed today, the nature of Archean tectonics is a subject of considerable controversy. Archean crust is characterized by generally low-grade granite-greenstone terranes and regions of high-grade gneisses. Some (for example, Kroner, 1981; Hargraves, 1981) have argued that higher heat flow in the Earth's early history precluded plate tectonics in the modern style, whereas others (for example, Burke and others, 1976) have argued that plate tectonics, not different in major details from those operating at present, has characterized the Earth's outer parts since the Early Archean. It is clearly beyond the scope and purpose of this article to summarize, much less to comment upon, the great diversity of opinion about Archean tectonics. Suffice it to say that although there is controversy because the geologic record is obscure, all agree that tectonic processes were a major factor in the evolution of even the earliest continental crust.

My hardly surprising conclusion is that at least since the Early Proterozoic, continental crust has evolved largely through accretion of crustal blocks, formed either by sea-floor processes (for example, hot-spot activity, mid-oceanic-ridge activity, formation of island arcs at subduction zones) or as rifted fragments from pre-existing continental regions. A major question is whether most of the continental crust was formed in the Earth's earliest history, and has simply been tectonically recycled ever since, or whether significant amounts of continental crust have formed by separation from the mantle throughout the Earth's history (for example, Patchett and Arndt, 1986). That question is not amenable to tectonic analysis. It is the stuff of geochemistry, particularly isotopic geochemistry, and will be addressed later.

Experimental Petrology and Geochemistry

My choices of articles as keynotes for discussion of the importance of experimental petrology and geochemistry are (1) "Origin of granite in the light of experimental studies in the system $KAlSi_3O_8$-$NaAlSi_3O_8$-SiO_2-H_2O," by O. F. Tuttle and N. L. Bowen (GSA Memoir 74, 1958), and (2) "Genesis of the calc-alkaline igneous rock suite," by T. H. Green and A. E. Ringwood (1968). Both papers are now classics in the literature of experimental petrology.

The work of Tuttle and Bowen is perhaps the most quoted in all of the petrologic literature. Published after years of experimental studies on parts of the "granite system," this paper provided a basis for understanding the behavior of granitic magmas and for interpreting the crystallization history and crustal setting of granitic rocks on the basis of petrographic features, such as the presence or absence of alkalic feldspar (or perthite), the stability of hydrous minerals such as biotite and hornblende, and the paragenetic relationships of quartz. A key part of this study was the role of water in both melting and crystallization of granitic compositions. It was shown that the minimum in the Q-Ab-Or-H_2O system (Fig. 7) is depressed and shifted toward more sodic compositions as P_{H_2O} increases. It was made clear that granitic liquids are the expected result of both (1) fractional crystallization of magmas of a wide range of compositions and (2) small degrees of partial melting of solid rocks of a wide range of compositions. Because the minimum in Q-Ab-Or-H_2O is depressed by increasing P_{H_2O}, it is clear that water is central to fluxing the melting process in the formation of most granitic magmas.

Green and Ringwood (1968) showed that partial melting of basaltic rocks under dry conditions can yield essentially andesitic liquids at pressures ranging from 27 to 36 kb because garnet and clinopyroxene are the stable liquidus phases. Under hydrous conditions, compositions shift toward granodioritic compositions. Quantitatively, Green and Ringwood showed that the compositions of liquids derived from dry melting of quartz eclogite by successive partial melting follow a broadly calc-alkaline trend as

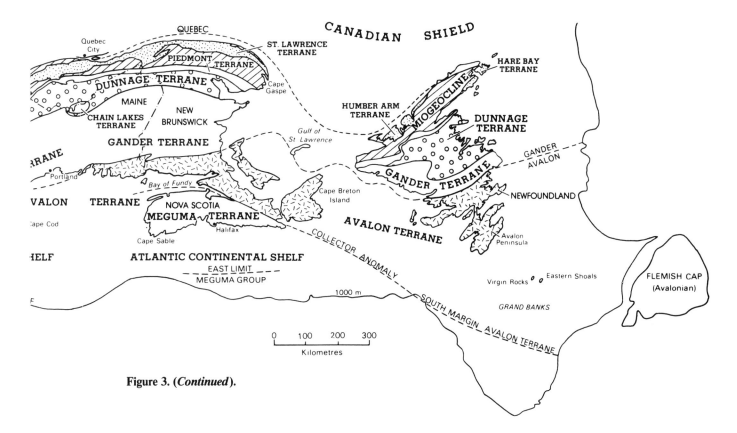

Figure 3. (*Continued*).

defined on an AFM plot, suggesting that calc-alkaline magmas may form by partial melting of basaltic rocks at mantle depths (70–80 km) or by fractional crystallization of basaltic magma under similar conditions.

In lower pressure experiments (9–10 kb) under hydrous conditions, Green and Ringwood found that amphibole and pyroxenes are near-liquidus phases, whereas calcic plagioclase and garnet occur nearer the solidus. Calculated liquid fractions also follow the calc-alkaline trend and suggest that such liquids could be derived from partial melting of basalt under wet conditions in the lower crust or by fractional crystallization of basaltic magma under similar conditions.

These classic studies illustrate the enormous impact experimental petrology has had upon our understanding of crustal genesis. Following the new tectonic understandings discussed earlier, it has become generally understood that continental crust is formed by a kind of three-stage fusion process. The first stage is the production of basaltic magma by partial melting of the mantle, a process that occurs mostly at mid-oceanic ridges with the formation of tholeiitic oceanic crust. Basalt, commonly alkalic, is also formed in isolated intraplate settings, presumably above mantle hot spots. Basalt formed beneath older continental crust is commonly intruded, with concomitant melting of the older silicic crust, forming the bimodal basalt-rhyolite suite that is associated with intracontinental rifting.

In the second stage, basaltic crust that is subducted during the formation of either oceanic island arcs or continental arcs apparently undergoes further partial melting to form andesitic or tonalitic magmas. The exact mechanism by which this process occurs is still the subject of intense debate and much observational and experimental study. It may be that the Green and Ringwood mechanism, that is, conversion of basaltic crust to quartz eclogite during subduction followed by partial melting yielding andesitic magmas, is what happens. The occurrence of garnet xenocrysts in many andesites (see Gill, 1981, p. 185–188, for a discussion) suggests that this mechanism may be important. The presence of amphibole phenocrysts or xenocrysts in many other andesites, however, indicates hydrous conditions, and some authors (for example, Ringwood, 1974, 1975; Mysen, 1982) believe that calc-alkaline magmas originate in subduction zones either through hydrous melting of subducted crust or through the introduction of water, perhaps carrying alkalis and other incompatible elements, into the overlying mantle wedge where it fluxes melting. It seems to me that subduction zones are unique in that they are the only places on Earth where rock material (basaltic oceanic crust) that has been in contact with sea water for perhaps tens of millions of years is taken to depths where melting of hydrated and chemically altered rock occurs. I thus favor the hypothesis that most calc-alkaline rocks of island arcs are the product of some kind of hydrous melting. A major problem is whether hydrous minerals are stable to the depth at which melting occurs (about 120 km; Marsh, 1979). Although detailed studies have not yet been done, I believe it is likely that the modest enrichment in incompatible trace elements and the slightly elevated $^{87}Sr/^{86}Sr$ ratios of the calc-alkaline rocks of island arcs can be traced to sea water.

The third stage in generation of continental crust evidently occurs after the calc-alkaline rocks of island arcs, continental arcs, or oceanic plateaus are accreted and stabilized. Then, anatexis of the newly accreted crust, variably contaminated and mixed with older continental material, forms later "post-kinematic" or "post-tectonic" granite. These rocks are commonly true granite, frequently high in K_2O, in contrast to the calc-alkaline volcanic rocks, tonalite, and granodiorite of accreted orogenic terranes. Post-orogenic granites most commonly occur in older terranes where uplift and erosion have exposed them beneath their cover of orogenic volcanic rocks and plutons. They are well known in the Early Proterozoic terranes of North America, for example in the Trans-Hudson orogen of northern Saskatchewan, where granitoids formed about 1830 Ma intrude 1880 Ma volcanics and 1860 Ma orogenic plutons (Van Schmus and others, 1987a), and in the Early Proterozoic terrane of central Colorado, where 1670 Ma granitoids intrude 1770–1740 Ma volcanic successions and 1750–1715 Ma tonalites (Bickford and Boardman, 1984; Bickford and others, 1988a).

The mechanism of formation of these post-orogenic magmas is still not clear. Their isotopic and chemical compositions (see below) indicate that they are products of partial melting of older crust. Van Schmus and I (Van Schmus and Bickford, 1981; Bickford and others, 1986; Van Schmus and others, 1987a) have argued that melting is triggered 20 to 30 m.y. after orogenic accretion by rifting of thickened orogenic crust

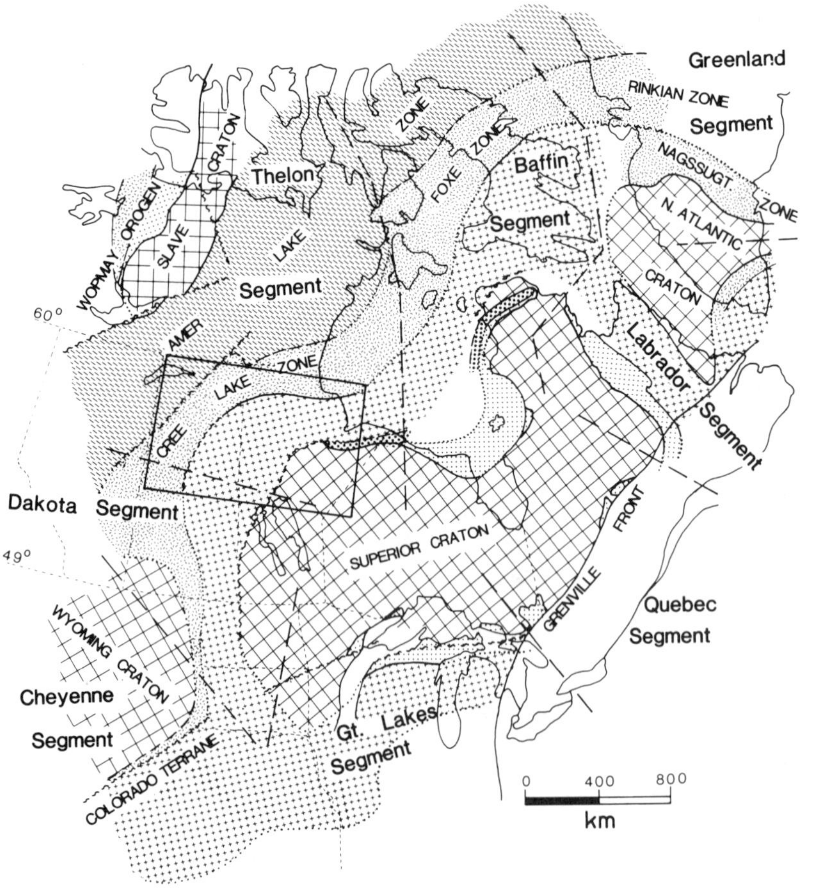

Figure 4. Geographic subdivisions and generalized lithotectonic elements of the Wopmay and Trans-Hudson orogens and other Early Proterozoic terranes in North America. Map is the work of J. F. Lewry as published in Van Schmus and others (1987b).

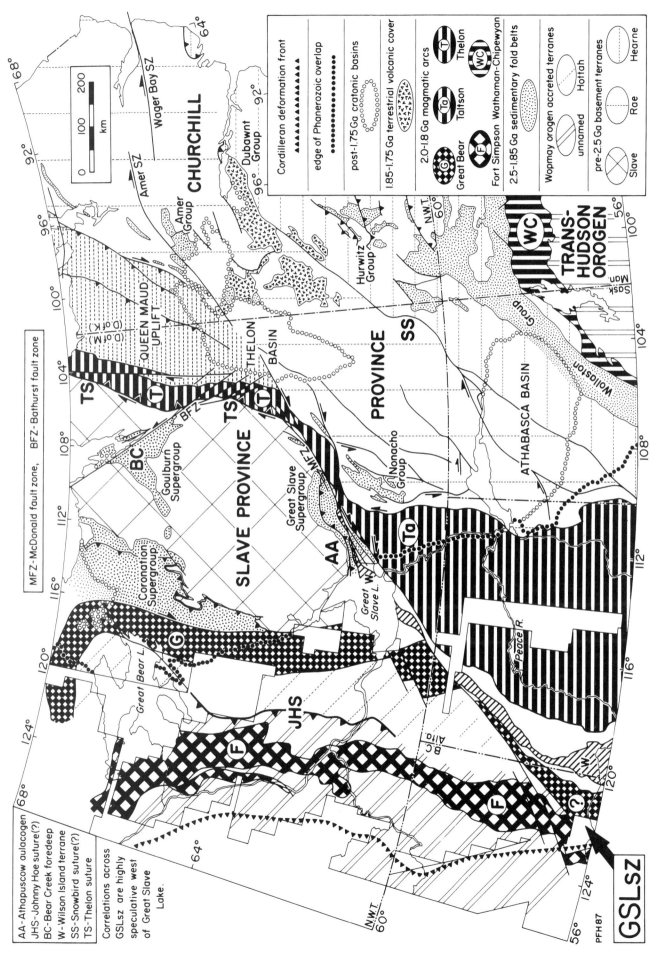

Figure 5. Tectonic elements in a portion of northwestern Canada, showing suggested correlations across Great Slave Lake shear zone. After Hoffman (1987).

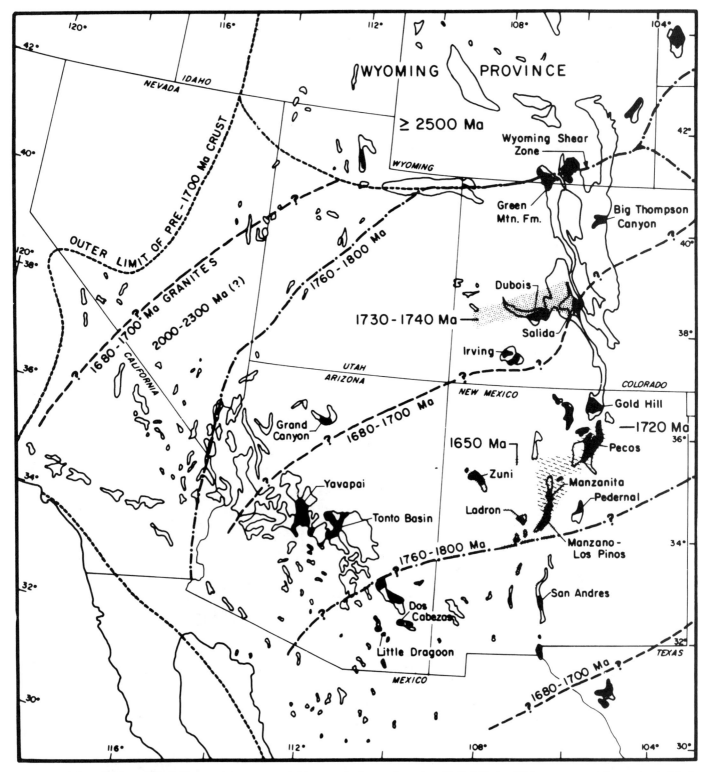

Figure 6. Distribution of Early Proterozoic supracrustal terranes in the southwestern United States. Major supracrustal areas are in black. Terrane ages are given inside boundaries, and those of limited areal extent are shown by stipple patterns. After Condie (1986).

with input of mantle-derived basalt as the heat source. In a review of this manuscript, however, P. F. Hoffman pointed out that accreted crust should not still be thick after this much time and that further, there is little evidence for rifting. This question will arise again in part 2 in a discussion of Middle Proterozoic "anorogenic" magmatism in southern North America.

The classic papers of England and Thompson (1984) and Thompson and England (1984) show clearly that anatexis and the production of granitic melts is the expected result of high-grade metamorphism following orogenesis and crustal thickening. It is not clear, however, whether large volumes of granite are formed by *in situ* anatexis during metamorphism, for where large granite plutons are present in high-grade metamorphic terranes, it is commonly arguable whether the plutons are the result or the cause of the metamorphism. Nevertheless, it is abundantly clear that granitic liquids are formed during migmatization in high-grade terranes, and this process must be a major factor in the formation of continental crust.

Continental crust thus appears to form through a combination of oceanic melting processes, in which basalt and the calc-alkaline series of island arcs are formed, accretion via plate movements, and later metamorphism and anatectic melting. I believe that the importance of anatexis and the formation of granite has been greatly exaggerated, for tonalitic gneisses—the ubiquitous "gray gneisses" of most Precambrian crystalline terranes—are very abundant relative to true granite. The continental crust may be characterized as a collage of accreted, and mostly strongly deformed, fragments of arcs and oceanic crust that have been modified through metamorphism and anatectic melting.

Geochronology and Isotope Systematics

As keynote articles for this section of my discussion, I have chosen "A low-contamination method for the hydrothermal decomposition of zircon and extraction of U and Pb for isotopic age determination" by T. E. Krogh (1973) and "Inferences about magma sources and mantle structure from variations of $^{143}Nd/^{144}Nd$" by D. J. DePaolo and G. J. Wasserburg (1976).

Tom Krogh's paper may seem a strange one to pick for great impact upon studies of the evolution of continental crust, for basically it described a technique for analyzing zircons for U and Pb isotopic composition and concentration. The first isotopic U-Pb age determinations were published by Nier (1939) and techniques for interpreting discordant U-Pb age patterns in zircons were developed by Wetherill (episodic lead loss, 1956), Tilton (continuous diffusion lead loss, 1960), Wetherill (combined lead and uranium diffusion, 1963), Wasserburg (continuous diffusion lead loss with time-dependent diffusion parameters, 1963), and Stern and others (dilatancy lead loss, 1966). Silver and Deutsch (1963) published careful studies of the nature of discordance in zircons and its relationship to U distribution and magnetic susceptibility. Nevertheless, prior to the development of microanalytical methods by Krogh, routine age determination of ordinary igneous rocks by U-Pb zircon methods was inhibited by the requirement that 500 to 1,000 mg of zircon be separated for analysis, a procedure necessitated by large laboratory contamination levels and often requiring processing of several hundred kilograms of rock.

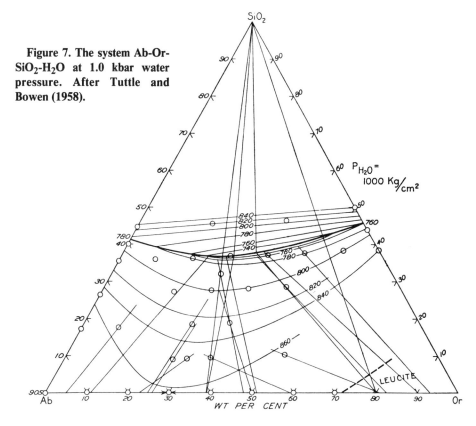

Figure 7. The system Ab-Or-SiO$_2$-H$_2$O at 1.0 kbar water pressure. After Tuttle and Bowen (1958).

Krogh's techniques, now in routine use in essentially all geochronological laboratories, allow high-precision analysis of zircon fractions of less than a milligram. Indeed, modern high-sensitivity, multicollector mass spectrometers permit analysis of single zircon grains (for example, Kober, 1986), and the ion-probe mass spectrometer (for example, the SHRIMP instrument of the Australian National University; Compston and others, 1982, 1984) allows U-Pb analysis of parts of single grains. Under ideal conditions, U-Pb ages of zircons can now be determined with about 0.25% uncertainty at the 2σ level, particularly if careful hand-picking and air-abrasion methods (Krogh, 1982) are employed to reduce discordancy. Geologic factors, such as lead or uranium migration, inherited xenocrystic components, metamorphic overgrowths, and so on, commonly detract from the ideal situation, however.

The implementation of routine, high-precision, accurate age determinations through analysis of zircons has had enormous impact upon understanding of the time dimension of continental evolution. Most geochronologists today accept the zircon method as the best technique for obtaining crystallization ages of igneous rocks older than about 100 m.y. because, despite problems caused by inheritance of xenocrystic zircon from magma source areas (for example, Grauert and Hoffman, 1973; Pidgeon and Aftalion, 1978; Bickford and others, 1981b; Wright and Haxel, 1982), the U-Pb system in zircons is far less readily disturbed than is the Rb-Sr or K-Ar system. Moreover, U-Pb analysis of baddeleyite (ZrO$_2$) has become increasingly important as a means of obtaining the ages of mafic rocks, and U-Pb analysis of monazite is being used to date metamorphic events. It is hard to overemphasize the importance of precise age determinations upon understanding of continental evolution, for the time dimension pervades analysis of paleomagnetic results, tectonic reconstructions, and calculations of rates of separation of continental crust from the mantle. I cite one major North American result as an example of the power of the use of zircon geochronology.

Zircon age data for major, Early Proterozoic orogenic belts has shown that the Penokean orogen of northern Michigan, Wisconsin, Minnesota, and adjacent southern Ontario (Van Schmus, 1976, 1980); the Wopmay orogen of the Canadian Northwest Territories (Hoffman and others, 1982; Hoffman and Bowring, 1984; Bowring, 1985); and the extensive Trans-Hudson orogen (Hoffman, 1981; Van Schmus and others, 1987b) were all formed between 1800 and 1900 Ma. Moreover, Table 1 (Van Schmus and others, 1987a) shows that the details of volcanic and plutonic history are remarkably similar. Figure 8 shows that these

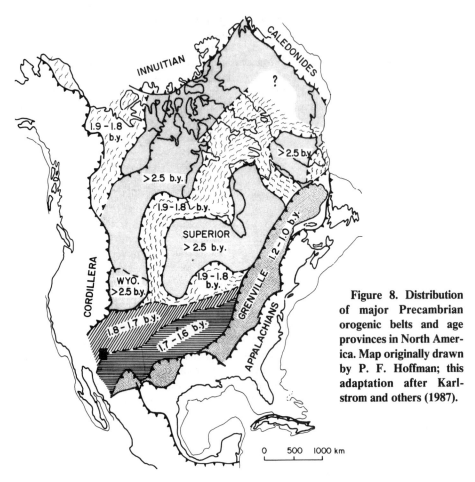

Figure 8. Distribution of major Precambrian orogenic belts and age provinces in North America. Map originally drawn by P. F. Hoffman; this adaptation after Karlstrom and others (1987).

orogenic belts border Archean cratons. Hoffman (1984, 1988) suggested that 1800–1900 Ma was the time of assembly of much of what is now North America. The orogenic belts, which consist of linear terranes of strongly deformed volcanic rocks, related immature sedimentary rocks, and plutons, were evidently formed during assembly of the Archean cratons as intervening oceanic crust was subducted. Clearly, reconstructions of this type, which are now being extended into the Sveccofennian shield, require precise and accurate age determinations that were not possible before the zircon method could be applied to a variety of igneous rocks.

The paper of DePaolo and Wasserburg (1976) was chosen because it marked the beginning of the application of the ^{147}Sm-^{143}Nd decay system to terrestrial rocks. Somewhat earlier, this system had been applied to meteorites and lunar basalt by Lugmair (1974) and Lugmair and others (1975). As indicated in Figure 9, the mantle Sm-Nd system may evolve along a path based upon the assumption that the bulk Earth has chondritic Sm-Nd ratios (CHUR = chondritic uniform reservoir). Many now feel, however, that this model is overly simplistic.

As pointed out by DePaolo (1981a), development of the Sm-Nd system marked the first time a radioactive parent-daughter system for which the geochemistry was relatively well understood was utilized both as a geochronometer and as a petrogenetic tracer. Both Sm and Nd are rare-earth elements and as a consequence, tend to share geochemical behavior. Fractionation of these elements thus should occur only when phases having significantly different distribution coefficients for Sm and Nd are present during melting, crystallization, or metamorphic events. Evidently, fractionation between Sm and Nd occurs during mantle melting events because garnet, with a strong tendency to incorporate heavy REE's, is present. The result of such melting events is that lighter Nd goes into the melt relative to heavier Sm. In contrast to the U-Pb and Rb-Sr systems, mantle melting events that lead to the formation of crustal rocks thus incorporate the *daughter* isotope in the melt in preference to the parent. The ^{143}Nd/^{144}Nd in the continental crust will therefore evolve more slowly (that is, along a trajectory of lesser slope; Fig. 10), whereas the residual "depleted mantle" will be enriched in Sm and evolve along a steeper trajectory (DM curve, Fig. 10). Subsequent melting, metamorphic, or weathering events in the continental crust, in which garnet is not involved, generally do not fractionate Sm from Nd. It is this conservative property that allows the calculation of "mantle separation ages," that is, the time when continental crustal material separated from the mantle. The contrast with the U-Pb and Rb-Sr systems is of great importance; in each of these commonly used radioactive parent-daughter systems, not only does mantle melting produce a liquid enriched in the *parent* element, but subsequent crustal melting or metamorphic events can also cause further fractionation.

The properties of the Sm-Nd isotopic system discussed above have led Patchett and Arndt (1986) to use Sm-Nd relationships to calculate the rate of growth of continental crust. These

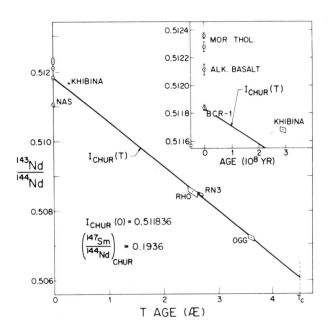

Figure 9. Observed initial ^{143}Nd/^{144}Nd versus time. $I_{CHUR}(T)$ represents evolution of ^{143}Nd/^{144}Nd in a reservoir with chondritic Sm/Nd (CHUR). After DePaolo and Wasserburg (1976).

authors have evaluated the Sm-Nd mantle separation ages of orogenic rocks formed during the Early Proterozoic (1.9–1.7 Ga) in North America, Greenland, and Europe in an attempt to determine how much material is juvenile, that is, derived directly from the mantle, and how much is recycled older continental crust. Patchett and Arndt found that terranes stabilized from 1.9–1.7 Ga make up 74% of pre–1.6 Ga crust but that only about 50% of this crust is juvenile, the other half being recycled Archean crust. These calculations lead to a crust production rate of about 1.2 km^3/yr, only slightly greater than the Phanerozoic island-arc accretion rate. It is clear that isotopic studies of this type are essential to evaluate Proterozoic and Archean tectonic styles and to compare Precambrian and Phanerozoic crustal growth mechanisms.

PART 2. PROTEROZOIC EVOLUTION OF CONTINENTAL CRUST IN SOUTHERN NORTH AMERICA

In this section, I will summarize current study of the Proterozoic evolution of the North American crust and show how the basic ideas developed above have been applied to understanding the history and mechanism of continental growth. Because I have already discussed the Early Proterozoic assembly of Archean cratons and the development of Wopmay, Trans-Hudson, and Penokean orogens as intervening oceanic crust was subducted, I shall concentrate here upon the accretionary and post-accretionary history of southern North America from about 1800 Ma until the Grenvillian event culminating about 1100 Ma. The discussion below is summarized in Figure 11.

The Mazatzal-Colorado Orogen

An extensive orogenic terrane is exposed across the southwestern United States. In Arizona, New Mexico, and Colorado, this terrane can be shown to consist of a northern belt, formed 1720–1780 Ma, and a southern belt,

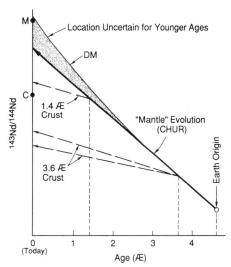

Figure 10. **Model for crust and mantle Nd isotopic evolution. Crust forms by extraction of chemically fractionated material from the mantle and evolves along lines of lower slope, reflecting low Sm/Nd ratios. The residual mantle, with increased Sm/Nd, must evolve away from CHUR in the opposite direction. Present values of ^{143}Nd/^{144}Nd in average crust (C) and in mid-ocean–ridge basalts (M) are shown. After DePaolo (1981a); DM = depleted mantle, added by author.**

formed 1630–1700 Ma, although recent work has shown that there is significant overlap between the two belts (Silver, 1968; Condie, 1986; Karlstrom and others, 1987). Both belts consist of volcanic rocks, commonly but not exclusively bimodal basalt-dacite-rhyolite assemblages, that are intercalated with thick sequences of volcanogenic, commonly turbiditic, metasedimentary rocks; in central Arizona and northern New Mexico, the younger belt includes thick sequences of fluvial to shelf metaquartzarenites (Mazatzal and Ortega Groups; for example, Trevena, 1979; Soegaard and Ericksson, 1985,

1986). These supracrustal rocks have been intruded by plutons, commonly tonalite to monzogranite, and by sills and stocks of gabbro. In central Colorado, Bickford and Boardman (1984) and Bickford and others (1988a) showed that the older terrane consists of a sequence of volcanic and sedimentary rocks formed about 1765 Ma that were intruded by plutons about 1755 Ma (Dubois Succession) and a second sequence (Cochetopa Succession) in which volcanic and sedimentary rocks were formed about 1735 Ma and were intruded about 1715 Ma. It seems likely that detailed zircon geochronology now in progress in Arizona and elsewhere (for example, Karlstrom and others, 1987) will show similar subdivisions.

Although rocks of these Early Proterozoic volcano-plutonic belts are strongly deformed and metamorphosed, primary features and chemical signatures are well enough preserved to make it clear that they are rocks of arc affinities, perhaps including back-arc assemblages as well as primary arc rocks (Bickford and Boardman, 1984; Condie and Shadel, 1984; Silver, 1984; Boardman, 1986; Boardman and Condie, 1986; Condie, 1986). Although structural aspects are still poorly known, Karlstrom and Bowring (1988) have made impressive progress in mapping and interpretation in Arizona and have already suggested that portions of the Arizona terranes represent tectonically assembled blocks.

The most easterly exposures of these Early Proterozoic orogenic rocks are in the Front Range of the Rocky Mountains. Studies of samples from the buried basement of the midcontinent region (Bickford and others, 1981c, 1986; Sims and Peterman, 1986; Van Schmus and others, 1987a), however, have shown that rocks of similar age (that is, 1600–1800 m.y.) occur as far east as Missouri and Wisconsin in the United States. Davidson (1986) has described 1700-m.y.-old rocks in the Georgian Bay region of Ontario, and Emslie and others (1983), Wardle and others (1986), Thomas and others (1986), and Gower and Ryan (1986) have described rocks of this age in Labrador. It thus is clear that igneous activity, and presumably the sedimentation and deformation that accompany it, occurred broadly across North America during the late Early Proterozoic. Unfortunately, very little is known about the petrography and chemistry of rocks in the buried basement because so few samples provide enough material to study in detail. In our recently published analysis (Bickford and others, 1986; Van Schmus and others, 1987a) and that of Sims and Peterman (1986), it has been assumed that the buried rocks of the northern midcontinent are broadly similar to those exposed in Colorado. Whereas both plutons, ranging from tonalite to granite, and

TABLE 1. COMPARISON OF OROGENIC CHRONOLOGIES (ROUNDED TO NEAREST 5 Ma)

Events	Trans-Hudson orogen*	Wopmay orogen†	Penokean orogen§	Svecofennian orogen**
Early volcanism	ca. 1910?	1895–1905	ca. 1910?	?
Early plutonism	?	1885–1890	?	1900–1940
Arc volcanism	1875–1890	1845–1875	1860–1880	1860–1900
Synvolcanic plutons	?	1845–1875	1860–1880	1860–1900
Postvolcanic plutons	1850–1865	?	1840–1860	1850–1860
Late plutons	1830–1840	?	1830–1840	ca. 1840

Note: table is from Van Schmus and others (1987b).
*Van Schmus and others (1987b); Baldwin and others (1985).
†Hoffman and Bowring (1984); Bowring (1985).
§Van Schmus (1980, 1984, unpub. data).
**Welin (1986).

metasedimentary rocks are common, however, metavolcanic rocks are rare.

Middle Proterozoic Granite-Rhyolite Terranes

Figure 11 shows that much of the basement of the midcontinent region is underlain by great terranes of rhyolite to dacite and epizonal plutons of similar composition. Within these vast terranes, which extend from the buried Grenville Front in eastern Ohio southwesterly across the continent at least to the Texas Panhandle, the highly silicic rocks are not deformed, and metamorphism is in greenschist facies at most; many samples appear to be unmetamorphosed. Sedimentary rocks, mafic igneous rocks, and igneous rocks of intermediate composition are exceedingly rare. A large number of U-Pb zircon age measurements (Bickford and others, 1981c; Van Schmus and Bickford, 1981; Hoppe and others, 1983; Thomas and others, 1984; Van Schmus and others, 1987a) indicate that rocks of the eastern granite-rhyolite province (Fig. 11) were formed 1480–1440 Ma, whereas those of the western granite-rhyolite province (Fig. 11) were formed 1400–1340 Ma.

The major exposure of the eastern province is in the St. Francois Mountains of southeastern Missouri, where about 3 km of rhyolitic to dacitic ash-flow tuff, rhyolitic flows, epizonal granitic plutons, and minor basaltic dikes and sills are associated with at least one major caldera (Anderson, 1970; Tolman and Robertson, 1969; Bickford and Mose, 1975; Sides and others, 1981; Bickford and others, 1981a; Cullers and others, 1981). Most of these rocks were formed at 1480 Ma, but two small plutons that intrude the older rocks are 1370 Ma. It thus is clear that magmas formed during the younger event (1400–1340 Ma) intrude at least some of the older (1480–1450 Ma) terrane. The only exposures of the western granite-rhyolite province are in Mayes County, Oklahoma, where the 1370 Ma granophyric Spavinaw granite occurs, and the eastern Arbuckle Mountains, where several major plutons and a gneissoid rock (Blue River gneiss) formed 1370 to 1400 Ma occur (Bickford and Lewis, 1979).

Plutons, formed 1480–1450 Ma and 1400–1340 Ma, occur to the north of the granite-rhyolite provinces as apparently isolated bodies intrusive into rocks of the 1600–1800 Ma belt. The major exposure of these isolated bodies is the 1480 Ma Wolf River batholith of central Wisconsin, which includes a rather wide range of compositions ranging from minor anorthosite to granite and syenite (Van Schmus and others, 1975; Anderson and Cullers, 1978; Anderson and others, 1980). In general, 1480–1440 Ma plutons occur to the north of both the eastern and western granite-rhyolite provinces, whereas 1400–1340 Ma plutons occur only north of the western granite-rhyolite province. To the west, numerous plutons with ages in the 1470- to 1420-m.y. range are known in exposures throughout Colorado and the southwestern United States, where they clearly intrude older rocks of the 1600–1800 Ma volcanogenic provinces. The only major occurrence of a pluton coeval with the younger western granite-rhyolite province is the 1360 Ma San Isabel batholith of the Wet Mountains of southern Colorado, which also intrudes rocks of the 1600–1800 Ma province (Thomas and others, 1984; Bickford and others, 1988b). These rocks form the well-known "Transcontinental Anorogenic Granite" suite (for example, Silver and others, 1977; Emslie, 1978; Anderson, 1983; Van Schmus and Bickford, 1981; Bickford and others, 1986; Van Schmus and others, 1987a).

The last major Proterozoic events affecting North America were the formation of the basaltic midcontinent rift system (for example, Van Schmus and Hinze, 1985) and the Grenville province of the eastern part of the continent. These events are of major importance but will

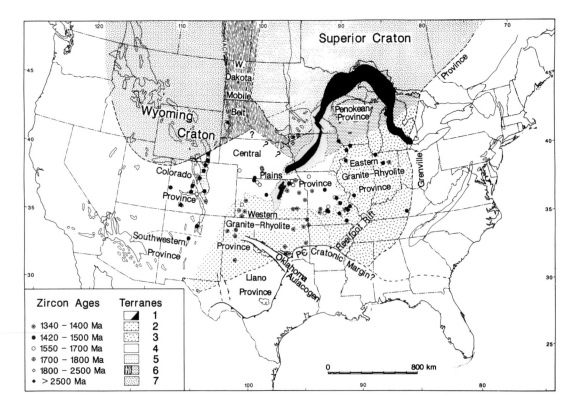

Figure 11. Generalized geologic map for Precambrian basement of central United States, showing major tectonic and petrologic provinces. Terrane 1 is midcontinent rift system; other terranes as named on map. After Bickford and others (1986).

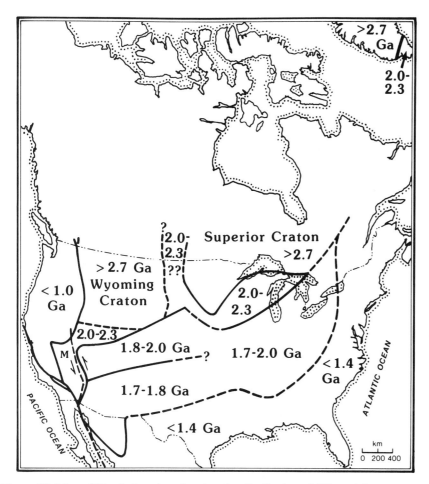

Figure 12. Map of North America, showing the distribution of Nd–model-age provinces. Based on data from Bennett and DePaolo (1987), Nelson and DePaolo (1985), Farmer and DePaolo (1983), Patchett and Arndt (1986), and Van Schmus and others (1987a). After Bennett and DePaolo (1987).

not be discussed herein. They may be thought of as terminating the continental evolution that occurred during the period 1800–1300 Ma. That period is the subject of the discussion to follow.

Evolution of Southern North American Crust from 1.8 to 1.3 Ga

The period between 1.8 and 1.3 Ga was a time of major crustal growth in southern North America. The amalgamation of Archean cratonal elements was complete, and the orogenic belts (for example, Wopmay, Trans-Hudson, Penokean, and numerous others; see Hoffman, 1988) that formed between them during ocean closure were stabilized. The southern portion of this continental mass is shown in Figure 12 (Wyoming craton, Superior craton, southern extension of Trans-Hudson orogen). The Mazatzal-Colorado orogen formed to the south of this continental mass.

Early Proterozoic Orogenic Terranes. Although the volcanic rocks of the Mazatzal-Colorado orogen are dominated by a bimodal basalt-dacite-rhyolite suite, the plutonic rocks are clearly calc-alkaline and the sedimentary rocks are mostly graywacke turbidites and other orogenic types. All students of this belt agree that these rocks were formed in island or continent-marginal arcs or in related back-arc settings (for example, Silver, 1984; Condie, 1982, 1986). This process was complex and protracted, beginning about 1.78 Ga and continuing until about 1.63 Ga. The details of terrane boundaries, tectonic settings, deformational histories, and petrogenetic processes are the subject of ongoing research.

That the 1.8–1.6 Ga terranes are essentially juvenile has been largely confirmed by isotopic and geochemical studies. Bickford and Boardman (1984), Boardman (1986), and Boardman and Condie (1986) have shown that the major- and trace-element geochemistry of mafic rocks in the Gunnison and Salida area of Colorado indicates juvenile derivation. The origin of abundant felsic volcanic rocks and plutons is less well constrained; these magmas may have resulted from melting of only slightly older juvenile crust or they may carry a significant component of recycled Archean crust. Condie (1986; see also Fig. 6 of this paper) has summarized his work and that of his students to show that the trace-element characteristics of volcanic rocks over most of the southwestern United States are compatible with formation in evolved oceanic arcs, continental-margin arcs, or associated back-arc basins.

Isotopic studies are also consistent with this interpretation. DePaolo (1981b) and Nelson and DePaolo (1984) have shown that the Sm-Nd and Sr isotopic characteristics of Proterozoic volcanic rocks in the southwestern United States are consistent with juvenile origin. Sm-Nd mantle separation ages (T_{DM}) are in the 1700- to 1900-m.y. range, essentially the same as crystallization ages determined from analysis of U-Pb in zircons, suggesting derivation directly from mantle sources (basalt) or perhaps from only slightly older continental crust (felsic volcanic rocks). Bennett and DePaolo (1987) have mapped Nd model ages (T_{DM}) in the southwestern United States, obtaining data either from dated Proterozoic rocks or from peraluminous Mesozoic and Tertiary plutons which they believe to have been derived from crustal sources. Their results confirm that an essentially juvenile crust of Early Proterozoic age underlies most of this region, but they found that Nd model ages in parts of eastern California, Nevada, and northern Utah are in the range 2.0–2.3 b.y. An earlier study by Nelson and DePaolo (1985) showed that Nd model ages of the granite-rhyolite provinces in the midcontinent region are in the range 1.7–2.0 b.y., despite the much younger crystallization ages of the rocks. Further, Nelson and DePaolo found that rocks in the 1.8–1.9 Ga Penokean orogen show Nd model ages of 2.0–2.3 b.y. Bennett and DePaolo (1987) summarized these results in a map presented here in Figure 12. The map clearly shows broad belts of Early Proterozoic crust with Nd model ages of 1.7–2.0 b.y., and within which there is no evidence for derivation from older Archean crust, trending northeasterly across the continent. The delineation of Archean and Proterozoic basement in the western United States on the basis of Pb isotopic data was also shown by Zartman (1974), and Stacey and Hedlund (1983) showed that Archean basement was absent in southwestern New Mexico through their study of Pb isotopic compositions of diverse rocks and mineral deposits.

Adjacent to the Archean cratons, Early Proterozoic crust has Nd model ages of 2.0–2.3 b.y., according to Bennett and DePaolo (1987) a consequence of mixing of Archean material with juvenile 1.7- to 1.9-b.y. material. It is interesting to note that the Nd model ages of the components of the Mazatzal-Colorado orogen shown in Figure 12 are consistently greater than the U-Pb zircon crystallization ages shown in Figure 6. This could have resulted from admixture of small amounts of Archean sediment from the cratons as the arc terranes were accreted to the south. An alternate possibility is that much of this region is underlain, at deep levels, by crust of Trans-Hudson–Penokean age, that is, about 1.85 b.y., and that this material has been the source of much of the volcanogenic rocks which crystallized in the 1.7–1.8 Ga period. The isotope systems available for study (Sm-Nd, Rb-Sr, U-Pb) are not sensitive enough to resolve this question; confirmation of the alternative model will require that 1.85–1.90 Ga rocks be identified at the surface.

Middle Proterozoic Granite-Rhyolite Provinces. A major subsequent event in the evolution of the southern North American continental crust has been the formation of the granite-rhyolite provinces and the coeval anorogenic plutons between 1.34 and 1.48 Ga. The origin of these rocks has been a puzzle since they were first described by Muehlberger and others (1966). The important facts are as follows.

(1) Igneous activity occurred during an earlier period, 1.48–1.45 Ga, and a later period, 1.34–1.40 Ga. Plutons of the younger episode intrude volcanic rocks of the older episode in the St. Francois Mountains. Nd model ages of many of these rocks (Nelson and DePaolo, 1985) are in the 1.7–1.9 range, indicating derivation of the younger silicic magmas from older crust of this age. Rocks with crystallization ages in the 1.7- to 1.9-b.y. range, however, have not been discovered anywhere within the granite-rhyolite provinces. Recently, however, Bowring and others (1988) have reported Nd model ages essentially equal to crystallization age (about 1.45 b.y.) for basement samples from eastern Missouri.

(2) The rocks are almost exclusively high-silica rhyolite, dacite, and high-level plutons of similar composition. Basalt and gabbro occur as sills and dikes in the St. Francois Mountains but make up a small fraction of the exposed rocks; a single andesite flow occurs. Mafic and intermediate igneous rocks are also rare in subsurface samples. Sedimentary rocks are rare throughout these terranes; where they occur, they are commonly volcaniclastic, tuffaceous rocks.

(3) Deformation is restricted mostly to block faulting, and metamorphism is absent or in lowest greenschist facies.

(4) The occurrence of coeval, epizonal plutons to the north of the granite-rhyolite terranes indicates that the igneous activity was broadly distributed across the older continental crust and suggests that supracrustal rocks may have extended far north of their present distribution.

(5) I am not aware that 1.34–1.48 Ga plutons have been identified as intrusions within Archean cratonic crust anywhere except in the Nain province of Labrador. There, isotopic data (Ashwal and others, 1986; Wooden and others, 1987) indicate that anorthosite, gabbro, and diabasic dikes that are probably 1450 m.y. (Krogh and Davis, 1973; Emslie, 1985) are contaminated with Early Archean crustal material. The absence of the 1.34–1.48 suite in most of the Archean of North America seems more significant than its occurrence in a limited area of Labrador. It is possible, however, that detailed isotopic study in the northern midcontinent region will reveal the presence of older Archean crust.

(6) Granite-rhyolite terranes in the midcontinent region are also known at 1.76 Ga (central Wisconsin; Van Schmus, 1978) and about 1.1 Ga (Franklin Mountains, west Texas; Wasserburg and others, 1962; Copeland and Bowring, 1988).

One of the major difficulties in formulating a model for the origin of the granite-rhyolite provinces has been the lack of a recent analogue. Where, in the Mesozoic to Recent Earth, do similar large terranes of high-silica rocks occur, and what is their tectonic setting? After years of reflection and conversation with numerous interested scientists, I must say that there do not appear to be any completely satisfying analogues. Below is a summary of those that come close.

(1) The Tertiary rocks of the Basin and Range province include great volumes of high-silica ash-flow tuff and epizonal granite that formed in an extensional environment beginning about 20 m.y. ago (for example, Eaton, 1982). These rocks, however, occur with a significant volume of basaltic rocks, forming the well-known bimodal suite that is characteristic of extended regions; as noted above, mafic rocks are rare in the midcontinent terranes. Further, remnants of older crust are abundantly exposed in tilted fault blocks, a feature evidently absent in the Middle Proterozoic of the midcontinent. Finally, the present Basin and Range province contains large volumes of alluvial sediment, and, as cooling and subsidence continue, this region should evolve into a major, sediment-filled basin. Neither sedimentary nor metasedimentary rocks are common in the midcontinent 1.34–1.48 Ma province.

(2) A large region in western Mexico in the Sierra Madre Occidental is underlain by a calc-alkaline magmatic suite, informally called the "upper volcanic supergroup" by McDowell and Clabaugh (1979). These rocks are of Tertiary age and are dominated by high-silica rhyolite and dacite that were erupted from numerous centers now marked by the presence of large calderas (Cameron and others, 1980; Swanson and McDowell, 1984); rocks of this terrane are alkaline to the east in trans-Pecos Texas (for example, Barker, 1977). Deep canyons in the Sierra Madre Occidental expose an older volcano-plutonic complex beneath the upper volcanic sequence. These rocks, which have yielded ages as old as Cretaceous in the western Sierra Madre Occidental but may be as young as Oligocene elsewhere, are predominantly intermediate in composition. An eastward-dipping subduction zone was active off the west coast of Mexico during the mid-Tertiary (Atwater and Molnar, 1973), and there is little question that the calc-alkaline rocks of at least the upper volcanic supergroup, including the voluminous high-silica rhyolites, were formed as a result. The origin of the more alkaline rocks to the east is still the subject of investigation, but it has been suggested that these were formed in a back-arc extensional setting (Barker, 1979; Parker and McDowell, 1979; Mauger, 1981).

The compositional and petrographic similarity of the high-silica rocks with those of the midcontinent Proterozoic granite-rhyolite terranes is striking and suggests that the midcontinent rocks may also have been formed in association with subduction, in this case along the southern and eastern margins of the continent. There is, however, an almost complete absence of mafic or intermediate members of the calc-alkaline series in the midcontinent Proterozoic rocks. The only occurrences of andesite are a single unit in the St. Francois Mountains and a few subsurface samples from a limited region in northeastern Oklahoma and southeastern Kansas (Denison, 1982). Unless there exists a significant volume of calc-alkaline rocks buried beneath the granite-rhyolite provinces, perhaps similar to the older volcanic complex in the Sierra Madre Occidental, the analogy breaks down.

(3) Ignimbritic sheets representing large volumes of rhyolitic magma occur in the central zone of the Andes (northern Chile, Bolivia), where they occur in association with andesitic and basaltic lavas (Thorpe and others, 1982). In contrast to the northern and southern zones of the Andes, the central zone is underlain by thick continental crust, including Precambrian crystalline rocks. The analogy of this region with the midcontinent granite-rhyolite terranes breaks down, as in 2 above, because of the lack of calc-alkaline rocks, particularly abundant andesite.

(4) Kisvarsanyi (1981) has called attention to the striking similarities between the volcanic and plutonic rocks of the St. Francois terrane of southeastern Missouri, which includes the only major exposure of the 1450–1480 Ma eastern granite-rhyolite province, and the Jurassic Nigeria-Niger province of western Africa (Jacobson and others, 1958; Turner, 1963; Bowden and van Breemen, 1972; Bowden, 1974; Bowden and Turner, 1974). The Nigerian occurrences evidently expose a somewhat deeper section of the crust than the St. Francois terrane, for volcanic rocks are not as abundant there. The rocks in both regions, however, are chemically and mineralogically quite similar; high-silica, alkaline to sub-alkaline, epizonal plutons and rhyolitic to trachytic volcanic rocks that may contain fayalite and Fe-rich pyroxenes are common, whereas sedimentary rocks, calc-alkaline andesite, and mafic igneous rocks are rare. The major ring complexes described in the Nigerian occurrences (Turner, 1963) are similar to structures interpreted in the subsurface by Kisvarsanyi (1981) that are perhaps related to the eroded calderas described in the surface exposures by Sides and others (1981).

The Nigerian complexes are not nearly so extensive and typically are more alkaline, compared to the Middle Proterozoic granite-rhyolite provinces. Nevertheless, the analogy seems a good one. The Nigerian complexes clearly formed in response to rifting as the Atlantic Ocean began to open.

My prejudice at present is that the midcontinent granite-rhyolite terranes, and the anorogenic plutons that occur to the north of them, have formed anatectically in response to broadly distributed crustal extension following (by about 100 m.y.) the accretion and stabilization of the 1.6–1.8 Ga orogenic terranes. The granite-rhyolite provinces are most likely veneers, probably only 3–10 km thick, that lie upon Early Proterozoic orogenic crust from which they were derived. This mechanism, rather than subduction-related magmatism, is favored because of the following arguments.

(1) It seems unlikely that as in the Sierra Madre Occidental, a suite of calc-alkaline, subduction-related rocks lies buried beneath the granite-rhyolite but is neither exposed nor encountered by the drill despite the occurrence of rocks of the 1.48–1.34 Ga suite in uplifted areas like the St. Francois Mountains (Ozark Dome) and the Wet Mountains of the southern Rocky Mountains. Further, if the isolated, coeval anorogenic plutons represent somewhat deeper crustal levels, their volcanic cover having been removed by erosion, one might then expect to find the calc-alkaline volcanic rocks exposed along the northern margin of the granite-rhyolite terranes. This is not the case. The absence of calc-alkaline rocks thus appears to be real.

(2) A related argument is that it is difficult to envision derivation of highly silicic rocks except by anatexis of pre-existing continental crust. The Nd model ages of Nelson and DePaolo (1985), indicating separation of this material from the mantle at 1.7–1.9 Ga, are consistent with remelting of the previously accreted orogenic crust. The recent determination of approximately 1.45-b.y. Nd model ages for rocks in eastern Missouri by Bowring and others (1988), however, indicates that at least some of the granite-rhyolite province rocks were separated from the mantle—or a region of the crust with mantle-like isotopic characteristics—at essentially the same time as their magmatic crystallization.

(3) Although basaltic rocks are rare in the granite-rhyolite provinces, they do occur. In the St. Francois Mountains, the principal exposure of these rocks, basalt occurs as late diabase dike swarms and gabbroic sills. This occurrence is consistent with an extensional setting. There may be much greater volumes of basalt, the mafic component of the extension-related bimodal suite, at depth.

(4) Whereas orogeny was clearly occurring in the southwestern United States at 1.76 Ga, a suite of high-silica granite and rhyolite formed at this time in Wisconsin to the south of the 1.85 Ga Penokean orogenic suite. These rocks are similar to the middle Proterozoic rocks in the southern midcontinent compositionally and petrographically and in the absence of penetrative deformation and metamorphism. Whatever the mechanism for forming these felsic suites, they seem to follow orogeny by about 100 m.y. in each case. It seems more likely that this mechanism involved extension following accretion rather than the initiation of a new subduction system.

The conclusion reached herein that the Middle Proterozoic granite-rhyolite provinces were formed in response to broadly distributed crustal extension following accretion of juvenile, orogenic rocks, clearly needs further testing. The obvious tests will be further detailed studies of the trace-element and isotopic chemistry of these rocks. An ultimate test will be to drill through the presumed veneer of granite and rhyolite and obtain samples of underlying crust for age determination and isotopic study.

ACKNOWLEDGMENTS

The research of the author and his students has been supported by the National Science Foundation since 1965, the most recent grants being EAR-8025257, EAR-8218463, EAR-8219137, EAR-8313167, EAR-8308311, EAR-8516550, and DOSECC contracts DSC 86-2 and DSC 87-1. Other support has come from the Nuclear Regulatory Agency through contracts to the Kansas Geological Survey (KGS), from the KGS, and from the General Research Fund of the University of Kansas. I have benefited from conversations with many colleagues over the past 25 years, but I am especially indebted to W. R. Van Schmus, with whom I work at the University of Kansas; P. F. Hoffman; L. T. Silver; S. A. Bowring; J. F. Lewry; R. Macdonald; A. Davidson; R. S. Hildebrand; H. W. Day; and K. C. Condie. Thorough reviews by P. F. Hoffman and R. E. Zartman were particularly helpful and are gratefully acknowledged. My graduate students, especially D. G. Mose, J. R. Sides, R. D. Shuster, and J. R. Chiarenzelli, have done most of the work and have greatly stimulated my thinking!

REFERENCES CITED

Anderson, D. L., 1984, The Earth as a planet: Paradigms and paradoxes: Science, v. 223, p. 347–355.
Anderson, J. L., 1983, Proterozoic anorogenic granite plutonism of North America, in Medaris, L. G., Jr., Mickelson, D. M., Byers, C. W., and Shanks, W. G., eds., Proterozoic geology: Selected papers from an international Proterozoic symposium: Geological Society of America Memoir 161, p. 133–154.
Anderson, J. L., and Cullers, R. L., 1978, Geochemistry and evolution of the Wolf River Batholith, a late Precambrian rapakivi massif in north Wisconsin, U.S.A.: Precambrian Research, v. 7, p. 287–324.
Anderson, J. L., Cullers, R. L., and Van Schmus, W. R., 1980, Anorogenic metaluminous and peraluminous granite plutonism in the mid-Proterozoic of Wisconsin, USA: v. 74, p. 311–328.
Anderson, R. E., 1970, Ash-flow tuffs of Precambrian age in southeast Missouri (Contributions to Precambrian Geology 2): Missouri Division Geological Survey and Water Resources Report of Investigations 46, 50 p.
Argand, E., 1924, La tectonique de l'Asie: International Geological Congress, 13th, Brussels, p. 171–372.
Ashwal, L. D., Wooden, J. L., and Emslie, R. F., 1986, Sr, Nd, and Pb isotopes in Proterozoic intrusives astride the Grenville Front in Labrador: Implications for crustal contamination and basement mapping: Geochimica et Cosmochimica Acta, v. 50, p. 2571–2585.

Atwater, Tanya, 1970, Implications of plate tectonics for the Cenozoic tectonic evolution of western North America: Geological Society of America Bulletin, v. 81, p. 3513–3536.

Atwater, T., and Molnar, P., 1973, Relative motion of the Pacific and North American plates deduced from sea-floor spreading in the Atlantic, Indian, and South Pacific Oceans, *in* Kovach, R. L., and Nur, A., eds., Proceedings of the conference on tectonic problems of the San Andreas fault system: Stanford University Publications, Geological Sciences, v. 13, p. 136–148.

Baadsgaards, H., Collerson, K. D., and Bridgwater, D., 1979, The Archean gneiss complex of N. Labrador, I. Preliminary U-Th-Pb geochronology: Canadian Journal of Earth Sciences, v. 16, p. 951–961.

Baldwin, D. A., Syme, E. C., Zwanzig, H. V., Gordon, T. M., Hunt, P. A., and Stevens, R. D., 1985, U-Pb zircon ages from the Lynn Lake and Rusty Lake metavolcanic belts, Manitoba: Two ages of Proterozoic magmatism: Geological Association of Canada Program and Abstracts, v. 10, p. A3.

Barker, D. S., 1977, Northern Trans-Pecos magmatic province: Introduction and comparison with the Kenya Rift: Geological Society of America Bulletin, v. 88, p. 1421–1427.

——— 1979, Cenozoic magmatism in the Trans-Pecos province: Relation to the Rio Grande Rift, *in* Rieker, R. E., ed., Rio Grande Rift: Tectonics and magmatism: Washington, D.C., American Geophysical Union, p. 382–392.

Ben-Avraham, Z., 1981, The movement of continents: American Scientist, v. 69, p. 291–299.

Ben-Avraham, Z., and Nur, A., 1982, An introductory overview to the concept of displaced terranes: Canadian Journal of Earth Sciences, v. 20, p. 994–999.

Bennett, V. C., and DePaolo, D. J., 1987, Proterozoic crustal history of the western United States as determined by neodymium isotopic mapping: Geological Society of America Bulletin, v. 99, p. 674–685.

Bickford, M. E., and Boardman, S. J., 1984, A Proterozoic volcano-plutonic terrane, Gunnison and Salida areas, Colorado: Journal of Geology, v. 92, p. 657–666.

Bickford, M. E., and Lewis, R. D., 1979, U-Pb geochronology of exposed basement rocks in Oklahoma: Geological Society of America Bulletin, v. 90, p. 540–544.

Bickford, M. E., and Mose, D. G., 1975, Geochronology of Precambrian rocks in the St. Francois Mountains, southeastern Missouri: Geological Society of America Special Paper 165, 48 p.

Bickford, M. E., Sides, J. R., and Cullers, R. L., 1981a, Chemical evolution of magmas in the Proterozoic terrane of the St. Francois Mountains, southeastern Missouri 1. Field, petrographic, and major element data: Journal of Geophysical Research, v. 86, p. 10365–10386.

Bickford, M. E., Chase, R. B., Nelson, B. K., Shuster, R. D., and Arruda, E. C., 1981b, U-Pb studies of zircon cores and overgrowths, and monazite: Implications for age and petrogenesis of the northwestern Idaho batholith: Journal of Geology, v. 89, p. 433–457.

Bickford, M. E., Harrower, K. L., Hoppe, W. J., Nelson, B. K., Nusbaum, R. L., and Thomas, J. J., 1981c, Rb-Sr and U-Pb geochronology and distribution of rock types in the Precambrian basement of Missouri and Kansas: Geological Society of America Bulletin, Part I, v. 92, p. 323–341.

Bickford, M. E., Van Schmus, W. R., and Zietz, I., 1986, Proterozoic history of the midcontinent region of North America: Geology, v. 14, p. 492–496.

Bickford, M. E., Shuster, R. D., and Boardman, S. J., 1988a, U-Pb geochronology of the Proterozoic volcano-plutonic terrane in the Gunnison and Salida areas, Colorado: Geological Society of America Special Paper (Tweto Volume) (in press).

Bickford, M. E., Cullers, R. L., Shuster, R. D., Premo, W. R., and Van Schmus, W. R., 1988b, U-Pb zircon geochronology of Proterozoic and Cambrian plutons in the Wet Mountains and southern Front Range, Colorado: Geological Society of America Special Paper (Tweto Volume) (in press).

Boardman, S. J., 1986, Early Proterozoic bimodal volcanic rocks in central Colorado, U.S.A., Part I: Petrography, stratigraphy, and depositional history: Precambrian Research, v. 34, p. 1–36.

Boardman, S. J., and Condie, K. C., 1986, Early Proterozoic bimodal volcanic rocks in central Colorado, U.S.A., Part II: Geochemistry, petrogenesis, and tectonic setting: Precambrian Geology, v. 34, p. 37–68.

Bowden, P., 1974, Oversaturated alkaline rocks: Granites, pantellerites, and comendites, *in* Sorensen, H., ed., The alkaline rocks: New York, John Wiley and Sons, p. 109–123.

Bowden, P., and Turner, D. C., 1974, Peralkaline and associated ring complexes in the Nigeria-Niger province, west Africa, *in* Sorensen, H., ed., The alkaline rocks: New York, John Wiley and Sons, p. 330–351.

Bowden, P., and van Breemen, O., 1972, Isotopic and chemical studies on younger granites from northern Nigeria, *in* Dessauvagie, T.F.J., and Whiteman, A. J., eds., African geology: Ibadan, Nigeria, University Press, p. 105–120.

Bowring, S. A., 1985, U-Pb zircon geochronology of Early Proterozoic Wopmay orogen, N.W.T. Canada: An example of rapid crustal evolution [Ph.D. dissert.]: Lawrence, Kansas, University of Kansas, 400 p.

Bowring, S. A., Arvidson, R. E., and Podosek, F. A., 1988, The Missouri gravity low: Evidence for a cryptic suture?: Geological Society of America Abstracts with Programs, v. 20 (in press).

Bridgwater, D., Collerson, K. D., and Myers, J. S., 1978, The development of the Archean gneiss complex of the North Atlantic region, *in* Tarling, D. H., ed., Evolution of Earth's crust: London, United Kingdom, Academic Press, p. 19–69.

Burke, K., Dewey, J. F., and Kidd, W.S.F., 1976, Dominance of horizontal movements, arc and microcontinental collisions during the later permobile regime, *in* Windley, B. F., ed., The early history of the Earth: New York, John Wiley and Sons, p. 113–129.

Cambray, F. W., 1978, Plate tectonics as a model for the environment of deposition and deformation of the Early Proterozoic of northern Michigan: Geological Society of America Abstracts with Programs, v. 10, p. 376.

Cameron, K. L., Cameron, M., Bagby, W. C., Moll, E. J., and Drake, R. E., 1980, Petrologic characteristics of mid-Tertiary volcanic suites, Chihuahua, Mexico: Geology, v. 8, p. 87–91.

Churkin, Michael, Jr., and Eberlein, G. D., 1977, Ancient borderland terranes of the North American Cordillera: Correlation and microplate tectonics: Geological Society of America Bulletin, v. 88, p. 769–786.

Collerson, K. D., and Bridgwater, D., 1979, Metamorphic development of Early Archean tonalitic and trondjhemitic gneisses: Saglek area, Labrador, *in* Barker, F., ed., Trondjhemites, dacites and related rocks: Amsterdam, the Netherlands, Elsevier, p. 205–273.

Compston, W., Williams, I. S., and Clement, S. W., 1982, U-Pb ages within single zircons using a sensitive high mass-resolution ion microprobe: American Society Mass Spectrometry Conference, 30th, Honolulu, Hawaii, Proceedings, p. 593–595.

Compston, W., Williams, I. S., and Meyer, C., 1984, U-Pb geochronology of zircons from lunar breccia 73217 using a sensitive high mass-resolution ion microprobe: Lunar Planetary Science Conference, 14th, Proceedings, Journal of Geophysical Research, v. 89, Supplement, p. B525–B534.

Condie, K. C., 1982, Plate tectonics model for Proterozoic continental accretion in the southwestern United States: Geology, v. 10, p. 37–42.

——— 1986, Geochemistry and tectonic setting of Early Proterozoic supracrustal rocks in the southwestern United States: Journal of Geology, v. 94, p. 845–864.

Condie, K. C., and Shadel, C., 1984, An early Proterozoic volcanic arc succession in southeastern Wyoming: Canadian Journal of Earth Sciences, v. 24, p. 415–427.

Coney, P. J., Jones, D. L., and Monger, J.W.H., 1980, Cordilleran suspect terranes: Nature, v. 188, p. 329–333.

Copeland, Peter, and Bowring, S. A., 1988, U-Pb zircon and $^{40}Ar/^{39}Ar$ ages for Proterozoic rocks, west Texas: Geological Society of America Abstracts with Programs, v. 20 (in press).

Cullers, R. L., Koch, R. J., and Bickford, M. E., 1981, Chemical evolution of magmas in the Proterozoic terrane of the St. Francois Mountains, southeastern Missouri, 2. Trace element data: Journal of Geophysical Research, v. 86, p. 10388–10401.

Davidson, A., 1986, Grenville front relationships near Killarney, Ontario, *in* Moore, J. M., Davidson, A., and Baer, A. J., eds., The Grenville province: Geological Association of Canada Special Paper 31, p. 107–117.

Day, H. W., Moores, E. M., and Tuminas, A., 1984, Structure and tectonics of the northern Sierra Nevada: Geological Society of America Bulletin, v. 96, p. 436–450.

Denison, R. E., 1982, Basement rocks in northeast Oklahoma: Oklahoma Geological Survey Circular 84, 84 p.

DePaolo, D. J., 1981a, Nd isotopic studies: Some new perspectives on Earth structure and evolution: EOS (American Geophysical Union Transactions), v. 62, p. 137–140.

——— 1981b, Neodymium isotopes in the Colorado Front Range and crustal evolution in the Proterozoic: Nature, v. 291, p. 193–196.

DePaolo, D. J., and Wasserburg, G. J., 1976, Inferences about magma sources and mantle structure from variations of $^{143}Nd/^{144}Nd$: Geophysical Research Letters, v. 3, p. 743–746.

Dietz, R. S., 1961, Continent and ocean basin evolution by spreading of the seafloor: Nature, v. 190, p. 854–857.

——— 1963, Collapsing continental rises: An actualistic concept of geosynclines and mountain building: Journal of Geology, v. 71, p. 314–333.

du Toit, A. L., 1937, Our wandering continents: Edinburgh, United Kingdom, Oliver and Boyd.

Eaton, G. P., 1979, Regional geophysics, Cenozoic tectonics, and geologic resources of the Basin and Range province and adjoining regions, *in* Newman, G. W., and Goode, H. D., eds., Basin and Range symposium: Denver, Colorado, Rocky Mountain Association of Geologists, p. 11–40.

——— 1982, The Basin and Range province: Origin and tectonic significance, *in* Wetherill, G. W., Albee, A. L., and Stehli, F. G., eds.: Palo Alto, California, Annual Reviews, Inc., Annual Reviews of Earth and Planetary Science, p. 409–440.

Eaton, G. P., Wahl, R. R., Prostka, H. J., Mabey, D. R., and Kleinkopf, M. D., 1979, Regional gravity and tectonic patterns—their relation to late Cenozoic epeirogeny and lateral spreading in the western Cordillera, *in* Smith, R. B., and Eaton, G. P., eds., Cenozoic tectonics and regional geophysics of the western Cordillera: Geological Society of America Memoir 152, p. 51–92.

Emslie, R. F., 1978, Anorthosite massifs, rapakivi granites, and late Proterozoic rifting of North America: Precambrian Research, v. 7, p. 61–98.

——— 1985, Proterozoic anorthosite massifs, *in* Touret, J., ed., The deep Proterozoic crust in the North Atlantic provinces: Proceedings, NATO Advisory Study Institute, Norway, July 1984: Amsterdam, the Netherlands, Reidel, p. 39–60.

Emslie, R. F., Loveridge, W. D., Stevens, R. D., and Sullivan, R. W., 1983, Igneous and tectonothermal evolution, Mealy Mountains, Labrador: Geological Association of Canada-Mineralogical Association of Canada Program with Abstracts, v. 8, p. A20.

England, P. C., and Thompson, A. B., 1984, Pressure-temperature-time paths of regional metamorphism I. Heat transfer during the evolution of regions of thickened continental crust: Journal of Petrology, v. 25, p. 894–928.

Farmer, G. L., and DePaolo, D. J., 1983, Origin of Mesozoic and Tertiary granites in the western U.S. and implications for pre-Mesozoic crustal structure, 1. Nd and Sr isotopic studies in the geocline of the northern Great Basin: Journal of Geophysical Research, v. 88, p. 3379–3401.

Froude, D. O., Ireland, T. R., Kinny, P. D., Williams, I. S., and Compston, W., 1983, Ion microprobe identification of 4,100–4,200 Myr-old terrestrial zircons: Nature, v. 304, p. 616–618.

Gibb, R. A., 1978, Slave-Churchill collision tectonics: Nature, v. 271, p. 50–52.

Gill, J. B., 1981, Orogenic andesites and plate tectonics: Berlin, Germany, Springer-Verlag, 390 p.

Glazner, A. F., and Bartley, J. M., 1984, Timing and tectonic setting of Tertiary low-angle normal faulting and associated magmatism in the southwestern United States: Tectonics, v. 3, p. 385–396.

Goldich, S. S., and Wooden, J. L., 1980, Origin of the Morton gneiss, southwestern Minnesota: Part 3. Geochronology: Geological Society of America Special Paper 182, p. 77–94.

Gower, C. F., and Ryan, A. B., 1986, Proterozoic evolution of the Grenville province and adjacent Makkovik province in eastern-central Labrador, *in* Moore, J. M. Davidson, A., and Baer, A. J. eds., The Grenville province: Geological Association of Canada Special Paper 31, p. 281–296.

Grauert, B., and Hoffman, A., 1973, Old radiogenic lead components in zircons from the Idaho batholith and its metasedimentary aureole: Carnegie Institution of Washington Year Book 72, p. 297–299.

Green, T. H., and Ringwood, A. E., 1968, Genesis of the calc-alkaline igneous rock suite: Contributions to Mineralogy and Petrology, v. 18, p. 105–162.

Hamilton, P. J., O'Nions, R. K., Evensen, N. M., Bridgwater, D., and Allaart, J. H., 1978, Sm-Nd isotopic investigations of Isua supracrustals and implications for mantle evolution: Nature, v. 272, p. 41–43.

Hamilton, Warren, 1969, Mesozoic California and the underflow of Pacific mantle: Geological Society of America Bulletin, v. 80, p. 2409–2430.

Hargraves, R. B., 1981, Precambrian tectonic style: A liberal uniformitarian interpretation, *in* Kroner, A., ed., Precambrian plate tectonics: Amsterdam, the Netherlands, Elsevier Scientific Publishing Co., p. 21–56.

Hatcher, R. D., Jr., 1978, Tectonics of the western Piedmont and Blue Ridge, southern Appalachians: Review and speculations: American Journal of Science, v. 278, p. 276–304.

Helwig, James, 1974, Eugeosynclinal basement and a collage concept of orogenic belts, *in* Dott, R. H., Jr., and Shaver, R. H., eds., Modern and ancient geosynclinal sedimentation: Society of Economic Paleontologists and Mineralogists Special Publication 19, p. 359–376.

Hoffman, P. F., 1980, Wopmay orogen: A Wilson cycle of Early Proterozoic age in the northwest of the Canadian Shield, *in* Strangway, D. E., ed., The continental crust and its mineral deposits: Geological Association of Canada Special Paper 20, p. 523–549.

——— 1981, Autopsy of Athapuscow Aulacogen: A failed arm affected by three collisions, *in* Campbell, F.H.A., ed., Proterozoic basins of Canada: Geological Survey of Canada Paper 81-10, p. 97–102.

——— 1984, Assembly and growth of the North American craton in Proterozoic times: International Geological Congress, 27th, Moscow, Abstracts, 9, Part 1, p. 156–157.

——— 1987, Continental transform tectonics: Great Slave Lake shear zone (ca. 1.9 Ga), northwest Canada: Geology, v. 15, p. 785–788.

——— 1988, United plates of America, the birth of a craton: Early Proterozoic assembly and growth of Laurentia: Annual Reviews of Earth and Planetary Sciences, v. 16, p. 543–603.

Hoffman, P. F., and Bowring, S. A., 1984, A short-lived 1.9 Ga continental margin and its destruction, Wopmay orogen, northwest Canada: Geology, v. 12, p. 68–72.

Hoffman, P. F., Bowring, S. A., McGrath, P. H., Thomas, M. D., and Van Schmus, W. R., 1982, Plate tectonic model for Wopmay orogen consistent with zircon chronology, gravity and magnetic anomalies east of Mackenzie Mountains: Canadian Geophysical Union, Annual Meeting, 9th, Abstracts, p. 13.

Holmes, Arthur, 1944, Principles of physical geology: London, United Kingdom, Thomas Nelson and Sons, p. 505–509.

Hoppe, W. J., Montgomery, C. W., and Van Schmus, W. R., 1983, Age and significance of Precambrian basement samples from northern Illinois and adjacent states: Journal of Geophysical Research, v. 88, p. 7276–7286.

Jacobson, R.R.E., MacLeod, W. N., and Black, R., 1958, Ring-complexes in the younger granite province of northern Nigeria: Geological Society of London Memoir 1, 72 p.

Jones, D. L., Silberling, N. J., and Hillhouse, J., 1977, Wrangellia—A displaced terrane in northwestern North America: Canadian Journal of Earth Sciences, v. 14, p. 2565–2577.

Karlstrom, K. E., and Bowring, S. A., 1988, Early Proterozoic assembly of tectonostratigraphic terranes in southwestern North America: Journal of Geology (in press).

Karlstrom, K. E., Bowring, S. A., and Conway, C. M., 1987, Tectonic significance of an Early Proterozoic two-province boundary in central Arizona: Geological Society of America Bulletin, v. 99, p. 529–538.

Kisvarsanyi, E. B., 1981, Geology of the St. Francois terrane, southeastern Missouri: Missouri Department of Natural Resources, Division of Geology and Land Survey, Report of Investigations 64, Contribution to Precambrian Geology 8, 58 p.

Kober, B., 1986, Whole-grain evaporation for $^{207}Pb/^{206}Pb$ age investigations on single zircons using a double-filament thermal ion source: Contributions to Mineralogy and Petrology, v. 93, p. 482–490.

Krogh, T. E., 1973, A low-contamination method for hydrothermal decomposition of zircon and extraction of U and Pb for isotopic age determination: Geochimica et Cosmochimica Acta, v. 37, p. 485–494.

——— 1982, Improved accuracy of U-Pb zircon ages by the creation of more concordant fractions using an air-abrasion technique: Geochimica et Cosmochimica Acta, v. 46, p. 637–649.

Krogh, T. E., and Davis, G. L., 1973, The significance of inherited zircons on the age and origin of igneous rocks—An investigation of the Labrador adamellites: Carnegie Institution of Washington Yearbook 72, p. 610–613.

Kroner, A., 1981, Precambrian plate tectonics, *in* Kroner, A., ed., Precambrian plate tectonics: Amsterdam, the Netherlands, Elsevier Scientific Publishing Co., p. 57–90.

Le Pichon, X., 1968, Seafloor spreading and continental drift: Journal of Geo-

physical Research, v. 73, p. 3611–3697.
Lewry, J. F., 1981, Lower Proterozoic arc-microcontinent collisional tectonics in the western Churchill province: Nature, v. 294, p. 69–72.
Lewry, J. F., Sibbald, T.I.I., and Schledewitz, D.C.P., 1985, Reworking of Archean basement in the western Churchill province and its significance, *in* Ayres, L. D., Thurston, P. C., Card, K. D., and Weber, W., eds., Archean crustal sequences: Geological Association of Canada Special Paper 28, p. 239–261.
Lugmair, G. W., 1974, Sm-Nd ages: A new dating method [abs.]: Meteoritics, v. 9, p. 369.
Lugmair, G. W., Scheinin, N. B., and Marti, K., 1975, Sm-Nd age and history of Apollo 17 basalt 75075: Evidence for early differentiation of the lunar exterior: Lunar Science Conference, 6th, Proceedings, p. 1419–1429.
Marsh, B. D., 1979, Island-arc volcanism: American Scientist, v. 67, p. 161–172.
Mauger, R. L., 1981, Geology and petrology of the central part of the Caldera Del Nido block, Chihuahua, Mexico, *in* Goodell, P. C., and Waters, A. C., eds., Uranium in volcanic and volcaniclastic rocks: American Association of Petroleum Geologists Studies in Geology, v. 13, p. 205–242.
McDowell, F. W., and Clabaugh, S. E., 1979, Ignimbrites of the Sierra Madre Occidental and their relation to the tectonic history of western Mexico, *in* Chapin, C. E., and Elston, W. E., eds., Ash-flow tuffs: Geological Society of America Special Paper 180, p. 113–124.
Michard-Vitrac, A., Lancelot, J., Allegre, C. J., and Moorbath, S., 1977, U-Pb ages on simple zircons from the early Precambrian rocks of W. Greenland and the Minnesota River Valley: Earth and Planetary Science Letters, v. 35, p. 449–453.
Moorbath, S., Allaart, J. H., Bridgwater, D., and McGregor, V. R., 1977, Rb-Sr ages of Early Archean supracrustal rocks and Amitsoq gneisses at Isua: Nature, v. 270, p. 43–45.
Morgan, W. J., 1968, Rises, trenches, great faults, and crustal blocks: Journal of Geophysical Research, v. 73, p. 1959–1982.
Muehlberger, W. R., Hedge, C. E., Denison, R. E., and Marvin, R. F., 1966, Geochronology of the midcontinent region, United States. Part 3. Southern area: Journal of Geophysical Research, v. 72, p. 5409–5426.
Mysen, B. O., 1982, The role of mantle anatexis, *in* Thorpe, R. S., ed., Andesites: Orogenic andesites and related rocks: New York, John Wiley and Sons, p. 489–522.
Nelson, B. K., and DePaolo, D. J., 1984, 1700 Myr greenstone volcanic successions in southwestern North America and isotopic evolution of the Proterozoic mantle: Nature, v. 312, p. 143–146.
——— 1985, Rapid production of continental crust 1.7–1.9 b.y. ago: Nd isotopic evidence from the basement of the North American continent: Geological Society of America Bulletin, v. 96, p. 746–754.
Nier, A. O., 1939, The isotopic constitution of uranium and the half-lives of the uranium isotopes I: Physical Review, v. 55, p. 150–153.
Parker, D. F., and McDowell, F. W., 1979, K-Ar geochronology of Oligocene volcanic rocks, Davis and Barrilla Mountains, Texas: Geological Society of America Bulletin, v. 90, Part I, p. 1100–1110.
Patchett, P. J., and Arndt, N. T., 1986, Nd isotopes and tectonics of 1.9–1.7 Ga crustal genesis: Earth and Planetary Science Letters, v. 78, p. 329–338.
Peterman, Z. E., Zartman, R. E., and Sims, P. K., 1980, Tonalitic gneiss of Early Archean age from northern Michigan: Geological Society of America Special Paper 182, p. 125–134.
Pidgeon, R. T., and Aftalion, M., 1978, Cogenetic and inherited zircon U-Pb systems in granites: Palaeozoic granites of Scotland and England: Geological Journal Special Issue 10, p. 183–220.
Ringwood, A. E., 1974, The petrological evolution of island arc systems: Geological Society of London Journal, v. 130, p. 183–204.
——— 1975, Composition and petrology of the Earth's mantle: New York, McGraw-Hill.
Saleeby, J. B., 1981, Ocean floor accretion and volcano-plutonic arc evolution of the Mesozoic Sierra Nevada, *in* Ernst, W. G., ed., The geotectonic evolution of California, Rubey Volume I: Englewood Cliffs, New Jersey, Prentice-Hall, p. 132–181.
——— 1982, Polygenetic ophiolitic belt of the California Sierra Nevada—Geochronological and tectonostratigraphic development: Journal of Geophysical Research, v. 87, p. 1803–1824.
Saleeby, J. B., Harper, G. D., Snoke, A. W., and Sharp, W. D., 1982, Time relations and structural-stratigraphic patterns in ophiolite accretion, west-central Klamath Mountains, California: Journal of Geophysical Research, v. 87, p. 3831–3848.
Sides, J. R., Bickford, M. E., Shuster, R. D., and Nusbaum, R. L., 1981, Calderas in the Precambrian terrane of the St. Francois Mountains, southeastern Missouri: Journal of Geophysical Research, v. 86, p. 10349–10364.
Silver, L. T., 1968, Precambrian batholiths of Arizona: Geological Society of America Abstracts for 1968, Special Paper 121, p. 558–559.
——— 1984, Observations on the Precambrian evolution of northern New Mexico and adjacent regions: Geological Society of America Abstracts with Programs, v. 16, p. 256.
Silver, L. T., and Deutsch, Sarah, 1963, Uranium-lead variations in zircons: A case study: Journal of Geology, v. 71, p. 721–758.
Silver, L. T., Bickford, M. E., Van Schmus, W. R., Anderson, J. L., Anderson, T. H., and Medaris, L. G., 1977, The 1.4–1.5 b.y. transcontinental anorogenic perforation of North America: Geological Society of America Abstracts with Programs, v. 9, p. 1176.
Sims, P. K., and Peterman, Z. E., 1986, The Early Proterozoic Central Plains orogen—A major buried structure in north-central United States: Geology, v. 14, p. 488–491.
Soegaard, K., and Ericksson, K. A., 1985, Evidence of tide, storm, and wave interaction on a Precambrian siliciclastic shelf: Journal of Sedimentary Petrology, v. 55, p. 672–684.
——— 1986, Transition from arc volcanism to stable shelf and subsequent convergent-margin sedimentation in northern New Mexico from 1.76 Ga: Journal of Geology, v. 94, p. 47–66.
Stacey, J. S., and Hedlund, D. C., 1983, Lead-isotopic compositions of diverse igneous rocks and ore deposits from southwestern New Mexico and their implications for Early Proterozoic crustal evolution in the western United States: Geological Society of America Bulletin, v. 94, p. 43–57.
Stern, T. W., Goldich, S. S., and Newell, M. F., 1966, Effects of weathering on the U-Pb ages of zircon from the Morton gneiss, Minnesota: Earth and Planetary Science Letters, v. 1, p. 369–371.
Swanson, E. R., and McDowell, F. W., 1984, Calderas of the Sierra Madre Occidental volcanic field, western Mexico: Journal of Geophysical Research, v. 89, p. 8787–8799.
Thomas, A., Nunn, G.A.G., and Krogh, T. E., 1986, The Labradorian orogeny: Evidence for a newly identified 1600 to 1700 Ma orogenic event in Grenville province crystalline rocks from central Labrador, *in* Moore, J. M., Davidson, A., and Baer, A. J., eds., The Grenville province: Geological Association of Canada Special Paper 31, p. 175–189.
Thomas, J. J., Shuster, R. D., and Bickford, M. E., 1984, A terrane of 1350–1400 m.y. old silicic volcanic and plutonic rocks in the buried Proterozoic of the midcontinent and in the Wet Mountains, Colorado: Geological Society of America Bulletin, v. 95, p. 1150–1157.
Thompson, A. B., and England, P. C., 1984, Pressure-temperature-time paths of regional metamorphism II. Their inference and interpretation using mineral assemblages in metamorphic rocks: Journal of Petrology, v. 25, p. 929–955.
Thorpe, R. S., Francis, P. W., Hammill, M., and Baker, M.C.W., 1982, The Andes, *in* Thorpe, R. S., ed., Andesites: Orogenic andesites and related rocks: New York, John Wiley and Sons, p. 187–205.
Tilton, G. R., 1960, Volume diffusion as a mechanism for discordant lead ages: Journal of Geophysical Research, v. 65, p. 2933–2945.
Tolman, C., and Robertson, F., 1969, Exposed Precambrian rocks in southeast Missouri: Missouri Division of Geological Survey and Water Resources Report of Investigations 44, 68 p.
Trevena, A. S., 1979, Studies in sandstone petrology: Origin of the Precambrian Mazatzal quartzite and provenance of detrital feldspar [Ph.D. dissert.]: Salt Lake City, Utah, University of Utah, 390 p.
Turner, D. C., 1963, Ring-structures in the Sara-Fier younger granite complex, northern Nigeria: Geological Society of London Quarterly Journal, v. 119, p. 345–366.
Tuttle, O. F., and Bowen, N. L., 1958, Origin of granite in the light of experimental studies in the system $NaAlSi_3O_8$-$KAlSi_3O_8$-SiO_2-H_2O: Geological Society of America Memoir 74, 153 p.
Van Schmus, W. R., 1976, Early and Middle Proterozoic history of the Great Lakes area, North America: Royal Society Philosophical Transactions, Series A, v. 280, p. 605–628.
——— 1978, Geochronology of southern Wisconsin rhyolites and granites: Geoscience Wisconsin, v. 2, p. 19–24.
——— 1980, Chronology of igneous rocks associated with the Penokean orogeny in Wisconsin, *in* Morey, G. B., and Hanson, G. N., eds., Selected studies of Archean gneisses and lower Proterozoic rocks, southern Canadian shield: Geological Society of America Special Paper 182, p. 159–168.
——— 1984, Recent contributions to the geochronology of the Precambrian of Wisconsin: Department of Geology, University of Wisconsin, Oskosh, Annual Institute of Lake Superior Geology, 30th, Wausau, Wisconsin, Program, p. 79.
Van Schmus, W. R., and Bickford, M. E., 1981, Proterozoic chronology and evolution of the midcontinent region, North America, *in* Kroner, A., ed., Precambrian plate tectonics: Amsterdam, the Netherlands, Elsevier Publishing Co., p. 261–296.
Van Schmus, W. R., and Hinze, W. J., 1985, The Midcontinent Rift system: Annual Reviews of Earth and Planetary Science, v. 13, p. 345–383.
Van Schmus, W. R., Medaris, L. G., and Banks, P. O., 1975, Geology and age of the Wolf River Batholith, Wisconsin: Geological Society of America Bulletin, v. 86, p. 907–914.
Van Schmus, W. R., Bickford, M. E., and Zietz, I., 1987a, Early and Middle Proterozoic provinces in the central United States, *in* Kroner, A., ed., Proterozoic lithospheric evolution: American Geophysical Union Geodynamics Series, v. 17, p. 43–68.
Van Schmus, W. R., Bickford, M. E., Lewry, J. F., and Macdonald, R., 1987b, U-Pb geochronology in the Trans-Hudson orogen, northern Saskatchewan, Canada: Canadian Journal of Earth Sciences, v. 24, p. 407–424.
Vine, F. J., and Matthews, D. H., 1963, Magnetic anomalies over ocean ridges: Nature, v. 199, p. 947–949.
Wardle, R. J., Rivers, T., Gower, C. F., Nunn, G.A.G., and Thomas, A., 1986, The northeastern Grenville province: New insights, *in* Moore, J. M., Davidson, A., and Baer, A. J., eds., The Grenville province: Geological Association of Canada Special Paper 31, p. 13–29.
Wasserburg, G. J., 1963, Diffusion processes in lead-uranium systems: Journal of Geophysical Research, v. 68, p. 4823–4846.
Wasserburg, G. J., Wetherill, G. W., Silver, L. T., and Flawn, P. T., 1962, A study of the ages of the Precambrian of Texas: Journal of Geophysical Research, v. 67, p. 4021–4047.
Wegener, A., 1915, Die Enstehung der Kontinente und Ozeane: Vieweg, Braunschweig, 94 p.
Welin, E., 1986, The depositional evolution of the Sveccofennian supracrustal sequence in Finland and Sweden: Precambrian Research (in press).
Wernicke, B., 1981, Low-angle normal faults in the Basin and Range province—Nappe tectonics in an extending orogen: Nature, v. 291, p. 645–648.
——— 1985, Uniform-sense normal simple shear of the continental lithosphere: Canadian Journal of Earth Sciences, v. 22, p. 108–125.
Wetherill, G. W., 1956, Discordant uranium-lead ages, 1: EOS (American Geophysical Union Transactions), v. 37, p. 320–326.
——— 1963, Discordant uranium-lead ages, II: Discordant ages resulting from diffusion of lead and uranium: Journal of Geophysical Research, v. 68, p. 2957–2965.
Williams, H., and Hatcher, R. D., Jr., 1982, Suspect terranes and accretionary history of the Appalachian orogen: Geology, v. 10, p. 530–536.
Wilson, J. T., 1965a, A new class of faults and their bearing on continental drift: Nature, v. 207, p. 343–347.
——— 1965b, Transform faults, oceanic ridges, and magnetic anomalies southwest of Vancouver Island: Science, v. 150, p. 482–485.
——— 1966, Did the Atlantic close and then reopen?: Nature, v. 207, p. 343–347.
——— 1968, Static or mobile Earth: The current scientific revolution, *in* Gondwanaland revisited: New evidence for continental drift: American Philosophical Society Proceedings, v. 112, p. 309–320.
Wooden, J. L., Ashwal, L. D., Wiebe, R. A., and Emslie, R. F., 1987, Regional Pb isotope systematics in Proterozoic intrusives, Nain province, Labrador [abs.]: EOS (American Geophysical Union Transactions), v. 68, p. 1519.
Wright, J. E., and Haxel, G., 1982, A garnet-two-mica granite, Coyote Mountains, southern Arizona: Geologic setting, uranium-lead systematics of zircon, and nature of granite source region: Geological Society of America Bulletin, v. 93, p. 1176–1188.
Zartman, R. E., 1974, Lead isotope provinces in the Cordillera of the western United States and their geologic significance: Economic Geology, v. 69, p. 792–805.

Manuscript Received by the Society January 12, 1988
Revised Manuscript Received May 3, 1988
Manuscript Accepted May 9, 1988

CENTENNIAL ARTICLE

The origin of granite: The role and source of water in the evolution of granitic magmas

JAMES A. WHITNEY *Department of Geology, University of Georgia, Athens, Georgia 30602*

ABSTRACT

Temperature and water content are the two most important parameters in the formation of granitic magmas. Evidence from volcanic and plutonic lithologies suggests that water contents of 2 to 4 wt. % are present in most silicic magmas. Calculations based on the stability of biotite yield water fugacities within the melt phase from about 500 to 2,000 bars, although these calculations determine the log f_{H_2O} and the effective uncertainty of ±0.5 log units yields a large absolute uncertainty. Comparison with crystallization experiments demonstrates that less than 2% water would require significant percentages of crystallization at temperatures above 900 °C with liquidus temperatures about 1000 °C. Water contents greatly in excess of 4 wt. % would mean that the magma would become vapor saturated at high pressures and would tend to crystallize during ascent to a fine-grained granite before reaching shallow depths.

The main sources of water for magma generation are the dehydration of hydrous silicates within the crust and volatiles transported into the crust from subducted oceanic crust and upper mantle in the form of hydrous basalts and andesites. Dehydration reactions of muscovite, biotite, and hornblende are of particular significance. Anatectic granites may be partially classified in terms of the probable dehydration reaction responsible for their generation.

Melts generated from muscovite dehydration are relatively cool, peraluminous in composition, high in K/Na ratio, and generally high in initial $^{87}Sr/^{86}Sr$ and delta ^{18}O. Biotite-generated melts tend to be higher in temperatures, peraluminous to meta-aluminous in composition, moderately high in K/Na ratio, and relatively high in strontium and oxygen ratios. Both types may contain metasedimentary enclaves. Granitoids generated by hornblende dehydration would be much higher in initial temperature, peralkaline to meta-aluminous in composition, and lower in K/Na ratio and in general would have lower strontium and oxygen ratios.

Volatiles deposited in hydrothermally altered oceanic crust and upper mantle will be released during subduction in the form of hydrous basalts and andesites. These melts are important energy-transfer mechanisms to drive anatexis within the crust. If these melts encounter a silicic magma chamber when intruding the crust, they will be trapped below the lower-density granitic melt. While partially quenching against the cooler melt, they may release volatiles and heat, which will aid in the melting process. Granites generated in this way will vary in their geochemical properties, depending on the relative importance of crustal versus mantle magma systems. In general, they will have moderate to high initial temperatures, be meta-aluminous in composition, be variable but somewhat low in K/Na ratio, and have lower initial strontium and delta ^{18}O ratios than will melts derived completely from the crust.

The fact that most terrestrial spreading centers are subaqueous thus may aid in the formation of a thick granitic continental crust.

INTRODUCTION

It has long been realized that studies of dry silicate systems do not yield exactly the conditions of formation of the igneous rocks because the salic rocks contain volatile ingredients, principally water. Hydrous minerals occur in nearly all granites, syenites, and nepheline syenites and as phenocrysts in many rhyolites, trachytes, and phonolites. Volatile materials emanating from volcanic rocks have been carefully analyzed chemically, and in them also water is the predominant volatile constituent. It is thus natural that water should be added to the feldspars and mixtures of feldspars and quartz, the principal constituents of granite. (Tuttle and Bowen, 1958, p. 5)

Thirty years ago, the Geological Society of America published Memoir 74, *Origin of granite in light of experimental studies* (Tuttle and Bowen, 1958). This book not only summarized many years of experimental work but has dominated our thinking about the formation of granitic magmas ever since. It was perhaps the final "battle" in the scientific "war" between the magmatists and metasomatists. Rereading it today, I am still amazed at the clarity of understanding represented by that book. The insight Bowen and his students demonstrated came from the careful integration of field, petrographic, and laboratory studies. Even today, we have not bettered much of the experimental work. We have, however, expanded our data base to higher pressures and lower water contents. We also have a great deal of new field, petrologic, and geochemical information, which has led to multiple models for magma genesis.

It is not possible in a single paper to summarize all recent advances in the studies of granite. Instead, this paper will concentrate on a central theme of Tuttle and Bowen (1958): the importance of water in the generation and evolution of granitic magmas. In the process, it will review magmatic conditions as evidenced by silicic volcanic analogues, present experimental studies dealing with water as a fundamental variable, and discuss sources of water as a controlling factor in silicic magma genesis.

VOLCANIC ANALOGUES OF GRANITIC BATHOLITHS

Field studies of granites deal with end products of processes long since completed, and the most detailed mapping and study of these products may fail to give convincing evidence concerning the exact nature of the processes responsible for the relations. (Tuttle and Bowen, 1958, p. 5)

The above statement is as true today as in 1958. Additional evidence can be gathered, however, from studies of volcanic analogues which were not available in 1958. Large-volume ash-flow–tuff eruptions (>500 km³) are the volcanic equivalent of silicic batholiths. Unlike plutons, however, these magmas have been quenched at various points in their development and therefore represent silicic batholiths frozen in time. Studies of these systems have yielded extensive data on magmatic parameters (for ex-

Geological Society of America Bulletin, v. 100, p. 1886–1897, 11 figs., 4 tables, December 1988.

ample, Smith, 1960, 1979; Smith and Bailey, 1966; Hildreth, 1977, 1979; Lipman, 1971; Fridrich and Mahood, 1987; Whitney and Stormer, 1985; Whitney and others, 1988). This effort has been made possible in the past 20 yr by the development of various thermodynamic models which allow us to quantify such variables as temperature, pressure, and the activities of gaseous components (for example, Buddington and Lindsley, 1964; Spencer and Lindsley, 1981; Andersen and Lindsley, 1985; Stormer, 1975; Brown and Parsons, 1981; Stormer and Whitney, 1985; Ghiorso, 1984; Price, 1985; Green and Usdansky, 1986; Wones, 1972; Bohlen and others, 1980; Whitney, 1984). These data, combined with experimental work on both naturally occurring and synthetic granitic materials, allow us to determine the probable water content of silicic magmas responsible for batholith-generating magma systems. Table 1 summarizes probable conditions for several such systems as determined by various authors.

Water content can be estimated two different ways. First, through the composition and stability data for hydrous phases such as biotite, the activity of water can be calculated (Wones, 1972; Wones and Eugster, 1965; Bohlen and others, 1980). Such calculations, however, have serious potential errors or uncertainties. It is difficult to determine the ferrous/ferric ratio in biotite at the time of magmatic equilibration, owing to oxidation and lack of micro-analytical techniques. The fluorine content in the hydroxyl site must also be carefully analyzed, and even with modern microprobes, the error is significant. Finally, the experimental data upon which the thermodynamic model is based have inherent uncertainties. Even given these uncertainties, several studies have yielded values that are probably good to 0.5 to 1.0 log units. Unfortunately, 1 log unit in the range of 2.0 to 3.0 is the difference between 100 and 1,000 bars water fugacity, which corresponds to approximately 0.5 to 4.0 wt. % water (Burnham, 1979; Day and Fenn, 1982).

A second approach, which is less numeric but at least as accurate, is the comparison of mineral assemblages and percent crystallization with the results of phase equilibria studies of similar compositions. Most large-volume sub-alkaline (including calc-alkaline) ash-flow tuffs have pre-eruption temperatures of between 750 and 850 °C (see Table 1; some fayalite-bearing alkaline systems may be higher in temperature and therefore dryer). Because these magmas are 60% to 90% liquid at this stage, analogies with laboratory experiments (for example, Whitney, 1975; Naney, 1977; Robertson and Wyllie, 1971) require between 2 and 6 wt. % water in the silicic melt phase (Whitney and Stormer, 1985; Whitney and others, 1988). Similar evaluation of less silicic systems, such as that erupted at Mount St. Helens (Melson and Hopson, 1981), at somewhat hotter temperatures requires less water, down to about 2 to 3 wt. % in the melt phase.

TABLE 1. ESTIMATED WATER CONTENT AND MAGMATIC CONDITIONS IN SOME ASH-FLOW TUFFS

	Temp. (°C)	% melt	log f_{O_2}	log f_{H_2O}	X_{H_2O}, melt	X_{H_2O}, bulk
Bishop Tuff	750 ± 30	85	−15 ± 1	2.7 ± 0.5	2.5	2.2
Fish Canyon Tuff	790 ± 20	60	−12 ± 1	3.2 ± 0.5	6.5	4.0
Carpenter Ridge Tuff	780 ± 30	90	−13.5 ± 1	2.9 ± 0.3	3.5	3.2
St. Helens dark pumice	950 ± 30	~85	−9.5 ± 0.5	~2.6	~2.3	~2

Note: values recalculated from the following authors: Bishop, Hildreth (1977); Fish Canyon, Whitney and Stormer (1985); Carpenter Ridge, Whitney and others (1988); St. Helens, Melson and Hopson (1981); and using solubility and thermodynamic data of Day and Fenn (1982); Burnham and Davis (1974); Burnham and others (1969). Values for water from St. Helens are approximations based on mineral assemblage and degree of crystallization as reported in Lipman and Mullinax (1981).

An important lesson from such volcanic studies is that the parameters of temperature, water content, and degree of crystallization are not independent parameters (Marsh, 1981). If a magma erupts with 10% crystals present and a known temperature, then the water content required can be estimated. Unless very large concentrations of fluorine, boron, or some other component which lowers the melting temperature of granite are present, the required water content is directly related to the temperature. If the magma is undersaturated with respect to water, this relationship is only moderately dependent on pressure.

Another important corollary is the possible contribution of CO_2 as a primary magmatic gas in such magma systems. Because CO_2 is quite insoluble in calc-alkaline granitic melts, it is strongly fractionated into the volatile phase and lowers the activity of water. If CO_2 is present in too great an abundance, the melt cannot have the low percentage of crystals present at the temperatures determined (Swanson, 1979). Therefore, although we do not have good thermodynamic measurements of CO_2 in such systems, the maximum amount possible can be estimated by analogy with synthetic systems using mixed gas phases (for example, Rutherford and others, 1985; Holloway, 1976).

It is difficult to do the same types of determinations of magmatic parameters in plutons because minerals re-equilibrate during cooling from magmatic temperatures. Tuttle and Bowen (1958) recognized this fact throughout their memoir and their experimental careers. Iron-titanium oxides appear to re-equilibrate at least down to solidus temperatures, and then continue to exsolve at lower temperatures. Some useful results have been obtained (Whitney and Stormer, 1976; Baldasari, 1981). Feldspars also exsolve, but they tend to re-equilibrate more slowly. Although calculations on co-existing feldspars have numerous uncertainties, including exsolution of the alkali feldspars, some results have been reported (Whitney and Stormer, 1977a, 1977b, 1976). The stability of the mafic phases may offer better indications of magmatic conditions. Biotite is limited in stability to below about 850 °C in the presence of quartz, or slightly higher in the presence of a high–silica-activity magma (Luth, 1967; Naney, 1977). Hornblende has a more complicated stability field highly dependent on bulk composition and pressure (Hammarstrom and Zen, 1986). Most studies have resulted in magmatic temperatures in the range of 850 to 650 °C for calc-alkaline granitic rocks. These temperatures again require the same range of water content as determined from volcanic analogues.

EXPERIMENTAL STUDIES OF SILICIC SYSTEMS

Laboratory studies, under controlled and measured conditions, of the common minerals of granites are natural avenues of attack on the granite problem. (Tuttle and Bowen, 1958, p. 5)

In the past 30 yr, many experimental studies have been published on haplogranite to haplogranodiorite systems. Experimentation on the "granite" system, $NaAlSi_3O_8$-$KAlSi_3O_8$-SiO_2-H_2O, has been carried to higher pressures (Luth and others, 1964). Other components have been added to the system, including Mg_2SiO_4 (Luth, 1968) and $CaAl_2Si_2O_8$ (James and Hamilton, 1969). In addition, a number of studies have been conducted with controlled water contents (Whitney, 1969; Robertson and Wyllie, 1971; Whitney, 1975a, 1975b; Naney, 1977) or with the activity of water reduced by an insoluble gas, in most cases CO_2 (for example, Swanson, 1979; Eggler, 1972; Eggler and Burnham, 1973). Both types of study allow the separation

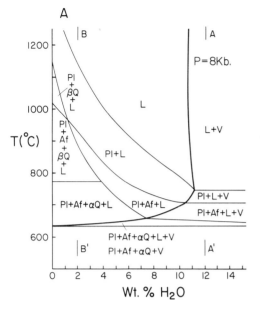
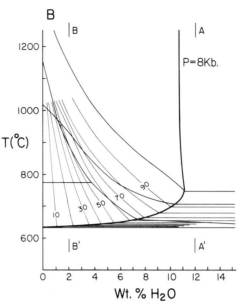

Figure 1. Temperature-X_{H_2O} diagrams for synthetic granite (compositions R4 of Whitney, 1972, 1975a, 1975b). See Whitney (1972, 1975a) for experimental procedures and results. Abbreviations are the same as in Table 3. (A) Stable phase assemblages at 8 kb confining pressure. (B) Phase-assemblage diagram contoured for volume-percent melt. Phase fields are the same as in A. Contours represent approximate percentage of melt present. Positions of contours are semiquantitative.

of water content and confining pressure as variables.

Perhaps the effect of water on melting relationships can best be illustrated through the use of phase diagrams which plot the stable phase assemblage with respect to two of the three variables: pressure, temperature, and total water content, with the third held constant. Because it is the phase assemblage that is being plotted, such diagrams may be referred to as "phase-assemblage diagrams" to distinguish them from more conventional phase diagrams (Whitney, 1972). Figure 1A is a temperature-X_{H_2O} diagram at 8 kb confining pressure for a synthetic composition within the system ternary feldspar-SiO_2 (R4 of Whitney, 1975a, 1975b). Also shown, in Figure 1B, are contours of approximate volume-percent melt. Similar diagrams at 2 kb pressure are shown in Figures 2A and 2B. The chemical compositions of R4 and other materials mentioned in this paper are summarized in Table 2. In the T-X diagrams, the dramatic effect of water content on degree of crystallization in the vapor-absent region is apparent.

Consider the temperature paths A-A' and B-B' in Figure 1. A-A' (12 wt. % water) is oversaturated with respect to a free volatile phase (V). Under these conditions, this granitic composition does not begin to crystallize until temperatures below about 750 °C. Crystallization is slow until below 700 °C, with the majority of crystallization occurring over a very short temperature range between 630 and 650 °C. In contrast, a composition with 2% water (B-B') has plagioclase stable to temperatures in excess of 1100 °C. Crystallization is slow, but continuous, between 1000 and 630 °C. The rapid crystallization of 50% of the magma over a small, 10–20 °C, temperature range occurs only under water-vapor-saturated conditions. The same phenomena can be observed at 2 kb (Fig. 2, A-A', B-B'), but now vapor saturation requires only 6 to 7 wt. % water. Therefore, in general, with other factors such as nucleation rate being

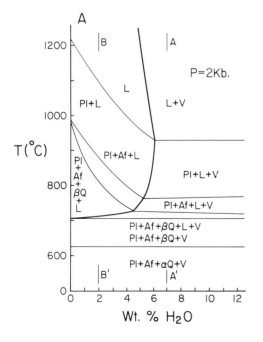
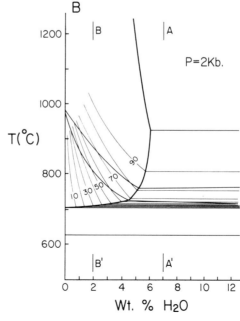

Figure 2. Temperature-X_{H_2O} diagrams for synthetic granite (R4) at 2 kb confining pressure. (A) Stable phase assemblage. (B) Diagram contoured for volume-percent melt. Positions of contours are semiquantitative.

TABLE 2. CHEMICAL COMPOSITION OF EXPERIMENTAL MATERIALS

Oxide	R4	Nockolds' average Hb-Bi qtz. monzonite	Cape Ann Granite	Westerly Granite	Mount Airy leucogranodiorite
SiO_2	70.99	65.88	77.61	72.34	71.03
TiO_2	..	0.81	0.25	0.26	N.D.
Al_2O_3	17.21	15.07	11.94	14.34	16.52
Fe_2O_3	..	1.74	0.55	0.68	..
FeO	..	2.73	0.87	1.13	1.75*
MnO	..	0.08	..	0.02	N.D.
MgO	..	1.38	Tr	0.37	Tr
CaO	2.83	3.36	0.31	1.52	1.93
Na_2O	4.00	3.53	3.80	3.37	5.53
K_2O	5.07	4.64	4.98	5.97	3.71
P_2O_5	..	0.26	N.D.	0.11	N.D.
H_2O	..	0.52	0.23	0.30	0.19
Total	100.0	100.0	100.54	100.30	100.66

Note: R4 and Nockolds' average, of which R4 represents the normative feldspar and quartz components, are taken from Whitney (1975a). The other analyses are as listed in Whitney (1969).
*Total iron reported as FeO.

equal, and assuming that heat flux out of the system is sufficient that latent heat will not arrest the cooling process, crystallization in the water-vapor–undersaturated region will lead to longer temperature ranges of crystallization and coarser grain size. Crystallization under vapor-saturated conditions will cause a shorter temperature range and finer grain size. The crossing of the vapor saturation surface during cooling will cause an increase in the rate of crystallization and in general smaller grain size (again assuming latent heat does not arrest the cooling process). An example of such a phenomenon may be the porphyritic, sub-porphyritic, and non-porphyritic phases of the Pinos Altos pluton, Silver City, New Mexico (Owen, 1983; Jones and others, 1967; El-Hindi, 1977). Other cases may be the fine-grained versus coarse-grained phases of the Siloam Granite, Georgia (Speer, 1977), and the Liberty Hill pluton, North Carolina (Speer and others, 1980).

Another way to reduce suddenly the confining pressure and volatile content is through venting from the surface. In this case, the pressure in the magma chamber is reduced to the pressure exerted by the magma column. In the case of a vesiculated magma, this pressure can be substantially lower than the lithostatic pressure (Whitney and Stormer, 1986). Such a phenomenon is probably responsible for the hiatal textures in many shallow porphyries.

Another diagram useful for understanding phase relations during upward migration of a magma is the isothermal phase-assemblage diagram that plots pressure and water content as independent variables. Figures 3A and 3B are such diagrams for the same synthetic R4 composition. Again, A–A' represents vapor-saturated conditions, whereas B–B' is vapor undersaturated throughout most of its path.

As pressure decreases along A–A', the magma crystallizes. This process is slow throughout the plagioclase field, with the majority of crystallization occurring within 1 kb of the solidus in a manner similar to classic eutectic crystallization. At pressures below the solidus, the melt is metastable. This phenomenon is the familiar pressure quench effect discussed by Tuttle and Bowen (1958). Such rapid crystallization due to decreasing confining pressure is probably the cause of many fine-grained aplites, aphanitic groundmass in some porphyries, and fine-grained homogeneous plutons such as the Elberton Batholith, Georgia (Stormer and Whitney, 1980). Because of this phenomenon, low-temperature granitic magmas are not stable under near-surface conditions. Granitic magmas with temperatures below about 650 °C cannot rise above the upper mid-crust before they tend to crystallize. Thus, it is rare to find extensive ash-flow sheets with magmatic temperatures below about 750 °C. Such magmas would tend to freeze before they reached shallow crustal depths.

At lower water contents (for example, 2 wt. %; B–B' in Fig. 3B), the crystallization history is quite different. As long as the magma remains isothermal and undersaturated with respect to water, the percentage of melt either increases or remains nearly the same. In the case of quartz, the degree of melting increases, causing resorption. This occurs because the quartz saturation surface within the granite system moves to higher silica content with decreasing pressure, as originally shown by Tuttle and Bowen (1958; see also Luth and others, 1964; Whitney, 1975a, 1975b; Nekvasil and Burnham, 1987). Alkali feldspar also appears to decrease in abundance

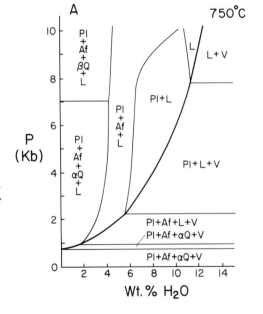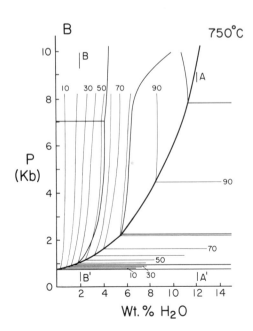

Figure 3. Pressure-X_{H_2O} diagrams for synthetic granite (R4) at 750 °C. (A) Phase-assemblage diagram showing experimental results and phase assemblages. (B) Same diagram as in A, contoured for volume-percent melt. Positions of contours are semiquantitative.

slightly for a number of granitic compositions (Whitney, 1975a). Plagioclase, however, either remains nearly the same or crystallizes somewhat, depending on the bulk composition. This occurs due to the shift in the plagioclase–alkali feldspar liquidus boundary within the granite system (Whitney, 1975a; Yoder and others, 1957) and the resulting shift of phase volumes in the water-undersaturated region. Eventually, as the pressure drops, the magma will become vapor saturated, and pressure quenching will again occur over a very short pressure range. Thus, the melt phase in most large ash flows and rhyolites is metastable by the time it reaches the surface. With the exception of very dry, high-temperature rhyolites, most silicic melts tend to crystallize rapidly to aphanitic products within 1 km of the surface.

The paragenesis of volcanic phenocrysts can also be understood in terms of isothermal decompression. Quartz phenocrysts are ubiquitously rounded or embayed. Such an occurrence is predicted by isothermal decompression in the water-vapor–undersaturated region. Such a phenomenon will not occur under vapor-saturated conditions. Similarly, early sanidine is in some cases rounded in silicic volcanics. Plagioclase, on the other hand, is rarely rounded or embayed. When plagioclase is seen to be resorbed, there are in many cases other signs of phenocryst disequilibrium involving the mafic phases. In such cases, field and petrographic evidence in many instances indicates disequilibrium processes such as magma mixing or assimilation. Under vapor-saturated decompression, resorption should not occur for most granitic compositions. (Note: There are exceptions to this rule in some alkaline granites or syenites due to the position of the "unique fractionation curve" in the granite system, but such a detailed phase-equilibrium discussion is beyond the scope of this general paper.)

We can further use experimental phase equilibrium to evaluate the crystallization history of some common granites. Figures 4, 5, and 6 show expanded 2-kb isobaric phase-assemblage diagrams for some common granite types. These are the Cape Ann Granite (Fig. 4, a body of the Quincy type) from Cape Ann, Massachusetts; the Westerly Granite from Westerly, Rhode Island (Fig. 5); and the Mount Airy leucogranodiorite from Mount Airy, North Carolina (Fig. 6). Two of these (the Quincy and Westerly) were chosen by Tuttle and Bowen (1958) themselves to represent alkaline (Quincy) and calc-alkaline (Westerly) compositions. The Mount Airy (Deitrich, 1961) is strongly peraluminous. Experimental data for these compositions are listed in Table 3. Percentage of glass was determined by modal analyses of powdered samples. Replicate analyses suggest that precision is ap-

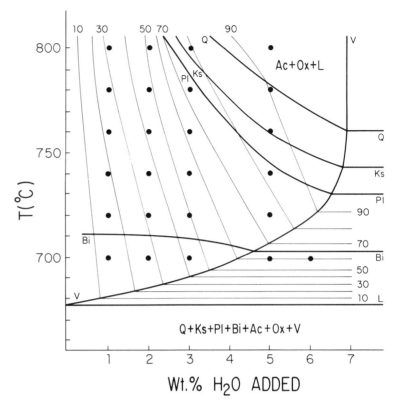

Figure 4. Temperature-X_{H_2O} phase-assemblage diagram for the Cape Ann Granite (Quincy type) at 2 kb confining pressure. Abbreviations are given in Table 2. Contours represent volume percentage of melt present. In each case, the sub-solidus assemblage is listed. The various boundaries are shown with the phase which appears or disappears along that boundary listed on the side of the curve on which it occurs. Percentage-melt contours are best-fits to the volumetric data in Table 3. See Whitney (1969) for experimental methods. All phase assemblages have been re-evaluated from the original work, using methods and criteria given in Whitney (1972, 1975a).

proximately ±5% glass; accuracy is thought to be about ±10%.

Again, note the contrast between vapor-absent and vapor-saturated crystallization. In the case of the Cape Ann Granite, 90% crystallization occurs between 720 and 680 °C under vapor-saturated conditions. If only 2% water is present, however, crystallization starts well above 800 °C (probably above 900° by comparison with other data) and continues to 680 °C. At 2 kb, this process is nearly constant until the composition is about 75% crystalline at 685 °C, at which point the composition becomes vapor saturated and finishes crystallization within 10 °C. The Westerly and Mount Airy compositions crystallize over a slightly greater temperature range under vapor-saturated conditions, but still the majority of the melt crystallizes within 20 to 30 °C of the solidus in a manner analogous to classic eutectic crystallization.

Thus, coarse-grained granites that evidence a long crystallization history with initial high-temperature paragenesis probably crystallized in the absence of a vapor phase with relatively low total water content. Fine-grained ones which have an abbreviated crystallization history starting at low temperature probably formed under vapor-saturated conditions with relatively higher water contents. The critical factor is the rate of cooling combined with the rate of crystallization. The rate of cooling is controlled by heat flow out of the body versus heat generated by latent heat of crystallization.

SOURCES OF WATER FOR THE GENERATION OF GRANITIC MELTS

If the water content of a segment of the earth's crust is on the order of 1–2 per cent by weight and the composition is approximately that of the average granite, a zone of melting at least 10 km thick will result. (Tuttle and Bowen, 1958, p. 123)

A critical factor in the generation of granitic rocks is therefore the availability of water. Water is required for the formation of granitic rocks under the conditions in which we see them

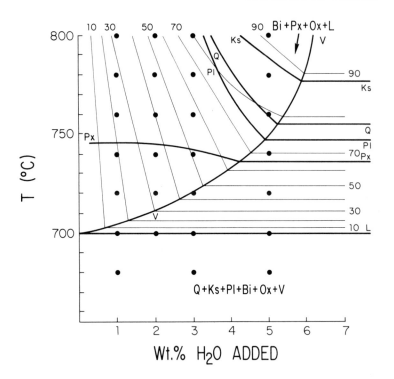

Figure 5. Temperature-X_{H_2O} phase-assemblage diagram for the Westerly Granite, Westerly, Rhode Island, at 2 kb confining pressure. Contours are volume-percent melt. See Figure 4 for symbols and explanation.

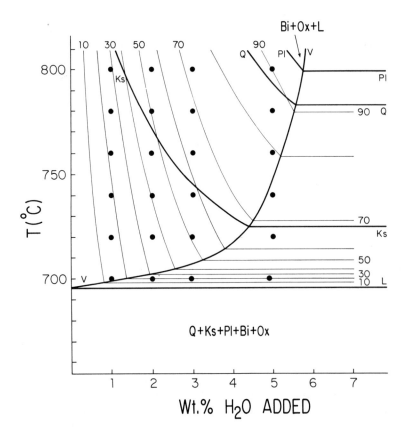

Figure 6. Temperature-X_{H_2O} phase-assemblage diagram for the Mount Airy leucogranodiorite, Mount Airy, Virginia, at 2 kb confining pressure. Contours are volume-percent melt. See Figure 4 for symbols and explanation.

(see Burnham, 1979, for an extended discussion). The granitic solidus is depressed by as much as 400 C° by the addition of water, as compared to only about 200 C° for basalt. Without water, granites do not begin to melt until 1000 °C at crustal pressures. To duplicate the paragenesis of mineral crystallization that we see in natural rocks requires about 2 wt. % water for dioritic rocks to as much as 4% or more for granitic rocks. Therefore, essential factors in the formation of granitic rocks are the sources and availability of water.

There are a number of sources of water for melting in the lower crust. First is the dehydration of hydrated phases which are present in high-grade metamorphic rocks (Hyndman, 1981). We also have very small amounts of intergranular volatile phases which may be called upon to help in that melting process. This intergranular water, however, is very low in abundance and yields only 0.1 to 0.3 wt. % of the 2 to 4 wt. % water we need to form most common granitic rocks. So, in anatexis of lower crustal rocks, the dehydration of hydrated phases is highly important and can be used to describe the nature of the melting process.

A second source of volatiles is subducted crust and upper mantle which have been hydrated during hydrothermal alteration at the ridge crest. These volatiles must be transported from the mantle into the crust, possibly by the formation of hydrated basalts and andesites from the subducted plate which eventually crystallize in the lower crust. Because many large granitic batholiths are associated with subduction-zone magmatism, such a mechanism could be highly important in those areas.

Dehydration Reactions

First, consider melting associated with the dehydration of hydrous crystalline phases, starting with muscovite. The dehydration melting of muscovite occurs along a series of incongruent melting reactions in which muscovite reacts with albite or plagioclase, K-feldspar, and quartz to form a liquid. These reactions originate at a series of invariant points generated by the intersection of the muscovite plus quartz reaction with the beginning of melting for the assemblage involved (see Thompson, 1982, and Thompson and Algor, 1977, for a more thorough discussion of Schreinemakers' constructions). For anatexis, the so-called "vapor-absent" reactions which do not require the presence of a hydrous volatile phase are particularly important. These reactions represent the melting of the hydrous assemblage without the input of additional water. Such reactions are the best experimental analogues to anatexis under crustal conditions.

For example, within the Inner Piedmont province of eastern Georgia, the reaction

Muscovite + Quartz + Na-Plag. = Sillimanite + K-spar + Liquid

appears to be responsible for the majority of migmatization (Potter, 1981). Layers which lack significant Na-rich plagioclase have not melted and remain competent even though they contain muscovite and quartz. Therefore, in this case, it appears as if the reaction

Muscovite + Quartz = Sillimanite + K-spar + Liquid

(Day, 1973) has not been exceeded. Thus, with rising temperatures, muscovite-generated migmatization will first begin in rocks containing quartz, albitic plagioclase, and muscovite and proceed at higher temperatures into rocks lacking plagioclase. The range of conditions for such muscovite-driven melting is shown in Figure 7.

The melts generated by muscovite anatexis are expected to have certain chemical characteristics. First, they tend to be relatively high in potassium, with K/Na atomic ratios of 1 or more. Second, they are granitic in composition. Third, they are strongly peraluminous. Finally, they have relatively low initial temperatures of below about 750 °C. Because the most common muscovite-rich lithologies are derived from continental pelitic sediments, most muscovite-generated granitoids will have high initial strontium ratios (0.705 or higher) and high delta ^{18}O values (in most cases, ~10 or more). These granites correspond to the ilmenite-series plutons described by Ishihara (1977). Because such sediments in many cases contain organic material which becomes graphite under metamorphic conditions, the initial oxygen activity may be low (Ishihara, 1977). Because the volatile system is within the ternary system C-O-H, and not simply C-O, however, a variety of oxygen fugacities are possible in the presence of graphite, and once the magma separates from the source area, these melts contain very little iron, owing to their low temperature, with biotite being the most common mafic phase. Therefore, during cooling, there is little control of oxygen activity in these silicic, peraluminous melts, and such granitoids may attain higher oxygen activities late in their crystallization history. The low iron content and initial low oxygen activity means that magnetite is rare, as most iron goes into silicate phases. Accessory minerals such as tourmaline and garnet are common. Schlieren or mafic enclaves carried within the magma may be formed from more-refractory sedimentary layers. Common metamorphic minerals inherited from the source area or assimilated crust would include biotite, garnet, and sillimanite. The Stone Mountain Granite (Whitney and others, 1975) may be an example of such a granite.

Melting reactions involving biotite occur along a similar set of reactions at higher temperatures (Fig. 7). The reactions involving the high-temperature stability of the magnesium end-member phlogopite have been studied by Luth (1967).

Phlogopite + Quartz = K-feldspar + Enstatite + Liquid.

Similar reactions occur for iron-rich biotites (Wones and Eugster, 1965), but at somewhat lower temperatures. Such reactions are also dependent on oxygen activity.

The melts generated by biotite dehydration will vary in geochemical characteristics, depending on the composition of the protolith. Melting

TABLE 3. EXPERIMENTAL DATA ON THE CAPE ANN, WESTERLY, AND MOUNT AIRY GRANITES

Run no.	Wt. % H_2O added	T (°C)	Time (hr)	Vol. % glass	Phase assemblage
Cape Ann Granite					
Q31	1.00	680	725	Tr	Q, Ks, Pl, Bi, Ox, Ac, L
Q32	3.04	680	725	Tr	Q, Ks, Pl, Bi, Ox, Ac, L, V
Q33	4.75	680	725	Tr	Q, Ks, Pl, Bi, Ox, Ac, L, V
Q34	9.96	680	725	Tr	Q, Ks, Pl, Bi, Ox, Ac, L, V
Q 1	1.00	700	746	ND	Q, Ks, Pl, Bi, Ox, Ac, L
Q 2	1.77	700	862	ND	Q, Ks, Pl, Bi, Ox, Ac, L
Q 3	2.85	700	862	ND	Q, Ks, Pl, Bi, Ox, Ac, L
Q 4	5.65	700	862	60	Q, Ks, Pl, Bi, Ox, Ac, L, V
Q 5	11.91	700	862	65	Q, Ks, Pl, Bi, Ox, Ac, L, V
Q 6	0.89	720	767	15	Q, Ks, Pl, Ox, Ac, L
Q 7	1.94	720	767	30	Q, Ks, Pl, Ox, Ac, L
Q 8	2.98	720	767	45	Q, Ks, Pl, Ox, Ac, L
Q 9	4.86	720	767	75	Q, Ks, Pl, Ox, Ac, L
Q10	9.97	720	767	90	Q, Ks, Pl, Ox, Ac, L, V
Q11	1.02	740	742	20	Q, Ks, Pl, Ox, Ac, L
Q12	2.00	740	725	40	Q, Ks, Pl, Ox, Ac, L
Q13	2.95	740	742	55	Q, Ks, Pl, Ox, Ac, L
Q14	4.94	740	742	80	Q, Ks, Pl, Ox, Ac, L
Q15	10.07	740	725	95	Q, Ks, Ox, Ac, L, V
Q16	1.00	760	897	20	Q, Ks, Pl, Ox, Ac, L
Q17	1.97	760	744	40	Q, Ks, Pl, Ox, Ac, L
Q18	2.97	760	744	60	Q, Ks, Pl, Ox, Ac, L
Q19	5.01	760	897	85	Q, Ox, Ac, L
Q20	9.72	760	744	95	Ox, Ac, L, V
Q21	1.00	780	778	25	Q, Ks, Pl, Ox, Ac, L
Q22	2.00	780	778	45	Q, Ks, Pl, Ox, Ac, L
Q23	2.92	780	778	70	Q, Ks, Pl, Ox, Ac, L
Q24	4.96	780	778	90	Q, Ox, Ac, L
Q25	9.91	780	897	97	Ox, Ac, L, V
Q26	0.99	800	756	30	Q, Ks, Pl, Ox, Ac, L
Q27	1.99	800	756	50	Q, Ks, Pl, Ox, Ac, L
Q28	2.99	800	756	75	Q, Ox, Ac, L
Q29	4.84	800	756	92	Ox, Ac, L
Q30	10.00	800	756	99	Ox, Ac, L, V
Westerly					
W54	0.99	680	725	..	Q, Ks, Pl, Bi, Ox, V
W55	2.91	680	725	..	Q, Ks, Pl, Bi, Ox, V
W56	4.98	680	725	..	Q, Ks, Pl, Bi, Ox, V
W57	10.00	680	725	..	Q, Ks, Pl, Bi, Ox, V
W19	1.00	700	746	Tr	Q, Ks, Pl, Bi, Ox, L
W20	2.04	700	791	Tr	Q, Ks, Pl, Bi, Ox, L, V
W21	2.95	700	791	Tr	Q, Ks, Pl, Bi, Ox, L, V
W22	5.00	700	791	Tr	Q, Ks, Pl, Bi, Ox, L, V
W23	9.40	700	791	Tr	Q, Ks, Pl, Bi, Ox, L, V
W24	1.04	720	767	ND	Q, Ks, Pl, Bi, Ox, L
W25	1.83	720	767	ND	Q, Ks, Pl, Bi, Ox, L
W26	3.02	720	767	ND	Q, Ks, Pl, Bi, Ox, L
W27	4.99	720	725	45	Q, Ks, Pl, Bi, Ox, L, V
W28	10.19	720	767	55	Q, Ks, Pl, Bi, Ox, L, V
W49	0.99	740	766	ND	Q, Ks, Pl, Bi, Ox, L
W50	1.99	740	766	ND	Q, Ks, Pl, Bi, Ox, L
W51	2.97	740	766	ND	Q, Ks, Pl, Bi, Ox, L
W52	4.99	740	766	ND	Q, Ks, Pl, Bi, Px, Ox, L, V
W53	9.93	740	766	70	Q, Ks, Pl, Bi, Px, Ox, L, V
W34	1.00	760	744	20	Q, Ks, Pl, Bi, Px, Ox, L
W35	1.93	760	744	35	Q, Ks, Pl, Bi, Px, Ox, L
W36	2.94	760	744	50	Q, Ks, Pl, Bi, Px, Ox, L
W37	5.00	760	897	75	Ks, Bi, Px, Ox, L
W38	10.85	760	744	80	Ks, Bi, Px, Ox, L, V
W39	1.00	780	897	20	Q, Ks, Pl, Bi, Px, Ox, L
W40	1.98	780	779	40	Q, Ks, Pl, Bi, Px, Ox, L
W41	3.00	780	779	60	Q, Ks, Pl, Bi, Px, Ox, L
W42	5.12	780	779	85	Ks, Bi, Px, Ox, L
W43	9.89	780	779	90	Bi, Px, Ox, L, V
W44	1.00	800	766	35	Q, Ks, Pl, Bi, Px, Ox, L
W45	1.88	800	766	60	Q, Ks, Pl, Bi, Px, Ox, L
W46	3.07	800	766	80	Q, Ks, Pl, Bi, Px, Ox, L
W47	4.97	800	766	90	Bi, Px, Ox, L
W48	9.94	800	897	95	Bi, Px, Ox, L, V

TABLE 3. (Continued)

Run no.	Wt. % H₂O added	T (°C)	Time (hr)	Vol. % glass	Phase assemblage
Mount Airy					
MA 1	0.81	700	766	ND	Q, Ks, Pl, Bi, Ox, L
MA 2	1.76	700	766	ND	Q, Ks, Pl, Bi, Ox, L, V
MA 3	3.24	700	766	ND	Q, Ks, Pl, Bi, Ox, L, V
MA 4	5.02	700	766	30	Q, Ks, Pl, Bi, Ox, L, V
MA 5	9.37	700	766	35	Q, Ks, Pl, Bi, Ox, L, V
MA 6	0.99	720	809	ND	Q, Ks, Pl, Bi, Ox, L
MA 7	1.97	720	725	ND	Q, Ks, Pl, Bi, Ox, L
MA 8	3.04	720	809	ND	Q, Ks, Pl, Bi, Ox, L
MA 9	4.96	720	809	65	Q, Ks, Pl, Bi, Ox, L, V
MA10	9.84	720	809	70	Q, Ks, Pl, Bi, Ox, L, V
MA11	0.94	740	725	20	Q, Ks, Pl, Bi, Ox, L
MA12	1.97	740	742	35	Q, Ks, Pl, Bi, Ox, L
MA13	2.98	740	742	50	Q, Ks, Pl, Bi, Ox, L
MA14	4.94	740	742	65	Q, Pl, Bi, Ox, L, V
MA15	9.74	740	742	75	Q, Pl, Bi, Ox, L, V
MA35	1.00	760	893	20	Q, Ks, Pl, Bi, Ox, L
MA36	2.01	760	893	45	Q, Ks, Pl, Bi, Ox, L
MA37	2.99	760	893	65	Q, Pl, Bi, Ox, L
MA38	4.99	760	893	75	Q, Pl, Bi, Ox, L
MA39	9.89	760	893	85	Q, Pl, Bi, Ox, L, V
MA21	0.98	780	771	25	Q, Ks, Pl, Bi, Ox, L
MA22	1.99	780	771	50	Q, Pl, Bi, Ox, L
MA23	2.99	780	771	70	Q, Pl, Bi, Ox, L
MA24	4.94	780	771	85	Q, Pl, Bi, Ox, L
MA25	9.51	780	771	90	Q, Pl, Bi, Ox, L, V
MA26	0.99	800	756	30	Q, Ks, Pl, Bi, Ox, L
MA27	1.99	800	756	55	Q, Pl, Bi, Ox, L
MA28	3.00	800	756	75	Q, Pl, Bi, Ox, L
MA29	4.96	800	756	90	Pl, Bi, Ox, L
MA30	9.97	800	756	95	Bi, Ox, L, V

Note: experimental procedure as described by Whitney (1969). Assemblages re-evaluated using criteria of Whitney (1972, 1975a), especially for the presence of a vapor phase. Abbreviations as follows: Tr = trace, ND = not determined, Q = quartz, Ks = K-feldspar, Pl = plagioclase, Bi = biotite, Px = pyroxene, Ac = acmite, Ox = Fe-Ti oxides, L = silicate melt, V = hydrous vapor. Glass percentages determined by point counting on powdered samples and are not accurate to better than ±5%. Oxygen fugacity was unbuffered but is thought to be somewhat above Ni-NiO. Lower oxygen activity may stabilize amphibole in the Cape Ann composition.

will again begin in lithologies containing Na-rich plagioclase through reactions of the form

Biotite + Na-Plag. + Quartz = K-feldspar + Pyroxene + Liquid.

The K/Na ratio of the melt will vary but in general will be near 1 or less. Depending on the abundance of sodium and calcium in the source region, the resulting plutons may be granitic to granodioritic in composition. The melt may be meta-aluminous to peraluminous, but would rarely be as rich in alumina as muscovite-generated melts unless muscovite had played an early roll in beginning anatexis. The initial temperature for biotite-generated melts would be about 750 to 850 °C. Because many protoliths contain biotite, other characteristics will be quite variable. If the source is metasedimentary, initial strontium and oxygen ratios may be high. If, on the other hand, the protolith was relatively young metavolcanic and volcaniclastic rocks, these ratios may be lower. Similarly, parameters such as the oxygen activity will depend on the protolith. Graphitic metasediments may yield low initial oxygen activity, whereas more oxidized metavolcanic units would yield higher values. The activity of oxygen is important in determining the abundance of magnetite as an accessory mineral (Ishihara, 1977). Under low oxygen activity, the iron remains in silicate phases, whereas under higher activity, it forms magnetite. Therefore, magnetite may or may not be important in biotite-generated granitoids. Mafic schlieren or enclaves may have a variety of mineralogies but are not extremely common. The Elberton Granite (Stormer and Whitney, 1980) and the so-called "New Hampshire series" plutons of northeastern Vermont may be of this type.

The most refractory of the common hydrous phases is hornblende. It is rare to see anatexis involving hornblende without the introduction of water. Migmatite within the aureole of the Bear Mountain Pluton on the Smith River of the Klamath Mountains (Snoke and others, 1981) may be an example. Such reactions may be more important in the deep crust where temperatures are higher. The approximate conditions of hornblende melting in gabbroic compositions determined by Wyllie (1971) are also shown in Figure 7.

The composition of hornblende-generated melts will depend on protolith and temperature (Helz, 1979). In general, the K/Na ratio is low, with the resulting magma being granodioritic to dioritic in composition. The magmas generated will be meta-aluminous to peralkaline in composition. The initial temperatures may be quite high and may range as high as 1000 °C. Because melting is occurring at such a high temperature, the initial magmas are quite dry. Crystallization therefore occurs over a long temperature range. Early accessory minerals such as pyroxene or even fayalitic olivine may be present, but the stability of magnetite versus fayalite will depend on oxygen activity. The most likely protoliths which would be high in amphibole content would be amphibolites and hornblende gabbros. Amphibole anatexis could be important in some alkaline granitic provinces such as the White Mountain Magma Series of New Hampshire or the Quincy-type granites in Massachusetts. It is also possible to generate both quartz-normative

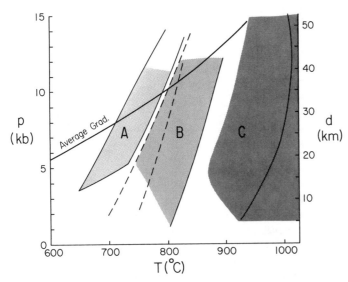

Figure 7. Pressure-temperature diagram showing the approximate conditions of dehydration melting of muscovite, biotite, and hornblende. See text for references. (A) Conditions of direct dehydration melting of muscovite-bearing assemblages. (B) Conditions of direct dehydration melting of biotite-bearing assemblages. (C) Conditions of direct dehydration melting of hornblende-bearing assemblages.

and nepheline-normative silicic magmas by stepwise fractional melting (Presnall and Bateman, 1973) of amphibole. At first melting, all available quartz and some feldspar would react with the amphibole to form a high-temperature, granitic melt. If this melt were removed, the residual amphibole and pyroxene would be peralkaline. Subsequent higher-temperature melting of the residue could yield nepheline-normative melts.

Thus, the origin of anatectic granites can be discussed in terms of the hydrous phases involved (Hyndman, 1981). Table 4 summarizes some of the characteristics expected from various melting reactions. Most of the granites classified as S-type in the original classification of Chappel and White (1974) would be generated by dehydration melting of muscovite and biotite. At low to moderate pressures, cordierite would be a product of the melting reaction at high temperatures. At lower temperatures and moderately higher pressures, however, cordierite would not form (Zen, 1987). Some of Chappel and White's I-types could be generated by biotite or biotite plus hornblende melting reactions. The type of protolith which could be involved is limited, however, to non-sedimentary materials by the low strontium and oxygen isotope ratios in such granites. Some I-types and some more alkaline granites could be generated by hornblende melting reactions. For the large-volume silicic batholiths so commonly associated with subduction zones, such as found in the Sierra Nevada range or the Andes of South America, however, it is very probable that other sources of volatiles exist that are related to the large volumes of andesites and basalts being generated from below the crust. In these terrains, there are very small percentages of true granites in the sense used by Tuttle and Bowen (1958; rocks in which two-thirds or more of the feldspar is alkali feldspar). The most common lithologies are intermediate granites (formerly termed "adamellite" or "quartz monzonite"), granodiorites, and tonalites.

Volatiles from Subducted Oceanic Crust and Mantle

One of the great advances since 1958 has been our understanding of plate tectonics. We now know that many of the great granitic batholiths are formed on compressive plate margins where they underlie basalt-andesite-rhyolite volcanic belts. Further, we now know from the study of hydrothermal processes at mid-ocean ridges and obducted ophiolites that the subducted oceanic crust and upper mantle contain significant quantities of volatiles in the form of hydrated alteration minerals (perhaps as much

TABLE 4. CHARACTERISTICS OF VARIOUS ANATECTIC GRANITES

Hydrated phase involved in melting	Muscovite	Biotite	Hornblende
Alkali/Al characteristics	Peraluminous	Peraluminous–meta-aluminous	Meta-aluminous–peralkaline
K/Na ratio	High, >1	Intermediate, ~1 or so	Low, <1
Initial temperature	650–750 °C	750–850 °C	>900 °C
Delta ^{18}O	High, ~10	Variable, as much as 10 or so	Lower, 7–8
$(^{87}Sr/^{86}Sr)_0$	High, >0.710	Variable, in many case >0.710	Lower, 0.703–0.710
Accessory minerals	Epidote, allanite, tourmaline, topaz, garnet, sillimanite, andalusite, ilmanite.	Garnet, allanite, sillimanite, cordierite, ilmenite, pyroxene	Pyroxene, magnetite, sphene

as 2 wt. % of the crust). Therefore, these volatiles are a potential source of water for the formation of granitic melts in the crust. The method of transport and exchange, however, is potentially complex (see Marsh, 1984, for a quantitative evaluation of the problem of magma ascent). Recently, numerous authors have recognized that mafic enclaves found within calc-alkaline batholiths represent magma mixing during which more-mafic magmas are quenched against the silicic host (Eichelberger, 1980; Gerlach and Grove, 1982; Van Bergen and others, 1983; Bacon and Metz, 1984; Bacon, 1986; Cantagrel and others, 1984; Dorais, 1987). Such a process could be important in the transport of volatiles.

Figure 8 illustrates the hydrothermal processes at mid-ocean ridges and surrounding sea-floor crust. During this process, water is fixed in the crust and upper mantle as hydrous minerals including chlorite, amphibole, and serpentine. Studies of alteration mineral chemistry as well as mass balance of hydrothermal fluids suggest that chlorine as well is present in hydrous minerals. Sulfur is also deposited in the form of sulfides in vein fillings and exhalative deposits. We know that this alteration system extends to rather great depths in the oceanic crust and upper mantle. Taylor (1983) and Gregory and Taylor (1981) have shown that pervasive hydrothermal exchange with sea water has occurred throughout the upper 6 km of oceanic crust and locally has penetrated into the tectonized peridotites of the upper mantle. Therefore, by the time the oceanic crust and upper mantle have cooled and moved away from the ridge, they are highly hydrated.

When oceanic crust and upper mantle are subducted, most of the sediments are left behind in the trench area. These materials are low in density, and seismic studies suggest that they are accumulated as a deformed sedimentary wedge. As the altered crust is subducted, it slowly heats as it interacts with the mantle. As it does, the hydrous minerals will begin to break down, first through dehydration reactions and then through melting. The products will be either tholeiitic basalts or andesites, depending on the availability of water (Green and Ringwood, 1968; Green, 1976; Mysen and Boettcher, 1975; Wyllie, 1979). Because the amount of water available is only 1% to 2%, however, most of the primary melts are probably basalt, with andesites being generated at shallower depths (for example, Maaloe and Petersen, 1981). These pass through the overlying continental lithosphere and perhaps part of the asthenosphere.

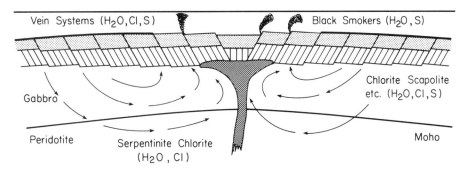

Figure 8. Hydrothermal circulation around a mid-ocean ridge. Alteration of the upper crust deposits water and chlorine in the alteration assemblages. Deeper circulation deposits water and chlorine through the formation of chlorite and serpentine. The crust is also relatively oxidized by the process. Hydrothermal vents on the ocean floor and vein fillings in the upper crust also fix sulfur. Circulation model after Gregory and Taylor (1981).

The products are the copious basalts and andesites associated with island arcs and compressive plate margins. Melting of peridotite, however, does not form true granite. Unless the continental lithosphere has a much higher silica activity (that is, it is made of quartz eclogite rather than peridotite), the generation of most granites must involve the lower crust. Thus, we find that true granites are absent in island arcs which do not involve continental crust.

When magmatism first begins in a compressive plate margin, the mafic magmas encounter lower crust at temperatures of around 500 to 700 °C or so. This process is schematically shown in Figure 9 (T_1). Under these conditions, much of the early mafic magma will tend to crystallize in the lower crust (Marsh, 1978). During this process, the crust may be thickened and partially underplated with hydrous gabbroic material. Although some of the water will be fixed as hornblende and/or biotite, part of the dissolved volatiles may be released to interact with the lower crust. As the lower crust heats, anatexis can begin where water is available. The melting relations in the Smith River area described by Snoke and others (1981) may be a small-scale analogue of what might happen around a crystallizing pluton in the lower crust. As the lower crust continues to heat, anatexis can begin over larger regions as the dehydration reactions of various phases are surpassed (Fig. 9, T_2). Thus, the passage of subduction-related magma through the crust may transfer heat and perhaps volatiles to the crust to cause melting if sufficient basalt traverses the crust (Marsh, 1984).

It is possible, however, that the mafic melts and their dissolved volatiles may interact more directly with granitic melts once silicic magmas begin to accumulate within the crust. Once a body of granitic magma begins to coalesce into a sizable coherent body, it forms a density and thermal barrier for upwelling mafic magmas. If a hydrous andesite or basalt penetrates a granitic magma chamber, the denser mafic magma will tend to underplate the less dense silicic magma (Fig. 10). Because the mafic magma has a higher crystallization temperature, it will also tend to partially quench against the cooler granitic pluton. Such a process has been hypothesized by a number of authors (for example, Eichelberger, 1980; Bacon, 1986; Dorais, 1987) to explain the mafic enclaves that are so common in the Sierra Nevada and other great calc-alkaline batholiths associated with compressive plate margins. Recently, Dorais (1987) has conducted a detailed geochemical study, including ion-probe data, on early phenocrysts of such enclaves in the Dinkey Creek pluton of the Sierra Nevada batholith and a volcanic analogue from the San Juan Moun-

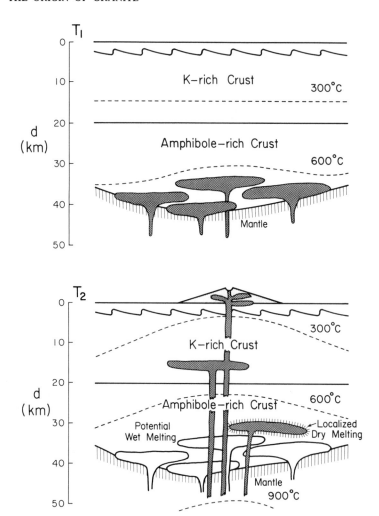

Figure 9. Thermal and magmatic development in the crust above an evolving subduction zone. T_1 = early in the volcanic history. T_2 = after substantial period of calc-alkaline magmatism.

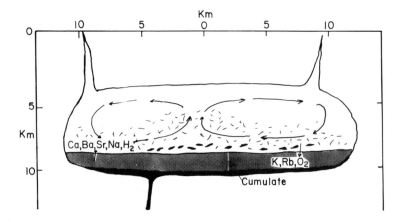

Figure 10. Underplating of a silicic magma chamber by a mafic magma of higher density. Modeled from the Carpenter Ridge Tuff, Whitney and others (1988). More-mafic magma underplates the chamber, partially quenches against the cooler silicic melt. Temperature and mass transfer are accomplished through mechanical and chemical diffusive methods. In the case of the Carpenter Ridge, the mafic enclaves become enriched in K, Rb, and oxygen, whereas the granitic magma becomes enriched in Ca, Ba, Sr, Na, and hydrogen.

tains, Colorado. Figure 11 is Dorais' model for such a system. Under these conditions, both thermal energy and volatiles can be transferred to the granitic system. Convection and mixing in the granitic magma will transport thermal energy throughout the body, aiding in additional assimilation, and at times disrupting the mixed, semi-solid boundary layer, strewing fragments of the mixed zone throughout the magma chamber. The chemistry and isotopic systematics of the resulting pluton will depend on the relative contributions of old continental crust, young hydrous gabbroic crust, and subduction-derived more-mafic magmas to the system, as well as subsequent fractionation and assimilation processes in the upper crust.

In this model, the volatiles responsible for large-scale melting of the lower crust in compressive plate margins would at least in part be derived from altered, subducted oceanic crust and mantle transported into the lower crust and released during crystallization of the hydrated mafic magmas, which would then directly interact with cooler, less dense silicic magmas. Such a system could explain why chlorine is so prominent in calc-alkaline plutons in compressive plate margins, while being far less common in tensional or anorogenic environments.

Granitic plutons formed by such a mixing process can be quite variable in composition because their characteristics depend on the nature of granitic melts formed in the crusts, the composition of more mafic magmas which interact with them, the relative quantity of such mafic magmas, and the degree of interaction. First, because basaltic and andesitic systems are rich in sodium, these systems will tend to have a lower K/Na ratio than will granitic magmas formed from crustal anatexis. If the degree of interaction is slight, they would still be intermediate granites, but with greater quantities of mafic component, they would be granodioritic to tonalitic or trondhjemitic. Second, they would tend to be meta-aluminous (that is, $K_2O + CaO + Na_2O > Al_2O_3 > Na_2O + K_2O$) in most cases. Their initial temperature would vary depending on the relative magnitude of mafic versus felsic magma, and thus the amount of heat transferred, but would in general be higher than in anatectic systems (that is, 800 °C or more) involving muscovite or biotite. The initial isotopic ratios would also be variable but in general, would be more primitive. Common mafic minerals would be hornblende and possibly pyroxene accompanying biotite. The activity of volatiles such as chlorine and sulfur might be expected to be higher than in many anatectic granites. The state of oxidation could also be variable. If the more mafic melts were derived from oxidized oceanic crust and upper mantle without the opportunity for reduction by primitive mantle, then the oxidation state could be high, yielding accessory minerals such as magnetite and sphene as is seen in many cases in arc-related rocks. If, on the other hand, the more mafic magmas had equilibrated with olivine-rich mantle, then they would be reduced, resulting in lower oxygen fugacity.

In most cases, such granites should contain mafic enclaves formed from the mafic magma which contain minerals such as hornblende and sphene with signs of interaction with the granitic melt. These enclaves may contain signs of quenching, such as acicular apatite (Wyllie and others, 1962; Vernon, 1983). They also contain a poikilitic mesostasis consisting of quartz, potassium feldspar, and sodic plagioclase. This material represents either residual liquids from partial quenching of the basaltic parent (Bacon, 1986) or commingling with the host silicic melt (Sparks and Marshall, 1986).

CONCLUSIONS

The origin of granitic melts in the crust requires several weight-percent water. The source and amount of water available have a direct effect on the temperatures at which melting occurs, the subsequent chemical composition of the magma, and the subsequent intrusion and crystallization history. Although all groups of rocks form a continuum and therefore can never be classified into neat pigeonholes, an alternate method of classification of associations of granites is one based on the sources of water. Hyndman (1981) suggested such a scheme for granites derived by anatexis involving hydrous phases. In cases in which volatiles from a subducted oceanic plate are involved, however, it is likely that the dehydration of crustal material will also occur, and so a mixture of magmatic characteristics may be encountered.

If the fluxing action of water carried down by subduction is important in the remelting of crustal materials to form granites, then the ultimate development of thick granitic continental crust may be directly related to the fact that the Earth is largely covered by water. Planetologists should investigate whether the differences in crustal development between Venus and the Earth could be affected by surface conditions which do not allow liquid water on our sister planet.

It is interesting to note that the Earth appears to be the only terrestrial planet with a thick continental crust and highly mobile crustal plates and is also the only one covered with water. As we begin to examine the planet as a complete system, we may increasingly discover that surface phenomena and conditions can affect the internal processes of the planet.

ACKNOWLEDGMENTS

Research was supported by National Science Foundation Grant EAR-8719700. Many of the ideas expressed have developed through interactions with many colleagues and students, including J. C. Stormer, Jr., M. J. Dorais, and numerous members of the U.S. Geological Survey. The manuscript was greatly improved by reviews from B. D. Marsh and P. C. Ragland. Errors in fact or concept, however, are solely the responsibility of the author.

REFERENCES CITED

Andersen, D. J., and Lindsley, D. H., 1985, New (and final!) models for the Ti-magnetite/ilmenite geothermometer and oxygen barometer [abs.]: EOS (American Geophysical Union Transactions), v. 66.
Bacon, C. R., 1986, Magmatic inclusions in silicic and intermediate volcanic rocks: Journal of Geophysical Research, v. 91, no. B6, p. 6091–6112.
Bacon, C. R., and Metz, J., 1984, Magmatic inclusions in rhyolites, contaminated basalts, and compositional zonation beneath the Coso volcanic field, California: Contributions to Mineralogy and Petrology, v. 85, p. 346–365.
Baldasari, A., 1981, Iron-titanium oxides in the Elberton Granite [M.S. thesis]: Athens, Georgia, University of Georgia, 72 p.
Bohlen, S. R., Pecor, D. R., and Essene, E. J., 1980, Crystal chemistry of metamorphic biotite and its significance in water barometry: American Mineralogist, v. 65, p. 55–62.
Brown, W. L., and Parsons, I., 1981, Toward a more practical two-feldspar geothermometer: Contributions to Mineralogy and Petrology, v. 76, p. 369–377.
Buddington, A. F., and Lindsley, D. H., 1964, Iron-titanium oxide minerals and synthetic equivalents: Journal of Petrology, v. 5, p. 310–377.

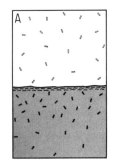

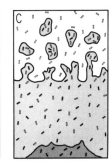

Figure 11. Detail of the partially quenched boundary layer between silicic and mafic magma. After Dorais (1987). Boundary layer develops (A), interacts through chemical diffusion (B), and is partially dismembered by convective movement in the overlying magma (C), yielding mafic enclaves which are themselves a chemical/mechanical mixture between the two parent magmas.

Burnham, C. W., 1979, Magmas and hydrothermal fluids, *in* Barnes, H. L., ed., Geochemistry of hydrothermal ore deposits (2nd edition): New York, John Wiley & Sons, p. 71–136.

Burnham, C. W., and Davis, N. F., 1974, The role of H_2O in silicate melts: II. Thermodynamic and phase relations in the system $NaAlSi_3O_8$-H_2O to 10 kilobars, 700° to 1000°C: American Journal of Science, v. 274, p. 902–940.

Burnham, C. W., Holloway, J. R., and Davis, N. F., 1969, Thermodynamic properties of water to 1000°C and 10,000 bars: Geological Society of America Special Paper 132, 96 p.

Cantagrel, J. M., Didier, J., and Gourgaud, A., 1984, Magma mixing: Origin of intermediate rocks and "enclaves" from volcanism to plutonism: Physics of Earth and Planetary Interiors, v. 35, p. 63–76.

Chappel, B. W., and White, A.J.R., 1974, Two contrasting granite types: Pacific Geology, v. 8, p. 173–174.

Day, H. W., 1973, The high temperature stability of muscovite plus quartz: American Mineralogist, v. 58, p. 255–262.

Day, H. W., and Fenn, P. W., 1982, Estimating the P-T-X_{H_2O} conditions during crystallization of low calcium granites: Journal of Geology, v. 90, p. 485–508.

Dietrich, R. V., 1961, Petrology of the Mount Airy "granite": Virginia Polytechnic Institute Engineering Station Bulletin Series 144, v. 54, no. 6, 63 p.

Dorais, M. J., 1987, The mafic enclaves of the Dinkey Creek granodiorite and the Carpenter Ridge Tuff: A mineralogical, textural and geochemical study of their origins with implications for the generation of silicic batholiths [Ph.D. dissert.]: Athens, Georgia, University of Georgia, 185 p.

Eggler, D. H., 1972, Water-saturated and undersaturated melting relations in a Paricutin andesite and an estimate of water content in the natural magma: Contributions to Mineralogy and Petrology, v. 34, p. 261–271.

Eggler, E. H., and Burnham, C. W., 1973, Crystallization and fractionation trends in the system andesite-H_2O-CO_2-O_2 at pressures to 10 kb: Geological Society of America Bulletin, v. 84, p. 2517–2532.

Eichelberger, J. C., 1980, Vesiculation of mafic magma during replenishment of silicic magma reservoirs: Nature, v. 288, p. 446–450.

El-Hindi, M. A., 1977, Geology and mineralization of the southern part of the Pinos Altos area, Fort Bayard quadrangle, Grant County, New Mexico [M.S. thesis]: Golden, Colorado, Colorado School of Mines.

Ernst, W. G.,1976, Petrologic phase equilibria: San Francisco, California, Freeman Press, 333 p.

Fridrich, C. J., and Mahood, G. A., 1987, Compositional layers in the zoned magma chamber of the Grizzly Peak Tuff: Geology, v. 15, p. 299–303.

Gerlach, D. C., and Grove, T. L., 1982, Petrology of Medicine Lake highland volcanics: Characterization of endmembers of magma mixing: Contributions to Mineralogy and Petrology, v. 80, p. 147–159.

Ghiorso, M. S., 1984, Activity/composition relations in the ternary feldspars: Contributions to Mineralogy and Petrology, v. 87, p. 282–296.

Green, D. H., 1976, Experimental testing of equilibrium partial melting of peridotite under water saturated, high-pressure conditions: Canadian Mineralogist, v. 14, p. 255–268.

Green, N. L., and Usdansky, S. I., 1986, Ternary-feldspar mixing relations and thermobarometry: American Mineralogist, v. 71, p. 1100–1108.

Green, T. H., and Ringwood, A. E., 1968, Genesis of the calc-alkaline igneous rock suite: Contributions to Mineralogy and Petrology, v. 18, p. 105–162.

Gregory, R. T., and Taylor, H. P., 1981, An oxygen isotope profile in a section of Cretaceous oceanic crust, Samail ophiolite, Oman: Evidence for delta ^{18}O buffering of the oceans by deep (>5 km) seawater-hydrothermal circulation at mid-ocean ridges: Journal of Geophysical Research, v. 86, p. 2737–2755.

Haggerty, S. E., 1976, Opaque mineral oxides in terrestrial igneous rocks: Mineralogical Society of America Short Course Notes, v. 3, p. 1–100.

Hammarstrom, J. M., and Zen, E-An, 1986, Aluminum in hornblende: An empirical igneous geobarometer: American Mineralogist, v. 71, p. 1297–1313.

Helz, R. T., 1979, Alkali exchange between hornblende and melt: A temperature-sensitive reaction: American Mineralogist, v. 64, p. 953–965.

Hildreth, E. W., 1977, The magma chamber of the Bishop Tuff: Gradients in temperature, pressure, and composition [Ph.D. dissert.]: Berkeley, California, University of California, 328 p.

——— 1979, The Bishop Tuff. Evidence for the origin of composition zonation in silicic magma chambers, *in* Chapin, C. E., and Elston, W., eds., Ash-flow tuffs: Geological Society of America Special Paper 180, p. 43–75.

Holloway, J. R., 1976, Fluids in the evolution of granitic magmas: Consequences of finite CO_2 solubility: Geological Society of America Bulletin, v. 87, p. 1512–1518.

Hyndman, D. W., 1981, Controls on source and depth of emplacement of granitic magma: Geology, v. 9, p. 244–249.

Ishihara, S., 1977, The magnetite series and ilmenite series granitic rocks: Mining Geology (Japan), v. 27, p. 293–305.

James, R. S., and Hamilton, D. L., 1969, Phase relations in the system $NaAlSi_3O_8$-$KAlSi_3O_8$-$CaAl_2Si_2O_8$-SiO_2 at 1 kilobar water vapour pressure: Contributions to Mineralogy and Petrology, v. 21, p. 111–141.

Jones, W. R., Hernon, R. M., and Moore, S. L., 1967, General geology of the Santa Rita quadrangle, Grant County, New Mexico: U.S. Geological Survey Professional Paper 555, 144 p.

Lindsley, D. H., 1976, The crystal chemistry and structure of oxide minerals as exemplified by the Fe-Ti oxides, *in* Oxide minerals: Mineralogical Society of America Short Course Notes, v. 3, p. L.1–L.88.

Lipman, P. W., 1971, Iron-titanium oxide phenocrysts in compositionally zoned ash-flow sheets from southern Nevada: Journal of Geology, v. 79, p. 438–456.

Lipman, P. W., and Mullinax, D. R., 1981, The 1980 eruption of Mount St. Helens, Washington: U.S. Geological Survey Professional Paper 1250.

Luth, W. C., 1967, Studies in the system $KAlSiO_4$-Mg_2SiO_4-SiO_2-H_2O: Inferred phase relations and petrologic applications: Journal of Petrology, v. 8, p. 372–416.

Luth, W. C., Jahns, R. H., and Tuttle, O. F., 1964, The granite system at pressures of 4 to 10 kilobars: Journal of Geophysical Research, v. 69, p. 759–773.

Maaloe, S., and Petersen, T. S., 1981, Petrogenesis of oceanic andesites: Journal of Geophysical Research, v. 86, p. 10273–10286.

Marsh, B. D., 1978, On the cooling of ascending andesitic magma: Royal Society of London Philosophical Transactions, v. A288, p. 611–625.

——— 1981, On the crystallinity, probability of occurrence, and rheology of lava and magma: Contributions to Mineralogy and Petrology, v. 78, p. 85–96.

——— 1984, Mechanics and energetics of magma formation and ascension, *in* Boyd, F. R., Jr., ed., Explosive volcanism: Inception, evolution, and hazards: Washington, D.C., National Academy Press, p. 67–83.

Melson, W. G., and Hopson, C. A., 1981, Pre-eruption temperatures and oxygen fugacities in the 1980 eruption sequence, *in* Lipman, P. W., and Mullinax, D. R., eds., The 1980 eruption of Mount St. Helens, Washington: U.S. Geological Survey Professional Paper 1250, p. 641–648.

Mysen, B. O., and Boettcher, A. L., 1975, Melting of a hydrous mantle: II. Geochemistry of crystals and liquids formed by anatexis of mantle peridotite at high pressures and high temperatures as a function of controlled activities of water, hydrogen, and carbon dioxide: Journal of Petrology, v. 16, p. 549–593.

Naney, M. T., 1977, Phase equilibria and crystallization in iron- and magnesium-bearing systems [Ph.D. thesis]: Stanford, California, Stanford University.

Naney, M. T., and Swanson, S. E., 1980, The effect of Fe and Mg on crystallization in granitic systems: American Mineralogist, v. 65, p. 639–653.

Nekvasil, H., and Burnham, C. W., 1987, The calculated individual effects of pressure and water content on phase equilibria in the granite system, *in* Mysen, B. O., ed., Magmatic processes: Physicochemical principles: Geochemical Society Special Publication 1, p. 433–445.

Owen, J. A., 1983, A lithogeochemical survey of the Pinos Altos pluton, Silver City, New Mexico [M.S. thesis]: Athens, Georgia, University of Georgia, 172 p.

Potter, P. M., 1981, The geology of the Carlton quadrangle, Inner Piedmont, Georgia [M.S. thesis]: Athens, Georgia, University of Georgia, 89 p.

Presnall, D. C., and Bateman, P. C., 1973, Fusion relations in the system $NaAlSi_3O_8$-$CaAl_2Si_2O_8$-$KAlSi_3O_8$-SiO_2-H_2O and generation of granitic magmas in the Sierra Nevada batholith: Geological Society of America Bulletin, v. 84, p. 3181–3202.

Price, J. G., 1985, Ideal site-mixing in solid solutions, with applications to two-feldspar geothermometry: American Mineralogist, v. 70, p. 696–701.

Robertson, J. K., and Wyllie, P. J., 1971, Rock-water systems with special reference to the water-deficient region: American Journal of Science, v. 271, p. 252–277.

Rutherford, M. J., Sigurdsson, H., Carey, S., and Davis, A., 1985, The May 18, 1980, eruption of Mount St. Helens. I. Melt composition and experimental phase equilibria: Journal of Geophysical Research, v. 90, p. 2929–2947.

Smith, R. L., 1960, Zones and zonal variations in welded ash-flows: U.S. Geological Survey Professional Paper 354-F, p. 149–159.

——— 1979, Ash-flow magmatism, *in* Chapin, C. E., and Elston, W., eds., Ash-flow tuffs: Geological Society of America Special Paper 180, p. 5–27.

Smith, R. L., and Bailey, R. A., 1966, The Bandelier Tuff: A study of ash-flow eruption cycles from zoned magma chambers: Bulletin of Volcanology, v. 29, p. 83–104.

Snoke, A. W., Quick, J. E., and Bowman, H. R., 1981, Bear Mountain igneous complex, Klamath Mountains, California: An ultrabasic to silicic calc-alkaline suite: Journal of Petrology, v. 22, p. 501–552.

Sparks, R.S.J., and Marshall, L. A., 1986, Thermal and mechanical contrasts on mixing between mafic and silicic magmas: Journal of Volcanology and Geothermal Research, v. 29, p. 99–124.

Speer, J. A., 1977, Description of the Siloam Pluton in evaluation and targeting of geothermal energy resources in the southeastern United States: U.S. Department of Energy Progress Report, Oct. 1977–Dec. 1977, 4648-1, p. A43–A63.

Speer, J. A., Becker, S. W., and Farrarr, S. S., 1980, Field relations and petrology of the post-metamorphic, coarse-grained granites and associated rocks in the Southern Appalachian Piedmont: IUGS Caledonide Volume, Memoir 2, p. 137–148.

Spencer, K. J., and Lindsley, D. H., 1981, A solution model for coexisting iron-titanium oxides: American Mineralogist, v. 66, p. 1189–1201.

Steiner, J. C., Jahns, R. H., and Luth, W. C., 1975, Crystallization of alkali feldspar and quartz in the haplogranite system $NaAlSi_3O_8$-$KAlSi_3O_8$-SiO_2-H_2O at 4 kb: Geological Society of America Bulletin, v. 86, p. 83–93.

Stormer, J. C., Jr., 1975, A practical two-feldspar geothermometer: American Mineralogist, v. 60, p. 667–674.

——— 1983, The effects of recalculation on estimates of temperature and oxygen fugacity from analyses of multicomponent iron-titanium oxides: American Mineralogist, v. 68, p. 586–594.

Stormer, J. C., Jr., and Whitney, J. A., eds., 1980, Geological, geochemical, and geophysical studies of the Elberton Batholith, eastern Georgia: Georgia Department of Natural Resources Guidebook 19, 134 p.

——— 1985, Two feldspar and iron-titanium oxide equilibria in silicic magmas and the depth of origin of large volume ash-flow tuffs: American Mineralogist, v. 70, p. 52–64.

Swanson, S. E., 1979, The effect of CO_2 on phase equilibria and crystal growth in the system $KAlSi_3O_8$-$NaAlSi_3O_8$-$CaAl_2Si_2O_8$-SiO_2-CO_2 to 8000 bars: American Journal of Science, v. 279, p. 703–720.

Thompson, A. B., 1982, Dehydration melting of pelitic rocks and the generation of H_2O-undersaturated granitic liquids: American Journal of Science, v. 282, p. 1567–1595.

Thompson, A. B., and Algor, J. R., 1977, Model systems for anatexis of pelitic rocks: Theory of melting reactions in the system $KAlO_2$-$NaAlO_2$-SiO_2-H_2O: Contributions to Mineralogy and Petrology, v. 63, p. 247–269.

Tuttle, O. F., and Bowen, N. L., 1958, Origin of granite in the light of experimental studies in the system $NaAlSi_3O_8$-$KAlSi_3O_8$-SiO_2-H_2O: Geological Society of America Memoir 74, 153 p.

Van Bergen, M. J., Ghezzo, C., and Ricci, C. A., 1983, Minette inclusions in the rhyodacite lavas of Mt. Amiata (central Italy): Mineralogical and chemical evidence of mixing between Tuscan and Roman type magmas: Journal of Volcanology and Geothermal Research, v. 19, p. 1–35.

Taylor, H. P., Jr., 1983, Oxygen and hydrogen isotope studies of hydrothermal interactions at submarine and subaerial spreading centers, *in* Rona, P. A., Bostrom, K., Laubier, Lucien, and Smith, K. L., Jr., eds., Hydrothermal processes at seafloor spreading centers: NATO Conference Series 4, v. 12, p. 83–139.

Vernon, R. H., 1983, Restite, xenoliths, and microgranitoid enclaves in granites: Royal Society N.S.W. Journal of Proceedings, v. 116, p. 77–103.

Whitney, J. A., 1969, Partial melting relationships of three granitic rocks [M.S. thesis]: Cambridge, Massachusetts, Massachusetts Institute of Technology, 142 p.

——— 1972, History of granodioritic and related magma systems: An experimental study [Ph.D. thesis]: Stanford, California, Stanford University, 192 p.

——— 1975a, The effects of pressure, temperature, and X_{H_2O} on phase assemblage in four synthetic rock compositions: Journal of Geology, v. 83, p. 1–27.

——— 1975b, Vapor generation in a quartz monzonite magma: A synthetic model with application to porphyry copper deposits: Economic Geology, v. 70, p. 346–358.

——— 1984, Fugacities of sulfurous gases in pyrrhotite-bearing silicic magmas: American Mineralogist, v. 69, p. 69–78.

Whitney, J. A., and Stormer, J. C., Jr., 1976, Geothermometry and geobarometry in epizonal granitic intrusions: A comparison of iron-titanium oxides and coexisting feldspars: American Mineralogist, v. 61, p. 751–761.

——— 1977a, Two-feldspar geothermometry, geobarometry in mesozonal granitic intrusions: Three examples from the Piedmont of Georgia: Contributions to Mineralogy and Petrology, v. 63, p. 51–64.

——— 1977b, The distribution of $NaAlSi_3O_8$ between coexisting microcline and plagioclase and its effect on geothermometric calculations: American Mineralogist, v. 62, p. 687–691.

——— 1985, Mineralogy, petrology, and magmatic conditions from the Fish Canyon Tuff, central San Juan volcanic field, Colorado: Journal of Petrology, v. 26, p. 726–762.

——— 1986, Model for the intrusion of batholiths associated with the eruption of large-volume ash-flow tuffs: Science, v. 231, p. 483–485.

Whitney, J. A., Jones, L. M., and Walker, R. L., 1976, Age and origin of the Stone Mt. Granite, Lithonia district, Georgia: Geological Society of America Bulletin, v. 87, p. 1067–1077.

Whitney, J. A., Dorais, M. J., Stormer, J. C., Jr., Kline, S. W., and Matty, D. J., 1988, Magmatic conditions and development of chemical zonation in the Carpenter Ridge Tuff, central San Juan volcanic field, Colorado: American Journal of Science Wones Memorial Volume (in press).

Wones, D. R., 1972, Stability of biotite: A reply: American Mineralogist, v. 57, p. 316–317.

Wones, D. R., and Eugster, H. P., 1965, Stability of biotite: Experiment, theory and application: American Mineralogist, v. 50, p. 1228–1272.

Wyllie, P. J., 1971, Experimental limits for melting in the earth's crust and upper mantle, *in* Heacock, J. G., ed., The structure and physical properties of the earth's crust: Geophysical Monograph 14, p. 279–301.

——— 1979, Magmas and volatile components: American Mineralogist, v. 64, p. 469–500.

Wyllie, P. J., Cox, K. G., and Biggar, G. M., 1962, The habit of apatite in synthetic systems and igneous rocks: Journal of Petrology, v. 3, p. 238–243.

Yoder, H. S., Jr., Stewart, D. B., and Smith, J. R., 1957, Ternary feldspars: Carnegie Institution of Washington Yearbook 56, p. 207–214.

Zen, E-An, 1987, Wet and dry AFM assemblages of peraluminous granites and the usefulness of cordierite as the prime indicator of S-type granite [abs.]: Geological Society of America Abstracts with Programs, v. 19, p. 904.

MANUSCRIPT RECEIVED BY THE SOCIETY MARCH 3, 1988
REVISED MANUSCRIPT RECEIVED JUNE 14, 1988
MANUSCRIPT ACCEPTED JUNE 15, 1988

CENTENNIAL ARTICLE

Crystal capture, sorting, and retention in convecting magma

BRUCE D. MARSH *Department of Earth & Planetary Sciences, The Johns Hopkins University, Baltimore, Maryland 21218*

ABSTRACT

The mystery of producing strong compositional diversity among suites of comagmatic igneous rocks is investigated by considering the dynamic evolution of basaltic magma in a sheet-like chamber. A central conclusion is that inward-progressing crystallization produces strong viscosity and temperature gradients that promote convection only near the leading edge of the upper thermal boundary layer. Convection is apparently confined to an essentially isoviscous, isothermal region that hugs the downward-growing roof zone. Strong changes in viscosity with crystallization divide the upper and lower thermal boundary layers into regions of decreasing viscosity and crystallinity (N) called "rigid crust" (N ≥ 0.5), "mush" (0.5 ≥ N ≥ 0.25), and "suspension" (N ≤ 0.25). The strong increase in viscosity near the mush-suspension interface acts as a capture front that overtakes and traps slowly settling crystals. Initial phenocrysts mostly escape capture, but crystals nucleated and grown in the suspension zone can escape only if the capture front slows to a critical rate attainable only in bodies thicker than about 100 m. Escaping crystals are redistributed and sorted by convection driven by the advance of the capture front itself. Crystal-laden plumes traverse the central, hot core of the body and deposit partially resorbed and sorted crystals within the lower suspension zone. Convection is never vigorous but is part of an overall intimate balance between roofward heat loss, rigid-crust growth, crystallization kinetics, and transport and sorting of sinking escaped crystals. There is a strong similarity between these processes and those producing both varves and saline pan deposits. It is clear that lavas, lava lakes, and sills are indeed examples of true magma chambers strongly exhibiting certain aspects of this over-all process. These aspects commonly also characterize the large mafic magmatic bodies. Because strong compositional changes in the residual melt occur largely outward (that is, at lower temperatures and higher crystallinities) of the capture front, which is immobile and mostly within rigid crust, the possible range in comagmatic compositions available for eruption anywhere within the active magma is very limited. This is in broad agreement with the compositional range observed in basaltic lava lakes, sills, and plutons like Skaergaard. The tuning of convection, crystallization kinetics, and phase equilibria in chambers of this type can produce a variety of textures and layering but not a diversity of compositions.

INTRODUCTION

The diversity of the igneous rocks gives petrologists their raison d'être. Every envisaged magmatic process, be it crystal fractionation, mixing, filter pressing, or zone refining, is an attempt to explain the variety of observed magmatic compositions. These processes rely on the fact that, without exception, igneous minerals are of a different chemical com-

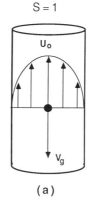

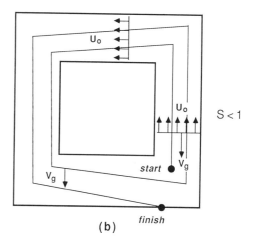

Figure 1. (a) A particle that settles with a velocity V_g in still fluid is suspended by fluid flowing upward with velocity U_o ($S = V_g/U_o = 1$). (b) A heavy particle is transported by flowing fluid. Along horizontal stretches of flow, the particle settles an amount proportional to S. The particle path forms a spiral that eventually intersects the wall. Appropriate adjustment of S and the flow field can make the particle circulate indefinitely.

Geological Society of America Bulletin, v. 100, p. 1720–1737, 10 figs., November 1988.

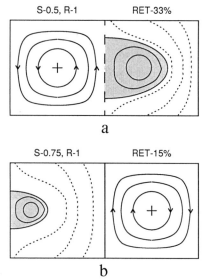

Figure 2. Retention of crystals (stippled areas) in a convecting magma chamber. The solid lines show the streamlines of fluid flow (which also takes place on the right side), and the dotted lines show the paths followed by sinking (a) and floating (b) crystals. Heavy crystals are retained in the center of the body, whereas floating crystals (b) are retained in a torus-shaped region about the midsection of the chamber (after Marsh and Maxey, 1985).

position from their host magma, and separation of liquid from solid gives rise to magmatic diversity.

The most popular of these processes is differentiation through crystal fractionation. Heavy or light crystals sink or float from their point of growth to yield a new magmatic composition. "The continuously changing mother liquor resulting from prolonged fractional crystallization can have every composition ranging from basaltic to granitic . . ." (Bowen, 1947). There is absolutely no doubt, whatsoever, that crystal fractionation is common, for there are systems for which the mass balances connecting distinct compositions seem just too close, in both mineral kind and amount, to believe otherwise.

Yet, just where it seems that diversity should be strong, it occurs hardly at all. Great masses of ocean-ridge and Hawaiian basaltic magma produce vanishing amounts of silicic differentiates. Thick diabase sills and gabbroic bodies like Skaergaard have made precious little progress along the liquid line of descent toward eruptable rhyolite. Perhaps, "cooling has been at such a rate that fractionation is . . . unable to effect the separation of liquid at intermediate stages," and "bodies . . . able to retain their volatile constituents . . . show the normal, natural differentiation series, basalt to rhyolite, as distinct from the laboratory (dry) series approached by the Greenland [Skaergaard] example" (Bowen, 1947). Even very large mafic, ultramafic, and more hydrous bodies, however, do not show much tendency to make silicic liquids. Why is it that the clearest examples of magma chambers, ones we can touch, show little differentiation, whereas the unseen chambers, those spawning suites of lavas, do fractionate?

A set of simple and fundamental questions must be answered: How does crystal fractionation physically happen? What has happened to all the separated (little seen) debris? What is the range of compositional diversity found in any single active magma chamber during fractional crystallization? What does the crystal size and sedimentary style of cumulates tell about the dynamic evolution of differentiating magma? Is only very limited differentiation possible in normal magma chambers? Finally, is the evidence of differentiation found in plutonic studies fundamentally different from that of volcanological studies?

To approach similar questions, Shaw (1965) once said, "we need to know the kind and size of crystals as a function of time, the number of crystals of various changing sizes, the effect of changing composition, temperature, and volume concentration of crystals on the bulk viscosity of the magma, together with a knowledge of temperature as a function of time. . . . If natural or forced convection occurs, . . . the prediction of differential movement of crystals and liquid is greatly complicated."

An approach to filling this tall order is attempted herein by elucidating what is now known of these separate subjects (that is, crystal retention, sedimentation, magmatic rheology and convection, and crystallization kinetics) and then integrating them into a generalized magmatic process applicable to lavas, lava lakes, sills, and other relatively small sheet-like magma chambers. The large layered intrusions are the opposite end member in size to which these processes are also applicable. All such bodies share certain aspects of this over-all process. In this sense, lavas, lava lakes, and sills indeed represent magma chambers accessible for careful study in order to provide insight into the dynamic behavior of magma in inaccessible, subterranean chambers.

CRYSTAL CAPTURE, RETENTION, SORTING, AND RAPID TRANSPORT

Convective Retention

Imagine a single spherical crystal settling in an infinite vertical tube containing upflowing fluid. The crystal settles under the action of gravity at the rate V_g, and the fluid flows upward everywhere at a velocity U. The position of the crystal within the column at any time is determined solely by its initial position and the ratio V_g/U. If $V_g < U$, the flow carries the crystal upward, whereas if $V_g > U$, the crystal settles downward despite the fluid motion. If $V_g = U$, the crystal position is invariant. Clearly, if $V_g/U >> 1$, the flow has little effect on the crystal as it settles, whereas if $V_g/U = 0$, the crystal never leaves the flow. The range of interest, where fluid flow is able to influence settling, is in the range $0 < S < 1$, where $S \equiv V_g/U$.

In a circulating flow, when the fluid moves horizontally, there is no vertical component of motion and the crystal settles unhindered, $S >> 1$. During each stretch of horizontal motion, the crystal settles a certain distance. For example, take the vertical pipe and bend it into the endless, square, vertical tube of Figure 1. Notice that if the horizontal stretches are too long, relative to pipe diameter, the settling crystal may reach the wall and be lost from the flow. Proper adjustment of the horizontal length, however, will always allow retention. The crystal is forever trapped in the flow.

Given a swarm of near-nuclei crystals growing in a circulating magma, it is clear that $S \simeq 0$, there is complete retention, and, temperature permitting, the crystals will be held throughout the flow field (Fig. 2). With continuous crystal growth, S increases, the retention zones decrease in size, and crystals escape the flow. Any distribution of crystal sizes will have a spread of S values and a similar distribution of retention zones. For the simple cellular flow of Figure 2, nuclei retention occurs throughout the chamber, whereas progressively larger crystals (larger S) have progressively smaller retention zones. Growing crystals eventually achieve a large enough value of S during circulation to escape retention and settle to the floor. Similarly, weakening or strengthening of the flow changes the degree of retention accordingly.

Light crystals will float to the roof unless they are held down by opposing fluid motion. In the chamber of Figure 2, the downward flows along the lateral walls produce a torus or doughnut-like retention zone about the midsection of the body. Larger crystals are closer to the walls, whereas smaller ones enjoy a wider range of circulatory freedom.

Iron-rich basaltic magmas, like icelandites, containing heavy opaques and pyroxenes and light plagioclase, could, upon circulation under these ideal conditions, tend to separate these phases into a plagioclase border phase and a central core or upwelling of pyroxenes and opaques. The initial crystal position within the chamber also influences its ultimate resting place. Crystals grown near the walls are mainly influenced by flow in that region. Heavy crystals in a strong downward flow are certain to reach the floor. Even crystals growing at the roof, because they enter the outer reaches of the flow field, are less likely to survive. Thus, whereas a single crystal would ordinarily take an inordinately long time to settle through still magma, convection can greatly hasten settling. Overall, light crystals nucleating and growing near the walls and heavy crystals in the core region are most likely to circulate and attain retention.

The simple, idealized cellular flow field of Figure 2 can be easily replaced by any desired flow by adding its stream function to that for crystal sinking (or floating). (The actual possible style of convection will be considered in a later section.) That is, by defining a particle stream function as

$$\phi = \psi(x,z) + V_g x \qquad (1)$$

where x and z are, respectively, the horizontal and vertical distances and $\psi(x,z)$ is the stream function describing the pattern of fluid motion, the zone of retention can be found for any arbitrary flow field. Details of the methods of calculation are given by Marsh and Maxey (1985) and Weinstein and others (1987). The behavior of crystals in magma flowing near a wall, as in dikes, which is a broadly related problem, has been considered by Komar (1972a, 1972b, 1976).

Capture

In studying Shonkin Sag laccolith (Weed and Pirsson, 1895, 1901), Osborne and Roberts (1931; see also Hurlbut, 1937) perceptively imagined that a downward-solidifying crust of magma might overtake slowly settling crystals, capture them, and incorporate them into the crust itself. Downward crust growth is a form of convection, and capture is again possible only if $S = V_g/U < 1$, where U is now the velocity of crust growth. Rapid crust growth ($S \to 0$) captures all crystals at their points of nucleation, or all pre-emplacement phenocrysts at their initial positions (Fig. 3). Slow crust growth or rapid crystal settling ($S \gg 1$) produces a layer of phenocrysts on the floor. Between these two extremes of straight and foot-shaped profiles lie any number of intermediate crystal distributions. Because, at the earliest instant of crust growth, the solidification velocity is very large, all but exceptionally large crystals are captured near the roof. Crust grows systematically slower (see later) with time (or distance downward), capturing fewer and fewer crystals. Pre-emplacement phenocryst concentration decreases downward and then increases, owing to basal accumulation (Fig. 3). This general interplay between crust growth and crystal settling produces the S-shaped modal profiles so commonly displayed by sills (Walker, 1940; Walker and Poldervaart, 1946; Hess, 1960) and Hawaiian lava lakes (Wright and Okamura, 1977). The overall shape of these profiles has been approximately treated by Gray and Crain (1969).

Consider first only those crystals initially in the body upon emplacement; the deepest point of capture in a Hawaiian lava lake, for example, is the depth at which the crust overtakes the crystal. As is well known from both the Stefan solidification problem and actual measurements of crust growth in Hawaii (for example, Peck, 1978; Wright and others, 1976; Turcotte and Schubert, 1982), crust thickens as

$$Z_2 = C_1 \sqrt{Kt} \qquad (2)$$

where Z_2 is crust thickness, t is time, K is thermal diffusivity, and C_1 is a numerical factor that depends on the rheological definition of the crust.

For reasons that will become clear later (see Convection), I define the leading edge of capturing crust (capture front) as a thermal front below which viscosity decreases downward by no more than a factor of 10 as the hottest temperature is reached. If one takes into account both liquid composition and concentration of solids, for typical Hawaiian basalt this is at a temperature of 1125 °C, whence $C_1 K^{1/2} = 0.158$ cm/s$^{1/2}$. Crust begins at zero thickness, thickens rapidly, and with time, slowly decreases in rate of thickening (that is, dZ_2/dt). Crystals beginning at a depth Z_o are assumed to settle according to Stokes' law. At any subsequent time, the crystal is found at a depth

$$Z_1 = V_g t + Z_o \qquad (3)$$

where changes in initial crystal size during settling are neglected (that is, no growth).

If at any time $Z_1 \leq Z_2$, crystal capture occurs. The deepest point of capture is where 2 and 3 lie tangent to one another (Fig. 4), which occurs when their slopes are equal. This gives the time of capture as

$$t_c = \left(\frac{C_1 K^{1/2}}{2 V_g}\right)^2 \qquad (4)$$

and the depth of maximum capture, from 2, is

$$Z_c = \frac{C_1^2 K}{2 V_g} \qquad (5)$$

The deepest initial position (Z_o) is obtained by substituting 4 and 5 into 3

$$Z_o = \frac{C_1^2 K}{4 V_g} \qquad (6)$$

which shows that $Z_o = Z_c/2$. That is, a captured crystal cannot have been initially deeper than half the capture depth. Of course, the faster a crystal sinks, the smaller are both Z_c and Z_o. Fast crystals are hard to capture.

An olivine crystal of radius a and density contrast ($\Delta\rho$) 0.6 g/cm^3 in a Hawaiian basaltic lava lake of viscosity 10^4 poises (Shaw and others, 1968; Murase and McBirney, 1973) at the capture front will be captured at a depth

$$Z_c = \frac{C_1^2 \, \mu K}{2(2/9) \Delta\rho g a^2} = \frac{95.5 \times 10^{-2}}{a^2} \qquad (7)$$

where V_g is given by Stokes' law and g = 980 cm/s^2. The deepest initial position of the same crystal is given by

$$Z_o = \frac{C_1^2 \, \mu K}{4(2/9) \Delta\rho g a^2} = \frac{47.8 \times 10^{-2}}{a^2} \qquad (8)$$

Thus, for a crystal of 1 mm (0.5 mm radius), deepest capture is at a depth of about 4 m, and its deepest initial position is about 2 m.

Wright and Okamura (1977; see also Helz, 1980) find, however, capture of olivines of this size to depths on the order of 10 m, which means that because the rate of isotherm advancement is well known, the viscosity estimate is too low by a factor of about 2.5. This is certainly reasonable, especially in light of the strong correlation of high vesicle concentrations and olivine occurrence in the lava lakes (see also Mangan and Helz, 1985; Helz, 1987). Bubbles act essentially as solids in terms of their effect on viscosity; their upward streaming causes a resistive flow, and their attachment to crystals lessens the crystal density contrast. The same over-all effect, however, could be produced if the fluid possessed a yield strength, which seems likely with increasing crystallinity (Shaw and others, 1968).

Figure 3. (a) Idealized modal distribution of phenocrysts in a sheet-like body as a function of rate of crystal settling relative to the rate of solidification; (b) the distribution of density, modal olivine, and magnesia in three sheet-like basaltic bodies. Note the over-all similarity in shape of these profiles to that of a when $0 \leqslant S \leqslant 1$.

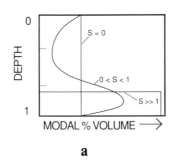

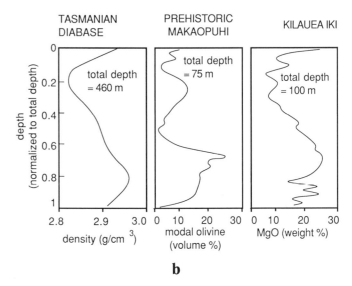

Captured crystals should also decrease in size with depth, which might be somewhat difficult to test if there is appreciable growth or resorption after capture. Careful measurement of crystal size with depth may, nevertheless, allow equation 7 to be fit to these data such that the group of constants $9C_1^2\mu/4\Delta\rho g$ can be found, whence $\mu/\Delta\rho$ can be estimated. On the other hand, if V_g is known with some certainty, the rate of crust growth (that is, C_1) can be found.

Sorting in Sedimentation

An initially homogeneous mixture of particles, differing in size or density, settling in a container, forms, both in the end and during all phases of settling itself, a layered sediment. The number of layers equals the number of distinct buoyancy groups (that is, size or density) within the particle ensemble. If each group is colored, a color-banded, layered sediment forms, as was originally demonstrated for magmas by Coats (1936). The theory of this process has been considered by Kynch (1952), Richardson and Zaki (1954), Smith (1965), Lockett and Al-Habbooby (1973, 1974), and Greenspan and Ungarish (1982).

An illustration of this effect is given in Figure 5. It shows an ensemble of two groups of particles, both of the same density but one consisting of large, and the other small, particles. Equal amounts of each are placed in a test tube of viscous (syrupy) fluid, mixed well (by inclusion of a ball bearing that is free to roll forth and back a few times), and left to settle under the action of gravity. Given that the initial mixture was homogeneous, at the first instant of settling, in a zone nearest the base, sediment begins collecting, for which the particle population is that of the initial mixture (Fig. 5b). Being so near the floor, no differential settling of large and small particles is possible; hence, this sediment is poorly sorted.

By contrast, at the top of the container, large particles soon settle away from the uppermost region, leaving only more slowly settling small particles. Because all of the small particles are identical in size, there will be

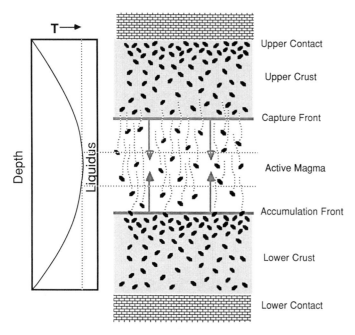

Figure 4a. Schematic depiction of the capture of settling phenocrysts in a sheet-like body by a downward-growing crust and the accumulation of phenocrysts within an upward-growing crust. The phenocryst content decreases downward from the upper contact, but it increases upward from the lower contact. The general variation in temperature shows strong gradients within the crusts and almost no gradient within the central, active region of magma.

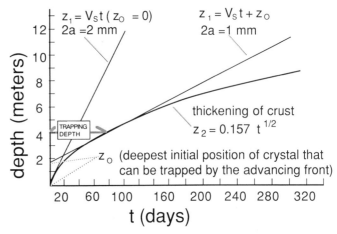

Figure 4b. Graphical depiction of the capture of sinking phenocrysts by downward-growing crust. The deepest point of capture or trapping of any phenocryst is where the crystal settling trajectory is tangent to the advance trajectory of the capture front. A phenocryst of olivine that has a diameter of 1 mm will be captured at no deeper than about 4 m (also see text).

a sharp interface separating the overlying perfectly clear, particle-free fluid from a settling ensemble, for which the top contains only small particles. The demarcation is sharp because mutual hindrance regulates the settling of each particle at the interface to be exactly the same. That is, consider a single particle left above the interface in the clear fluid. With no hindrance, it settles faster until encountering others, whence it slows and takes on the rate of its neighbors. In this regard, the interface is stable to such perturbations, although a local irregularity in particle concentration might produce instability (see later).

Deeper in the mixture, the large particles, having settled from the upper portions, find the going increasingly difficult. More and more large particles cause greater hinderance, due to increasing concentration, and at some level, a second sharp interface forms, above which large particles are absent. In essence, this interface is exactly like the first one except it can be thought of as forming in a dusty fluid. Further below this interface is the final interface marking the transition from settling particles to motionless sediment, which is characterized by a certain porosity. This interface propagates continually upward, meeting first the last large particle interface and finally the clear fluid interface, at which time all settling is complete. The final result is two layers, a lower one, for which the sorting (that is, size classification) improves upward, and a perfectly sorted upper layer (Fig. 5).

Any number (j) of *discrete* buoyancy groups will produce a sediment of j layers. The lowest layer is the least sorted, and the uppermost layer is perfectly sorted. Layering reflects discrete buoyancy groups within the particulate ensemble. A continuous spectrum of buoyancy, with equal numbers of all particles, will not show layering. Because any over-all size range contains (*sensu stricto*) an infinite number of discrete sizes, it is impossible to have equal numbers of a continuous particle size distribution. Hence, layering is the rule, and this very possibly, along with changing physical conditions, also explains the fundamental layered nature of all clastic rocks.

Understanding the temporal development of the layered assemblage in batch sedimentation amounts to being able to calculate settling velocities for any local concentration and size distribution of particles. In detail, it is virtually impossible to understand exactly the contribution to the total drag on a particle due to an arbitrary assemblage of neighbors. Yet, from measuring the settling rate of each boundary, simple and relatively accurate phenomenological functions (Fig. 5) have been found that predict settling velocity in virtually any mixture. For a binary mixture, for example, Lockett and Al-Habbooby (1973) found that the large and small particles settle, respectively, as

$$V_L = V_L^o (1 - N)^{e_L - 1} \quad (9)$$

$$V_S = V_S^o (1 - N)^{e_S - 1} \quad (10)$$

where V^o is the settling velocity in an otherwise particle-free fluid, N is the modal fraction of solids, and e is a number determined separately for each particle class by fluidization of a bed of such particles; e is in many cases in the range 3.0–3.5.

Theoretical formulations for many particle systems hold best at very small ($N \ll 0.1$) concentrations, but these equations, 9 and 10, hold well as high as $N \simeq 0.4$. At higher concentrations, the mixture approaches the point of maximum packing ($N_m \simeq 0.5$–0.6) where all particles move with similar (including zero) velocities (Fig. 5), and no universal formulae are available. The solid-liquid velocity of the packed assemblage is more akin to flow in a porous medium. This emphasizes the important feature that sorting is generally most efficient at low concentrations and slow (that is, small particle Reynolds number) settling rates (but see later).

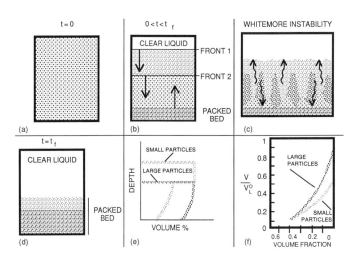

Figure 5. Dynamics of sedimentation (schematic) of two homogeneous populations of crystals of different buoyancy. The initial distribution (a) is homogeneously mixed, and kinematic shocks form and propagate downward (b), separating first clear fluid (top) from the less buoyant crystals, then these from a mixture containing also the more buoyant crystals; finally, an upwardly propagating sediment front separates the packed bed from settling crystals. The fronts may become unstable (c), forming particle fingers that hasten sedimentation into the final distribution of crystals (d and e). The difference in velocity of settling of the two solids during sedimentation (f) is large at small concentrations but converges to zero at concentrations near $N \simeq 0.40$ (also see text).

The sudden changes in particle concentration across an interface give rise to sudden discontinuities in settling velocity, and the attendant interfaces are thus termed "kinematic shocks" by Greenspan and Ungarish (1982). The full history of sedimentation amounts to monitoring the development and migration in time of the kinematic shocks or sedimentation fronts. Using the simple drag law $D = [1 - (N/N_m)]^{-2}$ with $N_m = 0.6$, which is similar to equations 9 and 10, and an unusually complete theoretical formulation, Greenspan and Ungarish (1982) computed a time-distance kinematic-shock map of a sedimenting mixture consisting of an even distribution of 4 particle sizes (diameters of 1.0, 0.8, 0.6, 0.4, in arbitrary units) initially in uniform concentration ($N_i = 0.05, 0.05, 0.05, 0.05; \Sigma N_i = 0.20$). Figure 6 has been constructed from their results.

The slopes of the slanting lines in this diagram give the rate of descent of the shocks

$$V_i = V_L(z^*/t^*) \quad (11)$$

where (z^*/t^*) is the actual numerical slope on the shock map and V_L is the settling velocity of the largest particle. The point of maximum packing, marking the developing sediment layer, propagates upward from the bottom ($z^* = 0$) at a stepwise continuous rate; its velocity is also given by equation 11 with the appropriate values of z^*/t^* as read from the figure. Thus, all trajectories on such a map are refracted when they meet one another. Once a downward-moving front meets the upward-moving sediment interface, sedimentation is complete for that zone or layer. The settling time for each layer is given by $t = (H/W)t^*$, where H is the container height and t^* is a number read from the figure. The total time of sedimentation, indicated by the circle, is when $t \simeq 7.0(H/V_L)$. That is, the time for the entire process is about seven times longer than the time for a single (largest) crystal to settle alone through the distance H. Speed of settling is sacrificed for order. The order is that of grading by hydraulic equivalence, with better sorting increasing upward (Fig. 6).

The over-all time of settling, at least for intermediate concentrations, is considerably shortened by the development of vertical streams of particles (Whitemore, 1955; Weiland and others, 1984). Within the approximate concentration range $0.10 \leqslant N \leqslant 0.30$, instabilities develop in the shock fronts in response to local perturbations in concentration, plumes form for each particle size (Fig. 5c), and sorting is remarkably hastened, although the total time for settling may not necessarily be shortened. Careful observation of the earlier-mentioned test-tube experiment will show such streams over a part of the total settling time. It appears that as long as the initial concentration of particles is less than ~0.3, the plumes will set in at some point. That is, even very dilute concentrations must pass through the critical concentration range of instability. It is still uncertain, however, if this instability is important when there is a large spectrum of crystal sizes.

When a crystal-laden density current travels downward along a magma chamber wall and slides across the floor, it is not difficult to imagine a cloud or suspension of crystals finally coming to rest in a fashion as described above (for example, Irvine, 1987). In a sheet-like body, however, it seems unlikely that single crystals can settle through a thick column of magma in any reasonably short time such that layer after layer of uniform crystals could be repeatedly deposited much as windless snowfalls. This has been well recognized by many people (for example, Wager and Deer, 1939; Jackson, 1961; Hess, 1960; Irvine, 1970; among others), and each has suggested some form of convective transport.

Rapid Transport

The mystery of the deposition in lakes of thin annual sediment layers or varves consisting of very fine particles of calcite and pollen from the air-water interface has been long recognized. Any single particle takes a substantial part of a year to reach the bottom, yet the purity of the layer itself appears to represent a burst of deposition reflecting rapid sediment transport from top to bottom. Escape from this riddle is that particles do not settle singly, but form rapidly descending particle-laden plumes (Bradley, 1965). A suspension of fine particles develops as a higher-density layer near the surface and becomes gravitationally unstable, forming a family of descending plumes. The plumes travel with a velocity increasing as the square of their radius and so reach the bottom in relatively short times. Upon approaching the bottom, the plumes spread, once again forming a particle-laden layer over the floor and depositing particles in the fashion described already. A sheet-like magma chamber, cooling and crystallizing from above, is apt to undergo similar transport and deposition. A series of experiments by Bradley (1965) reveals the fundamentals of this process.

Fine crystals of calcite sifted onto water first form a lens-like concentration from which a plume descends at more than 50 times the rate of fall of single crystals. When sifted onto a system of two layers, less dense and viscous over more dense and viscous, the particles first fall singly, then accumulate and spread into a thin layer near the interface, and finally give rise to a whole field of descending plumes. With a large field of closely spaced plumes, the return flows are localized and vigorous enough to disrupt the plumes, scattering the crystals. Adjustment of the fluid properties is also possible, such that although there is some concentration near the interface, plumes do not form and particles settle alone. When plumes do form, fluid is extracted along with the particles from whatever location the particles themselves come. That is, if the particles actually grow in the plume source itself, they travel with their original mother liquor and are protected from sudden changes in chemical equilibrium. (This will not be entirely true for magmas wherein temperature changes strongly influence equilibrium.) Last, the plumes themselves may be unstable and emanate from their surface yet smaller plumes, and so on, until the field of plumes is universally of a similar small size, yet much larger than individual crystals themselves.

The key to unsderstanding this whole process and applying it to crystal deposition in sheet-like magmas is recognizing the competition between the rate of settling of single crystals and the rate of instability development of a crystal-laden layer. There is no question whatsoever that single crystals settling in magma obey Stokes' law. That is,

$$V_g = C_g \frac{\Delta \rho_g \, g a^2}{\mu_1} \qquad (12)$$

where C_g is a constant dependent on particle shape (Komar and Reimers, 1978), g is gravity, $\Delta \rho_g$ is density contrast, a is radius, and μ_1 is fluid viscosity.

The instability of a higher-density layer develops over a characteristics time (t_i) of (for example, Whitehead and Luther, 1975; Marsh, 1979)

$$t_i = C_i \frac{\mu_1}{g h \Delta \rho_i} \qquad (13)$$

where C_i is a constant and h is the thickness of the unstable layer. The density contrast $(\Delta \rho_i)$ now is between the particle-laden layer and the underlying fluid, which is related to $\Delta \rho_g$ through $\Delta \rho_i = N(\Delta \rho_g)$, where N is the volume fraction of solids in the layer. Dividing each side of 13 by h gives a measure of the rate of rise of the instability on the scale of the layer thickness. The reciprocal of this rate (V_i) normalized by 12 gives a dimen-

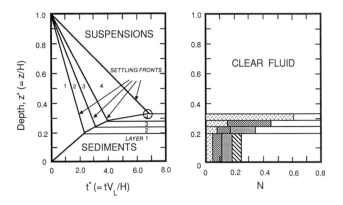

Figure 6. The history of sedimentation as calculated by Greenspan and Ungarish (1982) for four particles of the same initial modal fractions (0.05 each) and density but different sizes (1.0, 0.8, 0.6, 0.4; arbitrary units). The vertical axis is depth (dimensionless), and the horizontal axis is time (dimensionless, see text); H is the tank height and V_L is the unhindered settling velocity of the largest particle. Each line marks the progression of the kinematic shocks, the slope of which gives the rate of descent of the sediment interfaces. These shocks are refracted upon meeting the upward-propagating packed bed interface, the slope of which gives its rate of motion. The point of complete sedimentation is circled, which is at a time of about 7.0. The final distribution of sediment (right) shows progressive upward refinement in crystal size and increasing concentration of the smaller sizes.

sionless measure (B) of the competition between layer instability and settling by single crystals. Namely,

$$B \equiv \frac{V_i}{V_g} = C \left(\frac{h^2}{a^2} N \right) \quad (14)$$

where $C (\equiv 1/C_i C_g)$ is a new constant. Plumes form when $B \gg 1$, and single crystals settle when $B \ll 1$.

Consider a region of dispersed particles of thickness h settling without interaction. h is much larger than a, but if N is small enough, $B \ll 1$ and no plumes form. By holding N at this small value but increasing h, B arbitrarily can be made sufficiently large for equation 14 to indicate plume formation. Similarly, by letting the particles become as small as molecules, the layer simply becomes a more dense fluid overlying a less dense fluid and instability is inevitable. At the other extreme, for a thin layer (that is, small h), plumes form when N/a^2 is large, which is when there are many small particles shoulder to shoulder, with large mutual interaction.

Physical interaction between particles is only possible if the particles are highly concentrated, and only at maximum packing do all particles touch and settle at exactly the same rate (see Sorting). The layer itself then moves very slowly due to porous flow through the packed bed. There are therefore two limits that cause plume formation: (1) very small particles wherein $V_g \ll V_i$ and (2) large N when the bed, for any particle size, becomes packed. [It should also be clear that there is a functional relationship $h = f(N,a)$.] The limit of immediate interest is the instability arising from a mat or slowly settling packed layer of particles. A particle mat or suspended packed bed (that is, $N > N_m$) forms in any given region where the influx of particles at the top is much larger than the outflux from the bottom, which happens any time falling particles encounter a fluid in which they are sufficiently less (negatively) buoyant. Sifting particles from air to water and particles settling through tiers of increasingly dense or viscous fluids are obvious means of forming such mats, as shown by Bradley (1965).

The central lesson here is that plumes will form when particle settling is hindered by mutual interaction, which occurs when N exceeds some critical value or when the particles are so small as to be essentially motionless. Only particle mats produce particle plumes or density currents. The plumes themselves may also be unstable to smaller plume formation, although this is certainly only important at larger velocities (that is, larger Reynolds numbers based on plume diameter).

It is clear from this section that crystals growing about the margins of magma chambers may be transported, distributed, and sorted not only according to the mechanics of suspensions in otherwise still fluid magma but also according to the style and vigor of convection within the chamber. It is the mechanical and chemical (see later) interaction between sedimentation and convection that gives rise to ordered igneous bodies.

CONVECTIVE VIGOR AND STYLE

Convection and Rheology

A layer of constant-viscosity fluid heated from below and cooled from above convects once its thermal gradient reaches a point whereupon conduction alone is insufficient to carry away the supplied heat. Heat essentially piles up at the lower boundary, concentrating buoyancy and forming plumes that rise to the surface. Strong temperature gradients at the lower and upper boundaries enhance heat transfer, and the temperature at the center of the layer is the average of the boundary temperatures, that is, $(T_1 + T_2)/2$. How large the temperature gradient must be to offset viscous drag in the fluid and initiate convection is measured by the dimensionless Rayleigh number,

$$Ra = \frac{\alpha g \Delta T\, L^3}{\nu K} \quad (15)$$

where α is thermal expansion, L is layer thickness (Fig. 7a), ν is kinematic viscosity, and K is thermal diffusivity. Ra must rise beyond some critical value that is dependent on geometry and boundary conditions but most commonly is in the range 120–2,000. For tall, thin chambers cooled by cold walls there is no critical Ra, for any horizontal gradient in temperature (and hence density) promotes flow. With increases in Ra, both the

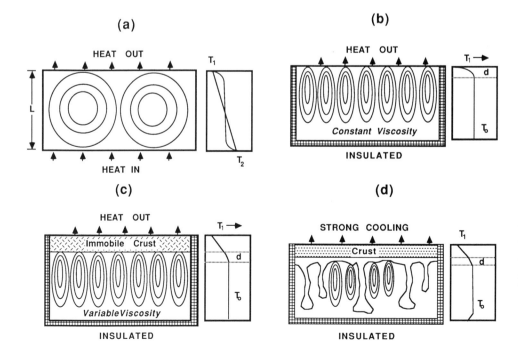

Figure 7. Possible styles of convection within a sheet of magma. Heated from below and cooled from above (a) with constant viscosity is greatly different from the same fluid cooled only from above (b) and otherwise insulated. With strongly variable viscosity (c), convection is confined to essentially an isoviscous and isothermal part of the layer, and most of the buoyancy of the cool upper thermal boundary layer is locked within the rigid crust. Under strong cooling from above (d), the form of convection may be more plume-like and intermittent.

vigor and complexity of convection also increase. All aspects of the dynamic behavior and longevity of convection in magma involves Ra.

Magma chambers are different from a simple heated layer in two important ways. They are not continually heated from below, but rather are cooled on at least their top and outer sides (Fig. 7b), and magma viscosity is extremely sensitive to temperature through the buildup of solid particles (Marsh, 1981). When magma acquires ~55% solids by volume, it is closest packed and becomes a dilatant material. Any shear then causes expansion of the body, and the magma may congest or plug the chamber. This rheological transition is clearly evidenced by the absence of lavas containing more than about 55% phenocrysts and by the sudden transition from hard, drillable rock to mushy, viscous magma in Hawaiian lava lakes (Shaw and others, 1968; Wright and Okamura, 1977). (Such rheological transitions are examined in more detail below.)

This rheological behavior makes all magma at temperatures less than that at some critical crystallinity part of the wall rock or container holding the actively convecting magma. In a layer heated from below, for example, the central temperature is still the average of the two boundary temperatures, but now the boundaries are those of only the actively flowing fluid, not of the original container (Fig. 7c). The crust locks up or isolates a great deal of the buoyancy normally available to drive the flow, for only the slightly cooled fluid inside the critical viscosity interface can fall back and stir the deeper fluid. The rate of convection is thus regulated by the rate of heat transfer through this rigid, partially molten crust and the original wall rock beyond.

An estimate of the maximum heat transfer for such a system is possible by comparing the heat flux from a continually (artificially) well-mixed, constant-viscosity layer to that of a stagnant layer cooling purely by conduction under exactly the same conditions (Marsh, 1988a, 1988c). Continual mixing with no crust formation allows a maximum of exactly twice as much heat transfer from the layer. This magical factor of 2 simply reflects the fact that the temperature at the contact is always that of the well-mixed (that is, isothermal) magma, whereas for a stagnant layer, the initial contact temperature is the average of the initial magma and wall-rock temperatures. This result greatly limits the vigor of convection, and it holds for any geometry of chamber, whether tall or sheet-like. Thus, the quandary in early thermal modeling (Lovering, 1935) over the possible importance of convection in greatly speeding solidification is now much less bothersome.

Vigor

The rate of magmatic convection can be measured by the magnitude of Ra. In convective systems (for example, Marsh, 1988a), the characteristic velocity (U) of convection is

$$U \simeq \frac{K}{L} Ra^{2b} \quad (16)$$

where b is a constant exponent that depends on the boundary conditions but typically is in the range 1/3 to 1/5. The total rate of heat transfer (Q_T) relative to a state of steady conduction alone (Q_{cd}) is measured by the dimensionless Nusselt (Nu) number.

$$Nu \equiv \frac{Q_T}{Q_{cd}} \simeq Ra^b \quad (17)$$

Combining these equations gives

$$U \simeq \frac{K}{L} Nu^2 \quad (18)$$

A common dimensionless measure of velocity, the Reynolds number (Re $\equiv UL/v$) can now be formed

$$Re \simeq Pr^{-1} Nu^2 \quad (19)$$

where Pr (= v/K), the Prandtl number, measures the rate of transfer of momentum relative to heat in the fluid. For magmas $v > \sim 10^2$ cm^2/s and K $\simeq 10^{-2}$ cm^2/s (Murase and McBirney, 1973), making Pr > 10^4. For the well-mixed body relative to stagnant, time-dependent cooling, Nu $\leq$ 2, but this is relative to a different basic state than that measured by 17. Converting Nu $\leq$ 2 to the same steady basic state as that of 17 increases the maximum Nu typically to no more than about 10 (Marsh, 1988c), whence

$$Re \leq \sim 10^{-2} \quad (20)$$

Such a low value suggests that the convective motion is slow and sluggish as long as the magma is not superheated.

If the magma is initially *superheated,* a possible though unlikely condition, Ra alone can be used to estimate vigor. Using equation 15 and choosing a characteristic length scale as a chamber of thickness of L $\simeq 10^5$ cm (1 km), plus $\alpha = 5 \times 10^{-5}$ °C^{-1}, $v = 10^2$ cm^2/s, K = 10^{-2} cm^2/s, g = 10^3 cm/s^2, and $\Delta T = 20$ °C, means that

$$Ra = 10^{15}$$

From 17, using b = 1/3,

$$Nu \simeq 10^5$$

and from 19,

$$Re \simeq 10^5$$

Turbulence generally sets in at Re $\simeq$ 2,000 (depending on Pr), and so this result implies highly vigorous convection, almost 10^{10} times faster than for Nu $\simeq$ 10. What is wrong here? What is inconsistent? How can the same equations yield two such vastly different estimates of Nu and Re?

The differences lie in the choice of the thermal regime of basic state of the system. For the bound that Nu $\leq$ 2, the basic state is the time-dependent conductive cooling of the body. For the result that Nu $\simeq 10^5$, the basic state is a linear temperature gradient ($\sim \Delta T/L$) throughout the body; the convective heat flux must be thus much larger than that of the basic state just to approach a heat flux of Nu $\simeq$ 2 using the time-dependent basic state. In terms of cooling history, both results give the correct result, but one mustn't construe any undue physical meaning, such as a state of extreme turbulent convection, from these large values of Ra and Nu. In fact, the vigor of convection is only slightly above the stable (that is, not stagnant) conductive state of cooling. Convection is thus necessarily a transient process linking stable initial and final thermal states.

This transient stage can be appreciated by imagining a tank of otherwise insulated fluid cooled from above (Fig. 7b–7d). A cooled layer grows downward from the top until it becomes unstable (just as for particle mats), sinks, and grows again. This process occurs regardless of the depth of the fluid; the process itself has no direct means of "knowing" the depth of the fluid. It is in essence a surface effect allowing more efficient cooling of the layer. Many experimental and analytical studies show that the form of convection is completely independent of layer depth (see review in Marsh, 1988a). The product of Ra_L (based on full layer thickness L) and the dimensionless time ($t_L = Kt/L^2$, to the 3/2 power) for convection to commence is found to be a constant.

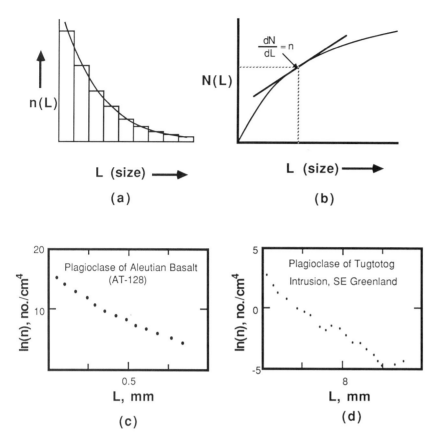

Figure 8. Population density, n(L), in a crystal size distribution (CSD) is related to the cumulative size distribution (b) through the derivative dN/dL. Measurements of N as a function of L also give measures of n. The measured CSD's of plagioclase in an Aleutian basalt (c) and in the Tugtotog pluton of southeast Greenland (d) are shown. Although there is a large difference in the over-all distribution of crystal size, each CSD is smooth and quasi linear. There is no bimodality, but a continuous range of crystal size.

$$Ra_L\, t_L^{3/2} = \text{Constant} \qquad (21)$$

or

$$\frac{\alpha g \Delta T\, L^3}{\upsilon K}\left(\frac{Kt}{L^2}\right)^{3/2} = \text{Constant} \qquad (22)$$

or

$$Ra_t \equiv \frac{\alpha g \Delta T\, [(Kt)^{1/2}]^3}{\upsilon K} = \text{Constant} \qquad (23)$$

The full layer depth L drops from the relation (because of its irrelevance), and the proper length scale appears as $(Kt)^{1/2}$, which, from diffusion theory, is clearly the thickness of the growing, cool boundary layer at the upper contact. The relevant Ra_t is time dependent, starting out small—not large as for Ra_L—and increasing to the point of instability, which is generally for $Ra_t \simeq 200$–300. The cool layer is probably only a few meters thick at initial instability. The vigor of convection is regulated by the rate of heat transfer through the roof rock, which is strongest at the start. Convection will increase in strength in proportion to the amount of superheat. Eventually, with warming of the roof rock, loss of superheat, and thickening of the crystallizing crust, heat flow will gradually drop, and convective vigor will again diminish systematically.

The additional effect of a strong variation in viscosity is now to lock up most of the cool margins of the magma chamber and allow only their leading edge, or sublayer, to go unstable. The sublayer is therefore only a fraction of the full thickness of the full cool boundary layer of the chamber, and so convective vigor is further diminished. To convection, the degree of superheat is not just the temperature increase above the liquidus, but it is a rheological boundary (that is, the sublayer critical crystallinity) beyond which convection is impeded. Addition of the effect of the rate of advancement of the solidification front itself also influences convection (Marsh, 1988a).

Style

The form of convection depends on the rate of cooling. Gentle cooling produces vertically elongate cells. Their strength lies in the cool, dense sublayer, but because of viscosity, they drive motion ahead of the advancing solidification front to a depth of about 7 times that of the cool sublayer (Smith, 1988). For strong cooling, the cool sublayer gives forth intermittent bursts of plumes, and it thus forms, destroys itself, and reforms, again and again. Depending on the total thickness of the magma body relative to the sublayer thickness, cool plumes may traverse its hot central region and reach the floor, forming a plume graveyard which stagnates the lowermost region of the chamber (Jaupart and others, 1984; Jaupart and Brandeis, 1986; Brandeis and Jaupart, 1986). Because convection occurs in the same region as crystallization, the flow will contain crystals subject to flow sorting, and the added thermal inertia and density of the solids may tend to favor plumes even for otherwise cellular flow. The over-all form of convection, then, is one of slow, small-scale circulation forward of the advancing crust, with plumes descending through the deeper, otherwise stagnant, reaches of the chamber. Because of the small scale and yet plume-like nature of the flow, it is apt to appear unsteady and chaotic, but this does not mean that it is turbulent. Small, low-density diapirs of differentiated magma may rise from the lower, upward-growing solidification front and collect against and be incorporated into the advancing upper crust (Helz, 1988).

When there is extensive melting of the roof rock, as has evidently happened over some larger ultramafic bodies like Bushveld (for example, Daly, 1928), there is no strong structural member for crust attachment, and the stability of such a unit itself becomes an issue. Cooling and crystallization still occur in the primary magma, but now all of the crystals accumulate according to their pattern of formation and deposition.

CRYSTAL SIZE DISTRIBUTION

Magmatic sedimentation is unique in that the particles nucleate and grow during deposition. A comprehensive analysis of sorting and deposition must of necessity not only appreciate and include crystallization kinetics, but must also be able to describe quantitatively the distribution of crystal sizes throughout the process. That is, modification of the initial crystal size distribution (CSD) is critical to predicting, for example, predominant crystal size, whether fronts will develop during sedimentation, longevity of crystals under adverse temperatures, and effective hydraulic equivalence. The relatively new field of CSD's allows direct access to such information.

CSD Theory

The understanding of particle size distributions in physical and chemical processes has been well developed by Randolph and Larson (for example, 1971). This approach has been applied to igneous and metamorphic rocks by Marsh (1988b), Cashman and Marsh (1988), and Cashman and Ferry (1988). The fundamental concept is the monitoring of crystal population density during any process. The population density (n) is defined as the number of crystals per unit size (for example, per unit size interval), per unit volume of magma. By establishing n on the basis of per unit size (that is, number density), the absolute size range over which the number of crystals is measured becomes immaterial. A histogram of n against crystal size (L, not to be confused with magma chamber thickness) is independent of the size range for each bar of the histogram. The total number of crystals (N) over some size interval 0 to L is then

$$N(L) = \int_0^L n(L)dL \quad (24)$$

which provides a strict definition of n, namely

$$\frac{dN(L)}{dL} = n(L) \quad (25)$$

On a plot of N against L, n is simply the slope of the curve (Fig. 8).

It happens that when ln(n) is plotted against L, the CSD's of many rocks plot as simple curves, in many cases even as straight lines (Fig. 8). This is not entirely surprising when one realizes that this equation (that is, ln(n) = mL + b; m and b are constants) is also a solution to the differential equation describing conservation of particles in many systems. To appreciate this, consider a school arranged such that the grade to which any pupil belongs is determined by her or his size, say, height. A certain number of students are born (nucleated) each year and enter grade 1; as they grow at various rates G_1, G_2, and so on, they move on to higher grades. At any time, the student size distribution of the school can be immediately learned by tallying the numbers in each grade. The net pupil accumulation in any grade is the difference between the rate of student growth into and out of that grade and also the difference between influx and outflux (that is, distinct transfers in and out) of students. Such reasoning leads to a governing equation of the form

$$\frac{\partial n}{\partial t} + G\frac{\partial n}{\partial L} + \frac{Qn}{V} = 0 \quad (27)$$

where it has been assumed that G is not dependent on size (L), the total volume of the system (school) is constant, and nucleation (birth) takes place within V such that only the outflux (Q) is nonzero. For a less accessible system, it is the outflux that furnishes a measure of the CSD. [More general conditions including batch systems are treated by Marsh (1988b).]

If the system can be considered to be at least in a quasi-steady state, the first term is zero, and if it is noted that V/Q ($\equiv \tau$) represents a recharge or residence time, equation 27 becomes

$$\frac{dn}{dL} + \frac{n}{G\tau} = 0 \quad (28)$$

which has the simple solution

$$\ln(n/n^o) = -\frac{L}{G\tau} \quad (29)$$

It may thus be no surprise that CSD's of many lavas are linear on a plot of ln(n) against L.

The slope of the CSD gives a direct measure of the product $G\tau$, from which growth rate itself can be found if τ is independently known. The intercept at L = 0 measures nucleation density (n^o), and the nucleation rate is given by (Randolph and Larson, 1971; Marsh, 1988b)

$$J = n^o G \quad (30)$$

Other useful measures (for example, total number, length, area, and volume or mass) can be found by finding the integral moments of $L^i n(L)$, where i is an integer. From the associated moment number densities, the predominant crystal size (L_d) in each moment distribution i yields the important relation

$$L_d^i = i\, G\tau \quad (31)$$

which says that the predominant crystal size is a simple multiple of $G\tau$. In considering the distribution of mass or volume in the system, for example, $L_d = 3\, G\tau$, whereas the distribution of length (L) gives $L_d = G\tau$. Such simple relations are especially useful in modeling.

The other unique value of this approach stems from the ability to study the effect of chemical and physical processes on the CSD through conservation equations like 27. For example, crystals growing at a rate dependent on size [that is, G = f(L)] can be easily modeled and found to cause curvature in the CSD. When crystal fractionation removes the larger crystals, the ensuing CSD is kinked downward (Fig. 9). A strong cooling event, such as eruption, kinks the CSD markedly upward at the smallest sizes. Annealing moves the grain size to an equilibrium range of small variation, which destroys the smaller crystals at the expense of larger ones.

Crystal Growth Rate

Petrologists have long recognized that variations in grain size across dikes are due to the interrelation of nucleation rate, growth rate, and cooling rate. Early on, Lane (1896, 1902) worked hard to quantify this by combining an accurate thermal conduction model with grain size observed in sills and dikes. The very same idea and technique persists to this day (for example, Gray, 1970; Kirkpatrick, 1977). From accurate observations of cooling rate, initial crystal size, and abundance in lava lakes, the CSD method, through image analysis, has yielded critical information on the crystallization kinetics of basaltic minerals (Cashman and Marsh, 1988).

Nucleation and growth rates are well known, especially in experimental systems, to depend on degree of undercooling, but the actual degree of undercooling is not well known in natural igneous systems. At least for lava lakes, the growth rates of common minerals—plagioclase, olivine, ilmenite, magnetite—all seem to be rather similar, $\sim 10^{-10}$ cm/s, which suggests very small ($<10^{-1}$ C°) undercoolings (Cashman and Marsh,

1988). Nucleation rates are also broadly similar for lava-lake minerals, but seem to vary considerably with cooling rate and/or melt composition. That is, nucleation rates of plagioclase are much smaller in the St. Helens dacite than in the lava lake, even though growth rates are not significantly smaller (Cashman, 1988). This also seems to be so (that is, similar growth rates and dissimilar nucleation rates) for minerals in contact-metamorphic aureoles (Cashman and Ferry, 1988). It is from such evidence that regardless of the exact mechanism of growth, practically speaking, it is reasonable to use a constant growth rate of $\sim 10^{-10}$ cm/s for crystallization during magmatic processes. Nucleation rates and densities, however, depend on the composition and rate of cooling of the system. For present purposes, those exhibited by the lava-lake data will be used. Because during transport and sorting, crystals may either grow or shrink, this growth rate—with appropriate sign—will be used for both processes.

THE COMBINED PROCESS

Introduction

Each of the processes already described is apt to be important within a crystallizing magma chamber. The final crystalline product, however, is not mainly the result of any single process, but is the net result of all these (and more) processes. The ultimate goal herein is thus to interweave these processes and sketch the integrated dynamics of magmatic crystallization, convection, and sedimentation. A general schematic illustration of this attempt is shown in Figure 10, and a brief description follows, which is then followed by a more detailed inspection of the major dynamic features.

Consider a relatively thin ($\leqslant \sim 1$ km) sheet-like magma chamber formed from a single batch of, for example, tholeiitic basalt magma. Examples might be the Palisade (F. Walker, 1940; K. Walker, 1969), Nipissing (Hriskevich, 1968), or Tasmanian (McDougall, 1962) sills, a Hawaiian lava lake, or perhaps even a thick lava flow (for example, Cornwall, 1951). The initial temperature is uniform at the liquidus temperature of 1200 °C, and cooling is primarily through the roof and floor with no significant influence from the walls. Rigid crusts begin forming at the roof and floor. The inner boundary of each crust is at a level where the viscosity is a factor of 10 over that of the deep interior, which is at the liquidus viscosity. [In terms of being eruptable, everything of crystallinity greater than about 50% (by volume) is rigid crust, which would also be so for drilling.] Crystallization commences a distance d inward from the crust, and in the intervening interval, olivine, clinopyroxene, and plagioclase appear in order. The presence of gas bubbles, which are probably often present in the earliest times, is ignored because they are not a general influence throughout the cooling history. The behavior of phenocrysts in the initial magma has already been treated, but the role of new-grown crystals in differentiation remains to be considered.

Convection occurs in response to cooling at the top and the downward advance of the upper crust. The cells are narrow and tall, extending downward from the crystallization interval to a depth of about 7d. The distributions of temperature and solids are both perturbed by this flow, encouraging plume formation near the downwellings. Crystal-laden plumes selectively sample the crystallization interval and traverse the hottest, central region of the layer (Brandeis and Jaupart, 1986). In transit, the crystals are reduced in size by partial resorption due to heating, and at large enough concentrations, they may be sorted by finger-like instabilities within the plumes themselves. Approaching the floor, the plumes, which carry a relatively well sorted assemblage, begin to slow and spread in response to the rise in viscosity associated with the lower crust. Further sorting and sedimentation occurs, producing, through interaction with neighboring plumes, a systematically layered series of cumulates. Further

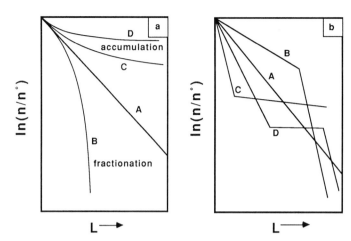

Figure 9. (a) Modification of original crystal size distribution (CSD) A by crystal fractionation B and crystal accumulation C and D. (b) Sudden eruption may produce a strong nucleation event C, which when coupled with crystal fractionation B, can give a complex CSD D relative to the original A.

crystal growth takes place during and after deposition. Early plumes are small and may not reach the floor, whereas late-stage plumes may be so numerous as to disrupt systematic crystal transport by vigorous return flows.

The entire process is one of crystal size refinement and sorting. The degree of order within the solidified body is strongly dependent on the extent to which each process runs to completion. Small bodies will be less ordered than large ones. Relative to gabbroic chambers, granitic bodies will be disordered or of more uniform texture and composition. Quantitative measures of these qualitative descriptions follow.

Rheology and Crystallinity

Crystallization at the top is marked by the downward propagation of a family of isotherms containing the liquidus as the leading isotherm (T_L = 1200 °C) and the solidus T_s at, say, 1000 °C. Between these isotherms, crystallization commences and finishes, viscosity rises, crystal size increases, and density increases (Fig. 10). It is these variations that drive convection and plume formation and, through the sequence of crystallization, furnish the initial sorting of solids. They are important to understand intimately.

Viscosity increases in response to cooling, the change in liquid composition, and the buildup of solids (Fig. 10), which is the strongest effect. The rigid, drillable portion of the crust extends downward to a crystallinity of about 50%–55% by volume (for example, Wright and Okamura, 1977); this is one way to define crust. Another is the level below which thermal convection in the underlying magma is able to deform the suspension of liquid and crystals. From experiments using variable-viscosity fluids (see Marsh, 1988a), it appears that convective shear deformation is possible as long as viscosity does not rise beyond a factor of 10 over the liquidus viscosity. Judging from the actual measurements of Shaw and others (1968) in Makaopuhi lava lake, this limit occurs at a crystallinity of about 20%–25%. Forward of this is a *suspension* (of thickness d) and backward is a *mush*, bordered by the *rigid crust*. Only in the suspension zone are convection and particle sorting possible.

Critical Time to Avoid Capture

The size range of *newly grown* crystals (as opposed to initial phenocrysts considered early on) available for sorting in the suspension zone is limited by the crystal residence or growth time (t_d) in this zone. (By definition, all crystals within and above the mush zone have been captured and are unable to participate further in differentiation.) This is the time required for the trailing delimiting isotherm (crystal capture front) to advance through the distance d, which itself increases with time. That is, as in equation 2, the position of the liquidus is given by

$$Z_1 = C_1(Kt)^{1/2} \qquad (32)$$

and the position of the crystal capture front (Z_2) is a large (~0.90) fraction of this, namely

$$Z_1 = 0.9C_1(Kt)^{1/2} \qquad (33)$$

The suspension-zone thickness is the difference of 32 and 33

$$d = 0.1C_1(Kt)^{1/2} \qquad (34)$$

From 33, the capture front advances at the rate

$$\frac{dZ_2}{dt} = V_2 = \frac{0.9C_1}{2}\left(\frac{K}{t}\right)^{1/2} \qquad (35)$$

The crystal residence time within the suspension zone (d) is then

$$t_d \equiv \frac{d}{V_2} = \frac{2}{9}t \qquad (36)$$

and from 31, the largest crystal (radius) captured is of the order

$$a = \left(\frac{1}{3}\right)Gt \qquad (37)$$

where i in equation 31 has been taken as 3 (that is, dominant size is based on crystal volume or mass) and $a = L_d^i/2$.

Because d increases with time, crystal size at the capture isotherm, which is the boundary between the suspension and mush zones, will also increase with time, but because crystallinity at the capture front remains, by definition, constant, the number of crystals decreases. This is in keeping with the CSD–crystallization kinetics, showing, at least in Makaopuhi, that nucleation rate decreases much more strongly with cooling rate than does growth rate, which is nearly constant. Because the rate of advancement of the capture front decreases with time (that is, equation 35), whereas crystal settling rate increases with time, there is a stage at which newly grown crystals can escape capture. This is found by equating settling rate to the advancement rate of the capture front (equation 35).

$$V_g = \frac{2}{9}\frac{\Delta\rho g(Gt/3)^2}{\mu_o} = 0.45C_1(K/t)^{1/2} \qquad (38)$$

The left side is simply Stokes' law (for a spherical crystal) with equation 37 substituted for crystal radius and μ_o being the viscosity at the capture front. Because this makes the simplifying assumption that no settling takes place until the crystal almost meets the capture front, the crystal size (a) and growth time are maximized. In solving for this critical escape time,

$$t_c = \left[\frac{18.23\ \mu_o\ C_1\ K^{1/2}}{\Delta\rho g G^2}\right]^{2/5} \qquad (39)$$

and using, for example, $C_1 = 1.6$ (see equation 2), $\mu_o = 10^4$ p, $K = 10^{-2}$ cm^2/s, $\Delta\rho = 10^{-1}$ g/cm^3, $g = 10^3$ cm/s^2, and $G = 10^{-10}$ cm/s, the critical time is about 50 yr; increasing the density contrast to 0.5 decreases t_c to about 25 yr. (These times are of course subject to changes in C_1 appropriate for the style of cooling, that is, whether the body is deeply buried, affected by hydrothermal circulation, and so on.) Beyond this time, crystals nucleated and grown within the advancing suspension will be large enough to escape capture. From equation 37, these crystals will have a radius of about 0.5–0.25 mm. They will be the largest crystals, having nucleated at the liquidus, whereas many others will be smaller, having nucleated later, at lesser temperatures. They too will be captured, but with time and slowing of the capture front, increasingly smaller crystals will also be able to escape capture. The size range of crystals escaping capture increases with time.

An added complexity is the fact that over the temperature interval of the suspension zone, more than a single solid phase may appear (Fig. 10). Instead of each phase having the full time t_d to grow before capture, each growth time is modified by the fractional part of d (see equation 36) over which growth is possible. Because each phase also has a different density contrast (and, strictly speaking, different growth rate), the critical capture time, beyond which escape is possible, is different for each phase. In basaltic magma, the critical time is relatively short with olivine as the liquidus phase, whereas because plagioclase appears later and has a small density contrast, it has a longer critical time. Cumulates of newly grown olivine and clinopyroxene are much more likely than of plagioclase. Phases appearing even later, behind the capture front and within the mush zone, have no chance at all of escape unless they are unusually dense for their size. Normally, in and above the mush zone, solidification is advancing much faster than any crystals can settle.

A critical time of about 50 yr implies a marked distinction between what is a small body and what is a large body. That is, using t = 50 in equation 32 gives a characteristic half-thickness of the cooling layer as 60 m; t = 25 gives 42 m. Although there may be some flexibility in these numbers, by this criterion the division between small and large basaltic sills, for example, is at an over-all thickness of, say, ~100 m. This critical time and thickness is not especially sensitive to sensible changes of any parameters in 39 except possibly crystal growth rate. Choosing a much larger growth rate (for example, 10^{-5} cm/s), as is sometimes considered realistic, decreases this time to about 170 days, whence the division between large and small bodies is at a thickness of about 10 m. It is also clear from this analysis that small diabase sills (for example, Hotz, 1953) showing strong internal order (for example, cumulate zones of bronzite) carried these crystals as phenocrysts in the original magma. This also may explain why thick (~100 m) lava flows show little internal order. Sills of highly viscous silicic magma, wherein μ_o may be as large as 10^9 p, would have to be more than ~1 km in thickness before capture could be averted.

Convective Sorting and Plumes

As crystals settle from the advancing capture front, they enjoy a continual decrease in viscosity and settle increasingly faster, ensuring escape. If these crystals are not influenced by convection, they may settle beyond the liquidus and suffer resorption in attempting to traverse the hot central core of the layer. Initial phenocrysts clearly can survive this traverse, but newly grown crystals can survive only if the lower crust is sufficiently near the upper crust. Mangan and Marsh (1988) found that only in the late stages of solidification can such a microcumulate form. Just as in Bradley's (1965) carbonate-varve experiments, crystals presumably can be transported to great depth in the layer if they descend as crystal-laden plumes. The added mass of a flock of such crystals gives a much faster descent, and the entrained liquid attending the crystals gives the plume an enhanced thermal inertia. Because there is no obvious means to produce a mat or layer of solids, the key here is to show that such plumes are a natural consequence of convection within the suspension zone.

As described earlier, the expected form of convection forward of the capture front is that of vertically elongated cells driven by cooling and advance of the suspension zone (Figs. 7 and 10). Combining this flow pattern with an initial uniform distribution of crystals suggests redistribution of crystals into discrete areas along the suspension zone (Fig. 10). Retention occurs at the upwellings, whereas at the downwellings, descending crystals will be concentrated and their descent hastened. The fluid circulates steadily into and ahead of the suspension zone, with the flow being very weak at its highest and deepest points of travel. Any crystals swept into the downwellings will find the flow too weak to recirculate them. The descending crystals will entrain some cool fluid, forming a crystal-laden plume. The size and detailed nature of these plumes is examined momentarily.

In the upwelling retention zones, crystals are trapped within a region the size of which is inversely proportional to S ($\equiv V_g/U$). These crystals continue growing, increasing their value of S and reducing their retention zone, until the flow can no longer support them, whence they sink. The size of the crystal when this happens can be estimated by equating the flow velocity to the settling velocity (that is, S = 1). The settling velocity is given by Stokes' law, and the flow velocity (U) is estimated from the condition that the heat transfer here is very nearly an equal balance between conduction and convection, neglecting latent heat, which is principally emitted at lower temperatures (Marsh, 1988a). That is, from conservation of energy $Ud/K \simeq 1$ or

$$V_g = U \simeq \frac{K}{d} \qquad (40)$$

or

$$a = \left[\frac{9}{2} \frac{\mu K^{1/2}}{\Delta\rho g 0.1 C_1 t^{1/2}} \right]^{1/2} \qquad (41)$$

where equation 34 has been used to remove the dependence on d. Given $\mu = 10^3$ p (that is, leading edge of suspension zone), $K = 10^{-2}$ cm^2/s, $\Delta\rho = 0.5$ g/cm^3, $C_1 = 1.6$, and for t, the critical time of ~50 yr, the size of the largest retained crystal is 0.12 mm; a critical time of 25 yr increases this size to 0.14 mm. Because d increases with time and the temperature gradients lessen, the flow weakens and is able to retain increasingly smaller crystals. It must be cautioned, nevertheless, that equation 40 is only an approximation, and a fuller knowledge of this relation (and also C_1) may change these estimates by as much as a factor of, say, two. But no matter how weak the flow becomes, its mere presence perturbs the distribution of crystals favoring plume formation.

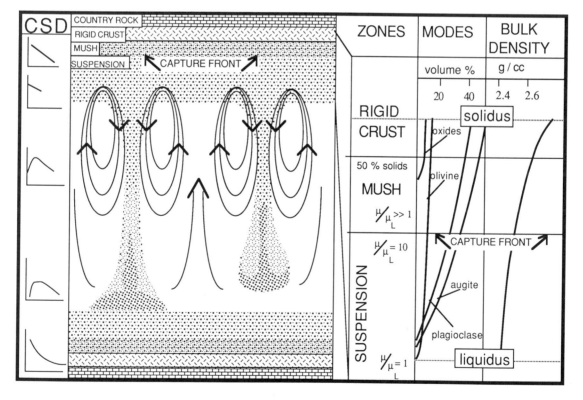

Figure 10. Schematic depiction of crystal fractionation and sorting due to convection beneath the advancing upper-crust capture front and within and forward of the suspension zone in a basaltic chamber. Convectively enhanced sorting during descent of crystal-laden plumes transports crystals from the upper suspension zone to the lower suspension zone. The effect of resorption during transit is reflected in the variation of the crystal size distribution (CSD) (left column) with depth. Fines are destroyed, and the lower suspension zone shows the CSD of a cumulate assemblage. The columns to the right show in detail the (1) rheological demarcations between suspension, mush, and rigid crust as a function of viscosity relative to the viscosity at the liquidus; (2) modal buildup of crystallizing phases; and (3) increase in density with progressive crystallization (see text for further details).

Plume Size and Life

The diameter (D) of plumes emanating from a layer is generally similar to the thickness of the layer itself (for example, Brandeis and Jaupart, 1986; Marsh, 1988a). At first guess, D ~ d, but this is probably an overestimate. Given a static, dense layer overlying a much thicker, less dense layer, there is a certain time (rise time) necessary before instability appears. At the forward edge of d, this rise time (t_1) cannot be exceeded because that portion of d is just newly formed. At the trailing edge of d, there has been plenty of time for instability, but it is easily overcome by the advancing, high-viscosity capture front. Instability, therefore, probably arises initially from a portion (h) of the midsection of the suspension zone (d) and also entrains some lower-viscosity fluid from the leading part of d. The exact fraction (h/d) need not be explicitly stated to develop the following formal argument, but for the sake of clarity, it is not unreasonable to assume h/d < ~½. All characteristics of d (that is, growth, physical properties, and so on) also describe h on a reduced scale. It is still not certain, however, how much of h will go to form plumes, for if there were no solids present, plumes might not form at all (see Fig. 7). In this context, the deep-going crystal plumes essentially represent "leaky" convection cells and will probably be scaled on a small fraction (≤ ~¼) of d. It is difficult to be more certain.

The longevity of these plumes depends on their rate of descent relative to their rate of heating, initial temperature, and amount of solids. For single, purely thermal plumes, Griffiths (1986) has shown that the temperature contrast (ΔT) is reduced to a tenth of its original value in a distance

$$Z_p \simeq 0.37\, D\, Ra_p^{1/2} \quad (42)$$

where Ra_p is the Rayleigh number based on ΔT and the volume of the plume head. Because D ~ h, and if a spherical plume head of radius h/2 is assumed,

$$Z_p \simeq 0.37 \left(\frac{4\pi}{24}\right)^{1/2} \left(\frac{\alpha g \Delta T}{\mu K}\right)^{1/2} h^2 \quad (43)$$

Because the present plumes contain perhaps 10%–20% (by volume) crystals, latent heat increases the effective ΔT, which altogether may be on the order of 50 C°. For basaltic magma,

$$Z_p \simeq 0.13 h^2 \quad (44)$$

such that if h = 10 cm, $Z_p \simeq$ 10 cm, but if h = 10^3 cm, $Z_p \simeq 10^5$ cm. In early times, plumes are not apt to go far, whereas with time in large bodies, as both d and h increase, descent might easily exceed a kilometer. In fact, as long as d grows large enough, there seems no way of prohibiting plume formation. At larger times, however, with thickening of the rigid crust, the head flux diminishes, crust growth slows, and convection itself dies. Plumes are thus prohibited at very short and very long times.

Plume Sorting

Once concentrated into descending plumes, crystals are free to sort themselves according to the general principles of sedimentation described earlier. The flow field within the plume itself will not hinder sorting, for it is broad and even, broadly analogous to, for example, flow in a pipe of the same size. This is mainly true within the trailing stalk of the plume. In the bulbous head, a vortex style flow, forced by the drag of the surrounding fluid, may keep the solids there well stirred. If the concentration in the stalk is in the appropriate range, perhaps 0.10 ≤ N ≤ 0.20, which is quite possible, sedimentation fingers may form, enhancing sorting. Sedimentation shocks will not form early in the descent because the size distribution is continuous. With time, the plume will warm, resorbing the smallest crystals, narrowing the size distribution, and continually refining the sorted assemblages. If there is more than a single solid phase, sedimentation fronts will probably now form. Crystals will not escape the plume, because they themselves hold much of the plume's buoyancy and, in effect, define the plume itself.

Approaching the upwardly propagating suspension zone on the floor, the plume will carry a sorted assemblage of crystals; the degree of sorting will depend on the concentration of solids and on the plume descent time relative to the time necessary for sorting. The time of sorting, based on the results of Greenspan and Ungarish (1982), is about 7 times the time for the largest crystal to settle through the sorting column, but in the optimum concentration range, this may be greatly reduced. The plume descent time depends on the distance to the lower suspension zone, and so as long as the descent distance is not so great as to fully resorb all the crystals, sorting will be best for long descents (where the descent distance is a large part of the resorption distance Z_p).

The original size distribution within the upper suspension zone—big on top—becomes inverted in the plume such that at the floor, the plume billows out as a suspension of already at least partly sorted crystals. These new crystals add to those of the lower suspension in the proper order, big on bottom. Further settling combines the two populations of crystals, but because the smaller crystals have been resorbed in transit, mainly larger crystals are added to the lower suspension zone. These crystals individually sink to their rightful (that is, hydraulic equivalence) positions within the zone. This deposition promotes rapid and discontinuous advancement of the mush-suspension zone interface, forming a layer of large crystals of fairly uniform size.

Phase Equilibria

The key ingredient of this process, dictating the sequence in which phases are added to the accumulating lower crust, is the sequence of hydraulic equivalence of the phases in the upper suspension zone. That is, the order of crystallization of phases in the suspension zone, their growth rates relative to their liquidus position, and their densities govern the nature of the assemblage carried by the plumes to the lower suspension zone. It is also conceivable that should a single phase (for example, chromite) have a much stronger negative buoyancy, for some, however short, time, only that single phase will be carried downward and deposited. The result may appear essentially as an extensive varve-like layer.

Much of the preceding implicitly assumes that the body is thin enough such that the liquidus is relatively unaffected by increasing pressure; a reasonable limiting thickness might be ≤ ~1 km, for which the pressure change is ≤ ~0.3 kbar. In thicker bodies, this effect may completely offset the tendency of heating to resorb crystals, and the plume crystals may grow during the later stages of descent, which may accentuate mode enhancement. (See more on this issue in Discussion.)

Postcumulus Processes

An enhanced modal variation will survive only if crystal size is also enhanced within the cumulate pile. That is, nutrients must be flushed through the mush zone to continue growth of the original suspension assemblage at the expense of new, later phases. Irvine (for example, 1980) has called on upward infiltration of intercumulus liquid in response to compaction of the cumulate pile, which is a common process in saline pans (Lowenstein and Hardie, 1985). Irvine also mentioned the possible concomitant downward infiltration of denser, more primitive liquid. Hess

(1972) gave the first quantitative consideration of intercumulus convection, and the process of light residual fluid loss has been experimentally and analytically modeled by Kerr and Tait (1986). Sparks and others (1985) have examined its application to layered intrusions. All of these works show the ease of adcumulus growth by melt transfusion within cumulates.

Compaction is a process that may enhance size without adding new phases; the solids actually deform to fill the interstitial pore space. The residual melt is driven upward, and conceivably, a monomineralic layer could be produced. Compaction is driven by the difference in pressure gradient between the contiguous solids and liquid. Resistance to deformation is supplied by the viscosity of the solid assemblage. The Irvine hypothesis can be modeled using equations developed by McKenzie (1984). The characteristic time for compaction is

$$t_{cp} = \frac{\left(\mu_b + \frac{4}{3}\mu_s\right)^{1/2}}{\Delta \rho g} \left(\frac{\mu}{K_p}\right)^{1/2} \frac{\epsilon}{(1-\epsilon)^2} \qquad (45)$$

where μ_b and μ_s are, respectively, the bulk (that is, compressive) and shear viscosity of the solids; μ is interstitial liquid viscosity; K_p is permeability; and ϵ is fractional porosity. This has nearly the same form and meaning as in equation 13, where $K_p^{1/2}$ plays the dimensional role of a length scale. Except for μ_b and μ_s, which are very nearly equal ($\sim 10^{19}$ p), and g, all of the quantities depend implicitly on porosity (ϵ). This is because in a closed system, decreasing porosity represents increasing fractionation of the residual melt such that density contrast and melt viscosity both in many cases (not all) increase significantly, whereas permeability decreases. Variations in μ and K_p increase compaction time, but increasing $\Delta \rho$ and decreasing ϵ decrease t_{cp}. The fact is, however, that increases in μ/K_p are far more substantial and greatly outweigh the effects of $\Delta \rho$ and ϵ. As ϵ gets very small, approaching zero, K_p also approaches zero and may, because of the dictates of phase equilibria on melt location, reach this condition before ϵ does.

Compaction is only possible if the time of the compaction is small relative to the time for the solidus to pass through the possible compaction zone, not the time for the whole body to solidify (Sparks and others, 1985). A good measure of this time is t_d from equation 36, which of course increases with thickness of crust Z_1, as in 32. Setting this time equal to that for compaction allows Z_1 to be found at commencement of compaction.

$$Z_1 = \left[\frac{9C_1^2}{2} K \frac{\left(\mu_b + \frac{4}{3}\mu_s\right)^{1/2}}{\Delta \rho g} \left(\frac{\mu}{K_p}\right)^{1/2} \frac{\epsilon}{(1-\epsilon)^2} \right]^{1/2} \qquad (46)$$

If one takes, for example, $C_1 = 1$, $(\mu_b + \frac{4}{3}\mu_s) = 10^{19}$ p, $\Delta \rho = 0.5$ g/cm^3, $\epsilon = 0.1$, $\mu = 10^4$ p, and K and g as usual, then

$$Z_1 = 1.9 \times 10^3 \, K_p^{-1/4} \qquad (47)$$

It is difficult to be precise about values for the permeability (K_p), but from examples given by Sparks and others (1985), a reasonable value might be 10^{-7} cm^2, whence $Z_1 \simeq 1$ km. There is clearly room, however, for a factor of 3 either way in porosity, making K_p ($\simeq 10^{-8}$) uncertain by 10^3, which means that compaction is ruled out with certainty only if $\mu \geq 10^6$ p, the body is thin, or cooling is in the early stages. It is clear that compaction is favored if the interstitial melt has low viscosity or if the body is large (that is, $Z_1 > 1$ km) and also that it will not be important during the early stages of cooling of any body. The compaction of ash flows is an analogous process (Riehle, 1973).

Interstitial viscosity can be kept small if the light residual melt continually escapes upward to be replaced by unfractionated melt. The main difficulty with compaction, however, is that it functions best at small porosities ($<<0.5$), at which point most of the secondary phases have already appeared and contaminated the original enhanced modal assemblage. The ideal sequence of events is thus continual loss of fractionated interstitial melt; leading to adcumulus growth due to infiltration of new, primitive melt; followed by compaction at the latest stages when infiltration flow becomes ineffective. Morse (1986) has presented a detailed discussion of the possible role of convection on adcumulus growth and the thermal effects of crystallization.

DISCUSSION

We see lava flows all the time, thick ones, and never think of them as magma chambers. They really are magma chambers, thin ones that have cooled quickly. Lava lakes, near-surface sills, gabbro sheets, and ultramafic layered intrusions are also all magma chambers. It is their degree of internal order that distinguishes them from one another.

Lava flows initially containing few, if any, crystals show no cumulates because newly grown crystals cannot escape the capture front. Phenocryst-rich lava may flow fast enough to suspend crystals until increasing crystallization (or topography) congests further flow, which dams and ponds the lava (Arndt, 1986). The suspended phenocryst load settles out, forming a cumulate layer, but newly grown crystals cannot normally escape the capture front. Cooling is so effective from the top, perhaps encouraged by rain and flood, that downward- and upward-propagating jointing may meet within about 20% of the bottom; that is, the ratio of upper to lower crust thickness (Z_u/Z_L) is ~ 4 (Long and Wood, 1986).

Many sheet-like gabbroic intrusions, like Bushveld and Muskox, seem to show essentially no upper crust ($Z_u/Z_L \simeq 0$) (Irvine, 1970). The lower crust appears to have grown until it reached the roof, with every crystal escaping the capture front. Some such bodies are a veritable garden of cumulate sorting, and the roof in many cases consists of granophyre apparently sweated out of the country rocks.

Lava lakes occupy the middle ground; their joints meet about 60% of the way down ($Z_u/Z_L \simeq 3/2$) (Evans and Moore, 1967). Their cumulates consist of intratellurically grown phenocrysts with little, if any, sign of escape of later grown crystals from the capture front (for example, Helz, 1988).

Order within these bodies is (inversely) measured by Z_u/Z_L: the more upper crust the less order. Thus, it may be the transport of crystals from roof to floor that produces order, and it is the dynamics of convection and crystallization near a downwardly moving planar, multiphase solidification front that stimulates the necessary style of transport and sorting.

Hess (1960) imagined the occasional instability of a roofward mat of crystals to produce the layering of Stillwater. In their study of Skaergaard, Wager and Brown (1968) saw the same need and agreed with Hess. (They also saw the additional and separate influence of flow down slanting side walls, which is not considered herein.) There is no easy means of making a mat of otherwise heavy crystals near the roof. But plume-like convection in response to cooling through the roof has all the necessary physical characteristics imagined by Hess. Furthermore, this process is by nature intermittent and can be tuned through phase relations and crystallization kinetics to be rhythmic (Brandeis and Jaupart, 1986). For example, when the upper crust is thin, but growing, plumes of only the very largest crystals may reach the floor, but with increasing time and d, plumes themselves become large and may periodically deplete the upper interface, retarding advancement and at the same time causing relatively vigorous stirring, due to the return flow, in the core of the layer. This stirring may be strong enough to disrupt the plumes, as Bradley (1965) saw experimentally, forcing settling by single crystals and producing a graded layer on the floor.

In general, depending on the phase relations, it might be possible that such a process could give rise to monomineralic varves in some intrusions. Irvine (1987) has suggested that large crystal-rich plumes were responsible for the macrorhythmic layering of the Skaergaard intrusion.

Convection, crystallization, and sorting, as envisaged here, are driven by roofward cooling. Concomitant upper crust growth is the central physical feature controlling crystal capture, defined by the phase equilibria, and encouraging initial sorting. Although its physical importance in regulating the flux of crystals to the lower crust is clear in lavas, lava lakes, and many sills, the upper crust may in full form appear only intermittently in certain deeply buried bodies. An upper crust can form only if it can stably attach itself to roof rock or can float on the underlying magma. Although floating would seem unreasonable, strong concentrations of gas bubbles could, as in Hawaiian lava lakes, make it possible. If vesiculation is unimportant and the roof is structurally weak, the upper crust is certain to detach, or not form at all, producing an intermittently unstable roof zone by growing a cool, crystal-rich zone that falls away in response to convection. The easiest means to weaken roof rock is to melt it, either *in situ* or as stoped blocks, and form a granophyre. The described style of convection (Fig. 10) and plume production still occurs, but it now takes place forward of the granophyre, which acts as the mush or rigid crust, depending on its viscosity. Granophyre is siliceous, highly viscous, and light. It is convectively stable, but too weak a structure on which to hang a basaltic crust. This has apparently happened over Muskox and Bushveld.

The presence of granophyre above a magmatic sheet furnishes a significant heat sink, via the needs of latent heat, to speed heat loss and crystallization (Irvine, 1970). As also noted by Irvine, however, the presence of granophyre itself does not necessarily mean that the magma was vigorously convecting nor superheated near its roof. Even with purely conductive cooling, melting is easily accomplished if the roof rock either has a low solidus temperature relative to the magma temperature or has an initial temperature near its own solidus temperature. Both conditions are satisfied if the magma is unusually hot, as ultramafic magma would be, and the roof rock is broadly granitic and lies at some depth in the crust. These seem only partly satisfied by the conditions of emplacement of the large ultramafic bodies, which in the cases of Bushveld and Muskox, were very shallow. Granophyre is also in many cases found associated with near-surface diabase sills (for example, Hotz, 1953) that *may* reflect either strong lateral flow and heat transfer during emplacement or melting of stoped blocks as shown by Walker and Poldervaart (1949).

Although the present investigation is primarily aimed at understanding relatively thin sheets of magma in which pressure has little effect on phase relations, it is reasonable to consider, at least briefly, the effect of pressure, for it has been a point of much discussion for many years (for example, Wager and Deer, 1939; Irvine, 1970; McBirney, 1985). When the sheet or sill is thick, say, more than ~3 km, the pressure increase downward through the body increases the liquidus temperature at about 2–3 times the adiabatic gradient. Cool plumes falling from the roof will at first heat up, fusing some crystals, but with increasing pressure (that is, depth), crystallization may again proceed and the plume will reach the floor still crystal laden. *If* the return flow is crystal free, which (*sensu stricto*) is unlikely, and ascends from the floor, it will follow an adiabat and reach the roof superheated. Cooling could now conceivably repeat the cycle or send crystal-free plumes downward, which may never strike the liquidus. These upward-moving plumes, however, may have lost crystals and density enough to remain at the roof, in spite of cooling, and be incorporated into the upper crust. Thus, the downward plumes may always be initially near their liquidus.

Although in the very earliest times, such a cycle may be possible, with cooling from both roof and floor, it is unlikely for the return flow fluid to come from the base. With plumes, the return flow may be chaotic, with no fluid rising more than a small part of the full chamber height. Also, any loss of crystals from plumes will change the local liquid composition, lessening the liquidus temperature. The liquidus may thus soon become as steep as the adiabat. Further, any crystals within sinking or rising fluid will buffer the temperature near the liquidus; 10° of superheat could be absorbed by fusing 2.5% (by mass) of olivine. Thus, even thick bodies are apt to be everywhere at the liquidus temperature.

The question of initial temperature distribution in thick bodies brings up another question that seems not to have been touched upon before. How are these bodies filled with the initial magma? Is is possible to fill any such layer, even for an instant, with fluid of constant composition and uniform temperature and crystallinity? Judging from the filling of lava lakes and sills, filling may be haphazard. Virtually all magma carries phenocrysts that would be impossible to hold at a homogeneous concentration during transport from the mantle and emplacement into a large sill; yet, exact knowledge of composition and phenocryst content is essential in considering the question in the previous paragraph. Furthermore, immediately after emplacement, whole-body convection is weak or absent, crystals cannot be suspended, and they must settle, perhaps as plumes, from everywhere within the body. For an initial phenocryst content of, say, 10% (by volume), a cumulate pile could form which occupies, considering packing porosity, some 20% or more of the full layer thickness. This assemblage will mainly reflect the process of filling the chamber and initial heterogeneities in crystallinity and not necessarily reflect phase equilibria and crystal growth after emplacement. Thus, the lower crust or cumulate pile may appear to grow much faster than the upper crust, or alternatively, during this time, the upper crust may appear not to have grown at all. The earlier ratio Z_u/Z_L may be strongly affected by initial phenocryst content. The point within a cumulate pile where initial phenocrysts give way to newly grown phases is important to identify.

All considerations herein hold tacitly to the conclusion that the vigor and style of convection within a sheet of magma are not governed by the usual measure of Rayleigh number (equation 15), which implicitly assumes heating from below. A much closer approach to magmatic convection is that due to cooling either from the roof and floor or from the roof alone. Convection occurs essentially when heat losses to the magma's wall rock are faster than the magma can supply by conduction alone. All motion is driven by the behavior of magma within and forward of the suspension zone. The coolest parts of the thermal boundary layer are held immobile by either crust or roof rock. Even under the most severe condition of steadily foundering upper crust, the heat transfer (without melting in the wall rock) is only twice what it would be for a purely stagnant body. If, instead, the definition in equation 15 is assumed true, the rate of heat transfer is so large that the predicted solidification time of 100-m-thick lava lakes—like Kilauea Iki—is about 1 month when in fact it has taken about 30 yr (Helz, 1988). There certainly is motion within sheets of magma, but it is so slight relative to what is needed to bring on rapid cooling that the thermal history is well approximated—especially with nonfoundering crust—by heat conduction alone with inclusion of latent heat.

It is this shift of emphasis from magma chambers controlled by vigorous, turbulent convection to the strong controls of solidification fronts that separates this investigation from earlier ones. Hess (1972) essentially fleshed out most all the dynamics of layered bodies with an insightful quantitative analysis, but he too did not consider any details of cooling and crystallization at the roof. He assumed that turbulent convection controls heat transfer within the layer, not realizing the inappropriateness of using equation 15 nor the effects of turbulence on sedimentation. This is not an uncommon position, for sedimentologists—excepting Bradley—also say precious little about how surface precipitates traverse lakes (for example, Hakanson and Jansson, 1983). Irvine (1970) has long recognized the role of the roof rock in controlling outward heat loss, which sets the over-all rate of crystallization, and his subsequent papers (for example, 1980, 1987) have thoughtfully sketched how crystals reach the floor in side-wall density currents or might form near the floor. The importance of his ideas is that they stem from field observations. Brandeis and Jaupart (1986) seem to have given the first semiquantitative description of plume transport of crystals and its possible role in forming basal cumulates.

An alternative to any of the above models is that of multiple injections, an old (see review by Jackson, 1961) but plausible idea by which repetitious layering is readily formed in ephemeral saline pans (Lowenstein and Hardie, 1985). As for single injections, any phenocrysts carried by the new magma can easily escape the capture front, especially if the new fluid underlies earlier magma, and will settle to form a cumulate layer. If the new magma is hotter and denser than the surrounding magma—having been injected, say, just over the lower mush zone—rapid cooling promotes crystallization and settling. The new magma's density decreases until it matches that of the overlying magma, whence instability and mixing commence (Huppert and Sparks, 1980). It is thus quite evident that there can be any number of models involving a combination of growing crusts and reinjection.

Any prolonged consideration of crystallization in magma chambers exposes its close similarity to chemical and mechanical sedimentation. Many early workers saw the strong role of physical sedimentation (see, for example, Grout, 1918; Wager and Deer, 1939; Turner and Verhoogen, 1960), but it seems to have been Jackson (1961) who first emphasized the role of chemical sedimentation even though others had also recognized the process itself (for example, Wager and Deer, 1939). In recognizing that convection is intimately coupled to crystallization, the role of chemical sedimentation emerges even more strongly. Halite precipitating on the surface of ephemeral saline pans sinks to form a porous deposit of syntaxially growing crystals. Desiccation encourages pore filling by saline cements and efflorescence. A new influx of water deposits mud that altogether produces a repetitious sequence of essentially monomineralic halite and mud layers (Lowenstein and Hardie, 1985). This is similar to a magmatic model of multiple injection, but different in being able to produce a full hiatus in deposition and nonindigenous layers—mud. Although it is doubtful that a similarly large number of hiatuses in deposition could occur in magma chambers (apparent hiatuses are known), magma chambers and perennial salt lakes in many cases have similar dimensions, and both can be multiply saturated, undergo reinjection, experience largely surface crystallization, and show vigorous postcumulus effects giving rise to monomineralic layers and intricate layered sequences. Lakes are not at all like some magmatic systems (for example, lavas and sills) in having no ever-thickening upper crust and thus no capture front. There is clearly a great deal for igneous petrologists to learn from chemical sediments, particularly because the dynamics of active salt-lake sedimentation can be studied first hand.

CONCLUSIONS

The evolution of simple magma chambers is characterized by an intricate and delicate balance between phase equilibria, crystallization kinetics, convection, and crystal sorting. Loss of heat, regulated by the wall rock, imposes this balance and gives impetus and time scale to the whole process. The shape of the chamber, whether it has been open to multiple injections or evacuations of liquids, and the composition of the liquids produce variations on this main theme. Because the vigor of convection, for example, is tied closely to heat transfer through the crusts and wall rock, it is no wonder that silicic plutons show differentiation so much less than do basic bodies.

Lava flows, lava lakes, sills, and plutons are solidified magma chambers. Although all have similar, unifying features, certain features are accentuated more in some than in others. Yet, by carefully studying all such bodies, using the same general principles and processes, a much clearer picture of how magma behaves in chambers will certainly emerge.

The single conclusion of foremost importance, however, is that chemical differentiation by crystal fractionation is extremely limited in such bodies. There is of course a great deal of liquid-solid separation involving the crystals already present upon emplacement, and this fractionation can lead to various degrees of evolution of essentially basaltic liquids, but there is little sign that this process continues and produces significant quantitites of andesitic, dacitic, or rhyolitic liquid. Why this is so may be in large part due to the presence of the solidification fronts. The dynamically active portion of the magma is inward of the mush zones which greatly limits, because of small crystallinity, any possible change in liquid composition by crystal separation. At large crystallinities, as in the mush zone, liquid compositions do certainly become highly fractionated (for example, Helz, 1982), but these liquids are generally inseparable because they reside in the high-crystallinity crust, behind the capture front, and are of small volume. A strong diversity of igneous rocks is not caused by such magma chambers. I hasten to add, however, that this view of diversity stems also from the dichotomy between pluton petrology and lava petrology. Small pockets or batches of highly evolved liquids can be found in plutons that because of their small volume or position within the chamber, may never have been viable as eruptible magma. A series of lavas from the same chamber spanning its period of solidification may thus show much less diversity than do the parts of the resulting pluton itself. It is on this basis that chemical diversity in these systems is found to be limited.

ACKNOWLEDGMENTS

Discussions with Lawrence Hardie, Geneviève Brandeis, George Bergantz, and Margaret Mangan have sharpened many of my ideas. Critical comments by Neil Irvine, Margaret Mangan, and David Yuen, as well as kind help with figures from George Bergantz and Mark Murphy, are appreciated. This work is supported by National Science Foundation Grant EAR-8418151-02. Hans Peter Eugster (1925–1987) introduced me to W. H. Bradley's work and to the mysteries of lake sedimentation. It is hoped that his (Eugster's) style and grace may somehow have affected this work.

REFERENCES CITED

Arndt, N. T., 1986, Differentiation of komatiite flows: Journal of Petrology, v. 27, p. 279–301.
Bowen, N. L., 1947, Magmas: Geological Society of America Bulletin, v. 58, p. 263–280.
Bradley, W. H., 1965, Vertical density currents: Science, v. 150, p. 1423–1428.
Brandeis, G., and Jaupart, C., 1986, On the interaction between convection and crystallization in cooling magma chambers: Earth and Planetary Science Letters, v. 77, p. 345–361.
Cashman, K. V., 1988, Crystallization of Mt. St. Helens 1980–1986 dacite: A quantitative textural approach: Bulletin of Volcanology (in press).
Cashman, K. V., and Ferry, J. M., 1988, Crystal size distribution (CSD) in rocks and the kinetics and dynamics of crystallization. III: Metamorphic crystallization: Contributions to Mineralogy and Petrology (in press).
Cashman, K. V., and Marsh, B. D., 1988, Crystal size distribution (CSD) in rocks and the kinetics and dynamics of crystallization. II: Makaopuhi lava lake: Contributions to Mineralogy and Petrology (in press).
Coats, R. R., 1936, Primary banding in basic plutonic rocks: Journal of Geology, v. 44, p. 407–419.
Cornwall, H. R., 1951, Differentiation in lavas of the Keweenawan Series and the origin of the copper deposits of Michigan: Geological Society of America Bulletin, v. 62, p. 159–202.

Daly, R. A., 1928, Bushveld igneous complex of the Transvaal: Geological Society of America Bulletin, v. 39, p. 703–768.

Evans, B. W., and Moore, J. G., 1967, Olivine in the prehistoric Makaopuhi tholeiitic lava lake, Hawaii: Contributions to Mineralogy and Petrology, v. 15, p. 202–223.

Gray, N. H., 1970, Crystal growth and nucleation in two large diabase dikes: Canadian Journal of Earth Sciences, v. 7, p. 366–375.

Gray, N. H., and Crain, I. K., 1969, Crystal settling in sills—A model for crystal settling: Canadian Journal of Earth Sciences, v. 6, p. 1211–1216.

Greenspan, H. P., and Ungarish, M., 1982, On hindered settling of particles of different sizes: International Journal of Multiphase Flow, v. 8, no. 6, p. 587–604.

Griffiths, R. W., 1986, Thermals in extremely viscous fluids, including the effects of temperature dependent viscosity: Journal of Fluid Mechanics, v. 166, p. 115–138.

Grout, F. F., 1918, Two phase convection in igneous magmas: Journal of Geology, v. 26, p. 481–499.

Hakanson, L., and Jansson, M., 1983; Principles of lake sedimentology: Berlin, Germany, Springer-Verlag, 316 p.

Helz, R. T., 1980, Crystallization history of Kilauea Iki lava lake as seen in drill core recovered in 1967-79: Bulletin Volcanologique, v. 43, p. 675–701.

——— 1987, Differentiation behavior of Kilauea Iki lava lake, Kilauea Volcano, Hawaii: An overview of past and current work, *in* Mysen, B. O., ed., Magmatic processes: Geochemical Society, Physicochemical Principles Special Publication 1, p. 241–258.

——— 1988, Diapiric transfer of melt in Kilauea Iki lava lake: A quick, efficient process of igneous differentiation: Geological Society of America Bulletin (in press).

Hess, G. B., 1972, Heat and mass transport during crystallization of the Stillwater igneous complex: Geological Society of America Memoir 132, p. 503–520.

Hess, H. H., 1960, Stillwater igneous complex: Geological Society of America Memoir 80, 230 p.

Hotz, P. E., 1953, Petrology of granophyre in diabase near Dillsburg, Pennsylvania: Geological Society of America Bulletin, v. 64, p. 675–704.

Hriskevich, M. E., 1968, Petrology of the Nipissing diabase sill of the Cobalt area, Ontario, Canada: Geological Society of America Bulletin, v. 79, p. 1387–1404.

Huppert, H. E., and Sparks, R.S.J., 1980, The fluid dynamics of a basaltic magma chamber replenished by influx of hot, dense ultrabasic magma: Contributions to Mineralogy and Petrology, v. 75, p. 279–289.

Hurlbut, C. S., 1939, Igneous rocks of the Highwood Mountains, Montana: Geological Society of America Bulletin, v. 50, p. 1043–1112.

Irvine, T. N., 1970, Heat transfer during solidification of layered intrusions. I. Sheets and sills: Canadian Journal of Earth Sciences, v. 7, p. 1031–1061.

——— 1980, Magmatic density currents and cumulus processes: American Journal of Science, v. 280-A, p. 1–58.

——— 1987, Layering and related structures in the Duke Island and Skaergaard intrusions: Similarities, differences, and origins, *in* Parsons, I., ed., Origins of igneous layering: Boston, Massachusetts, D. Reidel, p. 185–245.

Jackson, E. D., 1961, Primary textures and mineral associations in the ultramafic zone of the Stillwater Complex, Montana: U.S. Geological Survey Professional Paper 358, 106 p.

Jaupart, C., and Brandeis, G., 1986, The stagnant bottom layer of convecting magma chambers: Earth and Planetary Science Letters, v. 80, p. 183–199.

Jaupart, C., Brandeis, G., and Allegre, C. J., 1984, Stagnant layers at the bottom of convecting magma chambers: Nature, v. 308, p. 535–538.

Kerr, R. C., and Tait, S. R., 1986, Crystallization and compositional convection in a porous medium with application to layered igneous intrusion: Journal of Geophysical Research, v. 91, p. 3591–3608.

Kirkpatrick, R. J., 1977, Nucleation and growth of plagioclase, Makaopuhi and Alae lava lakes, Kilauea volcano, Hawaii: Geological Society of America Bulletin, v. 88, p. 78–84.

Komar, P. D., 1972a, Mechanical interactions of phenocrysts and flow differentiation of igneous dikes and sills: Geological Society of America Bulletin, v. 83, p. 973–988.

——— 1972b, Flow differentiation in igneous dikes and sills: Profiles of velocity and phenocryst concentration: Geological Society of America Bulletin, v. 83, p. 3443–3448.

——— 1976, Phenocryst interactions and the velocity profile of magma flowing through dikes or sills: Geological Society of America Bulletin, v. 87, p. 1336–1342.

Komar, P. D., and Reimers, C. E., 1978, Grain shape effects on settling rates: Journal of Geology, v. 86, p. 193–209.

Kynch, G. J., 1952, A theory of sedimentation: Faraday Society Transactions, v. 48, p. 166–176.

Lane, A. C., 1896, Igneous grain size: Geological Society of America Bulletin, v. 8, p. 403–407.

——— 1902, Studies of the grain of igneous intrusions: Geological Society of America Bulletin, v. 14, p. 369–384.

Lockett, M. J., and Al-Habbooby, H. M., 1973, Differential settling by size of two particle species in a liquid: Institute of Chemical Engineering Transactions, v. 51, p. 281–292.

——— 1974, Relative particle velocities in two-species settling: Powder Technology, v. 10, p. 67–71.

Long, P. E., and Wood, B. J., 1986, Structures, textures, and cooling histories of Columbia River basalt flows: Geological Society of America Bulletin, v. 97, p. 1144–1155.

Lovering, T. S., 1935, Theory of heat conduction applied to geological problems: Geological Society of America Bulletin, v. 46, p. 69–94.

Lowenstein, T. K., and Hardie, L. A., 1985, Criteria for the recognition of salt-pan evaporites: Sedimentology, v. 32, p. 627–644.

Mangan, M. T., and Helz, R. T., 1985, Vesicle and phenocryst distribution in Kilauea Iki lava lake, Hawaii [abs.]: EOS (American Geophysical Union Transactions), v. 66, p. 1133.

Mangan, M. T., and Marsh, B. D., 1988, Convection and crystallization of sheet-like magma bodies: III Crystal capture and basal cumulates [abs.]: EOS (American Geophysical Union Transactions), v. 69, p. 526.

Marsh, B. D., 1979, Island-arc developments: Some observation, experiments, and speculations: Journal of Geology, v. 87, p. 687–713.

——— 1981, On the crystallinity, probability of occurrence, and rheology of lava and magma: Contributions to Mineralogy and Petrology v. 78, p. 85–98.

——— 1988a, On convective style and vigor in sheet-like magma chambers: Journal of Petrology (in press).

——— 1988b, Crystal size distribution (CSD) in rocks and the kinetics and dynamics of crystallization I: Makaopuhi lava lake: Contributions to Mineralogy and Petrology (in press).

——— 1988c, Magma chambers: Annual Reviews of Earth and Planetary Science (in press).

Marsh, B. D., and Maxey, M. R., 1985, On the distribution and separation of crystals in convecting magma: Journal of Volcanology and Geothermal Research, v. 24, p. 95–150.

McBirney, A. R., 1985, Further considerations of double-diffusive stratification and layering in the Skaergaard intrusion: Journal of Petrology, v. 26, Part 4, p. 993–1001.

McDougall, I., 1962, Differentiation of the Tasmanian dolerites: Red Hill dolerite-granophyre association: Geological Society of America Bulletin, v. 73, p. 279–316.

McKenzie, D. P., 1984, The generation and compaction of partially molten rock: Journal of Petrology, v. 25, p. 713–765.

Morse, S. A., 1986, Convection in aid of adcumulus growth: Journal of Petrology, v. 27, p. 1183–1214.

Murase, T., and McBirney, A. R., 1973, Properties of some common igneous rocks and their melts at high temperatures: Geological Society of America Bulletin, v. 84, p. 3563–3592.

Osborne, F. F., and Roberts, E. J., 1931, Differentation in the Shonkin Sag laccolith: American Journal of Science, v. 22, p. 331–353.

Peck, D. L., 1978, Cooling and vesiculation of Alae lava lake, Hawaii: U.S. Geological Survey Professional Paper 935-B, 59 p.

Randolph, A. D., and Larson, M. A., 1971, Theory of particulate processes: New York, Academic Press, 251 p.

Richardson, J. F., and Zaki, W. N., 1954, Sedimentation and fluidization: Part I: Institute of Chemical Engineering Transactions, v. 32, p. 35–53.

Riehle, J. R., 1973, Calculated compaction profiles of rhyolitic ash-flow tuffs: Geological Society of America Bulletin, v. 84, p. 2193–2216.

Shaw, H. R., 1965, Comments on viscosity, crystal settling, and convection in granitic melts: American Journal of Science, v. 263, p. 120–152.

Shaw, H. R., Peck, D. L., Wright, T. L., and Okamura, R., 1968, The viscosity of basaltic magma: An analysis of field measurements in Makaopuhi lava lake, Hawaii: American Journal of Science, v. 266, p. 225–264.

Smith, M. K., 1988, Thermal convection during the directional solidification of a pure liquid with variable viscosity: Journal of Fluid Mechanics, v. 188, p. 547–570.

Smith, T. N., 1965, The differential sedimentation of particles of two different species: Institute of Chemical Engineering Transactions, v. 43, p. T69–T73.

Sparks, R.S.J., Huppert, H. E., Kerr, R. C., McKenzie, D., and Tait, S. R., 1985, Postcumulus processes in layered intrusions: Geology Magazine, v. 122, p. 555–568.

Turcotte, D. L., and Schubert, G., 1982, Geodynamics: New York, John Wiley & Sons, 450 p.

Turner, F. J., and Verhoogen, J., 1960, Igneous and metamorphic petrology: New York, McGraw-Hill, 694 p.

Wager, L. R., and Brown, G. M., 1968, Layered igneous rocks: San Francisco, W. H. Freeman, 588 p.

Wager, L. R., and Deer, W. A., 1939, Geological investigations in East Greenland: Part 3—The petrology of the Skaergaard intrusion, Kangerolugssuaq, East Greenland: Meddelelser om Grömland, v. 105, 335 p.

Walker, F., 1940, The differentiation of the Palisade diabase, New Jersey: Geological Society of America Bulletin, v. 51, p. 1059–1106.

Walker, F., and Poldervaart, A., 1949, Karroo dolerites of the Union of South Africa: Geological Society of America Bulletin, v. 60, p. 591–706.

Walker, K., 1969, The Palisades Sill, New Jersey: A reinvestigation: Geological Society of America Special Paper 111, 178 p.

Weed, W. H., and Pirsson, L. V., 1895, Highwood Mountains of Montana: Geological Society of America Bulletin, v. 6, p. 389–422.

——— 1901, Geology of the Shonkin Sag and Palisade Butte laccoliths in the Highwood Mountains of Montana: American Journal of Science, v. 12, p. 1–17.

Weiland, R. H., Fessas, Y. P., and Ramardo, B. V., 1984, On instabilities arising during sedimentation of two-component mixtures of solids: Journal of Fluid Mechanics, v. 142, p. 383–389.

Weinstein, S. A., Yuen, D. A., and Olson, P. L., 1987, Evolution of crystal settling in magma chamber convection: Earth and Planetary Science Letters, v. 87, p. 237–248.

Whitehead, J. A., and Luther, D. S., 1975, Dynamics of laboratory diapir and plume models: Journal of Geophysical Research, v. 80, p. 705–717.

Whitmore, R. L., 1955, The sedimentation of suspensions of spheres: British Journal of Applied Physics, v. 6, p. 239–245.

Wright, T. L., and Okamura, R. T., 1977, Cooling and crystallization of tholeiitic basalt, 1965 Makaopuhi lava lake, Hawaii: U.S. Geological Survey Professional Paper 1004, 78 p.

Wright, T. L., Peck, D. L. and Shaw, H. R., 1976, Kilauea lava lakes: Natural laboratories for study of cooling, crystallization, and differentiation of basaltic magma: American Geophysical Union Monograph 19, p. 375–390.

MANUSCRIPT RECEIVED BY THE SOCIETY DECEMBER 18, 1987
REVISED MANUSCRIPT RECEIVED JUNE 13, 1988
MANUSCRIPT ACCEPTED JUNE 14, 1988

The development of gravity and magnetic studies, emphasizing articles published in the *Geological Society of America Bulletin*

CENTENNIAL ARTICLE

G. R. KELLER *Department of Geological Sciences, University of Texas at El Paso, El Paso, Texas 79968-0555*

ABSTRACT

Gravity and magnetic studies have been published in the GSA *Bulletin* during most of its history. As demonstrated in the pages of the *Bulletin*, these techniques have evolved from crude measurements of only regional significance into mature research tools useful at any scale. Gravity and magnetic data have contributed to our understanding of geologic problems and features around the world. The equipment, data processing, and interpretation techniques, and, in many cases, the actual data needed to employ gravity and magnetic data in integrated studies are readily available. Thus, it is appropriate to anticipate that these techniques will find even greater use in the future. It is disappointing, however, to note that papers in the *Bulletin* involving gravity and magnetic data have decreased in number in recent years.

INTRODUCTION

Over the period during which the *Bulletin* has been published, gravity and magnetic surveying techniques have evolved into mapping tools which are used on a routine basis. Much of this evolution has been chronicled in the *Bulletin* and other Geological Society of America publications. For example, David White, in his 1923 Presidential address to the Society, mentioned the fact that there were 325 gravity measurements in the United States (Fig. 1) (White, 1924). Today the number of publicly available gravity measurements is near one million, and articles such as Lindreth Cordell's overview of the Rio Grande rift region (Cordell, 1978) presented maps (Fig. 2) produced from on the order of 100,000 readings. Because aeromagnetic surveying was not widely available until the early 1950s (for example, Hurley and Thompson, 1950; Vacquier and others, 1951), the evolution of this technique has been more recent. A magnetic map of the United States, however, has been compiled, and digital aeromagnetic data are available for the entire country. A significant contribution of the Decade of North American Geology effort is the compilation of gravity and magnetic maps for the continent. These maps have just been released and will provide many insights into the structural framework of North America as the scientific community evaluates their implications.

The purpose of this review is to discuss studies of the Earth's gravity and magnetic field in continental areas as published in the *Bulletin* over the years. The goal is to stress major developments, not every paper on this subject. This approach and the limited space available have probably resulted in some oversights, but it is fair to say that many important developments involving gravity and magnetic data have been reported in *Bulletin*.

STUDIES IN NORTH AMERICA

Because of interest in isostasy and the role it plays in the tectonic development of the Earth (McGee, 1891; Gilbert, 1893; Hobbs, 1907; Putnam, 1922; Reid, 1922), gravity measurements were of considerable interest during the early years of the *Bulletin*. The first measurements were painstakingly made with pendulums, and regional relationships were of primary interest. William Bowie, however, read a paper on "Gravity anomalies and geological formations" in 1912, and David White's presidential address (White, 1924) featured a long discussion about the correlation of gravity anomalies with individual geologic features. For example, he noted that a negative anomaly was associated with the sedimentary fill in the Newark (Triassic) basin in New Jersey. More than 50 years later, Sumner (1977) conducted a detailed study of this anomaly and basin.

White (1924) also noted that a positive anomaly was associated with the Black Hills uplift. The extent of the Black Hills anomaly suggested a high-density core, the boundaries of which spread well beyond the outcrop area. Chamberlin (1935) went on to further elucidate structural relations in this area. The midcontinent gravity high was reflected in positive values on his map (Fig. 1).

White (1924) also displayed considerable insight in suggesting that a large positive anomaly in the Texas Panhandle was due to dense rocks such as those found in the Arbuckle uplift area. These features are now considered to be related to the southern Oklahoma aulacogen which is associated with a major intracrustal mass excess, presumably a result of rifting processes. He also recognized the very large negative anomaly associated with the Arkoma basin, which has been the object of much recent interest (Lillie, 1985; Kruger and Keller, 1986).

George P. Woollard, however, for whom the award of the Geophysics Division of the Geological Society of America is named, deserves much credit for establishing gravity surveying as a geological tool. Much of his early work was published in the *Bulletin* (Woollard, 1943a, 1943b; Steenland and Woollard, 1952). The efforts of Woollard and those of Hersey (1944), Bean (1953), Joyner (1963), Simmons (1964), and Bothner (1974) employed gravity data to study structures in Pennsylvania, New York, and New England. Among other things, these studies established that granitic plutons usually cause gravity lows because granites are usually less dense than the "basement" rocks they intrude. The shape and mass distribution in granitic batholiths as delineated by gravity data was the subject of a significant study by Bott and Smithson (1967) which showed that plutons were denser at the bottom than at the top.

Aeromagnetic maps have been established as an important aid in geologic mapping in areas

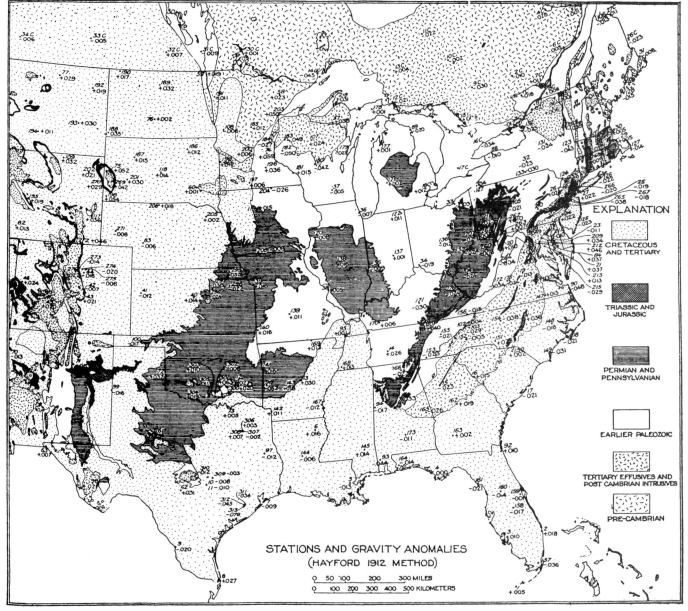

Figure 1. Map from White (1924) showing the distribution of gravity stations east of the Rocky Mountains. The numbers preceded by a plus or minus sign are isostatic anomalies based on the Hayford 1912 method. These values, when multiplied by 1,000, yield units of milligals, which are usually employed in gravity studies.

with poor exposures through studies that have often been published in the *Bulletin*. The early work of Hurley and Thompson (1950) in northwestern Maine is the first such study of which the author is aware. Further aeromagnetic studies in the Appalachian area directed toward geologic mapping were conducted by Nueschel (1970), Kane and others (1971), and Harwood and Zietz (1974). Memoir 47 (Vacquier and others, 1951) serves as a classic study on the interpretation of magnetic anomalies.

Even though the work was in an oceanic area, no discussion of magnetic studies published in the *Bulletin* would be complete without mentioning the work of Mason and Raff (Mason and Raff, 1961; Raff and Mason, 1961). Their magnetic survey off the west coast of North America provided early evidence for the magnetic lineations in the ocean floor which ultimately provided part of the foundation for ideas on sea-floor spreading and plate tectonics.

Long geophysical profiles have always attracted interest because of their ability to delineate relationships at all scales (that is, basement structure, lithospheric variations, and so on). Putnam (1922) mentioned that he had made a series of transcontinental gravity measurements

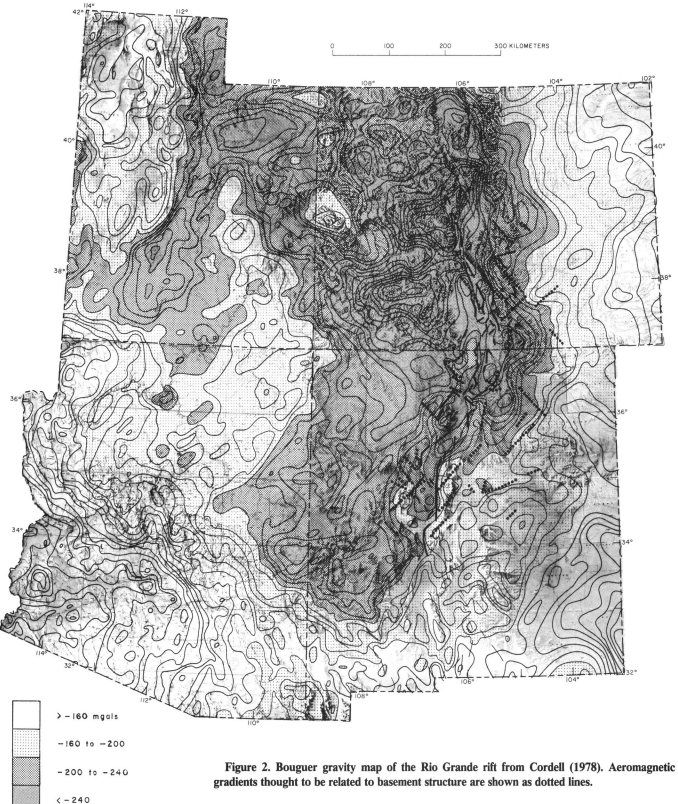

Figure 2. Bouguer gravity map of the Rio Grande rift from Cordell (1978). Aeromagnetic gradients thought to be related to basement structure are shown as dotted lines.

in 1894. Woollard, however, reported (1943a) on the first transcontinental gravity and magnetic profile which was sufficiently detailed to determine local relations (Fig. 3). For many years, this study was the standard for efforts to integrate gravity, magnetic, and geologic data. Isidore Zietz and his colleagues conducted a transcontinental aeromagnetic survey of a strip which spanned over 1° of latitude (Zietz and others, 1966, 1969, 1971). These studies were

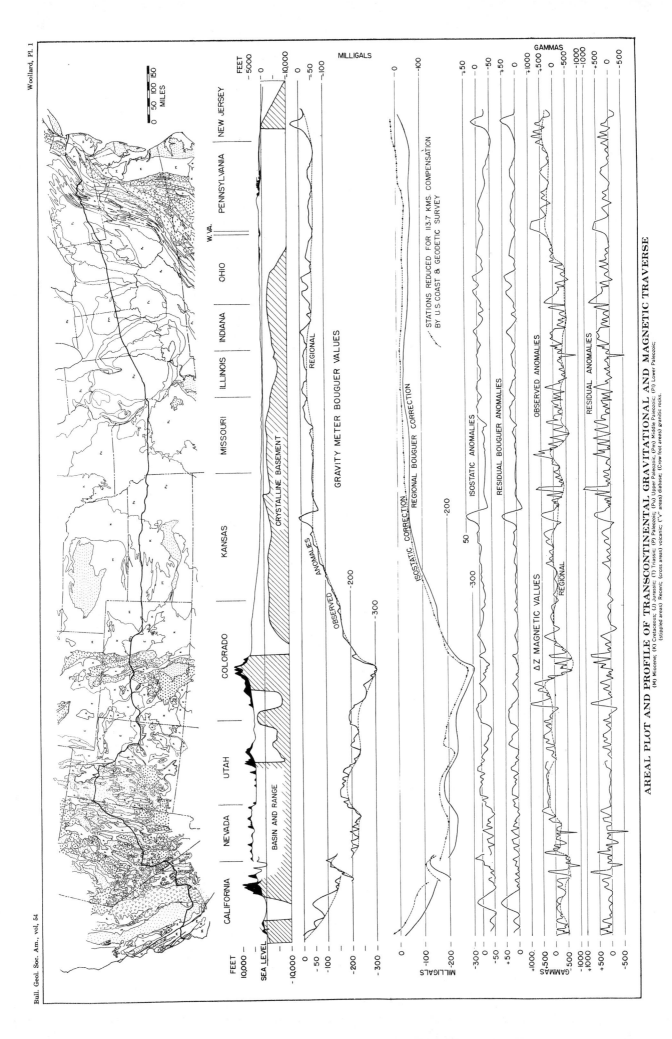

Figure 3. Transcontinental gravity and magnetic profile from Woollard (1943). Notice the Mid-continent gravity-high in Kansas.

part of the transcontinental survey sponsored primarily by the U.S. Geological Survey. The transects which are part of the Decade of North American Geology program are the most recent efforts to integrate geology and geophysics along strips across interesting features.

Such integrated efforts always yield interesting results. For example, the transcontinental profile of Woollard (1943a) delineated the mid-continent gravity high. This anomaly has since been recognized as the signature of the Keweenawan mid-continent rift system and has been studied primarily in efforts which have been reported in the *Bulletin* (Thiel, 1956; King and Zietz, 1971; Oray and others, 1973) and in Memoir 156 (Wold and Hinze, 1982). Gravity and magnetic data have primarily been used to delineate this feature which extends at least from Kansas through the Great Lakes into southern Michigan (Fig. 4). Recent geophysical studies in the Lake Superior portion of this rift system are tremendously interesting and indicate that "volcanic and interbedded and post-volcanic sedimentary rock extends to depths as great as 32 km" (Behrendt and others, 1988).

Studies of the mid-continent rift system laid the groundwork for the recognition of many other rift structures in the mid-continent region (Keller and others, 1983). The work of Ervin and McGinnis (1975) was published in the *Bulletin* and is particularly significant. They suggested that the Mississippi embayment formed by the reactivation of the late Precambrian-Early Cambrian Reelfoot rift (Fig. 5). This model motivated many studies in this area and has been confirmed by a variety of geological and geophysical data (McKeown and Pakiser, 1982; Hildenbrand, 1985). These studies have led to a much better understanding of the seismic activity in the New Madrid seismic zone (Braile and others, 1986).

Gravity and magnetic studies published in the *Bulletin* have generally shown that the basement of the mid-continent region is complex (Hase and Koch, 1968; Lidiak, 1971; O'Hara and Hinze, 1972, 1980; Klasner and Cannon, 1974). The repeated reactivation of old basement structures appears to be the rule, not the exception, and explain much of the intraplate seismicity in eastern North America (M. R. Keller and others, 1985; Braile and others, 1986). There is, therefore, renewed interest in basement structures and how they have reacted to changes in the stress regime through time. Gravity and magnetic data are an ideal aid to such studies.

Orogenic belts, continental margins and the sedimentary basins associated with them have been the subject of many papers in the *Bulletin* over the years. Walcott (1972) recognized the importance of lithospheric flexure at continental margins and how gravity anomalies can be analyzed to deduce the mechanical properties of the lithosphere. Gravity data have been used to

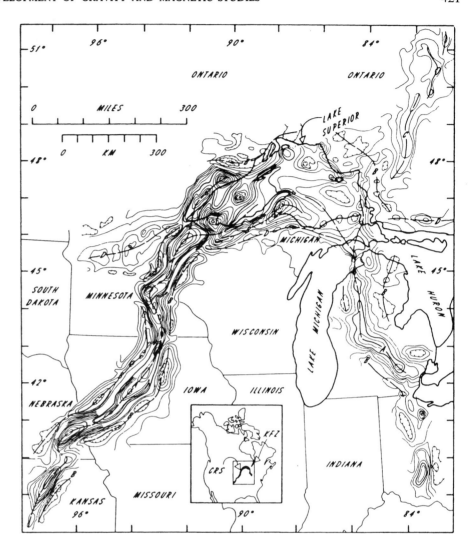

Figure 4. Bouguer gravity map of the mid-continent gravity high (CRS in inset). Heavy contour line = 0 mgals; contour interval = 10 mgals. Interpreted source of anomalies: A, Keweenawan mafics; B, marginal clastic wedges and/or crustal thickening; C, Archean granulites along Kapuskasing fault zone (KFZ in inset); D, middle Precambrian basins within Archean basement. From Halls (1978).

study the deep structure of the southern Appalachian (Best and others, 1973; Waskom and Butler, 1971; Dainty and Frazier, 1984; Pratt and others, 1985) and Ouachita orogenic belts (Keller and Cebull, 1973), as well as the large sedimentary basins associated with the adjacent continental margins (Horton, 1944; Nettleton, 1952; Keller and Shurbet, 1975).

Western North America has always fascinated geologists and geophysicists because of the good exposures, the recent tectonic history, and (if we dare to admit it) the scenery. The Sierra Nevada Mountains have long been of interest as a place to study isostasy (Lawson, 1936). Early studies of gravity (Johnston, 1940) and earthquake data (Byerly, 1937; Gutenberg, 1943) suggested that a substantial crustal root is associated with this mountain range. Some early seismic refraction results (Prodehl, 1970), however, could be interpreted to indicate that little crustal thickening is present. This apparent discrepancy was probably due to difficulties in interpreting lower crustal layers; but more modern gravity (Oliver, 1977) and seismic data (Pakiser and Brune, 1980) seem to clearly establish a substantial root for these mountains (Fig. 6). Better geophysical data will be needed to fully understand the deep structure of this mountain range, the uplift history of which is of continuing interest (Crough and Thompson, 1977; Chase and Wallace, 1986).

Other mountain ranges in the west coast region have also been of interest to workers who published in the *Bulletin*. The southern Cascade Range has been the object of much interest. Pakiser (1964) studied the large gravity low (70 mgals) in the Lassen Volcanic National Park area and interpreted it to indicate the presence of low-density rocks which, at least in part, isostatically compensate the topographic load above. LeFehr (1965) came to a similar conclusion in his analysis of the Mount Shasta area. Blakely and others (1985) employed both gravity and magnetic data to study the Southern Cascades. They found a broad magnetic low which encompassed the area of the Mount Shasta, Lassen Peak, and Medicine Lake volcanoes; they interpreted it as being due to an upwarp in the Curie isotherm. The volcanoes in the area were found to generally lie on the flanks of gravity lows which the authors thought were due to depressions filled with low-density magnetic material. LeFehr (1966) studied the eastern Klamath Mountains and delineated a large positive anomaly which he interpreted to be a deformed ultramafic intrusion. The central Coast Ranges and adjacent San Joaquin Valley of California were the object of a study by Byerly (1966). He found negative residual anomalies that delineated the Tertiary basins of the Coast Ranges. The region of the Coast Ranges was also the subject of Ruppel's study (1971) which employed both gravity and aeromagnetic data. He delineated several local anomalies associated with the Stony Creek fault, which acts as the boundary between the Coast Range and Great Valley provinces.

A regional gravity survey of Oregon was conducted by Thiruvathukal and others (1970). Their results indicated that neither the Coast nor Cascade Ranges possessed significant crustal roots. They suggested that significant crustal thickening occurred from west to east across the state and that plate movements played a role in the establishment of isostatic equilibrium. Twin Sisters Mountain in northern Washington was the object of a gravity and magnetic study by Thompson and Robinson (1975). The dunite body which forms these peaks was interpreted to be an alpine-type ultramafic body. This body was determined to be a thin (2 km) pod of peridotite which is sheathed by serpentine. Finally, an aeromagnetic survey of the San Gabriel Mountains was conducted by Cummings and Regan (1976). Their results and field results indicate that gabbro and norite underlie the anorthosite in this complex. They inferred a two-stage development for this Precambrian feature.

The various mountain ranges and basins in Wyoming have also been of keen interest over the years. The Wind River Mountains have long been considered to be a key to understanding the foreland deformation associated with the Laramide orogeny. A gravity study by Berg and Romberg (1966) confirmed that the southwest flank of this mountain range is underlain by a low-angle thrust fault. COCORP work about 10 years later went on to show that this fault probably extended well into the crust. The Laramide Mountains to the east were the object of studies by Hodge and Mayewski (1969) and Hodge and others (1973). These studies represent a good example of the integration of geophysics and petrology and were aimed at the Laramie anorthosite complex. The results suggested that the rock sequence present (anorthosite rimmed by norite-syenite) is probably not due to *in situ* differentiation of gabbroic magma.

From a geophysical point of view, the Great Basin is one of the most extensively studied tectonic provinces in North America. The results of many of these studies have been published in the *Bulletin*. George Thompson's long-standing interest in this area is documented by his early work (Thompson and Sandberg, 1958) in the Reno, Nevada, area which demonstrated the

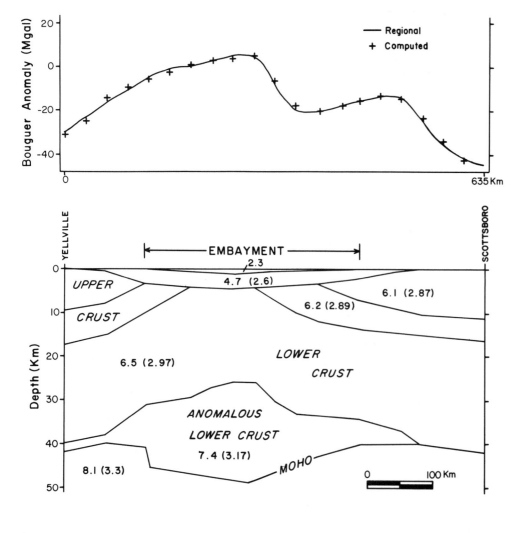

Figure 5. Regional gravity profile and inferred earth model for an east-west cross section across the Reelfoot rift. Yellville, Arkansas, is the western end of the profile. Density values are given in gm/cm^3 along with P-wave velocities in km/sec. Modified from Ervin and McGinnis (1975).

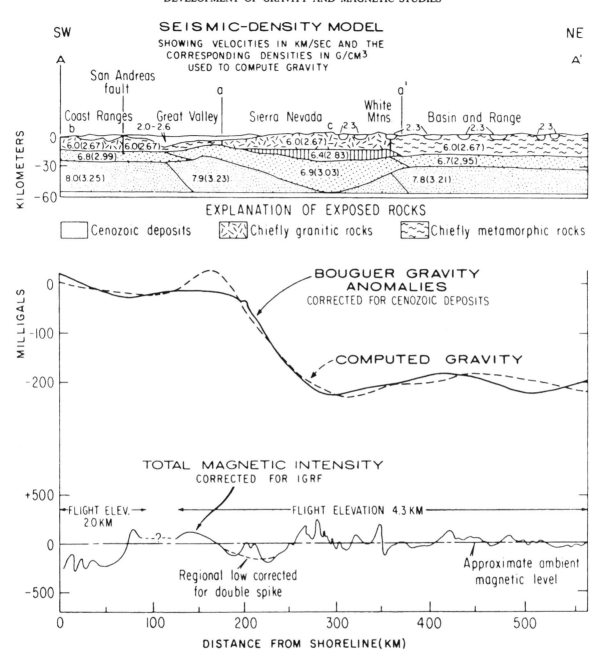

Figure 6. Seismic-density model (Oliver, 1977) for a northeast-southwest-trending profile across the Sierra Nevada Mountains. The interior portion of the profile (a-a') extends from near Fresno, California, to approximately the California-Nevada state line. The gravity profile is smoothed to remove the effects of Cenozoic basins. P-wave velocities are given in km/sec, and densities in gm/cm³ follow in parentheses. From Oliver (1977).

complex subsurface structure of the Cenozoic basins present. In the Mono Basin area to the south, Pakiser and others (1960) and Pakiser (1968) reported geophysical data suggesting that the area was a relatively deep volcano-tectonic depression. This interpretation was challenged by Gilbert and others (1968) and Christensen and others (1969), who believed that the basin was a shallow elongate warp. Pakiser (1976) presented seismic data which showed that the basin was intermediate in depth and was in fact a result of a combination of faulting and warping. This work has taken on renewed significance since the recognition that a volcanic eruption could occur in the area in the near future (Hill and Bailey, 1985).

Another study in the westernmost Great Basin area was the magnetic study of Lake Tahoe by Henyey and Palmer (1974). They established that this lake is in a graben typical of Basin and Range structure. They believed that volcanic activity associated with bounding faults was probably responsible for creating the closed Lake Tahoe basin. Other gravity and magnetic studies (Robinson, 1970; Speed and Cogbill, 1979; Green and Plouff, 1981) have done much to elucidate the structural framework of the area, as have studies in the eastern Great Basin in Utah (Cook and others, 1964; Cook and Hardman, 1967; Mikulich and Smith, 1974;

Halliday and Cook, 1980). Geological Society of America Memoir 152 (Smith and Eaton, 1978) also contains a very important collection of papers about the geophysics of the Great Basin.

In the past ten years, the Rio Grande rift has been accepted as one of the major Cenozoic examples of a continental rift. Cordell (1978) reviewed the geophysical data base for this feature. Gravity and magnetic data were shown to reflect control of rift structures by pre-existing zones of weakness. He also demonstrated that the rift has an anomalous lithospheric structure when compared to adjacent provinces such as the Great Plains (Shurbet, 1966). The rift in southern New Mexico was the subject of a thorough gravity study by Ramberg and others (1978). A broad, positive regional anomaly was delineated which was later shown to be due primarily to crustal thinning (Sinno and others, 1986). Towle (1980) measured geomagnetic field variations in the rift area and delineated conductivity anomalies in both the crust and upper mantle that correlate well with features determined from other data.

STUDIES IN CENTRAL AND SOUTH AMERICA

Although North America has been the subject of most gravity and magnetic studies published in the *Bulletin*, other areas of the western hemisphere have also received considerable attention. The Caribbean area was the focus of early attention because of the trench-island arc complexes present. The early gravity work of Bowie (1935) was followed by a series of studies by Maurice Ewing and his colleagues, G. L. Shurbet and J. Lamar Worzel (Ewing and Worzel, 1954; Worzel and Ewing, 1954; Shurbet and others, 1956; Shurbet and Ewing, 1956; Shurbet and Worzel, 1957a; Shurbet and Worzel, 1957b). These results established the basic structural framework of the West Indies. These studies and the work of F. A. Vening Meinesz (Vening Meinesz, 1954) did much to establish present understanding of island-arc systems world wide.

The continental areas adjacent to the Caribbean have also been of interest in the *Bulletin*. Colombia and Panama were the object of a series of gravity studies by J. E. Case and his colleagues (Case and others, 1971; Case and MacDonald, 1973; Case, 1974). These studies established that Panama is composed primarily of oceanic crust. The major structures of the Colombian Andes and their relations to the northern continental margin of South America were also delineated.

INTERNATIONAL STUDIES

The *Bulletin* is not exclusively for studies in the Americas, and this international flavor is reflected in subject matter of gravity and magnetic articles. For example, Hospers (1965) conducted a gravity study of the Niger delta of western Africa. He studied the sediment pile and inferred the amount of subsidence which has occurred, estimating that the maximum thickness of sediments in the delta was about 8 km. This feature is now widely thought to be a classic example of an aulacogen. Qureshy and others (1968) conducted a gravity study of the Godavari Valley in southern India. They interpreted the area to be a rift valley and pointed out that there were similar and related basins on the western and northern margins of Australia. They also suggested that the location of the rift was related to Precambrian zones of weakness. These conclusions were on the forefront of thinking at the time. Another study of a rift was conducted by Davis (1969). He constructed a gravity profile across southern Malawi and found anomalies similar to those farther north in the East African rift.

Compressional features around the world have also been of interest. The Jotun nappe in the Norwegian Caledonides was the subject of a gravity study by Smithson and Ramberg (1970). They were able to conclude that this major feature was probably due to locally derived folded nappes. Smithson (1972) also conducted a gravity study of the Trans-Antarctic Mountains, discovering that they separated regions of distinctly different crustal structure. He suggested that the Antarctic continent formed as a result of early Paleozoic continent-continent collision, with the suture presumably in the Trans-Antarctic Mountains region. Milson (1973) studied the ophiolite complex of the Papuan ultramafic belt in New Guinea. He observed that the large positive gravity anomalies present were similar to those in Cyprus, New Caledonia, and the Ivrea zone. He also recognized, however, that the structure of the area was more complex than that of a simple slice of obducted oceanic crust. The Himalayas were the subject of a gravity study by Qureshy and others (1974). They observed that the middle Himalayas were in isostatic equilibrium, indicating the presence of a thick root. More recent results would place crust of double normal continental thickness in this area.

SMALL-SCALE STUDIES

Finally, not all gravity and magnetic studies in the *Bulletin* have been of large scale. The eruption of the Paricutín volcano from a corn field in Mexico focused the attention of the world on that area. Barnes and Romberg (1948) conducted a gravity study of the area of this volcano but found no anomaly associated with it. Green and Lynch (1959) conducted a gravity study of a small basaltic intrusion in southwest Texas. There are many of these features which approximately follow the trend of the Ouachita system in the area. The authors showed that Mustang Hill was a laccolith, and they delineated the probable location for the feeder dike. Russell and others (1960) conducted a gravity study on the Salmon Glacier in British Columbia. Despite the fact that the glacier was moving at a rate sufficient to affect the readings, they concluded that gravity could very effectively delineate the shape of the underlying valley and the approximate thickness of the glacier. Onesti and Hinze (1970) conducted magnetic surveys over eskers in Michigan and observed distinct positive anomalies over these features. They concluded that magnetic materials are concentrated during the formation of eskers. Temporal variations of gravity in New Zealand's Wairakei geothermal field were the object of a study by Hunt (1970). Differences of up to 0.5 mgal over a 6-yr period were attributed to a net loss of water from the aquifer. Geothermal areas in Oregon and California were also the subject of studies by Cleary and others (1981) and Denlinger and Kovach (1981).

CONCLUSIONS

It is difficult to draw coherent conclusions from almost 100 years of research on a vast number of areas around the world. It is easy to see, however, that gravity and magnetic data have evolved into readily available tools which can be of use in a wide variety of geological applications. Large data bases are maintained by the National Geophysical Data Center and by several university and U.S. Geological Survey groups. The computer software needed to undertake first-rate analyses is also widely available. With this in mind, it is disappointing that gravity and magnetic studies have not been more prevalent in the *Bulletin* in recent years. Twenty such articles were published during the period 1960–1969. Forty-one were published during the period 1970–1979, but only eleven were published from 1980 to mid-1987. Within the geophysical community, potential field methods have suffered from benign neglect for some time. In fact, most geophysicists today are primarily seismologists, and they consider gravity and magnetic studies to be more in the realm of the structural geologist and explorationist. This situation does not belittle these powerful techniques, but it reflects the fact that many methods for data collection, processing, and interpretation are well established. The primary research interest thus has often not been on these techniques for their own sake but on their application to relevant problems. Recent instrumentation advances in both gravimetry and magnetometry, however, promise to revitalize interest in high-resolution and gradient measurements.

Perhaps the biggest barrier to the use of potential field techniques has been the perception that because they can be mathematically proven to be ambiguous, they are of little use. Many textbooks and courses stress mathematical developments rather than applications, and thus the fact that drilling results, geologic mapping, and plain common sense can greatly reduce the ambiguity in interpreting potential field data does not receive enough attention. For example, it seems reasonable to conclude that a linear gravity low between two mountain ranges in Nevada is not due to a buried body of salt. Likewise, a small circular gravity low in east Texas is probably due to salt. Another barrier may be the perception that seismic data are not ambiguous and that they thus render potential field methods obsolete. Seismic reflection is the most powerful (and expensive) geophysical technique for most studies in the upper crust, and a large volume of reflection data has become available in recent years. These data have indeed resolved some long-standing controversies, such as the extent of basement involvement in thrusting in the Valley and Ridge province of the Appalachians. Most reflection record sections, however, are subject to multiple interpretations (that is, they are ambiguous) and often provide as many new questions as answers for old questions. Many oil companies recognize this situation and regularly conduct gravity and magnetic surveys along their seismic profiles.

I do not wish to belittle other geophysical techniques but to stress that most geological and geophysical studies lead to results which are inherently ambiguous. This inexactness is what makes geological investigations so interesting to many of us. We are always striving to obtain an improved picture of the subsurface, and the integration of all available data is a key element to such efforts. My point is simply that gravity and magnetic data and techniques are widely available, relatively inexpensive, and would provide useful information in many situations. This fact has been documented over the years by many articles which have been published in the *Bulletin*.

ACKNOWLEDGMENTS

It is impossible to acknowledge the many individuals who have contributed to this paper through their efforts over the years. The long list of references is an attempt to recognize these workers. Reviews by William Hinze, Lindrith Cordell, Louis Pakiser, and George Thompson were very helpful, as were discussions with Thomas Hildenbrand, Carlos Aiken, and David Hill.

REFERENCES CITED

Aldrich, H. R., 1923, Magnetic surveying on the copper-bearing rocks of Wisconsin: Geological Society of America Bulletin, v. 34, p. 145.

Barnes, V. E., and Romberg, F. E., 1948, Observations of relative gravity at Paricutin volcano: Geological Society of America Bulletin, v. 59, p. 1019–1026.

Bean, R. J., 1953, Relation of gravity anomalies to the geology of central Vermont and New Hampshire: Geological Society of America Bulletin, v. 64, p. 509–538.

Behrendt, J. C., Green, A. C., Cannon, W. F., Hutchison, D. R., Lee, M. W., Milkereit, B., Agena, W. F., and Spencer, C., 1988, Crustal structure and deep rift basin of the mid-continent rift system—Results from GLIMPCE deep seismic reflection profiles: Geology, v. 16 (in press).

Berg, R. R., and Romberg, F. E., 1966, Gravity profile across the Wind River Mountains, Wyoming: Geological Society of America Bulletin, v. 77, p. 647–656.

Best, D. M., Geddes, W. H., and Watkins, J. S., 1973, Gravity investigation of the depth of source of the Piedmont gravity gradient in Davidson County, North Carolina: Geological Society of America Bulletin, v. 84, p. 1213–1216.

Bhattacharyya, D. S., and Bhattacharyya, T. K., 1970, Geological and geophysical investigations of a basaltic layer in Archean terrain of eastern India: Geological Society of America Bulletin, v. 81, p. 3073–3078.

Blakely, R. J., Jachens, R. C., Simpson, R. W., and Couch, R. W., 1985, Tectonic setting of the southern Cascade Range as interpreted from its magnetic and gravity fields: Geological Society of America Bulletin, v. 96, p. 43–48.

Bothner, W. A., 1974, Gravity study of the Exeter Pluton, southeastern New Hampshire: Geological Society of America Bulletin, v. 85, p. 51–56.

Bott, M.H.P. and Smithson, S. B., 1967, Gravity investigations of subsurface shape and mass distributions of granite batholiths: Geological Society of America Bulletin, v. 78, p. 859–878.

Bowie, W., 1912, Gravity anomalies and geological formations: Geological Society of America Bulletin, v. 23, p. 50.

——— 1935, Significance of gravity anomalies at stations in the West Indies: Geological Society of America Bulletin, v. 46, p. 869–878.

Braile, L. W., Hinze, W. J., Keller, G. R., Lidiak, E. G., and Sexton, J. L., 1986, Tectonic development of the New Madrid rift complex, Mississippi embayment, North America: Tectonophysics, v. 131, p. 1–21.

Byerly, P. E., 1937, Comment on the Sierra Nevada in the light of isostasy by A. C. Lawson: Geological Society of America Bulletin, v. 48, p. 2025–2031.

——— 1966, Interpretations of gravity data from the central Coast Ranges and San Joaquin Valley, California: Geological Society of America Bulletin, v. 77, p. 83–94.

Case, J. E., 1974, Oceanic crust forms basement of eastern Panama: Geological Society of America Bulletin, v. 85, p. 645–652.

Case, J. E., and MacDonald, W. D., 1973, Regional gravity anomalies and crustal structure in northern Colombia: Geological Society of America Bulletin, v. 84, p. 2905–2916.

Case, J. E., Duran S., L. G., Lopez R., A., and Moore, W. R., 1971, Tectonic investigations in western Colombia and eastern Panama: Geological Society of America Bulletin, v. 82, p. 2685–2712.

Chase, C. G., and Wallace, T. C., 1986, Uplift of the Sierra Nevada of California: Geology, v. 14, p. 730–733.

Christensen, M. N., Gilbert, C. M., Lajoie, K. R., and Al-Rawi, Y., 1969, Geological-geophysical interpretation of Mono basin, California-Nevada: Journal of Geophysical Research, v. 74, p. 5221–5239.

Clearly, J., Lange, I. M., Qamar, A. I., and Krouse, H. R., 1981, Gravity, isotope, and geochemical study of the Alvord Valley geothermal area, Oregon: Summary: Geological Society of America Bulletin, v. 92, p. 319–322.

Cook, K. L., and Hardman, E., 1967, Regional gravity survey of the Hurricane fault area and Iron Springs district, Utah: Geological Society of America Bulletin, v. 78, p. 1063–1076.

Cook, K. L., Halverson, M. O., Stepp, J. C., and Berg, J. W., Jr., 1964, Regional gravity survey of the northern Great Salt Lake desert and adjacent areas in Utah, Nevada, and Idaho: Geological Society of America Bulletin, v. 75, p. 715–740.

Cordell, L., 1978, Regional geophysical setting of the Rio Grande rift: Geological Society of America Bulletin, v. 89, p. 1073–1090.

Crough, S. T., and Thompson, G. A., 1977, Upper mantle origin of Sierra Nevada uplift: Geology, v. 5, p. 396–399.

Cummings, D., and Regan, R. D., 1976, Aeromagnetic survey of the San Gabriel anorthosite complex, San Gabriel Mountains, southern California: Geological Society of America Bulletin, v. 87, p. 675–680.

Dainty, A. M., and Frazier, J. E., 1984, Bouguer gravity in northeastern Georgia: A buried suture, a surface suture, and granites: Geological Society of America Bulletin, v. 95, p. 1168–1175.

Davis, R. W., 1969, A rift valley gravity profile in southern Malawi: Geological Society of America Bulletin, v. 80, p. 893–894.

Denlinger, R. P., and Kovach, R. L., 1981, Three-dimensional gravity modeling of the Geysers hydrothermal system and vicinity, northern California: Geological Society of America Bulletin, v. 92, p. 404–410.

Dennison, J. M., and Johnson, R. W., Jr., 1971, Tertiary intrusions and associated phenomena near the thirty-eighth parallel fracture zone in Virginia and west Virginia: Geological Society of America Bulletin, v. 82, p. 501–508.

De Witt, W., Jr., and Bayer, K. C., 1986, Seismicity, seismic reflection, gravity, and geology of the central Virginia seismic zone: Part 3. Gravity: Discussion and reply: Geological Society of America Bulletin, v. 97, p. 1285–1287.

Ervin, C. P., and McGinnis, L. D., 1975, Reelfoot rift: Reactivated precursor to the Mississippi embayment: Geological Society of America Bulletin, v. 86, p. 1287–1295.

Ewing, M., and Worzel, J. L., 1954, Gravity anomalies and structure of the West Indies, Part I: Geological Society of America Bulletin, v. 65, p. 165–174.

Gilbert, C. M., Christensen, M. N., Al-Rawi, Y., and Lajoie, K. R., 1968, Structural and volcanic history of Mono Basin, California-Nevada: Geological Society of America Memoir 116, p. 275–331.

Gilbert, G. K., 1893, Continental problems: Bulletin of the Geological Society of America, v. 4, p. 179–190.

Greene, R. C., and Plouff, D., 1981, Location of a caldera source for the Soldier Meadow Tuff, northwestern Nevada, indicated by gravity and aeromagnetic data: Summary: Geological Society of America Bulletin, v. 92, p. 4–6.

Greenwood, R., and Lynch, V. M., 1959, Geology and gravimetry of the Mustang Hill laccolith, Uvalde County, Texas: Geological Society of America Bulletin, v. 70, p. 807–826.

Gutenberg, B., 1943, Seismological evidence for roots of mountains: Geological Society of America Bulletin, v. 54, p. 473–498.

Halliday, M. E., and Cook, K. L., 1980, Regional gravity survey, northern Marysvale volcanic field, south-cenetral Utah: Geological Society of America Bulletin, v. 91, p. 502–508.

Halls, H. C., 1978, The late Precambrian central North American rift system — A survey of recent geological and geophysical investigations, *in* Ramberg, I. B., and Neuman, E. R., eds., Tectonics and geophysics of continental rifts: Dordrecht, Holland, Reidal Publishing Company, p. 111–123.

Harwood, D. S., and Zietz, I., 1974, Configuration of Precambrian rocks in southeastern New York and adjacent New England from aeromagnetic data: Geological Society of America Bulletin, v. 85, p. 181–188.

Hase, D. H., and Koch, D. L., 1968, Geophysical study of the Keota dome — Keokuk and Washington counties, Iowa: Geological Society of America Bulletin, v. 79, p. 935–940.

Henyey, T. L., and Palmer, D. F., 1974, Magnetic studies on Lake Tahoe, California-Nevada: Geological Society of America Bulletin, v. 85, p. 1907–1912.

Hersey, J. B., 1944, Gravity investigation of central-eastern Pennsylvania: Geological Society of America Bulletin, v. 55, p. 417–444.

Hildenbrand, T. G., 1985, Rift structure of the northern Mississippi embayment from the analysis of gravity and magnetic data: Journal of Geophysical Research, v. 90, p. 12,607–12,622.

Hill, D. P., and Bailey, R. A., 1985, Active tectonic and magmatic processes beneath Long Valley caldera, eastern California: An overview: Journal of Geophysical Research, v. 90, p. 11,111–11,120.

Hobbs, W. H., 1907, Origin of ocean basins in light of the seismology: Geological Society of America Bulletin, v. 18, p. 233–250.

Hodge, D. S., and Mayewski, P. A., 1969, Gravity study of a hypersthene syenite in the Laramie anorthosite complex, Wyoming: Geological Society of America Bulletin, v. 80, p. 705–714.

Hodge, D. S., Owen, L. B., and Smithson, S. B., 1973, Gravity interpretation of the Laramie anorthosite complex, Wyoming: Geological Society of America Bulletin, v. 84, p. 1451–1464.

Horton, C. W., 1944, Gravity anomalies due to extensive sedimentary beds: Geological Society of America Bulletin, v. 55, p. 1217–1228.

Hosphers, J., 1965, Gravity field and structure of the Niger delta, Nigeria, west Africa: Geological Society of America Bulletin, v. 76, p. 407–422.

Hunt, T. M., 1970, Gravity changes at Wairakei geothermal field, New Zealand: Geological Society of America Bulletin, v. 81, p. 529–536.

Hurley, P. M., and Thompson, J. B., 1950, Airborne magnetometer and geological reconnaissance survey in northwestern Maine: Geological Society of America Bulletin, v. 61, p. 835–842.

Johnston, W. D., Jr., 1940, Gravity section across the Sierra Nevada: Geological Society of America Bulletin, v. 51, p. 1391–1396.

Joyner, W. B., 1963, Gravity in north-central New England: Geological Society of America Bulletin, v. 74, p. 831–858.

Kane, M. F., Harwood, D. S., and Hatch, N. L., Jr., 1971, Continuous magnetic profiles near ground level as a means of discriminating and correlating rock units: Geological Society of America Bulletin, v. 82, p. 2449–2456.

Keller, G. R., and Cebull, S. E., 1973, Plate tectonics and the Ouachita system in Texas, Oklahoma, and Arkansas: Geological Society of America Bulletin, v. 84, p. 1659–1666.

Keller, G. R., and Shurbet, D. H., 1975, Crustal structure of the Texas Gulf Coastal Plain: Geological Society of America Bulletin, v. 86, p. 807–810.

Keller, G. R., Lidiak, E. G., Hinze, W. J., and Braile, L. W., 1983, The role of rifting in the tectonic development of the midcontinent, USA: Tectonophysics, v. 94, p. 391–412.

Keller, M. R., Robinson, E. S., and Glover, L., III, 1985, Seismicity, seismic reflection, gravity, and geology of the central Virginia seismic zone: Part 3. Gravity: Geological Society of America Bulletin, v. 96, p. 1580–1584.

King, E. R., and Zietz, I., 1971, Aeromagnetic study of the midcontinent gravity high of central United States: Geological Society of America

Bulletin, v. 82, p. 2187-2208.

Klasner, J. S., and Cannon, W. F., 1974, Geologic interpretation of gravity profiles in the western Marquette district, northern Michigan: Geological Society of America Bulletin, v. 85, p. 213-218.

Kodama, K. P., 1983, Magnetic and gravity evidence for a subsurface connection between the Palisades sill and the Ladentown basalts: Geological Society of America Bulletin, v. 94, p. 151-158.

Kruger, J. M., and Keller, G. R., 1986, Interpretation of crustal structure from regional gravity anomalies, Ouachita Mountains area and adjacent Gulf coastal plain: American Association of Petroleum Geologists Bulletin, v. 70, p. 667-689.

LaFehr, T. R., 1965, Gravity, isostasy, and crustal structure in the southern Cascade range: Journal of Geophysical Research, v. 70, p. 5881-5597.

———— 1966, Gravity in the eastern Klamath Mountains, California: Geological Society of America Bulletin, v. 77, p. 1177-1190.

Lawson, A. C., 1936, The Sierra Nevada in the light of isostasy: Geological Society of America Bulletin, v. 47, p. 1691-1712.

Lidiak, E. G., 1971, Buried Precambrian rocks of South Dakota: Geological Society of America Bulletin, v. 82, p. 1411-1420.

Lillie, R. J., 1985, Tectonically buried continent/ocean boundary, Ouachita Mountains, Arkansas: Geology, v. 13, p. 18-21.

Mason, R. G., and Raff, A. D., 1961, Magnetic survey off the west coast of North America, 32°N latitude to 42°N latitude: Geological Society of America Bulletin, v. 72, p. 1259-1266.

McGee, W. J., 1891, The Gulf of Mexico as a measure of isostasy: Bulletin of the Geological Society of America, v. 3, p. 501-504.

McKeown, F. A., and Pakiser, L. C., eds., 1982, Investigations of the New Madrid, Missouri, earthquake region: U.S. Geological Survey Professional Paper 1236, 201 p.

Mikulich, M. J., and Smith, R. B., 1974, Seismic reflection and aeromagnetic surveys of the Great Salt Lake, Utah: Geological Society of America Bulletin, v. 85, p. 991-1002.

Milsom, J., 1973, Papuan ultramafic belt: Gravity anomalies and the emplacement of ophiolites: Geological Society of America Bulletin, v. 84, p. 2243-2258.

Nettleton, L. L., 1952, Sedimentary volumes in Gulf coastal plain of the United States and Mexico, Part VI: Geophysical aspects: Geological Society of America Bulletin, v. 63, p. 1221-1228.

Neuschel, S. K., 1970, Correlation of aeromagnetics and aeroradioactivity with lithology in the Spotsylvania area, Virginia: Geological Society of America Bulletin, v. 81, p. 3575-3582.

O'Hara, N. W., and Hinze, W. J., 1972, Basement geology of the Lake Michigan area from aeromagnetic studies: Geological Society of America Bulletin, v. 83, p. 1771-1786.

———— 1980, Regional basement geology of Lake Huron: Geological Society of America Bulletin, v. 91, p. 348-358.

Oliver, H. W., 1977, Gravity and magnetic investigations of the Sierra Nevada batholith, California: Geological Society of America Bulletin, v. 88, p. 445-461.

Onesti, L. J., and Hinze, W. J., 1970, Magnetic observations over eskers in Michigan: Geological Society of America Bulletin, v. 81, p. 3453-3456.

Oray, E., Hinze, W. J., and O'Hara, N., 1973, Gravity and magnetic evidence for the eastern termination of the Lake Superior syncline: Geological Society of America Bulletin, v. 84, p. 2763-2780.

Pakiser, L. C., 1964, Gravity, volcanism, and crustal structure in the southern Cascade Range, California: Geological Society of America Bulletin, v. 75, p. 611-620.

———— 1968, Seismic evidence for the thickness of Cenozoic deposits in Mono Basin, California: Geological Society of America Bulletin, v. 79, p. 1833-1838.

———— 1976, Seismic exploration of Mono basin, California: Journal of Geophysical Research, v. 81, p. 3607-3618.

Pakiser, L. C., and Brune, J. N., 1980, Seismic models of the root of the Sierra Nevada: Science, v. 210, p. 1088-1094.

Pakiser, L. C., Press, F., and Kane, M. F., 1960, Geophysical investigation of Mono Basin, California: Geological Society of America Bulletin, v. 71, p. 415-448.

Pratt, T. L., Costain, J. K., Coruh, C., Glover, L., III, and Robinson, E. S., 1985, Geophysical evidence for an allochthonous Alleghanian(?) granitoid beneath the basement surface of the Coastal Plain near Lumberton, North Carolina: Geological Society of America Bulletin, v. 96, p. 1070-1076.

Prodehl, C., 1970, Seismic refraction study of crustal structure in the western United States: Geological Society of America Bulletin, v. 81, p. 2629-2646.

Putnam, G. R., 1922, Condition of the earth's crust and the earlier American gravity observations: Geological Society of America Bulletin, v. 33, p. 287-302.

Qureshy, M. N., Brahmam, N. K., Garde, S. C., and Mathur, B. K., 1968, Gravity anomalies and the Godavari rift, India: Geological Society of America Bulletin, v. 79, p. 1221-1230.

Qureshy, M. N., Venkatachalam, S., and Subrahmanyam, C., 1974, Vertical tectonics in the middle Himalayas: An appraisal from recent gravity data: Geological Society of America Bulletin, v. 85, p. 921-926.

Raff, A. D., and Mason, R. G., 1961, Magnetic survey off the west coast of North America, 40°N latitude to 52°N latitude: Geological Society of America Bulletin, v. 72, p. 1267-1270.

Ramberg, I. B., Cook, F. A., and Smithson, S. B., 1978, Structure of the Rio Grande rift in southern New Mexico and west Texas based on gravity interpretation: Geological Society of America Bulletin, v. 89, p. 107-123.

Reid, H. F., 1922, Isostasy and earth movements: Geological Society of America Bulletin, v. 33, p. 317-326.

Robinson, E. S., 1970, Relations between geological structure and aeromagnetic anomalies in central Nevada: Geological Society of America Bulletin, v. 81, p. 2045-2060.

Romberg, F. E., 1961, Exploration geophysics: A review: Geological Society of America Bulletin, v. 72, p. 883-932.

Rotstein, Y., Combs, J., and Biehler, S., 1976, Gravity investigation in the southeastern Mojave Desert, California: Geological Society of America Bulletin, v. 87, p. 981-993.

Ruppel, B. D., 1971, Geophysics of the Great Valley-Franciscan Junction between Paskenta and Stony Ford, California: Geological Society of America Bulletin, v. 82, p. 1717-1722.

Russell, R. D., Jacobs, J. A., and Grant, F. S. 1960, Gravity measurements on the Salmon glacier and adjoining snow field, British Columbia, Canada: Geological Society of America Bulletin, v. 71, p. 1223-1230.

Shurbet, D. H., 1966, Gravity field and isostatic equilibrium of the Llano Estacado of Texas and New Mexico: Geological Society of America Bulletin, v. 77, p. 213-222.

Shurbet, G. L., and Ewing, M., 1956, Gravity reconnaissance survey of Puerto Rico: Geological Society of America Bulletin, v. 67, p. 511-534.

Shurbet, G. L., and Worzel, J. L., 1957a, Gravity measurements in Oriente Province, Cuba: Geological Society of America Bulletin, v. 68, p. 119-124.

———— 1957b, Gravity anomalies and structure of the West Indies, Part III: Geological Society of America Bulletin, v. 68, p. 263-266.

Shurbet, G. L., Worzel, J. L., and Ewing, M., 1956, Gravity measurements in the Virgin Islands: Geological Society of America Bulletin, v. 67, p. 1529-1536.

Simmons, G., 1964, Gravity survey and geological interpretation, northern New York: Geological Society of America Bulletin, v. 75, p. 81-98.

Sinno, Y. A., Daggett, P. H., Keller, G. R., Morgan, P., and Harder, S. H., 1986, Crustal structure of the southern Rio Grande rift determined from seismic refraction profiling: Journal of Geophysical Research, v. 91, p. 6143-6156.

Smith, R. B., and Eaton, G. P., 1978, Cenozoic tectonics and regional geophysics of the western Cordillera: Geological Society of America Memoir 152, 388 p.

Smithson, S. B., 1972, Gravity interpretation in the Trans-Antarctic Mountains near McMurdo Sound, Antarctica: Geological Society of America Bulletin, v. 83, p. 3437-3442.

Smithson, S. B., and Ramberg, I. B., 1970, Geophysical profile bearing on the origin of the Jotun Nappe in the Norwegian Caledonides: Geological Society of America Bulletin, v. 81, p. 1571-1576.

Speed, R. C., and Cogbill, A. H., 1979, Deep fault trough of Oligocene age, Candelaria Hills, Nevada: Summary: Geological Society of America Bulletin, v. 90, p. 145-148.

Steenland, N. C., and Woollard, G. P., 1952, Gravity and magnetic investigation of the structure of the Cortland Complex, New York: Geological Society of America Bulletin, v. 63, p. 1075-1104.

Sumner, J. R., 1977, Geophysical investigation of the structural framework of the Newark-Gettysburg Triassic basin, Pennsylvania: Geological Society of America Bulletin, v. 88, p. 935-942.

Thiel, E., 1956, Correlation of gravity anomalies with the Keweenawan geology of Wisconsin and Minnesota: Geological Society of America Bulletin, v. 67, p. 1079-1100.

Thiruvathukal, J. V., Berg, J. W., Jr., and Heinrichs, D. F., 1970, Regional gravity of Oregon: Geological Society of America Bulletin, v. 81, p. 725-738.

Thompson, G. A., and Robinson, R., 1975, Gravity and magnetic investigation of the Twin Sisters dunite, northern Washington: Geological Society of America Bulletin, v. 86, p. 1413-1422.

Thompson, G. A., and Sandberg, C. H., 1958, Structural significance of gravity surveys in the Virginia City-Mount Rose area, Nevada and California: Geological Society of America Bulletin, v. 69, p. 1269-1282.

Towle, J. N., 1980, New evidence for magmatic intrusion beneath the Rio Grande rift, New Mexico: Geological Society of America Bulletin, v. 91, p. 626-630.

Vacquier, V., Steenland, N. C., Henderson, R. G., and Zietz, I., 1951, Interpretation of aeromagnetic maps: Geological Society of America Memoir 47, 151 p.

Vening Meinesz, F. A., 1954, Indonesian Archipelago: A geophysical study: Geological Society of America Bulletin, v. 65, p. 143-164.

Walcott, R. I., 1972, Gravity, flexure, and the growth of sedimentary basins at a continental edge: Geological Society of America Bulletin, v. 83, p. 1845-1848.

Waskom, J. D., and Butler, J. R., 1971, Geology and gravity of the Lilesville granite batholith, North Carolina: Geological Society of America Bulletin, v. 82, p. 2827-2844.

White, D., 1924, Gravity observations from the standpoint of the local geology: Geological Society of America Bulletin, v. 35, p. 207-278.

Wold, R. J., and Hinze, W. J., 1982, Geology and tectonics of the Lake Superior basin: Geological Society of America Memoir 156, 280 p.

Woollard, G. P., 1943a, Transcontinental gravitational and magnetic profile of North America and its relation to geologic structure: Geological Society of America Bulletin, v. 54, p. 747-790.

———— 1943b, Geologic correlation of areal gravitational and magnetic studies in New Jersey and vicinity: Geological Society of America Bulletin, v. 54, p. 791-818.

Worzel, J. L., and Ewing, M., 1954, Gravity anomalies and structure of the West Indies, Part II: Geological Society of America Bulletin, v. 65, p. 195-200.

Zietz, I., King, E. R., Geddes, W., and Lidiak, E. G., 1966, Crustal study of a continental strip from the Atlantic Ocean to the Rocky Mountains: Geological Society of America Bulletin, v. 77, p. 1427-1448.

Zietz, I., Bateman, P. C., Case, J. E., Crittenden, M. D., Jr., Griscom, A., King, E. R., Roberts, R. J., and Lorentzen, G. R., 1969, Aeromagnetic investigation of crustal structure for a strip across the western United States: Geological Society of America Bulletin, v. 80, p. 1703-1714.

Zietz, I., Hearn, B. C., Jr., Higgins, M. W., Robinson, G. D., and Swanson, D. A., 1971, Interpretation of an aeromagnetic strip across the northwestern United States: Geological Society of America Bulletin, v. 82, p. 3347-3372.

MANUSCRIPT RECEIVED BY THE SOCIETY JULY 29, 1987
REVISED MANUSCRIPT RECEIVED NOVEMBER 30, 1987
MANUSCRIPT ACCEPTED DECEMBER 4, 1987

CENTENNIAL ARTICLE

Three decades of geochronologic studies in the New England Appalachians

ROBERT E. ZARTMAN   *U.S. Geological Survey, M.S. 963, Denver Federal Center, Denver, Colorado 80225*

ABSTRACT

Over the past 30 years, both isotope geochronology and plate tectonics grew from infancy into authoritative disciplines in the geological sciences. Previously, mountain systems like the Appalachians had been viewed almost entirely in the context of the classical geosyncline, implying a gradualism in stratigraphic and structural change throughout the orogen. Age control, determined largely from distant fossiliferous strata, was unabashedly carried to high-grade metamorphic rocks based only on lithological correlations. With the new concepts in tectonics came the realization that abrupt breaks in stratigraphy and structure occur in many cases at the boundaries of lithotectonic zones. Fortunately, the new techniques of isotope geochronology could be brought to bear directly on the rocks of the immediate study area. This paper chronicles some of the major contributions to the geology of the New England Appalachians that resulted from these efforts during the past three decades.

In tracing the history of geochronologic research, one encounters an increasingly sophisticated approach to the analytical and interpretive aspects of the discipline. Today, the geochronologist can, under optimum conditions, constrain the age of stratigraphic units, igneous activity, deformation, and metamorphism with accuracy that is capable of resolving fine structure within individual orogenic pulses. He participates in full partnership with other colleagues of the science in unravelling the mysteries of mountain building. Several of the topical problems of New England geology in which geochronology played a key role include (1) the recognition and delineation of Avalonia as a Late Proterozoic eastern basement distinct from more western terranes, (2) the dating of the White Mountain Plutonic-Volcanic Suite, a Mesozoic igneous event spanning 100 m.y., and (3) the temporal and spatial separation of structural and metamorphic features imprinted by the Taconic and Acadian orogenies.

The existing geochronology is summarized into a map and table emphasizing the temporal construction of the New England Appalachians. By using lithotectonic zones as the building blocks of the orogen, seven such zones are defined in terms of pre-, syn-, and post-assembly geologic history. From west to east, these lithotectonic zones are (1) Berkshire–Green Mountain, (2) Rowe-Hawley, (3) Connecticut Valley, (4) Bronson Hill, (5) Kearsarge–Central Maine, (6) Tatnic Hill–Nashoba, and (7) Avalonia. Avalonia is further divided into three subzones, Hope Valley, Esmond-Dedham, and Penobscot Bay, which themselves may have had distinct origins and assembly histories. The boundaries between these zones are faults in most cases, some of which may have had recurring movement to further complicate any plate-tectonic scenario.

A delineation of underlying Grenvillian, Chain Lakes, and Avalonian basement is also attempted, which now can make use of isotopes in igneous rocks as petrogenetic indicators to supplement the rare occurrences of basement outcrop within mobile zones of the orogen. The belt of Permian thermal disturbance within the Kearsarge–Central Maine zone is hypothesized to reflect rapid rebound following compressional thickening of underlying Avalonian basement during the Alleghanian orogeny.

INTRODUCTION

Although past their prime in terms of active tectonism and alpine relief, the Appalachians remain an imposing mountain system. The erosionally dissected anatomy of this Late Proterozoic to Mesozoic geologic masterpiece dominates the physiography of eastern North America. Its rocks have been the subject of numerous classical studies in stratigraphy, paleontology, structural geology, and igneous and metamorphic petrology. Recent investigations of the Appalachians have added to these disciplines the revolutionary impact of plate tectonics. As a consequence of the evolving theories of sea-floor spreading and continental drift, virtually every aspect of Appalachian orogenesis has undergone reinterpretation during the past three decades.

This paper reviews the role of geochronology within the New England segment of the Appalachians in changing the way we perceive mountain-building processes. No single perspective holds a monopoly on the truth, and my task can only be accomplished within the broader context of what has happened to the whole science. Since being introduced to the New England Appalachians in 1963, it has been my good fortune to cross paths and associate with a number of the geologists who helped effect the changing concepts. Many of these colleagues are contributing to the Geological Society of America's Decade of North American Geology

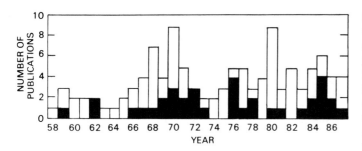

Figure 1. Histogram of number of papers containing radiometric ages from the six New England states published annually since 1958. Black boxes indicate Geological Society of America publications.

(DNAG) Appalachian-Ouachita region volume of The Geology of North America, and the reader seeking more detail and the latest ideas on New England geology is advised to consult this work. Reference to the Appalachian-Ouachita volume may be particularly helpful for those not familiar with geographic names and stratigraphic nomenclature of the area.

No attempt is made to cover even the restricted subject of this paper uniformly and completely. The emphasis placed on certain topical problems is not meant to slight other studies that cannot be treated in equal detail. Obviously, the complete story cannot be told yet, and this retrospection may well prove most useful in stimulating a renewed look at problems still to be resolved. To summarize the achievements already made, a map and table of the existing geochronology have been compiled to document current understanding about the temporal construction of the New England Appalachians. Recognizing lithotectonic terranes, or, using a more neutral term, zones, as the building blocks of the orogen has clearly become the key for deciphering orogenesis, and the role of time in identifying separate terranes—their pre-, syn-, and post-assembly history—cannot be overestimated.

I am aware of 132 published papers, excluding abstracts, unrefereed reports, or theses, that contain the primary documentation of radiometric ages for the 6 New England states through 1987.[1] In this centennial year of the Geological Society of America, perhaps special note should be made that 42 (32%) of these papers appeared in publications of the Society. Figure 1 is a histogram showing the distribution by year since 1958 of the published papers. A complete reference listing is contained in the U.S. Geological Survey's Radiometric Age Data Bank, where the interested reader can find and access the original literature. Internal reports of government agencies, companies, and universities and theses and abstracts not yet replaced by more formal publication provide additional radiometric ages. Furthermore, a familiarity with concurrent developments in the Maritime and the central/southern Appalachians ought not be neglected, although justice is not given in this paper to the literature spawned in these adjacent areas.

GEOSYNCLINES AND NEWFANGLED MACHINES

I beg the reader to allow me a digression to explain what has been a personal attachment to the Appalachians—going back to my youth in southeastern Pennsylvania. Not until undergraduate days at the Pennsylvania State University, however, did I begin to comprehend the structure and stratigraphy of the Valley and Ridge province, to which the topographically subdued Piedmont then seemed to be only an appendage. In the 1950s, the geosynclinal theory of mountain building reigned supreme (Kay, 1951), and one assumed stratigraphic continuity along the length and across the breadth of the entire orogen. These linear mobile belts, which lay along the margins of continents, were seen as crustal depressions receiving tremendous accumulations of sedimentary rocks. Subsequently, by a sort of rebound mechanism, the orogen would undergo an amazing accordion-type compression. The resulting 10-km–wavelength folds are revealed spectacularly in the truncated anticlines and synclines of central Pennsylvania.

An asymmetry to the orogen was recognized, with much of the sediment arriving from an eastern source, mysteriously called "Appalachia," the denuded roots of which now comprised the crystalline Piedmont. Several pulses of unrest in the east were recorded in the great Ordovician, Devonian, and Pennsylvanian deltaic deposits, but only the last pulse extended its tectonic deformation into the Valley and Ridge province. What chance was there of unravelling the polydeformed rocks of this eastern source where fossils are generally obliterated and metamorphism obscures the increasingly complex structure?

For me, the answer to that question had to wait 5 years while I pursued graduate studies across the continent at the California Institute of Technology. Under the tutorage of Professor G. J. Wasserburg, I was soon made aware of a whole new research discipline, which was to contribute substantially to the impending explosion of knowledge in the geological sciences. The mass spectrometer was already proving to be an awesome tool, and the potential for geochronologic and stable-isotope research seemed boundless. In rapid succession, the K-Ar, Rb-Sr, and U-Pb dating methods overcame their analytical barriers. Almost immediately, we had numbers for assignment to the geologic time scale.

I joined the newly organized Branch of Isotope Geology of the U.S. Geological Survey in 1962, where I was initially headquartered in Washington, D.C. Luckily, our Branch Chief, Samuel S. Goldich, had formed an alliance with William R. Shields of the National Bureau of Standards, which was for many years to assure our fledgling group of state-of-the-art instrumentation. These gentlemen also instilled into the Branch an uncompromising philosophy regarding analytical credibility. Reliable basic data are essential in both the field and laboratory, and a lasting respect is held for the geologist on whose careful mapping I depend for meaningful sample collection. One of my initial assignments was to provide geochronologic support for mapping projects in eastern Connecticut. In the Summer of 1963, I finally began my association with the eastern metamorphic and igneous rocks, albeit at some distance to the north of the Pennsylvania Piedmont.

During those first years, there was little hesitation to discuss a stratigraphy that relied on distant fossil evidence in New Hampshire and a tectonic framework that reached equally far. When the mobility of oceanic crust (Hess, 1962; Vine and Matthews, 1963) and its fate to be subducted beneath active continental margins (Guttenberg, 1959) was first receiving recognition, the geosyncline remained, for a short while longer, the secure model for Appalachian tectonics. With its basic tenet of gradualism, the geosyncline theory encouraged extrapolation of stratigraphic units, metamorphic isograds, and plutonic series. Not only were abrupt changes eschewed, but lithologic similarities wherever observed within the orogen were readily grasped as correlatives. To me, it appeared that exceptionally high-quality geologic maps often suffered in the frailty of some of their presumed age assignments. The geosyncline was about to make its acquaintance with the newfangled machine.

---

[1] Primary documentation is considered to be the first appearance of a radiometric age together with its supporting analytical data. The radiometric ages contained in these 132 published papers are on file in the Radiometric Age Data Bank (RADB) of the U.S. Geological Survey. U.S. Geological Survey Open-File Reports on the RADB by state are released periodically.

THE SIXTIES AND EARLIER—LEARNING BY TRIAL AND ERROR

The Spinelli pegmatite near Middletown, Connecticut, provided samarskite that Nier and others (1941) used for some of their first mass-spectrometric analyses. The resultant $^{207}Pb/^{206}Pb$ age of 280 m.y. is in remarkable agreement with modern analyses from the same pegmatite! Even before that, a number of gravimetric chemical analyses of uranium-bearing minerals from New England had allowed the calculation of lead-ratio ages. Without a lead isotopic measurement, however, the common lead content was not known, and no allowance could be made for age discordance among the decay schemes (for an evaluation of these early ages, see Rodgers, 1952). Likewise, occupying only a transitory niche among dating methods, because of analytical problems and lack of an isotopic measurement, were zircon lead-alpha ages—a variant of the lead-ratio method (Webber and others, 1956; Quinn and others, 1957; Lyons and others, 1957). Although occasionally giving good agreement with subsequent isotopic analyses, the general unreliability of the lead-ratio method relegates it to the realm of historical interest. Several U-Pb and K-Ar isotopic ages were determined in the 1950s, which, although becoming increasingly precise analytically, were directed as much toward development of technique as the solution of geologic problems (Wasserburg and others, 1956; Tilton and others, 1957; Aldrich and others, 1958).

The first topical studies employing K-Ar and Rb-Sr isotopic ages and directed primarily toward answering specific regional geologic questions were undertaken by Professor P. M. Hurley and his colleagues at the Massachusetts Institute of Technology. Obtaining minimum ages for a Lower Devonian slate near Jackman, Maine (Hurley and others, 1959), and for metamorphic Pennsylvanian rocks of the Narragansett basin (Hurley and others, 1960) represented important contributions at a time when the Phanerozoic time scale was still frought with major uncertainties. These papers provide constraints on the time scale that remain useful today, but the complicating effects of metamorphic overprinting would haunt other early efforts to establish a rock chronology over much of New England.

Geochronologic investigations in New England during the 1960s tended to fall into two categories. In the relatively unmetamorphosed rocks around the western and northern periphery of the orogen and/or of younger age, established paleostratigraphy invited a refining of the Paleozoic time scale (Hurley and others, 1959, 1960; Bottino and Fullagar, 1966; Fairbairn and others, 1967). In higher grade crystalline rocks largely devoid of fossils, attempts were made to determine stratigraphic age on the basis of radiometric age (Faul and others, 1963; Brookins and Hurley, 1965; Zartman and others, 1965). Neither approach was immune from problems of interpretation. The chief culprit turned out to be chemical disturbances in the various rocks and minerals, which could not be depended upon to yield the primary stratigraphic ages. Notably unreliable were many of the early K-Ar and Rb-Sr mica ages, which later work showed were partially or completely reset by subsequent metamorphism, and Rb-Sr whole-rock ages, which behaved as open chemical systems during diagenesis and metamorphic recrystallization. The successes and failures of this first sustained venture into dating the stratigraphy of the northern Appalachians are captured in the contemporaneous review of Lyons and Faul (1968).

Attention during the latter half of the 1960s turned to evaluating the relative responses of the various dating methods to heating and chemical alteration. The problem of disturbed ages was not unique to the New England Appalachians, and numerous workers had turned their attention to determining the activation energies for diffusion of daughter and parent isotopes within minerals of interest to geochronologists (see, for example, Dodson, 1973, and Hofmann and Giletti, 1970). Field and laboratory experiments eventually permitted semiquantitative calibration of several of the chronometers in terms of temperature above which the mineral or rock no longer behaved as a closed system. This concept of "blocking" or "closure" temperatures promised to make an important contribution to metamorphic petrology. In theory, one could now determine the time of passage through various metamorphic isograds, thereby measuring rates of uplift and cooling as well as fossil geothermal gradients. The complicating effects of recrystallization often invalidate the simple diffusive model, but we emerged with a better appreciation of the constraints these disturbed ages provided on the later metamorphic history.

An awareness of the complicated responses of the various dating methods to superimposed metamorphism is evident in the papers from this period (Brookins, 1967; Clark and Kulp, 1968; Bottino and others, 1970; Zartman and others, 1970). Ramifications of the work of Harper (1968), who was perhaps the first to use isotopic ages to establish a metamorphic chronology in New England, would reverberate long into the coming decade. Until his study demonstrating pervasive Ordovician Taconic metamorphism in Vermont, many workers had concluded that the major metamorphism everywhere in New England was of Devonian Acadian age. Of course, radiometric dating was still to be accepted as capable of yielding primary ages, but only after a strong case was built for preservation during any subsequent metamorphism of the chronometers (Brookins, 1968; Naylor, 1969; Brookins and others, 1969; Zartman and Marvin, 1971). Weathering, too, could disturb the isotopic systems of minerals and rocks (Bottino and Fullagar, 1968), and, thereafter, a premium was to be placed on sample freshness.

To overcome these difficulties, multiple dating methods involving minerals and rocks with varying chemical retentivity were frequently employed. Whenever possible, U-Pb zircon ages were included in the investigation of rocks that lie within the boundary of a thermally disturbed area. Because of this mineral's survival even under conditions of high-grade metamorphism, zircon often provides reliable primary ages where other methods fail. In the future, this refractory property would even pose a new problem—zircon xenocrysts in some igneous rocks actually retain a protolith age!

One important lesson was learned by the time we left the 1960s. An appreciation of the regional tectonics was paramount to avoiding the pitfalls associated with superimposed thermal overprinting. This shift to a more critical evaluation of what we were doing paved the way for a closer relationship between the field geologist and the geochronologist. Spurious results would no longer be accepted as valid in providing stratigraphic ages, but neither would they be ignored for the information they held about subsequent metamorphic disturbances. Mistakes were made during these formative years of isotopic dating, but we learned by trial and error and left the decade with greatly improved understanding.

THE SEVENTIES—FULL SPEED AHEAD

The decade of the 1970s was entered amidst the ferment created by the new visionaries. Some warning salvos had been heard earlier (Wilson, 1966; Kay, 1967), but now there appeared a paper by Bird and Dewey (1970) that brought plate tectonics right into the backyard of the New England Appalachians. We were introduced to passive and active continental margins, opening and closing oceanic basins, and the accretion of an intervening island arc and trench system. Major longitudinal sutures to the orogen marked the boundaries of distinct stratigraphic packages, which were assembled by the collision of rocks once lying hundreds and thousands of kilometers apart. No continuous stratigraphy across a geosyncline yielding only to gradual facies changes here! Admittedly, these megatectonicists sometimes did draw lines through a demonstratable map unit, setting off a lively exchange until the offending line was pushed safely to the east or west (Cady, 1972). But never again would we see in tectonic syntheses of the Appalachians a cross section that accordioned one continuous stratigraphy across the entire orogen.

Looking back, one recalls a time when simplistic ideas in geochronology and geology were rapidly giving way to a new sophistication. In the laboratory, innovations in chemistry and instrumentation were greatly increasing the accuracy and reducing the sample size of our analyses. Improved age resolution placed us on the threshold of differentiating fine structure within individual orogenic pulses. In the field, radical concepts of mountain building, which allowed for the juxtaposing of exotic lithotectonic terranes by long-distance convergence and by translation along strike-slip faults, had found their way into the Appalachians. Armed with these powerful new tools and concepts, we began to ask questions that only a few years earlier would have been considered brash indeed.

Consequently, the 1970s saw an explosion in the acquisition of geochronological data—if not in the actual number of publications (Fig. 1), then certainly in relevance to the regional geology. This new sophisitication brought the geochronologist into further alliance with the field geologist, as, together, they sought to define lithotectonic terranes and to delineate their boundaries. Such cooperation is perhaps nowhere more manifest than in Williams' (1979) compilation of the tectonic-lithofacies map of the Appalachian orogen. Now, let's look at a sampling of studies from this era that provided surprises by breaking with classical interpretation.

Avalonia—An Exotic Southeastern Terrane

The Early to Middle Devonian Acadian orogeny had historically been regarded as the predominant deformational event in New England, and rocks that escaped its metamorphic imprint were generally thought to be Middle Devonian or younger. On this basis and by lithologic similarity to rocks of the fossiliferous Carboniferous Narragansett basin, the Boston Bay Group of eastern Massachusetts was believed to comprise a Mississippian volcanic lower unit and a Pennsylvanian sedimentary upper unit (this view was espoused well into the decade by Billings, 1976). The crystalline basement upon which the group was deposited remained a candidate for Acadian magmatism and metamorphism, although Fairbairn and others (1967) had challenged this age for some of the plutonic rocks. Intruding the basement, there were a number of anorogenic alkalic granites and associated rocks, which because of their undeformed nature, had been given a Carboniferous or younger age. A possible correlation of these alkalic rocks with those of the Mesozoic White Mountain Plutonic-Volcanic Suite of New Hampshire, however, was already disputed on the basis of early lead-alpha and radiometric ages (Toulmin, 1961). Usually ignored in this discussion were several occurrences of fossiliferous Cambrian strata, which clearly had not fared badly during the Acadian orogeny.

A drastic revision to the entire scenario resulted from renewed field and geochronologic investigations. The crystalline basement, composed largely of calc-alkaline granitic rocks intruded into metavolcanic and metasedimentary rocks, was shown to be of unequivocal Late Proterozoic age (Hills and Dasch, 1972; Kovach and others, 1977; Olszewski, 1980). The Boston Bay Group actually lies with stratigraphic continuity below the Cambrian strata, and probably, its lower volcanic unit merges in age with the calc-alkaline granitic rocks of the basement as a volcanic-plutonic complex (Kaye and Zartman, 1980). Finally, the alkalic granites were found to be, in part, older than and unrelated to the Acadian orogeny—apparently emplaced during several episodes between Early Ordovician and Middle Devonian time (Zartman and Marvin, 1971; Lyons and Krueger, 1976; Zartman, 1977; Hermes and others, 1978).

We were coming to understand that Avalonia (the name extended to this coastal area from its counterpart in Newfoundland) had experienced an entirely different early history from the rest of New England. Either piecemeal or as a single unit, it only later became sutured—possibly already assembled with other lithotectonic terranes—to the rest of the Appalachians. Previously, Avalonia appears to have lain on the eastern side of an early Paleozoic seaway, perhaps the Iapetus Ocean, a separation that could readily account for its independent geologic evolution. After its attachment to North America, longitudinal movement along major strike-slip faults has further fragmented Avalonia and accentuated contrasts in degree of deformation at its contact with other terranes. More of the interesting saga of Avalonia is discussed when we look at the present decade.

White Mountain Plutonic-Volcanic Suite—An Igneous Event Spanning 100 m.y.

The post-tectonic character of the White Mountain Plutonic-Volcanic Suite in New Hampshire, western Maine, and east-central Vermont has never been questioned, but the vague stratigraphic control made age assignment difficult prior to the advent of geochronology. We knew only that the White Mountain Plutonic-Volcanic Suite intruded the New Hampshire Plutonic Suite, which, in turn, intruded metamorphosed, but fossiliferous, Silurian and Devonian rocks (Billings, 1956). It is not surprising that on the basis of similarities in petrography and post-tectonic setting, early workers had included the alkalic rocks of eastern Massachusetts with those of the White Mountain Suite. Because one of the Massachusetts alkalic granites lies unconformably below Pennsylvanian rocks of the Narragansett basin, Billings (1956) assigned them collectively a Mississippian age. Toulmin (1961) noted the inconsistency of this assignment and suggested that there must be at least two age groups of alkalic rocks. The Conway Granite from Redstone, New Hampshire, as a representative of the northern group, or White Mountain Plutonic-Volcanic Suite proper, had already been extensively dated, yielding good agreement among U-Pb zircon and K-Ar and Rb-Sr biotite methods (Tilton and others, 1957; Aldrich and others, 1958; Hurley and others, 1960). The resulting age of about 180 m.y. implied that this rock was Jurassic and younger than the pre-Pennsylvanian alkalic granite of the southern group.

Until a decade later, however, when interest in the White Mountain Plutonic-Volcanic Suite was revived, the range in ages to be found among the various intrusive centers was not anticipated. Although a few scattered results in the 1960s already had suggested a real temporal difference for some plutons and lamprophyric dikes of the suite (Faul and others, 1963; Hoefs, 1967; Zartman and others, 1967; Burke and others, 1969), the magnitude and geographic distribution of the age dispersion was first dramatically revealed in a systematic study of several different complexes by Foland and others (1971; see, also, Armstrong and Stump, 1971). Centered along a north-northwest line running the length of New Hampshire, there lie individual stocks and batholiths now unequivocally dated as spanning the Late Triassic to Early Cretaceous time interval.

Eventually, almost all of the approximately 30 complexes were studied (Foland and Faul, 1977; Foland and Friedman, 1977), and some generalities could be made about the distribution in ages. (1) There were three demonstratable peaks in magmatic activity at 220–240, 160–195, and 100–120 Ma, with the latter two having maxima at 180 and 120 Ma. (2) Emplacement of units within single differentiated complexes was brief and usually could not be resolved within analytical uncertainty. (3) The age pattern showed no consistent unidirectional trend along the north-

northwest alignment. Instead, if the 100 to 125 Ma alkalic Monteregian Hills intrusions of adjacent southern Quebec (Shafiqullah and others, 1970; Foland and others, 1986) are included, a roughly symmetrical younging pattern toward both the north and south from the central White Mountain batholith is observed, although notable exceptions to this generality also exist. This age distribution led Foland and Faul (1977) to prefer periodic emplacement along the continental extension of a transform fault, rather than the track of a migrating mantle hot spot or plume, as the mechanism for intrusion. Several distinctly older Triassic stocks in southwestern Maine even suggest a magmatic link back to late Paleozoic peraluminous plutonism, which may have sporadically perforated the country rocks in New Hampshire and Maine since the Acadian orogeny.

The Acadian Orogeny—Fact and Fiction

Few other aspects of the New England Appalachians have received more attention or have been the subject of more controversy than the Acadian orogeny—an Early to Middle Devonian episode of deformation, intrusion, and metamorphism. At no time has the existence of the Acadian orogeny or its pervasive influence on the tectonic development of the orogen been in doubt. What has caused considerable debate are the geographic extent, timing, and underlying mechanisms of this orogeny. As might be guessed, our understanding of the Acadian changed as our concept of mountain building changed to accommodate plate tectonics. Geochronological investigations have been in the vanguard of efforts to delineate this orogenic event in space and time and to determine to what extent it was responsible for the assembly of lithotectonic terranes.

Recognized by its impact on fossiliferous rocks in Quebec, New Brunswick, and northern and eastern Maine, the Acadian orogeny cannot be traced vigorously into adjacent areas of the orogen where structural complexity and metamorphic grade drastically increase. As biostratigraphic correlation gives way to lithologic correlation, as structural discontinuities interrupt an uncertain stratigraphy, and as increasing metamorphic grade further complicates identification of rock units, disagreements have often arisen more from presumptions than from evidence gained on the outcrop. Yet, despite the fragility of regional stratigraphic correlation, diligent field mapping and rare fossils (see Billings and Lyttle, 1981, for a summary of fossil localities) in the metasedimentary rock leave no doubt that an Acadian tectonic picture does extend far southward into central Massachusetts and Connecticut. On the other hand, the concurrent studies in Avalonia made it clear that some major orogenic activity had been mistakenly included in the Acadian sphere of influence. Complications also exist in defining the western limits of Acadian deformation and metamorphism. West of the Connecticut Valley is encountered the overlapping imprint of the Middle Ordovician Taconic orogeny, which cannot be easily differentiated in an area largely devoid of Silurian and Devonian rocks.

During the 1970s, a number of geochronological studies contributed to defining more precisely the timing and boundaries of the Acadian orogeny (Naylor, 1971; Lanphere and Albee, 1974; Moench and Zartman, 1976; Lyons and Livingston, 1977; Ashwal and others, 1979). By narrowly sandwiching the major deformational phase of the Acadian between the then-accepted youngest stratigraphic age of affected rocks and the age of syntectonic plutons, Naylor (1971) advanced his often quoted concept of "an abrupt and brief event." By recapturing for the Taconic orogeny and the closure of the Iapetus Ocean a swath of metamorphic rock along the western perimeter of the Appalachians, Lanphere and Albee (1974) confirmed the conclusion of Harper (1968) in his pioneering work. Somewhat diminished in size and restricted in time, the Acadian orogeny in New England nevertheless emerges from the decade, remaining one of the closing vises of the Appalachians. Hereafter, however, that honor must be shared more than previously with both the earlier, Middle Ordovician Taconic and the later, Pennsylvanian-Permian Alleghanian orogenies. Completely untangling the spheres of influence of these three orogenies remains today the greatest stumbling block in attempts to construct the temporal history of the Appalachians.

Within "the abrupt and brief" Acadian orogeny itself, age resolution was possible as progress in improving analytical precision continued. (1) A phase of major compression with nappe formation and emplacement of syntectonic bodies of granitic rock (Kinsman, Bethlehem, and Spaulding intrusive suites of New Hampshire, 385–410 Ma) terminates Early Devonian sedimentation and heralds the onset of the Acadian orogeny almost simultaneously throughout the New England Appalachians. (2) Uplift and erosion immediately begin to denude the resulting tectonic relief; diapiric rise of gneiss-cored domes occurs along the Bronson Hill (Oliverian domes) and eastern Connecticut Valley zones (380–390 Ma). (3) Discordant plutons (Concord and Barre intrusive suites of New Hampshire and Vermont, respectively, 350–390 Ma), which in some cases are accompanied by retrograde metamorphic aureoles, are emplaced during and following the dome-building phase. (4) Particularly in New Hampshire and western Maine, post-tectonic peraluminous plutonism (325–375 Ma) continues long after cessation of other Acadian deformation—in itself representing significant heat transport into the upper crust.

The big mystery continues to be the role played by the Acadian orogeny in assembling the lithotectonic terranes of the Appalachians. The classical scenario has had the Acadian orogeny as the time of major closure of the orogen, when most of the lithotectonic terranes were transported westward and sutured to the North American continent. Variations on this theme allow different combinations of partial assembly beforehand, such as conceded by the vestiges of the Iapetus Ocean incorporated into the Middle Ordovician rocks along the Taconian line in Vermont. Although not fully agreeing on the exact placement of the suture line, most recent proponents (for example, Osberg, 1978; Hall and Robinson, 1982) of this scenario envision a continent-continent type collision of alpine proportions to bring together the two sides of the orogen.

There is a rival scenario whereby all of the lithotectonic terranes, except perhaps for those of Avalonia, are already essentially joined together following the Middle Ordovician Taconic orogeny (Lyons and others, 1982; Boone and Boudette, 1988). In that case, subsequent sedimentation represents the local reopening of depositional basins, and subsequent tectonism results only in the jostling of adjacent structural blocks, presumably caused by the later arrival of Avalonian terranes. The role of the Acadian orogeny in effecting even this secondary reclosure of the orogen requires a better understanding of these enigmatic easternmost lithotectonic terranes. Questions of assembly between rocks originally so distant from recognizable North American stratigraphy are just now being broached. The possibility that some or all of Avalonia did not arrive until late Paleozoic time (Lux and Guidotti, 1985; O'Hara and Gromet, 1985) suggests that the Pennsylvanian-Permian Alleghanian orogeny may have been previously greatly underestimated in its effect on New England geology.

THE EIGHTIES—RECONCILIATION WITH THE NEW TECTONICS

Few papers relating to the geochronology of New England published in the decade of the 1980s fail to place the work in the context of modern plate tectonics. In fact, the primary purpose of most studies is now directed toward terrane identification and the structural relationships of the boundaries. Petrologically distinct rock suites are regarded as important indicators of tectonic environment, and, as such, are traced up and down the length of the orogen. Emphasis is placed on whether lithologic units are pre-, syn-, or post-assembly relative to their host and surrounding terranes. The geochronologist of the 1980s is constantly on the lookout for weak links in the chain of reasoning and for new approaches toward supplying definitive time constraints. Not surprisingly, attention often turns to problems where biostratigraphic control is absent or controversial and where structural breaks suggest terrane boundaries.

Avalonia Revisited

Half of the geochronological papers published on New England in this decade deal exclusively or in major part with the Avalonian terrane (Aleinikoff and others, 1980; Day and others, 1980; Olszewski, 1980; Dallmeyer, 1983; Lyons and others, 1982; Zartman and Naylor, 1984; Hermes and Zartman, 1985; O'Hara and Gromet, 1983, 1985; Wintsch and Aleinikoff, 1987; Zartman and Hermes, 1987; Zartman and others, 1988). From these studies, we have learned that one or more terranes composed of upper Proterozoic (650–750 Ma) sedimentary and volcanic rocks, which were extensively deformed, metamorphosed, and intruded by calc-alkaline granitic plutons soon after deposition (600–650 Ma), formed somewhere east(?) of the North American continent. Some of the plutons together with the sialic volcanic rocks they intrude appear to comprise a plutonic-volcanic complex, which grades upward into a thick sequence of *Bavlinella*-bearing Vendian (Lenk and others, 1982) clastic rocks of the Boston Bay Group, and, in turn, into several restricted occurrences of *Olenellus*-bearing Lower Cambrian and *Paradoxides*-bearing Middle Cambrian (Theokritoff, 1968) carbonate and clastic rocks. Because these Cambrian fossils have a close affinity with the fauna of the "Atlantic Province" found elsewhere in the coastal Maritime provinces of Canada, northwestern Europe, and western Africa, Avalonia is generally believed to have originated along the eastern side of the Late Proterozoic to early Paleozoic Iapetus Ocean.

Poorly known are the subsequent wanderings of Avalonia and whether the terrane traveled as a unit or as fragments accreting to North America at different times. A case has been made for at least two separate subterranes to Avalonia in southeastern New England—the Hope Valley subterrane to the west underlying other terranes as far west as the Bronson Hill anticlinorium, and already in place by Devonian or earlier time, and the Esmond-Dedham subterrane to the east, not involved in Appalachian tectonics prior to a late Paleozoic, probably Permian, accretion (O'Hara and Gromet, 1985). The Hope Valley subterrane shows evidence of having participated in Acadian and perhaps even older structural events, such as those that formed the Willimantic and Pelham domes and the Massabesic anticline revealing Avalonian rocks in their cores. By contrast, the Esmond-Dedham subterrane contains Ordovician to Middle Devonian alkalic granites that were first deformed, and then only locally, during the late Paleozoic. Naylor (1985) has drawn attention to contrasts in lithology and tectonism between the two subterranes, and he has suggested that the name "Avalonia" be retained only for the Esmond-Dedham subterrane. It is probably not advisable, however, to reject completely a possible kinship between the Hope Valley and Esmond-Dedham subterranes, attributing their disparities more to differences in erosional levels and strike-slip offset than to separate origins.

Although rocks significantly older than Late Proterozoic have not yet been identified at the surface within the Avalonian terrane, lead isotopes and zircon xenocrysts in Devonian granites of coastal Maine (Williams and others, 1986; Ayuso, 1986) and zircon xenocrysts in Permian granites of Rhode Island and Connecticut (Zartman and Hermes, 1987) suggest an underlying Late Archean basement. Zartman and Hermes (1987) specu-

lated that during the Permian, the southeastern part of Avalonia was obducted onto the West African shield, from whence the zircon xenocrysts were acquired by melting of overridden Archean crust. An even earlier acquisition of an ancient basement may have taken place in coastal Maine as revealed by an Archean signature in Devonian granites there. In contrast, no evidence exists for such old rocks west of Avalonia, although Middle Proterozoic zircon has been found in diamictite of the Chain Lake Massif (Naylor and others, 1973) and as xenocrysts in Concord-type granites of New Hampshire (Olszewski, 1980; Aleinikoff and others, 1980; Harrison and others, 1987; John Aleinikoff, 1987, personal commun.).

Unravelling the Taconic and Acadian Orogenies

Historically, the Middle Ordovician Taconic and the Early to Middle Devonian Acadian orogenies were defined on the basis of their structural and sedimentological record in fossiliferous rocks at low metamorphic grade wherein ages could be tightly constrained. In areas of higher grade metamorphic rocks, biostratigraphic resolution of the Taconic and Acadian orogenies is plagued not only by inadequate fossil control but also by structural ambiguities that prevent age assignment of superimposed deformation. Perhaps nowhere has controversy raged louder and longer than in a transitional tract across northern New England, where rocks experience a general southward increase in metamorphic grade perpendicular to the strike of the orogen. Here, despite the discovery of a number of fossil localities over the years and a regionally traceable unconformity at the base of the Silurian stratigraphic section, disagreement continues as to the effects of the Taconic and Acadian orogenies in pre-Silurian rocks (and, indeed, over what are pre-Silurian rocks; see, for example, Bothner and Finney, 1986). It is not surprising, then, that much of the geochronological effort at unravelling the two orogenies has dealt with these transitional rocks.

The story is complex, with multiple episodes of deformation, metamorphism, plutonism, and volcanism recorded within the span of a single orogeny (Rodgers, 1971; Robinson and Hall, 1980). With the superposition of multiple orogenies, the resulting structural complexities can be extremely challenging to unravel. The concept of terranes further complicates correlation because rocks that now lie adjacent to each other may have experienced very different histories while separated by great distances prior to assembly. In fact, unless working with rocks solidly anchored to the North American mainland where both the Taconic and Acadian orogenies can be stratigraphically defined, these very labels are misleading and should be replaced by numerical age assignments. Only after terranes are welded to the mainland can one view their future development in a coherent, regional "North American" context. And, of course, there is no guarantee that once joined together, terranes cannot later again split apart, as becomes very evident in the Mesozoic opening of the North Atlantic.

Nevertheless, there have been successes in placing radiometric age constraints on the pre-, syn-, and post-assembly stages of orogen development. Although the Taconic klippen of westernmost Massachusetts and Connecticut and adjacent eastern New York are the type locality of the Taconic orogeny, their presumed original location along the eastern flank of the Berkshire Highlands and Green Mountains now bears the combined metamorphic overprinting of the Taconic and Acadian orogenies. Separating the overlapping metamorphic isograds and other structural features observed in the Berkshires and lower Paleozoic(?) rocks to the east has been an important accomplishment (Sutter and others, 1985; Zartman and others, 1986). This same objective of distinguishing between Taconic and Acadian metamorphism was pursued successfully in northern Vermont (Laird and others, 1984). Delineating Taconic and Acadian plutonism has been carried as far east as the Bronson Hill anticlinorium (Foland and

Loiselle, 1981; Leo and others, 1984; Zartman and Leo, 1985; Lyons and others, 1986). All of these studies emphasize the growing caution against extrapolating metamorphic isograds and magmatic rock series beyond the immediate geologic setting under investigation. While refining pieces of the puzzle, we are made even more aware of how difficult it is to put the whole puzzle together correctly.

In Maine and southeastern New Hampshire, a number of studies have focused on the protracted period of plutonism, once entirely assigned to the Acadian orogeny, which now must be regarded as multiple thermal pulses that extend well beyond the Middle Devonian into the late Paleozoic (Dallmeyer and Van Breemen, 1981; Dallmeyer and others, 1982; Gaudette and others, 1982; Metzger and others, 1982; Aleinikoff and others, 1985; Harrison and others, 1987). Late- and post-Acadian plutons and batholiths have ages that merge with those of Permian granites associated with the Avalonian terrane and possibly with the oldest intrusive bodies of the White Mountain Plutonic-Volcanic Suite. The tectonic implications of these peraluminous rocks thus need re-examination. Collectively, they occupy approximately 50% of the surface area in western Maine and New Hampshire and represent tremendous transfer of heat from the deeper crust. Just what this sustained plutonism can tell us about the uplift and cooling history of the area is currently the subject of some debate, but the evidence supports a regional upwarping during the Permian of high-grade metamorphic rocks that previously had remained rather deeply buried following the Acadian orogeny. More isolated plutons intruding lower grade metamorphic rocks in eastern Maine suggest less initial burial and less subsequent uplift.

Isotopes as Petrogenetic Indicators

The direct dating of *in situ* rocks is not the only objective of radiogenic isotope measurements. The initial isotopic composition of an igneous rock provides valuable information about its petrogenesis. Likewise, zircon xenocrysts survive in some cases in silicate melts and can be used to date the source rock from which the magma was derived. Two decades ago, Lyons and Faul (1968) recognized the potential usefulness of initial $^{87}Sr/^{86}Sr$ when they tabulated all such ratios then available; however, the poor precision of most of those data prevents any meaningful interpretation by modern standards. Indeed, not until the 1980s have studies of initial Sr, Nd, and Pb been carried out in New England specifically to identify distinct crustal and mantle source rocks and to delineate basement boundaries. This decade also reports the first recognition of inherited zircon in plutons, which, in several instances, retains a remarkable record of source-rock age.

Indirect evidence about the subsurface was already mentioned in our revisit of Avalonia, where strong indications of an Archean basement are found. Two other applications of such isotopic criteria for characterizing lithotectonic zones can be cited. (1) The absence of older crustal material in the Ordovician volcanic and plutonic rocks of the Bronson Hill zone (Zartman and Leo, 1985) contrasts with the frequent occurrence of Middle Proterozoic inherited zircon in coeval and younger igneous rock of the Connecticut Valley and Kearsarge–Central Maine zones (Lyons and others, 1986; John Aleinikoff, 1988, personal commun.) and strongly suggests that these zones had different basements in Ordovician time. (2) The isotopic systematics of a mixed zircon population in a late kinematic granitic body cutting thrust faults in the Berkshire massif yielded an Ordovician crystallization age for the granite and identified its Grenvillian source rock (Zartman and others, 1986). Without careful unravelling of the two zircon populations, the hybrid results could easily have been mistaken for Late Proterozoic to Cambrian crystallization ages. Admittedly, isotopic probes may not always succeed in identifying crustal provenance, but positive indications of a long-standing crustal evolution are difficult to refute.

LITHOTECTONIC ZONES

What, then, has been the consequence of three decades of geochronologic study in the New England Appalachians? Let me draw from the accumulated data base of radiometric ages together with other geologic, geophysical, and geochemical evidence to characterize the main lithotectonic zones[2] that can now be recognized in the orogen. Such an undertaking is not new, but I believe that a compilation emphasizing age control is a useful exercise. In Table 1, for each zone, there are listed (1) preassembly unique characteristics of a terrane, (2) the time(s) and nature of collisional events with neighboring terranes, and (3) post-assembly history common to two or more sutured terranes. The tabulation by necessity is selective, but it attempts to fix in time key elements in the assembly of the Appalachians. Generally, I have borrowed freely from existing compilations. Especially helpful have been the recent tectonic models of Rodgers (1981), Hall and Robinson (1982), Hatch (1982), Lyons and others (1982), Zen (1983), Williams and Hatcher (1983), and Ando and others (1984).

The philosophy for defining the seven lithotectonic zones most nearly corresponds with that adopted by Williams (1979) in the adjacent Canadian Appalachians and extended into the United States by Williams and Hatcher (1983). These zones represent the fundamental building blocks of the orogen, which are constructed and assembled in the context of plate-tectonic environments (continental margins, island arcs, ocean basins, and so on) and processes (obduction, subduction, transcurrent faulting, and so on). The criteria for recognizing individual zones are a separate and distinct original geologic setting, subsequent tectonic joining with other terranes, and a common history thereafter. The seven lithotectonic zones so identified in Table 1 and Figure 2 for the New England Appalachians include, from west to east, (1) Berkshire–Green Mountain, (2) Rowe–Hawley, (3) Connecticut Valley, (4) Bronson Hill, (5) Kearsarge–Central Maine, (6) Tatnic Hill–Nashoba, and (7) Avalonia. The latter zone, Avalonia, consists of three subzones that reveal sufficient distinctions in their origins and/or assembly history to qualify as subdivisions of the building blocks. These components of Avalonia are the Hope Valley, Esmond–Dedham, and Penobscot subzones. A possible fourth component of Avalonia, the Merrimack trough, is herein still retained in the Kearsarge–Central Maine zone, with which it has been traditionally associated. That association, however, is questioned by Lyons and others (1982), who have argued persuasively that neither the stratigraphic age nor the subsequent metamorphism of rocks from the Merrimack trough are correlative with the rest of the Kearsarge–Central Maine zone.

Several special features of the map—subsurface basement boundaries, suture lines indicating time of suturing, and outline of the so-called "Permian disturbed belt"—need some amplification. Dated basement rock in domes and anticlinoria together with zircon xenocrysts from igneous rocks have been used to locate the Grenvillian-Avalonian boundary. This boundary is moved eastward into New Hampshire and Maine from that proposed by Lyons and others (1982), based on recent identification with the ion microprobe of 1.0 Ga (Grenvillian) zircon xenocrysts in plutons of central New Hampshire (Harrison and others, 1987; John Aleinikoff, 1988, personal commun.). Another departure from the map of Lyons and

---

[2]Although "lithotectonic terrane" and "zone" are interchangable terms as used in this paper, I will hereafter use the neutral term "zone," to avoid some of the genetic implications of "terranes" that may not be valid everywhere in the New England Appalachians.

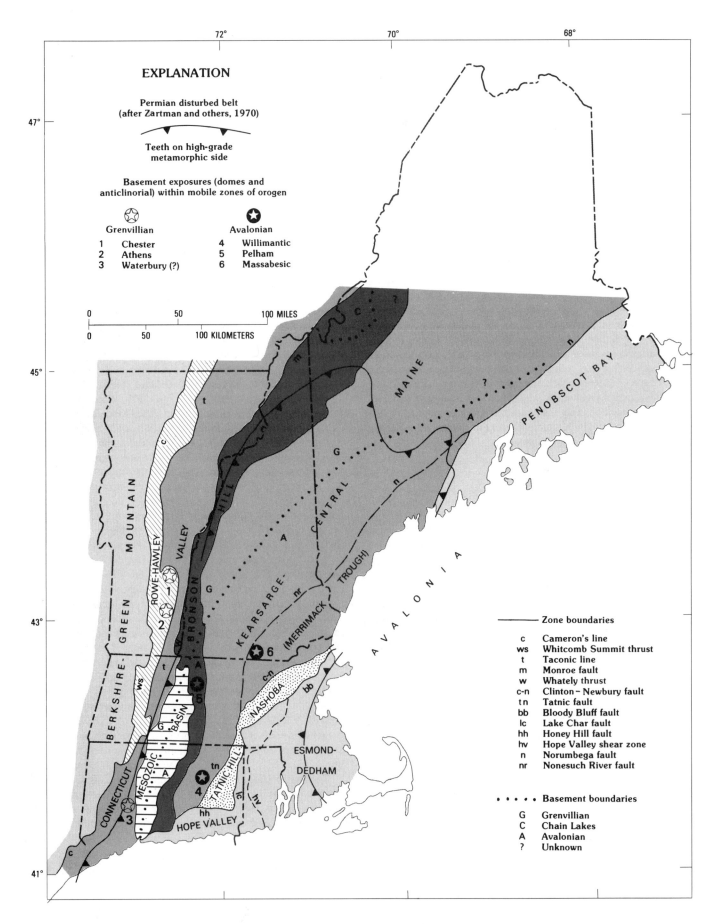

Figure 2. Generalized structural geology map of New England, showing location of lithotectonic zones given in Table 1.

TABLE 1. LITHOTECTONIC ZONES OF THE NEW ENGLAND APPALACHIANS

	West	Lithotectonic Zones		
	Berkshire–Green Mountain	Rowe-Hawley	Connecticut Valley	
Post-assembly	Metamorphism and deformation(?) during Acadian orogeny with subsequent uplift and cooling (405–355 Ma) (1–3); effects most pronounced within eastern flank of Berkshire Mountains and along Green Mountain anticlinorium in northern Vermont. Intrusion of late synkinematic plutons along basement thrust faults (440 ± 10 Ma) (4).	Intrusion of Middle Devonian late kinematic to postkinematic plutons (390–370 Ma) (8), especially in central and northern Vermont, with subsequent uplift and cooling (360–345 Ma) (2, 3). Diapiric rise of Athens and Chester domes, revealing Grenvillian basement (7). Low-grade retrograde metamorphism during the Acadian orogeny (390–355 Ma) (3, 9).	Intrusion of Mount Ascutney stock (120 Ma) (10, 11). Formation of Mesozoic rift basin (ca. 190 Ma) (12) at boundary of Connecticut Valley and Bronson Hill zones. Alleghanian thermal metamorphism (ca. 275 Ma) (13, 14) only peripherally affects this zone along southeastern margin. Intrusion of Middle Devonian late kinematic to postkinematic plutons (390–370 Ma) (8, 15), especially in northern Vermont, with subsequent uplift and cooling (360–345 Ma) (2, 3, 8, 15, 16).	
Syn-assembly	Metamorphism and thrusting of basement and cover rocks during the Taconic orogeny (470–440 Ma) (1, 3, 4). Although basement is autochthonous to North America, it undergoes significant structural telescoping at this time. Upper Proterozoic to Middle Ordovician rocks, including those of the Taconic allochthon, are transported westward at least 60 km from a source region east of the Berkshire and Green Mountains in the Rowe-Hawley zone.	Metamorphism and deformation of upper Proterozoic(?) to Middle Ordovician eugeoclinal rocks, and their tectonic attachment to North America during the Taconic orogeny (470–440 Ma) (3, 9). A belt of ultramafic rocks contained in Ordovician strata may represent slivers of oceanic crust derived outboard of North America.	Metamorphism and deformation of Silurian-Devonian and unconformably underlying older rocks during the Acadian orogeny (390–360 Ma) (3, 13). Seismic reflection evidence suggests major westward transport of an allochthon bearing the rocks of the Rowe-Hawley and Connecticut Valley zones onto the Grenvillian basement at this time. May have closed a Silurian-Devonian basin located between Rowe-Hawley and Bronson Hill zones, although evidence remains equivocal as to whether the Bronson Hill zone was already attached to North America following the Taconic orogeny.	Monroe fault—Whately thrust
Pre-assembly	Deposition of upper Proterozoic to Middle Ordovician quartzite and carbonate of the western shelf facies unconformably on Grenvillian basement, and time-equivalent eastern eugeoclinal-facies sedimentary rocks on rifted(?) continental slope. Intrusion of Tyringham Gneiss and igneous rocks of Mount Holly Complex (1120–950 Ma) (5–7) with accompanying metamorphism of older sedimentary and volcanic rocks. Deposition of percursor sedimentary and volcanic rocks of the Washington Gneiss, Lee Gneiss, Sherman Marble, and so on (>1100 Ma) (5).	Deposition of upper Proterozoic(?) to Middle Ordovician eugeoclinal rocks (including Lower Cambrian Hazens Notch Formation, Upper and Middle Cambrian Ottauquechee Formation, Lower Ordovician Stowe Formation, and Middle Ordovician Missisquoi Formation, probably as an accretionary prism contiguous with those being deposited on the adjacent continental slope of North America. Evidence for an original basement to these rocks is equivocal; where basement now appears in structures, such as the Athens and Chester domes, allochthonous relationships are suggested with the underlying Grenvillian rocks.	Deposition of Silurian and Lower Devonian sedimentary and volcanic rocks, presumably unconformably on deformed older rocks, although subsequent tectonism has obscured contact relationships. Recently discovered Ordovician graptolites (Bothner and Finney, 1986) in rocks of the Connecticut Valley zone bring into question some previous stratigraphic assignments and provide some hope of pre-Taconic correlation between the Rowe-Hawley and Bronson Hill zones.	

Cameron's line—Witcomb summit thrust separates Berkshire–Green Mountain from Rowe-Hawley. *Taconic line* separates Rowe-Hawley from Connecticut Valley.

	Lithotectonic zones			
	Bronson Hill	Kearsarge-Central Maine	Tatnic Hill-Nashoba	
	Intrusion of White Mountain Plutonic-Volcanic Suite (180–120 Ma) (10, 11). Low- to middle-grade Alleghanian thermal metamorphism decreasing to northeast (ca. 275 Ma) (14, 16, 17), followed by accelerated uplift, erosion, and cooling in southwest. Pegmatite emplacement (280–260 Ma) (18, 19) in the Middletown, Connecticut, area. Intrusion of Middle Devonian late kinematic plutons, (390–375 Ma) (20), including the Belchertown pluton in the Pelham dome. Diapiric rise of Oliverian gneisses and, in Massachusetts, underlying basement to form Oliverian domes and Pelham dome, respectively.	Intrusion of White Mountain Plutonic-Volcanic Suite (230–110 Ma) (10, 11, 26). Alleghanian thermal metamorphism and intrusion, decreasing to northeast (ca. 275 Ma) (14, 16, 27–29), followed by accelerated uplift, erosion, and cooling in southwest. Movement along the Norumbega and Nonesuch River faults. Deposition of Pennsylvanian Coal Mine Brook Formation and Harvard Conglomerate. Intrusion of Middle Devonian to Mississippian late kinematic to postkinematic plutons (390–325 Ma) (28–36). Deposition of Middle Devonian Mapleton Sandstone in eastern Maine adjacent to Norumbega fault.	Alleghanian thermal metamorphism (ca. 275 Ma) (14, 35, 40). Renewed movement on Lake Char and Honey Hill faults. Intrusion of Lower Devonian Canterbury Gneiss (395 ± 10 Ma) (35) as a stitching pluton along Tatnic fault. Intrusion of pegmatites and aplites (ca. 410 Ma) (35).	
Monroe fault—Whately thrust	Metamorphism and deformation occurred during the Acadian orogeny. Presumably, the Bronson Hill zone lay at the axis of the orogen and acted as a buttress, against which adjacent thick sedimentary rocks were thrown into large nappe structures. The location of the Bronson Hill zone relative to North America prior to the Acadian orogeny is still largely a matter of speculation. If it was already joined to the more western portions following the Taconic orogeny, then the Kearsarge-Central Maine zone, with which it shares a common sedimentary history since the Middle Ordovician (Robert Moench, 1988, personal commun.) must have accompanied it.	The Kearsarge-Central Maine and Bronson Hill zones, which share a common sedimentary history since at least Middle Ordovician time, were deformed together, presumably, by collision between North America and Avalonia—marking the early phase of the Acadian orogeny (405–390 Ma) (34–36). Extensive nappe formation and synkinematic plutonism occurred within a realtively brief inteval of time. As in the case of the Bronson Hill zone, it is not known where the Kearsarge-Central Maine zone was located relative to North America prior to the Acadian orogeny, and Silurian-Devonian sedimentation may have taken place in a secondary basin outboard of either North America or Avalonia.	Repeated movement, some associated with the Alleghanian orogeny, has occurred along its bounding faults with the Avalonia and Kearsarge-Central Maine zones. If the Esmond-Dedham subzone represents an independent unit of the Avalonia zone, it may have first impinged against the Tatnic Hill-Nashoba zone during the Alleghanian orogeny. Tatnic Hill-Nashoba zone incorporated as a tectonic slice above Avalonia zone and beneath Kearsarge-Central Maine zone. Initial assembly with adjacent zones took place either concurrently with the Acadian orogeny or possibly as early as Silurian time.	Bloody Bluff fault—Lake Char fault—Honey Hill fault
	Deposition of Silurian Clough and Fitch Formations and Lower Devonian Littleton Formation unconformably upon Cambrian-Ordovician rocks. Deposition of Cambrian Albee Formation (north) and upper Proterozoic(?) to Ordovician Monson Gneiss (south), Middle Ordovician Ammonoosuc Volcanics, and Middle Ordovician Partridge Formation. Calc-alkaline (Oliverian and Highlandcroft Suites) plutonism (445 ± 10 Ma) (21–25) occurred concurrently with and soon after Middle Ordovician volcanism. These rocks are stratigraphically continuous with those of the adjacent Kearsarge-Central Maine zone, implying an original structural continuity with at least the western part of that zone. If so, they may constitute an Ordovician paired island-arc and back-arc assemblage.	Deposition of thick sequence of Silurian and Lower Devonian sedimentary rocks. Ordovician and Silurian(?) deformation and metamorphism in Merrimack trough (37). Intrusion of Massabesic orthogneiss (ca. 475 Ma) (27), Newburyport Complex (450 ± 15 Ma) (35), and Ayer Granite (433 ± 5 Ma) (35). Deposition of Cambrian-Ordovician sedimentary and volcanic rocks, with the basal Boil Mountain Complex (510 ± 10 Ma) (38) overlying middle Proterozoic(?) Chain Lakes massif. Some strata in the Merrimack trough may be Proterozoic, and the trough itself may represent a separate lithotectonic zone. Rocks of apparent Late Proterozoic Avalonian association exposed in the Willimantic dome (39) and Massabesic anticlinorium (27).	Intrusion of late kinematic Preston Gabbro (424 ± 5 Ma) (35) and Sharpners Pond Diorite (430 ± 5 Ma) (35). High-grade metamorphism and anatectic melting to form gneissic phase of Andover Granite (450 ± 20 Ma) (35). Deposition of upper Proterozoic(?) Marlboro Formation (= Quinebaug Formation), Shawsheen Gneiss, Fish Brook Gneiss, Nashoba Formation (= Tatnic Hill Formation), and Tadmuck Brook Schist upon unidentified basement.	

Stratigraphic continuity since Middle Ordovician between Bronson Hill and Kearsarge-Central Maine. *Clinton-Newbury fault—Tatnic fault* separates Kearsarge-Central Maine from Tatnic Hill-Nashoba.

TABLE 1. (Continued)

Lithotectonic zones					East
	Avalonia				
	Hope Valley subzone		Esmond-Dedham subzone		Penobscot Bay subzone
Bloody Bluff fault—Lake Char fault—Honey Hill fault	Intrusion of Narragansett Pier and Westerly Granites (273 ± 2 Ma) (41). Metamorphism, deformation, and faulting during Alleghanian orogeny ($285-275$ Ma) (14, 39, 40, 42), including renewed movement along bounding faults. Intrusion of Devonian(?) or Carboniferous(?) alkalic Joshua Rock Granite (39).	Hope Valley shear zone	Intrusion of Narragansett Pier and Westerly Granites (273 ± 2 Ma) (41) with associated contact metamorphism of Narragansett Basin series. Faulting, deformation, and metamorphism during Alleghanian orogeny, especially along Hope Valley shear zone ($285-275$ Ma) (14, 39, 44). At this time, Avalonia may have been obducted eastward onto the West African craton (41). Deposition of Pennsylvanian Narragansett Basin series. Youngest unit of series is Rhode Island Formation of Stephanian (ca. 290 Ma) age, which was coincident with, or slightly preceded, the onset of the Alleghanian orogeny.	Norumbega fault (against Kearsarge–Central Maine zone)	Strike-slip dextral movement along Norumbega fault of 50+ km(?) since Middle Devonian time, with some Pennsylvanian or later movement. The Merrimack trough portion of the Kearsarge–Central Maine zone southeast of the Norumbega and Nonesuch River faults appears to be linked tectonically to the Penobscot Bay subzone since inception of strike-slip faulting. Intrusion of Middle Devonian to Mississippian late kinematic to postkinematic plutons ($390-325$ Ma) (52-54).
	Tectonic emplacement of the Hope Valley subzone beneath the Tatnic Hill–Nashoba, southern and eastern Kearsarge–Central Maine, and southern Bronson Hill zones is poorly constrained but is herein presumed to have occurred concurrently with the Acadian orogeny. A Silurian or even Ordovician assembly cannot be ruled out, however, in which case, the Kearsarge–Central Maine, Bronson Hill, and, possibly, Connecticut Valley zones may have accompanied the Hope Valley subzone during subsequent tectonic movements.		The Esmond-Dedham and Hope Valley subzones may be parts of a single structural unit, in which case, tectonic attachment of the Dedham subzone to North America presumably occurred simultaneously with the Hope Valley subzone. If so, the contrasts in structural style between the two subzones represent subsequent juxtapositioning of different crustal levels, most likely during the Alleghanian orogeny. Alternately, an independent arrival of the Dedham subzone could delay its initial collision with the Hope Valley subzone until the Alleghanian orogeny (44).		The Penobscot Bay subzone is separated from the Kearsarge–Central Maine zone by the Norumbega fault, which cuts the Middle Devonian (380 ± 54 Ma) (55) Lucerne pluton and Pennsylvanian rocks of the Fredericton basin in New Brunswick. Similarity in petrology of Middle Devonian and younger plutons across the Norumbega fault suggests proximity of the zones since the Acadian orogeny. Assembly age of individual tectonic blocks within the Penobscot Bay subzone is poorly constrained, but in part, precedes emplacement of the Lucerne pluton and follows Early Devonian (412 ± 14 Ma) (55) migmatization of Passagassawakeag Gneiss.
	Intrusion of Sterling Plutonic Suite (ca. 625 Ma) (39, 42-45), including Hope Valley Alaskite Gneiss, Potter Hill Granite Gneiss, Ponaganset Gneiss, and Stony Creek Granite Gneiss, and accompanying(?) metamorphism of older sedimentary and volcanic rocks. Deposition of Upper Proterozoic(?) Waterford Group (>625 Ma) (42), including Rope Ferry Gneiss, New London Gneiss, and Mamacoke Gneiss, and Plainfield Formation.		Intrusion of Ordovician to Middle Devonian alkalic granites ($480-370$ Ma) (43, 46-48), and extrusion of Upper Silurian to Lower Devonian Newbury Volcanic Complex. Deposition of upper Proterozoic to Middle Cambrian Boston Bay Group, Mattapan Volcanic Complex (602 ± 3 Ma) (49), Lower Cambrian Weymouth Formation, and Middle Cambrian Braintree Argillite. Intrusion of Dedham Granite, Milford Granite, and Esmond Granite ($630-600$ Ma) (35, 43, 50) and accompanying(?) metamorphism of older sedimentary and volcanic rocks. Deposition of Blackstone Group, Westboro Quartzite, Marlboro Formation, Middlesex Fells Volcanic Complex, and so on [$750(?)-650(?)$ Ma] (51).		Deposition of Upper Silurian to Lower Devonian Eastport Formation and Castine Volcanics (56). Deposition of Ordovician(?) Penobscot, Bucksport, and Appleton Formation (56). Deposition of Cambrian-Ordovician(?) quartzite, limestone, slate, and greenstone of Islesboro Formation. Deposition of precursor sedimentary and volcanic rocks to schist, quartzite, marble, and amphibolite of Seven Hundred Acre Island (>600 Ma) (57). Intrusion of protolith of Passagassawakeag Gneiss: the age of this polydeformed rock is poorly known but assumed to be Proterozoic.

Note: references to geochronological data: (1) Sutter and others, 1985, (2) Harper, 1968, (3) Laird and others, 1984, (4) Zartman and others, 1986, (5) Ratcliffe and Zartman, 1976, (6) Naylor, 1975, (7) Karabinos and Aleinikoff, 1988, (8) Naylor, 1971, (9) Lanphere and Albee, 1974, (10) Foland and others, 1971, (11) Foland and Faul, 1977, (12) Seidemann and others, 1984, (13) Clark and Kulp, 1968, (14) Zartman and others, 1970, (15) Arth and Ayuso, 1985, (16) Faul and others, 1963, (17) Brookins, 1967, (18) Brookins and others, 1969, (19) Brookins, 1970, (20) Ashwal and others, 1979, (21) Naylor, 1969, (22) Foland and Loiselle, 1981, (23) Leo and others, 1984, (24) Zartman and Leo, 1985, (25) Lyons and others, 1986, (26) Burke and others, 1969, (27) Aleinikoff and others, 1980, (28) Dallmeyer and Van Breemen, 1981, (29) Lux and Guidotti, 1985, (30) Moench and Zartman, 1976, (31) Dallmeyer and others, 1982, (32) Aleinikoff and others, 1980, 1985, (33) Harrison and others, 1987, (34) Lyons and Livingston, 1977, (35) Zartman and Naylor, 1984, (36) Gaudette and others, 1982, (37) Lyons and others, 1982, (38) Aleinikoff and Moench, 1985, (39) Zartman and others, 1988, (40) O'Hara and Gromet, 1983, (41) Zartman and Hermes, 1987, (42) Wintsch and Aleinikoff, 1987, (43) Hermes and Zartman, 1985, (44) O'Hara and Gromet, 1985, (45) Hills and Dasch, 1972, (46) Zartman and Marvin, 1971, (47) Zartman, 1977, (48) Hermes and others, 1978, (49) Kaye and Zartman, 1980, (50) Kovach and others, 1977, (51) Olszewski, 1980, (52) Brookins and Spooner, 1970, (53) Spooner and Fairbairn, 1970, (54) Metzger and others, 1982, (55) Marvin and Dobson, 1979, (56) Brookins, 1976, (57) Brookins, 1982.

others (1982) is the omission of an Iapetus remnant between the Grenvillian and Avalonian basements. Collisional events during the Acadian orogeny, and, perhaps, again during the Alleghanian orogeny, are likely to have destroyed any original continuity of Iapetus oceanic crust into the basement. If so, only the detached supracrustal remnants may now remain above a continent-continent suture. Substantial westward transport of the allochthonous Rowe-Hawley, Connecticut Valley, Bronson Hill, and western Kearsarge–Central Maine zones over Grenvillian basement is thereby implied, which agrees with the large thrust-ramp structure interpreted from deep seismic reflection to underlie the area (Ando and others, 1984).

Farther to the northeast, distinction in lead isotopic characteristics of Devonian plutons in the Kearsarge–Central Maine zone (Ayuso, 1986) shows that central Maine is not underlain by the same Avalonian basement as observed along the coast in the Penobscot Bay subterrane. The assignment of a basement to central Maine is thus left in doubt, although it may consist of thick upper Proterozoic and lower Paleozoic metasedimentary rocks derived from Avalonia(?) but without the older Proterozoic and Archean crystalline rocks associated with Avalonia itself (Williams and others, 1986). Also left to conjecture is the nature of the basement lying beneath northern Maine, which may be either Grenvillian (Ayuso, 1986) or the enigmatic Chain Lakes terrane (Boudette, 1982).

Where I am unable to decide among alternative age assignments for suturing events, or where assembly may have been accomplished during two or more episodes of movement between zones, such uncertainty is openly admitted. For example, many workers would follow Bird and Dewey (1970) in proposing collision of the Bronson Hill zone with North America in Ordovician time to explain the telescoping of the Berkshire–Green Mountain zone and westward obduction of the Taconic allochthon. The Acadian collision between North America and Avalonia (or some other landmass), then, results either from destruction of an ocean lying to the east of the Bronson Hill zone or from a second closure of some sedimentary basin that reopened to the west after the Taconic orogeny. Important to this interpretation is the polarity of the Bronson Hill zone, which is generally viewed as a converging-plate island arc, and which Bird and Dewey (1970) presumed was created above a west-dipping Middle Ordovician subduction zone peripheral to North America.

If the opposite polarity prevails with island-arc- (Bronson Hill zone) and back-arc- (Kearsarge-Central Maine zone) facies rocks essentially in depositional continuity rather than juxtaposed across any great distance, another story emerges. Recent mapping tends to confirm the proximity of these two zones since the Middle Ordovician, while suggesting a more profound break in pre-Acadian stratigraphy between the Connecticut Valley and Bronson Hill zones (Robert Moench, 1988, personal commun.). Although still allowing closure of an Ordovician Iapetus Ocean, and, perhaps, better explaining the obducted ultramafic remnants of that ocean, the arrival to North America already in the Ordovician of at least a part of the Kearsarge-Central Maine zone revises post-Taconic paleogeography. Now, if a more eastern location of a closing Silurian-Devonian ocean is sought, apparently we must look some distance into the Kearsarge-Central Maine zone, perhaps to the Norumbega-Nonesuch River fault, where another major break in pre-Acadian stratigraphy occurs. Separate Silurian-Devonian sedimentary basins on opposite sides of the Bronson Hill zone raise the question of two oceans, or an ocean and a back-arc basin (see, for example, Hatch, 1982). Provenance studies employing the dating of detrital zircon to identify sediment source and transport direction may help unravel the mystery.

Finally, let us turn our attention to the Permian disturbed belt of southeastern New England as originally defined on the basis of metamorphic mineral ages. Before the recognition of this discrete belt of thermal metamorphism in the field, a pattern of regional age resetting was already revealed by our earliest efforts to date the rocks radiometrically. Because the K-Ar and Rb-Sr systems in micas are easily disturbed even by mild heating, Faul and others (1963) and, later, Zartman and others (1970) were able to use mica ages to define a belt of Permian thermal disturbance over a broad area of southeastern New England (Fig. 2). Rocks within this disturbed belt, which roughly coincides with an area of Acadian sillimanite-grade metamorphism centered on the Kearsarge-Central Maine zone (then called the "Merrimack synclinorium"), must thus be recording temperature in excess of 200 °C long after their Devonian deformation. Attempts to locate better the perimeter of the belt and to reconstruct the thermal history of the affected rocks have produced two scenarios—one favoring a slow, delayed uplift and cooling of rocks deeply buried during the Acadian orogeny (Dallmeyer and Van Breemen, 1981) and the other favoring superimposed thermal events, which now finds support in extended post-Acadian plutonism that ranges at least into the Early Pennsylvanian (Dallmeyer and others, 1982; Lux and Guidotti, 1985; Aleinikoff and others, 1985). Plate tectonics prompts me to expand on these scenarios and offer a mechanism to explain this thermal anomaly.

The perimeter of the Permian disturbed belt is shown in Figure 2 to follow approximately the outline of presumed Avalonian basement in the subsurface. As previously discussed, there is evidence for the impingement of the Esmond-Dedham subzone of Avalonia at about the time recorded by the disturbed belt. To what extent newly arriving parts of Avalonia first collided with the orogen during the Alleghanian orogeny, and to what extent already-attached parts moved more or less parallel to the coast along transcurrent faults, such as is likely for the more northerly Penobscot Bay subzone, is not known. The indication of increasing compression southward, however, suggests some oblique component to the stress field that may have been transmitted to the Hope Valley subzone. It is not at all unlikely that a crust already softened by heat accompanying post-Acadian plutonism would thicken under such compressive stresses. If, indeed, the Hope Valley subzone does extend as a basement quite far to the northwest, as suggested in Figure 2, some underriding of Grenvillian basement could further thicken the continental crust. A rapid vertical rebound might be expected to follow cessation of these compressive stresses, imparting the tightly grouped 250-m.y. age pattern on the rocks as they passed through their mineral blocking temperatures.

CONCLUDING REMARKS

To single out any particular paper over the past three decades as having had unprecedented impact on unravelling the geochronology of the New England Appalachians is a difficult, if not impossible, task. As I recall, we worked together, argued together, and learned together toward common goals. Some papers have stood the test of time better than others, and occasionally, someone was in the right place at the right time to turn a few tables upside down. A stimulus to all of us was the Bird and Dewey (1970) paper, which, as I said, brought plate tectonics forever into our backyard. Yet, the plate-tectonics revolution was inevitable, and coming at us from many directions.

Progress in understanding the anatomy of this mountain system has always been a team effort. If there is a lesson to be learned in looking back over past accomplishments, it is that nature guards her secrets carefully, but, as in the case of the Purloined Letter, does not hesitate to put clues right in front of our noses. Only when we stop looking at the rocks and think we have a monopoly on the truth do we get sidetracked. Then we must hope that a colleague will come by to set straight our priorities. I have put together here a potpourri of some of the topical problems that confronted us in the past. The answers we fought for may appear obvious today, but so will the future answers to questions we now ask. I can't ever recall being bored by the challenges of the time, nor can unravelling the history of the lithotectonic zones as attempted in Table 1 and Figure 2 be regarded as a trivial exercise. Now let's get on with the job!

ACKNOWLEDGMENTS

As alluded to in the concluding remarks, the study of the New England Appalachians is a team effort, and few ideas can be claimed as solely one's own. In writing this paper, I became especially aware of major stratigraphic and structural revisions just now taking place in several of the lithotectonic zones. Discussions with John Aleinikoff, Robert Moench, John Lyons, and Craig Dietsch were particularly helpful in updating my own perceptions of the terranes and possible times of assembly. The effect has generally been to open up the limits of uncertainty to be attached to the timing of events throughout the orogen.

If I've not been remiss, collaborators in research over the years are duly acknowledged by co-authorship or otherwise in the resulting publications. In addition, however, there are a number of other colleagues who have, recurringly, enlightened me in the field and at scientific meetings and conferences. I take this opportunity, therefore, to give special recognition for the help and advice they offered me toward better understanding the magnificent Appalachians. My thanks and gratitude are extended to Lincoln Page, John Rodgers, Pete Robinson, Phil Osberg, Art Hussey, Jim Skehan, E-An Zen, Wally Cady, Jo Laird, and Dick Naylor. I hasten to add that none of these individuals is to be held responsible for any shortcoming of the present paper, the blame for which must lie squarely on my shoulders.

REFERENCES CITED

Aldrich, L. T., Wetherill, G. W., Davis, G. L., and Tilton, G. R., 1958, Radioactive ages of micas from granitic rocks by Rb-Sr and K-Ar methods: American Geophysical Union Transactions, v. 39, p. 1124–1134.

Aleinikoff, J. N., and Moench, R. H., 1985, Metavolcanic stratigraphy in northern New England—U-Pb zircon geochronology: Geological Society of America Abstracts with Programs, Northeastern Section, Annual Meeting, p. 1.

Aleinikoff, J. N., Zartman, R. E., and Lyons, J. B., 1980, U-Th-Pb geochronology of the Massabesic Gneiss and the granite near Milford, south-central New Hampshire—New evidence for Avalonian basement and Taconic and Alleghanian disturbances in eastern New England: Contributions to Mineralogy and Petrology, v. 71, p. 1–11.

Aleinikoff, J. N., Moench, R. H., and Lyons, J. B., 1985, Carboniferous U-Pb age of the Sebago batholith, southwestern Maine: Metamorphic and tectonic implications: Geological Society of America Bulletin, v. 96, p. 990–996.

Ando, C. J., Czuchra, B. L., Klemperer, S. L., Brown, L. D., Cheadle, M. J., Cook, F. A., Oliver, J. E., Kaufman, S., Walsh, T., Thompson, J. B., Jr., Lyons, J. B., and Rosenfeld, J. L., 1984, Crustal profile of mountain belt: COCORP deep seismic reflection profiling in New England Appalachians and implications for architecture of convergent mountain chains: American Association of Petroleum Geologists Bulletin, v. 68, p. 819–837.

Armstrong, R. L., and Stump, E., 1971, Additional K-Ar dates, White Mountain magma series, New England: American Journal of Science, v. 270, p. 331–333.

Arth, J. G., and Ayuso, R. A., 1985, The Northeast Kingdom batholith: Geochronology and isotopic composition of Sr, Nd, and Pb: Geological Society of America Abstracts with Programs, Annual Meeting, p. 515.

Ashwal, L. D., Leo, G. W., Robinson, P., Zartman, R. E., and Hall, D. J., 1979, The Belchertown quartz monzonite pluton, west-central Massachusetts: A syntectonic Acadian intrusion: American Journal of Science, v. 279, p. 936–969.

Ayuso, R. A., 1986, Lead-isotopic evidence for distinct sources of granite and for distinct basements in the northern Appalachians, Maine: Geology, v. 14, p. 322–325.

Billings, M. P., 1956, The geology of New Hampshire. Part 2, Bedrock geology: Concord, New Hampshire, New Hampshire State Planning and Development Commission, 203 p.

——— 1976, Geology of the Boston basin: Geological Society of America Memoir 146, p. 5–30.

Billings, M. P., and Lyttle, P., 1981, Paleontological control of Paleozoic stratigraphy, New England: U.S. Geological Survey Map MF-1302.

Bird, J. M., and Dewey, J. F., 1970, Lithosphere plate–continental margin tectonics and the evolution of the Appalachian orogen: Geological Society of America Bulletin, v. 81, p. 1031–1060.

Boone, G. M., and Boudette, E. L., 1988, Accretion of the Boundary Mountains terrane within the northern Appalachian orthotectonic zone, *in* Horton, J. W., and Rast, N. M., eds., Melanges and olistostromes of the U.S. Appalachians: Geological Society of America Special Paper (in press).

Bothner, W. A., and Finney, S. C., 1986, Ordovician graptolites in central Vermont: Richardson revived: Geological Society of America Abstracts with Programs, Annual Meeting, p. 548.

Bottino, M. L., and Fullagar, P. D., 1966, Whole-rock rubidium-strontium age of the Silurian-Devonian boundary in northeastern North America: Geological Society of America Bulletin, v. 77, p. 1167–1176.

——— 1968, The effects of weathering on whole-rock Rb-Sr ages of granitic rocks: Geological Society of America Bulletin, v. 79, p. 661–670.

Bottino, M. L., Fullagar, P. D., Fairbairn, H. W., Pinson, W. H., Jr., and Hurley, P. M., 1970, The Blue Hills igneous complex, Massachusetts: Whole-rock Rb-Sr open systems: Geological Society of America Bulletin, v. 81, p. 3739–3746.

Boudette, E. L., 1982, Ophiolite assemblage of early Paleozoic age in central western Maine, *in* St. Julien, P., and Beland, J., eds., Major structural zones and faults of the northern Appalachians: Geological Association of Canada Special Paper 24, p. 209–230.

Brookins, D. G., 1967, Rb-Sr age evidence for Permian metamorphism of the Monson gneiss, west central Massachusetts: Geochimica et Cosmochimica Acta, v. 31, p. 281–283.

——— 1968, Rb-Sr age of the Ammonoosuc volcanics, New England: American Journal of Science, v. 266, p. 605–608.

——— 1970, A summary of geochronological data for pegmatites of the Middletown, Connecticut, area accumulated mainly since 1952: Contributions to Geochronology in Connecticut, Connecticut Geology and Natural History Survey Report of Investigation 5, p. 10–18.

——— 1976, Geochronologic contributions to stratigraphic interpretation and correlation in the Penobscot Bay area, eastern Maine: Geological Society of America Memoir 148, p. 129–145.

——— 1982, Geochronologic studies in Maine—Part II: Precambrian rocks from Penobscot Bay: Isochron/West, v. 34, p. 13–15.

Brookins, D. G., and Hurley, P. M., 1965, Rb-Sr geochronological investigations in the Middle Haddam and Glastonbury quadrangles, eastern Connecticut: American Journal of Science, v. 263, p. 1–16.

Brookins, D. G., and Spooner, C. M., 1970, The isotopic age of the Oak Point and Stonington granites, eastern Penobscot Bay, Maine: Journal of Geology, v. 78, p. 570–576.

Brookins, D. G., Fairbairn, H. W., Hurley, P. M., and Pinson, W. H., 1969, A Rb-Sr geochronologic study of the pegmatites of the Middletown area, Connecticut: Contributions to Mineralogy and Petrology, v. 22, p. 157–168.

Burke, W. H., Otto, J. B., and Denison, R. E., 1969, Potassium-argon dating of basaltic rocks: Journal of Geophysical Research, v. 79, p. 1082–1086.

Cady, W. M., 1972, Are the Ordovician northern Appalachians and the Mesozoic Cordilleran system homologous?: Journal of Geophysical Research, v. 77, p. 3806–3815.

Clark, G. S., and Kulp, J. L., 1968, Isotopic age study of metamorphism and intrusion in western Connecticut and southeastern New York: American Journal of Science, v. 266, p. 865–894.

Dallmeyer, R. D., 1982, $^{40}Ar/^{39}Ar$ ages from the Narragansett basin and southern Rhode Island basement terrane: Their bearing on the extent and timing of Alleghanian tectonothermal events in New England: Geological Society of America Bulletin, v. 93, p. 1118–1130.

Dallmeyer, R. D., and Van Breemen, O., 1981, Rb-Sr whole-rock and $^{40}Ar/^{39}Ar$ mineral ages of the Togas and Hallowell quartz monzonite and Three Mile Pond granodiorite plutons, south-central Maine: Their bearing on post-Acadian cooling history: Contributions to Mineralogy and Petrology, v. 78, p. 61–73.

Dallmeyer, R. D., Van Breemen, O., and Whitney, J. A., 1982, Rb-Sr whole-rock and $^{40}Ar/^{39}Ar$ mineral ages of the Hartland stock, south-central Maine: A post-Acadian representative of the New Hampshire Plutonic Series: American Journal of Science, v. 282, p. 79–93.

Day, H. W., Brown, M. V., and Abraham, K., 1980, Precambrian(?) crystallization and Permian(?) metamorphism of hypersolvus granite in the Avalonian terrane of Rhode Island: Geological Society of America Bulletin, v. 91, Part II, p. 1669–1739.

Dodson, M. H., 1973, Closure temperature in cooling geochronological and petrological systems: Contributions to Mineralogy and Petrology, v. 40, p. 259–274.

Fairbairn, H. W., Moorbath, Stephen, Ramo, A. O., Pinson, H. W., Jr., and Hurley, P. M., 1967, Rb-Sr age of granitic rocks of southeastern Massachusetts and the age of the Lower Cambrian at Hoppin Hill: Earth and Planetary Science Letters, v. 2, p. 321–328.

Faul, Henry, Stern, T. W., Thomas, H. H., and Elmore, P.L.D., 1963, Ages of intrusion and metamorphism in the northern Appalachians: American Journal of Science, v. 261, p. 1–19.

Foland, K. A., and Faul, Henry, 1977, Ages of the White Mountain intrusives—New Hampshire, Vermont, and Maine: American Journal of Science, v. 277, p. 888–904.

Foland, K. A., and Friedman, I., 1977, Application of Sr and O isotope relations to the petrogenesis of the alkaline rocks of the Red Hill Complex, New Hampshire, USA: Contributions to Mineralogy and Petrology, v. 65, p. 213–225.

Foland, K. A., and Loiselle, M. C., 1981, Oliverian syenites of the Pliny region, northern New Hampshire: Geological Society of America Bulletin, v. 92, p. 179–188.

Foland, K. A., Quinn, A. W., and Giletti, B. J., 1971, K-Ar and Rb-Sr Jurassic and Cretaceous ages for intrusives of the White Mountain magma series, northern New England: American Journal of Science, v. 270, p. 321–330.

Foland, K. A., Gilbert, L. A., Sebring, C. A., and Chen, J.-F., 1986, $^{40}Ar/^{39}Ar$ ages for plutons of the Monteregian Hills, Quebec: Evidence for a single episode of Cretaceous magmatism: Geological Society of America Bulletin, v. 97, p. 966–974.

Gaudette, H. E., Kovach, A., and Hussey, A. M., II, 1982, Ages of some intrusive rocks of southwestern Maine: Canadian Journal of Earth Sciences, v. 19, p. 1350–1357.

Guttenberg, B., 1959, Physics of the Earth's interior: New York, Academic Press, 240 p.

Hall, L. M., and Robinson, P., 1982, Stratigraphic-tectonic subdivisions of southern New England, *in* St. Julien, P., and Beland, J., eds., Major structural zones and faults of the northern Appalachians: Geological Association of Canada Special Paper 24, p. 15–41.

Harper, C. T., 1968, Isotopic ages from the Appalachians and their tectonic significance: Canadian Journal of Earth Sciences, v. 5, p. 49–59.

Harrison, T. M., Aleinikoff, J. N., and Compston, W., 1987, Observations and controls on the occurrence of inherited zircon in Concord-type granitoids, New Hampshire: Geochimica et Cosmochimica Acta, v. 51, p. 2549–2558.

Hatch, N. L., Jr., 1982, The Taconic line in western New England and its implications to Paleozoic tectonic history, *in* St. Julien, P., and Beland, J., eds., Major structural zones and faults of the northern Appalachians: Geological Association of Canada Special Paper 24, p. 67–85.

Hermes, O. D., and Zartman, R. E., 1985, Late Proterozoic and Devonian plutonic terrane within the Avalon zone of Rhode Island: Geological Society of America Bulletin, v. 96, p. 272–282.

Hermes, O. D., Ballard, R. D., and Banks, P. O., 1978, Upper Ordovician peralkalic granites from the Gulf of Maine: Geological Society of America Bulletin, v. 89, p. 1761–1774.

Hess, H. H., 1962, History of the ocean basins, *in* Engel, A.E.J., and others, Petrological studies: A volume in honor of A. F. Buddington: Boulder, Colorado, Geological Society of America, p. 599–620.

Hills, R. A., and Dasch, E. J., 1972, Rb-Sr study of the Stony Creek granite, southern Connecticut. A case for limited remobilization: Geological Society of America Bulletin, v. 83, p. 3457–3464.

Hoefs, Joachim, 1967, A Rb-Sr investigation in the southern York County area, Maine, *in* 15th Annual Progress Report to U.S. Atomic Energy Commission: Massachusetts Institute of Technology, Department of Geology and Geophysics, p. 127–129.

Hofmann, A. W., and Giletti, B. J., 1970, Diffusion of geochronologically important nuclides in minerals under hydrothermal conditions: Eclogae geologicae Helvetiae, v. 63, p. 141–150.

Hurley, P. M., Boucot, A. J., Albee, A. L., Faul, Henry, Pinson, W. H., and Fairbairn, H. W., 1959, Minimum age of the Lower Devonian slate near Jackman, Maine: Geological Society of America Bulletin, v. 70, p. 947–950.

Hurley, P. M., Fairbairn, H. W., Pinson, W. H., and Faure, G., 1960, K-Ar and Rb-Sr minimum ages for the Pennsylvanian Section in the Narragansett Basin: Geochimica et Cosmochimica Acta, v. 18, p. 247–258.

Karabinos, Paul, and Aleinikoff, J. N., 1988, U-Pb zircon ages of augen gneisses in the Green Mountain massif and Chester dome, Vermont: Geological Society of America Abstracts with Programs, Northeastern Section, Annual Meeting, p. 29–30.

Kay, G. M., 1951, North American geosynclines: Geological Society of America Memoir 48, 143 p.

Kay, Marshall, 1967, Stratigraphy and structure of northeastern Newfoundland bearing on drift in North Atlantic: American Association of Petroleum Geologists Bulletin, v. 51, p. 579–600.

Kaye, C. A., and Zartman, R. E., 1980, A late Proterozoic Z to Cambrian age for the stratified rocks of the Boston Basin, Massachusetts, U.S.A., *in* Wones, D. R., ed., Proceedings of The Caledonides in the USA, I.G.C.P. project 27, Caledonide orogen: Virginia Polytechnic Institute and State University Memoir 2, p. 257–262.

Kovach, Adam, Hurley, P. M., and Fairbairn, H. W., 1977, Rb-Sr whole rock age determinations of the Dedham Granodiorite, eastern Massachusetts: American Journal of Science, v. 277, p. 905–912.

Laird, Jo, Lanphere, M. A., and Albee, A. L., 1984, Distribution of Ordovician and Devonian metamorphism in mafic and pelitic schists from northern Vermont: American Journal of Science, v. 284, p. 376–413.

Lanphere, M. A., and Albee, A. L., 1974, $^{40}Ar/^{39}Ar$ age measurements in the Worcester Mountains: Evidence of Ordovician and Devonian metamorphic events in northern Vermont: American Journal of Science, v. 274, p. 545–555.

Lenk, Cecilia, Strother, P. K., Kaye, C. A., and Braghoorn, E. S., 1982, Precambrian age of the Boston Basin: New evidence from microfossils: Science, v. 216, p. 619–620.

Leo, G. W., Zartman, R. E., and Brookins, D. G., 1984, Glastonbury Gneiss and mantling rocks (a modified Oliverian dome) in south-central Massachusetts and north-central Connecticut: Geochemistry, petrogenesis, and isotopic age: U.S. Geological Survey Professional Paper 1295, 23 p.

Lux, D. R., and Guidotti, C. V., 1985, Evidence for extensive Hercynian metamorphism in western Maine: Geology, v. 13, p. 696–700.

Lyons, J. B., and Faul, H., 1968, Isotope geochronology of the northern Appalachians, *in* Zen, E-An, and others, eds., Studies of Appalachian geology—Northern and Maritime: New York, Interscience Publishers, p. 305–318.

Lyons, J. B., and Livingston, D. E., 1977, Rb-Sr age of the New Hampshire Plutonic Series: Geological Society of America Bulletin, v. 88, p. 1808–1812.

Lyons, J. B., Jaffe, H. W., Gottfried, D., and Waring, C. L., 1957, Lead-alpha ages of some New Hampshire granites: American Journal of Science, v. 255, p. 527–546.

Lyons, J. B., Boudette, E. L., and Aleinikoff, J. N., 1982, The Avalonian and Gander zones in central eastern New England, *in* St. Julien, P., and Beland, J., eds., Major structural zones and faults of the northern Appalachians: Geological Association of Canada Special Paper 24, p. 43–66.

Lyons, J. B., Aleinikoff, J. N., and Zartman, R. E., 1986, Uranium-thorium-lead ages of the Highlandcroft Plutonic Suite, northern New England: American Journal of Science, v. 286, p. 489–509.

Lyons, P. C., and Krueger, H. W., 1976, Petrology, chemistry, and age of the Rattlesnake pluton and implications for other alkalic granite of southern New England: Geological Society of America Memoir 146, p. 71–102.

Marvin, R. F., and Dobson, S. W., 1979, Radiometric ages: Compilation B., U.S. Geological Survey: Isochron/West, no. 26, p. 3–32.

Metzger, W. J., Mose, D. G., and Nagel, M. S., 1982, Rb-Sr whole-rock ages of igneous rocks in the vicinity of Mt. Desert Island, coastal Maine: Northeastern Geology, v. 4, p. 33–38.

Moench, R. H., and Zartman, R. E., 1976, Chronology and styles of multiple deformation, plutonism, and polymetamorphism in the Merrimack synclinorium of western Maine: Geological Society of America Memoir 146, p. 203–238.

Naylor, R. S., 1969, Age and origin of the Oliverian domes, central-western New Hampshire: Geological Society of America Bulletin, v. 80, p. 405–428.

——— 1971, Acadian orogeny: An abrupt and brief event: Science, v. 172, p. 558–560.

——— 1975, Age provinces in the northern Appalachians: Annual Review of Earth and Planetary Sciences, v. 3, p. 387–400.

——— 1985, Acadian terranes in the northern Appalachians: Geological Society of America Abstracts with Programs, Northeastern Section, Annual Meeting, p. 56.

Naylor, R. S., Boone, G. M., Boudette, E. L., Ashenden, D. D., and Robinson, P., 1973, Pre-Ordovician rocks in the Bronson Hill and Boundary Mountain anticlinoria, New England, U.S.A. [abs.]: EOS (American Geophysical Union Transactions), v. 54, p. 495.

Nier, A. O., Thompson, R. W., and Murphey, B. F., 1941, The isotopic composition of lead and the measurement of geologic time: Physics Review, v. 60, p. 112–116.

O'Hara, K. D., and Gromet, L. P., 1983, Textural and Rb-Sr isotopic evidence for late Paleozoic mylonitization within the Honey Hill fault zone, southeastern Connecticut: American Journal of Science, v. 283, p. 762–779.

——— 1985, Two distinct late Precambrian (Avalonian) terranes in southeastern New England and their late Paleozoic juxtaposition: American Journal of Science, v. 285, p. 673–709.

Olszewski, W. J., Jr., 1980, The geochronology of some stratified metamorphic rocks in northeastern Massachusetts: Canadian Journal of Earth Sciences, v. 17, p. 1407–1416.

Osberg, P. H., 1978, Synthesis of the geology of the northeastern Appalachians, U.S.A., *in* Caledonian-Appalachian orogeny of the North Atlantic region: Geological Survey of Canada Paper 78-13, p. 137–147.

Quinn, A. W., Jaffe, H. W., Smith, W. L., and Waring, C. L., 1957, Lead-alpha ages of Rhode Island granitic rocks compared to their geologic age: American Journal of Science, v. 255, p. 547–560.

Ratcliffe, N. M., and Zartman, R. E., 1976, Stratigraphy, isotopic ages, and deformational history of basement and cover rocks of the Berkshire Massif, southwestern Massachusetts: Geological Society of America Memoir 148, p. 373-412.

Robinson, Peter, and Hall, L. M., 1980, Tectonic synthesis of southern New England, in Wones, D. R., ed., Proceedings of The Caledonides in the USA, I.G.C.P. project 27, Caledonide orogen: Virginia Polytechnic Institute and State University Memoir 2, p. 73-82.

Rodgers, John, 1952, Absolute ages of radioactive minerals from the Appalachian region: American Journal of Science, v. 250, p. 411-427.

——— 1971, The Taconic orogeny: Geological Society of America Bulletin, v. 82, p. 1141-1178.

——— 1981, The Merrimack synclinorium in northeastern Connecticut: American Journal of Science, v. 281, p. 176-186.

Seidemann, D. E., Masterson, W. D., Dowling, M. P., and Turekian, K. K., 1984, K-Ar dates and $^{40}Ar/^{39}Ar$ age spectra for Mesozoic basalt flows of the Hartford basin, Connecticut, and the Newark basin, New Jersey: Geological Society of America Bulletin, v. 95, p. 594-598.

Shafiqullah, M., Tupper, W. M., and Cole, T.J.S., 1970, K-Ar age of the carbonatite complex, Oka, Quebec: Canadian Journal of Earth Sciences, v. 10, p. 541-552.

Spooner, C. M., and Fairbairn, H. W., 1970, Relation of radiometric age of granitic rocks near Calais, Maine, to the time of Acadian orogeny: Geological Society of America Bulletin, v. 81, p. 3663-3670.

Sutter, J. F., Ratcliffe, N. M., and Mukasa, S. B., 1985, $^{40}Ar/^{39}Ar$ and K-Ar data bearing on the metamorphic and tectonic history of western New England: Geological Society of America Bulletin, v. 96, p. 123-136.

Theokritoff, George, 1968, Cambrian biogeography and biostratigraphy in New England, in Zen, E-An, and others, eds., Studies of Appalachian geology—Northern and Maritime: New York, Interscience Publishers, p. 9-22.

Tilton, G. R., Davis, G. L., Wetherill, G. W., and Aldrich, L. T., 1957, Isotopic ages of zircon from granites and pegmatites: American Geophysical Union Transactions, v. 38, p. 369-371.

Toulmin, Priestley, 3rd, 1961, Geological significance of lead-alpha and isotopic age determinations of "alkalic" rocks of New England: Geological Society of America Bulletin, v. 72, p. 775-780.

Vine, F. J., and Matthews, D. H., 1963, Magnetic anomalies over ocean ridges: Nature, v. 199, p. 947-949.

Wasserburg, G. J., Hayden, R. J., and Jensen, K. J., 1956, Ar^{40}-K^{40} dating of igneous rocks and sediments: Geochimica et Cosmochimica Acta, v. 10, p. 153-165.

Webber, G. R., Hurley, P. M., and Fairbairn, H. W., 1956, Relative ages of eastern Massachusetts granites by total lead ratios in zircon: American Journal of Science, v. 254, p. 574-583.

Williams, Harold, 1979, Appalachian orogen in Canada: Canadian Journal of Earth Sciences, v. 16, p. 792-807.

Williams, Harold, and Hatcher, R. D., Jr., 1983, Appalachian suspect terranes: Geological Society of America Memoir 158, p. 33-53.

Williams, I. S., Eriksson, S. C., and Compston, W., 1986, Application of inherited zircons to terrane accretion: An example of Archean zircons from the coastal belt of Maine: Geological Society of America Abstracts with Programs, Annual Meeting, p. 790.

Wilson, J. T., 1966, Did the Atlantic close and then re-open?: Nature, v. 211, p. 676-681.

Wintsch, R. P., and Aleinikoff, J. N., 1987, U-Pb isotopic and geologic evidence for late Paleozoic anatexis, deformation, and accretion of the Late Proterozoic Avalon terrane, south-central Connecticut: American Journal of Science, v. 287, p. 107-126.

Zartman, R. E., 1977, Geochronology of some alkalic rock provinces in eastern and central United States: Annual Review of Earth and Planetary Sciences, v. 5, p. 257-286.

Zartman, R. E., and Hermes, O. D., 1987, Archean inheritance in zircons from late Paleozoic granites from the Avalon zone of southeastern New England: An African connection: Earth and Planetary Science Letters, v. 82, p. 305-315.

Zartman, R. E., and Leo, G. W., 1985, New radiometric ages on gneisses of the Oliverian domes in New Hampshire and Massachusetts: American Journal of Science, v. 285, p. 267-280.

Zartman, R. E., and Marvin, R. F., 1971, Radiometric age (Late Ordovician) of the Quincy, Cape Ann, and Peabody Granites from eastern Massachusetts: Geological Society of America Bulletin, v. 82, p. 937-958.

Zartman, R. E., and Naylor, R. S., 1984, Structural implications of some radiometric ages of igneous rocks in southeastern New England: Geological Society of America Bulletin, v. 95, p. 522-539.

Zartman, R. E., Snyder, George, Stern, T. W., Marvin, R. F., and Bucknam, R. C., 1965, Implications of new radiometric ages in eastern Connecticut and Massachusetts: U.S. Geological Survey Professional Paper 525-D, p. D1-D10.

Zartman, R. E., Brock, M. R., Heyl, A. V., and Thomas, H. H., 1967, K-Ar and Rb-Sr ages of some alkalic intrusive rocks from central and eastern United States: American Journal of Science, v. 265, p. 848-870.

Zartman, R. E., Hurley, P. M., Krueger, H. W., and Giletti, B. J., 1970, A Permian disturbance of K-Ar radiometric ages in New England: Its occurrence and cause: Geological Society of America Bulletin, v. 81, p. 3359-3374.

Zartman, R. E., Kwak, L. M., and Christian, R. P., 1986, U-Pb systematics of a mixed zircon population—The granite at Yale Farm, Berkshire Massif, Connecticut, in Peterman, Z. E., and Schnabel, D. C., eds., Shorter contributions to isotope research: U.S. Geological Survey Bulletin 1622, p. 81-98.

Zartman, R. E., Hermes, O. D., and Pease, M. H., Jr., 1988, Zircon crystallization ages and subsequent isotopic disturbance events in gneissic rocks of eastern Connecticut and western Rhode Island: American Journal of Science, v. 288 (in press).

Zen, E-An, 1983, Exotic terranes in the New England Appalachians—Limits, candidates, and ages: A speculative essay: Geological Society of America Memoir 158, p. 55-81.

MANUSCRIPT RECEIVED BY THE SOCIETY OCTOBER 26, 1987
REVISED MANUSCRIPT RECEIVED APRIL 8, 1988
MANUSCRIPT ACCEPTED APRIL 11, 1988

100 years of economic geology and GSA

J. M. GUILBERT *Department of Geosciences, University of Arizona, Tucson, Arizona 85721*

Think what it must have been like to have been an economic geologist in 1888! Washington, Montana, and the Dakotas were still territories a year from statehood, and Idaho and Wyoming were two years from that status. Oklahoma would wait 19 years, and Arizona and New Mexico 24 years, before achieving statehood. Vast areas of the western United States were true frontiers. Railroads were single-track, east-west threads hundreds of miles apart, not the networks that were to come. More than 70% of the area of the western states was occupied by fewer than 2 people per square mile, including Indians. Transportation was slow. Travel away from railroads was by foot, boat, horseback, wagon, or coach, and so 50 miles a day was a real achievement. There were a few mining towns in the western interior, such as Butte and Virginia City, Montana; Silver City, Idaho; Virginia City, Nevada; Sacramento, California; Prescott and Tombstone, Arizona Territory; Silver City, New Mexico Territory; and Leadville, Colorado. There were a few inland mining support and transportation centers such as Tucson, Santa Fe, Salt Lake, and Denver, but the logistics of field geology must have been awful. Although Geronimo had surrendered a year earlier in 1887, Indian warfare would rage locally for another 20 years. Lawlessness was a general problem outside most city limits, and inside some. World population was about 1.5 billion, United States population had just topped 60 million, and there were fewer than 3 million dwellers in the City of New York.

There were few professional geological societies—The American Association of Geologists (1840) had broadened to become the American Association of Geologists and Naturalists (1842) and then the American Association for the Advancement of Science (1847). GSA was forming around a nucleus of people that we now identify as "economic" types, but the Society of Economic Geologists and the Mineralogical Society of America were 32 years away. National meetings were essentially unheard of—in fact, they had hardly been thought possible. The American Institute of Mining Engineers had formed 17 years earlier in 1871, but its meetings were still mainly regional ones. Several American journals were being published—the *American Journal of Science* (since 1818), the *Transactions of AIME* (since 1871), and *The American Geologist* (begun in January 1888!). About 700 minerals had been identified, but refined descriptions of stoichiometry suffered from primitive analytical procedures and from the fact that only 70 chemical elements had been discovered by 1888. Dana's *System of mineralogy* was in its 52nd year and 5th edition, and several modern treatises on ore deposits had been released. Von Cotta's *A treatise on ore deposits* (New York, 1870) was severely dated, but J. Arthur Phillips' *A treatise on ore deposits* (London, 1884) and D. C. Davies' *A treatise on metalliferous minerals and mining* (London, 4th ed., 1888) were hot off the presses. The U.S. Geological Survey (USGS) was by now 8 years old, with John Wesley Powell as director. Monographs on the Comstock Lode, Michigan copper, Leadville, and Eureka, Nevada, were out, and the series of USGS *Professional Papers* was about to begin. A dozen USGS *Bulletins* had been published, including one on phosphate deposits by R.A.F. Penrose, Jr., and the 5th annual USGS *Mineral resources of the United States* (for 1888) was in preparation. It would run to 651 pages and be sold for 50 cents.

The gold rush of 1849 had drawn tens of thousands of prospectors, miners, settlers, and merchants into the West, and the backwash from that wave of adventurers—and from Civil War peregrinations—had led to the discovery of many portentous districts. There was a clamor from entrepreneurs and engineers to know more about the geology of the western United States and of ore deposits in general. J. D. Whitney had written *The metallic wealth of the United States* (1854), and Browne and Taylor wrote a report to the Secretary of the Treasury in 1867 entitled *Reports upon the mineral resources of the United States*. Even with these documents, however, there was almost no backlog of published information on North American ore deposits, no significant communication between scientists except by hand-written letters, and only the barest stirrings of formal mineral-deposit instruction in the colleges and universities. H. C. Sorby had developed transmitted-light petrographic microscopy about 30 years earlier, but it was only now gaining rapidly in popularity. Reflected-light microscopy was far more advanced in theory than in practice. Quantitative physics and chemistry were enjoying great success. Lord Kelvin, and later Clarence King, was working on the determination of the age of the Earth by measurement of heat capacities and thermal conductivities of rocks. King's announcement 5 years later in 1893 that the Earth was only 24 million years old dismayed many but convinced many more of the effectiveness of such techniques. On another front, the Freiberg School in Germany was orchestrating the efforts of many practitioners and scholars of mining geology. The results of their practical work and their recognition of many of the details of upward structural guidance of hydrothermal ore-forming fluids would not become generally available until 1901, but its impact was being felt much earlier by word of mouth, both locally and abroad.

All in all, to be an economic geologist in 1888 required scholarship, patience, physical strength, robustness, and conviction. Field trips were measured in months, seasons, or even years, rather than in days or weeks. Preparation of reports on mining districts demanded that the authors describe not only the geology but also the culture and all aspects of natural history to the legislators and businessmen "back east" who had never visited the West. It must have been a lonely, isolated existence, with months at a time going by without substantial contact with one's peers.

1888–1912: THE FIRST QUARTER-CENTURY

Isolated labor and scholarship were not to last. With society's onrush from the Industrial Revolution into the Age of Electricity, the smokestack era was born, consumerism flourished, and the demand for commodities required an improvement in the science of economic geology. The base metals would in general prosper, but gold was held in the vise of the currency standard, and silver was crippled by the silver panics of 1893 and 1907, and humbled further by demonetization. While minerals research groups were organized in the U.S. Geological Survey, two other vital forces in mineral-deposits study emerged. Organization of mining geology and exploration groups by such new-born companies as Anaconda, Kennecott, U.S. Steel, and ASARCO would have a profound effect upon inquiry, discussion, and journal reporting, as would the emergence of powerful teaching and research facilities at several universities (Harvard, Yale, Columbia, Michigan, Chicago, Arizona, and Berkeley) and Schools of Mines (Columbia, Missouri, Montana, Colorado, New Mexico, Arizona, and Nevada). An important aspect of this emergence was the introduction of graduate instruction. Graduate student research augmentation of faculty output quickly earned recognition of the value of academic research.

The founding of the Geological Society of America was a landmark event in economic geology. Not only did it bring interested groups together in those early years—after all, annual national GSA meetings regularly had as many as two-score registrants by the end of the century!—but also it served to spawn subgroups of special-interest scientists. Early GSA *Bulletins* carried everything—meeting schedules, announcements, abstracts, programs, and papers. Even in these very early years, separate sections of regional and national meetings were allocated to "economic" geology. In view of the hindrances of distance and remoteness, of primitive chemistry and physics, and of religious doctrine in earth science, we were fortunate that so remarkable a cadre of men—many of them to be associated or identified with economic geology in the future—gathered in those early years. Volume 1 of the *Bulletin* carried papers by James Hall, T. C. Chamberlain, J. F. Kemp, G. K. Gilbert, Charles Walcott, J. W. Spencer, Andrew Lawson, William M. Davis, C. R. Van Hise, S. F. Emmons, Alexander Winchell, Nathaniel Shaler, and C. H. Hitchcock. The first economic geology paper of the hundreds that adorn subsequent *Bulletin* pages was by E. V. D'Invilliers (1891) on phosphate deposits of the Island of Navassa in the Caribbean. The Sudbury Complex had been discovered in 1882, and the first "definitive" description of it appeared in the GSA *Bulletin* (Bell, 1891). The last three pages of that article were astounding in that G. H. Williams described the amazing glass-sherd textures of the Onaping Formation (now thought to be an asteroid-impact fall-back breccia) as being provocatively unusual. He couldn't explain it, but he felt compelled to record its uncommon properties! Volume 3 (the last of three edited by W. J. McGee) carried one of the few articles by Richard Penrose. It was entitled "The Tertiary iron ores of Arkansas and Texas" (Penrose, 1893) and must have been written, submitted, and galley-proofed while he was launching his first successful mining venture at what he named the "Commonwealth Mine," at Pearce, Arizona. Earnings from that mine are estimated at 20 million dollars between 1893 and 1903, in spite of the lower silver prices after the Panic of 1893 and the Repeal of the Silver Standard in 1896. They constituted the nucleus of the fortune that "Penny" would later use to endow the American Philosophical Society and the Geological Society of America and to enrich the Society of Economic Geologists.

The first several issues of the *Bulletin* were dominated by "sed-strat-paleo" papers, as the journal was to be through most of its history. A series of reports by W. H. Weed, Arthur Winslow, J. J. Stevenson, W. C. Knight, and others established a tradition of coal-geology reporting that continues today. Elegant descriptions of coalfields in Montana, Wyoming, Missouri, New Mexico, and Mexico were interspersed with others on eastern coalfields and coal petrologic problems. Another area of obvious keen interest involved the Lake Superior province iron formations that were being funneled into the furnaces of the lower Great Lakes states. Their study first hit the *Bulletin* pages in 1894 with C. R. Van Hise's "The succession in the Marquette Iron District of Michigan." The breadth of coverage that would characterize the pages of the *Bulletin* under 40 years of editorship by Joseph Stanley-Brown (volumes 4–44, 1893–1933) was established early with articles on the Wisconsin Mississippi-Valley–type deposits (Blake, 1894), the California goldfields (Smith, 1894), nickel and titanium deposits (Kemp, 1894 and 1895), and Canadian gold veins (McKellar, 1898). A man described by Faul and Faul (1983) as "a Swede with a passion for the American West" broke into print in the *Bulletin* in 1895 with an article on "Characteristic features of California gold-quartz veins" (Waldemar Lindgren, 1895). During this first decade, a series of "areas of interest" were being developed in the *Bulletin*, including heavy involvement with nonmetallics such as coal, pegmatites and their products, phosphate rock, asbestos, and even diamonds. The fascinations of supergene enrichment were broached by W. H. Weed (1900) in his amazingly perceptive report on "Enrichment of mineral veins by later metallic sulfides." A few new textbooks came out during this period. Largely descriptive, they were welcome extensions of experience for a growing number of ore-deposit scholars. They included W. S. Tarr's *Economic geology of the United States* (1894) and Kemp's *Ore deposits of the United States* (1895) and *Ore deposits of the United States and Canada* (1903).

As the GSA *Bulletin* came into its early influence, speculation on ore-forming mechanisms, and therefore upon ore-deposit classifications, was just getting organized. A few scholars, such as Louis de Launay and J. A. Phillips, had made careful observations, reasoned with them, and correctly related basic physical principles to ore-deposit classification and genesis. Many field and mines geologists were grappling with the need to relate observations to principles of natural science, but there were still many quite fanciful explanations of ore-deposit genesis in circulation, some of them dogmatic. A group of cage-rattlers, however, had developed in and around the Freiberg Mining Academy. Incisive thinkers who were to become iconoclasts made *observations* in the mines of Saxony, Bohemia, and Slovakia that had to be answered. Professors such as the formidable Bernhard Von Cotta and Franz Posepny and their students, among whom were the irrepressible Raphael Pumpelly and the passionate Swede Waldemar Lindgren, both by then in the United States, were starting to close loops between observation, measurement, and explanation. Franz Posepny was an ill old man of 52 when the *Bulletin* was born. He was accorded Honorary Membership in AIME in GSA's first year, 1888, in recognition of his lifetime of practical studies on ore deposits. These studies culminated in his reading his book entitled *The genesis of ore deposits* to the annual AIME meeting assembled in Chicago in 1893. The paper (AIME Transactions, v. 23, 1893, p. 197–369, in English) carries the note "Translated by the Secretary," but it is not clear whether the paper was delivered in English or German. He died in 1895, and his 1893 text was published by AIME as a small book in English entitled *Ore deposits* that same year. Because it sold out immediately, and to ensure wide circulation, it was reprinted in *AIME Transactions* in 1901. His book was a true benchmark, a volume whose time had come. Its reasoned application of scientific method and observation that yielded a logical ore-deposit classification scheme rang true to many of his colleagues at home and abroad. It fulminated explosions of commentary, both laudatory and critical. The 1901 reprinting carried hefty responses from the luminaries of the day, including Van Hise, Kemp, Emmons, Weed, Rickard, Lindgren, and a young man named Vogt. Another volume of commentary called *Ore deposits—a discussion* was reprinted from the pages of *Engineering and Mining Journal* in 1905. Its authors include many of the same names, with comments made by prominent GSA members at two Geological Society of Washington meetings early in 1903.

Posepny's book naturally emphasized the Erzgebirge and other eastern European areas. Whether coincidentally or causally, a flood of texts by North Americans quickly followed: W. S. Tarr's (1905) *Economic geology of the United States* (2nd ed.), Merrill's (1905) *The nonmetallic minerals*, Beck and Weed's (1905) *The nature of ore deposits* (2 volumes),

Rickard's (1905) *Copper mines of Lake Superior*, and Weed's (1907) *Copper mines of the world*. Discussion spilled into the pages of the GSA *Bulletin*, inducing another important paper by Emmons (1904).

Another landmark event in these crossroad years was the inception in 1905 of the Economic Geology Publishing Company and the journal *Economic Geology*, which immediately absorbed *The American Geologist*. This new journal provided a prestigious pipeline for papers related to science, research, and the application of geological principles to ore deposits, and even the early volumes ran to 800 to 1,000 pages per year. Nonetheless, GSA *Bulletin* remained an important outlet for broader papers on secondary enrichment, iron ores, coal, diamonds, and phosphates and on specific areas such as Sudbury; Boulder County, Colorado; Wisconsin lead-zinc; and the Homestake district. Paige (1913) concluded that the Homestake ores were structurally controlled replacement deposits, starting an argument that would rage for 60 years. J. D. Irving, in discussion (1913), observed that there was no evident alteration and hence no evidence for replacement—the ores could not be distinguished from the country rocks, except by gold, silica, and iron content!

As the first quarter-century of GSA drew to a close, important new mines were being discovered, some old ones were closed down by economic shakedowns like the silver panics, and the emphasis in boardrooms was shifting toward businesslike management and scientific method. With Jackling's demonstration of the feasibility of open-pit mining and the development of flotation methods of extractive metallurgy in the first decade of the Twentieth Century, the stage was set for science and industry to burgeon.

1913–1937: THE SECOND QUARTER-CENTURY

This period started as it would end, with the darkening clouds of global war. World War I slowed the energy and rate of growth in economic geology into the western United States, even as the war effort put newly organized companies like Anaconda, ASARCO, Phelps Dodge, and USARCO to the test of accelerated production. The American involvement in WWI from 1914–1918 had no effect upon the thickness of volumes of *Economic Geology*, but the GSA *Bulletin* was noticeably thinner, especially in 1915. Several *Bulletin* articles on geology and the war effort, geopolitics, materials strategy and stockpiling, domestic development of critical minerals, and geologic knowledge as a tactical battlefield strategy appeared. Papers on Park City, Utah, coal petrography, hot-spring activity and geochemistry, the first article on Cripple Creek (Patton, 1915), on Kiruna, on Montana phosphate, on replacement mechanisms in Butte sulfides, on Muzo emeralds, and on many aspects of iron ores also appeared in the *Bulletin* during the war years. "Sericite, a low-temperature hydrothermal mineral" was defined by A. F. Rogers (1915), and the first truly experimental papers concerning water and magmas appeared with George Morey's "Importance of water as a magmatic constituent" (1916) and "The magmatic sulfides" (C. F. Tolman and A. F. Rogers, 1917). Overall emphasis in *Bulletin* pages continued strong in areas of observation and description, especially in stratigraphy and paleontology, with a strong mineralogy component.

1920 was an important year for both GSA and economic geologists for several reasons. Late in 1919, a group of GSA members with special interests met and decided to form a new society, the Society of Economic Geologists. R.A.F. Penrose, Jr., by then a wealthy 53-year-old philanthropist-scientist-business leader, was honored by his peers by election to the first presidency of SEG. The group of "activists" already listed above—Van Hise, Emmons, Weed, and the rest—had been joined by Reno Sales, Alan Bateman, J. E. Spurr, and Horace V. Winchell, but it was Penrose's efforts and leadership in forming the Society that earned him the honor of being its first president. The 16 founding members were quickly joined by hundreds of others, and the SEG immediately became the largest society affiliated with GSA, as it has remained. Dr. Penrose endowed the SEG Penrose Medal in 1923. Its first presentation to T. C. Chamberlain in 1924 preceded the first awarding of the GSA Penrose Medal by three years. Another sympathetic group was also spawned from GSA in 1920. *The American Mineralogist* had been in print since 1916, but it was not until 1920 that the Mineralogical Society of America was formed, after which time most of the detailed mineralogy papers were routed to *"Am. Min."* SEG and MSA have remained companions to GSA, and their paths have intertwined in many ways ever since. Penrose would become GSA president in 1930, just before his death from influenza the following year, and through the years, many officers and councilors have served two, even all three, Societies.

In the scientific arena, early *Bulletin* papers of J.H.L. Vogt, N. L. Bowen, J. W. Greig, E. Posnjak, and others were to start economic geology on a more quantitative, constrained path of physicochemical explanations. Ore deposits were still thought of mostly as flukes of nature, but the symbiosis of rocks and ores would now start to be developed. In retrospect, it is amazing how slowly physical chemistry pervaded economic geology, presumably because the latter was still an almost totally applied, field- and production-directed science.

GSA *Bulletin* papers in the Twenties were many and influential, then as now typically more of general than specific interest. Papers on the Foothills Copper Belt; the Adirondack and North Carolina magnetite ores; Cobalt, Ontario; Morococha, Peru; the Bushveld; Corocoro, Bolivia; and Mississippi Valley districts were supplemented by general notes on gypsum and anhydrite origin, collophane, jasperoid, coal, and pegmatites, the subject of W. T. Schaller's 1927 MSA presidential address. Waldemar Lindgren's 1925 GSA presidential address was titled simply "Metasomatism" (Lindgren, 1925), and it was thunderously received, following as it did another classic (Lindgren, 1924) on "Contact metamorphism at Bingham Canyon." The same volume carried papers by N. L. Bowen on Al_2O_3-SiO_2 and by Donald McLaughlin on the geology and ores of the Peruvian Cordillera. Bowen's paper was a harbinger, because it started economic geologists and petrologists alike on paths of exploration involving synthetic systems and of explaining hydrothermal processes and manifestations by physical chemical means. The Twenties must have been as heady and exciting for economic geologists as they were for atomic physicists, and as revolutionary as the plate-tectonics Seventies were for us all. Crystal chemistry, melts considered with vapor-phase components dissolved or exsolved, activities of dissolved species in fluids: all of these concepts began to become available through the pages of the GSA *Bulletin* and through GSA-SEG meetings.

As the Twenties drew to a close, reports of full-blown symposia on titles devised by economic geologists became even more important in the *Bulletin* with one symposium on Zonal Arrangement of Metalliferous Deposits (W. H. Emmons, Schofield, Berg, and Davison) and another on Lake Superior Banded Iron Formations (C. K. Leith, G. A. Thiel, and others) in 1927. 1928 saw R. A. Daly's epic description of the Bushveld, further reports on Tennessee phosphates and pegmatites, and first abstracts from three "heavyweights-to-be": James Gilluly, Tom Lovering, and Thomas Nolan. What may have been the first true joint sessions with GSA, MSA, and SEG in 1929 generated a number of articles on diamonds, the ores of Franklin Furnace, New Jersey; Llallagua, Bolivia; and Pachuca, Mexico.

The Thirties began with good news and bad. The Great Depression was bad enough, yet more personal tragedy befell both GSA and SEG in 1931 with the death of Richard Penrose. The American scientific world was soon electrified to learn, however, that Penrose had left $3,884,000 each to the American Philosophical Society and the Geological Society of America, probably the largest cash bequests to purely scientific societies to that date. Remember, those were 1931 dollars—a similar bequest today would be worth almost $30 million! The Geological Society was almost

50 years old and had 645 members. What an endowment, at $6,000 per member! GSA activities were expanded almost immediately following the bequest, with the initiation of the Memoir Series in 1934. The Memoirs were dedicated to topics too far-reaching and significant for single-paper presentations in the *Bulletin* pages. Although he published little in the *Bulletin*, it must be noted that this second quarter-century was the great era of Waldemar Lindgren and his thinking. Educated at Freiberg, his 31 years with the USGS (1884–1915; Chief Geologist, 1911–1912) and 21 at the Massachusetts Institute of Technology (1912–1933) afforded extensive travel, field study, and scholarly synthesis. His ore-deposit classification was based upon geologic habitat, geologic processes, and pressure-temperature in hydrothermal fluids. It guided thought and research for decades and is still widely applicable. The fourth edition of his book *Mineral Deposits* (1913, 1919, 1928, 1933) is still in print.

The end of the second quarter-century saw a spurt of interest in economic geology, although the greatest thrust in *Bulletin* pages was still toward stratigraphy, paleontology, and sedimentology. Volume 42 carried papers on ore-mineral assemblages (Merwin, 1931), chalcopyrite-sphalerite and chalcopyrite-bornite textures (Schwartz, 1931), and on lead-zinc deposits in Europe (Warren, 1931). SEG president L. C. Graton wrote on "Depth zones in ore deposition" (1932) in his official address, amending and commenting upon Lindgren's 4th and final edition of *Mineral Deposits* that would be published in 1933.

At the end of the second quarter-century, yet another massive GSA publication series funded by the Penrose bequest was launched. The Special Paper series, started in 1934, made large blocks of catalog-style material available at first, slipping later into regional studies, bibliographies, and even abstracts for a few years. At its outset, the series contained only paleontology monographs.

Volume 44 of the *Bulletin* saw the end of Joseph Stanley-Brown's heroic 40-year stint as editor. The *Bulletin* was soon redesigned to include only papers, not programs and abstracts. Important "economic" papers in the *Bulletin* were to include F. D. Adams (1934) on "Origin and nature of ore deposits—An historical study," W. H. Newhouse (1936) on "Opaque oxides and sulphides in common igneous rocks," and several short contributions on Grass Valley gold and the Front Range. Clearly, the GSA *Bulletin* was being read by economic geologists, and broader landmark papers were being published by it.

1938–1962: THE THIRD QUARTER-CENTURY

Global war started the third twenty-five years of GSA much as it had the second. *Bulletin* pages quickly reflected the stresses of wartime with thoughtful reports like C. K. Leith's "The role of minerals in the present international situation" (1939), and it was to be further felt in a group of papers from 1940–1945 on strategic element and mineral availability and procurement—W, Ni, Be, Co, and so on. Notwithstanding the dominance of the war effort, several classic papers appeared in the *Bulletin* during WWII. They included G. M. Schwartz' "The hydrothermal alteration of igneous rock"; Harrison Schmitt's epic on "The Pewabic Mine" (1939); P. F. Kerr's report on Golconda, Nevada, tungsten (1940); Bronson Stringham (1942) on West Tintic; and W. P. Hewitt's first report (1943) on Santa Eulalia. Even during the war years, the *Bulletin* published almost 2,000 pages a year, a healthy proportion of which were devoted to papers that were either on broad topics of interest to economic geologists or on limited, specific subjects (such as the skarn deposits of the Pewabic Mine at Silver City, New Mexico) that were of general interest in many geoscience sectors.

At least four advances stemming from the war years profoundly and positively affected economic geology after 1945. First, the globe had been shrunk both by our national preoccupation with world affairs and by the threshold of inexpensive, generally available air travel. Geologists could cover the globe in days rather than months, and the scenario set forth in the opening paragraphs of this paper was gone forever. Rapid information exchange, national meetings with casts of thousands, and national science funding had arrived. Second, the Age of Physical Chemistry had been augmented by the arrival of sophisticated X-ray methods—powder diffractometry, Debye-Scherrer and single-crystal techniques, and X-ray spectrofluorescence—spread like wildfire, and made petrology, mineralogy, and hybrid offshoots like the study of hydrothermal alteration effects intellectually and practically feasible for the first time. Combined with experimental studies of synthetic silicate and sulfide systems, particularly at the Carnegie Institute's Geophysical Laboratories and the USGS laboratories in Washington, the science was "on a roll." Application of the principles of petrology advanced by Lindgren and his associates would blossom in this same period through the comprehensive studies of Charles Meyer, Eugene Cameron, Paul Barton, Julian Hemley, Richard Stanton, and others. Thirdly and fourthly, the world had entered the Nuclear Age. The Manhattan Project (1942–1946) gave us not only the atomic bomb and the mad dash of the Forties and Fifties for uranium, but it also gave us the incredibly important insights of nuclear clocks—uranium-lead, potassium-argon, rubidium-strontium, and radiocarbon methods. Radiometric dating techniques were especially significant in Archean and Proterozoic terranes, made greenstone-belt studies productive, and thrust the economic geology of the Precambrian ahead at an expanding pace. Other outgrowths of the Second World War's industrial-military synergism would lead to the discovery of sea-floor spreading and consumption, plate motion, and to the unified theories of plate tectonics, but those really belong to the fourth quarter-century discussed below.

A hefty skein of papers appeared in the GSA *Bulletin* and Memoirs between 1946 and 1963. Topics most thoroughly treated were iron formations (Bruce, 1945; Alling, 1947; Tyler, 1949); James, 1951; and Goodwin, 1962), pegmatites and their minerals (Johnston, 1945; Johnston and Butler, 1946; Boos, 1954; Chadwick, 1958; Orville, 1960; and Wright, 1963), industrial minerals and their products (Mielenz, 1945), magmatism and ore minerals (Evrard, 1947; Lamey, 1961; and Robertson, 1962), precious-metal mineralization (Anderson, 1947; Noble, 1948, 1950; Noble and Harder, 1949; Noble and others, 1949), hydrothermal alteration and replacement (Kerr and others, 1950; Garrels and Dreyer, 1951; Kerr, 1951; Stringham, 1953; Leroy, 1954; Kelley and Kerr, 1957; and Buonorino, 1959), mineral economics (Lovering, 1953), uranium geology (Kerr, 1958; Kelley and Kerr, 1958; Malloy and Kerr, 1962; Miller and Kulp, 1963), and coal geology (Schultz, 1958; and Hedberg, 1961).

Among the classic, benchmark papers of the third quarter-century, these have to be listed, recognizing that the present writer's interests are not everyone's: Bowen and Tuttle's *The system MgO-SiO$_2$* (1947); Gustafson, Burrell, and Garrity on Broken Hill, NSW (1950); Noble's "Ore mineralization in the Homestake Gold Mine, Lead, South Dakota" (1950); Bronson Stringham (1953) on granitization and hydrothermal alteration at Bingham, Utah; "Uranium emplacement in the Colorado Plateau" by Paul Kerr (1958); and Hollis Hedberg's stratigraphic classification of coals (1961). As expected, a new crop of names that would become widely respected appeared on *Bulletin* pages, names such as Eugene Cameron (on Ni in Connecticut, 1943), John S. Brown (on porosity at Balmat-Edwards, 1947), E. C. Harder (on bauxite, 1947), Harold James (on BIF, 1951), Henry Cornwall (on Michigan copper, 1951), V. T. Allen (on bauxites, 1952, and skarns, 1957), E. L. Ohle (on southeast Missouri, 1954), Carl Lamey (on California iron, 1961), Bruce Doe (on Balmat sulfides, 1962), and Alan Goodwin (on Algoma iron, 1962). The districts most vividly illuminated with GSA *Bulletin* light were the Homestake district, South Dakota; the Lake Superior iron ranges; the southeast Missouri Lead Belt; Franklin Furnace–Sterling, New Jersey; the Colorado Plateau; and the Rocky Mountains and Great Plains.

In the third quarter-century, E. S. Bastin's Special Paper 24 (1939) on Mississippi Valley–type lead-zinc deposits, Joseph Singewald, Jr.'s bibliography of South American economic geology (Special Paper 50, 1943) and Poldevaart's *Crust of the Earth* (Special Paper 62, 1955) were of interest to economic geologists. The Colorado Plateau uranium boom prompted Kerr and others (1957) on Marysvale, Utah, uranium (Special Paper 64) and Margaret Cooper's uranium-thorium bibliography (Special Paper 67, 1958).

The third quarter-century (1938–1963) was the heyday of the GSA Memoir Series. Of the 139 Memoirs published between 1934 and 1974, a dozen were specifically devoted to economic geology. Most of them have become classic building blocks. John Ridge's Annotated Bibliographies (Memoirs 74 and 131, 1958 and 1972), the two on granites (Gilluly, Memoir 28, 1948; and Tuttle and Bowen, Memoir 74, 1958), Bastin on ore textures (Memoir 45, 1950; and Berry and Thompson on X-ray patterns of ore minerals (Memoir 85, 1962); Hess (Memoir 80, 1960) on the Stillwater; Kelly and Goddard (Memoir 109, 1969) on the Boulder County Telluride belt; Kerr on United States tungsten (Memoirs 15, 19); and Wisser on ore deposition and doming (Memoir 97, 1960) have all been profoundly influential.

The Memoirs also carried "experimental" tones. James Gilluly, in his foreword to Memoir 28 on the origin of granite (1948), noted that the book was a record of a new approach to scientific reporting—a full-day symposium, with discussants having exchanged papers beforehand to foster livelier debate from the floor and the podium. It worked. A young man named Donald E. White tentatively even suggested that oxygen isotopes might have something to say on the matter!

1963–1987: THE FOURTH QUARTER-CENTURY

The past 25-year period of GSA has been the most exciting period in its history to most of its members. It is even more remarkable, then, that its economic geology constituency has seen even *more* dramatic advances than has the profession at large! Three "core areas" of economic geology enjoyed spectacular development, only one of which—plate tectonics—was fully shared with others. The other two—the expansion of the concept of mineral deposit petrology and of the role of volcanism and "contemporary surfaces" in ore deposit formation—were singular to "economic."

Little need be said here about the melding of 1940–1960 advances in knowledge of sea-floor topography, marine magnetometry, and earthquake solutions that resulted in the marvelous revelations of plate tectonics and its mechanisms. One of the truly heartwarming aspects of a thoughtful trip through 100 years of the *Bulletin*—almost 130,000 pages!—is to see how the great minds of our men and women geoscientists observed and integrated billions of bits of information, most of which waited until the Seventies for mechanistic explanation. That all that information was patiently, honestly, and correctly stored for retrieval in modern times is a tribute to our antecedents. The impact of plate tectonics upon ore-deposit geology has been formidable in facilitation of understanding regional metallogenic zoning, in deciphering terrane histories by plate reconstructions, and most importantly by sharpening lithotectonic classifications of ore deposits that improve in many ways upon earlier commodity-based or phenomenological classifications (Sawkins, 1984; Guilbert and Park, 1986).

Almost at the outset of GSA's fourth quarter-century, a momentous conference was staged by the Canadian Institute of Mining and Metallurgy in Toronto, in 1965. At this meeting—and from its published proceedings—the early grapplings of a small group of enlightened geologists pushed concepts of volcanogenesis of ore deposits "over the hump" past conventional hydrothermal replacement doctrines as applied to what we know as massive sulfide deposits. Many major deposits such as Noranda, Cyprus, Jerome, Sullivan, and Broken Hill were effectively reclassified in a brilliant wrenching move that transformed exploration, lithotectonics, and genetic modeling in just a few years, opening new vistas as it happened. The 1966 start-up of the journal *Mineralium Deposita*—in many ways a voice of honest dissent from the then-staid *Economic Geology*—helped the transition. Not long thereafter, the publication of the textbook *Ore petrology* by Richard L. Stanton (1972) further emancipated thinking geologists and called for the more universal attitude that ore deposits more often than not are integral parts of the rocks that contain them, and that ore *systems* can most effectively be studied with petrologic tools and approaches.

These were obviously exciting and progressive times, of a tempo unrivaled since the excitement of Posepny's catalysis of thought at the turn of the century. That tempo was reflected in the GSA *Bulletin* pages. Topics treated included iron formations, a virtual tradition of inquiry (G.J.S. Govett, 1966; Chakraborty and Taron, 1968; Trendall, 1968; Faure and Kovach, 1969; Dimroth and Chauvel, 1973; Drever, 1974; and Lougheed, 1983); hydrothermal alteration (Jacobs and Kerr, 1965); pegmatites (C. V. Haynes, 1965); evaporites (Wardlaw and Schwerdtner, 1966; Wardlaw, 1968; Borchert, 1969; Ujueta, 1969; Kendall, 1969; and Jackson and Talbot, 1986); phosphates (Gibson, 1967; Rooney and Kerr, 1967; and Burnett, 1977); magmatism and ore minerals (Lindsley, 1973; Anderson, 1974; and Hudson and Arth, 1983); volcanism and ore deposits (Ishihara and others, 1975); uranium geology (Davidson and Kerr, 1968; Phair, 1979; and Reynolds and others, 1984); precious-metal geology (de Cserna, 1976; Lozy and Beales, 1977; and Thompson, Pierson, and Lyttle, 1982); industrial minerals and rocks (Winkler and Wilhelm, 1970; Carr and Murray, 1981; Heinrich, 1981; Murray, 1981; Goodwin and Baxter, 1981; Johnson and Sorenson, 1981; and Ault and Carr, 1981); coal geology (Hover and Davis, 1981; Palmer, 1981; Wood, 1981; Irvine, 1981; Glass, 1981; Wayland, 1981; Brant, 1981; Weir and McNulty, 1981; Carter and others, 1981; Trewargy and others, 1981; Flores and others, 1983; and Levine and Davis, 1983); oil shale (Bradley, 1970; Picard and High, 1972; Eugster and Surdam, 1973; Bradley, 1973; Eugster and Hardie and Surdam and Wolfbauer, 1975; and Desborough, 1978); manganese nodules (Schoettle and Freidman, 1971; Cronan and Thomas, 1972; and Toth, 1980); structural geology and ore deposits (Anderson, 1971; Cebull and Russell, 1979; and Fleet, 1979); plate tectonics and ore deposits (Keith, 1982; Clark, Foster, and Damon, 1982); sea-floor sulfide deposition (Dymond and others, 1973; Bonatti and others, 1976; Heath and Dymond, 1977; Czamanske and J. Moore, 1977; Leinen and Stakes, 1979; Honnorez and others, 1981; Mottl, 1983; Zierenberg, Shanks, and Bischoff, 1984; and Koski, Clague, and Oudin, 1984); sedimentary base metals (Harrison, 1972; Gibbs, 1977; and Wavra, Isaacson, and Hall, 1986); carbonatites (Nash, 1972); metallogenic zoning (Noble, 1970; Stewart, W. J. Moore, and Zeitz, 1977; and Stacey and Hedlund, 1983); and minerals policy and economics (Flawn, 1979).

Again, new names were abundantly revealed in the *Bulletin* pages, including G.J.S. Govett (1966), Donald Davidson (1968), Carl Anhaeusser, the Viljoen twins Morris and Richard, and Gunter Faure (1969), E. M. Winkler and William Bradley (1970), Jack Harrison (1972), Shunso Ishihara and Chikao Nishiwaki (1974), Enrico Bonatti (1976), Bill Moore and Gerry Czamanske (1977), George Desborough (1978), Michael Fleet (1979), Tommy Thompson and Ken Clark (1982), and Wayne Shanks and James Bischoff (1984).

Districts emphasized—again almost traditionally—included Archean terranes (Anhaeusser, Mason, and M. J. and R. P. Viljoen, 1969), Broken Hill (Anderson, 1971), the Green River Basin (Bradley, 1971; Richard and High, 1972), and Sudbury (Broucoum and Dalziel, 1974; Fleet, 1979).

The fourth quarter-century saw only three of the Special Paper series that were of immediate appeal to economic geologists, all of them in the category of regional studies. They were R. B. Mattox, ed., on saline deposits (no. 88, 1968); E. C. Dapples and M. E. Hopkins, eds., on environments of coal deposition (no. 114, 1969); and Eldridge Moores, Jr., on the Vourinos Complex (no. 118, 1969). The Special Papers continue on a strong note at the end of the quarter with Christiansen, Sheridan, and Burt on topaz rhyolites (no. 205, 1986) and Lyons and Rice on United States coal-forming basins (no. 210, 1987).

The reverberation of R.A.F. Penrose's bequest has been felt again during this fourth quarter-century. The Memoir series that was underwritten by the Penrose endowment continued, with many titles directly or indirectly profitable to economic geologists—for example, on the Nazca Plate and the Lake Superior region. GSA wisely chose also to fund the Penrose Conference program. The First Penrose Conference was convened by Brian Skinner and Wayne Burnham in February 1969 in Tucson, Arizona, a fitting place to discuss the geology, geochemistry, and genetic models of porphyry-copper systems. Information exchanged at that conference transformed our collective understanding of that deposit type and facilitated the advances of the Seventies in porphyry-copper geology. Several Penrose Conferences since then have focused on a wide variety of economic-geology topics, including limestone replacement deposits, brine-hydrothermal fluid interfacing, sea-floor processes and mechanisms, and silicic volcanism and ore genesis. The Penrose torch was profitably extended from the Memoir series to the Conferences, both series that continue dynamically into GSA's second century.

This past quarter-century saw another major turning point, namely, the watershed from "There just isn't enough information on my area" to "How will I find time to *read* all the stuff on my area?" With the gigantic printing schedules of the specific journals (*Economic Geology* has about doubled its annual output from 1963 to 1983) and the keen competition for major articles, it is not surprising that no real benchmark papers for the economic geologist appeared in the *Bulletin* during its fourth quarter-century, although there were strong contenders, such as that of James Noble (1970) on metal provinces in the western United States and A. F. Trendall's 1968 article on regional banded iron-formation comparisons. Two explanatory points, however, must be made that have ensured that most economic geologists stay current with GSA publications. First is the fact that the *Bulletin* has always carried the best published selection of ancillary or complementary papers. None of those has been cited in this review for reasons of length (and the writer's sanity), but an example is David Boden's paper on "Eruptive history and structural development of the Toquima caldera complex, central Nevada" (1986). The paper is not overtly economic, and yet the Toquima caldera volcanic history is intimately intertwined with the genesis of the epithermal disseminated gold deposits at Round Mountain, Nevada, one of the world's largest such deposits, and no student of epithermal precious-metal systems could be "complete" without having read Boden's report. One can immediately perceive that other areas as visceral as this one have been inconspicuously but vitally served by the GSA *Bulletin*—areas such as volcanotectonics, regional geology, petrology, granitoid genesis, marine geology, deep structure of the Earth, and the philosophy of science, to name but a few. Economic geology embraces all disciplines of geology, and the GSA *Bulletin* has served it well.

The second point is that the magazine-format *Geology* was begun in 1973. Because economic geology is evolving so rapidly, many papers that might have been "saved" for the *Bulletin* in more measured times have recently been published there. Scarcely an issue goes by that does not carry at least one jolt of economic character. The precept of *Geology* is that the papers be short, newsworthy, fast-paced, and therefore somewhat ephemeral, and so they have not been evaluated in detail here. The well-informed economic geologist, however, is well aware of the formidable intellectual-scientific content of the pages of *Geology* and how they relate to the *Bulletin*.

The publications and activities of the Geological Society of America have played major roles both in the support mode and in direct contribution to the affairs of the economic geologist during its first hundred years. Just as GSA and SEG have walked the same path since their birth, individual Fellows and Members have shared many common interests, read many articles of mutual interest and value, and contributed in major ways to one another's scientific progress. It can be hoped that the same cooperation, sharing of interests, and harmony will continue through GSA's entire second century. Think how distant the economic geologist of 1888 will appear by then! With luck, the same will be said of those of us fortunate to enjoy this Centennial in 1988.

1988–2012: THE FIFTH QUARTER-CENTURY

As we enter the second century of GSA, it is certain that we are growing and changing prodigiously. It is easy to say that the rate of change has been greatest during the past quarter-century, but that may be only because we have actually experienced that period. It must have been equally exciting to have watched the developments of that first quarter-century unfold as Weed, Tarr, the Emmons, and others must have.

Where is the field of economic geology today? What issues are stirring it? What issues will stir it in another 25 years? The present domestic status of the field is somewhat muddled and ambivalent as these words are written. Intellectually and scientifically, we are riding the crest of a wave. National SEG meetings send ripples of excitement through the audience, the journals of economic geology flourish, and excellent science is certainly being carried out in many sectors of our Society. The funding to continue that science is faltering, however, and many of the great engines that drove the machinery of economic geology during the past century seem to be sputtering to a ghastly halt. The USGS has recently been hamstrung by hiring freezes, low travel budgets, and disproportionate involvement in repetitive inventory-style activities. Domestic industry has been even more eclipsed by increased environment protection regulation and cost in both exploration and operation, real and threatened land withdrawals, unprecedentedly low metals prices, and general lack of governmental and public understanding of minerals problems. University programs in mineral-deposit research and mining-related activity are undermined by low levels of National Science Foundation support and almost non-existent corporate funding. University administrations are focused more and more on pure research, whereas industry and government call for more and more basic, applied studies. Several of the once-proud mining academies of the United States have been almost completely dismantled. The combined effects of a national mineral policy vacuum filled by harassment-style regulations, of land withdrawal, of foreign competition, and of American economic failure at the global scale have seriously darkened and chilled the theatre of economic geology as the second century opens.

Let us not hang too much crêpe. As stated so well by James Cook in an article entitled "The molting of America," the nation, and to some extent the world, has undergone a transition in the Eighties, not a conventional recession or depression. I wrote in mid-1983 that "industrial activity (is) near historically low levels, and the doldrums in metal-consuming sectors such as automobile manufacture and commercial-residential construction (have) dragged metal prices to all-time inflation-adjusted lows. Three years earlier, base- and precious-metal prices were soaring at the highest relative levels in decades" (p. 4, Guilbert and Park, 1986). As these words are written in late 1987, many metal prices are soaring again, but it seems clear that irrevocably changed consumer buying habits will not soon favor the supply-side for conventional metals such as iron, copper, lead, and zinc; with the near-disappearance of America's "smokestack industries," the demand for those basic metals has markedly softened. The fact that many mines have closed suggests that prices in the next decades will probably fluctuate wildly between times of oversupply and undersupply, as economic fortunes wax and wane. Precious-metal prices have been strong for a decade now and probably will continue. Keen interest is blossoming in rare elements for space-age materials, including yttrium in superconductors, niobium, tantalum, and lanthanum in advanced ceramics, and a host of elements in supermagnets, laser technology, high-tech electronics and optics, and superalloys. As these elements emerge as key technologic commodities, economic geologists and mineral-deposit petrologists will be called upon to supply information on their occurrence and aid in their discovery. Data bases and bibliographies will continue to swell into the fifth quarter-century, but the methodologies of the academic, industrial, and government economic geologists will become less distinctive as they depend ever more upon geochemistry, geophysics, and petrology.

It is difficult to describe today's "average" economic geologists and to dimension their dependence upon their professional societies and their journals. Most of us in universities and companies are probably less specialized than we were, and so we are probably more journal-dependent in servicing those broader activities. My interests a decade ago were more riveted upon porphyry base-metal systematics. My principal personal research now is in three broader areas: (1) the lithogeochemistry and metallogeny of two-mica granites, involving their petrology, geostatistical treatment of multielement analyses, and determination of their plate-tectonic settings through radioisotope measurements; (2) the lithotectonic occurrence and classification of ore deposits; and (3) preliminary evaluation of chlorine-isotope separations in natural environments upon sustained unidirectional flow through semipermeable systems as in brine migration in MVD source-basins or in convective flow patterns producing phyllic overprints. Interests being carried through student affiliates range from Great Basin "micron gold" deposits and tectonic-lithologic controls of Andean deposits along and near the West Fissure of Chile to the geology and characteristics of diatomite deposits. I suspect that my broader activities are typical. My needs for continual refurbishment require that I read not only my specific journals but also the GSA *Bulletin*, *Geology*, the Memoirs, and the Special Papers. As Associate Editor of the *Bulletin*, I strive to see the traditional high quality of GSA *Bulletin* papers perpetuated in those of interest to the economic geologist.

Let's conclude by speculating, as we started. How will economic geologists appear at the end of the fifth quarter-century of GSA, in the year 2013? They will certainly be fully computer-literate, and while in the office will communicate, acquire data and information, and create texts, files, and maps with computers, their networkings, and their video imagery. Conventional journals as we know them today will have yielded to electronic storage accessed by instant look-up and retrieval. If instrumentation development continues even at a fraction of the rate that it has in the Eighties, analytical data will be at their fingertips in avalanche proportions. They will be computer-literate and instrumentation-literate as never before. In spite of signal improvements in remote sensing and image processing, field data will still be essential. Global, perhaps even lunar, transportation will have become faster and cheaper. Mapping methods will involve automatic and continuous location monitoring from satellite systems, digitized data recording from electronic compasses and geophysical devices, and the analysis of optical, physical, and electrical properties of rocks in the field that may have enormously eased field identification. Samples will still be taken with a hammer, and boot leather will still scrape off on outcrops. Perception, experience, and wisdom will still be required of the explorationist, but information—for example, from drill core—will be gathered, digitized, and made available for maximum use to computer graphics, simulation-comparison, and genesis-characteristics modeling in hours rather than months. Gone will be the days of "we have as many opinions as we have geologists" and the maddening aptness of "there's gold in them thar hills!"

Although we can hardly speculate upon the economic environment in which they will operate, it is safe to predict that economic geologists of 2013 will have grown from another quarter-century of merging the collective advances of the Geological Society of America with the more specific progress of their immediate associates, just as they have done for the past century.

ACKNOWLEDGMENTS

This paper profited from comments by William A. Thomas, Brian J. Skinner, and an anonymous reviewer. I thank them all—may they live to be a hundred.

BIBLIOGRAPHY

(Excluding references to the GSA authors and publications sufficiently identified in text)

Beck, R., and Weed, W. H., 1905, The nature of ore deposits (2 vols.): New York, Engineering and Mining Journal, 695 p.
Browne, J. R., and Taylor, J. W., 1867, Reports upon the mineral resources of the United States: Washington, D.C., U.S. Government Printing Office, 360 p.
Cook, J., 1982, The molting of America: Forbes, November 22, p. 161–167.
Dana, J. D., 1868, The system of mineralogy (5th edition): New York, John Wiley and Sons, 600 p.
Davies, D. C., 1888, A treatise on metalliferous minerals and mining: London, England, Crosby, Lockwood and Son, 438 p.
Day, D. T., 1890, Mineral resources of the United States—Calendar year 1888: Washington, D.C., U.S. Government Printing Office, 652 p.
Faul, H., and Faul, C., 1983, It began with a stone: New York, Wiley Interscience, 270 p.
Guilbert, J. M., and Park, C. F., Jr., 1986, The geology of ore deposits: New York, W. H. Freeman and Company, 985 p.
Kemp, J. F., 1895, Ore deposits of the United States: New York, The Scientific Publishing Company, 343 p.
—— 1903, Ore deposits of the United States and Canada: New York, McGraw-Hill Book Company, 481 p.
Lindgren, W., 1913, 1919, 1928, 1933, Mineral deposits (four editions): New York, McGraw-Hill, 883, 957, 1,049, 930 p.
Merrill, G. P., 1905, The nonmetallic minerals: New York, John Wiley and Sons, 414 p.
Phillips, J. A., 1884, A treatise on ore deposits: London, MacMillan and Company, 651 p.
Posepny, F., 1895, Ore deposits: New York, American Institute of Mining Engineers, 265 p.
Rickard, T. A., ed., 1905a, Ore deposits—A discussion: New York, Engineering and Mining Journal, 97 p.
—— 1905b, The copper mines of Lake Superior: New York, Engineering and Mining Journal, 164 p.
Sawkins, F. J., 1984, Metal deposits in relation to plate tectonics: New York, Springer, 325 p.
Stanton, R. L., 1972, Ore petrology: New York, McGraw-Hill Book Company, 713 p.
Tarr, W. S., 1894, Economic geology of the United States: New York, MacMillan and Company, 509 p.
—— 1905, Economic geology of the United States (2nd edition): New York, The MacMillan Company, 525 p.
Von Cotta, B., 1870, A treatise on ore deposits (trans. by F. Prime): New York, D. Van Nostrand, 575 p.
Weed, W. H., 1907, Copper mines of the world: New York, Hill Publishing Company, 375 p.
Whitney, J. D., 1854, The metallic wealth of the United States: Philadelphia, Pennsylvania, Lippincott, Grambo, and Company, 510 p.

MANUSCRIPT RECEIVED BY THE SOCIETY AUGUST 28, 1987
REVISED MANUSCRIPT RECEIVED JANUARY 12, 1988
MANUSCRIPT ACCEPTED JANUARY 13, 1988

Printed in U.S.A.

CENTENNIAL ARTICLE

The *Bulletin of the Geological Society of America* and Charles Doolittle Walcott

ELLIS L. YOCHELSON   *U.S. Geological Survey (retired)* and *Research Associate, Department of Paleobiology, National Museum of Natural History, Washington, D.C. 20560*

ABSTRACT

Charles Doolittle Walcott, who became the third Director of the U.S. Geological Survey and the fourth Secretary of the Smithsonian Institution, was author of a paper in volume 1, number 1, of the *Bulletin of the Geological Society of America*. From 1890 through 1906, he published six scientific papers, one abstract, eight discussions, and a presidential address in that journal. Examination of these four categories of publication helps trace the history of the Society and the *Bulletin* through their early years.

Walcott made a very few errors of fact and of judgment in the six papers. Notwithstanding those, the quality and breadth of the papers demonstrate that he was a geologist of wide-ranging interests and confirm his importance in American geology; only part of his scientific activities during this 16-year interval were published in the *Bulletin*. The subsequent impact of Walcott's scientific papers is included in this historical review.

INTRODUCTION

Charles Doolittle Walcott was an eminent administrator in science, the only man to be both Director of the U.S. Geological Survey and Secretary of the Smithsonian Institution. He was also *de facto* head of the Forest Service, Chief of the Reclamation Service, and founder of the National Advisory Committee on Aeronautics. He was equally an eminent scientist. Although Walcott is best known as a paleontologist, his publications in the *Bulletin of the Geological Society of America* document a breadth of geologic investigations.

WHO WAS WALCOTT AND WHERE WAS HE IN 1890?

Chas. D. Walcott, as he often wrote his name, was born March 31, 1850, in New York Mills, New York, just west of Utica; he was the youngest in his family, and his father died when Walcott was two years old (Yochelson, 1967). Very early in life, he began picking up fossils from the Utica Shale (Middle Ordovician). In 1863, he spent a summer at Trenton Falls, New York, collecting more and different fossils from the Trenton Limestone (Middle Ordovician). Interest in fossils became a passion which lasted all his life.

Walcott began public school in Utica at age 8 and finished all formal education a decade later. His uncle wanted him to study for the ministry, for the sickly youngster who was addicted to rocks was deemed a poor risk to manage the family knitting mills. Walcott refused and worked as a clerk in a hardware store until 1871, when he had had a bellyfull of hardware and of clerking.

The next five years were spent at Trenton Falls with William P. Rust, "farmer and paleontologist." Rust was a skilled collector and, in return for his room and board, Walcott helped with farm chores and did much collecting. Trenton fossils were a profitable sideline to farming; in 1873, the Rust-Walcott collection was sold to Louis Agassiz for $5,000, a most impressive sum. Walcott, inspired by a few days with Agassiz, began to make the transition from purveyor of fossils to actually studying them. He wrote his first paper in 1875, a two-page description of a new Ordovician trilobite; this transition to science was against a backdrop of his wife's death (she was one of Rust's sisters) after only 16 months of marriage.

In November 1877, James Hall of Albany offered Walcott a position as special assistant (Yochelson, 1987). Walcott received $75.00 a month in his first paid geologic position. He did a variety of jobs for Hall, yet he was the only one of Hall's assistants who published under his own name while in Hall's employ. What stands out among these papers is Walcott's discovery of the limbs of trilobites, which he found at the Rust farm by sectioning enrolled specimens from the Trenton Limestone and which he pursued independently of his work for Hall. James Hall was a complex and tough-minded individual addicted to the acquisition and description of fossils; Walcott was equally tough-minded, and ultimately there was friction. Late in 1878, Hall did not renew Walcott's contract, but for months Walcott continued to study fossils and to write in Albany, position or no position. Prospects for employment in geology were so abysmal that in the spring of 1879, he returned to Trenton Falls.

On March 3, 1879, the United States Geological Survey was founded. Quite unexpectedly, in July, Walcott was hired as a temporary geologist at $600 a year and was sent off to the Colorado Plateau. He was a good field man and, in one year earned a permanent position at $1,200. After four years in the West and in the office, which culminated in *Paleontology of the Eureka District* (Walcott, 1884), Walcott began investigations in eastern New York and New England, originally to help clarify the issue of the Taconic System. The year 1888 was particularly fine for Walcott, for he remarried and, during a working honeymoon in Newfoundland, discovered that the Cambrian trilobite sequence there had been misinterpreted and was essentially the same as the Scandinavian series. The Walcotts continued on to the International Congress of Geologists in London, where he reported this new finding. During 1889, Walcott plunged more deeply into work on the Cambrian, but he took time to visit Toronto for the first summer meeting of the new Geological Society of

America (GSA). On August 29, 1889, he presented a talk, "Study of a line of displacement in the Grand Cañon, in northeastern Arizona." The last line of his published introduction indicates that it was the manuscript of the speech; the formal term "reading a paper" certainly applied.

WALCOTT'S FIRST *BULLETIN* PAPER

In the first volume of the *Bulletin*, a fair amount of space is devoted to organizational matters. This is followed by James Hall's presidential address and some abstracts. Walcott's is the second paper printed, following one by James Dwight Dana. To have been on the first scientific program and to have his paper published in full demonstrates that Walcott was well known to his peers. For whatever the honor is worth, Walcott published the first text figures to appear in the *Bulletin*.

Walcott began his paper,

During the summer and fall of 1882 I was engaged in studying the Paleozoic rocks of southern Utah and northern Arizona, north of the Grand Cañon of the Colorado River, and in the winter of 1882–'83 in a detailed study of a portion of the Grand Cañon. The area under investigation in the Cañon, included its head, at the foot of Marble Cañon, and the Grand Cañon with its lateral cañon valleys on the west, from Nun-ko-weap valley outlet to the westward turn of the cañon, where it cuts through the Kaibab plateau and exposes the Archean rocks in the depths of the inner cañon. A partial account of the notable sections of Algonkian and Paleozoic strata has been published, but nothing has yet appeared relating to a line of displacement whose early history was mainly determined by the study of the stratigraphy within the cañon. To-day I wish to describe this displacement and also to call your attention to certain conclusions drawn from the consideration of the phenomenon presented by it. (Walcott, 1890a, p. 49)

At the end, he summarized,

The history of the displacement is briefly stated as follows: The East Kaibab movement began in the region of the Grand Cañon as a pre-Cambrian fault displacing the older Algonkian strata, with a downthrow to the west of from 400 to 4,000 feet. A period of rest then ensued that was broken, in the latter part of Paleozoic time, by the formation of an eastward-facing monoclinal fold of a few hundred feet. So far as known this movement ceased with the close of the Paleozoic, and was not resumed until Tertiary time. It then began and continued until the East Kaibab fold and the accompanying fault had developed; the displacement aggregating over 2,700 feet in the vicinity of the Grand Cañon. This occurred before the removal, by erosion, of the Permian and probably more or less superjacent strata over the Grand Cañon area. (Walcott, 1890a, p. 64)

In the body of the paper, which included a dozen cross sections and a clearly written text, he named and described in detail the Butte fault and reconstructed the history of its movements. Because there are effectively no fossils below the Middle Cambrian Tonto (now Tapeats) Sandstone, much of this work was based on matching parts of the Algonkian (now middle Proterozoic) section on opposite sides of the fault. During the early Tertiary, lavas in the inner river canyon accompanied the faulting and were affected by later vertical movements. Having analyzed the various times and directions of movement on the fault, Walcott indicated how the fault at depth had affected development of the great monocline, how the Butte fault compared in magnitude with the Hurricane fault, and the possible role that this fault might have played in directing the course of the Colorado River.

This is a succinct paper, building from the particular to the general. Nowhere is there a hint that John Wesley Powell thought the beds in the canyon to be Lower Silurian, as the term was then used, and that Walcott spent many a futile day searching for fossils in the Algonkian shales. Likewise, nowhere is there any hint that when his field companion became depressed with the gloomy depths of the canyon and left, and when the cook went for supplies, for days Walcott was alone. The structural work entailed hazardous climbs and was not what he had been sent into the canyon to investigate (Yochelson, 1967). His study is a fine example of initiative, finding an alternative when the original problem fails.

The time between submission of manuscript and publication early in 1890 was short. The numerous cross sections required some weeks of drafting time. Perhaps Walcott used large cross sections to illustrate his lecture, thereby having the drafting already completed when the manuscript was submitted. A display of lantern slides is mentioned in connection with the following meeting, but for years they were an uncommon event.

The subject is unexpected in that Walcott commonly published the results of his field work promptly, having gotten out a brief note on Paleozoic stratigraphy of the canyon in 1883. It is a reasonable assumption that the problems of finishing his monograph on fossils from Nevada, and his subsequent focus on eastern investigations, prevented his writing more on the Grand Canyon for a few years.

As to the long-term results of this paper, no one followed in Walcott's footsteps for nearly half a century. Apart from the work of Edwin McKee in the 1930s, the Grand Canyon, and particularly the north side, was neglected by most geologists. It was not until the uranium boom of the 1950s that structural and stratigraphic investigations of the plateau country began in earnest. Although Walcott's paper is almost never cited, it was a useful contribution and, in retrospect, one that was correct. The rocks exposed in the Grand Canyon have finally been mapped in detail (Huntoon and others, 1976), and, fittingly, on the east side of the Walhalla Plateau near where the Colorado River makes almost a right-angle bend, a structural feature is designated as "Walcott graben."

WALCOTT'S SECOND PAPER

At the second annual meeting, held at the American Museum of Natural History, Walcott again spoke; volume I of the *Bulletin* also contains "The value of the term 'Hudson River Group' in geologic nomenclature," an 18-page work followed by 2 pages of discussion (Walcott, 1890b). In contrast to the original work reported in his first paper, this is in large measure a literature survey. Faunal lists are given to support conclusions, but the paper is essentially a summary of physical stratigraphy and correlation.

The Hudson River beds or group was delineated in the 1830s by the Natural History Survey of New York for strata near the Hudson River. In today's terminology, it would be a series term within the Ordovician; "Ordovician" was not used at that time, and the notion of named intervals of time, as distinguished from named formations, was not generally appreciated. James Hall subsequently extended use of this term, presumably in a time sense, throughout the state and, later, as far west as Iowa. In 1862, Hall changed his mind and restricted "Hudson River group" to older beds; as a consequence, in 1865 Meek and Worthen proposed "Cincinnati" for the younger interval. In 1877, before Walcott arrived in Albany, Hall reversed himself and revived the geographically widespread use of "Hudson River" to designate what has come to be called "Late Ordovician" time.

Walcott reviewed the older literature and added his 1879 work on the Utica Shale and later discoveries of faunas which correlated shales near Albany to those in the Mohawk Valley. He also reported new discoveries of graptolites in northeast New York, forms which occur low in the section. Walcott recognized that in general, the shales on the east side of the Hudson were older than those to the west in the Mohawk Valley, and he also recognized the over-all westward thinning of the section.

Walcott suggested that "Hudson" be retained for the "series of strata between the Trenton limestone and the superjacent Upper Silurian rocks" and added "The second part of the name is dropped in order to bring it into harmony with the names Trenton, Chazy, Niagara, etc., and to adapt to its position as a generic term" (Walcott, 1890b, p. 352 and footnote). He also emphasized that terms such as "Maqoketa shale," "Cincinnati shale and limestone," and "Frankfort shale" should be retained and used locally for rocks of Hudson age. Walcott was clear on the distinction of time versus strata.

> In reply, then to the question, "What is the value of the term Hudson River in the light of recent geologic research," I think we may say that its essential part is established by the rules of geological nomenclature, except against the prior use of the name Salmon River. In relation to this, I think all geologists will agree that the interests of geology will be subserved by leaving the term Salmon River in the obscurity in which it has so long remained. The term Hudson has a clear and distinct meaning. It is known in the geologic nomenclature of America and Europe, and it is sustained by the testimony of the rocks in the valley of the Hudson. (Walcott, 1890b, p. 353)

It is difficult to understand why Walcott felt impelled to advocate the term "Hudson," for, after twenty-five years, "Cincinnati" had come into fairly common use as a time term. Cincinnati was more accurate, as the typical Hudson River beds were older, and it was only by extending the name to younger shales to the west in New York that the sequence reaches to the Silurian limestones. Walcott, the paleontologist, noted in his summary that "Hudson" had priority, although he rejected the still earlier name "Salmon River" because it had not come into common usage. Many paleontologists are strong advocates of the principle of priority, but not every paleontologist is or was in favor of strict priority; obviously, Walcott was not in this instance. Legalistic as these arguments are, priority was not an established concept at that time, only one proclaimed by individuals. Indeed, this paper was written some years before the formal adopting of an international code of zoological nomenclature which supported priority as a basic principle. It was nearly half a century before a code of stratigraphic nomenclature, which includes some of the concepts of the zoological code, was formally prepared.

Walcott's paper did not have the intended impact. On this nomenclatural issue, the concept of accepted usage triumphed over the principle of priority. For decades, students memorized that in the Ordovician, the Trentonian was followed by the Cincinnatian, which in turn was divided into Eden, Maysville, and Richmond. Precise correlation between the New York and the Ohio River sections continued to provoke argument among paleontologists and stratigraphers. Eventually, the U.S. Geological Survey began mapping the state of Kentucky in detail, approaching Cincinnati from the south. An ancillary activity was renewal of interest in correlation. Independent lines of evidence from several different fossil groups now confirm that the lower part of the Cincinnati section is the correlative of the upper part of the Trenton Limestone (Sweet and Bergström, 1976). Those who supported "Cincinnati" and those who chose "Hudson" were both partly right and partly wrong. Perhaps this is true of many geological controversies, especially when the issue is as much nomenclatural as scientific.

There is no clue from its context why the Hudson nomenclatural paper was written or when the manuscript was prepared, but some guesses may be made. During his 1888 trip to the International Congress of Geologists, Walcott had learned details of the newly proposed Ordovician System. Furthermore, by that time, most of his field work on the Taconic problem had been completed, and discovery of the correct trilobite zonation in Newfoundland had clarified many of the remaining stratigraphic problems. There still remained the point of correct terminology for rocks in the upper portion of the Taconic terrane, and a paper on nomenclature would have assisted resolution of this issue.

THE THIRD *BULLETIN* PAPER OF WALCOTT

Walcott's third publication by the Geological Society of America was based on a remarkable discovery. In December 1890, he was examining some invertebrates collected by T. W. Stanton from the Harding Sandstone of southern Colorado to more precisely date that formation (Yochelson, 1983). The material was mainly internal molds of Ordovician pelecypods, but intermixed with them were a few phosphatic pieces. Walcott immediately recognized these as fragments of fish, like those that he had seen in the Devonian of the Grand Canyon in 1879. They were far older than the classic Old Red Sandstone fishes of Europe. This identification was an ideal example both of serendipity and of fortune favoring the prepared mind. In a few days, Walcott set off for Colorado to collect more specimens and to examine the local stratigraphy in detail. Early in the new year, his investigations had progressed sufficiently that he reported his find to a local scientific society in Washington, D.C. Rumors of these very old vertebrates spread, and several geologists came to Washington to see the specimens.

As an aside, one may wonder why there are a larger number of GSA meetings than there are years of the GSA, but the explanation is simple. Originally, the new Society was closely associated with the American Association for the Advancement of Science and held both a winter annual meeting and a summer meeting, generally in connection with the Association (Eckel, 1982, p. 104–106). On August 24, 1891, the Geological Society of America met in Washington, D.C., for its third summer meeting. Walcott spoke of his dramatic discovery; in contrast to his other papers, the title of the talk, "The Lower Silurian (Ordovician) ichthyic fauna and its mode of occurrence," differed slightly from that of the publication. This meeting was unique in that it immediately preceded the Fifth International Congress of Geologists; 240 persons attended the Congress, 73 from outside the United States and about one-third of the remainder from Washington, D.C. (Rabbitt, 1980, p. 291). It is apparent, considering that the GSA membership was only a bit more than 200, that a large percentage of the American geological community either was present when Walcott spoke or heard about his discovery from others during the next few days. It was also the second time that he announced a dramatic discovery before an international audience.

Walcott described local stratigraphy and gave lists of marine invertebrate fossils to establish the Trenton (Middle Ordovician) age of the Harding Sandstone. After getting the age and the environment of the rocks firmly in hand, Walcott concluded by naming three new genera of fish, each based on a single species. Walcott had never worked on vertebrates and, being aware of the limitations of his knowledge, entitled the publication "Preliminary notes on the discovery of a vertebrate fauna in Silurian (Ordovician) strata" (Walcott, 1892).

There was considerable discussion from the audience on taxonomic placement of the fish, although no one really argued with the basic conclusion that they were more similar to the Devonian armored fish than to the scraps known from the Silurian. Professor James Hall supported Walcott's age interpretation. Professor Otto Jaekel from Germany looked at thin sections, and, as a result, an extra plate and several pages of text were added to the manuscript to confirm the vertebrate features. This information and a footnote added in March 1892 establish that the talk was modified for publication. The fossils are illustrated by drawings, even though the *Bulletin* had included photographs of plant fossils the previous year. Likewise, photomicrographs had already been published, but Walcott used drawings; this may have been done to insure that the illustrations met with the approval of Professor Jaekel before he left America.

The long-range scientific effect of this paper is disappointing. Fish remains, far more fragmentary and slightly older, had been described from Baltic Russia, although that work was generally ignored. For about ninety years, Walcott held the record as having described the oldest vertebrates, but these Ordovician forms did little to influence thought on the development of the vertebrates. Because the material was mainly individual plates or scales, it was given little attention, but even a partial skull waited a quarter of a century for its first description.

For about fifty years, vertebrate paleontologists argued that for physiological reasons, fish must have evolved in fresh water. As a consequence, they surmised that the fragments in the Harding Sandstone must have been transported great distances. Had they examined the delicate material of *Dictyorhabdus*, one of Walcott's genera, they would have seen that extensive transport was unlikely. The evidence suggests that the Harding Sandstone fish disarticulated more or less *in situ*; their construction may not have been so rigid as in the Devonian armored fish.

The key point was not just the new fish, but their antiquity. Walcott went to great lengths to document the associated invertebrates. He wrote several pages on his locality for the Congress guidebook (Walcott, 1893a). A German geologist at the meeting later visited the Canyon City area, made some collections, and, in a short, vitriolic note, insisted that the beds were Devonian, infaulted into the area. Everyone else, however, accepted Walcott's evidence on age for the careful work that it was.

The red herring of transport from fresh water inhibited ideas about the early vertebrates and tended to negate the importance of this discovery. Perhaps the most significant effect of this paper was the prominence it gave to Charles Walcott. This is not a frivolous remark, for in 1891 and particularly 1892, Director Powell was in grave political difficulty with Congress, and the U.S. Geological Survey budget was slashed. Walcott was picked to head the Geologic Division in 1894, followed Powell as Director. It is not clear why he was selected in 1892, but his fine scientific reputation, which was burnished by the finding of these early fish, could well have been the deciding factor.

THE DISCUSSIONS

For the first few years, some, although by no means all, papers in the *Bulletin* included "Discussion by . . .," when there were matters of substance. This interchange between speaker and audience always added to those reports and, for a few, was more interesting and informative than the report itself. Even abstracts carried audience comments. In the early days of the Society, few geologists attended the meetings. For the first decade of its scientific sessions, attendance at the annual meeting ranged from 29 to 77; attendance at the summer meetings was always small. Those present apparently thought it almost a duty to comment on the material presented for their benefit.

Charles Walcott was not a shrinking violet in the GSA audience. At the Toronto meeting, he commented on Dana's paper, although the remark is not recorded. At the second annual meeting, held in New York City, he discussed two papers. To the first, on Ordovician stratigraphy around Lake Champlain, he added some details on thickness and extended observations southward into Tennessee (Walcott, 1890c). To a talk on Devonian correlations in central New York, Walcott (1890d) noted the importance of this work to the worrisome issue of homotaxis in correlation. He also tossed a compliment to a talk by W J ("no stop") McGee, who, among other activities, had been the first editor of the *Bulletin*.

At the third annual meeting, in Washington, D.C., Walcott made a mark by commenting on *four* papers. He amplified on the overthrusts of the southern Appalachians by emphasizing how unrecognized thrust faults in Vermont had given a spurious stratigraphic section—confusing the geologists—and how thrusting was a common phenomenon along the length of the Appalachians (Walcott, 1891a). He compared rocks of the Blue Ridge near Harpers Ferry to those of the Adirondack area (Walcott, 1891b). Both additions were helpful and favorable, not surprisingly, as he had just been in the field with the authors of these papers. Several years earlier, he had investigated around Quebec, and, thus, he made remarks about H. M. Ami's research there (Walcott, 1891c); most of the comments centered on terminology as to whether "Quebec" should be used for younger beds in the sequence and "Levis" for older ones. It is a bit unexpected to find an inquiry by Walcott (1891d) about the Comanchean of Texas, but he had been in the Llano region in 1884 to examine the older rocks exposed in the domal uplift, and he wanted confirmation that the Cretaceous was removed by erosion. Certainly, Mr. Walcott had a variety of interests.

In addition, he commented on G. M. Dawson's work in the Selkirk Range in western Canada. This is a curious case, for remarks by J. C. Spencer and G. K. Gilbert are appended to Dawson's paper, whereas Walcott's (1891e) remarks appear hundreds of pages later in the volume. In summary, Dawson noted that Walcott incorrectly dated the *Olenellus*-bearing Bow River Series as Algonkian rather than Early Cambrian. Walcott agreed with Dawson, although he was not quite up to actually stating that he was wrong in his earlier interpretation. This is the only *Bulletin* discussion by Walcott which was separately printed and distributed as a reprint.

There is no discussion by Walcott listed in the next volume, but, in the summer of 1893, in spite of difficulties within the U.S. Geological Survey, he traveled to the Madison meeting of the Geological Society of America. Walcott had been elected to the council, and he treated elective office as a responsibility, not an honor; simultaneously, he was vice-president of section E of the American Association for the Advancement of Science. At the meeting, he made remarks following J. W. Spencer's talk on submergence in the southeastern United States (Walcott, 1893b). Although this talk was mainly on Pleistocene epeirogenic movement, Walcott reiterated that Paleozoic sedimentation came from an eastern source. That was the last published comment by Walcott in the *Bulletin*. Within a few years, the tradition of printing remarks made by the audience lapsed. For a few more years, discussants were named, although what they said was not recorded; that is not very helpful information.

WALCOTT'S FOURTH *BULLETIN* PAPER

Walcott was examining rocks in Pennsylvania during October 1893, and, the following month, he worked with Arthur Keith on the Ocoee rock sequence in Tennessee (Rabbitt, 1980, p. 233). He did not attend the Boston meeting of the Society, but "Paleozoic intraformational conglomerates" was read by title and published February 9, 1894. By anyone's standards, that is quick time from field observation to printed page.

Walcott began,

Usually the presence of a conglomerate in a stratigraphic series of rocks is a matter of considerable importance to the geologist. He naturally infers the presence of a break in the continuity of sedimentation; an orographic movement of greater or less extent; erosion of a prëexisting formation. He sees in his mental review, the waves sorting and depositing sand, pebbles, and bowlders derived from the uplifted land. The idea of a lapse of a period of time of considerable and often long duration is formed as he recalls orographic movement, erosion and unconformity of deposition. If the conglomerate is near the base of formation or series of formations, he views it as almost conclusive evidence of the marked change that introduced the new deposits. This is all fair induction from observed facts, and it is general and approved experience of geologists. When I ventured to describe to a veteran geologist the peculiarities of a formation of conglomerate that occurs in the Lower Cambrian rocks of the eastern United States he advised my reviewing my field work and opinions, as the latter were unusual. This has been done and observations extended, with the result that I find the presence of intra-formational conglomerates a not uncommon phenomenon, one that must receive the attention of every field geologist working in the Appalachian region, from the Saint Lawrence valley in the northeast to the Cretaceous boundary of the Paleozoic of the far southwest in Georgia and Alabama. (Walcott, 1894, p. 191–192)

Now that is nice writing! Apart from the introduction, this short paper consists of three parts. First, Walcott defined his concept of consolidated limestone that was broken and then moved into matrix, commonly also limestone, of the same age. The Lower Cambrian at Schodack Landing, New York, is his type locality.

Second, Walcott reviewed various occurrences of this type of rock and differentiated between limestone breccia and conglomerate. He briefly noted localities in Quebec and New England. A longer section is on York, Pennsylvania, where these breccias are well developed; Walcott included three photographs to supplement his text. A few occurrences in Virginia are mentioned. Most space is devoted to Tennessee localities, for conglomerates had been cited as evidence that part of the Ocoee Series was Ordovician, rather than older. By emphasizing that these particular conglomerates did not automatically indicate a long hiatus, Walcott was able to counter that interpretation.

The third part, the last page, is devoted to origin of the deposits. Walcott observed that the breccias indicated fairly rapid consolidation of lime mud. He proposed modest uplift as a method for breaking and transporting the limestone; this is quite a plausible interpretation. For moving large limestone boulders, however, the only mechanism he could evoke was the old standby of rafting on sea ice. In that respect, he followed Sir William Dawson's interpretation of the Levis conglomerates.

By distinguishing between conglomerates at unconformities and those which indicated essentially no time break, Walcott both clarified ideas on stratigraphy and contributed data to carbonate sedimentation. The descriptive work is good, but again, for decades, his observations were not extended by other geologists. Today, some of the edgewise breccias would be ascribed to storm ripup, or tidal effects, not too different from the modest uplift Walcott suggested. In contrast, many of the conglomerates are olistostromes moved down submarine slopes, possibly during earthquakes. Long after Walcott died, the beds around York, Pennsylvania, stimulated the concept of a steep foreslope to the Lower Cambrian limestone platform of eastern North America (Rodgers, 1968). Turbidity currents and other transport mechanisms, which we consider plausible today, are notions that no one had in the 19th century.

NON-GSA PAPERS OF THIS DECADE

For a generation, Society activities were published in the *Bulletin*. They provide a chronicle of who was working on what in geology. From this source, one can determine that at the seventh annual meeting in Baltimore, Walcott presented two talks. Although he had been Director of the U.S. Geological Survey for a scant six months, having succeeded John Wesley Powell, he still found time to prepare "The Appalachian type of folding in the White Mountain Range of Inyo County, California," and "Lower Cambrian rocks in eastern California"; the latter "was presented by the author in a few words." Walcott had been able to get into the field in California for two weeks during August 1894, and he made the most of this opportunity. Then, as now, not all talks presented at a GSA meeting were published in the *Bulletin*, and both papers appeared in 1895 in other journals. Subsequent non-GSA publications of Walcott are not discussed, even though some were presented at meetings of the Society.

THE PENULTIMATE WALCOTT *BULLETIN* PAPER

After a hiatus of five years in GSA publication, Walcott produced "Pre-Cambrian fossiliferous formations" (Walcott, 1899). This *magnum opus* for the *Bulletin* was his longest work in that journal, with the most varied illustrations, including plates of fossils, cross sections, and correlation tables. It also contains his most prominent error in paleontology.

Walcott introduced the subject by reviewing the various areas of Algonkian outcrop. He discussed the Belt rocks of Montana, emphasizing the physical unconformity between them and the overlying Middle Cambrian. He redescribed the strata in the Grand Canyon and touched briefly on the rocks of the Llano uplift in Texas. Walcott used the literature and letters from G. F. Matthew to supplement his hasty look at Newfoundland in 1888; he discussed the Lake Superior region, again relying mainly on literature, especially the work of Charles Van Hise. A few comments about Utah, Nevada, and California concluded the review of lithostratigraphy. Walcott established that in all regions, a profound unconformity separates the Cambrian from the underlying rocks.

Next, Walcott considered presumed Precambrian fossils described by others. He touched on "*Eozoon*" and on the presence of graphite, concluding that neither necessarily indicated organic origin. A long-questioned form, "*Palaeotrochus*" was given its final push into the realm of pseudofossils. The Lake Superior area had yielded some fossils preserved in iron ore, but Walcott was able to show that they were Cambrian, not older. He also disposed of "*Aspidella*" from Newfoundland. Walcott ultimately noted that authentic Algonkian fossils occurred only in the Grand Canyon, in the Belt terrane, and "in the Etcheminian terrane of New Brunswick and Newfoundland, if the latter proves to be truly pre-Cambrian" (Walcott, 1899, p. 227). In hindsight, it is evident that Matthew included fossiliferous pre-*Olenellus* beds in the Etcheminian stage. Matthew had made some miscorrelations in the past and was suspect, but in considering the Etcheminian, Walcott missed the significance of Matthew's find. Strictly speaking, these rocks contain Tommatian fossils, earliest Cambrian, not Precambrian. Perhaps, if Walcott had spent more time in the area, he might have seen what Matthew saw.

Part of the paper presented new data. "When collecting material from the Chuar terrane in 1883, I was strongly impressed with their resemblance to the forms occurring in the upper Cambrian rocks of Saratoga county, New York, which Professor James Hall subsequently described as *Cryptozoan proliferum*" (Walcott, 1899, p. 235). Sir William Dawson saw the stromatolite in 1883, shortly after Walcott returned from the Grand Canyon, but did not describe it until 1897. Dawson's original comments are reproduced in Walcott's paper, along with illustration and description. This may be the first authentic Precambrian fossil described from North America.

Walcott described a new genus, *Chuaria*, also from the Grand Canyon. It is a nondescript, small, carbonaceous sphere, crushed flat. During the last two decades, specimens have been found in Scandinavia, India, China, and elsewhere in North America. Within broad limits, *Chuaria* seems to be a good index fossil in the Middle Proterozoic.

From the Belt rocks in Montana, Walcott had two sorts of fossils. The first were presumed trails, most of which have turned out to be filamentous algae (see Yochelson, 1979, p. 274–275). The second was *Beltina*, described as a eurypterid. It was an error, but an understandable one, as the forms do resemble fragments of Silurian eurypterids. Walcott had looked hard and long for fossils; he deserved to find some! Despite the fact that *Beltina* was not an arthropod, it has biostratigraphic utility in the Belt of Montana.

Australia and, later, other regions have yielded the Ediacara fauna of larger invertebrates. This pre-trilobite fauna is far younger than and different from fossils in the Belt rocks. Had radioactive dating been available to Walcott, so that the great age of the Belt, relative even to the Ediacara fossils was known, he might not have been so keen to interpret the scraps as animals as complex as eurypterids.

No *apologia* is needed. Even with misinterpretations of some of the Belt material, this paper is exceptionally important. It took Precambrian fossils out of the area of myth and moved them to reality. The stage was set for great advances. Walcott did more work on the Precambrian and even endowed a medal to encourage research (Yochelson, 1979), but no one followed his lead. A full generation of geologists elapsed between Walcott's death and renewed interest in such early fossils

THE ABSTRACT

Ten percent of the Society membership, 23 persons, attended the summer meeting at Columbus, Ohio, in August 1899. Walcott was not among those present; however, "Random, a pre-Cambrian Upper Algonkian terrane" was printed as part of the proceedings (Walcott, 1900). The structure of meetings and use of abstracts was far different then. Typically, in the report of a meeting, the title of a talk was published, followed by where it was published in the *Bulletin* or, more rarely, in another journal. Occasionally, when a speaker was absent, the Secretary would read the paper, but more common was the comment "by title only." In effect, "title only" announced what research was in progress and, in a sense, staked a claim to that problem. It was assumed that in the near future, a paper would be forthcoming.

Abstracts were published in the "Proceedings" section of the *Bulletin* and included either informal remarks or exceedingly short notes which the author did not consider worthy of full publication. Walcott's three pages is printed in six-point type and would be a respectable contribution in a modern journal. He judged it as merely a follow-up to his 1899 paper, however. In his eyes, it was just a report on field work, not a full-fledged study.

The 1899 season in Newfoundland was a busy one for Walcott. Several years earlier, G. F. Matthew had proposed "Etcheminian" for a Paleozoic series below the Cambrian. For a start, Walcott and S. W. Loper found *Olenellus* within those beds, thus moving them back to the Cambrian. Next, with J. P. Hawley, they measured sections and searched for fossils between the dated Cambrian and the underlying Signal Hill Conglomerate. The beds between were named by Walcott and Hawley the "Random terrane," for Random Island in Trinity Bay.

> The Random terrane is probably 1,000 feet, and possibly more in thickness. It fills in a portion, if not all, of the gap between the Signal Hill conglomerate and the Cambrian. The erosion preceding the deposition of the Cambrian about Trinity bay appears to have been slight as the conglomerate resting on the Random is rarely over 18 inches in thickness, and usually much less. This however, is not a safe deduction, as great erosion may leave but slight trace, either in conglomerates or in apparent nonconformity in the dip and strike of the strata.
> The Random terrane is considered to be the upper member of the Avalon series [named by Walcott in 1899]. Animal life existed during the deposition of a portion of it, as is evidenced by clearly marked annelid trails. (Walcott, 1900, p. 5)

It was not until the 1950s, when Martin Glaessner began detailed investigation of the Ediacara fossils and new finds were made in Siberia, that the notion of pre-trilobite faunas again became a subject of study. It was only in the late 1960s that unconfirmed reports of body fossils in the upper Precambrian of southern Newfoundland began to circulate by word of mouth. It was not until the 1970s that the idea of trace fossils older than early body fossils was seriously considered as a working hypothesis.

After years of neglect of this part of the Newfoundland section, geologists and paleontologists have finally recognized that the area might contain both Ediacaran and Tommotian, earliest Cambrian, faunas. Because of this, in 1987, Newfoundland was seriously considered as a candidate for the Precambrian-Cambrian boundary stratotype. Who can say how rapidly studies of this boundary might have progressed if Walcott had had the time to spend another season examining the Random terrane, or if others had read his abstract and followed him in the field?

THE PRESIDENTIAL ADDRESS

Walcott was an original fellow of GSA, that is, a member of section E of the American Association for the Advancement of Science, but he was among the last to join the new Society during its first year (Fairchild, 1932, p. 101). After three years on the Council, beginning in 1893, he was elected second vice-president for 1899 and first vice-president for 1900. In 1901, Walcott succeeded G. M. Dawson, to serve as the thirteenth president. An address by the retiring president is one of the nicer GSA traditions. Walcott gave his address on December 31, 1901, at the Rochester, New York, meeting, on the "Outlook of the geologist in America." A few days earlier, Walcott had helped Andrew Carnegie launch the Carnegie Institution of Washington, and he must have been in an optimistic frame of mind when he spoke.

The rationale for the subject was to indicate the prospects of a young man [sic] who was considering geology as his career. The first part of this paper surveyed the various organizations employing geologists and paleontologists in 1901. Walcott noted that the U.S. Geological Survey gave continuous employment to 36 geologists and paleontologists and temporarily employed about another 50. He then examined the museums, academic institutions, and state surveys of the United States.

> Taking account of all these agencies, whether surveys, museums, or educational institutions, I estimate that seventy geologists are enabled by financial support to devote themselves wholly to professional research work; that fifty geologists, mining engineers and technologists, though occupied chiefly in other ways, receive pay for special work in the field of research and that seventy other geologists, employed and salaried as teachers either are urged or permitted without prejudice to devote part of their time to scientific investigations. (Walcott, 1902, p. 103)

Since that time, geology in America has boomed in terms of research positions. Walcott would have been pleased with the number and variety of opportunities today. In particular, because he was of a practical turn of mind so far as pure research is concerned—not really an oxymoron if you think about it—he would have been delighted with the research laboratories of the petroleum industry which so influenced mid-century geology in America. Walcott's remark concerning teachers "permitted without prejudice" to engage in research stands in stark contrast to the "publish or perish" pressure cooker of current campus life.

Walcott commented on the product of geologic effort, pointing out that of the 21,000 pages on American geology published in 1899, the GSA supported 500 pages. Even though the journal

> ... does not contribute to the support of the student, it is sometimes the factor which turns the doubtful scale and makes a contribution to geologic science possible. The *Bulletin* ... [is] supported by the scientific men themselves, and from one point of view may seem to give no aid to the needy investigator, but it is really the readers who pay the printer, and the investigator is called on to pay only because he is also a reader. (Walcott, 1902, p. 103)

Almost half of Walcott's address is devoted to indicating who was working on what problems during 1900. It is a who's who of geology, and most people should be able to recognize many of the names, if not the research problems. Walcott then moved to the present and, in four pages, summarized outstanding problems and interesting areas for investigation. More than 80 years later, some of these problems still remain, and anyone in need of ideas is referred to it. This section is a tribute to an interest in research, regardless of the specialty, and might be read with profit by those who are a bit jaded in their outlook.

In the final pages, Walcott explored the future.

> ... geology, although affording occupation to a somewhat limited number of persons, is nevertheless a well established profession—a profession which flourishes in so many places and under such a variety of conditions that it may well be assumed to have altogether passed the experimental stage. If its recent history were reviewed in connection with its present status, its development as a profession would seem to have fully kept pace not only with population but with the general development of culture factors. There is no reason to doubt that its expansion will continue. (Walcott, 1902, p. 116)

As one might hope, in a presidential address, there are a few good quotes. "In the interaction between applied geology and pure geologic science lies the charm and recompense of every-day routine geologic work" (Walcott, 1902, p. 115). "Every investigation undertaken to solve some geologic problem, whether it proves successful or not, is sure to develop other problems, and the geologic Alexander will never lack worlds to conquer" (Walcott, 1902, p. 116). "It is impossible to forecast the problems of the future" (Walcott, 1902, p. 116). "The support of the geologist depends on public appreciation of the value of his services" (Walcott, 1902, p. 117). "It is more and more understood by men whose ability puts them in positions of responsibility that material progress depends, in the ultimate analysis, on the growth of knowledge, and from this increasing confidence in the ultimate utility of pure science research is reaping a generous harvest of endowment" (Walcott, 1902, p. 117). Words to reflect upon.

Walcott concluded with advice that still seems appropriate.

> In closing I wish to say a word about the training of the men who will probably reap the largest results from the great opportunities in geology that will be offered during the century. The practical economic geologist will undoubtedly receive the largest financial returns, but in this field the man with the broadest, most thorough training will win out as competition becomes more and more active. In the more purely scientific lines a broad, general culture should be the groundwork for the special geological training. A few months of business training will be almost invaluable to any student who aspires to be more than a directed assistant throughout his career. Business method and habit must underlie all successful administrative work, whether it be of a small party or of a great survey. It is needless to say that, as in modern business life so in science, character of the highest standard is essential to permanent success. The outlook of the well balanced, well trained student of geology in America is most encouraging—far more so than when I began work with an honored leader, James Hall, a quarter of a century ago. (Walcott, 1902, p. 118)

Currently, opportunity does not seem to be knocking for geologists, and young students [sic!] tend to be discouraged. To regain perspective, one measure is comparison of research geologists to population size. In 1900, the United States population was 76,212,168, and by 1980 it tripled to 226,548,861. Even at the nadir of the worst cycle in geologic employment, research positions are still many-fold above triple what were available at the start of the twentieth century.

WALCOTT'S FINAL GSA *BULLETIN* PAPER

"Algonkian formations of northwestern Montana" (Walcott, 1906) was presented by title only at the 1905 meeting. This is the second longest of Walcott's *Bulletin* papers and is the one most profusely illustrated by photographs, with 10 plates containing 13 photographs. These show Walcott's skill as an outdoor photographer and include several taken using a panorama camera, the first such photographs published in the *Bulletin*. The paper is in part a continuation of Walcott's 1899 work and incorporates observations made during several field seasons. The quickest way to approach this study is to repeat the summary.

> The Algonkian rocks which form the subject of this paper represent a total thickness of 37,000 feet and occupy an area extending from the Little Belt mountains on the southeast to the vicinity of Coeur d'Alene on the west and northward into British Columbia. The Camp Creek, Mission Range section occurs near the center of this area and is taken as the type because of the great vertical extent (24,770 feet) and the fact that it is capped by Cambrian strata.
>
> In the four sections measured by the writer the Algonkian or Belt terrane is overlain unconformably by massive coarse grained sandstones referred to the Middle Cambrian. The unconformity is usually indicated by great changes in the volume of the underlying strata and represents a considerable time interval. From the presence of Lower Cambrian fossils in the Bow river Series of McConnell, it is believed that this series was laid down during the erosion interval between the Algonkian and the Middle Cambrian in Montana.
>
> The physical conditions under which the Belt terrane was deposited are very clearly indicated by the change in the character of the sedimentation from the conglomerates, grits, and coarse sandstones on the northwest to the limestones, fine sandstones, and shales on the southeast. The land area from which these sediments were mainly derived must have been to the west and northwest of the Kootenay valley at Porthill, Idaho. The sediments which extend 300 miles or more to the eastward betray frequent evidence of shallow water deposition, and in the Little Belt mountains indicate that the eastern land area was of low relief and situated still further to the east, although the presence of a limited land area is shown by the conglomerates at the base of the Algonkian section near Neilhart. (Walcott, 1906, p. 27–28)

Walcott covered a huge area in this report. He had crossed the Belt Mountains in 1900 and observed to the west, below the Cambrian, rock unlike those of the Little Belt Mountains. The following year, Bailey Willis, F. L. Ransome, and Frank Calkins concentrated on other areas of Algonkian. In 1905, Walcott was back in western Montana and Idaho, where he was able to tie the various observations together. Considerably more is now known of stratigraphy of the Belt Supergroup. Still, it seems fair to say that Walcott's basic units and his notion of an east-west facies change are essentially correct. Sections are long, and the rocks are recalcitrant; knowledge of this part of geologic time has grown slowly.

Close reading of the paper brings out two aspects. First, several pages are devoted to long quotes from Canadian reports concerning the Bow River Series. Walcott had made an error, which was partially corrected in his discussion of Dawson's paper years earlier. Here, he developed the concept that in Montana, the Middle Cambrian lapped onto the Precambrian. Only to the west, and particularly the west-northwest, did one find Lower Cambrian rocks. He was still explaining away his original view.

A more significant point is the use of fossils for correlation. Walcott identified the Siyeh Limestone as Algonkian because it contained stromatolites. "Typical fragments of *Beltina danai* of the Newland limestone occur in the Altyn formation; also Cryptozoon" (Walcott, 1906, p. 19). On a correlation chart, Walcott noted that the "fossiliferous" Newland Limestone is the principal horizon for correlation. Walcott was using fossils for biostratigraphy in the Precambrian. He was wrong on the nature of *Beltina*, and he had the notion that stromatolites were indicative of fresh water—his only model was lake balls (Yochelson, 1979, p. 280–284). Even without a clear understanding of their origin, he trusted the stromatolites as indicators of time. It was not until half a century later, when Russian workers in Siberia demonstrated the utility of stromatolites for Precambrian correlations, that Walcott's remarkable insight was vindicated.

As to Scapegoat Mountain in the Mission Range, Montana, there should be a tangible item resulting from this paper, but no map records it. Walcott wrote, "The southeast point is Scapegoat (9,185 feet elevation) and the northwest elevation (9,000 feet) I shall call Cambria" (Walcott, 1906, p. 3). Far more important than a name, Glacier National Park, itself, may be attributed in part to the paper. Walcott was an advocate of setting aside this portion of the majestic scenery for the enjoyment of all, and he was one of a group of geologists who campaigned to add the area to the national park system.

MISCELLANEOUS OBSERVATIONS

In our present-day scientific climate, many papers are joint or multi-authored, and "interdisciplinary" and "team research" are the current buzz words. In contrast, all of Walcott's GSA papers were written solely by him, and, indeed, throughout his long career, he was involved in only one joint publication.

Nowadays, it is automatically assumed that an author's address is given. For his first paper, the authorship is by "Charles Doolittle Walcott of the U.S. Geological Survey." On all subsequent papers, only the name is given, for it was then the custom of GSA not to include addresses. *Bulletin* papers were written by Fellows for Fellows, and if one was not personally known, it was easy enough for others to consult the membership list.

One minor curiosity is a footnote on his 1984 paper, "Printed by permission of the Director of the U.S. Geological Survey," which does not appear on the three earlier papers. Presumably, after Walcott became Chief Geologist in 1893, he instituted the practice of channeling all manuscripts through the Director's office to bring some order into Powell's untidy administration. Walcott then set an example of obtaining permissions, for he always played by the rules. Inasmuch as Walcott was Director from July 1904 onward, no permission is given for his presidential address or his last two papers; after all, the Director ought to know what the Director is doing.

In none of Walcott's *Bulletin* publications is there a section on acknowledgements, for such was not the custom. Within the body of the text, where appropriate, Walcott mentions those he visited in the field or those who assisted him with collecting. A footnote in his presidential address acknowledges G. K. Gilbert and F. B. Weeks for advice, as well as those who furnished data. There is no indication of what draftsmen made the numerous cross sections, but, again, this is not strange; it is still not the custom within the U.S. Geological Survey to single out such effort. Likewise, there is no indication of who drew the pictures of fish for plates or of who photographed the fossils for the 1899 paper.

In contrast, the few photographs that Walcott published which were not taken by him are acknowledged. Walcott was an outstanding photographer, and his landscape pictures are as good as any published in the *Bulletin*. G. K. Gilbert may have been the one who taught him the value of a camera in the field. The panorama camera Walcott packed around Montana and Canada for two decades was an enormous load, yet he saw a scientific advantage to it. (Pictures may show a bit more than just how the outcrops looked. Plate 2 of Walcott, 1894, and plate 8 of Walcott, 1906, each contain a long-handled hammer, but not the same one. This was long before the Estwing era, and these hammers are what one would expect of a hard-rock man. The limestones of the Appalachians are hard, but the Belt Series rocks are harder still, and Walcott's later hammer is heavier.)

Years ago, *Bulletin* papers did not include an abstract. Whether long or short, each had a table of contents following the title. An abstract is a relatively new idea, and key words at the end of the abstract is an even newer concept. In Walcott's day and for some time thereafter, references were given as footnotes in highly abbreviated form. Anyone who has searched the library for the likes of "Jb. Nassau Natur. 1847" will appreciate the list of references at the end of a modern paper.

Walcott's spellings, like "bowlder" and "pre-Cambrian," appear strange to our eyes. Words such as "limestone," "formation," or a geographic locality that we expect to see capitalized are not. Diacritical marks on English words have now almost entirely disappeared. Even if the conventions are different, however, the editing of the *Bulletin* from volume 1 onward has always been impressively uniform.

Finally, it may be worthwhile to direct attention to the back of the earlier volumes of the *Bulletin*. The index included all papers, all talks, and all discussions, and these were indexed in depth. Walcott is quoted on fossils that he identified from California. Walcott is cited in regard to the age of the Brigham Quartzite. Walcott is mentioned in connection with the formation of natural bridges. The examples go on and on, of nuggets of information that otherwise might not be found. Anyone who has a hero or a historical subject to pursue should not forget the *Bulletin* index.

AFTERWARD

Walcott's last *Bulletin* publication also signaled a major career change. In 1907, he became the fourth Secretary of the Smithsonian Institution and immediately started a research program in western Canada. For nearly twenty years he was in the field each summer. Discovery and description of the Burgess Shale fauna were only one of his accomplishments. Almost all of his investigations were published in the *Smithsonian Miscellaneous Collections*, and he filled six entire volumes.

Walcott expected fewer administrative duties in heading the Smithsonian. He accumulated new organizations and new responsibilities just as a rolling snowball accumulates snow, however. It is a wonder that he still found time to come to some of the annual meetings, yet he continued to present his findings to the geologic community fairly regularly. In 1919, he gave his final talk on Middle Cambrian sponges. In 1920, he did not appear, but "The Secretary then presented for the author, in his absence, an account of the wonderful anatomical structures preserved in the Middle Cambrian Burgess shale branchiopod crustacea near Field, British Columbia. Illustrated by lantern slides and specimens" (Bassler, 1921, p. 127).

After his seventieth birthday, Walcott's health was slowly failing, but he pushed himself and his research for nearly seven more years. Considering his long association with the Geological Society of America, it is appropriate that the longest and finest obituary of Charles Doolittle Walcott appeared in the *Bulletin* (Darton, 1928).

ACKNOWLEDGMENTS

Several persons generously gave their time in helping me form my judgment of Walcott's scientific contributions discussed above. They include Arly Allen, R. H. Flower, H. J. Hofmann, C. B. Hunt, Nicholas Rast, and Donald Winston. For dramatic improvement of the original manuscript, I am greatly indebted to Tom Dutro, Bill Thomas, and A. R. Palmer.

REFERENCES CITED

Bassler, R. S., 1921, Proceedings of the twelfth annual meeting of the Paleontological Society, held at Chicago, Illinois, December 28–30, 1928: Geological Society of America Bulletin, v. 32, p. 119–158.

Darton, N. H., 1928, Memorial of Charles Doolittle Walcott: Geological Society of America Bulletin, v. 39, p. 80–116, 1 fig.

Eckel, E. B., 1982, The Geological Society of America, life history of a learned society: Geological Society of America Memoir 155, 167 p., numerous illustrations.

Fairchild, H. L., 1932, The Geological Society of America 1888–1930, a chapter in earth science history: New York, Geological Society of America, 232 p.

Huntoon, P. W., and others, 1976, Geologic map of Grand Canyon National Park, Arizona: Grand Canyon Natural History Association and Museum of Northern Arizona, scale 1:62,500.

Rabbitt, M. C., 1980, Minerals, lands, and geology for the common defence and general welfare, Volume 2, 1879–1904, United States Geological Survey: Washington, D.C., U.S. Government Printing Office, 407 p., numerous illustrations.

Rodgers, John, 1968, The eastern edge of the North American continent during the Cambrian and Early Ordovician, *in* Zen, E-An, White, W. S., Hadley, J. B., and Thompson, J. B., eds., Studies of Appalachian geology: Northern and maritime: New York, Interscience Publishers, p. 141–150.

Sweet, W. C., and Bergström, S. M., 1976, Conodont biostratigraphy of the Middle and Upper Ordovician of the United States Midcontinent, *in* Bassett, M. G., ed., The Ordovician System: Cardiff, Wales, University of Wales Press and National Museum of Wales, p. 121–152.

Walcott, C. D., 1884, Paleontology of the Eureka District: U.S. Geological Survey Monograph 8, 292 p., 24 pls., 7 figs.

———1890a, Study of a line of displacement in the Grand Cañon of the Colorado, in northern Arizona: Geological Society of America Bulletin, v. 1, p. 49–64, 12 figs.

———1890b, The value of the term "Hudson River Group" in geologic nomenclature: Geological Society of America Bulletin, v. 1, p. 335–355, 1 fig.

———1890c, Discussion of paper by Ezra Brainard and Henry M. Seely on the Calciferous formation in the Champlain valley: Geological Society of America Bulletin, v. 1, p. 512–513.

———1890d, Discussion of paper by H. S. Williams on the *Cuboides* zone and its fauna: A discussion of methods of correlation: Geological Society of America Bulletin, v. 1, p. 499.

———1891a, Discussion of paper by C. Willard Hayes on the overthrust faults of the southern Appalachians: Geological Society of America Bulletin, v. 2, p. 153.

———1891b, Discussion of paper by H. R. Geiger and Arthur Keith on the structure of the Blue Ridge near Harper's Ferry: Geological Society of America Bulletin, v. 2, p. 163–164.

———1891c, Discussion of paper by H. M. Ami on the geology of Quebec and environs: Geological Society of America Bulletin, v. 2, p. 501–502.

———1891d, Discussion of paper by Robert T. Hill on the Comanche series of the Texas-Arkansas region: Geological Society of America Bulletin, v. 2, p. 526–527.

———1891e, Discussion of paper by George M. Dawson on the geological structure of the Selkirk Range: Geological Society of America Bulletin, v. 2, p. 611.

———1892, Preliminary notes on the discovery of a vertebrate fauna in Silurian (Ordovician) strata: Geological Society of America Bulletin, v. 3, p. 153–172, pls. 3–5, 1 fig.

———1893a, Silurian vertebrate life at Canyon City, Colorado: Congrès Géologique International, Compte Rendu de la 5ME session, Washington, D.C., 1891, p. 427–428.

———1893b, Discussion of paper by J. W. Spencer on terrestrial subsidence southeast of the American continent: Geological Society of America Bulletin, v. 4, p. 22.

———1894, Paleozoic intraformational conglomerates: Geological Society of America Bulletin, v. 5, p. 191–198, pls. 6–7.

———1899, Pre-Cambrian fossiliferous formations: Geological Society of America Bulletin, v. 10, p. 199–214, pls. 22–28, 7 figs.

———1900, Random, a pre-Cambrian Upper Algonkian terrane: Geological Society of America Bulletin, v. 11, p. 3–5.

———1902, Outlook of the geologist in America: Geological Society of America Bulletin, v. 13, p. 99–118.

———1906, Algonkian formations of northwestern Montana: Geological Society of America Bulletin, v. 17, p. 1–28, pls. 1–11.

Yochelson, E. L., 1967, Charles Doolittle Walcott 1850–1927: National Academy of Sciences Biographical Memoir, v. 39, p. 471–540, 1 fig.

———1979, Charles D. Walcott—America's pioneer in Precambrian paleontology and stratigraphy, *in* Kupsch, W. O., and Sarjeant, W.A.S., eds., History of concepts in Precambrian geology: Geological Association of Canada Special Paper 19, p. 261–292, 6 figs.

———1983, Walcott's discovery of Middle Ordovician vertebrates: Earth Sciences History, v. 2, p. 66–75, 2 figs.

———1987, C. D. Walcott, James Hall's "special assistant": Earth Sciences History, v. 6, p. 84–92.

MANUSCRIPT RECEIVED BY THE SOCIETY MARCH 19, 1987
REVISED MANUSCRIPT RECEIVED JUNE 18, 1987
MANUSCRIPT ACCEPTED JUNE 25, 1987

CENTENNIAL ARTICLE

Paleontology and The Geological Society of America: The first 100 years

J. THOMAS DUTRO, JR. *U.S. Geological Survey, Washington, D.C. 20560*

ABSTRACT

Since its early years of publication, the GSA *Bulletin* has been an outlet for paleontologists who want their results to reach a broad scientific audience. During the first 40 years, paleontologic papers featured a wide variety of subjects, including systematics, biostratigraphy, correlation, and paleogeography. After the *Journal of Paleontology* and other specialized journals began publication, applied paleontology characterized many of the *Bulletin* articles. From about 1930 to 1960, a series of correlation papers on the Phanerozoic systems dominated the *Bulletin*. Systematic studies were generally published as *Special Papers* or *Memoirs*, starting in the 1930s. Beginning in the 1950s and continuing to the present, the GSA and the University of Kansas Press have produced 36 volumes of the *Treatise on Invertebrate Paleontology*, arguably the most significant paleontologic enterprise of the century. In the past 20 years, the *Bulletin* has contained many papers in which paleontology was applied to solving general geologic and biologic problems. Doubtless, this trend will continue, in both the *Bulletin* and in *Geology*, for many years to come.

INTRODUCTION

The GSA *Bulletin* has been a place where paleontologists alerted the rest of the profession to the general geologic significance of new paleontologic hypotheses and spectacular new fossil finds, although its role has varied during the first 100 years of its existence. This changing role of the *Bulletin* can be characterized in three phases: (1) 1888–1928, the 40-year period before the *Journal of Paleontology* became an effective medium for publishing detailed research on fossils and strata; (2) 1928–1968, a period of major biostratigraphic synthesis in North America; and (3) 1968–1988, the modern era of detailed analyses and documentation of the plate-tectonic theory.

All subdisciplines of paleontology are represented in each period, but the focus has shifted gradually through the years as paleontologists have developed new ways of looking at fossils and applying them to the solution of geologic and biologic problems. From the early years before the turn of the century, biostratigraphy and correlations were prominent topics for articles in the *Bulletin*; these subjects were especially emphasized during the second period. Speculations on the origin of life on Earth and the meaning of Precambrian fossils were published early on. Ideas on the environmental significance of trace fossils and their use in facies analysis appeared sporadically. Evolution and the systematics of major fossil groups were features in the first period. Paleoecology, paleoclimatology, and statistical analysis quite expectedly have had greater exposure in the *Bulletin* during the modern period. Paleogeography and paleobiochemistry have also been well represented in the pages of the *Bulletin*.

THE FIRST FORTY YEARS

Before the appearance of the *Journal of Paleontology* and other specialized paleontological publications, the *Bulletin* was one of several general outlets for paleontological papers. Announcements of new finds and new faunas, correlations, and biostratigraphic syntheses appeared not only in the *Bulletin* but also in the *Journal of Geology*, the *American Journal of Science*, and other North American general journals.

Among the earlier papers, two that deal with the origins of life stand out. C. D. Walcott (1899) provided a comprehensive summary of what was known to that time of fossils in Precambrian rocks. In 1909, R. A. Daly discussed the earliest calcareous fossils and their significance in the origin of certain kinds of limestones. These two papers pointed the way for subsequent work with organic remains in Precambrian strata, but it was many years before new techniques and new hypotheses were applied to deciphering the enigma of the origin of life.

During the first three decades of *Bulletin* publication, classification, and correlation of the Phanerozoic systems were undergoing explosive investigation. Several papers discussed the principles that should be applied to these problems, probably triggered by E. O. Ulrich's "Revision of the Paleozoic Systems" (1911). F. H. Knowlton (1916) published a short paper on the use of fossil plants in correlation, W. D. Matthew (1913) discussed theoretical problems affecting the use of fossil vertebrates in correlation and in phylogenetic studies, H. S. Williams (1913) considered correlation problems from the perspective of his work in the Eastport area, Maine; Charles Schuchert evaluated correlations and chronology on the basis of paleogeographic reconstructions (1916). Schuchert initiated his paleogeographic syntheses during this period; in addition to the 1916 paper, two earlier studies had laid the foundations for his methods. In 1910, Schuchert's monumental "Paleogeography of North America" had created quite a stir. Lengthy reviews appeared in both the *American Journal of Science* and *Science*, and both his methods and the resulting maps were under heated discussion for at least the next two decades. His 1911 paper on the paleogeographic and geologic significance of recent Brachiopoda indicated how he relied on biologic (fossil) evidence in his reconstructions, especially as modern biogeographies might give clues to past geographies.

H. S. Williams was perhaps the keenest geologist of his time when it came to applying fossil data to constructions of biostratigraphies and then using them for practical geologic purposes. Papers published in the *Bulletin* in 1900, 1905, and 1910 dealt, respectively, with the Silurian-Devonian boundary in North America, bearing of fossil data on nomenclature and classification of sedimentary formations, and the migration and shifting of Devonian faunas with respect to facies.

Specific fossil groups were being applied to more limited problems in paleogeography. For example, David White considered the value of floral evidence in marine strata as an indicator of distance from the shoreline (1911); F. H. Knowlton (1918) analyzed the relations between the Mesozoic floras of North and South America; T. W. Stanton (1918) sketched the Mesozoic history of Mexico, Central America, and the West Indies mainly on the basis of mollusks; and T. W. Vaughan (1918) presented a comparable story for the Cenozoic of Central America and the West Indies.

Descriptive systematic studies of particular fossil groups or assemblages were relatively sparse in the early days of the *Bulletin*. Most studies of this kind, with their requirements for fossil plates and detailed documentation beyond the scope of the *Bulletin*, were published elsewhere—either by state surveys, in museum proceedings, or as monographs or Professional Papers of the U.S. Geological Survey. Nevertheless, several outstanding contributions of this kind were published in the *Bulletin*, and the fact that most of them have stood the test of time indicates that the original decision to publish was a wise one.

Two papers by E. R. Cumings on bryozoans are still considered significant seminal studies by workers in the field. The earlier one (1912) discussed the development and systematic position of the monticuliporoid bryozoans, and a subsequent study with J. J. Galloway (1915) is a more detailed morphologic and histologic analysis of the group. T. W. Vaughan (1924) described American and European Tertiary foraminifers in some depth. R. T. Jackson's classic study of Palaeechinoida [sic] was actually the earliest work of this kind to appear in the *Bulletin* (1896). Stuart Weller published one of his Kinderhook faunal studies in the *Bulletin* (1909) in which he described the important Fern Glen fauna; this work is still a useful reference to megafaunas of early Osage age in North America. An early taxonomic coral study by W. H. Sherzer was published in 1892; this paper on the Silurian-Devonian genus *Chonophyllum*, although subsequently revised, remains a valuable contribution to coral systematics.

Contemplation of the meaning of trace fossils has been a continuing theme in the *Bulletin*. Two papers in 1892 led the way. J. F. James discussed *Scolithus* [sic], its possible origin, and its environmental significance; E. H. Barbour elaborated on the nature and structure of the giant burrow *Daimonelix*. Somewhat later, J. J. Galloway (1922) discussed the nature of *Taonurus* and its potential use in estimating geologic time. A. W. Grabau (1913) also examined trace fossils in his detailed exposition on deltaic deposits and environments of North America.

More localized biostratigraphic studies were published by Charles Schuchert and W. H. Twenhofel (1910) on the Ordovician and Silurian sequence on Anticosti and the Mingan Islands. An analysis of similar scope on the Laramie and related formations in Wyoming was prepared by F. H. Knowlton and T. W. Stanton (1897).

From this brief and selective summary of the first 40 years of paleontologic contributions to the *Bulletin*, it is clear that paleontologists were discovering, describing, and developing hypotheses on many of the same fossil problems that intrigue us today. Of course, the evidence was much sparser 75–100 years ago. Many of these geologists were the men who started paleontology onto its modern course. Research tools were not as sophisticated, many paleobiological ideas had not yet been developed, and the rudimentary statistical approaches left much to be desired.

THE SECOND FORTY YEARS

This period of *Bulletin* publication was dominated by a nearly 30-year project summarizing the biostratigraphy and correlation of the Phanerozoic systems in North America. Several lines of activity in the earth science community during the early 1930s converged to produce this intensive attack that ultimately involved nearly 100 stratigraphers and paleontologists. The idea was initiated by W. H. Twenhofel, who was chairman of the Division of Geology and Geography of the National Research Council in 1932. A Committee on Stratigraphy was established, with C. O. Dunbar as its chairman. Original plans were to produce 12 charts covering the Phanerozoic systems; the Triassic and Jurassic were to be combined, and the Cretaceous and Cenozoic were each to be displayed in two charts. Ultimately, 16 charts were published. Not only were the Triassic and Jurassic separated, but they each had two charts because the Canadian sequences were treated separately. A third Cretaceous chart dealt with Greenland and Alaska. Nearly ten years of work preceded the publication of the first chart in 1942 (L. W. Stephenson and others), and the last chart (Permian) was not completed until 1960 (C. O. Dunbar and others).

A major advance in North American biostratigraphy was the general acceptance of the "stage" as a fundamental time-stratigraphic unit for correlation between regions and on an international scale. Most of the charts represent the first consistent North American use of the stage, ushering in a period of close international cooperation which has continued to the present-day activities by International Subcommissions for every system.

In his introduction to the publication of the first chart (1942), Dunbar acknowledged the group's philosophical debt to W. J. Arkell, whose great book on the Jurassic of Great Britain was published in 1933, just as the North American activity began. Another spur was provided by the appearance of the so-called Ashley Code (G. H. Ashley and others, 1933) for the classification and naming of rock units. A major reference supplement appeared in 1938, M. Grace Wilmarth's *Lexicon of Geologic Names*. No doubt, the general ferment of the two International Geological Congresses of the decade also added an incentive to codify and correlate the North American sequences to make them available for international use.

The format of the charts themselves provided a template for all subsequent publications of this kind. Most charts have, in addition to a standard section for the covered region, a column showing the European stages for correlation purposes. Depending on the complexity of the sequences and the facies, each chart has a series of columns arranged in geographic order and aligned by correlations based mainly on fossil occurrences, generic or specific ranges of key forms, or zones of various kinds. Some of the charts include columns showing major zones of certain kinds of fossils or ranges of key fossils. In a few of the charts, stratigraphic diagrams indicate complex facies relationships. This major effort by North American geologists helped push the international community into the current intensive effort to refine worldwide correlations and to establish internationally useful boundary stratotypes. The first of these was the Silurian-Devonian boundary committee initiated in 1960 just as the Permian chart was published. The history of this committee and the philosophical basis for its work have been nicely summarized by D. J. McLaren (1977).

The introduction of time-stratigraphy in the original charts, and the ensuing turmoil among stratigraphers, produced a number of philosophical discussions of the subject. Two of the more significant of these appeared as *Bulletin* articles. H. D. Hedberg (1948) analyzed the time-stratigraphic classification of sedimentary rocks in some detail and carefully delineated the differences between rock and time-rock concepts. The relationships of both kinds of stratigraphic units to time, itself, were also discussed. Hedberg, however, failed to appreciate the contribution of detailed faunal analyses to the delineation of stages, as pointed out a decade later by Curt Teichert (1958). Teichert emphasized the unique characteristics of fossils as time indicators because of the unidirectional nature of evolution. It is true that all aspects of the geological record, including its fossil contents, are incomplete and will always remain so. As Teichert contended, however, the fossils in the rocks provide the single trustworthy clue to the relative sequence of events in geologic history. Although radiometric means of dating single, or multiple, events in Earth history have become much more precise and sophisticated during the 30 years since Teichert's paper, it is clear that fossils remain the unifying method for the ordering of events in later Earth history.

After a long drought, two papers by C. L. and M. A. Fenton (1936, 1937), one by P. E. Raymond (1935), and a fourth by K. Rankama (1948) continued the documentation of evidence of life in the Precambrian. Raymond's paper, his address as retiring president of the Paleontological Society, summarized the evidence for Precambrian life and examined the six prevalent theories to explain the scarcity of fossil evidence. Rejecting Walcott's idea that the available Precambrian sequences were mainly nonmarine, consequently low in calcium carbonate, and Daly's notion that the simple lack of calcium in the sea meant that organisms had no skeletons, Raymond suggested that the reason lay in the life style of early organisms. If most were mobile animals, they had no need for heavy mineralized skeletons; therefore, we have found none. The Fentons based their conclusions on field work on the Belt Series [sic] in western Montana. Their 1936 paper described the algae and pseudo-algae of the Newland Limestone; the 1937 paper went into more detail to explain the sedimentary features of the Belt, including many stromatolitic forms that were attributed to algae. Rankama's study of carbon isotopes in Precambrian samples provided the first convincing geochemical evidence for the organic origin of a number of carbon-bearing samples from Fennoscandia.

The GSA provided new outlets for book-length studies of all kinds, including a number of systematic paleontologic and biostratigraphic works, through its Memoir and Special Paper series supported by the Penrose bequest starting in the early 1930s. During the next two decades, magnificent systematic studies were published in both series. Selected Special Papers are H. G. Schenck's description of the nuculid bivalve *Acila* (4; 1936), A. K. Miller's monograph of Devonian ammonoids of America (14; 1938), F. M. Anderson's description of the geology and mollusks of the Lower Cretaceous of California and Oregon (16; 1938), A. K. Miller and W. M. Furnish's monograph of Permian ammonoids of west Texas (26; 1940), A. S. Romer and L. W. Price's review of the Pelycosauria (28; 1940), J. B. Knight's synthesis of Paleozoic gastropod genotypes (32; 1941), P. E. Cloud's monograph of Silurian and Devonian terebratuloid brachiopods (38; 1942), and T. W. Vaughan and J. W. Wells' revision of the scleractinid corals (44; 1943).

Some major Memoirs include Julia Gardner's description of Tertiary mollusks from northeastern Mexico (11; 1945), Rudolph Ruedemann's monograph of the graptolites of North America (19; 1947), A. K. Miller's description of Tertiary nautiloids of the Americas (23; 1947), N. D. Newell and others' *Upper Paleozoic faunas of Peru* (58; 1953), and Helen Muir-Wood and G. A. Cooper's monograph on the productoid brachiopods (81; 1960).

Nevertheless, the *Bulletin* continued to publish a few systematic papers. Typical of these is the study of the Tully Formation of New York by G. A. Cooper and J. Stewart Williams (1935). This is a detailed analysis of the stratigraphy and systematic paleontology of this Middle Devonian formation that provides a key link in the international correlation of the New York sequence. Although the lithic stratigraphy has been recently revised, and new fossils found and older ones reinterpreted, this paper remains an essential reference for all Devonian geologists. Two other papers of this kind are Newell's description of the dominantly molluscan faunas of the Upper Permian Whitehorse Sandstone (1940) and Halka Chronic's systematic description of the molluscan fauna of the Kaibab Formation in Arizona (1952). A systematic study of ostracodes from the Middle Devonian Onondaga Limestone of central Pennsylvania by F. M. Swartz and F. M. Swain (1941) remains the classic faunal study of ostracodes of this age in eastern North America. During this period, also, R. W. Imlay produced three paleontologic papers on the Mesozoic of Latin America that are still standard references for the biostratigraphy of that region. These include systematic descriptions of Late Jurassic ammonites from Mexico (1939), Neocomian faunas of northern Mexico (1940), and Late Jurassic fossils from Cuba (1942). Papers of this kind remain useful because they document biostratigraphic relationships and correlations that are still used in paleotectonic and paleogeographic syntheses relating to the plate-tectonic interpretation of various regions.

In the 1950s, the GSA helped the University of Kansas Press launch the most significant paleontologic project of this century—the *Treatise on Invertebrate Paleontology*. This continuing effort by hundreds of paleontologists has now produced 36 volumes that synthesize, at the generic level, most of the invertebrate phyla. Several of the volumes have been revised and updated, and the work continues. The role of the GSA as co-publisher and financial supporter of this monumental publishing undertaking leaves all paleontology in its debt.

Despite the fact that most systematic studies had other publishing outlets during this period (in addition to the *Journal of Paleontology*, the British journal *Palaeontology* was initiated in 1957), a relatively large number of significant papers continued to be published in the *Bulletin*. A sampling of these is listed, to give an idea of the breadth of coverage across several phyla.

Two papers by J.R.P. Ross (1963, 1964) provided modern systematic descriptions of Ordovician cryptostome bryozoans from the standard Chazyan Series and a synthesis of the morphology and phylogeny of early Ectoprocta. Carboniferous and Permian ctenostomatous bryozoans were described by G. E. Condra and M. K. Elias (1944).

Two papers by W. A. Oliver, Jr. (1954, 1956) are prime examples of detailed biostratigraphic analyses, particularly of corals, applied to regional stratigraphic synthesis of a major geologic unit—in this case, the Onondaga Limestone of New York.

Coral studies began to concentrate on the role that this group played in the development of reefs, with several papers comparing modern reef environments to those of the past. The early paper by E. R. Cumings and R. R. Shrock on the Niagaran reefs of Indiana and adjacent states (1928) is a truly basic study which gave rise to subsequent research on limestone depositional environments. Many of these results, because of their economic significance in the search for oil and gas, were published in the AAPG *Bulletin* and similar journals.

E. R. Cumings' presidential address before the Paleontological Society provided a superb summary of reefs and their relationships with fossil organisms (1932). It was here that the term "bioherm" was coined, and the various kinds of organic build-ups compared. Shortly after this, P. B. King (1934) published his grand synthesis of the Permian stratigraphy of Trans-Pecos Texas. This regional study not only exhaustively described the Permian geology of the region but also treated the "reef" problem in some detail and placed the build-ups in a depositional framework. These three papers are, in my mind, the forerunners of the massive outpouring of research on carbonate rocks and their depositional settings that has flooded the geologic literature during the past 50 years. King's paper also was a fine example of bio- and lithostratigraphic synthesis based on detailed field work and geologic mapping. It was a model for subsequent studies.

Mollusks received a large share of attention during this period. In addition to the papers by Imlay, Newell, and Chronic, mentioned above, W. J. Arkell (1946) provided a discussion of ammonites in the standard sequence of the European Jurassic. Trilobites were the center of attention in papers by P. E. Raymond (1937) on the Upper Cambrian and Lower Ordovician of Vermont, C. E. Resser and B. F. Howell (1938) on the Lower Cambrian *Olenellus* Zone in the Appalachians, and Christina Lochman-Balk (1956) on the *Elliptocephalus asaphoides* strata of New York. Brachiopod biostratigraphy highlights the study of the lower Middle Ordovician of the Shenandoah Valley by B. N. Cooper and G. A. Cooper (1946).

As techniques for preparing and studying microfossils advanced, their use in solving a wide variety of geologic problems became dominant in paleontology. C. R. Stauffer described and illustrated conodonts from the Glenwood beds (Ordovician) of Minnesota (1935); this paper, along with his studies of the Decorah conodonts (1930, 1935), placed Stauffer among the first American paleontologists who saw the tremendous usefulness of conodonts for biostratigraphic studies. Stauffer also described polychaetes from Minnesota (1933). Another descriptive conodont paper of this period was by L. A. Thomas (1949) on Devonian and Mississippian assemblages from Iowa.

Graptolite research advanced rapidly during the 1960s, and three papers in the *Bulletin* reflect this activity. Foremost among these was D. E. Jackson's (1964) analysis of the sequence and correlation of Early and Middle Ordovician graptolite faunas in North America. Also appearing at about this same time were W.B.N. Berry's paper on the stratigraphy, zonation, and age of the classic eastern New York Ordovician graptolites (1962) and G. M. Kay's study of shelly and graptolitic Chazyan sequences from central Nevada (1962).

The study of trace fossils, mainly as environmental indicators, grew steadily after the pioneering work of the earlier period. In addition to contributions in the Precambrian papers mentioned above, C. E. Van Gundy (1951) discussed traces in the Nankoweap Group of the Grand Canyon. G. E. Condra and M. K. Elias (1944) published an informative paper on the borings left by ctenostomatous bryozoans, and Bradford Willard and A. B. Cleaves (1930) described amphibian footprints from the Pennsylvanian of the Narragansett Basin.

The interest in analyzing fossil communities (assemblages), from a paleoecologic point of view, developed during the 1960s, spurred by the publication of the monumental *Treatise on Marine Ecology and Paleoecology*, edited by J. W. Hedgpeth and H. S. Ladd, as Memoir 67 (1957). A landmark paper on models and methods of analysis of the formation of different kinds of fossil assemblages was published by R. G. Johnson (1960). This was quickly followed, in 1964, by J. A. Fagerstrom's paper on the recognition of fossil communities and their significance in paleoecology. D. R. Lawrence (1968) published an important discussion of taphonomy and information loss in fossil communities, and P. W. Bretsky (1969) documented Late Ordovician communities in the central Appalachians.

THE MODERN PERIOD

Five trends that developed rapidly during the past two decades drastically changed not only the content and tone of paleontologic articles published in the *Bulletin* but also the whole course of paleontologic research. These trends are interwoven into the complex fabric of modern geology. First, the general acceptance of the plate-tectonic theory directed some paleontologic efforts toward paleobiogeographic and paleoclimatologic research aimed at testing and augmenting the theory. Second, the massive deep-sea drilling program, itself an outgrowth of plate tectonics because of the need to know more about the geology of the ocean basins, has produced a mass of new data on the later part of the geologic record. The explosive growth of Cretaceous and Tertiary microfossil studies, together with integrated paleomagnetic, sedimentologic, isotopic, and geochemical research, has produced a detailed geologic history that is unrivaled in its precision. Third, the growth in statistical applications to geologic problems, aided by the computer power to attack massive data accumulations like those of the deep-sea program, has reoriented many aspects of paleontologic research. Fourth, the burgeoning international effort to codify geologic history by determining boundary stratotypes for time-stratigraphic units has concentrated the world's paleontologic community on detailed systematic and phylogenetic studies in fossil groups that will provide the desired biostratigraphic precision. Fifth, the appearance of a number of new publishing outlets has siphoned off most lengthy paleontological contributions from general journals such as the GSA *Bulletin*.

The *Bulletin* remains a significant publishing medium for paleontologists who apply their talents to the solution of general scientific problems. The multidisciplinary nature of most earth-science research today results in many multi-authored papers that include more than a dash of paleontologic tabasco. Even the more traditional research on fossils is influenced by the trends outlined above. Paleontology, along with all of the other subdisciplines of the earth sciences, has changed inevitably.

The late 1960s are a good place to begin this review of the change. My discussion is highly subjective, partly because the past two decades are still fresh in the minds of most of us, but also because time has not tested the usefulness of many of the new ideas.

Perhaps most visible are the results of the deep-sea drilling. Among the first of these papers was the study of Pliocene and Pleistocene sediments of the equatorial Pacific, documenting the paleomagnetic, biostratigraphic, and climatic record, by J. D. Hays and others (1969). A few years later (1973), J. S. Gartner elaborated on the absolute chronology of the late Neogene nannofossil succession in the same region. In 1971, W. F. Ruddiman published an analysis of the Pleistocene stratigraphy and faunal paleoclimatology in the equatorial Atlantic. Already in 1970, hypotheses were being developed to explain some of the results of the deep-sea work; one of these was W. H. Berger's idea that biogenous deep-sea sediments could be explained by fractionation due to deep-sea circulation (1970). As the drilling efforts spread to all of the oceans, a number of papers appeared in the *Bulletin* and in many other outlets. More recent important contributions in the *Bulletin* are the 1984 paper by K. J. Hsü and others on the numerical ages of Cenozoic biostratigraphic datums, and the summary of Cenozoic geochronology by W. A. Berggren and others (1985). Two more general papers are the study of late Pliocene climate changes in the North Atlantic as they relate to the onset of glaciation in the Northern Hemisphere by Paul Loubere and others (1986) and the analysis of late Quaternary deep-sea circulation as revealed by distribution patterns of foraminifers by B. H. Corliss (1986).

The international effort to refine stratigraphic correlations and boundaries has resulted in several papers in the *Bulletin*; among them are the restating of the Morrowan-Atokan (Pennsylvanian) boundary problem by P. K. Sutherland and others (1983); the correlation of Gulf Coast Paleogene stages with European standard stages by W. G. Siesser and others (1985); a revision of the geochronology of the Jurassic and Cretaceous, using mainly ammonites, by D. V. Kent and others (1985); and the revision of the time-stratigraphy of North American Middle and Upper Ordovician rocks by W. C. Sweet and S. M. Bergstrom (1971). In addition, A. J. Boucot and W.B.N. Berry were instrumental in the preparation of a series of new Silurian correlation charts that appeared as GSA Special Papers (102, 1970; 133, 1972; 137, 1972; 147, 1973; 150, 1975; 154, 1974; and 202, 1986).

The application of statistical methods is now nearly universal. A landmark paper, appearing in 1970, was J. E. Hazel's analysis of the use of binary coefficients and clustering in biostratigraphy. A. J. Rowell and others (1973) applied quantitative techniques in their study of latest Cambrian trilobite distributions in North America. Species diversity patterns of Miocene and modern foraminifers were compared by T. G. Gibson and M. A. Buzas (1973).

Paleontologic testing and documentation of the new paleogeographies implied by the plate-tectonic theory are reflected in a number of papers, listed in no particular order, which show the diversity of interests and applications of modern paleontological research. R. J. Ross, Jr., and J. K. Ingham (1970) presented a synthesis of the biogeography of the Whiterock (Middle Ordovician) faunal realm in the Northern Hemisphere. W.B.N. Berry and A. J. Boucot (1973) implied glacio-eustatic control of Upper Ordovician and Lower Silurian platform sediments and faunal changes. In two papers, Christina Lochman-Balk summarized Late Cambrian faunal patterns on the North American craton (1970) and Late Cambrian biostratigraphy of North America (1974). J. G. Johnson (1970) discussed the paleogeographic and faunal implications of the Middle Devonian (Taghanic) onlap of the North American craton. In 1984, Johnson and others extended these ideas to present a time-rock model for the Silurian and Devonian shelf deposition in the western United States. A year later, Johnson and others (1985) extended these eustatic arguments to the Devonian of Euramerica. R. B. Neuman (1984) reviewed the geology and paleontology of islands in the Ordovician Iapetus Ocean and discussed the plate-tectonic implications of these ideas. Wang Yu and others (1984) presented the first comprehensive English-language paper on the Silurian and Devonian biogeography of China. M. E. Johnson and others (1985) made intercontinental correlations between North America and China, on the basis of graptolites and analysis of coordinated land-sea events. Two papers in 1986, one by J. L. Hannah and E. M. Moores and the other by R. E. Hanson and R. A. Schweickert, documented the puzzling tectonic history of Paleozoic island-arc sequences in the northern Sierra Nevada of California.

Problems concerning evolution and extinction of fossil groups continue to receive attention, particularly as a result of the meteorite impact hypothesis to explain the mass extinction at the end of the Cretaceous. D. J. McLaren's presidential address (1983) is a thoughtful analysis of the "bolide hypothesis" and the methods by which paleontologists and biostratigraphers can test its validity. P. W. Bretsky (1973) documented evolutionary patterns in Paleozoic pelecypods and discussed theoretical implications of these phylogenies. In 1970, Bretsky and D. M. Lorenz suggested the role played by genetic and adaptive strategies in mass extinctions in the fossil record. Also in 1970, T.J.M. Schopf analyzed taxonomic diversity gradients of some bryozoans and pelecypods and discussed some of the geologic implications of these patterns. In 1982, D. E. Schindel analyzed punctuation as the mode of evolution of some Pennsylvanian gastropods.

Paleoenvironments and paleoclimatology, in various guises, were the subject of a number of recent papers. H. J. Hofmann and G. L. Snyder (1985) discussed the paleoenvironmental significance of Archean stromatolites from the Hartville Uplift of Wyoming. F. R. Ettensohn and others (1985) provided an intriguing peek at paleo-oceans with their definition and location of the Late Devonian–Early Mississippian pycnocline in eastern Kentucky. D. K. Watkins (1986) developed a picture of the paleoceanography of the Cretaceous Greenhorn Sea using calcareous nannofossils. C. H. Stevens (1986) discussed the evolution of the Ordovician through Middle Pennsylvanian carbonate shelf in east-central California. A. L. Saltsman (1986) reconstructed the paleoenvironment of the Upper Pennsylvanian Ames Limestone and associated strata in western Pennsylvania. C. A. Ross (1986) synthesized the Paleozoic evolution of the southern margin of the Permian Basin in a major contribution that effectively closes the circle begun more than half a century ago by the King brothers' work on the West Texas Permian.

Finally, five magnificent papers published last year in the *Bulletin* epitomize how far paleontology has traveled during the 100 years of this rapid review. The paper by Nancy Chow and N. P. James (1987) presented a fine synthesis of Cambrian grand cycles from a northern Appalachian perspective. J. H. Craft and J. S. Bridge (1987) presented a definitive study of the shallow-marine sedimentary processes at work in the Late Devonian Catskill Sea in New York State. R. A. Armin (1987) developed a regional picture of the sedimentology and tectonic significance of Lower Permian conglomerates in the Pedregosa Basin of southeastern Arizona, southwestern New Mexico, and northern Mexico. E. K. Wright (1987) described a paleoceanographic model for the Late Cretaceous Western Interior Seaway using paleotemperature and salinity profiles derived from carbon- and oxygen-isotope studies of fossil material. V. A. Rejebian, A. G. Harris, and J. S. Huebner (1987) carried conodont color and textural alteration studies into thermal regions above 300 degrees Celsius and discussed their application to problems of thermal and regional metamorphism and hydrothermal alteration. All of these papers attack major geologic problems; all use a multidisciplinary approach in which the role of paleontology is highlighted.

What have we learned from this look at paleontology as published by the GSA during its first 100 years? Although the role of the *Bulletin* has shifted during the century, throughout its existence it has published significant paleontologic contributions, especially those that tackled broader geologic problems. As the philosophical bases and technological tools of paleontology have evolved, so have its products become more widely useful in both geology and biology. We can look forward to the next 100 years with anticipation that paleontology will play an increasingly important role. There are many questions now being asked in the earth or biological sciences to which paleontologists can contribute critical data and fresh insights.

REFERENCES CITED

(Excluding references to GSA authors and publications sufficiently identified in text)

Arkell, W. J., 1933, The Jurassic system in Great Britain: Oxford, England, The Clarendon Press, 681 p.
Ashley, G. H., and others, 1933, Classification and nomenclature of rock units: Geological Society of America Bulletin, v. 44, p. 423–459.
McLaren, D. J., 1977, The Silurian-Devonian Boundary Committee—A final report, *in* Martinsson, A., ed., The Silurian-Devonian boundary, International Union of Geological Sciences, ser. A, No. 5, p. 1–34: Stuttgart, Germany, E. Schweizerbart'sche Verlagsbuchhandlung (Nagele u. Obermiller).
Stauffer, C. R., 1930, Conodonts from the Decorah shale: Journal of Paleontology, v. 4(2), p. 121–128.
—— 1935, The conodont fauna of the Decorah shale: Journal of Paleontology, v. 9(7), p. 596–620.
Wilmarth, M. G., compiler, 1938, Lexicon of geologic names of the United States: U.S. Geological Survey Bulletin 896, 2,396 p.

MANUSCRIPT RECEIVED BY THE SOCIETY MARCH 7, 1988
REVISED MANUSCRIPT RECEIVED MAY 16, 1988
MANUSCRIPT ACCEPTED MAY 16, 1988

ISBN 0-8137-2253-5